THE DIGITAL IC GALLERY

By
Clement S. Pepper

THE DIGITAL IC GALLERY

By
Clement S. Pepper

PROMPT®
PUBLICATIONS

PROMPT© Publications is an imprint of Howard W. Sams & Company, A Bell Atlantic Company, 2647 Waterfront Parkway, E. Dr., Indianapolis, IN 46214-2041.

International Standard Book Number: 0-7906-1167-8
Library of Congress Catalog Card Number: 98-067315

Acquisitions Editor: Loretta Yates
Editor: Pat Brady
Assistant Editors: J.B. Hall
Typesetting: Pat Brady
Cover Design: Christy Pierce
Graphics Conversion: William Skinner
Illustrations and Other Materials: Courtesy of the Author

Trademark Acknowledgments:

All product illustrations, product names and logos are trademarks of their respective manufacturers. All terms in this book that are known or suspected to be trademarks or services have been appropriately capitalized. PROMPT® Publications, Howard W. Sams & Company, and Bell Atlantic cannot attest to the accuracy of this information. Use of an illustration, term or logo in this book should not be regarded as affecting the validity of any trademark or service mark.

PRINTED IN THE UNITED STATES OF AMERICA

9 8 7 6 5 4 3 2 1

Dedication

To family and those many friends who have been understanding of the time taken from them as I concentrated on what had to be done.

Acknowledgments

Several people have been especially helpful in times of need as I worked my way through the preparation of this work. Among these are Jim Davison of Fairchild Semiconductor; Karen Bosco of Motorola, Semiconductor Products Sector; Mr. Richard Young, Motorola Technical Publications; Karen Matzen and Thomas Wyatt of Philips Semiconductors; Sue of Micro-Tex, Inc.; Shantha Natarajan of National Semiconductor; and Mr. Gary Honeycutt, Mike He, and Don Anderson of Texas Instruments.

The Mororola Distribution Center, Texas Instruments Literature Fulfillment Center, and Micro-Tex, Inc. were generous in meeting my requests for product data books.

Contents

Introduction

The Purpose of This Book

My objective in writing this book is to aid the Experimenter, Hobbyist, and Engineer in identifying digital integrated circuits of interest in two primary logic families: 7400 series TTL and the several series of CMOS. TTL was preceded by several short lived logic families such as RTL and DTL that we are less likely to be involved with in these times and therefore not included.

Unless we have access to a large number of manufacturer's data books it can be rather difficult to learn the full range of devices that are or, in many cases, have been available for any given category of digital logic. Catalogs accessible to experimenters such Digi-Key, Jameco, Mouser and others provide only a partial listing of what is current. In part these may be whatever is selling well at the time. If we are working with older equipment it can be a challenge to track down needed information.

The descriptions provided are necessarily limited in detail. Included are connection diagrams for the dual-in-line (DIP) package, logic diagrams, truth and function tables and/or logic symbol diagrams plus a brief functional summary. Package pinning for SSOP, TSSOP and SOIC is generally the same as for the dual-in-line and is not defined individually. Performance tables provide a limited selection of frequencies and/or time performance appropriate to the device. There are many ways in which this data is defined, and not only between manufacturers. The data in the tables should be used for comparisons only; go to the manufacturer's data for precise information.

The appendix contains a listing of references and device manufacturer's data books referenced by the author along with others that may be of interest to those of us who enjoy doing things in our own home labs as well as on the job. These should be referred to whenever greater detail is required.

A Look At What's Inside

Glossary

A review of manufacturer's data books reveals a differing terminology for many of the same parameters. The purpose of the glossary is to assist in relating the definitions found in the various manufacturer's sources.

Chapter One. Digital IC Basics

This chapter is to get ourselves grounded in the logic family essentials beginning with standard 7400 TTL and following through to the currently available high performance types within the TTL and CMOS families. Early TTL logic types other than standard were High Speed (H) and Low Power (L). The High Speed and Low Power logic has been superseded by the Schottky (S), Low-Power Schottky (LS), Advanced Low-Power Schottky (ALS), Advanced Schottky (AS), and Fast (F).

The original CD4000 series of CMOS was offered by RCA in the early '70s. Shortly thereafter National Semiconductor introduced a CMOS line based on the existing 7400 series. Both CMOS lines have been considerably extended by several manufacturer's with inclusion of the CD4500, CD4700, 74C/HC/HCT, 74AC/ACT and 74AHC/AHCT series.

A recent trend is the introduction of CMOS devices duplicating the functions and package pinouts of existing logic but operating at reduced levels of supply voltage. These include voltage tolerant devices capable of interfacing between five and three volt logic. Examples are the LV, LVC, LVCH, LCX, LVX, LVQ, LVT, ALVCH, ALVT and others.

Chapter Two. Gate Logic

The initial TTL IC offerings were a variety of logic gates, which we can readily verify by a quick glance at the data listings of any IC data book. Gate logic occupies the initial two pages of the Signetics 1978 TTL Data Manual listing for instance. After that the flood gates open on flip-flops, monostables, latches, counters, and comparators. This is less true for the early CMOS releases which included a mix of flip-flops, latches, and counters along with the gates. The CD4000/4500 offerings are on a much smaller scale than their TTL cousins. The later 74HC/HCT high speed logic duplicate many of the TTL devices.

Chapter Three. Monostables and Timers

The monostable represents the next level of complexity. The TTL 74121, 74122, and 74123 were early arrivals. The CD4047, CD4528/38 and the 74C221 are CMOS standard bearers. The TTL 555 Timer was a welcome addition when introduced around 1974. The device functions in both mono- and astable modes. Capable of driving TLL loads, the 555 is rated at up to 18 volts. *Intersil* introduced the CMOS 7555 which duplicates the pinout and functions of the 555 with an added square wave mode.

Chapter Four. Flip-Flops

The first of the TTL flip-flops were the "D" and variations on the "J-K." Later versions, in particular all the quad, hex, octal, nine- and ten-bit devices are "D" logic. The total TTL and CMOS flip-flop variety is very large and seems to be still growing. Virtually all of the spread in device features arise from bus operation requirements, in particular input enabling and 3-state output provisions with set and reset options.

The flip-flop in one of its several variations is the building block for all sequential logic.

Chapter Five. The Data Latch

In many respects the latch is a flip-flop under a new name, but the applications and distinctions warrant a chapter of its own. Latches find application whenever a time related save and hold function is required. A common use is as a storage element between a counter and the count display. Latch circuits are integrated into many multifunction devices we discover in later chapters.

Chapter Six. Shift and Data Storage Registers

The shift register is a popular application of the flip-flop for conversion from serial to parallel mode and its opposite. Data shifting is a common means for multiplying and dividing: a right shift performs a binary multiplication; a left a binary division.

Chapter Seven. Counters and Dividers

The first of Medium Scale Integration (MSI) devices were the TTL 7490 series of ripple counters. In these the output of each stage is the input to the succeeding, hence the "ripple" designation. These were soon followed by synchronous types with features such as presettability, preset and clear, and up/down count options. Similar counters have become available within the enhanced TTL logic offerings as well the range of CMOS logic families.

Chapter Eight. Decoders and Encoders, Multiplexers and Demultiplexers

A major task of decoders is to function as translators between number bases. That is, we have BCD-to-Decimal, BCD-to-Binary, BCD-to-7-segment, binary to 1-of-4, and so on. Among the first of these were the TTL 7442 BCD-decimal, 7443 Excess 3-to-decimal, 7445 BCD-to-decimal/driver with open collector output.

Primary encoders are the 74147 10-bit and 74148 8-bit Priority Encoders.

The priority encoder monitors its input lines to determine which is to have priority. The 74148 8-to-3 Priority Encoder looks at eight parallel inputs and assigns a 3-bit code representation to the outputs. The "priority" arises from the ability to make assignments between inputs having the same active status. The 74148 is found in the TTL 7400 standard/LS/F series. CMOS versions include the MC14532 and 74HC/HCT148.

The CMOS 74C922 and 74C923 are 16- and 20-Key encoders for arrays of SPST switches. I have not found these in the HC and HCT logic series.

The digital multiplexer is the logical replica of the rotary switch. Just as with the rotary switch, these may include multiple decks and/or poles. The determination of which input is to appear at the output is made through a 3- or 4-bit input code. Thus an 8-input multiplexer will have three selector, or addressing, bits. Though multiplexers are digital devices many, such as the CD4016 and others provide analog switching.

Analog multiplexers, or switches, are a combination offering digital control for the transmission of uni- or bipolar analog signals. Switches are availability for TTL and CMOS logic interfacing. When used with computer buses a logic latch may be required to maintain the circuit while the computer is away on other activities.

The digital demultiplexer functions as the opposite of the multiplexer. In these devices the selector inputs decide which output of a group will respond by a change of state. An example is the popular 74154 1-of-16 Decoder/Demultiplexer. This device features four address lines. The binary value of these determines which of 16 outputs will respond by an exclusive high-to-low transition.

Chapter Nine. Arithmetic and Logical Functions

These include the half- and full-adders and the 4-bit Arithmetic Logic Unit (ALU). The variety of device types is rather small, but distributed throughout all levels of TTL and CMOS logic. This chapter also describes the Lookahead Carry Generator. The functions of the devices in this chapter have been taken over in large part by the increased use of programmable logic.

The TTL logic comparator made an early appearance along with the flip-flop, monostable and counter. There are applications for only so many comparator types, 4-bit and 8-bit magnitude comparators are available in the TTL and CMOS families. This chapter includes parity generators/checkers and parity encoders, which utilize bit comparison in their operation.

Error Detection/Correction and Binary/BCD conversion is included also.

Chapter Ten. Buffers and Line Drivers/Receivers

Standard TTL logic provides a fanout to ten inputs. Low power and CMOS logic is severely limited in driving power. Drivers with 3-state outputs serve as buffers and bus interfacing. Often these not provide enhanced driving power and 3-state operation, but provide protection for more vulnerable and/or expensive devices in the system.

Chapter Eleven. Drivers and Bus Transceivers

Transceivers enable two-way transmission between source and destination. Devices are available to meet a wide variety of requirements. This group of devices has grown from a basic two-way communication device to those including storage registers and parity generators/checkers.

GLOSSARY

The tables and drawings here provide comprehensive definitions of the terminology employed in semiconductor manufacturer literature. Over time these have become pretty much standardized, though some variations may still be found. The four tables define the terms and definitions you will need to follow the data in the chapters which follow. Logic symbols and truth tables are not included here, as they appear with the descriptions of those devices to which they relate.

Glossary Table and Figure Listing:

Abbreviations For Logic Forms:

CML	Current Mode Logic
CTL	Complementary-Transistor Logic
DCL	Designer's-Choice Logic[1]
DCTL	Direct-Coupled Logic
DTL	Diode-Transistor Logic
ECL	Emitter-Coupled Logic
RCTL	Resistor-Capacitor-Transistor Logic
RTL	Resistor-Transistor Logic
TTL (T^2L)	Transistor-Transistor Logic
UTILOGIC	Signetics developed industrial logic form[1]
CMOS	Complementary MOS Logic
HCMOS	High-Speed CMOS Logic
ABT	BiCMOS bus interface products
ABTE	Enhanced BiCMOS bus interface products
LVC	Low Voltage CMOS Logic
LVT	Low Voltage Transistor Logic
ALVC	Advanced Low Voltage CMOS
LVTZ	Low Voltage Transistor Logic, High Impedance
FB	Future Bus (TTL, BTL translation)
GTL	Reduced voltage switching standard
CBT	Computer Bus Interface logic

[1] Trademark, Signetics Corporation.

Table G.1. Logic Terms

Accumulator	A register in an arithmetic unit for performance of arithmetical and logic functions such as addition and shifting.
Adder	Full adder: A device for performing the sum with carry of two binary digits and carry of a previous operation. Half adder: A device for performing the addition and carry of two binary digits without a facility to include a previous carry.
Alphanumeric	A combination of alphabetic characters and digital numbers.
AND	Logic implementation generating a ONE output when all inputs are ONE. It is sufficient, if any input is ZERO to generate a ZERO.
Astable Multivibrator	An oscillator generally with complementary outputs, no trigger is necessary.
Asynchronous Operation	Generally an operation that is started by a completion signal from a previous operation. It then proceeds at the maximum speed of the circuits until finished and generates its own completion signal.
Bar	Used to symbolize the inverse, or complement, of a function.
Binary Coded Decimal	Code that covers the numbers 1, 2, 3, 4, 5, 6, 7, 8, 9, and 0 in binary notation using four binary digits. Examples: 8-4-2-1 BCD, 2-4-2-1 code. Note that only the 8-4-2-1 code is defined as BCD; all other codes are spelled out.
Binary Number	Each digit of a binary number has only one of two possible states, defined as 0 and 1. Each digit represents a number with the radix 2.
Bistable Multivibrator	Flip-Flop, or Binary. A circuit having two stable states.
Bit	Binary Digit. The smallest part of information in a binary notation system. A bit is either a ONE or a ZERO.
Boolean Algebra	Algebraic rules in manipulating logic equations.
Buried Layer	A layer of heavily doped material (high conductivity) under the collector region generally to reduce the collector saturation resistance. This layer is diffused on the wafer before the epitaxial growth.
Buffer	Usually an isolation stage between input and output of a digital circuit. Also: An intermediate storage device in a computer, or;

	A term sometimes used for a line driving element with high capacity loading capability and/or high fanout.
Buss (Line)	A common connection line for distribution of ground, power supply or signals.
Byte	An eight-bit grouping of bits. May contain two 4-bit characters or some other combinations of bit sub-grouping to facilitate information handling.
Carry	When adding two binary digits, a carry is generated when the sum exceeds the radix 2.
Character	A 4-bit, 6-bit or 8-bit alphanumeric code.
Clear	Setting a number of memory elements or binaries to the ZERO state. Same as Reset.
Clock	Basic timing device in a system, usually providing a continuous chain of timing pulses (clock pulses).
Clock Skew	A phase shift in a single clock distribution system. Can be introduced by different delays in clock driving elements and/or by distribution paths. Excessive skew will result in malfunction.
Clock Stagger	A term generally applied to the time separation between clock pulses in a multiphase clock system. It may be applied to the voltage separation of the clock thresholds in a flip-flop such as the common DC clock Master-Slave devices.
Coder	A device for converting data from one notation system to another.
Collector Logic	See Wired AND.
Comparator	Digitally, an arrangement of gates that checks two signals or numbers for equality and usually indicates "equality," "less than," "greater than." An analog comparator compares two input levels and indicates "less than" or "greater than."
Complement	The logical inversion of a Boolean function. (See De Morgan's Theorem). Used to express negative numbers to facilitate subtraction.
Core Memory	Storage device consisting of ferromagnetic cores with provisions to enter data and to read data. A ferromagnetic core has two stable states which are used to store a binary digit.
Darlington Pair	Two transistors connected together to obtain a higher gain and higher input impedance.
Debugging	Act of checking system operation to identify malfunctions.
Decimal Numbers	Decimal notation, a system using numbers with the radix 10.

De Morgan's Theorem
The theorem states that the inversion of a series of AND implications is equal to the same series of inverted OR implications, or the inversion of a series of OR implications is equal to the same series of AND implications.

$$\overline{A \bullet B \bullet C} = \bar{A}+\bar{B}+\bar{C} \quad \overline{A+B+C} = \bar{A} \bullet \bar{B} \bullet \bar{C}$$

Die (Dice) or Chip
Tiny silicon bar containing one or more circuit components such as transistors, diodes, resistors, etc. A die may contain a quantity of circuits of varying complexity.

Die Attach
Process of attaching by alloying the tiny silicon bars to the mechanical support.

Die Sort
An electrical evaluation of the dice while in the wafer form.

Dielectric Isolation
A process in which the individual components are separated through dielectric isolating layers.

Double Diffusion
See Isolation Diffusion.

EEPROM
Electrically erasable Programmable Read Only Memory (PROM)

EPROM
Erasable (ultraviolet) Programmable Read Only Memory (PROM)

Epitaxial Growth
A depositing process of a single crystal P or N material or intrinsic silicon on a substrate.

Excess Three Code
A self-complementing BCD code. Any 8-4-2-1 BCD code number might easily be transformed to an excess three number by adding a binary 3 (0011) to each number. If all zeros in a number are changed to ones and all ones to zeros, the nines complement of the number is obtained.

Exclusive-OR
The function, either one, but not both. $F = A\bar{B} + \bar{A}B$

False
Statement for a ZERO in the Boolean Algebra.

Fan In
Total number of inputs to a particular gate or function.

Fan Out
The number of loads connected to the output of a gate or function.

FIFO
First In, First Out memory register.

Flip-Flop
A circuit with two and only two stable states.

RS Flip-Flop
A binary with Set and Reset inputs and the restriction that both inputs cannot be set high simultaneously as the resultant state will be indeterminate.

T Flip-Flop
A binary with a synchronous T-input: if T input is high, the flip-flop will trigger synchronously.

RST Flip-Flop	A binary with the features of an RS and a T flip-flop.
RS/T Flip-Flop	An RS flip-flop which may be connected to operate in a toggling mode.
JK Flip-Flop	A binary with synchronous set and reset inputs; all input combinations permitted.
D Flip-Flop	Delay binary. The output shows the input signal at the next clock pulse (one clock pulse delayed).
Gray Code	A binary code which is useful in input/output mechanisms, for example, shaft position encoders, because each number differs from the preceding by one digit.
Hexadecimal Numbers	Hexadecimal notation, a system using numbers with the radix 16 (0 -9, A,B,C,D,E,F).
Holding Time	The period of time that the input states must remain after activation of the clock pulse.
Hybrid	A mating of two or more technologies: 1. A class of integrated circuits where two or more silicon chips are interconnected within the package. 2. A combination of the monolithic and thin film methods of manufacture.
Input Latch	Minimum voltage between two emitters (multiple emitters) after breakdown.
Inverse Beta	Resulting gain of a transistor when the emitter and collector loads are physically reversed in a circuit's operation.
Inverter	A device or circuit to complement a Boolean Function.
Isolation Diffusion	A technique to separate the individual components within the monolithic silicon N-structure. P-diffused isolation zones form the P-N junctions which act as reverse-biased diodes.
Isolation Diode	PN junctions, which surround all components, and act as biased diodes.
Latch	Usually a feedback loop in a symmetrical digital circuit (such as a flip-flop) for retaining a state. A very simple RS flip-flop derived from two inverting gates.
LIFO	Last In, First Out memory register.
Logic Depth/Logic Levels	Generally the number of driven gates in a synchronous system between an output of a memory element binary and the enabling input of the next memory element binary. The maximum allowable logic depth is determined through the switching delays of the gates and the input/output delays of the memory element binaries.

Logic Levels	One of two possible states, ZERO or ONE.			
	Definition:	for positive logic	"UP"	level 1 = TRUE
			"DOWN"	level 0 = FALSE
		for negative logic	"DOWN"	level 1 = TRUE
			"UP"	level 0 = FALSE

LSB — Least Significant Bit. The lowest weighted digit of a binary number.

Master-Slave — A binary element containing two independent storage stages a definite separation of the clock function to enter information to the master and to transfer it to the slave.

Monolithic — Elements or circuits formed within a single semiconductor substrate.

Monostable Multivibrator — One shot. A circuit with one stable state and one quasi-stable state. When triggered, the monostable changes its state and falls back to the stable state after a certain time determined by its RC time constant.

MOS — Metal Oxide Semiconductor. A field effect transistor whose gate is isolated from its channel through an oxide film. Also a capacitor formed by using similar techniques; semiconductor material forms one plate, metal (Al) the other plate, with oxide form ing the dielectric.

Multiplex — Commutate: sequentially connect a central unit or system to one of several channels.

MSB — Most Significant Bit. The highest weighted digit of a binary number.

NAND — Logic implication that produces the inverted AND function.

Noise Margin (DCM) — The DC difference between the levels required at an input and available at an output. (Refer to *Figure 1.7*, Chapter 1).

NOR — Logic implication that produces the inverted OR function.

Octal Numbers — Octal notation, a system using numbers with the radix 8.

OR (Inclusive) — Logical implication that produce at the output a ONE if at least one input is ONE.

Pair Delay — Propagation delay over two inverting stages.

Parallel Mode — The operation is performed for all binary digits simultaneously. For example, in a register, read-in and read-out occur simultaneously for all digits contained in the register. Usually parallel mode requires more equipment than series mode, but speeds up the operation.

Parity Check	A simple kind of error-detecting scheme that requires an additional parity bit for each set of binary data. The binary set is always brought to an even (or odd) number of one digits with the parity bit before transfer operations and storage. The parity is tested after data transfers and storage readouts to check for errors that may have occurred during these operations.
Propagation Delay (tpd)	Average propagation time per stage.
Pull-Down Resistor	A resistor connected to ground or a negative voltage.
Pull-Up Resistor	A resistor connected to the positive supply voltage.
Pull-Up (Active)	A transistor replaces the pull-up resistor to effect low output impedance without high power consumption.
Race	The condition that exists whenever a signal propagates through two or more memory elements during one clock period. Also: A condition that occurs whenever changing the state of a system requires changing two or more variables. If the final state is dependent up which variable changes first, the condition is a critical race.
RAM	Random-Access Memory. Volatile semiconductor memory.
Register	A storage device for binary data generally used to store numbers for arithmetic operations or their results.
Reset	Changing the state of a binary from ONE to ZERO condition.
Ring Counter	A feedback shift register with only one stage in the ONE condition at any time. The shift register's output is fed back to the input to keep this single binary digit circulating. (See also Twisted Ring Counter.)
Ripple Counter	An asynchronously controlled counter; the clock is derived from an output of a previous stage.
ROM	Read-Only Memory. Non-volatile semiconductor memory.
Scratch Pad Memory	Fast memory unit to store temporary data between operations in computers.
Serial Mode	The operation is performed bit by bit, generally beginning with the least significant bit. Read-in and read-out occurs bit after bit by shifting the binary data through the register.

Set	Changing the state of a binary from ZERO to ONE condition.
Set-Up Time	The periods of time that logic levels must be presented at a binary's logic inputs before the clock input may be activated.
Shift	Transfer data from one memory cell to another.
Shift Register	A storage device with the ability to shift the content in one or the other direction.
Single-Rank Binary	A flip-flop which requires not more than one full clock pulse of a single clock system to transfer the logic from a synchronous input to the binary's output. Contains but a single memory stage.
Sink Load	A load with a current flow out of its input. A sink load must be driven by a current sink.
Source Load	A load with a current flow into its input. A source load must be driven by a current source.
Standard Unit Load	Within each logic series there is a standard load defined. Maximum fanout is determined with this standard load or an equivalent circuit.
Synchronous Operation	Operation controlled by a clock pulse.
Triple Diffused	The transistors are fabricated within the monolithic substrate by three different diffusion steps.
Turn-off Time (T_{off})	Switching time delay of a driven digital element measured between input and output at specified voltage levels, usually the switching threshold voltages.
True	True condition, the statement for a ONE in the Boolean Algebra.
Truth Table	A tabular list with all possible input logic combinations and the resulting output logic for all these combinations.
Twisted Ring Counter	A feedback shift register with 2N states for N stages. The input to the first stage is the complement of the last stage.
Up-Down Counter	(Reversible Counter) A counter counting in ascending or descending order depending upon the logic at the up-down inputs.
Wafer	Semiconductor slice for dice manufacture.

Wired AND/Dot AND Collector Logic	An AND function formed by wiring two or more outputs together. An AND will result from the connection of NPN collectors as in a conventional DTL or RTL logic. Equal to wired-OR if notation is negative logic.
Wired OR/Dot OR	An OR function formed by wiring two or more outputs employing active pull up and passive pull down together. Frequently applied to wired AND.
Word	A term for the largest grouping of bits, characters, or bytes.

Table G.2. Symbols Commonly Used with Function and Truth Tables

H/1	Indicates a HIGH logic level. Use of “H” or “1” varies between manufacturers. In this book “1” is used in truth and function tables.
h	When used indicates a HIGH logic level one setup time prior to the HIGH-to-LOW clock transition.
L/0	Indicates a LOW logic level. Use of “L” or “0” varies between manufacturers. In this book “0” is used in truth and function tables.
l	When used indicates a LOW logic level one setup time prior to the LOW-to-HIGH clock transition.
↑	Indicates the transition from a logic LOW to a logic HIGH level.
↓	Indicates the transition from a logic HIGH to a logic LOW level.
X	Irrelevant. Usually designated as “Don’t Care.” The logic state of this function has no effect on the response.
Z	Indicates a high impedance state. Typical of the output of a 3-state device in the high output state condition.
Q0/q	Output level prior to the referenced steady state condition. The output prior to a clock transition.
$\overline{Q}0/\overline{q}$	The complement of Q0/q.
$\overline{Q}n/Qn$	Level of the output, $Q/\overline{Q}$, prior to the most recent clock transition indicated by clock transitions ↑ or ↓.
Toggle	The output changes to the complement of the previous level on each ↑ or ↓ transition.

Table G.3. Letter Symbols, Terms and Definitions

The following symbols, terms, and definitions are in accordance with currently agreed upon by the JEDEC Council of the Electronics Industries Association (EIA) for use in the USA and the International Electrotechnical Commission (IEC) for international use. The definitions are grouped by section as applying to voltages, currents, switching characteristics and classification of circuit complexity.

CAPACITANCES:

C_i	Input capacitance. The internal capacitance at the input of the device.
C_{pd}	Power dissipation capacitance. Used to determine the no-load dynamic power dissipation per logic function. (See individual circuit pages.) $P_D = C_{pd}V_{CC}^2f + I_{CC}V_{CC}$.

VOLTAGES:

V_{IH}	High-level input voltage. An input voltage level within the more positive (less negative) of the two ranges of values used to represent the binary variables. A minimum value is specified which is the least-positive (most negative) value of high-level input voltage for which operation of the logic element within specification limits is guaranteed.
V_{IL}	Low-level input voltage. An input voltage level within the less positive (more negative) of the two ranges of values used to represent the binary variables. A minimum value is specified which is the most-positive (least negative) value of low-level input voltage for which operation of the logic element within specification limits is guaranteed.
V_{T+}	Positive-going threshold voltage. The voltage level at a transition-operated input that causes operation of the logic element according to specification as the input voltage rises from a level below the negative-going threshold voltage, V_{T-}.
V_{T-}	Negative-going threshold voltage. The voltage level at a transition-operated input that causes operation of the logic element according to specification as the input voltage falls from a level above the positive-going threshold voltage, V_{T+}.
V_{OH}	High-level output voltage. The voltage at an output terminal for a specified output current I_{OH} with input conditions applied that according to the product specification will establish a high level at the output.
V_{OL}	Low-level output voltage. The voltage at an output terminal for a specified output current I_{OL} with input conditions applied that according to the product specification will establish a low level at the output.
$V_{O(on)}$	On-state output voltage. The voltage at an output terminal for a specified output current with input conditions applied that according to the product specification will cause the output switching element to be in the on state.

Note: This characteristic is usually specified only for outputs not having internal pull-up elements.

$V_{O(off)}$ — Off-state output voltage. The voltage at an output terminal for a specified output current with input conditions applied that according to the product specification will cause the output switching element to be in the off state.

Note: This characteristic is usually specified only for outputs not having internal pull-up elements.

CURRENT:

I_{IH} — High-level input current. The current flowing into an input * when a specified high-level voltage is applied to that input.

I_{IL} — Low-level input current. The current flowing into an input * when a specified low-level voltage is applied to that input.

I_{OH} — High-level output current. The current flowing into the output ** with a specified high-level output voltage V_{OH} applied.

$I_{O(off)}$ — Off-state output current. The current flowing into an output ** with a specified output voltage applied and input conditions applied that according to the product specification will cause the output switching element to be in the off state.

I_{OS} — Short-circuit output current. The current flowing into * an output when that output is short-circuited to ground (or other specified potential) with input conditions applied to establish the output logic level farthest from ground potential (or other specified potential).

I_{CCH} — Supply current, output(s) high. The current flowing into * the V_{CC} supply terminal of a circuit when the reference output(s) is (are) at a high-level voltage.

I_{CCL} — Supply current, output(s) low. The current flowing into * the V_{CC} supply terminal of a circuit when the reference output(s) is (are) at a low-level voltage.

* Current flowing out of a terminal is negative value.

** This parameter is usually specified for open collector outputs intended to drive other logic circuits.

DYNAMIC CHARACTERISTICS:

f_{max} — Maximum clock frequency. The highest rate at which the clock input of a bistable circuit can be driven through its required sequence while maintaining stable transitions of logic level at the output with input conditions established that should cause a change of output state with each clock pulse.

tHZ	Output disable time (of a three-state output) from high level. The time between the specified reference points on the input and output voltage waveforms with the three-state output changing from the defined high level to a high-impedance (off) state.
tLZ	Output disable time (of a three-state output) from low level. The time between the specified reference points on the input and output voltage waveforms with the three-state output changing from the defined low level to a high-impedance (off) state.
tPLH	Propagation delay time, low-to-high level output. The time between the specified reference points on the input and output voltage waveforms with the output changing from the defined low level to the defined high level.
tPHL	Propagation delay time, high-to-low level output. The time between the specified reference points on the input and output voltage waveforms with the output changing from the defined high level to the defined low level.
tTLH	Transition time, low-to-high level output. The time between a specified low-level voltage and a specified high-level voltage on a waveform that is changing from the defined low level to the defined high level.
tTHL	Transition time, high-to-low level output. The time between a specified high-level voltage and a specified low-level voltage on a waveform that is changing from the defined high level to the defined low level.
ta	Access time. The time interval between the application of a specified input pulse and the availability of valid signals at an output.
tdis	Disable time (of a three-state output). The time interval between the specified reference points on the input and output voltage waveforms, with the three-state output changing from either of the defined active levels (high or low) to a high impedance (off) state. (tdis = tPHZ or tPLZ).
ten	Enable time (of a three-state output). The time interval between the specified reference points on the input and output voltage waveforms, with the three-state output changing from a high impedance (off) state to either of the defined active levels (high or low). (ten = tPHZ or tPLZ).
tf	Fall time. The time interval between two reference points (90% and 10% unless otherwise specified) on a waveform that is changing from the defined high level to the defined low level.
th	Hold time. The time interval for which a signal or pulse is retained at a specified input terminal after an active transition occurs at another specified input terminal.
tpd	Propagation delay time. The time between the specified reference points on the input and output voltage waveforms with the output changing from one defined level (high or low) to the other defined level.

tr	Rise time. The time interval between two reference points (90% and 10% unless otherwise specified) on a waveform that is changing from the defined low level to the defined high level.
tw	Average pulse width. The time between 50% amplitude points (or other specified reference points) on the leading and trailing edges of a pulse.
trelease	Release time. The time interval between the release from a specified input terminal of data intended to be recognized and the occurrence of an active transition at another specified input terminal. Note: When specified, the interval designated "release time" falls within the setup interval and constitutes, in effect, the negative hold time.
tsu	Setup time. The time interval for which a signal is applied and maintained at a specified input terminal before an active transition occurs at another specified input terminal.
tZH	Output enable time (of a three-state output) to a high level. The time between the specified reference points on the input and output voltage waveforms with the three-state output changing from a high-impedance (off) state to the defined high level.
tZL	Output enable time (of a three-state output) to a low level. The time between the specified reference points on the input and output voltage waveforms with the three-state output changing from a high-impedance (off) state to the defined low level.

CLASSIFICATION OF CIRCUIT COMPLEXITY

Gate equivalent circuit	A basic unit-of-measure of relative digital-circuit complexity. The number of gate equivalent is that number of individual logic gates that would have to be interconnected to perform the same function.
LSI	Large Scale Integration. A concept whereby a complete major subsystem or system function is fabricated as a single microcircuit. In this context a major subsystem or system, whether logical or linear, is considered to be one that contains 100 or more equivalent gates or circuitry of similar complexity.
MSI	Medium Scale Integration. A concept whereby a complete major subsystem or system function is fabricated as a single microcircuit. The subsystem or system is smaller than for LSI, but whether digital or linear, os considered to be one that contains 12 or more equivalent gates or circuitry of similar complexity.
SSI	Small Scale Integration. Integrated circuits of less complexity than medium-scale integration (MSI).

DC POWER:

Iin	DC input current, per pin. Expressed in milliamperes (mA) or microamperes (mA).

I_{CC}	DC supply current, V_{CC} and GND pins. Expressed in milliamps (mA) or microamps (mA).
I_{out}	DC output current, per pin. Expressed in milliamperes (mA) or microampers (mA).
GND	Power supply common; the reference from which voltages and currents are measured.
P_D	Power dissipation in still air. Expressed in milliwatts (mW).
V_{CC}	Positive supply voltage referenced to ground (GND). Expressed in positive volts (+V)
V_{DD}	Supply voltage referenced to ground. Originally defined for MOS devices and negative power supply (ref 1602 UART). Originally used with CMOS (ref Motorola CMOS Data Book, DL-105, 1978).
V_{in}	DC input voltage range. Typically expressed relative to V_{CC}; i.e., -1.5 to VCC +1.5.
V_{out}	DC output voltage range. Typically expressed relative to V_{CC}; i.e., -0.5 to VCC +0.5.
V_{SS}	Substrate voltage. Employed as ground designation on early CD4000 CMOS devices. (ref Motorola CMOS Data Book, DL-105, 1978).

Table G.4. Low-Voltage Logic Definitions
(that differ or add to those of Table G.3)

DC CHARACTERISTICS

I_{CCH}	The current flowing into the V_{CC} supply terminal when the outputs are in the high state.
I_{CCL}	The current flowing into the V_{CC} supply terminal when the outputs are in the low state.
⇑ I_{CC}	Additional I_{CC} due to TTL high levels forced on CMOS inputs.
I_{CCZ}	The current flowing into the V_{CC} supply terminal when the outputs are disabled (high impedance).
I_I, I_{IN}	Input current. The current flowing into or out of an input when a specified low or high voltage is applied to that input.
I_{OZ}	Output OFF current. The current flowing into or out of a disabled three-state output when a specified low or high voltage is applied to that output.

$I_{(HOLD)}$ — Input hold current. Input current that holds the input at the previous state when the driving device goes to a high impedance state.

$I_{(OD)}$ — Input overdrive current. Input current that is specified to switch a logic level which is held at a previous state.

I_{OFF} — Input/Output power-off leakage current. The maximum leakage current into or out of the input/output transistors when forcing the input/output from 0V to 5.5V with V_{CC} = 0V.

V_{JK} — Input Clamp Diode voltage. The voltage on an input (-) when a specified current is pulled from that input.

V_{IH} — Input high voltage. The minimum input voltage that is recognized as a DC high level.

V_{IHD} — Dynamic input high voltage. The minimum input voltage that is recognized as a high level during a Multiple Output Switching (MOS) operation.

V_{IL} — Input low voltage. The maximum input voltage that is recognized as a DC low level.

V_{ILD} — Dynamic input low voltage. The maximum input voltage that is recognized as a low level during a Multiple Output Switching (MOS) operation.

V_{OLP} — Maximum (peak) voltage induced on a static low output during switching of other outputs.

V_{OLV} — Minimum (valley) voltage induced on a static low output during switching of other outputs.

AC CHARACTERISTICS

f_{max} — Toggle Frequency/Operating Frequency. The maximum rate at which clock pulses may be applied to a sequential circuit. Above this frequency the device may refuse to function.

t_{PHZ} — Same definition as for t_{HZ}.

t_{PLZ} — Same definition as for t_{LZ}.

t_{PZH} — Output enable time (of a 3-state output) to a high level.

t_{PZL} — Output enable time (of a 3-state output) to a low level.

t_{rec} — Recovery time. The time between the specified level on the trailing edge of an asynchronous input control pulse and the same level on a synchronous input (clock) pulse such that the device will respond to the synchronous input.

AC DYNAMIC (NOISE) CHARACTERISTICS

VOLP, VOLV — Ground Bounce (Quiet Output Switching). Measured parameters with 50 pF loading relate the amount that a static conditioned output will change in voltage under multiple outputs switching condition with outputs operating in phase . They are heavily influenced by the magnitude that VCC and Ground move internal to the IC.

VILD, VIHD — Dynamic Threshold. Dynamic threshold measures the shift of an IC's input threshold due to noise generated while under multiple outputs switching condition with outputs operating phase. This test is package and test environment sensitive.

Input Edge Rate — This test is performed to determine what minimum edge rate can be applied to an input to have the corresponding output transition with no abnormalities such as glitches or oscillations.

SERIES	V CC(V)	V IL(MAX)	V IH(MIN)	VT (V)	dt/dv (nS/V)
SN74	4.75–5.25	0.8	2	1.4	15
SN74LS	4.75–5.25	0.8	2	1.4	15
SN74S	4.75–5.25	0.8	2	1.4	15
SN74ALS	4.5–5.5	0.8	2	1.4	15
SN74AS	4.5–5.5	0.8	2	1.4	15
SN74F	4.5–5.5	0.8	2	1.4	15
SN74HC	2	0.3	1.5	1.4	625
	4.6	0.9	3.15	2.25	110
	6	1.2	4.2	3	80
SN74HCT	4.5–5.5	0.8	2	1.4	125
74AC	3	0.9	2.1	1.5	10
	4.5	1.35	3.15	2.25	10
	5.5	1.65	3.85	2.75	10
74ACT	4.5–5.5	0.8	2	1.4	10
SN74BCT	4.5–5.5	0.8	2	1.4	10
SN74ABT	4.5–5.5	0.8	2	1.4	5/10
SN74LV	2.7–3.6	0.8	2	~1.5	100
SN74LVC	2.7–3.6	0.8	2	~1.5	5/10
SN74LVT	3.0–3.6	0.8	2	1.4	10

(a) Required minimum input Rise/Fall time rates for logic families.

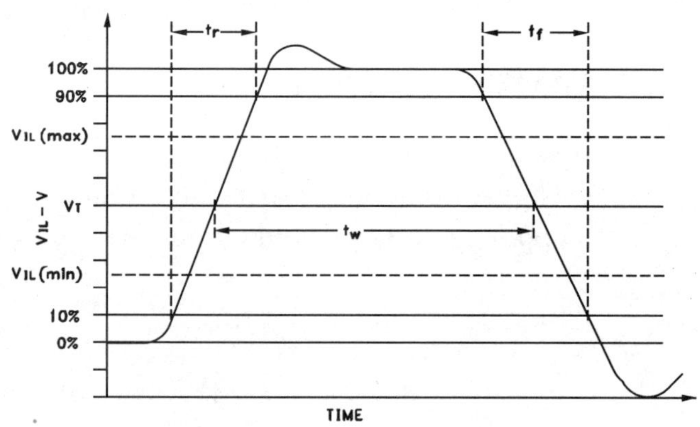

(b) Definition of signal amplitude and pulse width.

Figure G.1. (a) A tabulation of minimum rise and fall time rates for the currently available logic families. (b) Graphical display of pulse timing definitions.

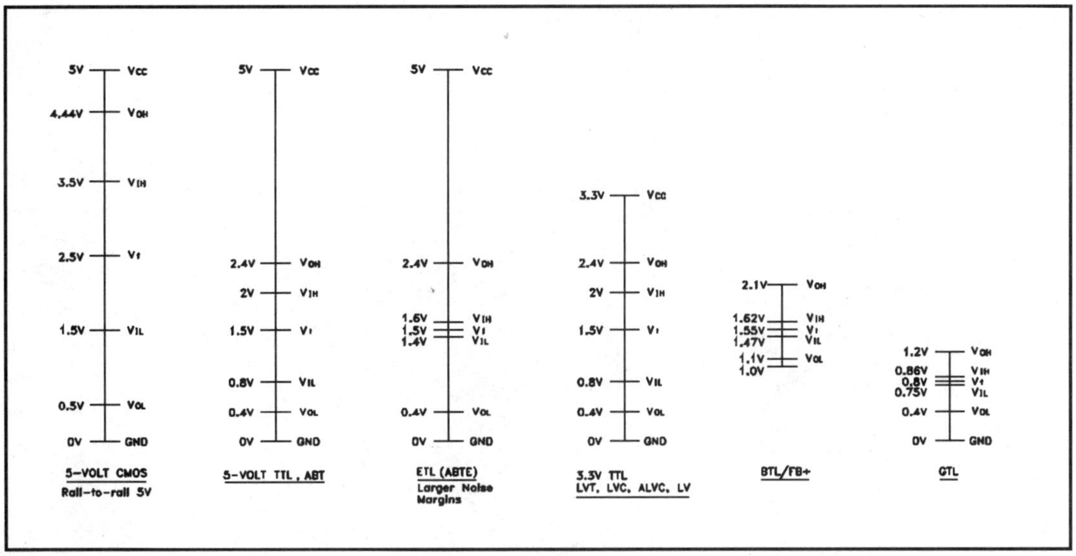

Figure G.2. Logic voltage definitions for the currently available logic families.

Chapter 1

Digital IC Basics

Introduction

In just a touch over half a century ago an infant semiconductor technology emerged from the laboratory. In consequence, to a considerable degree we have witnessed in these few short years an unparalleled revolution in all areas of education, science, engineering, and manufacturing. Our lives have been affected to an extent few of us can even begin to comprehend.

It seems appropriate, therefore, to begin this book with a brief review of the origins of the integrated circuit. The role of the semiconductor in electronics actually goes back somewhat further. There are still old-timers around whose experience in electronics got its send-off with the crystal radio. The crystal was Galena (PbS). Germanium crystals functioned as microwave detectors in World War II. It was however, the invention of the transistor by J. Bardeen and W. H. Brattain in 1948 that launched the semiconductor on its path to dominance in electronics we see in these times.

The function of the crystal radio is that of a semiconductor diode rectifying the radio's carrier wave to reveal the intelligence embedded within its amplitude modulation. It was the study of this phenomenon that led to the initial *point contact transistor*.

The point contact transistor is now relegated to its brief moment in history, but it was an important milestone on the road that lead to its successor: the *junction transistor*. Both transistor types, the point contact and the junction, obtained their names through the process of their manufacture.

The point transistor consists of a small germanium crystal mounted on a supporting fixture with two rectifying point contacts (whiskers) attached close to each other and a single non-rectifying contact at a further distance. This is shown in *Figure 1.1.* A power gain of 20 dB and an upper frequency limit of 10 MHz are typical.

The junction transistor underwent an evolutionary progress that led up to the integrated circuit. Early devices were of germanium, which were less demanding in their manufacture, but also limited in their performance when compared to silicon. Improvements in silicon refining combined with improved construction methods led to higher performance with lowered costs. In the early 1960s silicon transistors were expensive—I recall paying twenty-five dollars for devices now available for around twenty-five cents. *Table 1.1* lists the sequence of manufacturing methods that in time enabled production of the integrated circuit (IC).

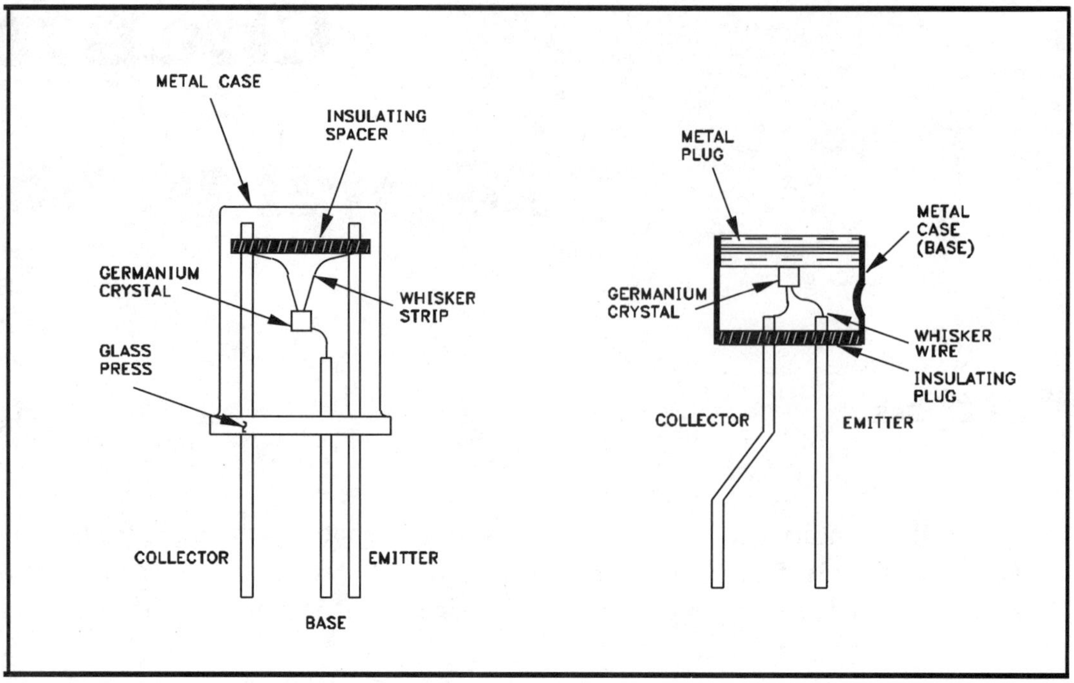

Figure 1.1. Point contact transistor construction examples.

At this point a brief description of semiconductor properties is in order. The prefix *semi-* accurately describes the electrical conductivity of the material. It is neither an insulator nor a good carrier conductor. *Intrinsic semiconductor* material is the absolutely pure element in a rigidly structured crystal lattice. The slightest imperfection in the lattice, referred to as *dislocations*, renders the material useless. The material also exhibits a significant temperature coefficient.

Charge carriers within the crystal structure move about at random. We think of electrons as charge carriers. However the "holes" left behind by wandering electrons are also charge carriers. Electrons carry a negative charge; holes a positive. Since the holes and electrons are equal in number their effects cancel: there is no net charge. This random movement is referred to as a *diffusion.*

To be useful for other than temperature-sensitive resistors, the material must be modified to create one of the two carrier types. This is accomplished by introducing a small amount of a doping element into the lattice structure. If the doping contributes an excess of electrons, the material is said to be of type *n.* If the doping contributes an excess of holes, that is, a shortage of electrons, the material is said to be type *p.*

A small portion of the periodic table is shown below:

Group III	**Group IV**	**Group V**
B (Boron)		
Al (Aluminum)	Si (Silicon)	P (Phosphorous)
Ga (Gallium)	Ge (Germanium)	As (Arsenic)
In (Indium)		Sb (Antimony)

A useful semiconductor device requires a junction between P-type and N-type semiconductor material. This must be mechanically strong with a continuous, undistorted crystal lattice. Here are several manufacturing processes that have been employed:

Alloy Junctions:

A crystal of one material type is grown and the required impurity added. This is crystallized on a prepared crystal base by dipping the base in the melted material and slowly withdrawing and freezing. The grown crystal is then cut into slabs. A button of opposite material type is placed on the slab and the assembly heated in an oven. The button melts and forms an alloy with the slab. On cooling, the alloy crystallizes as an extension of the slab to form a uniform junction.

Grown Junctions:

Initially as with the alloy junction, supposing the material being grown is N-type, P-type material in a greater amount than the N-type is added to the melt. At some point the material makes a changeover to P-type. The junction thus formed is more gradual than with the alloy procedure.

Diffused Junctions:

The procedure starts with a crystal slab of one type, or extremely pure intrinsic material. The impurity is diffused into the crystalline solid. Junctions are forming by changing the diffused impurity type. The process provides for a variety of options, based on the impurities used and the rate of the diffusion.

Deposited PN Junctions:

First developed and patented in 1954, this procedure is a technique for deposition of thin high-quality material films on previously prepared surfaces. The resulting product is one of dramatic improvement over previous techniques. This led the way to the development of the integrated circuit in the early 1960s.

Epitaxial Junctions:

Epitaxy is a Greek derivation meaning "arranged upon." In this process, material is deposited on a prepared substrate in such a way as to build up a crystal structure of the same orientation. The process usually involves exposing a prepared single crystal substrate to a hot gas containing the required impurities. The term *vapor growth* is often applied to this procedure. Technologies based on evaporation and sputtering have also been developed.

Table 1.1. Summary of manufacturing processes developed for semiconductor device fabrication.

Group III elements have three electrons in their outer shell, Group IV have four, and Group V, five. Thus if an impurity such as Boron is infused, holes will be created in the lattice and the material will be type P. With Phosphorous, there will be a surplus of electrons, and the material will be type N. These elements are sufficiently close enough in their physical properties to fit in the crystal structure without dislocations.

As noted in *Table 1.1,* vapor growth is suited to the fabrication of integrated circuits. Understandably, the manufacturing processes are challenging with the requirements for high temperatures, extremely close physical tolerances, and a material quality of an ultra-high purity approaching absolute perfection of the crystal lattice.

Semiconductor devices are described by the number of PN junctions. Thus diodes are single-junction devices, bipolar transistors are two-junction, junction field-effect transistors, two-junctions, and so forth.

Figure 1.2 shows the electrical characteristics of the two-junction PN diode. Note that when the polarity of the battery terminal and the junction are the same—there is a repulsion between the charges that narrows the junction. When the polarities differ, the resulting attraction widens the gap between the two regions.

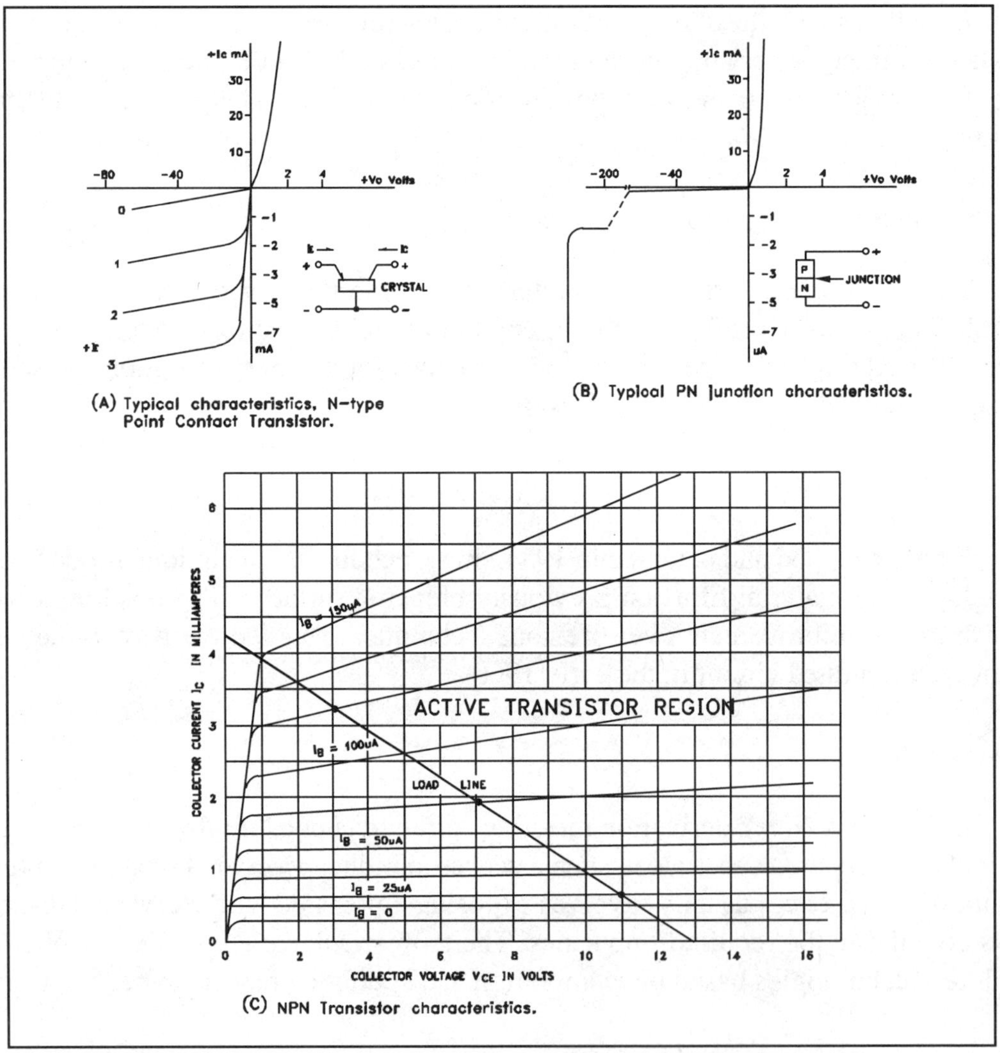

Figure 1.2. (a) Typical characteristics, N-type point contact transistor.
(b) Typical PN junction characteristics. (c) NPN transistor characteristics.

As its name suggests, the junction transistor depends on the junctions of three regions of semiconductor material. The regions are called the *collector*, the *base*, and the *emitter*. Two of the regions, the emitter and the collector are of the same donor type, P or N, but of differing charge densities. The third region, the base, is of the opposite type. *Figure 1.3* illustrates the characteristics and construction of a typical NPN junction transistor. The diode concept shown is illustrative of the action taking place within, but is not a literal reality with discrete devices—we cannot make our own transistors in this fashion.

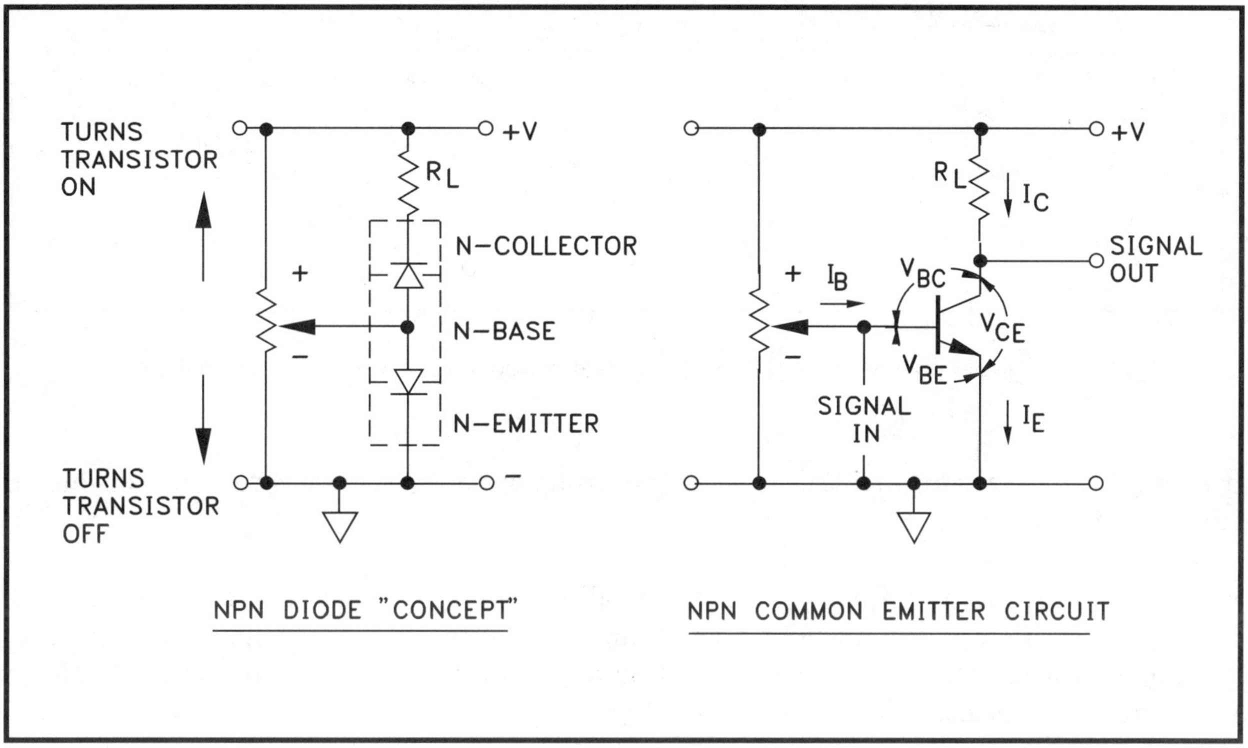

Figure 1.3. Bipolar transistor basic concepts. Though the NPN only is shown, the PNP model is identical with diode polarities and voltages reversed.

Before moving on to the integrated circuit we need to make the acquaintance of the Field-Effect Transistor (FET), shown in *Figure 1.4*, as it occupies an important role in the formation of NMOS, PMOS, and CMOS devices.

The junction FET (JFET) consists of a doped semiconductor bar with a single PN junction formed at its side. The polarity and properties of the junction are such that in the absence of an external potential the junction has no effect on the conductivity of the bar. However, if a repulsive charge is provided, the widening junction narrows the slab column, with a resultant decrease in its conductivity. In this manner, a signal applied to the gate modulates the current flow in the bar, providing amplification. Because the junction is reverse biased in its operation it presents a high impedance to the driving source. In this respect it somewhat resembles the vacuum tube.

The insulated gate, or Metal-Oxide Semiconductor FET (MOSFET) transistor, also shown in *Figure 1.2*, consists of a bar with two differing material types positioned so as to isolate the drain and the source.

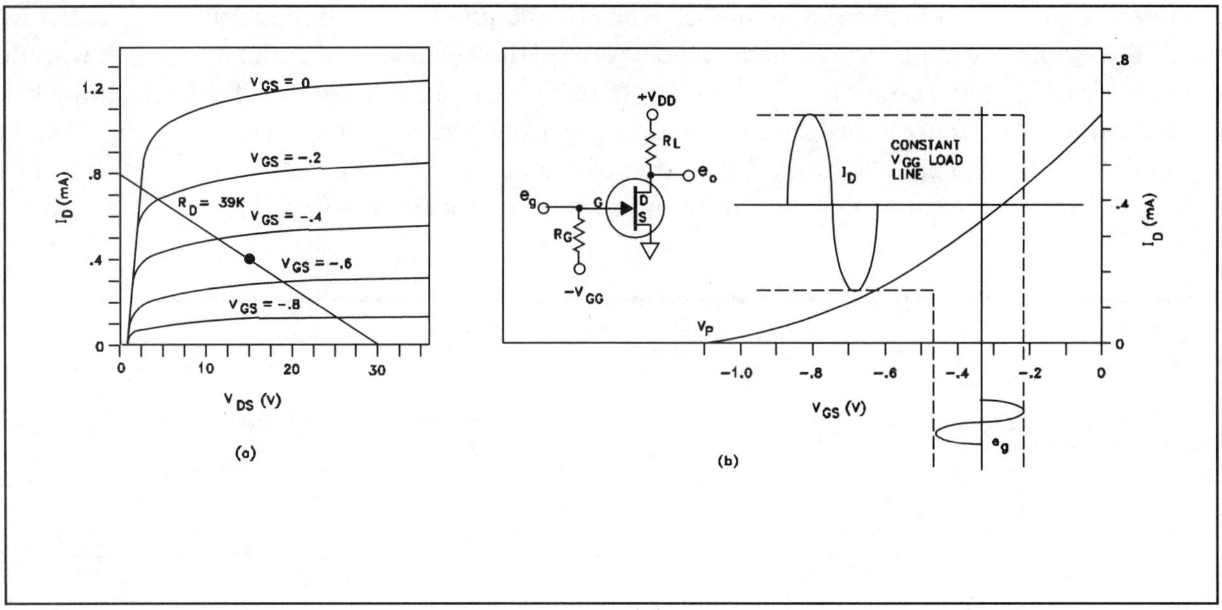

Figure 1.4. JFET constant-voltage biased amplifier, output characteristics and transfer curve.

The bar surface is oxidized to provide an insulated covering. A metal electrode is deposited on this surface to form the gate connection.

In contrast to the JFET the MOSFET is typically normally off, though depletion devices are also available. The insulated gate offers an extremely high input impedance. The frequency response is enhanced by a reduced input capacitance. *Figure 1.5* illustrates an application of the enhancement mode N-channel MOSFET as a low-resistance analog switch.

TTL Families

74 Standard /H/L/S/LSxxx Series Logic

The 54/7400 series of Transistor Transistor Logic (TTL), introduced in 1965, was preceded by several earlier IC families that, with a couple exceptions, did not enjoy a long lived existence. These include *Direct- Coupled Transistor Logic (DCTL), Resistor-Transistor Logic (RTL), Current Mode Logic (CML), Diode-Transistor Logic (DTL),* and *Emitter-Coupled Logic (ECL).* The two significant exceptions were DTL and ECL. *Figure 1.6* shows dual-input gate examples of these early series. Because the families other than TTL are seldom found in current use they are not included in this book.

TTL, derived from the earlier experience with DTL, had several things going for it that enabled its take over of the competition. Two features of importance are the multiple-emitter structure of its input transistor and the totem pole output stage. In its conception, the TTL design was modeled from the DTL with an emphasis on reducing parasitic capacitance at the input. Passive parasitics are effectively reduced by mini-

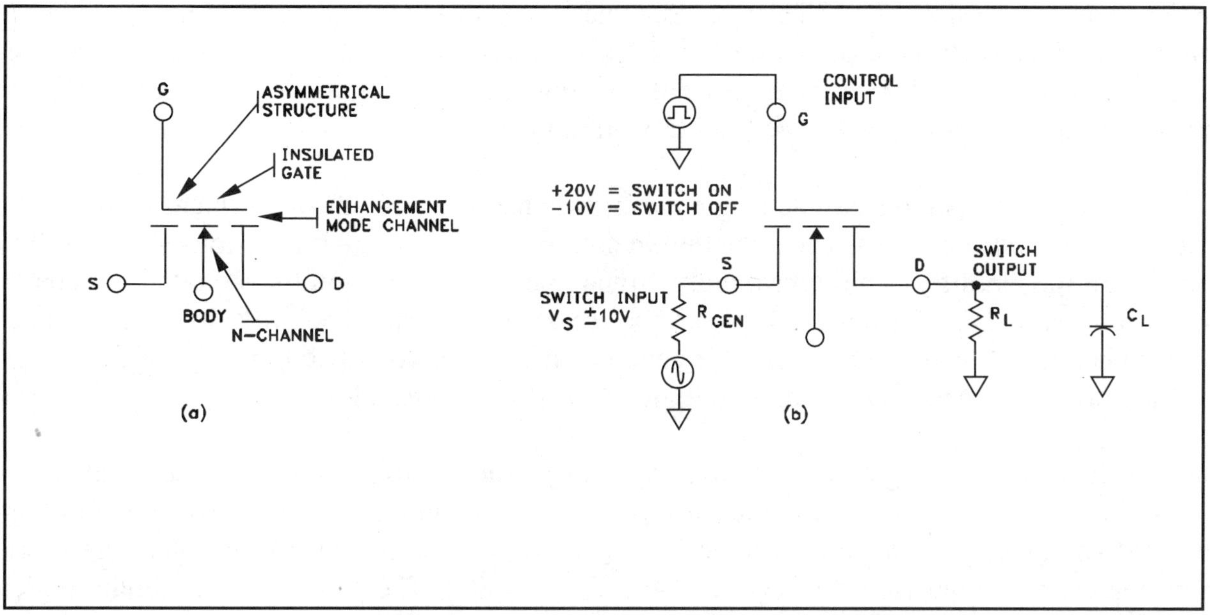

Figure 1.5. DMOS FET switch. (a) Electrical symbol. (b) Application as a ±10 volt analog switch.

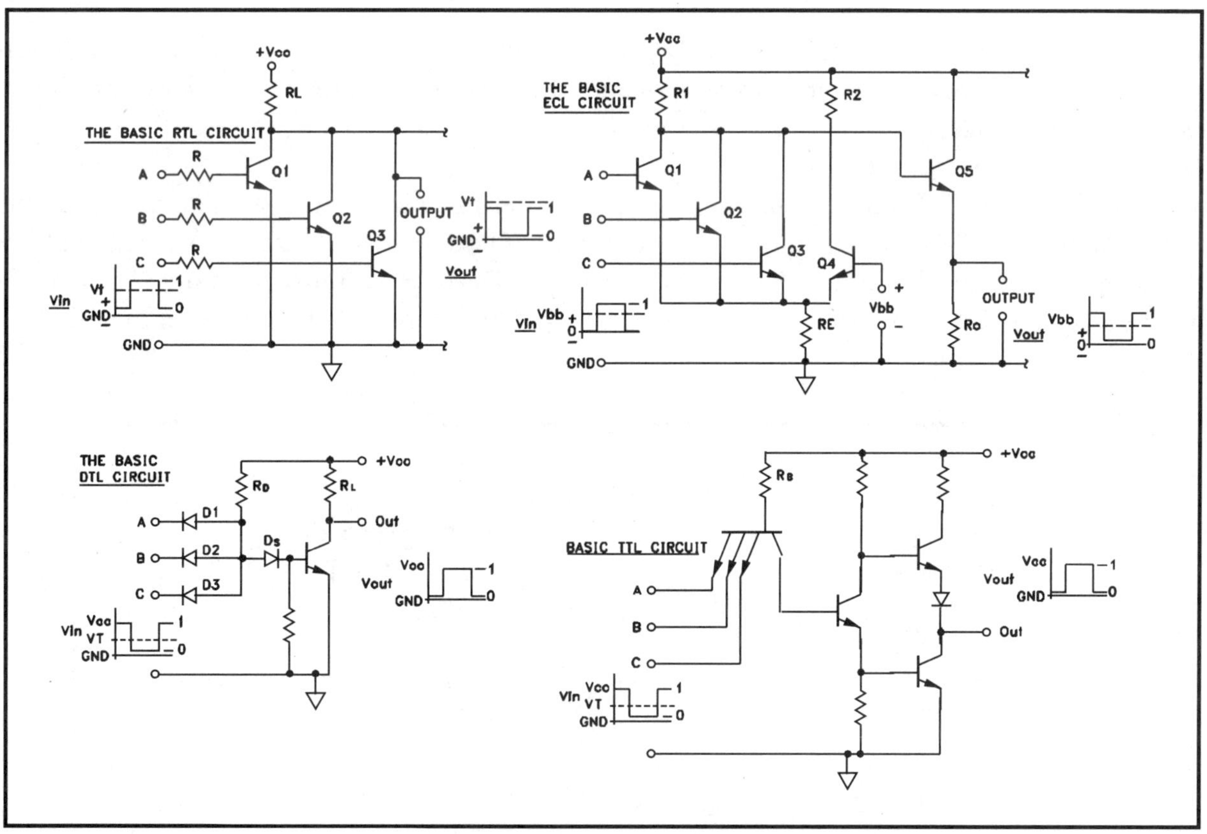

Figure 1.6. Resistor-Transistor Logic (RTL), Emitter-Coupled Logic, and Diode-Transistor Logic paved the way to Transistor-Transistor Logic (TTL).

mizing the device dimensions. The TTL input characteristics closely follow those of the DTL without the pull-up resistor. The totem pole output enables high load current in both current sourcing and sinking modes. Two associated disadvantages are current hogging in the sinking mode and current spiking when both transistors are in the on state for the brief transition time.

Noise immunity and generation assumed a greater importance with the advent of integrated circuits. This arose from several considerations. The reduction in both individual circuit dimensions and overall system size brought all the circuitry in close proximity. Signal, power, and ground lines on densely constructed printed-circuit boards raise the potential for serious crosstalk concerns. The provision of low-impedance ground lines is difficult with the necessity for generous use of bypass capacitors. *Figure 1.7* is a graphic representation of noise margins in relation to logic one and zero thresholds.

To assure a safe operating margin, devices are tested and guaranteed to specific datasheet limits. For 5400/7400 logic, these are specified for a fanout of ten over the recommended ranges of supply voltage and temperature. The guaranteed maximum logical zero output voltage is 400 mV less than the guaranteed minimum logical zero input voltage ($V_{OL} \leq 0.4V$, $V_{IL} \leq 0.8V$). The guaranteed minimum logical one output voltage is 400 mV greater than the guaranteed maximum logical one input voltage ($V_{OH} \geq 2.4V$, $V_{IH} \geq 2.0V$). With these guaranteed levels the gates will operate correctly. The 400 mV margin of safety in the two logic states is termed *the guaranteed dc noise margin.* This means a 54/74 gate output is compatible with a 54/74 gate input when driving as many as ten inputs, with a guaranteed safety margin of 400 mV. Knowledge of the input and output capabilities are essential when interfacing this logic to the logic of another system.

This section groups the standard, high-speed (H), low-power (L), Schottky (S), and low-power Schottky because they have features largely in common. Where they differ is in areas of power consumption and higher speed performance.

The circuitry of the 54H/74H devices differs from the standard by having lower resistance values and the inclusion of damping diodes on each multiple-emitter input. These reduce the transmission line effects that arise at the higher rise and fall times. The output stages are Darlington transistor pairs. These result in a slightly higher speed (6 nS per gate), at a cost of some what higher power consumption over the standard TTL series.

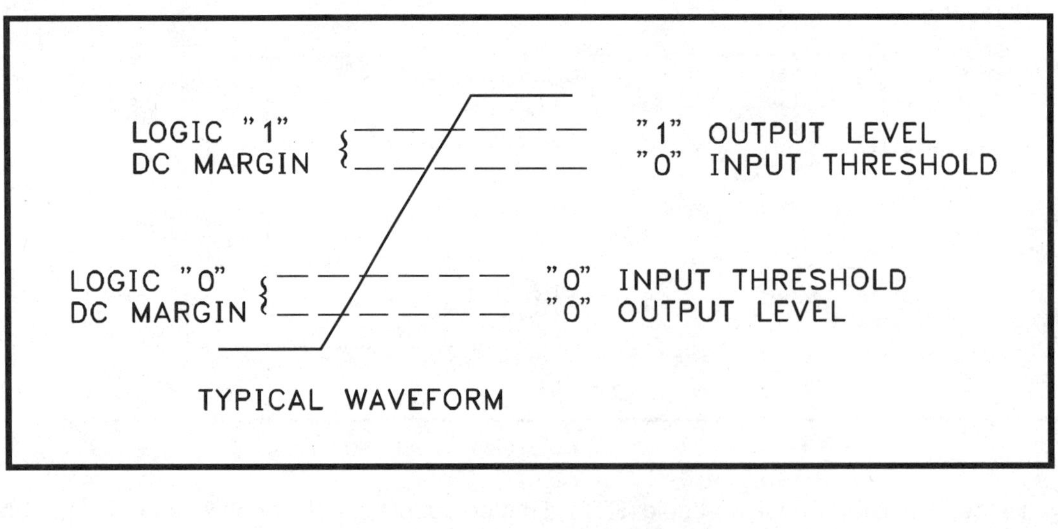

Figure 1.7. Logic level thresholds and DC margins.

The low-power 54L/74L provide the best speed-power product of the logics available at the time of their introduction. The basic gate is essentially the same gate as the standard series, except the resistance values are increased. Power requirements are less than one-tenth that of the standard, with 1 mW per gate. Speeds of 33 nS per gate are said to be typical.

Note that the 54H/74H and 54L/74L are not currently being offered, having been largely supplanted by Schottky clamped (S) and Low-Power Schottky devices (LS). *Figure 1.8* provides a comparison of four gate varieties discussed in this section: standard, low-power, Schottky, and Low-Power Schottky. Of these, Schottky logic has the highest speed, achieved by combining the high speed of unsaturated logic (ECL) with the lower power consumption of TTL. *Figure 1.9* describes the composition of the Schottky transistor. Outstanding features of the series are described as 3 nS average propagation delay time and average power dissipation of 20 mW per gate.

In contrast to the other 54/74 series devices, 54LS/74LS logic does not employ multiple-emitter inputs, as we see in *Figure 1.8 (c)*. This family enjoys a surprisingly five-fold reduction in current and power when compared to the standard TTL, making it truly low powered. The gold doping used with other TTL types reduces switching time at the expense of current gain. In its place the LS family uses the Baker clamp (*Figure 1.9*), which prevents the transistor from going into saturation. By avoiding full saturation the transistor makes a faster recovery. The diode-transistor inputs are faster and provide a higher breakdown voltage. The output structure is a Darlington transistor to increase speed and drive capabilities. Typical device power is 2 mW per gate; typical propagation delay is 10 nS while driving a 15 pF, 2 Kohm load.

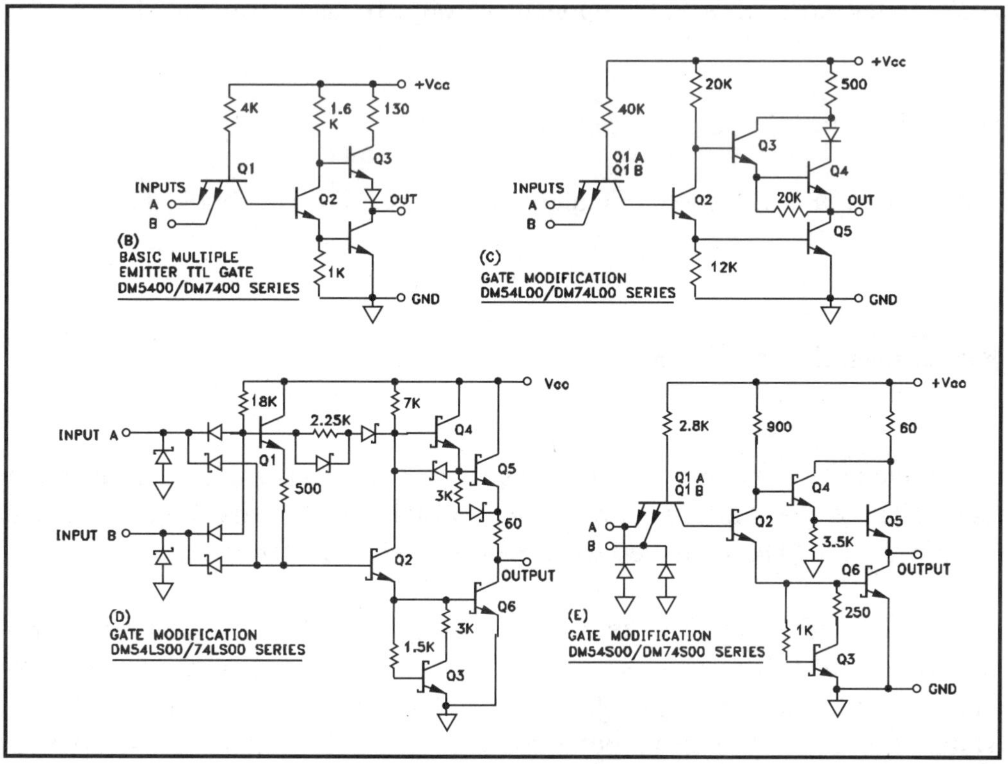

Figure 1.8. 7400 series Standard, Low Power (L), Low Power Schottky (LS), and Schottky (S) TTL gate circuit examples.

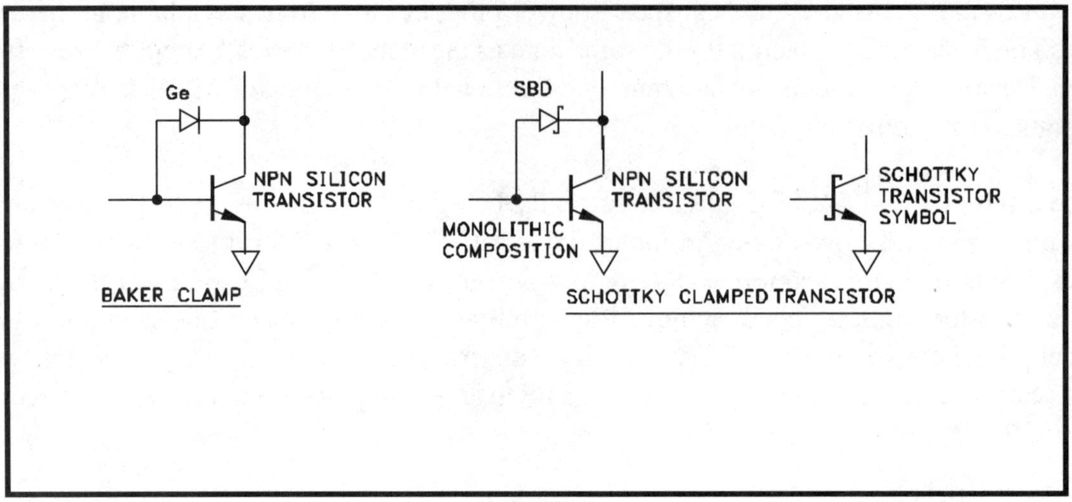

Figure 1.9. Derivation of the Schottky clamped transistor.

References:

Analysis and Design of Integrated Circuits, Engineering Staff, Motorola Inc., Semiconductor Products Division, McGraw-Hill Book Co., 1967

Designing With TTL Integrated Circuits, Components Group, Texas Instruments Inc., McGraw-Hill Book Co., 1971

Application Memos, Signetics Corporation, September 1969

Guide to Bipolar Logic Families, National Semiconductor, ALS/AS Databook, 1987.

TTL Design Considerations

Since TTL is constructed with bipolar transistors, handling precautions are less stringent than with CMOS devices. The *FAST* TTL devices, however, are subject to Electrostatic Discharge (ESD) damage. We and our handling equipment should be grounded when working with this family.

TTL design rules require unused inputs to be tied to either logic high or low, depending on the device and the situation. A 1 Kohm isolating resistor is advisable when connection is to the positive power supply. A number of inputs can be connected to a single resistor. Some suggested connections are: unused active NAND or AND inputs to Vcc, to the used input of the same gate if loading rules will not be exceeded, or to the unused output of another gate that is held high; unused active NOR or OR to ground.

Mixing the slower family members with more recent faster devices can be done, but must be carefully thought out. Each family of TTL devices has certain unique features designed to optimize the desired speed and/or power performance. Higher-speed devices achieve their improved performance in part through internal rise and fall enhancements. The fast transitions can cause an increase in power and ground noise as well as signal noise throughout the system. These can be reduced by shorter signal lines, good ground planes, use of bypass capacitors, and care in PC board layout to avoid parallel lines that can result in crosstalk, and impedance matching to reduce transmission reflections. The best way to protect against these is to prevent them in the first place.

The logic levels of all TTL family members are fully compatible. However, input and output loading rules do vary. *Table 1.2* is a comparison of load driving capabilities. Input and fanout loads in TTL are specified in unit loads as well as actual loads. Standard TTL fanout is ten unit loads; 74F TTL fanout to standard inputs is 12.5 unit loads. The upper row shows the actual value of the unit load for the types listed.

Totem pole outputs cannot be connected to each other. Only open-collector and 3-state outputs can be so connected. Open-collector outputs require a pull-up. The pull-up can be calculated from:

$$R(min) = (Vcc(max) - V_{OL})/(I_{OL} - N2(I_{IL}))$$

$$R(max) = (Vcc(min) - V_{OH})/(N1(I_{OH}) + N2(I_{IH}))$$

where:

I_{OL} = Minimum I_{OL} guarantee for OR-tied elements.

$N2(I_{IL})$ = Cumulative maximum input low current for all OR-tied inputs.

$N1(I_{OH})$ = Cumulative maximum output high leakage currents for all OR-tied outputs.

$N2(I_{IH})$ = Cumulative maximum high leakage current for all OR-tied inputs.

DRIVEN DEVICE FAMILY		74F	74F(NPN)	74LS	74	74S	8200 9300	82S00
Driving Device Family	I_{OL} (MIN)	I_L (MAX)						
		0.6mA	20uA	0.4mA	1.6mA	2.0mA	1.6mA	0.4mA
		Maximum Number of Loads Driven						
74F	20mA	33	1000	50	12.5	10	12	50
74F(NPN)	64mA	100	3200	160	40	32	40	160
74LS	8mA	13	400	20	5	4	5	20
74LS BUFFER	24mA	40	1200	60	15	12	15	60
74	16mA	26	800	40	10	8	10	40
74 BUFFER	40mA	78	2400	120	30	24	30	120
74S	20mA	33	1000	50	12.5	10	12	50
74S BUFFER	60mA	100	3000	150	37.5	30	37	150
8200/9300	16mA	26	800	40	10	8	10	40
82S00	20mA	33	1000	50	12	10	12	50

Table 1.2. TTL family member loading comparisons.

3-state outputs are designed to be tied together, but are not intended to be active simultaneously; only one 3-state output should be active at one time. Output enable signals should not be overlapping. Most output enable inputs are active low, shift registers or edge-triggered storage registers are good output enable buffers. Shift registers such as the 74x164, '194 with one circulating low bit are ideal for sequential enable signals.

Reference:

Signetics Corporation, *FAST Data Manual*, 1987. "Design Considerations," pp 4-12 - 14.

74ALS/AS Series Logic

Advanced Low Power Schottky (ALS)

One of the most advanced of the TTL series, 74ALS devices deliver twice the data handling efficiency while saving up to 50% in power usage compared to the 74LS series. This is possible because of newer fabrication processes, whereby component isolation is obtained by a selectively grown thick-oxide rather than the P-N junction used in conventional manufacturing. A contributing factor is improved circuit design yielding smaller component geometries, shallower diffusions, and lower junction capacitances. This enables a doubling of the switching speeds while offering lower operating currents.

In addition to the pin-to-pin compatibility with corresponding family devices, many of the MSI and LSI devices are offered in high-density, 24-pin 300 mil DIP. This enhances the cost effectiveness with advantages of reduced component count, reduced PC board real estate, and increased power-speed performance.

The basic ALS gate is similar circuit-wise to the LS gate. It consists of either PNP transistor or diode inputs, Darlington transistor pair pull-up and active pull-down (squaring network) at the output. ESD protection is provided by a base-emitter shorted transistor at the input for rapid discharge of high-voltage static electricity. Also the inputs and outputs are clamped by Schottky diodes to prevent excessive swinging below ground level. A buried N+ guard ring all around input and output structures prevents crosstalk. The ALS family features yields a typical power dissipation of 1 mW per gage and a typical propagation delay of 4 nS into a 50 pF/2 Kohm load.

Figure 1.10 provides circuit schematics of the ALS and the AS modified gates.

Advanced Schottky (AS)

The AS family has virtually the same circuitry as the ALS. It features PNP transistor or diode inputs with electrostatic protection base-emitter shorted transistors. The output totem-pole consists of a Darlington transistor pull-up and an active pull-down squaring network. Inputs and outputs are Schottky clamped to

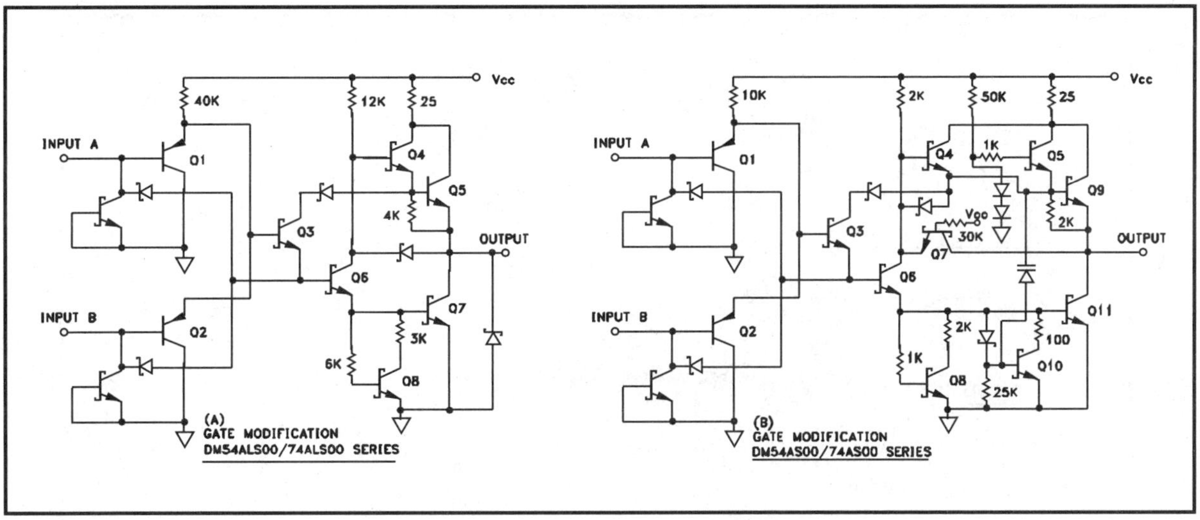

Figure 1.10. 7400ALS/AS gate circuit examples.

attenuate transmission reflections. The AS family yields typical power dissipation of 7 mW per gate and propagation delay time of 1.5 nS when driving a 50 pF/2 Kohm load.

Reference:

Guide to Bipolar Logic Families, National Semiconductor, ALS/AS Databook, 1987, p1-4.

74Fxx Series Logic

There are three different circuit types used in the input of FAST logic devices: these are diffusion diode; PNP vertical transistor, and NPN transistor. These are shown in *Figure 1.11.*

The diffusion diode input is the most often used. Though one diode is shown in the drawing, there may be more, as with the NAND gates. In the oxide-isolated processes these are base-collector diffusions. Each input pin also has a Schottky clamp diode. This diode, standard for most TTL circuits, is provided to limit negative input voltage excursions arising frequently from undershoots. With the input low, current flows from the positive supply through R1 and D1 to the terminal. With a logic one at the input, the current flow is through R1 to Q3, Q2, and Q1. At this level current from the source is limited to the very minute leakage through the diodes.

The PNP vertical transistor has been widely used in low-power Schottky (74LS) because it presents a high impedance to the source. It was not initially used with the FAST devices due to manufacturing difficulties which have since been resolved. As seen in the figure, Q3 functions as an emitter-follower with the input pin at logic low. On the low-to-high transition Q3 is cut off; the internal current is then directed to Q2 via diode D1. Schottky diode D3 speeds up the high-to-low transition if the pin voltage falls more rapidly than the base of Q2, otherwise D3 is off.

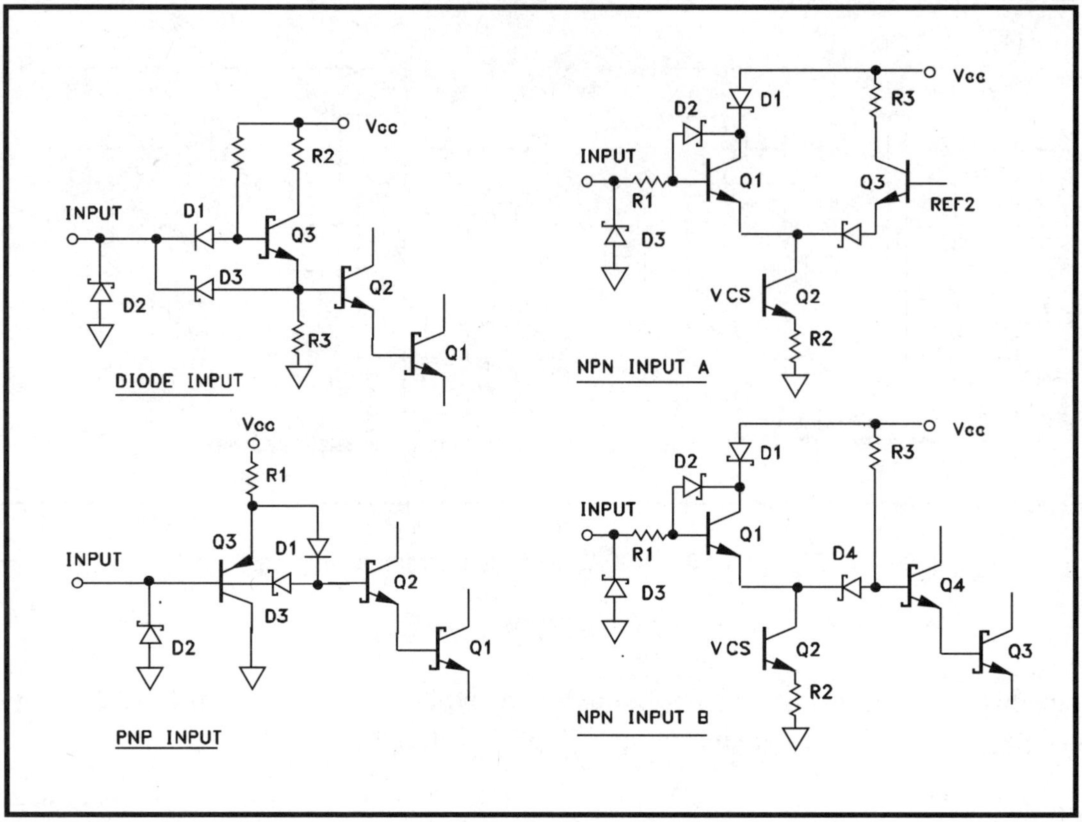

Figure 1.11. 7400 FAST series input circuit options.

The NPN input is shown with two versions. This circuit has limited use in standard TTL circuits. It is used with FAST logic for its superior high input impedance where this is of importance. When compared with the diffusion diode there is a much reduced input current with a logic low. From zero volts to the switching threshold transistor Q1 is off and the input current is very low. As the input rises Q1 is turned on, back-biasing Schottky diode D4 and shutting off Q3 (input a) or Q4 (input b).

Figure 1.12 shows the output stage for two versions: that which provides 3-state control, and that which does not. The basic totem-pole structure appears in both versions, with the difference appearing in the preceding region.

All the design considerations described previously must be scrupulously observed with this very high-speed sensitive logic series.

Reference:

Signetics Corporation, *FAST Data Manual*, 1987. "Design Considerations," pp 3-3, 3-4, p 3-8.

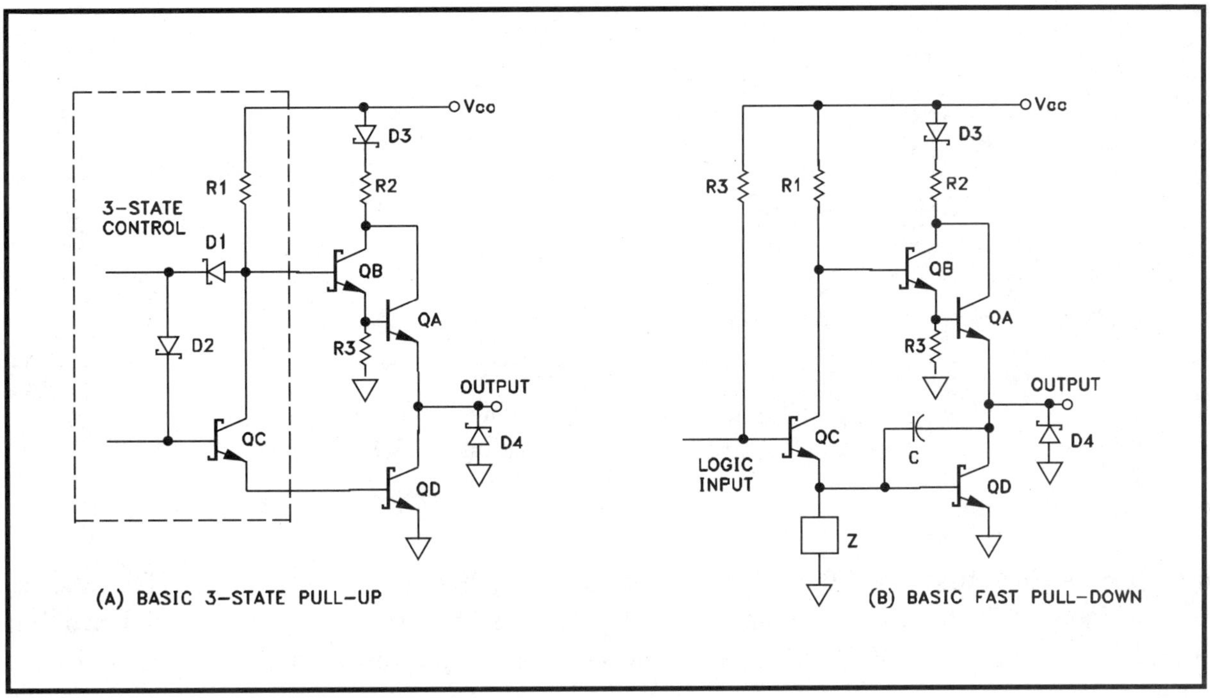

Figure 1.12. 7400 FAST series output circuit options.

CMOS Families

Introduction to CMOS

The metal-oxide-silicon FET (MOSFET) design is based on the fact that it is not necessary to form a semiconductor junction on the channel of a FET to achieve gate control of the device. It turns out that a metallic or polysilicon gate may be isolated from the channel by a thin layer of silicon dioxide (i.e., glass). Using an N-channel device for a model, the insulating material is in contact with the P-type silicon substrate. Free electrons accumulate in the interface, inverting the P-type to form an N-type channel. The effect created is a conduction channel between the diffused N-type channel source and drain regions.

JFETs are inherently depletion devices; that is, they operate in a pinched-off mode with conductivity modulation from the signal source. A MOSFET can be manufactured to function either in this, or its opposite, the enhancement, mode.

The most elemental CMOS circuit, the inverter, consists of one P-channel and one N-channel enhancement mode devices connected in a series arrangement, as seen in *Figure 1.13.* In this configuration the lower NMOS transistor, Q1, is switched while the upper PMOS, Q2, performs as an active fixed load. When the signal turns Q1 on, current flows through Q2 to the load. Because Q1 includes a small resistance the output can approach zero, but not actually drop to it. When the input signal is low (ground), Q1 is off, and Q2 turns on. If the load is another MOS gate, the current drain is minute, allowing the output to closely approach the positive supply.

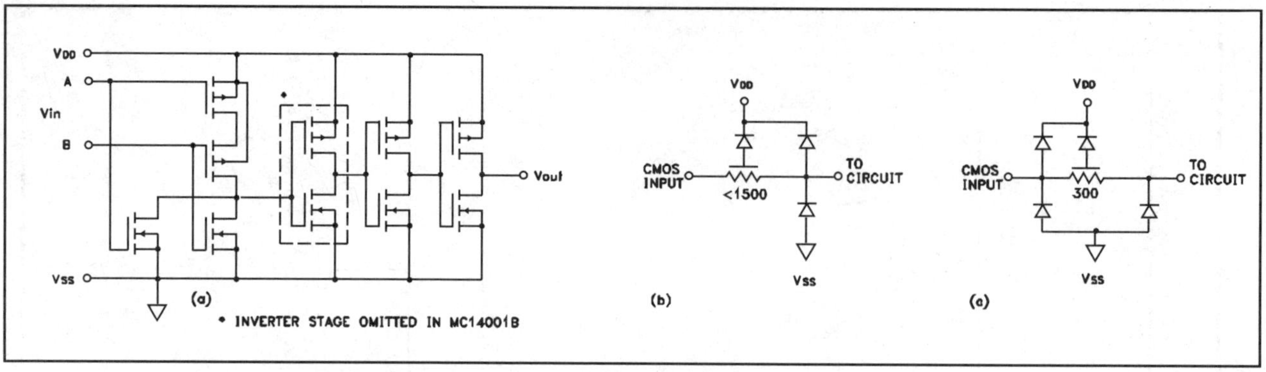

Figure 1.13. (a) The MC14001 Buffered NOR gate (one of four). (b) Two examples of Electrostatic Breakdown (ESD) Protection.

When CMOS devices first made their appearance the positive supply was labelled VDD; the negative, ground, was labelled VSS. Texts and data books of the '70s and '80s use this notation. Later publications use VCC and Ground. Most CMOS ICs employ the same power pins as the popular TTL devices.

The high current logic families frequently require switching regulators to maintain their regulation. CMOS with lightly loaded devices is less dependent on high-current power supplies. The devices function well with batteries and even unregulated power supplies, provided the maximum supply limits are not exceeded.

Electrostatic Protection Needs

The very high input impedance of MOS devices makes them vulnerable to destruction from electrostatic voltages. Walking across a carpeted room on a day with low humidity can generate a static potential of thousands of volts. Even a gate designed to withstand a one hundred volt potential will cry uncle. The early CD4000 "A" series did not feature input protection and were particularly vulnerable. Current "B" and "UB" devices do provide protection. A couple Motorola examples are shown in *Figure 1.13(b)*. Manufacturers are also providing output circuit protection.

The basic protection against ESD is careful grounding—of ourselves, in handling the parts, in storage, of our bench-top equipment, in assembly and testing operations. Unfortunately, a device does not necessarily show catastrophic failure when exposed. It may be weakened, continue to function for a time, then fail.

"B and "UB" CMOS Devices

CMOS ICs with these designations meet the EIA/JEDEC CMOS industry specifications. These include the following features:

- 3- to 18-volt operational limits.

- Capable of driving two low-power TTL loads or one low-power Schottky load over the rated temperature range. "B" series devices are guaranteed to sink a minimum of 0.36 mA over the temperature range with a five-volt supply.

- Direct interface to high-speed CMOS.

- Maximum input current of ±1 mA with a 15-volt supply over the temperature range.

- Parameters specified at 5, 10, and 15 volt supply. Note that some devices, notably 74C/HC, specify 4.5 volts rather than five. CMOS speed improves significantly at the higher voltages. The device data provided in this book for CD4000/4500 and 74C/HC FAMILIES use the 4.5/5.0-volt speed data throughout.

Noise margins have been described in the Glossary.

Figure 1.14 illustrates the primary difference between "B" and "UB" devices. Buffered devices, as shown in *1.14(b)* have two inverter stages preceding the output pin. The unbuffered do not have this circuitry. This is a good news and bad news situation. The bad news is that the unbuffered provide less gain and a less than ideal transfer characteristic The good news is they perform better when used in a "linear" mode for oscillators, monostables, or amplifiers. The decreased gain yields better stability and a cleaner output wave. Also, "UB" gates and inverters offer higher speed since there is but a single stage.

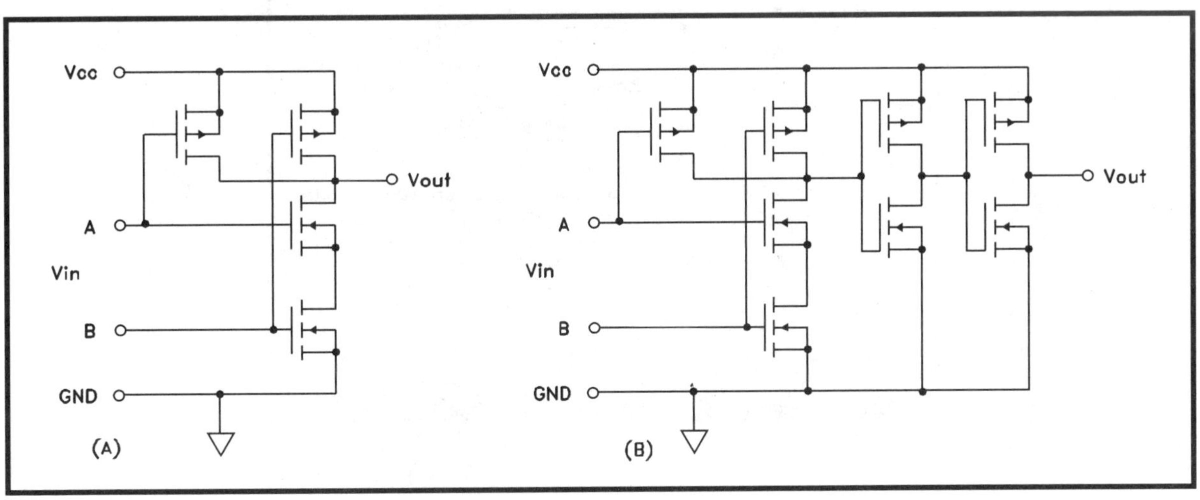

Figure 1.14. (a) Unbuffered NAND gate. (b) Buffered NAND gate.

Latch-Up

Figure 1.15(a) identifies the presence of a bipolar parasitic associated with the CMOS inverter. *Figure 1.15(b)* shows a typical cross-section in which the parasitics reside. Those familiar with this circuit recognize it as a breakdown structure similar to the Silicon-Controlled Rectifier (SCR). Which means that once triggered, there is no way to continue operation with this device: it is locked up. The device may or may not be destroyed, depending on power supply current limiting. Factors that may lead to latch-up include:

- Exceeding the maximum rated limits on input and output signals. For instance, we should never connect external powered-up instrumentation such as pulse generators to a CMOS input until its circuit is energized.

- If severe input or output transients are a possibility, employ voltage clamps such as external diodes or resistors to reduce the threat.

- Sequence power supply turn-on so that inputs or outputs are not activated before the power is applied.

- Use on-board regulating and filtering to hold down noise from outside interference.

- Employ power supply current limiting within the range at which latch-up may occur.

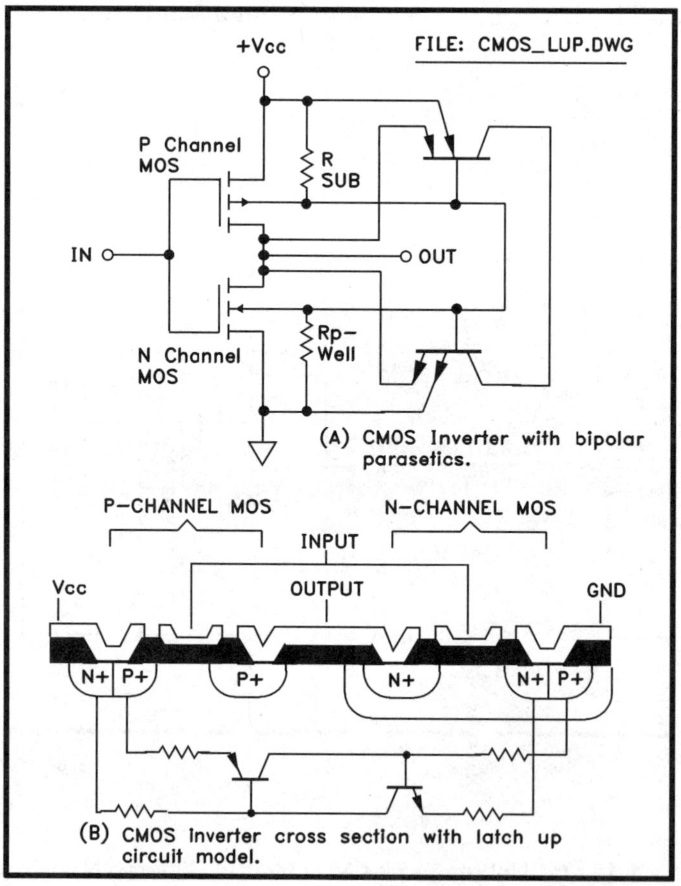

Figure 1.15. (a) CMOS inverter with bipolar parasitics revealed.
(b) The CMOS inverter layup cross-section with the latch-up model circuit.

CD40xx/45xx Series Logic

The CMOS 4000 series of digital integrated circuits was introduced by RCA in 1968. It rapidly gained a large following with a wide selection of functions and the rapid availability of alternate sources. Several manufacturers introduced proprietary devices under the 4000 designation; Motorola, for example, with its MC140xx series, its MC145xx series of MSI devices, and MC144xx series of monolithic MSI and LSI communications devices. Philips markets their series as HEF/HEC4000. Where the HEF series has a temperature range of -40 to +85°C, the HEC series offers the full range of -55 to 125°C.

The family has enjoyed rapid improvement in its early years in its input, output, and internal transfer characteristics. A major step was the introduction of the buffered, “B” series as described in the previous sections.

Another CD4000/4500 series source is National Semiconductor Corporation, which provides these along with their proprietary 54C/74C family.

Unlike TTL logic, all CMOS flip-flops are edge-sensitive; information is transferred only on the positive leading edge.

The 74C/HC/HCT Series

In the early 1970s National Semiconductor introduced their 54/74C family of CMOS ICs. These were modeled on the existing 54/74TTL families. An intent was to duplicate the LSTTL pinouts to facilitate interchangeability. (For the most part this is true of the standard series; even so, check it out. Don’t plug a 74C86 into a socket occupied by a 7486—if a 74LS86, it’s OK.)

National claims a fifty percent speed and sink current advantage for the 74C series over the CD4000 series. The series is spec’d at TTL levels as well as CMOS. The basic circuitry appears to be pretty much identical between the two series. As with the CD series, both buffered and unbuffered gates and inverters are offered.

There are some differences when interfacing 4000 series and 74C series to other logic systems. 74C devices can be driven directly by TTL and DTL over the commercial temperature range without the need for pull-up resistors. 4000 series have no guarantee that they will operate correctly in this mode. If in doubt in any situation, use a pull-up resistor. Interfacing to PMOS or NMOS series can be particularly tricky. It is best in situations such as this to refer to the data manual for the manufacturer’s devices.

The 54/74HC family of high-speed logic devices is uniquely characterized for speed and power consumption in comparison to any bipolar or other CMOS families. Operating speeds are similar to low-power Schottky TTL (54/74LS) technology. Speeds are about half as fast as the 54/74ALS and 54/74S families. Compared to 4000 and 54/74C this is an order of magnitude improvement. The devices provide the functions and pinouts of the 54/74LS logic components. Many of the CD4000 metal gate series have been implemented in this technology as well. Although there are many similarities with the 54/74LS series, differences do exist. The manufacturer’s data manual should always be consulted where these could be critical to the performance. This data manual also provides a thorough analysis of the latch-up problem with 54HC/74HC logic.

Figure 1.16 illustrates the assembly of basic HC devices from assemblies of inverters, transmission gates, and logic gates.

The 54/74HCT logic family is a subgroup of the 54/74HC logic series. The two are nearly identical except that the HCT devices are tailored to be compatible with TTL logic levels. One precaution stands out: maximum power supply voltages are those of TTL, not CMOS. The HCT series was created to resolve interfacing difficulties with other logic families such as TTL, NMOS, and various other recent proliferations. Many of these difficulties arise with NMOS microprocessor and bipolar peripheral interfacing.

Figure 1-17 illustrates some interfacing approaches. A fanout table for MM54HC/MM74HC devices to TTL family logic is included.

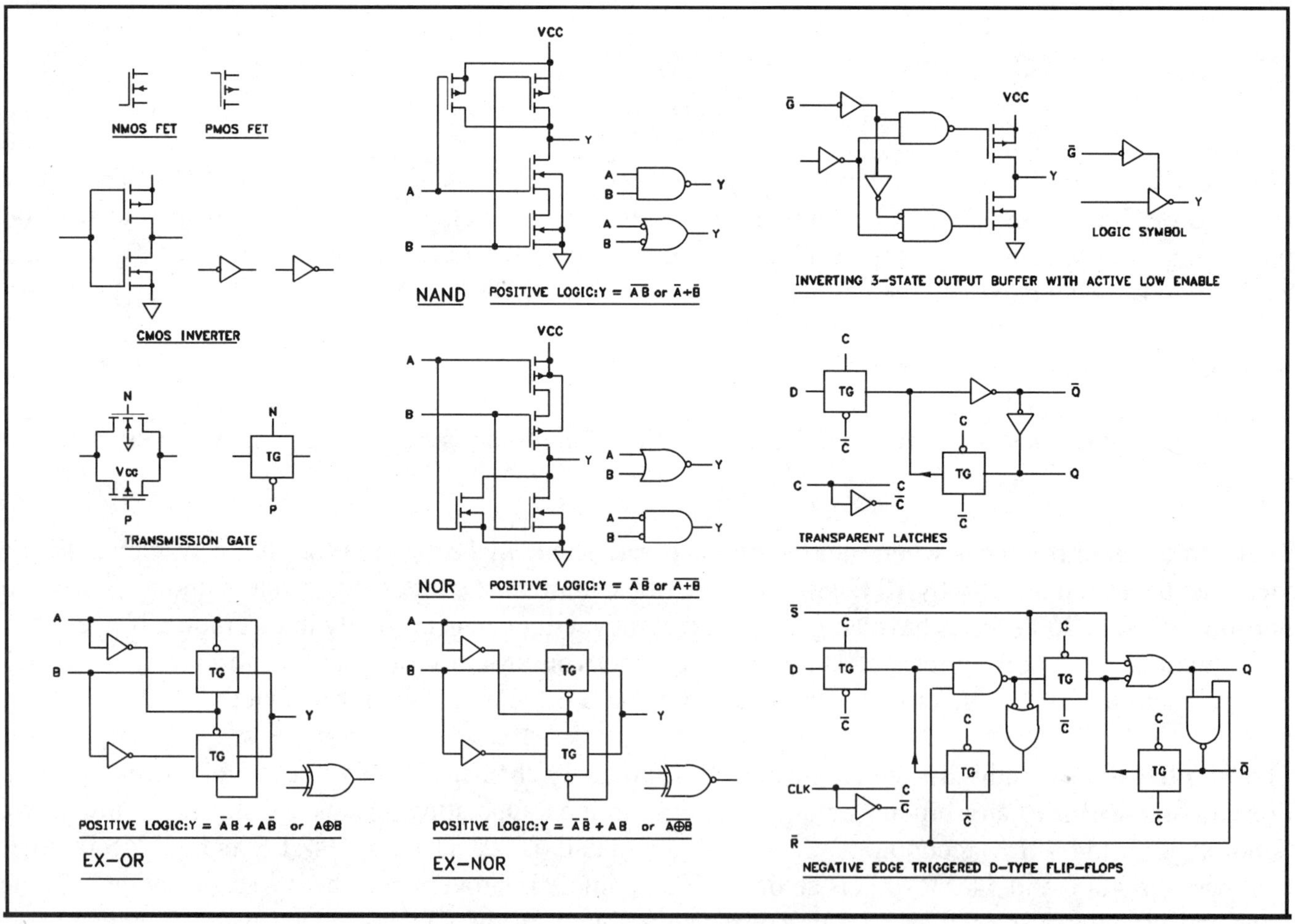

Figure 1.16. 74HC CMOS circuit construction examples.

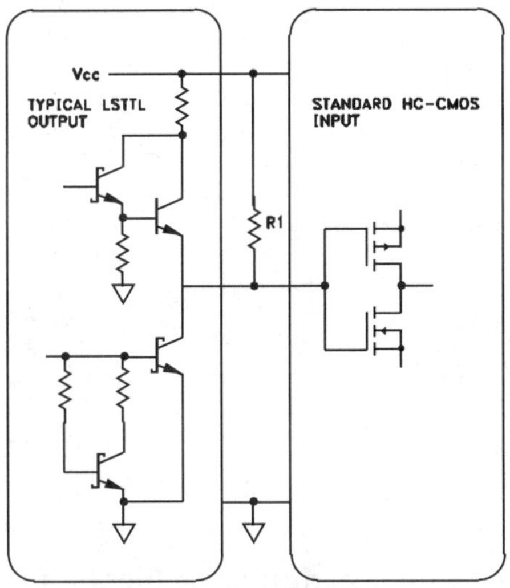

(A) Interfacing LSTTL outputs to standard HC-CMOS inputs using a pull-up resistor.
p 5-89

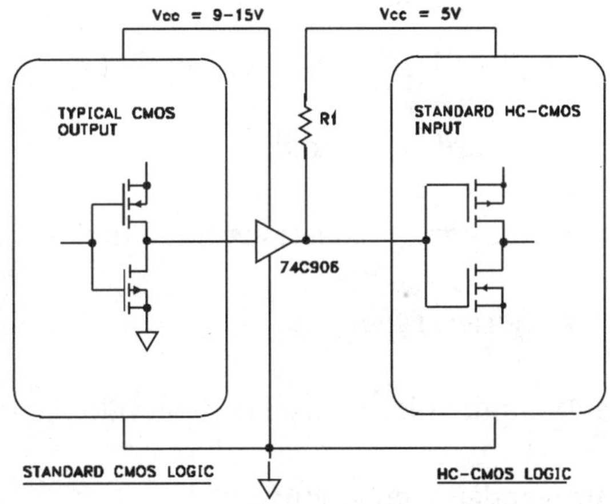

(C) CD4000 or 74C series CMOS to HC-CMOS connection with logic level conversion using an open drain CMOS.
p 5-95, FIG 11(b)

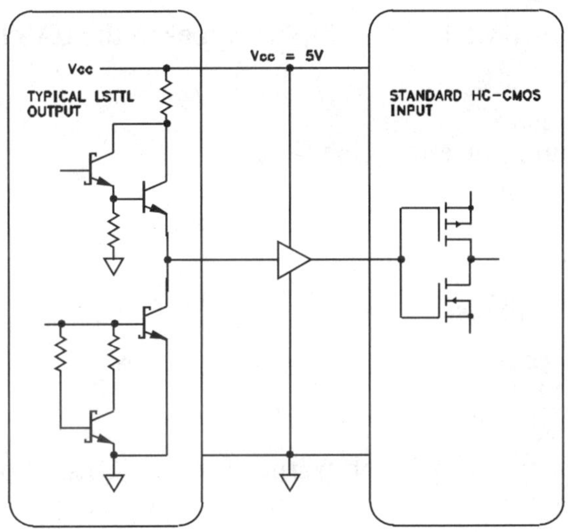

(B) LSTTL output directly drives 54HCT/74HCT logic which can interface to 54HC/74HC.
p5-90, FIG 4

EQUIVALENT FANOUT CAPABILITIES OF HIGH SPEED CMOS LOGIC

HC-CMOS EQUIVALENT FANOUTs	LSTTL		TTL		S-TTL		ALS-TTL	
	MIN	TYP	MIN	TYP	MIN	TYP	MIN	TYP
Standard Output MM54HC/MM74HC	10	20	2	4	2	4	20	40
Bus Driver Output MM54HC/MM74HC	15	30	4	8	3	6	30	60

P5-91, FIG. 6

Figure 1.17. Some 74HC/HCT LSTTL/HC-CMOS and standard CMOS/HC-CMOS circuit interfacing examples.

Advanced CMOS Logic ICs (54/74AC/ACT)

This advanced series of CMOS devices is characterized as comprising a broad range of logic types comparable in performance and speed with FAST, Advanced Schottky (AS), and Schottky (S) bipolar types. Superiority is claimed with respect to requiring substantially less power. Features of the two logic types include:

CD54/74ACTxxx Series Types:

- TTL input voltage-level compatibility.
- Same standardized pinouts.
- Significant power reductions compared to similar bipolar families.

CD54/74ACxxx Series Types:

- CMOS input voltage-level compatibility.
- Same standardized pinouts.
- Enhanced system performance at speeds similar to the high-performance bipolar families.

AC/ACT Family Features:

- Functionally and pin compatibility with industry 54 and 74 bipolar types in the FAST, AS, and S series.
- CMOS rail-to-rail output swing for maximum noise margins.
- Fanout over-temperature:

  2400 AC/ACT loads; 15 FAST loads; 48 AS loads.
- Balanced propagation and output transition times.
- Outputs reliably drive 50-ohm lines (at 85°C) and 75-ohm lines (at 125°C) without need for terminations.
- CMOS input compatibility.
- Octal types typically feature a 1-volt peak (DIP package) simultaneous switching voltage transient, similar to FAST series. Peak is typically 0.8 volt in the small-outline packages.

Electrostatic breakdown is an even greater concern with these very high-speed advanced devices. *Figure 1-18* shows an approach to providing input and output protection. The primary protection device is a low-voltage-triggered silicon rectifier (LVTSCR). In an ESD occurrence, most of the current is diverted through this. Additional protection is afforded by clamp transistors. Texas Instruments claims immunity to over 2000 volts in the human-body-model test, over 200 volts in the machine-model test, and over 1000 volts in the charge-device model test.

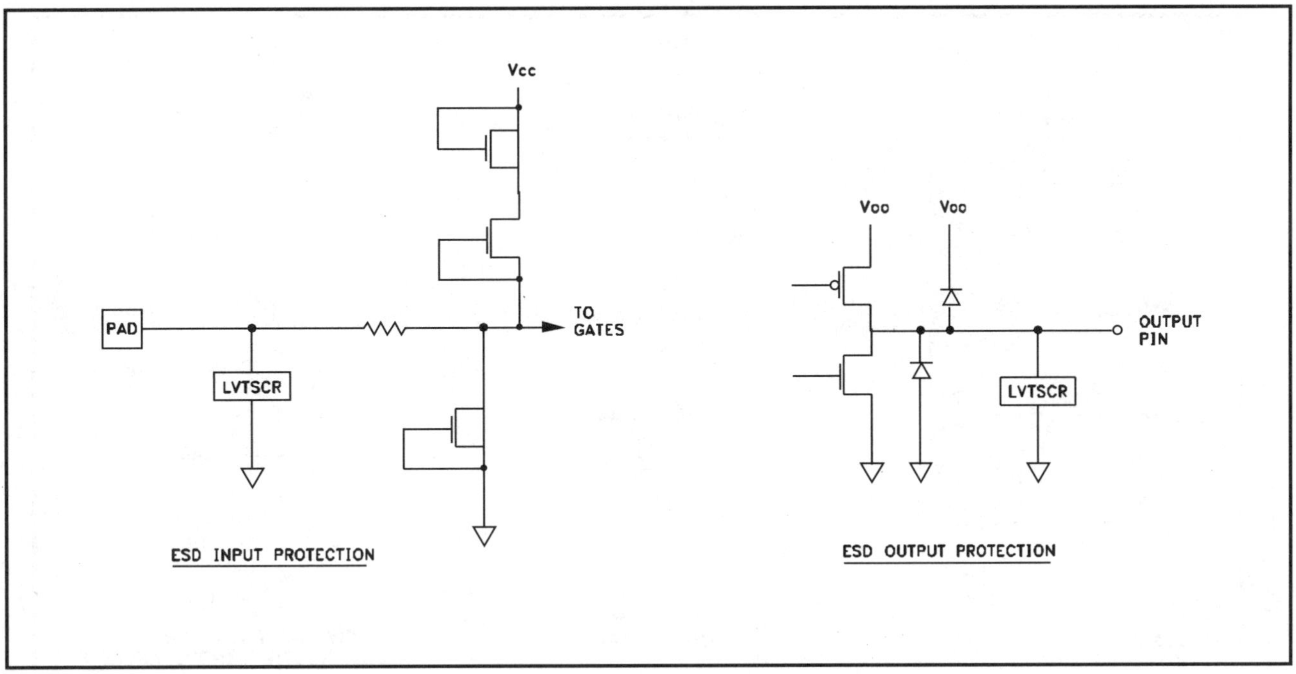

Figure 1.18. Input and output circuit ESD protection for advance high-speed CMOS.

Glossary Figure 3.1 provides a tabulation of high and low input and transition voltages for the various logic families. For the AC/ACT series, these are given for three values of supply voltage. The input circuitry of these devices are sensitive to voltage levels. The construction is such that the only input current is the reverse diode leakage, on the order of a few picoamperes, of the input electrostatic protection. The actual shift of the input voltage over the temperature range of -55 to +125°C is an impressive 100 millivolts.

Figure 1-19 is a comparison of two types of AC/ACT output circuitry with a TTL FAST version. The very fast switching capability of these series introduces a potential for "ground bounce" potentials. The source for these are the inductances shown in *Figure 1.19*. The effect of the bounce is to reduce the input noise immunity with the result that data stored in flip-flops or latches could be upset.

Advanced High-Speed CMOS (AHC/AHCT)

This series is a Texas Instruments advanced high-speed logic family intended to meet the need for more speed with lower power, low-drive applications. The family consists of gates, medium-scale integrated circuits, and octal functions fabricated using EPICTM process featuring higher performance over the HCMOS HC product family.

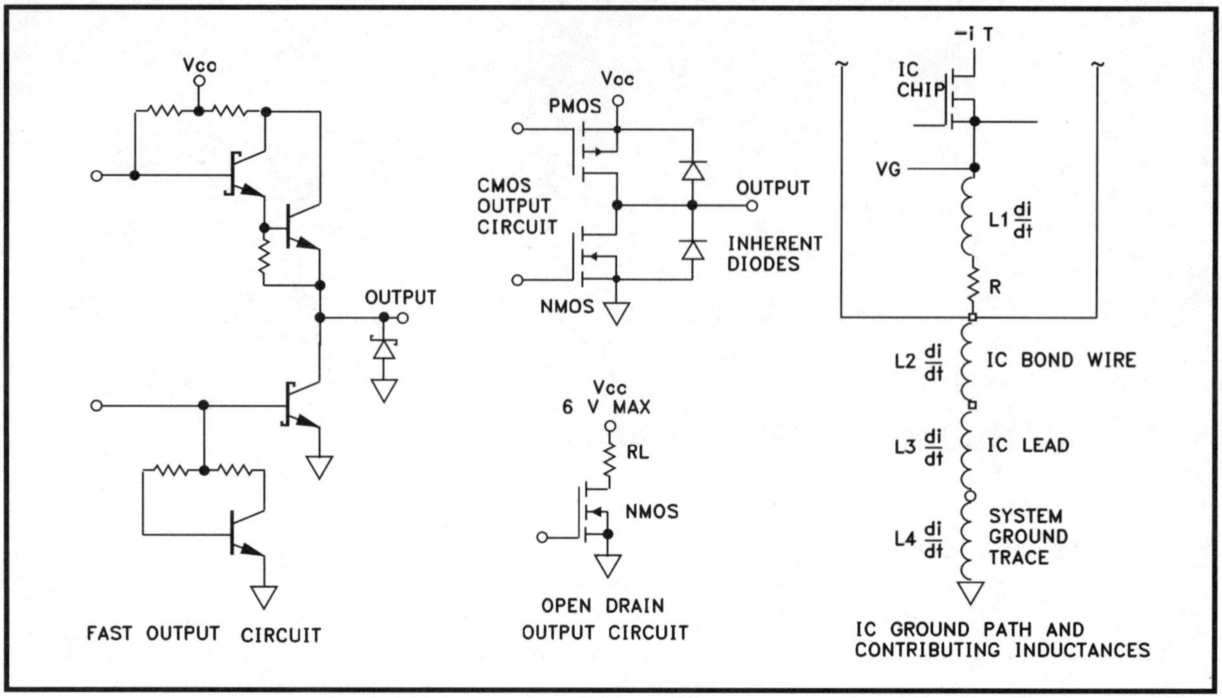

Figure 1.19. Comparison of CMOS AC/ACT output circuitry with TTL FAST.

Some advantages claimed for AHC over HC are:

- Almost three times faster than the HC family. A typical propagation delay of about 5.2 nS
- Combines the low-noise characteristics of HCMOS devices with its performance levels without the under/overshoot problems typical of higher-drive devices.
- Lower power consumption.
- Output drives of ±8 mA at 5VCC and ±4 mA at 3.3VCC.
- Wide range of packaging options.
- Availability of Microgate Logic (single-gate) versions that simplify routing.

Switching characteristics of AHC versus AHCT devices are similar in terms of operating terms and limits, except for AHCT input TTL compatibility. AHCT devices are used primarily to interface TTL output signals to CMOS inputs. The need for compatibility required changes in input circuitry that increase power consumption over the AHC devices.

Figure 1.18 shows the input and output circuit provisions for ESD protection. This can be compared with AC/ACT output circuitry in *Figure 1.19.*

All AHC devices are 5-volt input tolerant when operated at 3.3V. To partially power-down a device, no paths exist from the input or output pins to Vcc. The AHC/AHCT devices have rail-to-rail swings that make them compatible with the HC/HCT family. They do not have paths from the output pins to VCC and can be partially powered down. Special consideration may have to be given to the output circuit termination. For a trace length in excess of four inches, reflections begin to appear on the line leading to undesirable effects. The TI AHC/AHCT databook referenced provides common termination techniques.

The Microgate Logic devices are single gates that simplify PCB routing. Typical propagation delay is 3.5 nS.

Advanced High-Speed VHC\VHCT CMOS Logic

This logic family is designed for operation from Vcc = 2.5 - 5.5V. In operation at less than five volts the family features five-volt-tolerant inputs to aid 3V - 5V mixed systems. Low power, low switching noise and fast switching speeds make this family well suited to low-power, low-cost portable applications.

The VHCT devices accept TTL level inputs and delivers full swing outputs of 4.5 - 5.5 volts, the VHCT supply range. Inputs tolerate voltages from 0 - 7 volts. When the voltage at the input exceeds the supply, no current path to the supply exists. This feature facilitates 3-volt-to-5-volt interfacing. A level shifter may be constructed with a combination of VHC and VHCT devices.

Low/Cross-Voltage CMOS Families

Introduction

The requirements for ever faster speeds coupled with reduced power consumption has led to the development of CMOS and BiCMOS devices capable of providing high performance at supply voltages ranging over a low of 2.7 volts to 3.6 volts, with 3.3 volts nominal. Translation devices capable of bridging the 3.3 - 5.0 volt span have been developed to accommodate those applications utilizing both levels of supply power. Because many of these devices are in the development stage, databooks define the status as *Production, Advance Information,* or *Product Review.*

Because manufacturers products differ in some respects, there exists a diversity of logic family designations that may be confusing. For example, National Semiconductor defines the following families:

- LCX Low-voltage, high-speed, with 5-volt tolerant inputs and outputs
- LVX Translator, low-voltage dual supply translating transceivers. Extended range from 2.7 to 3.6 volts
- LVX Bus switch, low-voltage bus switches

- LVX Low-voltage, with 5-volt tolerant inputs, extended range from 2.7 to 3.6 volts
- LVQ Low-voltage quiet CMOS, extended range from 2.7 to 3.3 volts
- LVT Low-voltage, high-speed BiCMOS logic, extended range from 2.7 to 3.3 volts

Texas Instruments defines four independent logic families as:

- ALVC Advanced low-voltage CMOS
- LV Low-voltage HCMOS
- LVC Low-voltage CMOS
- LVT Low-voltage technology

Within these categories are:

- ALB Widebus
- ALVC Widebus
- GTL Transceivers and Backplane Drivers
- LV MSI and Octals
- LVC MSI and Octals
- LVC Widebus
- LVT JTAG/IEEE 1149.1
- LVT Octals
- LVT Widebus
- LVTZ Octals

The need to move toward power supply voltages of 3.3 volts and below became an issue in 1984 when two JEDEC standards were developed. *Standard 8.0* was directed toward regulated (3.3V to 3.6V) while *8.1* was for unregulated (2V to 3.6V) battery usage. A concern was the need to interface to standard 5V TTL devices as well as low-voltage devices. It has been determined these standards fall short and a review is in process.

Because of the diversity, only general features of these logic families can be discussed here. More details will be found in the chapters dealing with the respective devices; i.e., gates, flip-flops, transceivers, and so forth.

The LVT devices feature an even different, BiCMOS, output design. Because of the evolving mixed voltage requirements the semiconductor industry has been confronted with the interfacing of low and traditional 5-volt systems. The source of the dilemma can be realized by a review of *Figure 1.20*.

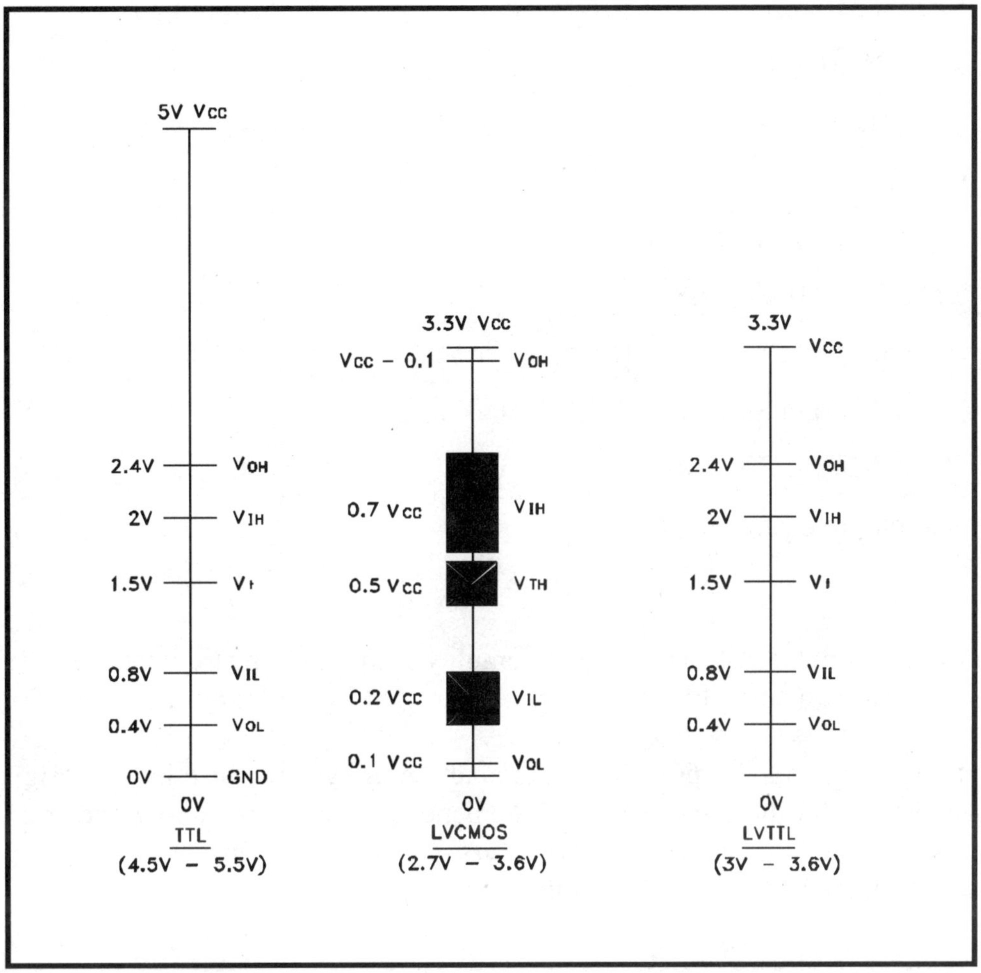

Figure 1.20. Comparison of 3.3V and 5V interfaces.

The Texas Instruments LVT series of devices rely on a state-of-the-art submicron BiCMOS process to provide up to a 90% reduction in static power dissipation over ABT devices. The family characteristics are as follows:

- 5.5-volt maximum input voltage
- Specified 2.7V to 3.6V supply range
- I/O structures that support power-on (live) insertion
- Standard TTL output drives of:

  $V_{OH} = 2V$ at $I_{OH} = -32$ mA

  $V_{OL} = 0.55V$ at $I_{OL} = 64$ mA
- Rail-to-rail switching for driving CMOS
- Maximum supply currents of:

  $V_{CCL} = 15$ mA

V_{CCH} = 250 mA

V_{CCZ} = 250 mA

- Propagation delays of:

  t_{pd} < 4.6 nS

  t_{pd}(LE to Q) < 5.1 nS

  t_{pd}(CLK to Q) < 6.3 nS

- Surface-mount package support including fine-pitch packages:

  48- and 56-pin SSOP for LVT Widebus13*

  20- and 24-pin TSSOP for standard LVT

- LVT input/output characteristics.

Figure 1.21 shows the simplified output structure employed by Texas Instruments LVT devices. This figure also shows a similar structure, differing in detail, provided by National Semiconductor.

National has its own proprietary approach to the reduction of system noise in the design of high-speed, low- voltage line drivers for the preservation of the benefits of reduced supply and compressed voltage swings. This approach employs their *Graduated Turn-On (GTO)*** circuitry to meet specified ground bounce, undershoot, and dynamic input threshold requirements.

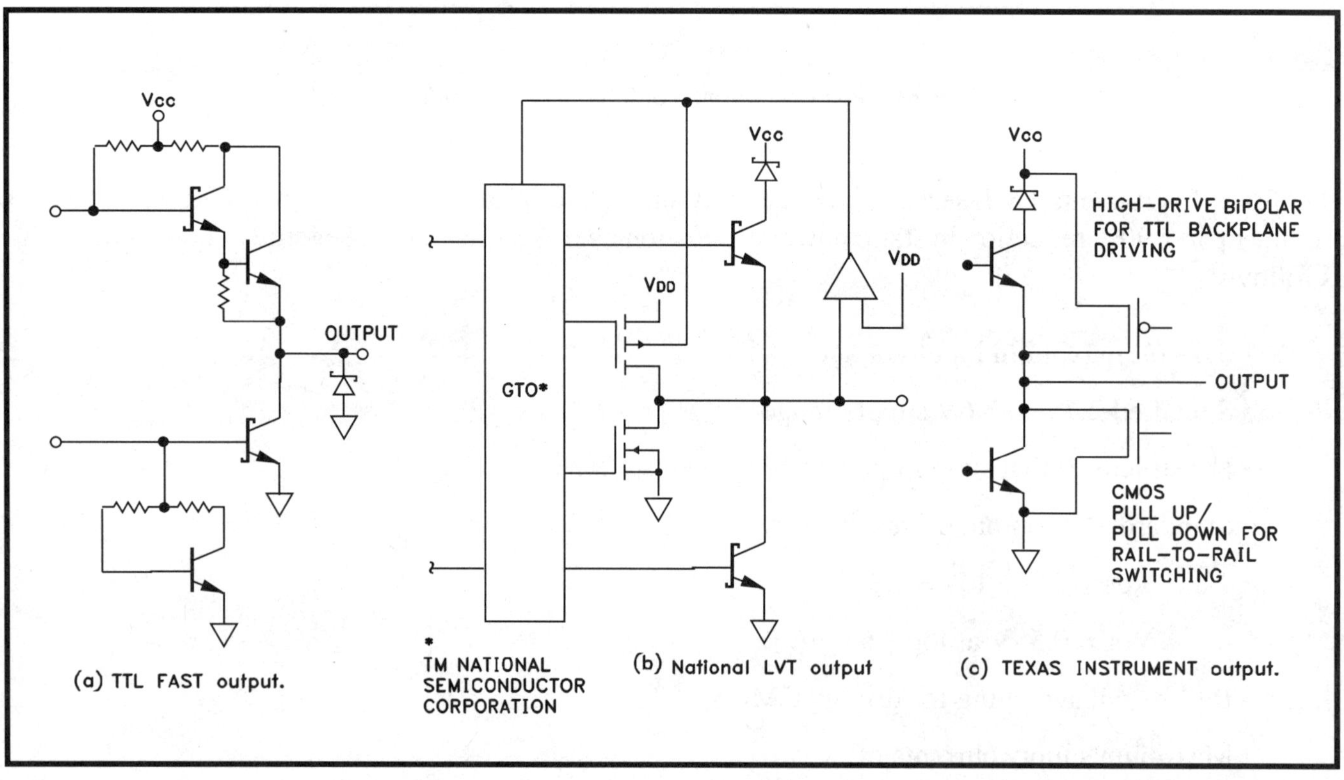

Figure 1.21. Comparison of CMOS LVT output circuitry with TTL FAST.

The National Crossvolt** Product line features the following:

- LVQ Low-Voltage Quiet CMOS

  Targeted for 3.3-volt-only applications where interfacing to 5-volt or other interface levels is not a requirement.

- LVX Low-Voltage CMOS Logic with 5V Tolerant Inputs

  This family is made up of three product groupings, each possessing a unique set of interfacing capabilities and characteristics optimized for specific applications.

  - Gates, octals, and MSI types
  - Dual-supply translating transceivers
  - Bus switches

- LVX Low-Voltage Dual-Supply CMOS Translating Transceivers

  These devices are true translators, meaning they receive 3V signals and output 5V signals and receive 5V signals and output 3V signals.

- LVX Low-Voltage CMOS Bus Switches

  Featuring a low "on" resistance, these devices offer a high-speed, bidirectional interface between mixed supply busses.

- LCX Low-Voltage High-Speed CMOS with 5V Tolerant Inputs and Outputs

  These devices offer mixed 3V-5V capability for applications where 3V and 5V subsystems interface with one another and where low power consumption is a necessity.

- LVT Low-Voltage High-Speed BiCMOS Logic

  These include all the mixed-supply interface features of the LCX family in addition to enhanced propagation delays. 5V tolerant, they are recommended for applications where 3.3V and 5V subsystems interface with one another with high-speed and high-drive requirements.

Further details will be found in the chapters on specific devices.

Summary

This chapter is intended to provide an overview of current TTL and CMOS family capabilities. It is not possible to provide the detailed comprehensive descriptions required for other than casual design information. Please refer to the manufacturers' literature referenced in this and following chapters for detailed design information.

References:

Siliconix Incorporated, *Low Power Discretes Data Book,* p 9.5.

Motorola Incorporated, *CMOS Logic Data,* Series C, 1991. pp 5-2 to 5-6.

National Semiconductor Corporation, *CMOS Logic Databook,* 1988 Edition, p 2-3.

Texas Instruments Inc., *High-Speed CMOS Logic Data Book, 1988, "Designer's Information," pp 4-5 to 4-38.*

National Semiconductor Corporation, CMOS Logic Databook, 1988 Edition, , "Comparison of MM54HC/ MM74HC to 54LS/74LS, 54S/74S, and 54ALS/74ALS Logic," pp 2-106 tp 2-111.

Ibid, "National's Process Enhancements Eliminate the CMOS SCR Latch-Up Problem in 54HC/74HC Logic," pp 2-112 to 2-119.

Ibid, "An Introduction to and Comparison of 54HCT/74HCT Compatible CMOS Logic," pp 2-135 to 2-140.

GE/RCA/Intersil Semiconductors, *Advanced CMOS Logic ICs AC/ACT,* ssd-283A, "Technical Overview," pp 14 -32, "System Design," pp 33 - 47.

National Semiconductor Corporation, *VHC Advanced CMOS Logic Databook,* 1996, pp 5-88 to 5-95.

Texas Instruments Inc., *AHC/AHCT Advanced High-Speed CMOS Logic Data Book,* 1997.

Texas Instruments Inc., *Low-Voltage Logic ALB/ALVC/LV/LVT,LVTZ,GTL Families Data Book,* 1996.

National Semiconductor Corporation, *Crossvolt Low Voltage Logic Series Databook,* 1994 Edition.

* Trademark, Texas Instruments, Inc.

** Trademark, National Semiconductor Corporation.

Chapter 2

Gate Logic

Logic Gates are a system's decision makers, which is a reason for there being so many of them. *Figure 2.1* defines the symbols and function table for the commonly used gates. One gate not illustrated is the Exclusive-NOR. It is simply the complement of the Exclusive-OR.

Figure 2.2 illustrates some basics of switching logic. Almost any function can be constructed through the appropriate selection of AND and OR gates, with or without inverters. That is, the NAND gate is simply an AND followed by an inverter, or an OR with its inputs preceded by an inverter. Because of its versatility the NAND is far and away the most widely used, followed by the NOR.

When we look at an individual gate it appears to be quite straightforward in its functioning. For a given input condition we have a defined output status. The complications arise when we begin interconnecting a number of gates. A system of mathematics has been developed to deal with the problem. The system is called *Boolean Algebra.* The "Algebra" in the name can be confusing, as the math does not conform to the algebraic rules as we normally understand them. Thus $A + B$ is not the sum of the two variables. It simply defines the relationship as one or the other, *A or B.* Similarly, $A \bullet B$ are not the product obtained by multiplying A by B, but rather a requirement that *A and B* must function as a unit.

Even so, certain algebraic procedures do work with Boolean Algebra. Logic equations can be simplified through factoring. For instance, the equation $A \bullet B + A \bullet C$ can be written as $A \bullet (B + C)$. Truth tables are a useful tool for verifying the results of our math manipulations. We are assisted in this by the following fifteen identities:

1. $A \bullet B = B \bullet A$
2. $A + B = B + A$

These are known as "commutative laws"

3. $(A + B) + C = A + (B + C)$
4. $(A \bullet B) \bullet C = A \bullet (B \bullet C)$

These are known as "associative laws"

5. $A(B + C) = AB + AC$
6. $A + B \bullet C = (A + B) \bullet (A + C)$

These are known as "distributive laws"

7. $A + \overline{A} = 1$
8. $A \bullet \overline{A} = 0$

9. A + A = A

10. A•A = A

11. (A) = A (The complement of the complement remains unchanged.)

12. A + 1 = 1 (If one input to an OR function is always true, so is the output.)

13. A•0 = 0 (If one input to an AND function is always false, so is the output.)

14. $\overline{(A+B)} = \overline{A} \bullet \overline{B}$

15. $\overline{(A \bullet B)} = \overline{A} + \overline{B}$

Numbers 14 and 15 are known as *De Morgan's Theorem.* The theorem tells us the same physical gate can perform both a NAND and a NOR function, depending on how the logic levels are defined; that is, with positive true logic the theorem holds for the NAND, and with negative true logic the theorem holds for the NOR. We see this in the NAND and NOR symbols shown in *Figure 2.1.*

In the early days of digital circuit design, individual logic gates had to be constructed and interconnected as a system development. Mathematical procedures for optimizing the gate requirements were developed. With advances in large-scale integrated circuit manufacturing the requirement for extensive combinations of gate logic was largely eliminated. So unless we are engaged in the design of complex integrated circuits we will have little need of these.

Our real need is awareness of where the digital IC industry is taking us with new products. In general, these are produced to meet requirements of higher speed combined with reduced power dissipation and the ability to function at lower voltage levels. This chapter provides drawing details of dual in-line packaging, logic diagrams, and function tables for the majority of TTL and CMOS logic gates, past and present. The accompanying tables describe propagation delays as a basis for performance comparisons. For this the Low-to-High (↑) and High-to-Low (↓) transitions seemed like the best choice.

Chapter Organization

Nine tables are provided. *Tables 2.1 - 2.4* provide a device source listing by manufacturer. The figure number for each device is provided in these tables. There are sources beyond those listed, but those given provide a realistic picture of what has been and is currently available.

Tables 2.5 - 2.22 provide timing references; the LOW-to-HIGH propagation delay (t_{PLH}) and HIGH-to-LOW (t_{PHL}) propagation delay as a comparative figure of merit. Because of the manner in which manufacturers define these quantities, the data in the tables is not necessarily precise. The intent is to provide insights helpful for performance comparisons. In particular, we can see the spectacular gains in speed and reduced power dissipation, as well as complexity, achieved with the latest devices.

The drawings of *Figures 2.1 - 2.19* describe the Dual In-Line (DIP) package pinning, a Truth Table and Logic Diagrams. To simplify comparisons as well as to conserve space, all devices with matching features, specifically pinouts and truth tables, are included in the same drawing.

The relationship between the chapter drawings and associated tables are as follows:

NAND gates:	Figures 2.3 - 2.5	Tables 2.5 (TTL) and 2.6 (CMOS)
NOR gates:	Figures 2.6, 2.7	Tables 2.7 (TTL) and 2.8 (CMOS)
AND gates:	Figures 2.8, 2.9	Tables 2.9 (TTL) and 2.10 (CMOS)
OR gates:	Figure 2.10	Tables 2.11 (TTL) and 2.12 (CMOS)
EX-OR/NOR gates:	Figure 2.11	Tables 2.13 (TTL) and 2.14 (CMOS)
HEX inv/non-inv.	Figure 2.12	Tables 2.15 (TTL) and 2.16 (CMOS)
AND-OR-INV gates:	Figures 2.13, 2.14	Tables 2.17 (TTL) and 2.18 (CMOS)
Mixed Functions	Figure 2.15, 2.16	Table 2.19 (TTL), (CMOS)
CMOS UB/HCU/AHCU	Unbuffered logic	Table 2.20 (CMOS) (No specific drawing)
Low, Dual-Voltages	Figures 2.17, 2.18	Table 2.21 (CMOS)
Single Gate, all	Figure 2.19	Table 2.22 (CMOS)

Summary

This chapter provides tables and drawings describing logic gates, inverters, and buffers of the 7400 TTL series and the many varieties of the 4000 and 7400 series of CMOS devices. Some points to keep in mind are:

- Timing information in the tables is for comparison purposes only. There is a general lack of consistency in how these parameters are defined, and not just between manufacturers. Where timing is critical consult the manufacturer's data manual.
- All timing data is shown for the industrial/commercial or recommended free-air operating temperature range. Where data is for 25°C it is given here.

- The specifications for CMOS devices may cover voltage ranges from 2 to 15 volts. Data is typically provided for VCC's of 5, 10, and 15 volts, but not always. Some manufacturers use 5.0 volts; others 4.5. There is a significant speed factor—in general, the higher the voltage the faster the performance. Again, where this is important to your application consult the data manual.

- A similar situation exists for the more recent low-voltage devices. Typical VCC ranges given are 2.7 - 3.6 volts, but values may be found for 2, 2.6, 3.0, 3.3 or 3.6 volts. (Glossary *Figures G.1* and *G.2* may provide helpful insights on performance.)

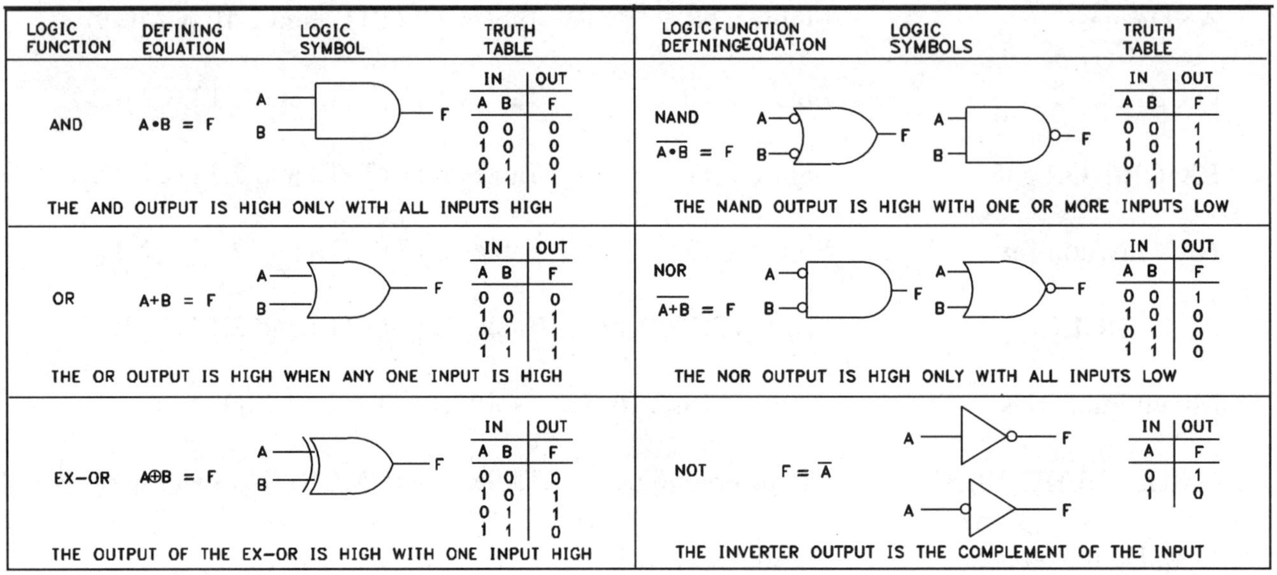

Figure 2.1. Logic definitions, defining equations, and symbols for the six basic logic gates.

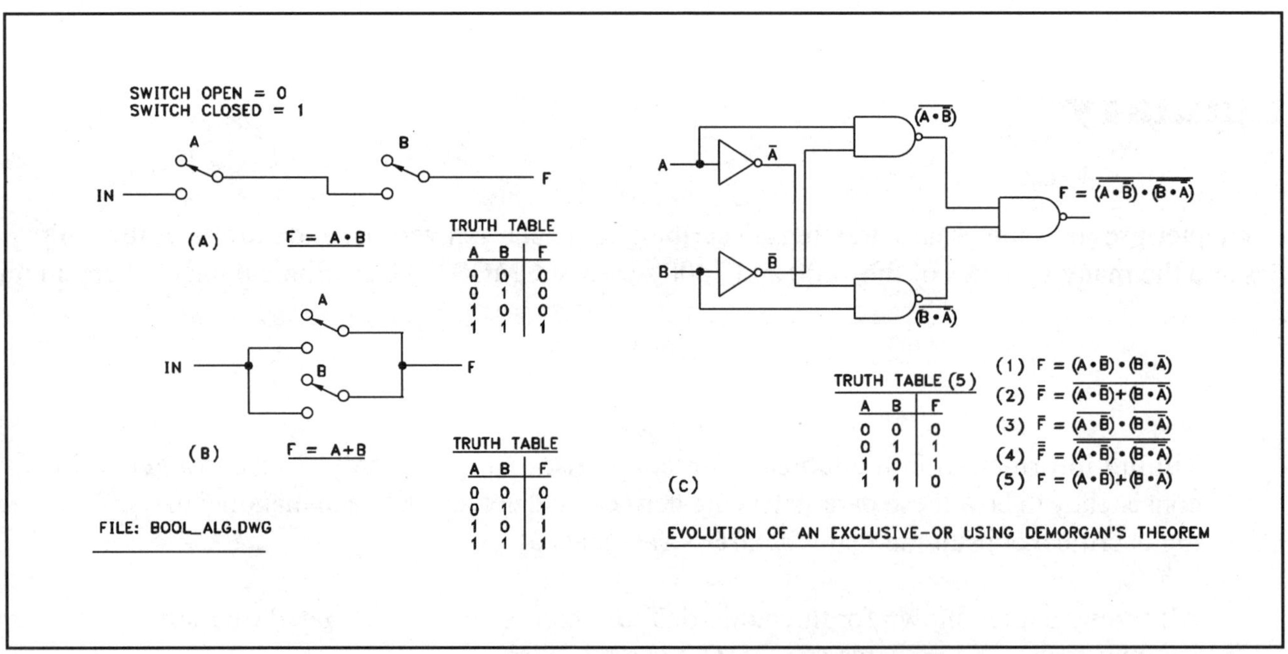

Figure 2.2. Switching logic basics: (a) AND logic. (b) OR logic. (c) derivation of a 2-input Exclusive -OR (EX_OR) from inverter and NAND components.

DEVICE 74---	STD		L		H		S		LS		AS		ALS		F		Figure No.	DEVICE 74---
	NM	ST	NM	ST	NM	ST	NM	ST	NM	ST	NM	ST	NM	ST	NM	ST		
7400	x	xx	x	x		xx	x	xx	xx	xx	x	x	x	x	x	xx	2.3	7400
7401	x	xx		x		xx			x	xx			x	x			2.3	7401
7402	x	xx	x	x			x	xx	xx	xx	x	x	x	x	x	xx	2.6	7402
7403	x	xx		x			x	xx	xx	xx			x	x			2.3	7403
7404	x	xx	x	x		xx	x	xx	xx	xx	x	x	x	x	x	xx	2.12	7404
7405	x	xx				xx	x	xx	xx	xx			x	x			2.12	7405
7406	x	xx														x	2.12	7406
7407	x	xx														x	2.12	7407
7408	x	xx				xx	x	xx	xx	xx	x	x	x	x		xx	2.8	7408
7409	x	xx				xx	x	xx	xx	xx			x	x			2.9	7409
7410	x	xx	x	x		xx	x	xx	xx	xx	x	x	x	x	x	xx	2.3	7410
7411		xx				xx	x	xx	xx	xx	x	x	x	x	x	xx	2.8	7411
7412		xx							xx	xx			x	x			2.3	7412
7413		xx							x	xx			x		x		2.4	7413
7414	x	xx							xx	xx			x		x		2.12	7414
7415	x	xx						xx	x	xx			x	x			2.8	7415
7416	x	xx															2.12	7416
7417	x	xx															2.12	7417
7419										x							2.12	7419
7420	x	xx	x	x		xx	x	xx	xx	xx	x	x	x	x	x	xx	2.14	7420
7421		x				xx			xx	xx	x	x	x	x	x	x	2.8	7421
7422		x				xx		xx	x	xx			x	x			2.4	7422
7423		xx															2.6	7423
7424										x							2.3	7424
7425		xx															2.6	7425
7426	x	xx							xx	xx							2.3	7426
7427	x	xx							xx	xx	x	x	x	x	x	xx	2.6	7427
7428		xx							x	xx			x	x			2.6	7428
7430	x	xx		x		xx	x	xx	xx	xx	x	x	x	x	x	xx	2.4	7430
7432	x	xx				xx	x	xx	xx	xx	x	x	x	x	x	xx	2.10	7432
7433		xx							x	xx				x		x	2.6	7433
7434												x		x			2.12	7434
7436														x			2.12	7435
7436	x	xx														x	2.6	7436
7437	x	xx						xx	xx	xx		x		x	x	x	2.3	7437
7438	x	xx						xx	xx	xx		x		x	x	xx	2.3	7438
7439		xx															2.3	7439
7440		xx				xx	x	xx	xx	xx		x		x	x	x	2.4	7440

Key to Manufacturer: N – National, M – Motorola,
S – Signetics/Philips, T – Texas Instruments

Table 2.1. 7400 TTL source summary by manufacturer.

DEVICE 74---	STD NM	STD ST	L NM	L ST	H NM	H ST	S NM	S ST	LS NM	LS ST	AS NM	AS ST	ALS NM	ALS ST	F NM	F ST	Figure No.	DEVICE 74---
7450		xx				xx											2.13	7450
7451		xx		x		xx	x	xx	xx	xx						xx	2.13	7451
7452						xx											2.13	7452
7453		xx				xx											2.13	7453
7454		xx		x		xx			x	xx							2.13	7454
7455				x		xx			x	xx							2.13	7455
7460		x				xx											2.13	7460
7461						xx											2.13	7461
7462						xx											2.13	7462
7464							x	xx								x	2.13	7464
7465								xx									2.13	7465
7486	x	xx					x	xx	xx	xx	x	x	x	x		xx	2311	7486
74132	x	xx						x	xx	xx			x			x	2.3	74132
74133							x	x					x	x		x	2.4	74133
74134								x									2.4	74134
74135		x						x									2.11	74135
74136		x							x	xx	x	x	x	x			2.11	74136
74260								xx	x	x						xx	2.6	74260
74266									x	xx							2.11	74266
74386										x							2.11	74386
74804											x	x	x	x		x	2.5	74804
74805											x	x	x	x		x	2.7	74805
74808											x	x	x	x		x	2.9	74808
74810											x	x	x	x			2.11	74810
74811											x	x	x	x			2.11	74811
74832											x	x	x	x		x	2.10	74832
741000											x	x	x	x			2.3	741000
741002													x	x			2.6	741002
741003													x	x			2.3	741003
741004											x	x	x	x			2.12	741004
741005													x	x			2.12	741005
741008											x	x	x	x			2.8	741008
741010											x	x	x	x			2.3	741010
741011											x	x	x	x			2.8	741011
741020												x	x	x			2.4	741020
741032											x	x	x	x			2.10	741032
741034											x	x	x	x			2.12	741034
741035													x	x			2.12	741035
741036											x	x					2.6	741036
741804											x	x		x		x	2.5	741804
741805											x	x		x		x	2.7	741805
741808											x	x		x		x	2.9	741808
741832											x	x		x		x	2.10	741832

Key to Manufacturer: N – National, M – Motorola,
S – Signetics/Philips, T – Texas Instruments

Table 2.1a. 7400 TTL source summary by manufacturer (continuation).

DEVICE	4000 NM		4000 ST	4100 NM		4100 ST	4500 NM		4500 ST	4700 NM		4700 ST	Figure No.	DEVICE
4000	x		x										2.15	4000
4000UB	x		x										2.15	4000UB
4001	x	x	x										2.6	4001
4001UB		x	x										2.6	4001UB
4002	x	x	x										2.6	4002
4002UB		x											2.6	4002UB
4007	x												2.15	4007
4007UB		x	x										2.15	4007UB
4009	x												2.12	4009
4010	x												2.12	4010
4011	x	x	x										2.3	4011
4011UB		x	x										2.3	4011UB
4012	x	x	x										2.4	4012
4012UB		x	x										2.4	4012UB
4019	x												2.14	4019
4023	x	x	x										2.3	4023
4023UB		x											2.3	4023UB
4025	x	x	x										2.6	4025
4025UB		x											2.6	4025UB
4030	x		x										2.11	4030
4041			x										2.15	4041
4041UB	x												2.15	4041UB
4048	x												2.15	4048
4049	x	x	x										2.12	4049
4049UB	x	x											2.12	4049UB
4050	x	x	x										2.12	4050
4068		x	x										2.4	4068
4069UB	x	x	x										2.12	4069UB
4070	x	x	x										2.11	4070
4071	x	x	x										2.10	4071
4072	x	x	x										2.10	4072
4073	x	x	x										2.8	4073
4075	x	x	x										2.10	4075
4077		x	x										2.11	4077
4078		x	x										2.6	4078
4081	x	x	x										2.8	4081
4082	x	x	x										2.8	4082
4085			x										2.13	4085
4086			x										2.13	4086
4093	x	x											2.3	4093
40097			x										2.16	40097
40098			x										2.16	40098
40106	x												2.12	40106
4106					x								2.12	4106
4501UB								x					2.15	4501UB
4502								x	x				2.16	4502
4503							x	x					2.16	4503
4506								x					2.14	4506
4519							x	x					2.14	4519
4572UB								x					2.15	4572UB
4583								x					2.15	4583
4584							x	x					2.12	4584

Key to Manufacturer:

	Prefix		Prefix
N – National,	CD	S – Signetics/Philips,	HEC/HEF
M – Motorola,	MC1	T – Texas Instruments	--

Table 2.2. 4000 series CMOS source summary by manufacturer.

DEVICE	C		HC		HCT		AHC (5V)		AHCT		VHC (5V)		VHCT		AC (5V)		ACT		Figure	DEVICE
74---	NM	ST	NM	ST	NM	ST	NM	ST	NM	ST	NM	ST	NM	ST	NM	ST	NM	ST	No.	74---
7400	x		xx	xx	xx	xx		x		x	xx		xx			x		x	2.3	7400
7401				x															2.3	7401
7402	x		xx	xx		x		x		x	xx		xx						2.6	7402
7403			xx	xx	x	x													2.3	7403
7404	x		xx	xx	xx	xx		x		x	xx		xx			x		x	2.12	7404
74U04	x		xx	xx															2.12	74U04
7405			xx	x	xx			x		x									2.12	7405
7408	x		xx	xx	xx	x		x		x	xx		xx			x		x	2.8	7408
7409				x															2.9	7409
7410	x		xx	xx		x		x		x						x		x	2.3	7410
7411			xx	xx		x		x		x						x		x	2.8	7411
7414	x		xx	x		x					xx		xx			x		x	2.12	7414
7420	x		xx	xx		x													2.14	7420
7421				xx		x													2.8	7421
7427			xx	xx		x		x		x	x								2.6	7427
7430	x		xx	xx		x					x		xx						2.4	7430
7432	x		xx	xx	xx	x		x		x	xx		xx			x		x	2.10	7432
7434			xx		xx														2.12	7434
7435			x		x														2.12	7435
7436				x															2.6	7436
7451			xx	x															2.13	7451
7458			xx	x															2.13	7458
7486	x		xx	xx		x													2311	7486
74132			xx	x															2.3	74132
74133			xx	xx		x													2.4	74133
74266			x	x															2.11	74266
74386				x															2.11	74386
74804				x															2.5	74804
74805				x															2.7	74805
74808				x															2.9	74808
74832				x															2.10	74832
744002			xx	xx		x													2.6	744002
744049			x	x															2.12	744049
744050			x	x															2.12	744050
744075			x	xx		x													2.10	744075
744078			x	x															2.11	744078
747001				x															2.8	747001
747002				x															2.6	747002
747006				x															2.15	747006
747008				x															2.15	747008
747032				x															2.10	747032
747266			x	xx															2.11	747266
7411000																x			2.3	7411000
7411004																x		x	2.12	7411004
7411008																x		x	2.8	7411008
7411032																x		x	2.10	7411032
7411086																x			2.11	7411086

Key to Manufacturer: N – National, M – Motorola, S – Signetics/Philips, T – Texas Instruments

Table 2.3. 7400 series 5V CMOS source summary by manufacturer.

DEVICE	LV		LVC		LVT		LVX		LVQ		LCX		AC (3V)		ALVCH		HL		AHC (3V)		VHC (3V)		Figure	DEVICE
74---	NM	ST	NM	ST	NM	ST	NM	ST	NM	ST	NM	ST	NM	ST	NM	ST	NM	ST	NM	ST	NM	ST	No.	74---
7400		xx		xx		x	x		x		x			x						x	xx		2.3	7400
7402		xx		xx		x	x		x											x	xx		2.6	7402
7403		x		xx																			2.3	7403
7404		xx		xx		x	x		x		x			x						x	xx		2.12	7404
74U04		xx		xx																x	xx		2.12	74U04
7405		xx		xx																x			2.12	7405
7408		x		xx		x	x		x		x			x						x	xx		2.8	7408
7410		x		xx		x								x						x			2.3	7410
7411		x		xx							x			x						x			2.8	7411
7414		xx		xx		x	x		x		x			x						x	xx		2.12	7414
7420		xx		xx		x	x		x		x			x						x	xx		2.14	7420
7427		x		x																x			2.6	7427
7430		x		xx										x						x			2.4	7430
7432		xx		xx		x	x		x		x			x						x	xx		2.10	7432
7486		xx		xx		x	x		x		x			x						x	xx		2311	7486
74132		x																		x	xx		2.3	74132
7411000														x									2.3	7411000
7411004														x									2.12	7411004
7411008														x									2.8	7411008
7411032														x									2.10	7411032
7411086														x									2.11	7411086

Key to Manufacturer: N – National, M – Motorola, S – Signetics/Philips, T – Texas Instruments

Table 2.4. 7400 series 3V CMOS source summary by manufacturer.

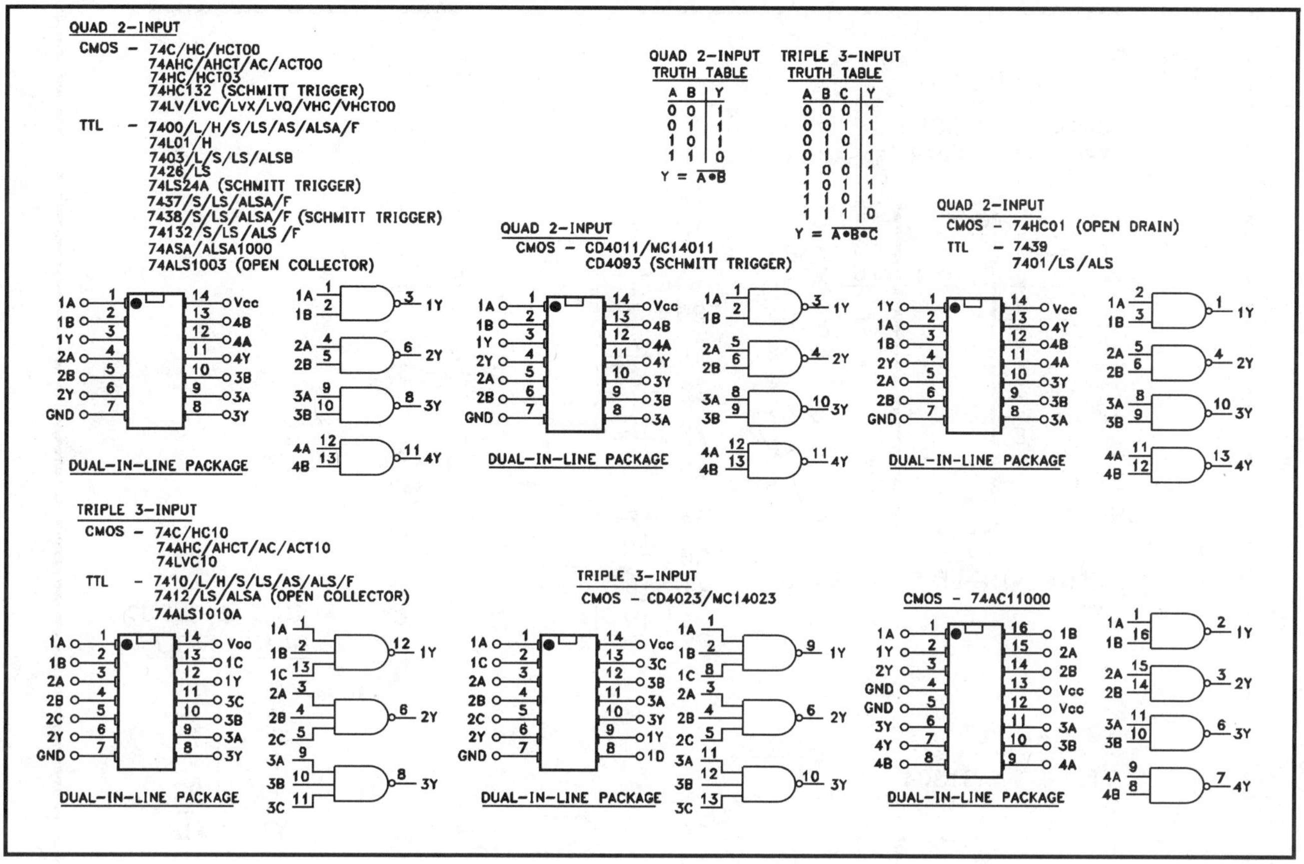

Figure 2.3. Quad 2-input and triple 3-input NAND logic gates, five-volt TTL, three- and five-volt CMOS series devices.

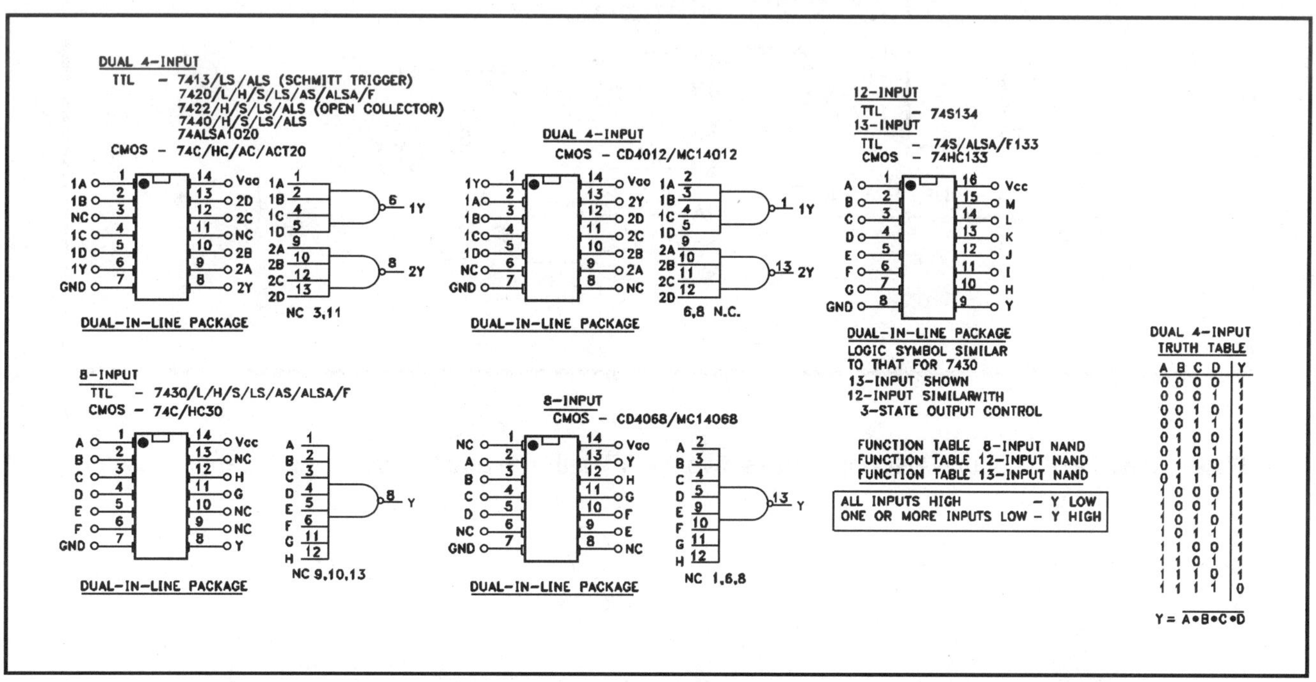

Figure 2.4. Dual 4-input, 8-, 12- and 13-input NAND logic gates, five-volt TTL, three- and five-volt CMOS series devices.

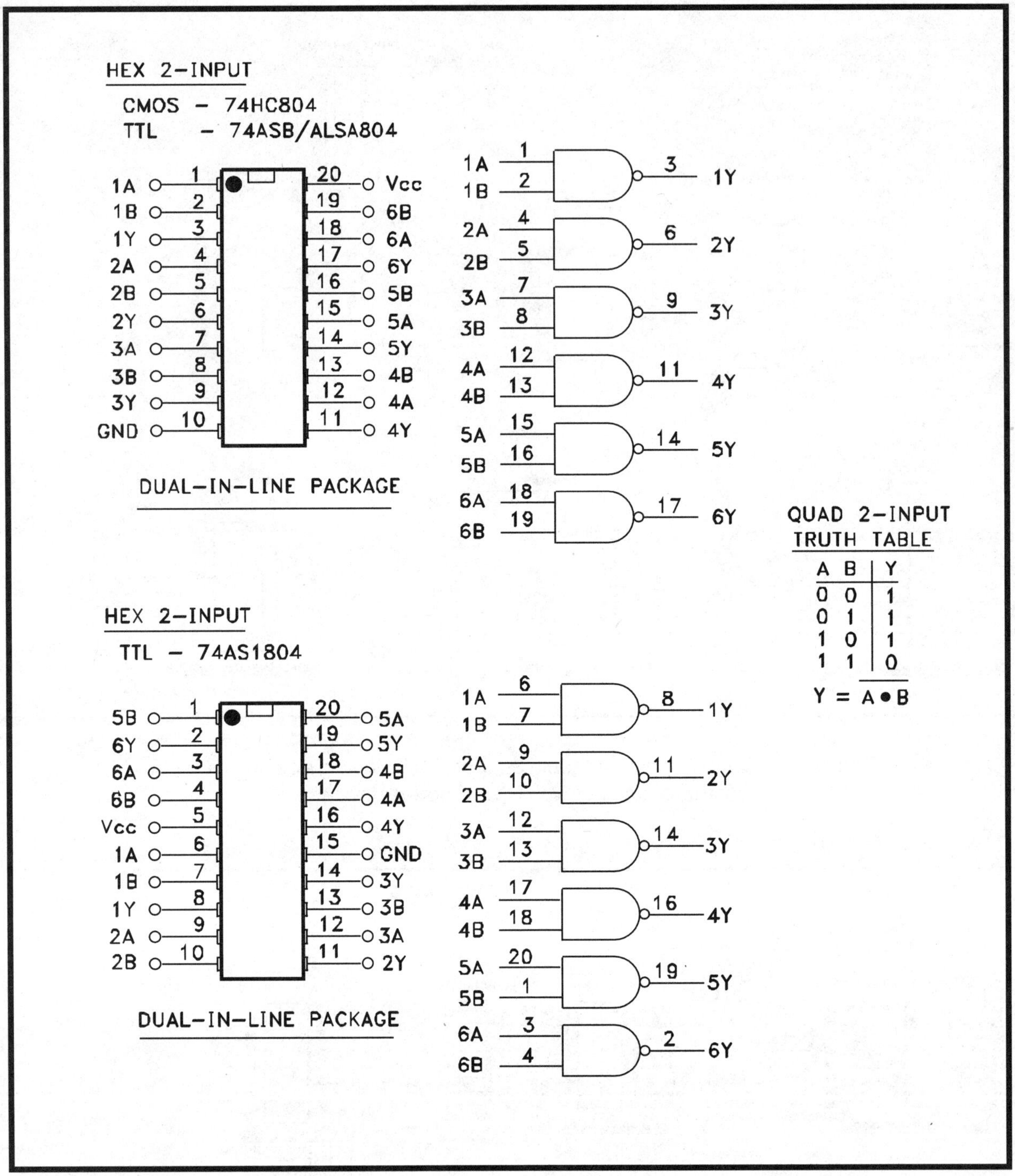

Figure 2.5. HEX 2-input NAND logic gates, five-volt TTL, three- and five-volt CMOS series devices.

DEVICE	STD	H	L	S	LS	AS/A/B	ALS/A/B	F	DELAY
7400	11-22	10	31-60	3-4.5	9-15	1-4.5	3.7-5	3.7-5	tPLH↑
——	7-15	10	35-60	3-5.0	10-15	1-4.0	2-8	3.2-4.3	tPHL↓
7401	35-55	15	90	—	17-32	—	23-54	—	tPLH↑
——	8-15	12	60	—	15-28	—	8-28	—	tPHL↓
7403	35-45	—	90	5-7.5	17-32	—	—	—	tPLH↑
——	8-15	—	60	4.5-7.0	15-28	—	—	—	tPHL↓
7410	11-22	5.9-10	35-60	3-4.5	9-15	1-4.5	—	—	tPLH↑
——	7-15	6.3-10	31-60	3-5.0	10-15	1-4.5	—	—	tPHL↓
7412	35-45	—	—	—	17-32	—	—	—	tPLH↑
——	8-15	—	—	—	15-28	—	—	—	tPHL↓
7413	18-27	—	—	—	15-22	—	—	—	tPLH↑
——	15-22	—	—	—	18-27	—	—	—	tPHL↓
7420	12-22	6-10	35-60	3-4.5	9-15	1-5.0	3-11	3.7-5.0	tPLH↑
——	15-22	6-10	31-60	3-5.0	10-15	1-4.5	3-10	3.2-4.3	tPHL↓
7422	35-45	10-15	—	2-7.5	17-32	—	23-45	—	tPLH↑
——	8-15	7.5-12	—	2-7.0	15-28	—	4-18	—	tPHL↓
7424	—	—	—	—	13-20	—	—	—	tPLH↑
——	—	—	—	—	25-40	—	—	—	tPHL↓
7426	16-24	—	—	—	—	—	—	—	tPLH↑
——	11-17	—	—	—	—	—	—	—	tPHL↓
7430	13-22	6.8-10	35-60	4.0-6.0	8-15	1-5.0	3-10	1.5-5.0	tPLH↑
——	8-15	8.9-12	70-100	4.5-7.0	13-20	1-4.5	3-12	1.0-3.0	tPHL↓
7437	13-22	—	—	4.0-6.5	12-24	—	2-8	2.5-5.5	tPLH↑
——	8-15	—	—	4.0-6.5	12-24	—	2-7	1.5-4.5	tPHL↓
7438	14-22	—	—	6.5-10	20-32	—	—	7.5-12.5	tPLH↑
——	11-18	—	—	6.5-10	18-28	—	—	1.5-5.0	tPHL↓
7439	22	—	—	—	—	—	—	—	tPLH↑
——	18	—	—	—	—	—	—	—	tPHL↓
7440	13-22	8.5-12	—	4-6.5	12-24	—	2-8	—	tPLH↑
——	8-15	6.5-12	—	6.5-6.5	12-24	—	2-7	—	tPHL↓
74132	15-22	—	—	7-10.5	15-22	2-12	—	—	tPLH↑
——	15-22	—	—	8.5-13	15-22	2-11	—	—	tPHL↓
74133	—	—	—	4-6.0	—	—	3-11	—	tPLH↑
——	—	—	—	4.5-7.0	—	—	5-25	—	tPHL↓
74134	—	—	—	4-6.0	—	—	—	—	tPLH↑
——	—	—	—	5-7.5	—	—	—	—	tPHL↓
74140	—	—	—	6.5	—	—	—	—	tPLH↑
——	—	—	—	6.5	—	—	—	—	tPHL↓
74804	—	—	—	—	—	1-4(B)	2-7	—	tPLH↑
——	—	—	—	—	—	1-4(B)	2-8	—	tPHL↓
741000	—	—	—	—	—	1-4(A)	2-8(A)	—	tPLH↑
——	—	—	—	—	—	1-4(A)	2-7(A)	—	tPHL↓
741003	—	—	—	—	—	—	10-33(A)	—	tPLH↑
——	—	—	—	—	—	—	2-12(A)	—	tPHL↓
741010	—	—	—	—	—	—	2-8(A)	—	tPLH↑
——	—	—	—	—	—	—	2-8(A)	—	tPHL↓
741020	—	—	—	—	—	—	2-8(A)	—	tPLH↑
——	—	—	—	—	—	—	2-7(A)	—	tPHL↓
741804	—	—	—	—	—	1-4	—	—	tPLH↑
——	—	—	—	—	—	1-4	—	—	tPHL↓
748003	—	—	—	—	—	—	3-11	—	tPLH↑
——	—	—	—	—	—	—	2-8	—	tPHL↓

Table 2.5. TTL NAND Logic gate propagation delays (nS). Vcc = 5 volts, 25°C.

DEVICE	74C	74HC	74HCT	74AHC	74AHCT	74AC	74ACT	CD4n	DELAY
74n00	50-90	9-18	18-23	3.7-5.5	5-6.9	1.5-8.5	1-9.5	—	tPLH↑
——	50-90	9-18	18-23	3.7-5.5	5/6.9	1-7	1-8	—	tPHL↓
74n01	—	13-25	—	—	—	—	—	—	tPLH↑
——	—	10-20	—	—	—	—	—	—	tPHL↓
74n03	—	10-20	10-20	—	—	—	—	—	tPLH↑
——	—	10-20	10-20	—	—	—	—	—	tPHL↓
74n10	60-100	10-18	—	3.9-5.9	3.9-5.9	1-8	1-10	—	tPLH↑
——	60-100	10-18	—	3.9-5.9	3.9-5.9	1-6.5	1-9.5	—	tPHL↓
74n20	70-115	9-18	—	—	—	3.1-11	3.5-12	—	tPLH↑
——	70-115	9-18	—	—	—	3.1-11	3.5-12	—	tPHL↓
74n30	125-180	23-35	—	—	—	—	—	—	tPLH↑
——	125-180	23-35	—	—	—	—	—	—	tPHL↓
74n132	—	13-25	—	—	—	—	—	—	tPLH↑
——	—	13-25	—	—	—	—	—	—	tPHL↓
74n133	—	23-35	—	—	—	—	—	—	tPLH↑
——	—	23-35	—	—	—	—	—	—	tPHL↓
7411000	—	—	—	—	—	1.5-6.5	—	—	tPLH↑
——	—	—	—	—	—	1.5-6.1	—	—	tPHL↓
4011	—	—	—	—	—	—	—	125-250	tPLH↑
——	—	—	—	—	—	—	—	125-250	tPHL↓
4012	—	—	—	—	—	—	—	160-300	tPLH↑
——	—	—	—	—	—	—	—	160-300	tPHL↓
4023	—	—	—	—	—	—	—	160-300	tPLH↑
——	—	—	—	—	—	—	—	160-300	tPHL↓
4068	—	—	—	—	—	—	—	200-350	tPLH↑
——	—	—	—	—	—	—	—	200-350	tPHL↓
4093	—	—	—	—	—	—	—	125-250	tPLH↑
——	—	—	—	—	—	—	—	125-250	tPHL↓

Table 2.6. CMOS NAND Logic gate propagation delays (nS). Vcc = 5 volts, 25°C.

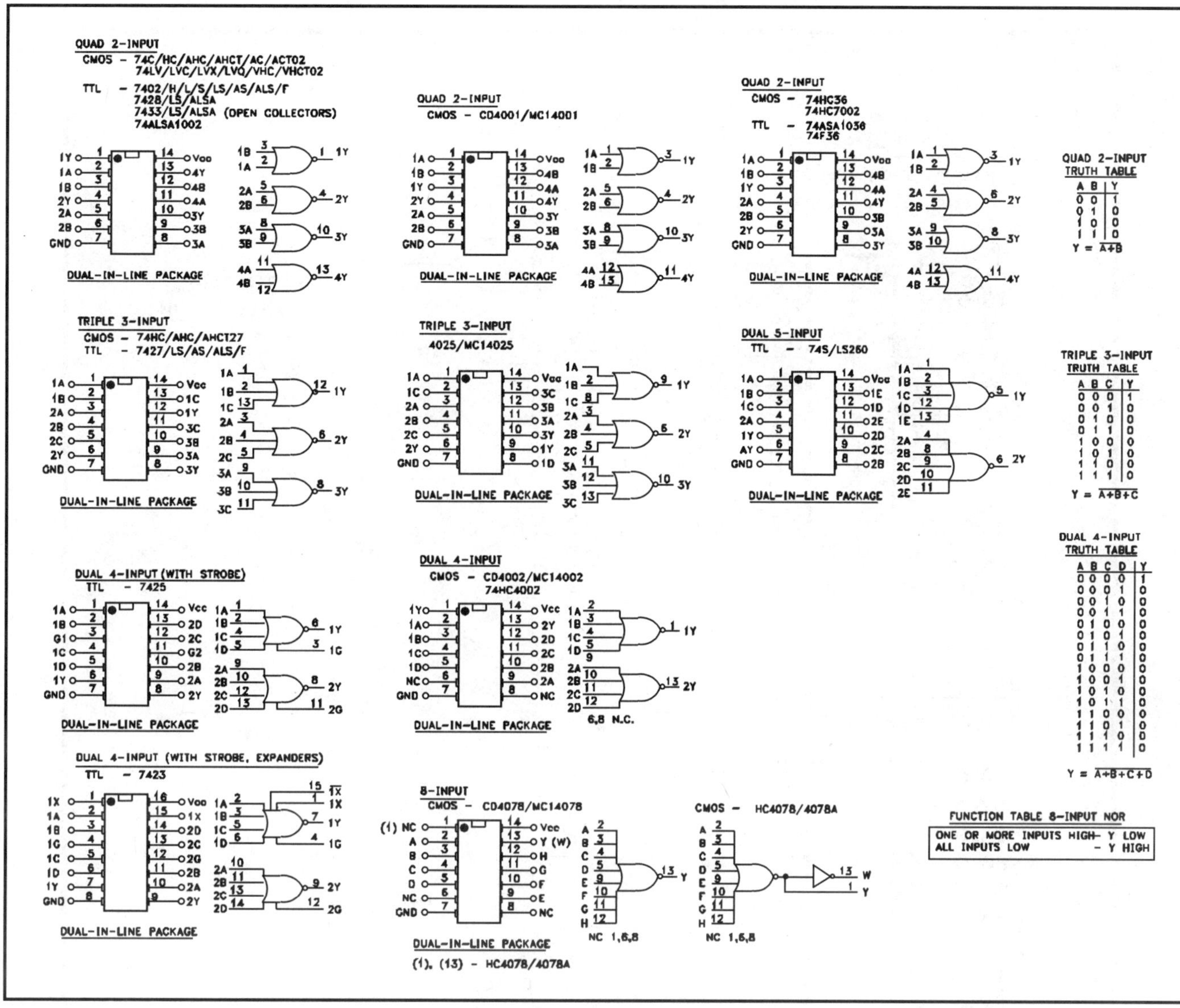

Figure 2.6. Quad 2-input, triple 3-input, dual 4-input, and 8-input NOR logic gates, five-volt TTL, three- and five-volt CMOS series devices.

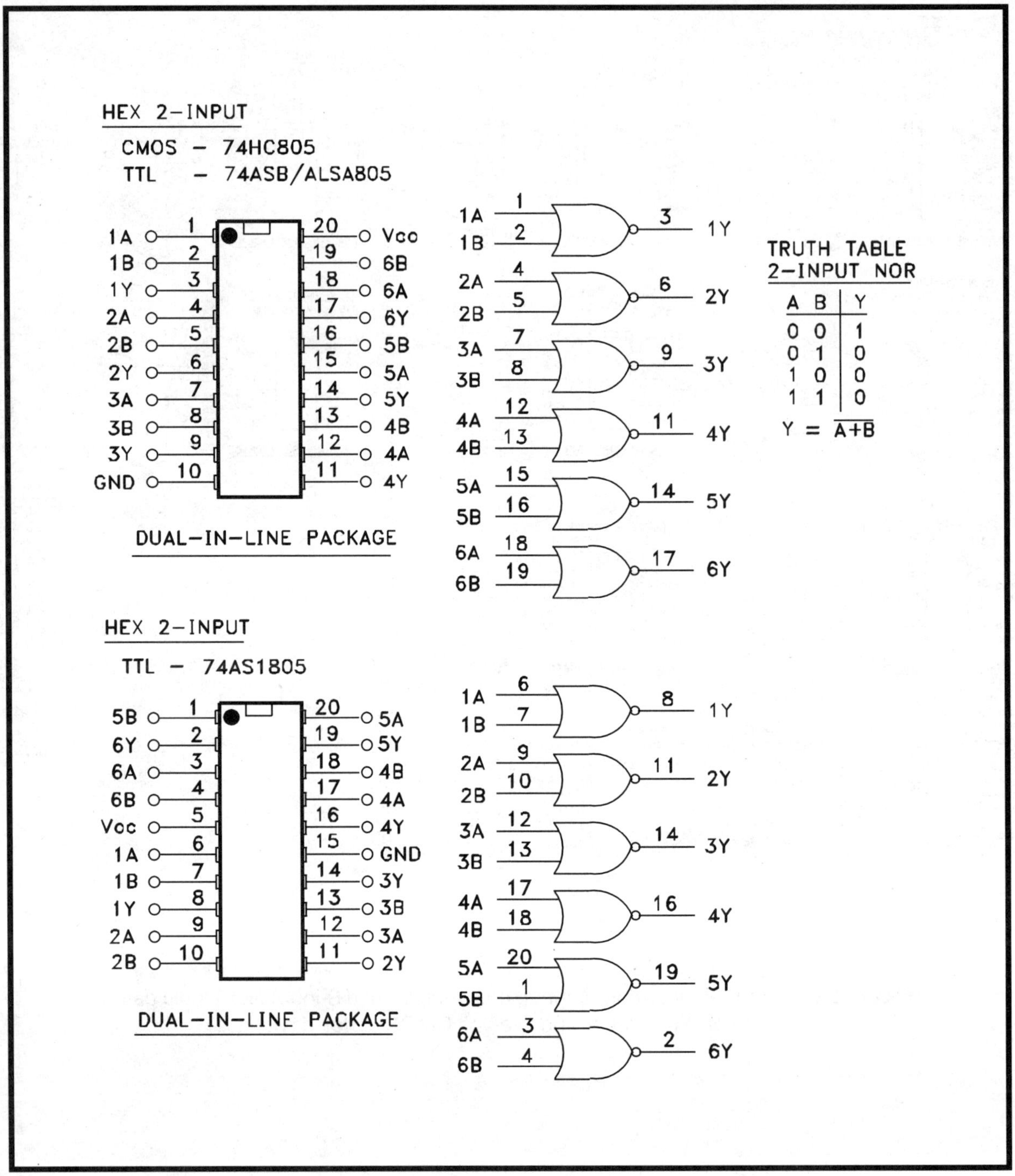

Figure 2.7. HEX 2-input NOR logic gates, five-volt TTL and CMOS series devices.

DEVICE	STD	H	L	S	LS	AS/A/B	ALS/A/B	F	DELAY
7402	12/15	10	31-60	3.5-5.5	10-15	1-4.5	3-12	2.5-4.4	tPLH↑
——	12/19	10	31-60	3.5-5.5	10-15	1-4.5	3-10	2.5-4.4	tPHL↓
7423	13-22	—	—	—	—	—	—	—	tPLH↑
——	8-15	—	—	—	—	—	—	—	tPHL↓
7425	13-22	—	—	—	—	—	—	—	tPLH↑
——	8-15	—	—	—	—	—	—	—	tPHL↓
7427	10-15	—	—	—	10-15	1-5.5	4-15	2-5	tPLH↑
——	7-11	—	—	—	10-15	1-4.5	3-9	1-4.5	tPHL↓
7428	6-9	—	—	—	—	—	2-8	—	tPLH↑
——	8-12	—	—	—	—	—	2-7	—	tPHL↓
7433	10-16	—	—	—	20-32	—	10-33	—	tPLH↑
——	12-18	—	—	—	18-28	—	2-12	—	tPHL↓
7436	—	—	—	—	—	—	—	NA	tPLH↑
——	—	—	—	—	—	—	—	NA	tPHL↓
74260	—	—	—	4-5.5	15	—	—	—	tPLH↑
——	—	—	—	4-6.0	15	—	—	—	tPHL↓
74805	—	—	—	—	—	1-4.3	2-7	—	tPLH↑
——	—	—	—	—	—	2-4.3	2-8	—	tPHL↓
741002	—	—	—	—	—	2-8	2-7	—	tPLH↑
——	—	—	—	—	—	3-7	2-8	—	tPHL↓
741036	—	—	—	—	—	1-4.3	—	—	tPLH↑
——	—	—	—	—	—	1-4.3	—	—	tPHL↓
741805	—	—	—	—	—	1-4.3	—	—	tPLH↑
——	—	—	—	—	—	1-4.3	—	—	tPHL↓

Table 2.7. TTL NOR Logic gate propagation delays (nS). Vcc = 5 volts, 25°C.

DEVICE	74C	74HC	74HCT	74AHC	74AHCT	74AC	74ACT	CD4n	DELAY
74n02	50-90	9-23	—	3.6-5.5	2.4-5.5	3-10.4	3.1-11.1	—	tPLH↑
——	50-90	9-23	—	3.6-5.5	3.5-5.5	3-10.4	3.1-11.1	—	tPHL↓
74n27	—	10-18	—	4.1-5.9	4.1-5.9	—	—	—	tPLH↑
——	—	10-18	—	4.1-5.9	4.1-5.9	—	—	—	tPHL↓
74n36	—	10-25	—	—	—	—	—	—	tPLH↑
——	—	10-25	—	—	—	—	—	—	tPHL↓
74n805	—	10-19	—	4.1-5.9	4.1-5.9	—	—	—	tPLH↑
——	—	10-19	—	4.1-5.9	4.1-5.9	—	—	—	tPHL↓
74n4002	—	12-24	—	—	—	—	—	—	tPLH↑
——	—	12-24	—	—	—	—	—	—	tPHL↓
74n7002	—	18-33	—	—	—	—	—	—	tPLH↑
——	—	18-33	—	—	—	—	—	—	tPHL↓
4001	—	—	—	—	—	—	—	125-250	tPLH↑
——	—	—	—	—	—	—	—	125-250	tPHL↓
4002	—	—	—	—	—	—	—	160-300	tPLH↑
——	—	—	—	—	—	—	—	160-300	tPHL↓
4025	—	—	—	—	—	—	—	160-300	tPLH↑
——	—	—	—	—	—	—	—	160-300	tPHL↓
4078	—	—	—	—	—	—	—	200-350	tPLH↑
——	—	—	—	—	—	—	—	200-350	tPHL↓

Table 2.8. CMOS NOR Logic gate propagation delays (nS). Vcc = 5 volts, 25°C.

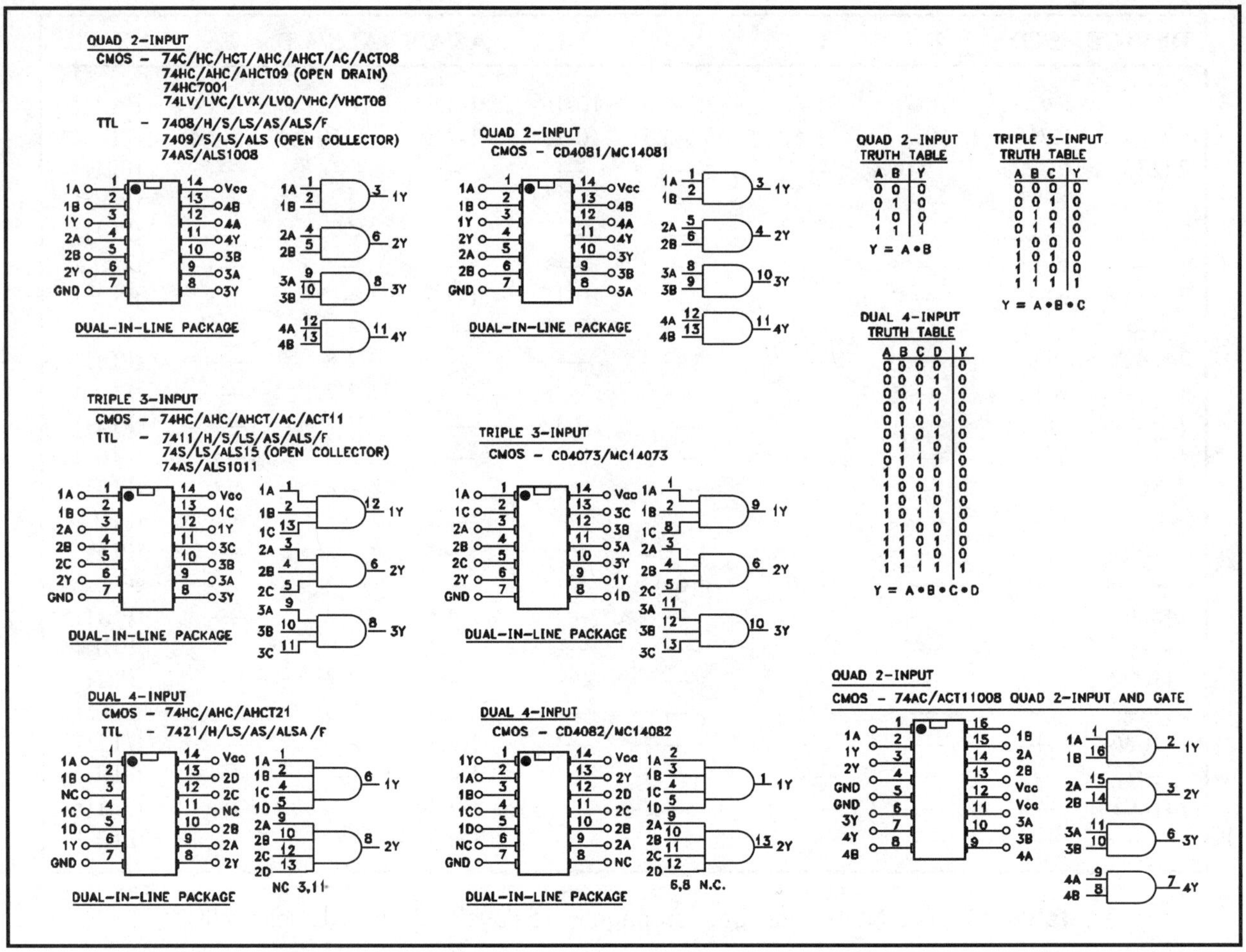

Figure 2.8. Quad 2-input, triple 3-input, and dual 4-input AND logic gates, five-volt TTL, three- and five-volt CMOS series devices.

DEVICE	STD	H	L	S	LS	AS/A/B	ALS/A/B	F	DELAY
7408	17-27	12	—	4.5-7.0	8-15	1-5.5	4-14	4.2-5.6	tPLH↑
——	12-19	12	—	5-7.5	10-20	1-5.5	3-10	4.0-6.3	tPHL↓
7409	21-32	—	—	6.5-10	20-35	—	23-54	—	tPLH↑
——	16-24	—	—	6.5-10	17-35	—	5-15	—	tPHL↓
7411	27	8.8-12	—	4.5-7	8-15	1-6	2-13	4.2-5.6	tPLH↑
——	19	7.6-12	—	5-7.5	10-20	1-5.5	2-10	4.1-5.5	tPHL↓
7415	—	—	—	5.5-8.5	20-35	—	20-45	—	tPLH↑
——	—	—	—	6-9	17-35	—	6-20	—	tPHL↓
7421A	15	8.8-12	—	—	8-15	1-6	4-15	1-5.3	tPLH↑
——	20	7.6-12	—	—	10-20	1-6	2-10	1.5-5.5	tPHL↓
74808	—	—	—	—	—	1-6	2-9	—	tPLH↑
——	—	—	—	—	—	1-6	1-8	—	tPHL↓
741008	—	—	—	—	—	1-6	2-9	—	tPLH↑
——	—	—	—	—	—	1-6	3-9	—	tPHL↓
741011	—	—	—	—	—	1-6	2-9	—	tPLH↑
——	—	—	—	—	—	1-6	3-9	—	tPHL↓
741808	—	—	—	—	—	1-6	—	—	tPLH↑
——	—	—	—	—	—	1-6	—	—	tPHL↓

Table 2.9. TTL AND Logic gate propagation delays (nS). Vcc = 5 volts, 25°C.

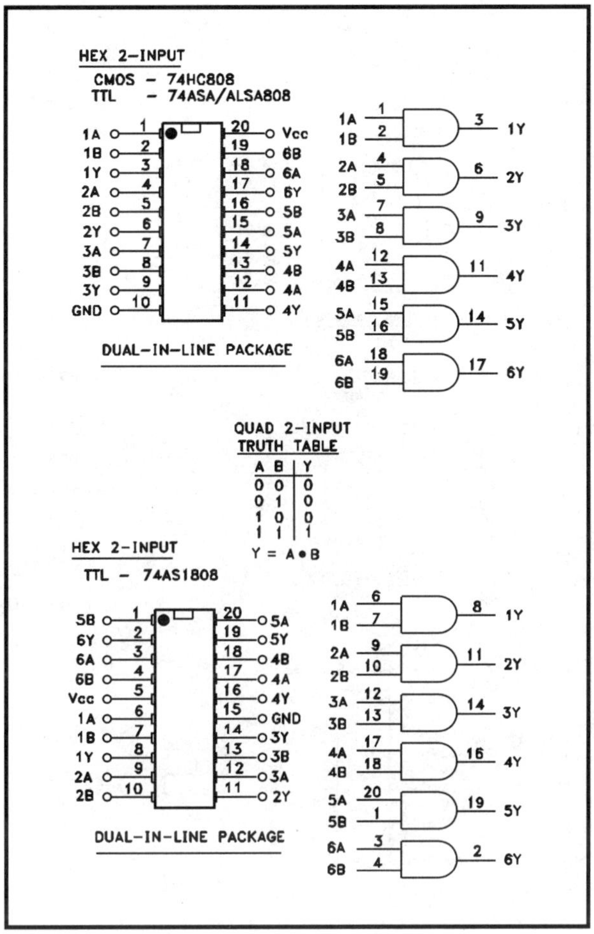

Figure 2.9. HEX 2-input AND logic gates, five-volt TTL and CMOS series devices.

DEVICE	74C	74HC	74HCT	74AHC	74AHCT	74AC	74ACT	CD4n	DELAY
74n08	80-140	10-18	9-15	4.3-5.9	5-6.9	1-8.5	1-10	—	tPLH↑
——	80-140	10-18	9-15	4.3-5.9	5-6.9	1-7.5	1-10	—	tPHL↓
74n09	—	13-25	—	4.1-5.9	4.1-5.9	—	—	—	tPLH↑
——	—	13-25	—	4.1-5.9	4.1-5.9	—	—	—	tPHL↓
74n11	—	10-20	—	4.1-5.9	4.1-5.9	1-8.5	1-10.5	—	tPLH↑
——	—	10-20	—	4.1-5.9	4.1-5.9	1-7.5	1-10.5	—	tPHL↓
74n21	—	14-22	—	4.1-5.9	4.1-5.9	—	—	—	tPLH↑
——	—	14-22	—	4.1-5.9	4.1-5.9	—	—	—	tPHL↓
74n808	—	10-20	—	—	—	—	—	—	tPLH↑
——	—	10-20	—	—	—	—	—	—	tPHL↓
74n11008	—	—	—	—	—	1.5-6.2	1.5-8	—	tPLH↑
——	—	—	—	—	—	1.5-5.9	1.5-7.7	—	tPHL↓
4081	—	—	—	—	—	—	—	160-300	tPLH↑
——	—	—	—	—	—	—	—	160-300	tPHL↓
4073	—	—	—	—	—	—	—	160-300	tPLH↑
——	—	—	—	—	—	—	—	160-300	tPHL↓
4082	—	—	—	—	—	—	—	160-300	tPLH↑
——	—	—	—	—	—	—	—	160-300	tPHL↓

Table 2.10. CMOS AND Logic gate propagation delays (nS). Vcc = 4.5/5.0 volts, 25°C.

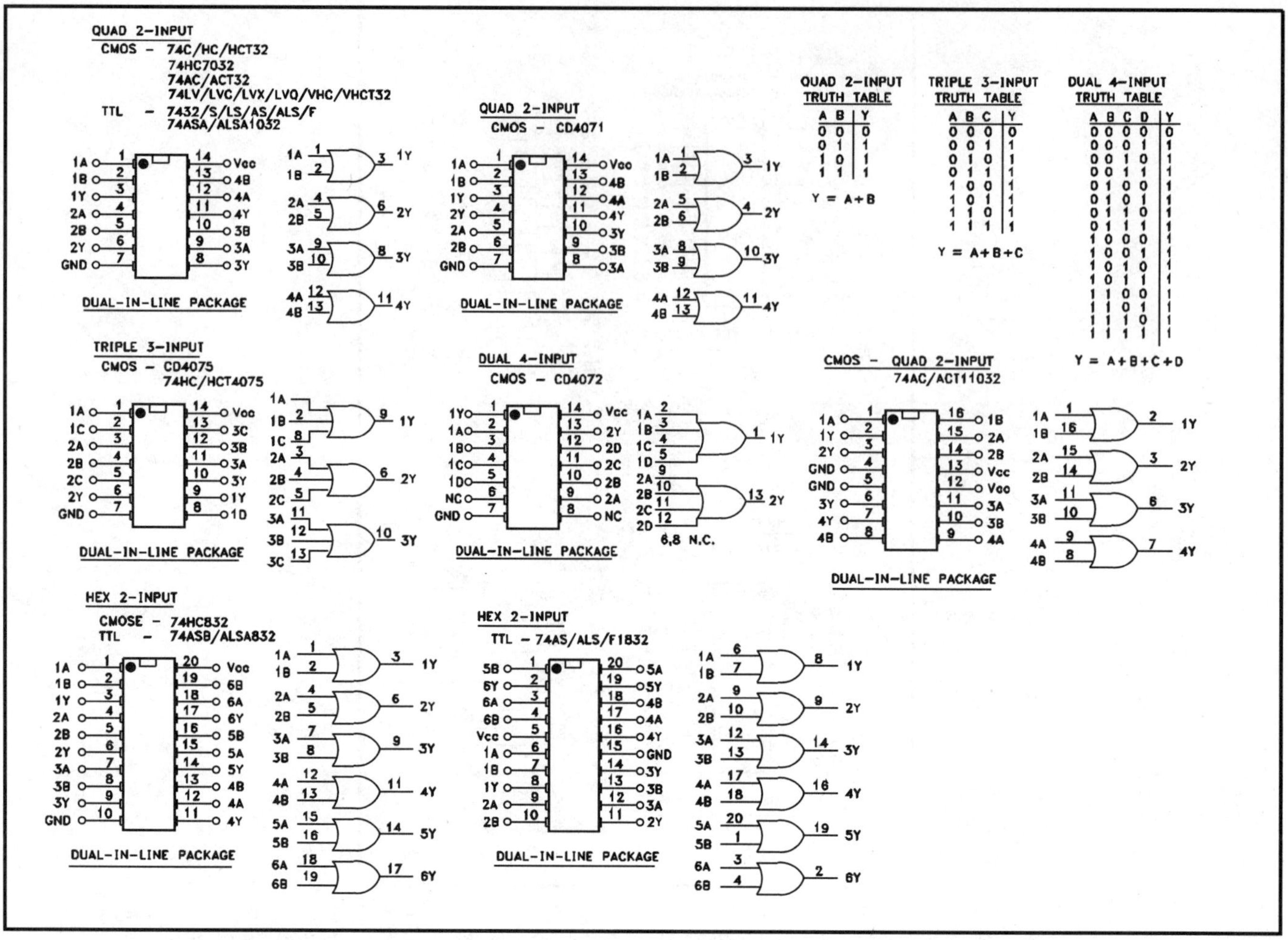

Figure 2.10. Quad 2-input, triple 3-input, dual 4-input, and HEX 2-input OR logic gates, five-volt TTL, three- and five-volt CMOS series devices.

DEVICE	STD	H	L	S	LS	AS/A/B	ALS/A/B	F	DELAY
7432	10-15	—	—	4-7.0	14-22	1-5.8	3-14	3-5.6	tPLH↑
——	12-19	—	—	4-7.0	14-22	1-5.8	3-12	3-5.3	tPHL↓
74832	—	—	—	—	—	1-6.3	2-9	—	tPLH↑
——	—	—	—	—	—	1-6.3	1-8	—	tPHL↓
741032	—	—	—	—	—	1-6.3	2-9	—	tPLH↑
——	—	—	—	—	—	1-6.3	3-12	—	tPHL↓
741832	—	—	—	—	—	1-6.3	—	—	tPLH↑
——	—	—	—	—	—	1-6.3	—	—	tPHL↓

Table 2.11. TTL OR Logic gate propagation delays (nS). Vcc = 5 volts, 25°C.

DEVICE	74C	74HC	74HCT	74AHC	74AHCT	74AC	74ACT	CD4n	DELAY
74n32	12-20	10-18	12-20	3.8-5.5	5-6.9	1.5-7.5	1-9	—	tPLH↑
——	12-20	10-18	9-15	3.8-5.5	5-6.9	1.5-7	1-9	—	tPHL↓
74n11032	—	—	—	—	—	1.5-6.2	1.5-8.1	160-300	tPLH↑
——	—	—	—	—	—	1.5-5.5	1.5-7.4	160-300	tPHL↓
4071	—	—	—	—	—	—	—	160-300	tPLH↑
——	—	—	—	—	—	—	—	160-300	tPHL↓
4075	—	—	—	—	—	—	—	160-300	tPLH↑
——	—	—	—	—	—	—	—	160-300	tPHL↓
4072	—	—	—	—	—	—	—	160-300	tPLH↑
——	—	—	—	—	—	—	—	160-300	tPHL↓

Table 2.12. CMOS OR Logic gate propagation delays (nS). Vcc = 4.5/5.0 volts, 25°C.

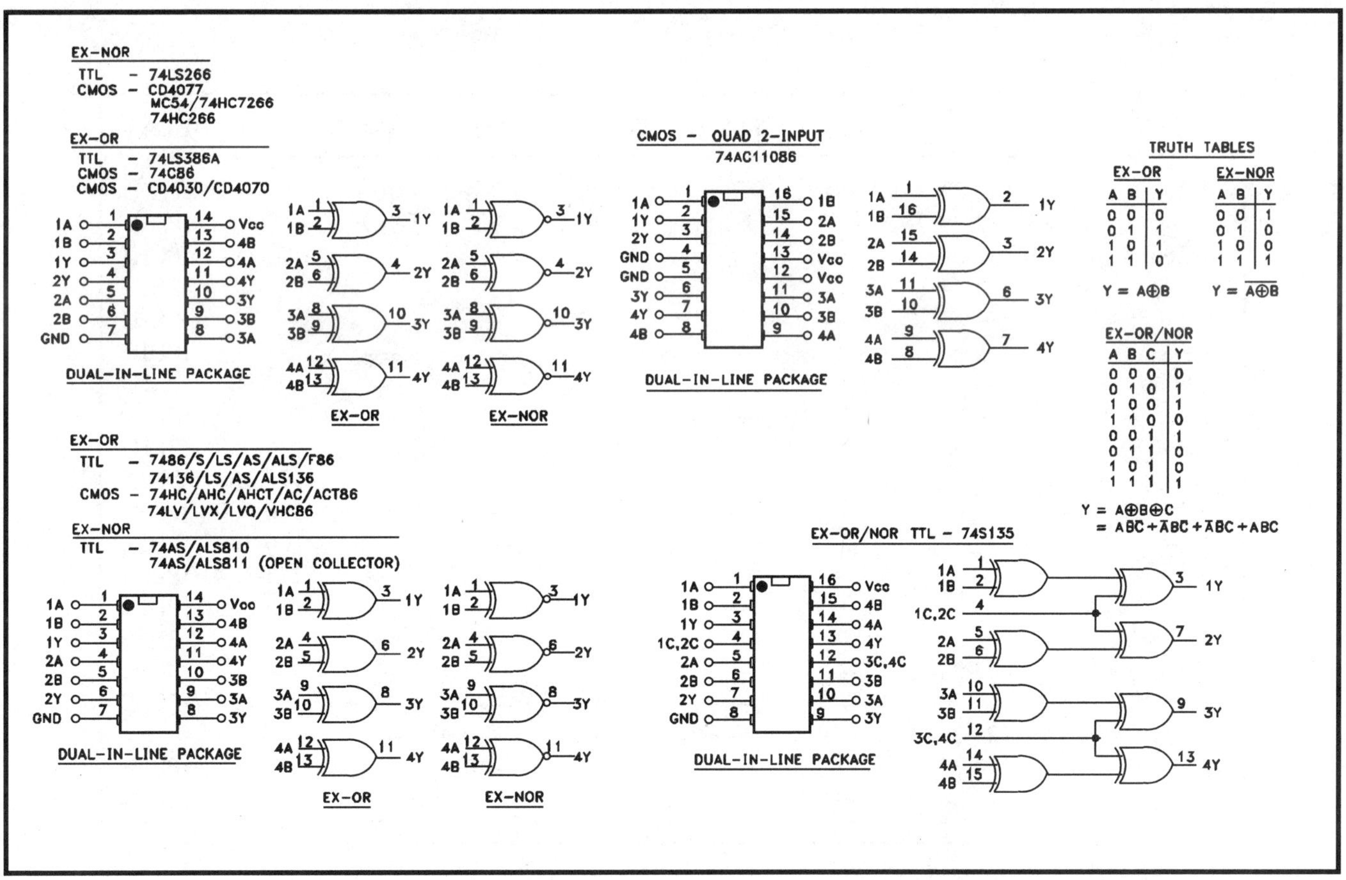

Figure 2.11. Quad 2-input Exclusive-OR/NOR logic gates, five-volt TTL, three- and five-volt CMOS series devices.

DEVICE	STD	H	L	S	LS	AS/A/B	ALS/A/B	F	DELAY
7486	15-23	—	—	7-10.5	12-23	3.6	3-17	3-5.5	tPLH↑
—	12-19	—	—	6.5-10	10-17	3.5	2-12	3-5.5	tPHL↓
74135	—	—	—	8-14.5	—	—	—	—	tPLH↑
—	—	—	—	8-15	—	—	—	—	tPHL↓
74136	18-30	—	—	—	12-22	10.5	20-50	—	tPLH↑
—	18-30	—	—	—	38-55	4.3	3-15	—	tPHL↓
74266	—	—	—	—	18-30	—	—	—	tPLH↑
—	—	—	—	—	18-30	—	—	—	tPHL↓
74386	—	—	—	—	12-23	—	—	—	tPLH↑
—	—	—	—	—	10-17	—	—	—	tPHL↓
74810	—	—	—	—	—	1-6.5	5-20	—	tPLH↑
—	—	—	—	—	—	1-6.5	3-14	—	tPHL↓
74811	—	—	—	—	—	5-45	25-55	—	tPLH↑
—	—	—	—	—	—	1-8.5	5-28	—	tPHL↓

Table 2.13. TTL EX-OR/NOR Logic gate propagation delays (nS). Vcc = 5 volts, 25°C.

DEVICE	74C	74HC	74HCT	74AHC	74AHCT	74AC	74ACT	CD4n	DELAY
74n86	110-185	12-24	—	4.8-6.8	5-6.9	1-9	1-10	—	tPLH↑
—	110-185	12-24	—	4.8-6.8	5-6.9	1-9.5	1-10.5	—	tPHL↓
74n266	—	12-24	—	—	—	—	—	—	tPLH↑
—	—	12-24	—	—	—	—	—	—	tPHL↓
74n11086	—	—	—	—	1.5-6.8	—	—	—	tPLH↑
—	—	—	—	—	1.5-6.2	—	—	—	tPHL↓
CD4030	—	—	—	—	—	—	—	100-300	tPLH↑
—	—	—	—	—	—	—	—	100-300	tPHL↓
MC14070	—	—	—	—	—	—	—	175-350	tPLH↑
—	—	—	—	—	—	—	—	175-350	tPHL↓
MC14072	—	—	—	—	—	—	—	160-300	tPLH↑
—	—	—	—	—	—	—	—	160-300	tPHL↓
MC14077	—	—	—	—	—	—	—	175-350	tPLH↑
—	—	—	—	—	—	—	—	175-350	tPHL↓
MC74HC	—	24-30	—	—	—	—	—	—	tPLH↑
86	—	24-30	—	—	—	—	—	—	tPHL↓
MC74HC	—	24-30	—	—	—	—	—	—	tPLH↑
266	—	24-30	—	—	—	—	—	—	tPHL↓
MC74HC	—	24-30	—	—	—	—	—	—	tPLH↑
7266	—	24-30	—	—	—	—	—	—	tPHL↓

Table 2.14. CMOS EX-OR/NOR Logic gate propagation delays (nS). Vcc = 4.5/5.0 volts, 25°C.

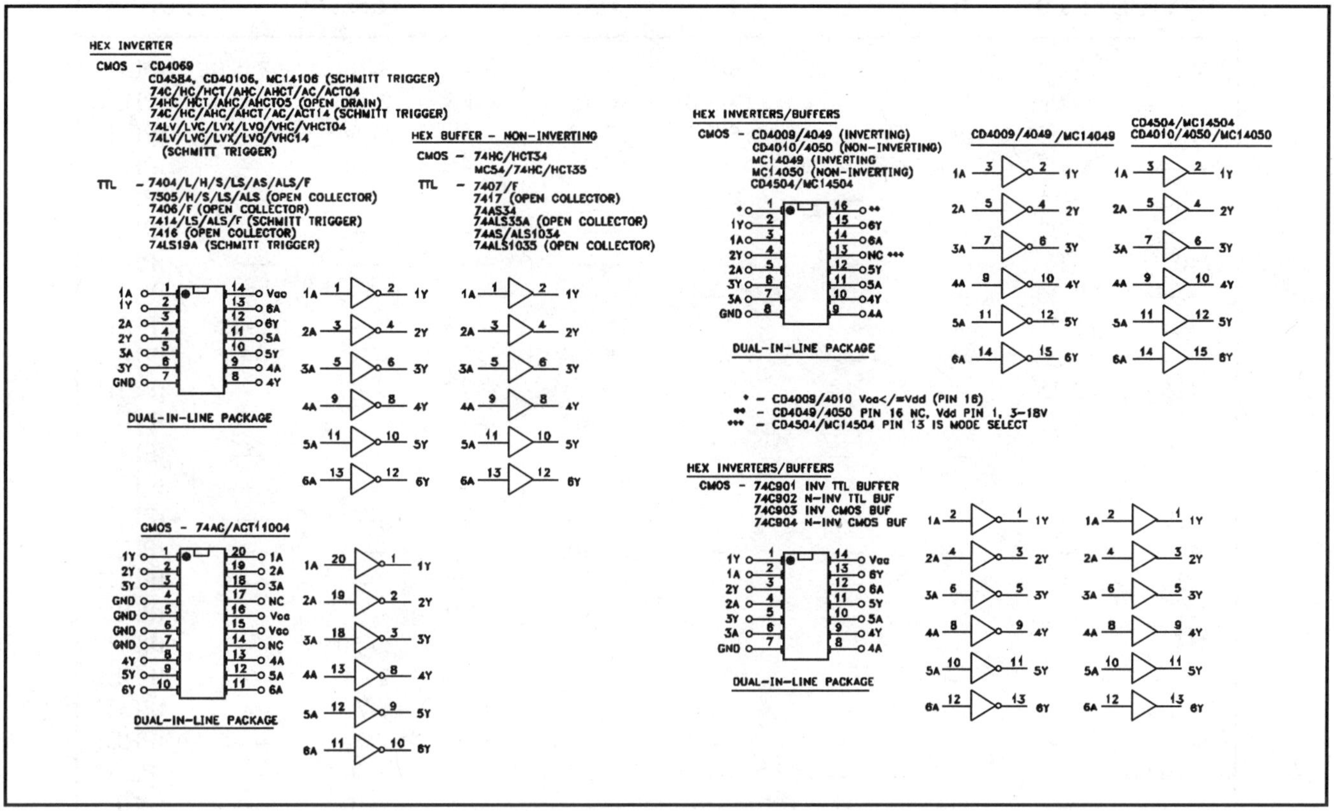

Figure 2.12. HEX inverter and non-inverter five-volt TTL, three- and five-volt CMOS series devices.

DEVICE	STD	H	L	S	LS	AS/A/B	ALS/A/B	F	DELAY
7404	12-22	10	31-60	3-4.5	9-15	1-6	3-14	3.7-5.0	tPLH↑
——	8-15	10	35-60	3-5.0	10-15	1-4.5	2-12	3.2-4.3	tPHL↓
7405	40-55	15	—	5-7.5	17-32	—	23-54	—	tPLH↑
——	8-15	12	—	4.5-7.0	15-28	—	4-14	—	tPHL↓
7406	10-15	—	—	—	—	—	—	—	tPLH↑
——	15-23	—	—	—	—	—	—	—	tPHL↓
7407	6-15	—	—	—	—	—	—	—	tPLH↑
——	20-26	—	—	—	—	—	—	—	tPHL↓
7414	15-22	—	—	—	15-22	—	—	6.5-8.5	tPLH↑
——	15-22	—	—	—	15-22	—	—	5.0-6.5	tPHL↓
7416	10-15	—	—	—	—	—	—	—	tPLH↑
——	15-23	—	—	—	—	—	—	—	tPHL↓
7417	6-15	—	—	—	—	—	—	—	tPLH↑
——	20-26	—	—	—	—	—	—	—	tPHL↓
7419	—	—	—	—	13-20	—	—	—	tPLH↑
——	—	—	—	—	25-40	—	—	—	tPHL↓
7434	—	—	—	—	—	1-5.5	—	—	tPLH↑
——	—	—	—	—	—	1-6.0	—	—	tPHL↓
7435	—	—	—	—	—	—	20-50	—	tPLH↑
——	—	—	—	—	—	—	2-14	—	tPHL↓
741004	—	—	—	—	—	1-4	1-7	—	tPLH↑
——	—	—	—	—	—	1-4	1-6	—	tPHL↓
741005	—	—	—	—	—	—	5-30	—	tPLH↑
——	—	—	—	—	—	—	2-10	—	tPHL↓
741034	—	—	—	—	—	1-6	1-8	—	tPLH↑
——	—	—	—	—	—	1-6	1-8	—	tPHL↓
741035	—	—	—	—	—	—	5-30	—	tPLH↑
——	—	—	—	—	—	—	2-12	—	tPHL↓

Table 2.15. TTL Inverter/Buffer propagation delays (nS). Vcc = 5 volts, 25°C.

DEVICE	74C	74HC	74HCT	74AHC	74AHCT	74AC	74ACT	CD4n	DELAY
74n04	50-90	8-15	10-18	3.8-5.5	4.7-6.7	1.5-7	1-8.5	—	tPLH↑
——	50-90	8-15	10-18	3.8-5.5	4.7-6.7	1.5-6.5	1-8	—	tPHL↓
74n05	—	8-19	8-15	3.8-5.5	3.8-5.5	—	—	—	tPLH↑
——	—	8-19	9-16	3.8-5.5	3.8-5.5	—	—	—	tPHL↓
74n14	220-400	13-25	—	5.5-8.6	4-7	1.5-10	1.5-11.5	—	tPLH↑
——	220-400	13-25	—	5.5-8.6	4-7	1.5-8.5	1.5-10	—	tPHL↓
74n34	—	10-16	10-20	—	—	—	—	—	tPLH↑
——	—	10-16	10-20	—	—	—	—	—	tPHL↓
MC74n35	—	22-28	24-30	—	—	—	—	—	tPLH↑
——	—	22-28	24-30	—	—	—	—	—	tPHL↓
74n901	38-70	—	—	—	—	—	—	—	tPLH↑
——	38-70	—	—	—	—	—	—	—	tPHL↓
74n902	38-70	—	—	—	—	—	—	—	tPLH↑
——	38-70	—	—	—	—	—	—	—	tPHL↓
74n903	38-70	—	—	—	—	—	—	—	tPLH↑
——	38-70	—	—	—	—	—	—	—	tPHL↓
74n904	38-70	—	—	—	—	—	—	—	tPLH↑
——	38-70	—	—	—	—	—	—	—	tPHL↓
74n4049	—	10-17	—	—	—	—	—	—	tPLH↑
——	—	10-17	—	—	—	—	—	—	tPHL↓
74n4050	—	10-17	—	—	—	—	—	—	tPLH↑
——	—	10-17	—	—	—	—	—	—	tPHL↓
74n11004	—	—	—	—	—	1.5-6.3	1.5-9	—	tPLH↑
——	—	—	—	—	—	1.5-5.5	1.5-8.7	—	tPHL↓
4049	—	—	—	—	—	—	—	80-140	tPLH↑
——	—	—	—	—	—	—	—	40-80	tPHL↓
4050	—	—	—	—	—	—	—	80-140	tPLH↑
——	—	—	—	—	—	—	—	40-80	tPHL↓
4106	—	—	—	—	—	—	—	125-250	tPLH↑
——	—	—	—	—	—	—	—	125-250	tPHL↓
4502	—	—	—	—	—	—	—	135-270	tPLH↑
——	—	—	—	—	—	—	—	135-270	tPHL↓
4503	—	—	—	—	—	—	—	75-150	tPLH↑
——	—	—	—	—	—	—	—	75-150	tPHL↓
4584	—	—	—	—	—	—	—	125-250	tPLH↑
——	—	—	—	—	—	—	—	125-250	tPHL↓

Table 2.16. CMOS Inverter/Buffer propagation delays (nS). Vcc = 4.5/5.0 volts, 25°C.

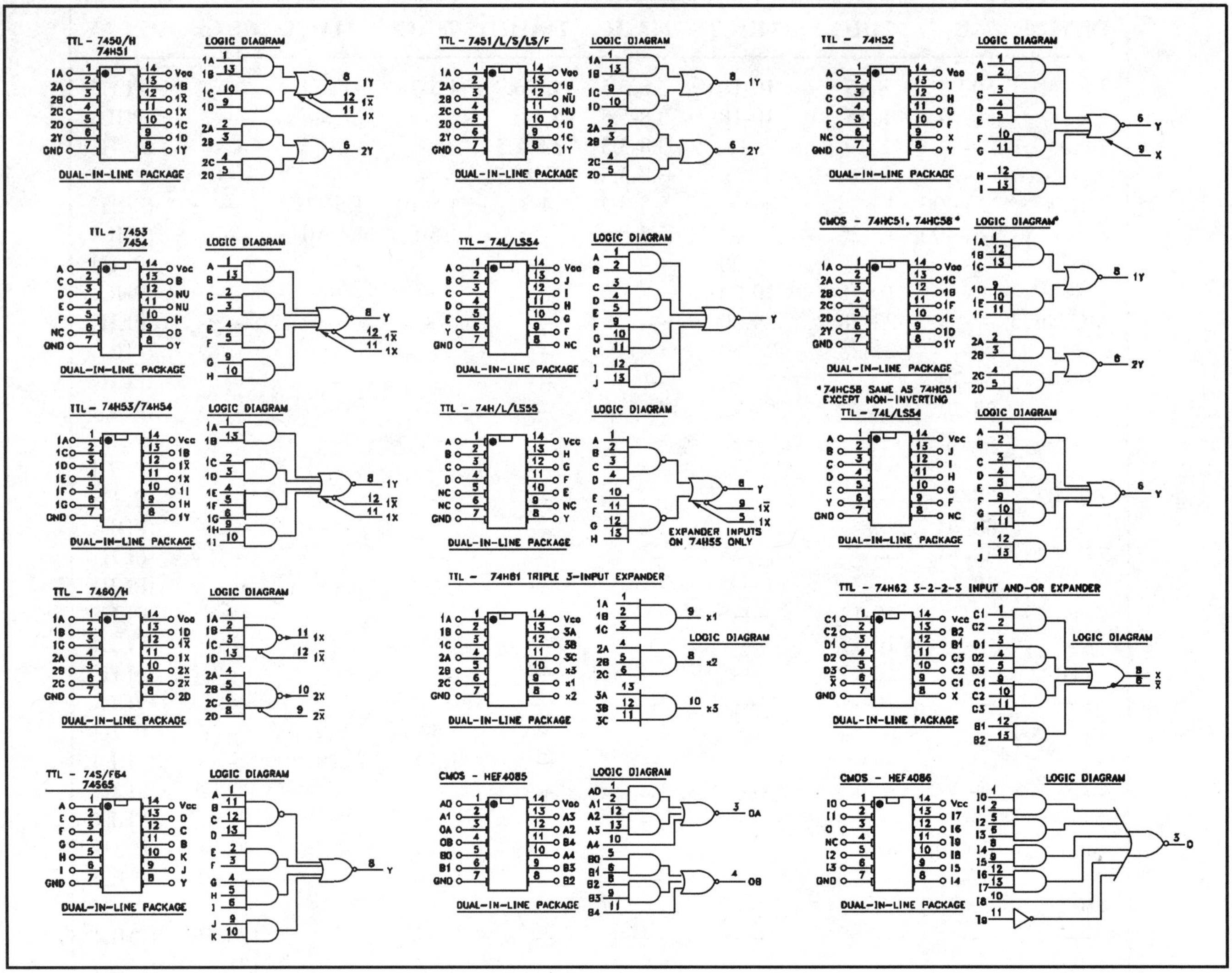

Figure 2.13. AND-OR-INVERT logic gates, five-volt TTL, three- and five-volt CMOS series devices.

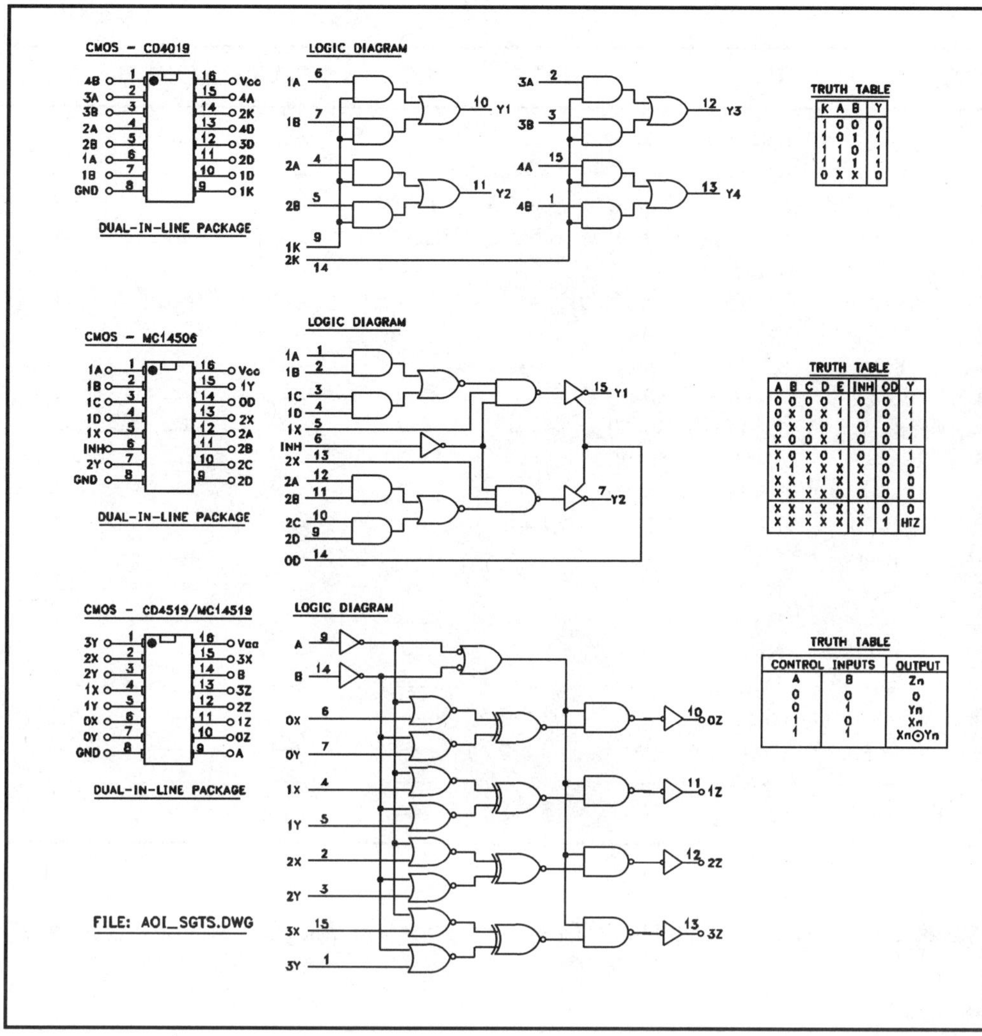

Figure 2.14. AND-OR select logic gates, five-volt CMOS series devices.

DEVICE	STD	H	L	S	LS	AS/A/B	ALS/A/B	F	DELAY
7450	13-22	6.8-11	—	—	—	—	—	—	tPLH↑
——	8-15	6.2-11	—	—	—	—	—	—	tPHL↓
7451	13-22	6.8-11	50-90	3.5-5.5	12-20	—	—	1.5-6.5	tPLH↑
——	8-15	6.2-11	50-90	3.5-5.5	12.5-20	—	—	1-4.5	tPHL↓
7452	—	10.6-15	—	—	—	—	—	—	tPLH↑
——	—	9.2-15	—	—	—	—	—	—	tPHL↓
7453	13-22	7-11	—	—	—	—	—	—	tPLH↑
——	8-15	6.2-11	—	—	—	—	—	—	tPHL↓
7454	13-22	7-11	50-90	—	—	—	—	—	tPLH↑
——	8-15	6.2-11	35-60	—	—	—	—	—	tPHL↓
7455	—	7-11	50-90	—	15-22	—	—	—	tPLH↑
——	—	6.5-11	35-60	—	15-22	—	—	—	tPHL↓
7460	10-30	No	—	—	—	—	—	—	tPLH↑
——	10-30	Data	—	—	—	—	—	—	tPHL↓
7461	—	No	—	—	—	—	—	—	tPLH↑
——	—	Data	—	—	—	—	—	—	tPHL↓
7462	—	No	—	—	—	—	—	—	tPLH↑
——	—	Data	—	—	—	—	—	—	tPHL↓
7464	—	3.5-5.5	—	3.5-5.5	—	—	—	2.5-7.0	tPLH↑
——	—	3.5-5.5	—	3.5-5.5	—	—	—	2.0-5.5	tPHL↓
7465	—	5-7.5	—	5-7.5	—	—	—	—	tPLH↑
——	—	5.5-8.5	—	5.5-8.5	—	—	—	—	tPHL↓

Table 2.17. TTL AND-OR-INVERT (AOI) propagation delays (nS). Vcc = 5 volts, 25°C.

DEVICE	74C	74HC	74HCT	74AHC	74AHCT	74AC	74ACT	CD4n	DELAY
74n51	—	25-31	—	—	—	—	—	—	tPLH↑
——	—	25-31	—	—	—	—	—	—	tPHL↓
74n58	—	25-31	—	—	—	—	—	—	tPLH↑
——	—	25-31	—	—	—	—	—	—	tPHL↓
4019	—	—	—	—	—	—	—	100-300	tPLH↑
——	—	—	—	—	—	—	—	100-300	tPHL↓
4085	—	—	—	—	—	—	—	65-135	tPLH↑
——	—	—	—	—	—	—	—	75-155	tPHL↓
4086	—	—	—	—	—	—	—	80-155	tPLH↑
——	—	—	—	—	—	—	—	30-65	tPHL↓
4506	—	—	—	—	—	—	—	180-580	tPLH↑
——	—	—	—	—	—	—	—	200-480	tPHL↓
4519	—	—	—	—	—	—	—	250-500	tPLH↑
——	—	—	—	—	—	—	—	250-500	tPHL↓

Table 2.18. CMOS AND-NOR/OR Logic gate propagation delays (nS). Vcc = 4.5/5.0 volts, 25°C.

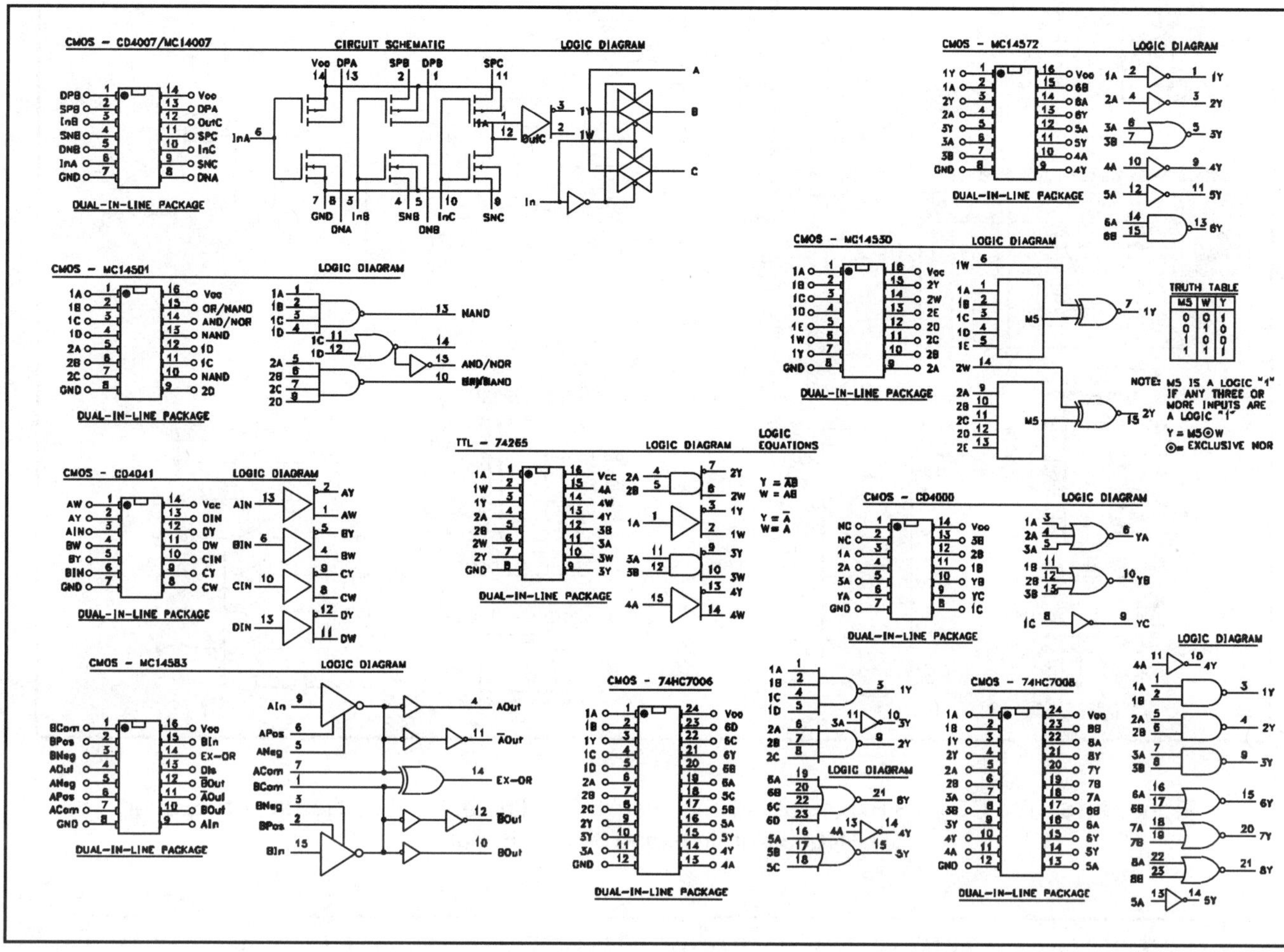

Figure 2.15. Mixed function logic gates, five-volt TTL and CMOS series devices.

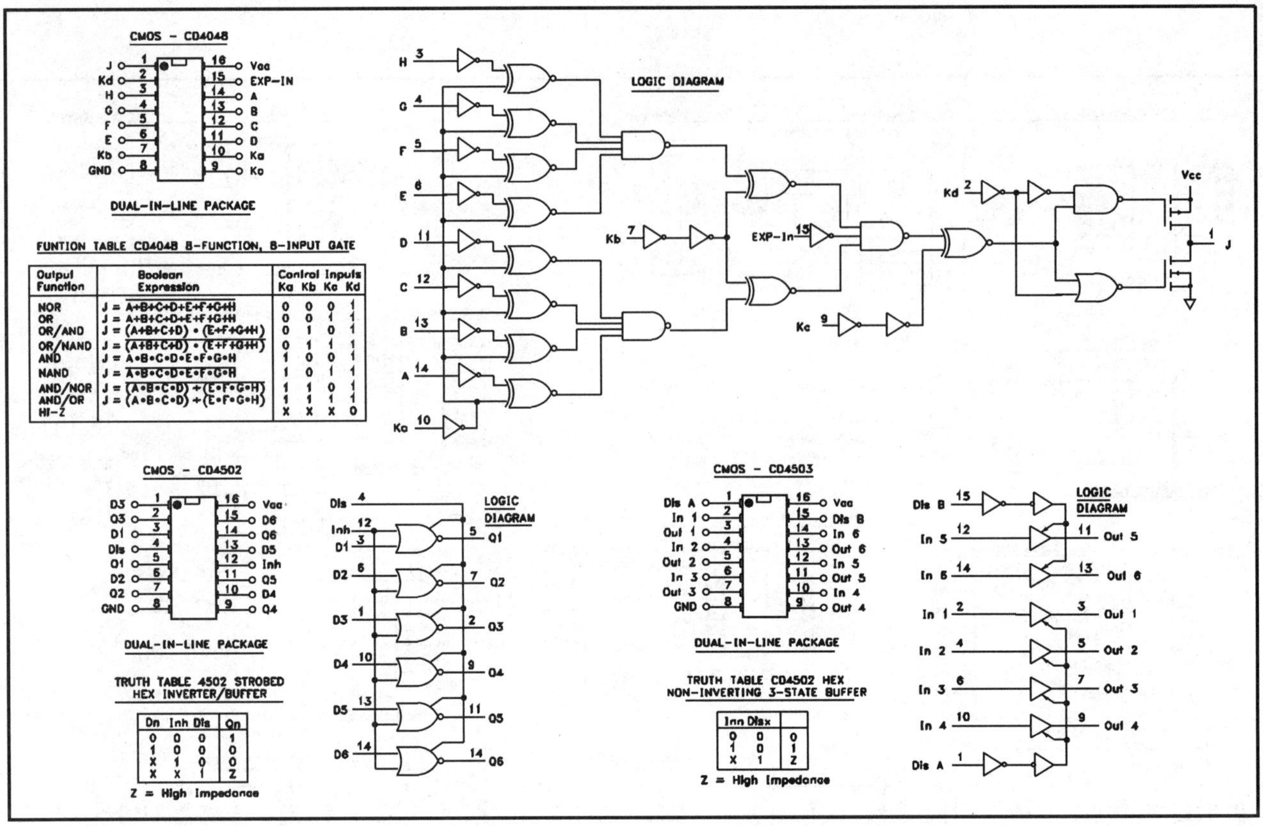

Figure 2.16. The CD4048 Expandable 8-Function, 8-Input gate; CD4502 Hex Inverter/Buffer; CD4503 Non-Inverting 3-State Buffer.

DEVICE	74C	74HC	74HCT	74AHC	74AHCT	74AC	74ACT	CD4n	DELAY
74265	9.8-18	—	—	—	—	—	—	—	tPLH↑
——	11.3-18	—	—	—	—	—	—	—	tPHL↓
747006	—	9-24	—	—	—	—	—	—	tPLH↑
——	—	9-24	—	—	—	—	—	—	tPHL↓
747008	—	9-24	—	—	—	—	—	—	tPLH↑
——	—	9-24	—	—	—	—	—	—	tPHL↓
4000	—	—	—	—	—	—	—	115-230	tPLH↑
——	—	—	—	—	—	—	—	115-230	tPHL↓
4007	—	—	—	—	—	—	—	60-125	tPLH↑
——	—	—	—	—	—	—	—	60-125	tPHL↓
4041	—	—	—	—	—	—	—	60-150	tPLH↑
——	—	—	—	—	—	—	—	60-150	tPHL↓
4048	—	—	—	—	—	—	—	425-850	tPLH↑
——	—	—	—	—	—	—	—	425-850	tPHL↓
4501	—	—	—	—	—	—	—	115-260	tPLH↑
——	—	—	—	—	—	—	—	115-260	tPHL↓
4530	—	—	—	—	—	—	—	375-960	tPLH↑
——	—	—	—	—	—	—	—	375-960	tPHL↓
4572	—	—	—	—	—	—	—	90-180	tPLH↑
——	—	—	—	—	—	—	—	90-180	tPHL↓

Table 2.19. CMOS/TTL Mixed Gate Logic propagation delays (nS). Vcc = 4.5/5.0 volts, 25°C.

DEVICE	B	UB	HC	HCU	3V_AHCU	5V_AHCU	DELAY
MC14001	125-300	90-180	—	—	—	—	tPLH↑
——	125-300	90-180	—	—	—	—	tPLH↑
MC14002	125-300	90-180	—	—	—	—	tPLH↑
——	125-300	90-180	—	—	—	—	tPLH↑
MC14007	—	60-125	—	—	—	—	tPLH↑
——	—	60-125	—	—	—	—	tPLH↑
MC14011	125-300	90-180	—	—	—	—	tPLH↑
——	125-300	90-180	—	—	—	—	tPLH↑
MC14012	125-300	90-180	—	—	—	—	tPLH↑
——	125-300	90-180	—	—	—	—	tPLH↑
MC14023	125-300	90-180	—	—	—	—	tPLH↑
——	125-300	90-180	—	—	—	—	tPLH↑
MC14025	125-300	90-180	—	—	—	—	tPLH↑
——	125-300	90-180	—	—	—	—	tPLH↑
MC14049	40-80	30-60	—	—	—	—	tPLH↑
——	40-80	30-60	—	—	—	—	tPLH↑
MC14069	—	65-125	—	—	—	—	tPLH↑
——	—	65-125	—	—	—	—	tPLH↑
MC14501	—	115-260	—	—	—	—	tPLH↑
——	—	115-260	—	—	—	—	tPLH↑
MC14506	—	180-580	—	—	—	—	tPLH↑
——	—	180-580	—	—	—	—	tPLH↑
MC14572	—	90-180	—	—	—	—	tPLH↑
——	—	90-180	—	—	—	—	tPLH↑
74HCU04	—	—	—	—	5-8.9	3.5-5.5	tPLH↑
——	—	—	—	—	5-8.9	3.5-5.5	tPLH↑

Table 2.20. CMOS UB/HCU/AHCU Unbuffered Logic delays (nS). Vcc =4.5/5.5, 2.7/3.3 volts, 25°C.

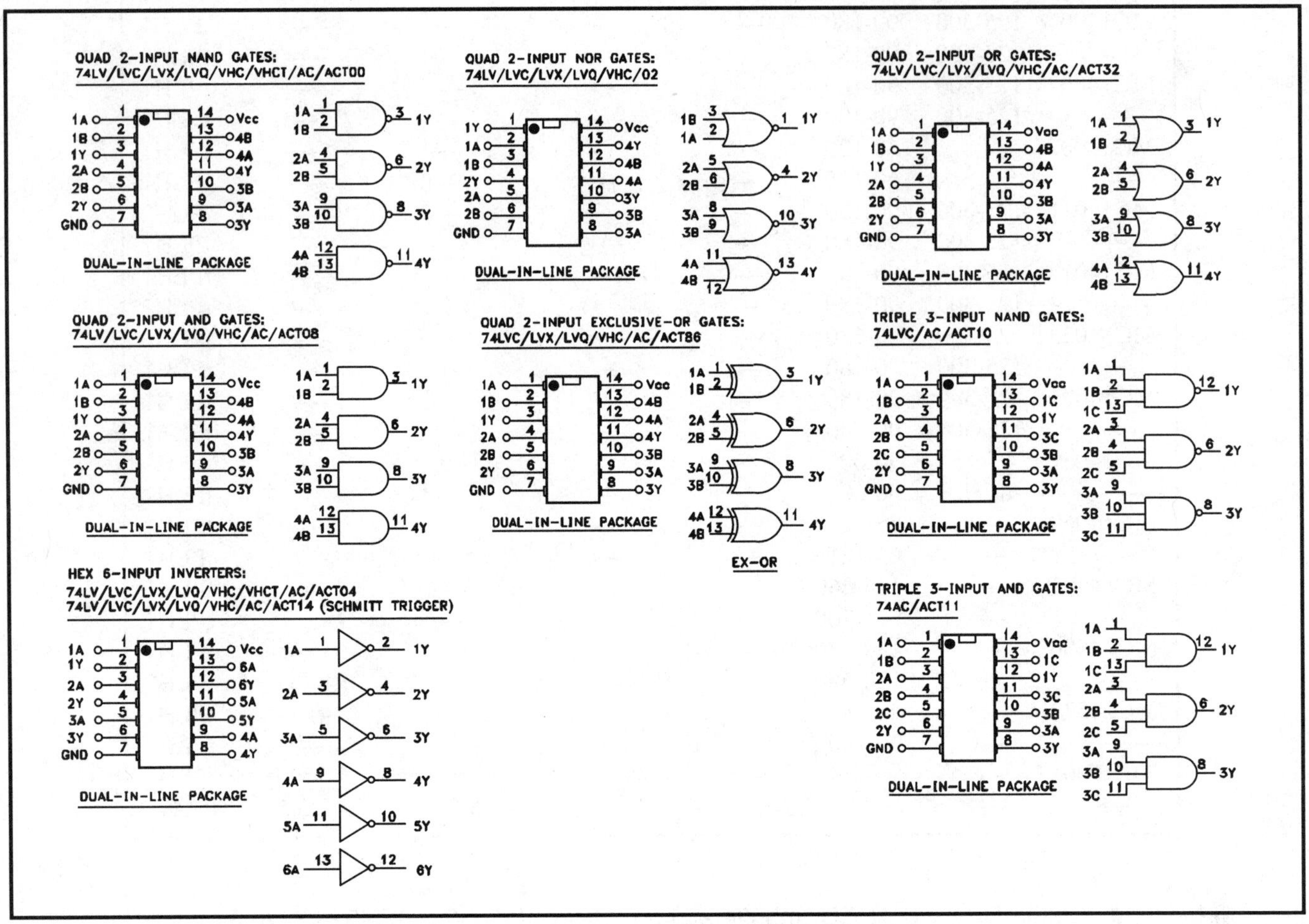

Figure 2.17. Low-voltage, (2.7 - 3.6V), and dual supply (2.7 - 3.6, 4.5 - 5.5V) translator CMOS logic gates and hex inverter.

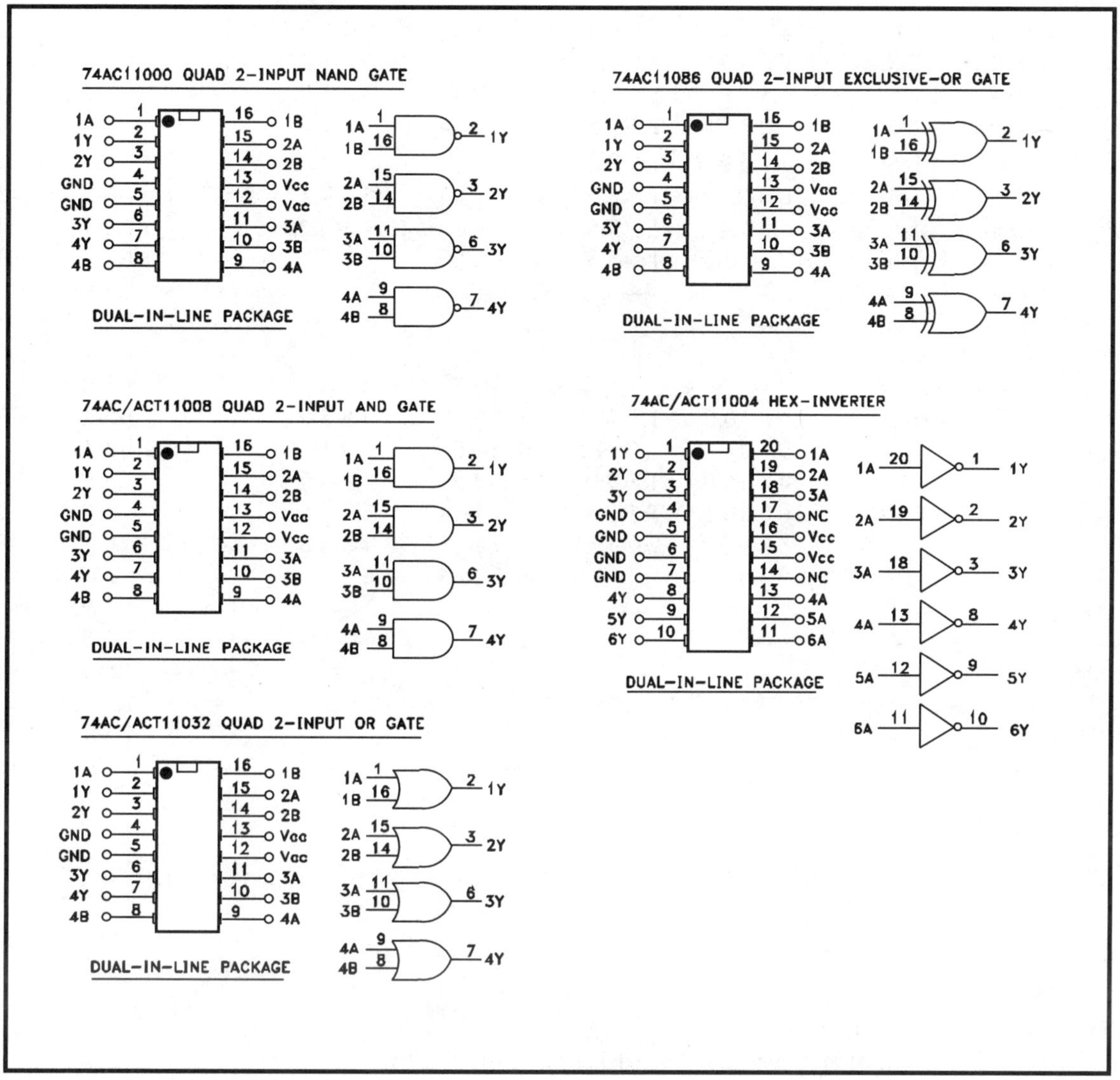

Figure 2.18. Dual-voltage (3.3V ±0.3V, 5V ±0.5V) AC/ACT series CMOS logic gates and hex inverter. While these devices are shown on their respective function drawings, we can see them here as a set.

DEVICE	LV	LVC	LVX	LVQ	3V VHC	5V VHC	5V VHCT	3V AC	DELAY
74x00	9-18	1-6	4.1-6.2	2.0-9.5	5.5-7.9	5.0-6.9	5.0-6.9	2-9.5	tPLH↑
——	9-18	1-6	6.6-9.7	1.5-8.0	5.3-7.9	5.3-7.9	5.5-7.9	1.5-8	tPHL↓
74x02	9-16	1-6	4.5-6.6	1.5-7.5	5.6-7.9	3.0-5.5	—	—	tPLH↑
——	9-16	1-6	7.0-10.1	1.5-7.5	8.1-11.4	5.1-7.5	—	—	tPHL↓
74x04	7-15*	1-6•	4.1-6.2	1.5-9.0	5.0-7.1■	3.8-5.5	4.7-6.7	1.5-9	tPLH↑
——	7-15	1-6	6.6-9.7	1.5-8.5	7.5-10.6	5.8-7.5	5.5-7.7	1.5-8.5	tPHL↓
74x08	10-18	1-6	4.8-7.1	1.5-9.5	6.2-8.8	4.3-5.9	—	1.5-9	tPLH↑
——	10-18	1-6	7.3-10.6	1.5-8.5	8.7-12.3	5.8-7.9	—	1.5-8.5	tPHL↓
74x10	—	1-6	—	—	—	—	—	1.5-9.5	tPLH↑
——	—	1-6	—	—	—	—	—	1.5-8.5	tPHL↓
74x11	—	—	—	—	—	—	—	1.5-9.5	tPLH↑
——	—	—	—	—	—	—	—	1.5-8.5	tPHL↓
74x14	13-30	1-7	6.8-10.6	1.5-13.5	8.3-12.8	5.5-8.6	—	1.5-13.5	tPLH↑
——	13-30	1-7	9.3-14.1	1.5-11.5	10.8-16.3	7.0-10.6	—	1.5-11.5	tPHL↓
74x32	8-17	1.5-6	4.4-16.6	1.5-9.0	5.5-7.9	3.8-5.5	—	1.5-9	tPLH↑
——	8-17	1.5-6	6.9-10.1	1.5-8.5	8.0-11.4	5.3-7.5	—	1.5-8.5	tPHL↓
74x86	—	1-6.5	5.8-9.3	2.0-11.5	7.0-11.5	4.8-6.8	—	2-11.5	tPLH↑
——	—	1-6.5	8.3-12.8	2.0-11.5	9.5-14.5	6.3-8.8	—	2-11.5	tPHL↓
74x11000	—	—	—	—	—	—	—	1.3-9.8	tPLH↑
——	—	—	—	—	—	—	—	1.5-8.6	tPHL↓
74x11004	—	—	—	—	—	—	—	1.5-9	tPLH↑
——	—	—	—	—	—	—	—	1.5-7.4	tPHL↓
74x11008	—	—	—	—	—	—	—	1.5-9	tPLH↑
——	—	—	—	—	—	—	—	1.5-7.8	tPHL↓
74x11032	—	—	—	—	—	—	—	1.5-8.7	tPLH↑
——	—	—	—	—	—	—	—	1.5-7.4	tPHL↓
74x11086	—	—	—	—	—	—	—	1.5-9.4	tPLH↑
——	—	—	—	—	—	—	—	1.5-7.4	tPHL↓

* 74LVU04 = 6/14,6/14 nS
• 74LCVU04 = 1/5,1/54 nS
■ 74VHCU04 = 5.0/8.9,7.5/11.4 nS
5V 74AC/ACT times provided in Tables 2.6, 2.8, 2.10, 2.12, 2.14, 2.16, 2.18.

Table 2.21. CMOS Low-Voltage Gate, Inverter delays (nS). Vcc = 3.3 volts, 25°C.

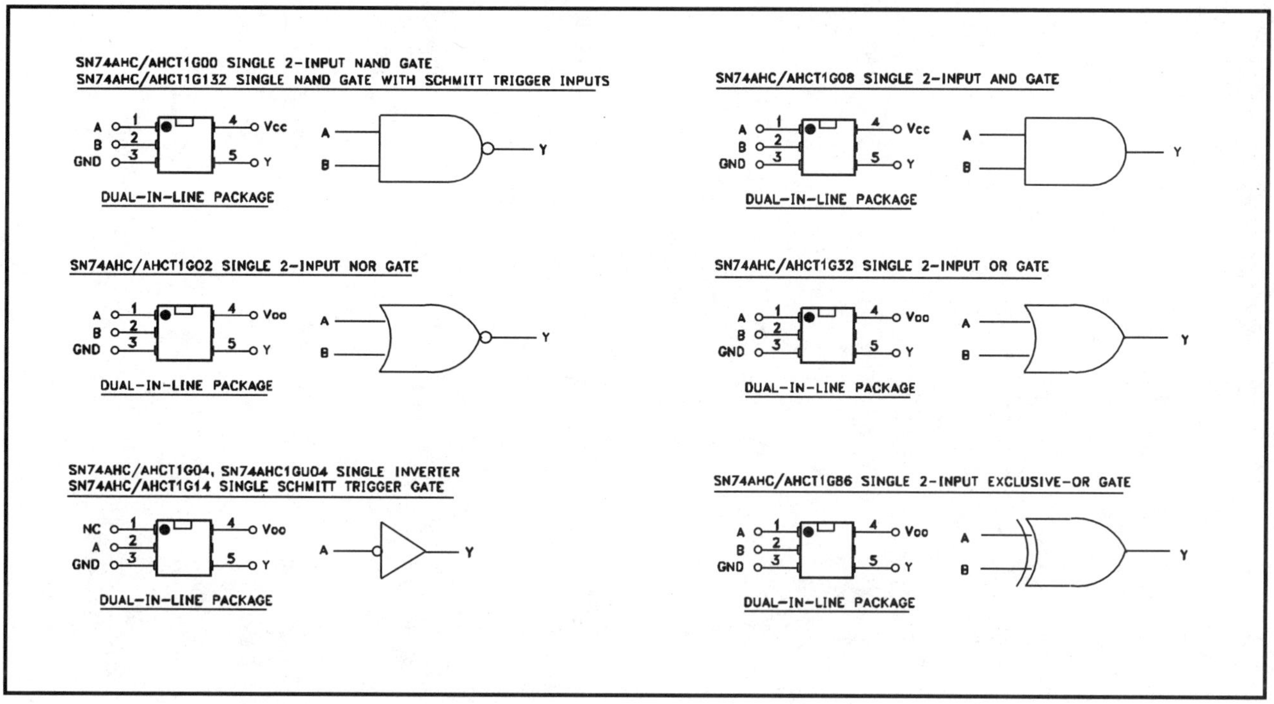

Figure 2.19. Single-gate, dual-voltage (3.3V ±0.3V, 5V ±0.5V) AHC/AHCT series CMOS logic gates and hex inverter.

DEVICE	3V AHC	5V AHC	5V AHCT	DELAY
74x1G00	5.5-7.9	3.7-5.5	5.0-6.9	tPLH↑
——	5.5-7.9	3.7-5.5	5.0-6.9	tPHL↓
74x1G02	5.6-7.9	3.6-5.5	2.4-5.5	tPLH↑
——	5.6-7.9	3.6-5.5	3.5-6.9	tPHL↓
74x1G04	5.0-7.1	3.8-5.5*	4.7-6.7	tPLH↑
——	5.0-7.1	3.8-5.5	4.7-6.7	tPHL↓
74x1G08	6.2-8.8	4.3-5.9	4.7-6.7	tPLH↑
——	6.2-8.8	4.3-5.9	4.7-6.7	tPHL↓
74x1G14	8.3-12.8	5.5-8.6	4.0-7.0	tPLH↑
——	8.3-12.8	5.5-8.6	4.0-7.0	tPHL↓
74x1G32	5.6-7.9	3.8-5.5	5.0-6.9	tPLH↑
——	5.6-7.9	3.8-5.5	5.0-6.9	tPHL↓
74x1G86	7.0-11.0	4.8-6.8	5.6-9.0	tPLH↑
——	7.0-11.0	4.8-6.8	5.6-9.0	tPHL↓
74x1G132	5.5-7.9	3.7-5.5	5.0-6.9	tPLH↑
——	5.5-7.9	3.7-5.5	5.0-6.9	tPHL↓

* 74AHC1U04 - 3.3V: 5.0/8.9, 5.0/8.9; 5V: 3.5/5.5, 3.5,5.5

Table 2.22. CMOS single 2-input gate and inverter delays (nS). Vcc = 3.3/5.0V, 25°C.

Chapter 3

Monostables and Timers

We often employ a monostable, frequently called "one-shots," or, more formally, "multivibrators," when the need arises for a single pulse of a predetermined width. In the pre-solid state era a monostable consisted of two capacitor-coupled triode vacuum tubes such that the circuit rested in a stable quiescent state, logical "0," until disturbed by a trigger that placed the circuit in a momentary logical "1" condition. After a predetermined time interval the circuit reverted to logical "0" to await the arrival of the next trigger. Then the transistor came on the scene and the vacuum tube circuit gave way to its solid state equivalent. There were many variations of these: *Figure 3.1* illustrates a typical circuit. With the values shown the circuit provides a 750 mS pulse which comes out to about 0.7R3C2, or a single time constant of these two timing components.

If a pulse is all that is required, this circuit, or any number of variants, will do the job. But the modern monostable IC has features obtainable only with significant effort from a circuit of discrete components. Prominent of these are multiple inputs, both positive-edge and negative-edge triggering, retriggerability and resettability. They are easy to use and cost very little.

Table 3.1 summarizes the features of available TTL and CMOS monostables. Their operations are described with some detail in the figures which show the dual in-line package terminations, the logic diagram, an input table, and a summary of features.

Table 3.2 is a summary of circuit monostable and timer integrated circuit sources referenced in this book. The table also relates the device to its figure number.

The 9600 series, shown in *Figure 3.2*, while not part of the TTL 7400 series, were among the first of the IC monostables. Though not found in the current catalog listings, these were versatile with features carried over into the later devices. The 74121 single one-shot, *Figure 3.3*, features a Schmitt input that gives it jitter-free response to slow-rising triggers—similarly for the 74122/LS122, which offers more triggering options as well as the Low Power Schottky version.

Possibly the most popular mono is the 74123 in its TTL and CMOS versions, *Figure 3.4*. Several devices, both TTL and CMOS, shown in the figure, follow its package pinning. The 74C221 and its HC version are also very popular, differing primarily from the '123 in not being retriggerable.

The CD/MC14528 and 4538 series of CMOS monos, *Figure 3.5*, differ in their pinouts and triggering. 74HC/HCT versions of the 4538 are offered by Philips. Two Motorola devices, the MC14536B and MC14541, are more complex monostables, doubling as timers.

Space limitations prevent showing the many capabilities of these; the manufacturer's data books should be consulted before making use of them. The same is true for the versatile CD4047, shown in *Figure 3.6*.

Figure 3.7 is a selection guide that compares the timing ranges of the Motorola monostables. This can be helpful when deciding on a device for a specific timing span.

Figure 3.8 describes features of the TTL 555/CMOS 7555 Timers. These are also provided in a dual timer package, the TTL 556/CMOS 7556. These highly useful devices feature very high timing stability over their full range of supply voltages and temperatures. Both a free running (astable) mode and a triggered single pulse mode are possible.

Figure 3.9 describes the TTL 558 Timer. The drawing includes a simplified logic diagram and a long-duration timer application. Note that the device consists of four independent timers with a common power supply and voltage reference. Since each stage is brought out there may be opportunities to utilize intermediate times in a sequence.

In *Figure 3.10* we see another complex mono/timer, the HC/HCT5555, with many unique features. Again, the Philips reference should be consulted for details not provided here.

Although monostables are generally used as a source of single pulses on demand, their use as a continuous pulse source is feasible. *Figure 3.11* shows a 74LS123 dual-mono in a feedback circuit with a 74LS04 inverter.

The Texas Instrument 74S124 Dual-Voltage Controlled Oscillator is a rather fascinating device. Unfortunately, the databook provides no function table or logic diagram. Some readers may question the presence of the Philips HEF4753, seen in *Figure 3.13*. It is included in this chapter as a timer because it has great potential for that as well as its single or multiple pulse output.

Summary

In comparison to the devices described in the preceding and following chapters, the quantity and variety of monostables and timers is limited, which would simply appear to suggest that what exists is meeting the need. Can't ask for much more than that.

Some points to keep in mind as we review the data provided in the figures are:

- Frequencies and times are provided for comparison purposes only. Because of individual distinctions in how sources describe their products, this book can only provide an overview. Where performance requirements are rigorous, always consult the manufacturer's data books.

- Data shown is for the industrial/commercial or recommended free-air operating temperature range.

- The specifications for CMOS devices may cover voltage ranges from 2 to 15 volts. Data is typically provided for VCC's of 5, 10, and 15 volts, but not always. Some sources use 5.0 volts; others 4.5. There is a significant speed factor. The 4.5 or 5.0 volt data is provided here.

- A similar situation exists for the more recent low-voltage devices. Only a few monostables are provided, Vcc ranges are typically 2.7 - 3.6 volts, but VCC values may be given for 2, 2.6, 3.0, 3.3, or 3.6 volts. Data shown is typically for 2.7 - 3.3 volts.

Device Number	Sngl/ Dual	Retrig	Reset	Capac Min Max	Resis Min Max	Timing Eqn Cext > 1000 pF
74121	Sngl	No	No	0 1000	1.4 40	tW = .55RC(1+0.7/R)
74LS122	Sngl	Yes	Yes	No Limit	5 260	tW = .45RC
74123	Dual	Yes	Yes	No Limit	5 50	tW = .34RC(1+0.7/R)
74LS123	Dual	Yes	Yes	No Limit	5 260	tW = .45RC
74LS221	Dual	No	Yes	0 1000	1.4 100	tW = .7RC
DM9601	Sngl	Yes	No	No Limit	5 50	tW = .32RC(1+0.7/R)
DM9602	Dual	Yes	Yes	No Limit	No Limit	tW = .31RC(1+1/R)

Table 3.1 (a). TTL family monostable features.

Device Number	Sngl/ Dual	Retrig	Reset	Capac Min Max	Resis Min Max	Timing Eqn Cext > 1000 pF
74HC123	Dual	Yes	Yes	No Limit	2 1M	tW = RC
74HCT123	Dual	Yes	Yes	No Limit	2 1M	tW = RC
74HC423	Dual	Yes	Yes	No Limit	2 1M	tW = .45RC
74HCT423	Dual	Yes	Yes	No Limit	2 1M	tW = .45RC
74VHC123	Dual	Yes	Yes	No Limit	1 No Lim	tW = RC
74C221	Dual	No	Yes	No Limit	1.4 100	tW = RC
74C221	Dual	No	Yes	No Limit	1 No Lim	tW = RC
4528	Dual	Yes	Yes	No Limit	5 1M	tW = .2RCln(VDD-VSS)
4538	Dual	Yes	Yes	No Limit	5 1M	tW = .2RCln(VDD-VSS)
74HC4538	Dual	Yes	Yes	No Limit	5 1M	tW = .7RC
74HCT4538	Dual	Yes	Yes	No Limit	2 1M	tW = .7RC

Table 3.1 (b). CMOS family monostable features.

Notes:

1. Resistance in Kohms.
2. CMOS timing equations based on 4.5/5.5 Vcc. 2.0/3.0 - 5.5 VCC for VHC monos.
3. 74HC/HCT423 same as 74HC/HCT123 but cannot be triggered via the reset line.

	TTL						5V CMOS				HI-SPEED CMOS				3V CMOS					
	STD		S		LS		4000		C		HC		HCT		LV		VHC		Figure	
DEVICE	NM	ST	NM	ST	NM	ST	NM	ST	NM	ST	NM	ST	NM	ST	NM	ST	NM	ST	No.	DEVICE
9600	x																		3.2	9600
9601	x																		3.2	9601
9602	x																		3.2	9602
74121	x	xx																	3.3	74121
74122		xx			xx														3.3	74122
74123	x	xx			xx	x						x		x		x	x		3.4	74123
74124				x															3.12	74124
74130		x				x													3.4	74130
74221		xx			x	xx			x		x	x		x			x		3.4	74221
74422						x													3.4	74422
74423						x						x		x					3.4	74423
744538											x	x		x					3.5	744538
745555												x		x					3.10	745555
4047							x	x											3.6	4047
4098 *							--	--											3.5	4098
4528							x	x											3.5	4528
4536							x												3.5	4536
4538							x	x											3.5	4538
4541																			3.5	4541
4753							x	x											3.13	4753
4938								x											3.5	4938
555								x											3.8	555
556								x											3.8	556
7555 **							--	--											3.8	7555
7556 **							--	--											3.8	7556
558								x											3.8	558
5555								x											3.10	5555

Key to Manufacturer: N – National, M – Motorola,
S – Signetics/Philips, T – Texas Instruments

Manufacturer's prefix for 4000 series CMOS

Manufacturer	Prefix
N – National,	CD
M – Motorola,	MC1
S – Signetics/Philips,	HEC/HEF
* – RCA, Harris	CD
** – Intersil,Harris	ICM

Table 3.2. TTL and CMOS monostable and timer source summary by manufacturer.

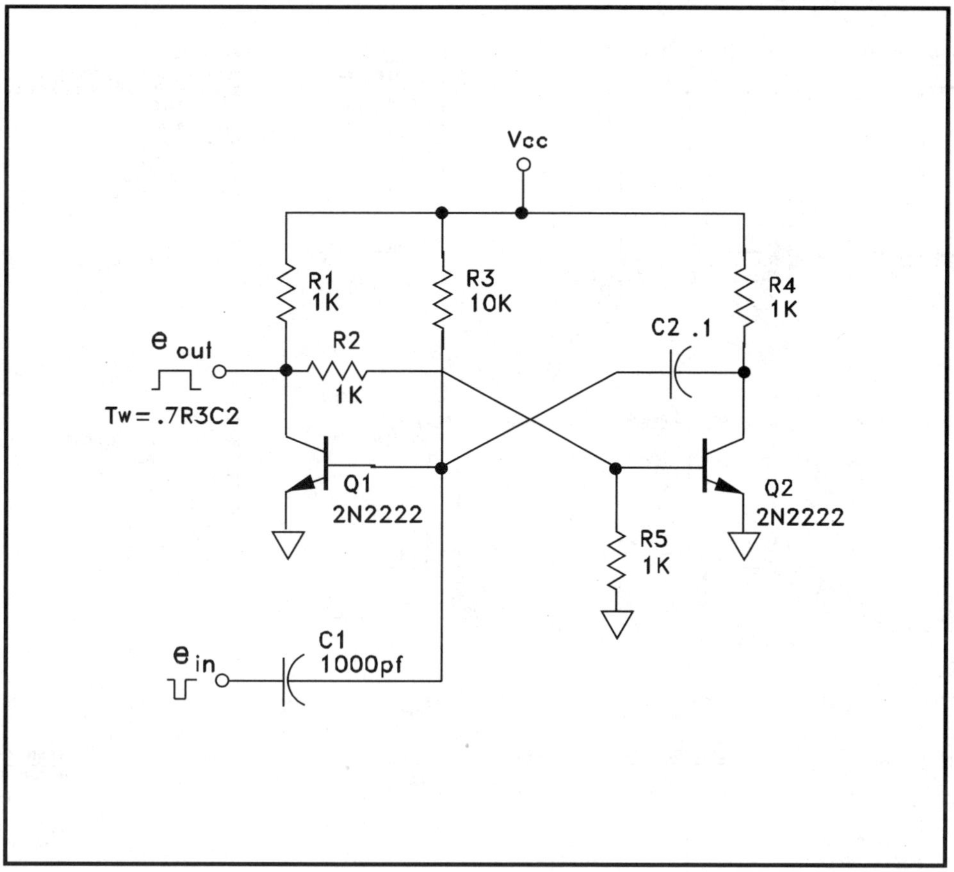

Figure 3.1. A discrete component transistor one-shot. With the values shown the output pulse width is approximately 750 mS.

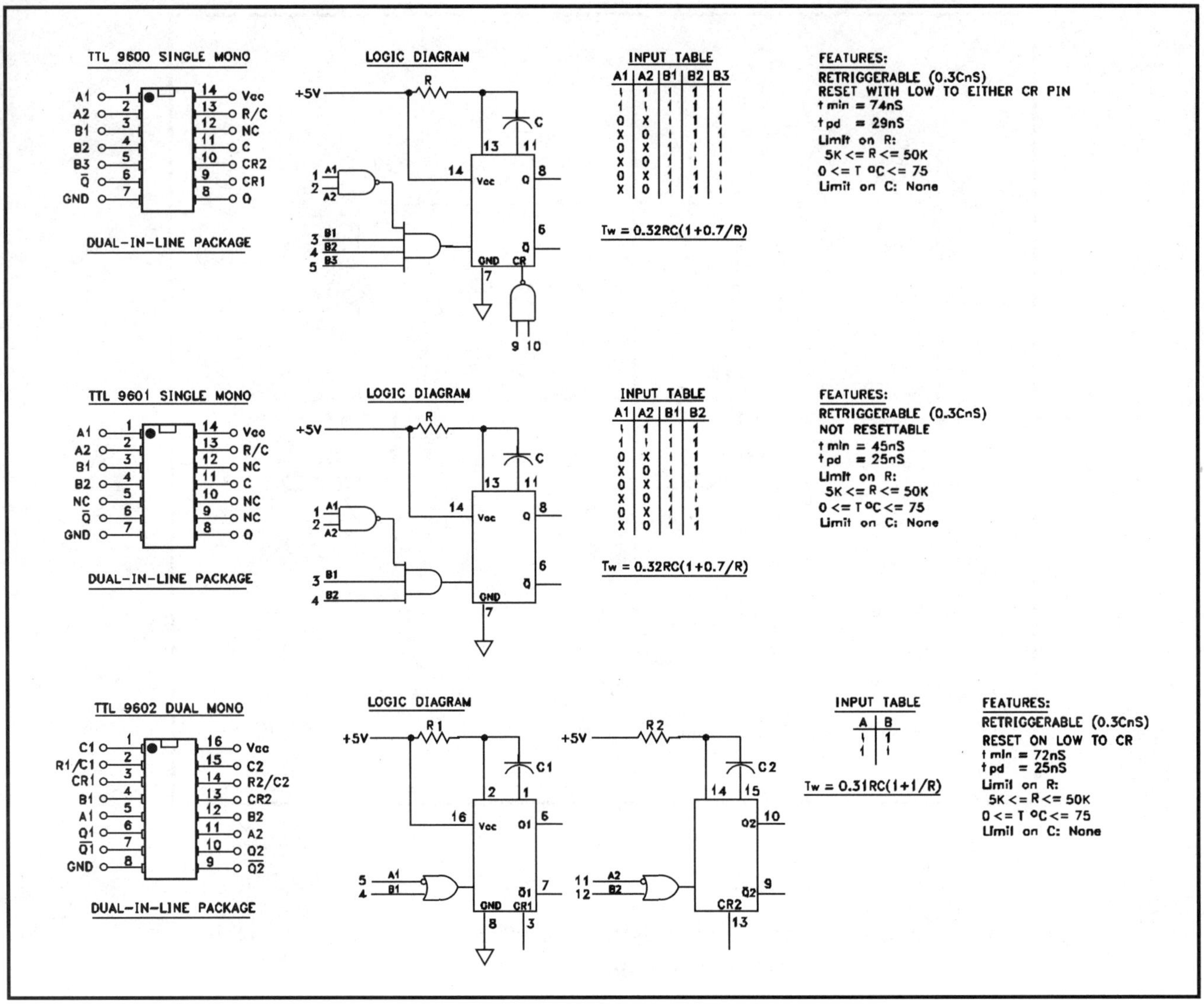

Figure 3.2. The 9600 series TTL single- and dual-IC Monostables.

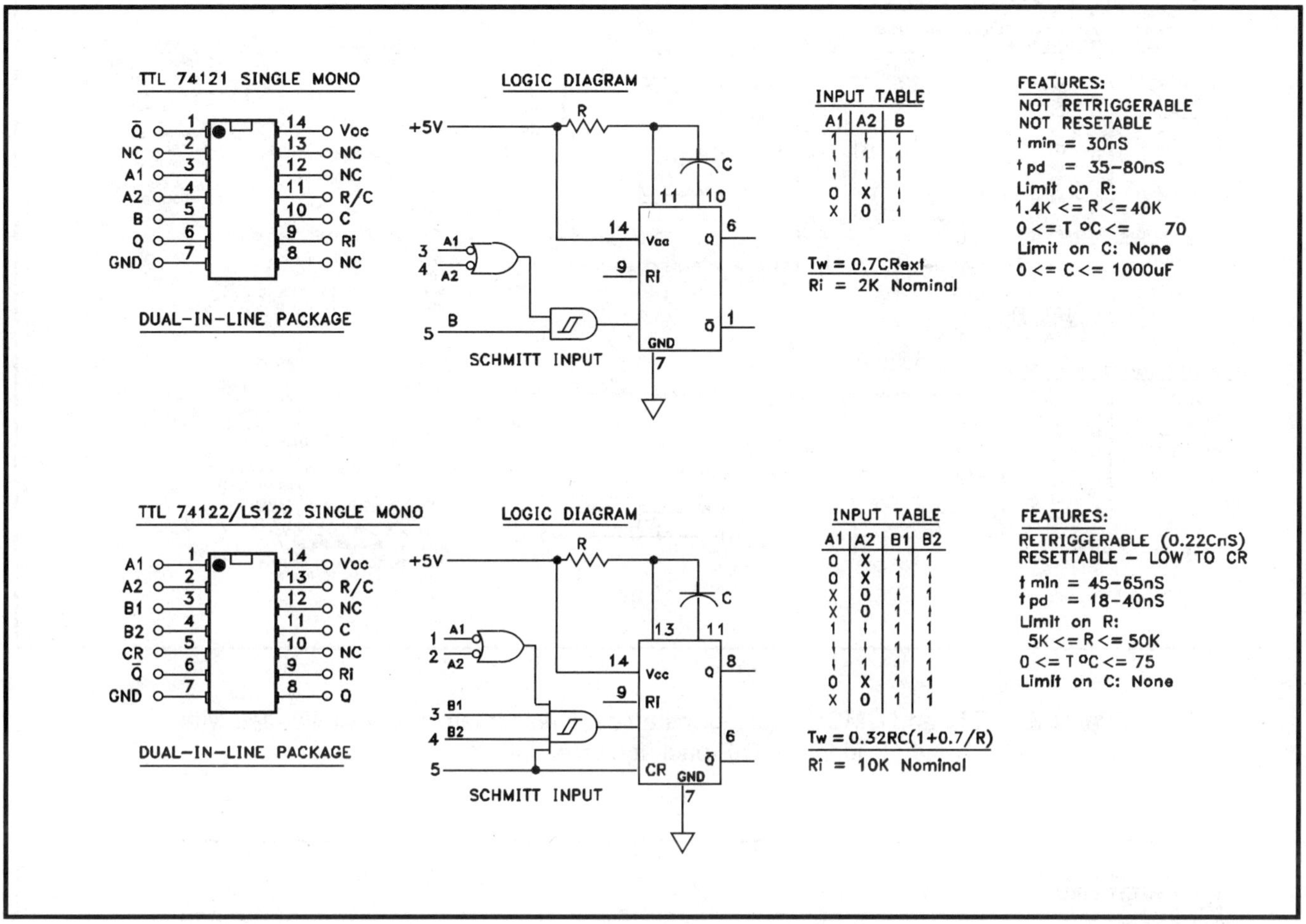

Figure 3.3. The 74121 and 74122/LS122 series of single TTL monostables.

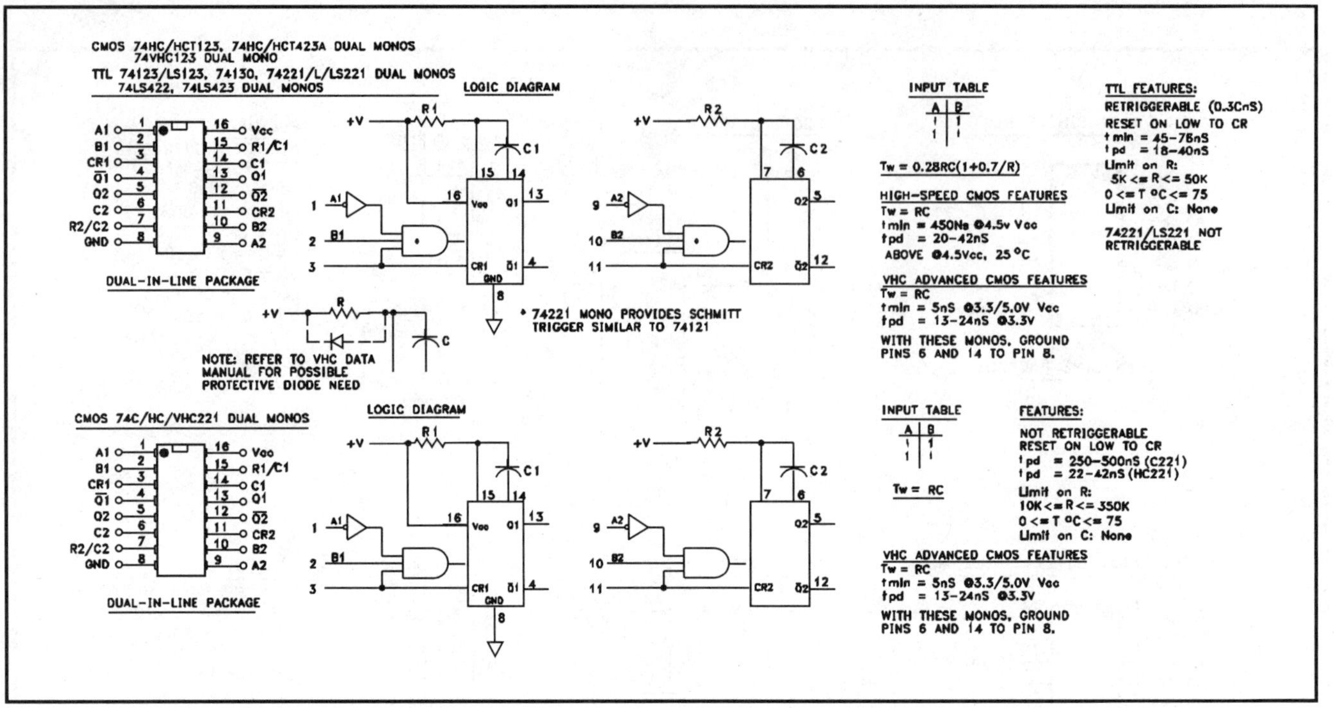

Figure 3.4. TTL and CMOS dual monostables. These share many similarities with identical DIP package terminations.

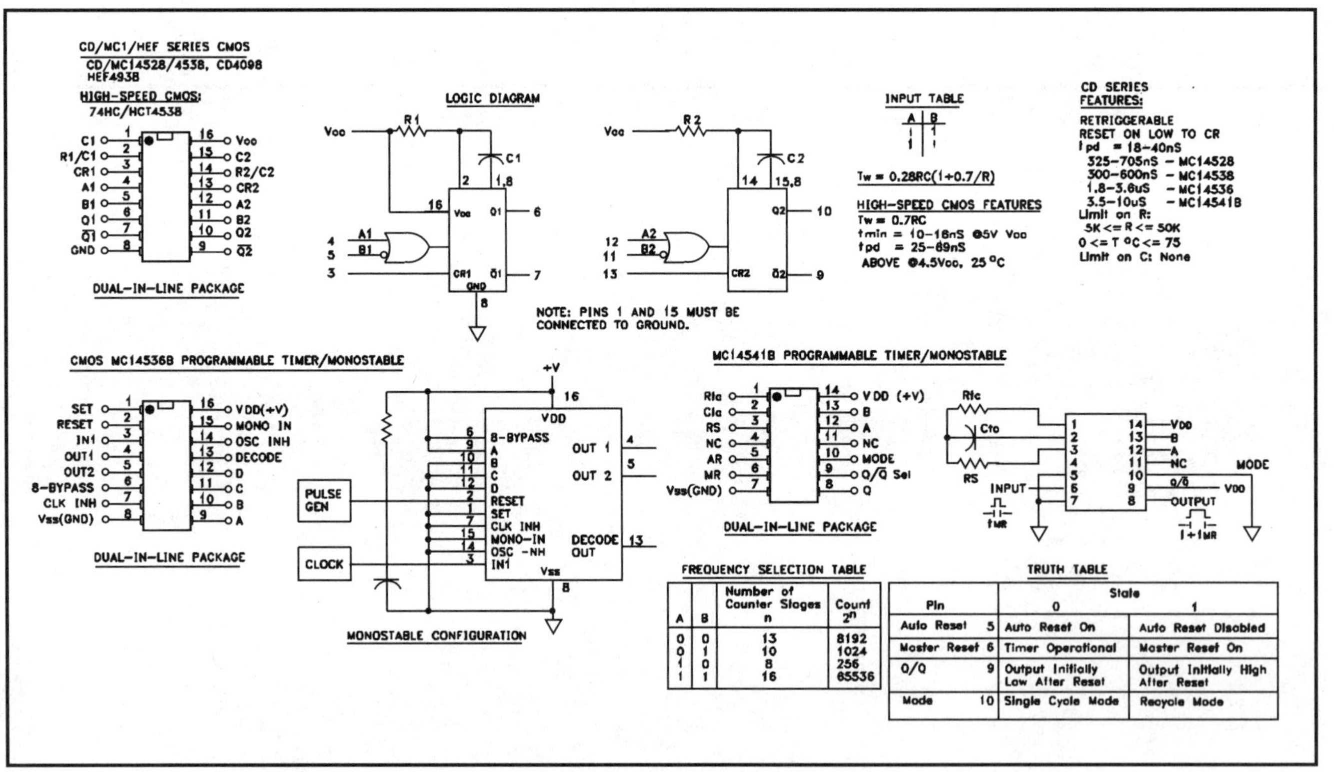

FREQUENCY SELECTION TABLE

A	B	Number of Counter Stages n	Count 2^n
0	0	13	8192
0	1	10	1024
1	0	8	256
1	1	16	65536

TRUTH TABLE

Pin		State 0	State 1
Auto Reset	5	Auto Reset On	Auto Reset Disabled
Master Reset	6	Timer Operational	Master Reset On
Q/Q	9	Output Initially Low After Reset	Output Initially High After Reset
Mode	10	Single Cycle Mode	Recycle Mode

Figure 3.5. The 4528/4538 CMOS and High-Speed CMOS monostable series, which includes the CD4098. Also included are two versatile programmable timer/monostable ICs: the MC14536B and MC14541.

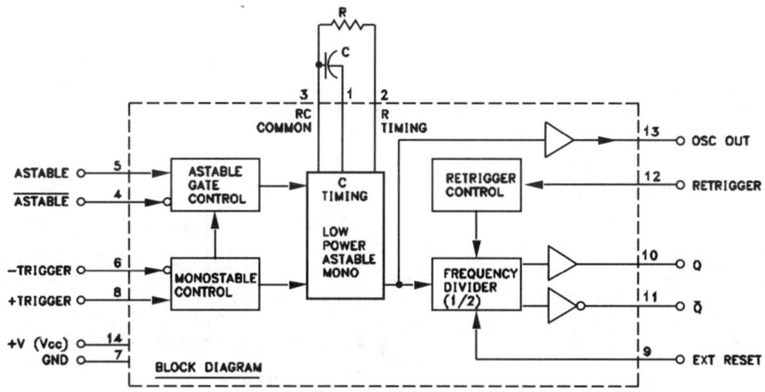

(b)

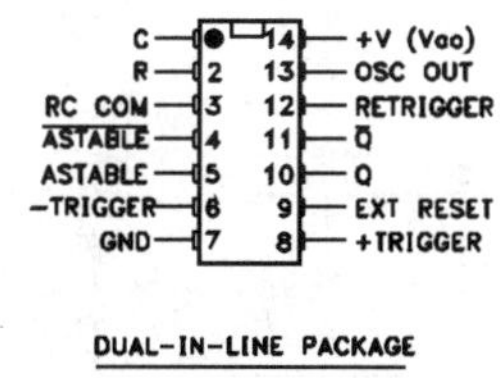

(a)

TRUTH TABLE

FUNCTION	TERMINAL CONNECTIONS TO Vcc	TO GND	INPUT PULSE TO	OUTPUT PULSE FROM	TYPICAL OUTPUT PERIOD OR PULSE WIDTH
ASTABLE MULTIVIBRATOR					
FREE-RUNNING	4,5,6,14	7,8,9,12		10,11,13	tA(10,11)=4.40RC
TRUE GATING	4,6,14	7,8,9,12	5	10,11,13	
COMPLEMENT GATING	6,14	5,7,8,9,12	4	10,11,13	tA(13)=2.20RC
MONOSTABLE MULTIVIBRATOR					
POSITIVE-EDGE TRIGGER	4,14	5,6,7,9,12	8	10,11	
NEGATIVE-EDGE TRIGGER	4,8,14	5,7,9,12	6	10,11	tM=(10,11)=2.48RC
RETRIGGERABLE	4,14	5,6,7,9	8,12	10,11	
EXTERNAL COUNTDOWN	14	5,6,7,8,9,12	REF FIGURE	REF FIGURE	REF FIGURE

NOTE: EXTERNAL RESISTOR BETWEEN TERMINALS 2 AND 3
EXTERNAL CAPACITOR BETWEEN TERMINALS 1 AND 3

(c)

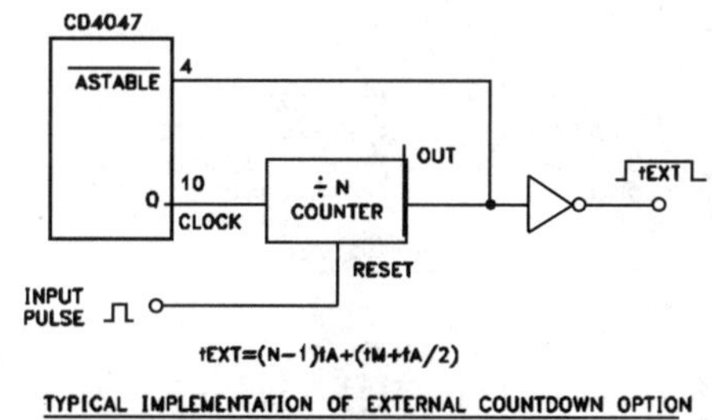

(d)

Figure 3.6. The CD4047 low-power monostable/astable multivibrator.

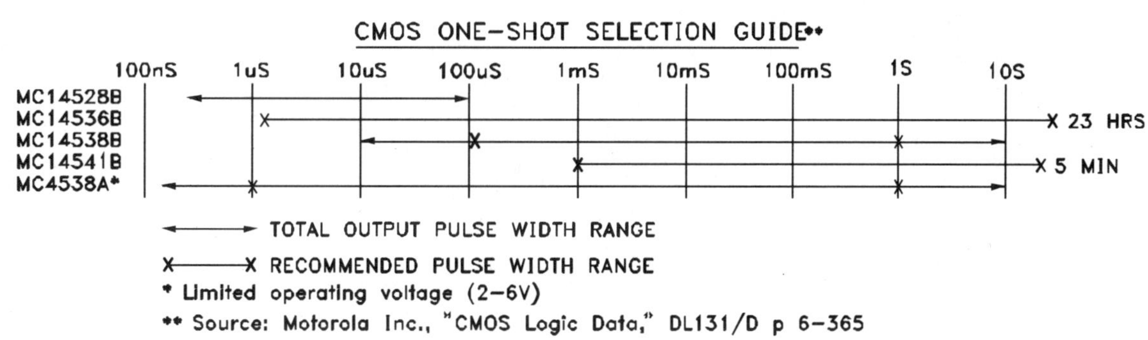

Figure 3.7. A performance selection guide for Motorola CMOS monostables.

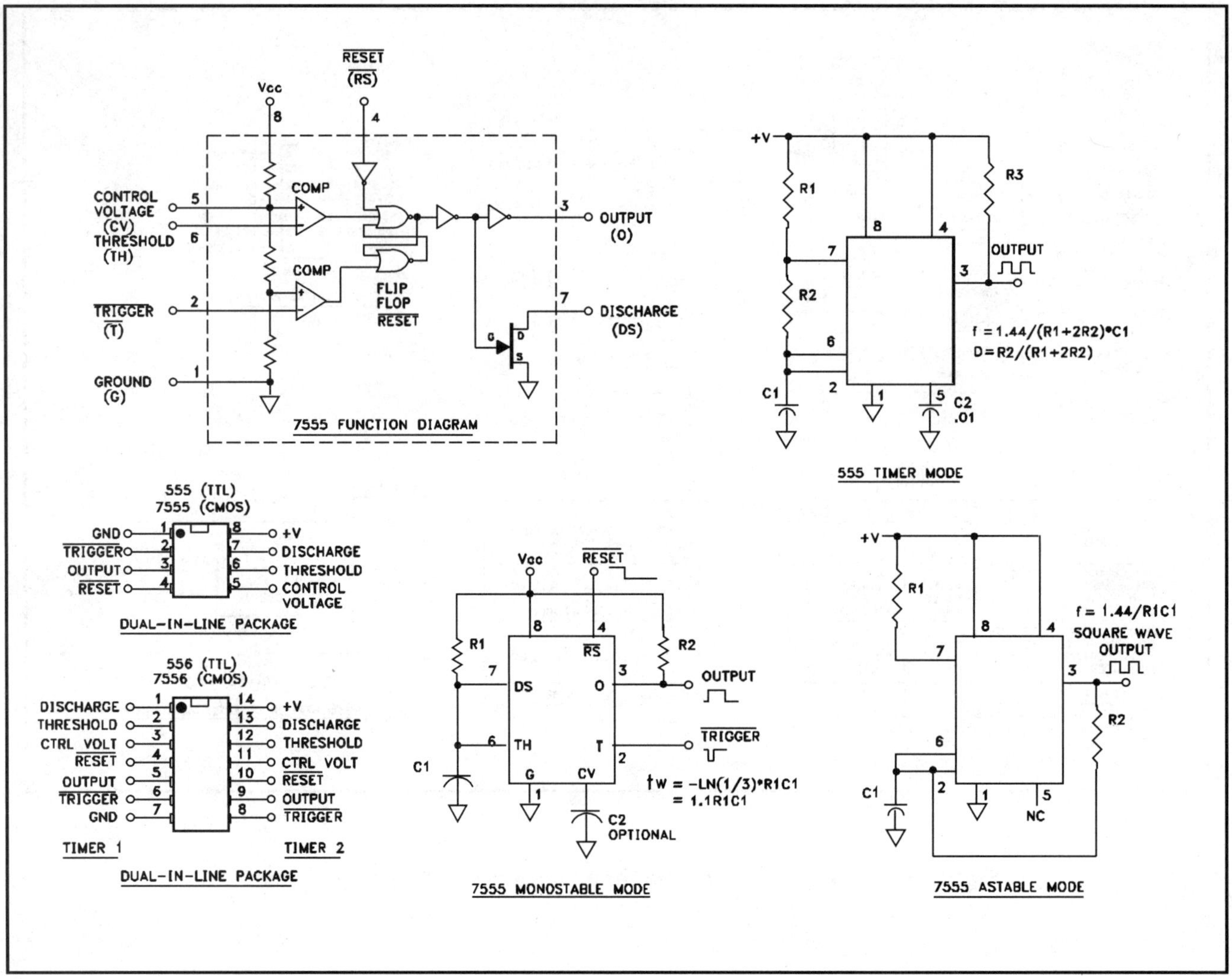

Figure 3.8. The 555/556(TTL) and 7555/7556 (CMOS) IC monostable/astable timer.

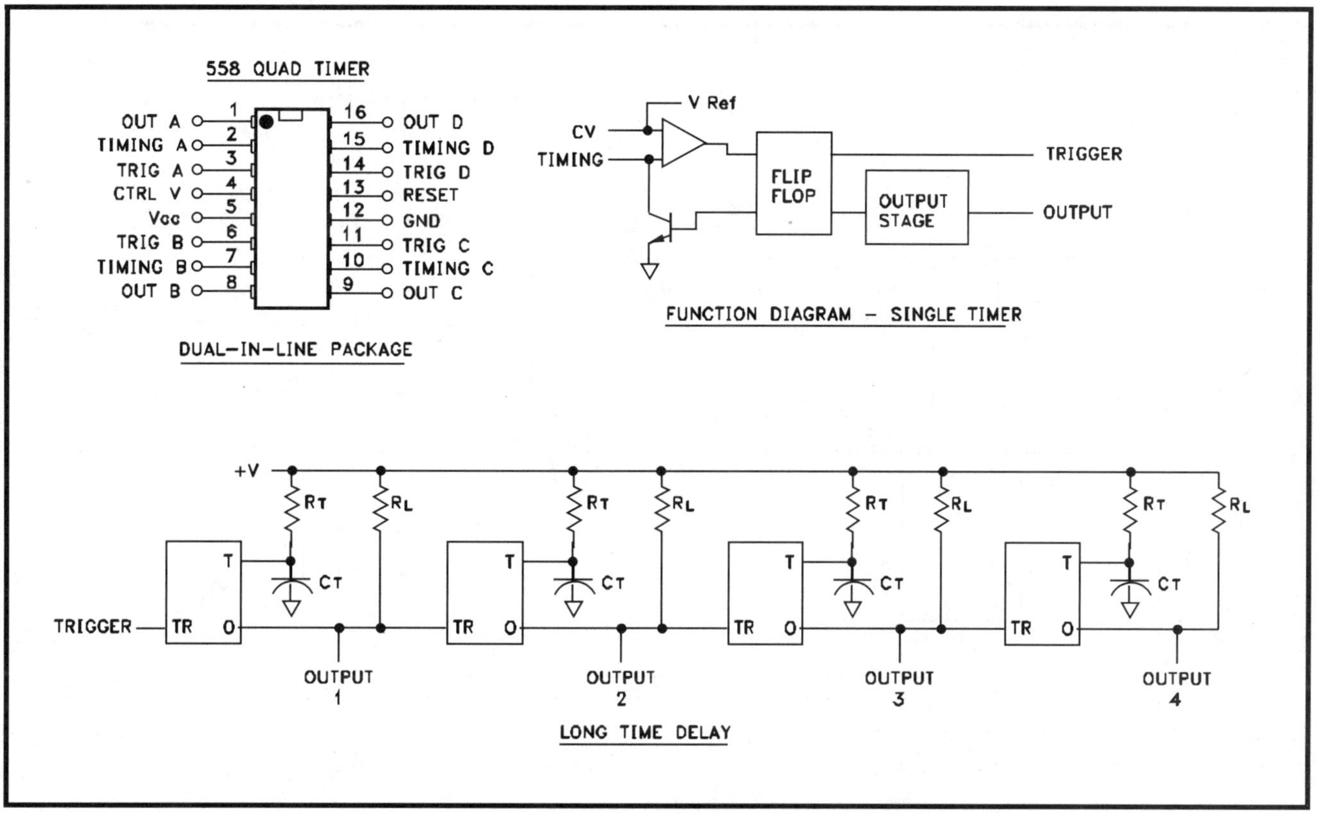

Figure 3.9. The 558 Quad IC monostable/astable timer.

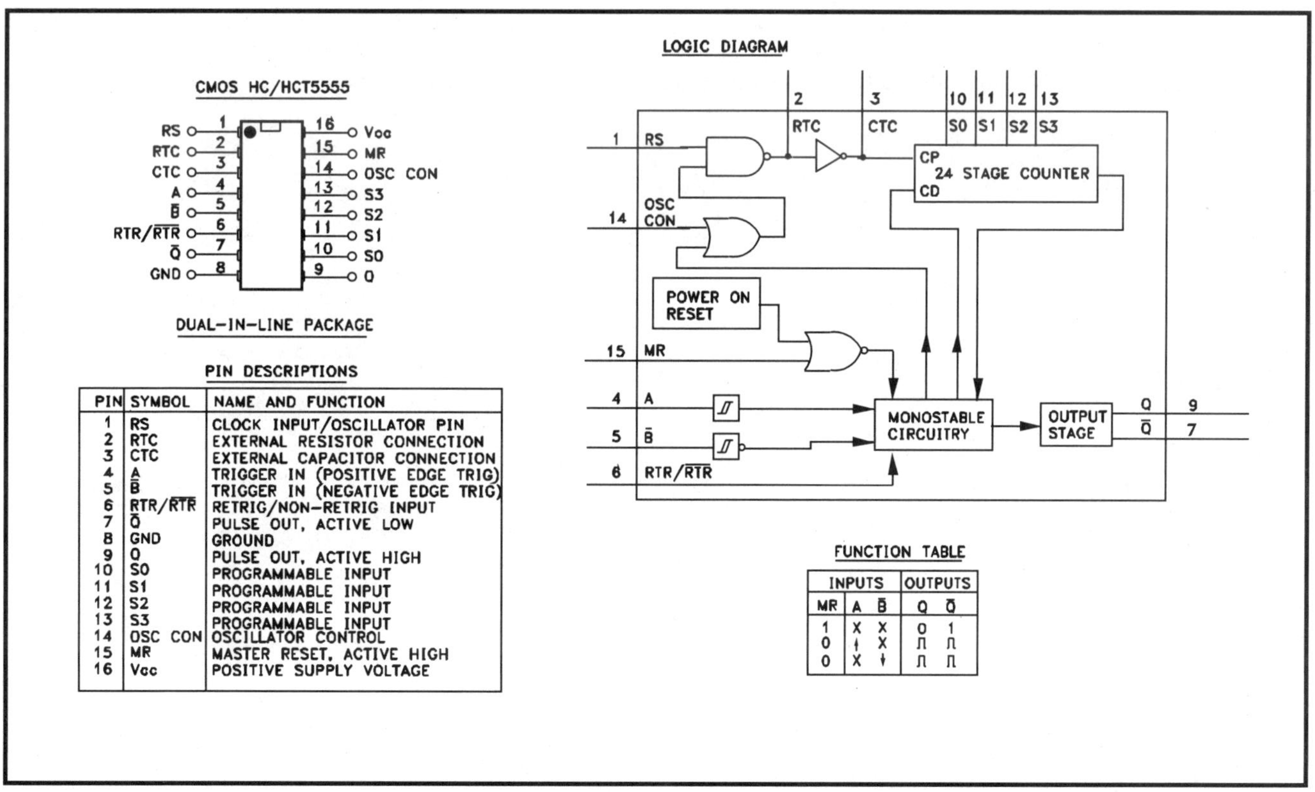

PIN	SYMBOL	NAME AND FUNCTION
1	RS	CLOCK INPUT/OSCILLATOR PIN
2	RTC	EXTERNAL RESISTOR CONNECTION
3	CTC	EXTERNAL CAPACITOR CONNECTION
4	A	TRIGGER IN (POSITIVE EDGE TRIG)
5	B̄	TRIGGER IN (NEGATIVE EDGE TRIG)
6	RTR/R̄T̄R̄	RETRIG/NON-RETRIG INPUT
7	Q̄	PULSE OUT, ACTIVE LOW
8	GND	GROUND
9	Q	PULSE OUT, ACTIVE HIGH
10	S0	PROGRAMMABLE INPUT
11	S1	PROGRAMMABLE INPUT
12	S2	PROGRAMMABLE INPUT
13	S3	PROGRAMMABLE INPUT
14	OSC CON	OSCILLATOR CONTROL
15	MR	MASTER RESET, ACTIVE HIGH
16	Vcc	POSITIVE SUPPLY VOLTAGE

INPUTS			OUTPUTS	
MR	A	B̄	Q	Q̄
1	X	X	0	1
0	↑	X	⊓	⊓
0	X	↓	⊓	⊓

Figure 3.10. The Signetics HC/HCT5555 programmable delay timer with oscillator.

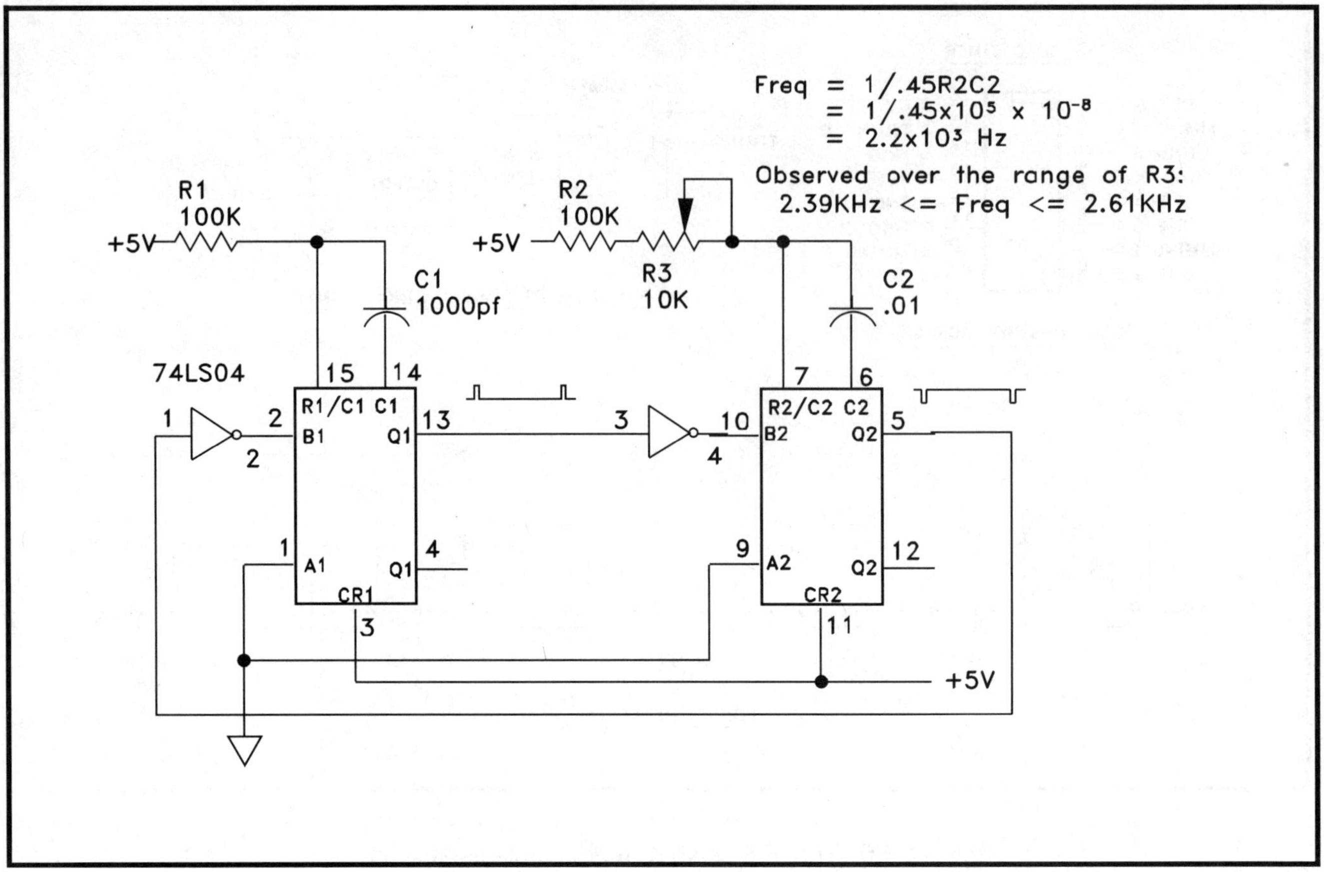

Figure 3.11. A monostable repetitive pulse oscillator example using the 74LS123 dual-mono.

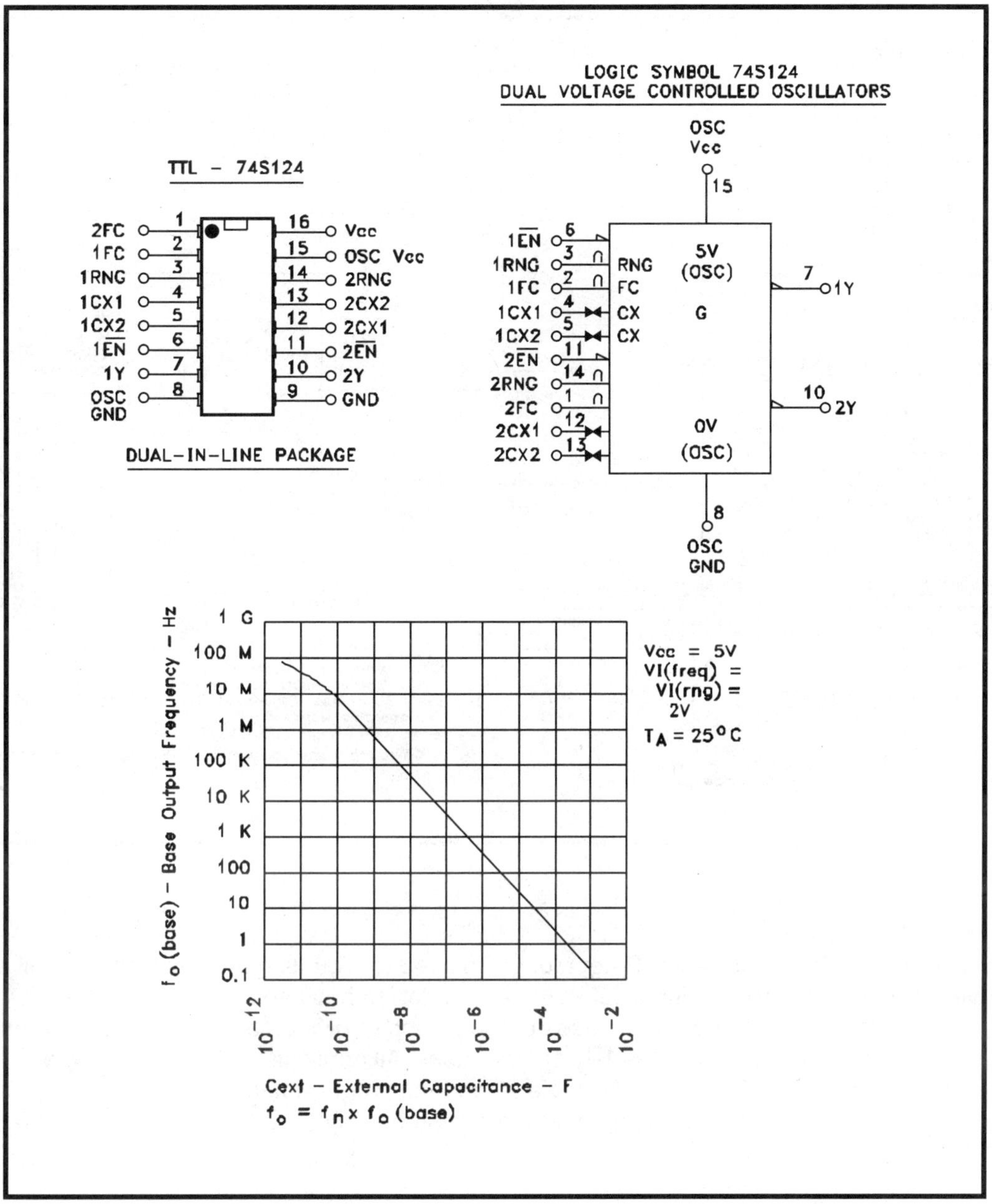

Figure 3.12. The 74S124 Dual-Voltage Controlled Oscillators feature two independent VCOs in a single package. Output frequency is set by a single external capacitor for fixed or variable frequency operation. It is highly stable with a typical upper frequency approaching 100 MHz. The frequency spectrum is defined as 1 Hz to 60 MHz. The Texas Instruments data book is somewhat lacking in details, so you may need to contact the company for more information.

5V CMOS – HEF4753

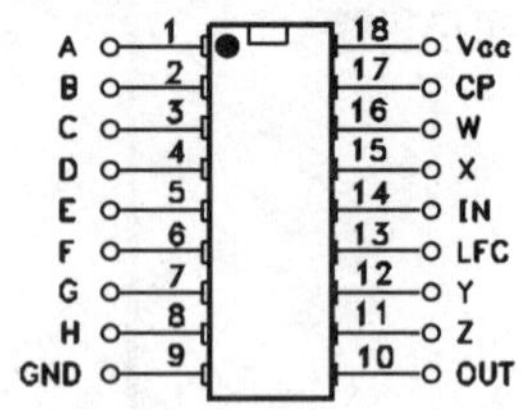

DUAL-IN-LINE PACKAGE

FUNCTIONAL DIAGRAM HEF4753 UNIVERSAL TIMER

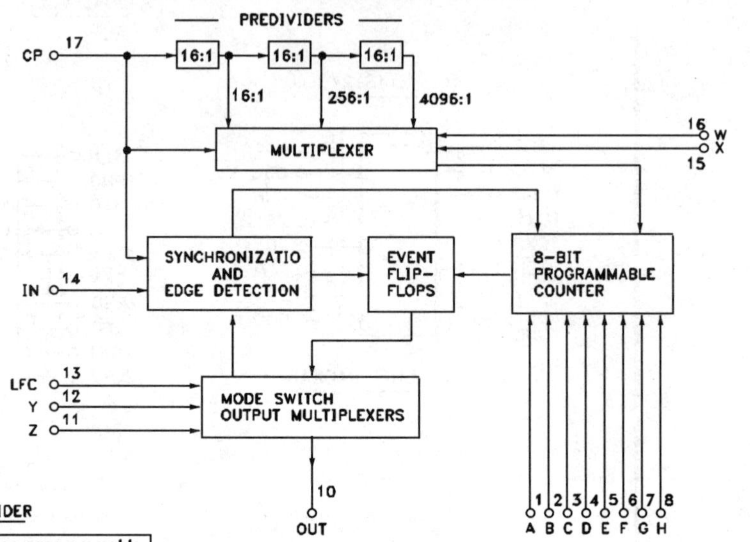

FUNCTION TABLES 4753 UNIVERSAL TIMER MODULE

INPUTS			
LFC	Y	Z	OPERATING MODE
0	0	1	Counter
0	1	0	Divider
1	1	0	Delayed LOW-to-HIGH edge
1	0	1	Delayed HIGH-to-LOW edge
1	1	1	Transient pulse suppression
0	1	1	Frequency recognition
LFC	0	0	Digital pulse duration selector

PROGRAMMABLE 8-BIT COUNTER

Inputs * Active LOW	VALUE
A	1
B	2
C	4
D	8
E	16
F	32
G	64
H	128

* All inputs A – H active HIGH is not allowed.

12-Bit PREDIVIDER

W	X	Clock for programmable counter CP/X
0	0	X = 1
0	1	X = 16
1	0	X = 256
1	1	X = 4096

DEVICE FUNCTIONS:

Clock Divider and Decoder
8-Bit Programmable Counter
Event Flip-Flops, Synchronization and Edge Detector
Retriggerable Monostable
Mode Switch and Output Multiplexer
6 Operating Modes activated by the Logic States of LFC, Y, and Z.

Figure 3.13. The HEF4753 Universal Timer Module is used for counting and dividing, as well as for event recognition and manipulation of input sequences. The device has uses as a presettable 8-bit counter with transient pulse suppression, pulse duration selector divider, positive- or negative-edge delaying module, and a variety of other uses. All applications depend on time scaling, adjustable by an 8-bit programmable counter and the system clock.

Chapter 4

Flip-Flops

We can best define the flip-flop as a *basic unit of memory.* With the logic gate we saw how the gate output status related to an external combination of input states. With the monostable we observed the state of the output to be transitory, remaining changed only for a time duration defined by an external RC combination. In contrast to these the output state of the flip-flop is determined by an internal "lock," so to speak, which once set externally, can only be changed by an opposing signal. A common example for describing flip-flop action is the wall switch. Flip it up to turn on the light, and the light remains on until the switch is flipped down.

But strictly speaking, this does not define the flip-flop. With the wall switch the "up" that provides the "set" is washed out with a "down" to execute the "reset." This is not the case with the flip-flop. In general, there are two distinct input conditions to effect a change in the output. *Toggling* an IC flip-flop refers to the output, which is toggled by the inputs. The trigger pulse that does the toggling is like the "push on/push off" switch we find on many of our VCRs, Stereos, and the like.

Figure 4-1 is the circuitry for a discrete transistor Set-Reset (S-R) flip-flop. From the circuit we see that a positive-going pulse switches transistor Q1 to the ON state. Q1 is going to remain in this state until we find a way to turn it off. A way to turn it off is to pull the base potential down to the point where the base-to-emitter conduction is cut off. The simplest approach to turning Q1 *OFF* is to turn Q2 *ON*.

Back in the early 1960s I constructed a lot of flip-flops for a variety of needs. Most of them looked pretty much like the circuit of *Figure 4-1*. Then I acquired a copy of *The Integrated Circuits Catalog For Design Engineers* from our local rep for Texas Instruments. (I am happy to say I still have this catalog, which I treasure very highly.) The catalog showed the schematic for the complete flip-flop and I thought "Wow!!—what for with all those transistors?" The 7473/107 has about thirty.

Out of sight.

But it didn't take long to realize that those thirty could do much the simple discrete circuits could do, only with lots more work; for example, clocking, gated inputs, and overriding preset and clear (or reset) inputs.

The basic flip-flop designations are the RS, D, T, JK, and JK Master-Slave (MS). The RS is more latch than flip-flop, so it will be described in the next chapter. The "D" stands for "Data," the "T" for "Triggered," and as for the "JK"—your guess is as good as mine. The "J" and "K" are logic inputs that influence the outcome of the clock trigger.

As a point of interest, the flip-flop is the most complex of the small-signal integrated circuits (SSI). The next level, medium-scale integration (MSI), begins with the ripple counter.

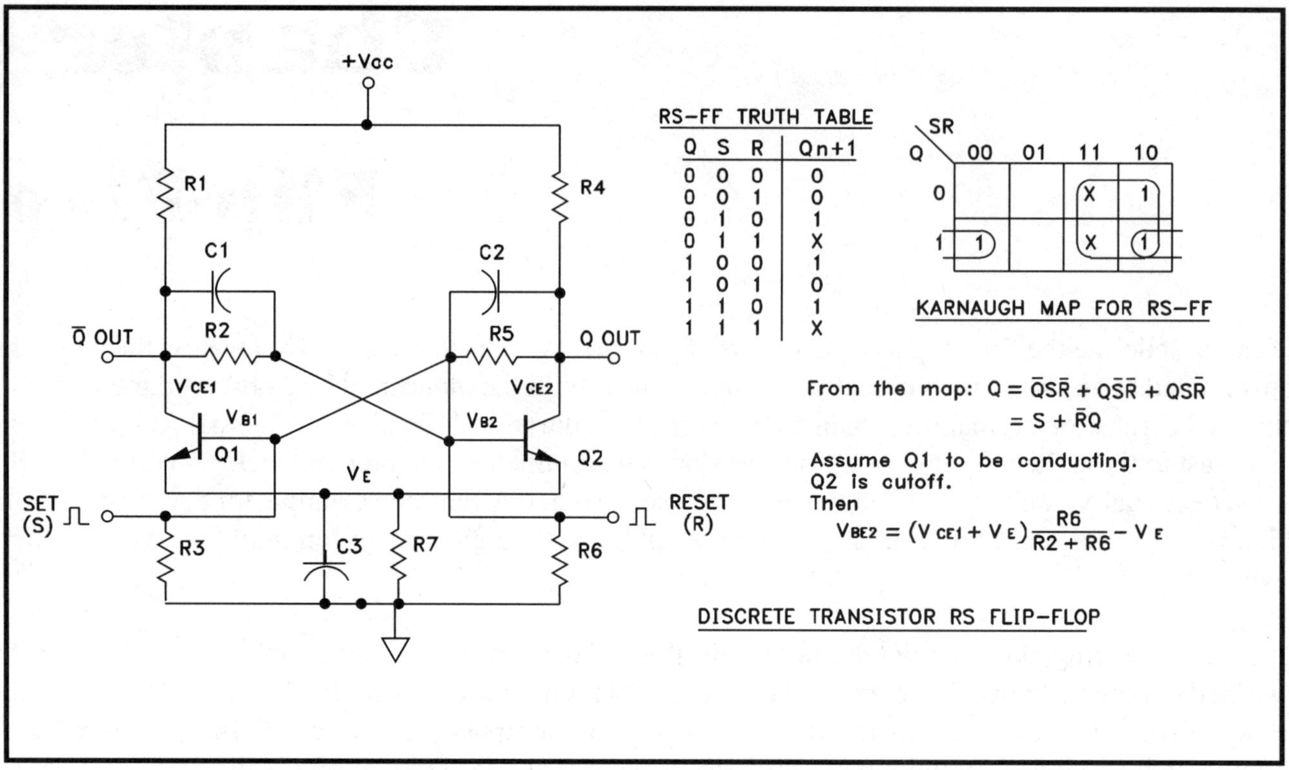

Figure 4.1. The discrete Eccles-Jordan bistable flip-flip. The capacitors are for speed-up only, not timing. Emitter resistor R7 provides turn-off action on the transistor being driven into cutoff.

Earlier, I stated two input conditions are required for toggling a flip-flop. In general, the first consists of data (J, K, D) and the second a clock pulse. Changes to the data are made while the clock is low and the output responds to the input logic in synchronism with some aspect of the clock signal. This may be an entire high-low cycle, or the leading or trailing edge. There is a system advantage to this, as all elements will simultaneously respond to the clock—well, almost. Because of time delays through the various devices, some clock skew takes place. One of the challenges to the logic designer is to ensure the skew remains within acceptable performance limits.

Chapter Organization

In this chapter there are ten tables. *Tables 4.1 - 4.4* provide a flip-flop listing by manufacturer. Four manufacturers are included: National Semiconductor Corporation, Motorola Incorporated, Signetics/Philips Semiconductors, and Texas Instruments. Harris, Fairchild, and Toshiba also offer many of the devices described here.

Tables 4.5 - 4.10 provide frequency and timing references. Because of the manner by which manufacturers define these quantities, the data in these tables is not precise. But they provide insight helpful for performance comparisons. In particular, we can see the spectacular gains in speed and reduced power dissipation achieved with the latest devices.

The drawings of *Figures 4.2 - 4.40* describe the Dual In-Line package, truth tables, and logic diagrams. Again, to simplify comparisons as well as to conserve space all devices with matching features are included in the same drawing.

The D Flip-Flop

Just about all we need to know about this flip-flop type is shown in *Figure 4-2*. This is a clocked flip-flop. The TTL and CMOS devices shown are positive-edge triggered, which is typical for this type. The truth tables for these commercial devices describe their characteristic performance.

There are no single D flip-flop ICs - we have to buy at least two in a package. And the only CD4000 CMOS JK, the CD4027, is a dual. Rather like the packaging in modern merchandizing. The details shown *Figure 4-2* include each TTL and CMOS dual D flip-flop IC available.

A point we can make here in that the versatile D flip-flop can be configured to emulate any other flip-flop type through appropriate data gating.

The T Flip-Flop

Figure 4-2 includes the logic for this flip-flop type. The simplest construction approach is to tie the /Q output to the D input. Then each clock cycle toggles the output. Insertion of a gate—a 2-input AND is shown—provides the means for asserting gate logic in the operation.

The T flip-flop is a do-it-yourself type—an IC for it does not exist. It is frequently found in divider and counter circuits because of its inherent divide-by-two feature.

The JK Flip-Flop

The JK flip-flop has two data inputs. A glance at *Figure 4-3* shows that from the standpoint of the FF this is true, but the IC may provide additional logic expansion. The JK truth table resembles that of the RS FF in *Figure 4-1* with one exception: The JK will accept simultaneous high data inputs which are a no-no with the RS.

A variation of the JK FF is the JK MS (JK Master-Slave). The master-slave type is essentially two latches in a serial connection. The first latch is the master and the second the slave. Clocking action takes place in four steps:

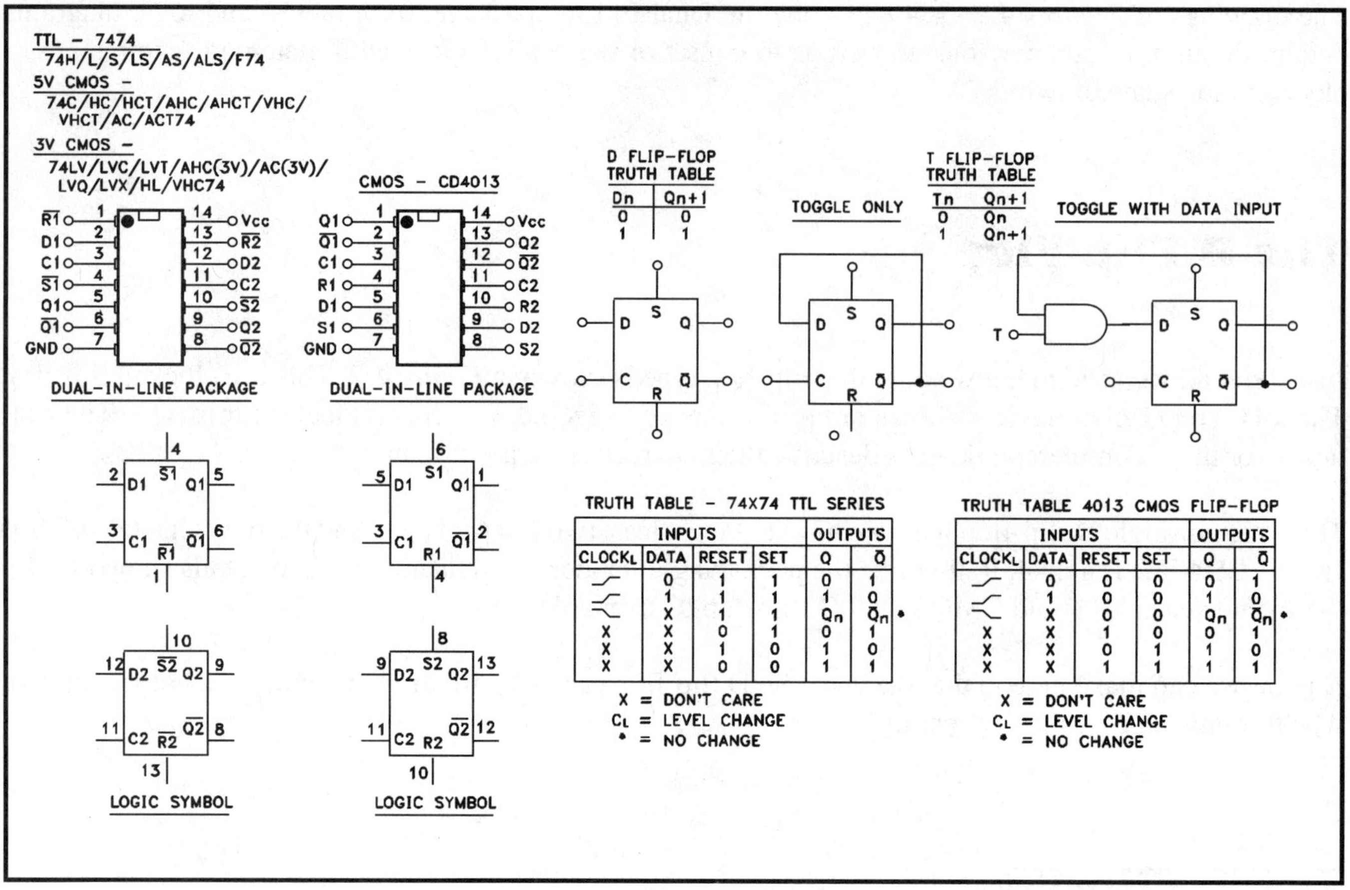

TRUTH TABLE - 74X74 TTL SERIES

INPUTS				OUTPUTS	
CLOCK$_L$	DATA	RESET	SET	Q	$\bar{Q}$
↑	0	1	1	0	1
↑	1	1	1	1	0
↓	X	1	1	Q_n	$\bar{Q}_n$ •
X	X	0	1	0	1
X	X	1	0	1	0
X	X	0	0	1	1

X = DON'T CARE
C$_L$ = LEVEL CHANGE
• = NO CHANGE

TRUTH TABLE 4013 CMOS FLIP-FLOP

INPUTS				OUTPUTS	
CLOCK$_L$	DATA	RESET	SET	Q	$\bar{Q}$
↑	0	0	0	0	1
↑	1	0	0	1	0
↓	X	0	0	Q_n	$\bar{Q}_n$ •
X	X	1	0	0	1
X	X	0	1	1	0
X	X	1	1	1	1

X = DON'T CARE
C$_L$ = LEVEL CHANGE
• = NO CHANGE

Figure 4.2. These TTL, 5V 4000 and High-Speed series, and Low-Voltage series, CMOS dual-D flip-flop series feature positive-edge triggering with Data, Clock, Set, and Reset inputs; complementary Q and /Q outputs are provided.

1. Isolate the slave from the master, which is the normal condition with the clock low.

2. Enable the data inputs to the master. This takes place on the low-to-high clock transition.

3. Disable the data inputs. This occurs with the clock at its high level.

4. Transfer the data from the master to the slave. This takes place with the clock descent from high-to-low.

The advantage is total isolation of the data inputs from the FF output.

The SN7470 is the first flip-flop listed in *The Integrated Circuits Catalog For Design Engineers,* section 6, series 54/74 SSI Circuits. This edge-triggered JK Flip-Flop features gated inputs with direct (asynchronous) preset and clear inputs. These should be asserted with the clock low. Clock triggering occurs at a specific voltage level of the clock pulse when the input status is transferred to the outputs. Once the clock input threshold has been reached the gated inputs are locked out.

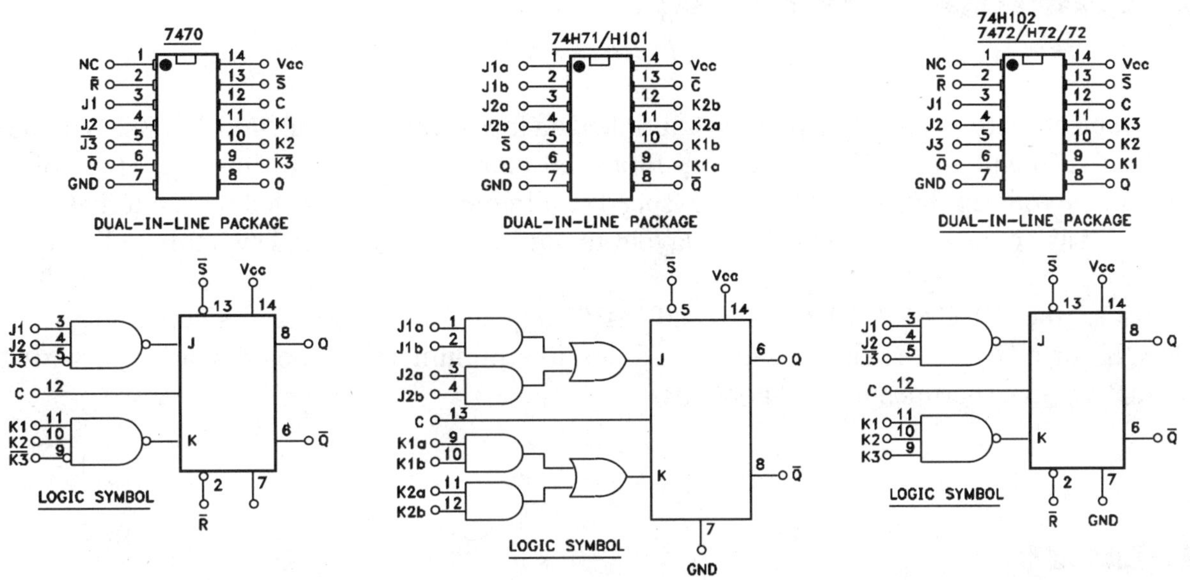

TRUTH TABLE TTL 7470 JK FLIP-FLOP

	INPUTS					OUTPUTS	
	C	J	K	$\bar{S}$	$\bar{R}$	Q	$\bar{Q}$
1	0	X	X	0	1	1	0
2	0	X	X	1	0	0	1
3	X	X	X	0	0	0	0
4	↑	h	h	1	1	$\bar{q}$	q
5	↑	l	h	1	1	0	1
6	↑	h	l	1	1	1	0
7	↑	l	l	1	1	q	$\bar{q}$

$J = J \cdot J \cdot \bar{J}$ $K = K \cdot K \cdot \bar{K}$

LOWER CASE INDICATES LOGIC STATUS PRIOR TO CLOCK TRANSITION.

TRUTH TABLE TTL 74H71/H101 JK FLIP-FLOP

	INPUTS				OUTPUTS	
	C	J	K	$\bar{S}$	Q	$\bar{Q}$
1	X	X	X	0	1	0
4	⎍	h	h	1	$\bar{q}$	q
6	⎍	h	l	1	1	0
7	⎍	l	l	1	q	$\bar{q}$

$J=(J1a \cdot J1b)+(J2a \cdot J2b)$
$K=(K1a \cdot K1b)+(K2a \cdot K2b)$

LOWER CASE INDICATES LOGIC STATUS PRIOR TO CLOCK TRANSITION.

OPERATING MODE NOTES:

1. Asynohronous Set
2. Asynchronous Reset (Clear)
3. Not predictable
4. Toggle
5. Load "0" (Reset)
6. Load "1" (Set)
7. Hold "No Change"

TRUTH TABLE TTL 7472/H72/L72/H102 JK FLIP-FLOP

	INPUTS					OUTPUTS	
	C	J	K	$\bar{S}$	$\bar{R}$	Q	$\bar{Q}$
1	0	X	X	0	1	1	0
2	0	X	X	1	0	0	1
3	X	X	X	0	0	1	1
4	⎍	h	h	1	1	$\bar{q}$	q
5	⎍	l	h	1	1	0	1
6	⎍	h	l	1	1	1	0
7	⎍	l	l	1	1	q	$\bar{q}$

$J = J \cdot J \cdot J$ $K = K \cdot K \cdot K$

LOWER CASE INDICATES LOGIC STATUS PRIOR TO CLOCK TRANSITION.

Figure 4.3. The JK series in this figure are very early types. The 74104 and 74105 are gated JK Master-Slave (MS) types with direct preset and clear. They have multiple J and K inputs, plus a unique JK input. The 74105 differs from the 74104 only in /J2 and /K2 inputs. No CMOS equivalents exist.

The other two JK types are positive-pulse triggered master-slave flip-flops with gated inputs as shown. The input logic is loaded into the master with the clock high, then transferred to the slave on the high-to-low transition. The logic inputs need to be stable throughout this time. The 74H71 lacks a reset provision. This can be achieved through the input logic.

Note that none of these flip-flops is available in a CMOS version.

Multiple-Input Flip-Flops

A considerable variety of packages provide multiple data inputs ranging from 4 to 20 bits. With the exception of the 74276 and 74376 JK QUAD Flip-Flops, all employ the D-type flip-flop. Many of these are tailored for bus-wide applications where high-speed requirements dictate both device internal features and external, such as pinouts favoring PC board layout to minimize crosstalk and switching glitches.

Three devices: the 74HC7074, 74HC7075, and 74HC7076, are multipurpose ICs containing inverters, 2-input NAND or NOR gates. The versatility and possible minimizing of board space and wiring makes these attractive for experimenters and hobbyists.

Summary

In this chapter we find forty drawings describing over 100 flip-flops, from the single 7470 to the impressive 16821 3.3-volt 20-bit flip-flop with 3-state outputs. Many devices comprise a series, providing an impressive array of options. Some points to keep in mind are:

- Frequencies and times are for comparison purposes only. There is a general lack of consistency in defining these parameters, and not just between manufacturers. Where timing is critical, go to the manufacturer's data manual.

- In general, 3-state devices are not affected internally when in the high output impedance state.

- Frequency and time data shown is for the industrial/commercial or recommended free-air operating temperature range.

- The specifications for CMOS devices may cover voltage ranges from 2 to 15 volts. Data is typically provided for Vcc's of 5, 10, and 15 volts, but not always. Some manufacturers use 5.0 volts, others 4.5 volts. There is a significant speed factor, so again, where timing is important or your application is at other than 5 volts VCC, consult the data manual.

- A similar situation exists for the more recent low-voltage devices. Typical VCC ranges are 2.7 - 3.6 volts, but values may be given for 2, 2.6, 3.0, 3.3, or 3.6.

DEVICE 74---	STD NMST	H NMST	L NMST	S NMST	LS NMST	AS/A/B NMST	ALS/A/B NMST	F NMST	Figure No.	DEVICE 74---
7470	xx								4.3	7470
7471		xx	x						4.4	7471
7472	xx	xx	x x						4.4	7472
7473	x xx	xx	x x		xxxx				4.6	7473
7474	x xx	xx	x x	x x	xxxx	x	x	x	4.2	7474
7476	x xx	x			xxx				4.6	7476
7478		x	x		xx				4.7	7478
74101		xx							4.4	74101
74102		xx							4.4	74102
74103		xx							4.6	74103
74104	x								4.3	74104
74105	x								4.3	74105
74106		xx							4.6	74106
74107	x xx				xxxx				4.8	74107
74108		x							4.7	74108
74109	x xx				xxxx	x	x	x	4.8	74109
74110	x								4.5	74110
74111	x								4.5	74111
74112				x xx	xxxx		x	xxx	4.9	74112
74113				x xx	xxx			x	4.8	74113
74114				xx	xxx		x	x	4.7	74114
74173	x				xx			x	4.10	74173
74174	x xx			x xx	xxx	x x	x x	x x	4.11	74174
74175	x xx			x xx	xxx	x x	x x	x x	4.12	74175
74273	x			x	xx		x	x	4.13	74273
74276	x								4.14	74276
74364					x				4.15	74364
74374				x x	x xx	x x	x	xxx	4.16	74374
74376	x								4.17	74376
74377					xx			xx	4.18	74377
74378					xxx			x	4.19	74378
74379					xx			x	4.20	74379
74534						x x	x x	xxx	4.21	74534
74564							x x	xx	4.22	74564
74574						x	x	xxx	4.22	74574
74575						x x	x		4.23	74575
74576						x x	x x		4.22	74576
74577						x x	x		4.23	74577
74821						x		x	4.24	74821
74822						x		x	4.24	74822
74823						x		x	4.25	74823
74824						x		x	4.25	74824
74825						x		x	4.26	74825
74826						x		x	4.26	74826
74874						x	x		4.27	74874
74876						x	x		4.28	74876
74878						x	x		4.29	74878
74879						x	x		4.29	74879
744374						x			4.30	744374
7429821						x	x		4.24	7429821
7429822						x	x		4.24	7429822
7429823						x	x		4.25	7429823
7429824						x	x		4.25	7429824
7429825						x	x		4.26	7429825
7429826						x	x		4.26	7429826

Key to Manufacturer: N – National, M – Motorola, S – Signetics/Philips, T – Texas Instruments

Table 4.1. TTL flip-flop source summary by manufacturer.

DEVICE	4000 NMST	4100 NMST	4500 NMST	Figure No.
4013	x xx			4.2
4027	x xx			4.9
40174	x x			4.11
40175	x x			4.12
14174		x		4.11
14175		x		4.12
40374	x			4.16

Key to Manufacturer: Prefix
N – National, CD
M – Motorola, MC1
S – Signetics/Philips, HEC/HEF
T – Texas Instruments --

Table 4.2. CMOS 4000 series flip-flop source summary by manufacturer.

DEVICE 74---	C NMST	HC NMST	HCT NMST	AHC(5V) NMST	AHCT NMST	VHC(5V) NMST	VHCT NMST	AC (5V) NMST	ACT NMST	ABT NMST	Figure No.	DEVICE 74---
7473	x	xxxx	x								4.6	7473
7474	x	x	xxxx	x	x	xx	x	x	x		4.2	7474
7476	x	x x	x								4.6	7476
7478		x									4.7	7478
74107	x	xxx	x								4.8	74107
74109		xxxx	x x								4.8	74109
74112		xxxx	x x			x	x				4.9	74112
74113		x x									4.8	74113
74173	x	xxxx	x								4.10	74173
74174	x	xxxx	xx								4.11	74174
74175	x	xxxx	x			x					4.12	74175
74273		xxxx	xxxx	x	x					x	4.13	74273
74374		xxxx	xxxx	x	x	xx	x	x	x	x	4.16	74374
74377		xx	xx							x	4.18	74377
74378		x									4.19	74378
74379		x									4.20	74379
74534		xxxx	x x					x	x		4.21	74534
74564		xxx	x					x	x		4.22	74564
74574		xxxx	xxx	x	x	xx	x	x	x	x	4.22	74574
74821										x	4.24	74821
74823										x	4.25	74823
747074		x									4.31	747074
747075		x									4.32	747075
747076		x									4.33	747076
7411074								x	x		4.34	7411074
7411374								x	x		4.35	7411374
7416374				x	x			x	x	x	4.36	7416374
7416821								x	x	x	4.39	7416821
7416823					x			x	x	x	4.40	7416823

Key to Manufacturer: N – National, M – Motorola, S – Signetics/Philips, T – Texas Instruments

Table 4.3. 5V High Speed CMOS flip-flop source summary by manufacturer.

DEVICE 74---	LV NMST	LVC NMST	LVT/ABT NMST	LVX NMST	LVQ NMST	LCX NMST	AC(3V) HE NMST	LVCH/ ALVCH NMST	GTL NMST	VHC NMST	Figure No.	DEVICE 74---
7474	x	x		x	x		x			xx	4.2	7474
74107	x										4.8	74107
74109	x	x									4.8	74109
74112		x								xx	4.9	74112
74174	x			x	x						4.11	74174
74175										xx	4.12	74175
74273	x			x	x					xx	4.13	74273
74374	x	x	x	x	x	x	x	x		xx	4.16	74374
742374		x						x		x	4.16	742374
74377	x	x						x			4.18	74377
74534							x				4.21	74534
74564							x				4.22	74564
74574	xx	xx				x	x	x		xx	4.22	74574
742574		x					x	x			4.22	742574
74821		xx						x			4.24	74821
74823		xx						x			4.25	74823
742821		x						x			4.24	742821
742823		x						x			4.25	742823
7411074							x				4.34	7411074
7411374							x				4.35	7411374
7416374		x	x					xx			4.36	7416374
74162374		x						x			4.36	74162374
7416721								x			4.37	7416721
7416820								x			4.38	7416820
7416821								x			4.39	7416821
7416823								x			4.40	7416823
7416921									x		4.41	7416921
7433374							x				4.35	7433374
7433534							x				4.35	7433534
74162820								x			4.38	74162820

Key to Manufacturer: N – National, M – Motorola, S – Signetics/Philips, T – Texas Instruments

Table 4.4. 2.7-3.3V High-Speed CMOS flip-flop source summary by manufacturer.

DEVICE 74---	STD	H	L	S	LS	AS/A/B	ALS/A/B	F	DATA	Figure	DEVICE 74---
7470	20-35	--	--	--	--	--	--	--	f MHz	4.3	7470
----	27-50	--	--	--	--	--	--	--	tpd	--	----
----	20/5	--	--	--	--	--	--	--	ts/th	--	----
7471	--	25	3	--	--	--	--	--	f MHz	4.4	7471
----	--	13-27	10-200	--	--	--	--	--	tpd	--	----
----	--	12/0	100/0	--	--	--	--	--	ts/th	--	----
7472	15	25	3	--	--	--	--	--	f MHz	4.4	7472
----	25-40	13-27	10-200	--	--	--	--	--	tpd	--	----
----	20/0	12/0	200/0	--	--	--	--	--	ts/th	--	----
7473	15	30	3	--	30	--	--	--	f MHz	4.6	7473
----	25-40	6-27	10-200	--	20-30	--	--	--	tpd	--	----
----	20/0	>tp(clk)	>tp(clk)	--	20/0	--	--	--	ts/th	--	----
7474	15	43	3	110	25	105	34	125	f MHz	4.2	7474
----	14-40	4-30	10-150	5-13.5	13-40	3-10.5	3-15	3.2-10.5	tpd	--	----
----	40/25	15/0	30/0	3/2	20/5	2/0	10/0	3/1	ts/th	--	----
7476	15	25	--	--	30	--	--	--	f MHz	4.6	7476
----	25-40	21-70	--	--	20-30	--	--	--	tpd	--	----
----	20/0	12/0	--	--	20/0	--	--	--	ts/th	--	----
7478	--	--	--	--	30-45	--	--	--	f MHz	4.7	7478
----	--	--	--	--	15-20	--	--	--	tpd	--	----
----	--	--	--	--	20/0	--	--	--	ts/th	--	----
74101	--	40	--	--	--	--	--	--	f MHz	4.4	74101
----	--	12-35	--	--	--	--	--	--	tpd	--	----
----	--	13/0	--	--	--	--	--	--	ts/th	--	----
74102	--	40	--	--	--	--	--	--	f MHz	4.4	74102
----	--	12-35	--	--	--	--	--	--	tpd	--	----
----	--	13/0	--	--	--	--	--	--	ts/th	--	----
74103	--	40	--	--	--	--	--	--	f MHz	4.6	74103
----	--	12-35	--	--	--	--	--	--	tpd	--	----
----	--	13/0	--	--	--	--	--	--	ts/th	--	----
74104	--	--	--	--	--	--	--	--	f MHz	4.3	74104
----	9-25	--	--	--	--	--	--	--	tpd	--	----
----	35/10	--	--	--	--	--	--	--	ts/th	--	----
74105	--	--	--	--	--	--	--	--	f MHz	4.3	74105
----	9-25	--	--	--	--	--	--	--	tpd	--	----
----	35/10	--	--	--	--	--	--	--	ts/th	--	----
74106	--	40	--	--	--	--	--	--	f MHz	4.6	74106
----	--	12-35	--	--	--	--	--	--	tpd	--	----
----	--	13/0	--	--	--	--	--	--	ts/th	--	----
74107	15	--	--	--	30	--	--	--	f MHz	4.8	74107
----	25-40	--	--	--	20-30	--	--	--	tpd	--	----
----	20/0	--	--	--	20/0	--	--	--	ts/th	--	----
74108	--	40	--	--	--	--	--	--	f MHz	4.7	74108
----	--	12-35	--	--	--	--	--	--	tpd	--	----
----	--	13/0	--	--	--	--	--	--	ts/th	--	----
74109	25	--	--	--	25	--	--	--	f MHz	4.8	74109
----	15-35	--	--	--	25-40	--	--	--	tpd	--	----
----	10/6	--	--	--	20/5	--	--	--	ts/th	--	----
74110	20-25	--	--	--	--	--	--	--	f MHz	4.5	74110
----	12-30	--	--	--	--	--	--	--	tpd	--	----
----	5/20	--	--	--	--	--	--	--	ts/th	--	----
74111	20-25	--	--	--	--	--	--	--	f MHz	4.5	74111
----	12-30	--	--	--	--	--	--	--	tpd	--	----
----	30/0	--	--	--	--	--	--	--	ts/th	--	----
74112	--	--	--	80	30	--	30	85-100	f MHz	4.9	74112
----	--	--	--	7.0	20-30	--	3.0/19	2.0/6.5	tpd	--	----
----	--	--	--	3.0/0	20/0	--	22/0	5.0/0	ts/th	--	----
74113	--	--	--	80	30	--	--	110-135	f MHz	4.8	74113
----	--	--	--	7.0	20-30	--	--	2.0/4.5	tpd	--	----
----	--	--	--	3.0/0	20/0	--	--	4.0/0	ts/th	--	----
74114	--	--	--	80	30	--	30	110/125	f MHz	4.7	74114
----	--	--	--	7.0	20-30	--	5-18	3-5	tpd	--	----
----	--	--	--	3/0	20/0	--	20/0	4/0	ts/th	--	----

Table 4.5. TTL single/dual flip-flop frequency (MHz), typical delays (tpd), and setup/hold times (ts/th).

DEVICE 74---	STD	S	LS	AS/A/B	ALS/A/B	F	DATA	Figure	DEVICE 74---
74173	30	--	30	--	--	100-125	f_{MHz}	4.10	74173
----	14-43	--	13-36	--	--	1.5-8.5	t_{pd}	--	----
----	17/10	--	17/10	--	--	7.0/0	t_s/t_h	--	----
74174	25	75	30	100	50	80-100	f_{MHz}	4.11	74174
----	30-35	13-15	24-35	3.5-14	3-23	4.5-10	t_{pd}	--	----
----	17/0	5/3	12/0	-/-	-/-	4/0	t_s/t_h	--	----
74175	25	50	35	100	50	100-140	f_{MHz}	4.12	74175
----	30-35	12-22	22-28	4-13	3-23	4.0/9.0	t_{pd}	--	----
----	12/0	8/2	6/0	-/-	-/-	3/1	t_s/t_h	--	----
74273	30-40	75	30-40	--	35	130-145	f_{MHz}	4.13	74273
----	17-27	15	17-27	--	2-18	4.0-9.5	t_{pd}	--	----
----	25/5	5/3	15/0	--	10/0	3/0	t_s/t_h	--	----
74276	35-50	--	--	--	--	--	f_{MHz}	4.14	74276
----	17-30	--	--	--	--	--	t_{pd}	--	----
----	10/10	--	--	--	--	--	t_s/t_h	--	----
74364	35	--	--	--	--	--	f_{MHz}	4.15	74364
----	28-38	--	--	--	--	--	t_{pd}	--	----
----	10/3	--	--	--	--	--	t_s/t_h	--	----
74374	--	75-100	35	125	35	70	f_{MHz}	4.16	74374
----	--	7-18	3-18	2-10	3-18	2.0-13	t_{pd}	--	----
----	--	5/2	10/0	2/2	10/0	2/2	t_s/t_h	--	----
74376	30	--	--	--	--	--	f_{MHz}	4.17	74376
----	17-35	--	--	--	--	--	t_{pd}	--	----
----	10/20	--	--	--	--	--	t_s/t_h	--	----
74377	--	--	30-40	--	--	110-120	f_{MHz}	4.18	74377
----	--	--	17-27	--	--	4-9	t_{pd}	--	----
----	--	--	15/5	--	--	4/0	t_s/t_h	--	----
74378	--	--	30-40	--	--	110-120	f_{MHz}	4.19	74378
----	--	--	17-27	--	--	4-9	t_{pd}	--	----
----	--	--	15/5	--	--	4/0	t_s/t_h	--	----
74379	--	--	30-40	--	--	110-120	f_{MHz}	4.20	74379
----	--	--	17-27	--	--	4-9	t_{pd}	--	----
----	--	--	15/5	--	--	4/0	t_s/t_h	--	----

Table 4.6. TTL QUAD and HEX input flip-flop frequency (MHz), typical delays (tpd), and setup/hold times (ts/th).

DEVICE 74---	STD	S	LS	AS/A/B	ALS/A/B	F	DATA	Figure No.	DEVICE 74---
74534	--	--	--	125	35	70	f MHz	4.21	74534
----	--	--	--	2-10	2-18	3-13	tpd	--	----
----	--	--	--	2/0	10/0	2/2	ts/th	--	----
74564	--	--	--	30-50	30-50	70	f MHz	4.22	74564
----	--	--	--	4-18	4-18	2-12.5	tpd	--	----
----	--	--	--	15/0	15/0	2/2	ts/th	--	----
74574	--	--	--	30-50	30-50	70	f MHz	4.22	74574
----	--	--	--	4-18	4-18	2-12.5	tpd	--	----
----	--	--	--	15/0	15/0	2/2	ts/th	--	----
74575	--	--	--	125	30	--	f MHz	4.23	74575
----	--	--	--	2-10	2-18	--	tpd	--	----
----	--	--	--	2/0	15/0	--	ts/th	--	----
74576	--	--	--	125	30-50	--	f MHz	4.22	74576
----	--	--	--	2-10	4-18	--	tpd	--	----
----	--	--	--	2/0	15/0	--	ts/th	--	----
74577	--	--	--	125	30	--	f MHz	4.23	74577
----	--	--	--	2-10	2-18	--	tpd	--	----
----	--	--	--	2/0	15/0	--	ts/th	--	----
74821	--	--	--	125	--	--	f MHz	4.24	74821
----	--	--	--	2-12	--	--	tpd	--	----
----	--	--	--	6/0	--	--	ts/th	--	----
74822	--	--	--	125	--	--	f MHz	4.24	74822
----	--	--	--	2-12	--	--	tpd	--	----
----	--	--	--	6/0	--	--	ts/th	--	----
74823	--	--	--	--	125	110-120	f MHz	4.25	74823
----	--	--	--	--	2-12	4-9	tpd	--	----
----	--	--	--	--	6/0	4/0	ts/th	--	----
74824	--	--	--	--	125	110-120	f MHz	4.25	74824
----	--	--	--	--	2-12	4-9	tpd	--	----
----	--	--	--	--	6/0	4/0	ts/th	--	----
74825	--	--	--	--	125	110-120	f MHz	4.26	74825
----	--	--	--	--	2-12	4-9	tpd	--	----
----	--	--	--	--	6/0	4/0	ts/th	--	----
74826	--	--	--	--	125	--	f MHz	4.26	74826
----	--	--	--	--	2-12	--	tpd	--	----
----	--	--	--	--	6/0	--	ts/th	--	----
74874	--	--	--	125	30	--	f MHz	4.27	74874
----	--	--	--	2-10.5	2-18	--	tpd	--	----
----	--	--	--	2/0	15/0	--	ts/th	--	----
74876	--	--	--	125	30	--	f MHz	4.28	74876
----	--	--	--	2-10.5	2-18	--	tpd	--	----
----	--	--	--	2/0	15/0	--	ts/th	--	----
74878	--	--	--	125	30	--	f MHz	4.29	74878
----	--	--	--	2-10.5	2-18	--	tpd	--	----
----	--	--	--	2/0	15/0	--	ts/th	--	----
74879	--	--	--	125	30	--	f MHz	4.29	74879
----	--	--	--	2-10.5	2-18	--	tpd	--	----
----	--	--	--	2/0	15/0	--	ts/th	--	----
744374	--	--	--	125	--	--	f MHz	4.30	744374
----	--	--	--	8-8	--	--	tpd	--	----
----	--	--	--	4/1	--	--	ts/th	--	----
7429821	--	--	--	--	-/-	--	f MHz	4.24	7429821
----	--	--	--	--	2-16	--	tpd	--	----
----	--	--	--	--	4/2	--	ts/th	--	----
7429822	--	--	--	--	-/-	--	f MHz	4.24	7429822
----	--	--	--	--	2-16	--	tpd	--	----
----	--	--	--	--	4/2	--	ts/th	--	----
7429823	--	--	--	--	-/-	--	f MHz	4.25	7429823
----	--	--	--	--	2-16	--	tpd	--	----
----	--	--	--	--	6/2	--	ts/th	--	----

Table 4.7. TTL Octal, 9-, 10-, 16-bit input flip-flop frequency (MHz), typical delays (tpd), and setup/hold times (ts/th).

DEVICE	4000	4100	4500	--	DATA	Figure	DEVICE
4013	4	--	--	--	f MHz	4.2	4013
----	175–350	--	--	--	tpd	--	----
----	20/20	--	--	--	ts/th	--	----
4027	1.5–3.0	--	--	--	f MHz	4.9	4027
----	175–350	--	--	--	tpd	--	----
----	70/70	--	--	--	ts/th	--	----
40174	5–10	--	--	--	f MHz	4.11	40174
----	75/110	--	--	--	tpd	--	----
----	45/0	--	--	--	ts/th	--	----
40175	31–50	--	--	--	f MHz	4.12	40175
----	75/110	--	--	--	tpd	--	----
----	45/0	--	--	--	ts/th	--	----
14174	--	2–7	--	--	f MHz	4.11	14174
----	--	210–500	--	--	tpd	--	----
----	--	20/40	--	--	ts/th	--	----
14175	--	5–11	--	--	f MHz	4.12	14175
----	--	220–500	--	--	tpd	--	----
----	--	60/40	--	--	ts/th	--	----
40374	6–17	--	--	--	f MHz	4.16	40374
----	250–250	--	--	--	tpd	--	----
----	20/20	--	--	--	ts/th	--	----

(a)

DEVICE 74---	C	HC	HCT	AHC	AHCT	VHC(5V)	VHCT	AC (5V)	ACT	DATA	Figure	DEVICE 74---
7473	2.5–4	25	30–72	--	--	--	--	--	--	f MHz	4.6	7473
----	180–300	19–32	7–38	--	--	--	--	--	--	tpd	--	----
----	175/110	30/0	12/–2	--	--	--	--	--	--	ts/th	--	----
7474	3.5	25	22	90–170	80–160	130–170	100–160	140/160	145–210	f MHz	4.2	7474
----	180/400	20–25	20–23	4.6/9.7	5.8/11.4	10.1–9.7	5.8–11.4	3.0/10	3.0/11	tpd	--	----
----	50	25/0	0/0	3.0/0.5	3.5/0	5.0/0.5	5.0/0	0/0.5	3/1	ts/th	--	----
7476	4	25	27	--	--	--	--	--	--	f MHz	4.6	7476
----	200–300	19–39	22–35	--	--	--	--	--	--	tpd	--	----
----	110/0	25/0	35/0	--	--	--	--	--	--	ts/th	--	----
7478	--	31–50	--	--	--	--	--	--	--	f MHz	4.7	7478
----	--	16–31	--	--	--	--	--	--	--	tpd	--	----
----	--	2.5/0	--	--	--	--	--	--	--	ts/th	--	----
74107	2.5–4	25	30–66	--	--	--	--	--	--	f MHz	4.8	74107
----	70–300	16–39	7–38	--	--	--	--	--	--	tpd	--	----
----	175/0	25/0	20/–2	--	--	--	--	--	--	ts/th	--	----
74109	--	30–50	22–27	--	--	--	--	--	--	f MHz	4.9	74109
----	--	20–30	25–44	--	--	--	--	--	--	tpd	--	----
----	--			--	--	--	--	--	--	ts/th	--	----
74112	--	21–27	22–27	--	--	NO	--	--	--	f MHz	4.9	74112
----	--	20–39	20–44	--	--	SPEED	--	--	--	tpd	--	----
----	--	25/0	25/0	--	--	DATA	--	--	--	ts/th	--	----
74113	--	21	--	--	--	--	--	--	--	f MHz	4.8	74113
----	--	25–41	--	--	--	--	--	--	--	tpd	--	----
----	--	25/0	--	--	--	--	--	--	--	ts/th	--	----

(b)

Table 4.8. (a) CMOS 4000 series flip-flop frequency (MHz), typical delays (tpd), and setup/hold times (ts/th). (b) 5V High-Speed CMOS series flip-flop frequency (MHz), typical delays (tpd), and setup/hold times (ts/th).

DEVICE 74---	C	HC	HCT	AHC (5V)	AHCT	VHC (5V)	VHCT	AC (5V)	ACT	ABT	DATA	Figure	DEVICE 74---
74173	3-4	45-27	30-80	--	--	--	--	--	--	--	fMHz	4.10	74173
----	220-400	23-45	16-35	--	--	--	--	--	--	--	tpd	--	----
----	80/0	20/0	16/1	--	--	--	--	--	--	--	ts/th	--	----
74174	2.0-6.5	27	30-90	--	--	--	--	--	--	--	fMHz	4.11	74174
----	150-300	18-33	16-33	--	--	--	--	--	--	--	tpd	--	----
----	75/0	12/1	2/3	--	--	--	--	--	--	--	ts/th	--	----
74175	2.0-3.5	60-30	30-75	--	--	150-210	--	--	--	--	fMHz	4.12	74175
----	180-300	14-25	18-35	--	--	4.8-9.3	--	--	--	--	tpd	--	----
----	100/0	20/0	16/5	--	--	4.0/1.0	--	--	--	--	ts/th	--	----
74273	--	30-60	19-33	--	--	120-165	--	--	--	150-200	fMHz	4.13	74273
----	--	18-30	18-35	--	--	8.7/11.0	--	--	--	2.0-6.8	tpd	--	----
----	--	12/3	16/3	--	--	4.5/1.0	--	--	--	2.0/1.2	ts/th	--	----
74374	--	30-70	26-44	--	--	130-185	90-140	--	--	150-200	fMHz	4.16	74374
----	--	18-33	16-32	--	--	5.4-10.1	5.6-10.4	--	--	2.0-5.0	tpd	--	----
----	--	12/5	12/5	--	--	4.5/2.0	2.5/2.5	--	--	1.5/1.0	ts/th	--	----
74377	--	30-70	27-48	--	--	--	--	--	--	150-200	fMHz	4.18	74377
----	--	16/32	17-32	--	--	--	--	--	--	2.2-6.8	tpd	--	----
----	--	12/3	12/2	--	--	--	--	--	--	3.0/1.0	ts/th	--	----
74378	--	25-44	--	--	--	--	--	--	--	--	fMHz	4.19	74378
----	--	15-32	--	--	--	--	--	--	--	--	tpd	--	----
----	--	20/5	--	--	--	--	--	--	--	--	ts/th	--	----
74379	--	25-44	--	--	--	--	--	--	--	--	fMHz	4.20	74379
----	--	15-32	--	--	--	--	--	--	--	--	tpd	--	----
----	--	20/5	--	--	--	--	--	--	--	--	ts/th	--	----
74534	--	30-55	22-36	--	--	--	--	--	--	--	fMHz	4.21	74534
----	--	15-33	16-30	--	--	--	--	--	--	--	tpd	--	----
----	--	12/5	12/5	--	--	--	--	--	--	--	ts/th	--	----
74564	--	30-115	27-56	--	--	--	--	85-95	75-90	--	fMHz	4.22	74564
----	--	18-33	19-35	--	--	--	--	10.5-9.5	10.5-9.5	--	tpd	--	----
----	--	12/5	12/3	--	--	--	--	2.5/2.0	3/1	--	ts/th	--	----
74574	--	30-112	30-69	--	--	130-180	90-140	85-153	85-110	--	fMHz	4.22	74574
----	--	17-30	18-33	--	--	5.6-10.6	5.6-10.4	9.5-8.5	11-10	--	tpd	--	----
----	--	12/5	12/5	--	--	3.5/1.5	2.5/2.5	2/1.5	2.5/1	--	ts/th	--	----
74821	--	--	--	--	--	--	--	--	--	|	fMHz	4.24	74821
----	--	--	--	--	--	--	--	--	--	ADVANCE	tpd	--	----
----	--	--	--	--	--	--	--	--	--	INFOR-	ts/th	--	----
74823	--	--	--	--	--	--	--	--	--	MATION	fMHz	4.25	74823
----	--	--	--	--	--	--	--	--	--	|	tpd	--	----
----	--	--	--	--	--	--	--	--	--		ts/th	--	----
747074	--	28-50	--	--	--	--	--	--	--	--	fMHz	4.31	747074
----	--	15-35	--	--	--	--	--	--	--	--	tpd	--	----
----	--	5/5	--	--	--	--	--	--	--	--	ts/th	--	----
747075	--	28-50	--	--	--	--	--	--	--	--	fMHz	4.32	747075
----	--	15-35	--	--	--	--	--	--	--	--	tpd	--	----
----	--	5/5	--	--	--	--	--	--	--	--	ts/th	--	----
747076	--	28-50	--	--	--	--	--	--	--	--	fMHz	4.33	747076
----	--	15-35	--	--	--	--	--	--	--	--	tpd	--	----
----	--	5/5	--	--	--	--	--	--	--	--	ts/th	--	----
7411074	--	--	--	--	--	--	--	125-150	100-125	--	fMHz	4.34	7411074
----	--	--	--	--	--	--	--	1.5-8.2	1.5-11.3	--	tpd	--	----
----	--	--	--	--	--	--	--	3.5/1	4.5/2	--	ts/th	--	----
7411374	--	--	--	--	--	--	--	95-110	55-70	--	fMHz	4.35	7411374
----	--	--	--	--	--	--	--	1.5-11	1.5-12.7	--	tpd	--	----
----	--	--	--	--	--	--	--	2.5/3.5	3/5.5	--	ts/th	--	----
7416374	--	--	--	85-185	85-140	--	--	100	65	--	fMHz	4.36	7416374
----	--	--	--	5.4-8.1	5.6-9.4	--	--	3.8-9.5	5.1-10.9	--	tpd	--	----
----	--	--	--	5/3	2.5/2.5	--	--	5/0	6.5/1	--	ts/th	--	----
7416821	--	--	--	--	--	--	--	--	70	--	fMHz	4.39	7416821
----	--	--	--	--	--	--	--	--	4.5-12.6	--	tpd	--	----
----	--	--	--	--	--	--	--	--	7.5/12.5	--	ts/th	--	----
7416823	--	--	--	--	--	--	--	115	90	--	fMHz	4.40	7416823
----	--	--	--	--	--	--	--	3.1-11.4	4.2-11.5	--	tpd	--	----
----	--	--	--	--	--	--	--	7.2/1	3.5/2.5	--	ts/th	--	----

Table 4.9. 5V High-Speed CMOS series flip-flop frequency (MHz), typical delays (tpd), and setup/hold times (ts/th).

DEVICE 74---	LV	LVC	LVT/ABT	LVX	LVQ	LVCH ALVCH	LCX	AC (3V) HL	GTL	VHC	DATA	Figure No.	DEVICE 74---
7474	24-100	83-100	175-200	45-145	100-125	--	--	100	--	80-125	f_{MHz}	4.2	7474
----	16-28	1/9	2.8-3.5	7.3-13.2	8.0-10.5	--	--	12-13.5	--	6.7-15.4	t_{pd}	----	----
----	13/3	3/2	1.7/0	8/0.5	4.0/0.5	--	--	4.5/0.5	--	6.0/0.5	ts/th	----	----
74107	24-70	--	--	--	--	--	--	--	--	--	f_{MHz}	4.8	74107
----	18-26	--	--	--	--	--	--	--	--	--	t_{pd}	----	----
----	15/5	--	--	--	--	--	--	--	--	--	ts/th	----	----
74109	24-70	150-225	--	--	--	--	--	--	--	--	f_{MHz}	4.8	74109
----	36-33	8.0-9.0	--	--	--	--	--	--	--	--	t_{pd}	----	----
----	16/5	2.5/2.0	--	--	--	--	--	--	--	--	ts/th	----	----
74112	--	--	--	--	--	--	--	--	--	ADV	f_{MHz}	4.9	74112
----	--	10	--	--	--	--	--	--	--	INFOR-	t_{pd}	----	----
----	--	--	--	--	--	--	--	--	--	MATION	ts/th	----	----
74174	24-80	--	--	45-180	60-100	--	--	--	--	--	f_{MHz}	4.11	74174
----	12-33	--	--	5.9-18.5	2.0-10.8	--	--	--	--	--	t_{pd}	----	----
----	5/5	--	--	7.5/0	2.5/8.0	--	--	--	--	--	ts/th	----	----
74175	--	--	--	--	--	--	--	--	--	90-140	f_{MHz}	4.12	74175
----	--	--	--	--	--	--	--	--	--	7.5-15	t_{pd}	----	----
----	--	--	--	--	--	--	--	--	--	5.0/1.0	ts/th	----	----
74273	19-90	80-100	150	45-150	50-90	--	--	--	--	75-120	f_{MHz}	4.13	74273
----	12-31	1.5-9.5	5.5-5.5	7.1-20.4	4.0-17.6	--	--	--	--	8.7-17.1	t_{pd}	----	----
----	5/5	1.5/1.5	2.3/0	5.5/1.0	5.0/0	--	--	--	--	5.5/1.0	ts/th	----	----
74374	19-75	175-200	150	45-160	55-75	175-200	-/-	60	--	80-130	f_{MHz}	4.16	74374
----	14-38	3.2-3.5	5.5-5.5	6.7-19.8	3.0-18.3	3.2-3.5	2.7-3.6	13.5-12.5	--	8.1-16.2	t_{pd}	----	----
----	16/5	2.0/.3	2.3/0	4.5/2.0	4.0/1.5	2.0/.3	3.6/3.6	5.5/1	--	4.5/2.0	ts/th	----	----
742374	--	75-150	--	--	--	75-150	--	--	--	--	f_{MHz}	4.16	742374
----	--	1.5-9.5	--	--	--	1.5-9.5	--	--	--	--	t_{pd}	----	----
----	--	-/1.0	--	--	--	-/1.0	--	--	--	--	ts/th	----	----
74377	24-70	200	--	--	--	--	--	--	--	--	f_{MHz}	4.18	74377
----	15-30	5.5-5.5	--	--	--	--	--	--	--	--	t_{pd}	----	----
----	16/5	3/3	--	--	--	--	--	--	--	--	ts/th	----	----
74534	--	--	--	--	--	--	--	70	--	--	f_{MHz}	4.21	74534
----	--	--	--	--	--	--	--	14-13	--	--	t_{pd}	----	----
----	--	--	--	--	--	--	--	6.5/1.5	--	--	ts/th	----	----
74564	--	--	--	--	--	--	--	60-75	--	--	f_{MHz}	4.22	74564
----	--	--	--	--	--	--	--	14-12.5	--	--	t_{pd}	----	----
----	--	--	--	--	--	--	--	3/2	--	--	ts/th	----	----
74574	19-70	80-100	150	--	--	75-150	150	60-75	--	80-125	f_{MHz}	4.22	74574
----	14-38	1.5-9.5	5.4-5.9	--	--	4.8-4.8	8.5-8.5	13.5-12	--	8.5-5.6	t_{pd}	----	----
----	16/5	1.5/1.5	2.0/0.3	--	--	1.0/1.0	2.5/1.5	3/1.5	--	3.5/1.5	ts/th	----	----
742574	--	75-150	--	--	--	75-150	--	--	--	--	f_{MHz}	4.22	742574
----	--	4.8-4.8	--	--	--	4.8-4.8	--	--	--	--	t_{pd}	----	----
----	--	1.0/1.0	--	--	--	1.0/1.0	--	--	--	--	ts/th	----	----
74821	--	75-150	--	--	--	75-150	--	--	--	--	f_{MHz}	4.24	74821
----	--	7.0-7.0	--	--	--	7.0-7.0	--	--	--	--	t_{pd}	----	----
----	--	1.0/1.0	--	--	--	1.0/1.0	--	--	--	--	ts/th	----	----
74823	--	75-150	--	--	--	75-150	--	--	--	--	f_{MHz}	4.25	74823
----	--	8.5-8.5	--	--	--	8.5-8.5	--	--	--	--	t_{pd}	----	----
----	--	1.0/1.0	--	--	--	1.0/1.0	--	--	--	--	ts/th	----	----
742821	--	75-150	--	--	--	75-150	--	--	--	--	f_{MHz}	4.24	742821
----	--	8.0-8.0	--	--	--	8.0-8.0	--	--	--	--	t_{pd}	----	----
----	--	1.0/1.0	--	--	--	1.0/1.0	--	--	--	--	ts/th	----	----
742823	--	75-150	--	--	--	75-150	--	--	--	--	f_{MHz}	4.25	742823
----	--	9.5-9.5	--	--	--	9.5-9.5	--	--	--	--	t_{pd}	----	----
----	--	1.0/1.0	--	--	--	1.0/1.0	--	--	--	--	ts/th	----	----
7433374	--	--	--	--	--	--	--	250-350	--	--	f_{MHz}	4.35	7433374
----	--	--	--	--	--	--	--	3.2-4.3	--	--	t_{pd}	----	----
----	--	--	--	--	--	--	--	0.5/0.5	--	--	ts/th	----	----
7433534	--	--	--	--	--	--	--	250-350	--	--	MHz	4.35	7433534
----	--	--	--	--	--	--	--	3.2-4.3	--	--	t_{pd}	----	----
----	--	--	--	--	--	--	--	0.5/0.5	--	--	ts/th	----	----
7416374	--	125-350	350	--	--	200-350	170	--	--	--	f_{MHz}	4.36	7416374
----	--	3.5-6.0	3.0-4.0	--	--	1-5.9	6.2-6.1	--	--	--	t_{pd}	----	----
----	--	2.0/1.5	1.5/1.0	--	--	2.1/0.6	2.5/1.5	--	--	--	ts/th	----	----
74162374	--	240		--	--	240	--	--	--	--	f_{MHz}	4.36	74162374
----	--	8.0-8.0		--	--	8-8	--	--	--	--	t_{pd}	----	----
----	--	2.0/2.0		--	--	2.0/2.0	--	--	--	--	ts/th	----	----
7416721	--	--	--	--	--	0-150	--	--	--	--	f_{MHz}	4.37	7416721
----	--	--	--	--	--	1-5.7	--	--	--	--	t_{pd}	----	----
----	--	--	--	--	--	4/0	--	--	--	--	ts/th	----	----
7416820	--	--	--	--	--	0-150	--	--	--	--	f_{MHz}	4.38	7416820
----	--	--	--	--	--	1-6.5	--	--	--	--	t_{pd}	----	----
----	--	--	--	--	--	1.7/1.1	--	--	--	--	ts/th	----	----
7416821	--	--	--	--	--	0-150	--	--	--	--	f_{MHz}	4.39	7416821
----	--	--	--	--	--	1-6.4	--	--	--	--	t_{pd}	----	----
----	--	--	--	--	--	4.4/0	--	--	--	--	ts/th	----	----
7416823	--	--	--	--	--	0-150	--	--	--	--	f_{MHz}	4.40	7416823
----	--	--	--	--	--	1-6.4	--	--	--	--	t_{pd}	----	----
----	--	--	--	--	--	1.9/0.8	--	--	--	--	ts/th	----	----
7416921	--	--	--	--	--	--	--	--	NO	--	f_{MHz}	4.41	7416921
----	--	--	--	--	--	--	--	--	SPEED	--	t_{pd}	----	----
----	--	--	--	--	--	--	--	--	DATA	--	ts/th	----	----
74162820	--	--	--	--	--	0-150	--	--	--	--	f_{MHz}	4.38	74162820
----	--	--	--	--	--	1-7	--	--	--	--	t_{pd}	----	----
----	--	--	--	--	--	1.8/1.1	--	--	--	--	ts/th	----	----

Table 4.10. Low voltage high-speed CMOS series flip-flop frequency (MHz), typical delays (tpd), and setup/hold times (ts/th).

TTL – 74H71/H101

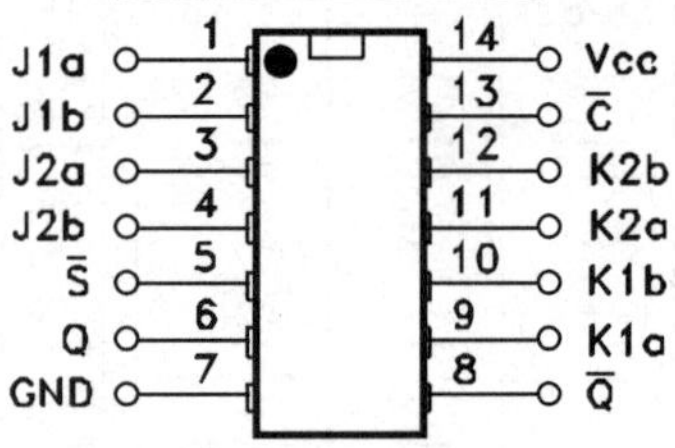

DUAL-IN-LINE PACKAGE

TTL – 74H102
TTL – 7472/H72

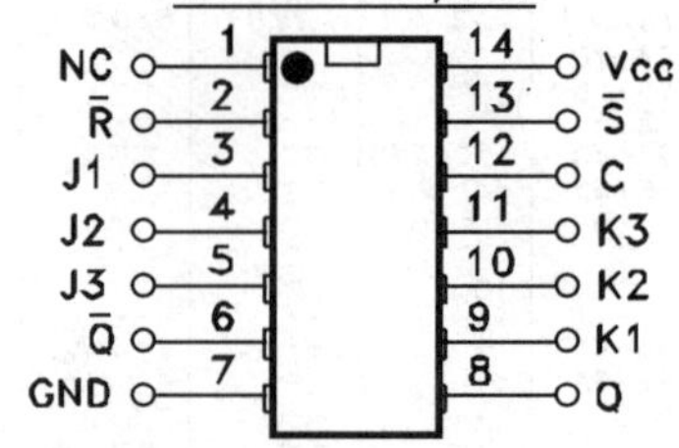

DUAL-IN-LINE PACKAGE

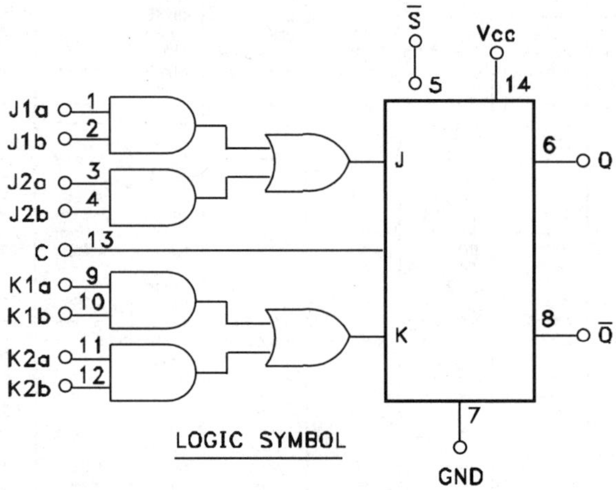

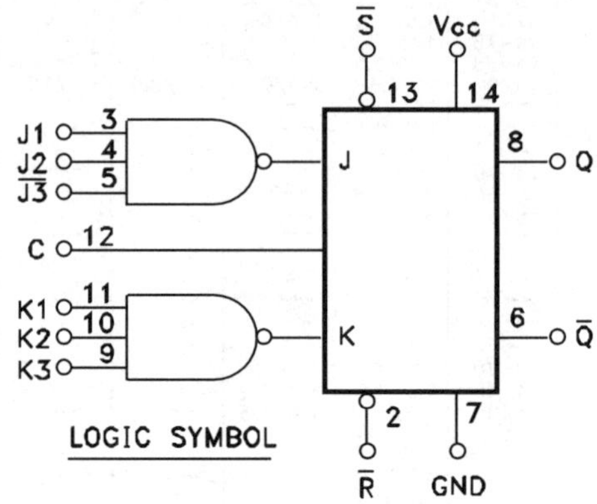

TRUTH TABLE TTL 74H71/H101 JK FLIP-FLOP

	INPUTS				OUTPUTS	
	C	J	K	$\overline{S}$	Q	$\overline{Q}$
1	X	X	X	0	1	0
4	⎍	h	h	1	$\overline{q}$	q
6	⎍	h	l	1	1	0
7	⎍	l	l	1	q	$\overline{q}$

$J=(J1a \bullet J1b)+(J2a \bullet J2b)$
$K=(K1a \bullet K1b)+(K2a \bullet K2b)$

LOWER CASE INDICATES LOGIC STATUS PRIOR TO CLOCK TRANSITION.

TRUTH TABLE TTL 7472/H72/L72/H102 JK FLIP-FLOP

	INPUTS					OUTPUTS	
	C	J	K	$\overline{S}$	$\overline{R}$	Q	$\overline{Q}$
1	0	X	X	0	1	1	0
2	0	X	X	1	0	0	1
3	X	X	X	0	0	1	1
4	⎍	h	h	1	1	$\overline{q}$	q
5	⎍	l	h	1	1	0	1
6	⎍	h	l	1	1	1	0
7	⎍	l	l	1	1	q	$\overline{q}$

$J = J1 \bullet J2 \bullet J3$ $K = K1 \bullet K2 \bullet K3$

LOWER CASE INDICATES LOGIC STATUS PRIOR TO CLOCK TRANSITION.

OPERATING MODE NOTES:

1 Asynchronous Set
2 Asynchronous Reset (Clear)
3 Not predictable
4 Toggle
5 Load "0" (Reset)
6 Load "1" (Set)
7 Hold "No Change"

Figure 4.4. These early high-speed JK series with multiple inputs feature positive-pulse triggering with Data, Clock, Set, and Reset inputs; complementary Q and /Q outputs. Note that the 74H71/74H101 have the preset only. No CMOS equivalents exist.

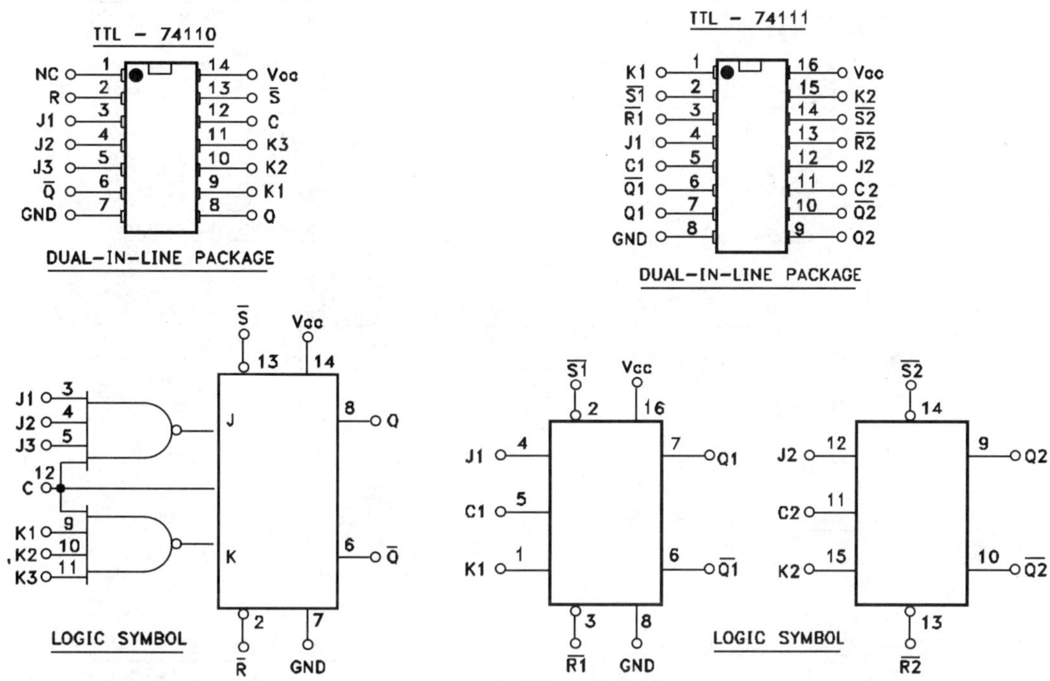

TRUTH TABLE TTL 74110 JK FLIP-FLOP

	INPUTS					OUTPUTS	
	C	J	K	$\bar{R}$	$\bar{S}$	Q	$\bar{Q}$
	X	X	X	0	X	0	1
1▷	X	X	X	X	0	1	0
4▷	⎍	h	h	1	1	$\bar{q}$	q
6▷	⎍	h	l	1	1	1	0
7▷	⎍	l	l	1	1	q	$\bar{q}$

LOWER CASE INDICATES LOGIC STATUS PRIOR TO CLOCK TRANSITION.

TRUTH TABLE 74111 JK FLIP-FLOP

	INPUTS					OUTPUTS	
	C	J	K	$\bar{S}$	$\bar{R}$	Q	$\bar{Q}$
1▷	X	X	X	0	1	1	0
2▷	X	X	X	1	0	0	1
3▷	X	X	X	0	0	1	1
4▷	⎍	l	l	1	1	q	$\bar{q}$
5▷	⎍	h	l	1	1	0	1
6▷	⎍	l	h	1	1	1	0
7▷	⎍	h	h	1	1	q	$\bar{q}$

LOWER CASE INDICATES LOGIC STATUS PRIOR TO CLOCK TRANSITION.

OPERATING MODE NOTES:
1▷ Asynchronous Set
2▷ Asynchronous Reset (Clear)
3▷ Not predictable
4▷ Toggle
5▷ Load "0" (Reset)
6▷ Load "1" (Set)
7▷ Hold "No Change"

Figure 4.5. The 74110 is a gated JK MS type with data lockout and variable skew having a short enable time. The clock input is logically AND'd with the J and K inputs. Direct preset and clear are provided with complementary outputs. The 74111 is a dual JK MS with data lockout. Direct preset and clear are provided with complementary outputs.These two flip-flops are early varieties from Texas Instruments. No CMOS equivalents exist.

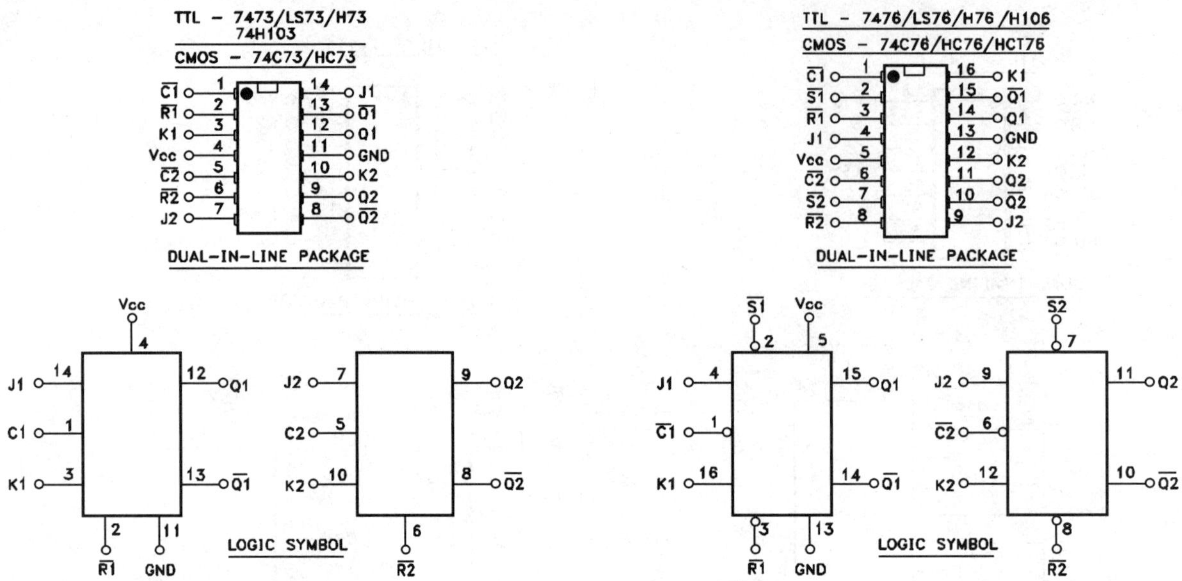

TRUTH TABLE 74x73/H103 JK FLIP-FLOP

	INPUTS				OUTPUTS	
	$\overline{C}$	J	K	$\overline{R}$	Q	$\overline{Q}$
2▷	X	X	X	0	0	1
4▷	⎍	h	h	1	$\overline{q}$	q
5▷	⎍	h	l	1	0	1
6▷	⎍	h	l	1	1	0
7▷	⎍	l	l	1	q	$\overline{q}$

LOWER CASE INDICATES LOGIC STATUS PRIOR TO CLOCK TRANSITION.

OPERATING MODE NOTES:

1▷ Asynchronous Set
2▷ Asynchronous Reset (Clear)
3▷ Not predictable
4▷ Toggle
5▷ Load "0" (Reset)
6▷ Load "1" (Set)
7▷ Hold "No Change"

TRUTH TABLE 74x76/H106 JK FLIP-FLOP

	INPUTS					OUTPUTS	
	$\overline{C}$	J	K	$\overline{S}$	$\overline{R}$	Q	$\overline{Q}$
1▷	X	X	X	0	1	1	0
2▷	X	X	X	1	0	0	1
3▷	X	X	X	0	0	1	1
4▷	⎍	h	h	1	1	$\overline{q}$	q
5▷	⎍	l	h	1	1	0	1
6▷	⎍	h	l	1	1	1	0
7▷	⎍	l	l	1	1	q	$\overline{q}$

LOWER CASE INDICATES LOGIC STATUS PRIOR TO CLOCK TRANSITION.

Figure 4.6. These TTL and CMOS dual-D flip-flop series feature positive-edge triggering with Data, Clock, Set, and Reset inputs; complementary Q and /Q outputs are provided. Observe the nonstandard power connections.

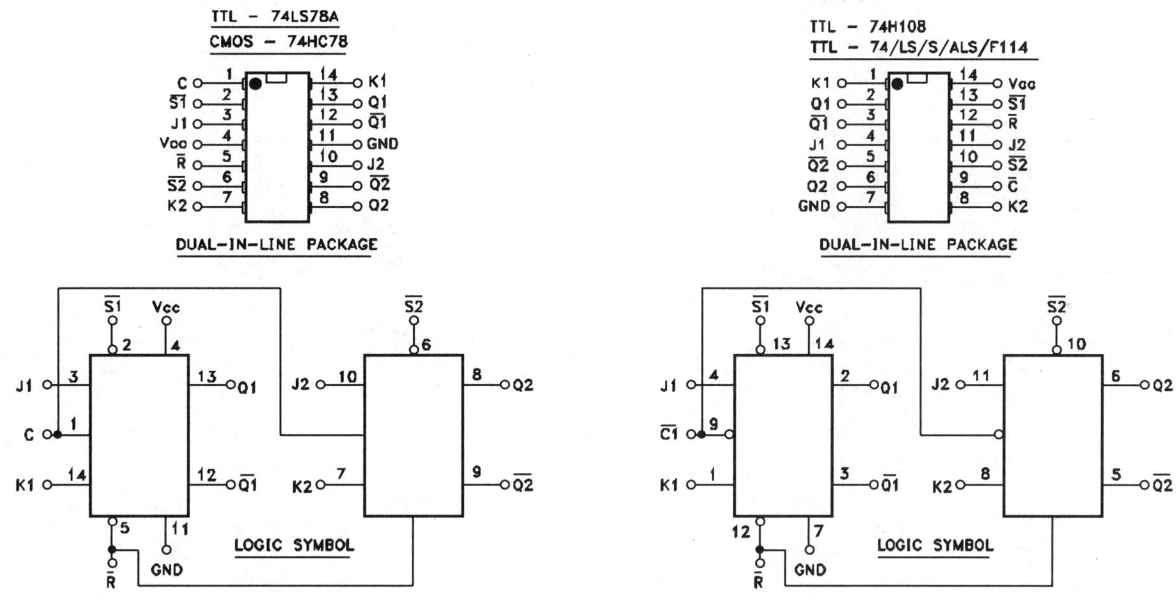

TRUTH TABLE 74LS78A/HC78 DUAL JK FLIP-FLOPS

	INPUTS					OUTPUTS	
	C	J	K	$\overline{S}$	$\overline{R}$	Q	$\overline{Q}$
1	X	X	X	0	1	1	0
2	X	X	X	1	0	0	1
3	X	X	X	0	0	1	1
7	⌐_	0	0	1	1	Qo	$\overline{Qo}$
4	⌐_	1	0	1	1	1	0
5	⌐_	0	1	1	1	0	1
6	⌐_	1	1	1	1	TOGGLE	
7	1	X	X	1	1	Qo	$\overline{Qo}$

OPERATING MODE NOTES:

1. Asynchronous Set
2. Asynchronous Reset (Clear)
3. Not predictable
4. Toggle
5. Load "0" (Reset)
6. Load "1" (Set)
7. Hold "No Change"

TRUTH TABLE 74H108/x114 JK FLIP-FLOP

	INPUTS					OUTPUTS	
	$\overline{C}$	J	K	$\overline{S}$	$\overline{R}$	Q	$\overline{Q}$
1	X	X	X	0	1	1	0
2	X	X	X	1	0	0	1
3	X	X	X	0	0	1	1
4	⌐_	h	l	1	1	$\overline{q}$	q
5	⌐_	l	h	1	1	0	1
6	⌐_	h	l	1	1	1	0
7	⌐_	l	h	1	1	q	$\overline{q}$

LOWER CASE INDICATES LOGIC STATUS PRIOR TO CLOCK TRANSITION.

Figure 4.7. These Dual JK series feature negative-edge triggering with Data, Common Clock, and Common Reset inputs; complementary Q and /Q outputs are provided. Observe the nonstandard power connections with the 74 x 78.

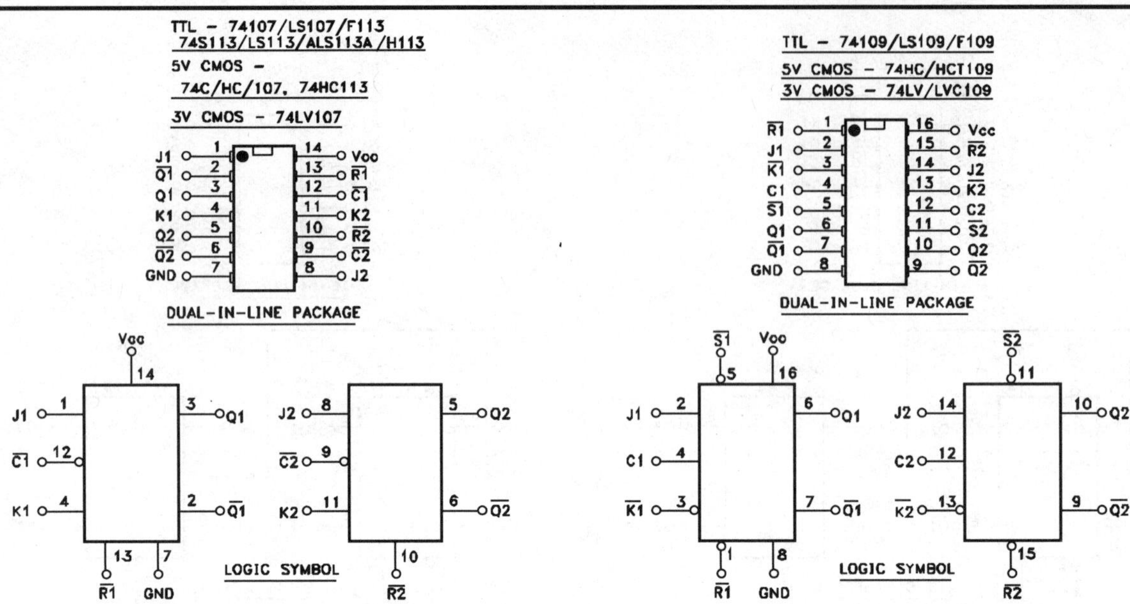

TRUTH TABLE 74x107/x113 JK FLIP-FLOP

	INPUTS				OUTPUTS	
	$\overline{C}$	J	K	$\overline{R}$	Q	$\overline{Q}$
2	X	X	X	0	0	1
4	⎍	h	h	1	$\overline{q}$	q
5	⎍	l	h	1	0	1
6	⎍	h	l	1	1	0
7	⎍	l	l	1	q	$\overline{q}$

LOWER CASE INDICATES LOGIC STATUS PRIOR TO CLOCK TRANSITION.

OPERATING MODE NOTES:
1. Asynchronous Set
2. Asynchronous Reset (Clear)
3. Not predictable
4. Toggle
5. Load "0" (Reset)
6. Load "1" (Set)
7. Hold "No Change"

TRUTH TABLE 74x109 JK FLIP-FLOP

	INPUTS					OUTPUTS	
	C	J	$\overline{K}$	$\overline{S}$	$\overline{R}$	Q	$\overline{Q}$
1	X	X	X	0	1	1	0
2	X	X	X	1	0	0	1
3	X	X	X	0	0	1	1
4	↑	h	l	1	1	$\overline{q}$	q
5	↑	l	h	1	1	0	1
6	↑	h	l	1	1	1	0
7	↑	l	h	1	1	q	$\overline{q}$

LOWER CASE INDICATES LOGIC STATUS PRIOR TO CLOCK TRANSITION.

Figure 4.8. These TTL and CMOS Dual JK series feature positive-pulse/edge triggering with Data, Clock, Set, and Reset inputs; complementary Q and /Q outputs are provided. The 74 x 107 and 74 x 113 series have a Reset input, but not Set.

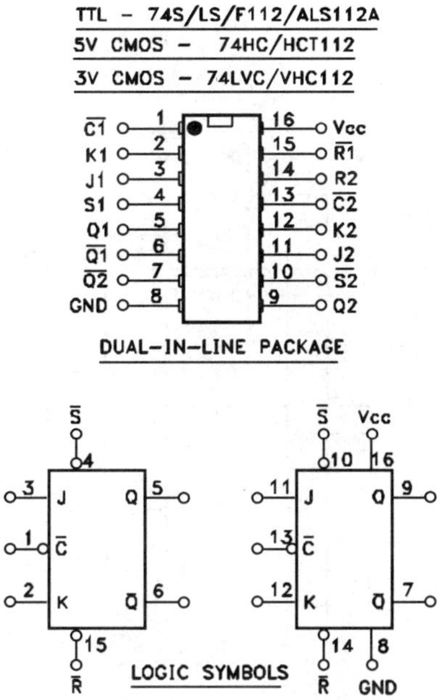

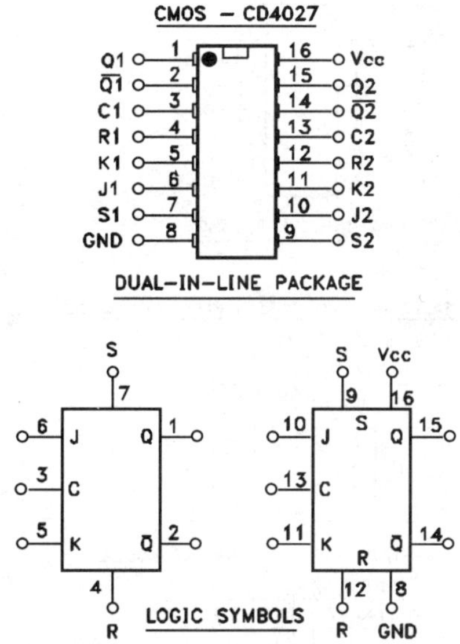

TRUTH TABLE TTL 74x112 JK FLIP-FLOP

	INPUTS					OUTPUTS	
	$\bar{C}$	J	K	$\bar{S}$	$\bar{R}$	Q	$\bar{Q}$
1	X	X	X	0	1	1	0
2	X	X	X	1	0	0	1
3	X	X	X	0	0	1	1
4	↓	h	h	1	1	$\bar{q}$	q
5	↓	l	h	1	1	0	1
6	↓	h	l	1	1	1	0
7	↓	l	l	1	1	q	$\bar{q}$

LOWER CASE INDICATES LOGIC STATUS PRIOR TO CLOCK TRANSITION.

OPERATING MODE NOTES:

1. Asynchronous Set
2. Asynchronous Reset (Clear)
3. Not predictable
4. Toggle
5. Load "0" (Reset)
6. Load "1" (Set)
7. Hold "No Change"

TRUTH TABLE - CD4027 JK FLIP-FLOP

INPUTS						OUTPUTS *	
Cι	J	K	S	R	Qn **	Qn+1	$\overline{Qn+1}$
↑	1	X	0	0	0	1	0
↑	X	0	0	0	1	1	0
↑	0	X	0	0	0	0	1
↑	X	1	0	0	1	0	1
↑	1	1	0	0	Qo	$\bar{Q}o$	Qo
↓	X	X	0	0	X	Qn	$\bar{Q}n$
X	X	X	1	0	X	1	0
X	X	X	0	1	X	0	1
X	X	X	1	1	X	1	1

X = DON'T CARE ** = PRESENT STATE
ι = LEVEL CHANGE * = NEXT STATE

Figure 4.9. These TTL and CMOS Dual JK series feature edge triggering: negative for the TTL, positive for the CD4027, with Data, Clock, Set, and Reset inputs; complementary Q and /Q outputs are provided.

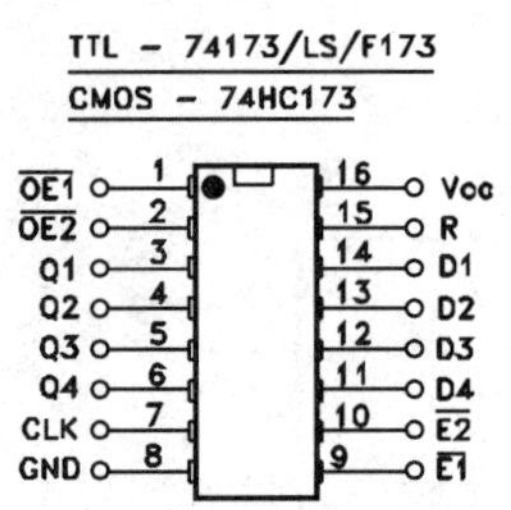

DUAL-IN-LINE PACKAGE

TRUTH TABLE QUAD-D X173 SERIES FLIP-FLOPS

REGISTER OPERATING MODES

	INPUTS					
	CLK	D	MR	$\overline{E1}$	$\overline{E2}$	Q
5	1	X	X	X	X	0
1	⎍	1	0	0	0	1
2	⎍	0	0	0	0	0
3	X	X	0	1	X	Qo
4	X	X	0	X	1	Qo

3-STATE BUFFER OPERATING MODES

	INPUTS			
	Qn (Register)	$\overline{OE1}$	$\overline{OE2}$	Q_{1-4}
6	0 1	0 0	0 0	0 1
7	X X	1 X	X 1	Z Z

OPERATING MODE NOTES:

1. Load "1" (Set)
2. Load "0" (Reset)
3. Hold "No Change"
4. Hold "No Change"
5. Clear
6. Read
7. Disabled (High Impedance)

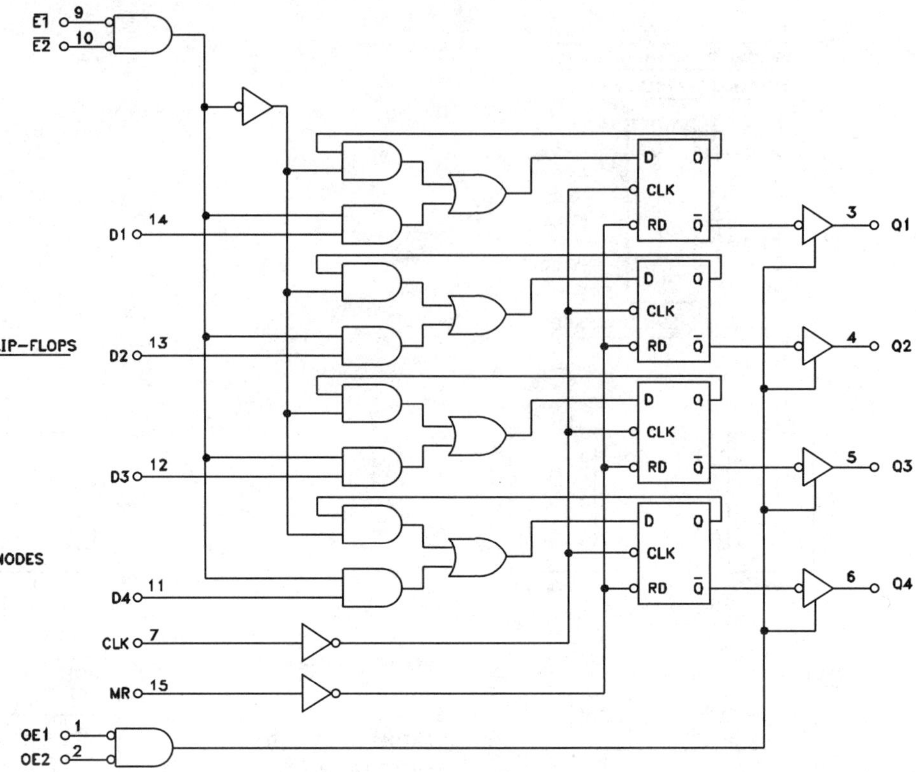

LOGIC DIAGRAM

Figure 4.10. This versatile Quad D flip-flop series features positive-edge triggering with Data and Clock inputs. Q outputs only available. No set provision. This is a 3-state output IC with dual active low input enables and dual active low output high Z enables.

TTL – 74174/LS/S/F/ALS/AS174

5V CMOS – CD40174, MC14174
74C/HC/HCT174

3V CMOS – 74LV/LVX/LVQ174

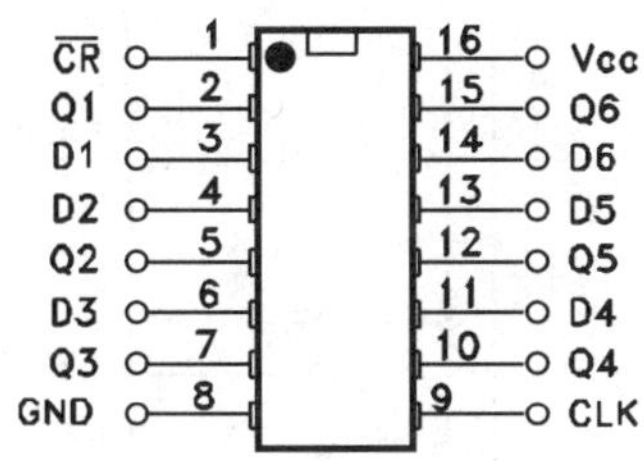

DUAL-IN-LINE PACKAGE

TRUTH TABLE HEX-D X174 SERIES FLIP-FLOPS

	INPUTS			
	C	D	$\overline{CR}$	Q
1▷	X	X	0	0
2▷	⎍	1	1	1
3▷	⎍	0	1	0
4▷	0	X	1	Qo

OPERATING MODE NOTES:

1▷ Asynchronous Reset (Clear)
2▷ Load "1" (Set)
3▷ Load "0" (Reset)
4▷ Hold "No Change"

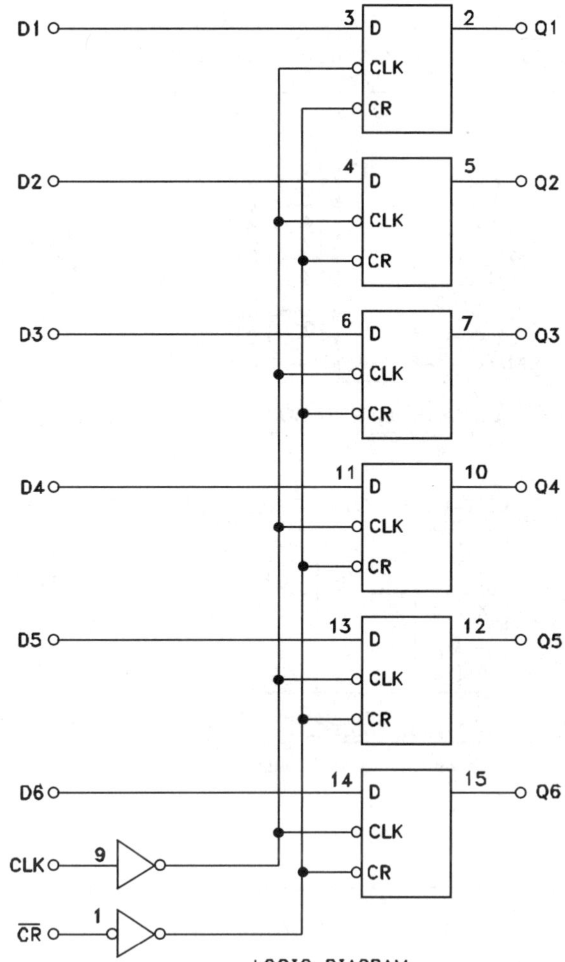

LOGIC DIAGRAM

Figure 4.11. Of the two HEX D flip-flops available, this TTL and CMOS series has the most speed and power options. Features include positive-edge triggering with Data, Clock, and Clear inputs. Q outputs only available. No Set provision. This is not a 3-state IC.

TTL – 74175/LS/S/F/ALS/AS175

5V CMOS – 74C/HC/HCT/VHC(5V)175
CD40175, MC14175

3V CMOS – 74VHC175

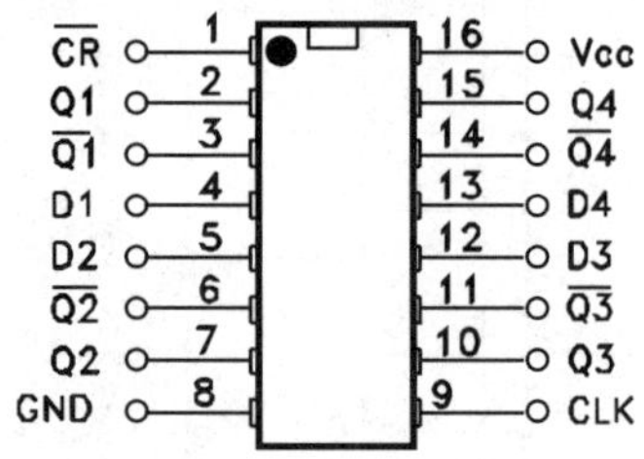

DUAL-IN-LINE PACKAGE

TRUTH TABLE QUAD-D X175 SERIES FLIP-FLOPS

	INPUTS			OUTPUTS	
	C	D	$\overline{CR}$	Q	$\overline{Q}$
1	X	X	0	0	1
2	↑	1	1	1	0
3	↑	0	1	0	1
4	0	X	1	Qo	$\overline{Qo}$

OPERATING MODE NOTES:

1 Asynchronous Reset (Clear)
2 Load "1" (Set)
3 Load "0" (Reset)
4 Hold "No Change"

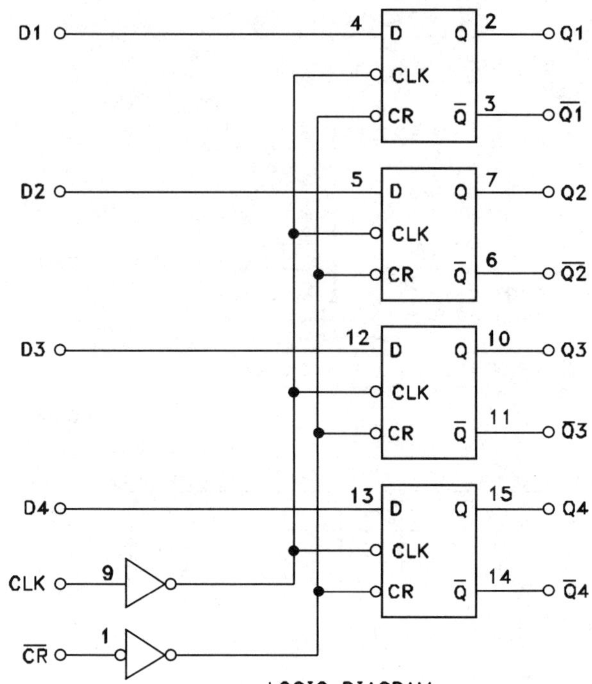

LOGIC DIAGRAM

Figure 4.12. This TTL and CMOS QUAD D flip-flop series features positive-edge triggering with Data, Clock, and Clear inputs. Complementary outputs available. No set provision. This is not a 3-state IC.

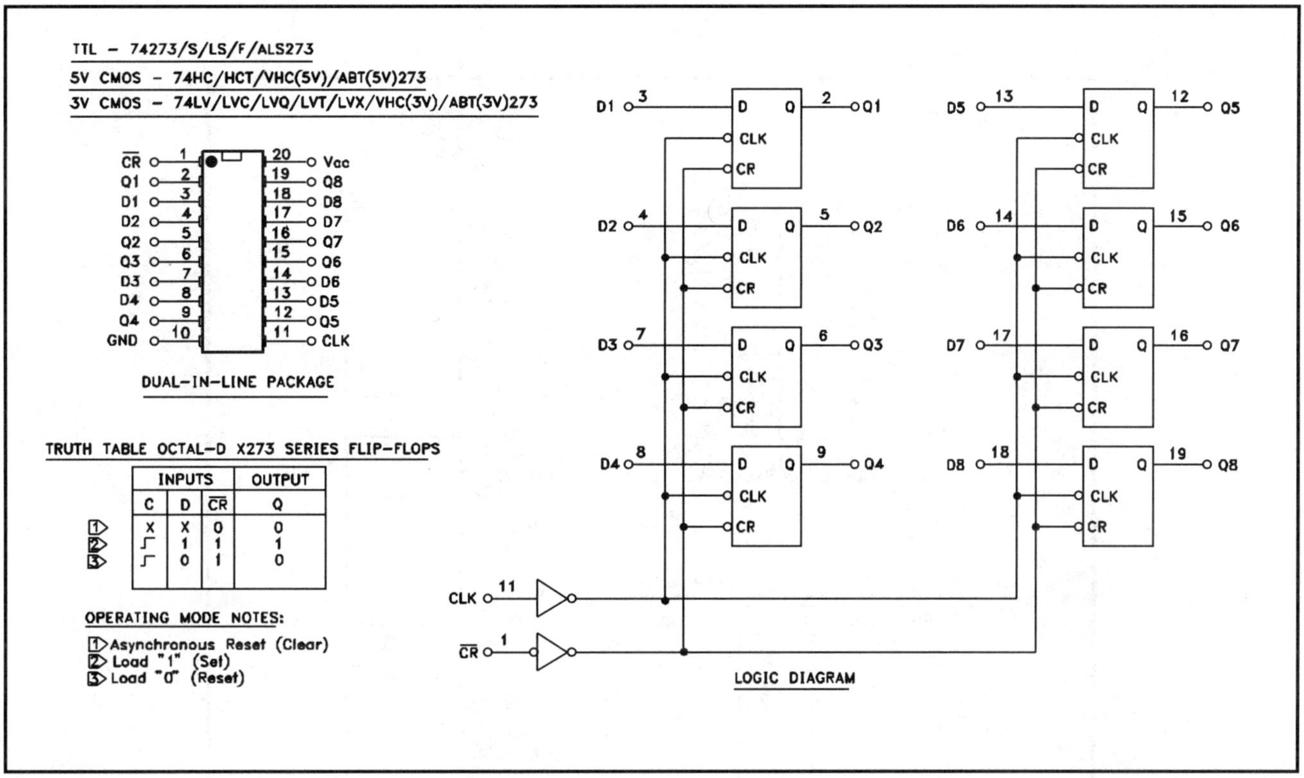

Figure 4.13. This Octal D flip-flop series is available with a wide range of TTL and CMOS speed and power options. Features are positive-edge triggering with Data, Clock and active Low Clear inputs. Q outputs only provided. No set provision. This is not a 3-state IC.

TTL - 74276

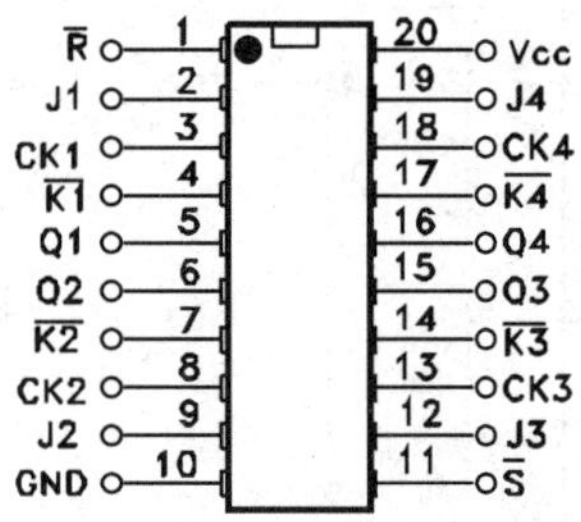

DUAL-IN-LINE PACKAGE

TRUTH TABLE 74276 QUAD JK̄ FLIP-FLOP

	INPUTS			COMMON		
	CK	J	K̄	S̄	R̄	Q
1	X	X	X	0	1	1
2	X	X	X	1	0	0
3	X	X	X	0	0	1
7	⌊	0	1	1	1	Qo
5	⌊	1	1	1	1	0
6	⌊	0	0	1	1	1
4	⌊	1	0	1	1	/Q
7	1	X	X	1	1	Qo

OPERATING MODE NOTES:

1. Asynchronous Set
2. Asynchronous Reset (Clear)
3. Not predictable
4. Toggle
5. Load "0" (Reset)
6. Load "1" (Set)
7. Hold "No Change"

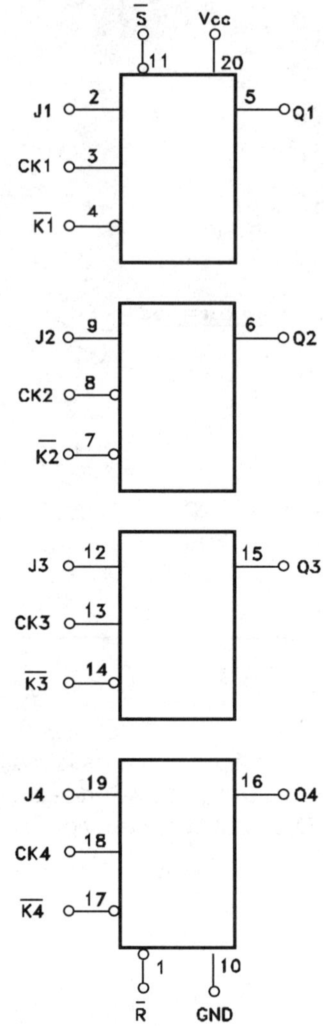

LOGIC SYMBOL

Figure 4.14. This Quad J/K flip-flop features negative-edge triggering with Data and Clock inputs. Q output only. The four independent flip-flops share a common Active Low Set and Clear input. This is not a 3-state IC. A CMOS version does not exist.

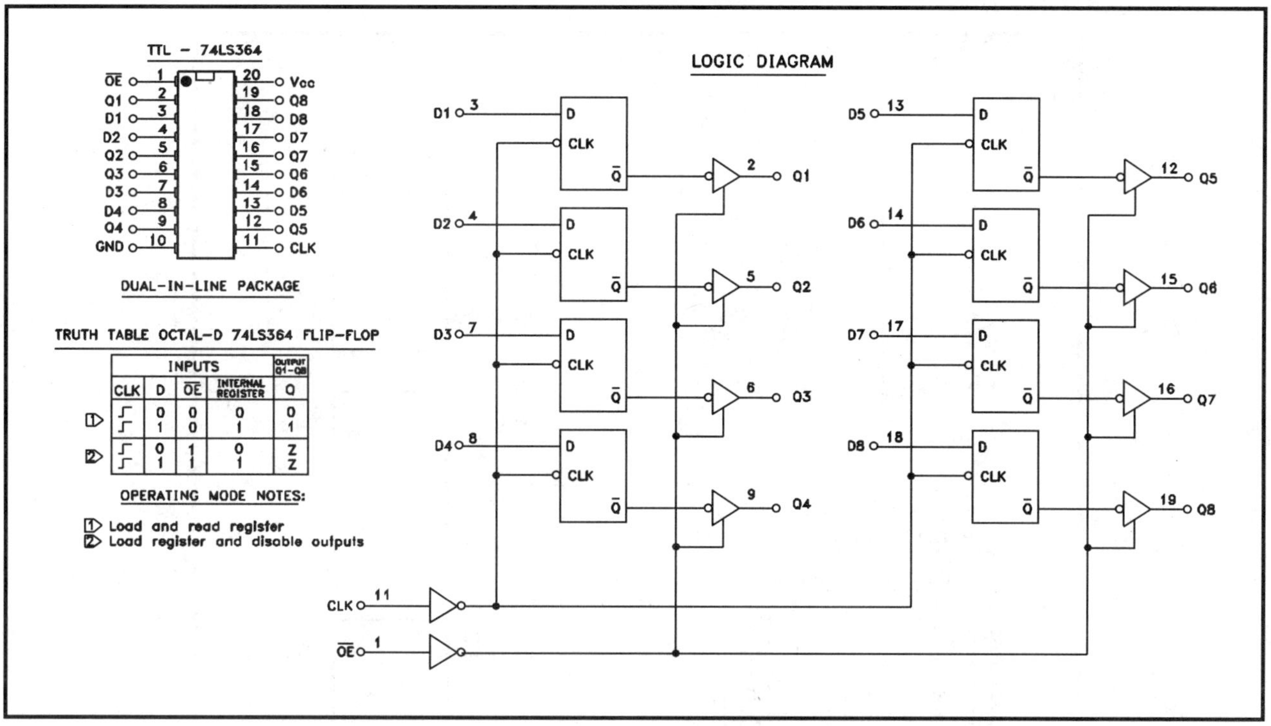

Figure 4.15. This Octal D flip-flop features positive-edge triggering with Data and Clock inputs. Q output only provided. The eight flip-flops feature a common clock; there are no Set and Clear functions. This is a 3-state IC designed with MOS compatible output buffers. A CMOS version does not exist.

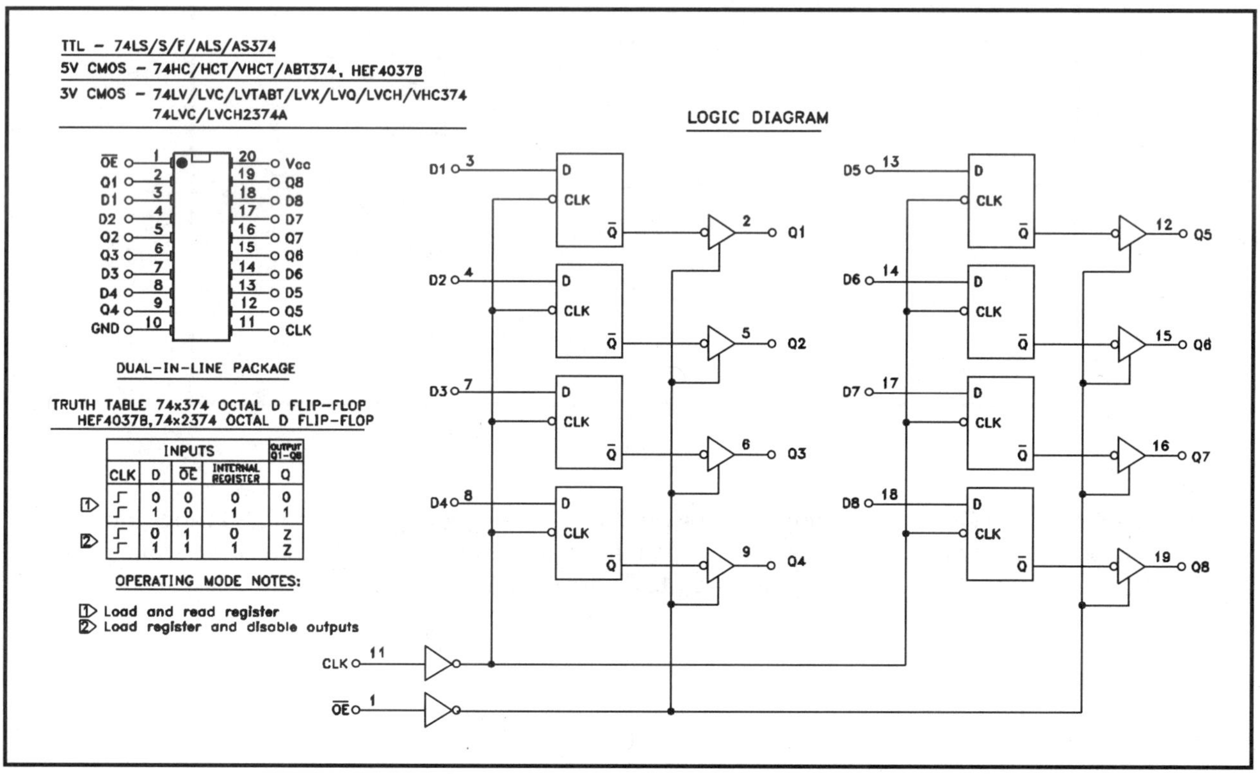

Figure 4.16. This popular Octal D flip-flop series is available with a wide range of TTL and CMOS speed and power options. The device features positive-edge triggering with Data and Clock inputs. Q output only provided. No Set or Clear provision. This is a 3-state output IC with active Low output enable.

TTL - 74376

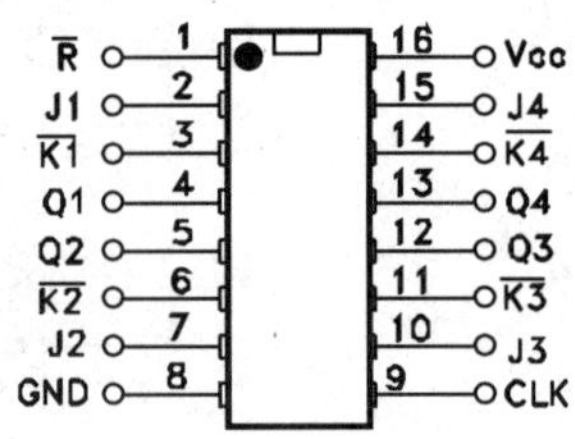

TRUTH TABLE 74376 QUAD J$\overline{K}$ FLIP-FLOP

	INPUTS		COMMON		
	J	$\overline{K}$	CLK	$\overline{R}$	Q
1	X	X	X	0	0
2	0	1	↑	1	Qo
3	1	1	↑	1	1
4	0	0	↑	1	0
5	1	0	↑	1	/Q
2	X	X	0	1	Qo

OPERATING MODE NOTES:

1. Asynchronous Reset (Clear)
2. Hold "No Change"
3. Load "1" (Set)
4. Load "0" (Reset)
5. Toggle

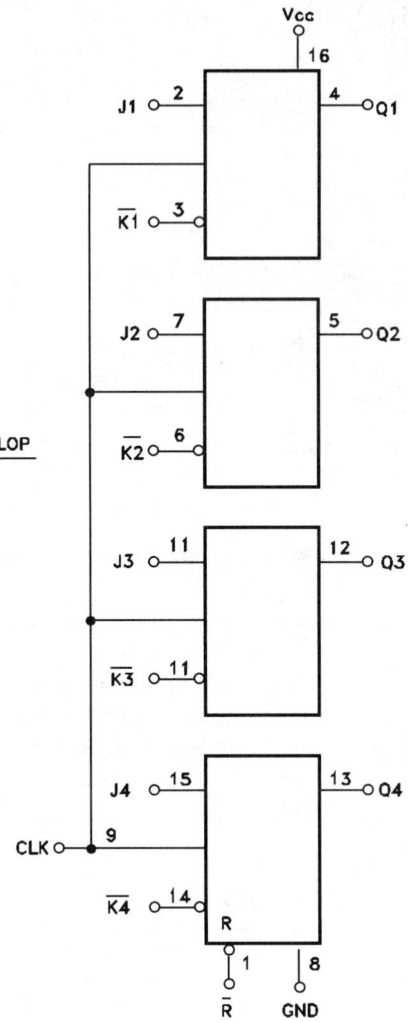

LOGIC SYMBOL

Figure 4.17. This Quad J/K flip-flop features positive-edge triggering with Data and Clock inputs. Q output only provided. The four flip-flops feature a common clock; there is a common Clear function. A CMOS version does not exist.

TTL - 74LS/F377
5V CMOS - 74HC/HCT/ABT377
3V CMOS - 74LV/LVC/ABT377

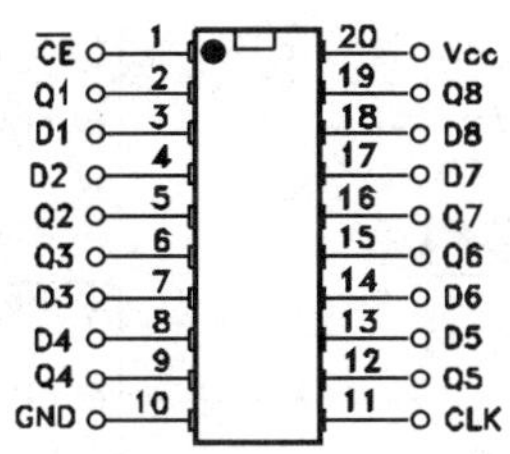

DUAL-IN-LINE PACKAGE

TRUTH TABLE OCTAL-D 74x377 FLIP-FLOP

	INPUTS			OUTPUT Q1-Q8
	CLK	D	$\overline{CE}$	Q
1	↑	0	0	0
2	↑	1	0	1
3	↑	X	1	No Change
3	X	X	1	No Change

OPERATING MODE NOTES:

1. Load "0"
2. Load "1"
3. Hold (Do nothing)

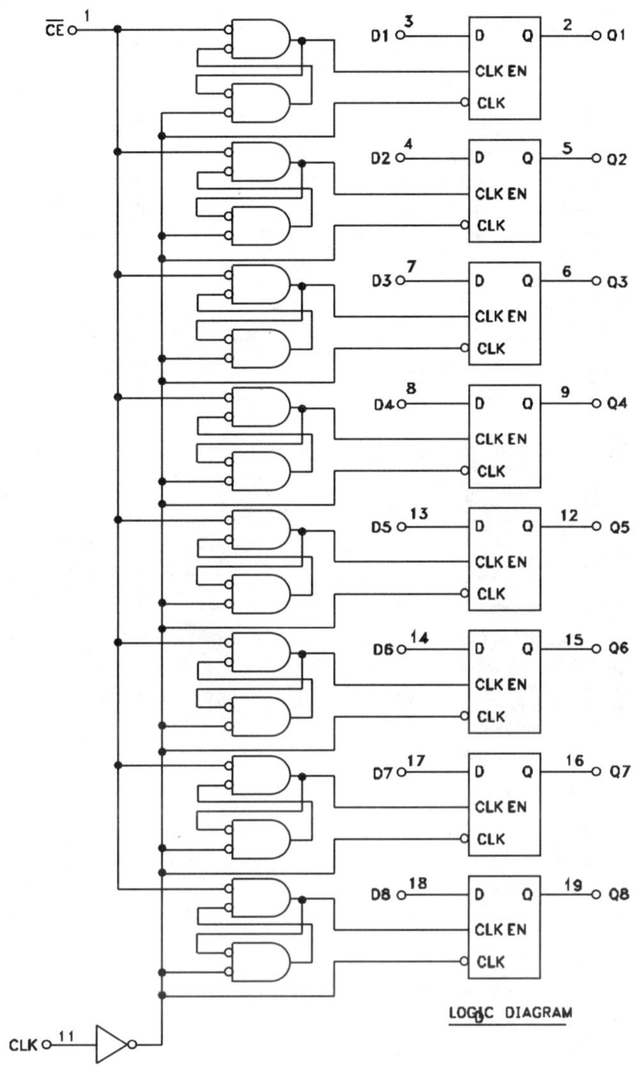

LOGIC DIAGRAM

Figure 4.18. This Octal D flip-flop features a latched input enable designed to prevent false clocking at the active Low (/CE) input. The device is available with a wide range of TTL and CMOS speed and power options. Clock triggering occurs at a particular voltage level not directly related to the input pulse. Q output only provided. The eight flip-flops feature a common clock; there is no Set or Clear function. The output is not 3-state. The 74x377, 74x378, and 74x379 are similar to the 74x273, 74x174, and 74x175 respectively, but feature a latched clock enable in place of a common clear.

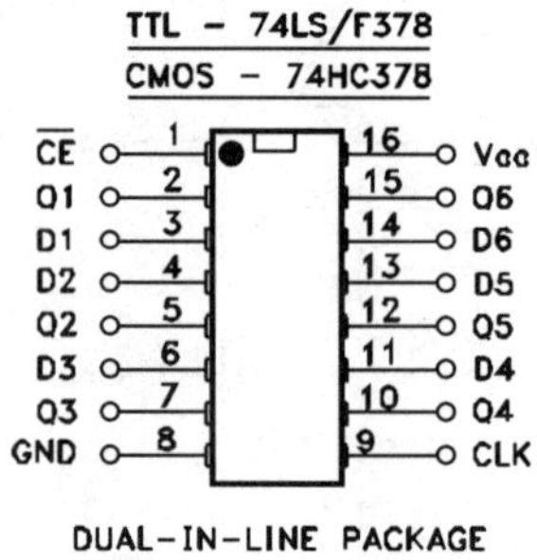

TRUTH TABLE HEX D X378 SERIES FLIP-FLOPS

	INPUTS			
	CLK	D	$\overline{CE}$	Q
1	↑	X	l	0
2	↑	1	i	1
3	↑	0	h	Qo
4	X	X	1	Qo

OPERATING MODE NOTES:

1 Load "1" (Set)
2 Load "0" (Reset)
3 Hold "No Change"
4 Hold "No Change"

l, h = low, high voltage level one setup time prior to the low-to-high clock transition.

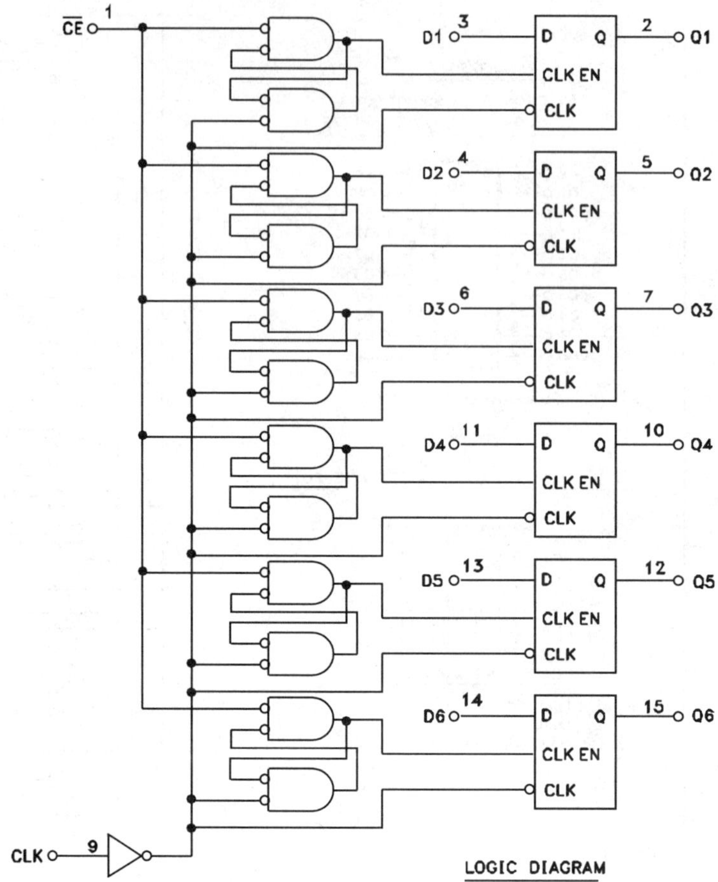

Figure 4.19. This Hex D flip-flop with limited TTL and CMOS availability features positive-edge triggering with Data, an active Low latched Clock Enable with common Clock inputs. Q output only provided. The six flip-flops feature a common clock; there is no Set or Clear function. The output is not 3-state. The 74x377, 74x378, and 74x379 are similar to the 74x273, 74x174, and 74x175 respectively, but feature a latched clock enable in place of a common clear.

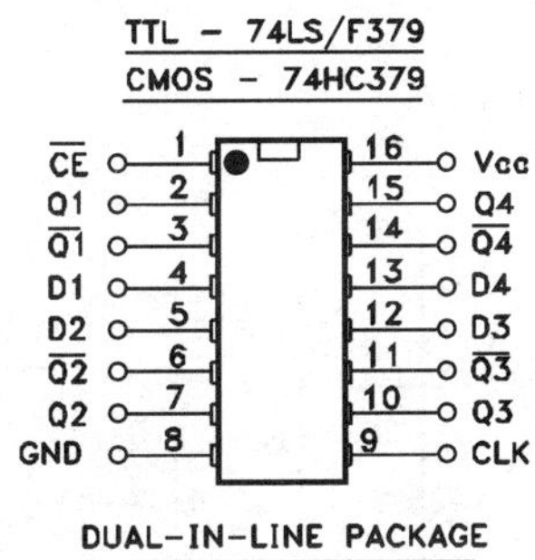

TRUTH TABLE QUAD-D X379 SERIES FLIP-FLOPS

	INPUTS			OUTPUTS	
	C	D	$\overline{CE}$	Q	$\overline{Q}$
1	↑	1	0	1	0
2	↑	0	0	0	1
3	↑	X	1	Qo	$\overline{Qo}$
4	X	X	1	Qo	$\overline{Qo}$

OPERATING MODE NOTES:

1 Load "1" (Set)
2 Load "0" (Reset)
3 Hold "No Change"
4 Hold "No Change"

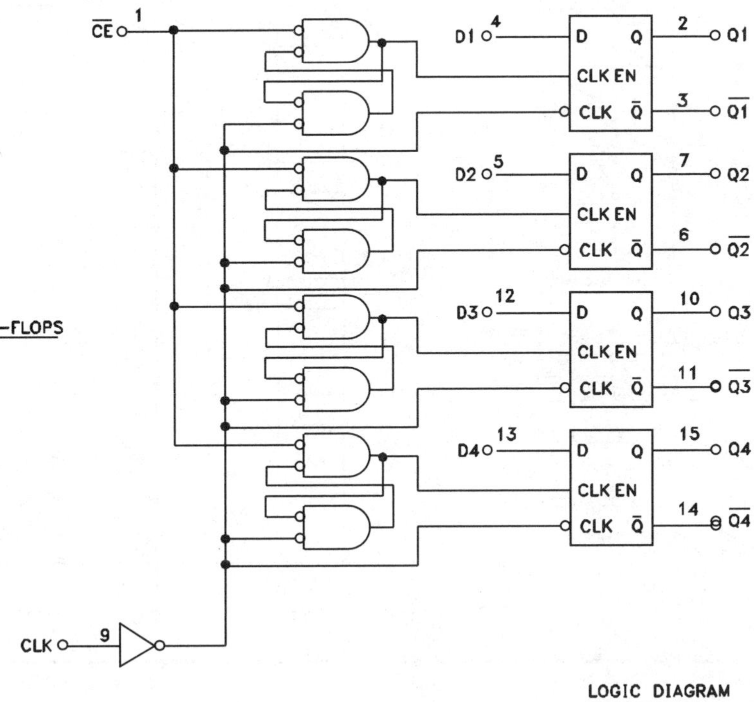

Figure 4.20. This TTL and CMOS Quad D flip-flop features positive-edge triggering with Data and Clock inputs. Q and /Q outputs are provided. The four flip-flops feature a common clock; there is no Set or Clear function. A buffered active Low input enable is provided. The enable is latched to avoid false clocking. Register data is retained with the input enable High. The output is not 3-state. The 74x377, 74x378, and 74x379 are similar to the 74x273, 74x174, and 74x175 respectively, but feature a latched clock enable in place of a common clear.

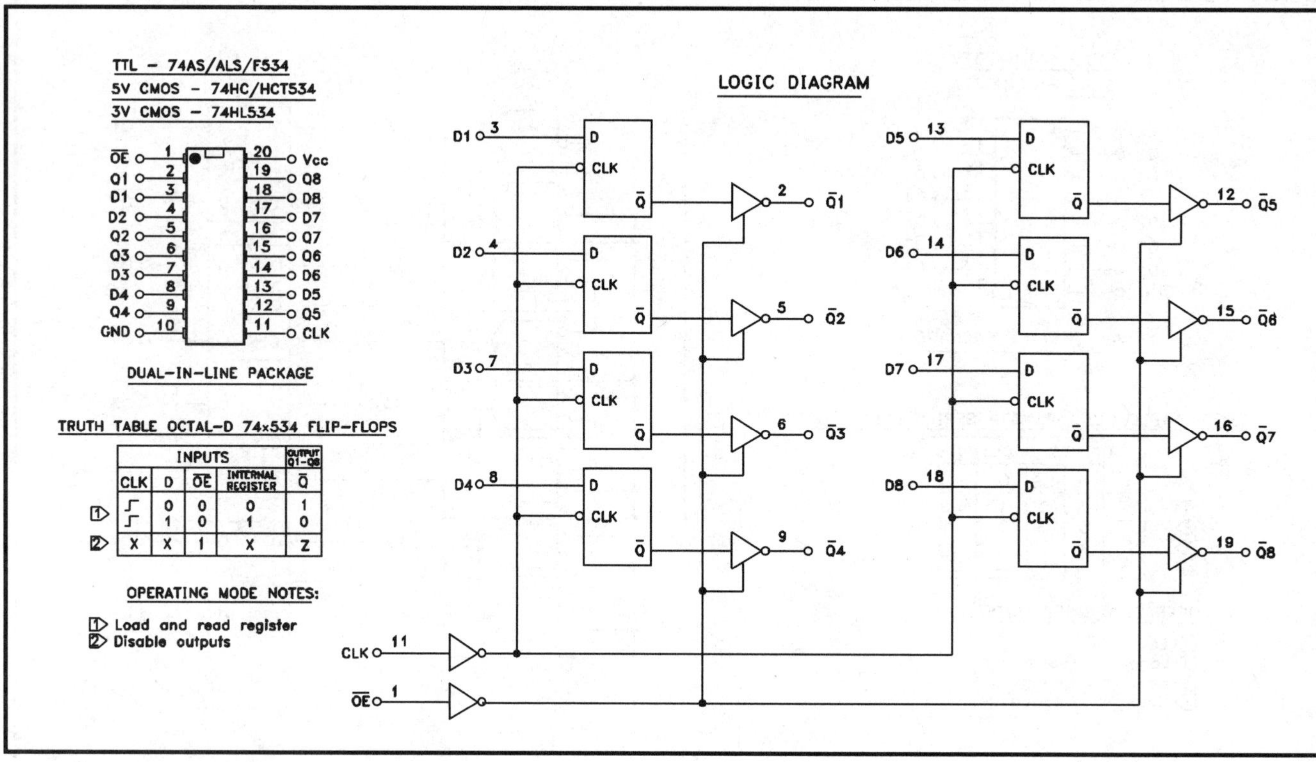

Figure 4.21. This TTL and CMOS Octal D flip-flop features positive-edge triggering with Data and Clock inputs. Q output only. The eight flip-flops feature a common clock; there is no Set or Clear function. The /Q output only is provided. An active low enabled 3-state high current output is provided. The 74HCT534 is capable of driving 15 LSTTL loads.

TTL – 74AS/ALS/F564
TTL – 74ALS574A, AS/F574
TTL – 74ALS576A, AS576

5V CMOS – 74HC/HCT/AC(5V)/ACT564
74HC/HCT/VHC(5V)/VHCT/AC(5V)/ACT574

3V CMOS – 74AC564 (3V)
74LV/LVC/LVT/LVCH/LCX/AC(3V)/LCX/VHC574

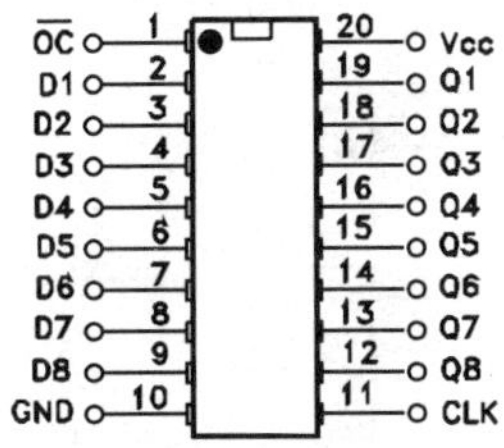

DUAL-IN-LINE PACKAGE

TRUTH TABLE OCTAL-D 74x564 FLIP-FLOPS
TRUTH TABLE OCTAL-D 74x574 FLIP-FLOPS
TRUTH TABLE OCTAL-D 74x576 FLIP-FLOPS

74x564 – TRUE IN COMPLEMENTRY OUT
74x574 – TRUE IN TRUE OUT
74x576 – TRUE IN COMPLEMENTRY OUT
TRUE IN TRUE OUT SHOWN

	INPUTS				OUTPUT Q1–Q8	OUTPUT $\overline{Q1}$–$\overline{Q8}$
	CLK	D	$\overline{OC}$	INTERNAL REGISTER	Q	$\overline{Q}$
1	↑	0	0	0	0	1
	↑	1	0	1	1	0
2	X	0	X	X	Z	Z

OPERATING MODE NOTES:

1 Load and read register
2 Disable outputs

LOGIC DIAGRAM

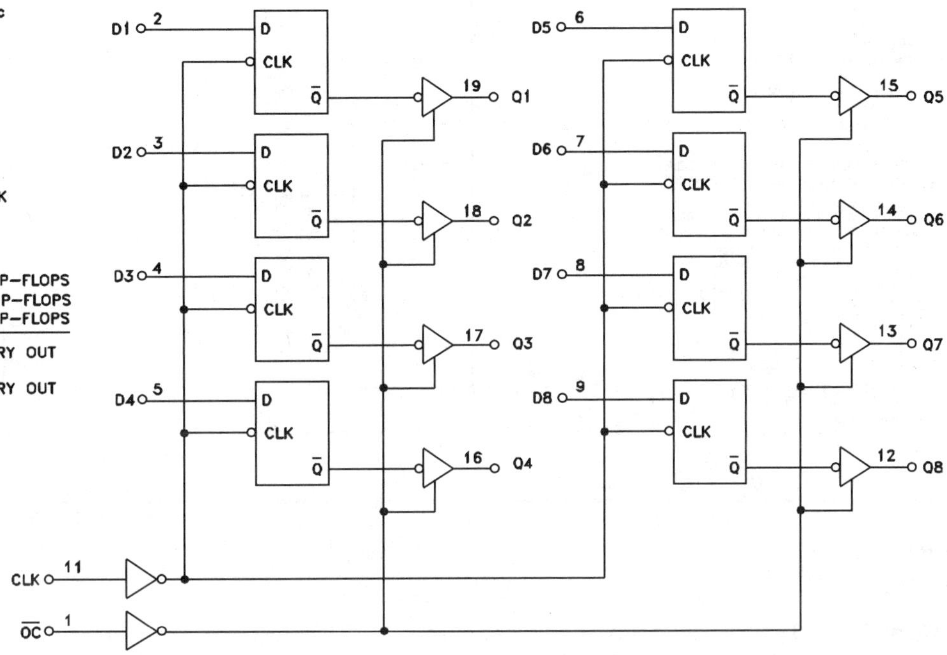

Figure 4.22. The 74x564, 74x574, and 74x576 series of 3-state Octal D flip-flops are essentially identical in their performance. The 74x564 and 74x576 feature /Q outputs only; the 74x574 provide true outputs only. Each of the series feature positive-edge triggering with Data and Clock inputs. No Set or Clear functions are provided. The 3-state output enable is active low. The package pinout is bus structured. The devices are available with a variety of TTL and CMOS high speed and low power options. The 74HC/HCT564/574 are capable of driving bus lines directly or up to 15 LSTTL loads.

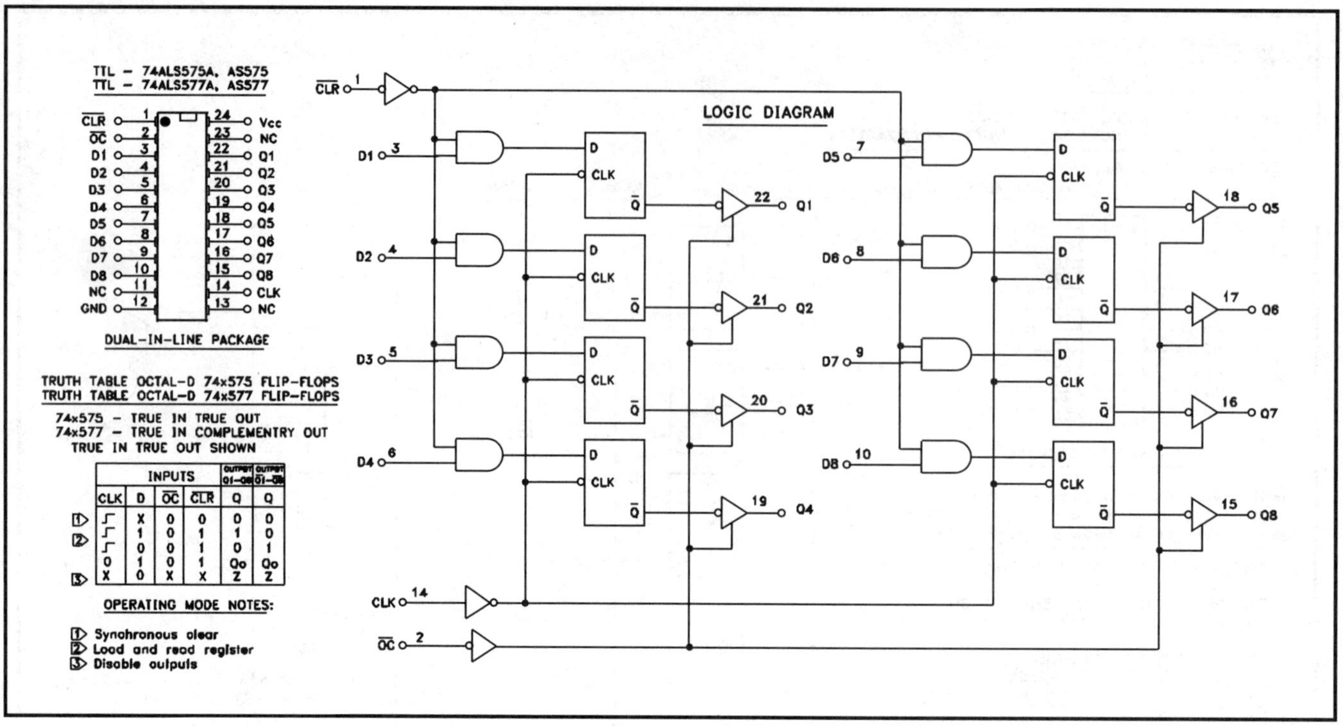

	INPUTS				OUTPUT Q1–Q8	OUTPUT /Q1–/Q8
	CLK	D	/OC	/CLR	Q	/Q
1	↑	X	0	0	0	0
2	↑	1	0	1	1	0
	↑	0	0	1	0	1
	0	1	0	1	Qo	Qo
3	X	0	X	X	Z	Z

Figure 4.23. These advanced 74ALS577A and 74AS577 3-state Octal D flip-flops are essentially identical in their performance to the 74ALS575A and 74AS575, respectively. The '755 feature true outputs; the '577 /Q outputs only. The pinout is bus structured; outputs drive buses directly. Positive-edge triggering with Data and Common Clock inputs. These have a Synchronous Clear function. Control inputs are buffered. These devices are tailored for use as buffer registers, I/O ports, bi-directional bus drivers and working registers. No CMOS devices are available.

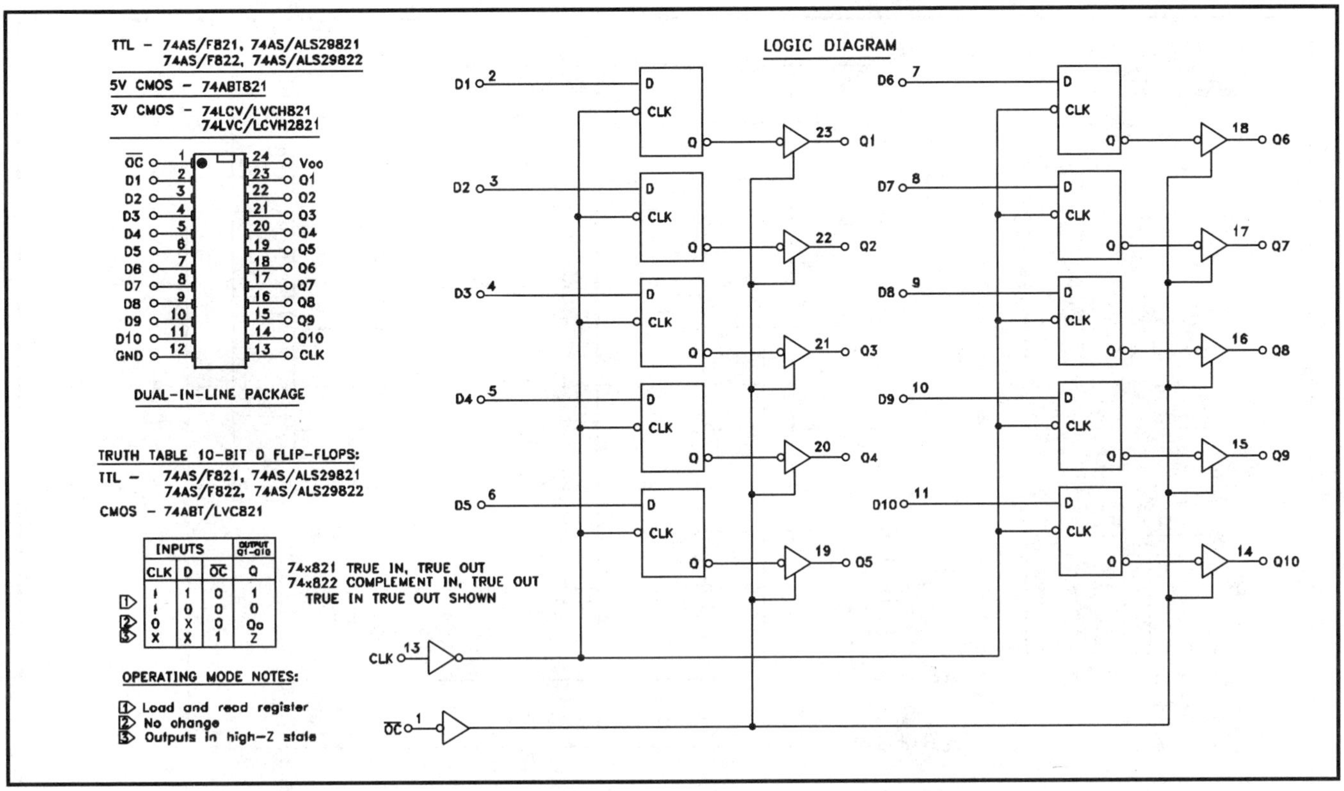

Figure 4.24. The 74AS821 and 74AS822 3-state 10-input D flip-flops are identical in their speed and time performance. The 74ALS29821/29822 feature similar performance. On a positive clock transition the outputs of the 'AS821/'29821 will be true; complementary to the inputs with the 'AS822/'29822. The drawing shows the true in/true out provision. The package pinout is bus structured with the extra width for wider address/data needs. Buffered control inputs reduce DC loading effects. Power up is in the high-impedance state. The active low 3-state output enable is designed for driving highly capacitive or relatively low-impedance loads. These are tailored for use as wider buffer registers, I/O ports, bidirectional bus drivers with parity and working registers. The 74F821/822 feature high-impedance NPN base input structure to minimize bus loading. The devices are also available with a variety of CMOS high-speed and low-power options.

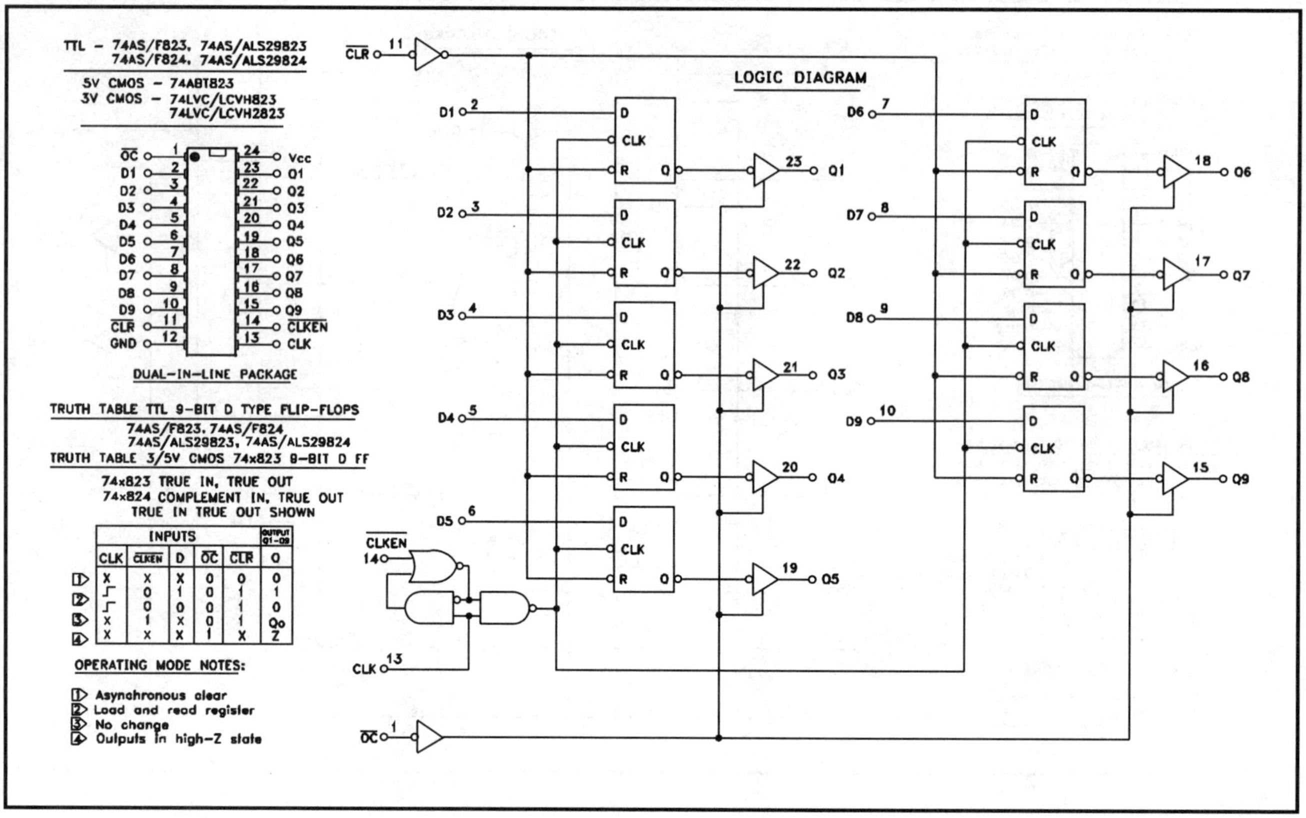

Figure 4.25. The 74AS823 and 74AS824 3-state 9-input D flip-flops are identical in their speed and time performance. The devices feature an active Low Clock enable. With /CLKEN low and a positive clock transition the outputs of the 'AS823 will be true; complementary to the inputs with the 'AS824. With the enable high the clock buffer is disabled, latching the outputs. The package pinout is bus structured. Buffered control inputs reduce DC loading effects. The active low 3-state output enable is designed for driving highly capacitive or relatively low-impedance loads. Power-up is in the high-impedance state. The 74AS/ALS29823 and 74AS/ALS29824 feature similar performance. The 74F823/824 feature high-impedance NPN base input structure to minimize bus loading. The devices are also available with a variety of CMOS high-speed and low-power options.

TTL – 74AS/ALS/F825, 74AS/ALS29825
74AS/ALS/F826, 74AS/ALS29826

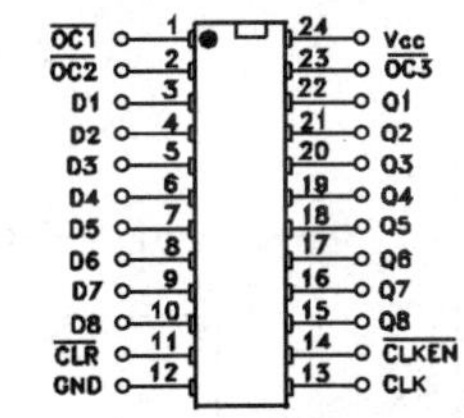

DUAL-IN-LINE PACKAGE

TRUTH TABLE OCTAL D-TYPE FLIP-FLOPS:
74AS/ALS/F825, 74AS/ALS29825
74AS/ALS/F826, 74AS/ALS29826

74AS/ALS/F825, 74AS/ALS29825 TRUE IN, TRUE OUT
74AS/ALS/F826, 74AS/ALS29826 COMPLEMENT IN, TRUE OUT
TRUE IN TRUE OUT SHOWN

	INPUTS					OUTPUT Q1–Q8
	CLK	$\overline{CLKEN}$	D	$\overline{OC}$	$\overline{CLR}$	Q
1	X	X	X	0	0	0
2	↑	0	1	0	1	1
2	↑	0	0	0	1	0
3	X	1	X	0	1	Q_0
4	X	X	X	1	X	Z

OPERATING MODE NOTES:

$\overline{OC}$ = 1 If any of $\overline{OC1}$, $\overline{OC2}$, or $\overline{OC3}$ are High

$\overline{OC}$ = 0 If any of $\overline{OC1}$, $\overline{OC2}$, or $\overline{OC3}$ are Low

1 Asynchronous clear
2 Load and read register
3 No change
4 Outputs in high-Z state

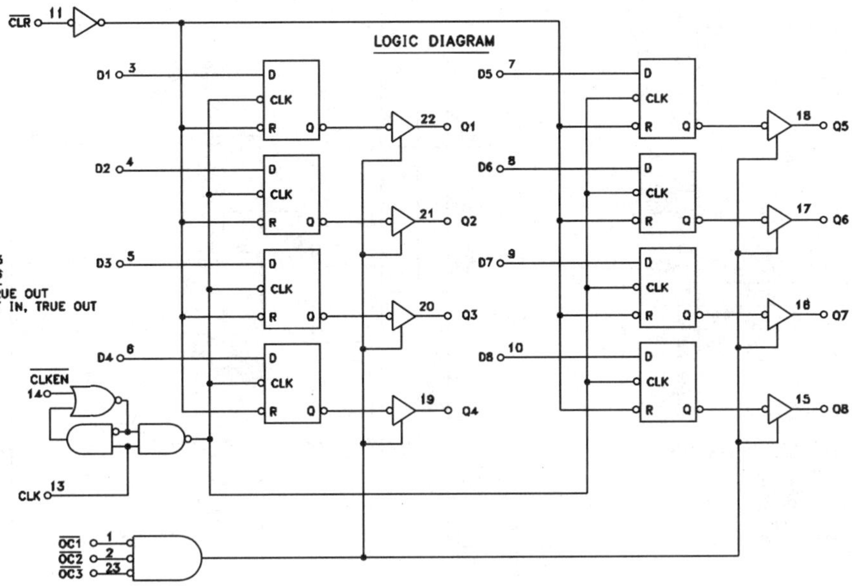

Figure 4.26. The 74AS825 and 74AS826 3-state 8-input D flip-flops are identical in their speed and time performance. The devices feature an active Low Clock enable. With /CLKEN low and a positive clock transition the outputs of the 'AS823 will be true; complementary to the inputs with the 'AS824. With the enable high the clock buffer is disabled, latching the outputs. The package pinout is bus structured. Buffered control inputs reduce DC loading effects. The multi-user active low buffered 3-state output enable is designed for driving highly capacitive or relatively low-impedance loads. Power-up is in the high-impedance state. The 74AS/ALS29825 and 74AS/ALS29826 feature similar performance. The 74F825/826 feature high-impedance NPN base input structure to minimize bus loading. No CMOS devices are available.

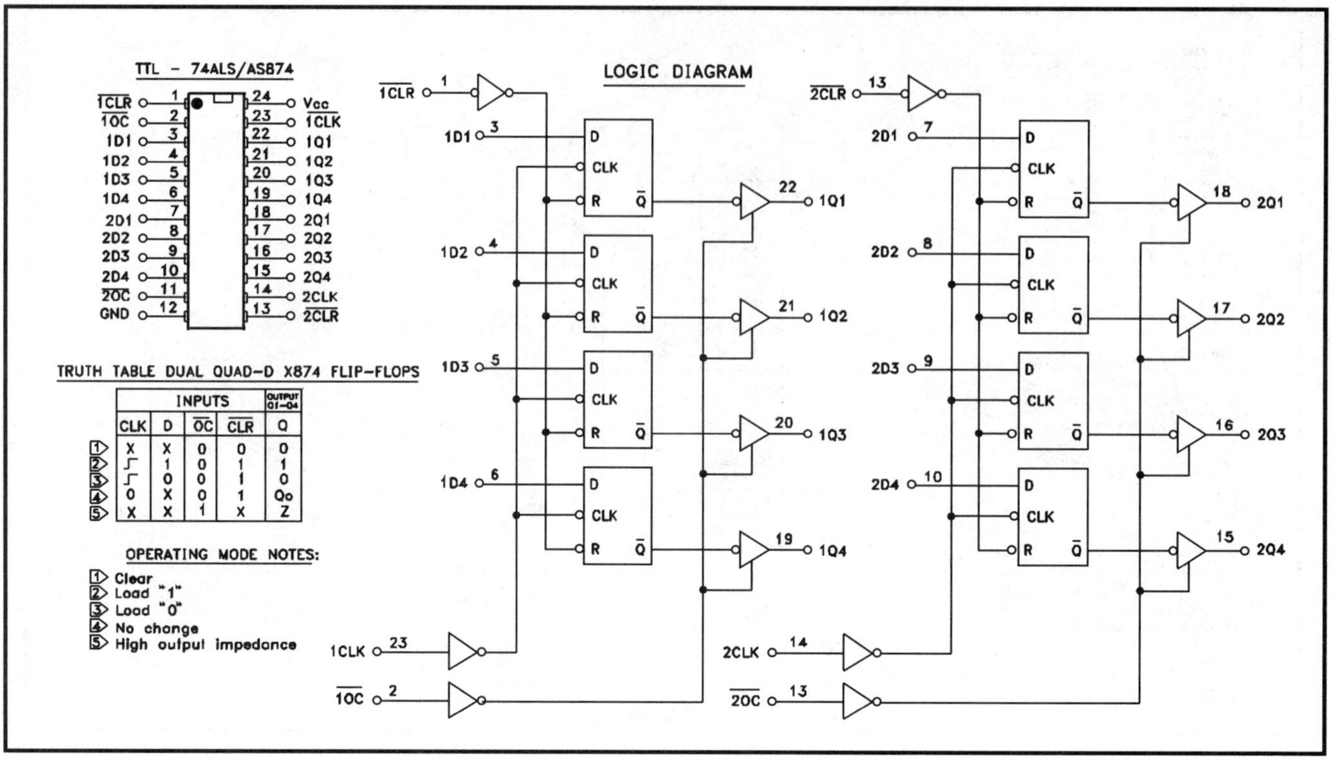

Figure 4.27. The 74ALS874B and 74AS874A are dual 4-bit D flip-flops. Each set features a common Clock and active Low Asynchronous Clear. With a positive clock transition the output is true. The active low 3-state output is designed specifically for bus driving. The package pinout is bus structured. No CMOS equivalents exist.

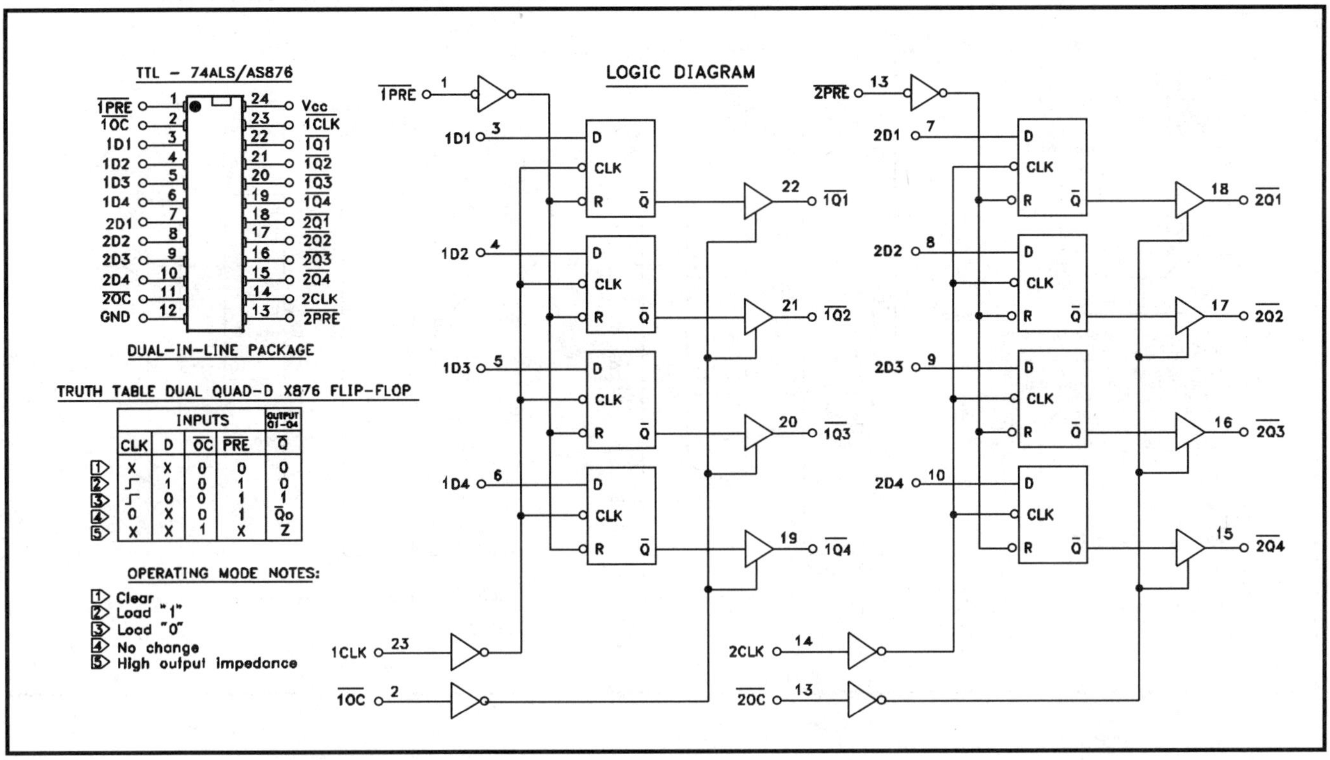

	INPUTS				OUTPUT Q1-Q4
	CLK	D	$\overline{OC}$	$\overline{PRE}$	$\overline{Q}$
1	X	X	0	0	0
2	↑	1	0	1	0
3	↑	0	0	1	1
4	0	X	0	1	$\overline{Q}_0$
5	X	X	1	X	Z

Figure 4.28. The 74ALS876A and 74AS876 are dual 4-bit D flip-flops. Each set features a common Clock and active Low Preset. With a positive clock transition the output is complementary. The active low 3-state output is designed specifically for bus driving. The package pinout is bus structured. No CMOS equivalents exist.

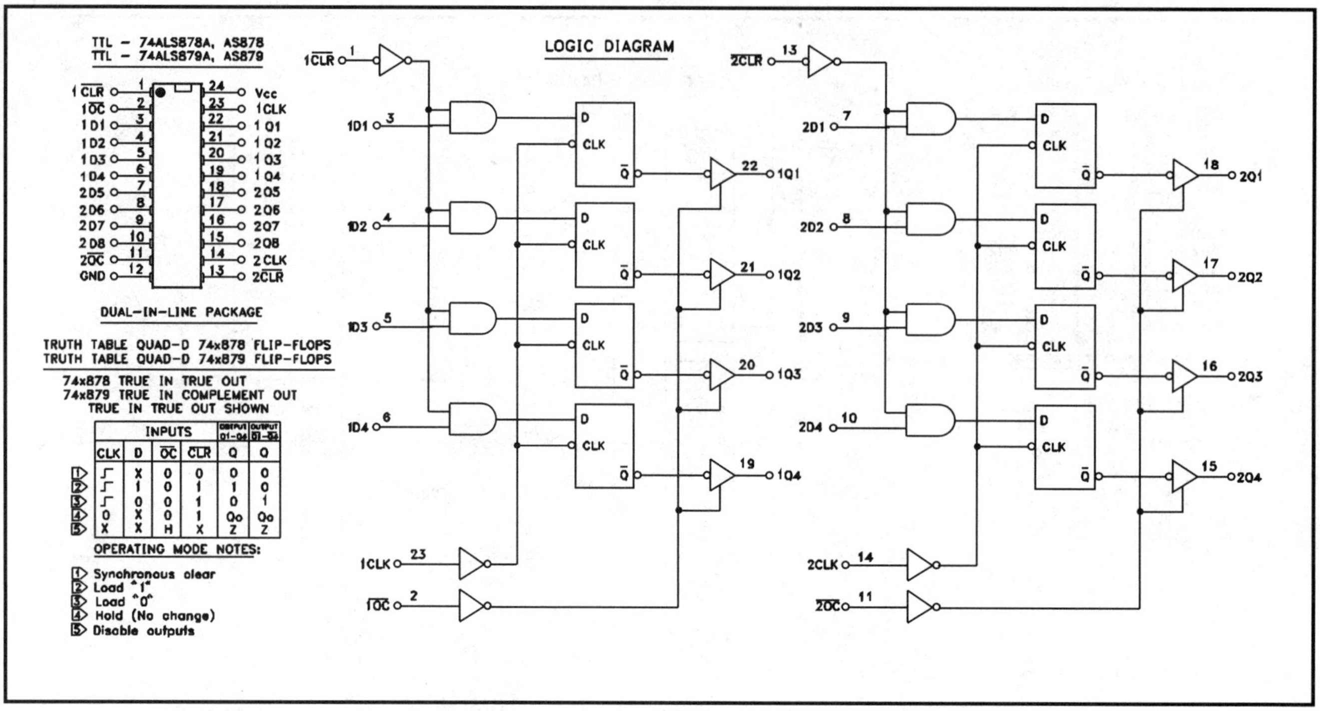

	INPUTS				OUTPUT Q1–Q4	OUTPUT $\overline{Q1}$–$\overline{Q4}$
	CLK	D	$\overline{OC}$	$\overline{CLR}$	Q	$\overline{Q}$
1	↑	X	0	0	0	0
2	↑	1	0	1	1	0
3	↑	0	0	1	0	1
4	0	X	0	1	Qo	Qo
5	X	X	H	X	Z	Z

Figure 4.29. The 74ALS878A/879A and 74AS878/879 are dual 4-bit D flip-flops. Each register features a common Clock and active Low Synchronous Clear. With a positive clock transition the 'ALS/AS878 outputs are true; the 'ALS/AS879 outputs are complementary. The active low 3-state output is designed specifically for bus driving. The package pinout is bus structured. No CMOS equivalents exist.

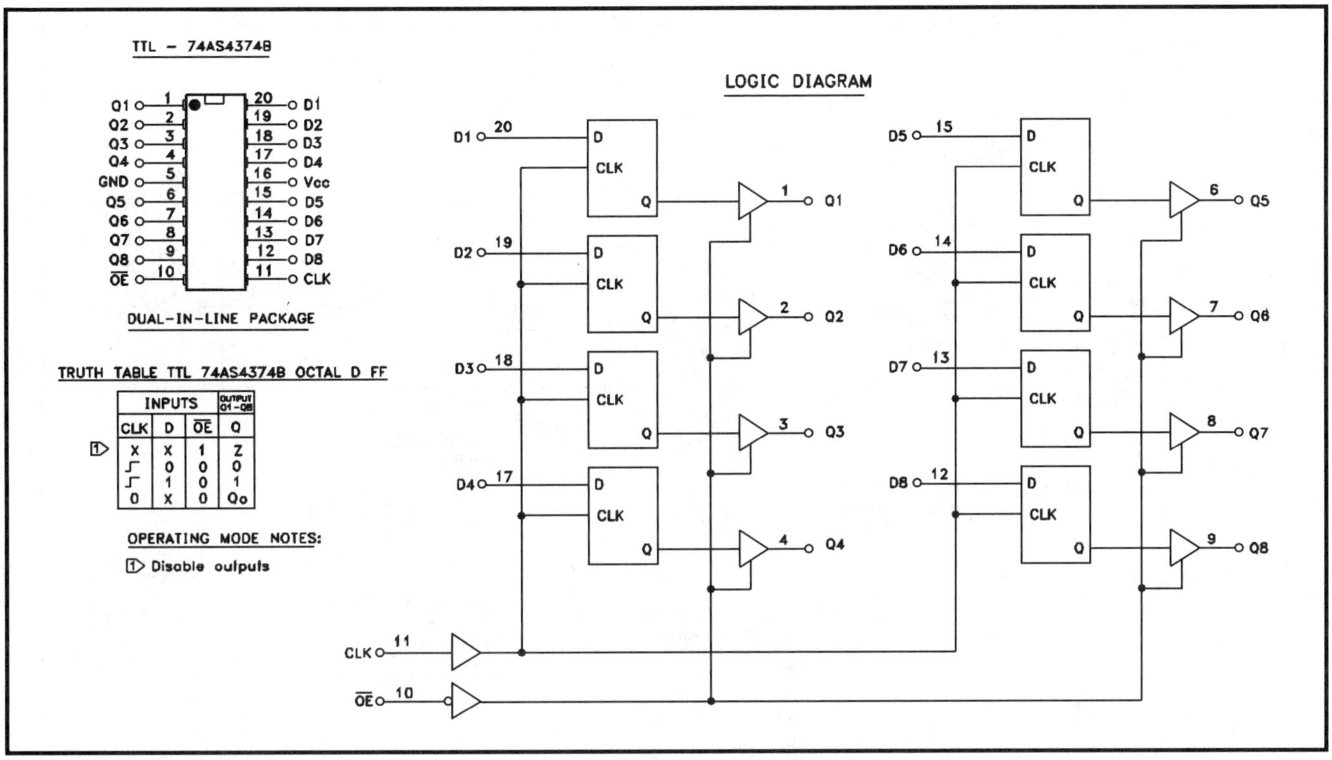

Figure 4.30. The 74AS4374B is an 8-bit edge-triggered D type flip-flop designed specifically for driving highly capacitive or relatively low impedance loads. Its 3-state outputs can drive bus lines directly. The output enable does not affect the internal operation. The device is unique in that two clock pulses are required to effect an output transition. On the 2nd clock positive transition the status of the D inputs are transferred to the Q outputs. There is no CMOS equivalent.

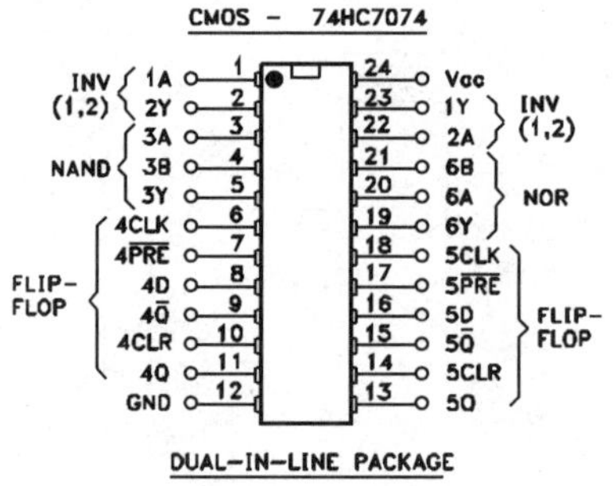

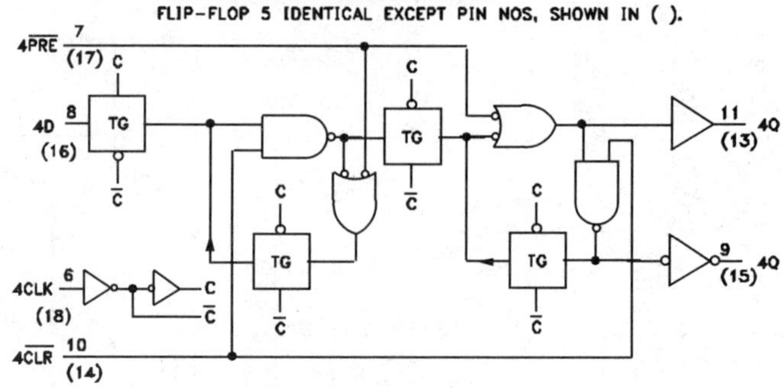

FUNCTION TABLE
(EACH D FLIP-FLOP)

INPUTS				OUTPUTS	
PRE	CLR	CLK	D	Q	Q̄
0	1	X	X	1	0
1	0	X	X	0	1
0	0	X	X	1*	1*
1	1	↑	1	1	0
1	1	↑	0	0	1
1	1	0	X	Q0	Q̄0

* Unstable; will not persist when either PRE or CLR returns to inactive (high) level.

INVERTERS

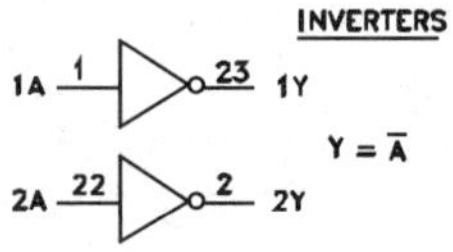

FUNCTION TABLE
(EACH INVERTER)

INPUT	OUTPUT
A	Y
1	0
0	1

2-INPUT NAND GATE

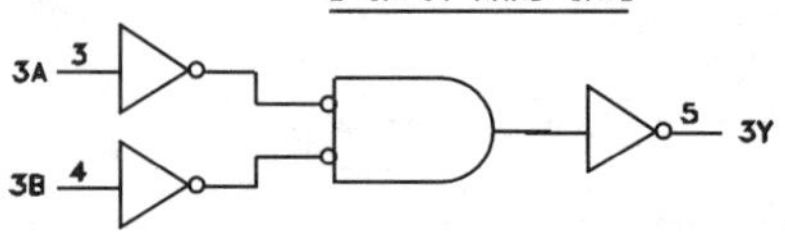

FUNCTION TABLE

INPUTS		OUTPUT
A	B	Y
1	1	0
0	X	1
X	0	1

$Y = \overline{A \cdot B}$ or $Y = \bar{A} + \bar{B}$

2-INPUT NOR GATE

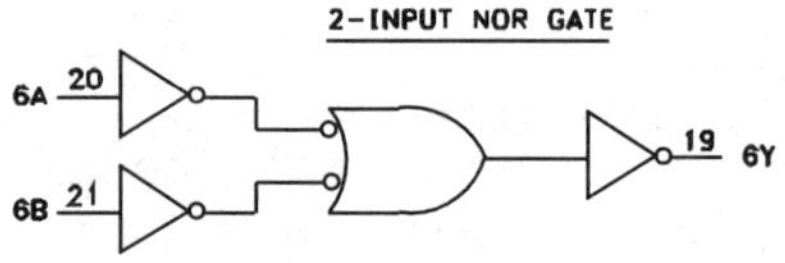

FUNCTION TABLE

INPUTS		OUTPUT
A	B	Y
1	X	0
X	1	0
0	0	1

$Y = \bar{A} \cdot \bar{B}$ or $Y = \overline{A + B}$

Figure 4.31. The High-Speed CMOS 74HC7074 is a multifunction device incorporating two inverters, a 2-input NAND gate, and a 2-input NOR gate along with 2 D-type Flip-Flops. Each flip-flop is provided with independent asynchronous Preset and Clear and complementary outputs. The flip-flops are functionally equivalent to the 74HC74. There is no TTL equivalent.

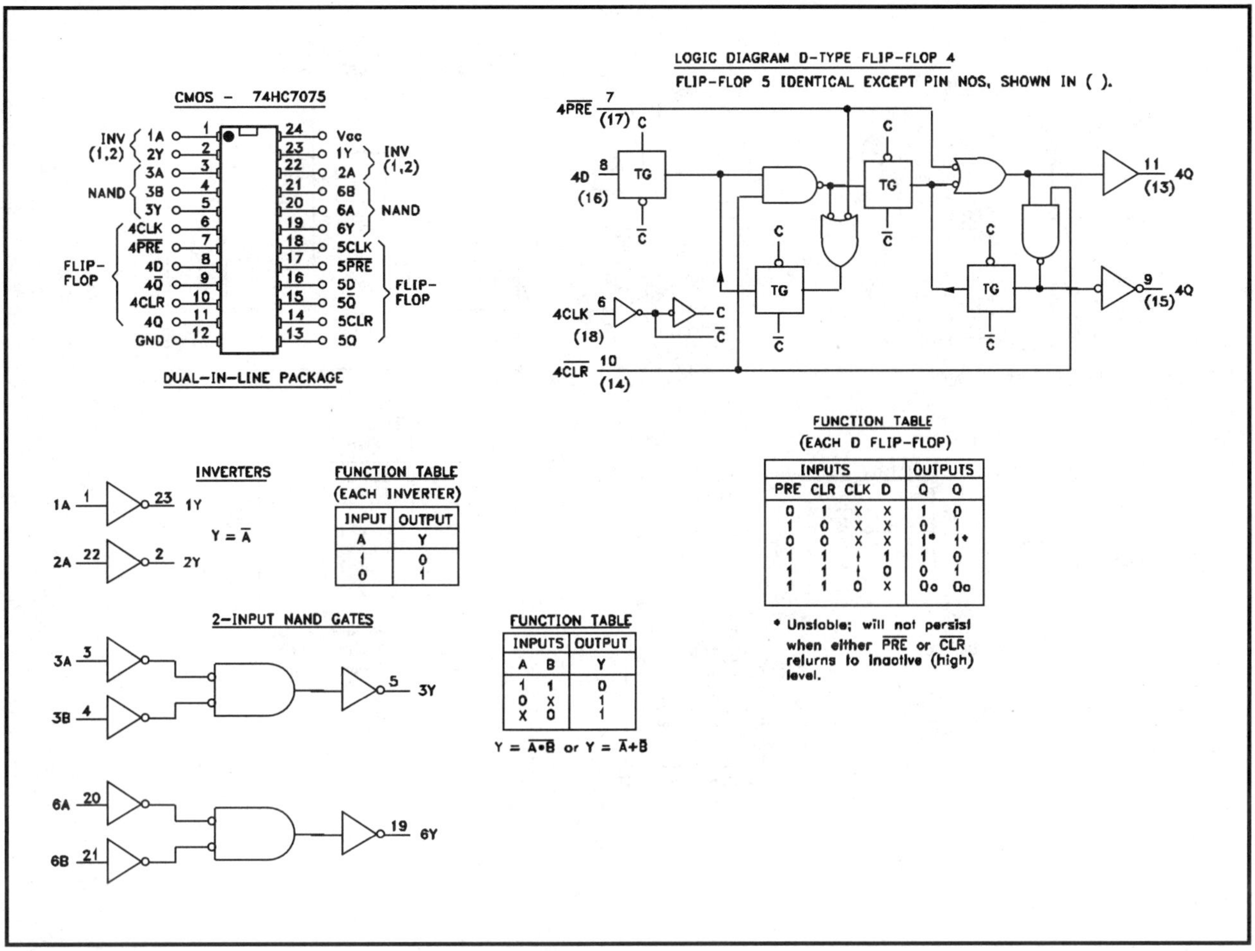

FUNCTION TABLE (EACH D FLIP-FLOP)

PRE	CLR	CLK	D	Q	Q̄
0	1	X	X	1	0
1	0	X	X	0	1
0	0	X	X	1*	1*
1	1	↑	1	1	0
1	1	↑	0	0	1
1	1	0	X	Q0	Q̄0

FUNCTION TABLE (EACH INVERTER)

INPUT A	OUTPUT Y
1	0
0	1

FUNCTION TABLE

INPUTS A	B	OUTPUT Y
1	1	0
0	X	1
X	0	1

Figure 4.32. The High-Speed CMOS 74HC7075 is a multifunction device incorporating two inverters, and two 2-input NAND gates, along with two D-type Flip-Flops. Each flip-flop is provided with independent asynchronous Preset and Clear and complementary outputs. The flip-flops are functionally equivalent to the 74HC74. There is no TTL equivalent.

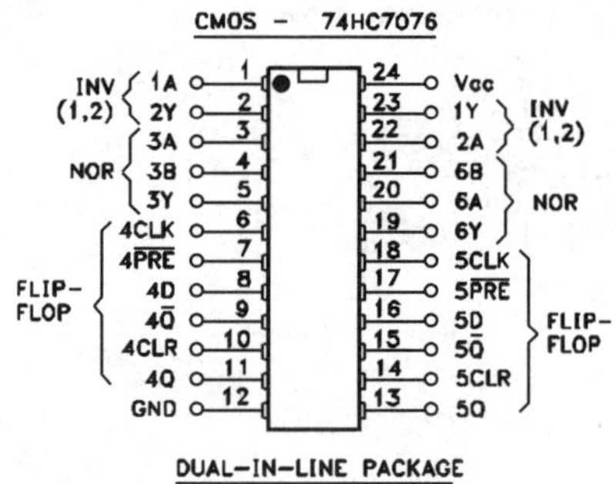

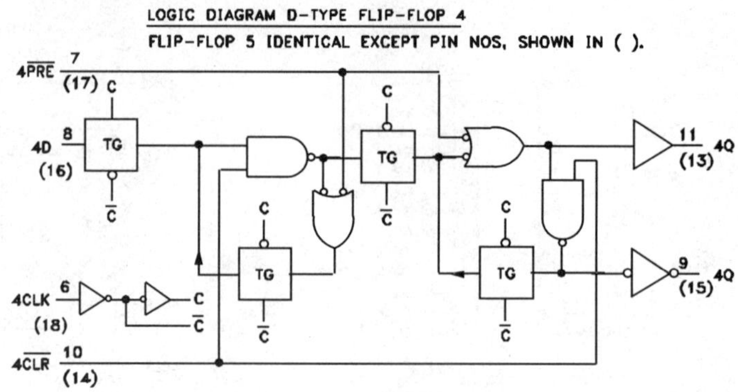

FUNCTION TABLE
(EACH D FLIP-FLOP)

INPUTS				OUTPUTS	
PRE	CLR	CLK	D	Q	Q̄
0	1	X	X	1	0
1	0	X	X	0	1
0	0	X	X	1*	1*
1	1	↑	1	1	0
1	1	↑	0	0	1
1	1	0	X	Q_0	$\bar{Q}_0$

* Unstable; will not persist when either $\overline{PRE}$ or $\overline{CLR}$ returns to inactive (high) level.

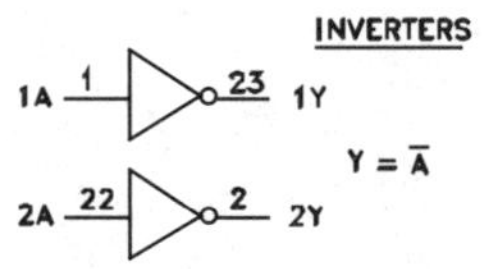

FUNCTION TABLE
(EACH INVERTER)

INPUT	OUTPUT
A	Y
1	0
0	1

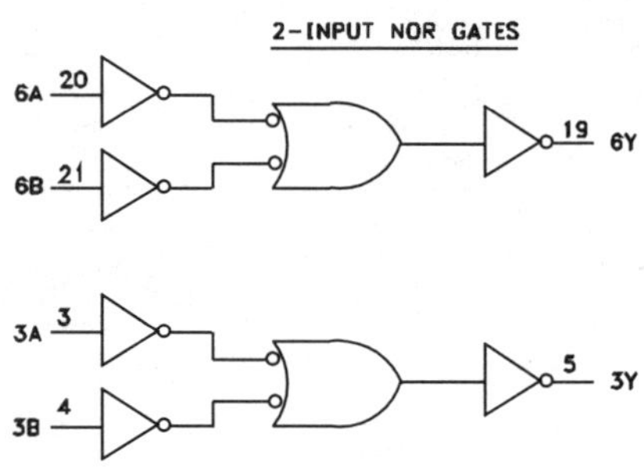

FUNCTION TABLE

INPUTS		OUTPUT
A	B	Y
1	X	0
X	1	0
0	0	1

$Y = \bar{A} \bullet \bar{B}$ or $Y = \overline{A+B}$

Figure 4.33. The High-Speed CMOS 74HC7076 is a multifunction device incorporating two inverters and two 2-input NOR gates, along with 2 D-type Flip-Flops. Each flip-flop is provided with independent asynchronous Preset and Clear and complementary outputs. The flip-flops are functionally equivalent to the 74HC74. There is no TTL equivalent.

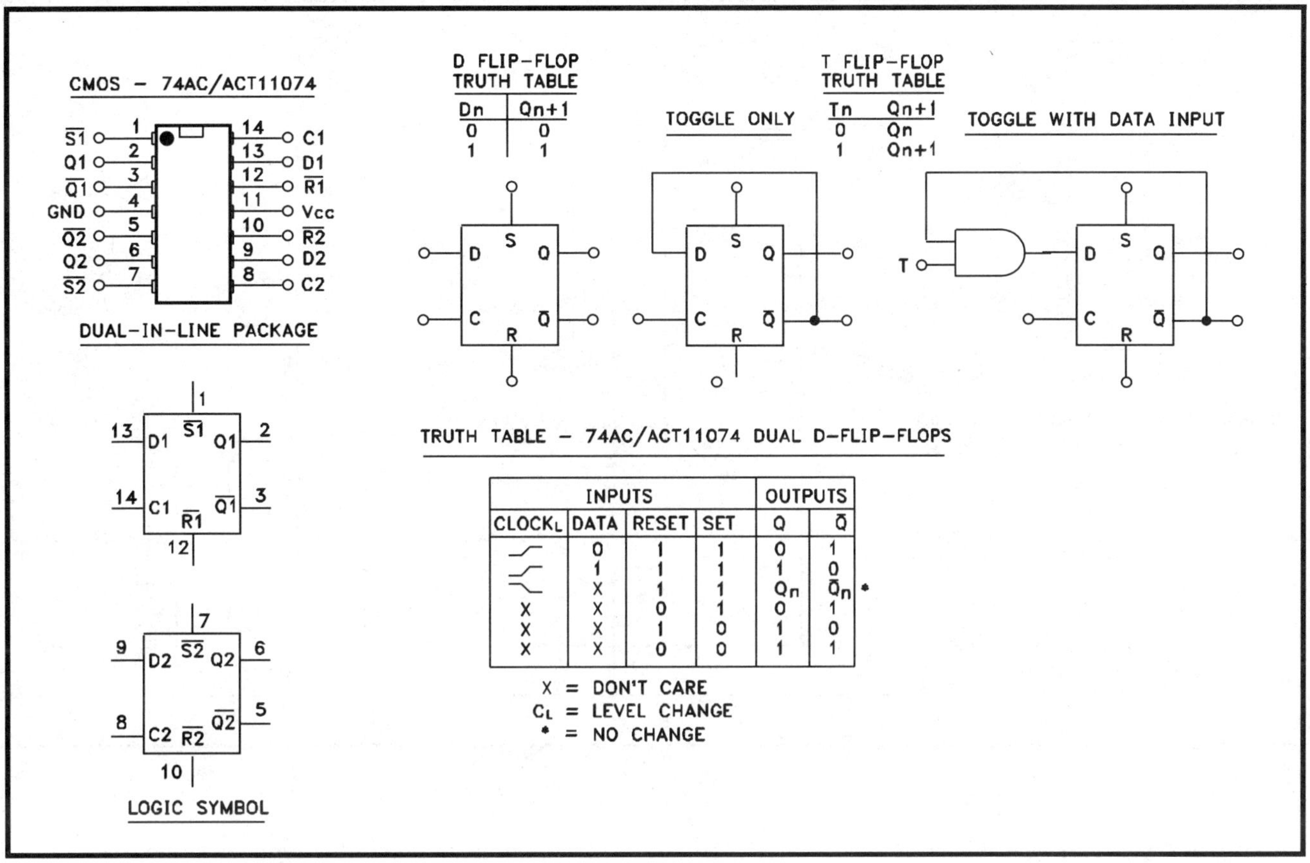

INPUTS				OUTPUTS	
CLOCK$_L$	DATA	RESET	SET	Q	$\overline{Q}$
rising	0	1	1	0	1
rising	1	1	1	1	0
falling	X	1	1	Q_n	$\overline{Q}_n$ *
X	X	0	1	0	1
X	X	1	0	1	0
X	X	0	0	1	1

Figure 4.34. The 3V/5V High-Speed CMOS 74AC/ACT11074 are dual D-type flip-flops. Each flip-flop is provided with independent asynchronous Preset and Clear and complementary outputs. Center pin Vcc and GND configurations minimize high-speed switching noise. There is no TTL equivalent.

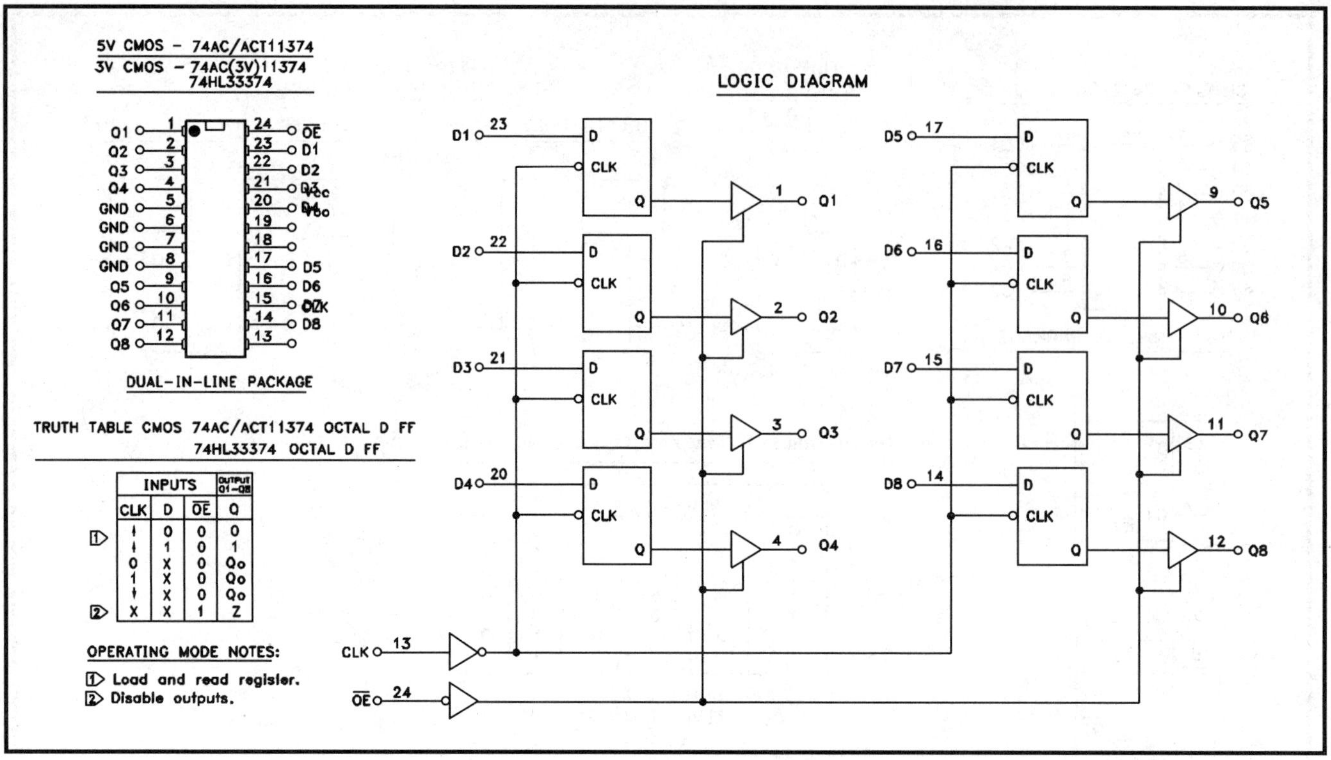

Figure 4.35. The 3V/5V High-Speed CMOS 74AC/ACT11374 consists of eight D-type flip-flops. This is a 3-state device, active Low enable, with bus driving true outputs. Center pin Vcc and GND configurations minimize high-speed switching noise. There is no TTL equivalent.

5V CMOS – 74AHC/AHCT/AC/ACT16374
3V CMOS – 74AHC(3V)/AC(3V)/LVTABT/LCX/ LVCH/ALVCH16374 74LVC/LVCH162374A

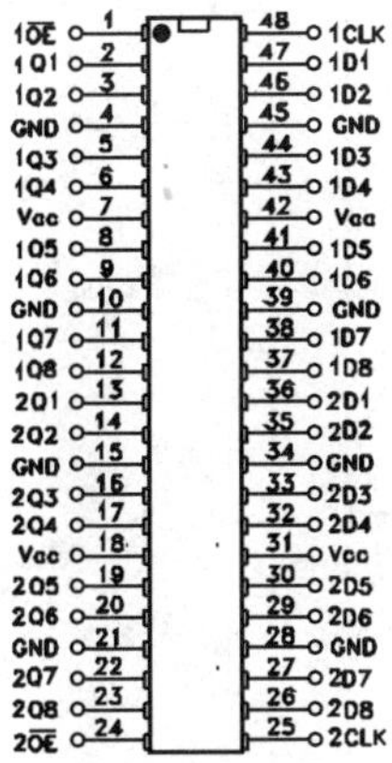

DUAL-IN-LINE PACKAGE

TRUTH TABLE CMOS 3V/5V 74x16374 DUAL OCTAL D FF
74x162374A DUAL OCTAL D FF

	INPUTS			OUTPUT Q1–Q8
	CLK	D	$\overline{OE}$	Q
	↑	0	0	0
	↑	1	0	1
	X	X	0	Qo
	↑	X	0	Qo
1▷	X	X	1	Z

OPERATING MODE NOTES:
1▷ Disable outputs

LOGIC DIAGRAM

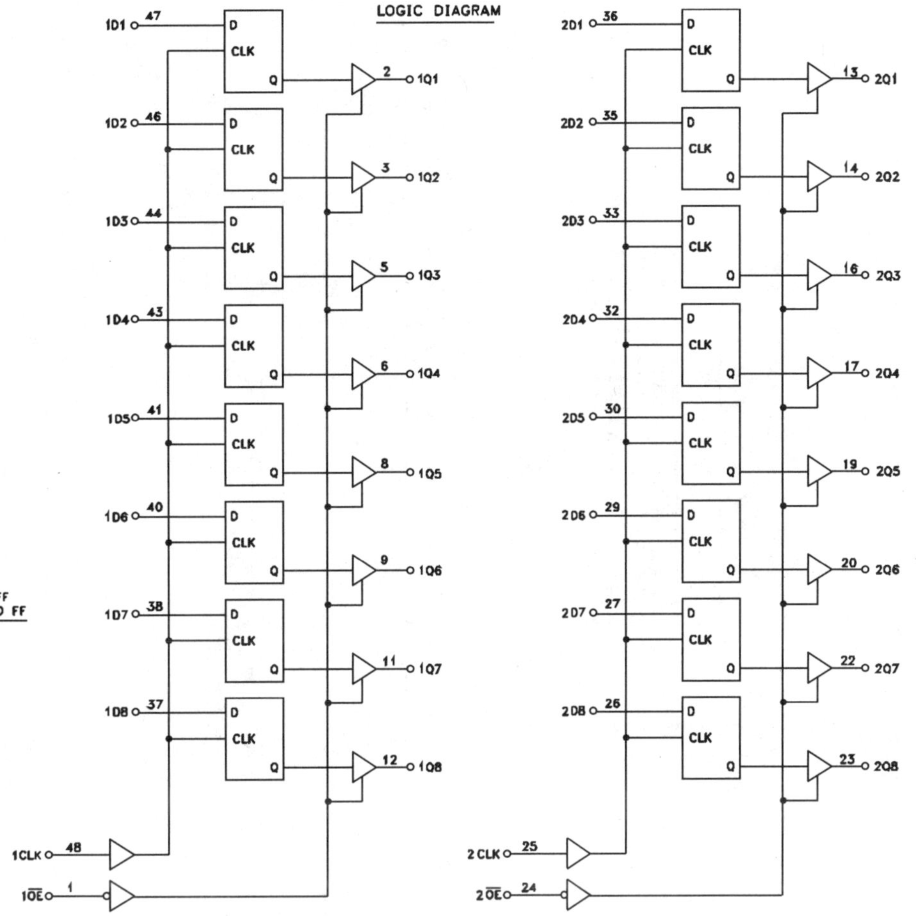

Figure 4.36. The 5V/3V CMOS 74x16374 series consists of sixteen non-inverting D flip-flops structured as two distinct groups of eight. Each group has its individual Clock and Output Enable input, but by shorting these the two may be combined into a single group of sixteen. The output enable does not affect the internal operation. Distributed power pinouts minimizes high-speed switching noise. Flow-through architecture facilitates PC board layout. The active low 3-state output enable is designed for driving highly capacitive or relatively low-impedance loads. Inputs are TTL voltage compatible.

3V - CMOS - 74ALVCH16721

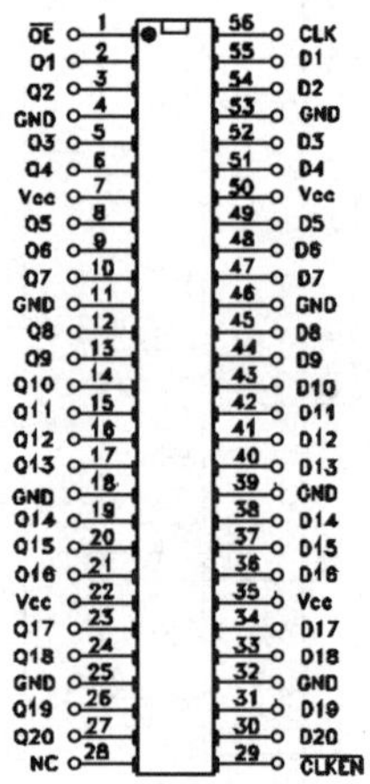

DUAL-IN-LINE PACKAGE

TRUTH TABLE CMOS 74ALVCH16721 20-BIT FF

	INPUTS				OUTPUT
	$\overline{CLKEN}$	CLK	D	$\overline{OE}$	Q
	1	X	X	0	Qo
	0	↑	0	0	0
	0	↑	1	0	1
	0	0/1	X	0	Qo
1▷	X	X	X	1	Z

OPERATING MODE NOTES:

1▷ Disable outputs

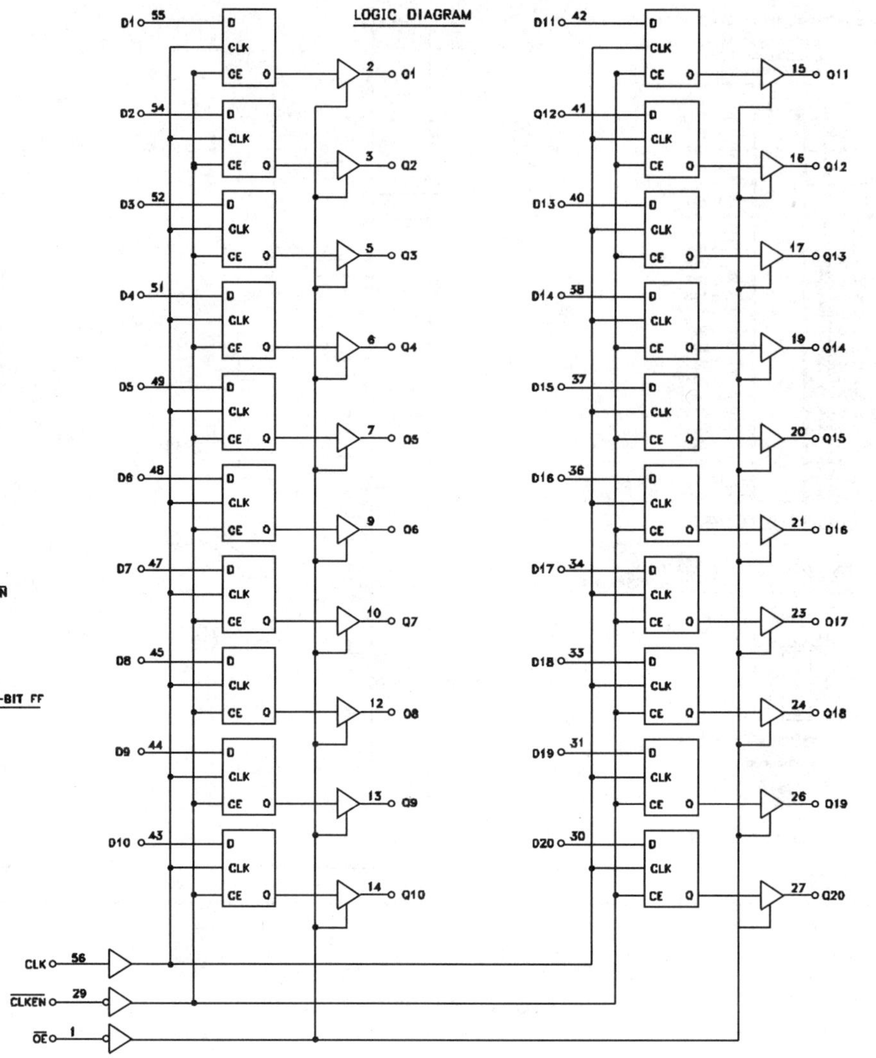

Figure 4.37. The 3V High-Speed CMOS 74ALVCH16721 is a 20-bit D-type flip-flop with 3-state outputs designed for 2.3 - 3.6 volt operation. ESD protection exceeds 2000V per MIL-STD-883C, Method 3015; exceeds 200V using machine model. Latch-up performance exceeds 250 mA per JEDEC Standard JESD-17. Bus hold on data inputs eliminates the need for external pull-up/pull-down resistors. Distributed pin Vcc and GND configurations minimize high-speed switching noise. To ensure the high-impedance state during power up/down /OE should be tied to Vcc through a pull-up resistor. Active bus-hold circuitry is provided to hold unused or floating data inputs at a valid logic level. There is no TTL equivalent.

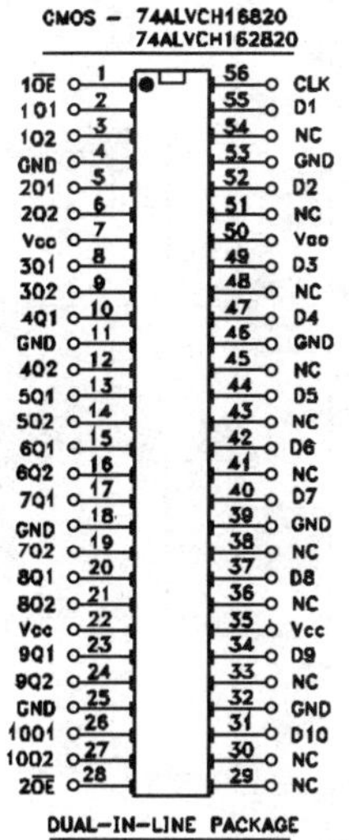

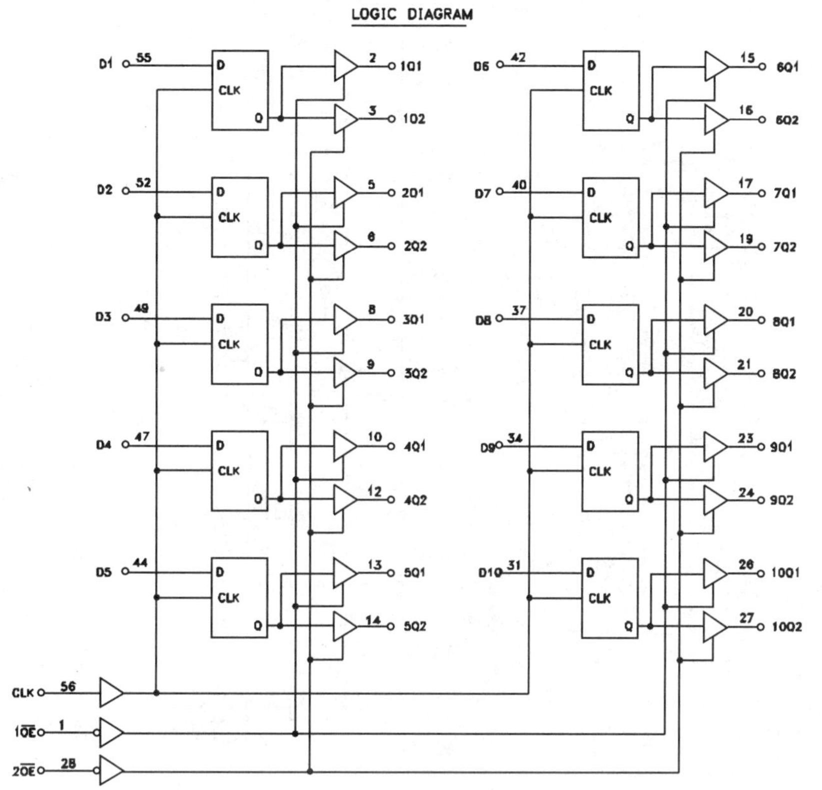

TRUTH TABLE CMOS 74ALVCH168220 10-BIT FF
74ALVCH16820 10-BIT FF

INPUTS			OUTPUT
CLK	D	$\overline{OE}$n (2)	Q
↑	0	0	0
↑	1	0	1
0	X	0	Qo
X (1)	X	1	Z

OPERATING MODE NOTES:

(1) Disable outputs

(2) n = 1,2

Figure 4.38. The 3V High-Speed CMOS 74ALVCH16820/162820 is a 10-bit D-type flip-flop with 3-state outputs designed for 2.3 - 3.6 volt operation. ESD protection exceeds 2000V per MIL-STD-883C, Method 3015; exceeds 200V using machine model. Latch-up performance exceeds 250 mA per JEDEC Standard JESD-17. Bus hold on data inputs eliminates the need for external pull-up/pull-down resistors. Distributed pin Vcc and GND configurations minimize high-speed switching noise. To ensure the high impedance state during power up/down /OE should be tied to Vcc through a pull-up resistor. Output ports, designed to sink up to 12mA, have equivalent 26-ohm resistors; no external resistors are required. Active bus-hold circuitry is provided to hold unused or floating data inputs at a valid logic level. There is no TTL equivalent.

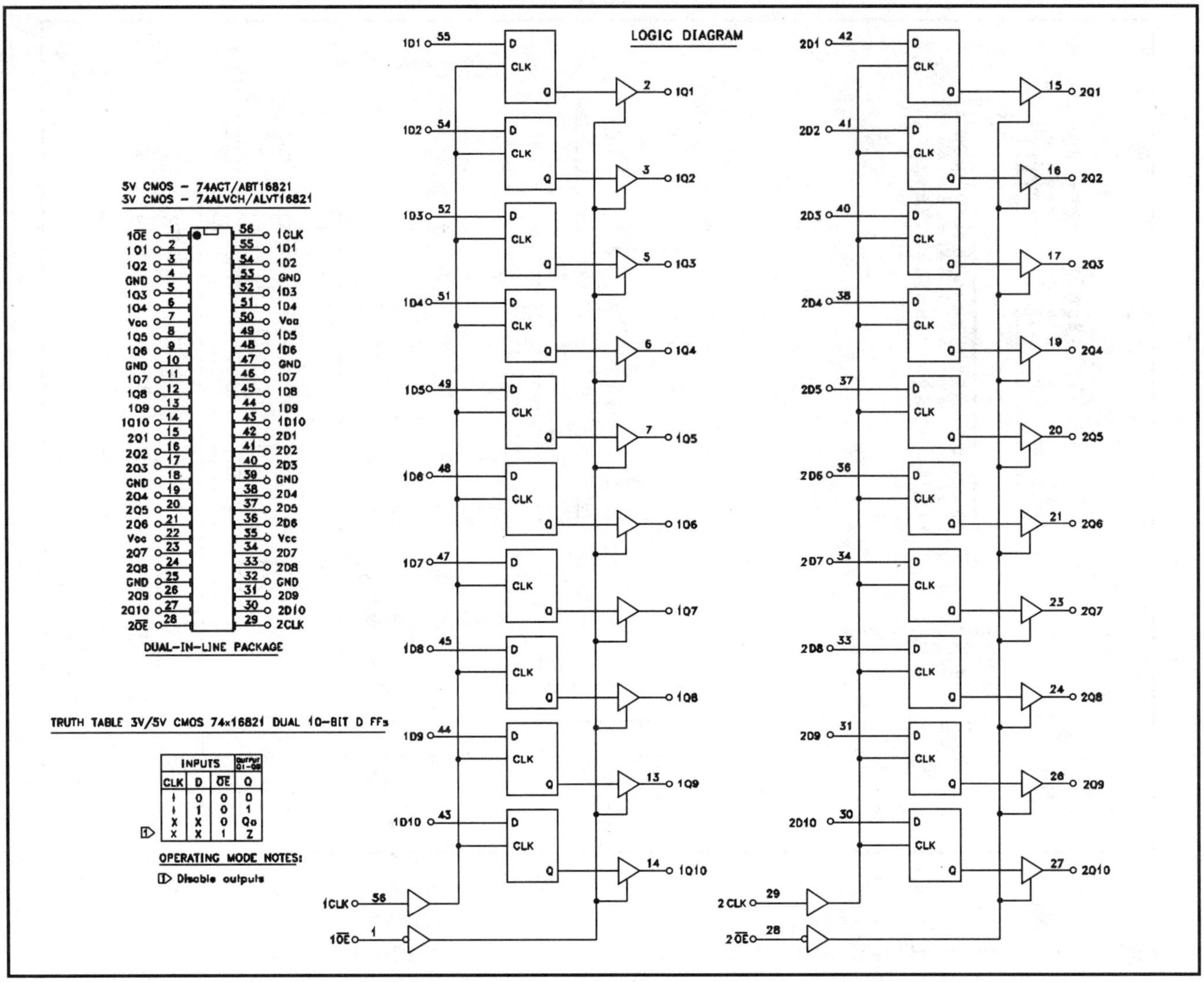

Figure 4.39. The 3V High-Speed CMOS 74ALVCH16821/162821 is a 20-bit D-type flip-flop with 3-state outputs designed for 2.3 - 3.6 volt operation. It may be configured as two 10-bit flip-flops. ESD protection exceeds 2000V per MIL-STD-883C, Method 3015; exceeds 200V using machine model. Latch-up performance exceeds 250 mA per JEDEC Standard JESD-17. Bus hold on data inputs eliminates the need for external pull-up/pull-down resistors. Distributed pin Vcc and GND configurations minimize high-speed switching noise. To ensure the high-impedance state during power up/down /OE should be tied to Vcc through a pull-up resistor. Output ports, designed to sink up to 12mA, have equivalent 26-ohm resistors; no external resistors are required. Active bus-hold circuitry is provided to hold unused or floating data inputs at a valid logic level. The 74ACT16821 is a 5V High-Speed CMOS similar to the above. Inputs are TTL voltage compatible. The output is capable of sinking 24mA; no internal resistors are provided. There are no TTL equivalents.

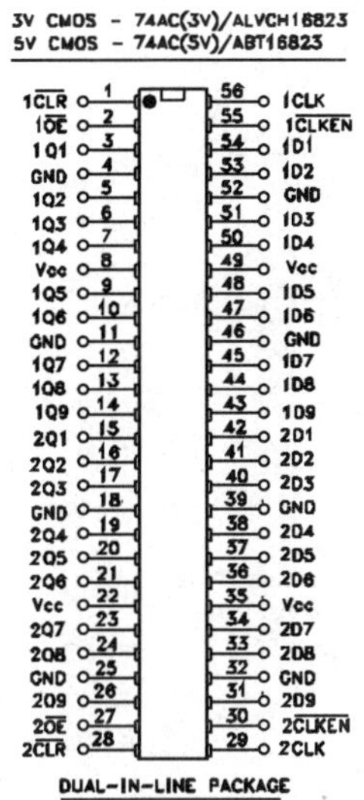

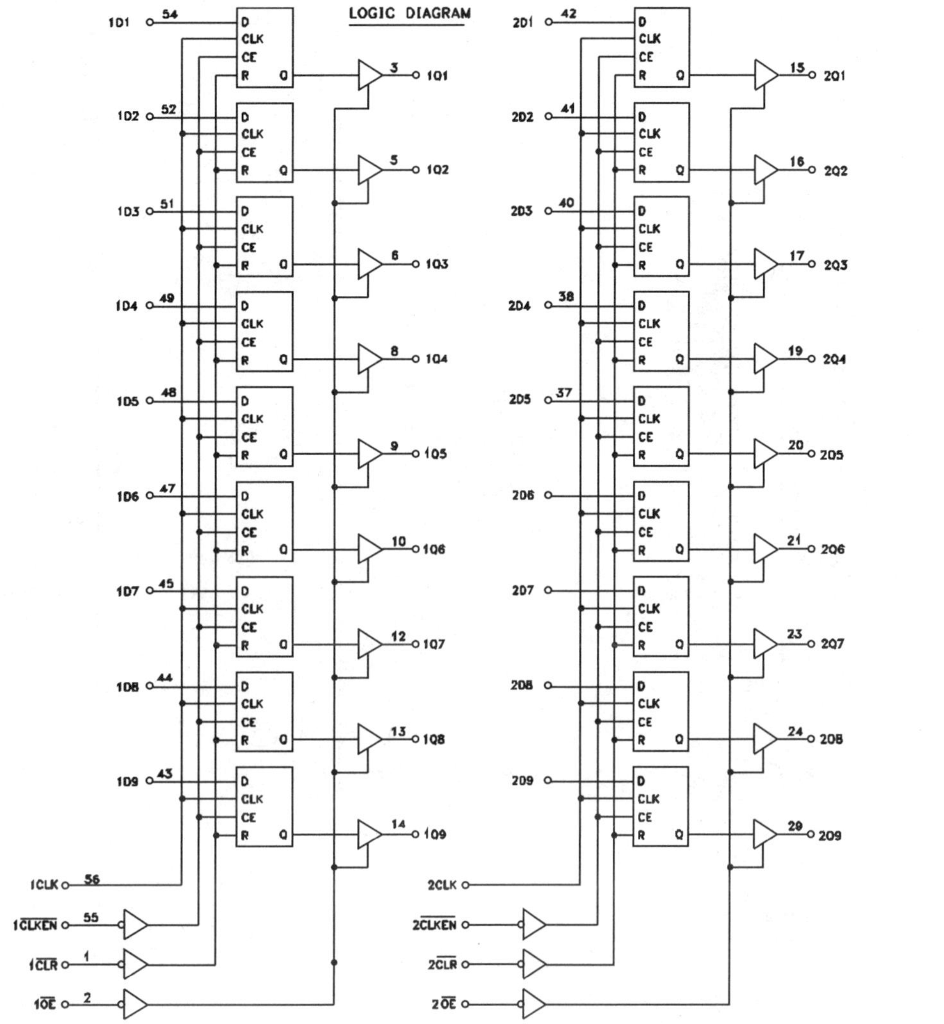

TRUTH TABLE 3V/5V CMOS 74x16823
DUAL 9-BIT D FLIP-FLOPS

	INPUTS					Output Q1-Q9
	$\overline{OE}$	$\overline{CLR}$	$\overline{CLKEN}$	CLK	D	Q
	0	0	X	X	0	0
	0	1	0	↑	1	1
	0	1	0	↑	0	0
	0	1	0	0	X	Q_0
	0	1	1	X	X	Q_0
(1)	1	X	X	X	X	Z

OPERATING MODE NOTES:

(1) Disable outputs

Figure 4.40. The 3V High-Speed CMOS 74ALVCH16823 is an 18-bit D-type flip-flop with 3-state outputs designed for 2.3 - 3.6 volt operation. It may be configured as two 9-bit flip-flops. ESD protection exceeds 2000V per MIL-STD-883C, Method 3015; exceeds 200V using machine model. Latch-up performance exceeds 250 mA per JEDEC Standard JESD-17. Bus hold on data inputs eliminates the need for external pull-up/pull-down resistors. Distributed pin Vcc and GND configurations minimize high-speed switching noise. To ensure the high-impedance state during power up/down /OE should be tied to Vcc through a pull-up resistor. Output ports, designed to sink up to 12mA, have equivalent 26 ohm resistors; no external resistors are required. Active bus-hold circuitry is provided to hold unused or floating data inputs at a valid logic level. The 74AC/ACT16823 is a 3V/5V High Speed CMOS similar to the above. Inputs are TTL voltage compatible. The output is capable of sinking 24mA; no internal resistors are provided. There are no TTL equivalents.

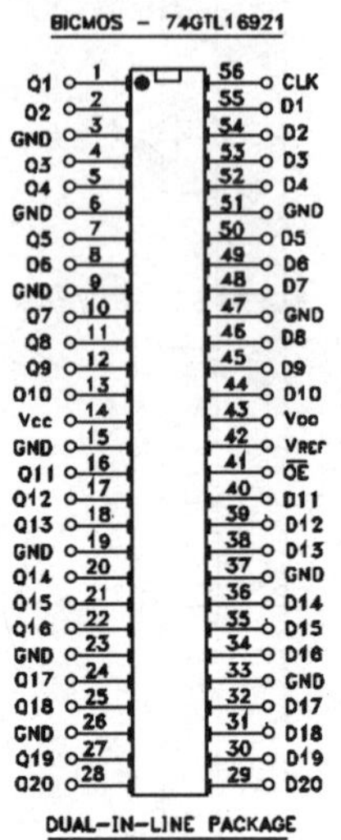

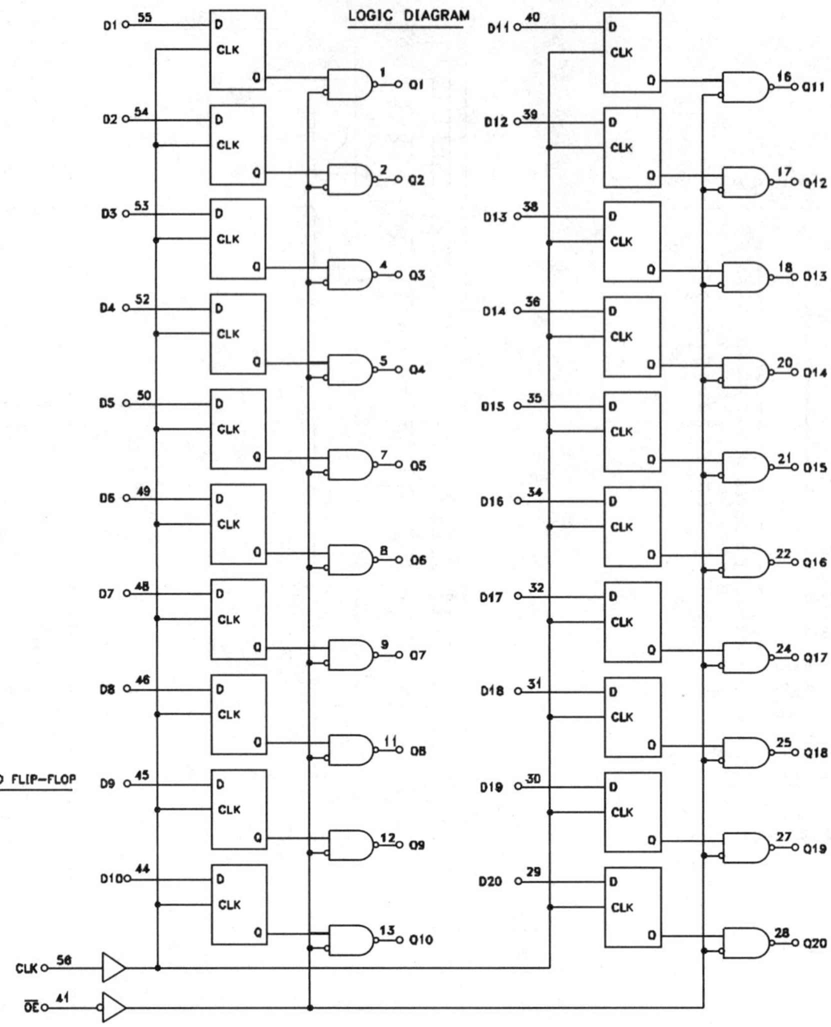

TRUTH TABLE BICMOS 74GTL16921 20-BIT D FLIP-FLOP

	INPUTS			OUTPUT Q1-Q20
	CLK	D	$\overline{OE}$	Q
	↑	0	0	0
	↑	1	0	1
	X	X	0	Q_0
1▷	X	X	1	Z

OPERATING MODE NOTES:

1▷ Disable outputs

Figure 4.41. The 3V 74GTL16921 is a 20-bit Input D-type flip-flop with 3-state outputs designed for 3 - 3.6 volt operation. The output enable has no effect on the internal operation. Distributed Vcc and GND pin configurations minimize high-speed switching noise. Distributed power pinouts minimize high-speed switching noise. Flow-through architecture optimizes PC Board layout. Input and output provide GTL signal levels.

Chapter 5

The Data Latch

In Chapter 4 we gained an understanding of the Flip-Flop. The flip-flop, and in particular the D-type, is an extremely versatile memory device which provides the basis for just about any application involving memory. This is certainly true for the Data Latch, which in most respects is an unclocked flip-flop at heart. A significant distinction for the latch is the presence of a Latch Enable in lieu of a Clock input.

The simplest of latch circuits is designated Set-Reset, or simply, S-R. S and R are standard designators for Set and Reset, respectively. Thus, when the Set input is active the output is set to the True, HIGH, or Logic "1" state. Conversely, when the Reset is active the output is reset to the False, LOW, or Logic "0" state. Typically the Set and Reset inputs are momentary, and for the majority of circuits they are active low.

A very basic latch is the NAND, illustrated with the 74X279 in *Figure 5.6*. The circuit shown enjoys popular usage as a bounceless switch. Although the drawing shows an ON-ON switch, even with an ON-OFF-ON version the circuit would hold its Set state until reset due to the LOW provided by the opposing NAND.

A pair of NOR gates can also function as a latch circuit. This circuit becomes a little tricky to implement, as the NOR inputs reside in the LOW state as opposed to the NAND. It is difficult to pull up a grounded input, so an inverter is needed, illustrated here with NPN transistors.

Many of the design considerations for flip-flop circuits apply to the latch as well. In particular this relates to high-speed applications where timing becomes critical. As with the flip-flop, we see that exotic devices have been developed for extended bus applications, such as microprocessors and storage registers.

The data latch is analogous to the camera in many respects; that is, just as the camera captures a scene as it exists in a moment of time, the latch captures the state of its data inputs at the instant of activation. Of importance to latch circuits are the Setup Time, t_{su}, the Pulse Width, t_w, the Hold Time, t_h, and the propagation times, t_{PLH} and t_{PHL}. *Tables 5.5 -5.9* list representative setup, hold, and propagation times.

Chapter Organization

Nine tables are provided. *Tables 5.1 - 5.4* provide a listing of latch devices by manufacturer. There are sources beyond those shown in the tables but the four shown provide a realistic picture of what has been and is currently available.

Tables 5.5 - 5.8 provide timing references. Because of the manner by which manufacturers define these quantities the data in the tables is not precise. The intent is to provide insight helpful for performance comparisons. In particular, we can see the spectacular gains in speed and reduced power dissipation achieved with the latest devices.

The drawings of *Figures 5.1 -5.39* describe the Dual In-Line (DIP) package pinouts, a Function Table, a Logic Diagram and/or Logic Symbol. Again, to simplify comparisons as well at to conserve space, all devices with matching features, specifically pinouts and function tables, are included in the same drawing.

Summary

This chapter provides tables and drawings describing over 100 variations of data latches from the early 7475 to the very sophisticated such as the 74ALVCH16260, a 12- to 24-Bit Multiplexed D-Type Latch with 3-State Outputs fitted into a 56 DIP package. Many devices comprise a series with an impressive array of options. Some points to keep in mind are:

- Timing information in the tables is for comparison purposes only. There is a general lack of consistency in how these parameters are defined, and not just between manufacturers. Where timing is critical consult the manufacturer's data manual.

- In general, 3-state output devices are not affected internally when in the high impedance output state.

- All timing data shown is for the industrial/commercial or recommended free-air operating temperature range. Where data is for 25°C it is given here.

- The specifications for CMOS devices may cover voltage ranges from 2 to 15 volts. Data is typically provided for VCC's of 5, 10, and 15 volts, but not always. Some manufacturers use 5.0 volts, others 4.5. There is a significant speed factor—in general the higher the voltage, the faster the performance. Again, where this is important to your application consult the data manual.

- A similar situation exists for the more recent low-voltage devices. Typical VCC ranges given are 2.7 - 3.6 volts, but values may be given for 2, 2.6, 3.0, 3.3 or 3.6 volts.

DEVICE 74---	STD NM	STD ST	S NM	S ST	LS NM	LS ST	AS/A/B NM	AS/A/B ST	ALS/A/B NM	ALS/A/B ST	F NM	F ST	Figure No.	DEVICE 74---
7475	x	xx			x	xx							5.1	7475
7477					x								5.1	7477
74100		x											5.2	74100
74116		xx											5.3	74116
74256					x	x					x	x	5.4	74256
74259	x	x			xx	xx				x	x	x	5.5	74259
74279		xx			xx	x							5.6	74279
74363						x							5.7	74363
74373			x	x	xx	x	x	x	x	x	x	xx	5.8	74373
74375					x	xx							5.9	74375
74412												x	5.10	74412
74432												x	5.10	74432
74533							x	x	x	x	x		5.11	74533
74563									x	x			5.12	74563
74573							x	x	x	x		xx	5.12	74573
74580							x	x	x	x			5.12	74580
74604						x						x	5.13	74604
74605												x	5.13	74605
74606						x							5.13	74606
74607						x							5.13	74607
74666										x			5.14	74666
74667										x			5.14	74667
74841								x		x		x	5.15	74841
74842								x		x		x	5.15	74842
74843								x		x		x	5.16	74843
74844								x		x		x	5.16	74844
74845								x		x		x	5.17	74845
74846								x		x		x	5.17	74846
74873							x	x	x	x			5.18	74873
74880							x	x	x	x			5.19	74880
74990										x			5.20	74990
74991										x			5.20	74991
74992										x			5.21	74992
74993										x			5.21	74993
74994										x			5.22	74994
74995										x			5.22	74995
74996										x			5.23	74996
7429841								x		x			5.15	7429841
7429842								x		x			5.15	7429842
7429843								x		x			5.16	7429843
7429844								x		x			5.16	7429844
7429845								x		x			5.17	7429845
7429846								x		x			5.17	7429846

Key to Manufacturer: N — National, M — Motorola,
S — Signetics/Philips, T — Texas Instruments

Table 5.1. TTL latch source summary by manufacturer.

DEVICE	4000 NM	4000 ST	4500 NM	4500 ST	4700 NM	4700 ST	Figure No.	DEVICE
4042	x x	x					5.24	4042
4043	x x	x					5.25	4043
4044	x x	x					5.26	4044
4076	x						5.27	4076
4099	x x						5.28	4099
40373		x					5.8	40373
4508				x			5.29	4508
4597			x				5.30	4597
4598			x				5.31	4598
4599			x				5.32	4599
4723					x		5.33	4723
4724					x	x	5.33	4724

Key to Manufacturer:	Prefix
N – National,	CD
M – Motorola,	MC1
S – Signetics/Philips,	HEC/HEF
T – Texas Instruments	--

Table 5.2. 5V CMOS 4000 series latch source summary by manufacturer.

DEVICE 74---	C NM	C ST	HC NM	HC ST	HCT NM	HCT ST	AHC (5V) NM	AHC (5V) ST	AHCT NM	AHCT ST	VHC (5V) NM	VHC (5V) ST	VHCT NM	VHCT ST	AC (5V) NM	AC (5V) ST	ACT NM	ACT ST	Figure No.	DEVICE 74---
7475			x x	x x		x													5.1	7475
7477				x															5.1	7477
74259			x x	x x		x													5.5	74259
74373	x		x x	x x	x x	x x		x		x	x x		x x			x	x	x	5.8	74373
74375				x															5.9	74375
74533			x x	x x	x	x x										x		x	5.11	74533
74563			x x	x x		x x										x		x	5.12	74563
74573			x x	x x	x	x x		x		x	x x		x x			x	x	x	5.12	74573
744724				x															5.33	744724
7411373																x		x	5.34	7411373
7416373								x		x						x	x	x	5.35	7416373

Key to Manufacturer: N – National, M – Motorola, S – Signetics/Philips, T – Texas Instruments

Table 5.3. 5V CMOS 74xxx series latch source summary by manufacturer.

DEVICE 74---	LV NM	LV ST	LVCH LVC NM	LVCH LVC ST	ALVT LVT NM	ALVT LVT ST	LVX NM	LVX ST	LVQ NM	LVQ ST	LCX NM	LCX ST	AC NM	AC ST	ALVCH NM	ALVCH ST	HL NM	HL ST	AHC NM	AHC ST	VHC NM	VHC ST	Figure No.	DEVICE 74---
74259					x	x																	5.5	74259
74373		xx		xx	x	x	x		x		x									x	x		5.8	74373
742373		x																					5.8	742373
74533														x									5.11	74533
74563														x									5.12	74563
74573		xx		xx		xx			x					x						x	x		5.12	74573
74841				xx																			5.15	74841
742841				x																			5.15	742841
74843				x																			5.16	74843
7416260				x												x							5.36	7416260
7416373				x	x	x					x			x		x							5.35	7416373
74162373				x																			5.35	74162373
7416841						x										x							5.37	7416841
7416843																x							5.38	7416843
7433373																		x					5.39	7433373
7433533																		x					5.39	7433533

Key to Manufacturer: N – National, M – Motorola, S – Signetics/Philips, T – Texas Instruments

Table 5.4. Low-voltage CMOS 74xxx series latch source summary by manufacturer.

DEVICE 74---	STD	S	LS	AS/A/B	ALS/A/B	F	DATA	Figure	DEVICE 74---
-7475	20	--	20	--	--	--	t_w	-5.1	-7475
----	16-40	--	12-30	--	--	--	t_{PLH}↕	---	----
----	20/5	--	20/5	--	--	--	t_s/t_h	---	----
-7477	--	--	20	--	--	--	t_w	-5.1	-7477
----	--	--	15-30	--	--	--	t_{PLH}↕	---	----
----	--	--	20/0	--	--	--	t_s/t_h	---	----
-74100	20	--	--	--	--	--	t_w	-5.2	-74100
----	30	--	--	--	--	--	t_{PLH}↕	---	----
----	20/5.0	--	--	--	--	--	t_s/t_h	---	----
-74116	15-30	--	--	--	--	--	t_w	-5.3	-74116
----		--	--	--	--	--	t_{PLH}↕	---	----
----	14.0/8.0	--	--	--	--	--	t_s/t_h	---	----
-74256	--	--	15	--	--	--	t_w	-5.4	-74256
----	--	--	32-38	--	--	--	t_{PLH}↕	---	----
----	--	--	20/10	--	--	--	t_s/t_h	---	----
-74259	15	--	10-17	--	10-15	3.0-7.5	t_w	-5.5	-74259
----	11-17	--	12-17	--	4-22	4.0-10.5	t_{PLH}↕	---	----
----	15/0	--	17/0	--	20/0	2.0/0	t_s/t_h	---	----
-74279	20	--	20	--	--	--	t_w	-5.6	-74279
----	12-22	--	12-22	--	--	--	t_{PLH}↕	---	----
----	NA	--	NA	--	--	--	t_s/t_h	---	----
-74363	--	--	15	--	--	--	t_w	-5.7	-74363
----	--	--	18-30	--	--	--	t_{PLH}↕	---	----
----	--	--	3.0/10	--	--	--	t_s/t_h	---	----
-74373	--	6-7.3	15	4.5	10	3.5-4.0	t_w	-5.8	-74373
----	--	12-14	28-32	3.5-11.5	2-29	3-11.5	t_{PLH}↕	---	----
----	--	5/2	20/0	2/3	10/7	2.0/3.0	t_s/t_h	---	----
-74375	--	--	20	--	--	--	t_w	-5.9	-74375
----	--	--	12-30	--	--	--	t_{PLH}↕	---	----
----	--	--	20/0	--	--	--	t_s/t_h	---	----
-74412	--	--	--	--	--	8.0-9.0	t_w	-5.10	-74412
----	--	--	--	--	--	3.5-13	t_{PLH}↕	---	----
----	--	--	--	--	--	0/8.0	t_s/t_h	---	----
-74432	--	--	--	--	--	3.0-13	t_w	-5.10	-74432
----	--	--	--	--	--	2.5-9.5	t_{PLH}↕	---	----
----	--	--	--	--	--	0/9.0	t_s/t_h	---	----
-74533	--	--	--	2	15	5.0-6.0	t_w	-5.11	-74533
----	--	--	--	4-9	4-9	4.0-11.0	t_{PLH}↕	---	----
----	--	--	--	2/3	15/7	2.0/2.0	t_s/t_h	---	----
-74563	--	--	--	--	3-29	--	t_w	-5.12	-74563
----	--	--	--	--	15	--	t_{PLH}↕	---	----
----	--	--	--	--	10/10	--	t_s/t_h	---	----
-74573	--	--	--	4.5	10	3.0-3.5	t_w	-5.12	-74573
----	--	--	--	3-16.5	2-33	3-11.5	t_{PLH}↕	---	----
----	--	--	--	2/3	10/7	2.0/3.0	t_s/t_h	---	----
-74580	--	--	--	2	15	--	t_w	-5.12	-74580
----	--	--	--	3-9	3-29	--	t_{PLH}↕	---	----
----	--	--	--	2/3	10/10	--	t_s/t_h	---	----
-74604	--	--	20	--	--	5.0-6.0	t_w	-5.13	-74604
----	--	--	15-45	--	--	5.0-10	t_{PLH}↕	---	----
----	--	--	20/0	--	--	2/0	t_s/t_h	---	----
-74605	--	--	--	--	--	5.0-6.0	t_w	-5.13	-74605
----	--	--	--	--	--	7.5-11.0	t_{PLH}↕	---	----
----	--	--	--	--	--	3/1	t_s/t_h	---	----
-74606	--	--	20	--	--	--	t_w	-5.13	-74606
----	--	--	22-50	--	--	--	t_{PLH}↕	---	----
----	--	--	20/0	--	--	--	t_s/t_h	---	----
-74607	--	--	20	--	--	--	t_w	-5.13	-74607
----	--	--	28-70	--	--	--	t_{PLH}↕	---	----
----	--	--	20/0	--	--	--	t_s/t_h	---	----

Table 5.5a. TTL latch representative propagation, setup, and hold times (nS).

DEVICE 74---	STD	S	LS	AS/A/B	ALS/A/B	F	DATA	Figure	DEVICE 74---
-74666	--	--	--	--	10	--	tw	-5.14	-74666
----	--	--	--	--	3-21	--	tPLH	---	----
----	--	--	--	--	10/5	--	ts/th	---	----
-74667	--	--	--	--	10	--	tw	-5.14	-74667
----	--	--	--	--	6-28	--	tPLH	---	----
----	--	--	--	--	10/5	--	ts/th	---	----
-74841	--	--	--		20	3.0-3.5	tw	-5.15	-74841
----	--	--	--	1-13	8.5-18	2.0-10	tPLH	---	----
----	--	--	--	2.5/2.5	10/5	0/3.0	ts/th	---	----
-74842	--	--	--	4	20	3.0-3.5	tw	-5.15	-74842
----	--	--	--	1-13	11-23	2.0-10	tPLH	---	----
----	--	--	--	2.5/2.5	10/5	0/3.0	ts/th	---	----
-74843	--	--	--	4	20-35	3.0-3.5	tw	-5.16	-74843
----	--	--	--	1-13	7-18	3.5-10	tPLH	---	----
----	--	--	--	2.5/2.5	10/5	0/4.0	ts/th	---	----
-74844	--	--	--	4	7-18	3.0-4.0	tw	-5.16	-74844
----	--	--	--	1-12.5	11-24	3.5-10	tPLH	---	----
----	--	--	--	2.5/2.5	10/5	0/4.0	ts/th	---	----
-74845	--	--	--	4	7-18	3.0-4.0	tw	-5.17	-74845
----	--	--	--	1-14	7-26	3.5-10	tPLH	---	----
----	--	--	--	2.5/2.5	10/5	0/4.0	ts/th	---	----
-74846	--	--	--	4	7-18	3.0-4.0	tw	-5.17	-74846
----	--	--	--	1-14.5	11-19	3.5-10	tPLH	---	----
----	--	--	--	2.5/2.5	10/5	0/4.0	ts/th	---	----
-74873	--	--	--	3.5-4.5	10-15	--	tw	-5.18	-74873
----	--	--	--	3-9.5	2-22	--	tPLH	---	----
----	--	--	--	2/4.5	10/7	--	ts/th	---	----
-74880	--	--	--	2.5-3.5	15	--	tw	-5.19	-74880
----	--	--	--	4-11.5	14-19	--	tPLH	---	----
----	--	--	--	2/1	10/10	--	ts/th	---	----
-74990	--	--	--	--	10	--	tw	-5.20	-74990
----	--	--	--	--	8-22	--	tPLH	---	----
----	--	--	--	--	10/5	--	ts/th	---	----
-74991	--	--	--	--	10	--	tw	-5.20	-74991
----	--	--	--	--	12-21	--	tPLH	---	----
----	--	--	--	--	10/5	--	ts/th	---	----
-74992	--	--	--	--	10	--	tw	-5.21	-74992
----	--	--	--	--	7-15	--	tPLH	---	----
----	--	--	--	--	10/5	--	ts/th	---	----
-74993	--	--	--	--	10	--	tw	-5.21	-74993
----	--	--	--	--	11-20	--	tPLH	---	----
----	--	--	--	--	10/5	--	ts/th	---	----
-74994	--	--	--	--	10	--	tw	-5.22	-74994
----	--	--	--	--	7-10	--	tPLH	---	----
----	--	--	--	--	10/5	--	ts/th	---	----
-74995	--	--	--	--	10	--	tw	-5.22	-74995
----	--	--	--	--	12-16	--	tPLH	---	----
----	--	--	--	--	10/5	--	ts/th	---	----
-74996	--	--	--	--	10	--	tw	-5.23	-74996
----	--	--	--	--	16-24	--	tPLH	---	----
----	--	--	--	--	15/0	--	ts/th	---	----
-7429841	--	--	--	|	6	--	tw	-5.15	7429841
----	--	--	--	|	2-16	--	tPLH	---	----
----	--	--	--	|	2.5/4.5	--	ts/th	---	----
-7429842	--	--	--			--	--	-5.15	7429842
-7429843	--	--	--	NOT	NOT	--	--	-5.16	7429843
-7429844	--	--	--	AVAIL-	AVAIL-	--	--	-5.16	7429844
-7429845	--	--	--	ABLE	ABLE	--	--	-5.17	7429845
-7429846	--	--	--	|	|	--	--	-5.17	7429846
	--	--	--	|	|	--	--		

Table 5.5b. TTL Latch representative propagation, setup, and hold times (nS) (continuation from Table 5.5a).

DEVICE CD\MC1	4000	4500	4700	DATA	Figure	DEVICE CD\MC1
-4042	100	--	--	tw	- 5.24	-4042
----	220-440	--	--	tPLH↑	---	----
----	50/100	--	--	ts/th	---	----
-4043	80-100	--	--	tw	- 5.25	-4043
----	100-200	--	--	tPLH↑	---	----
----	NA	--	--	ts/th	---	----
-4044	80-100	--	--	tw	- 5.26	-4044
----	100-200	--	--	tPLH↑	---	----
----	NA	--	--	ts/th	---	----
-4076	260	--	--	tw	- 5.27	-4076
----	300-600	--	--	tPLH↑	---	----
----	30/130	--	--	ts/th	---	----
-4099	40-150	--	--	tw	- 5.28	-4099
----	200-400	--	--	tPLH↑	---	----
----	100/150	--	--	ts/th	---	----
-40373	30-160	--	--	tw	- 5.8	-40373
----	125-250	--	--	tPLH↑	---	----
----	15/25	--	--	ts/th	---	----
-4508	--	70-140	--	tw	- 5.29	-4508
----	--	220-440	--	tPLH↑	---	----
----	--	50/50	--	ts/th	---	----
-4597	--	100-320	--	tw	- 5.30	-4597
----	--	160-320	--	tPLH↑	---	----
----	--	100/100	--	ts/th	---	----
-4598	--	100-320	--	tw	- 5.31	-4598
----	--	160-320	--	tPLH↑	---	----
----	--	100/100	--	ts/th	---	----
-4599	--	75-150	--	tw	- 5.32	-4599
----	--	200-400	--	tPLH↑	---	----
----	--	100/150	--	ts/th	---	----
-4723	--	--	100-200	tw	- 5.33	-4723
----	--	--	225-450	tPLH↑	---	----
----	--	--	40/60	ts/th	---	----
-4724	--	--	100-200	tw	- 5.33	-4724
----	--	--	225-450	tPLH↑	---	----
----	--	--	40/60	ts/th	---	----

Table 5.6. 5V CMOS 4000 series latch representative propagation, setup, and hold times (nS).

DEVICE 74---	C	HC	HCT	AHC (5V)	AHCT	VHC(5V)	VHCT	AC (5V)	ACT	DATA	Figure	DEVICE 74---
-7475	--	6-20	4-20	--	--	--	--	--	--	tw	-5.1	-7475
----	--	12-24	15-28	--	--	--	--	--	--	tPLH↓	---	----
----	--	5/-3	4/-2	--	--	--	--	--	--	ts/th	---	----
-7477	--	16	--	--	--	--	--	--	--	tw	-5.1	-7477
----	--	12-30	--	--	--	--	--	--	--	tPLH↓	---	----
----	--	20/5	--	--	--	--	--	--	--	ts/th	---	----
-74259	--	16		--	--	--	--	--	--	tw	-5.5	-74259
----	--	21-37	23-39	--	--	--	--	--	--	tPLH↓	---	----
----	--	7/-6	7/-6	--	--	--	--	--	--	ts/th	---	----
-74373		16	17-20	5	6.5	5	6.5	5.5-6	7-8	tw	-5.8	-74373
----	155-310	15-30	17-30	1-8.5	1-9.5	4.9-7.2	7.7-12.3	1.5-10.5	1.5-11.5	tPLH↓	---	----
----	140/0	5/-3	6/-1	4/1	1.5/3.5	4.0/1.0	1.5/3.5	4.5/1	8/1	ts/th	---	----
-74375	--	16	--	--	--	--	--	--	--	tw	-5.9	-74375
----	--	14-30	--	--	--	--	--	--	--	tPLH↓	---	----
----	--	20/5	--	--	--	--	--	--	--	ts/th	---	----
-74533	--	16	17-20	--	--	--	--	--	--	tw	-5.11	-74533
----	--	26-38	28-40	--	--	--	--	--	--	tPLH↓	---	----
----	--	10/10	9/5	--	--	--	--	--	--	ts/th	---	----
-74563	--	17-29	18-30	--	--	--	--	4	3	tw	-5.12	-74563
----	--	17-29	18-30	--	--	--	--	2-9.5	3-11.5	tPLH↓	---	----
----	--	4/-2	3/-1	--	--	--	--	2/2	4/0	ts/th	---	----
-74573	--	17-30	20-35	4.5-6.8	4.2-6	5	6.5	5	3.5	tw	-5.12	-74573
----	--	17-30	20-35	5-7.7	5.6-7.5	5.0-7.7	7.7-12.3	2.5-10	2.5-10.5	tPLH↓	---	----
----	--	5/4	7/4	3.5/1.5	3.5/1.5	3.5/1.5	1.5/3.5	3/2	3.5/0	ts/th	---	----
-744724	--	18-30	--	--	--	--	--	--	--	tw	-5.33	744724
----	--	18-30	--	--	--	--	--	--	--	tPLH↓	---	----
----	--	15/5	--	--	--	--	--	--	--	ts/th	---	----
-7411373	--	--	--	--	--	--	--	4	5	tw	-5.34	7411373
----	--	--	--	--	--	--	--	1.5-8.9	1.5-10.3	tPLH↓	---	----
----	--	--	--	--	--	--	--	3.5/2	3.5/3.5	ts/th	---	----
-7416373	--	--	--	5	6.5	--	--	4	4	tw	-5.35	7416373
----	--	--	--	5-7.2	5.1-8.5	--	--	3.1-8.5	3.8-9.4	tPLH↓	---	----
----	--	--	--	4/1	1.5/3.5	--	--	1.5/2.5	1/5	ts/th	---	----

Table 5.7. 5V CMOS 74xxx series latch representative propagation, setup, and hold times (nS).

DEVICE 74---	LV	LVC LVCH	ALVT LVT	LVX	LVQ	LCX	AC(3V)	ALVCH	HL	AHC (3V)	VHC(3V)	DATA	Figure No.	DEVICE 74---
-74259	6-20	--	--	--	--	--	--	--	--	--	--	tw	-5.5	-74259
----	20-29	--	--	--	--	--	--	--	--	--	--	tPLH↓	---	----
----	7/6	--	--	--	--	--	--	--	--	--	--	ts/th	---	----
-74373	6-20	3.0-4.0	3.3	5.0	2.0	4.0	5.5-6	--	--	5	5	tw	-5.8	-74373
----	13-22	1.5-7.5	1.0-4.2	6.0-9.7	2.5-9.6	6.0-9.7	1.5-13.5	--	--	7.3-11.4	7.0-11.0	tPLH↓	---	----
----	10/5	2.0/2.0	0.7	4.0/1.0	4.0/1.5	4.0/1.0	5.5/1	--	--	4/1	4.0/1.0	ts/th	---	----
-742373	--	4.0	--	--	--	--	--	—	--	--	--	tw	-5.8	742373
----	--	1.5-9.5	--	--	--	--	--	--	--	--	--	tPLH↓	---	----
----	--	2.0/2.0	--	--	--	--	--	--	--	--	--	ts/th	---	----
-74533	--	--	--	--	--	--	5.5	--	--	--	--	tw	-5.11	-74533
----	--	--	--	--	--	--	2-14.5	--	--	--	--	tPLH↓	---	----
----	--	--	--	--	--	--	5.5/1.5	--	--	--	--	ts/th	---	----
-74563	--	--	--	--	--	--	6	--	--	--	--	tw	-5.12	-74563
----	--	--	--	--	--	--	3-11.5	--	--	--	--	tPLH↓	---	----
----	--	--	--	--	--	--	4/0	--	--	--	--	ts/th	---	----
-74573	20	3.0	3.3	--	2.0-4.0	--	6	--	--	5	5.0	tw	-5.12	-74573
----	14-23	1.5-4.3	1-4.2	--	2.5-10.5	--	2.5-13	--	--	7-11.9	7.6-11.9	tPLH↓	---	----
----	10/8	1.0/1.0	.7/1.6	--	3.0/1.5	--	3.5/2	--	--	3.5/1.5	3.5/1.5	ts/th	---	----
-74841	--	3.0	--	--	--	--	--	--	--	--	--	tw	-5.15	-74841
----	--	1.5-8.5	--	--	--	--	--	--	--	--	--	tPLH↓	---	----
----	--	1.0/1.0	--	--	--	--	--	--	--	--	--	ts/th	---	----
-742841	--	3.0	--	--	--	--	--	--	--	--	--	tw	-5.15	-742841
----	--	1.5-8.5	--	--	--	--	--	--	--	--	--	tPLH↓	---	----
----	--	1.0/1.0	--	--	--	--	--	--	--	--	--	ts/th	---	----
-74843	--	Speeds	--	--	--	--	--	--	--	--	--	tw	-5.16	-74843
----	--	Not	--	--	--	--	--	--	--	--	--	tPLH↓	---	----
----	--	Defined	--	--	--	--	--	--	--	--	--	ts/th	---	----
-7416260	--	--	--	--	--	--	--	3.3	--	--	--	tw	-5.36	7416260
----	--	--	--	--	--	--	--	1.2-6	--	--	--	tPLH↓	---	----
----	--	--	--	--	--	--	--	1.4/1.6	--	--	--	ts/th	---	----
-7416373	--	4.0	--	--	--	--	--	3.3	--	--	--	tw	-5.35	7416373
----	--	3.1-5.5	--	--	--	--	--	1-5.1	--	--	--	tPLH↓	---	----
----	--	2.0/1.5	--	--	--	--	--	1/1.5	--	--	--	ts/th	---	----
-74162373	--	3.0	--	--	--	--	--	--	--	--	--	tw	-5.35	74162373
----	--	8.0-8.5	--	--	--	--	--	--	--	--	--	tPLH↓	---	----
----	--	2.2/2.2	--	--	--	--	--	--	--	--	--	ts/th	---	----
-7416841	--	--	--	--	--	--	--	3.3	--	--	--	tw	-5.37	7416841
----	--	--	--	--	--	--	--	1.2-3.9	--	--	--	tPLH↓	---	----
----	--	--	--	--	--	--	--	1.1/1.1	--	--	--	ts/th	---	----
-7416843	--	--	--	--	--	--	--	Speeds	--	--	--	tw	-5.38	7416843
----	--	--	--	--	--	--	--	Not	--	--	--	tPLH↓	---	----
----	--	--	--	--	--	--	--	Defined	--	--	--	ts/th	---	----
-7433373	--	--	--	--	--	--	--	--	2.0-2.5	--	--	tw	-5.39	7433373
----	--	--	--	--	--	--	--	--	4.0-44.4	--	--	tPLH↓	---	----
----	--	--	--	--	--	--	--	--	0.5/0.5	--	--	ts/th	---	----
-7433533	--	--	--	--	--	--	--	--	2.0-2.5	--	--	tw	-5.39	7433533
----	--	--	--	--	--	--	--	--	4.0-44.4	--	--	tPLH↓	---	----
----	--	--	--	--	--	--	--	--	0.5/0.5	--	--	ts/th	---	----

Table 5.8. Low-Voltage CMOS 74xxx series latch representative propagation, setup, and hold times (nS).

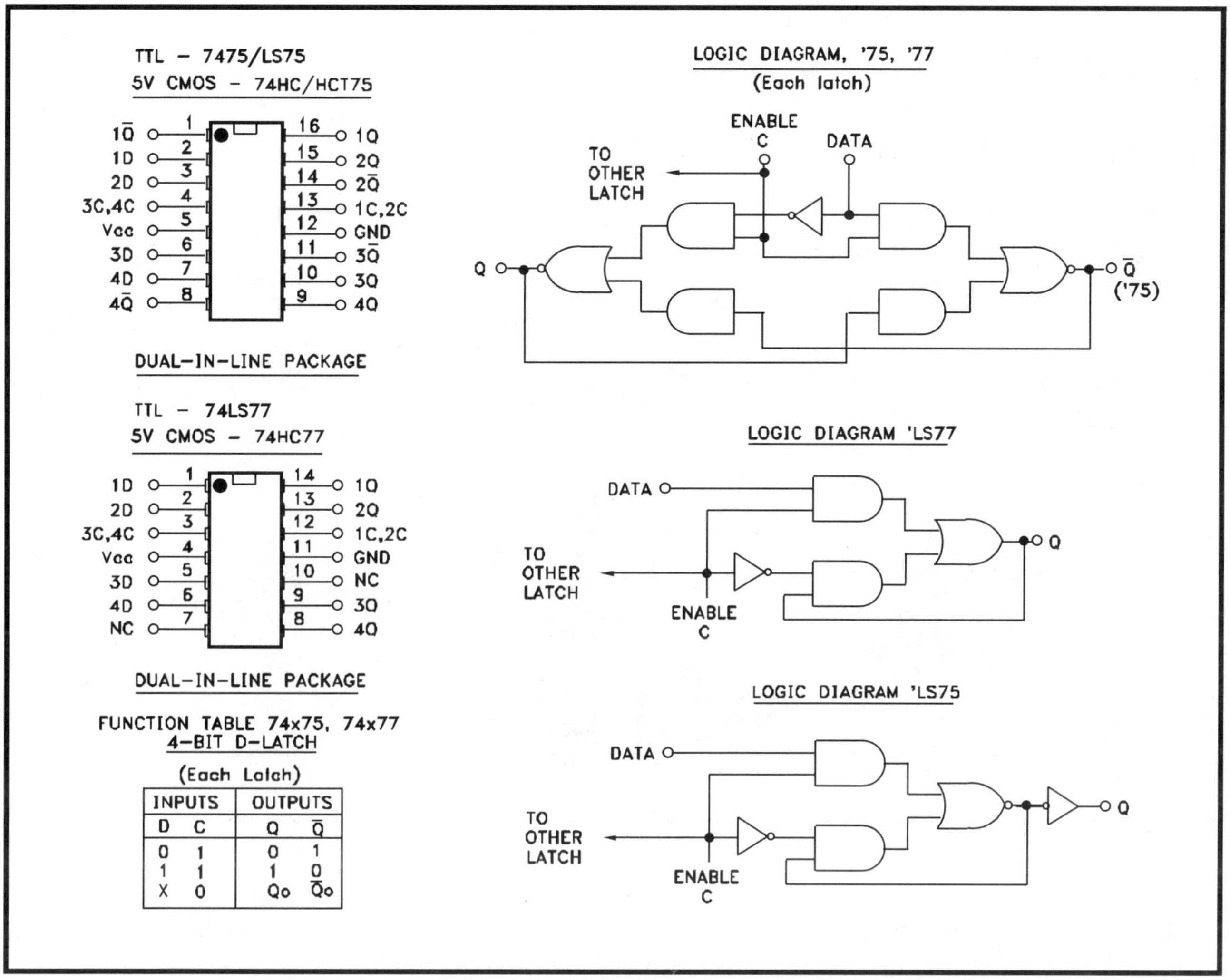

INPUTS		OUTPUTS	
D	C	Q	Q̄
0	1	0	1
1	1	1	0
X	0	Qo	Q̄o

Figure 5.1. These two latch types, available in TTL and CMOS versions, are used for temporary storage of binary data. Information present at a Data input is transferred to the Q output when the Enable is HIGH. The Q output follows the Data in while the Enable is HIGH. When the Enable drops to the LOW state the data is retained unchanged in the latch until the Enable returns to the HIGH state. Note the logic diagram for the LS versions differs from the original standard devices.

TTL – 74100

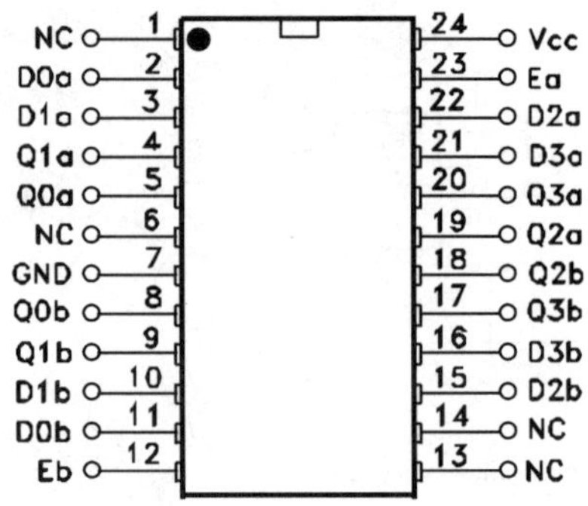

DUAL-IN-LINE PACKAGE

LOGIC DIAGRAM

(Each latch)

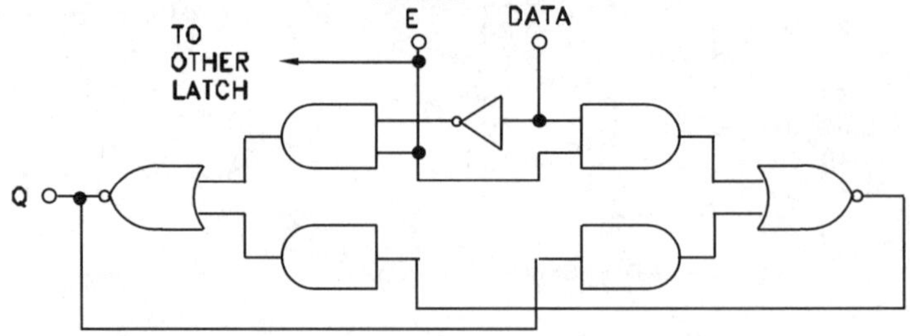

FUNCTION TABLE 74100 DUAL 4-BIT TRANSPARENT LATCH

(Each Latch)

INPUTS		OUTPUTS
D	E	Q
0	1	0
1	1	1
X	0	Qo

LOGIC SYMBOL

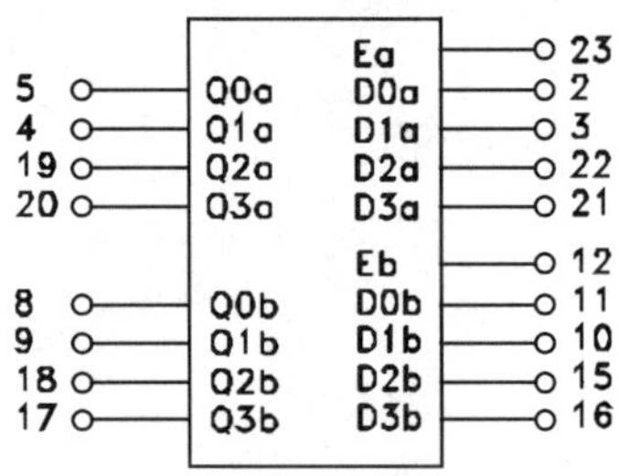

Figure 5.2. The 74100 consists of two independent 4-bit transparent latches. Each latch has its own active HIGH enable input. With Enable HIGH data entering the latch flows through and appears at the output. Data present at the "D" inputs one setup time prior to the HIGH-to-LOW Enable transition will be stored in the latch. Latch content remains stable with the Enable in the LOW state.

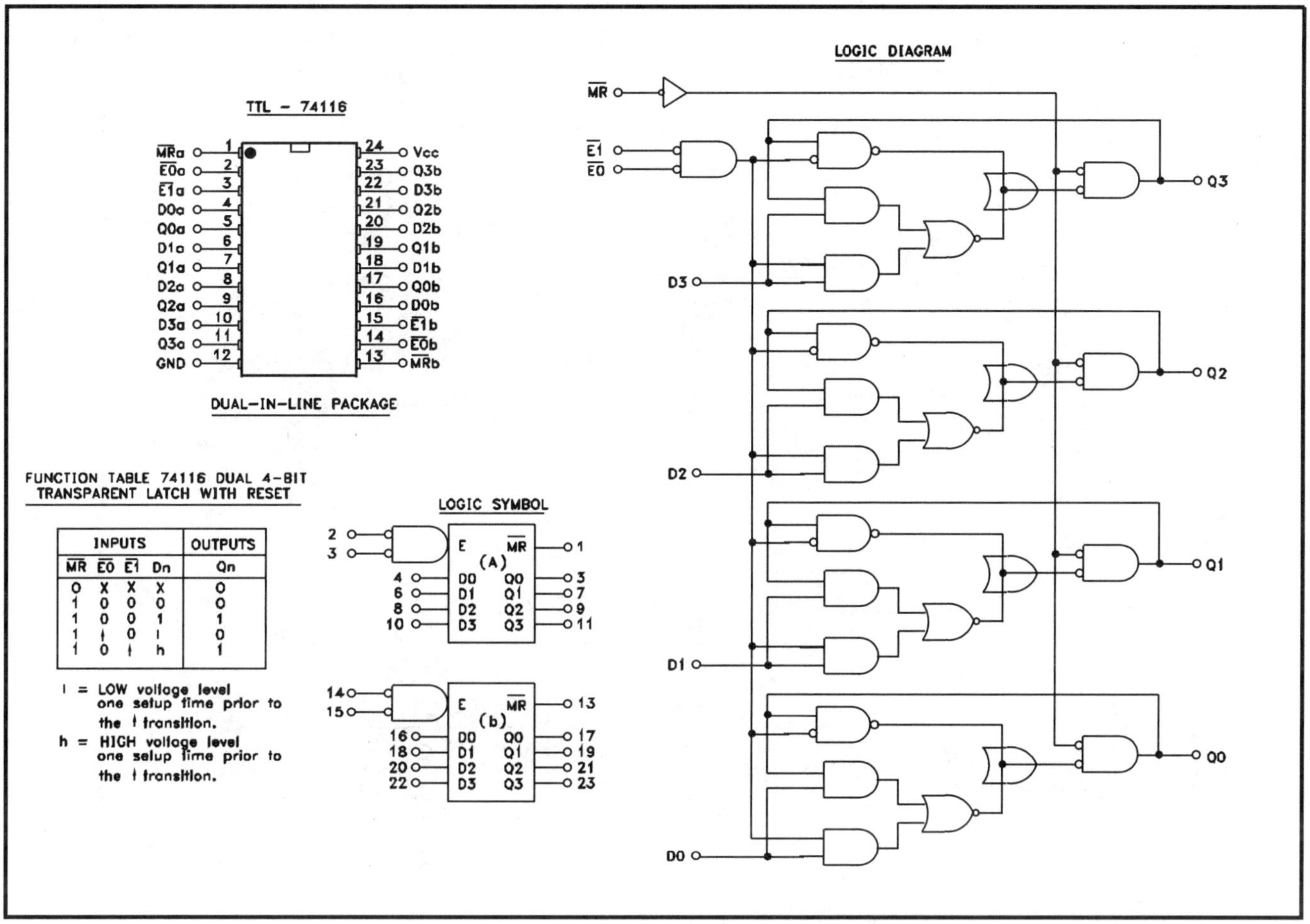

Figure 5.3. The 74116 consists of two independent 4-bit transparent latches. Each latch is enabled by a 2-input active LOW enable gate. With both Enables LOW data entering the latch flows through and appears at the output. Data present at the "D" inputs one setup time prior to the LOW-to-HIGH transition of either Enable will be stored in the latch. Latch content remains stable with either Enable in the HIGH state. Each latch has an active LOW asynchronous Master Reset. The reset overrides the data and enable inputs, setting the four latches to the LOW state.

TTL – 74LS/F256

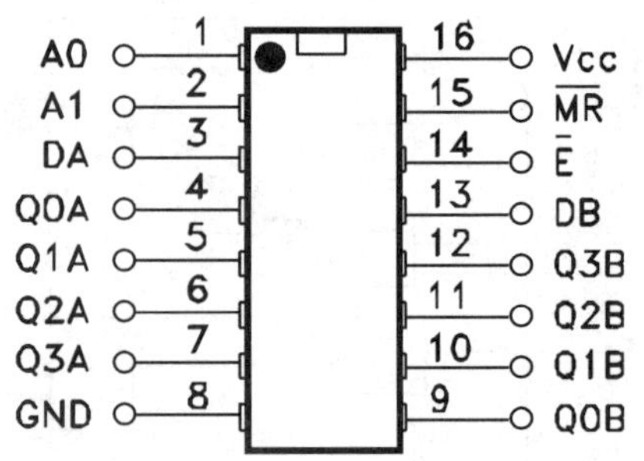

DUAL–IN–LINE PACKAGE

LOGIC SYMBOL

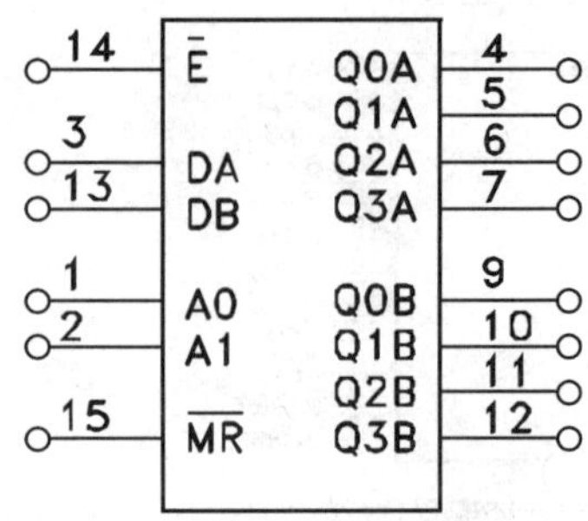

FUNCTION TABLE 74x256 DUAL 4–BIT ADDRESSABLE LATCH

OPERATING MODE	INPUTS					OUTPUTS			
	$\overline{MR}$	$\bar{E}$	Dn	A0	A1	Q0	Q1	Q2	Q3
Master Reset	0	1	X	X	X	0	0	0	0
Demultiplex (Active–High decoder when D = H)	0	0	D	0	0	Q=d	0	0	0
	0	0	d	1	0	0	Q=d	0	0
	0	0	d	0	1	0	0	Q=d	0
	0	0	D	1	1	0	0	0	Q=d
Store (Do nothing)	1	1	X	X	X	q0	q1	q2	q3
Addressable Latch	1	0	d	0	0	Q=d	q1	q2	q3
	1	0	d	1	0	q0	Q=d	q2	q3
	1	0	d	0	1	q0	q1	Q=d	q3
	1	0	D	1	1	Q0	Q1	Q2	Q=D

d = HIGH or LOW data one setup time prior to the LOW–to–HIGH Enable transistion.

q = the state of the referenced output established during the last cycle in which it was addressed or cleared.

Figure 5.4. The 74256 provides two 4-bit addressable latches. The latches respond to common address and control inputs. These are two address inputs, an active LOW Enable, and an active LOW Clear input. Each latch has a single "D" input and four outputs. The device combines the features of a dual 1-in-4 demultiplexer and an 8-bit transparent latch. It can be used as a dual 1-of-4 active HIGH decoder. The operation is shown in the function table. The device has four distinct modes of operation selectable by controlling the Clear and Enable inputs. In the Addressable Latch mode, data at the inputs is written into the addressed latches. These follow the data while the unaddressed latches remain in their existing states. In the Memory mode all latches remain in their previous states unaffected by changes at the data inputs. The Enable should be held HIGH while the address lines are changing. In the Decoding (demultiplexing) mode (/CLR=/E=LOW) addressed outputs willfollow the level of the "D" inputs with all other outputs LOW. In the Clear mode all outputs are LOW and unaffected by address and data inputs.

TTL – 74259/LS/ALS/F
5V CMOS – 74HC/HCT259
3V CMOS – 74LV259

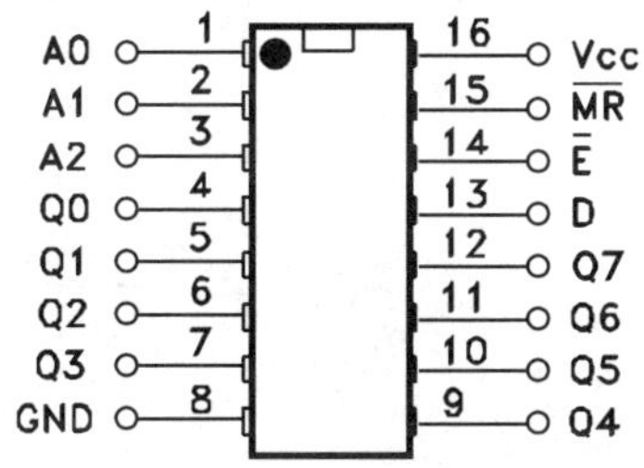

DUAL-IN-LINE PACKAGE

LOGIC SYMBOL

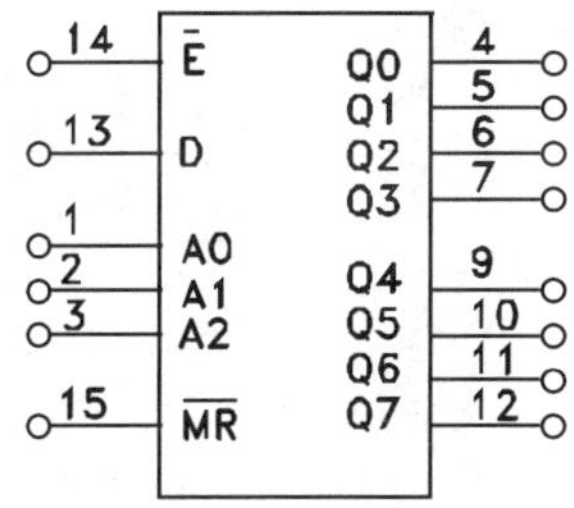

FUNCTION TABLE 74x259 8-BIT ADDRESSABLE LATCH

OPERATING MODE	INPUTS						OUTPUTS							
	$\overline{MR}$	$\overline{E}$	Dn	A0	A1	A2	Q0	Q1	Q2	Q3	Q4	Q5	Q6	Q7
Master Reset	0	1	X	X	X	X	0	0	0	0	0	0	0	0
Demultiplex (Active-High decoder when D = H)	0	0	d	0	0	0	Q=D	0	0	0	0	0	0	0
	0	0	d	1	0	0	0	Q=d	0	0	0	0	0	0
	0	0	d	0	1	0	0	0	Q=d	0	0	0	0	0
	.	.	.	.	.	.	.	.	.	.	.	.	.	.
	.	.	.	.	.	.	.	.	.	.	.	.	.	.
	.	.	.	.	.	.	.	.	.	.	.	.	.	.
	0	0	d	1	1	1	0	0	0	0	0	0	0	Q=d
Store (Do nothing)	1	1	X	X	X	X	q0	q1	q2	q3	q4	q5	q6	q7
Addressable Latch	1	0	d	0	0	0	Q=d	q1	q2	q3	q4	q5	q6	q7
	1	0	d	1	0	0	q0	Q=d	q2	q3	q4	q5	q6	q7
	1	0	d	0	1	0	q0	q1	Q=d	q3	q4	q5	q6	q7
	.	.	.	.	.	.	.	.	.	.	.	.	.	.
	.	.	.	.	.	.	.	.	.	.	.	.	.	.
	.	.	.	.	.	.	.	.	.	.	.	.	.	.
	1	0	d	1	1	1	Q0	Q1	Q2	q3	q4	q5	q6	Q=d

d = HIGH or LOW data one setup time prior to the LOW-to-HIGH Enable transition.

q = the state of the referenced output established during the last cycle in which it was addressed or cleared.

Figure 5.5. The 74259 is an 8-bit addressable latch with three control inputs. These are three address inputs, an active LOW Enable, and an active LOW Clear input. Each latch has a single "D" input and individual outputs. The device combines the features of a dual 1-in-8 demultiplexer and an 8-bit transparent latch. It can be used as a 1-of-8 active HIGH decoder. The operation is shown in the function table. The device is similar to the 74256 in having four distinct modes of operation selectable by controlling the Clear and Enable inputs. In the Addressable Latch mode data at the inputs is written into the addressed latches. These follow the data while the unaddressed latches remain in their existing states. In the Memory mode all latches remain in their previous states unaffected by changes at the data inputs. The Enable should be held HIGH while the address lines are changing. In the Decoding (demultiplexing) mode (/CLR=/E=LOW) addressed outputs will follow the level of the "D" inputs with all other outputs LOW. In the Clear mode all outputs are LOW and unaffected by address and data inputs.

TTL – 74279
74LS279

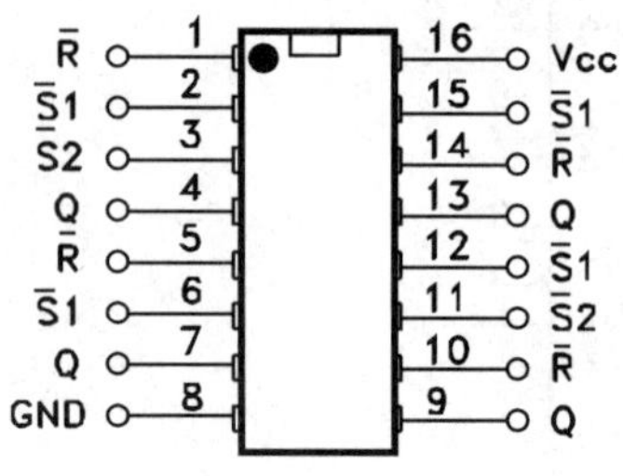

DUAL–IN–LINE PACKAGE

FUNCTION TABLE 74x279
QUAD SET–RESET LATCH

$\overline{S}1$	$\overline{S}2$	$\overline{R}$	Q
0	0	0	h
0	X	1	1
X	0	1	1
1	1	0	0
1	1	1	NO CHANGE

h = The output is HIGH as long as S1 or S2 is LOW. If all inputs go HIGH simultaneously the output state is indeterminate.

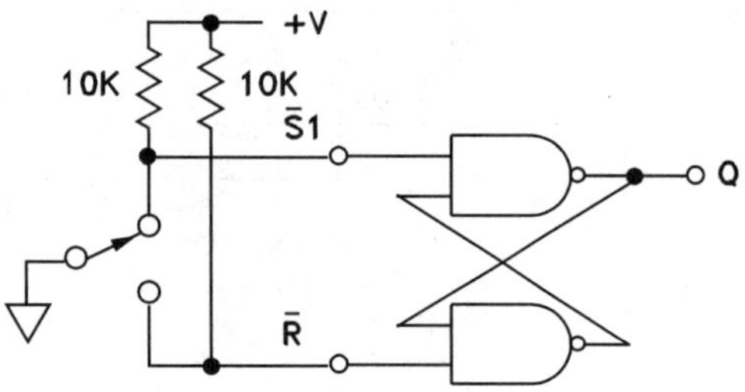

NAND R–S LATCH APPLICATION
Q IS HIGH WITH SWITCH AS SHOWN

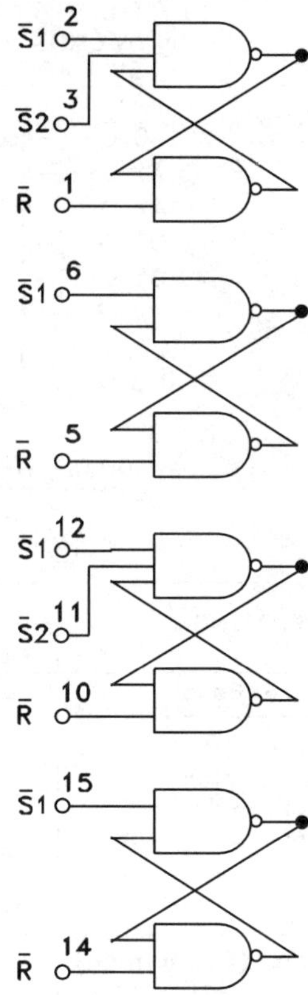

LOGIC DIAGRAM

Figure 5.6. The 74279 provides four individual active low Set-Reset latches in a single 16-pin package. In operation the Latch inputs are normally held HIGH. Momentarily dropping the Set input sets the output HIGH. Note that two of the four latches feature two Set inputs.

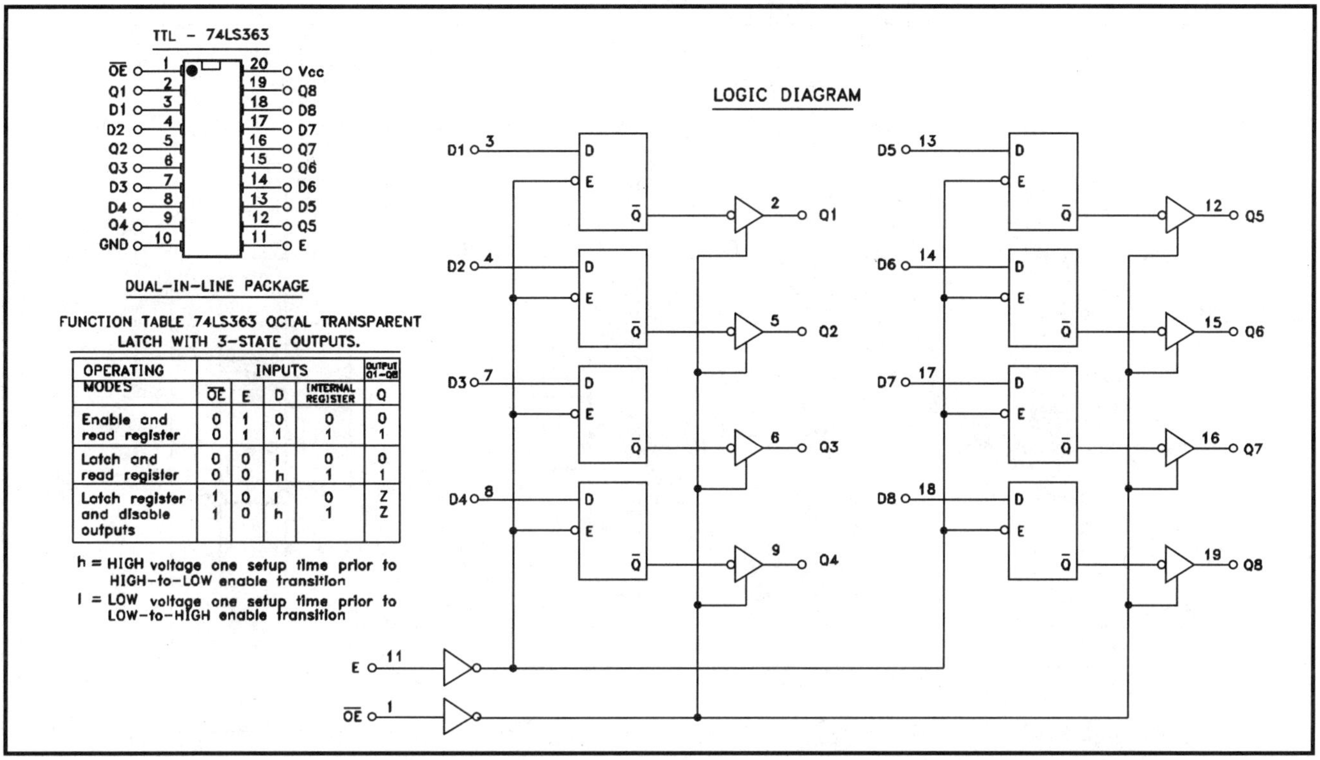

Figure 5.7. The 74LS363 is an 8-bit transparent latch with 3-state buffered outputs. This latch is mostly used for driving MOS memories or microprocessors which require higher than normal VIH levels. The outputs follow the inputs when the Enable is HIGH. With the Enable LOW the outputs are stable. The outputs are controlled by an active LOW Output Enable (/OE). A high on this line forces the outputs to a high impedance state.

TTL - 74LS/S/F/ALS/AS373

5V CMOS - 74C/HC/HCT/AC/ACT/AHC/AHCT/VHC/VHCT373
HEF40373

3V CMOS - 74LV/LVC/LVT-ABT/LVX/LVQ/LCX/LVC
/LVCH373

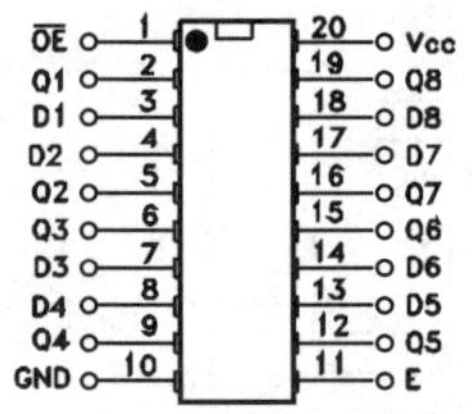

DUAL-IN-LINE PACKAGE

FUNCTION TABLE 74x373, HEF40373 OCTAL TRANSPARENT LATCH WITH 3-STATE OUTPUTS

OPERATING MODES	INPUTS				OUTPUT Q1-Q8
	$\overline{OE}$	E	D	INTERNAL REGISTER	Q
Enable and read register	0 0	1 1	0 1	0 1	0 1
Latch and read register	0 0	0 0	l h	0 1	0 1
Latch register and disable outputs	1 1	0 0	l h	0 1	Z Z

h = HIGH voltage one setup time prior to HIGH-to-LOW enable transition

l = LOW voltage one setup time prior to LOW-to-HIGH enable transition

LOGIC DIAGRAM

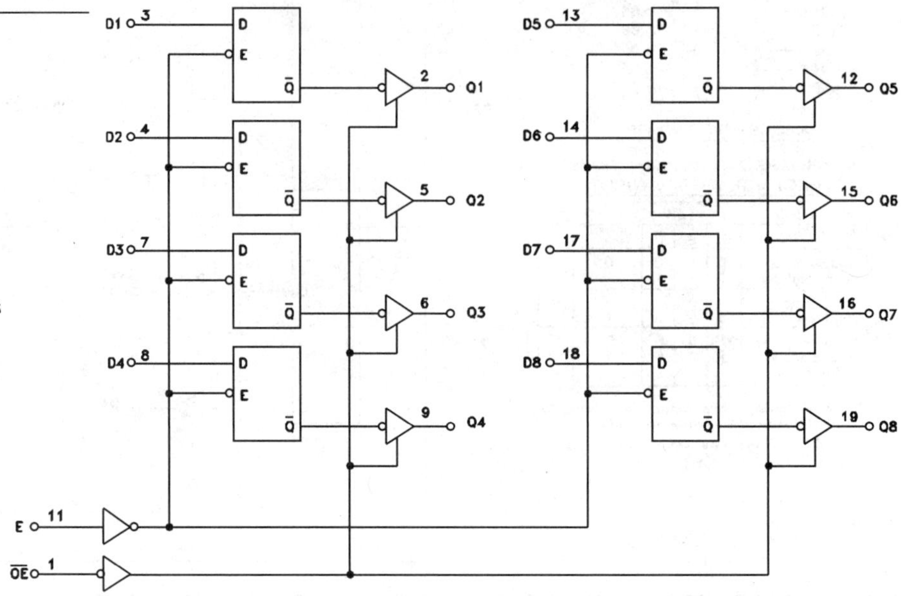

Figure 5.8. The popular 74373 series is an Octal transparent latch with 3-state buffered outputs. Observe that it is essentially identical in performance to the 74363. The outputs follow the inputs when the Enable is HIGH. With the Enable LOW the outputs are stable. The outputs are controlled by an active LOW Output Enable (/OE). A high on this line forces the outputs to a high impedance state.

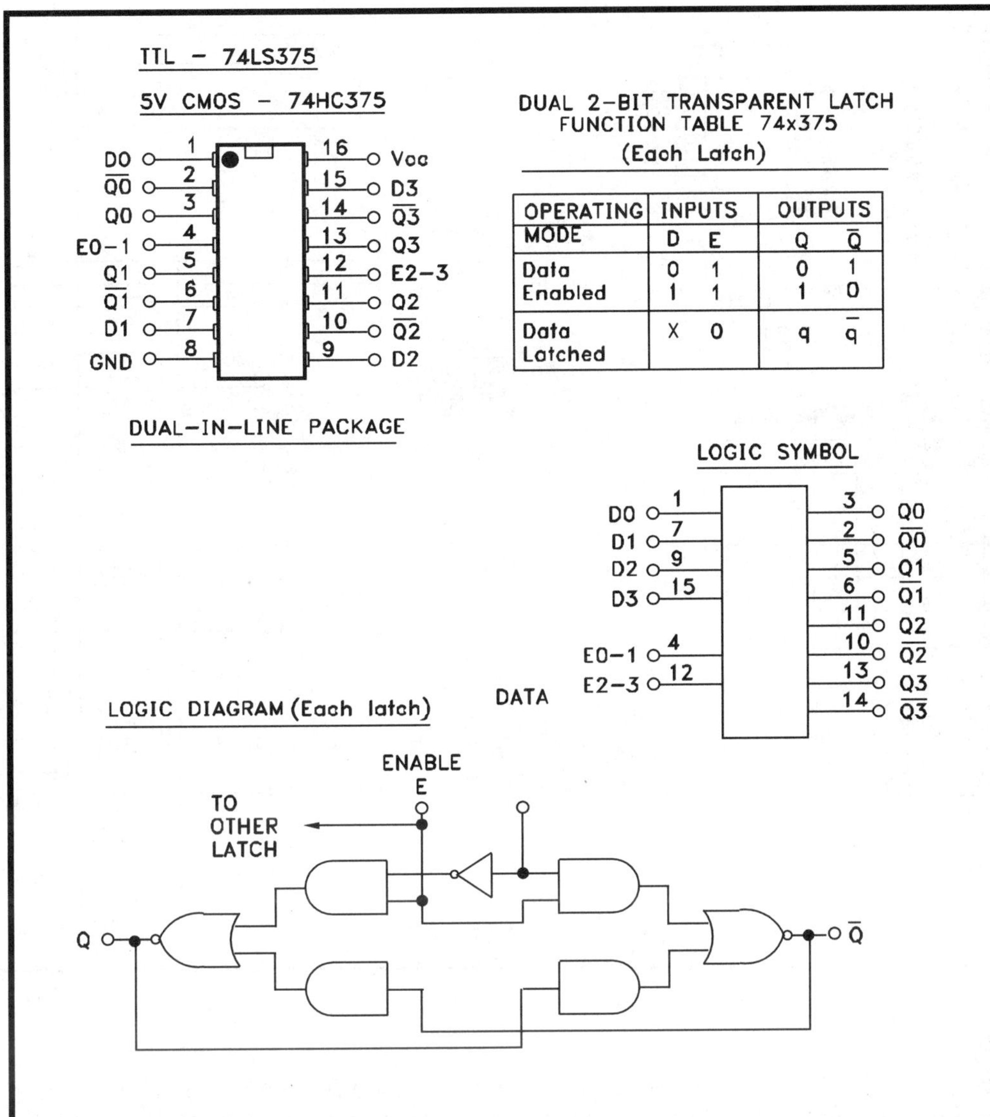

Figure 5.9. The 74375 is a dual 2-bit transparent latch very similar to the earlier 7475. A look back at *Figure 5.1* will verify the similarity of the Logic Diagram and Function Table. Two Enable inputs are provided, each controls two latches. With the Enable HIGH data present at the input flows through to the Q and /Q outputs. Data present one setup time prior to the HIGH-to-LOW Enable transition is stored in the latch until return of the Enable to the HIGH state.

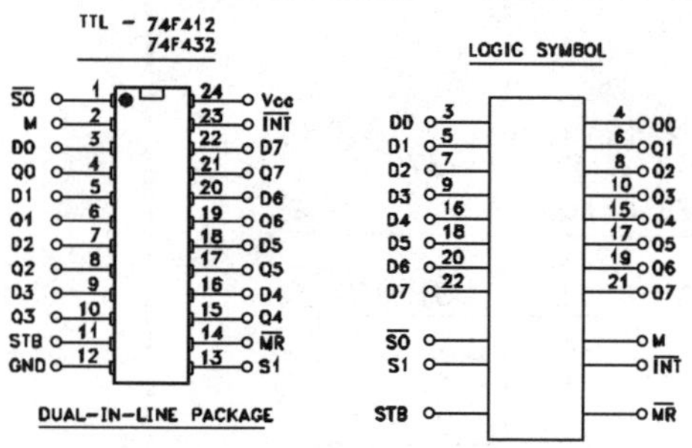

NOTE: 74F412 LATCH, NONINVERTING, 3-STATE
74F432 LATCH, INVERTING, 3-STATE

FUNCTION TABLE (For Data Latches)
74F412, 74F432 MULTI-MODE BUFFERED LATCHES

OPERATING MODE	INPUTS					DATA IN	DATA OUT	
	MR	M	S0	S1	STB		'F412	'F432
Clear	0	1	1	X	X	X	0	1
	0	0	0	1	0	X	0	1
De-select	X	0	X	0	X	X	Z	Z
	X	0	1	X	X	X	Z	Z
Hold	1	1	1	0	X	X	Q0	$\overline{Q0}$
	1	0	0	1	0	X	Q0	$\overline{Q0}$
Data Bus	1	1	0	1	X	0	0	1
	1	1	0	1	X	1	1	0
Data Bus	1	0	0	1	↑	0	0	1
	1	0	0	1	↑	1	1	0

FUNCTION TABLE (For Status Flip-Flop)

INPUTS				$\overline{INT}$
MR	S0	S1	STB	
0	1	X	X	1
0	X	0	X	1
1	X	X	↑	0
1	0	1	X	0

↑ = LOW-to-HIGH clock transition.

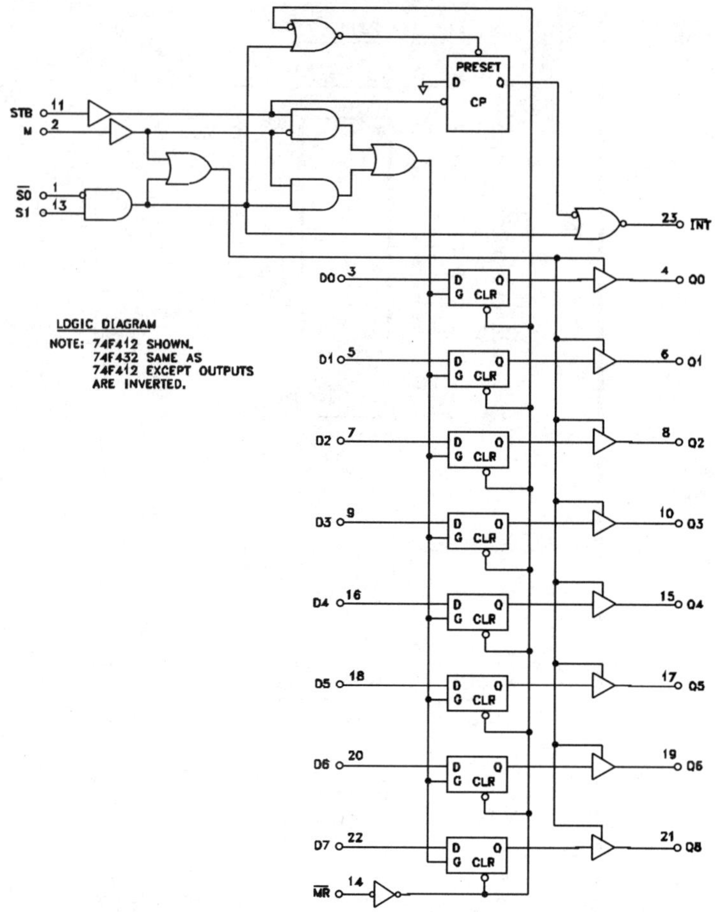

Figure 5.10. The FAST 74F412/432 are described as Multi-Mode Buffered Latches. The '412 provides true outputs; the '432 inverted. In other respects the devices are identical. The devices are 8-bit latches with 3-state output buffers. A status flip-flop (/INT) provides "device busy" or "request interrupt" commands. Separate Mode (M) and Select (/S0, S1) inputs allow data to be stored with either output status. These may also be operated in a fully transparent mode.

TTL – 74AS/ALS/F533
5V CMOS – 74HC/HCT/AC/ACT533
3V CMOS – 74AC533

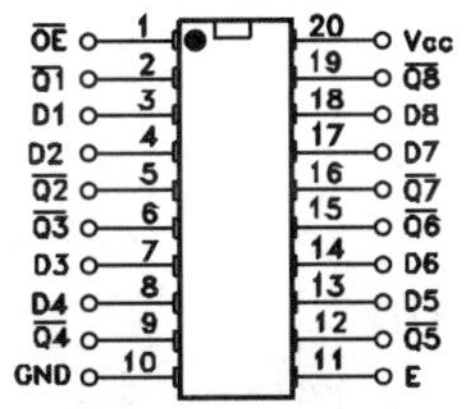

DUAL-IN-LINE PACKAGE

FUNCTION TABLE 74x533 OCTAL TRANSPARENT LATCH WITH 3-STATE INVERTING OUTPUT BUFFERS

OPERATING MODES	INPUTS				OUTPUT Q1–Q8
	$\overline{OE}$	E	Dn	INTERNAL REGISTER	$\overline{Q}$
Enable and read register	0	1	0	0	1
	0	1	1	1	0
Latch and read register	0	↓	l	0	1
	0	↓	h	1	0
Hold	0	0	X	NC	NC
Disable Outputs	1	X	X	X	Z

↓ = HIGH-to-LOW clock transition.
h = HIGH voltage level one setup time priot to the ↓ clock transition.
l = LOW voltage level one setup time priot to the ↓ clock transition.

LOGIC DIAGRAM

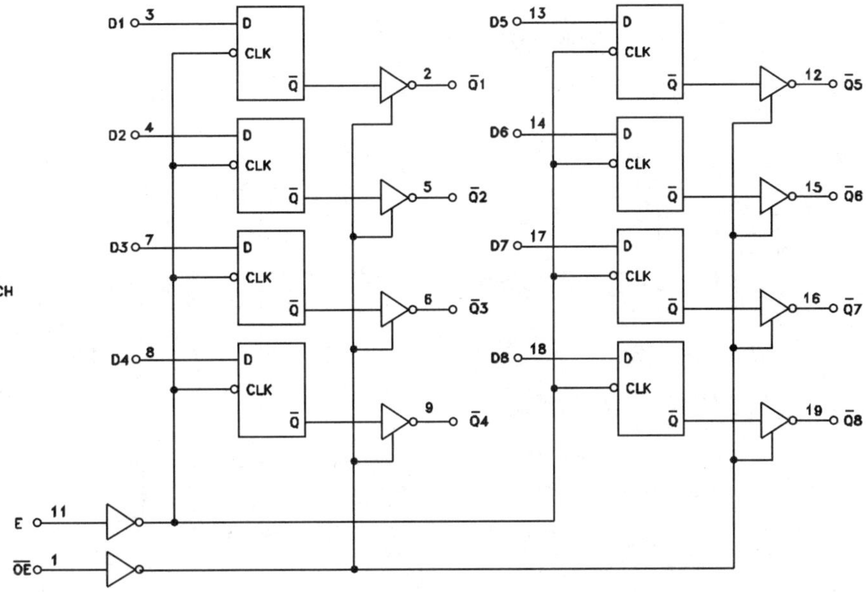

Figure 5.11. The 74x533 family of TTL and CMOS devices are 8-bit transparent latches featuring outputs designed for driving highly capacitive or relatively low-impedance loads. With eight latches, full parallel loading access, buffered control inputs, and 3-state driving outputs they are well suited for implementing buffer registers, I/O ports, bi-directional bus drivers, and working registers. While the Enable is HIGH the /Q outputs follow the complements of the data inputs. When the Enable is taken LOW the outputs are latched at the inverse of the input. The device is functionally equivalent to the 74x373 other than for the inverted outputs.

TTL – 74ALS/F563
74AS/ALS/F573
74AS/ALS580

5V CMOS – 74HC/HCT/AC(5V)/ACT563
74HC/HCT/AHC(5V)/AHCT/VHC/VHCT/
AC(5V)/AHCT573

3V CMOS – 74AHC(3V)/VHC(3V)/AC(3V)/LV/LVC/
LVQ/LVCH573

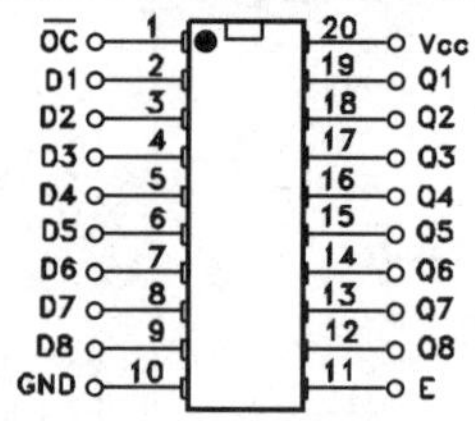

DUAL-IN-LINE PACKAGE

FUNCTION TABLE 74x563, 74x573, 74x580
OCTAL TRANSPARENT LATCH, 3-STATE OUTPUTS

74563 – TRUE IN, COMPLEMENTRY OUT
74573 – TRUE IN, TRUE OUT
74580 – TRUE IN, COMPLEMENTRY OUT
TRUE IN TRUE OUT SHOWN

OPERATING MODES	INPUTS				OUTPUT $\overline{O1}$–$\overline{O8}$
	$\overline{OE}$	E	Dn	INTERNAL REGISTER	$\overline{Q}$
Enable and read register	0 0	1 1	0 1	0 1	1 0
Latch and read register	0 0	↓ ↓	l h	0 1	1 0
Hold	0	0	X	NC	NC
Disable Outputs	1	X	X	X	Z

↓ = HIGH-to-LOW clock transition.
h = HIGH voltage level one setup time prior to the ↓ clock transition.
l = LOW voltage level one setup time prior to the ↓ clock transition.

LOGIC DIAGRAM

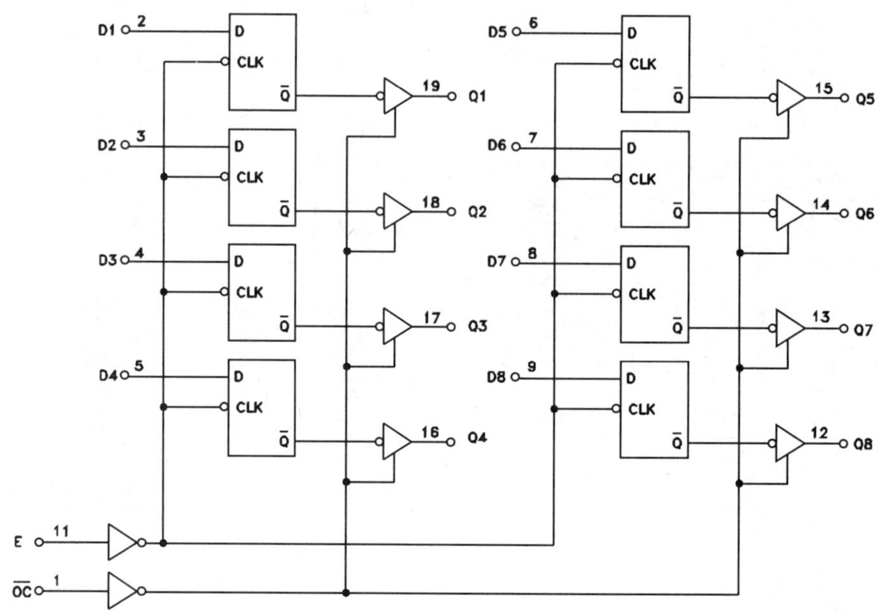

Figure 5.12. The 74x563/573/580 family of TTL and CMOS devices are 8-bit D-type transparent latches featuring outputs designed for driving highly capacitive or relatively low-impedance loads. Like the 74x533 with eight latches, full parallel loading access, buffered control inputs, and 3-state driving outputs they are well suited for implementing buffer registers, I/O ports, bi-directional bus drivers, and working registers. Additionally, the package pinout is bus structured. While the Enable is HIGH the /Q outputs follow the complements of the data inputs. When the Enable is taken LOW the outputs are latched to retain the existing data.

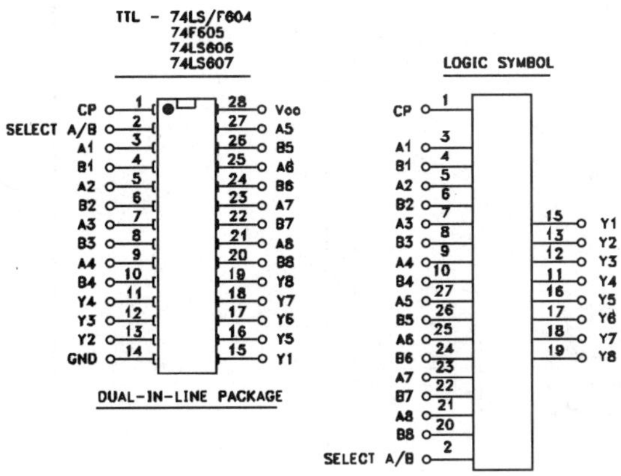

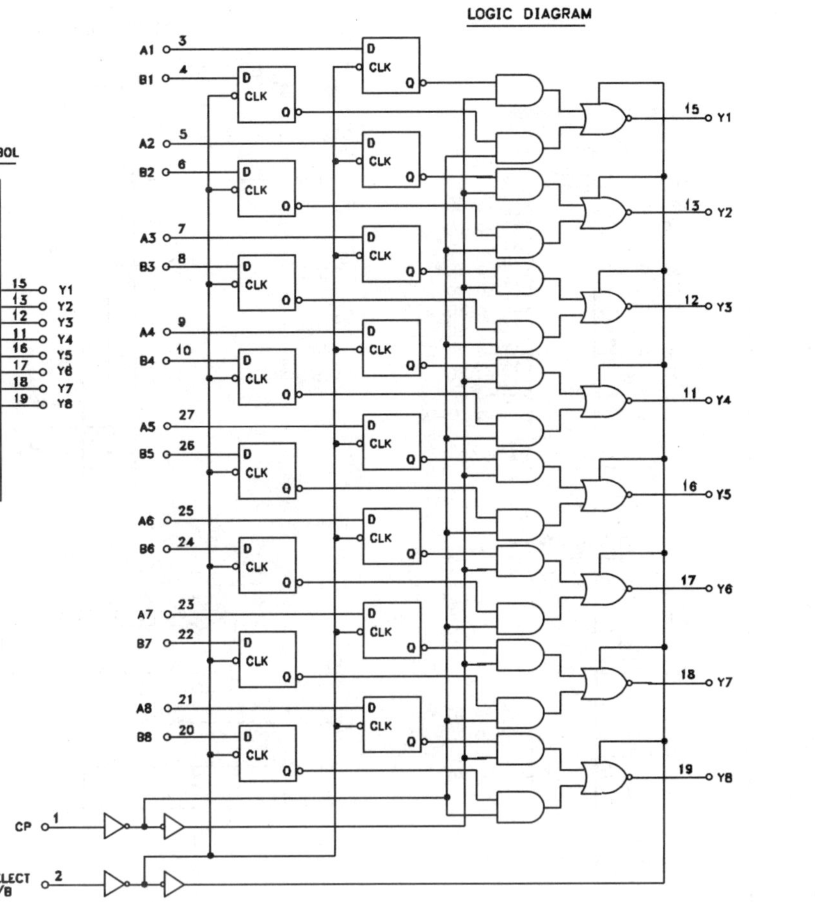

FUNCTION TABLE 74x604, '605, '606, '607
DUAL OCTAL REGISTER, 3-STATE OUTPUTS

INPUTS			OUTPUTS
A1–A8 B1–B8	SEL A/B	CP	Y1–Y8
A data B data	0	↑	B data
A data B data	1	↑	A data
X X	X	0	Z
X X	0	1	B Reg stored data
X X	1	1	A Reg stored data

↑ = LOW-to-HIGH clock transition.

Figure 5.13. The 74x604/605/606/607 are multiplexed latches useful for operation with two data sources, providing an output from either of two internal storage registers. With these devices the clock has a dual role: the normal one of loading data on the positive going transition, but also controlling the active and high-impedance states of the output. When the clock is low the outputs are in the high impedance OFF state, when HIGH the outputs are enabled. The '604 is optimized for speed, the '605 is open collector, the '606/'607 are designed to eliminate decoding spikes. The devices are suited for simultaneous 16-bit inputs transmitted as two sequential eight-bit output words.

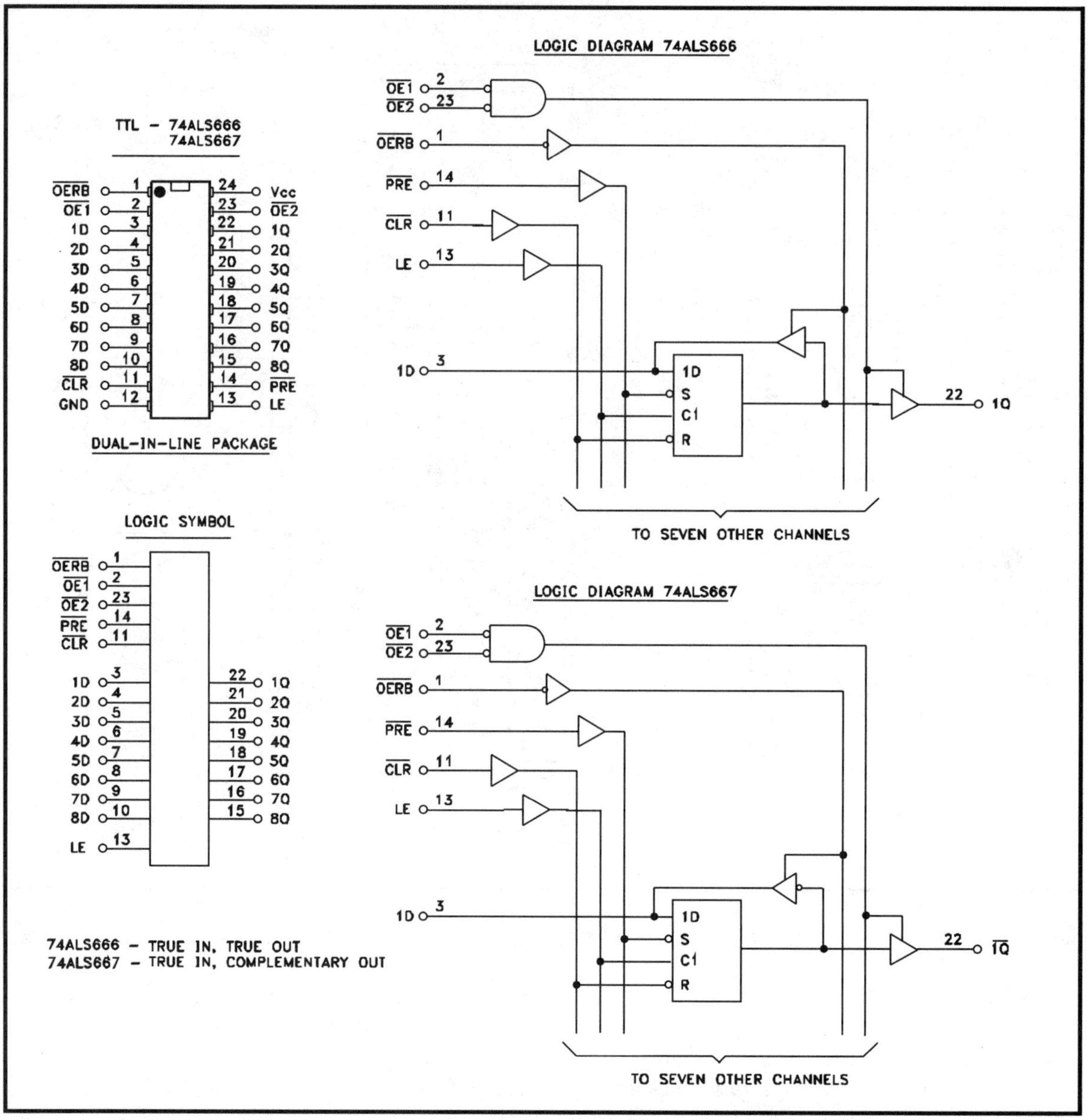

Figure 5.14. The 74x666/667 are 8-bit transparent "read-back" latches designed for the purpose of storing the content of the input data bus plus reading the content back onto the data bus when so directed. Primary device features are this I/O type of read-back, a bus-structured package pinout, preset and clear inputs, and choice of true or complement outputs. With the Latch Enable HIGH the '666 outputs follow the data; the '667 the complement. The output is in the high impedance state if either of the two output enables is at a HIGH logic level. Read-back is controlled by the /OERB input. When this input is taken LOW the data present at the output is relayed back onto the input data bus.

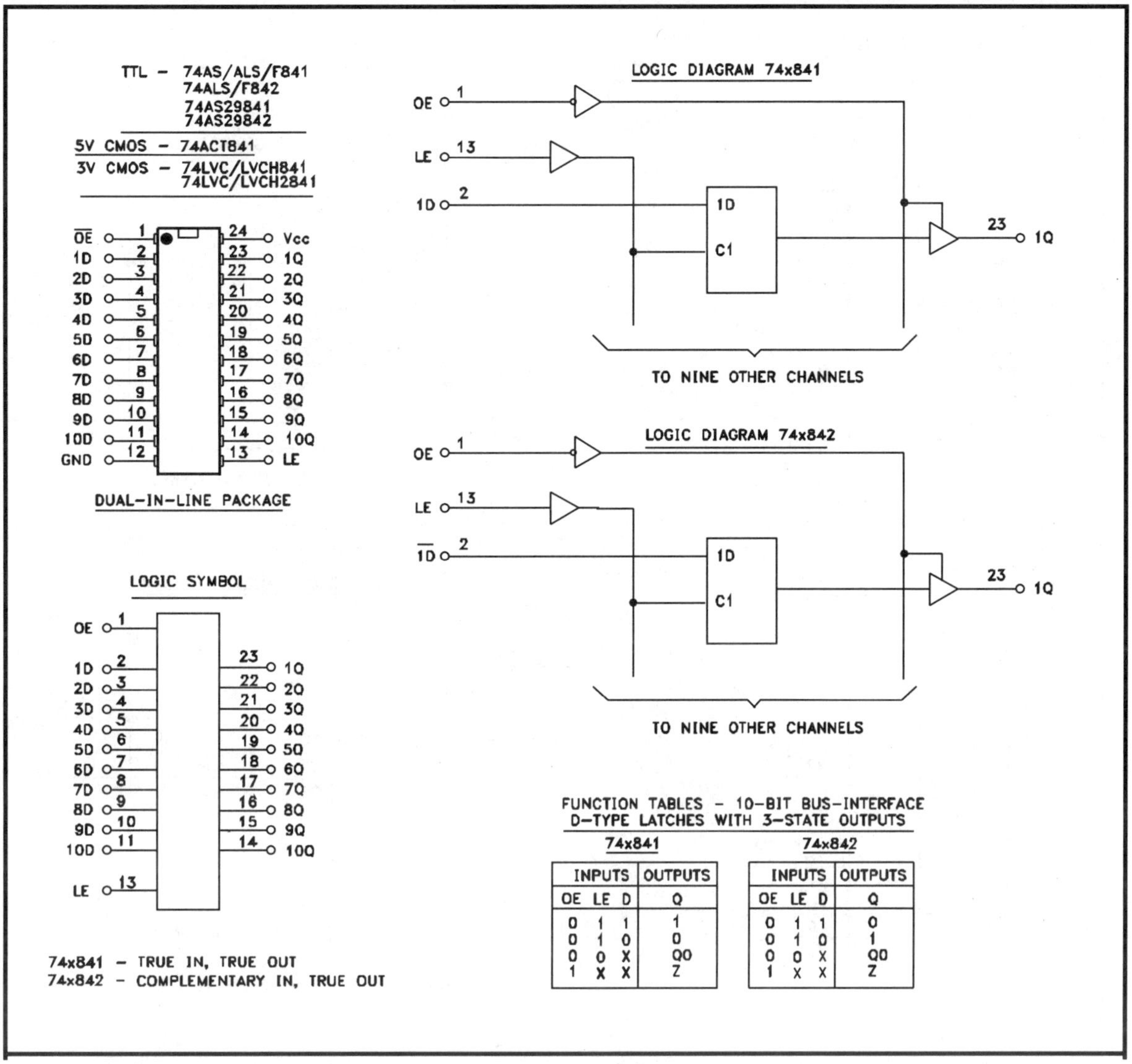

74x841

INPUTS			OUTPUTS
OE	LE	D	Q
0	1	1	1
0	1	0	0
0	0	X	Q0
1	X	X	Z

74x842

INPUTS			OUTPUTS
OE	LE	D	Q
0	1	1	0
0	1	0	1
0	0	X	Q0
1	X	X	Z

Figure 5.15. The 74x841/842 TTL and CMOS devices are 10-bit wide bus interface D-type latches featuring outputs designed for driving highly capacitive or relatively low-impedance loads. With ten latches, full parallel loading access, buffered control inputs, and 3-state driving outputs they are well suited for implementing buffer registers, I/O ports, bi-directional bus drivers, and working registers. Increased drive capability allows bus line driving without the need for interface or pull-up components. The added width allows for wider address/data paths or buses with parity. Powering up is to the high impedance state. The '841 provides true outputs with true inputs, the '842 the inverse; otherwise these are identical.

TTL – 74AS/ALS/F843
74AS/ALS/F844
74AS29843
74AS29844

3V CMOS – 74LVC843

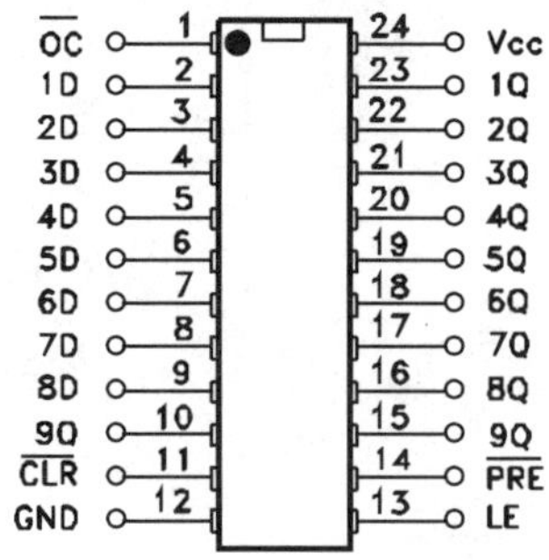

DUAL-IN-LINE PACKAGE

LOGIC SYMBOL

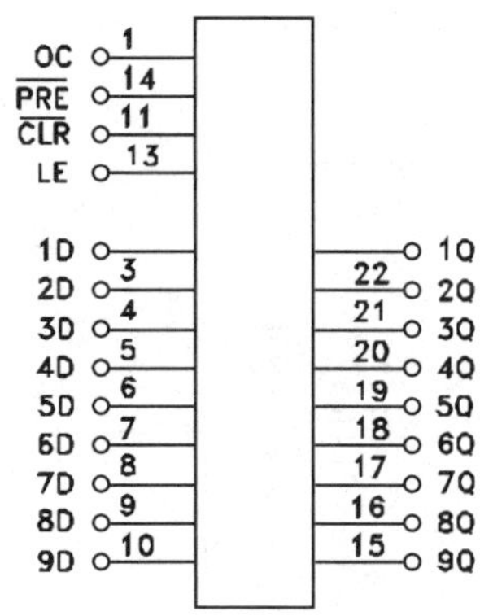

74x843 – TRUE IN, TRUE OUT
74x844 – COMPLEMENT IN, TRUE OUT
TRUE IN, TRUE OUT SHOWN

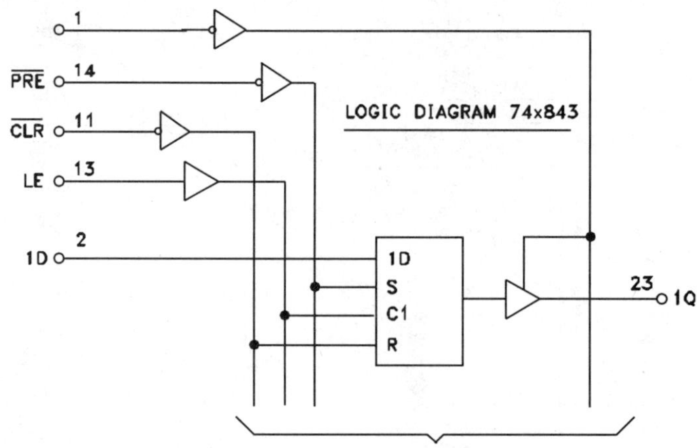

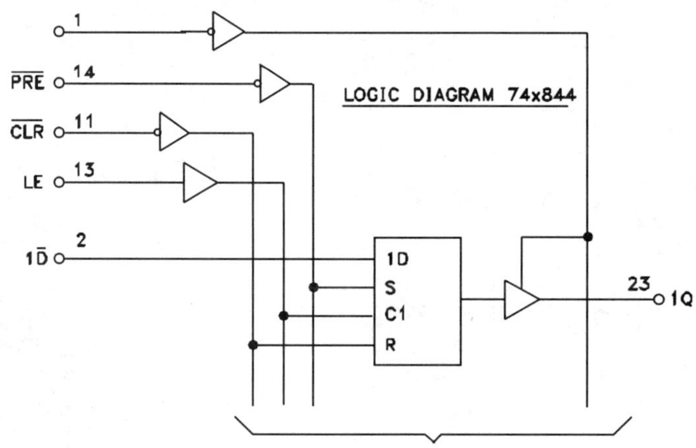

FUNCTION TABLE 9-BIT INTERFACE D-TYPE LATCH WITH 3-STATE OUTPUTS

74x843

INPUTS					OUTPUT
PRE	CLR	OE	LE	D	Q
0	X	0	X	X	1
1	0	0	X	X	0
1	1	0	1	0	0
1	1	0	1	1	1
1	1	0	0	X	Q0
X	X	1	X	X	Z

74x844

INPUTS					OUTPUT
PRE	CLR	OE	LE	D	Q
0	X	0	X	X	0
1	0	0	X	X	1
1	1	0	1	0	1
1	1	0	1	1	0
1	1	0	0	X	Q0
X	X	1	X	X	Z

Figure 5.16. The 74x843/844 TTL and CMOS devices are 9-bit wide bus interface D-type latches featuring outputs designed for driving highly capacitive or relatively low-impedance loads. With nine latches, full parallel loading access, buffered control inputs, and 3-state driving outputs they are well suited for implementing buffer registers, I/O ports, bi-directional bus drivers, and working registers. Increased drive capability allows bus line driving without the need for interface or pull-up components. The added width allows for wider address/data paths or buses with parity. Powering up is to the high impedance state. The '843 provides true outputs with true inputs, the '844 the inverse; otherwise these are identical.

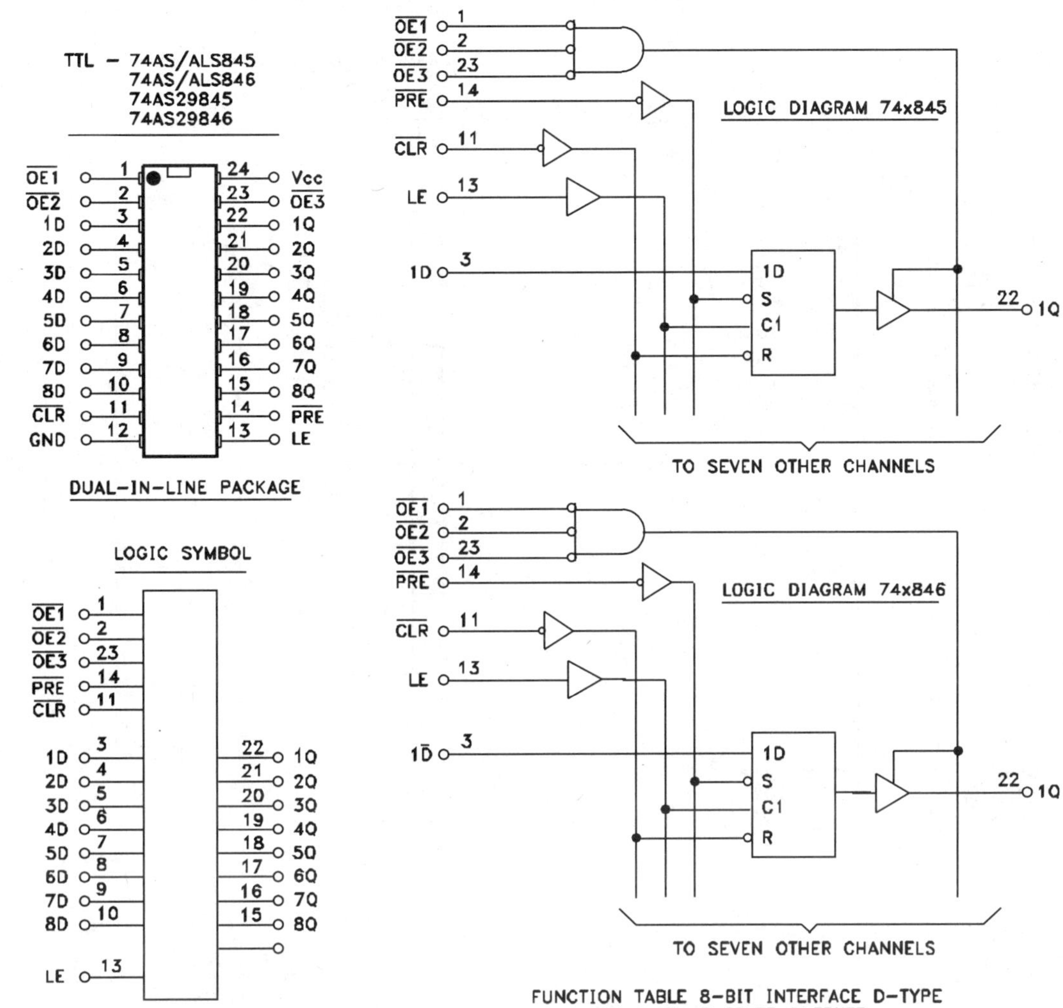

74x845 – TRUE IN, TRUE OUT
74x846 – COMPLEMENT IN, TRUE OUT
TRUE IN, TRUE OUT SHOWN

FUNCTION TABLE 8-BIT INTERFACE D-TYPE LATCH WITH 3-STATE OUTPUTS

74x845

INPUTS							OUTPUT
/PRE	/CLR	/OE1	/OE2	/OE3	LE	D	Q
0	1	0	0	0	X	X	1
1	0	0	0	0	X	X	0
0	0	0	0	0	X	X	1
1	1	0	0	0	1	0	0
1	1	0	0	0	1	1	1
1	1	0	0	0	0	X	Q0
X	X	X	X	X	X	X	Z
X	X	X	X	X	X	X	Z
X	X	1	1	1	X	X	Z

74x846

INPUTS							OUTPUT
/PRE	/CLR	/OE1	/OE2	/OE3	LE	D	Q
0	1	0	0	0	X	X	1
1	0	0	0	0	X	X	1
0	0	0	0	0	X	X	1
1	1	0	0	0	1	0	1
1	1	0	0	0	1	1	0
1	1	0	0	0	0	X	Q0
X	X	X	X	X	X	X	Z
X	X	X	X	X	X	X	Z
X	X	1	1	1	X	X	Z

Figure 5.17. The TTL 74x845/846 are 8-bit wide bus interface transparent D-type latches featuring outputs designed for driving highly capacitive or relatively low-impedance loads. With eight latches, full parallel loading access, buffered control inputs, and 3-state driving outputs they are well suited for implementing buffer registers, I/O ports, bi-directional bus drivers, and working registers. The /CLR and /PRE inputs are independent of the clock; taking /CLR LOW causes the 8 outputs to go LOW. Taking /PRE LOW causes the eight outputs to go HIGH. When both /CLR and /PRE are taken LOW the outputs follow the PRESET condition. Increased drive capability allows bus line driving without the need for interface or pull-up components. Powering up is to the high impedance state. The '845 provides true outputs with true inputs, the '846 the inverse; otherwise these are identical.

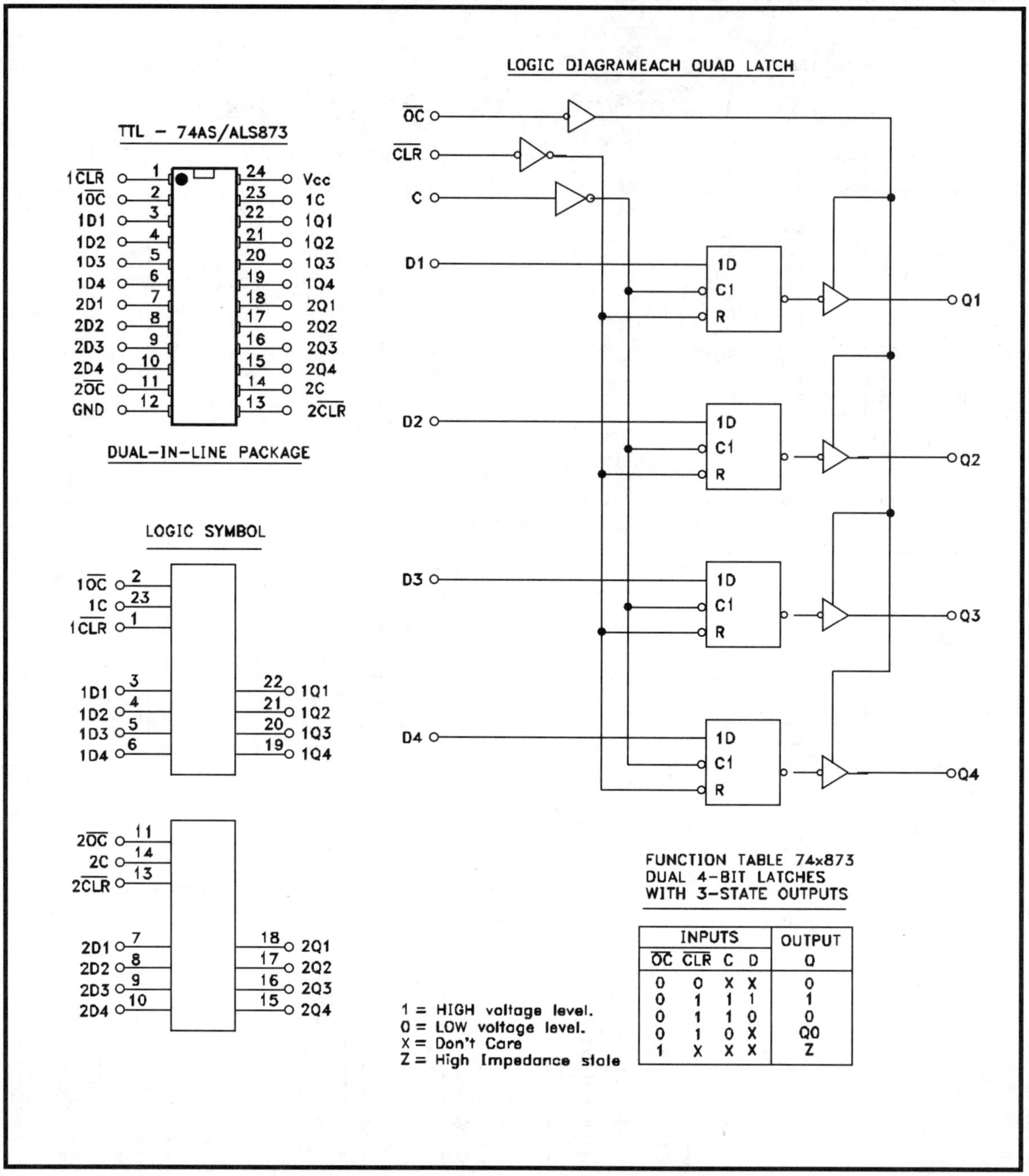

Figure 5.18. The TTL 74ALS/AS873 are dual 4-bit wide bus structured transparent D-type latches featuring outputs designed for bus driving. With dual latch pairs, buffered control inputs, and 3-state driving outputs they are well suited for implementing buffer registers, I/O ports, bi-directional bus drivers, and working registers. While the Latch Enable (LE) is HIGH the Q outputs follow the input data. The /CLR input is independent of the Latch Enable; taking /CLR LOW causes the 4 outputs to go LOW.

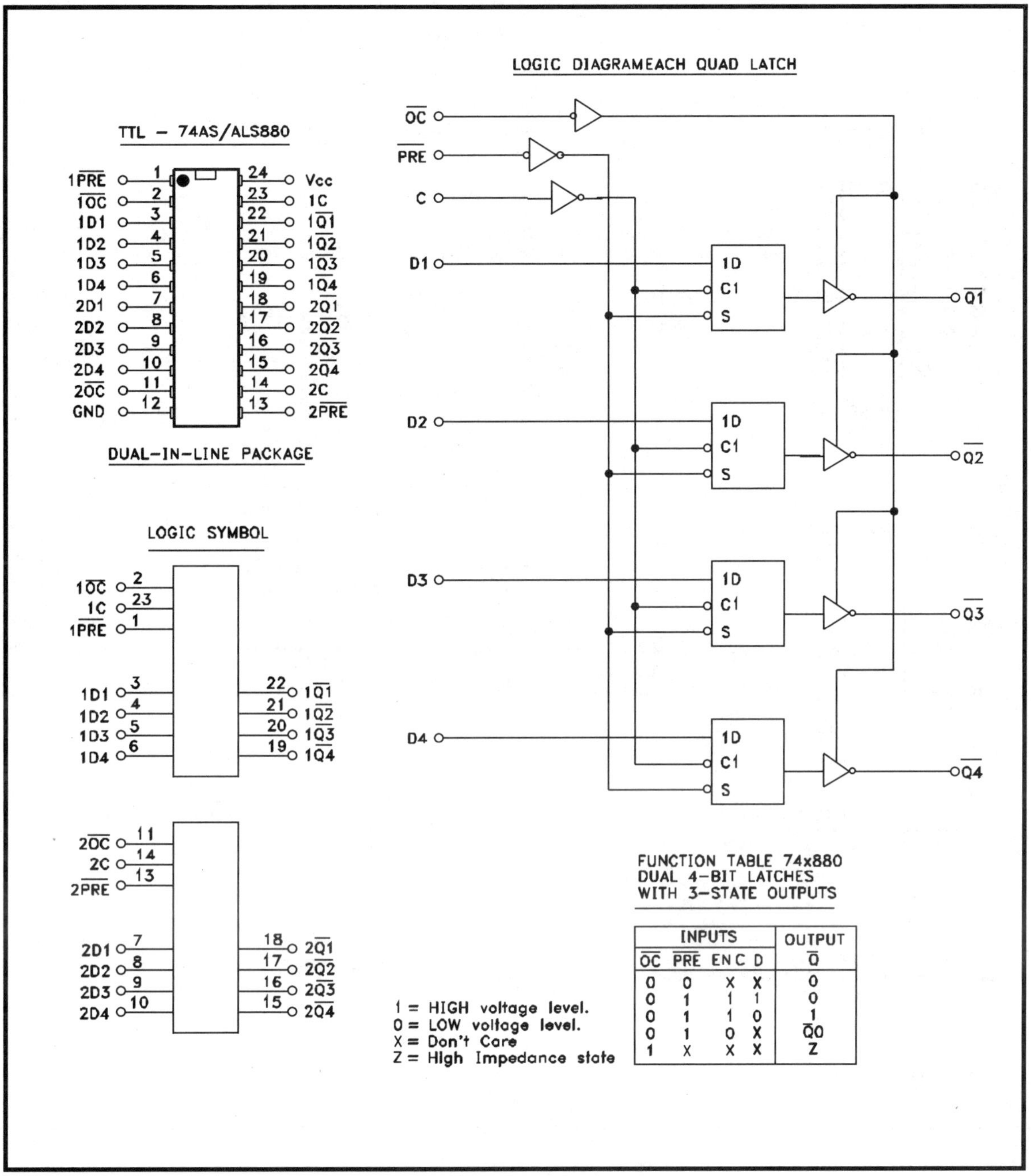

Figure 5.19. The TTL 74ALS/AS880 is an alternative version of the 74ALS/AS873 (Figure 5.18) with inverting outputs and a /PRE input replacing the /CLR. Like the '873 the '880 are dual 4-bit wide bus structured transparent D-type latches featuring outputs designed for bus driving. With dual latch pairs, buffered control inputs, and 3-state driving outputs they are well suited for implementing buffer registers, I/O ports, bi-directional bus drivers, and working registers. While the Latch Enable (1CLK or 2CLK)is HIGH the /Q outputs follow the input data in inverted form. The /PRE input is independent of the clock; taking /PRE LOW causes the 4 outputs to go LOW.

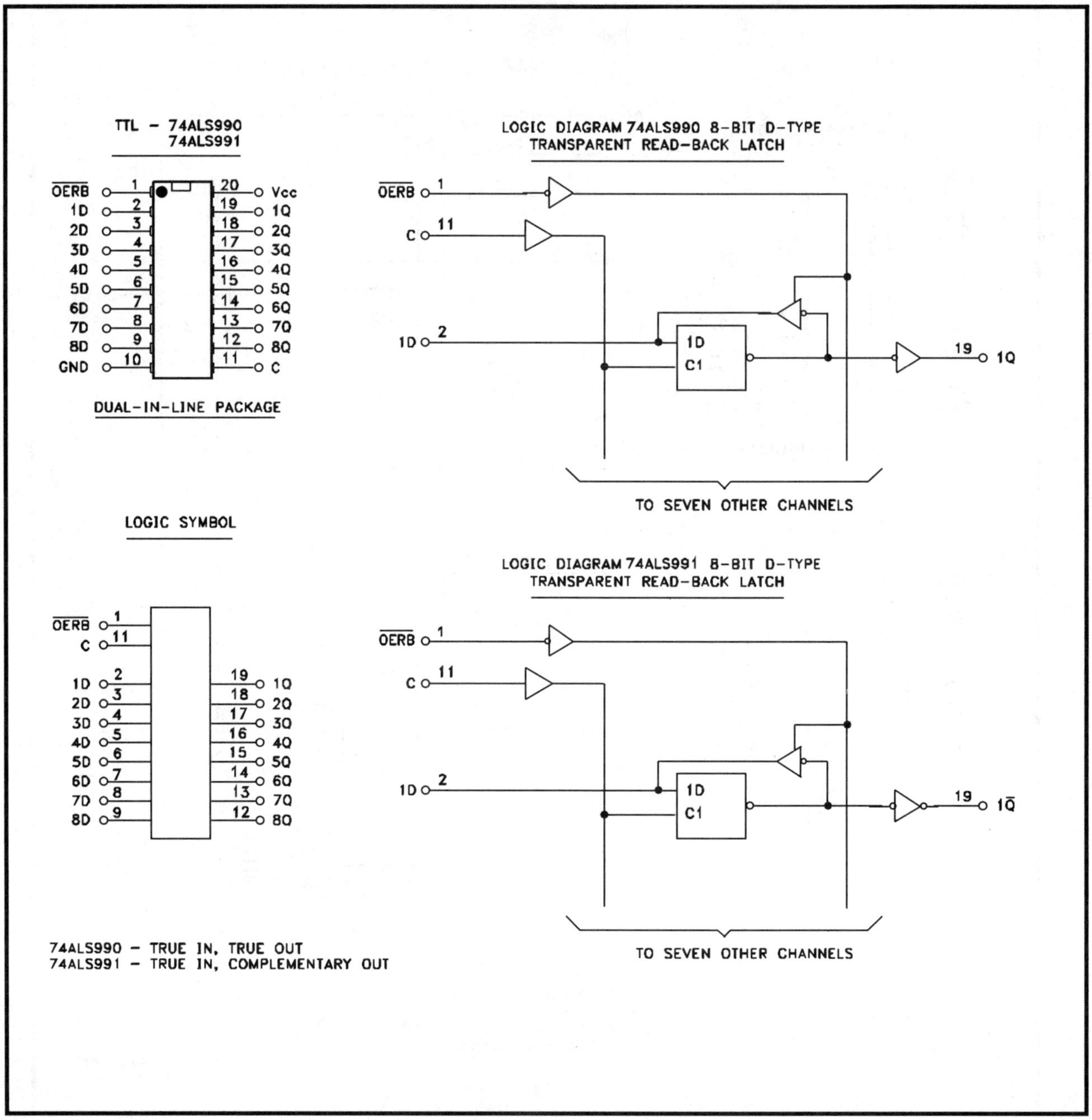

Figure 5.20. The TTL 74ALS990/991 are 8-Bit D-type Transparent Read-Back Latches featuring 3-State I/O type Read-Back inputs and bus structured pinout. These devices are designed for storing the content of the data bus with the capability of reading back the stored content to the input data bus. With the Enable HIGH the '990 outputs will follow the input data; the '991 provides the complement. Read-Back is effected by control input /OERB. When the control is taken LOW the data present at the output is allowed back onto the input data bus. Isolation occurs with the control HIGH. This control has no effect on the internal operation though precautions against a bus conflict must be observed.

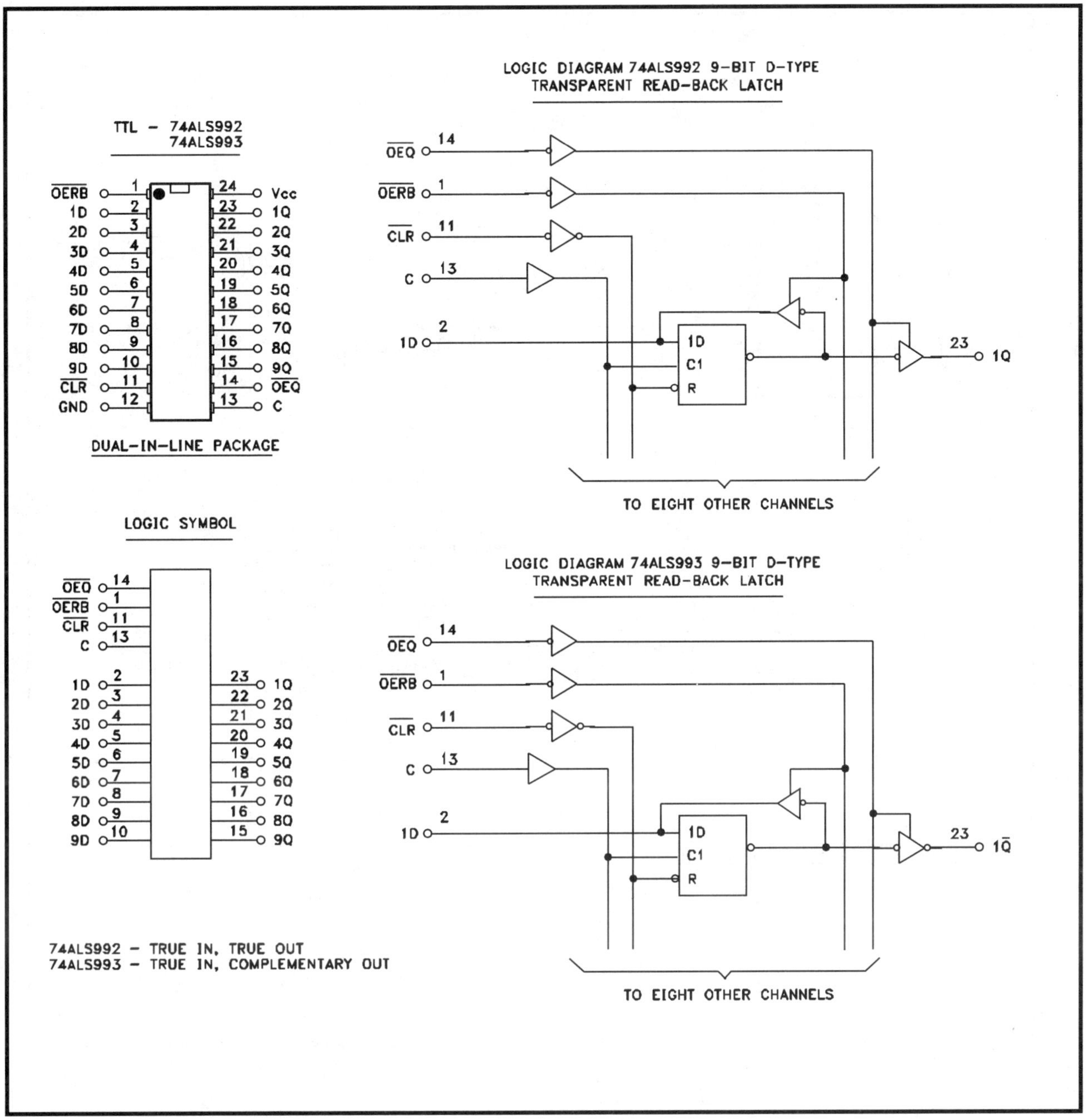

Figure 5.21. The TTL 74ALS992/993 are 9-Bit D-type Transparent Read-Back Latches featuring 3-State I/O type Read-Back inputs and bus structured pinout. This pinout enhances parity applications. These devices are designed for storing the content of the data bus with the capability of reading back the stored content to the input data bus. With the Enable HIGH the '992 outputs will follow the input data; the '993 provides the complement. Read-Back is effected by control input /OERB. When the control is taken LOW the data present at the output is allowed back onto the input data bus. Isolation occurs with the control HIGH. This control has no effect on the internal operation though precautions against a bus conflict must be observed.

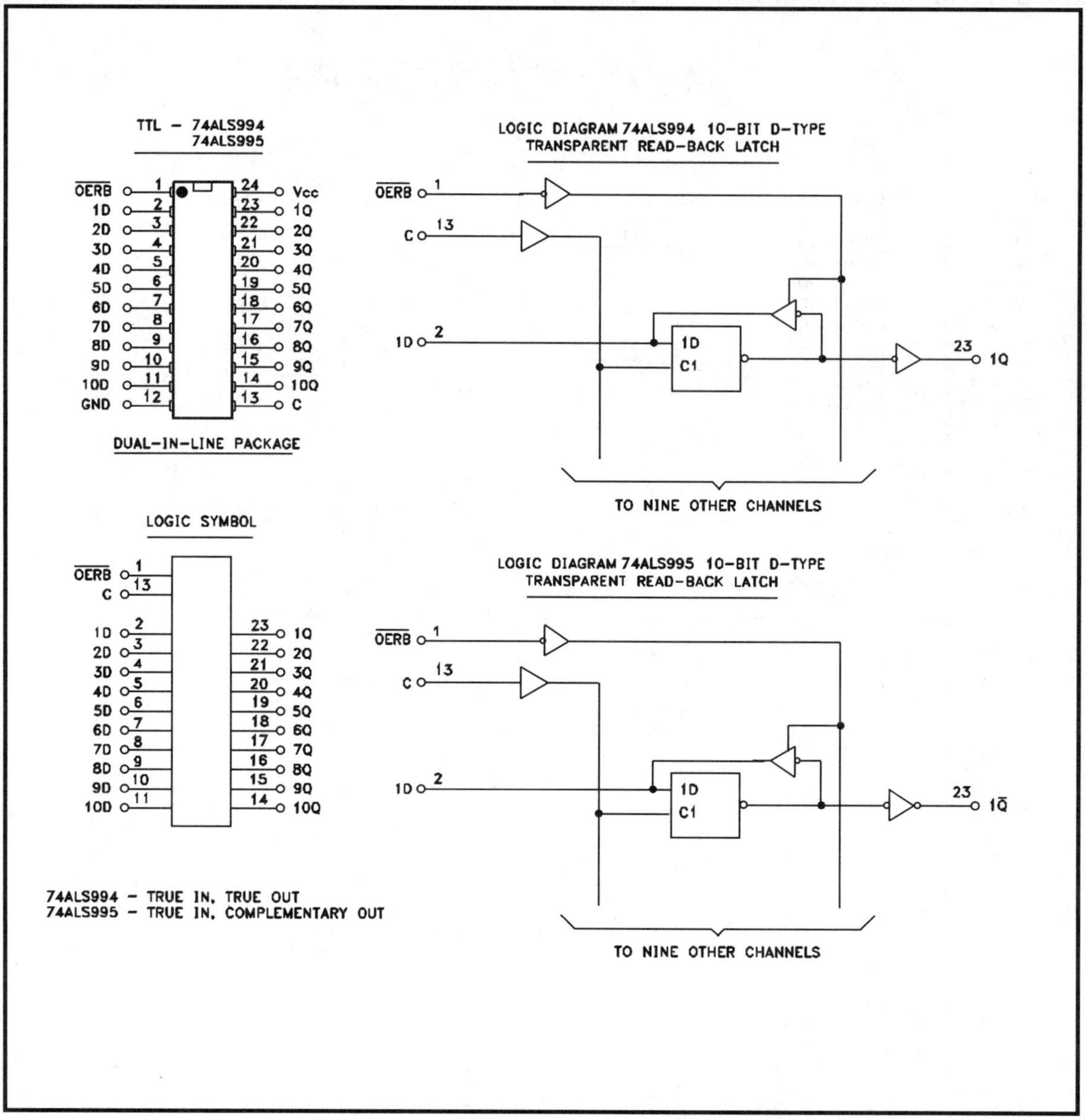

Figure 5.22. The TTL 74ALS994/995 are 10-Bit D-type Transparent Read-Back Latches featuring 3-State I/O type Read-Back inputs and bus structured pinout. This pinout enhances parity and wide-bus applications. These devices are designed for storing the content of the data bus with the capability of reading back the stored content to the input data bus. With the Enable HIGH the '994 outputs will follow the input data; the '995 provides the complement. Read-Back is effected by control input /OERB. When the control is taken LOW the data present at the output is allowed back onto the input data bus. Isolation occurs with the control HIGH. This control has no effect on the internal operation though precautions against a bus conflict must be observed.

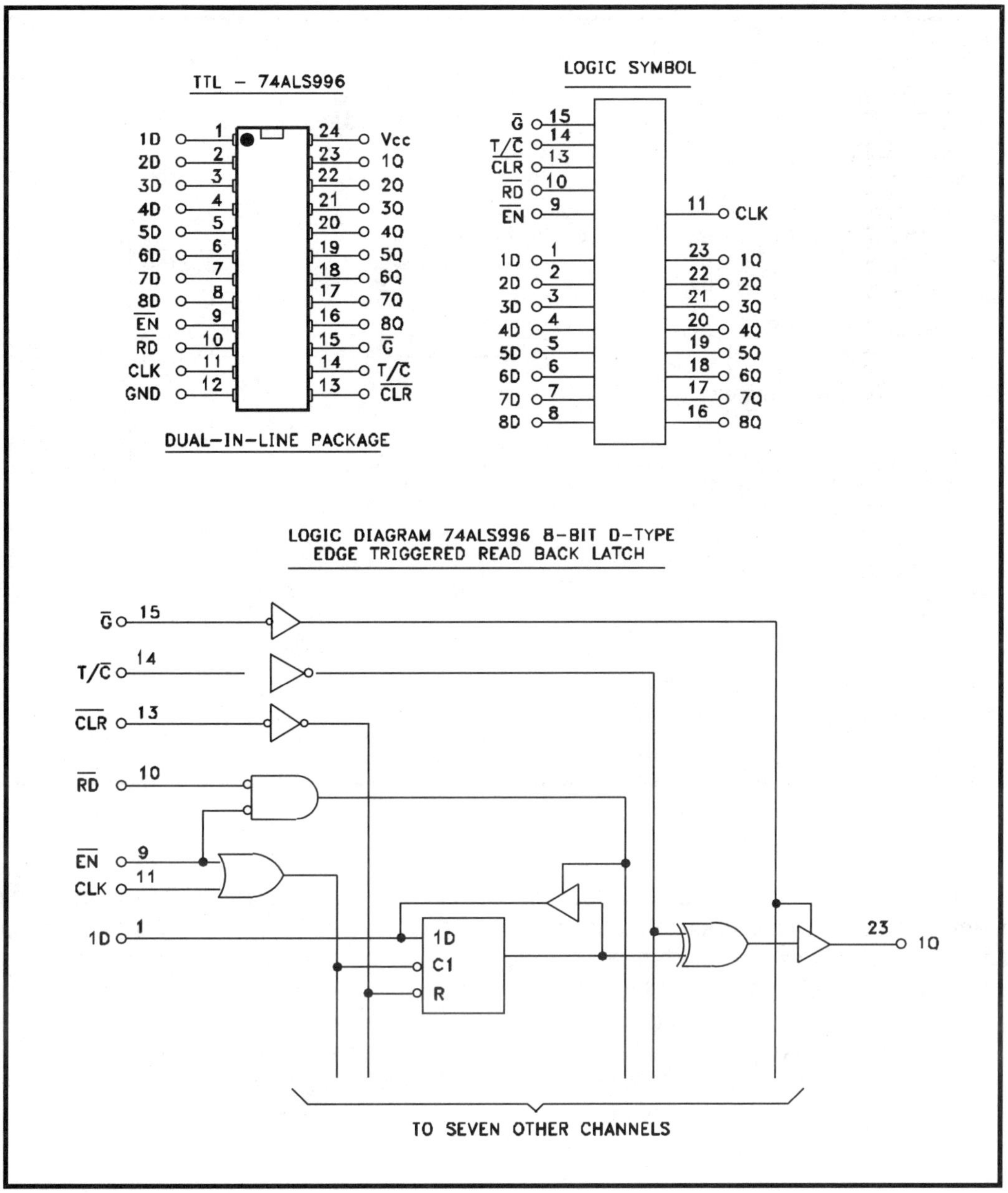

Figure 5.23. The TTL 74ALS996 is an 8-Bit D-type Edge-Triggered Read-Back Latch featuring 3-State I/O type Read-Back inputs and bus structured pinout. Control T//C determines True or Complement data at the Q outputs. These devices are designed for storing the content of the data bus with the capability of reading back the stored content to the input data bus. The edge-triggered flip-flops enter the data on the LOW-to-HIGH clock transition with the Enable LOW. With the Enable HIGH both the Read-Back and Write modes are disabled. Read-Back is effected by taking control input /RB LOW in addition to having the Enable LOW. Whenever EN is HIGH both the Read-Back and Write modes are disabled. /EN transitions should be made only when the Clock is high to avoid false clocking.

5V CMOS – CD4042

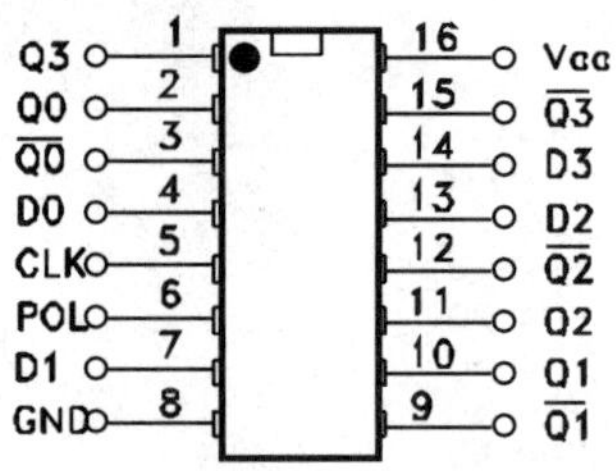

DUAL-IN-LINE PACKAGE

LOGIC DIAGRAM

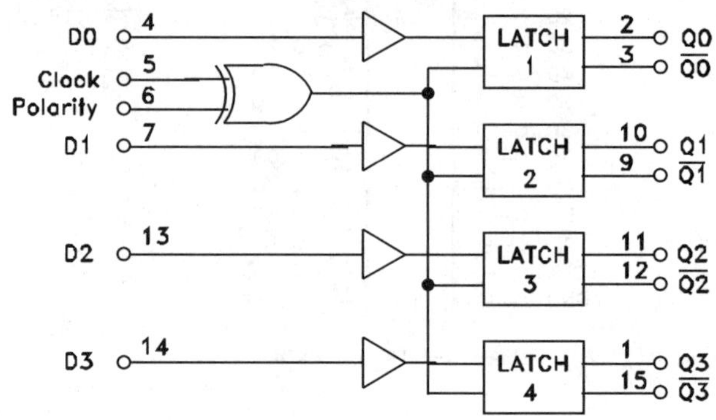

FUNCTION TABLE CD4042
QUAD TRANSPARENT LATCH

(Each latch)

CLOCK	POLARITY	Q
0	0	Data
1	0	Latch
1	1	Data
0	1	Latch

Figure 5.24. The CD4042 CMOS Quad Transparent Latch is constructed with MOS P-channel and N-channel enhancement mode devices. Each latch has a separate data input, but all four share a common clock. The clock polarity used to strobe data through the latch is reversible using the polarity input (observe the EXOR gate in the logic diagram). Information at the data inputs is transferred to the latch Q and /Q outputs during the clock level determined by the Polarity input. With the Polarity input at logic "0" data is transferred with the clock level LOW. Conversely, with the Polarity at the logic "1" level the transfer takes place with a HIGH clock level.

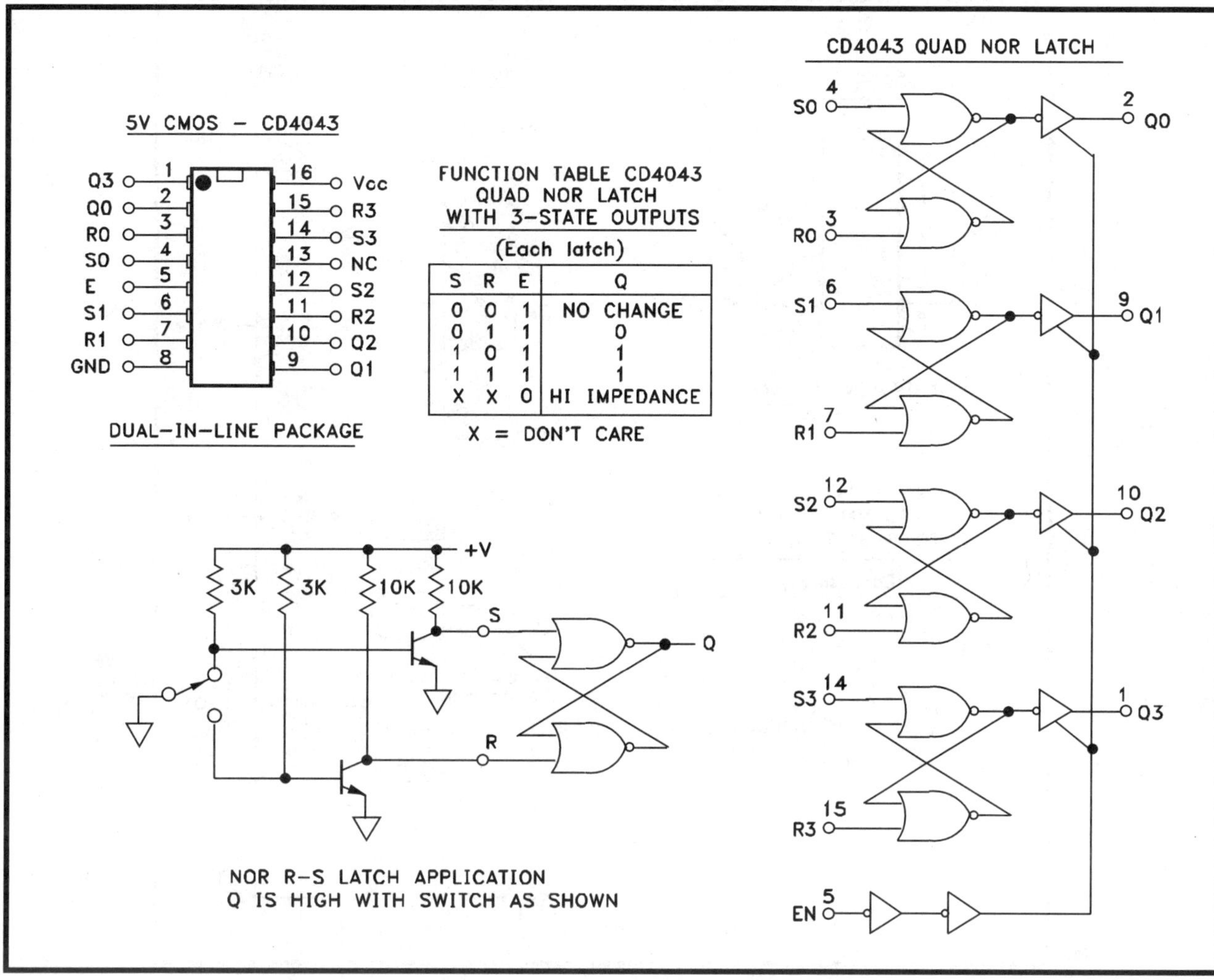

Figure 5.25. The CD4043 CMOS Quad R-S Latch is constructed with MOS P-channel and N-channel enhancement mode NOR logic circuitry and common 3-state outputs. Each latch has an independent Q output with Set and Reset inputs. With the enable LOW the outputs are at a high impedance. Note that the NOR inputs are active HIGH.

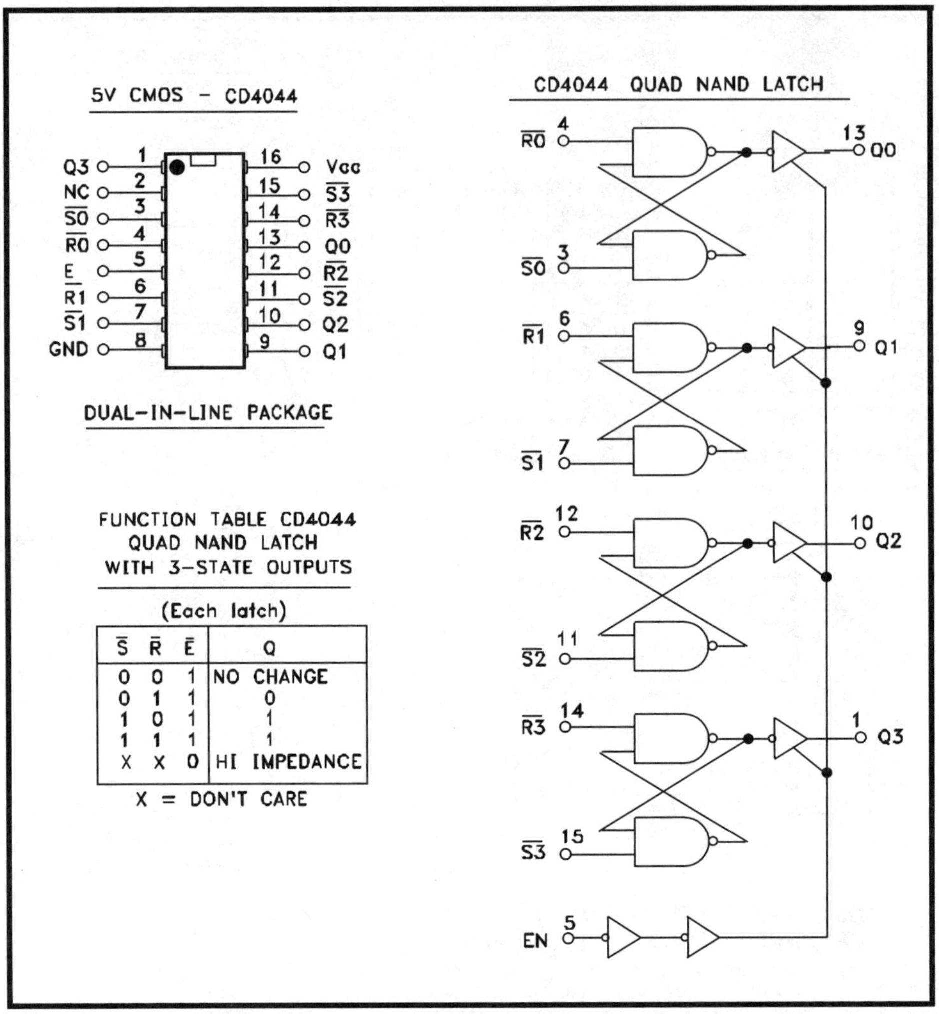

Figure 5.26. The CD4044 CMOS Quad R-S Latch is constructed with MOS P-channel and N-channel enhancement mode NAND logic circuitry and common 3-state outputs. Each latch has an independent Q output with Set and Reset inputs. With the enable LOW the outputs are at a high impedance. Note that the NAND inputs are active LOW.

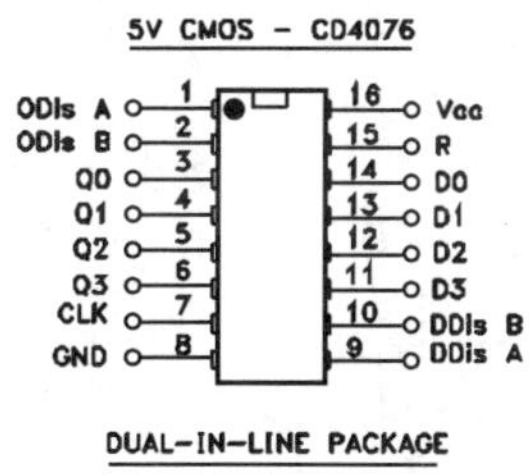

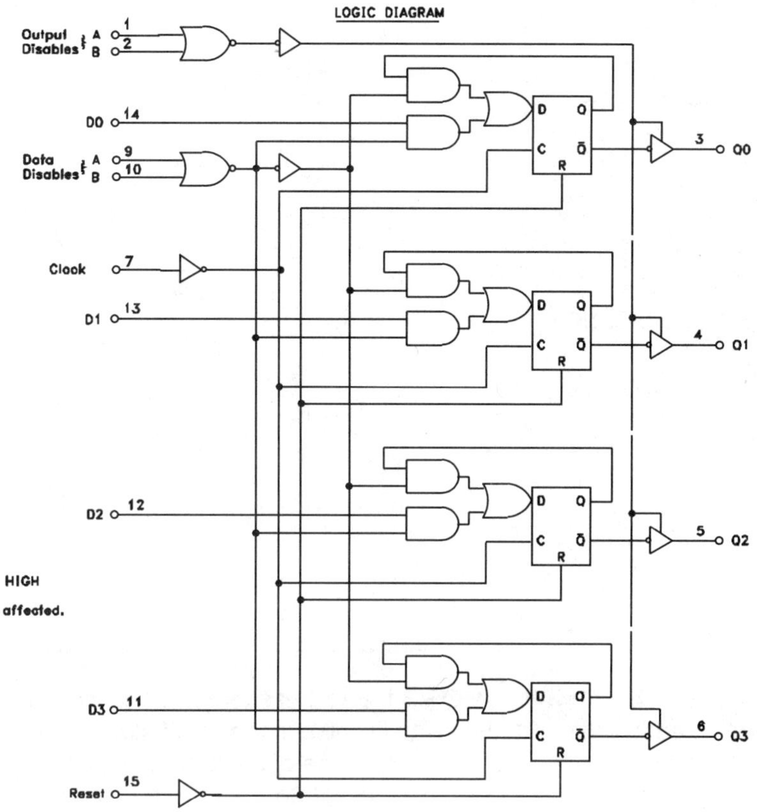

FUNCTION TABLE CD4076 4-BIT D-TYPE REGISTER WITH 3-STATE OUTPUTS

INPUTS					
		DATA DISABLE		DATA	OUTPUT
RESET	CLOCK	A	B	D	Q
1	X	X	X	X	0
0	0	X	X	X	Qn
0	⎍	1	X	X	Qn
0	⎍	X	1	X	Qn
0	⎍	0	0	0	0
0	⎍	0	0	1	1

When either output disable A or B (or both) is (are) HIGH the output is disabled to the high impedance state; however sequential operation of the flip-flops is not affected.

Figure 5.27. The CD4076 CMOS 4-Bit D-Type Register consists of four D-type flip-flops operating synchronously from a common clock. NOR gated output disable logic designed for bus organized systems force the output into a high impedance state. Similar input disable control logic causes the Q outputs to be fed back to the D inputs of the flip-flops. This feature inhibits a change of state with a stable clock input. An asynchronous Master Reset clears all four flip-flops simultaneously independent of the clock or disable inputs.

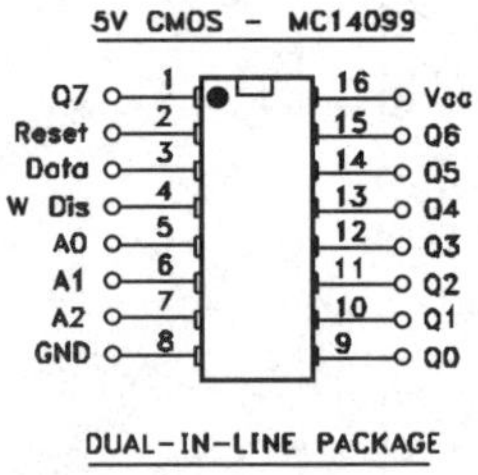

FUNCTION TABLE MC14099
8-BIT ADDRESSABLE LATCH

Write Disable	Reset	Addressed Latch	Unaddressed Latches
0	0	Data	On*
0	1	Data	Reset
1	0	On*	On*
1	1	Reset	Reset

* is previous state of latch
Reset is to zero state

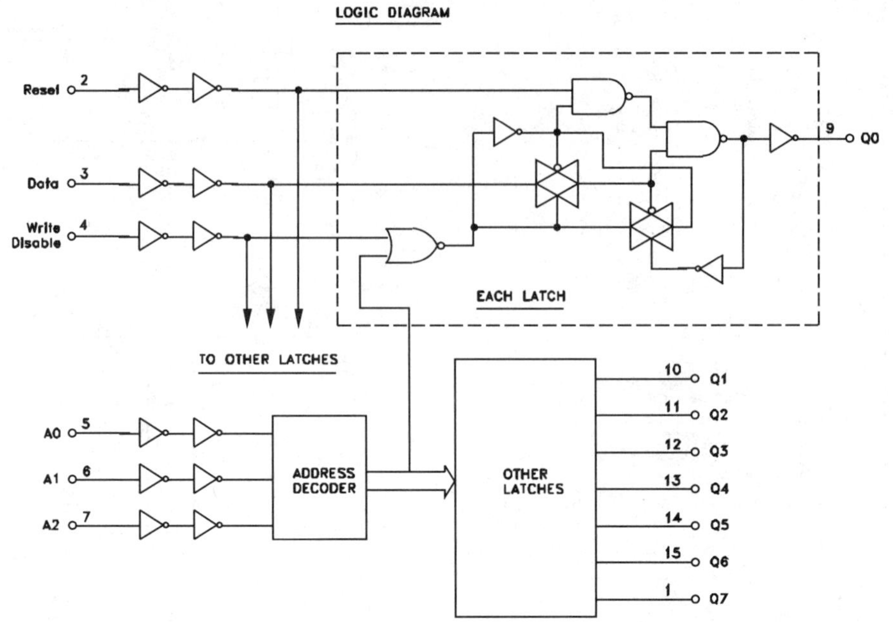

Figure 5.28. The CD4099 CMOS 8-Bit Addressable Latch provides for serial data entry when the appropriate latch is addressed with the Write Disable in the LOW state. The data input is unidirectional. The data is presented in parallel format at the outputs of the eight latches independently of the Write Disable logic state. The Master Reset is active HIGH.

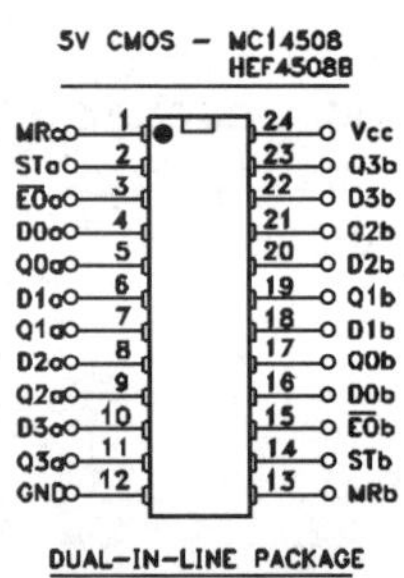

FUNCTION TABLE 4508
DUAL 4-BIT LATCH
(Each Latch)

INPUTS				OUTPUT
MR	ST	EO	Dn	On
0	1	0	1	1
0	1	0	0	0
0	0	0	X	latched
1	X	0	X	0
X	X	1	X	Z

1 = HIGH voltage level.
0 = LOW voltage level.
X = Don't Care
Z = High Impedance state

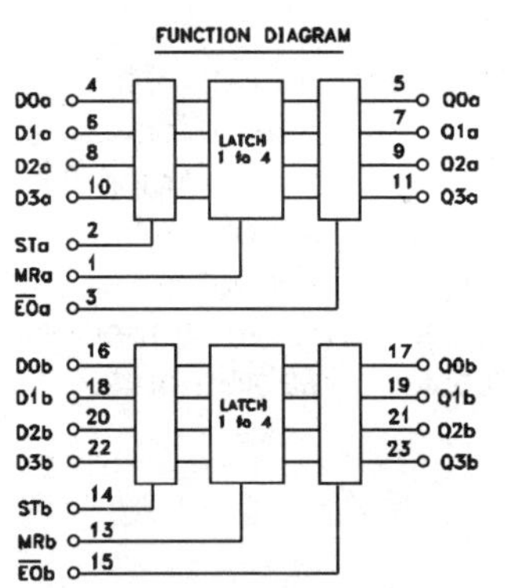

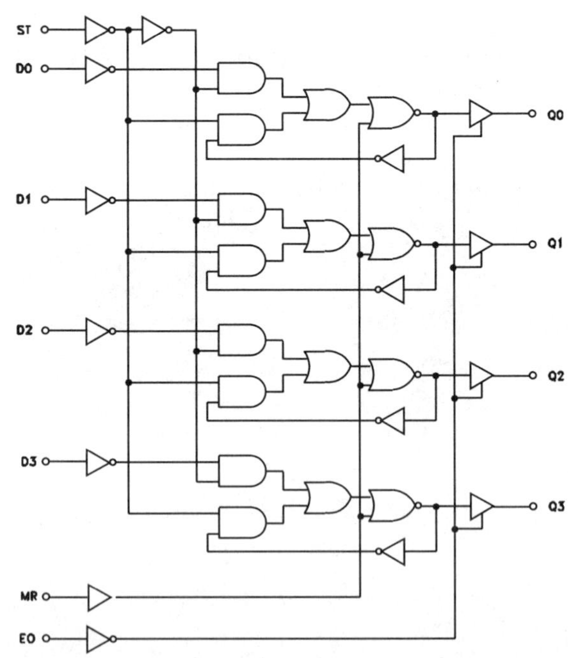

Figure 5.29. The CD4508 CMOS Dual 4-Bit Latch consists of two identical 4-bit latches with separate Strobe and Master Reset controls. Separate Disable inputs force the outputs into a high impedance state for use with time sharing logic. Applications include buffer storage, holding register, or general logic use where high noise immunity and low power dissipation are an asset.

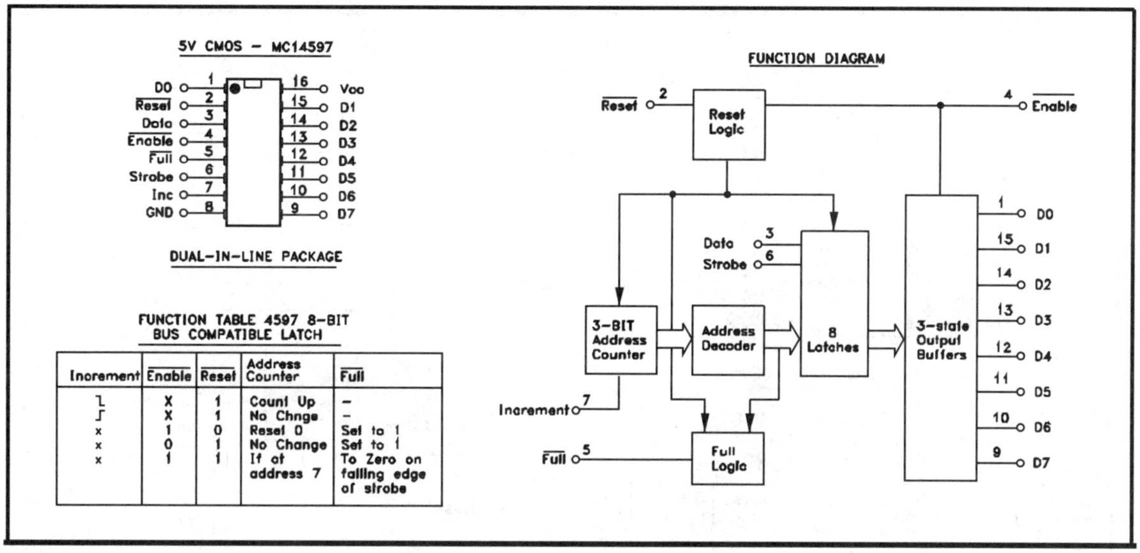

Increment	Enable	Reset	Address Counter	Full
⌊	X	1	Count Up	–
⌈	X	1	No Chnge	–
x	1	0	Reset 0	Set to 1
x	0	1	No Change	Set to 1
x	1	1	If at address 7	To Zero on falling edge of strobe

Figure 5.30. The CD4597 CMOS 8-Bit Bus-Compatible Latch features addressing from an internal counter employing an Increment input activated on the input's falling edge. A /FULL flag indicates the position of the embedded address counter. All eight latch outputs are available in parallel when the /Enable is in the LOW state. Data is entered into a specified latch from the Data pin when the Strobe is HIGH. The 3-state Enable control input is TTL compatible. Capable of driving one TTL or four LSTTL loads over the rated temperature range.

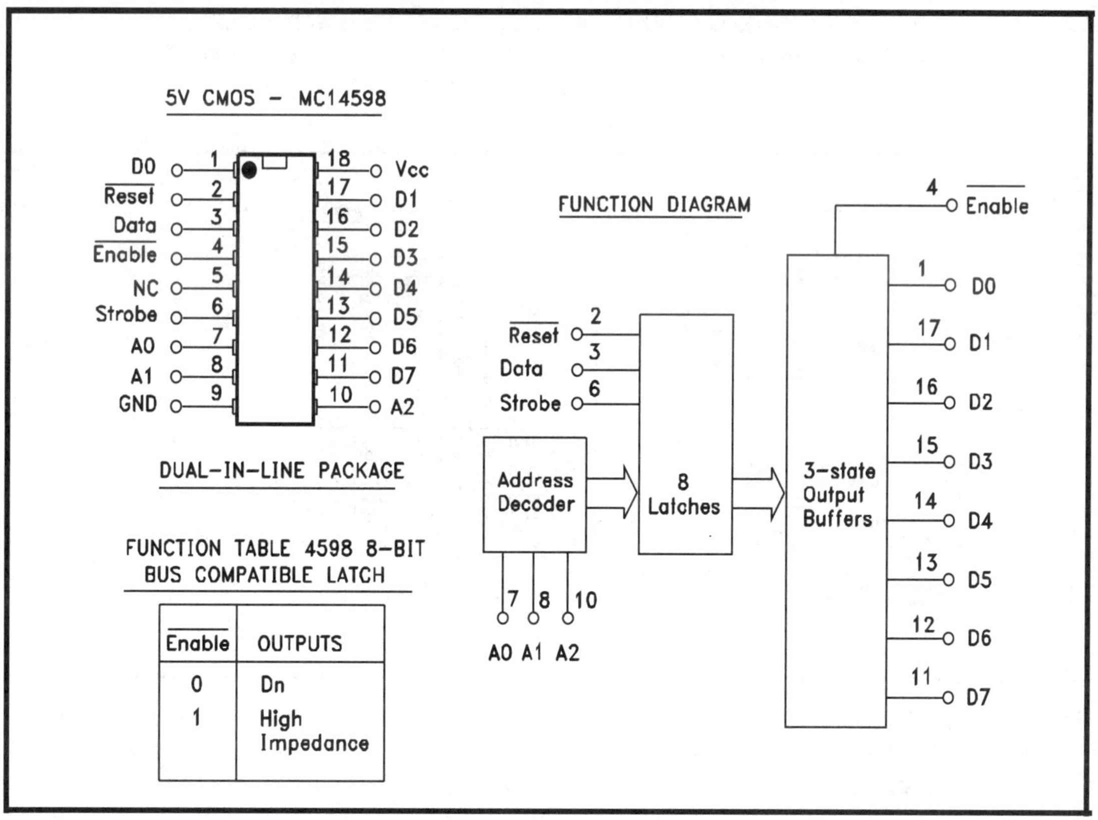

Enable	OUTPUTS
0	Dn
1	High Impedance

Figure 5.31. The CD4598 CMOS 8-Bit Bus-Compatible Latch, in contrast to the CD4597, features external addressing with three address inputs. As with the 4597, all eight latch outputs are available in parallel when the /Enable is in the LOW state. Data is entered into a specified latch from the Data pin when the Strobe is HIGH. The 3-state Enable control input is TTL compatible. Capable of driving one TTL or four LSTTL loads over the rated temperature range.

5V CMOS – MC14599

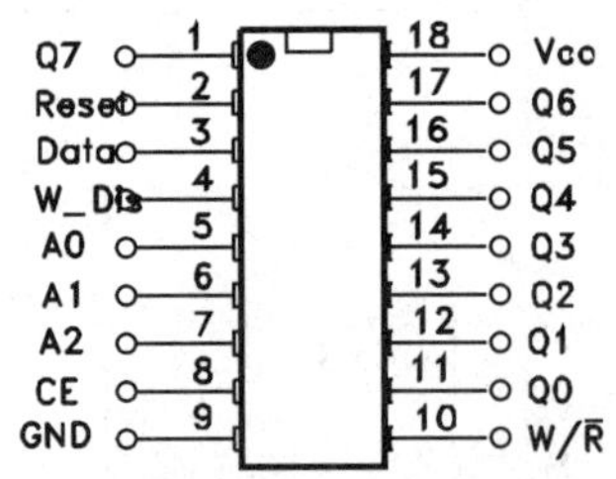

DUAL–IN–LINE PACKAGE

FUNCTION DIAGRAM

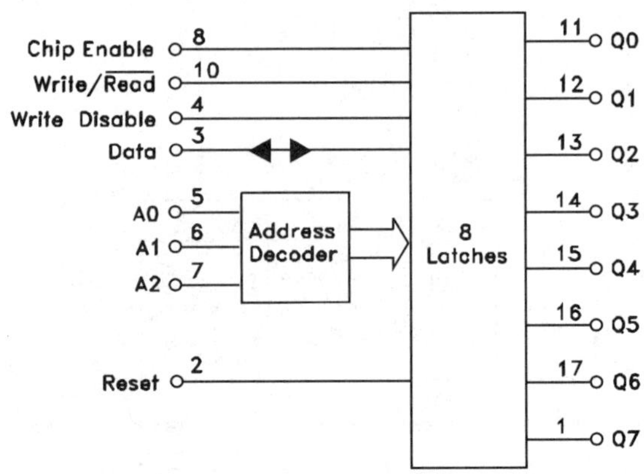

FUNCTION TABLE 4599 8–BIT ADDRESSABLE LATCH WITH BIDIRECTIONAL DATA PORT

Chip Enable	Write/Read	Write Disable	Reset	Addressed Latch	Other Latches	Data Pin
0	X	X	0	•	•	Z
1	1	0	0	Data	•	Input
1	1	1	0	•	•	Z
1	0	X	0	•	•	Qn
X	X	X	1	0	0	Z/0

• = No change in state of latch
Z = High impedance
Qn= State of addressed latch

Figure 5.32. The CD4599 CMOS 8-Bit Addressable Latch admits data in a serial form when the appropriate latch is addressed and the Write Disable input is in the LOW state with the Chip Enable at a HIGH logic state. The data pin is a bi-directional data port controlled by the Write//Read line. Output data is presented in parallel independently of the three control inputs, but not the Master Reset, which resets all latches to logic "0."

5V CMOS – CD4723

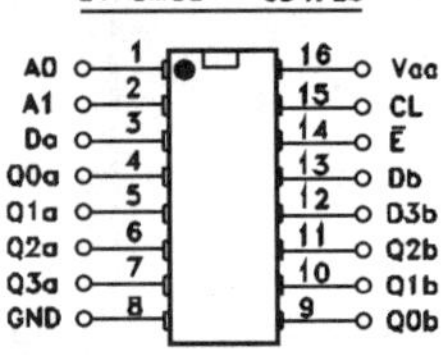

DUAL–IN–LINE PACKAGE

5V CMOS – CD4724
HEF4724

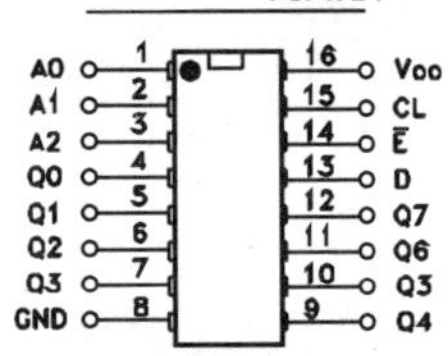

DUAL–IN–LINE PACKAGE

FUNCTION TABLE CD4723 DUAL 4–BIT ADDRESSABLE LATCH
CD4724 8–BIT ADDRESSABLE LATCH

Ē	CL	Addressed Latch	Unaddressed Latch	Mode
0	0	Follows Data	Holds Previous Data	Addressable Latch
1	0	Holds Previous Data	Holds Previous Data	Memory
0	1	Follows Data	Reset to "0"	Demultiplexer
1	1	Reset to "0"	Reset to "0"	Clear

FUNCTION TABLE CD4724 8–BIT ADDRESSABLE LATCH

CL	Ē	D	A0	A1	A2	Q0	Q1	Q2	Q3	Q4	Q5	Q6	Q7	Mode
1	1	X	X	X	X	0	0	0	0	0	0	0	0	Clear
1	0	D1	0	0	0	D1	0	0	0	0	0	0	0	demultiplexer; unaddressed latch is cleared
1	0	D1	1	0	0	0	D1	0	0	0	0	0	0	
1	0	D1	0	1	0	0	0	D1	0	0	0	0	0	
1	0	D1	1	1	0	0	0	0	D1	0	0	0	0	
1	0	D1	0	0	1	0	0	0	0	D1	0	0	0	
1	0	D1	1	0	1	0	0	0	0	0	D1	0	0	
1	0	D1	0	1	1	0	0	0	0	0	0	D1	0	
1	0	D1	1	1	1	0	0	0	0	0	0	0	D1	
0	1	X	X	X	X	Qn–1	Qn–1	Qn–1	Qn–1	Qn–1	Qn–1	Qn–1	Qn–1	memory
0	0	D1	0	0	0	D1	Qn–1	Qn–1	Qn–1	Qn–1	Qn–1	Qn–1	Qn–1	addressable latch; unaddressed latch holds previous state
0	0	D1	1	0	0	Qn–1	D1	Qn–1	Qn–1	Qn–1	Qn–1	Qn–1	Qn–1	
0	0	D1	0	1	0	Qn–1	Qn–1	D1	Qn–1	Qn–1	Qn–1	Qn–1	Qn–1	
0	0	D1	1	1	0	Qn–1	Qn–1	Qn–1	D1	Qn–1	Qn–1	Qn–1	Qn–1	
0	0	D1	0	0	1	Qn–1	Qn–1	Qn–1	Qn–1	D1	Qn–1	Qn–1	Qn–1	
0	0	D1	1	0	1	Qn–1	Qn–1	Qn–1	Qn–1	Qn–1	D1	Qn–1	Qn–1	
0	0	D1	0	1	1	Qn–1	Qn–1	Qn–1	Qn–1	Qn–1	Qn–1	D1	Qn–1	
0	0	D1	1	1	1	Qn–1	Qn–1	Qn–1	Qn–1	Qn–1	Qn–1	Qn–1	D1	

Qn–1 = state before the positive transition of E.
D1 = either 1 (HIGH) OR 0 (low).

Figure 5.33. The CD4723 CMOS Dual 4-Bit Addressable Latch features common control inputs including two address lines (A0, A1), an active LOW Enable, and an active HIGH Clear. Each latch is provided with a Data input and four outputs. The CD4724 8-Bit Addressable Latch is similar other than for three address lines and eight outputs. Data for each is entered into the latch designated by the address inputs with the Enable LOW. Data entry is inhibited when the Enable line is HIGH. With both the Clear and Enable lines HIGH all outputs are LOW. When Clear is HIGH with Enable LOW channel demultiplexing takes place. The bit that is addressed has an active output which follows the data input while all unaddressed outputs are held LOW. While operating in the Addressable Latch Mode, Clear and Enable both LOW, changing more than one bit of the address could introduce a transient error. This should only be done in the Memory Mode, Enable HIGH, Clear LOW.

5V CMOS – 74AC(5V)/ACT11373

3V CMOS – 74AC(3V)11373

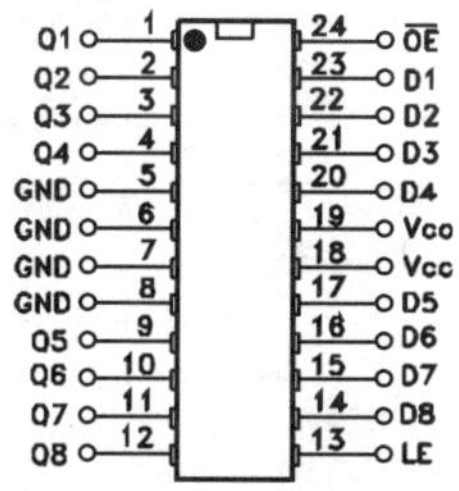

DUAL-IN-LINE PACKAGE

FUNCTION TABLE 74AC/ACT11373 OCTAL TRANSPARENT LATCH WITH 3-STATE OUTPUTS

OPERATING MODES	INPUTS				OUTPUT 01–08
	$\overline{OE}$	LE	D	INTERNAL REGISTER	Q
Enable and read register	0 0	1 1	0 1	0 1	0 1
Latch and read register	0 0	0 0	l h	0 1	0 1
Latch register and disable outputs	1 1	0 0	l h	0 1	Z Z

h = HIGH voltage one setup time prior to HIGH–to–LOW enable transition

l = LOW voltage one setup time prior to LOW–to–HIGH enable transition

LOGIC DIAGRAM

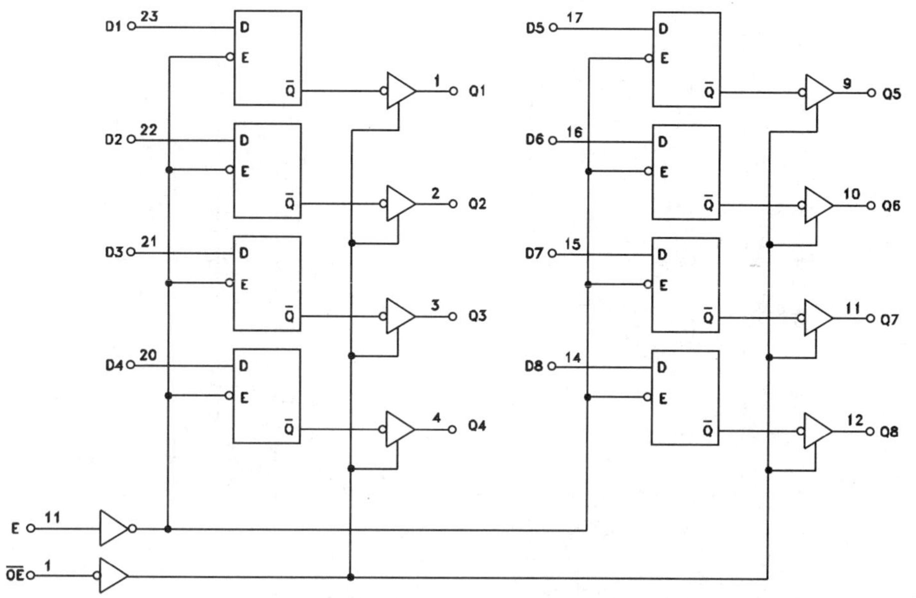

Figure 5.34. The eight latches of the 74AC/ACT11373 are transparent D-type. With the Latch Enable HIGH, the outputs follow the Data (D) inputs. When LE goes LOW, the existing inputs are latched. The device features 3-state outputs designed for driving highly capacitive or relatively low-impedance loads. It is useful for implementing buffer registers, I/O ports, bidirectional bus drivers and working registers. Entering the high-impedance state at the outputs has no effect on the internal latch operations. Old data is retained and new data entered while the outputs are OFF. The package pinout facilitates parallel access for loading and PCB layout. The center-pin power connections minimize high-speed switching noise. Latch-up immunity is typically 500 mA at 125°C.

5V CMOS – 74AHC(5V)/AHCT/AC(5V)ACT16373
3V CMOS – 74AHC(3V)/AC(3V)/LVTABT/LCX/
LVCH/ALVCH16373
74;VC/LVCH162373

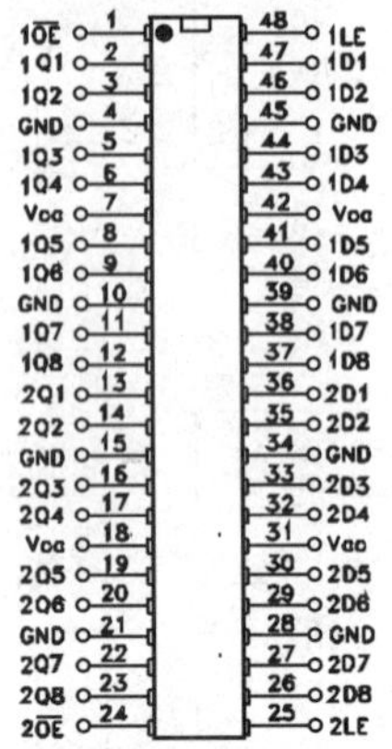

DUAL-IN-LINE PACKAGE

FUNCTION TABLE CMOS 3V/5V 74x16373, 16x162373 16-BIT
D-TYPE DUAL OCTAL TRANSPARENT LATCH

OPERATING MODES	INPUTS			INTERNAL LATCHES	OUTPUTS Q1–Q8
	$\overline{OE}$	LE	D		
Enable and read reg. (transparent mode)	0	1	0	0	0
	0	1	1	1	1
Latch and read reg.	0	0	l	0	0
	0	0	h	1	1
Latch register and disable outputs	1	0	l	0	Z
	1	0	h	1	Z

h = HIGH voltage level one setup time prior to the LE transition
l = LOW voltage level one setup time prior to the LE transition

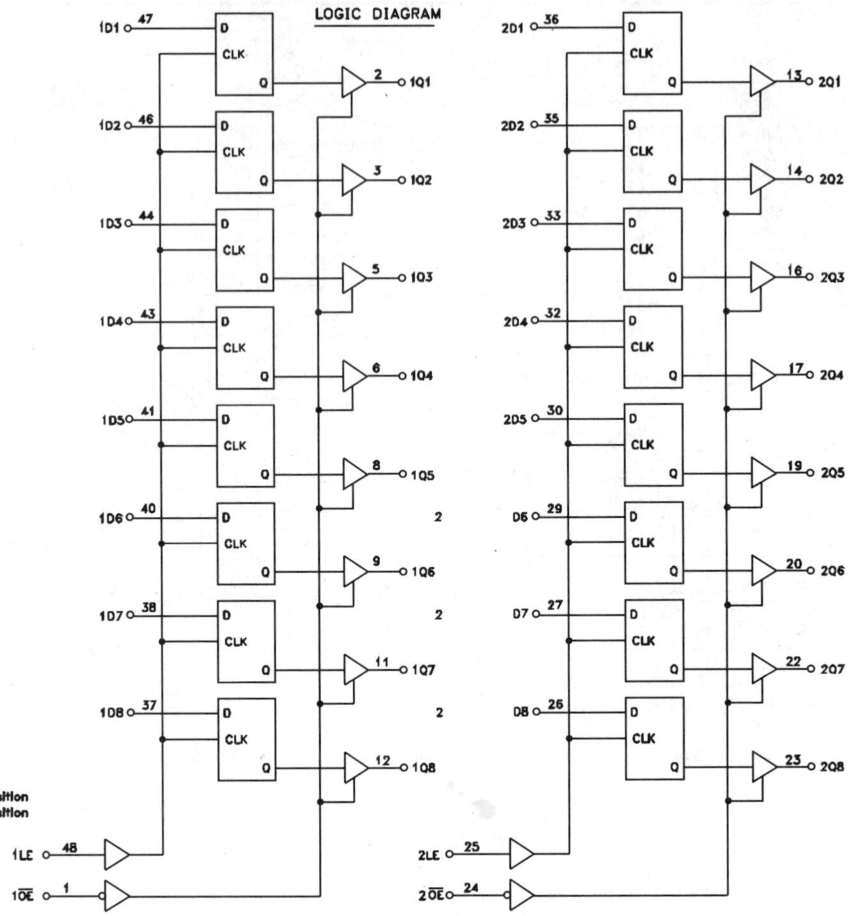

Figure 5.35. The 16 latches of the 74x16373 family are transparent D-type. These may be configured as a single 16-bit or two 8-bit latches. With the Latch Enable HIGH, the outputs follow the Data (D) inputs. When LE goes LOW, the existing inputs are latched. The device features 3-state outputs designed for driving highly capacitive or relatively low-impedance loads. It is useful for implementing buffer registers, I/O ports, bidirectional bus drivers and working registers. Entering the high-impedance state at the outputs has no effect on the internal latch operations. Old data is retained and new data entered while the outputs are OFF. The package pinout facilitates parallel access for loading and PCB layout. The distributed pin power connections minimize high-speed switching noise. Latch-up immunity is typically 500 mA at 125°C.

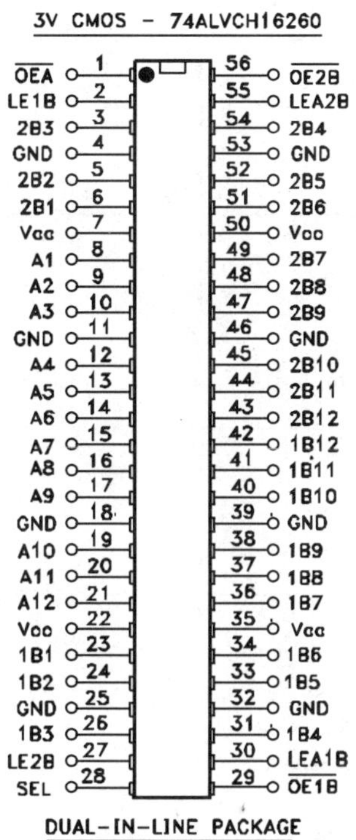

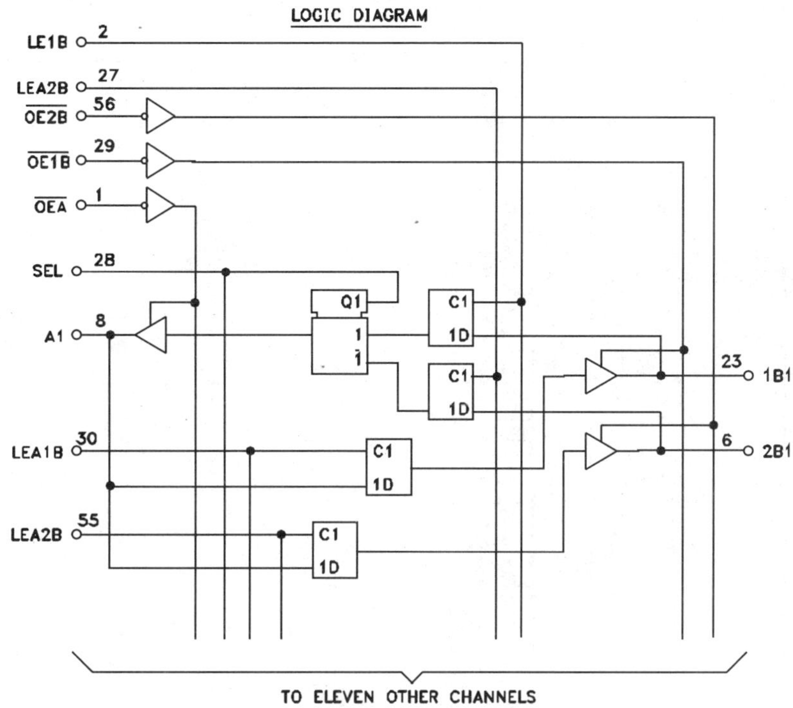

FUNCTION TABLES 74ALVCH16260 12-BIT TO 24-BIT MULTIPLEXED D-TYPE LATCH

A TO B (OEA=H)

INPUTS					OUTPUT	
A	LEA1B	LEA2B	OE1B	OE2B	1B	2B
1	1	1	0	0	1	1
0	1	1	0	0	0	0
1	1	0	0	0	1	2B0
0	1	0	0	0	0	2B0
1	0	1	0	0	1B0	1
0	0	1	0	0	1B0	0
X	0	0	0	0	1B0	2B0
X	X	X	1	1	Z	Z
X	X	X	0	1	Active	Z
X	X	X	1	0	Z	Active
X	X	X	0	0	Active	Active

B TO A (OEB=H)

INPUTS						OUTPUT
1B	2B	SEL	LE1B	LE2B	OEA	A
1	X	1	1	X	0	1
0	X	1	1	X	0	0
X	X	1	0	X	0	A0
X	1	0	X	1	0	1
X	0	0	X	1	0	0
X	X	0	X	0	0	A0
X	X	X	X	X	1	Z

Figure 5.36. The 74ALVCH16260 is a 12-Bit-to-24-Bit Multiplexed D-Type Latch with 3-state outputs. It is used in applications where two separate data paths must be multiplexed onto, or demultiplexed from, a single datapath. Applications include multiplexing/demultiplexing address and data information in microprocessor and bus-interface applications. It is also useful in memory interleaving. Three 12-Bit I/O ports (A1 - A12, 1B1 - 1B12, 2B1 - 2B12) are available for address and/or data transfer. The output enables (/OE1B, /OE2B, /OEA) control the transceiver functions. /OE1B and /OE2B control also allow bank control in the A-to-B direction. With the Latch Enable HIGH, input data flows through to the output; with the enable LOW, the data present is latched and held. To ensure the high-impedance state during power-up or -down, /OE should be tied to Vcc through a pull-up resistor. Active bus-hold circuitry is used to hold unused or floating inputs at a valid logic level. ESD protection exceeds 2000 volts per MIL-STD-883C, Method 3015; exceeds 200 volts using machine model. Latch-up exceeds 250 mA per JEDEC Standard JESD-17.

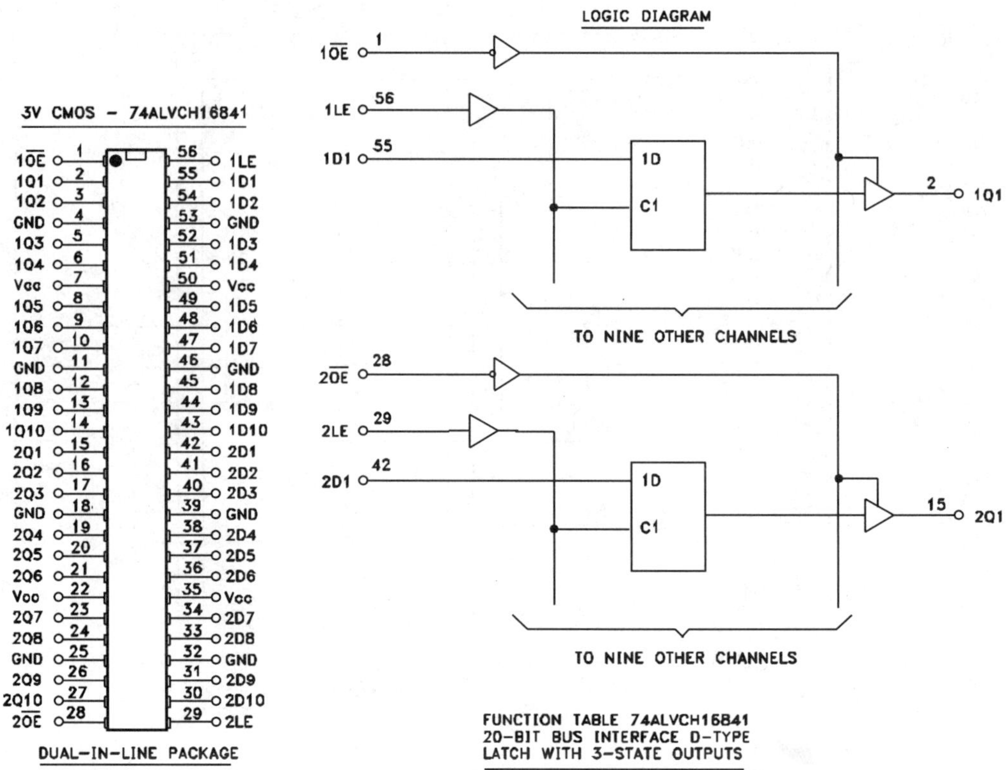

FUNCTION TABLE 74ALVCH16841
20-BIT BUS INTERFACE D-TYPE
LATCH WITH 3-STATE OUTPUTS

(Each 10-bit latch)

INPUTS			OUTPUTS
OE	LE	D	Q
0	1	1	1
0	1	0	0
0	0	X	Q0
1	X	X	Z

Figure 5.37. The 74ALVCH16841 is a 20-Bit Bus Interface D-Type Latch with 3-state outputs. The device can be configured as two 10-bit or a single 20-bit latch. With the Latch Enable HIGH, the outputs follow the Data (D) inputs. When LE goes LOW, the existing inputs are latched. The device features 3-state outputs designed for driving highly capacitive or relatively low-impedance loads. It is useful for implementing buffer registers, I/O ports, bidirectional bus drivers and working registers. Entering the high-impedance state at the outputs has no effect on the internal latch operations. Old data is retained and new data entered while the outputs are OFF. The package pinout facilitates parallel access for loading and PCB layout. The distributed pin power connections minimize high-speed switching noise. To ensure the high-impedance state during power-up or -down, /OE should be tied to Vcc through a pull-up resistor. Active bus-hold circuitry is used to hold unused or floating inputs at a valid logic level. ESD protection exceeds 2000 volts per MIL-STD-883C, Method 3015; exceeds 200 volts using machine model. Latch-up exceeds 250 mA per JEDEC Standard JESD-17.

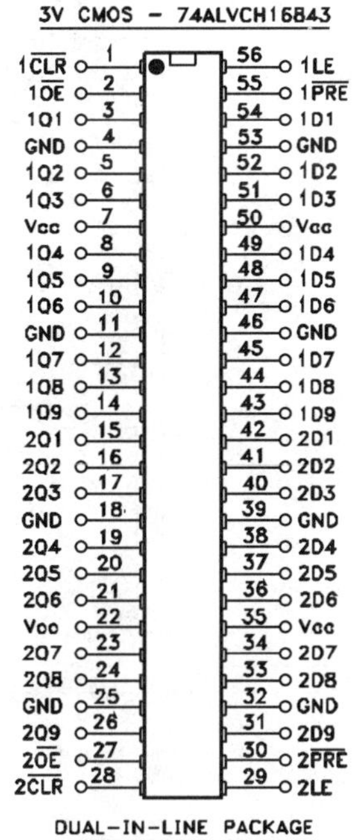

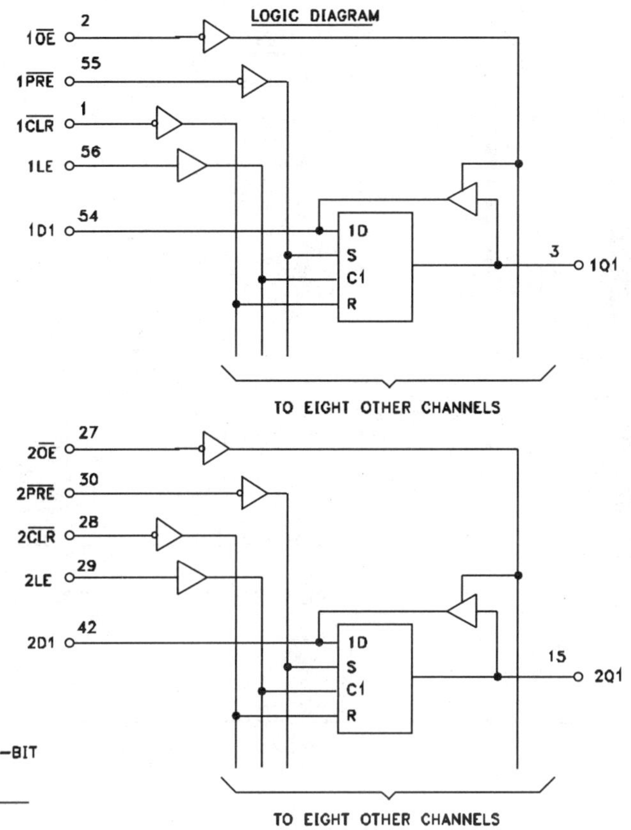

FUNCTION TABLE 74ALVCH16843 18-BIT BUS INTERFACE D-TYPE LATCH WITH 3-STATE OUTPUTS

(each 9-bit latch)

INPUTS					OUTPUT
$\overline{PRE}$	$\overline{CLR}$	$\overline{OE}$	LE	D	O
0	X	0	X	X	1
1	0	0	X	X	0
1	1	0	1	0	0
1	1	0	1	1	1
1	1	0	0	X	O0
X	X	1	X	X	Z

Figure 5.38. The 74ALVCH16843 is an 18-Bit Bus Interface D-Type Latch with 3-state outputs. The device can be configured as two 9-bit or a single 18-bit latch. With the Latch Enable HIGH, the outputs follow the Data (D) inputs. When LE goes LOW, the existing inputs are latched. The device features 3-state outputs designed for driving highly capacitive or relatively low-impedance loads. It is useful for implementing buffer registers, unidirectional bus drivers and working registers. Entering the high-impedance state at the outputs has no effect on the internal latch operations. Old data is retained and new data entered while the outputs are OFF. The package pinout facilitates parallel access for loading and PCB layout. The distributed pin power connections minimize high-speed switching noise. To ensure the high-impedance state during power-up or -down, /OE shold be tied to Vcc through a pull-up resistor. Active bus-hold circuitry is used to hold unused or floating inputs at a valid logic level.

3V CMOS – 74HL33373
74HL33533

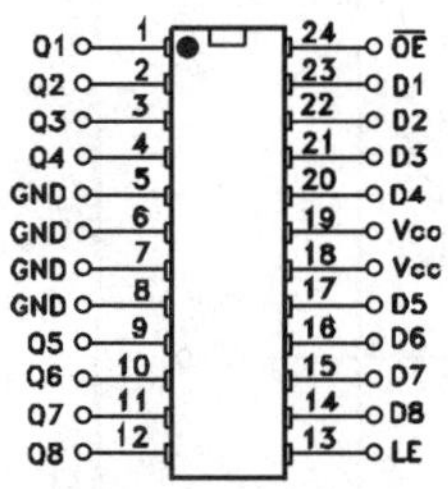

DUAL-IN-LINE PACKAGE

FUNCTION TABLE 74HL33373, '33533 OCTAL D-TYPE TRANSPARENT LATCH WITH 3-STATE OUTPUTS

OPERATING MODES	INPUTS			OUTPUTS
	$\overline{OE}$	LE	D	Q1 TO Q8
Enable and read register	0 0	1 1	0 1	0 1
Latch and read register	0 0	0 0	l h	0 1
Latch register and disable outputs	1	X	X	Z

h = HIGH voltage one setup time prior to HIGH–to–LOW enable transition
l = LOW voltage one setup time prior to LOW–to–HIGH enable transition

LOGIC DIAGRAM

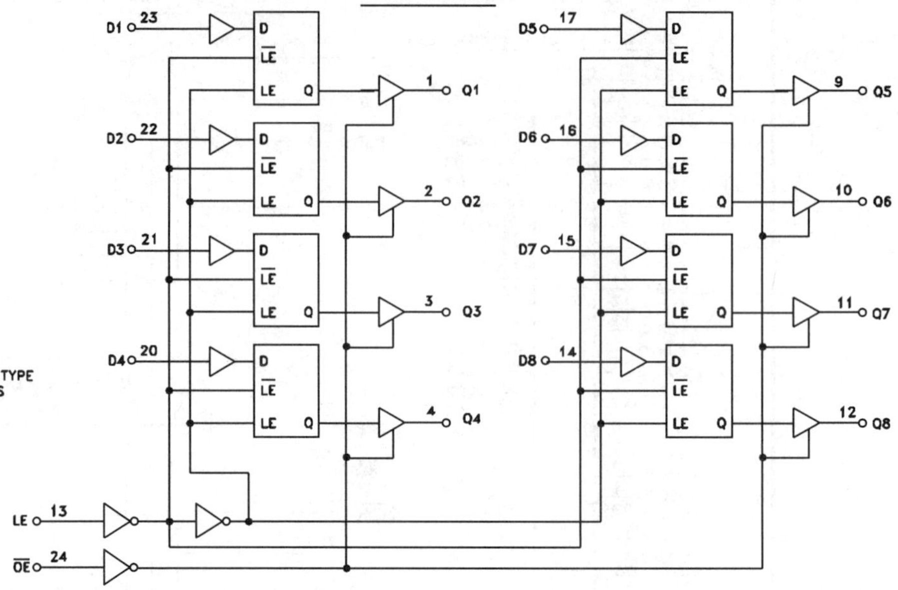

NOTE: THE 74HL33533 IS FUNCTIONALLY IDENTICAL TO THE 74HL33373 EXCEPT FOR INVERTED OUTPUTS.

Figure 5.39. The 74HL33373 and 74HL33533 Octal D-Type Transparent Latches with 3-state outputs. The two are identical in function; the '533 has inverted outputs. Separate D-type inputs are provided for each latch. With the Latch Enable HIGH, the outputs follow the Data (D) inputs. When LE goes LOW, the existing inputs are latched. Entering the high-impedance state at the outputs has no effect on the internal latch operations. Old data is retained and new data entered while the outputs are OFF. The center power distribution provides a low inductance for minimum noise and ground bounce. These provide direct interface with TTL levels of 5V to 3.3-volt level shifting.

Chapter 6

Shift and Data Storage Registers

The Shift Register, as with the Data Latch, consists basically of a number of flip-flops preceded by selection logic. The complexity varies over a broad range of relative simplicity to some of the most obtuse Function Tables we will ever encounter.

The basic concept is simple. We have a series of flip-flops in which the output of each stage functions as the input for the next. Each stage is connected to a clock input such that each changes state synchronously. The first stage input, whether a logic "0" or "1," is transferred to the second stage on succeeding clock inputs. Eventually the initial input arrives at the output of the final stage. If the output of each stage is brought out to a package pin we do not have to wait the entire length of the chain. With many types a provision is made to enter and/or retrieve data at each stage individually.

To introduce some kind of order, shift registers are identified by the type of input and output. Thus, if the device can only accept data in a serial form and deliver a serial output the device is designated serial-in/serial-out. If provided with individual stage outputs the designation might be serial-in/parallel-out, and possibly include serial-out as well. Where each stage accepts individual inputs the register is parallel-in and could be parallel-out, serial-out, or a combination.

A frequent application of the shift register is the *ring counter*. Construction is simple—the output of the last stage is connected to the first stage input. However complexity may enter in with presetting and decoding requirements.

As we will see, there is a surprisingly large variety of shift registers. The majority are TTL, many of which are available in CMOS equivalent packaging.

The first shift register types were fairly simple, consisting of chains of flip-flops for serial transmission of the data from stage to stage with each input from a clock source. We tend to think of a clock as a uniform train of pulses on some kind of a time basis. That is not necessarily so with the shift register. Clock inputs can occur randomly or in any kind of a sequence up to the maximum allowable input frequency.

D type flip-flops are frequently employed—these may be a master-slave combination with high-speed systems susceptible to timing difficulties. Several devices, the 74HC/HCT7402 and '7404, are high speed First-In-First-Out (FIFO) registers. These feature great flexibility with independent reading and writing operations due to separate shift-in (SI) and shift-out (SO) controls. Data transfers may be synchronous or asynchronous. Data-in-Ready (DIR) and Data-out-Ready (DOR) flags provide device status information.

Chapter Organization

Nine tables are provided. *Tables 6.1 - 6.4* provide a listing of shift registers by manufacturer. There are sources beyond those listed but those given provide a realistic picture of what has been and is currently available.

Tables 6.5 -6.8 provide timing references; namely, the latch pulse width (t_W), the LOW-to-HIGH propagation delay (t_{PLH}), and the setup/hold times. Because of the manner in which manufacturers define these quantities, the data in the tables is not precise. The intent is to provide insight helpful for performance comparisons. In particular, we can see the spectacular gains in speed and reduced power dissipation, as well as complexity, achieved with the latest devices.

The drawings of *Figures 6.1 - 6.55* describe the Dual In-Line (DIP) package pinning, a Function Table, Logic Diagrams and/or Logic Symbol. To simplify comparisons as well as to conserve space, all devices with matching features, specifically pinouts and function tables, are included in the same drawing.

Summary

This chapter provides tables and drawings describing over 100 variations of shift registers, from the early 7491 to the sophisticated FIFO registers. Many devices display an impressive array of capabilities. Some points to keep in mind are:

- Timing information in the tables is for comparison purposes only. There is a general lack of consistency in how these parameters are defined, and not just between manufacturers. Where timing is critical consult the manufacturer's data manual.
- In general, 3-state devices are not affected internally when in the high impedance output state.
- All timing data is shown for the industrial/commercial or recommended free-air operating temperature range. Where data is for 25°C it is given here.
- The specifications for CMOS devices may cover voltage ranges from 2 to 15 volts. Data is typically provided for V_{CC}'s of 5, 10, and 15 volts, but not always. Some manufacturers use 5.0 volts; others 4.5. There is a significant speed factor—in general, the higher the voltage the faster the performance. Again, where this is important to your application consult the data manual.
- A similar situation exists for the more recent low-voltage devices. Typical V_{CC} ranges given are 2.7 - 3.6 volts, but values may be found for 2, 2.6, 3.0, 3.3 or 3.6 volts. (Glossary *Figures G.1 and G.2* may provide helpful insights on performance.)

DEVICE	STD		L		H		S		LS		AS		ALS		F		Figure	DEVICE
74---	NM	ST	NM	ST	NM	ST	NM	ST	NM	ST	NM	ST	NM	ST	NM	ST	No.	74---
7491		xx								x							6.1	7491
7494		xx															6.2	7494
7495		xx	x						x	xx		x					6.3	7495
7496		xx								xx							6.4	7496
7498			x														6.5	7498
74164		xx							x	xx				x	x	x	6.6	74164
74165		xx							x	x				x			6.7	74165
74166		xx							x	x				x		x	6.8	74166
74170		xx							x	xx							6.9	74170
74172		x						x									6.10	74172
74194		xx															6.11	74194
74x194							x	xx	xx	xx		x			x	x	6.12	74x194
74195		xx					x	xx		xx		x			x	x	6.13	74195
74198		xx														x	6.14	74198
74199		xx														x	6.15	74199
74295										xx							6.16	74295
74298		xx								xx		x					6.17	74298
74299									x	xx				x		xx	6.18	74299
74322									x	x						x	6.19	74322
74323									x	xx				x	x	xx	6.18	74323
74350								x								x	6.20	74350
74395									x	xx						x	6.21	74395
74396										x							6.22	74396
74398									x	x					x	x	6.23	74398
74399									x	xx					x	x	6.24	74399
74594										x							6.26	74594
74595										x						x	6.27	74595
74596										x							6.27	74596
74597										x						x	6.28	74597
74598										x						x	6.28	74598
74599										x							6.26	74599
74670									x	x						x	6.29	74670
74671										x							6.30	74671
74672										x							6.30	74672
74673										x							6.31	74673
74674										x						x	6.32	74674
74676																x	6.33	74676
74835																x	6.34	74835
74870												x		x			6.35	74870
74871												x		x			6.35	74871
74963														x			6.36	74963
74964														x			6.37	74964

Key to Manufacturer: N – National, M – Motorola,
S – Signetics/Philips, T – Texas Instruments

Table 6.1. TTL Shift and Data Storage Register source summary by manufacturer.

DEVICE	4000 NM	4000 ST	4500 NM	4500 ST	4100 NM	4100 ST	4700 NM	4700 ST	Figure No.	DEVICE
4006	xx	x							6.38	4006
4014	xx	x							6.39	4014
4015	xx	x							6.40	4015
4021	xx	x							6.39	4021
4031	x	x							6.41	4031
4034	xx								6.42	4034
4035	xx	x							6.43	4035
4094	xx	x							6.44	4094
40194		x							6.12	40194
40195		x							6.13	40195
4194					x				6.12	4194
4517			x	x					6.45	4517
4557			x	x					6.46	4557
4562			x						6.47	4562
4731								x	6.48	4731

Key to Manufacturer:	Prefix
N – National,	CD
M – Motorola,	MC1
S – Signetics/Philips,	HEC/HEF
T – Texas Instruments	--

Table 6.2. 4000 Series CMOS Shift and Data Storage Register source summary by manufacturer.

DEVICE 74---	C NM	C ST	HC NM	HC ST	HCT NM	HCT ST	AHC (5V) NM	AHC (5V) ST	AHCT NM	AHCT ST	VHC (5V) NM	VHC (5V) ST	VHCT NM	VHCT ST	AC (5V) NM	AC (5V) ST	ACT NM	ACT ST	Figure No.	DEVICE 74---
7495	x																		6.3	7495
74164	x		xx	xx	xx	x					x								6.6	74164
74165	x		xx	xx	x	x		x		x									6.7	74165
74166			x	xx	x	x													6.8	74166
74194			xx	xx	x	x													6.12	74194
74195	x		xx	xx	x	x													6.13	74195
74299			xx	xx		x					x								6.18	74299
74589			xx																6.25	74589
74594				xx		x													6.26	74594
74595			xx	xx		x		x		x	x								6.27	74595
74597			xx	x		x													6.28	74597
74670				x		x													6.29	74670
744015				x		x													6.40	744015
744094				x		x													6.44	744094
747030				x		x													6.49	747030
747403				x		x													6.50	747403
747404				x		x													6.51	747404
747597				x		x													6.52	747597
747731				x		x													6.53	747731
7440104				x		x													6.54	7440104
7440105				x		x													6.55	7440105

Key to Manufacturer: N – National, M – Motorola, S – Signetics/Philips, T – Texas Instruments

Table 6.3. 7400 Series 5V CMOS Shift and Data Storage Register source summary by manufacturer.

DEVICE 74---	LV NM	LV ST	LVCH LVC NM	LVCH LVC ST	ALVT LVT NM	ALVT LVT ST	LVX NM	LVX ST	LVQ NM	LVQ ST	LCX NM	LCX ST	AC (3V) NM	AC (3V) ST	ALVCH NM	ALVCH ST	HL NM	HL ST	AHC (3V) NM	AHC (3V) ST	VHC (3V) NM	VHC (3V) ST	Figure No.	DEVICE 74---
74164		xx																			x		6.6	74164
74165		xx																		x			6.7	74165
74299																					x		6.18	74299
74595		x																		x	xx		6.27	74595
744094		x																					6.44	744094

Key to Manufacturer: N – National, M – Motorola, S – Signetics/Philips, T – Texas Instruments

Table 6.4. 7400 Series 3V CMOS Shift and Data Storage Register source summary by manufacturer.

DEVICE 74---	STD	L	H	S	LS	AS/A/B	ALS/A/B	F	DATA	Figure	DEVICE 74---
-7491	10-18	--	--	--	10-18	--	--	--	f_{MHz}	-6.1	-7491
----	24-40	--	--	--	24-40	--	--	--	t_{pd}	---	----
----	25/0	--	--	--	25/0	--	--	--	t_s/t_h	---	----
-7494	10	--	--	--	--	--	--	--	f_{MHz}	-6.2	-7494
----	25-40	--	--	--	--	--	--	--	t_{pd}	---	----
----	35/0	--	--	--	--	--	--	--	t_s/t_h	---	----
-7495	25-36	6	--	--	25-36	100	--	--	f_{MHz}	-6.3	-7495
----	18-32	90-90	--	--	18-32	2-10	--	--	t_{pd}	---	----
----	15/0	50/0	--	--	20/20	2.5/3.5	--	--	t_s/t_h	---	----
-7496	10	--	--	--	25	--	--	--	f_{MHz}	-6.4	-7496
----	25-40	--	--	--	25-40	--	--	--	t_{pd}	---	----
----	30/0	--	--	--	30/0	--	--	--	t_s/t_h	---	----
-7498	--	6	--	--	--	--	--	--	f_{MHz}	-6.5	-7498
----	--	80-100	--	--	--	--	--	--	t_{pd}	---	----
----	--	100/150	--	--	--	--	--	--	t_s/t_h	---	----
-74164	25-36	--	--	--	25-36	--	60	80-100	f_{MHz}	-6.6	-74164
----	8-37	--	--	--	17-36	--	10-12	3.0-10.5	t_{pd}	---	----
----	20/5	--	--	--	20/5	--	-/0	7/1	t_s/t_h	---	----
-74165	20-26	--	--	--	25-35	--	25	--	f_{MHz}	-6.7	-74165
----	11-40	--	--	--	13-35	--	13-26	--	t_{pd}	---	----
----	45/0	--	--	--	45/0	--	-/0	--	t_s/t_h	---	----
-74166	25-35	--	--	--	25-35	--	60	135-175	f_{MHz}	-6.8	-74166
----	17-35	--	--	--	5-30	--	10-13	4.0-10.0	t_{pd}	---	----
----	20/0	--	--	--	30/0	--	-/0	5/0	t_s/t_h	---	----
-74170	25 *	--	--	--	25 *	--	--	--	t_w	-6.9	-74170
----	10-45	--	--	--	20-45	--	--	--	t_{pd}	---	----
----	15/5	--	--	--	15/5	--	--	--	t_s/t_h	---	----
-74172	20	--	--	20	--	--	--	--	f_{MHz}	-6.10	-74172
----	33-50	--	--	30-38	--	--	--	--	t_{pd}	---	----
----	45/0	--	--	30/0	--	--	--	--	t_s/t_h	---	----
-74194	25	--	--	--	--	--	--	--	f_{MHz}	-6.11	-74194
----	7-30	--	--	--	--	--	--	--	t_{pd}	---	----
----	20/0	--	--	--	--	--	--	--	t_s/t_h	---	----
-74x194	--	--	--	70	25	--	80	105-150	f_{MHz}	-6.12	-74x194
----	--	--	--	4.0-18.5	22-30	--	3-12	3.5-7.0	t_{pd}	---	----
----	--	--	--	8.0/3.0	30/0	--	3/0	8/0	t_s/t_h	---	----
-74195	30	--	--	70	30	--	70	165-190	f_{MHz}	-6.13	-74195
----	22-30	--	--	12-18.5	22-30	--	2.5-11.5	2.5-9.5	t_{pd}	---	----
----	20/0	--	--	5.0/3.0	25/0	--	4/1	3.0/0	t_s/t_h	---	----
-74198	25	--	--	--	--	--	--	80-95	f_{MHz}	-6.14	-74198
----	26-35	--	--	--	--	--	--	5.0-11.0	t_{pd}	---	----
----	30/0	--	--	--	--	--	--	9/0	t_s/t_h	---	----
-74199	25	--	--	--	--	--	--	80-95	f_{MHz}	-6.15	-74199
----	26-35	--	--	--	--	--	--	5.5-12.5	t_{pd}	---	----
----	30/0	--	--	--	--	--	--	1.5/4.5	t_s/t_h	---	----
-74295	--	--	--	--	30	--	--	--	f_{MHz}	-6.16	-74295
----	--	--	--	--	23-30	--	--	--	t_{pd}	---	----
----	--	--	--	--	10/5.0	--	--	--	t_s/t_h	---	----
-74298	20 *	--	--	--	20	62	--	--	f_{MHz}	-6.17	-74298
----	27-32	--	--	--	27-32	2-11	--	--	t_{pd}	---	----
----	15/5	--	--	--	15/5	4.5/3.5	--	--	t_s/t_h	---	----
-74299	--	--	--	--	35	--	30	70-115	f_{MHz}	-6.18	-74299
----	--	--	--	--	25-39	--	5-19	4.0-10.0	t_{pd}	---	----
----	--	--	--	--	20/3.0	--	16/0	6.5/0	t_s/t_h	---	----
-74322	--	--	--	--	20-35	--	--	110-125	f_{MHz}	-6.19	-74322
----	--	--	--	--	16-35	--	--	4.0-7.0	t_{pd}	---	----
----	--	--	--	--	20/2	--	--	12.5/0	t_s/t_h	---	----
-74323	--	--	--	--	25-35	--	17	70-115	f_{MHz}	-6.18	-74323
----	--	--	--	--	17-39	--	4-19	3.5-6.5	t_{pd}	---	----
----	--	--	--	--	20/3	--	18/0	6.5/0	t_s/t_h	---	----

* frequency not given, write enable pulse, tw, shown.

Table 6.5A. TTL Shift and Data Register representative frequency (MHz), delays (t_{pd}), and setup/hold (t_s/t_h) times in nS.

DEVICE 74---	STD	L	H	S	LS	AS/A/B	ALS/A/B	F	DATA	Figure	DEVICE 74---
-74350	--	--	--	NA	--	--	--	NA	f MHz	-6.20	-74350
----	--	--	--	5-20	--	--	--	2.5-10	tpd	---	----
----	--	--	--	NA	--	--	--	NA	ts/th	---	----
-74395	--	--	--	--	30-45	--	--	105-120	f MHz	-6.21	-74395
----	--	--	--	--	15-35	--	--	3.5-11.0	tpd	---	----
----	--	--	--	--	40/10	--	--	2.5/1.5	ts/th	---	----
-74396	--	--	--	--	30	--	--	--	f MHz	-6.22	-74396
----	--	--	--	--	20-30	--	--	--	tpd	---	----
----	--	--	--	--	20/5	--	--	--	ts/th	---	----
-74398	--	--	--	--	20 *	--	--	100-120	f MHz	-6.23	-74398
----	--	--	--	--	18-32	--	--	3.0-8.5	tpd	---	----
----	--	--	--	--	45/0	--	--	3.0/1.0	ts/th	---	----
-74399	--	--	--	--	20 *	--	--	100-120	f MHz	-6.24	-74399
----	--	--	--	--	18-32	--	--	3.0-8.5	tpd	---	----
----	--	--	--	--	45/0	--	--	3.0/1.0	ts/th	---	----
-74594	--	--	--	--	20-25	--	--	--	f MHz	-6.26	-74594
----	--	--	--	--	12-57	--	--	--	tpd	---	----
----	--	--	--	--	40/0	--	--	--	ts/th	---	----
-74595	--	--	--	--	25	--	--	115-135	f MHz	-6.27	-74595
----	--	--	--	--	12-35	--	--	2.5-10	tpd	---	----
----	--	--	--	--	40/0	--	--	4.5/0	ts/th	---	----
-74596	--	--	--	--	25	--	--	--	f MHz	-6.27	-74596
----	--	--	--	--	14-42	--	--	--	tpd	---	----
----	--	--	--	--	40/0	--	--	--	ts/th	---	----
-74597	--	--	--	--	20-35	--	--	120-135	f MHz	-6.28	-74597
----	--	--	--	--	15-60	--	--	6.0-12.0	tpd	---	----
----	--	--	--	--	40/0	--	--	6.0/0	ts/th	---	----
-74598	--	--	--	--	20-35	--	--	85-160	f MHz	-6.28	-74598
----	--	--	--	--	11-48	--	--	5.0-11.5	tpd	---	----
----	--	--	--	--	40/0	--	--	7.0/0	ts/th	---	----
-74599	--	--	--	--	20-25	--	--	--	f MHz	-6.26	-74599
----	--	--	--	--	12-60	--	--	--	tpd	---	----
----	--	--	--	--	40/0	--	--	--	ts/th	---	----
-74670	--	--	--	--	25 *	--	--	--	tw	-6.29	-74670
----	--	--	--	--	23-50	--	--	--	tpd	---	----
----	--	--	--	--	15/5	--	--	--	ts/th	---	----
-74671	--	--	--	--	30 *	--	--	--	f MHz	-6.30	-74671
----	--	--	--	--	10-45	--	--	--	tpd	---	----
----	--	--	--	--	35/0	--	--	--	ts/th	---	----
-74672	--	--	--	--	30 *	--	--	--	f MHz	-6.30	-74672
----	--	--	--	--	10-45	--	--	--	tpd	---	----
----	--	--	--	--	35/0	--	--	--	ts/th	---	----
-74673	--	--	--	--	20-28	--	--	--	f MHz	-6.31	-74673
----	--	--	--	--	21-45	--	--	--	tpd	---	----
----	--	--	--	--	35/0	--	--	--	ts/th	---	----
-74674	--	--	--	--	20-28	--	--	80-95	f MHz	-6.32	-74674
----	--	--	--	--	21-45	--	--	6.0-12.5	tpd	---	----
----	--	--	--	--	35/0	--	--	8.0/0	ts/th	---	----
-74676	--	--	--	--	--	--	--	100-110	f MHz	-6.33	-74676
----	--	--	--	--	--	--	--	4.5-12.5	tpd	---	----
----	--	--	--	--	--	--	--	10.0/2.0	ts/th	---	----
-74835	--	--	--	--	--	--	--	130-150	f MHz	-6.34	-74835
----	--	--	--	--	--	--	--	5.0-9.5	tpd	---	----
----	--	--	--	--	--	--	--	4.5/0	ts/th	---	----
-74870	--	--	--	--	--	12 *	10 *	--	tw	-6.35	-74870
----	--	--	--	--	--	5-22	12.5-16	--	tpd	---	----
----	--	--	--	--	--	15/0	4/0	--	ts/th	---	----
-74871	--	--	--	--	--	12 *	10 *	--	tw	-6.35	-74871
----	--	--	--	--	--	5-26	12.5-16	--	tpd	---	----
----	--	--	--	--	--	4/0	4/0	--	ts/th	---	----
-74963	--	--	--	--	--	--	25-30	--	f MHz	-6.36	-74963
----	--	--	--	--	--	--	10-14	--	tpd	---	----
----	--	--	--	--	--	--	NA	--	ts/th	---	----
-74964	--	--	--	--	--	--	25-30	--	f MHz	-6.37	-74964
----	--	--	--	--	--	--	10-14	--	tpd	---	----
----	--	--	--	--	--	--	NA	--	ts/th	---	----

* Frequency not given, write enable pulse, tw, shown.

Table 6.5b. TTL Shift and Data Register representative frequency (MHz), delays (tpd), and setup/hold (ts/th) times in nS. (Continuation)

DEVICE CD\MC1	4000	4500	4100	4700	DATA	Figure	DEVICE CD\MC1
-4006	9-18	--	--	--	fMHz	-6.38	-4006
----	90-180	--	--	--	tpd	---	----
----	10/-5	--	--	--	ts/th	---	----
-4014	6-13	--	--	--	fMHz	-6.39	-4014
----	130-260	--	--	--	tpd	---	----
----	5/10	--	--	--	ts/th	---	----
-4015	7-15	--	--	--	fMHz	-6.40	-4015
----	130-260	--	--	--	tpd	---	----
----	-15/20	--	--	--	ts/th	---	----
-4021	6-13	--	--	--	fMHz	-6.39	-4021
----	125-250	--	--	--	tpd	---	----
----	-15/20	--	--	--	ts/th	---	----
-4031	2.5-5	--	--	--	fMHz	-6.41	-4031
----	180-380	--	--	--	tpd	---	----
----	0/10	--	--	--	ts/th	---	----
-4034	1.2-2.5	--	--	--	fMHz	-6.42	-4034
----	525-1050	--	--	--	tpd	---	----
----	35	--	--	--	ts/th	---	----
-4035	5-10	--	--	--	fMHz	-6.43	-4035
----	170-340	--	--	--	tpd	---	----
----	25/10	--	--	--	ts/th	---	----
-4094	2-5	--	--	--	fMHz	-6.44	-4094
----	220-440	--	--	--	tpd	---	----
----	10/10	--	--	--	ts/th	---	----
-40194	6-12	--	--	--	fMHz	-6.12	-40194
----	100-205	--	--	--	tpd	---	----
----	40/-30	--	--	--	ts/th	---	----
-40195	60-120	--	--	--	fMHz	-6.13	-40195
----	105-215	--	--	--	tpd	---	----
----	40/-25	--	--	--	ts/th	---	----
-4194	--	--	1.8-3.6	--	fMHz	-6.12	-4194
----	--	--	275-550	--	tpd	---	----
----	--	--	10/180	--	ts/th	---	----
-4517	--	2-5	--	--	fMHz	-6.45	-4517
----	--	220-440	--	--	tpd	---	----
----	--	30/45	--	--	ts/th	---	----
-4557	--	2.5-5	--	--	fMHz	-6.46	-4557
----	--	240-480	--	--	tpd	---	----
----	--	180/-110	--	--	ts/th	---	----
-4562	--	1.1-1.9	--	--	fMHz	-6.47	-4562
----	--	600-1.2K	--	--	tpd	---	----
----	--	-20/350	--	--	ts/th	---	----
-4731	--	--	--	2.25-6	fMHz	-6.48	-4731
----	--	--	--	115-230	tpd	---	----
----	--	--	--	25/50	ts/th	---	----

Table 6.6. 4000 Series CMOS Shift and Data Register representative frequency (MHz), delays (t_{pd}), and setup/hold (t_s/t_h) times in nS.

DEVICE 74---	C	HC	HCT	AHC	AHCT	VHC(5V)	VHCT	AC (5V)	ACT	DATA	Figure	DEVICE 74---
-7495	3-5	--	--	--	--	--	--	--	--	f MHz	-6.3	-7495
----	200-400	--	--	--	--	--	--	--	--	tpd	---	----
----	60/25	--	--	--	--	--	--	--	--	ts/th	---	----
-74164	2.0-3	30-71	27-55	--	--	85-175	--	--	--	f MHz	-6.6	-74164
----	230-310	15-34	17-36	--	--	5.8-11	--	--	--	tpd	---	----
----	200/0	8/-6	6/-2	--	--	4.5/1.0	--	--	--	ts/th	---	----
-74165	2.5-6	30-51	26-44	NO	NO	--	--	--	--	f MHz	-6.7	-74165
----	200-400	19-33	17-34	DATA	DATA	--	--	--	--	tpd	---	----
----	150/50	8/2	10/-7	GIVEN	GIVEN	--	--	--	--	ts/th	---	----
-74166	--	30-57	25-45	--	--	--	--	--	--	f MHz	-6.8	-74166
----	--	18-30	23-40	--	--	--	--	--	--	tpd	---	----
----	--	12/-10	15/-13	--	--	--	--	--	--	ts/th	---	----
-74194	--	30-93	30-70	--	--	--	--	--	--	f MHz	-6.12	-74194
----	--	17-29	18-32	--	--	--	--	--	--	tpd	---	----
----	--	8/-5	10/-7	--	--	--	--	--	--	ts/th	---	----
-74195	2.0-3.0	30-52	27-52	--	--	--	--	--	--	f MHz	-6.13	-74195
----	150-300	18-30	18-32	--	--	--	--	--	--	tpd	---	----
----	110/-10	12/-3	12/-5	--	--	--	--	--	--	ts/th	---	----
-74299	--	25-45	25-42	--	--	115-160	--	--	--	f MHz	-6.18	-74299
----	--	24-40	22-37	--	--	12.2-20.7	--	--	--	tpd	---	----
----	--	14/-10	18/-17	--	--	7.0/1.5	--	--	--	ts/th	---	----
-74589	--	24-30	--	--	--	--	--	--	--	f MHz	-6.25	-74589
----	--	20-45	--	--	--	--	--	--	--	tpd	---	----
----	--	20/5	--	--	--	--	--	--	--	ts/th	---	----
-74594	--	30-92	30-92	--	--	--	--	--	--	f MHz	-6.26	-74594
----	--	16-30	18-32	--	--	--	--	--	--	tpd	---	----
----	--	6/-3	7/-3	--	--	--	--	--	--	ts/th	---	----
-74595	--	30-91	30-52	95-185	95-185	--	--	--	--	f MHz	-6.27	-74595
----	--	19-32	25-42	5.4-7.4	5.4-7.4	--	--	--	--	tpd	---	----
----	--	8/-2	8/-2	5/2	5/2	--	--	--	--	ts/th	---	----
-74597	--	30-87	30-75	--	--	--	--	--	--	f MHz	-6.28	-74597
----	--	20-35	23-40	--	--	--	--	--	--	tpd	---	----
----	--	4/-2	5/-2	--	--	--	--	--	--	ts/th	---	----
-74670	--	5 *	9 *	--	--	--	--	--	--	tw	-6.29	-74670
----	--	21-39	21-40	--	--	--	--	--	--	tpd	---	----
----	--	2/0	4/-1	--	--	--	--	--	--	ts/th	---	----
-744015	--	30-100	30-67	--	--	--	--	--	--	f MHz	-6.40	-744015
----	--	19-35	21-35	--	--	--	--	--	--	tpd	---	----
----	--	3/0	4/-2	--	--	--	--	--	--	ts/th	---	----
-744094	--	30-87	30-80	--	--	--	--	--	--	f MHz	-6.44	-744094
----	--	18-30	23-39	--	--	--	--	--	--	tpd	---	----
----	--	10/-5	9/-4	--	--	--	--	--	--	ts/th	---	----
-747030	--	23-30	22-26	--	--	--	--	--	--	f MHz	-6.49	-747030
----	--	25-42	30-51	--	--	--	--	--	--	tpd	---	----
----	--	-13/16	-16/18	--	--	--	--	--	--	ts/th	---	----
-747403	--	23-30	18-30	--	--	--	--	--	--	f MHz	-6.50	-747403
----	--	25-42	30-51	--	--	--	--	--	--	tpd	---	----
----	--	-13/16	-5/30	--	--	--	--	--	--	ts/th	---	----
-747404	--	23-30	23-30	--	--	--	--	--	--	f MHz	-6.51	-747404
----	--	25-42	30-51	--	--	--	--	--	--	tpd	---	----
----	--	-13/16	-16/18	--	--	--	--	--	--	ts/th	---	----
-747597	--	30-90	30-72	--	--	--	--	--	--	f MHz	-6.52	-747597
----	--	18-35	20-35	--	--	--	--	--	--	tpd	---	----
----	--	4/-3	5/-4	--	--	--	--	--	--	ts/th	---	----
-747731	--	30-78	30-80	--	--	--	--	--	--	f MHz	-6.53	-747731
----	--	18-31	24-42	--	--	--	--	--	--	tpd	---	----
----	--	8/-3	6/-3	--	--	--	--	--	--	ts/th	---	----
-7440104	--	30-56	27-52	--	--	--	--	--	--	f MHz	-6.54	-7440104
----	--	16-34	18-34	--	--	--	--	--	--	tpd	---	----
----	--	8/-5	9/-5	--	--	--	--	--	--	ts/th	---	----
-7440105	--	30-36	28	--	--	--	--	--	--	f MHz	-6.55	-7440105
----	--	19-35	18-35	--	--	--	--	--	--	tpd	---	----
----	--	-14/16	-14/16	--	--	--	--	--	--	ts/th	---	----

* Frequency not given, write enable pulse width shown.

Table 6.7. 7400 Series 5V CMOS Shift and Data Register representative frequency (MHz), delays (tpd), and setup/hold (ts/th) times in nS.

DEVICE 74---	LV	LVC	ALVT LVT/ABT	LVX	LVQ	LCX	AC(3V)	LVCH/ ALVCH	HL	AHC (3V)	VHC(3V)	DATA	Figure No.	DEVICE 74---
-74164	24-70	--	--	--	--	--	--	--	--	--	50-125	f MHz	-6.6	-74164
----	14-23	--	--	--	--	--	--	--	--	--	8.4-16.3	tpd	---	----
----	3/2	--	--	--	--	--	--	--	--	--	5.0/1.0	ts/th	---	----
-74165	24-70	--	--	--	--	--	--	--	--	NO	--	f MHz	-6.7	-74165
----	21-44	--	--	--	--	--	--	--	--	DATA	--	tpd	---	----
----	8/2	--	--	--	--	--	--	--	--	GIVEN	--	ts/th	---	----
-74299	--	--	--	--	--	--	--	--	--	--	55-100	f MHz	-6.18	-74299
----	--	--	--	--	--	--	--	--	--	--	12.7-20.7	tpd	---	----
----	--	--	--	--	--	--	--	--	--	--	14.5/0.5	ts/th	---	----
-74595	24-70	--	--	--	--	--	--	--	--	55-150	55-150	f MHz	-6.27	-74595
----	18-38	--	--	--	--	--	--	--	--	7.7-13	7.7-15.4	tpd	---	----
----	8/2	--	--	--	--	--	--	--	--	2.5/2	8/1.5	ts/th	---	----
-744094	24-87	--	--	--	--	--	--	--	--	--	--	f MHz	-6.44	-744094
----	14-34	--	--	--	--	--	--	--	--	--	--	tpd	---	----
----	10/2	--	--	--	--	--	--	--	--	--	--	ts/th	---	----

Table 6.8. 7400 Series 3V CMOS Shift and Data Register representative frequency (MHz), delays (tpd), and setup/hold (ts/th) times in nS.

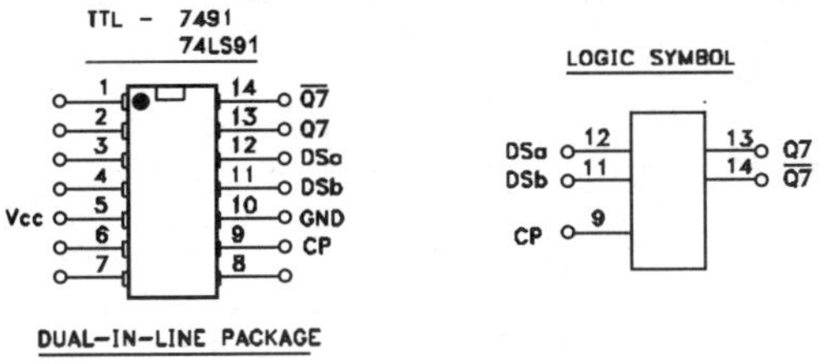

FUNCTION TABLE 7491, 'LS91 8-BIT SHIFT REGISTER

OPERATING MODES	INPUTS CP	DSa	DSb	1st STAGE Q0	$\overline{Q0}$	OUTPUTS Q7	$\overline{Q7}$
Shift, Reset First Stage	↑	l	x	0	1	q6	$\overline{q6}$
	↑	x	l	0	1	q6	$\overline{q6}$
Shift, Set First Stage	↑	h	h	1	0	q6	$\overline{q6}$

↑ = LOW-to-HIGH clock transition
h = HIGH voltage level one setup time prior to the ↑ clock transition.
l = LOW voltage level one setup time prior to the ↑ clock transition.
q = lower case letters indicate the state of the referenced input one setup time prior to the ↑ clock transition.

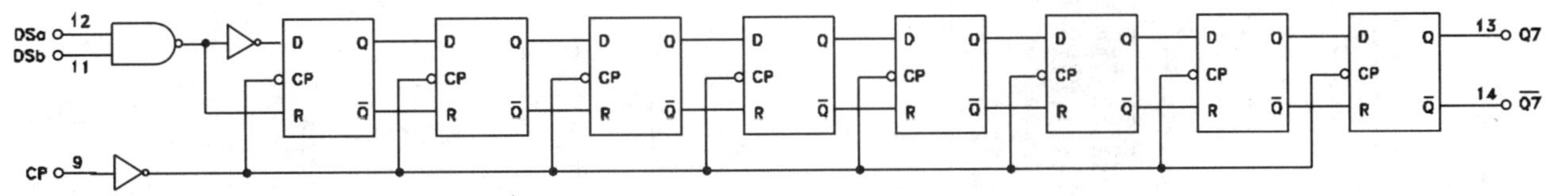

Figure 6.1. The 7491 and 'LS91 are serial-in, serial-out 8-bit shift registers composed of eight R-S master-slave flip-flops, input gating, and a clock driver. The clock pulse inverter/driver causes a shift on the positive edge of the clock pulse.

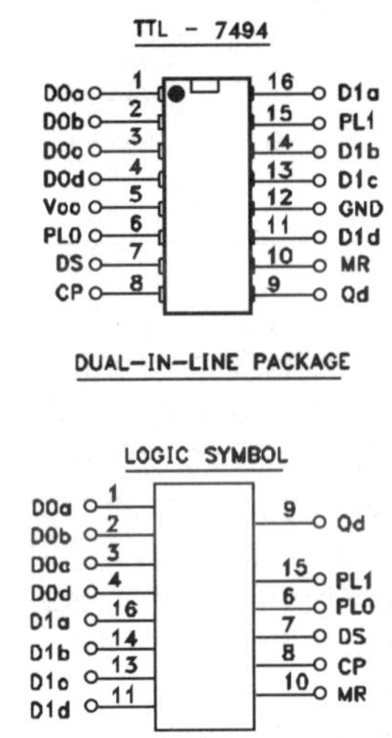

FUNCTION DIAGRAM 7494 4-BIT SHIFT REGISTER

OPERATING MODE	INPUTS PL0	PL1	D0n	D1n	MR	CP	DS	OUTPUTS Qa	Qb	Qc	Qd
Parallel Load	1	0	0	X	X	X	X	Qa	Qb	Qc	Qd
	1	0	1	X	X	X	X	1	1	1	1
	0	1	X	0	X	X	X	Qa	Qb	Qc	Qd
	0	1	X	1	X	X	X	1	1	1	1
Reset (Clear)	0	0	X	X	1	X	X	0	0	0	0
Shift Right	0	0	X	X	0	↑	l	0	qa	qb	qc
	0	0	X	X	0	↑	h	1	qa	qb	qc

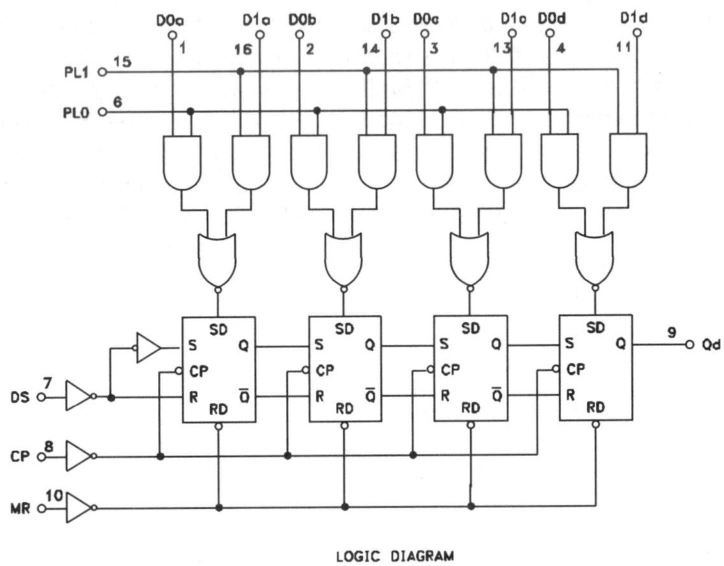

Figure 6.2. The 7494 is a parallel-to-serial converter composed of four R-S master-slave flip-flops, four AND-OR-INVERT gates, and four inverter-drivers. The device performs right-shift operations as a serial-in, serial-out register or as a dual-source, parallel-to-serial converter. Devices may be series-connected to form an n-bit register.

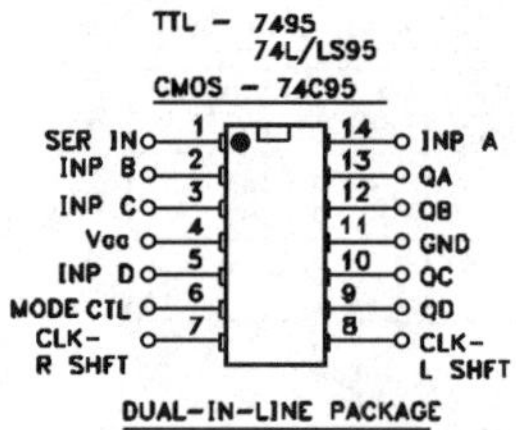

FUNCTION DIAGRAM 74x95 4-BIT SHIFT REGISTER

MODE CONTROL	CLOCKS 2(L)	1(R)	INPUTS SER	A	B	C	D	OUTPUTS QA	QB	QC	QD
1	1	X	X	X	X	X	X	QA0	QB0	QC0	QD0
1	↓	X	X	a	b	c	d	a	b	c	d
1	↓	X	X	QB	QC	QD	d	QBn	QCn	QDn	d
0	0	1	X	X	X	X	X	QA0	QB0	QC0	QD0
0	X	↓	1	X	X	X	X	1	QAn	QBn	QCn
0	X	↓	0	X	X	X	X	0	QAn	QBn	QCn
↑	0	0	X	X	X	X	X	QA0	QBn	QC0	QD0
↓	0	0	X	X	X	X	X	QA0	QB0	QC0	QD0
↓	0	1	X	X	X	X	X	QA0	QB0	QC0	QD0
↑	1	0	X	X	X	X	X	QA0	QB0	QC0	QD0
↑	1	1	X	X	X	X	X	QA0	QB0	QC0	QD0

Shifting LEFT requires ext connection of QB to A, QC to B, QD to C.
Serial data is entered at input D.
Steady state: 1 = HIGH, 0 = LOW, X = Don't Care
↓ = HIGH-to-LOW transitions. ↑ = LOW-to-HIGH transitions.
a, b, c, d, = the level of steady state at inputs A, B, C, D respectively
QAn, QBn, QCn, QDn = The level of QA, QB, QC, QD respectively before the most recent ↓ transition of the clock

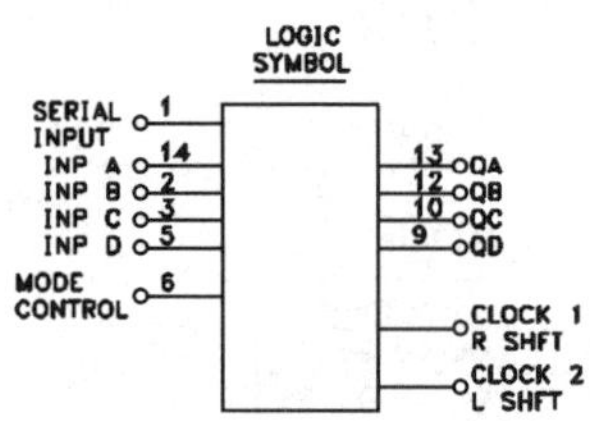

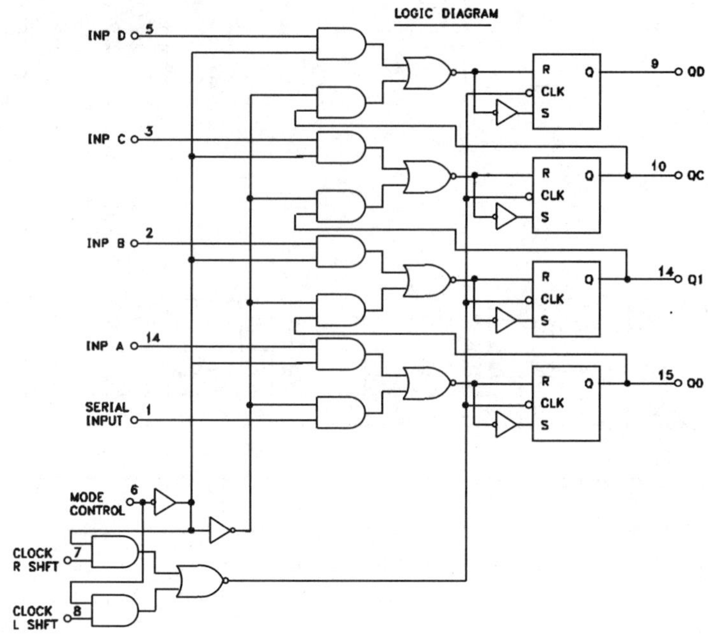

Figure 6.3. The 7495, 'L95, 'LS95, and 'C95 are 4-bit registers featuring parallel and serial inputs, parallel outputs, mode control, and two clock inputs. The registers have three modes of operation: Parallel load, Shift Right (QA toward Qd), and Shift Left (QD toward QA). Parallel loading is accomplished by taking the mode control input high. The data is loaded, appearing at the output following the high-to-low transition of the clock 2 input. Shift Right takes place on the high-to-low transition of clock 1 when the mode control is low. Shift left is done on the high-to-low transition of clock 2 when the mode control is high by connecting the output of each flip-flop to the parallel input of the previous flip-flop. The clock input may be applied commonly to both clocks if both modes can be clocked from the same source.

TTL - 7496
74LS96

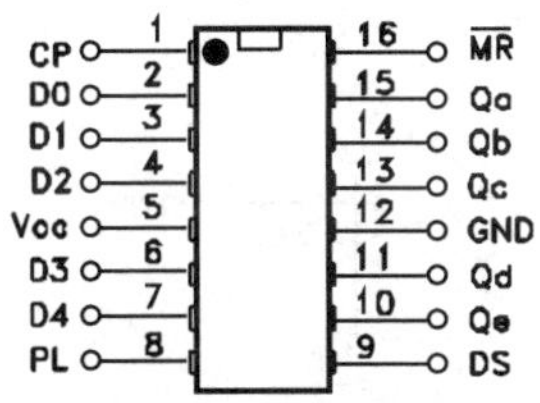

DUAL-IN-LINE PACKAGE

FUNCTION DIAGRAM 7496 5-BIT SHIFT REGISTER

OPERATING MODE	INPUTS					OUTPUTS				
	PL	Dn	MR	CP	DS	Qa	Qb	Qc	Qd	Qe
Parallel Load	1	0	X	X	X	Qa	Qb	Qc	Qd	Qe
	1	1	X	X	X	1	1	1	1	1
Reset (Clear)	0	X	0	X	X	0	0	0	0	0
Shift Right	0	X	1	↑	l	0	qa	qb	qc	qd
	0	X	1	↑	h	1	qa	qb	qa	qd

↑ = LOW-to-HIGH clock transition.
h = HIGH voltage level one setup time prior to the ↑ clock transition.
l = LOW voltage level one setup time prior to the ↑ clock transition.
dn, qn = Lower case indicates state of referenced input (or output) one setup time prior to the ↑ clock transition.

LOGIC SYMBOL

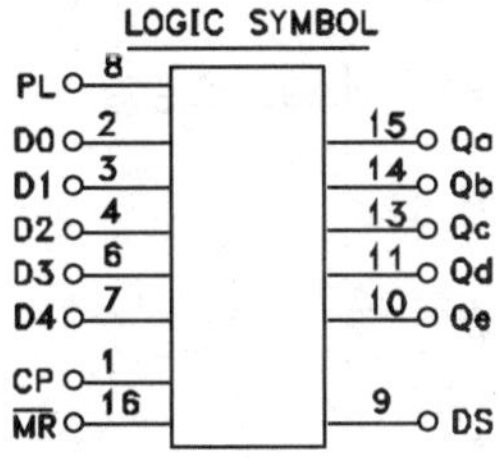

LOGIC DIAGRAM

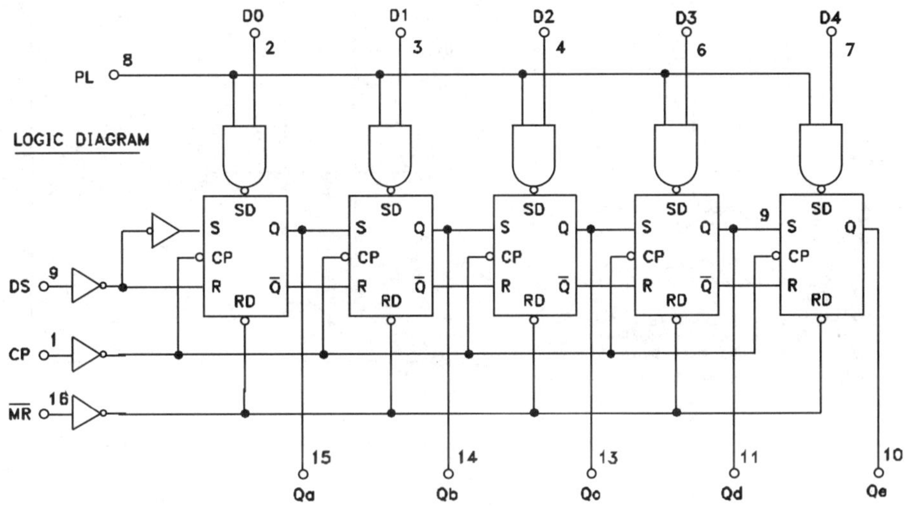

Figure 6.4. The 7496 and 'LS96 consist of five R-S master-slave flip-flops connected to perform parallel-to-serial or serial-to-parallel conversions. Inputs and outputs of each flip-flop are accessible, facilitating parallel-in/parallel-out or serial-in/serial-out operation. Applying a low level to the Master Reset clears all flip-flops when the preset is low. Clearing is independent of the clock.

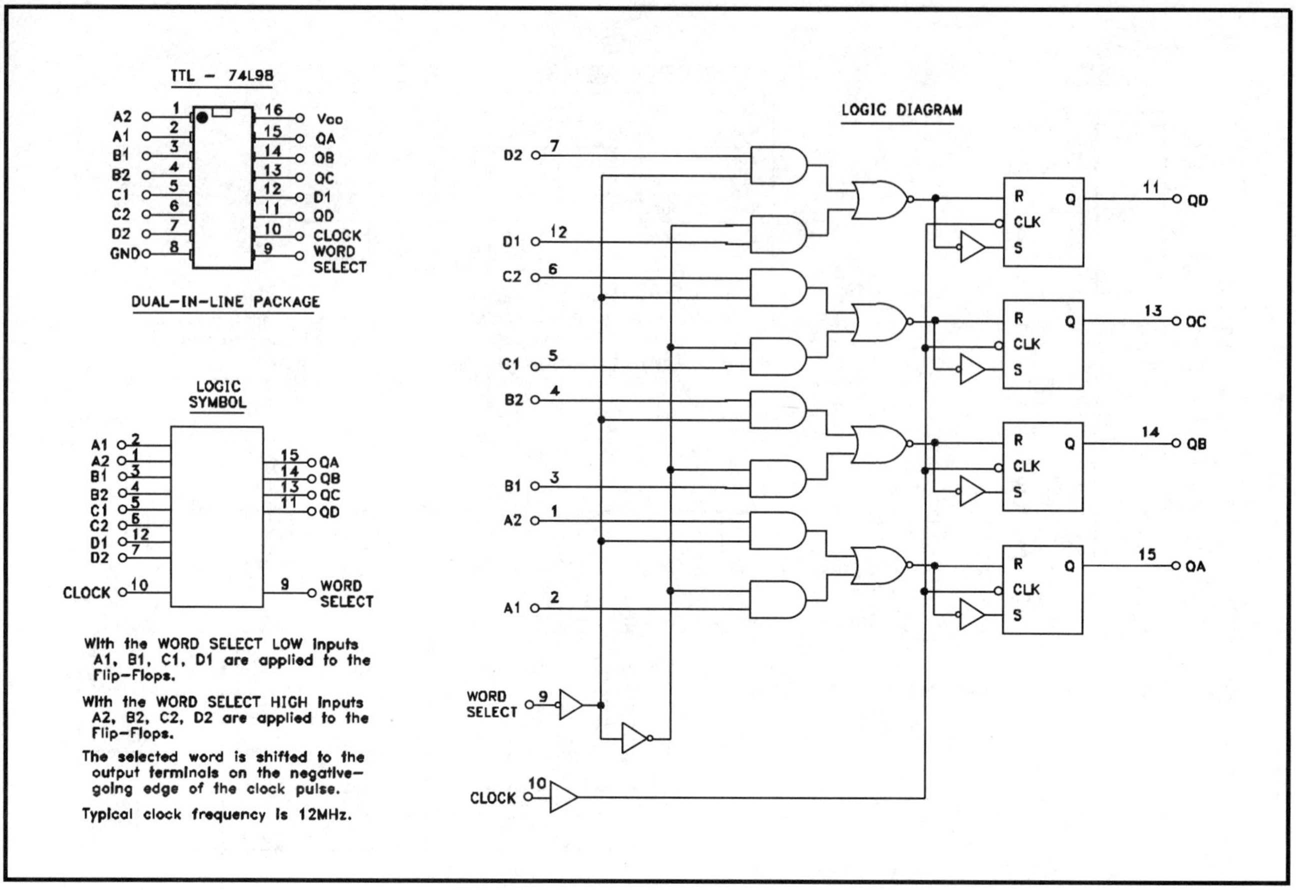

Figure 6.5. The 74L98 is composed of four R-S master-slave flip-flops, four AND-OR-INVERT gates, a buffer, and six inverter-drivers. With the word-select input low word 1 is applied to the flip-flops. A high-level word-select causes the selection of word 2. The selected word is shifted to the output on the negative-going edge of the clock pulse.

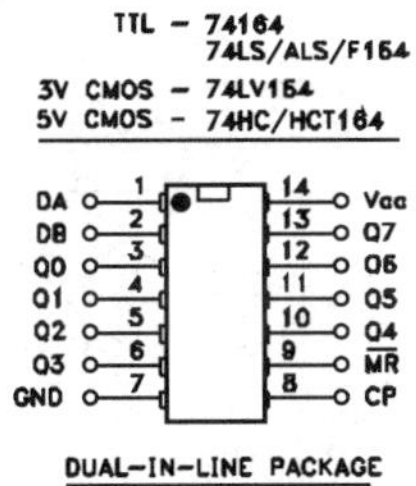

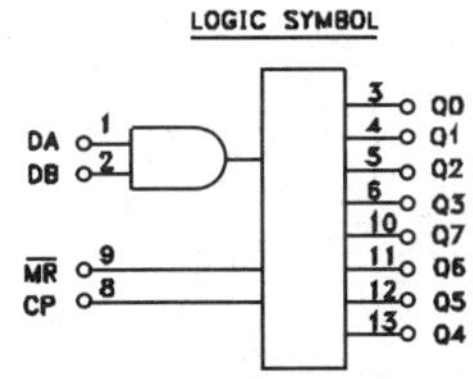

FUNCTION TABLE 74x164 8-BIT SERIAL-IN/PARALLEL OUT SHIFT REGISTER

OPERATING MODES	INPUTS				OUTPUTS	
	MR	CP	DA	DB	Q0	Q1–Q7
Reset (Clear)	X	X	X	X	0	0 – 0
Shift	↑	↑	l	l	0	q0 – q6
	↑	↑	l	h	0	q0 – q6
	↑	↑	h	l	1	q0 – q6
	↑	↑	h	h	1	q0 – q6

↑ = LOW-to-HIGH clock transition
h = HIGH voltage level one setup time prior to the ↑ clock transition.
l = LOW voltage level one setup time prior to the ↑ clock transition.
q = lower case letters indicate the state of the referenced input one setup time prior to the ↑ clock transition.

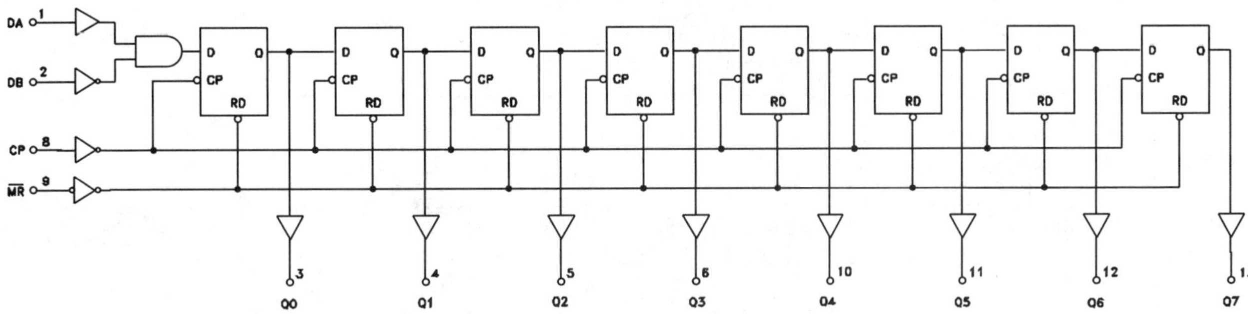

Figure 6.6. The 74x164 series are 8-bit shift registers featuring gated serial inputs and an asynchronous clear. The gated inputs permit complete control over incoming data as a low at either input inhibits entry of the new data, resetting the first flip-flop to the low level at the next clock pulse. A high level input enables the other input which then determines the state of the first flip-flop. Data at the serial inputs may be changed while the clock is high or low, but only information meeting the setup time will be entered. Clocking occurs on the low-to-high transition of the clock. Clock and serial inputs are fully buffered.

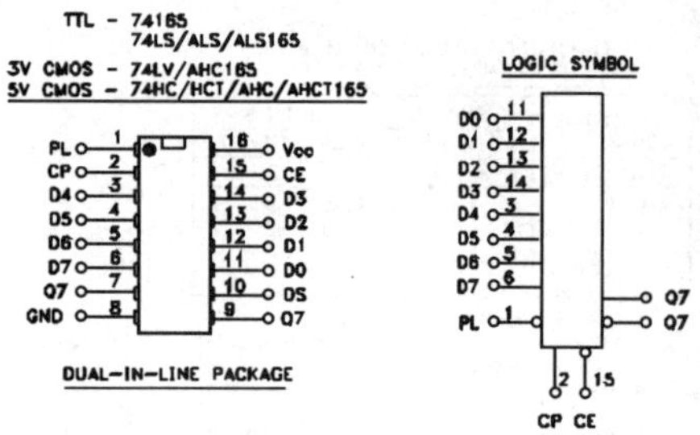

FUNCTION TABLE 74x165 8-BIT PARALLEL-IN/SERIAL-OUT SHIFT REGISTER

OPERATING MODES	INPUTS					Qn REGISTERS		OUTPUTS	
	PL	CE	CP	DS	D0-D7	Q0	Q1-Q6	Q7	Q7
Parallel Load	0	X	X	X	0	0	0 - 0	0	0
	0	X	X	X	1	1	1 - 1	1	1
Serial Shift	1	0	↑	l	X	0	q0-q5	q6	q6
	1	0	↑	h	X	0	q0-q5	q6	q6
Hold "Do Nothing"	1	1	X	X	X	q0	q1-q6	q7	q7

↑ = LOW-to-HIGH clock transition
h = HIGH voltage level one setup time prior to the ↑ clock transition.
l = LOW voltage level one setup time prior to the ↑ clock transition.
q = lower case letters indicate the state of the referenced input one setup time prior to the ↑ clock transition.

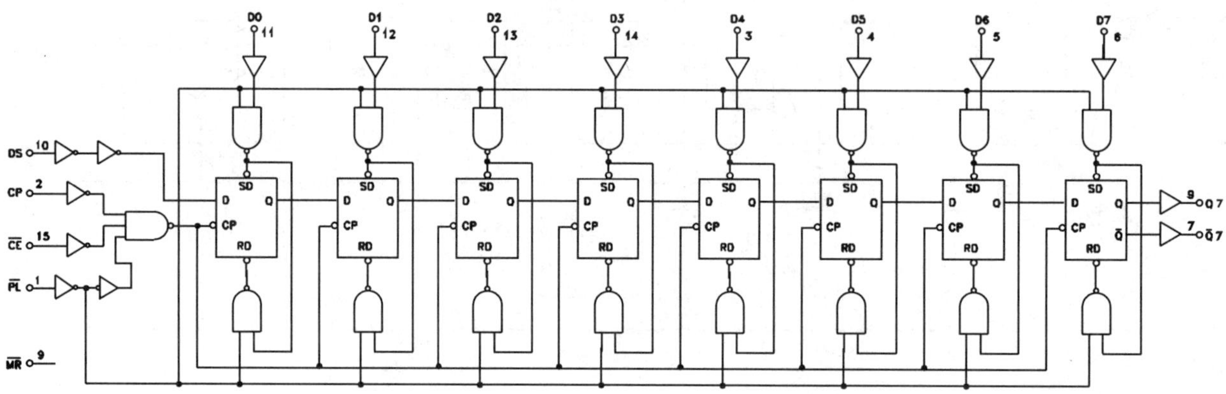

Figure 6.7. The 74x165 series are 8-bit serial shift registers which shift the data in the direction QA to QH when clocked. Parallel-in access to each stage is made available by eight individual data inputs enabled by a low level at the shift/load input. The registers also feature gated clock inputs and a complementary output from the eighth bit. Clocking via a 2-input NOR gate permits one input to be employed as a clock inhibit. Either input held high inhibits clocking and either input low with the shift/load high enables the other clock input. Data at the parallel inputs are loaded directly into the register when the shift/load is low independently of the clock, clock inhibit, or serial inputs.

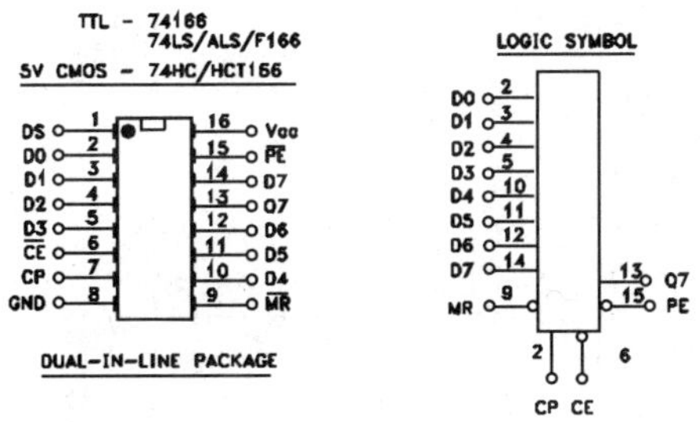

FUNCTION TABLE 74x166 8-BIT
PARALLEL-IN/SERIAL-OUT SHIFT REGISTER

OPERATING MODES	INPUTS						Qn REGISTERS		OUTPUT
	MR	PE	CE	CP	DS	D0–D7	Q0	Q1–Q6	Q7
Reset	0	X	X	X	X	X	0	0 – 0	0
Parallel Load	1	l	l	↑	X	l	0	0 – 0	0
	1	l	l	↑	X	h	1	1 – 1	1
Serial Shift	1	h	l	↑	l	X	0	q0 – q5	q6
	1	h	l	↑	h	X	1	q0 – q5	q6
Hold "Do Nothing"	1	X	h	X	X	X	q0	q1 – q6	q7

↑ = LOW-to-HIGH clock transition
h = HIGH voltage level one setup time prior to the ↑ clock transition.
l = LOW voltage level one setup time prior to the ↑ clock transition.
q = lower case letters indicate the state of the referenced input one setup time prior to the ↑ clock transition.

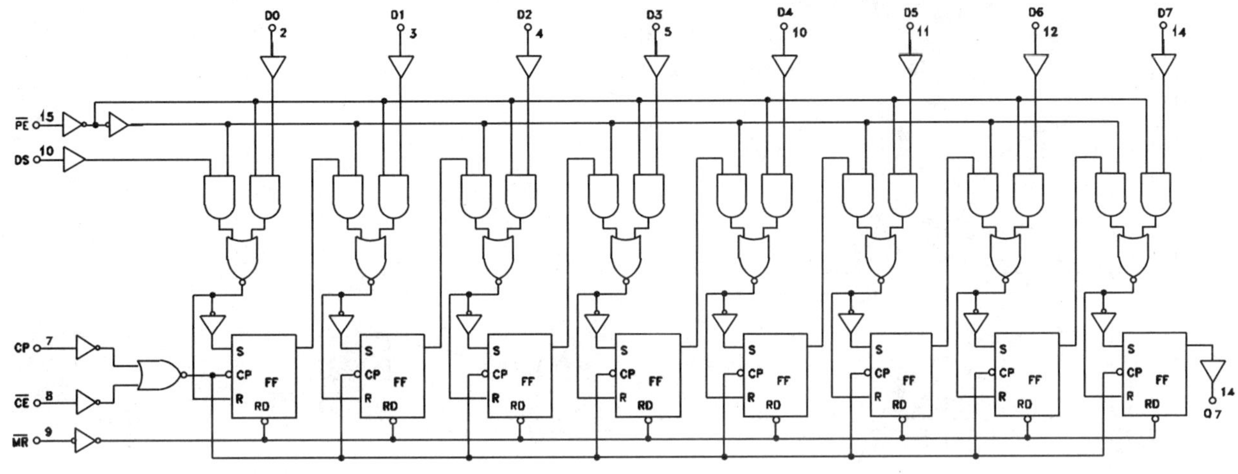

Figure 6.8. The 74x166 series are parallel load 8-bit shift registers with buffered inputs to lower the drive requirements to one standard load (TTL devices). The devices feature gated clock inputs and an overriding clear input. The parallel-in or serial-in modes are set by the shift/load input. When high, this enables the serial data input and couples the eight flip-flops for serial shifting with each clock pulse. When low, the parallel inputs are enabled and synchronous loading occurs on the next clock pulse. During parallel loading serial data flow is inhibited. Clocking occurs on the low-to-high edge of the clock pulse through a 2-input NOR gate permitting one input to be used as a clock inhibit or enable. Either input high inhibits clocking; either low enables the other's clocking. The inhibit high transition should occur only while the clock input is high.

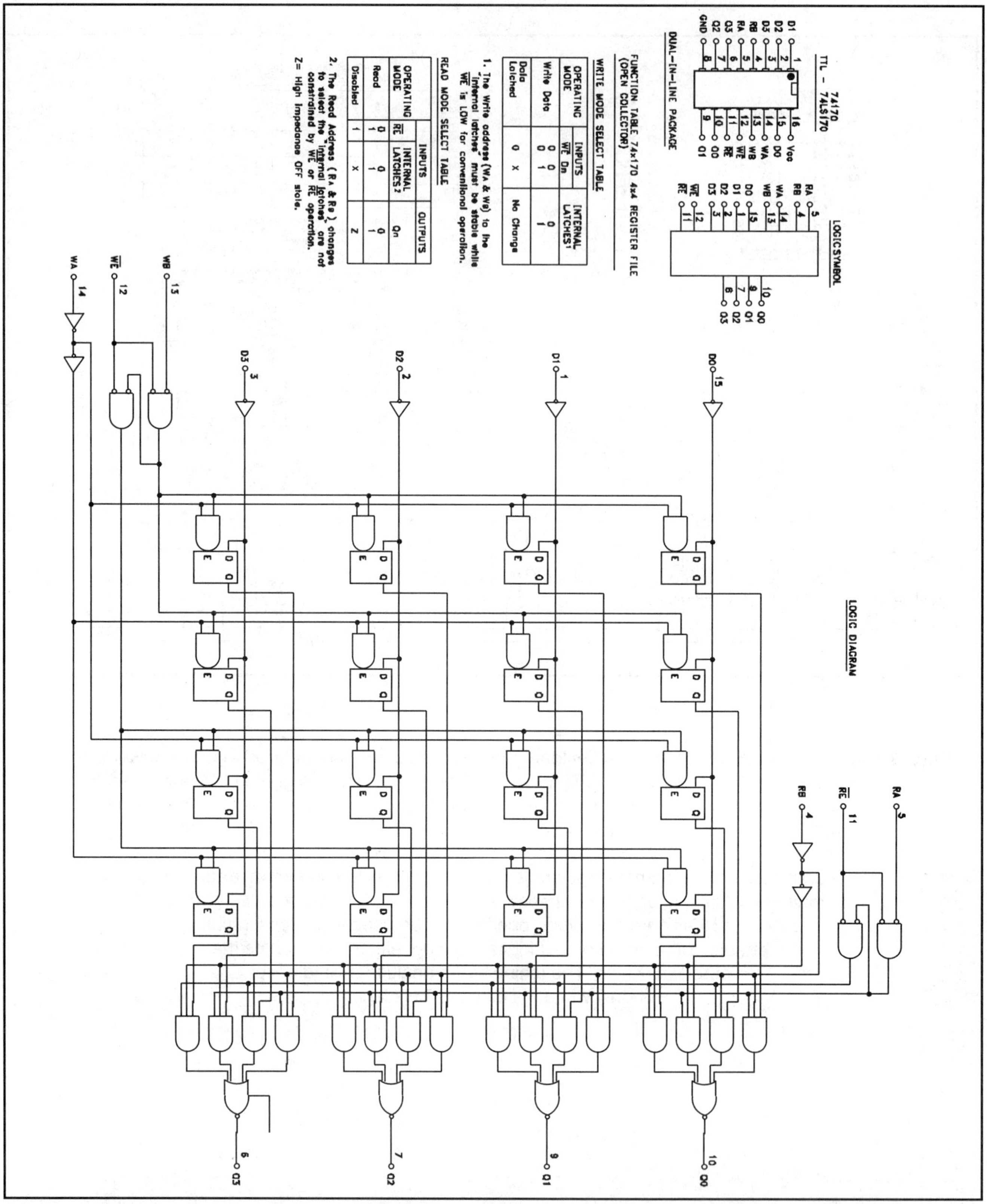

Figure 6.9. The 74170 and 'LS170 are organized as 4 words of 4 bits each in a file equivalent to 98 gates. On-chip decoding is provided for addressing the locations for either write-in or data retrieval. This permits writing to one location while reading from another word location. Location of the word is determined by the write-address A and B inputs in conjunction with an active-low write-enable. Data at the inputs should be present in its true form. Latch inputs are such that data willbe accepted only if both internal address gates are high. In this condition, data at the D input is transferred to the latch output. Data entry is inhibited with the write-enable input high. Data outputs are inhibited and remain high with the read-enable input high.

TTL – 74172
74S172

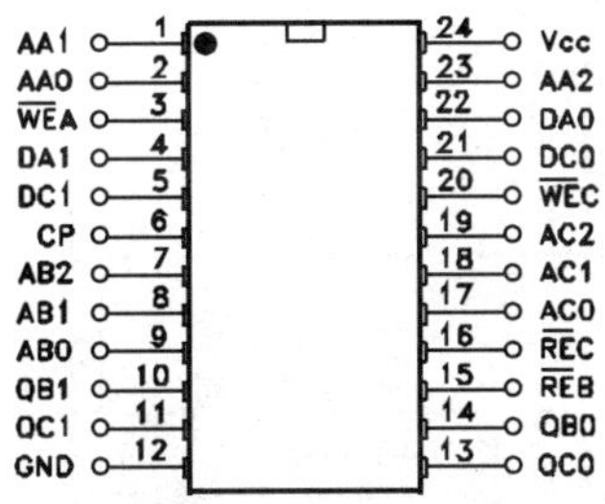

DUAL-IN-LINE PACKAGE

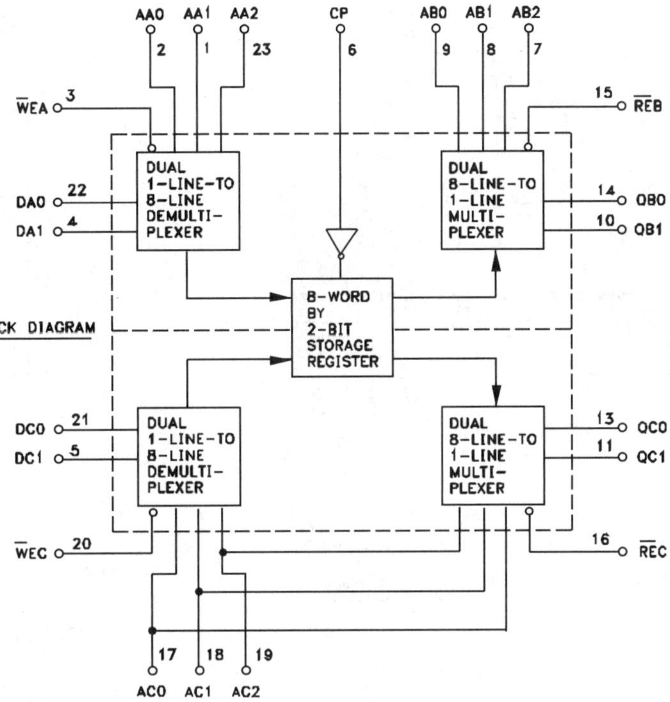

BLOCK DIAGRAM

FUNCTION TABLE 74x172 16-BIT MULTIPLE PORT REGISTER FILE

WRITE MODE SELECT TABLE

OPERATING MODE	INPUTS CP	$\overline{WE}$	Dn	ADDRESSED REGISTER
Write Data[b]	↑	l	l	0
	↑	l	h	1
Hold[c]	↓		h	X

READ MODE SELECT TABLE

OPERATING MODE	INPUTS $\overline{RE}$	ADDRESSED REGISTER	OUTPUTS On
Read	0	0	0
	0	1	1
Disabled	1	X	Z

↓ = HIGH-to-LOW clock transition.
↑ = LOW-to-HIGH clock transition.
h = HIGH voltage level one setup time prior to the ↓ clock transition.
l = LOW voltage level one setup time prior to the ↓ clock transition.
b = The Write Address (AA & AC) to the "internal register" must be stable whiel $\overline{WE}$ and CP are LOW for conventional operation
c = The Write Enable must be HIGH before the ↓ clock transition to ensure that the data in the register is not changed.

Figure 6.10. The 74172 and 'S172 are organized as 8 words of 2 bits each in a file incorporating the equivalent of 201 gates. On-chip decoding is provided for addressing the locations for either write-in or data retrieval. Multiple address decoding is such as to permit simultaneous read and write operations independently on two locations. The file essentially consists of two distinct sections. 3-state outputs simplify application to bus-organized systems. Applications include stacked data registers, scratch-pad memory, buffer storage between processors, and fast multiplication.

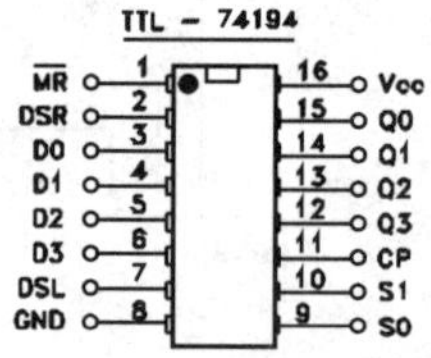

DUAL-IN-LINE PACKAGE

FUNCTION TABLE 74194 4-BIT BIDIRECTIONAL UNIVERSAL SHIFT REGISTER

OPERATING MODE	INPUTS							OUTPUTS			
	CP	MR	S1	S0	DSR	DSL	Dn	Q0	Q1	Q2	Q3
Reset	x	0	x	x	x	x	x	L	L	L	L
Hold	x	1	l(b)	l(b)	x	x	x	q0	q1	q2	q3
Shift Left	↑	1	h	l(b)	x	l	x	q1	q2	q3	L
	↑	1	h	l(b)	x	h	x	q1	q2	q3	H
Shift Right	↑	1	l(b)	h	l	x	x	L	q0	q1	q2
	↑	1	l(b)	h	h	x	x	H	q0	q1	q2
Parallel Load	↑	1	h	h	x	x	dn	d0	d1	d2	d3

h = HIGH voltage level one setup time prior to the ↑ clock transition.
l = LOW voltage level one setup time prior to the ↑ clock transition.
dn, qn = Lower case indicates state of referenced input (or output) one setup time prior to the ↑ clock transition.
(b) = The ↑ transition of the S0, S1 inputs should occur only when the CP is HIGH for conventional operation.

LOGIC SYMBOL

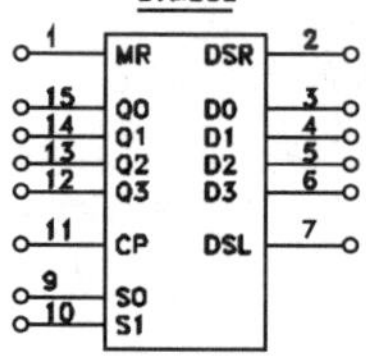

LOGIC DIAGRAM

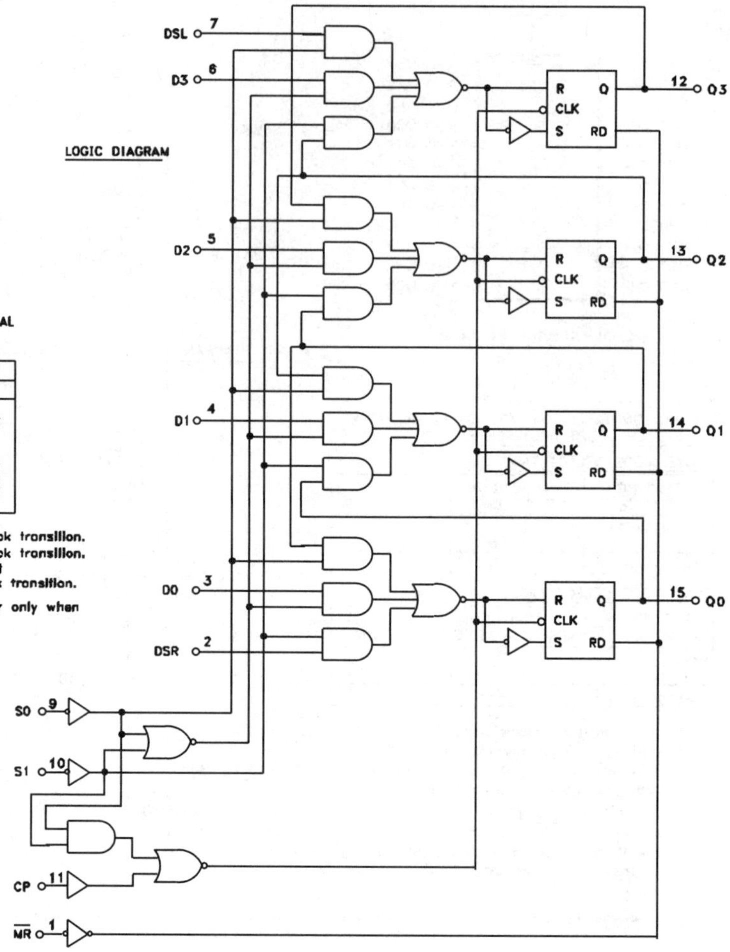

Figure 6.11. The 74194 is a bidirectional shift register designed to accommodate a broad variety of shift register tasks. Features include parallel inputs and outputs, right shift and left shift serial inputs, operating mode control, and a direct overriding clear. Four distinct modes of operation are: Clock inhibit (do nothing), Shift right in the direction QA > QD, Shift left in the direction QD > QA, and parallel (broadside) load. Synchronous parallel loading is achieved by applying the four data bits and taking both mode control inputs high. Data is loaded and appears at the outputs following the positive transition of the clock input. Serial data flow is inhibited during loading. Shift right is performed synchronously with the rising edge of the clock when S0 is high, S1 low. Data shifts left synchronously with S0 low, S1 high.Clocking is inhibited with both mode controls low.

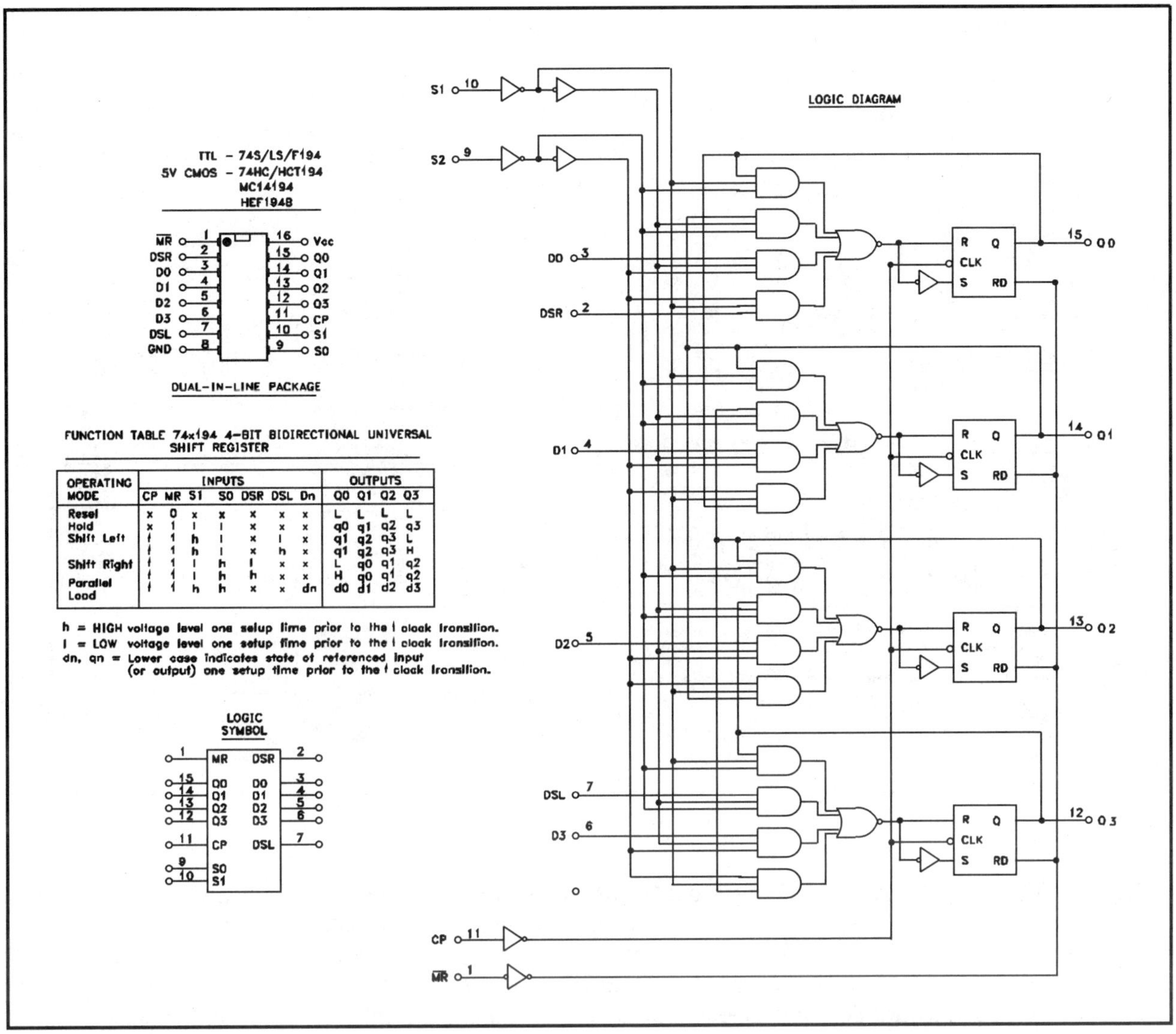

FUNCTION TABLE 74x194 4-BIT BIDIRECTIONAL UNIVERSAL SHIFT REGISTER

OPERATING MODE	INPUTS							OUTPUTS			
	CP	MR	S1	S0	DSR	DSL	Dn	Q0	Q1	Q2	Q3
Reset	x	0	x	x	x	x	x	L	L	L	L
Hold	x	1	l	l	x	x	x	q0	q1	q2	q3
Shift Left	↑	1	h	l	x	l	x	q1	q2	q3	L
	↑	1	h	l	x	h	x	q1	q2	q3	H
Shift Right	↑	1	l	h	l	x	x	L	q0	q1	q2
	↑	1	l	h	h	x	x	H	q0	q1	q2
Parallel Load	↑	1	h	h	x	x	dn	d0	d1	d2	d3

h = HIGH voltage level one setup time prior to the ↑ clock transition.
l = LOW voltage level one setup time prior to the ↑ clock transition.
dn, qn = Lower case indicates state of referenced input (or output) one setup time prior to the ↑ clock transition.

Figure 6.12. The 74x194 are identical in performance to the 74194. However, they feature a differing internal structure, for which it seemed worthwhile to provide an individual figure.

TTL – 74195
74S/LS/AS/F195
5V CMOS – 74C/HC/HCT195
HEF40195B

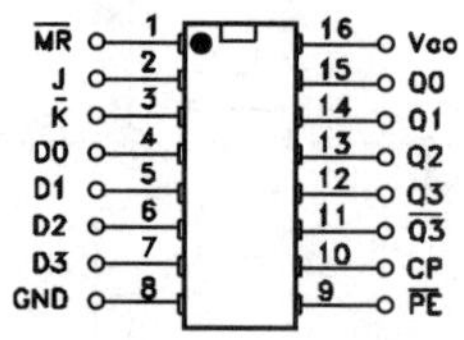

DUAL-IN-LINE PACKAGE

LOGIC SYMBOL

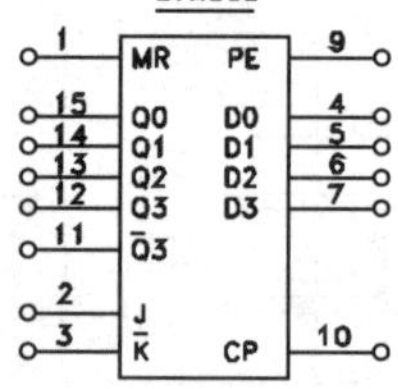

FUNCTION TABLE 74x195 4-BIT PARALLEL ACCESS SHIFT REGISTER

OPERATING MODE	INPUTS						OUTPUTS				
	MR	CP	$\overline{PE}$	J	$\bar{K}$	Dn	Q0	Q1	Q2	Q3	$\bar{Q}3$
Asynoh Reset	0	x	x	x	x	x	L	L	L	L	H
Shift,Set 1st Stage	1	↑	h	h	h	x	H	q0	q1	q2	$\bar{q}2$
Shift,Rst 1st Stage	1	↑	h	l	l	x	L	q0	q1	q2	$\bar{q}2$
Shift,Tgl 1st Stage	1	↑	h	h	l	x	$\bar{q}0$	q0	q1	q2	$\bar{q}2$
Shift Retain 1st Stg	1	↑	h	l	h	x	q0	q0	q1	q2	$\bar{q}2$
Parallel Load	1	↑	l	x	x	dn	d0	d1	d2	d3	$\bar{d}3$

h = HIGH voltage level one setup time prior to the ↑ clock transition.
l = LOW voltage level one setup time prior to the ↑ clock transition.
dn, qn = Lower case indicates state of referenced input (or output) one setup time prior to the ↑ clock transition.

LOGIC DIAGRAM

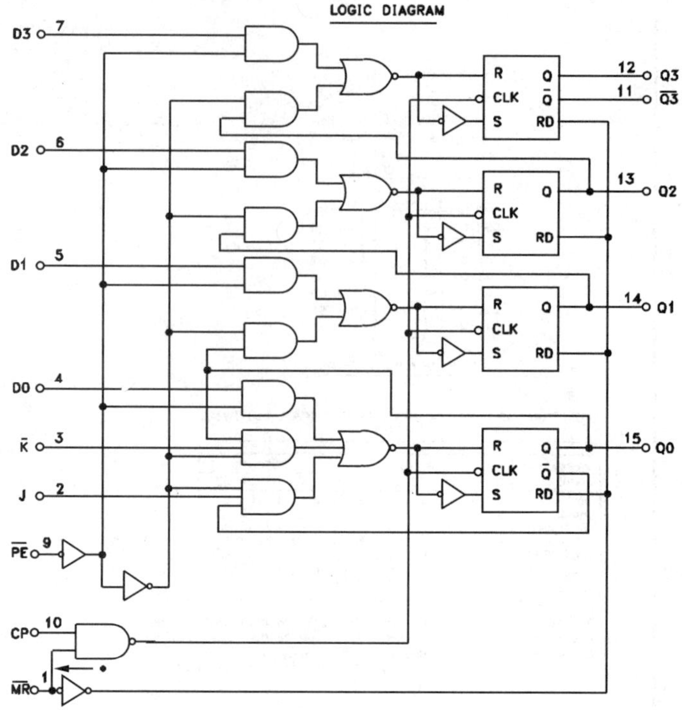

Figure 6.13. The 74x195 are 4-bit registers featuring parallel inputs and outputs, J and /K logic inputs, shift/load control inputs, and a direct overriding clear. All inputs are buffered. The register's two modes of operation are: Parallel (broadside) load, and shift in the direction QA > QD. Parallel loading occurs with shift/load input low. Data is loaded and appears at the outputs following the positive transition of the clock. Serial flow is inhibited during loading. Shifting takes place synchronously with the shift/load input high. Serial data is entered at the J, /K inputs. These inputs allow the first stage to operate in the J, /K, D, or T-flip-flop modes.

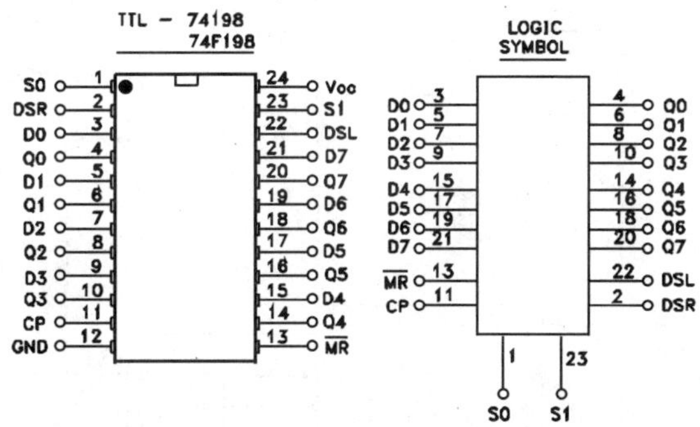

FUNCTION TABLE 74198 8-BIT BIDIRECTIONAL UNIVERSAL SHIFT REGISTER
74F198 8-BIT BIDIRECTIONAL UNIVERSAL SHIFT REGISTER

OPERATING MODE	INPUTS							OUTPUTS		
	CP	$\overline{MR}$	S1	S0	DSR	DSL	Dn	Q0	Q1 . . . Q6	Q7
Reset	x	0	x	x	x	x	x	0	0 · 0 · 0	0
Hold (Do Nothing)	↑	1	l	l	x	x	x	q0	q1 · · · q6	q7
Shift Left	↑	1	h	l	x	l	x	q1	q2 · · · q7	0
	↑	1	h	l	x	h	x	q1	q2 · · · q7	1
Shift Right	↑	1	l	h	l	x	x	0	q0 · · · q5	q6
	↑	1	l	h	h	x	x	1	q0 · · · q5	q6
Parallel Load	↑	1	h	h	h	x	dn	d0	d1 · · · d6	d7

↑ = LOW–to–HIGH clock transition.
h = HIGH voltage level one setup time prior to the
l = LOW voltage level one setup time prior to the ↑ clock transition.
dn, qn = Lower case indicates state of referenced input (or output) one setup time prior to the ↑ clock transition.

The HIGH–to–LOW transition of S0, S1 inputs should only take place while CP is HIGH for conventional operation.

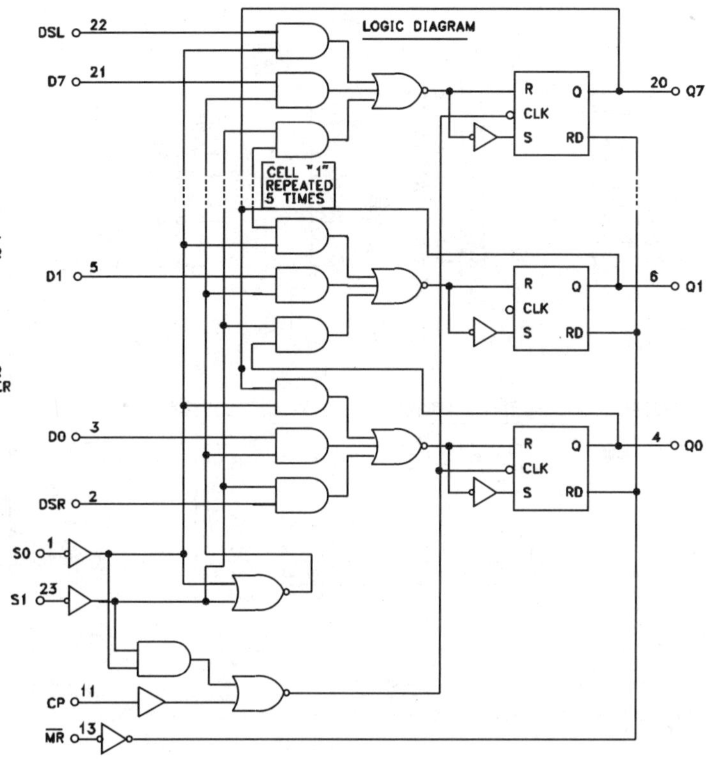

Figure 6.14. The 74x198 are 8-bit registers featuring parallel inputs and outputs, right- and left-shift serial inputs, operating mode control inputs, and a direct overriding clear. All inputs are buffered. The register's four modes of operation are: Inhibit clock (do nothing), shift right (QA > QH), shift left (QH > QA), and Parallel (broadside) load. Parallel loading occurs with taking mode control inputs S0 and S1 high. Data is loaded and appears at the outputs following the positive transition of the clock. Serial flow is inhibited during loading. Shift right takes place synchronously with the rising edge of the clock when S0 is high, S1 low. Serial data is entered at the shift-right data input. Shift left occurs synchronously when S0 is low, S1 high. Serial data in this mode is entered at the shift-left serial input. Clocking is inhibited with both mode controls low.

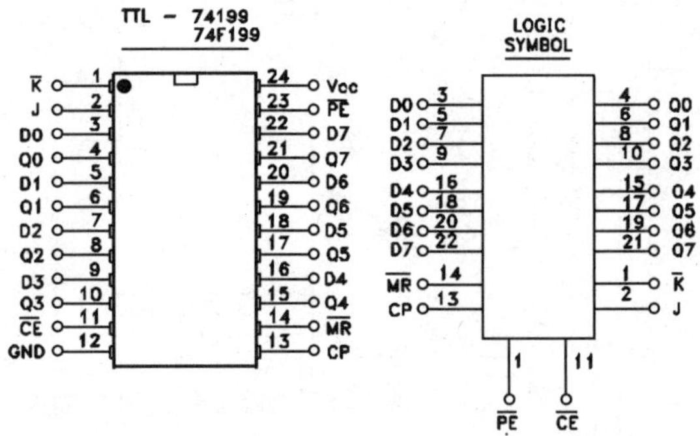

FUNCTION TABLE 74199/F199 8-BIT PARALLEL ACCESS SHIFT REGISTER

OPERATING MODE	INPUTS							OUTPUTS				
	$\overline{MR}$	CP	$\overline{PE}$	$\overline{CE}$	J	$\bar{K}$	Dn	Q0	Q1	…	Q6	Q7
Asynch Reset	0	x	x	x	x	x	x	0	0	…	0	0
Shift,Set 1st Stage	1	↑	h	l	h	h	x	1	q0	…	q5	q6
Shift,Rst 1st Stage	1	↑	h	l	l	l	x	0	q0	…	q5	q6
Shift,Tgl 1st Stage	1	↑	h	l	h	l	x	$\bar{q}0$	q0	…	q5	q6
Shift Retain 1st Stg	1	↑	h	l	l	h	x	q0	q0	…	q5	q6
Parallel Load	1	↑	l	l	x	x	dn	d0	d1	…	d6	d7
Hold (Do Nothing)	1	↑	x	h	x	x	x	q0	q1	…	q6	q7

↑ = LOW-to-HIGH clock transition.
h = HIGH voltage level one setup time prior to the ↑ clock transition.
l = LOW voltage level one setup time prior to the ↑ clock transition.
dn, qn = Lower case indicates state of referenced input (or output) one setup time prior to the ↑ clock transition.
The LOW-to-HIGH transition of CE should only occur when CP is HIGH for conventional operation.

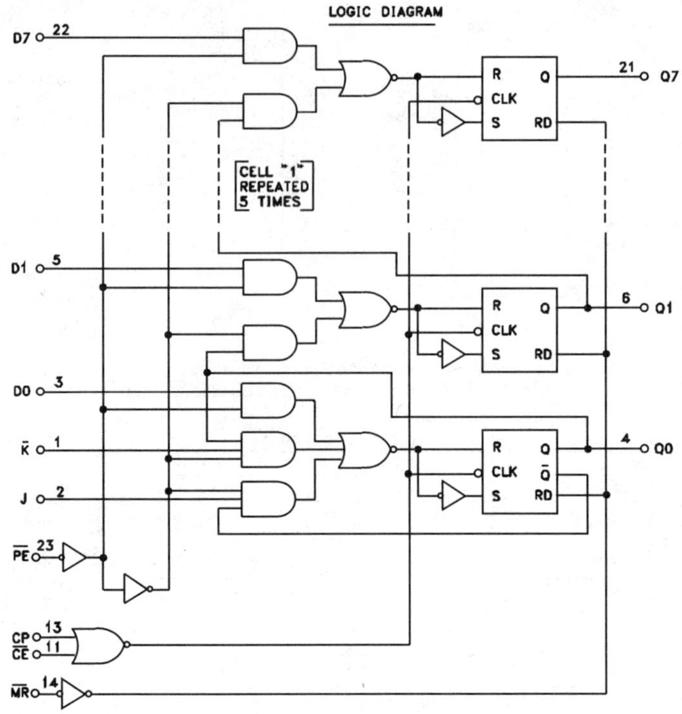

Figure 6.15. The 74x199 are 8-bit registers featuring parallel inputs and outputs, J, /K serial inputs, operating mode control inputs, a direct overriding clear and gated clock inputs. All inputs are buffered. The register's three modes of operation are: Inhibit clock (do nothing), shift QA > QH, and parallel (broadside) load. Parallel loading occurs with taking the shift/load control input low. Data is loaded and appears at the outputs following the positive transition of the clock. Serial flow is inhibited during loading. Shifting takes place synchronously with the shift/load high and the clock not inhibited. The two clock inputs are identical and may be used interchangeably as clock or clock inhibit inputs. Clocking is inhibited with either input held high.

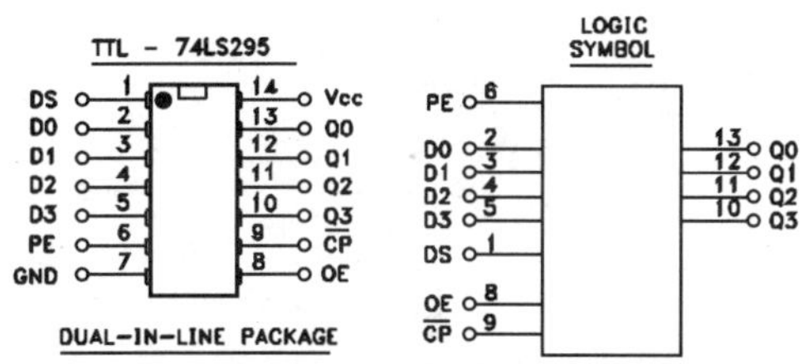

FUNCTION TABLE – 74LS295 4-BIT SHIFT REGISTER WITH 3-STATE OUTPUTS

REGISTER OPERATING MODES	INPUTS				OUTPUTS			
	$\overline{CP}$	PE	DS	Dn	Q0	Q1	Q2	Q3
Parallel Load	↓	l	l	x	0	q0	q1	q2
	↓	l	h	x	1	q0	q1	q2
Shift Right	↓	h	x	x	0	0	0	0
	↓	h	x	x	1	1	1	1

↓ = HIGH-to-LOW clock transition.
h = HIGH voltage level one setup time prior to the ↓ clock transition.
l = LOW voltage level one setup time prior to the ↓ clock transition.
dn, qn = Lower case indicates state of referenced input (or output) one setup time prior to the ↓ clock transition.

FUNCTION TABLE

3-STATE OPERATING MODES	INPUTS		OUTPUTS
	OE	Qn(Reg)	Q0 Q1 Q2 Q3
Read	1	0	0
	1	1	1
Disabled	0	X	Z

Z = High impedance OFF state

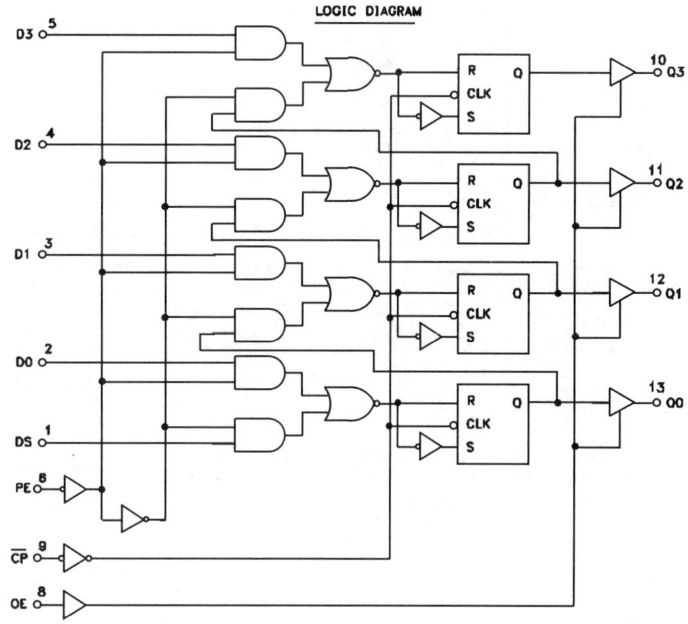

Figure 6.16. The 74LS295 is a 4-bit right-shift, left-shift register with 3-state outputs. The device features parallel inputs and outputs, clock, mode, and output control inputs. The register's three modes of operation are: shift right (QA > QD), shift left (QD > QA), and parallel (broadside) load. Parallel loading occurs with taking the mode control input high. Data is loaded and appears at the outputs following the high-to-low transition of the clock. Serial flow is inhibited during loading. Shift right takes place with the mode control low. Shift left is achieved when the mode control is high by connecting the output of each flip-flop to the parallel input of the previous flip-flop with serial data entered at input D3. Normal logic levels are available at the output with the output control high. Taking the control low causes the output to be a high impedance which neither load nor drive the bus.

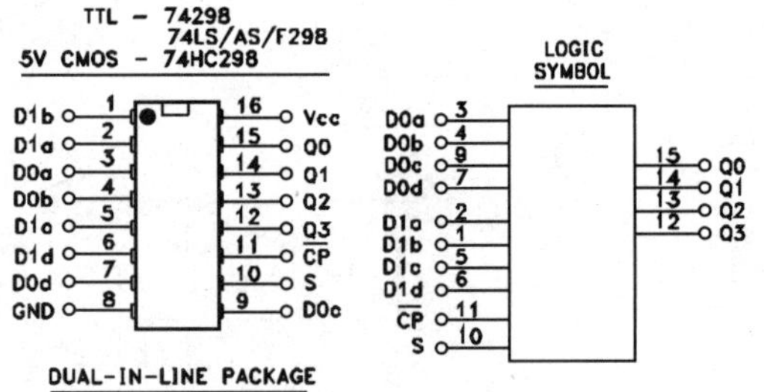

FUNCTION TABLE 74x298 QUAD 2-PORT REGISTER

OPERATING MODES	INPUTS				OUTPUTS
	$\overline{CP}$	S	D0	D1	Qn
Load Source "0"	↓	l	l	x	0
	↓	l	h	x	1
Load Source "1"	↓	h	x	l	0
	↓	h	x	h	1

↓ = HIGH-to-LOW clock transition.
h = HIGH voltage level one setup time prior to the ↓ clock transition.
l = LOW voltage level one setup time prior to the ↓ clock transition.

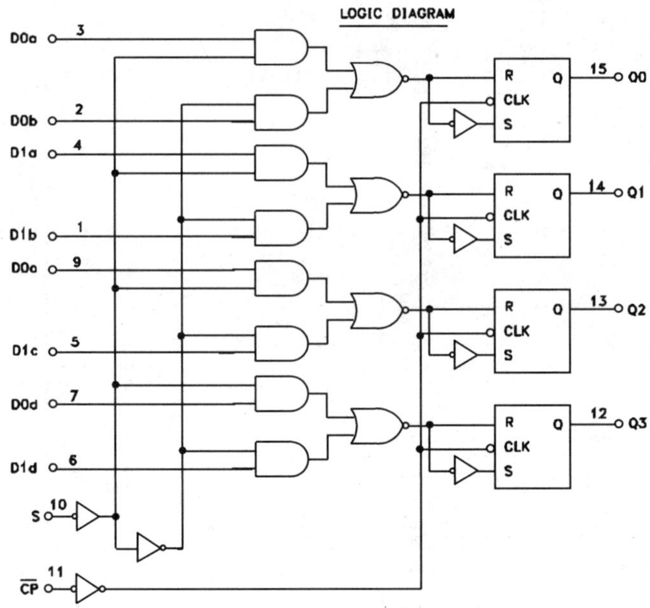

Figure 6.17. The 74x298 are quadruple 2-input multiplexers with storage. These implement separate registers capable of parallel exchange of contents, while retaining external load capability. Their application is as a universal type register for implementing various shift patterns, including compound left-right capabilities. With the word-select input low word 1 is applied to the flip-flops. A high on the select causes the selection of word 2. The selected word is clocked to the output on the negative-going edge of the clock pulse.

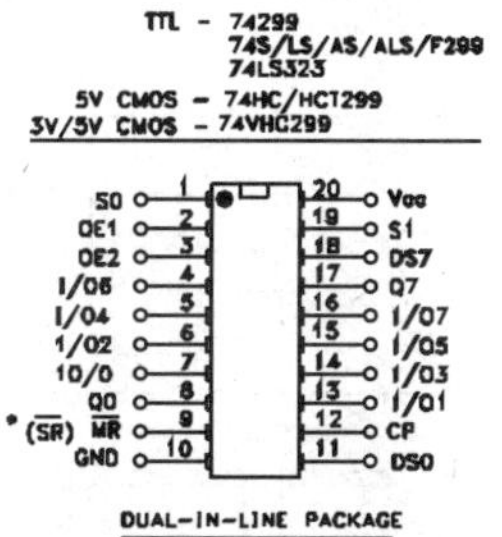

* SYNCHRONOUS RESET IN PLACE OF THE ASYNCHRONOUS MASTER RESET FOR THE 74LS323.

FUNCTION TABLE 74x299 8-BIT UNIVERSAL SHIFT/STORAGE RESISTER

OPERATING MODE	INPUTS							OUTPUTS				
	MR	CP	S0	S1	DS0	DS7	I/On	Q0	Q1	...	Q6	Q7
Reset	0	x	x	x	x	x	x	L	L	...	L	L
Shift Right	1	↑	h	l	l	x	x	L	q0	...	q5	q6
	1	↑	h	l	h	x	x	H	q0	...	q5	Q6
Shift Left	1	↑	l	h	x	l	x	q1	q2	...	q7	L
	1	↑	l	h	x	h	x	q1	q2	...	q7	H
Hold	1	x	l	l	x	x	x	q0	q1	...	q6	q7
Parallel Load	1	↑	h	h	x	x	l	L	L	...	L	L
	1	↑	h	h	x	x	h	H	H	...	H	H

↑ = LOW-to-HIGH clock transition.
h = HIGH voltage level one setup time prior to the ↑ clock transition.
l = LOW voltage level one setup time prior to the ↑ clock transition.
qn = Lower case indicates the state of the referenced output one setup time prior to the ↑ clock transition.

FUNCTION TABLE 3-STATE I/O PORT OPERATING MODES

OPERATING MODE	$\overline{OE1}$	$\overline{OE2}$	S0	S1	Qn (Register)	INPUTS/OUTPUTS I/O0 ... I/O7
Read Register	0	0	0	x	0	0
	0	0	0	x	1	1
	0	0	x	0	0	0
	0	0	x	0	1	1
Load Register	x	x	1	1	Qn = I/On	I/On = Inputs
Disable I/O	1	x	x	x	x	Z
	x	1	x	x	x	Z

Z= High impedance OFF state.

LOGIC SYMBOL

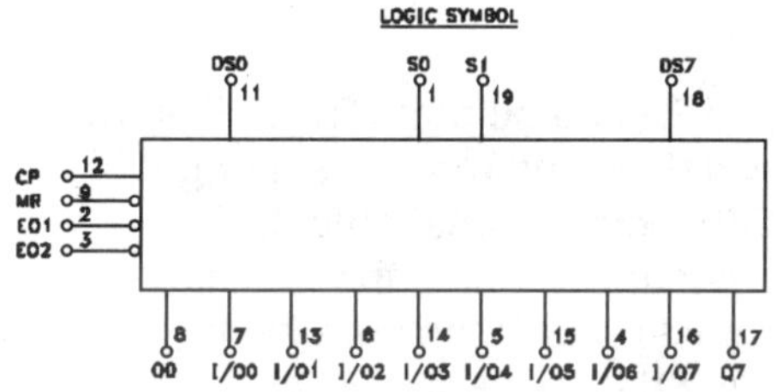

LOGIC DIAGRAM

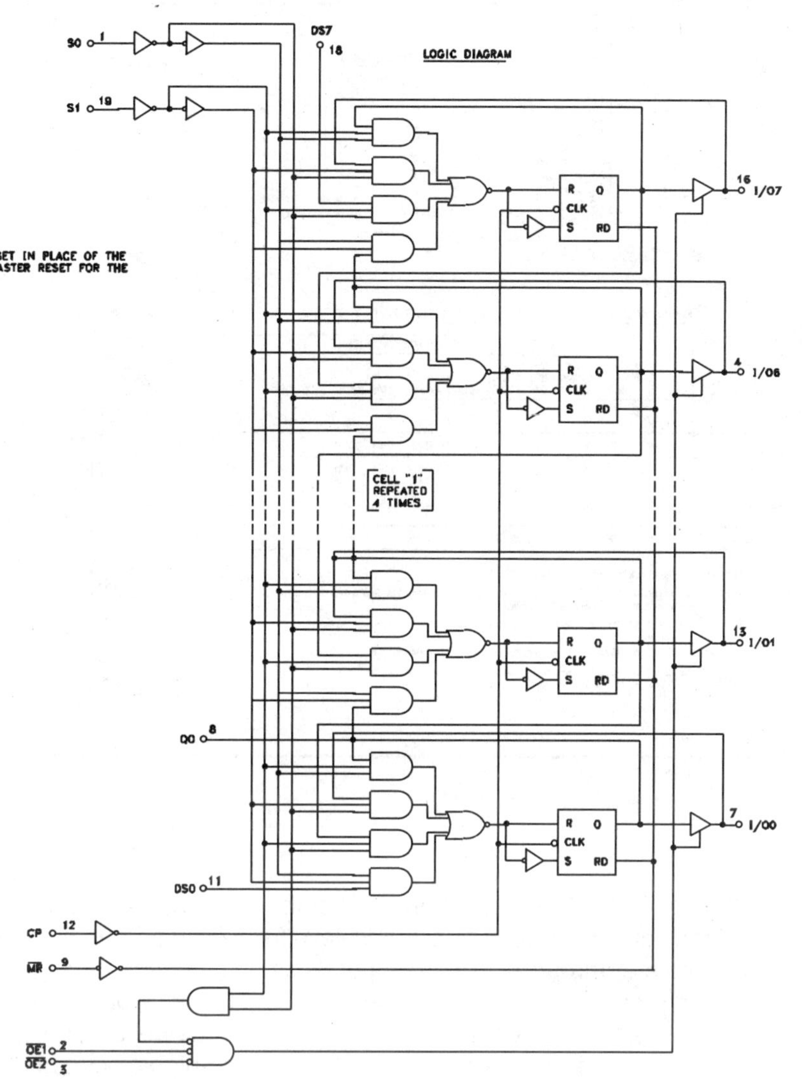

Figure 6.18. The 74x299 and 'LS323 are 8-bit universal shift/storage registers with 3-state outputs. These feature multiplexed input/outputs to achieve full 8-bit data handling. The device is provided with two function select inputs and two output control inputs for use in selecting the desired operating mode from the function table. Synchronous parallel loading occurs with both select lines S0, S1 high. This also places the outputs in the high impedance state. The register content can be read out in any mode in which the outputs are enabled. The direct overriding clear functions whether the outputs are enabled or off.

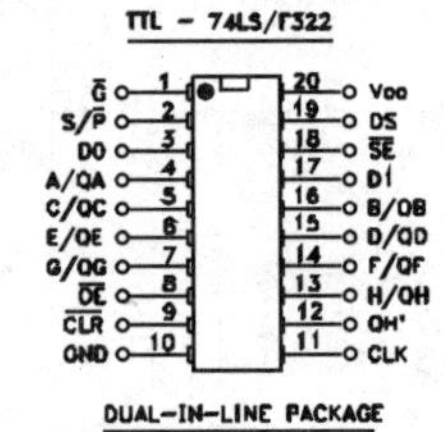

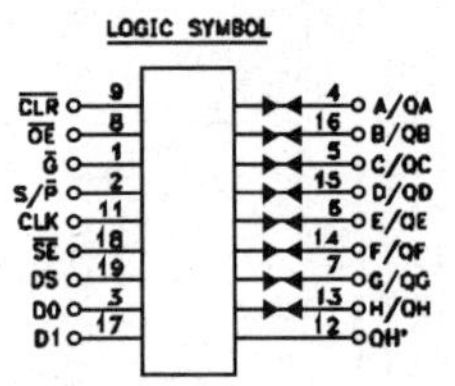

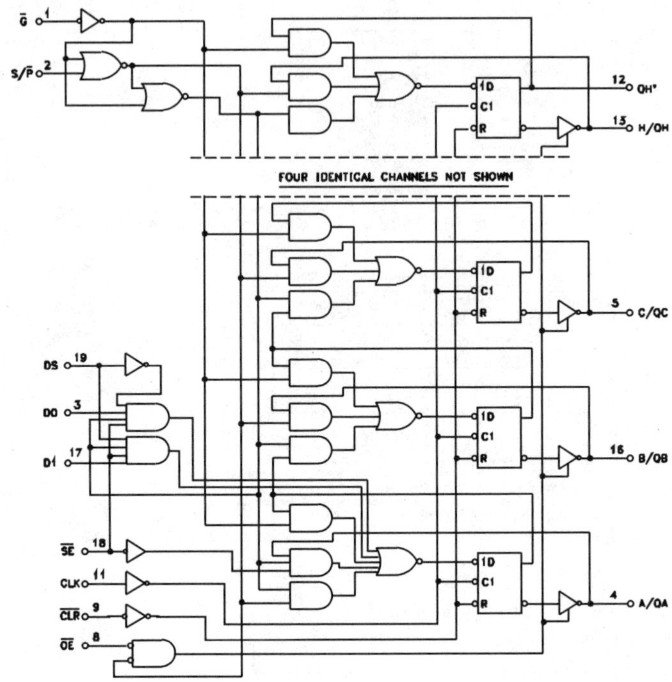

FUNCTION TABLE 74x322 8-BIT SHIFT REGISTERWITH SIGN EXTEND

OPERATION	INPUTS CLR	REG ENB G̅	S/P̅	SIGN EXTND S̅E̅	DATA SEL DS	OUT ENB O̅E̅	CLK	INPUTS/OUTPUTS A/QA	B/QB	C/QC ...	H/QH	OUTPUT QH'
Clear	0	1	X	X	X	0	X	0	0	0	0	0
	0	x	1	X	X	0	X	0	0	0	0	0
Hold	1	1	X	X	X	0	X	QA0	QB0	QC0	QH0	QH0
Shift Right	1	0	1	1	1	0	↑	D0	QAn	QBn	QGn	QGn
	1	0	1	1	0	0	↑	D1	QAn	QBn	QGn	QGn
Sign Extend	1	0	1	0	X	0	↑	QAn	QAn	QBn	QGn	QGn
Load	1	0	0	X	X	X	↑	a	b	c	h	h

↑ = LOW-to-HIGH clock transition.

QA0–QH0 = the level of QA through QH, respectively, before the indicated steady state conditions were established.

QAn–QHn = the level of QA through QH, respectively, before the most recent ↑ transitions of the clock.

D0, D1 = the level of steady state inputs D0 and D1 respectively.

a...h = the level of steady inputs at inputs A through H respectively.

Figure 6.19. The 74x322 are 8-bit shift registers with sign extend and 3-state outputs. These feature multiplexed input/outputs to achieve full 8-bit data handling. Serial data is entered into shift-right register through D0 or D1 input as selected by the data select (DS) input. A serial output, QH', is provided to facilitate expansion. Synchronous parallel loading occurs with both the register enable and S//P lines low. This also places the 3-state input/output ports in the data input mode. Data are entered on the low-to-high clock transition. The data extend function repeats the sign in the QA flip-flop during shifting. The direct overriding clear functions whether the outputs are enabled or off. The output enable does not interfere with synchronous operation of the register.

TTL – 74S/F350

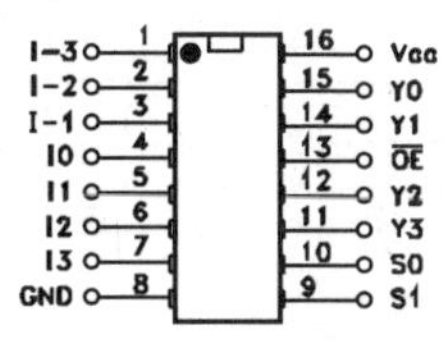

DUAL-IN-LINE PACKAGE

LOGIC SYMBOL

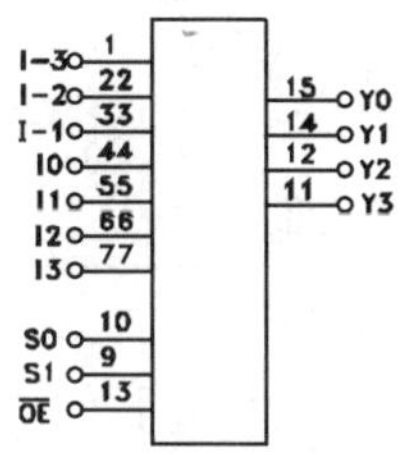

LOGIC DIAGRAM

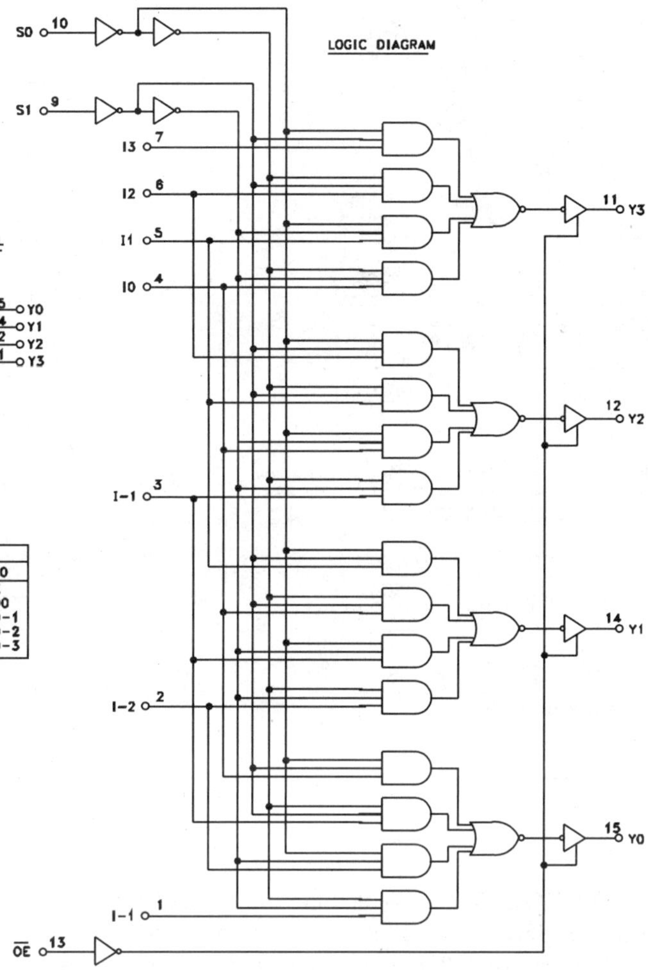

FUNCTION TABLE 74x350 4-BIT SHIFTER

INPUTS										OUTPUTS			
$\overline{OE}$	S1	S0	I3	I2	I1	I0	I-1	I-2	I-3	Y3	Y2	Y1	Y0
1	X	X	X	X	X	X	X	X	X	Z	Z	Z	Z
0	0	0	D3	D2	D1	D0	X	X	X	D3	D2	D1	D0
0	0	1	X	D2	D1	D0	D-1	X	X	D2	D1	D0	D-1
0	1	0	X	X	D1	D0	D-1	D-2	X	D1	D0	D-1	D-2
0	1	1	X	X	X	D0	D-1	D-2	D-3	D0	D-1	D-2	D-3

Dn = HIGH (1) or LOW (0) slate of referenced In Inpul.

Figure 6.20. The 74x350 are a 4-bit shifter with 3-state outputs. Operationally these are equivalent to a 4-input multiplexer with the inputs connected such that the select code causes shifts of the data word. It is possible to achieve shifts of 0, 1, 2, or 3 places on words of any length with appropriate interconnections. A 7-bit data word introduced at the D inputs is shifted according to the code applied to the select inputs S0 and S1. When the output enable is low, the outputs follow the selected data inputs. With the output enable high, the output is in the high impedance state. Shifters may be cascaded on the same output lines or to a common bus. The shift function can be logical with zeros pulled in at either or both ends of the shifting field, arithmetic with the sign bit repeated during a shift down, or end-around with the data word forming a continuous loop.

TTL - 74LS/F395

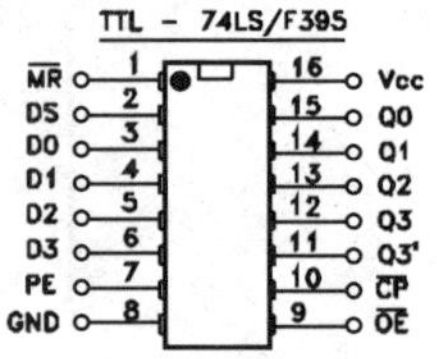

DUAL-IN-LINE PACKAGE

LOGIC SYMBOL

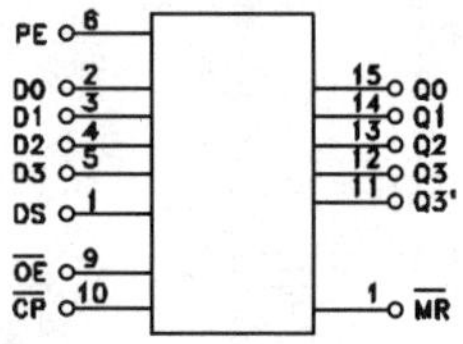

LOGIC DIAGRAM

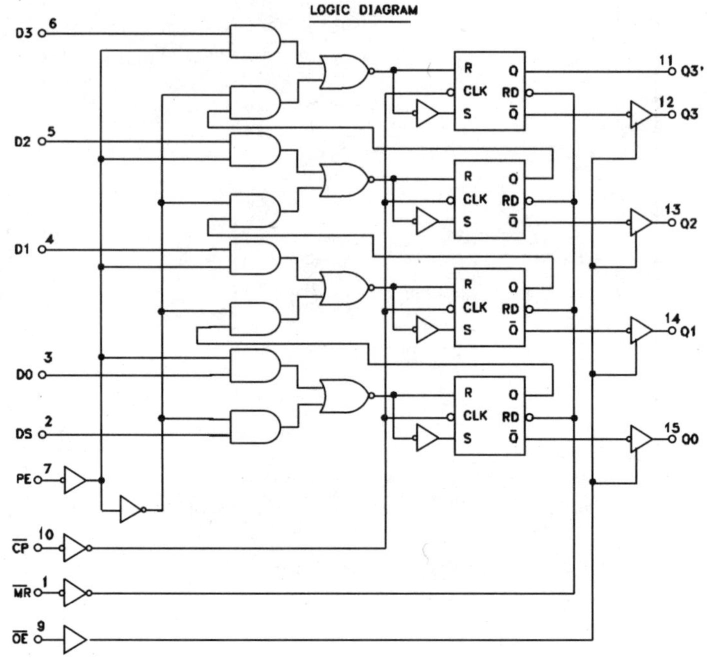

FUNCTION TABLE - 74LS/F395 4-BIT CASCADABLE SHIFT REGISTER WITH 3-STATE OUTPUTS

REGISTER OPERATING MODE	INPUTS					OUTPUTS			
	$\overline{MR}$	$\overline{CP}$	PE	DS	Dn	Q0	Q1	Q2	Q3
Reset (Clear)	0	x	x	x	x	0	0	0	0
Shift Right	1	↓	l	l	x	0	q0	q1	q2
	1	↓	l	h	x	1	q0	q1	q2
Parallel Load	1	↓	h	x	l	0	0	0	0
	1	↓	h	x	h	1	1	1	1

↓ = HIGH-to-LOW clock transition.
h = HIGH voltage level one setup time prior to the ↓ clock transition.
l = LOW voltage level one setup time prior to the ↓ clock transition.
dn, qn = Lower case indicates state of referenced input (or output) one setup time prior to the ↓ clock transition.

FUNCTION TABLE - OPERATING MODES

3-STATE OPERATING MODES	INPUTS		OUTPUTS	
	$\overline{OE}$	Qn(Reg)	Q0 Q1 Q2 Q3	Q3'
Read	0	0	0	0
	0	1	1	1
Disabled	1	0	Z	0
	1	1	Z	1

Z = High impedance OFF state

Figure 6.21. The 74x395 are a 4-bit cascadable shift register with 3-state outputs. These feature parallel input and outputs, and clock, serial, load shift, output control, and direct overriding clear inputs. Shifting takes place with the load/shift control low. Parallel loading takes place by applying the four bits of data and taking the load/shift control high. The data is loaded into the flip-flops, appearing at the outputs after the high-to-low transition of the clock. Serial data is inhibited during parallel loading. The outputs are disabled independently of the clock level by a high logic level at the output control input. During the high-impedance mode the Q3' output is still available for cascading.

TTL – 74LS396

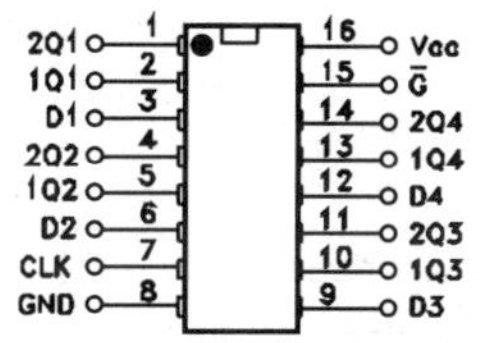

DUAL-IN-LINE PACKAGE

FUNCTION TABLE 74LS396 OCTAL STORAGE REGISTER

INPUTS						OUTPUTS							
STROBE		DATA				BYTE 1				BYTE 2			
G	CLK	D1	D2	D3	D4	1Q1	1Q2	1Q3	1Q4	2Q1	2Q2	2Q3	2Q4
1	X	X	X	X	X	0	0	0	0	0	0	0	0
0	↑	a	b	c	d	a	b	c	d	$1Q1_n$	$1Q2_n$	$1Q3_n$	$1Q4_n$

↑ = LOW-to-HIGH clock transition.

LOGIC SYMBOL

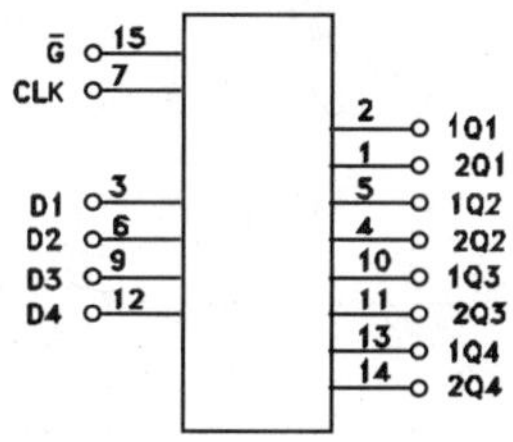

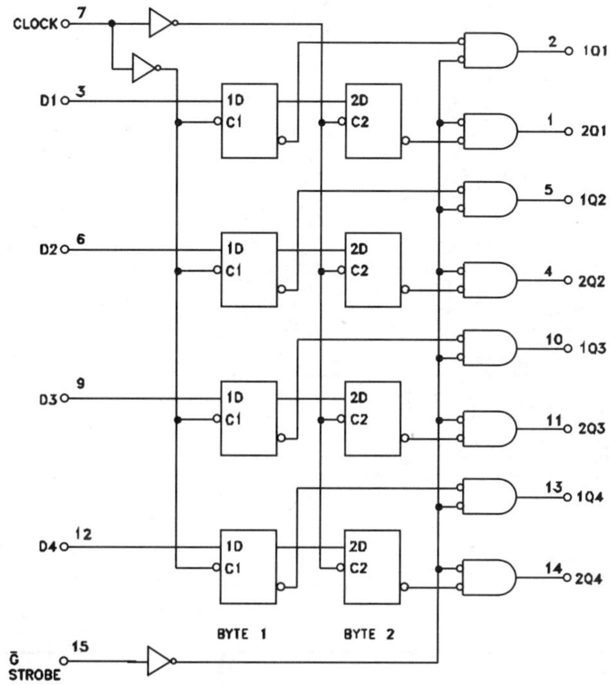

Figure 6.22. The 74LS396 is an octal register organized as two 4-bit bytes of storage. With the input of a positive-going clock signal information stored in byte 1 is transferred into byte 2 as a new 4-bit byte is loaded into the byte 1 location from the four data lines. The full 8-byte data word is available at the outputs after two clock cycles. The clock and the strobe lines are fully buffered.

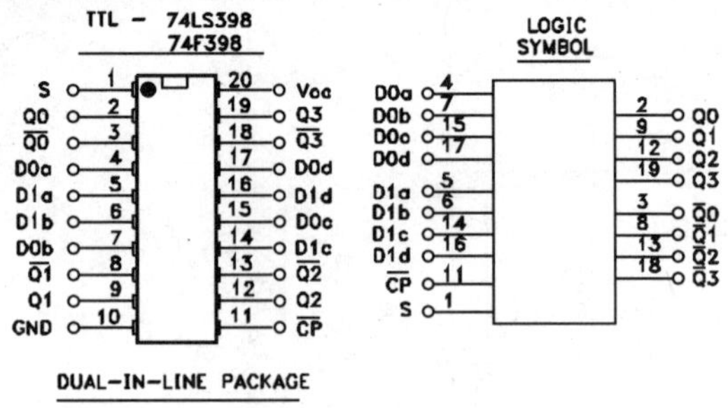

FUNCTION TABLE 74x398 QUAD 2-PORT REGISTER

OPERATING MODES	INPUTS $\overline{CP}$	S	D0	D1	OUTPUTS Qn	$\overline{Qn}$
Load Source "0"	↑	l	l	x	0	1
	↑	l	h	x	1	0
Load Source "1"	↑	h	x	l	0	1
	↑	h	x	h	1	0

↑ = LOW-to-HIGH clock transition.
h = HIGH voltage level one setup time prior to the ↑ clock transition.
l = LOW voltage level one setup time prior to the ↑ clock transition.

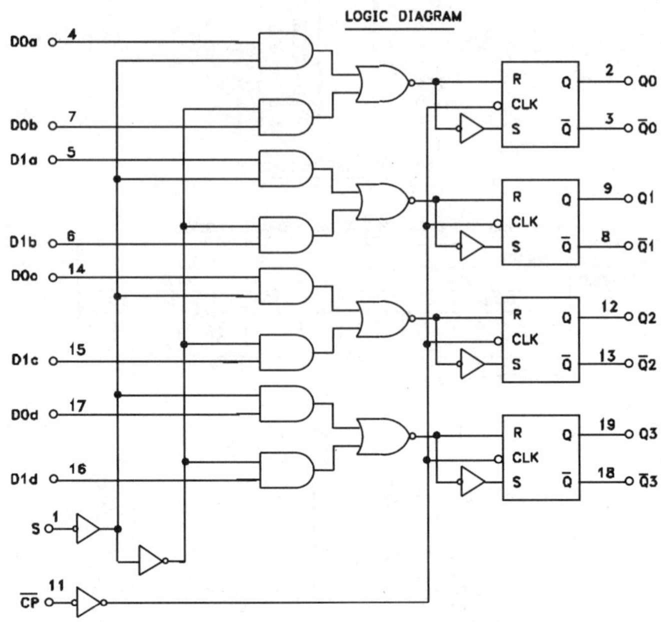

Figure 6.23. The 74x398 are quad 2-port registers with true and complementary outputs; the logical equivalent of quad 2-input multiplexer feeding into four edge-triggered flip-flops. A common select determines which of the two four bit words is accepted. The selected data enters the flip-flops on the rising edge of the clock. The data is transferred to a 4-bit output register synchronous with the low-to-high transition of the clock input.

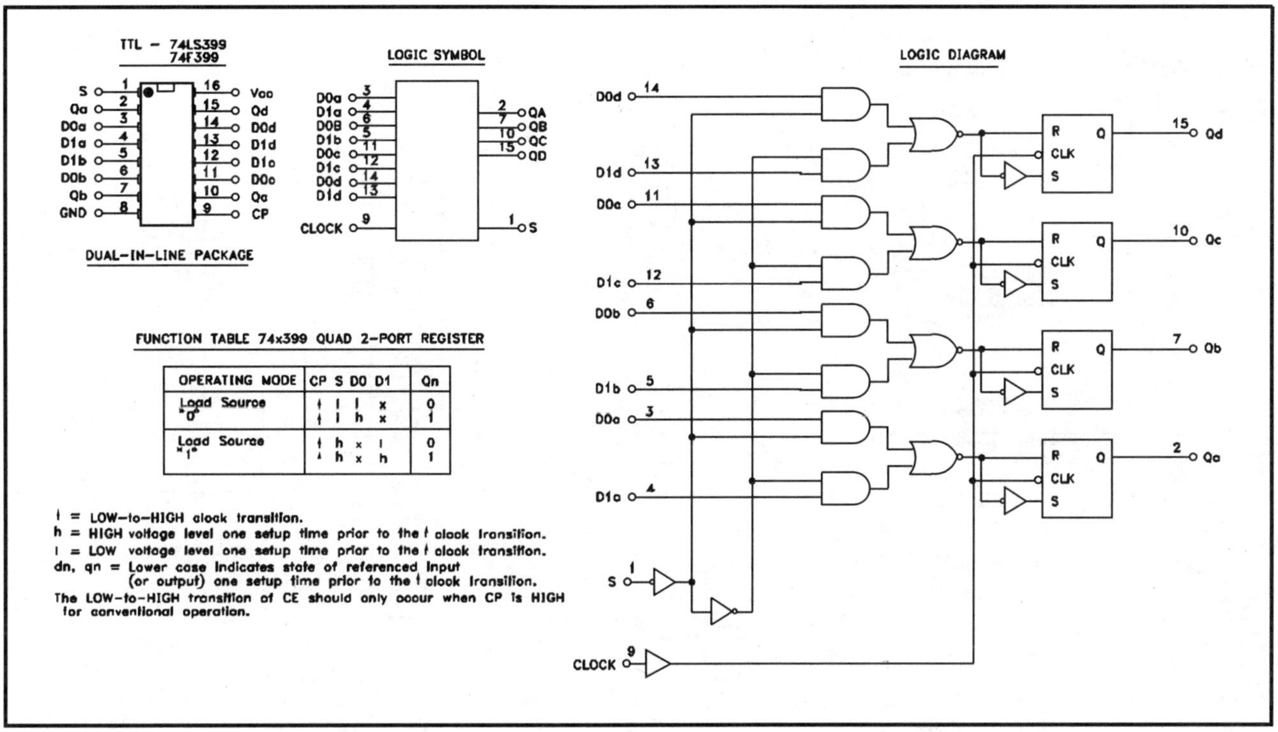

OPERATING MODE	CP	S	D0	D1	Qn
Load Source "0"	↑	l	l	x	0
	↑	l	h	x	1
Load Source "1"	↑	h	x	l	0
	↑	h	x	h	1

Figure 6.24. The 74x399 are quad 2-port registers with true outputs only. In other respects they are 16-pin equivalents of the 20-pin 74x398.

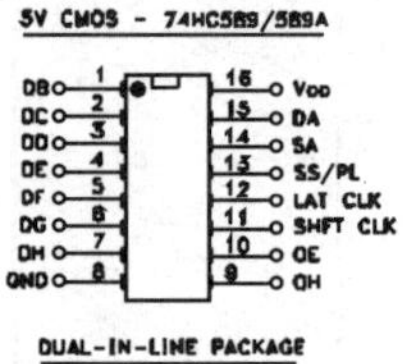

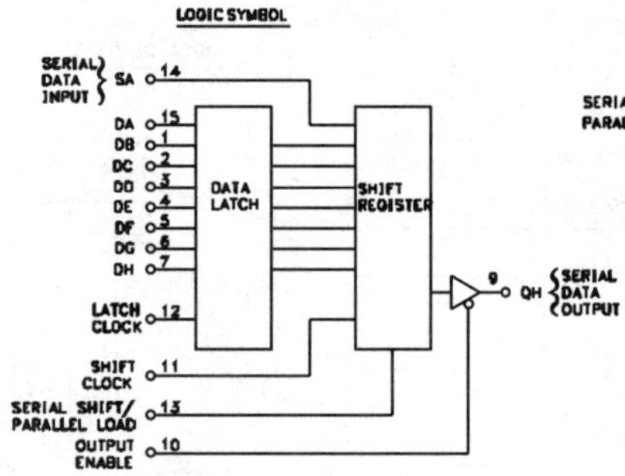

FUNCTION TABLE 74589/589A 8-BIT SERIAL OR PARALLEL-INPUT SERIAL-OUTPUT SHIFT REGISTER WITH 3-STATE OUTPUT

OPERATION	Output Enable	Serial Shift/ Parallel Load	Latch Clock	Shift Clock	Ser Inp SA	Par Inps A–H	Data Lat Contents	Shf Reg Contents	Output QH
Force output into high impedance state	1	X	X	X	X	X	X	X	Z
Load parallel data into data latch	0	1	↑	0,1,↓	X	a–h	a–h	U	U
Transfer latch contents to shift register	0	0	0,1,↓	X	X	X	U	LRn–SRn	LRH
Contents of input latch and shift register are unchanged	0	1	0,1,↓	0,1,↓	X	X	U	U	U
Load parallel data into data latch and shift register	0	0	↑	X	X	a–h	a–h	a–h	h
Shift serial data into shift register	0	1	X	↑	D	X	.	SRA=D, SRn–SRn+1	SRG–SPH
Load parallel data in data latch and shift serial data into shift register	0	0	↑	↑	D	a–h	a–h	SRA=D, SRn–SRn+1	SRG–SPH

LR = latch register contents
SR = shift register contents
a–h = data at parallel data inputs A–H
D = dat (L,H) at serial data input SA

U = Unchanged
X = Don't Care
Z = High Impedance
. = depends on Latch Clock input

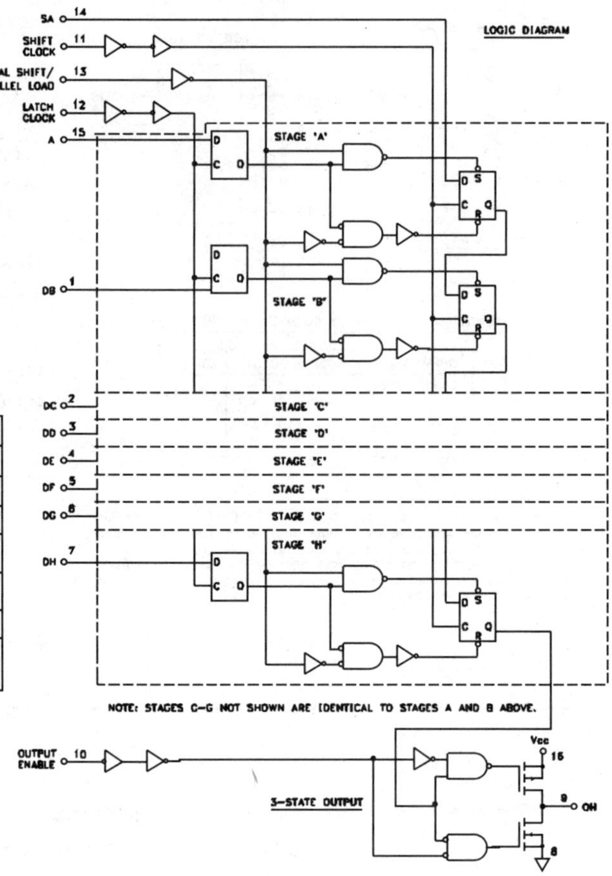

Figure 6.25. The 74HC589 is a high-speed CMOS 8-bit shift register with input latches and 3-state serial output. The device consists of an 8-bit storage latch feeding a parallel-in, serial-out 8-bit shift register. Data can also be serially entered to the register via the SA input (pin 14). The storage and shift registers both have positive edge-triggered clocks. The serial shift/parallel load input (pin 13) controls parallel load or serial shift operations for the shift register.

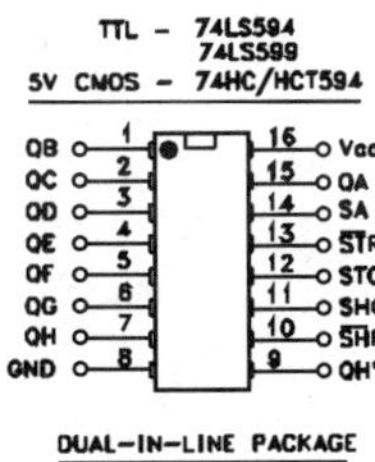

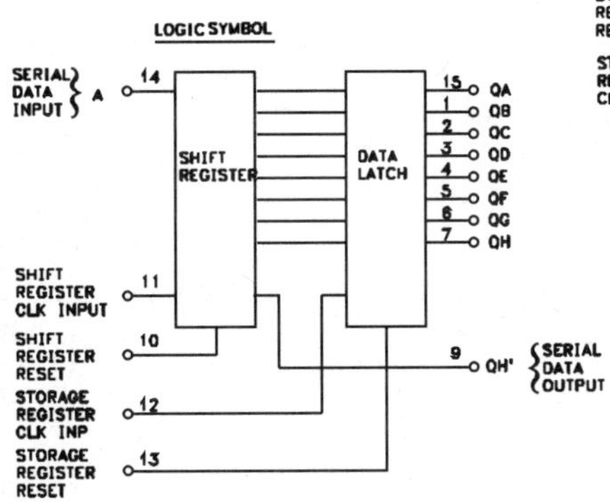

FUNCTION TABLE 74x594, 74LS599 8-BIT SHIFT REGISTER WITH OUTPUT REGISTER

OPERATION	Shift Clock	Storage Reg Clock	Shift Reg Reset	Storage Reg Reset	Serial Input A	Serial Output Q7'	Parallel Outputs QA–QH
Reset Shift Register Active Low	X	X	0	X	X	0	NC
Storage Register Reset Active Low	X	X	X	0	X	NC	0
Shift Register Content loaded into storage register	X	↑	0	1	X	0	0
Content of shift register stages shifted one stage	↑	X	1	X	1	Q6'	NC
Content of shift register stages transferred to storage registers and parallel output stages	X	↑	1	1	X	NC	Qn'
Content of shift registers shifted through. Previous content transferred to storage registers and parallel output stages	↑	↑	1	1	X	Q6n	Qn'

NC = No change ↑ = LOW-to-HIGH transition
X = Don't Care

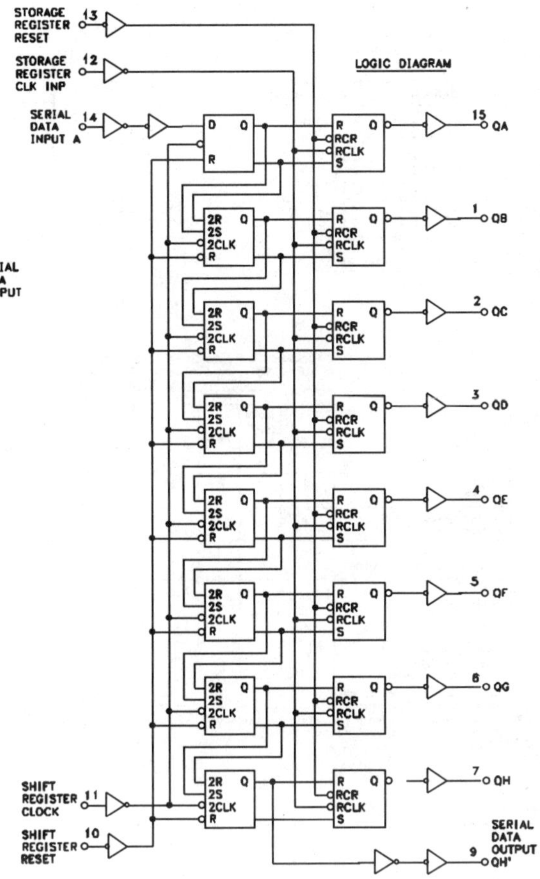

Figure 6.26. The 74x594 is a high-speed 8-bit serial-in, parallel-out shift register with output storage. The device provides an 8-bit storage register with buffered outputs. Separate clocks and direct overriding clear inputs are provided on both the shift and storage registers. A shift output, QH', is provided for cascading devices. The shift and storage registers have separate clock inputs; both are positive-edge triggered. If the two clock inputs are tied together the shift register will always be on clock pulse ahead of the storage register.

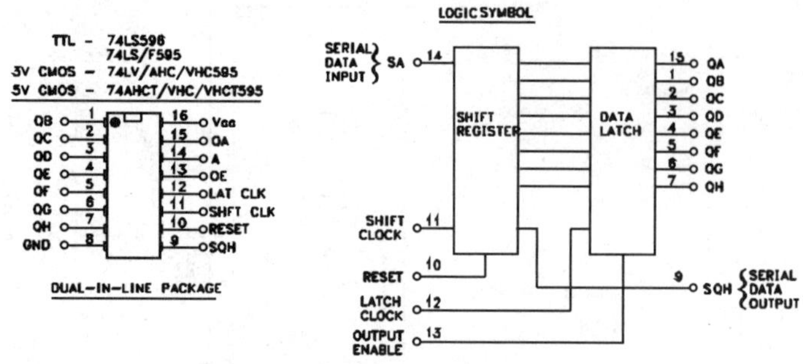

FUNCTION TABLE 74x595/LS596 8-BIT SERIAL-INPUT SERIAL OR PARALLEL-OUTPUT SHIFT REGISTER WITH LATCHED 3-STATE OUTPUTS

OPERATION	Reset	Serial Input A	Shift Clock	Latch Clock	Output Enable	Shift Register Contents	Latch Register Contents	Serial Output SQH	Parallel Outputs QA–QH
Reset Shift Register	0	X	X	0,1,↓	0	0	U	0	U
Shift Data into Shift Register	1	D	↑	0,1,↓	0	D—SRA, SRN — SRN+1	U	SRG—SRH	U
Shift Register remains unchanged	1	X	0,1,↓	0,1,↓	0	U	U	U	U
Transfer Shift Register contents to Latch Register	1	X	0,1,↓	↑	0	U	SRN—LRN	U	SRN
Latch Register remains unchanged	X	X	X	0,1,↓	0	•	U	•	U
Enable Parallel Outputs	X	X	X	X	0	•	••	•	Enabled
Force Outputs into High Impedance state	X	X	X	X	1	•	••	•	Z

LR = latch register contents
SR = shift register contents
• = Depends on Reset and Shift Clock Inputs.
•• = Depends on Latch Clock Input

U = Unchanged
X = Don't Care
Z = High Impedance

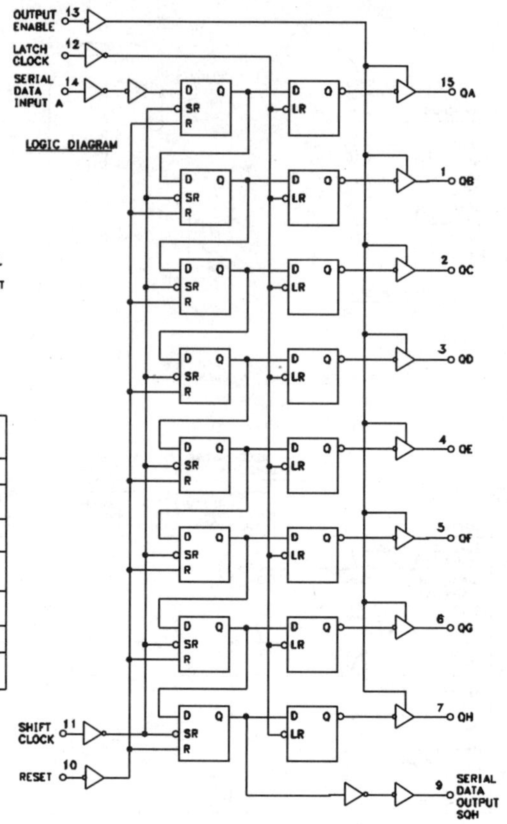

Figure 6.27. The 74x595, 'LS596 are an 8-bit serial-in, parallel-out shift register with output storage under 3-state control. Other than for the 3-state control, the device is functionally equivalent to the 74x594. The device provides an 8-bit storage register with buffered outputs. Separate clocks and direct overriding clear inputs are provided on both the shift and storage registers. A shift output, QH', is provided for cascading devices. The shift and storage registers have separate clock inputs; both are positive-edge triggered. If the two clock inputs are tied together the shift register will always be on clock pulse ahead of the storage register. The 74LS596 is identical to the 74x595 but has open collector outputs.

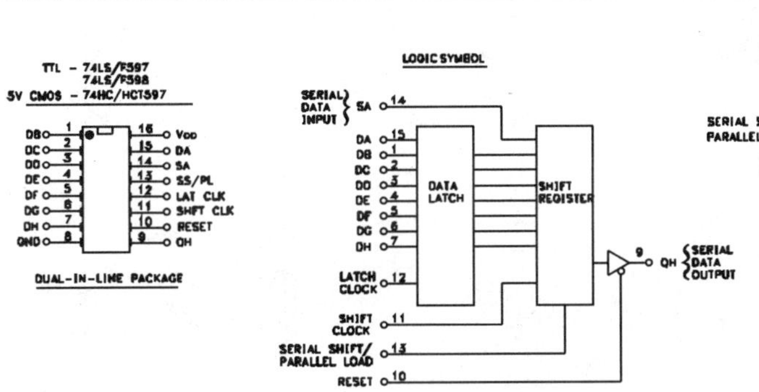

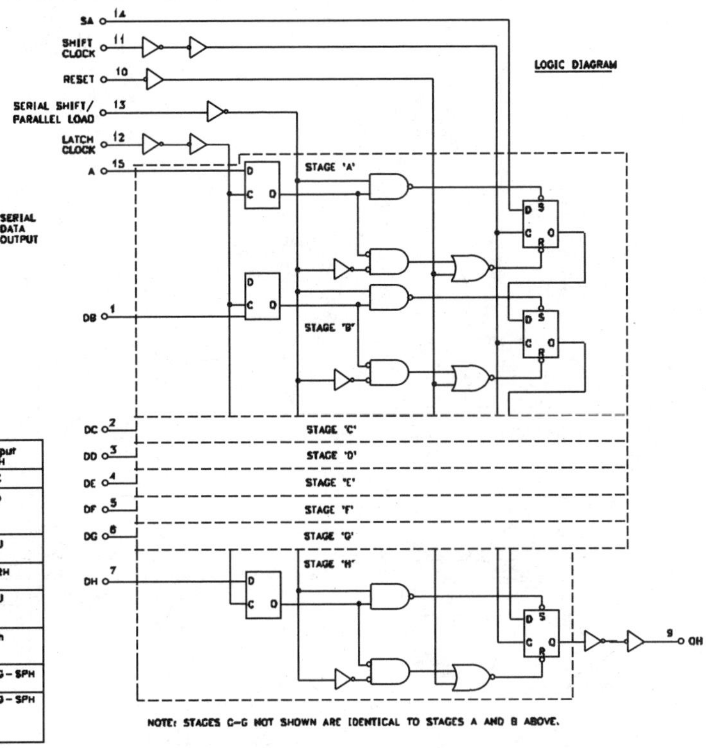

FUNCTION TABLE 74x597, 74x598 8-BIT SERIAL OR PARALLEL INPUT SERIAL-OUTPUT SHIFT REGISTER WITH INPUT LATCH

OPERATION	Reset	Serial Shift/ Parallel Load	Latch Clock	Shift Clock	Ser Inp SA	Par Inps A–H	Data Lat Contents	Shf Reg Contents	Output QH
Reset Shift Register	0	X	X	X	X	X	X	X	Z
Reset Shift Register; Load parallel data into data latch	0	X	↑	X	X	a–h	a–h	0	0
Load parallel data into data latch	1	1	↑	0,1,↓	X	a–h	a–h	U	U
Transfer latch contents to shift register	1	0	0,1,↓	X	X	X	U	LRn–SRn	LRH
Contents of input latch and shift register are unchanged	1	1	0,1,↓	0,1,↓	X	X	U	U	U
Load parallel data into data latch and shift register	1	0	↑	X	X	a–h	a–h	a–h	h
Shift serial data into shift register	1	1	X	↑	D	X	.	SRA=D, SRn–SRn+1	SRG–SPH
Load parallel data in data latch and shift serial data into shift register	1	0	↑	↑	D	a–h	a–h	SRA=D, SRn–SRn+1	SRG–SPH

LR = latch register contents
SR = shift register contents
a–h = data at parallel data inputs A–H
D = data (L,H) at serial data input SA

U = Unchanged
X = Don't Care
Z = High Impedance
. = depends on Latch Clock input

Figure 6.28. The 74x597, 'x598 are an 8-bit shift register with input storage register and parallel 3-state control. The '597 has a 16-pin package. The '598 offers 3-state I/O ports providing parallel shift register outputs and multiplexed serial data inputs in addition to the all the '597 features. It comes in a 20-pin package, not shown in the drawing. The devices provide an 8-bit storage latch feeding a parallel-in, serial-out 8-bit shift register. The shift and storage registers have separate clock inputs; both are positive-edge triggered. The shift register also has direct load from storage and clear inputs.

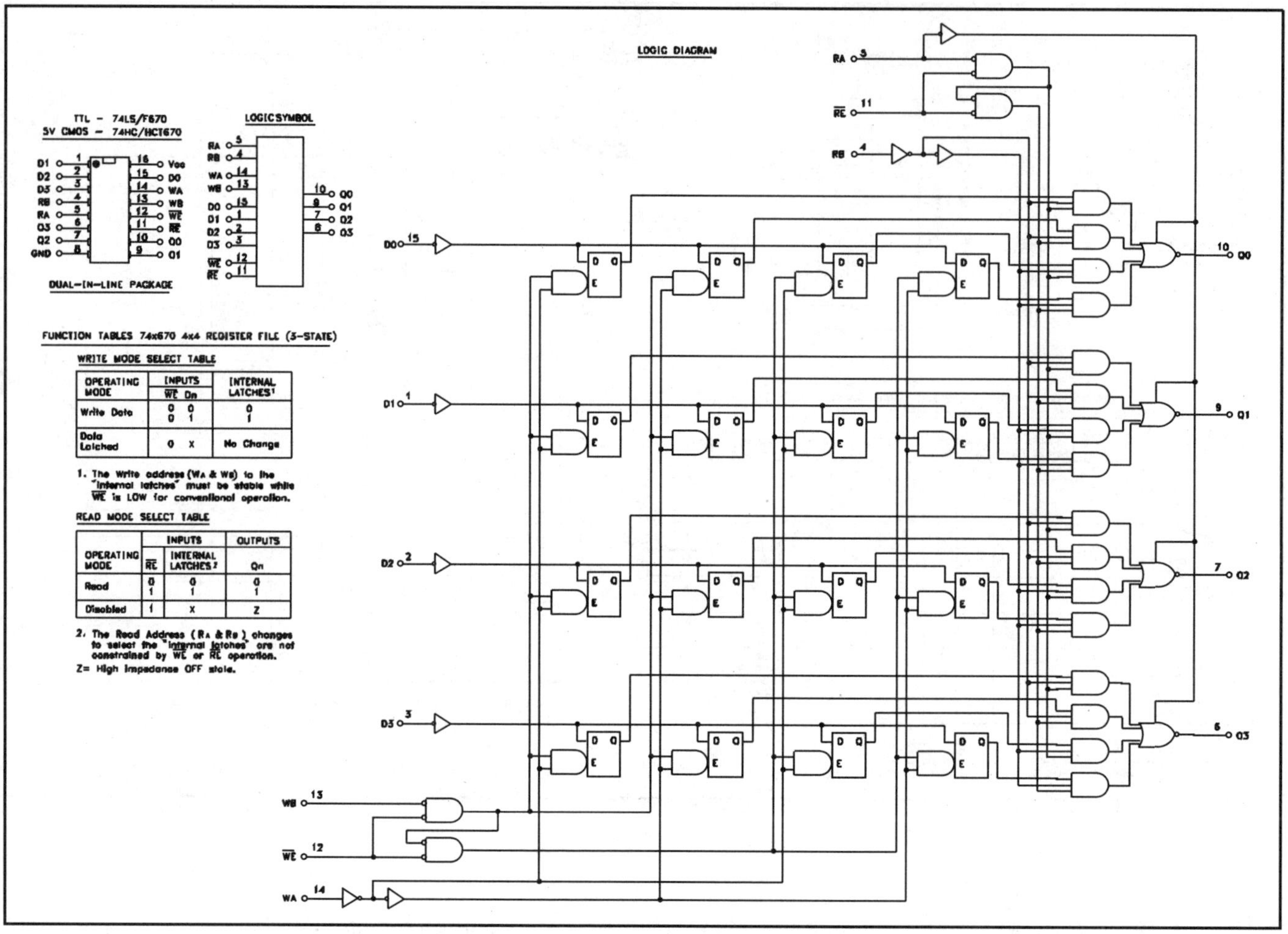

WRITE MODE SELECT TABLE

OPERATING MODE	INPUTS $\overline{WE}$	INPUTS Dn	INTERNAL LATCHES[1]
Write Data	0 0	0 1	0 1
Data Latched	0	X	No Change

READ MODE SELECT TABLE

OPERATING MODE	INPUTS $\overline{RE}$	INPUTS INTERNAL LATCHES[2]	OUTPUTS Qn
Read	0 1	0 1	0 1
Disabled	1	X	Z

Figure 6.29. The 74x670 is a 16-bit 3-state register file organized as 4 words of 4 bits each. Individual read, write, and enable inputs permit simultaneous writing into one word location while reading from another. The 4-bit data to be stored is presented to four inputs. The write address inputs (WA, WB) determine the location of the stored word. Storage occurs with the write enable input low. Stored data is read out in true form from the 3-state outputs. Data and address inputs are inhibited when the write enable is high. Direct acquisition of data stored in any of the four registers is made possible by the individual read address inputs (RA, RB). The addressed word appears at the output when the read enable (/RE) is low. The output is in the high-impedance state with the enable high.

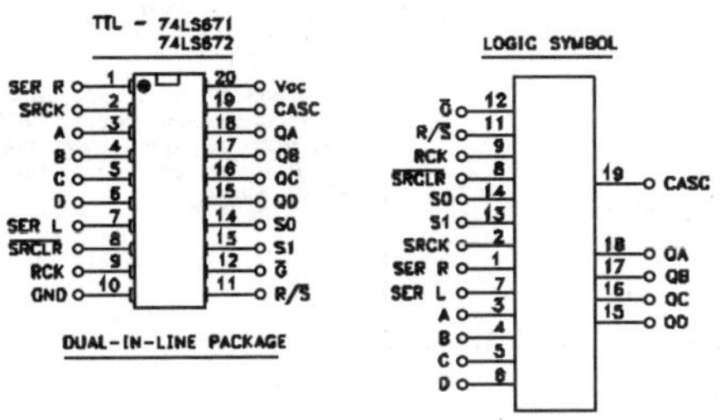

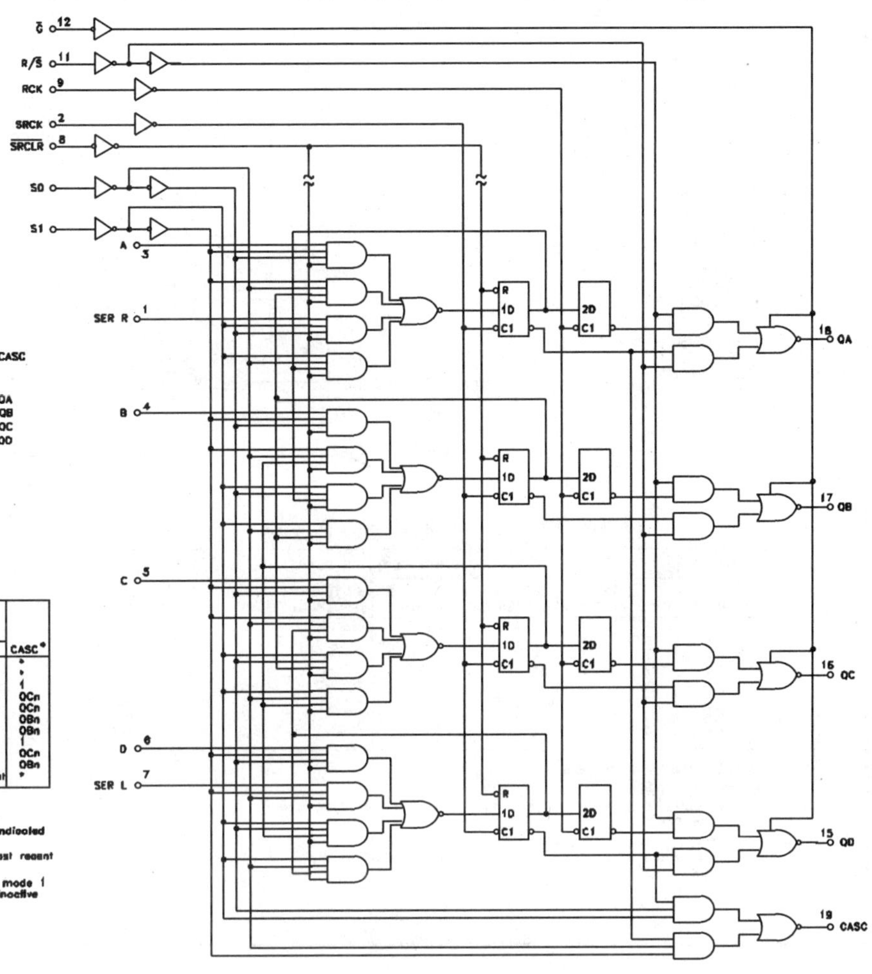

FUNCTION TABLE 74LS671, 74LS672 4-BIT UNIVERSAL SHIFT REGISTERS/LATCHES WITH 3-STATE OUTPUTS

G	R/S	SRCLR	SR MODE S1	SR MODE S0	SRCK 'LS671	SRCK 'LS672	SERIAL INPUTS SL	SERIAL INPUTS SR	PAR'EL INPUTS A	B	C	D	PARALLEL OUTPUTS QA	QB	QC	QD	CASC*
0	0	0	X	X	X	↑	X	X	X	X	X	X	0	0	0	0	*
0	0	1	X	X	0	0	X	X	X	X	X	X	QA0	QB0	QC0	QD0	*
0	0	1	0	0	X	X	X	X	X	X	X	X	QA0	QB0	QC0	QD0	1
0	0	1	0	1	↑	↑	X	1	X	X	X	X	1	QAn	QBn	QCn	QCn
0	0	1	0	1	↑	↑	X	0	X	X	X	X	0	QAn	QBn	QCn	QCn
0	0	1	1	0	↑	↑	1	X	X	X	X	X	QBn	QCn	QDn	1	QBn
0	0	1	1	0	↑	↑	0	X	X	X	X	X	QBn	QCn	QDn	0	QBn
0	0	1	1	1	↑	↑	X	X	a	b	c	d	a	b	c	d	1
1	X	X	0	1	↑	↑	X	X	X	X	X	X	Z	Z	Z	Z	QCn
1	X	X	1	0	↑	↑	X	X	X	X	X	X	Z	Z	Z	Z	QBn
0	1	X	X	X	X	X	X	X	X	X	X	X	Internal reg content				*

↑ = LOW-to-HIGH clock transition.
a,b,c,d = the level of steady state input at A, B, C, D respectively.
QA0 – QD0 = the level of QA, QB, QC, QD respectively before the indicated steady state input conditions were established.
QAn, QBn, QCn = the level of QA, QB, QC respectively before the most recent transition of the clock.
* = the cascade output displays the D bit of the shift register in mode 1 (S1, S0 = 0,1), the A bit in mode 2 (S1, S0 = 10) and is inactive (1 level) in modes 0 and 3 (S1, S0 = 00, 11).
Z = the high impedance state.

Figure 6.30. The 74LS671, 'LS672 are 4-bit universal shift register/latches with 3-state outputs. Each device contains a 4-bit universal shift register similar to the 'LS194A and a 4-bit storage register multiplexed to a 3-state output stage. Register selection is by register/shift select input (R//S). The 'LS671 has a direct overriding shift register clear while the LS672 features a synchronous shift register clear. The shift register has four modes of operation: inhibit clock (do nothing), shift right (QA > QD), shift left (QD > QA), and parallel load. A cascade output for the shift register is provided so that full shift functionality is provided even while the outputs are in the high-impedance mode. QA data is output in the shift-left mode, QD in the shift-right.

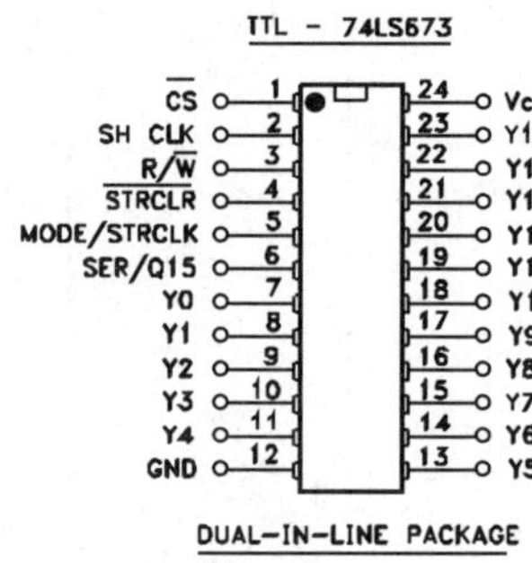

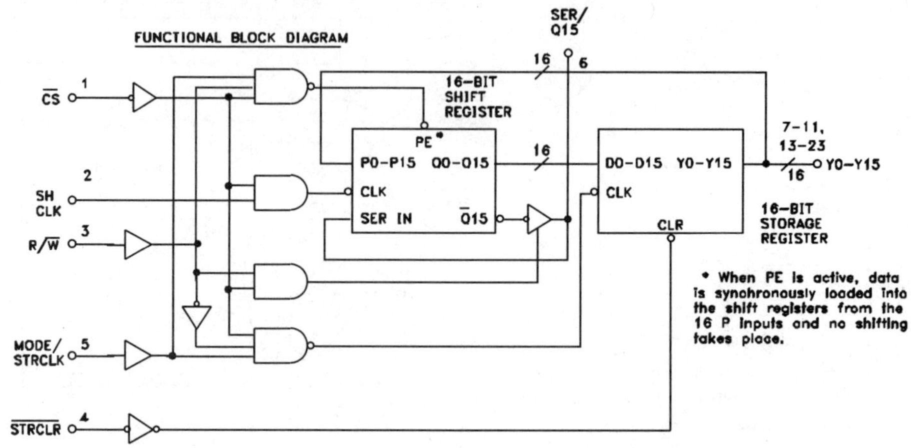

FUNCTION TABLE 74LS673 16-BIT SHIFT REGISTER

INPUTS							SHIFT REGISTER FUNCTIONS			STORAGE REGISTER FUNCTIONS	
/CS	R/W	SH CLK	/STRCLR	MODE/ STRCLK	SER Q15	SHIFT	READ FROM SER OUTPUT	WRITE INTO SER INPUT	PARALLEL LOAD	CLEAR	LOAD
1	X	X	X	X	Z	NO	NO	NO	NO		NO
X	X	X	0	X						YES	
0	0	↓	X	X	Z	YES	NO	YES	NO		
0	1	X	X	X	Q15		YES	NO			NO
0	1	↓	X	0	Q14n	YES	YES	NO	NO		NO
0	1	↓	0	1	0	NO	YES		YES	YES	NO
0	1	↓	1	1	Y15n	NO	YES		YES	YES	NO
0	0	X	1	↑	Z		NO		NO	NO	YES

↓ = HIGH-to-LOW clock transition.
↑ = LOW-to-HIGH clock transition.
Q14n = content of 14th bit of the shift register before the most recent ↓ of the clock.
Q15 = present content of the 15th bit of the shift register.
Y15n = content of the 15th bit of the storage register before the most recent ↓ of the clock.
P15 = level of input P15.

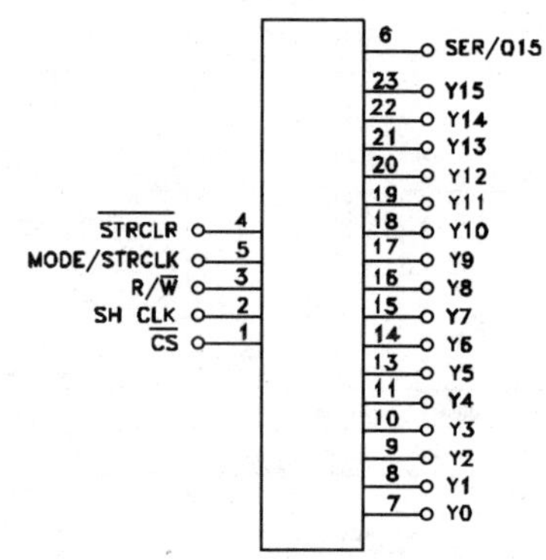

Figure 6.31. The 74LS673 is a 16-bit shift register and 16-bit storage register. A 3-state I/O port to the shift register provides for serial entry and/or reading of data. The storage register is connected in a parallel data loop with the shift-register and may be asynchronously cleared by taking the store-clear input low. The storage register may be parallel loaded with shift-register data to provide shift register status via the paralleled outputs. The shift register can be parallel loaded with the storage register data upon command. A high logic level at the chip-select (/CS) input disables both the shift- and storage-register clocks and places SER/Q15 in the high-impedance state. The store-clear function is not disabled by the chip-select.

TTL – 74LS674
74F674

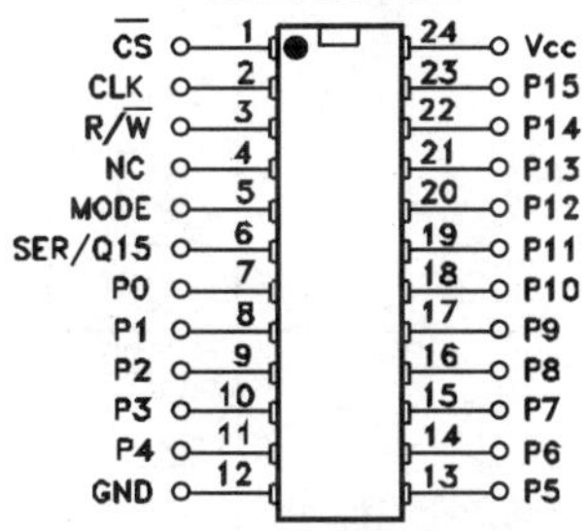

DUAL-IN-LINE PACKAGE

FUNCTIONAL BLOCK DIAGRAM

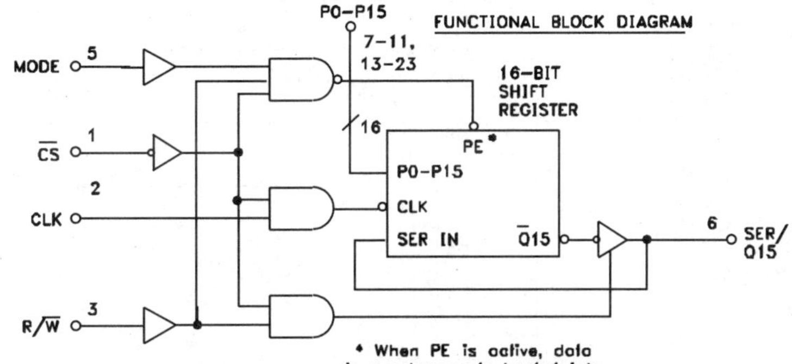

* When PE is active, data is synchronously loaded into the shift registers from the 16 P inputs and no shifting takes place.

LOGIC SYMBOL

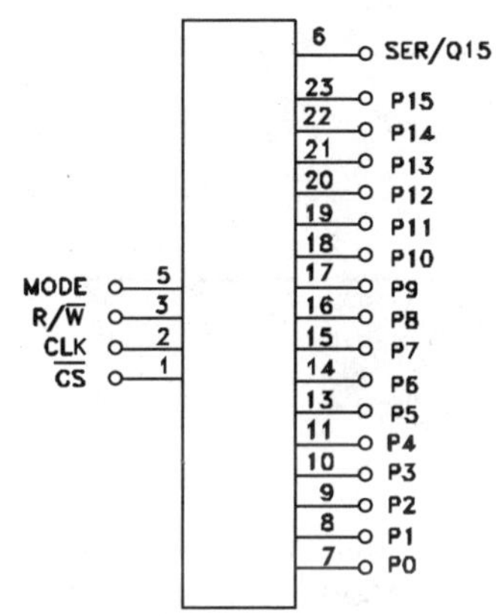

FUNCTION TABLE 74x674 16-BIT SHIFT REGISTER

INPUTS				SER/	
CS	R/W	MODE	CLK	Q15	OPERATION
1	X	X	X	Z	Do nothing
0	0	X	↓	Z	Shift and write (ser load)
0	1	0	↓	Q14n	Shift and read
0	1	1	↓	P15	Parallel load

↓ = HIGH-to-LOW clock transition.
Q14n = content of 14th bit of the shift register before the most recent ↓ of the clock.
Q15 = present content of the 15th bit of the shift register.
Y15n = content of the 15th bit of the storage register before the most recent ↓ of the clock.
P15 = the level of Input P15

Figure 6.32. The 74x674 is a 16-bit parallel-in, serial-out shift register. A 3-state I/O port provides access for serial data entry or reading the shift register word in a circulating loop. The device features four basic modes of operation: hold (do nothing), write, serially via input/output, read serially, parallel load via the data inputs. Low-to-high-level changes at the chip select input should be made only when the clock input is low to prevent false clocking.

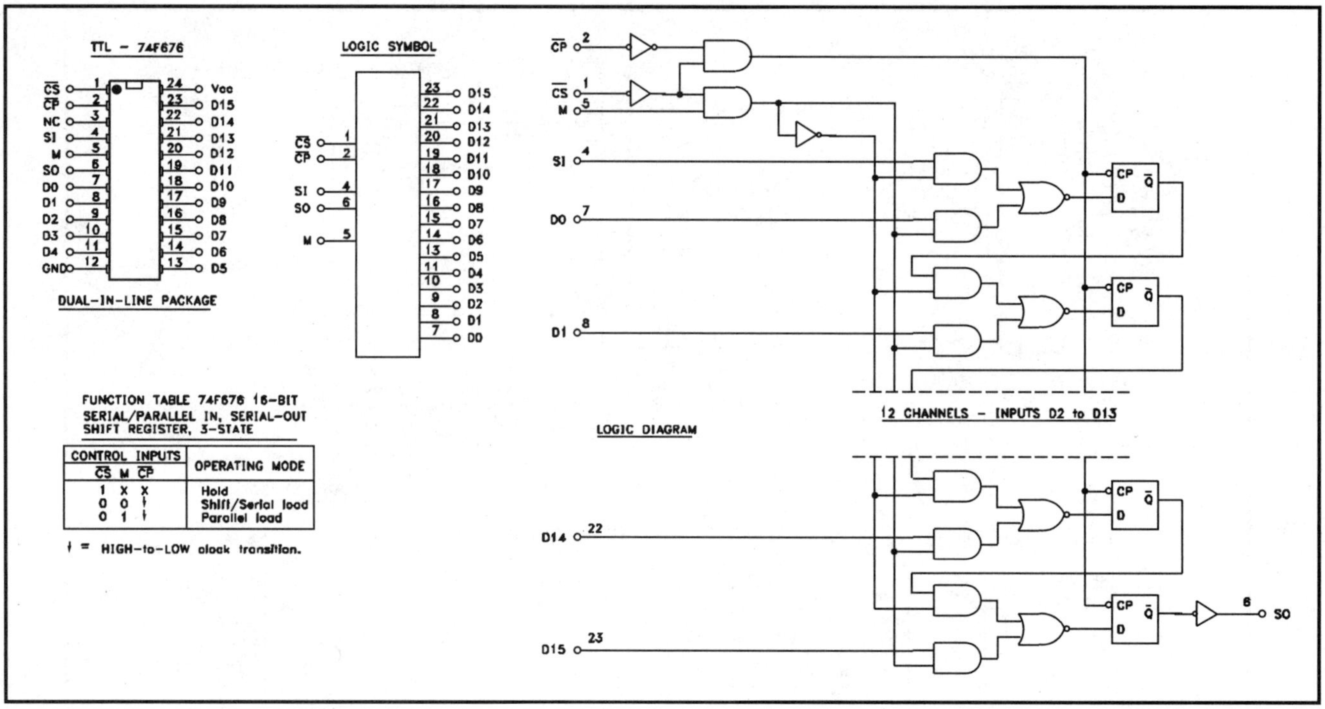

Figure 6.33. The 74F676 is a 16-bit serial/parallel-in, serial-out shift register with 3-state output control. The device contains sixteen flip-flops with provision for synchronous parallel or serial entry and serial output. With the mode (M) input high, information present at the parallel data inputs is entered on the falling edge of the clock pulse. When M is low, data is shifted out of the most significant bit position while information present on the serial input shifts into the least significant bit position. A high level on the chip select (/CS) input prevents both serial and parallel operations. The device operates in one of three modes as seen in the function table.

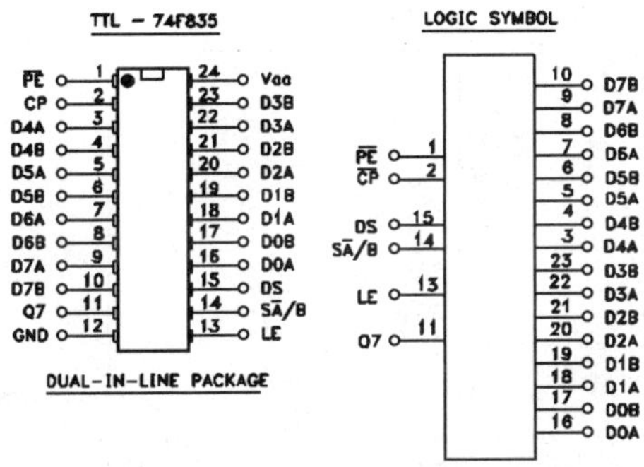

FUNCTION TABLE 74F835 8-BIT SHIFT REGISTER WITH 2:1 MUX-IN, LATCHED "B" INPUTS AND SERIAL OUT

OPERATING MODE	INPUTS							INTERNAL			OUTPUT
								B	SERIAL REG		
	PE̅	CP	LE	SA̅/B	DnA	DnB	DS	LATCH	Q0	Q1–6	Q7
Parallel Load A Data	0	↑	X	0	h l	X X	X X	X X	1 0	1 0	1 0
Latch B Data	X	X	0	X	X X	h l	X X	1 0	X X	X X	X X
Parallel Load B Data (From Latch)	0	↑	0	1	X X	X X	X X	h l	1 0	1 0	1 0
Parallel Load B Data (Transparent Mode)	0	↑	1	1	X X	h l	X X	h l	1 0	1 0	1 0
Serial Shift	1	↑	X	X	X X	X X	h l	X X	1 0	qn–1 qn–1	q6 q6

↑ = LOW-to-HIGH clock transition.
h = HIGH voltage level one setup time prior to ↑ clock transition.
l LOW voltage level one setup time prior to ↑ clock transition.
qn Lower case letters indicate the state of the referenced flop cell one cycle prior to the ↑ clock transition.

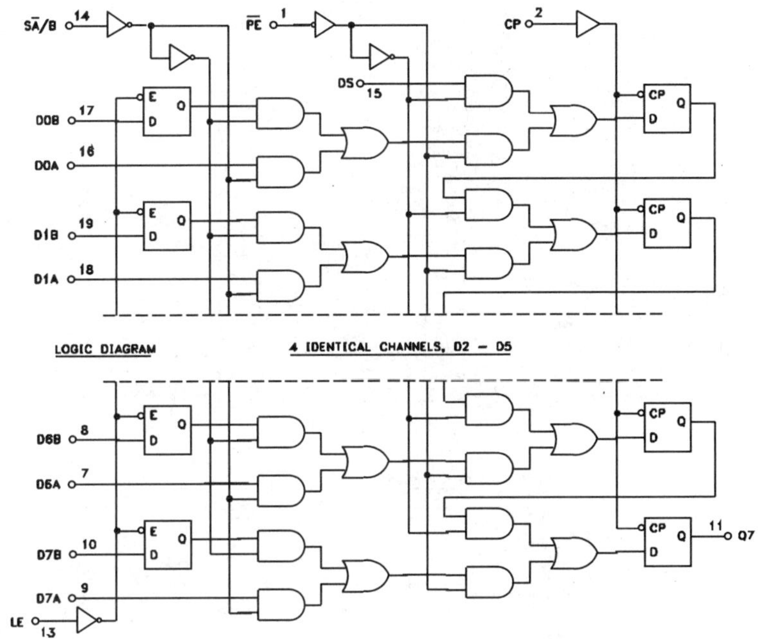

Figure 6.34. The 74F835 is an 8-bit shift register with 2:1 mux-in, latched "B" inputs, and serial out. The device was specifically designed for video applications where interleaved loading is desired and parts count is critical. The parallel-inputs have been connected to an internal octal two-to-one multiplexer with all the "B" inputs connected to an octal latch.

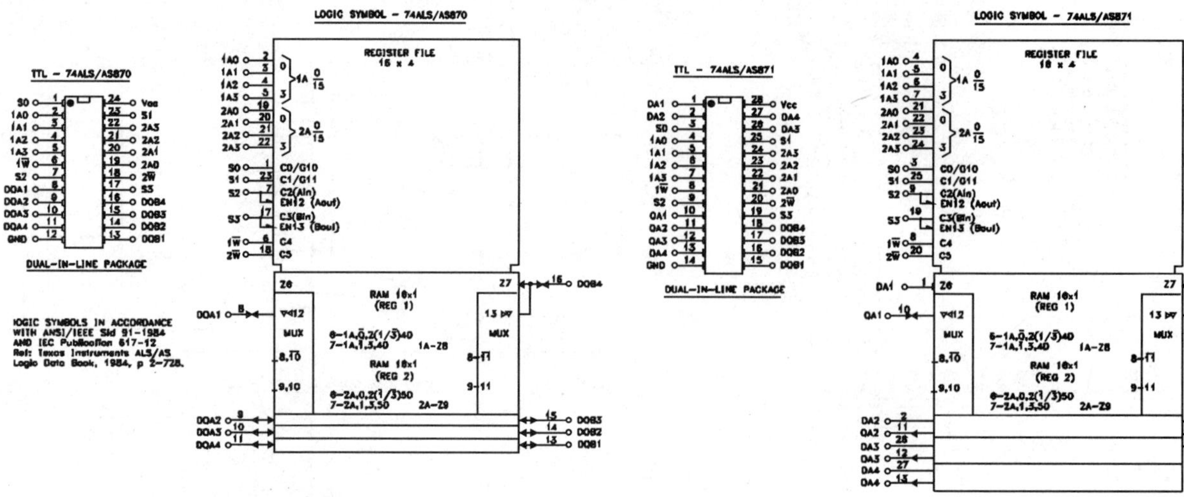

FUNCTION TABLE 74ALS/AS 870, '871
DUAL 16-BY-4 REGISTER FILES

FILE SELECT			INPUT/OUTPUT		
S0	S1	FILE SEL	S2	S3	I/O SEL
0 1 0 1	0 0 1 1	1R TO A, 1R TO B 2R TO A, 1R TO B 1R TO A, 2R TO B 2R TO A, 2R TO B	0	0	A OUT, B OUT
0 1 0 1	0 0 1 1	A TO 1R, 1R TO B A TO 2R, 1R TO B A TO 1R, 2R TO B A TO 2R, 2R TO B	1	0	A IN, B OUT
0 1 0 1	0 0 1 1	1R TO A, B TO 1R 2R TO A, B TO 1R 1R TO A, B TO 2R 2R TO A, B TO 2R	0	1	A OUT, B IN
0 1 0 1	0 0 1 1	B TO 1R A TO 2R, B TO 1R A TO 1R, B TO 2R B TO 2R	1	1	A IN, B IN

Figure 6.35. The 74AS/ALS870 and 74AS/ALS871 are dual 16-by-4 register files with 3-state buffer-type outputs. Each device has individual write enable controls and address lines. The '870 has two 4-bit data I/O ports; the '871 has one 4-bit data I/O port with the other port having individual data inputs and outputs. The ports can output to Bus A and Bus B, receive input from Bus A and Bus B, or output to Bus A and receive input from Bus B. To prevent conflicts in the dual-input mode the B input port takes priority. Select lines S0 and S1 control which port has access to which register. S2 determines whether the A ports are in the input or output modes and S3 does likewise for the B ports. The address lines are decoded by an internal 1-of-16 decoder to select which register word is to be accessed.

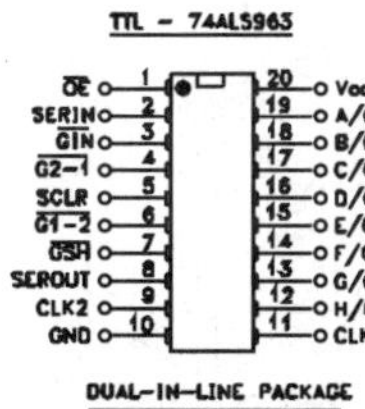

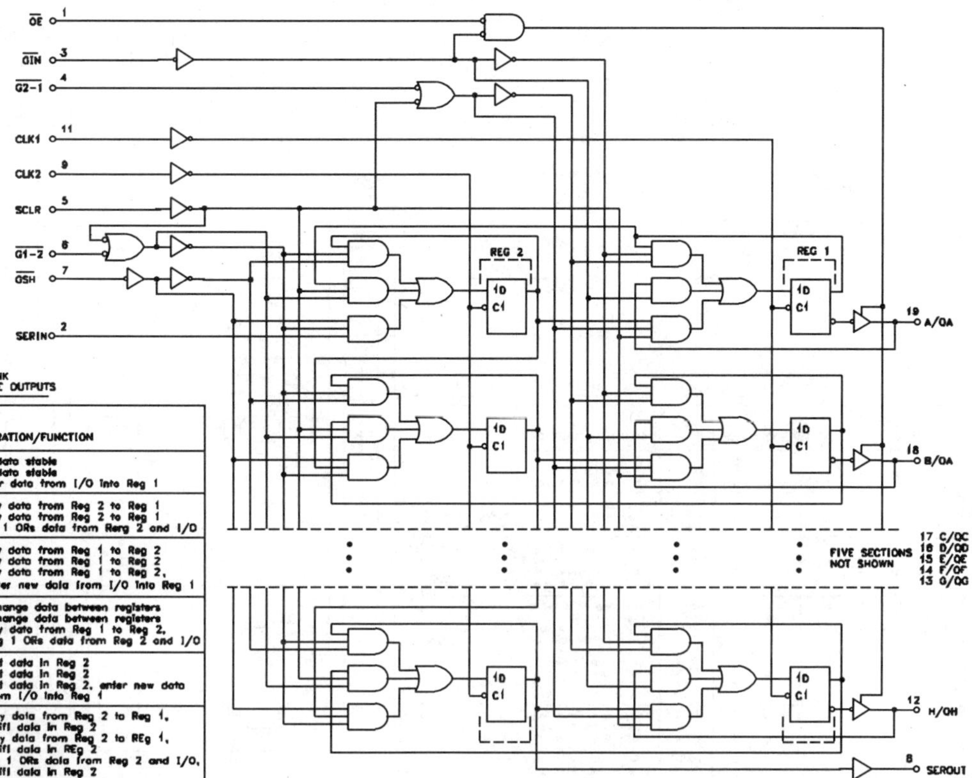

FUNCTION TABLE 74ALS963 DUAL-RANK 8-BIT SHIFT REGISTER WITH 3-STATE OUTPUTS

INPUTS								A/QA THRU H/QH	OPERATION/FUNCTION
$\overline{OE}$	$\overline{GIN}$	$\overline{GE-1}$	$\overline{G1-2}$	$\overline{GSH}$	CLK1	CLK2	SCLR		
1	1	1	1	1	X	X	0	HI-Z	All data stable
0	1	1	1	1	X	X	0	OUTPUT	All data stable
X	0	1	1	1	↑	X	0	INPUT	Enter data from I/O into Reg 1
1	1	0	1	1	↑	X	0	HI-Z	Copy data from Reg 2 to Reg 1
0	1	0	1	1	↑	X	0	OUTPUT	Copy data from Reg 2 to Reg 1
X	0	0	1	1	↑	↑	0	INPUT	Reg 1 ORs data from Rerg 2 and I/O
1	1	1	0	X	X	↑	0	HI-Z	Copy data from Reg 1 to Reg 2
0	1	1	0	X	X	↑	0	OUTPUT	Copy data from Reg 1 to Reg 2
X	0	1	0	X	↑	↑	0	INPUT	Copy data from Reg 1 to Reg 2, enter new data from I/O into Reg 1
1	1	0	0	X	↑	↑	0	HI-Z	Exchange data between registers
0	1	0	0	X	↑	↑	0	OUTPUT	Exchange data between registers
X	0	0	0	X	↑	↑	0	INPUT	Copy data from Reg 1 to Reg 2, Reg 1 ORs data from Reg 2 and I/O
1	1	1	1	0	X	↑	0	HI-Z	Shift data in Reg 2
0	1	1	1	0	X	↑	0	OUTPUT	Shift data in Reg 2
X	0	1	1	0	↑	↑	0	INPUT	Shift data in Reg 2, enter new data from I/O into Reg 1
1	1	0	1	0	↑	↑	0	HI-Z	Copy data from Reg 2 to Reg 1, shift data in Reg 2
0	1	0	1	0	↑	↑	0	OUTPUT	Copy data from Reg 2 to REg 1, shift data in REg 2
X	0	0	1	0	↑	↑	0	INPUT	Reg 1 ORs data from Reg 2 and I/O, shift data in Reg 2
X	1	X	X	X	↑	X	1		Synchronously clear Reg 1
X	X	X	X	X	X	↑	1		Synchronously clear Reg 2
X	1	X	X	X	↑	↑	1		Synchronously clear both registers
X	0	X	X	X	↑	↑	1	INPUT	Enter data from I/O into REg 1 and synchronously clear Reg 2
X	0	X	X	X	↑	X	1	INPUT	Enter data from I/O into Reg 1

Figure 6.36. The 74ALS963 is a dual-rank 8-bit shift register with 3-state outputs. The device contains an 8-bit register in parallel with an 8-bit I/O register. In addition to serial-to-parallel and parallel-to-serial conversions, the device is capable of exchanging data between the shift and I/O registers. Control lines determine the mode of operation, as seen in the function table. The '963 features individual shift and I/O register clock inputs; clocking is by positive transitions at the clock input. Active high synchronous clearing is provided.

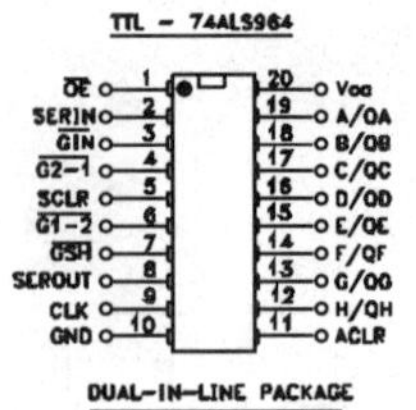

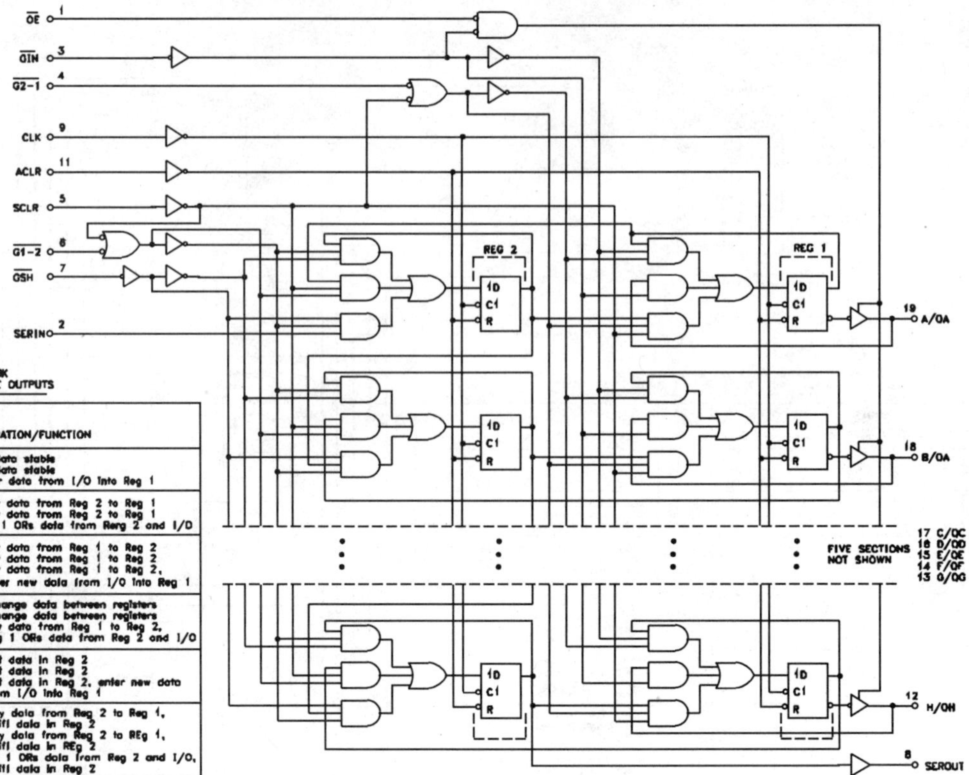

FUNCTION TABLE 74ALS964 DUAL-RANK 8-BIT SHIFT REGISTER WITH 3-STATE OUTPUTS

INPUTS								A/QA THRU H/QH	OPERATION/FUNCTION
$\overline{OE}$	$\overline{GIN}$	$\overline{GE-1}$	$\overline{G1-2}$	$\overline{GSH}$	CLK	ACLR	SCLR		
1	1	1	1	1	X	0	0	HI-Z	All data stable
0	1	1	1	1	X	0	0	OUTPUT	All data stable
X	0	1	1	1	↑	0	0	INPUT	Enter data from I/O into Reg 1
1	1	0	1	1	↑	0	0	HI-Z	Copy data from Reg 2 to Reg 1
0	1	0	1	1	↑	0	0	OUTPUT	Copy data from Reg 2 to Reg 1
X	0	0	1	1	↑	0	0	INPUT	Reg 1 ORs data from Rerg 2 and I/D
1	1	1	0	X	X	0	0	HI-Z	Copy data from Reg 1 to Reg 2
0	1	1	0	X	X	0	0	OUTPUT	Copy data from Reg 1 to Reg 2
X	0	1	0	X	↑	0	0	INPUT	Copy data from Reg 1 to Reg 2, enter new data from I/O into Reg 1
1	1	0	0	X	↑	0	0	HI-Z	Exchange data between registers
0	1	0	0	X	↑	0	0	OUTPUT	Exchange data between registers
X	0	0	0	X	↑	0	0	INPUT	Copy data from Reg 1 to Reg 2, Reg 1 ORs data from Reg 2 and I/O
1	1	1	1	0	X	0	0	HI-Z	Shift data in Reg 2
0	1	1	1	0	X	0	0	OUTPUT	Shift data in Reg 2
X	0	1	1	0	↑	0	0	INPUT	Shift data in Reg 2, enter new data from I/O into Reg 1
1	1	0	1	0	↑	0	0	HI-Z	Copy data from Reg 2 to Reg 1, shift data in Reg 2
0	1	0	1	0	↑	0	0	OUTPUT	Copy data from Reg 2 to REg 1, shift data in REg 2
X	0	0	1	0	↑	0	0	INPUT	Reg 1 ORs data from Reg 2 and I/O, shift data in Reg 2
X	1	X	X	X	↑	0	1		Synchronously clear Reg 1 and Reg 2
X	X	X	X	X	X	1	X		Asynchronously clear Reg and Reg 2
X	0	X	X	X	↑	0	1	INPUT	Enter data from I/O into REg 1 and synchronously clear Reg 2

Figure 6.37. Like the 74ALS963, the 74ALS964 is a dual-rank 8-bit shift register with 3-state outputs. The device contains an 8-bit register in parallel with an 8-bit I/O register. In addition to serial-to-parallel and parallel-to-serial conversions, the device is capable of exchanging data between the shift and I/O registers. Control lines determine the mode of operation, as seen in the function table. The '964 features simultaneous register clocking through a single clock input; clocking is by positive transitions at the clock input. Active high asynchronous and synchronous clearing is provided.

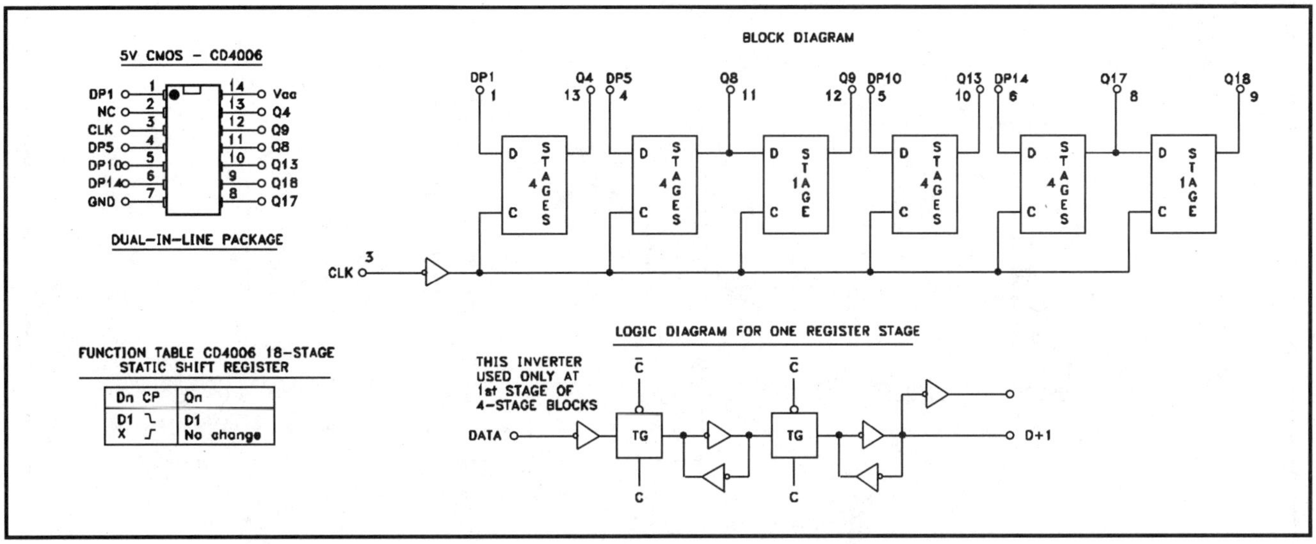

Figure 6.38. The CD4006 CMOS 18-bit static shift register is comprised of four separate shift register sections sharing a common clock. Two sections have four stages, and two sections have five stages with an output on both the 4th and 5th. With this configuration it is possible to obtain a register of 4, 5, 8, 9, 10, 12, 13, 14, 16, 17, or 18 bits by input and output selections. Output transitions occur on the falling edge of the clock pulse. Devices can be cascaded for longer register lengths.

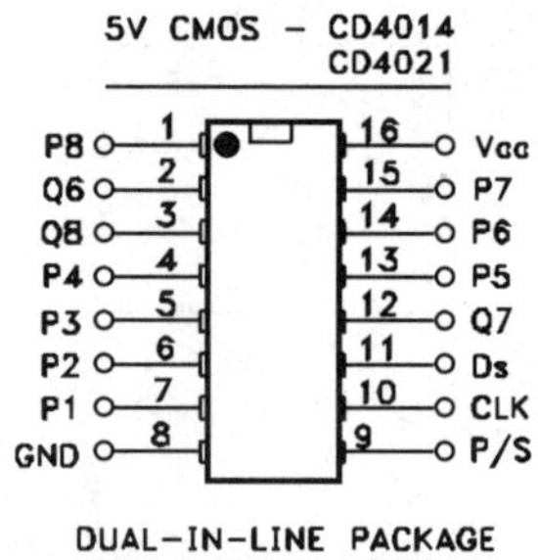

FUNCTION TABLES CD4014, 4021 8-STAGE
STATIC SHIFT REGISTER

SERIAL OPERATION

t	CLK	Ds	P/S	Q6 t=n+6	Q7 t=n+7	Q8 t=n+8
n	↑	0	0	0	?	?
n+1	↑	1	0	1	0	?
n+2	↑	0	0	0	1	0
n+3	↑	1	0	1	0	1
	↑	x	0	Q6	Q7	Q8

PARALLEL OPERATION

CLOCK					
4014	4021	Ds	P/S	Pn	*Qn
↑	X	X	1	0	0
↑	X	X	1	1	1

*Q6, Q7, and Q8 are available externally
X = DON'T CARE

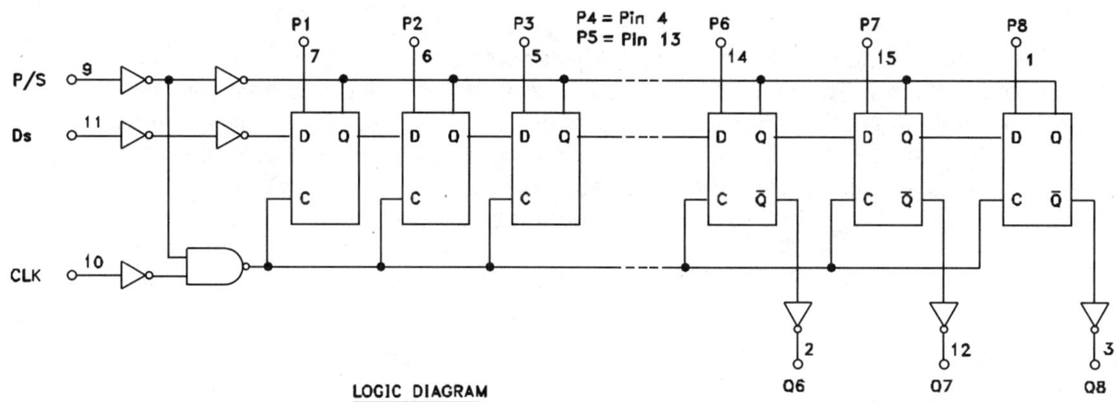

Figure 6.39. The CD4014 and 4021 are CMOS 8-bit static shift registers, finding primary use for parallel-to-serial data conversion, serial output data queuing, and various shift register functions. The 4014 features synchronous parallel input/serial output; the 4021 asynchronous parallel input/serial output. Both provide synchronous serial input/serial output. Operation is fully static. "Q" outputs are available from the 6th, 7th, and 8th stages.

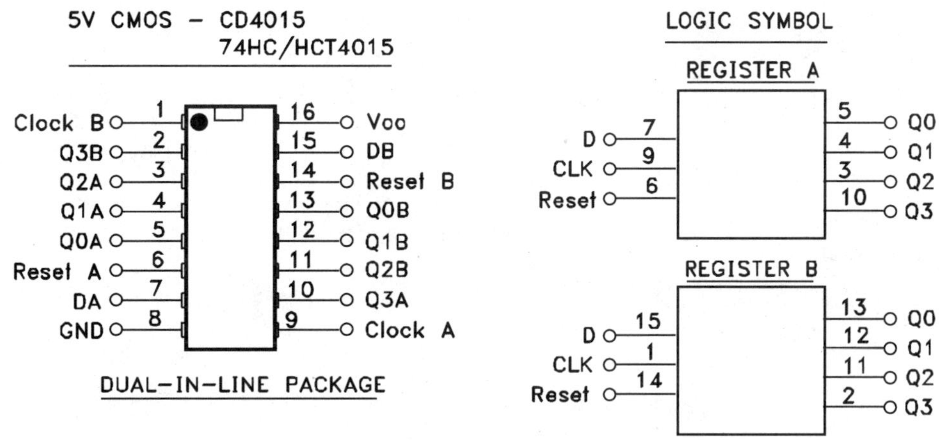

FUNCTION TABLE CD4015, 74x4015
DUAL 4-BIT SERIAL-IN/PARALLEL-OUT
SHIFT REGISTER

Clock	D Input	Reset	Q0	Qn
↑	0	0	0	Qn−1
↑	1	0	1	Qn−1
↓	X	0	NC	NC
X	X	1	0	0

X = DON'T CARE. NC = No Change
Qn = Q0, Q1, Q2, or Q3 as applicable
Qn−1 = Output of prior stage

Figure 6.40. The CD4015 is a CMOS dual 4-bit static shift register consisting of two identical, independent 4-stage serial-input/parallel output registers. Each register provides independent clock and reset inputs with a single serial data input. The register stages are D-type master-slave flip-flops. Data is shifted from one stage to the next during the positive-going clock transition. Each register is cleared by a high level input on the reset line. The existing logic state is retained indefinitely with the clock at either level. Information is transferred only on the positive-going edge of the clock pulse.

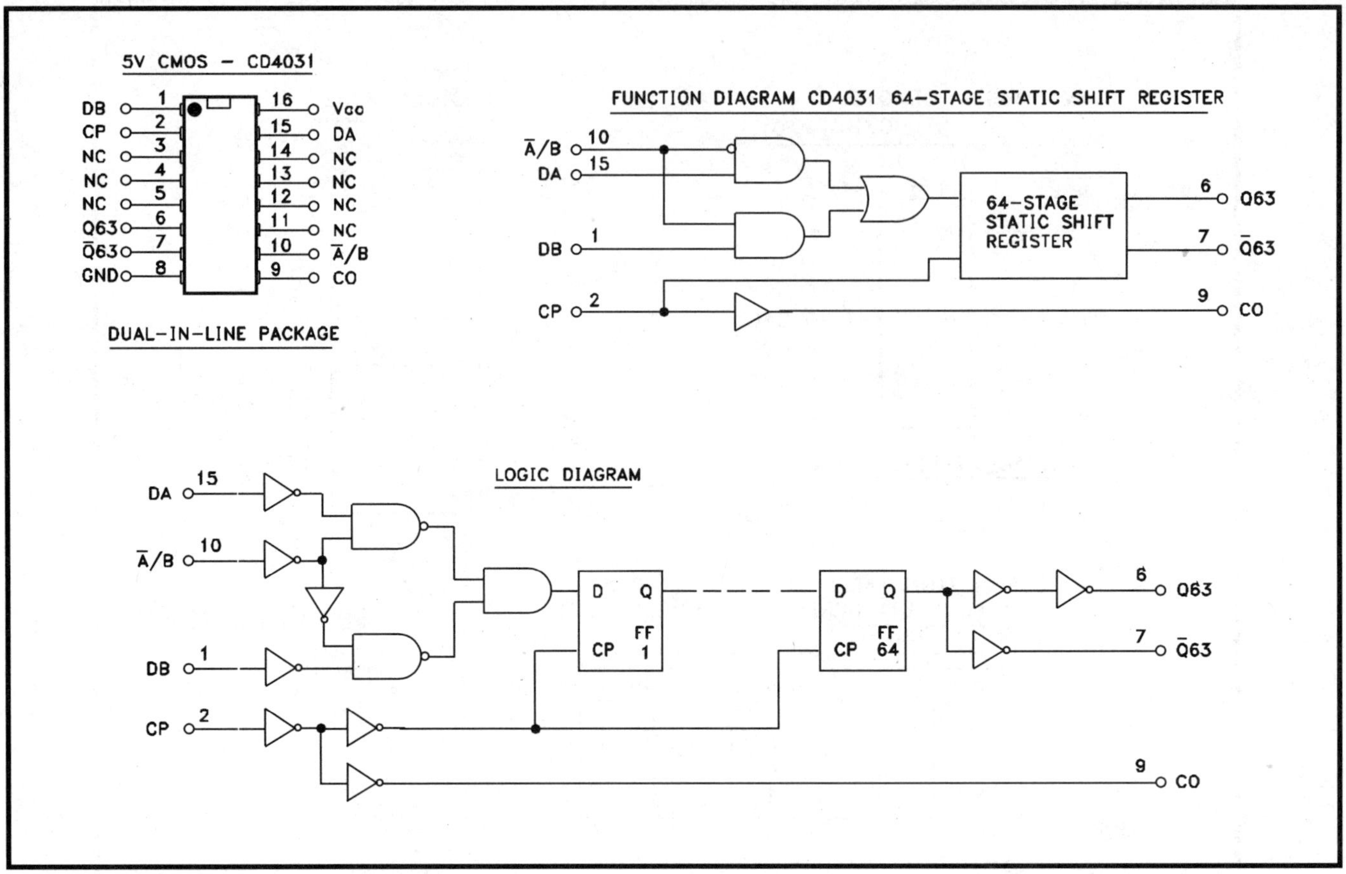

Figure 6.41. The CD4031 is a CMOS dual 64-stage fully static shift register. Two data inputs, Data-in and Recirculate-in and a Mode control input are provided. With the Mode control low, data present at the Data-in input which meets the setup and hold time requirements is entered into the register's first stage. With the Mode control high, data present at the Recirculate-in input is entered. Data is shifted one stage with each positive transition of the clock pulse. Data is available in both true and complementary forms at the output of the 64th stage. A delayed clock output has been provided to allow reduced clock drive fanout and transition time requirements when cascading stages.

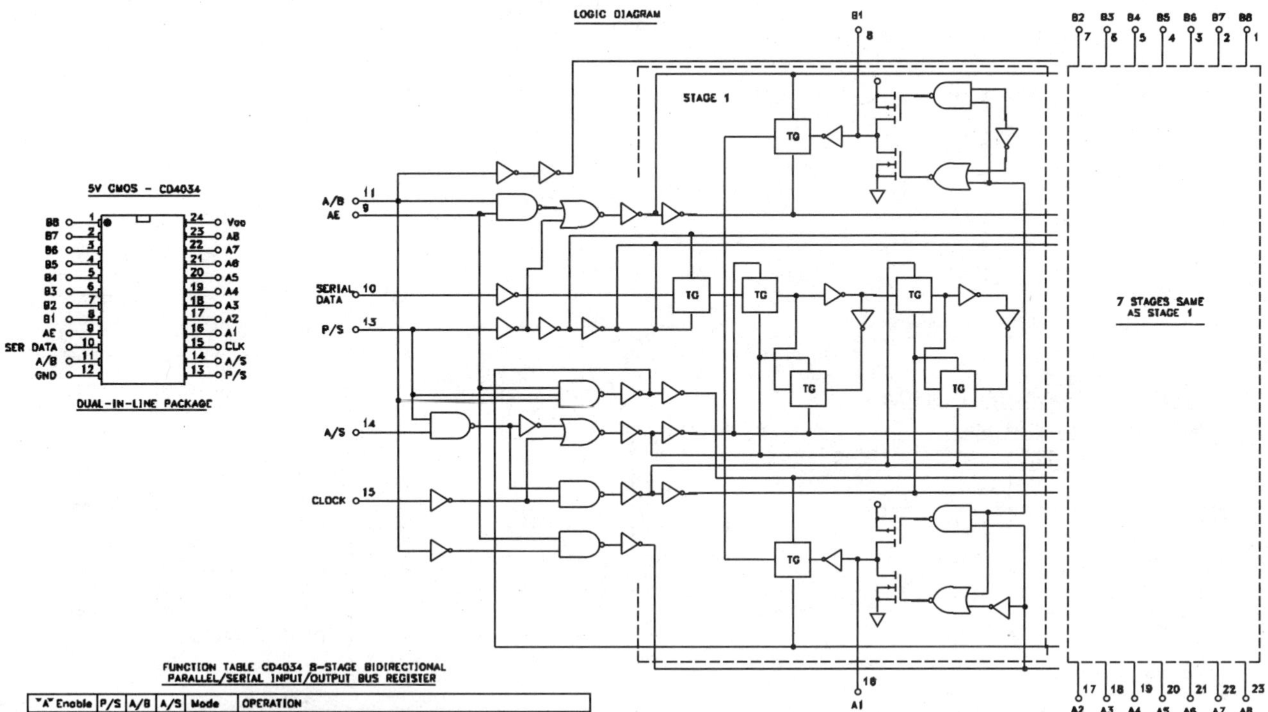

FUNCTION TABLE CD4034 8-STAGE BIDIRECTIONAL PARALLEL/SERIAL INPUT/OUTPUT BUS REGISTER

"A" Enable	P/S	A/B	A/S	Mode	OPERATION
0	0	0	X	Serial	Synchronous Serial data input, A- and B- Parallel data outputs disabled.
0	0	1	X	Serial	Synchronous Serial data input, B Parallel data output.
0	1	0	0	Parallel	B Synchronous Parallel data inputs, A-Parallel data outputs disabled.
0	1	0	1	Parallel	B Asynchronous Parallel data inputs, A-Parallel data outputs disabled.
0	1	1	0	Parallel	A-Parallel data inputs disabled, B-Parallel data outputs, synchronous data recirculation.
0	1	1	1	Parallel	A-Parallel data inputs disabled, B-Parallel data outputs, asynchronous data recirculation
1	0	0	X	Serial	Synchronous Serial data input, A-Parallel data output.
1	0	1	X	Serial	Synchronous Serial data input, B-Parallel data output.
1	1	0	0	Parallel	B synchronous Parallel data input, A-Parallel data output.
1	1	0	1	Parallel	B asynchronous Parallel data input, A-Parallel data output.
1	1	1	0	Parallel	A synchronous Parallel data input, B-Parallel data output.
1	1	1	1	Parallel	A asynchronous Parallel data input, B-Parallel data output.

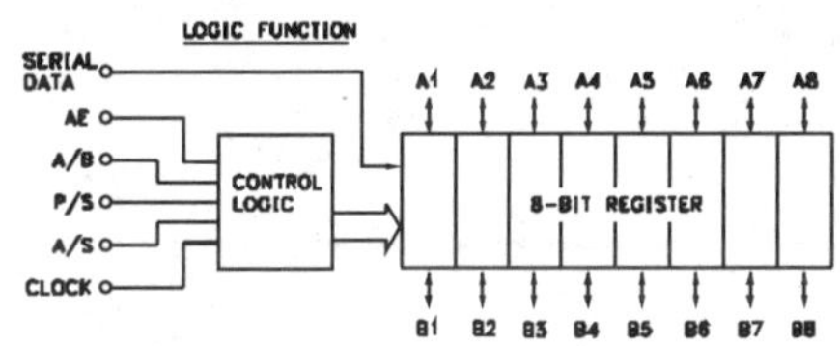

Figure 6.42. The CD4034 is a CMOS 8-stage bidirectional parallel/serial static shift I/O register with 3-state outputs. The two bidirectional ports, A and B, when combined with serial shifting operations, can be used to (1) bidirectionally transfer parallel data between two buses, (2) convert serial data to parallel form and direct it to either of two buses, (3) store (recirculate) parallel data, or (4) accept parallel data from either of two buses and convert it to serial form. These operations are controlled by five inputs as defined in the function table.

5V CMOS - CD4035

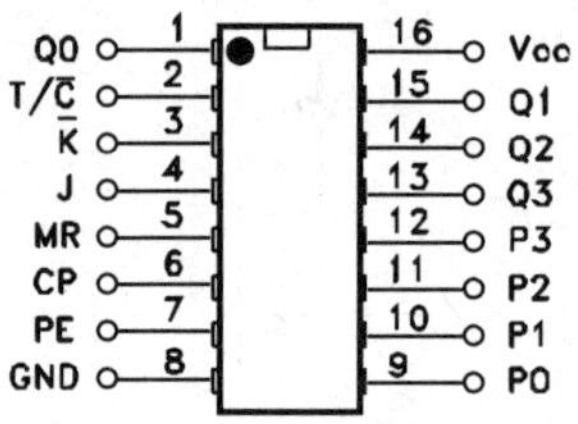

DUAL-IN-LINE PACKAGE

FUNCTION TABLES CD4035
4-BIT UNIVERSAL SHIFT REGISTER

Serial Operation First Stage

INPUTS				OUTPUT	
CP	J	K̄	MR	Q0+1	OPERATING MODE
⌠	1	1	0	1	D Flip-Flop
⌠	0	0	0	0	D Flip-Flop
⌠	1	0	0	Q̄0	Toggle
⌠	0	1	0	Q0	No Change
X	X	X	1	0	Reset

T/C̄ = HIGH, PE = LOW

Parallel Operation

	INPUTS				OUTPUTS			
CP	P0	P1	P2	P3	Q0	Q1	Q2	Q3
⌠	1	1	1	1	1	1	1	1
⌠	0	0	0	0	0	0	0	0

T/C̄ = HIGH, PE = HIGH, MR = LOW

FUNCTION DIAGRAM CD4035 4-BIT UNIVERSAL SHIFT REGISTER

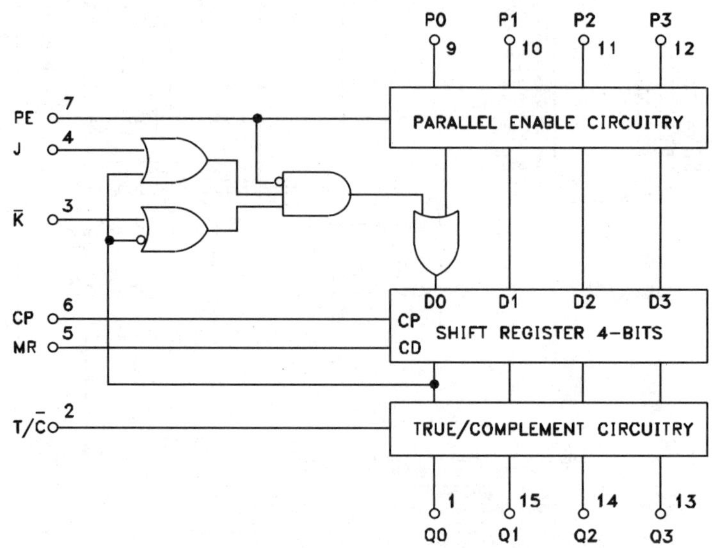

Figure 6.43. The CD4035 is a CMOS 4-bit parallel-in/parallel-out serial clocked shift register. The register has provisions for synchronous parallel inputs to each stage and serial input to the first via J/K logic. Register stages 2,3, and 4 are coupled in a serial D-type flip-flop configuration when the register is in the serial mode, parallel/serial control input low. Parallel entry via the "D" line of each stage only when this control is high. In either mode data is transferred on positive clock transitions. With the true/complement control input high, the true register content is available at the output terminals. With this control low, the outputs are the complement of the register content. This control functions asynchronously with the clock. J/K logic is provided on the first stage serial input. With the J/K inputs tied together the first stage becomes a D-flip-flop. A common asynchronous reset is provided.

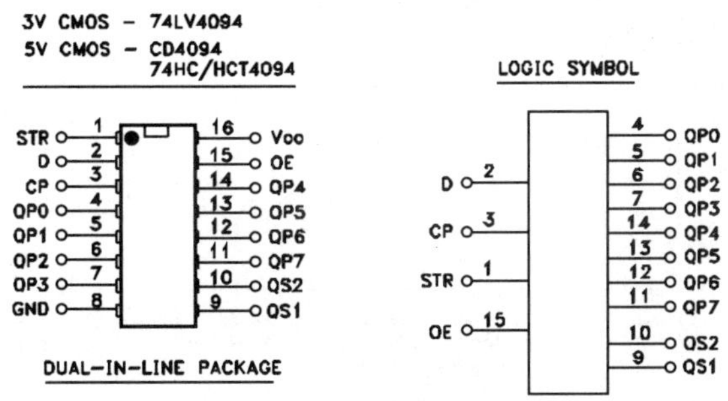

FUNCTION TABLE CD4094, 74o4094 8-STAGE SHIFT-AND-STORE BUS REGISTER

INPUTS				PARALLEL OUTPUTS		SERIAL OUTPUTS	
CP	OE	STR	D	QP0	QPn	QS1	QS2
↑	0	X	X	Z	Z	Q'6	NC
↓	0	X	X	Z	Z	NC	QP7
↑	1	0	X	NC	NC	Q'6	NC
↑	1	1	0	0	QPn−1	Q'6	NC
↑	1	1	1	1	QPn−1	Q'6	NC
↓	1	1	1	NC	NC	NC	QP7

↓ = HIGH-to-LOW clock transition.
↑ = LOW-to-HIGH clock transition.
Z = high impedance OFF state
Q'6 = the information in the seventh register stage is transferred to the 8th register stage and QSn output at the ↓ clock edge.

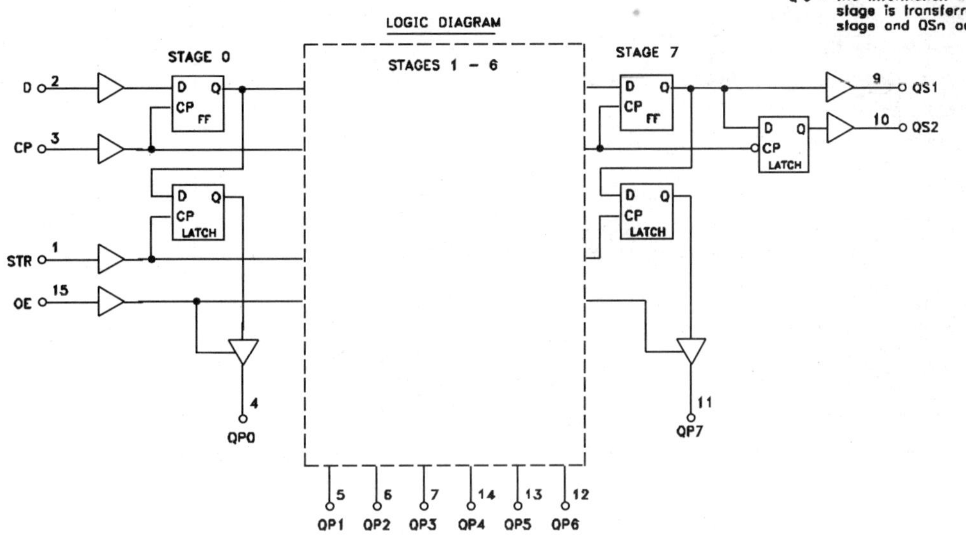

Figure 6.44. The CD4094 is a CMOS 8-bit shift register featuring a 3-state latch. Data is shifted serially through the register on positive transitions of the clock. The last stage (QS1) output can be used for cascading several devices. Data on the QS1 output is transferred to a second output, QS2, on the following negative transition of the clock. Each register stage feeds a latch which functions on the negative edge of the strobe input. With the strobe input high, data propagates through the latch to the 3-state output gates. The gates are enabled with the output enable input high.

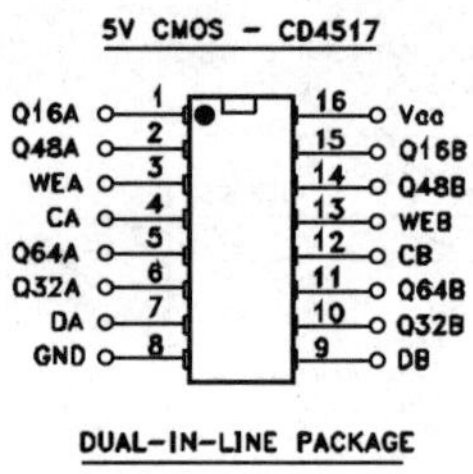

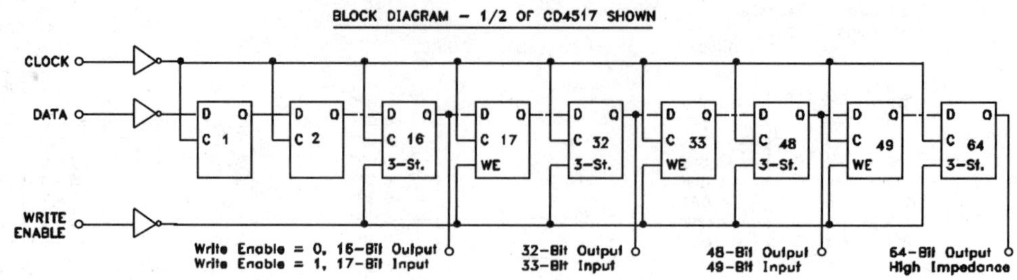

FUNCTION TABLE CD4517 DUAL 64-BIT STATIC SHIFT REGISTER

16-BIT TAP	32-BIT TAP	48-BIT TAP	64-BIT TAP	CLOCK	WRITE ENABLE	DATA
Content of 16-Bit Displayed	Content of 32-Bit Displayed	Content of 48-Bit Displayed	Content of 64-Bit Displayed	0	0	X
High Impedance	High Impedance	High Impedance	High Impedance	0	1	X
Content of 16-Bit Displayed	Content of 32-Bit Displayed	Content of 48-Bit Displayed	Content of 64-Bit Displayed	1	0	X
High Impedance	High Impedance	High Impedance	High Impedance	1	1	X
Content of 16-Bit Displayed	Content of 32-Bit Displayed	Content of 48-Bit Displayed	Content of 64-Bit Displayed	↑	0	Data Entered Into 1st Bit
Data at Tap entered Into 17-Bit	Data at Tap entered Into 33-Bit	Data at Tap entered Into 49-Bit	High Impedance	↑	1	Data Entered Into 1st Bit
Content of 16-Bit Displayed	Content of 32-Bit Displayed	Content of 48-Bit Displayed	Content of 64-Bit Displayed	↓	0	X
High Impedance	High Impedance	High Impedance	High Impedance	↓	1	X

Figure 6.45. The CD4517 is a CMOS 64-bit static shift register consisting of two identical, independent 64-bit registers, each having separate clock and enable inputs. Outputs are available at bits 16, 32, 48, and 64. Data at the data input is entered by clocking regardless of the state of the write enable input. The 3-state output is disabled with the write enable line high. During this time data entry may be continued. The device may be loaded with 64 bits in 16 clock cycles. Output transitions occur on the rising edge of the clock pulse.

5V CMOS – CD4557

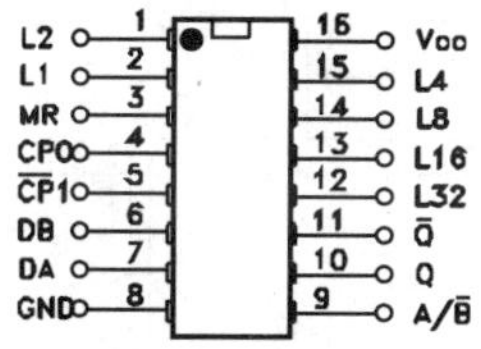

DUAL-IN-LINE PACKAGE

FUNCTION DIAGRAM

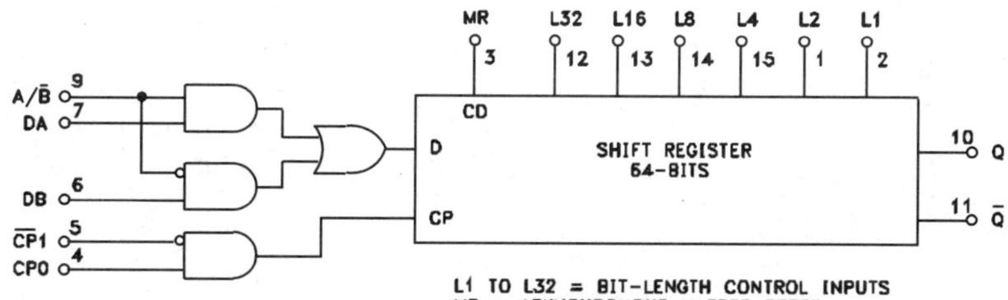

L1 TO L32 = BIT-LENGTH CONTROL INPUTS
MR = ASYNCHRONOUS MASTER RESET
CP0 = CLOCK INPUT
CP1 = CLOCK ENABLE
DA, DB = DATA INPUTS
A/B̄ = SELECT DATA INPUT
Q, Q̄ = BUFFERED OUTPUTS

FUNCTION TABLE CD4557 1-to64-BIT VARIABLE LENGTH SHIFT REGISTER

INPUTS						OUTPUT
MR	A/B̄	DA	DB	CP0	C̄P̄1	Q*
0	0	D1	D2	↑	0	D2
0	1	D1	D2	↑	0	D1
0	0	D1	D2	1	↓	D2
0	1	D1	D2	1	↓	D1
1	X	X	X	X	X	0

* The moment Dn appears at Q depends on the bit-length shown in the Bit-Length Select table.

BIT-LENGTH SELECT TABLE

L32	L16	L8	L4	L2	L1	Register Length
0	0	0	0	0	0	1-Bit
0	0	0	0	0	1	2-Bits
0	0	0	0	1	0	3-Bits
0	0	0	0	1	1	4-Bits
0	0	0	1	0	0	5-Bits
0	0	0	1	0	1	6-Bits
0	0	0	1	1	0	7-Bits
0	0	0	1	1	1	8-Bits
↓	↓	↓	↓	↓	↓	↓
0	1	1	1	1	1	32-Bits
1	0	0	0	0	0	33-Bits
1	0	0	0	0	1	34-Bits
↓	↓	↓	↓	↓	↓	↓
1	1	1	1	0	0	61-Bits
1	1	1	1	0	1	62-Bits
1	1	1	1	1	0	63-Bits
1	1	1	1	1	1	64-Bits

INTERPOLATION TABLE

LENGTH CONTROL INPUTS						MINIMUM NUMBER OF BITS SELECTED	SETUP, HOLD RECOVERY TIMES
L1	L2	L4	L8	L16	L32		
0	0	0	0	0	0	1	Specified
1	0	0	0	0	0	2	↓
X	1	0	0	0	0	3	
X	X	1	0	0	0	5	6 equal steps
X	X	X	1	0	0	9	
X	X	X	X	1	0	17	↓
X	X	X	X	X	1	33	Specified

The set-up, hold and recovery times vary with the minimum number of bits selected. For other values as specified one may interpolate as shown in the interpolation table.

Figure 6.46. The CD4557 is a CMOS 1-to-64-bit variable length static clocked serial shift register whose length may be programmed to be any number between 1 and 64. The number of bits selected is equal to the sum of the subscripts of the enabled Length Control inputs. (Refer to the function table.) Serial data may be selected from the A or B input using the A//B select input. The clock enable input allows for clock gating from either edge of the clock pulse.

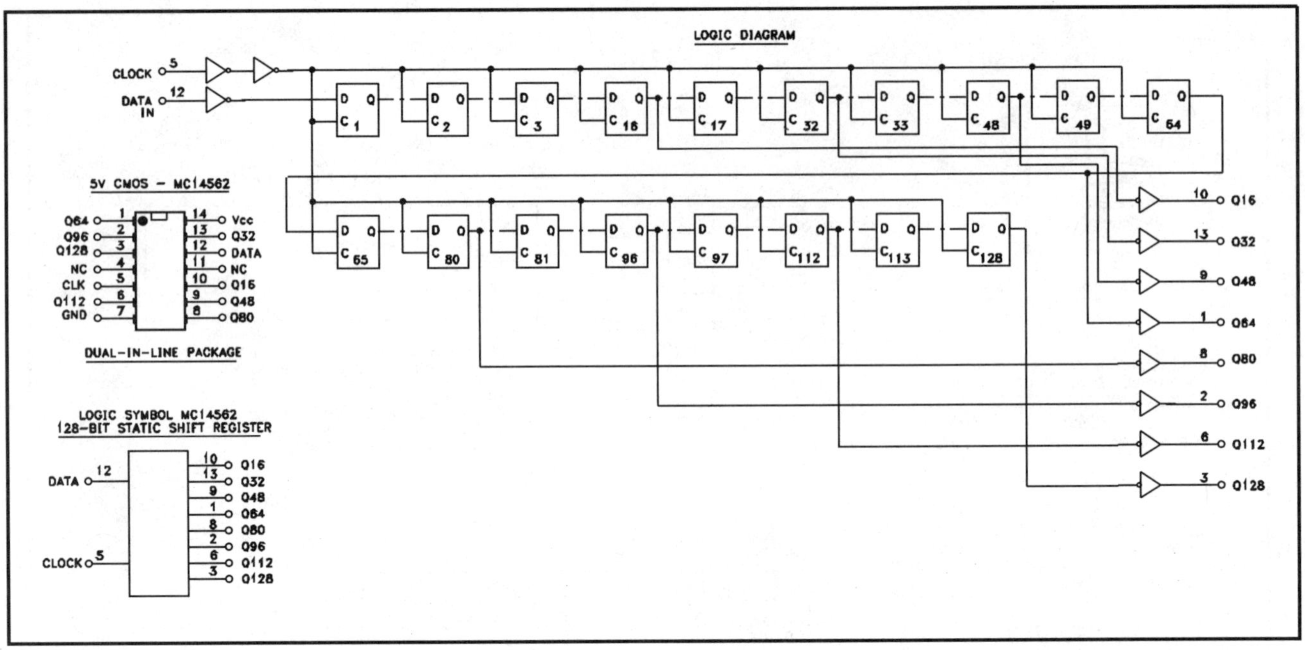

Figure 6.47. The MC14562 is a CMOS 128-bit static shift register. Data is clocked in and out of the register on the positive edge of the clock pulse. Data outputs are available at intervals of every 16 bits from 16 to 128. The device is cascadable to provide for extended lengths

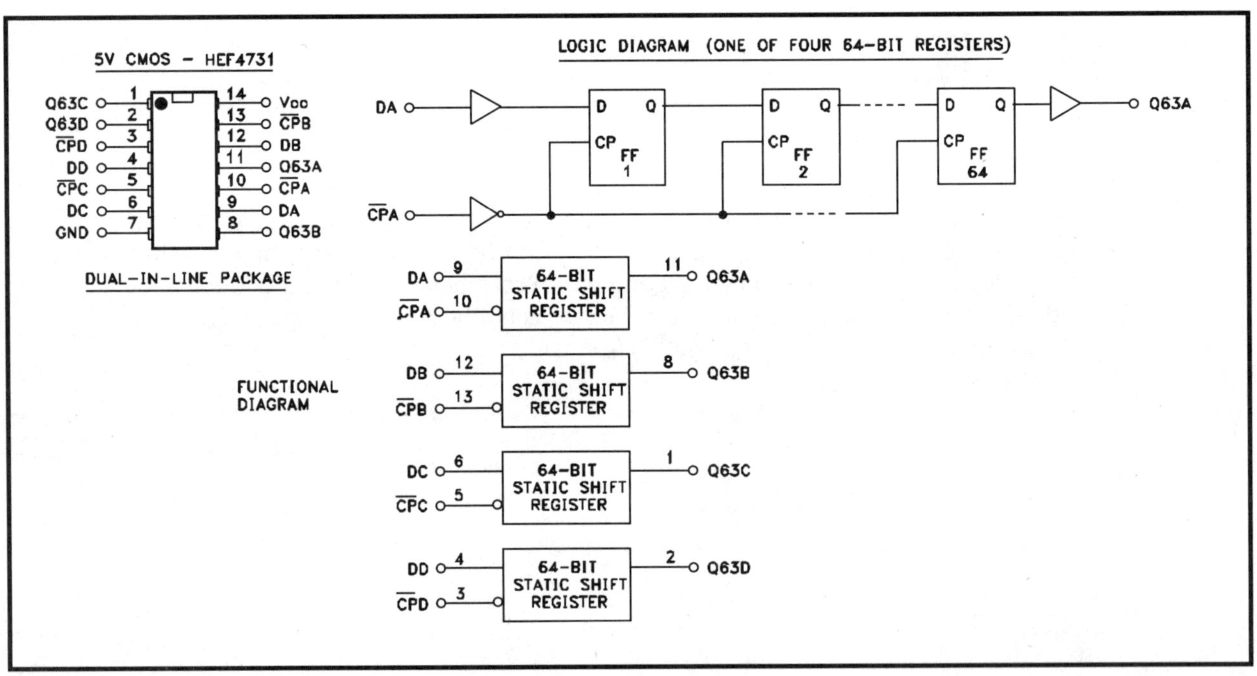

Figure 6.48. The HEF4731 is a CMOS quadruple 64-bit static shift register featuring separate serial data inputs (DA - DD), clock inputs (/CPA - /CPD) and data outputs (Q63A - Q63D) from the 64th register position. Data shifts take place on the negative-going edge of the clock. Low-impedance outputs are provided for direct connection to TTL.

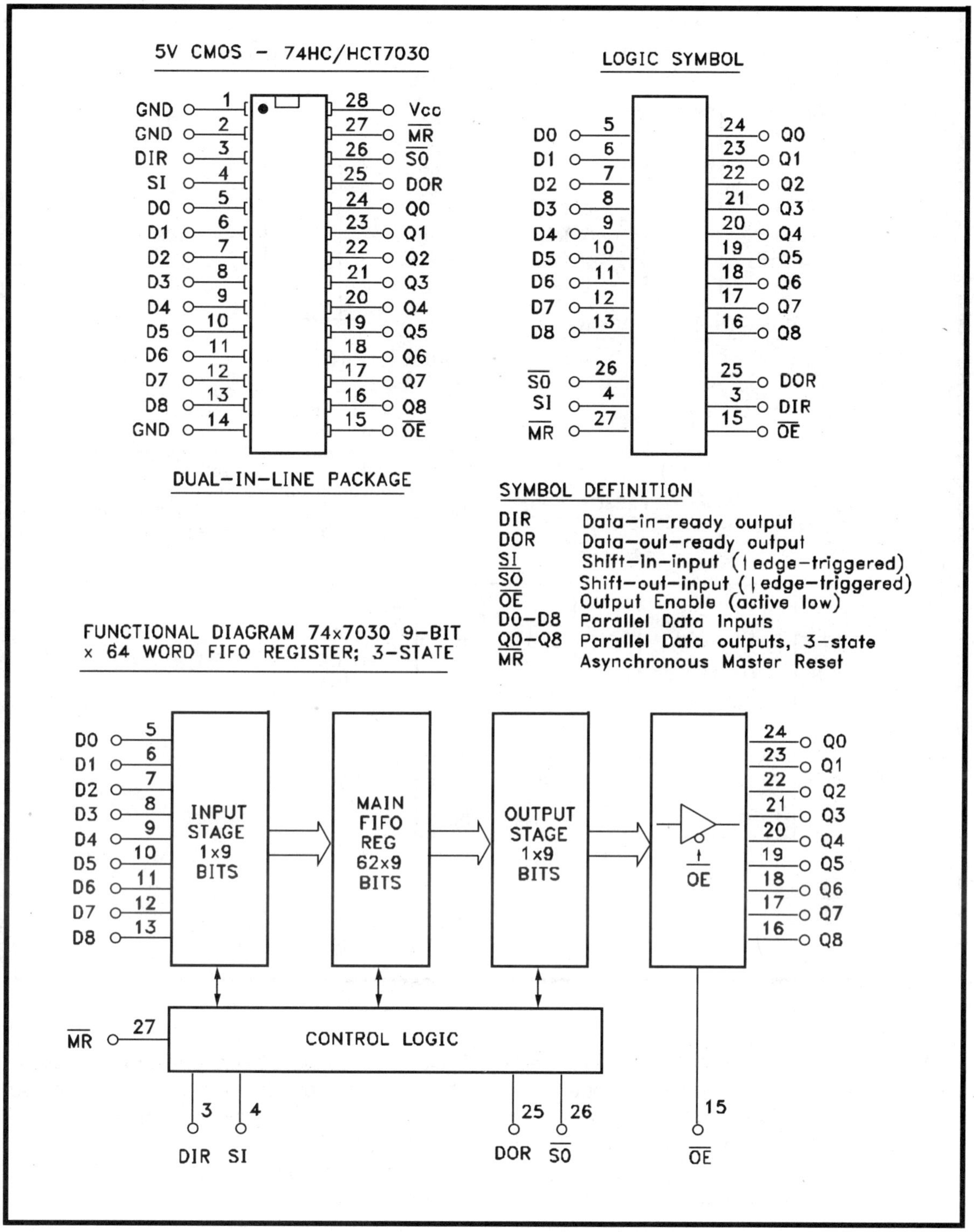

Figure 6.49. The 74HC/HCT7030 is a high-speed CMOS 9-bit-by-64-word expandable FIFO register with 3-state outputs. Reading and writing operations are fully independent due to separate shift-in (SI) and shift-out (/SO) controls. Data transfers may be asynchronous or synchronous. A Master Reset (/MR) and Output Enable (/OE) input are provided. Data-in-Ready (DIR) and Data-out-Ready (DOR) flags indicate device status. The device is expandable in bit and word dimensions. The package pinning is designed for simplified PC board layout.

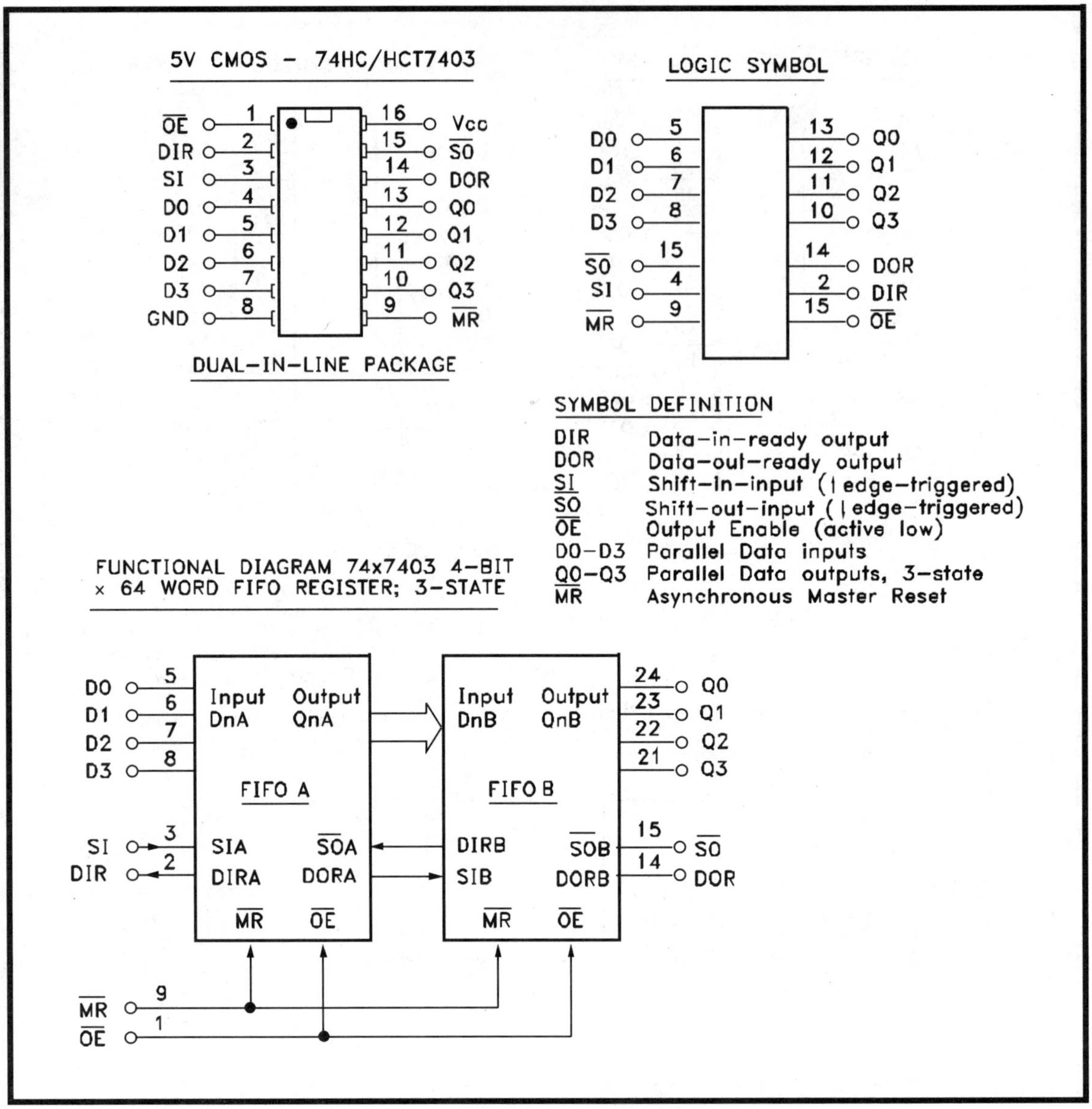

Figure 6.50. The 74HC/HCT7403 is a high-speed CMOS 4-bit-by-64-word expandable FIFO register with 3-state outputs. Reading and writing operations are fully independent due to separate shift-in (SI) and shift-out (/SO) controls. Data transfers may be asynchronous or synchronous. A Master Reset (/MR) and Output Enable (/OE) input are provided. Data-in-Ready (DIR) and Data-out-Ready (DOR) flags indicate device status. The device is expandable in bit and word dimensions. The package pinning is designed for simplified PC board layout.

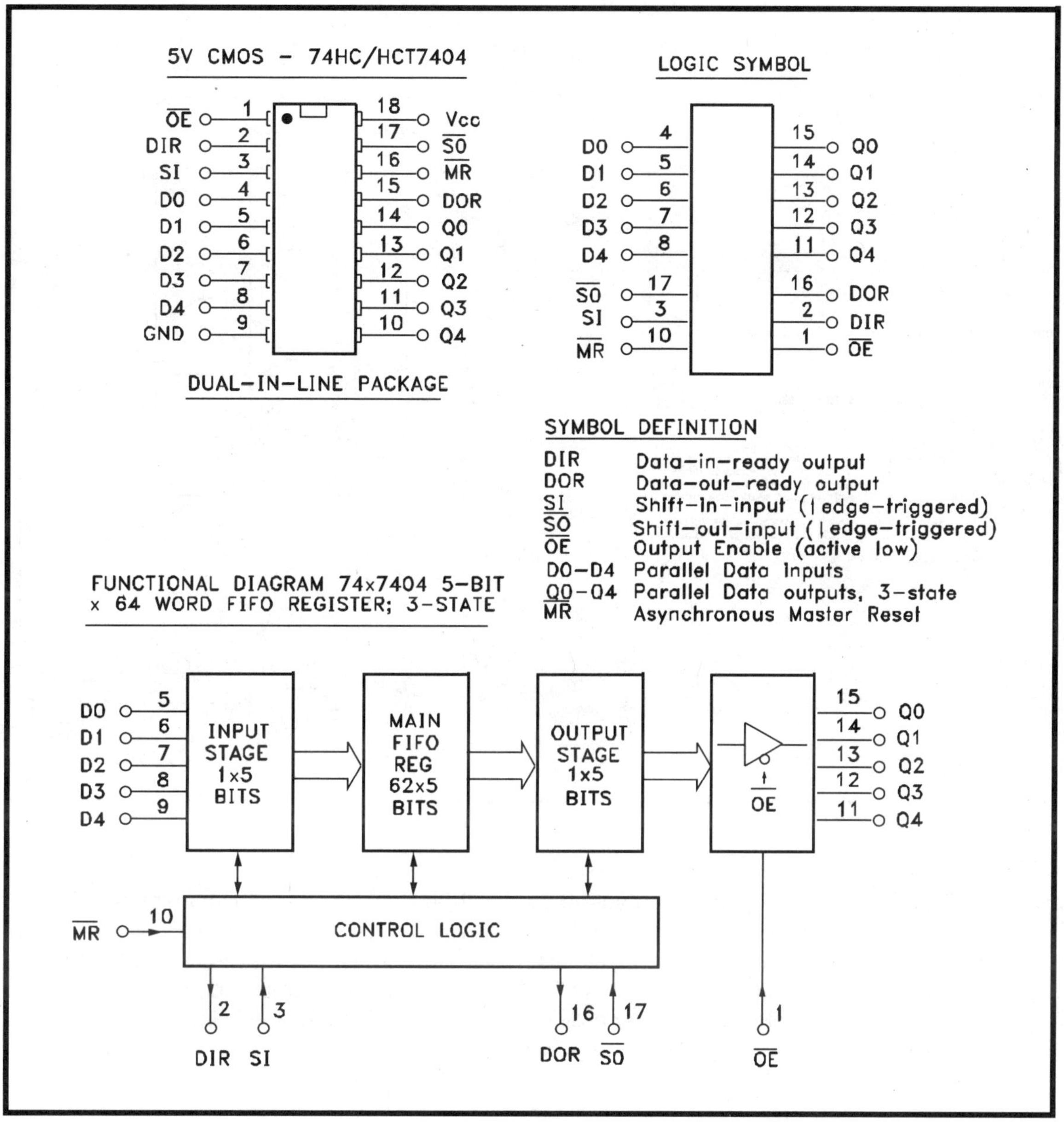

Figure 6.51. The 74HC/HCT7404 is a high-speed CMOS 5-bit-by-64-word expandable FIFO register with 3-state outputs. Reading and writing operations are fully independent due to separate shift-in (SI) and shift-out (/SO) controls. Data transfers may be asynchronous or synchronous. A Master Reset (/MR) and Output Enable (/OE) input are provided. Data-in-Ready (DIR) and Data-out-Ready (DOR) flags indicate device status. The device is expandable in bit and word dimensions. The package pinning is designed for simplified PC board layout.

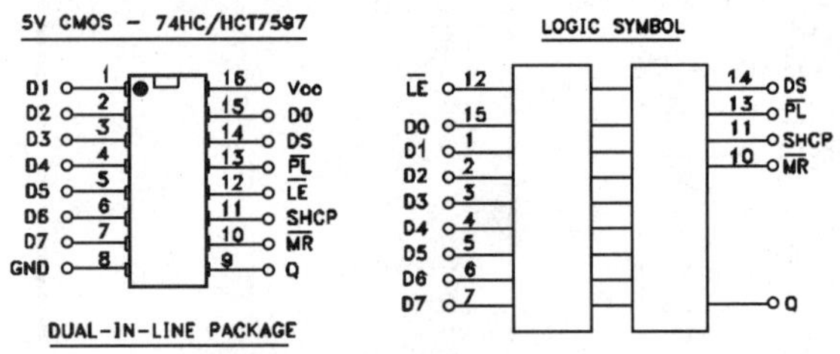

FUNCTION TABLE 74x7597 8-BIT
SHIFT REGISTER WITH INPUT LATCHES

LE	SHCP	PL	MR	FUNCTION
0	X	X	X	Data enabled to input latches (transparent)
1	X	X	X	Data stored into latches (non-transparent)
X	X	0	1	Data transferred from latches to shift register
X	X	0	0	Invalid logic; shift register state indeterminate
X	X	1	0	Shift register cleared
X	↑	1	1	Shift register clocked: Qn=Qn−1, Q0=DS

↑ = LOW-to-HIGH clock transition.

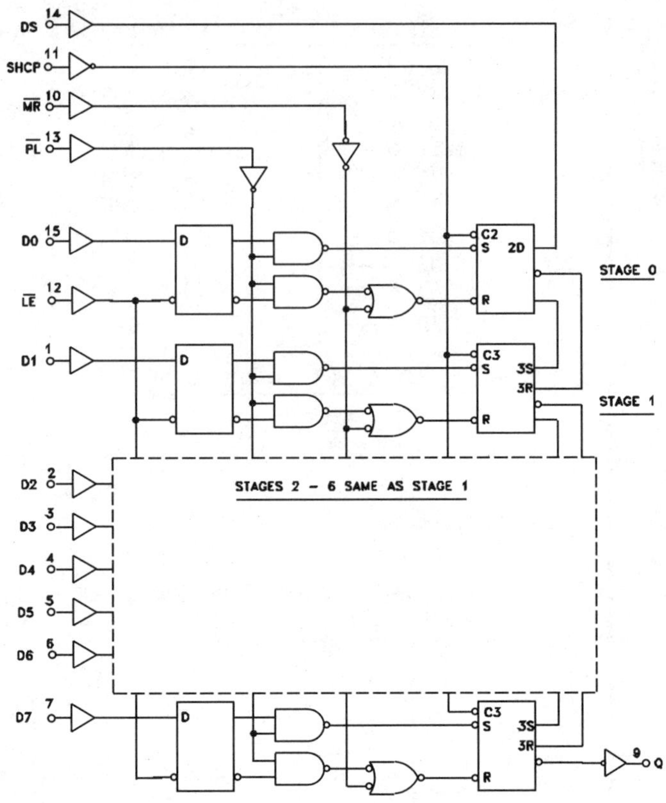

Figure 6.52. The 74HC/HCT7597 is a high-speed CMOS 8-bit shift register with input latches. The device consists of an 8-bit parallel-in/serial-out shift register. With the Latch Enable (/LE) low, data at the Dn inputs enter the latches, which, in this condition, are transparent. With /LE high, the latches store the data present at the Dn inputs meeting the setup and hold time requirements. The register provides a positive-edge triggered clock, direct load (from storage) and clear inputs.

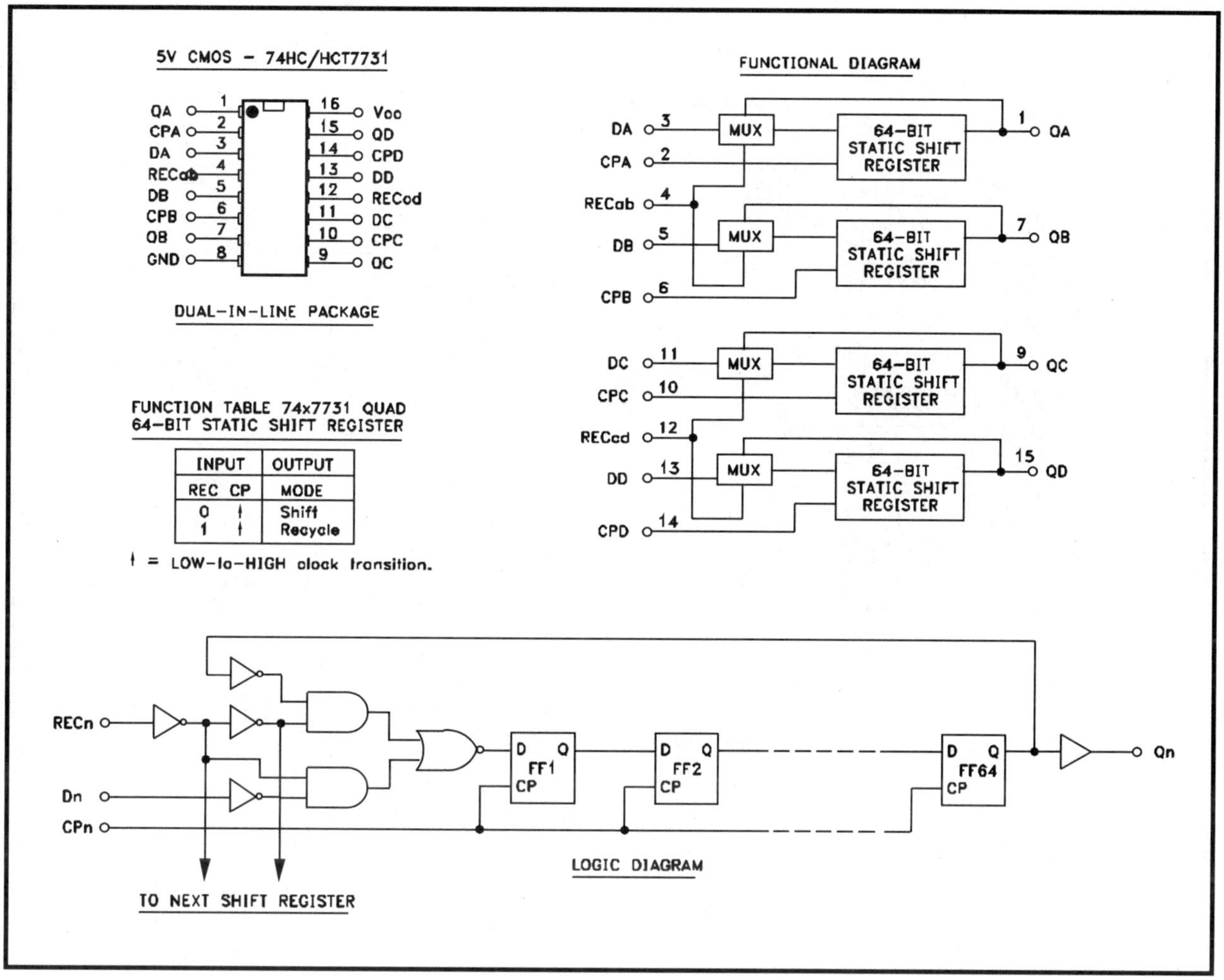

Figure 6.53. The 74HC/HCT7731 is a high-speed CMOS quad 64-bit static shift register with a recycling mode. Each register has its own data and clock inputs and data outputs. With each low-to-high clock transition, data is shifted one stage towards the output. Each Recycling Mode input controls two registers: RECab for registers A and B; RECcd for registers C and D. With the REC line high, the device is in the recycling mode and data at the output is shifted back into the input of the register. Thus, after 64 clock pulses the register content is returned to its original status. Data may be tapped off at any position. External data is shifted in with the REC input low.

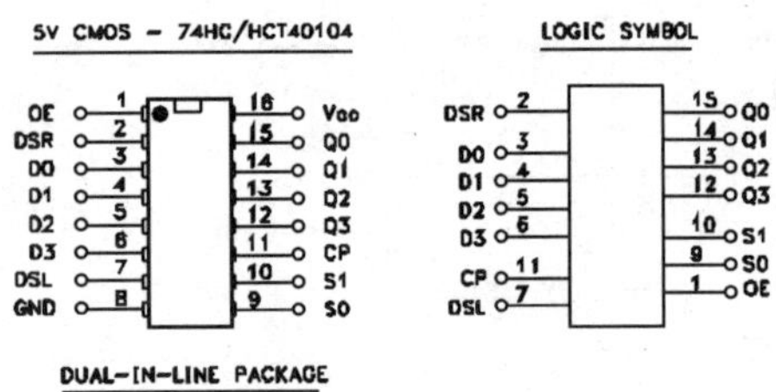

FUNCTION TABLE 74x40104 4-BIT BIDIRECTIONAL UNIVERSAL SHIFT REGISTER. 3-STATE

OPERATING MODES	INPUTS (OE=HIGH)					OUTPUTS at tn+1			
	S1	S0	DSR	DSL	D0 to D3	Q0	Q1	Q2	Q3
Reset	0	0	X	X	X	0	0	0	0
Shift Left	1	0	X	0	X	Q1	Q2	Q3	0
	1	0	X	1	X	Q1	Q2	Q3	1
Shift Right	0	1	0	X	X	0	Q0	Q1	Q2
	0	1	1	X	X	1	Q0	Q1	Q2
Parallel Load	1	1	X	X	0	0	0	0	0
	1	1	X	X	1	1	1	1	1

tn+1 = state after next LOW-to-HIGH transition of CP.

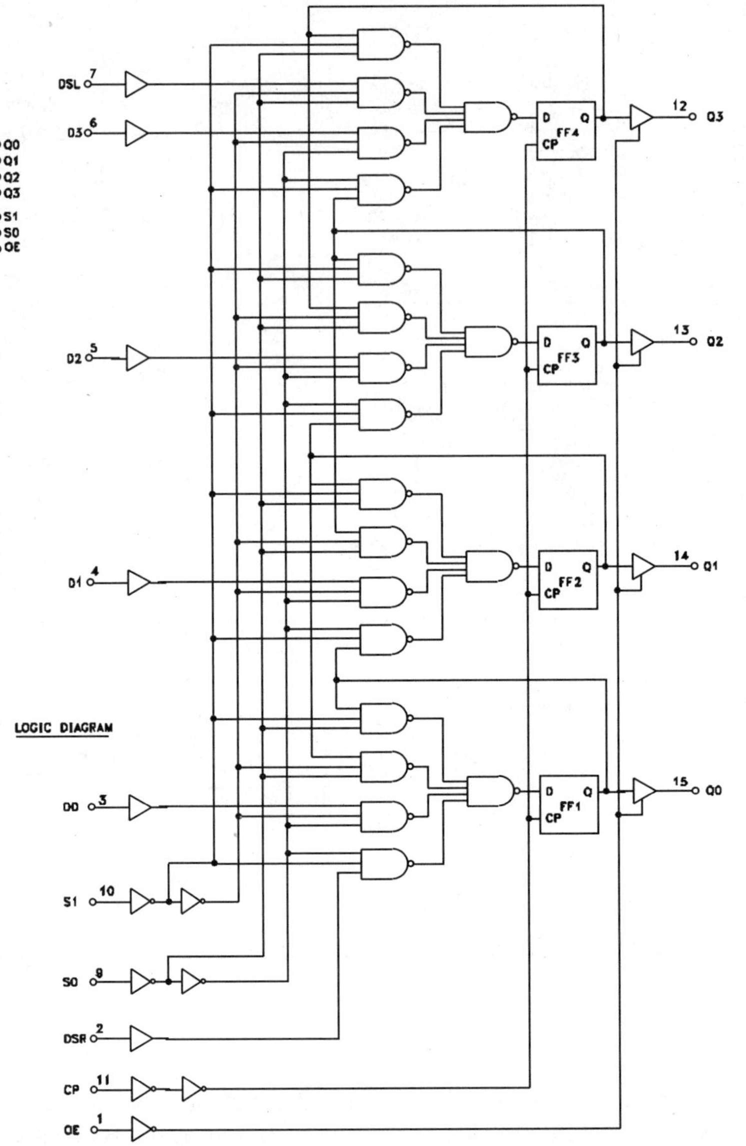

Figure 6.54. The 74HC/HCT40104 is a high-speed CMOS 4-bit bidirectional universal shift register featuring parallel inputs and outputs, shift-right and shift-left serial inputs, and 3-state outputs. With S0 and S1 high, the device is in the parallel mode, data is loaded into the associated flip-flops and appears at the output following the positive transition of the clock input. During this loading serial data flow is inhibited. Shifts left and right are accomplished synchronously on the positive edge of the clock with data entered at the shift-right (DSR) and shift-left (DSL) serial inputs respectively. Clearing occurs with S0 and S1 both low and clocking the register. With the output enable (OE) low, the outputs are in the high-impedance state.

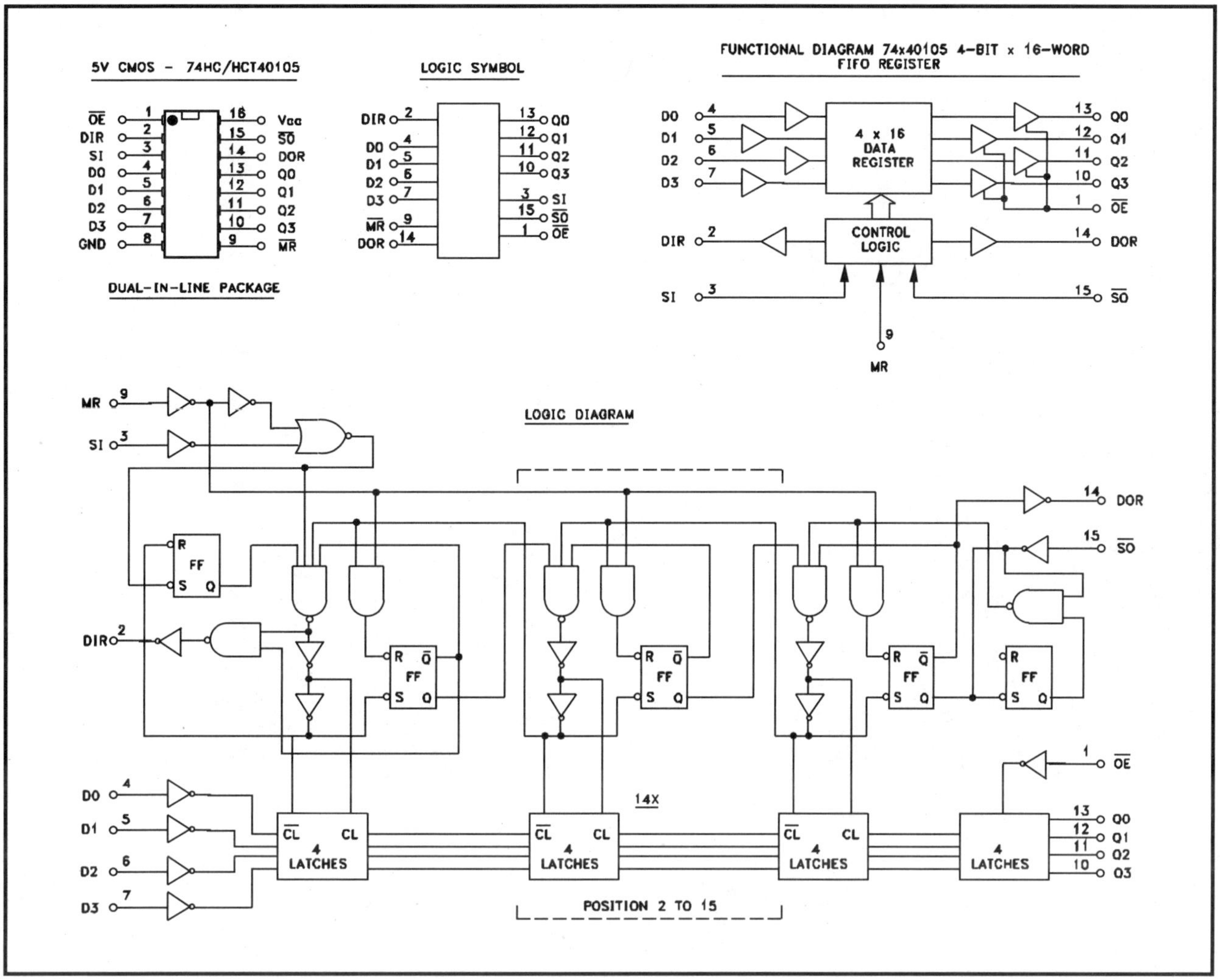

Figure 6.55. The 74HC/HCT40105 is a high-speed CMOS 4-bit-by-16-word FIFO "elastic" storage register. The "elastic" refers to an ability to handle input and data at differing shifting rates which makes the device useful as a buffer in asynchronous data systems. Each word position in the register is clocked by a control flip-flop which stores a marker bit. A "1" indicates the position's data is filled; a "0" that it is vacant. The control flip-flop detects the status of the preceding flip-flop and passes its own status on to the following flip-flop. Status flags DIR and DOR indicate if the register is full or able to accept data. As earliest data is removed at the output later data ripples forward.

Chapter 7

Counters and Dividers

Frequency division and data counting represent a vital application of the flip-flop. Frequency division is, in general, performed by a rather simplistic circuit consisting of a chain of flip-flops in which the changes migrate from stage to stage. The term commonly used for this mode of state transfer is *ripple*. Not all frequency dividers are simple: several described in this chapter include self-contained oscillators.

Counters may follow this same pattern as dividers, and often do. But the counter is frequently called upon to perform tasks other than frequency division and can become very complex in its inner workings. Counters are classified as *ripple* or *synchronous*. As with dividers, in ripple counters the stage states progress from one flip-flop to the next, with the data entering at the first stage. With the synchronous counter, each flip-flop stage is provided with a clock input. State changes then occur simultaneously in all stages, consistent with internal gating and enables. The output spikes often associated in asynchronous ripple counters are avoided. High-speed counters also often feature a carry look-ahead feature that facilitates cascading counters. Typically, the true output of each stage is brought out to a package pin.

Two basic counter types are the *binary* and the *decade* or BCD. Binary counters simply follow the powers-of-2 division to as many stages as required: mostly four, in some eight. Decade counters employ internal decoding to recognize the number nine, at which point the counter is reset to zero. Several CMOS devices employ a counter scheme named *Johnson* (after its inventor), employing internal decoding such that an output pulse is provided for each count over the designed range. Decoding is not restricted to the number nine—the 7492 divides by twelve for 12- or 24-hour clock applications.

Advances in technology raised demands for additional features: a Master Reset for clearing the stages; presetting individual stages to an initial state; selection of an up or down counting mode; the ability to divide by any number, n, within a specific range; multiple counters in a single package with multiplexing for output selection; internal display drivers; and other features worked their way into the design. Examples of these will be found in the detail descriptions which follow.

Figure 7.1 illustrates do-it-yourself binary and decade ripple counters. These circuits employ the CMOS 4027 dual JK flip-flops. The design includes an RC circuit for power-up reset and four LED display circuits so the operation can be observed as the data is entered. A bounceless switch is an excellent clock source.

Chapter Organization

Ten tables are provided. *Tables 7.1 - 7.4* list the counters described here by their manufacturer. There are sources other than those included, but the four provided show a realistic picture of what has been and is now currently available.

Tables 7.5A - 8.0 provide timing reference: these are clock frequency, f_{MHz}; propagation delay, t_p; and clock pulse width, t_w in nS. Because of variations between manufacturers and databook descriptions, this data is not precise. The intent is to provide insight helpful for performance comparisons. In particular, we can see the spectacular gains made in speed and power reduction achieved in the more recent devices.

The drawings of *Figures 7.2 - 7.54* depict device dual in-line packaging pinouts, a Function Table, Logic Symbol, and Logic Diagrams. Some of these do not appear in every drawing and many drawings include additional information. Several drawings are not of devices, but the timing diagrams for counters selected as representative of their class and operating modes.

Summary

This chapter provides tables and drawings describing several hundred variations of frequency counters and dividers. Many devices display an impressive array of capabilities. Some points to keep in mind are:

- Timing information in the tables is for comparison purposes only. There is a general lack of consistency in how these parameters are defined and not just between manufacturers. Where timing is critical consult the manufacturer's data manual.

- In general, 3-state devices are not affected internally when in the high impedance output state.

- All timing data is shown for the industrial/commercial or recommended free-air operating temperature range. Where data is for 25°C it is given here.

- The specifications for CMOS devices may cover voltage ranges from 2 to 15 volts. Data is typically provided for V_{CC}'s of 5, 10, and 15 volts, but not always. Some manufacturers use 5.0 volts; others 4.5. The data shown in Tables 7.5a - 7.8 is as defined for 4.5/5.0 volts. There is a significant speed factor—in general the higher the voltage the faster the performance. Again, where this is important to your application consult the data manual.

- A similar situation exists for the more recent low voltage CMOS devices. Typical V_{CC} ranges given are 2.7 - 3.6 volts. But values may be found for 2, 2.8, 3.0, 3.3 or 3.6 volts. Data shown in the tables is for the range of 2.7 - 3.3 volts. (Glossary *Figures G.1 and G.2* may provide helpful insights on performance.)

DEVICE	STD		L		H		S		LS		AS		ALS		F		Figure	DEVICE
74---	NM	ST	NM	ST	NM	ST	NM	ST	NM	ST	NM	ST	NM	ST	NM	ST	No.	74---
7468										x							7.2	7468
7469										x							7.2	7469
7490	x	xx		x					xx	xx							7.3	7490
7492		xx							x	xx							7.4	7492
7493	x	xx		x					xx	xx							7.5	7493
7497		x															7.7	7497
74160		xx							x	xx	x	x	x	x	xx	xx	7.8	74160
74161	x	xx					x		xx	xx	x	x	x	x	xx	xx	7.9	74161
74162	x	xx						x	x	xx	x	x	x	x	xx	xx	7.8	74162
74163	x	xx					x	x	xx	xx	x	x	x	x	xx	xx	7.9	74163
74167		x															7.11	74167
74168									x	x	x	x	x	x	x	xx	7.12	74168
74169								x	x	xx	x	x	x	x	x	xx	7.12	74169
74176		x															7.13	74176
74177		x															7.13	74177
74190		xx							xx	xx			x	x	x	x	7.14	74190
74191		xx							xx	xx			x	x	x	x	7.14	74191
74192		xx							x	xx			x	x	x	x	7.15	74192
74193		xx							xx	xx			x	x	x	x	7.15	74193
74196		xx						x	x	xx							7.13	74196
74197		xx						x	x	xx							7.13	74197
74269															x		7.17	74269
74290		x							x	xx							7.18	74290
74293		x							x	xx							7.18	74293
74390		x							x	xx			x				7.19	74390
74393		x							x	xx						x	7.20	74393
74490									x	xx							7.21	74490
74560														x			7.22	74560
74561														x			7.22	74561
74568										x				x	x	x	7.23	74568
74569									x	x				x	x	xx	7.23	74569
74579																x	7.24	74579
74590										x			x				7.25	74590
74591										x							7.25	74591
74592										x							7.25	74592
74593										x							7.25	74593
74668										x							7.12	74668
74669									x	x							7.12	74669
74690										x							7.26	74690
74691										x							7.26	74691
74693										x							7.26	74693
74696										x							7.27	74696
74697										x							7.27	74697
74699										x							7.27	74699
74779																x	7.24	74779
741779																x	7.24	741779

Key to Manufacturer: N – National, M – Motorola,
S – Signetics/Philips, T – Texas Instruments

Table 7.1. TTL Counter and Divider source summary by manufacturer.

DEVICE	4000 NM		4000 ST		4100 NM		4100 ST		4500 NM		4500 ST		4700 NM		4700 ST		Figure No.	DEVICE
	N	M	S	T	N	M	S	T	N	M	S	T	N	M	S	T		
4017	x	x	x														7.28	4017
4018	x	x	x														7.30	4018
4020	x	x	x														7.31	4020
4022	x	x	x														7.33	4022
4024	x	x	x														7.34	4024
4029	x	x	x														7.35	4029
4040	x	x	x														7.36	4040
4059			x														7.37	4059
4060	x	x	x														7.38	4060
40102																	7.52	40102
40103																	7.52	40103
40160	x	x	x														7.8	40160
40161	x	x	x														7.9	40161
40162	x	x	x														7.8	40162
40163	x	x	x														7.9	40163
40192	x		x														7.15	40192
40193	x		x														7.15	40193
4160						x											7.8	4160
4161						x											7.9	4161
4162						x											7.8	4162
4163						x											7.9	4163
4510									x	x	x						7.39	4510
4516									x	x	x						7.40	4516
4518									x	x	x						7.41	4518
4520									x	x	x						7.42	4520
4521										x	x						7.43	4521
4522									x	x	x						7.44	4522
4526									x	x	x						7.45	4526
4534										x	x						7.46	4534
4553										x							7.47	4553
4568										x							7.48	4568
4569										x							7.48	4569
4737															x		7.49	4737

Key to Manufacturer:	Prefix
N – National,	CD
M – Motorola,	MC1
S – Signetics/Philips,	HEC/HEF
T – Texas Instruments	--

Table 7.2. 4000 Series CMOS Counter and Divider source summary by manufacturer.

DEVICE	C		HC		HCT		AHC(5V)		AHCT		VHC(5V)		VHCT		AC (5V)		ACT		Figure	DEVICE
74---	NM	ST	NM	ST	NM	ST	NM	ST	NM	ST	NM	ST	NM	ST	NM	ST	NM	ST	No.	74---
7490	x																		7.3	7490
7493	x			x		x													7.5	7493
74160	x		x x	x x		x													7.8	74160
74161	x		x x	x x	x	x		x		x	x				x		x		7.9	74161
74162	x		x x	x x		x													7.8	74162
74163	x		x x	x x	x	x		x		x	x				x		x		7.9	74163
74169															x		x		7.12	74169
74190			x	x x	x														7.14	74190
74191			x	x x	x										x				7.14	74191
74192	x		x	x x	x														7.15	74192
74193	x		x	x x	x														7.15	74193
74390			x x	x x		x					x x								7.19	74390
74393			x x	x x		x					x								7.20	74393
74490				x															7.21	74490
74590				x															7.25	74590
74925	x																		7.50	74925
74926	x																		7.50	74926
74927	x																		7.50	74927
74928	x																		7.50	74928
744017			x x	x x		x													7.28	744017
744020			x x	x x		x													7.32	744020
744024			x	x x		x													7.34	744024
744040			x x	x x		x													7.36	744040
744059			x x	x x		x													7.37	744059
744060			x x	x x		x													7.38	744060
744510				x		x													7.39	744510
744516				x		x													7.40	744516
744518				x		x													7.41	744518
744520				x		x													7.42	744520
746323				x		x													7.51	746323
747022				x															7.52	747022
747060															x*		x*		7.53	747060
747061															x*		x*		7.53	747061
7440102				x		x													7.54	7440102
7440103				x		x													7.54	7440103

Key to Manufacturer: N – National, M – Motorola, S – Signetics/Philips, T – Texas Instruments
* RCA CD54/74AC and CD54/74ACT Series

NOTE: 74AC/ACT161, '163, '169 and 74AC191 data is from "FACT Advanced Logic Databook," National Semiconductor #400022, 1993.

74AC7060, '7061 data is from RCA "Advanced CMOS Logic Databook" GE Corporation, 1988

Table 7.3. 7400 Series 5V CMOS Counter and Divider source summary by manufacturer.

DEVICE	LV		LVCH LVC		ALVT LVT		LVX		LVQ		LCX		AC (3V)		ALVCH		HL		AHC (3V)		VHC (3V)		Figure	DEVICE
74---	NM	ST	NM	ST	NM	ST	NM	ST	NM	ST	NM	ST	NM	ST	NM	ST	NM	ST	NM	ST	NM	ST	No.	74---
74161		x		x									x								x		7.9	74161
74163		x		x									x								x		7.9	74163
74169				x									x										7.12	74169
74191		x											x										7.14	74191
74393		x																					7.20	74393
744020		x																					7.32	744020
744040		x																					7.36	744040
744060		x																					7.38	744060
747060													x *										7.53	747060
747061													x *										7.53	747061

Key to Manufacturer: N – National, M – Motorola, S – Signetics/Philips, T – Texas Instruments, * RCA CD54/74AC and CD54/74ACT Series

NOTE: 74AC/ACT161, '163, '169 and 74AC191 data is from "FACT Advanced Logic Databook," National Semiconductor #400022, 1993.

74AC7060, '7061 data is from RCA "Advanced CMOS Logic Databook" GE Corporation, 1988

Table 7.4. 7400 Series 3V CMOS Counter and Divider source summary by manufacturer.

DEVICE 74---	STD	L	H	S	LS	AS/A/B	ALS/A/B	F	DATA	Figure	DEVICE 74---
-7468	--	--	--	--	50-70	--	--	--	f MHz	- 7.2	-7468
----	--	--	--	--	7-40	--	--	--	tpd	---	----
----	--	--	--	--	10-25	--	--	--	tw	---	----
-7469	--	--	--	--	50-70	--	--	--	f MHz	- 7.2	-7469
----	--	--	--	--	7-40	--	--	--	tpd	---	----
----	--	--	--	--	10-25	--	--	--	tw	---	----
-7490	10-18	3	--	--	32-42	--	--	--	f MHz	- 7.3	-7490
----	60-100	340	--	--	10-50	--	--	--	tpd	---	----
----	15-30	200	--	--	15-30	--	--	--	tw	---	----
-7492	10-18	--	--	--	32-42	--	--	--	f MHz	- 7.4	-7492
----	60-100	--	--	--	10-50	--	--	--	tpd	---	----
----	15-30	--	--	--	15-30	--	--	--	tw	---	----
-7493	10-18	3	--	--	32-42	--	--	--	f MHz	- 7.5	-7493
----	75-135	280-450	--	--	10-70	--	--	--	tpd	---	----
----	15-30	200	--	--	15-30	--	--	--	tw	---	----
-7497	25-32	--	--	--	--	--	--	--	f MHz	- 7.7	-7497
----	15-23	--	--	--	--	--	--	--	tpd	---	----
----	15-20	--	--	--	--	--	--	--	tw	---	----
-74160	25-32	--	--	--	25-32	75	40	100-130	f MHz	- 7.8	-74160
----	11-38	--	--	--	9-35	1-16.5	3-24	2.0-9.0	tpd	---	----
----	20-25	--	--	--	20-25	6.7-8	12.5-15	4.0-5.5	tw	---	----
-74161	25-32	--	--	40-70	25-32	75	40	100-130	f MHz	- 7.9	-74161
----	11-38	--	--	8-25	9-35	1-16.5	3-24	1.5-8.5	tpd	---	----
----	20-25	--	--	10	20-25	6.7-8	12.5-15	4.0-6.0	tw	---	----
-74162	25-32	--	--	--	25-32	75	40	100-130	f MHz	- 7.8	-74162
----	11-38	--	--	--	9-35	1-16.5	3-20	2.0-9.0	tpd	---	----
----	20-25	--	--	--	20-25	6.7-8	12.5-15	4.0-5.5	tw	---	----
-74163	25-32	--	--	40-70	25-32	75	40	100-130	f MHz	- 7.9	-74163
----	11-38	--	--	8-25	9-35	1-16.5	3-20	1.5-8.5	tpd	---	----
----	20-25	--	--	10	20-25	6.7-8	12.5-15	4.0-6.0	tw	---	----
-74167	25-32	--	--	--	--	--	--	--	f MHz	- 7.11	-74167
----	6-39	--	--	--	--	--	--	--	tpd	---	----
----	25/0	--	--	--	--	--	--	--	tw	---	----
-74168	--	--	--	--	25	75	40	100-115	f MHz	- 7.12	-74168
----	--	--	--	--	14-35	1-16.5	2-20	2.5-12.5	tpd	---	----
----	--	--	--	--	30/0	6.7	12.5	5.0	tw	---	----
-74169	--	--	--	--	25	75	40	100-115	f MHz	- 7.12	-74169
----	--	--	--	--	14-35	1-16.5	2-20	2.5-12.5	tpd	---	----
----	--	--	--	--		6.7	12.5	5.0	tw	---	----
-74176	35-50	--	--	--	--	--	--	--	f MHz	- 7.13	-74176
----	8-51	--	--	--	--	--	--	--	tpd	---	----
----	14-28	--	--	--	--	--	--	--	tw	---	----
-74177	35-50	--	--	--	--	--	--	--	f MHz	- 7.13	-74177
----	8-51	--	--	--	--	--	--	--	tpd	---	----
----	14-28	--	--	--	--	--	--	--	tw	---	----
-74190	20-25	--	--	--	20-25	--	25	100-125	f MHz	- 7.14	-74190
----	14-52	--	--	--	13-52	--	3-45	2.0-16.0	tpd	---	----
----	25-35	--	--	--	25-35	--	16.5-20	3.5-6.0	tw	---	----
-74191	20-25	--	--	--	20-25	--	30	100-125	f MHz	- 7.14	-74191
----	14-52	--	--	--	13-52	--	3-45	2.0-16.0	tpd	---	----
----	25-35	--	--	--	25-35	--	16.5-20	3.5-6.0	tw	---	----
-74192	25-32	--	--	--	25-32	--	25	100-125	f MHz	- 7.15	-74192
----	16-47	--	--	--	15-47	--	4-30	2.0-9.5	tpd	---	----
----	20	--	--	--	20	--	10-20	3.5-10.0	tw	---	----
-74193	25-32	--	--	--	25-32	--	30	100-125	f MHz	- 7.15	-74193
----	16-47	--	--	--	15-47	--	4-30	2.0-9.5	tpd	---	----
----	20	--	--	--	20	--	10-20	3.5-10.0	tw	---	----
-74196	50-70	--	--	100-140	30-40	--	--	--	f MHz	- 7.13	-74196
----	10-63	--	--	5-37	8-62	--	--	--	tpd	---	----
----	10-20	--	--	5-30	15-30	--	--	--	tw	---	----
-74197	50-70	--	--	100-140	30-40	--	--	--	f MHz	- 7.13	-74197
----	10-63	--	--	5-37	8-62	--	--	--	tpd	---	----
----	10-20	--	--	5-30	15-30	--	--	--	tw	---	----
-74269	--	--	--	--	--	--	--	100	f MHz	- 7.17	-74269
----	--	--	--	--	--	--	--	3.0-10.0	tpd	---	----
----	--	--	--	--	--	--	--	4.0-4.5	tw	---	----
-74290	32-42	--	--	--	32-42	--	--	--	f MHz	- 7.18	-74290
----	10-50	--	--	--	10-50	--	--	--	tpd	---	----
----	15-30	--	--	--	15-30	--	--	--	tw	---	----
-74293	32-42	--	--	--	32-42	--	--	--	f MHz	- 7.18	-74293
----	10-70	--	--	--	10-70	--	--	--	tpd	---	----
----	15-30	--	--	--	15-30	--	--	--	tw	---	----

Table 7.5A. TTL Counter and Divider representative frequency (MHz), delays (tpd), and clock pulse width (tw) times in nS.

DEVICE 74---	STD	L	H	S	LS	AS/A/B	ALS/A/B	F	DATA	Figure	DEVICE 74---
-74390	25-35	--	--	--	25-35	--	--	--	f MHz	-7.19	-74390
----	12-60	--	--	--	12-60	--	--	--	tpd	---	----
----	20-25	--	--	--	20-40	--	--	--	tw	---	----
-74393	25-35	--	--	--	25-35	--	--	--	f MHz	-7.20	-74393
----	12-60	--	--	--	12-60	--	--	--	tpd	---	----
----	20-25	--	--	--	20-40	--	--	--	tw	---	----
-74490	--	--	--	--	25-35	--	--	--	f MHz	-7.21	-74490
----	--	--	--	--	12-54	--	--	--	tpd	---	----
----	--	--	--	--	20	--	--	--	tw	---	----
-74560	--	--	--	--	--	--	20	--	f MHz	-7.22	-74560
----	--	--	--	--	--	--	2-55	--	tpd	---	----
----	--	--	--	--	--	--	15-25	--	tw	---	----
-74561	--	--	--	--	--	--	30	--	f MHz	-7.22	-74561
----	--	--	--	--	--	--	2-55	--	tpd	---	----
----	--	--	--	--	--	--	15-25	--	tw	---	----
-74568	--	--	--	--	25	--	20	50-115	f MHz	-7.23	-74568
----	--	--	--	--	14-35	--	1-28	1.5-7.5	tpd	---	----
----	--	--	--	--	20	--	15-25	4.5-7.0	tw	---	----
-74569	--	--	--	--	25	--	30	50-115	f MHz	-7.23	-74569
----	--	--	--	--	14-35	--	1-28	1.5-7.5	tpd	---	----
----	--	--	--	--	20	--	15-25	4.5-7.0	tw	---	----
-74579	--	--	--	--	--	--	--	100-115	f MHz	-7.24	-74579
----	--	--	--	--	--	--	--	3.5-6.0	tpd	---	----
----	--	--	--	--	--	--	--	3.0-4.5	tw	---	----
-74590	--	--	--	--	20-35	--	NO	--	f MHz	-7.25	-74590
----	--	--	--	--	12-45	--	DATA	--	tpd	---	----
----	--	--	--	--	20-25	--	GIVEN	--	tw	---	----
-74591	--	--	--	--	20-35	--	--	--	f MHz	-7.25	-74591
----	--	--	--	--	12-45	--	--	--	tpd	---	----
----	--	--	--	--	20-25	--	--	--	tw	---	----
-74592	--	--	--	--	20-35	--	--	--	f MHz	-7.25	-74592
----	--	--	--	--	15-53	--	--	--	tpd	---	----
----	--	--	--	--	20-40	--	--	--	tw	---	----
-74593	--	--	--	--	20-35	--	--	--	f MHz	-7.25	-74593
----	--	--	--	--	14-63	--	--	--	tpd	---	----
----	--	--	--	--	20-40	--	--	--	tw	---	----
-74668	--	--	--	--	25-32	--	--	--	f MHz	-7.12	-74668
----	--	--	--	--	11-60	--	--	--	tpd	---	----
----	--	--	--	--	20	--	--	--	tw	---	----
-74669	--	--	--	--	25-32	--	--	--	f MHz	-7.12	-74669
----	--	--	--	--	11-60	--	--	--	tpd	---	----
----	--	--	--	--	20	--	--	--	tw	---	----
-74690	--	--	--	--	20	--	--	--	f MHz	-7.26	-74690
----	--	--	--	--	12-40	--	--	--	tpd	---	----
----	--	--	--	--	20-25	--	--	--	tw	---	----
-74691	--	--	--	--	20	--	--	--	f MHz	-7.26	-74691
----	--	--	--	--	12-40	--	--	--	tpd	---	----
----	--	--	--	--	20-25	--	--	--	tw	---	----
-74693	--	--	--	--	20	--	--	--	f MHz	-7.26	-74693
----	--	--	--	--	12-40	--	--	--	tpd	---	----
----	--	--	--	--	20-25	--	--	--	tw	---	----
-74696	--	--	--	--	20	--	--	--	f MHz	-7.27	-74696
----	--	--	--	--	12-40	--	--	--	tpd	---	----
----	--	--	--	--	20-25	--	--	--	tw	---	----
-74697	--	--	--	--	20	--	--	--	f MHz	-7.27	-74697
----	--	--	--	--	12-40	--	--	--	tpd	---	----
----	--	--	--	--	20-25	--	--	--	tw	---	----
-74699	--	--	--	--	20	--	--	--	f MHz	-7.27	-74699
----	--	--	--	--	12-40	--	--	--	tpd	---	----
----	--	--	--	--	20-25	--	--	--	tw	---	----
-74779	--	--	--	--	--	--	--	125-145	f MHz	-7.24	-74779
----	--	--	--	--	--	--	--	3.0-5.5	tpd	---	----
----	--	--	--	--	--	--	--	4.0	tw	---	----
-741779	--	--	--	--	--	--	--	115-130	f MHz	-7.24	-741779
----	--	--	--	--	--	--	--	2.0-5.0	tpd	---	----
----	--	--	--	--	--	--	--	3.0-4.5	tw	---	----

Table 7.5b. TTL Counter and Divider representative frequency (MHz), delays (tpd), and clock pulse width (tw) times in nS (continuation).

DEVICE CD\MC1	4000	4100	4500	4700	DATA	Figure	DEVICE CD\MC1
-4017	6-12	--	--	--	f_{MHz}	- 7.28	-4017
----	115-260	--	--	--	t_{pd}	---	----
----	40-80	--	--	--	t_w	---	----
-4018	2-4	--	--	--	f_{MHz}	-7.30	-4018
----	205-350	--	--	--	t_{pd}	---	----
----	20-50	--	--	--	t_w	---	----
-4020	5-10	--	--	--	f_{MHz}	- 7.32	-4020
----	180-210	--	--	--	t_{pd}	---	----
----	25-50	--	--	--	t_w	---	----
-4022	3-6	--	--	--	f_{MHz}	- 7.33	-4022
----	245-485	--	--	--	t_{pd}	---	----
----	35-75	--	--	--	t_w	---	----
-4024	5-10	--	--	--	f_{MHz}	-7.34	-4024
----	120-210	--	--	--	t_{pd}	---	----
----	30-60	--	--	--	t_w	---	----
-4029	2-4	--	--	--	f_{MHz}	- 7.35	-4029
----	280-385	--	--	--	t_{pd}	---	----
----	55-110	--	--	--	t_w	---	----
-4040	10-20	--	--	--	f_{MHz}	-7.36	-4040
----	90-170	--	--	--	t_{pd}	---	----
----	25-50	--	--	--	t_w	---	----
-4059	3-7	--	--	--	f_{MHz}	- 7.37	-4059
----	90-200	--	--	--	t_{pd}	---	----
----	---	--	--	--	t_w	---	----
-4060	4-8	--	--	--	f_{MHz}	-7.38	-4060
----	210-420	--	--	--	t_{pd}	---	----
----	60-120	--	--	--	t_w	---	----
-40102	REFER	--	--	--	f_{MHz}	- 7.52	-40102
----	TO	--	--	--	t_{pd}	---	----
----	7440102	--	--	--	t_w	---	----
-40103	AND	--	--	--	f_{MHz}	-7.52	-40103
----	7440103	--	--	--	t_{pd}	---	----
----		--	--	--	t_w	---	----
-40160 [1]	2.5-5	1-2	--	--	f_{MHz}	- 7.8	-40160
----	145-230	440-880	--	--	t_{pd}	---	----
----	50-100	125-250	--	--	t_w	---	----
-40161 [1]	2.5-5	1-2	--	--	f_{MHz}	- 7.9	-40161
----	145-230	440-880	--	--	t_{pd}	---	----
----	50-100	125-250	--	--	t_w	---	----
-40162 [1]	2.5-5	1-2	--	--	f_{MHz}	-7.8	-40162
----	130-230	440-880	--	--	t_{pd}	---	----
----	50-100	125-250	--	--	t_w	---	----
-40163 [1]	2.5-5	1-2	--	--	f_{MHz}	- 7.9	-40163
----	13-230	440-880	--	--	t_{pd}	---	----
----	50-100	125-250	--	--	t_w	---	----
-40192	2.5-5	--	--	--	f_{MHz}	-7.15	-40192
----	365-340	--	--	--	t_{pd}	---	----
----	75-150	--	--	--	t_w	---	----
-40193	2.5-5	--	--	--	f_{MHz}	- 7.15	-40193
----	365-340	--	--	--	t_{pd}	---	----
----	75-150	--	--	--	t_w	---	----
-4510	--	--	5-10	--	f_{MHz}	-7.39	-4510
----	--	--	260-500	--	t_{pd}	---	----
----	--	--	45-95	--	t_w	---	----
-4516	--	--	3-6	--	f_{MHz}	- 7.40	-4516
----	--	--	260-500	--	t_{pd}	---	----
----	--	--	45-95	--	t_w	---	----
-4518	--	--	8-16	--	f_{MHz}	-7.41	-4518
----	--	--	120-240	--	t_{pd}	---	----
----	--	--	30-60	--	t_w	---	----
-4520	--	--	8-16	--	f_{MHz}	- 7.42	-4520
----	--	--	110-220	--	t_{pd}	---	----
----	--	--	30-60	--	t_w	---	----
-4521	--	--	6-12	--	f_{MHz}	-7.43	-4521
----	--	--	950-1.9K	--	t_{pd}	---	----
----	--	--	35-80	--	t_w	---	----
-4522	--	--	6-12	--	f_{MHz}	- 7.44	-4522
----	--	--	200-420	--	t_{pd}	---	----
----	--	--	40-80	--	t_w	---	----
-4526	--	--	6-12	--	f_{MHz}	-7.45	-4526
----	--	--	210-420	--	t_{pd}	---	----
----	--	--	40-80	--	t_w	---	----
-4534	--	--	2.5-5	--	f_{MHz}	- 7.46	-4534
----	--	--	550-1.1K	--	t_{pd}	---	----
----	--	--	35-70	--	t_w	---	----
-4553	--	--	1.5/C1*	--	f_{MHz}	-7.47	-4553
----	--	--	600-1.8K	--	t_{pd}	---	----
----	--	--	275-550	--	t_w	---	----
-4568	--	--	1.2-1.8	--	f_{MHz}	-7.48	-4568
----	--	--	450-450	--	t_{pd}	---	----
----	--	--	75-250	--	t_w	---	----
-4569	--	--	2.1-3.5	--	f_{MHz}	- 7.48	-4569
----	--	--	380-700	--	t_{pd}	---	----
----	--	--	100-300	--	t_w	---	----
-4737	--	--	--	3-6	f_{MHz}	-7.49	-4737
----	--	--	--	490-700	t_{pd}	---	----
----	--	--	--	80-160	t_w	---	----

NOTES:

* Frequency in Hertz, C1 in uF

1. Also Motorola series MC14160B - MC14163B.

Table 7.6. 4000 Series CMOS Counter and Divider representative frequency (MHz), delays (t_{pd}), and clock pulse width (t_w) times in nS.

DEVICE 74---	C	HC	HCT	AHC (5V)	AHCT	VHC(5V)	VHCT	AC (5V)	ACT	DATA	Figure	DEVICE 74---
-7490	2	--	--	--	--	--	--	--	--	f_{MHz}	-7.3	-7490
----	150-1.2K	--	--	--	--	--	--	--	--	t_{pd}	---	----
----	100-600	--	--	--	--	--	--	--	--	t_w	---	----
-7493	2	30-91	30-70	--	--	--	--	--	--	f_{MHz}	-7.5	-7493
----	150-1.2K	15-155	17-58	--	--	--	--	--	--	t_{pd}	---	----
----	100-600	5-16	5-16	--	--	--	--	--	--	t_w	---	----
-74160	2.0-3	30-55	16-28	--	--	--	--	--	--	f_{MHz}	-7.8	-74160
----	70-450	17-44	9-50	--	--	--	--	--	--	t_{pd}	---	----
----	90-170	8-16	8-20	--	--	--	--	--	--	t_w	---	----
-74161	2.0-3	23-40	23-41	95-185	95-185	95-185	--	110-167	115-125	f_{MHz}	-7.9	-74161
----	70-450	12-44	17-51	4.9-12.3	4.9-12.3	4.9-12.3	--	1.5-8.5	1.5-8.0	t_{pd}	---	----
----	90-170	7-16	7-20	5	5	5	--	2.0-4.5	2.0-3.0	t_w	---	----
-74162	2.0-3	30-57	17-29	--	--	--	--	--	--	f_{MHz}	-7.8	-74162
----	70-450	14-43	12-51	--	--	--	--	--	--	t_{pd}	---	----
----	90-170	8-16	7-16	--	--	--	--	--	--	t_w	---	----
-74163	2.0-3	27-46	26-45	95-185	95-185	95-185	--	110-140	120-140	f_{MHz}	-7.9	-74163
----	70-450	13-43	17-49	4.9-12.3	4.9-12.3	4.9-12.3	--	1.5-8.5	1.5-11.5	t_{pd}	---	----
----	90-170	6-16	6-20	5	5	5	--	2.0-3.0	2.0-3.5	t_w	---	----
-74169	--	--	--	--	--	--	--	100-154	90	f_{MHz}	-7.12	-74169
----	--	--	--	--	--	--	--	1.5-13	2.0-11.5	t_{pd}	---	----
----	--	--	--	--	--	--	--	2.0-3.0	4.0	t_w	---	----
-74190	--	15-25	16-27	--	--	--	--	--	--	f_{MHz}	-7.14	-74190
----	--	16-51	18-58	--	--	--	--	--	--	t_{pd}	---	----
----	--	9-31	12-25	--	--	--	--	--	--	t_w	---	----
-74191	--	20-33	20-33	--	--	--	--	90-133	--	f_{MHz}	-7.14	-74191
----	--	12-51	19-51	--	--	--	--	1.5-12.5	--	t_{pd}	---	----
----	--	8-25	9-22	--	--	--	--	2.0-3.0	--	t_w	---	----
-74192	2.5-4.0	20-36	20-41	--	--	--	--	--	--	f_{MHz}	-7.15	-74192
----	120-400	12-63	16-64	--	--	--	--	--	--	t_{pd}	---	----
----	35-80	14-28	6-20	--	--	--	--	--	--	t_w	---	----
-74193	2.5-4.0	20-41	20-43	--	--	--	--	--	--	f_{MHz}	-7.15	-74193
----	120-400	14-58	15-58	--	--	--	--	--	--	t_{pd}	---	----
----	35-80	7-20	7-25	--	--	--	--	--	--	t_w	---	----
-74390	--	30-60	27-55	--	--	--	--	--	--	f_{MHz}	-7.19	-74390
----	--	17-42	21-51	--	--	--	--	--	--	t_{pd}	---	----
----	--	8-17	8-17	--	--	--	--	--	--	t_w	---	----
-74393	--	30-90	30-90	--	--	--	--	--	--	f_{MHz}	-7.20	-74393
----	--	5-28	5-28	--	--	--	--	--	--	t_{pd}	---	----
----	--	6-16	6-16	--	--	--	--	--	--	t_w	---	----
-74490	--	31	--	--	--	--	--	--	--	f_{MHz}	-7.21	-74490
----	--	19-47	--	--	--	--	--	--	--	t_{pd}	---	----
----	--	16	--	--	--	--	--	--	--	t_w	---	----
-74590	--	16	--	--	--	--	--	--	--	f_{MHz}	-7.25	-74590
----	--	24-60	--	--	--	--	--	--	--	t_{pd}	---	----
----	--	20-25	--	--	--	--	--	--	--	t_w	---	----
-74925	1.5-4	--	--	--	--	--	--	--	--	f_{MHz}	-7.50	-74925
----	NA	--	--	--	--	--	--	--	--	t_{pd}	---	----
----	15	--	--	--	--	--	--	--	--	t_w	---	----
-74926	1.5-4	--	--	--	--	--	--	--	--	f_{MHz}	-7.50	-74926
----	NA	--	--	--	--	--	--	--	--	t_{pd}	---	----
----	15	--	--	--	--	--	--	--	--	t_w	---	----
-74927	1.5-4	--	--	--	--	--	--	--	--	f_{MHz}	-7.50	-74927
----	NA	--	--	--	--	--	--	--	--	t_{pd}	---	----
----	15	--	--	--	--	--	--	--	--	t_w	---	----
-74928	1.5-4	--	--	--	--	--	--	--	--	f_{MHz}	-7.50	-74928
----	NA	--	--	--	--	--	--	--	--	t_{pd}	---	----
----	15	--	--	--	--	--	--	--	--	t_w	---	----

NOTE: 74AC/ACT161, '163, '169 and 74AC191 data is from "FACT Advanced Logic Databook," National Semiconductor #400022, 1993.

Table 7.7a. 7400 Series 5V CMOS Counter and Divider representative frequency (MHz), delays (t_{pd}), and clock pulse width (t_w) times in nS.

DEVICE 74---	C	HC	HCT	AHC (5V)	AHCT	VHC(5V)	VHCT	AC (5V)	ACT	DATA	Figure	DEVICE 74---
-744017	--	30-70	30-61	--	--	--	--	--	--	f MHz	-7.28	-744017
----	--	19-50	20-50	--	--	--	--	--	--	tpd	---	----
----	--	6-16	4-16	--	--	--	--	--	--	tw	---	----
-744020	--	30-92	25-47	--	--	--	--	--	--	f MHz	-7.32	-744020
----	--	8-34	8-56	--	--	--	--	--	--	tpd	---	----
----	--	4-16	7-20	--	--	--	--	--	--	tw	---	----
-744024	--	30-82	30-64	--	--	--	--	--	--	f MHz	-7.34	-744024
----	--	9-40	9-40	--	--	--	--	--	--	tpd	---	----
----	--	6-16	6-16	--	--	--	--	--	--	tw	---	----
-744040	--	30-82	30-72	--	--	--	--	--	--	f MHz	-7.36	-744040
----	--	10-37	10-45	--	--	--	--	--	--	tpd	---	----
----	--	5-16	6-16	--	--	--	--	--	--	tw	---	----
-744059	--	21-36	21-36	--	--	--	--	--	--	f MHz	-7.37	-744059
----	--	18-40	24-46	--	--	--	--	--	--	tpd	---	----
----	--	6-18	7-20	--	--	--	--	--	--	tw	---	----
-744060	--	30-80	30-80	--	--	--	--	--	--	f MHz	-7.38	-744060
----	--	8-60	8-66	--	--	--	--	--	--	tpd	---	----
----	--	6-16	6-16	--	--	--	--	--	--	tw	---	----
-744510	--	30-52	30-53	--	--	--	--	--	--	f MHz	-7.39	-744510
----	--	13-52	17-68	--	--	--	--	--	--	tpd	---	----
----	--	7-20	4-20	--	--	--	--	--	--	tw	---	----
-744516	--	30-49	30-52	--	--	--	--	--	--	f MHz	-7.40	-744516
----	--	13-60	18-68	--	--	--	--	--	--	tpd	---	----
----	--	7-16	5-20	--	--	--	--	--	--	tw	---	----
-744518	--	30-55	25-50	--	--	--	--	--	--	f MHz	-7.41	-744518
----	--	16-42	17-53	--	--	--	--	--	--	tpd	---	----
----	--	9-24	11-20	--	--	--	--	--	--	tw	---	----
-744520	--	30-58	30-58	--	--	--	--	--	--	f MHz	-7.42	-744520
----	--	16-48	16-53	--	--	--	--	--	--	tpd	---	----
----	--	8-24	10-20	--	--	--	--	--	--	tw	---	----
-746323	--	50-85	50-85	--	--	--	--	--	--	f MHz	-7.51	-746323
----	--	22-67	24-70	--	--	--	--	--	--	tpd	---	----
----	--	6.0-16	6-16	--	--	--	--	--	--	tw	---	----
-747022	--	31-50	--	--	--	--	--	--	--	f MHz	-7.52	-747022
----	--	23-50	--	--	--	--	--	--	--	tpd	---	----
----	--	16	--	--	--	--	--	--	--	tw	---	----
-747060	--	--	--	--	--	--	--	114	114	f MHz	-7.53	-747060
----	--	--	--	--	--	--	--	1.5-21	1.5-22.1	tpd	---	----
----	--	--	--	--	--	--	--	4.4	3.5-4.4	tw	---	----
-747061	--	--	--	--	--	--	--	114	114	f MHz	-7.53	-747061
----	--	--	--	--	--	--	--	1.5-21	1.5-22.1	tpd	---	----
----	--	--	--	--	--	--	--	4.4	3.5-4.4	tw	---	----
-7440102	--	15-27	15-27	--	--	--	--	--	--	f MHz	-7.54	7440102
----	--	18-68	25-83	--	--	--	--	--	--	tpd	---	----
----	--	8-33	11-43	--	--	--	--	--	--	tw	---	----
-7440103	--	15-29	15-28	--	--	--	--	--	--	f MHz	-7.54	7440103
----	--	18-63	23-75	--	--	--	--	--	--	tpd	---	----
----	--	8-33	10-38	--	--	--	--	--	--	tw	---	----

NOTE: 74AC/ACT7060, '7061 data is from RCA "Advanced CMOS Logic Databook" GE Corporation, 1988.

Table 7.7b. 7400 Series 5V CMOS Counter and Divider representative frequency (MHz), delays (tpd), and clock pulse width (tw) times in nS (continuation).

DEVICE 74---	LV	LVC	ALVT LVT/ABT	LVX	LVQ	LCX	AC(3V)	LVCH/ ALVCH	HL	AHC (3V)	VHC(3V)	DATA	Figure No.	DEVICE 74---
- 74161	24-70	125-160	--	--	--	--	70-111	--	--	--	55-130	f_{MHz}	-7.9	- 74161
----	10-44	4.0-9.5	--	--	--	--	1.5-8.5	--	--	--	7.5-20.7	t_{pd}	---	----
----	6-20	1.2-4.0	--	--	--	--	2.0-5.5	--	--	--	5.0	t_w	---	----
- 74163	24-70	125-160	--	--	--	--	70-95	--	--	--	55-130	f_{MHz}	-7.9	- 74163
----	10-44	4.0-9.5	--	--	--	--	1.5-11	--	--	--	7.5-20.7	t_{pd}	---	----
----	6-20	1.2-4.0	--	--	--	--	3.0-4.0	--	--	--	5.0	t_w	---	----
- 74169	--	125-160	--	--	--	--	75-118	--	--	--	--	f_{MHz}	-7.12	- 74169
----	--	4.0-10	--	--	--	--	2.5-18	--	--	--	--	t_{pd}	---	----
----	--	1.2-4.0	--	--	--	--	2.0-3.0	--	--	--	--	t_w	---	----
- 74191	--	--	--	--	--	--	70-105	--	--	--	--	f_{MHz}	-7.14	- 74191
----	--	--	--	--	--	--	2.0-18	--	--	--	--	t_{pd}	---	----
----	--	--	--	--	--	--	2.0-3.5	--	--	--	--	t_w	---	----
- 74393	24-90	--	--	--	--	--	--	--	--	--	--	f_{MHz}	-7.20	- 74393
----	5-29	--	--	--	--	--	--	--	--	--	--	t_{pd}	---	----
----	6-20	--	--	--	--	--	--	--	--	--	--	t_w	---	----
- 744020	24-92	--	--	--	--	--	--	--	--	--	--	f_{MHz}	-7.32	- 744020
----	8-38	--	--	--	--	--	--	--	--	--	--	t_{pd}	---	----
----	4-20	--	--	--	--	--	--	--	--	--	--	t_w	---	----
- 744040	24-82	--	--	--	--	--	--	--	--	--	--	f_{MHz}	-7.36	- 744040
----	9-41	--	--	--	--	--	--	--	--	--	--	t_{pd}	---	----
----	5-20	--	--	--	--	--	--	--	--	--	--	t_w	---	----
- 744060	24-90	--	--	--	--	--	--	--	--	--	--	f_{MHz}	-7.38	- 744060
----	8-48	--	--	--	--	--	--	--	--	--	--	t_{pd}	---	----
----	5-20	--	--	--	--	--	--	--	--	--	--	t_w	---	----
- 747060	--	--	--	--	--	--	81	--	--	--	--	f_{MHz}	-7.54	- 747060
----	--	--	--	--	--	--	2.2-29	--	--	--	--	t_{pd}	---	----
----	--	--	--	--	--	--	4.4-6.1	--	--	--	--	t_w	---	----
- 747061	--	--	--	--	--	--	81	--	--	--	--	f_{MHz}	-7.54	- 747061
----	--	--	--	--	--	--	2.2-29	--	--	--	--	t_{pd}	---	----
----	--	--	--	--	--	--	4.4-6.1	--	--	--	--	t_w	---	----

NOTE: 74AC/ACT161, '163, '169 and 74AC191 data is from "FACT Advanced Logic Databook," National Semiconductor #400022, 1993.

74AC7060, '7061 data is from RCA "Advanced CMOS Logic Databook" GE Corporation, 1988

Table 7.8. 7400 Series 3V CMOS Counter and Divider representative frequency (MHz), delays (t_{pd}), and clock pulse width (t_w) times in nS.

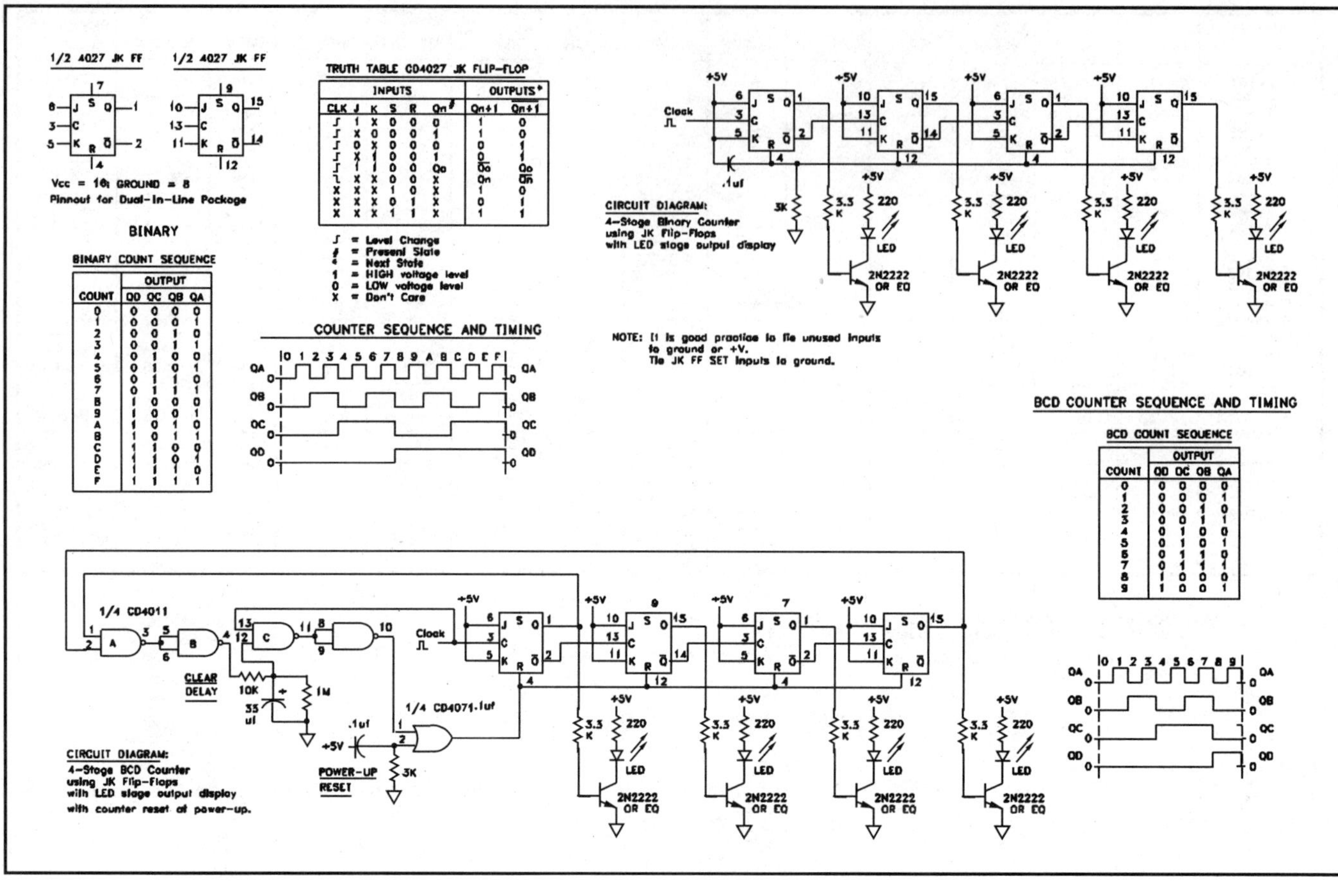

Figure 7.1. Binary and Decade counters constructed using the CD4027 dual JK Flip-Flop. LED displays provide for observing the stage-by-stage transitions. An RC reset circuit initializes the counter to zero on power-up.

TTL - 74LS68 (DECADE)
74LS69 (BINARY)

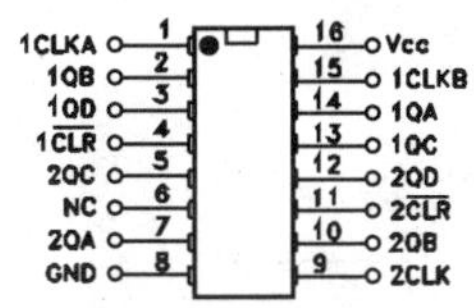

DUAL-IN-LINE PACKAGE

FUNCTION TABLES 7468 DUAL 4-BIT DECADE COUNTER

BCD COUNT SEQUENCE

COUNT	OUTPUT			
	QD	QC	QB	QA
0	0	0	0	0
1	0	0	0	1
2	0	0	1	0
3	0	0	1	1
4	0	1	0	0
5	0	1	0	1
6	0	1	1	0
7	0	1	1	1
8	1	0	0	0
9	1	0	0	1

Output QA connected to Clook-2 Input.

BIQUINARY COUNT SEQ

COUNT	OUTPUT			
	QD	QC	QB	QA
0	0	0	0	0
1	0	0	0	1
2	0	0	1	0
3	0	0	1	1
4	0	1	0	0
5	1	0	0	1
6	1	0	1	0
7	1	0	1	1
8	1	0	0	0
9	1	1	0	1

Output QD connected to Clock-1 Input.

BINARY COUNT SEQUENCE

COUNT	OUTPUT			
	QD	QC	QB	QA
0	0	0	0	0
1	0	0	0	1
2	0	0	1	0
3	0	0	1	1
4	0	1	0	0
5	0	1	0	1
6	0	1	1	0
7	0	1	1	1
8	1	0	0	0
9	1	0	0	1
A	1	0	1	0
B	1	0	1	1
C	1	1	0	0
D	1	1	0	1
E	1	1	1	0
F	1	1	1	1

FUNCTION TABLE 74LS69 DUAL BINARY COUNTER

LOGIC DIAGRAM 74LS68 DUAL 4-BIT DECADE COUNTER

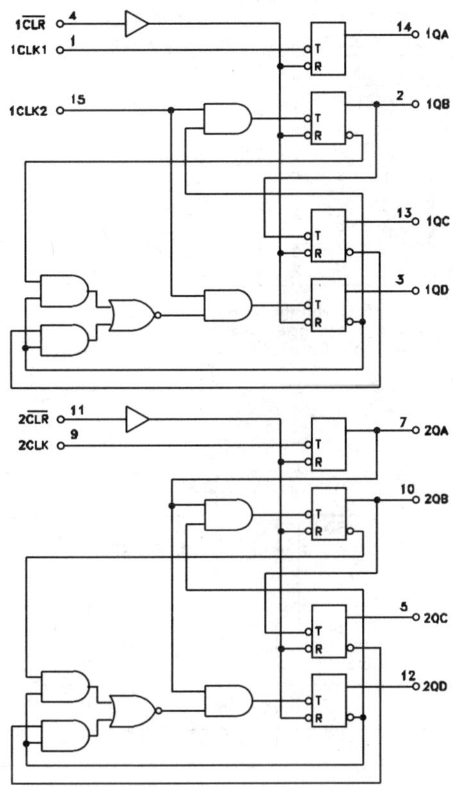

LOGIC DIAGRAM 74LS69 DUAL 4-BIT BINARY COUNTER

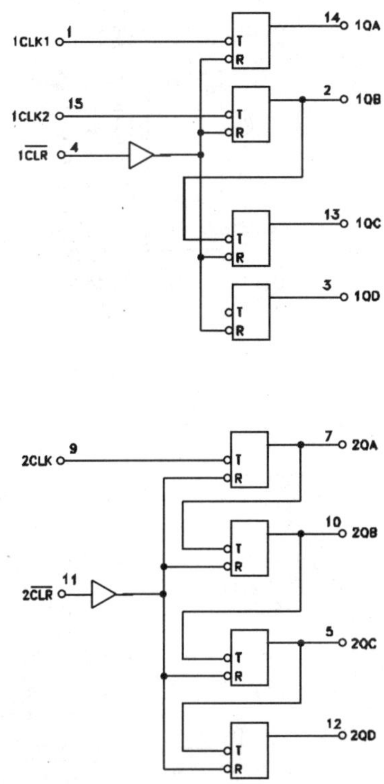

Figure 7.2. The 74LS68 Decade and 74LS Binary counters are dual counters each containing two four-bit counting circuits. Counter one of each device has two clock pins, as shown in the logic diagram. Counter one of the 'LS68 can perform bi-quinery counting. Counter one provides adequate drive for clock two while maintaining full fanout capability. Function tables for each counter are included. Triggering occurs on the high-to-low clock transition.

TTL – 7490/L/LS
CMOS – 74C/HC90

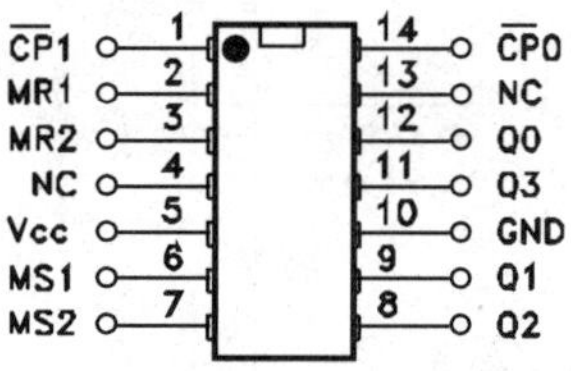

DUAL-IN-LINE PACKAGE

FUNCTION TABLE 74x90 4-BIT RIPPLE
DECADE COUNTER – MODE SELECTION

RESET/SET INPUTS				OUTPUTS			
MR1	MR2	MS1	MS2	Q0	Q1	Q2	Q3
1	1	0	X	0	0	0	0
1	1	X	0	0	0	0	0
X	X	1	1	1	0	0	1
0	X	0	X	Count			
X	0	X	0	Count			
0	X	X	0	Count			
X	0	0	X	Count			

1 = HIGH voltage level
0 = LOW voltage level
x = Don't Care

LOGIC SYMBOL

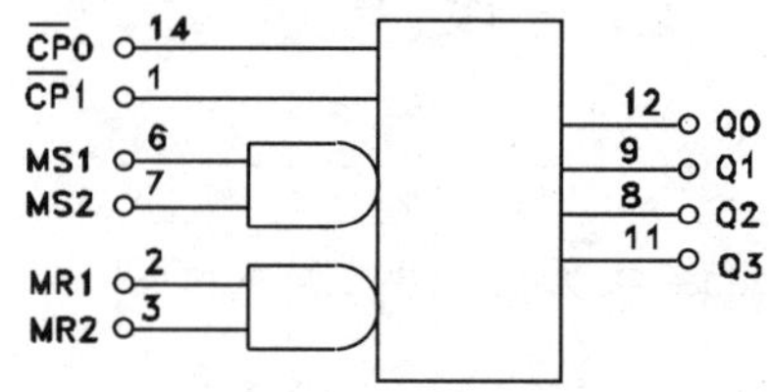

LOGIC DIAGRAM

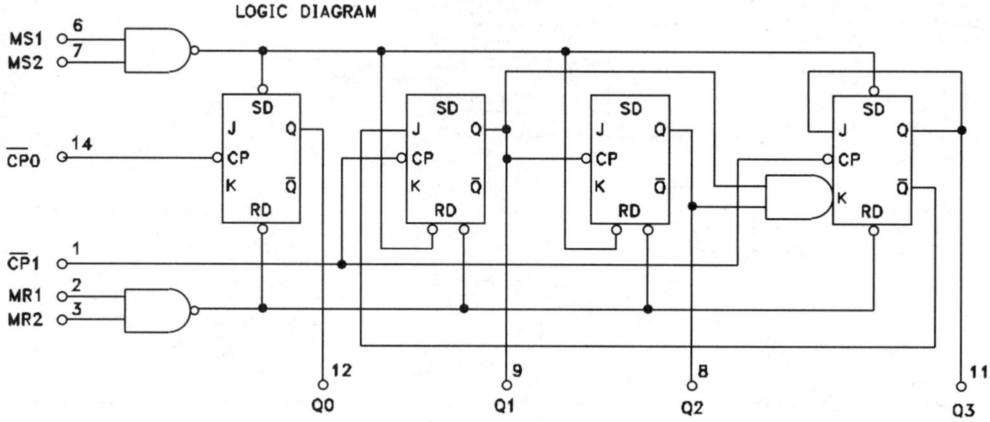

Figure 7.3. The 7490 series of TTL and CMOS decade counters feature four master-slave flip-flops with gating to provide a divide-by-two counter preceding three additional stages. The counter features a gated zero reset and also gated set-to-nine inputs. *Figure 7.6* provides a function table and timing diagram for this counter.

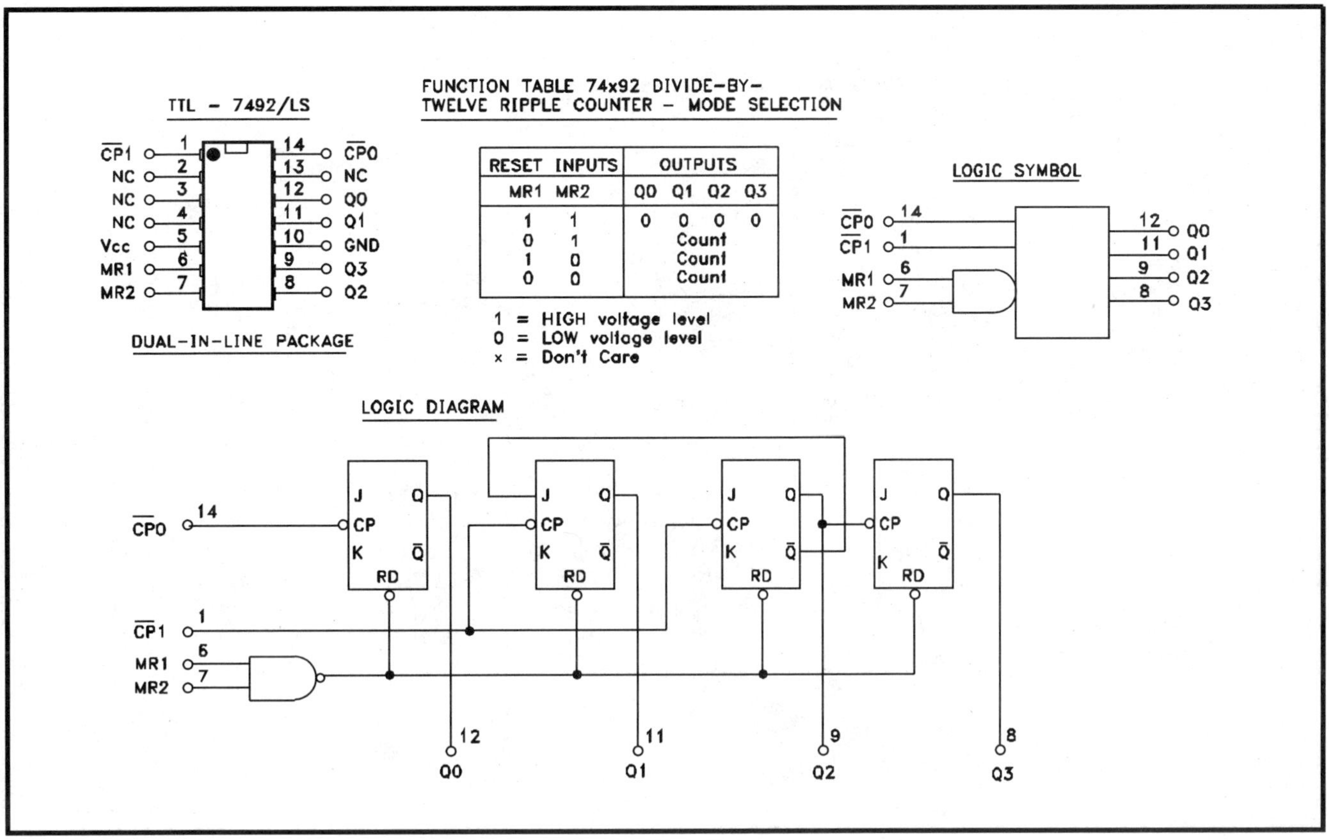

Figure 7.4. The 7492 and 'LS92 TTL divide-by-twelve counters feature four master-slave flip-flops with gating to provide a divide-by-two counter preceding three additional stages. The counter features a gated zero reset.

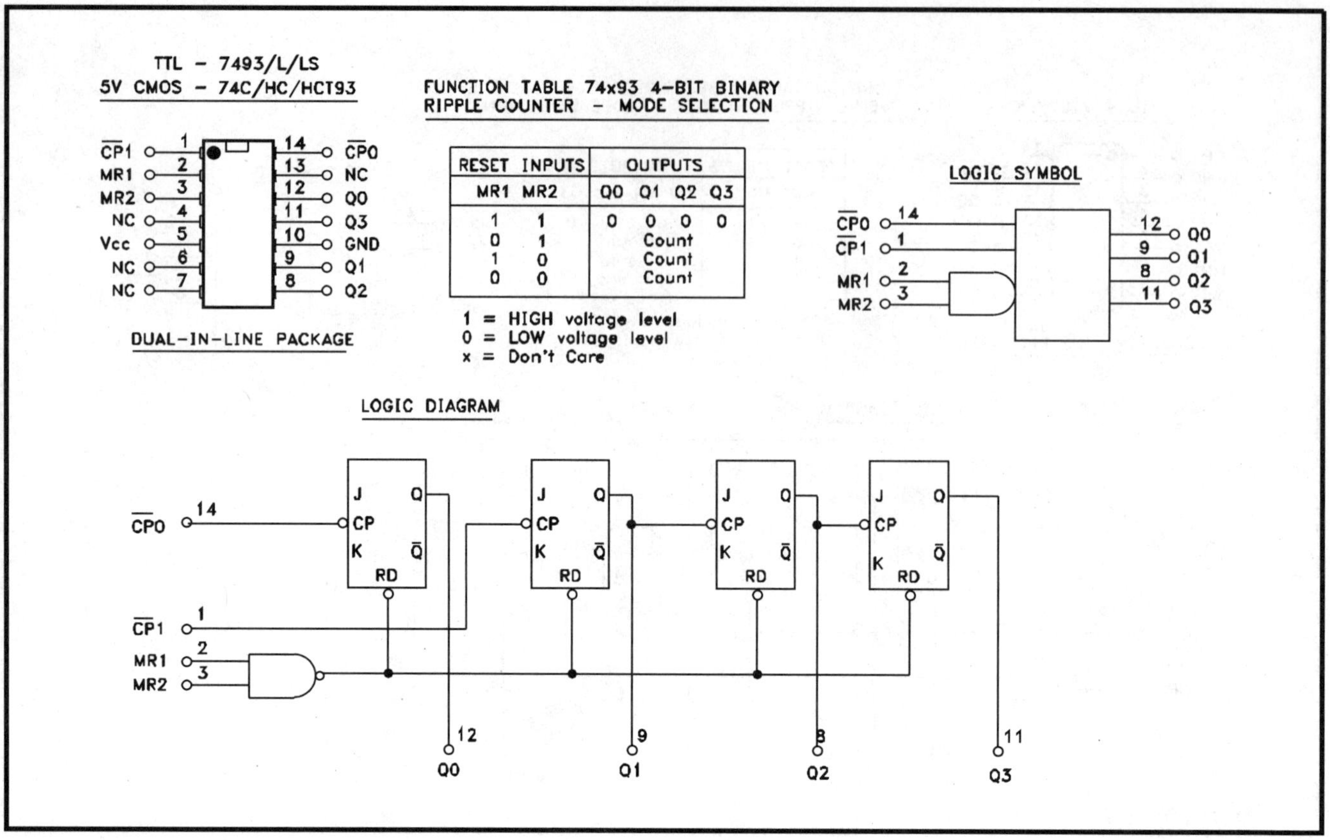

RESET INPUTS		OUTPUTS			
MR1	MR2	Q0	Q1	Q2	Q3
1	1	0	0	0	0
0	1	Count			
1	0	Count			
0	0	Count			

Figure 7.5. The 7493 series of TTL and CMOS binary counters feature four master-slave flip-flops with gating to provide a divide-by-two counter preceding three additional stages. The counter features a gated zero reset. *Figure 7.6* provides a function table and timing diagram for this counter.

BCD COUNTER SEQUENCE AND TIMING

BCD COUNT SEQUENCE

COUNT	OUTPUT QD	QC	QB	QA
0	0	0	0	0
1	0	0	0	1
2	0	0	1	0
3	0	0	1	1
4	0	1	0	0
5	0	1	0	1
6	0	1	1	0
7	0	1	1	1
8	1	0	0	0
9	1	0	0	1

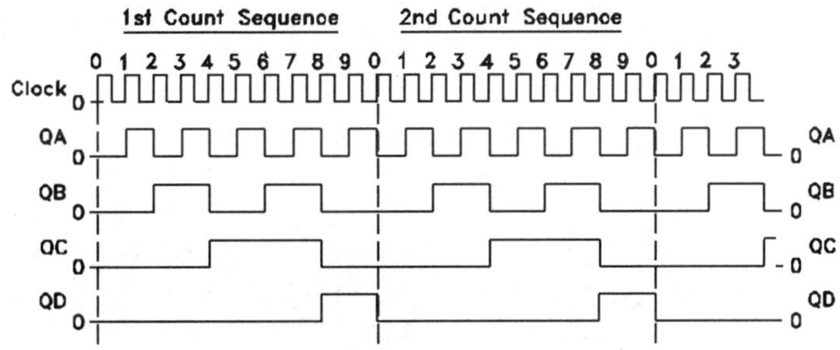

BINARY COUNTER SEQUENCE AND TIMING

BINARY COUNT SEQUENCE

COUNT	OUTPUT QD	QC	QB	QA
0	0	0	0	0
1	0	0	0	1
2	0	0	1	0
3	0	0	1	1
4	0	1	0	0
5	0	1	0	1
6	0	1	1	0
7	0	1	1	1
8	1	0	0	0
9	1	0	0	1
A	1	0	1	0
B	1	0	1	1
C	1	1	0	0
D	1	1	0	1
E	1	1	1	0
F	1	1	1	1

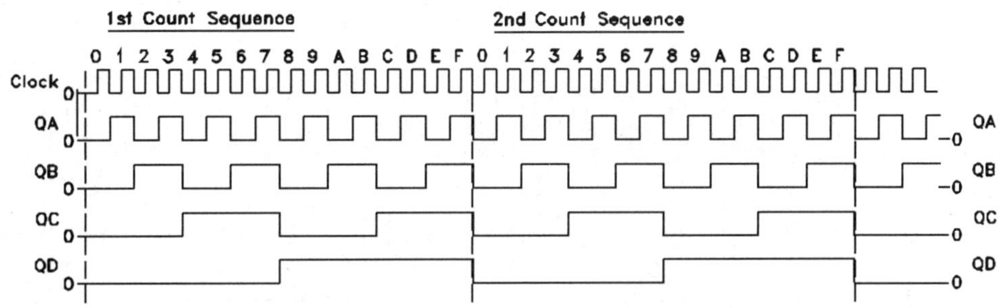

Figure 7.6. Decade and binary counter function tables and timing diagrams. Although provided for the 7490 decade and 7493 counters, these have application to any basic ripple or synchronously clocked counter.

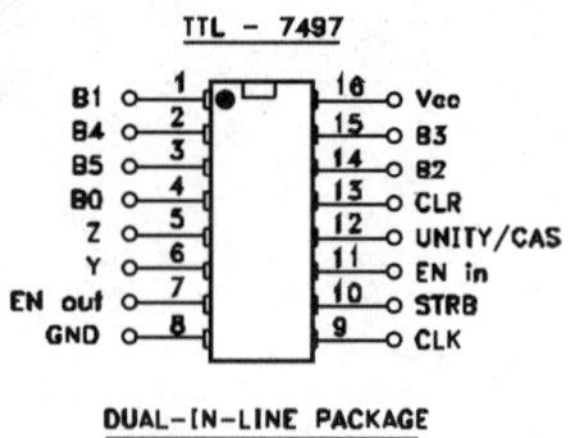

STATE AND/OR RATE FUNCTION TABLE
SYNCHRONOUS 6-BIT BINARY RATE MULTIPLIER

INPUTS											OUTPUTS			
			BINARY RATE						NO. OF CLK	UNITY/	LOGIC LEVEL OR NO OF PULSES			
CLR	EN	STRB	B5	B4	B3	B2	B1	B0	PULSES	CASCADE	Y	Z	ENABLE	NOTES
1	X	1	X	X	X	X	X	X	X	1	L	H	H	B
0	0	0	0	0	0	0	0	0	64	1	L	H	1	C
0	0	0	0	0	0	0	0	1	64	1	1	1	1	C
0	0	0	0	0	0	0	1	0	64	1	2	2	1	C
0	0	0	0	0	0	1	0	0	64	1	4	4	1	C
0	0	0	0	0	1	0	0	0	64	1	8	8	1	C
0	0	0	0	1	0	0	0	0	64	1	16	16	1	C
0	0	0	1	0	0	0	0	0	64	1	32	32	1	C
0	0	0	1	1	1	1	1	1	64	1	63	63	1	C
0	0	0	1	1	1	1	1	1	64	0	H	63	1	D
0	0	0	1	0	1	0	0	0	64	1	40	40	1	E

NOTES: A. H, 1 = HIGH LEVEL; L, 0 = LOW LEVEL. X = IRRELEVANT.
B. A simplified illustration of the clear functioin. The clock and strobe status can affect logic level of Y and Z. A LOW Unity/Cascade will cause output Y to remain HIGH.
C. Each rate illustrated assumes a constant value of rate inputs; however, these illustrations in no way prohibit variable-rate inputs.
D. Unity/cascade is used to inhibit output Y.
E. f out = 40f in / 64 = 0.625 f in.

LOGIC DIAGRAM

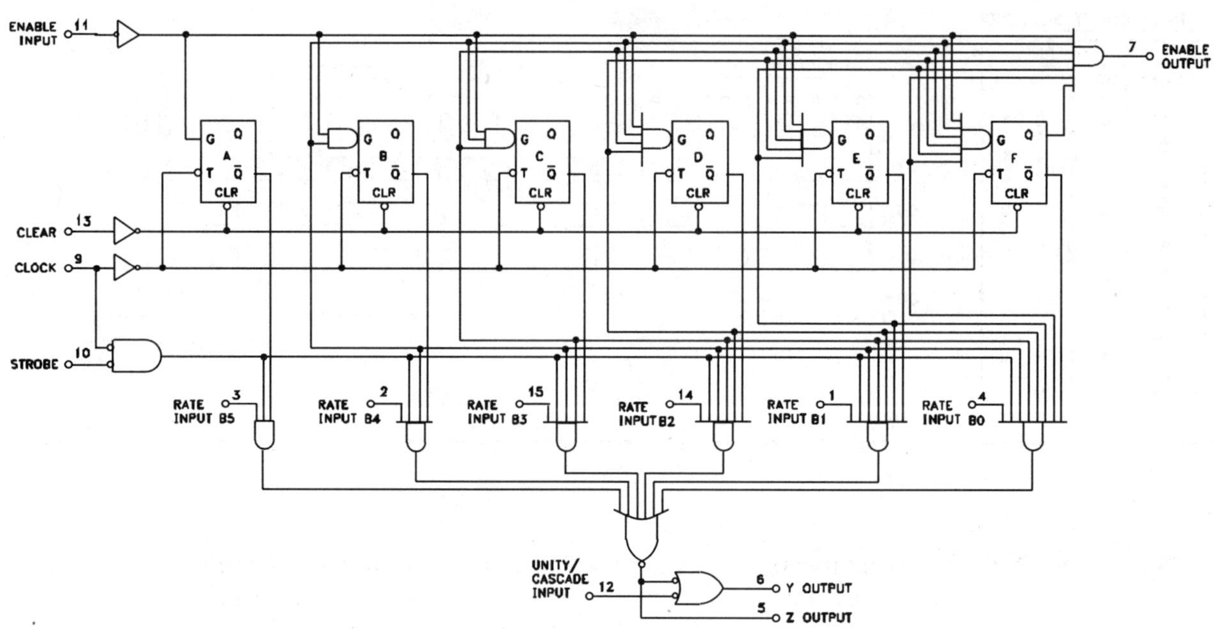

Figure 7.7. The 7497 is a six-bit serial binary counter featuring buffered clock, clear and enable inputs for its control and a strobe input to enable or inhibit the rate input/decoding AND-OR-INVERT gates. The counter is enabled when the clear, strobe and enable inputs are low. When enabled, the count frequency is equal to the input frequency multiplied by the rate input M and divided by 64. When the rate input is binary 0, the Z output remains high. The device may perform fixed or variable-rate frequency division.

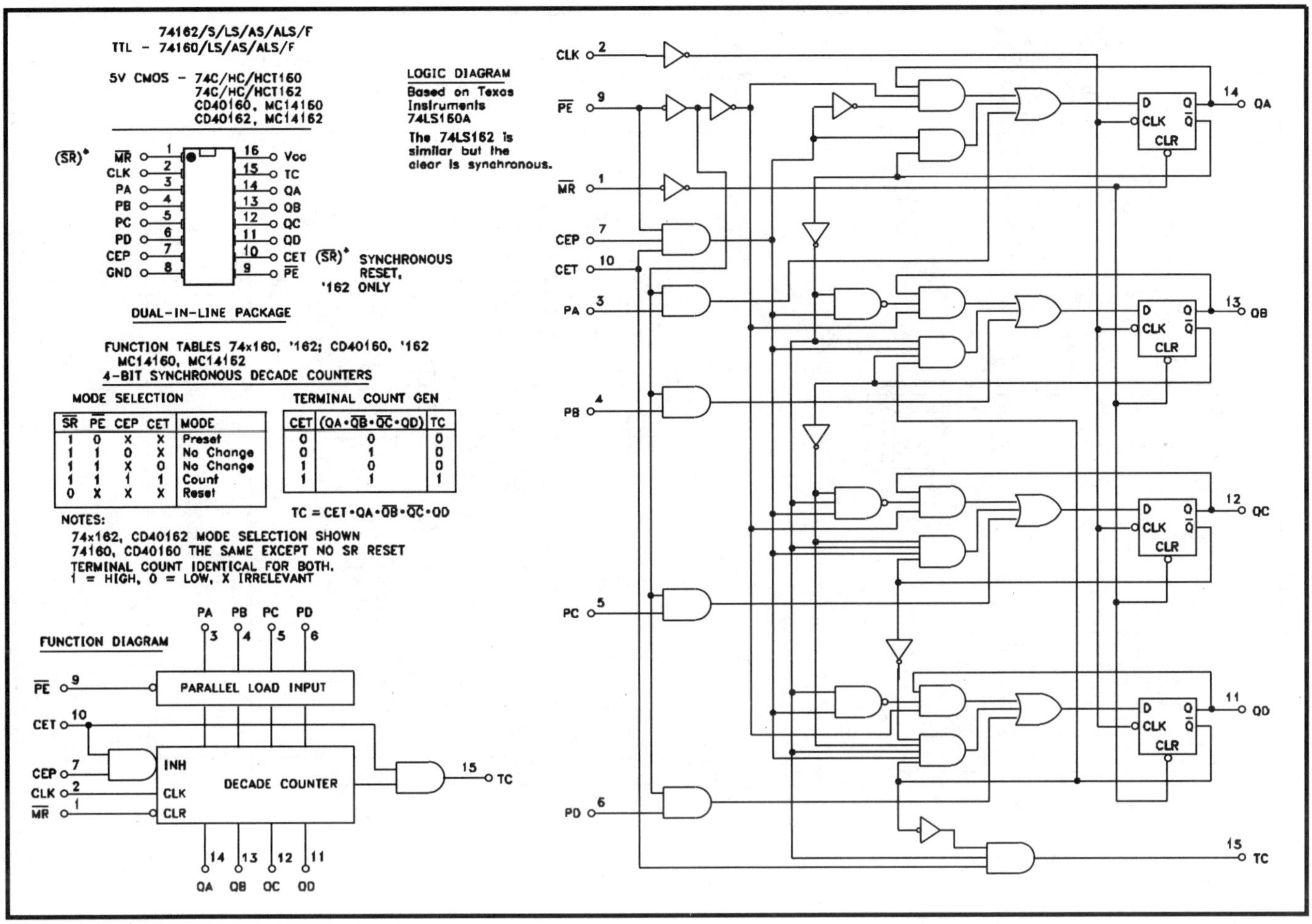

Figure 7.8. The 74x160 and 74x162 are synchronous presettable decade counters featuring internal look-ahead. TTL and CMOS devices are available. With a maximum clock frequency rating above 30 MHz these are well suited to high-speed applications. The counters are fully presettable—outputs may be preset to either level. The preset is synchronous; that is, the preset does not take effect until the next clock pulse. Clearing is asynchronous.

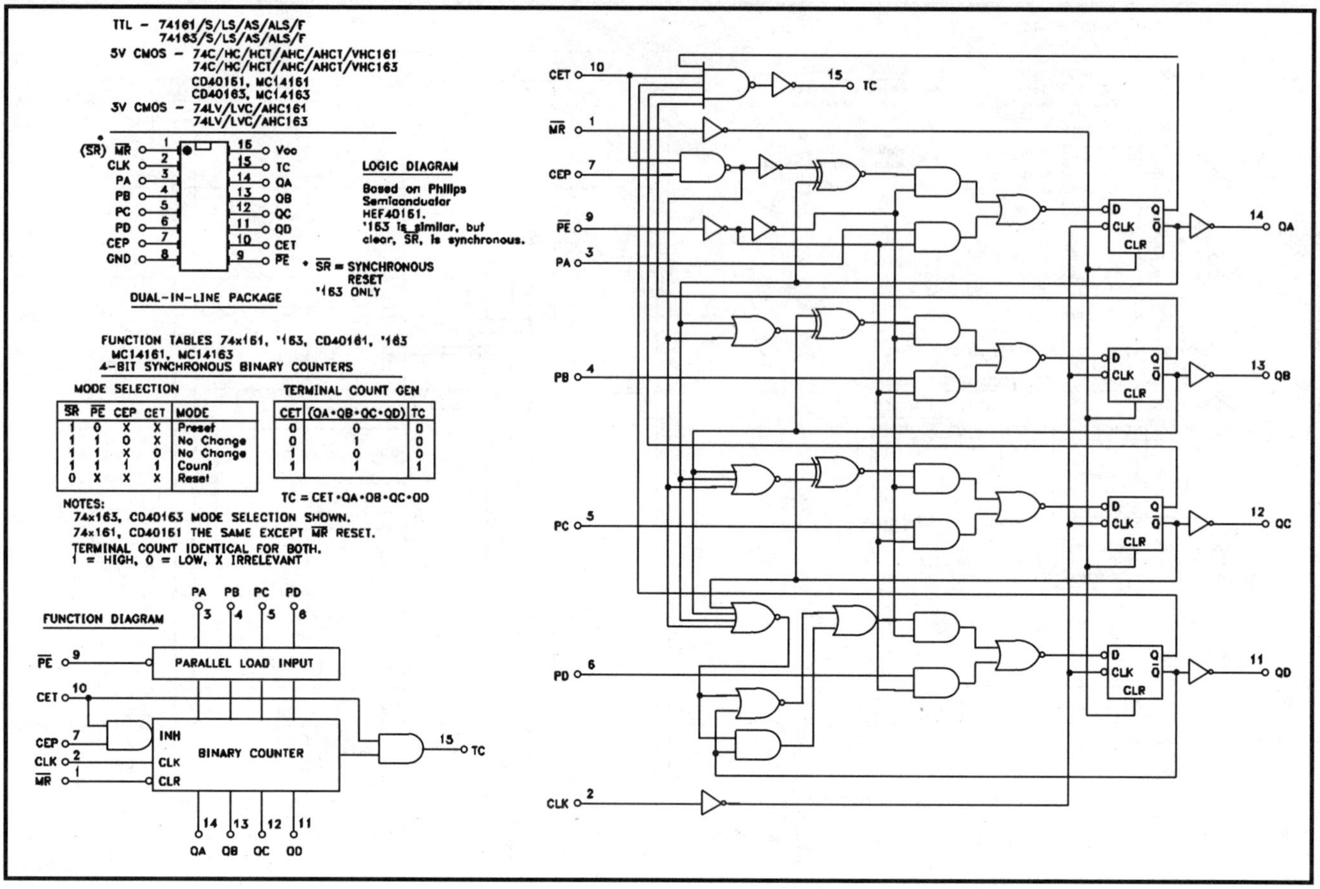

MODE SELECTION

SR	PE	CEP	CET	MODE
1	0	X	X	Preset
1	1	0	X	No Change
1	1	X	0	No Change
1	1	1	1	Count
0	X	X	X	Reset

TERMINAL COUNT GEN

CET	(QA•QB•QC•QD)	TC
0	0	0
0	1	0
1	0	0
1	1	1

Figure 7.9. The 74x161 and 74x163 are synchronous presettable binary counters featuring internal look-ahead. TTL and CMOS devices are available. With a maximum clock frequency rating above 30 MHz, these are well suited to high-speed applications. The counters are fully presettable—outputs may be preset to either level. The preset is synchronous; that is, the preset does not take effect until the next clock pulse. Clearing is synchronous, active low, effective after the following clock pulse..

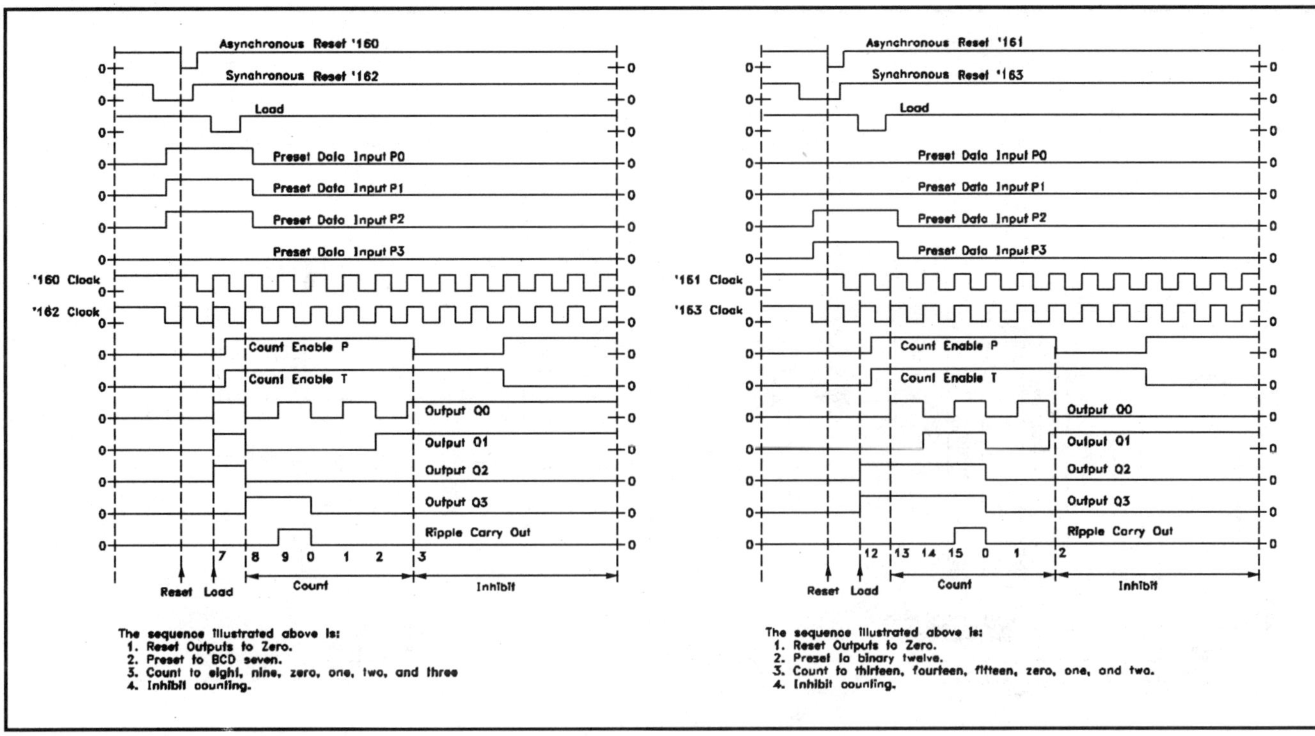

Figure 7.10. Timing diagrams for the 74x160 - 74x163 synchronous counters.

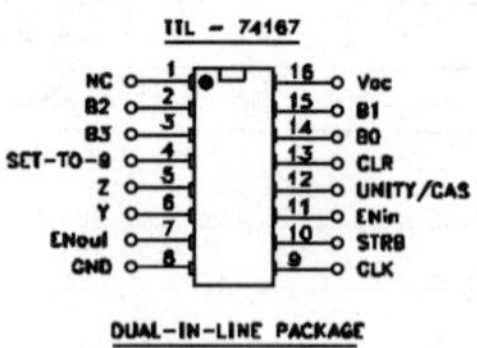

STATE AND/OR RATE FUNCTION TABLE
74167 DECADE RATE MULTIPLIER

INPUTS									OUTPUTS			
CLR	EN	STRB	BCD RATE B3	B2	B1	B0	No. of CLK Pls	UN/CAS	LOGIC LEVEL or No. of CLK Pls Y	Z	EN	NOTES
1	X	1	X	X	X	X	X	1	L	H	H	B
0	0	0	0	0	0	0	10	1	L	H	1	C
0	0	0	0	0	0	1	10	1	1	1	1	C
0	0	0	0	0	1	0	10	1	2	2	1	C
0	0	0	0	0	1	1	10	1	3	3	1	C
0	0	0	0	1	0	0	10	1	4	4	1	C
0	0	0	0	1	0	1	10	1	5	5	1	C
0	0	0	0	1	1	0	10	1	6	6	1	C
0	0	0	0	1	1	1	10	1	7	7	1	C
0	0	0	1	0	0	0	10	1	8	8	1	C
0	0	0	1	0	0	1	10	1	9	9	1	C
0	0	0	1	0	1	0	10	1	8	8	1	C,D
0	0	0	1	0	1	1	10	1	9	9	1	C,D
0	0	0	1	1	0	0	10	1	8	8	1	C,D
0	0	0	1	1	0	1	10	1	9	9	1	C,D
0	0	0	1	1	1	0	10	1	8	8	1	C,D
0	0	0	1	1	1	1	10	1	9	9	1	C,D
0	0	0	1	0	0	0	10	1	H	9	1	E

NOTES:

A. 1, H = HIGH level, 0,L = LOW level. Other entries numerical counts.
B. Simplified clear function. Clock and Strobe states can affect Y and Z logic levels. A LOW Unity/Cascode causes output Y to remain HIGH.
C. Each illustrated rate assumes a constant value at at rate inputs; however these in no way prohibit variate rate inputs.
D. These input conditions exceed the range of the decimal rate inputs.
E. Unity/Cascade can be used to inhibit output Y.

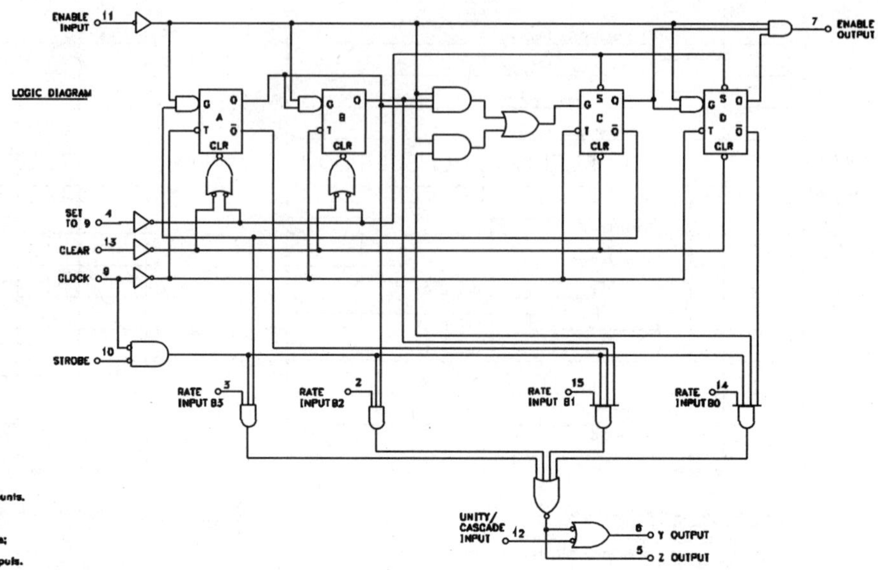

Figure 7.11. The 74167 is a synchronous programmable decade counter featuring buffered clock, clear and enable inputs for its control and a strobe input to enable or inhibit the rate input/decoding AND-OR-INVERT gates. The counter is enabled when the clear, strobe set-to-nine, and enable inputs are low. When enabled, the count frequency is equal to the input frequency multiplied by the rate input M and divided by 10. The outputs have additional gating for cascading and transferring unity-count rates. When the rate input is binary 0, the Z output remains high. The device may perform fixed or variable-rate frequency division.

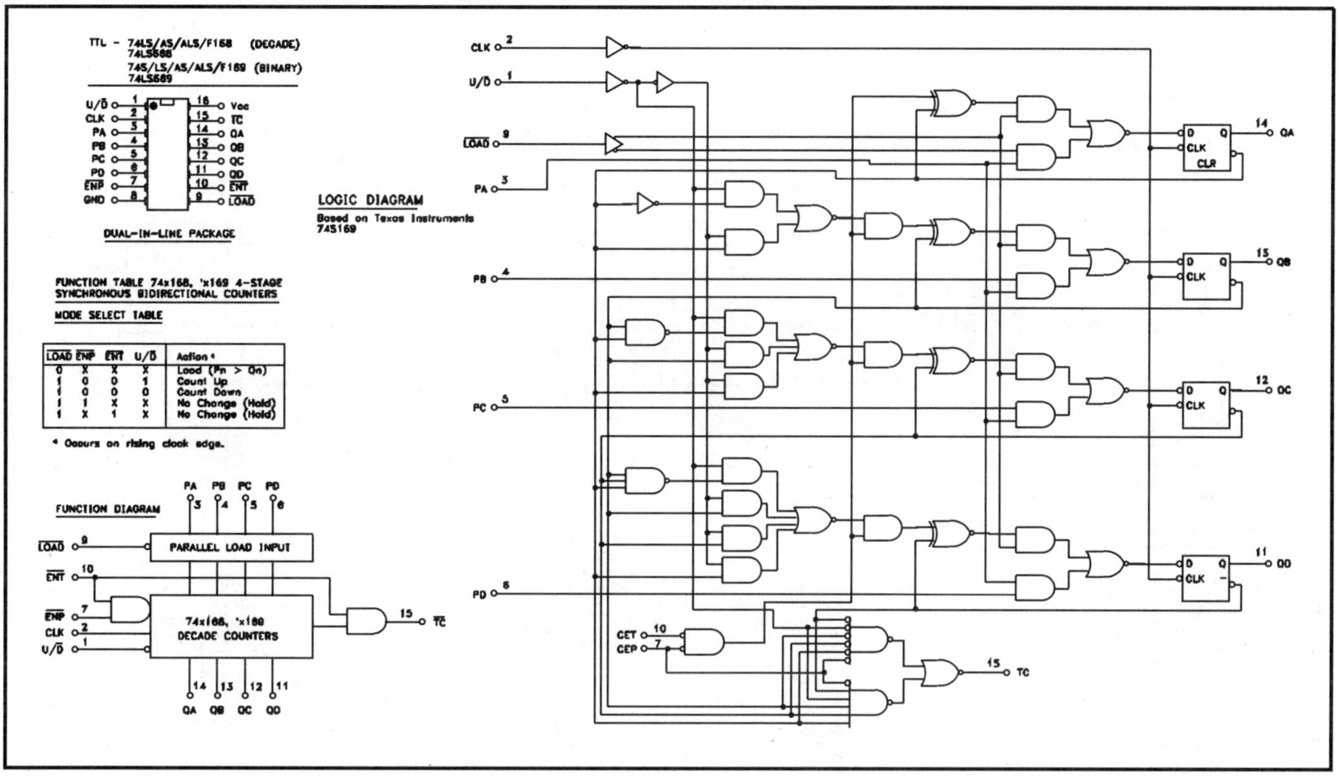

Figure 7.12. The 74x168 and 74x169 are synchronous programmable decade and binary 4-bit up/down counters, respectively, featuring an internal look-ahead for high-speed cascading. A buffered clock input triggers four master-slave flip-flops on the rising edge of the clock input. The counters are programmable to either output logic level. Loading is synchronous. Count direction is set by the Up/Down input—both count enable inputs must be low to count. Input /ENT is fed forward to enable the carry output. The enable carry output produces a low-level output pulse of approximate duration of the QA output when counting up, and the low portion of the QA output when counting down.

TTL – 74176, 74196/S/LS (DECADE)
74177, 74197/S/LS (BINARY)

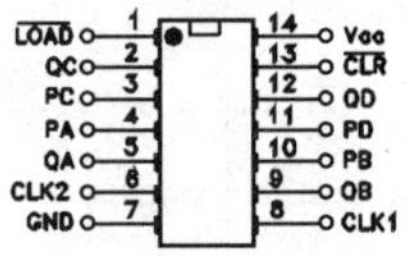

DUAL-IN-LINE PACKAGE

NOTE: DIAGRAMS AND TABLES SHOWN FOR THE 74176, '177 APPLY TO THE 74196, '197 RESPECTIVELY.

FUNCTION TABLES 74176 PRESETTABLE DECADE COUNTER

BCD COUNT SEQUENCE

COUNT	OUTPUT			
	QD	QC	QB	QA
0	0	0	0	0
1	0	0	0	1
2	0	0	1	0
3	0	0	1	1
4	0	1	0	0
5	0	1	0	1
6	0	1	1	0
7	0	1	1	1
8	1	0	0	0
9	1	0	0	1

Output QA connected to Clock–2 input.

BIQUINARY COUNT SEQ

COUNT	OUTPUT			
	QD	QC	QB	QA
0	0	0	0	0
1	0	0	0	1
2	0	0	1	0
3	0	0	1	1
4	0	1	0	0
5	1	0	0	1
6	1	0	1	0
7	1	0	1	1
8	1	0	0	0
9	1	1	0	1

Output QD connected to Clock–1 input.

BINARY COUNT SEQUENCE

COUNT	OUTPUT			
	QD	QC	QB	QA
0	0	0	0	0
1	0	0	0	1
2	0	0	1	0
3	0	0	1	1
4	0	1	0	0
5	0	1	0	1
6	0	1	1	0
7	0	1	1	1
8	1	0	0	0
9	1	0	0	1
A	1	0	1	0
B	1	0	1	1
C	1	1	0	0
D	1	1	0	1
E	1	1	1	0
F	1	1	1	1

FUNCTION TABLE 74177 PRESETTABLE BINARY COUNTER.

LOGIC DIAGRAM 74176 DECADE PRESETTABLE COUNTER

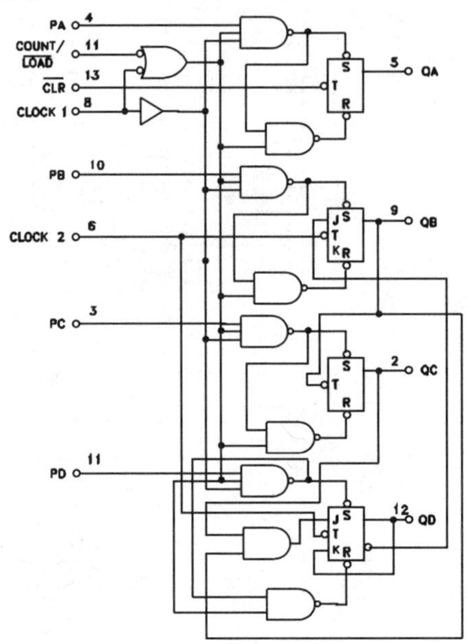

LOGIC DIAGRAM 74177 BINARYPRESETTABLE COUNTER

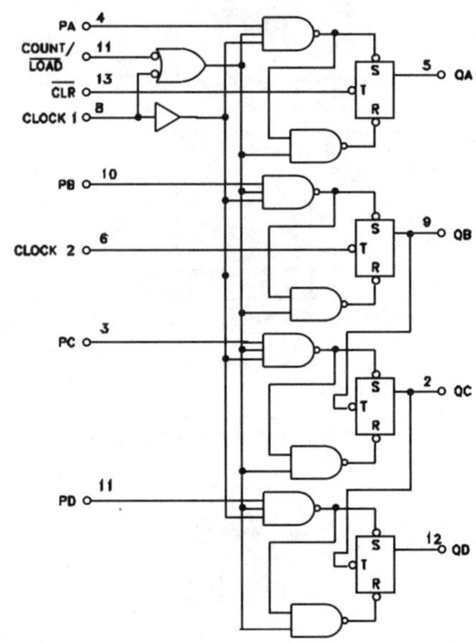

Figure 7.13. The 74x196, 74x197 50/30/100 MHz decade/binary counter/latches consist of four DC coupled master-slave flip-flops internally connected to provide a divide-by-two or divide-by-ten counter ('x196), or divide-by-two or divide-by-eight counter ('x197). The counters are fully programmable. Outputs change state on the negative transition of the clock input. The direct clear is independent of the clock. The counters may be used as latches by using the count/load input as a strobe with data presented at the four data inputs. The outputs will follow the data inputs while the count/load input is low. With this input high and the clock inputs inactive, the outputs will remain unchanged. The 74176 and 74177 are 35 MHz reduced-power versions of the 50 MHz 74196 and 74197 decade and binary counter/latches.

TTL – 74190/LS/ALS/F (DECADE)
74191/LS/ALS/F (BINARY)
5V CMOS – 74HC/HCT190 (DECADE)
74HC/HCT191 (BINARY)
3V CMOS – 74LV/LVC191 (BINARY)

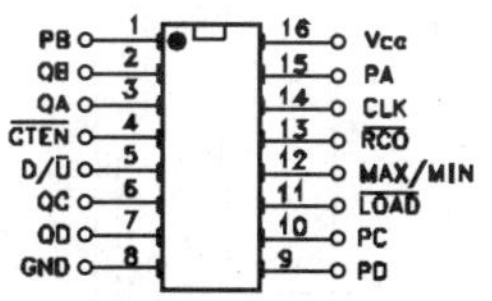

DUAL-IN-LINE PACKAGE

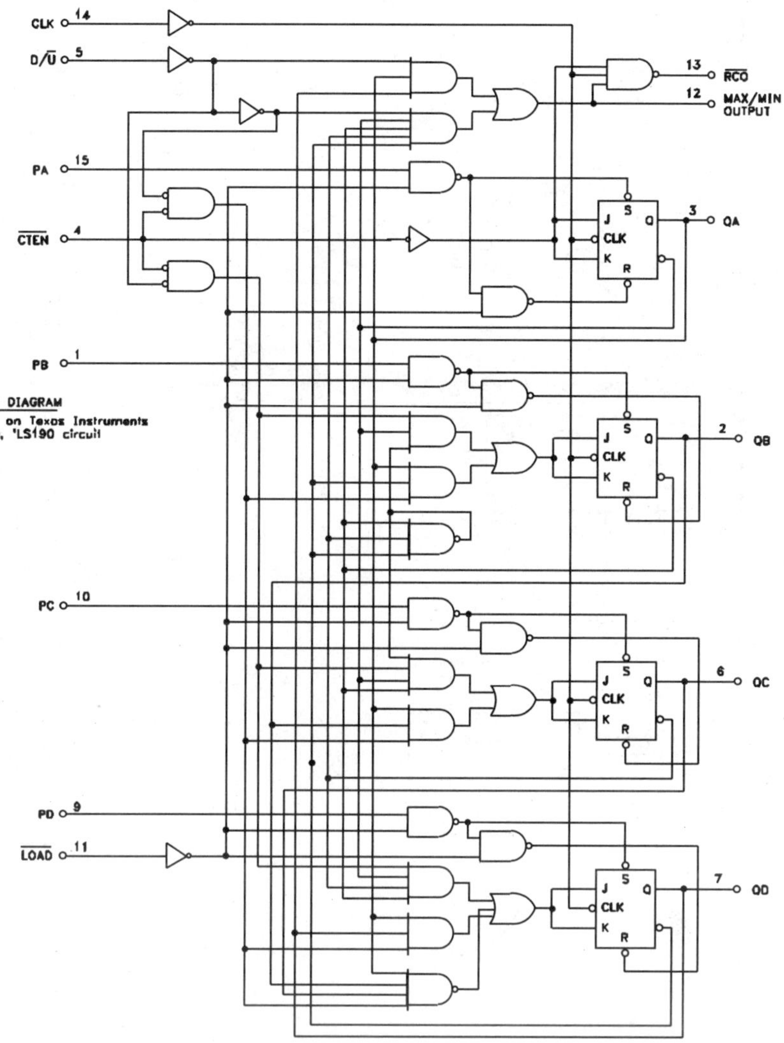

LOGIC DIAGRAM
Based on Texas Instruments 74190, 'LS190 circuit

FUNCTION TABLES 74x190, '191
PRESETTABLE SYNCHRONOUS UP/DOWN COUNTERS

MODE SELECT TABLE

OPERATING MODE	INPUTS					OUTPUTS
	$\overline{LOAD}$	D/$\overline{U}$	$\overline{CTEN}$	CLK	Dn	On
Parallel Load	0	X	X	X	0	0
	0	X	X	X	1	1
Count Up	1	0	l	↑	X	C_UP
Count Down	1	1	l	↑	X	C_DN
Hold	1	X	1	X	X	NC

l = LOW voltage level one setup time prior to the ↑ transition.
↑ = LOW-to-HIGH clock transition.
1 = HIGH, 0 = LOW, X = Irrelevant, NC = no change
C_UP = Count Up, C_DN = Count Down

OUTPUT TABLE

INPUTS			TERM COUNT				OUTPUT	
D/$\overline{U}$	$\overline{LOAD}$	CLK	QA	QB	QC	QD	M/M	$\overline{RCO}$
1	X	X	1	X	X	1	0	1
0	1	X	1	X	X	1	1	1
0	0	⊔	1	X	X	1	1	⊔
0	X	X	0	0	0	0	0	1
1	1	X	0	0	0	0	1	1
1	0	⊔	0	0	0	0	1	⊔

FUNCTION DIAGRAM

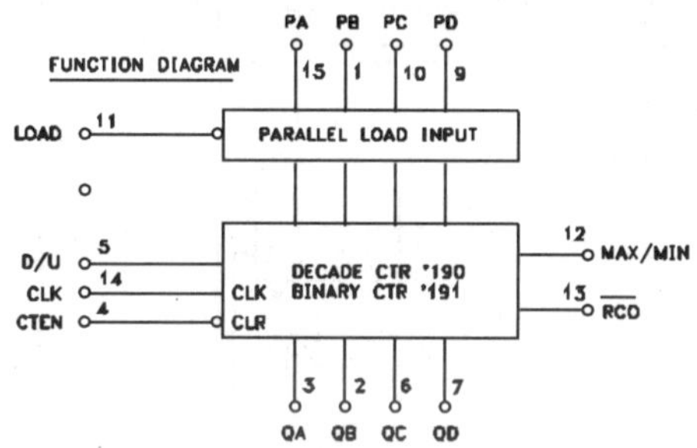

Figure 7.14. The 74x190/74x191 are synchronous up/down decade/binary counters, respectively, employing four master-slave flip-flops. The counters are fully programmable; the outputs are asynchronously presettable to either logic level by entering a low on the load input with the data present at the four inputs. This enables utilizing the counters as modulo-n dividers by using the preset inputs to modify the count length. All inputs are buffered. Two outputs are provided to perform the cascading function: ripple clock and maximum/minimum count. The counters are easily cascaded by coupling the ripple carry out to the enable input of the succeeding counter with parallel clocking.

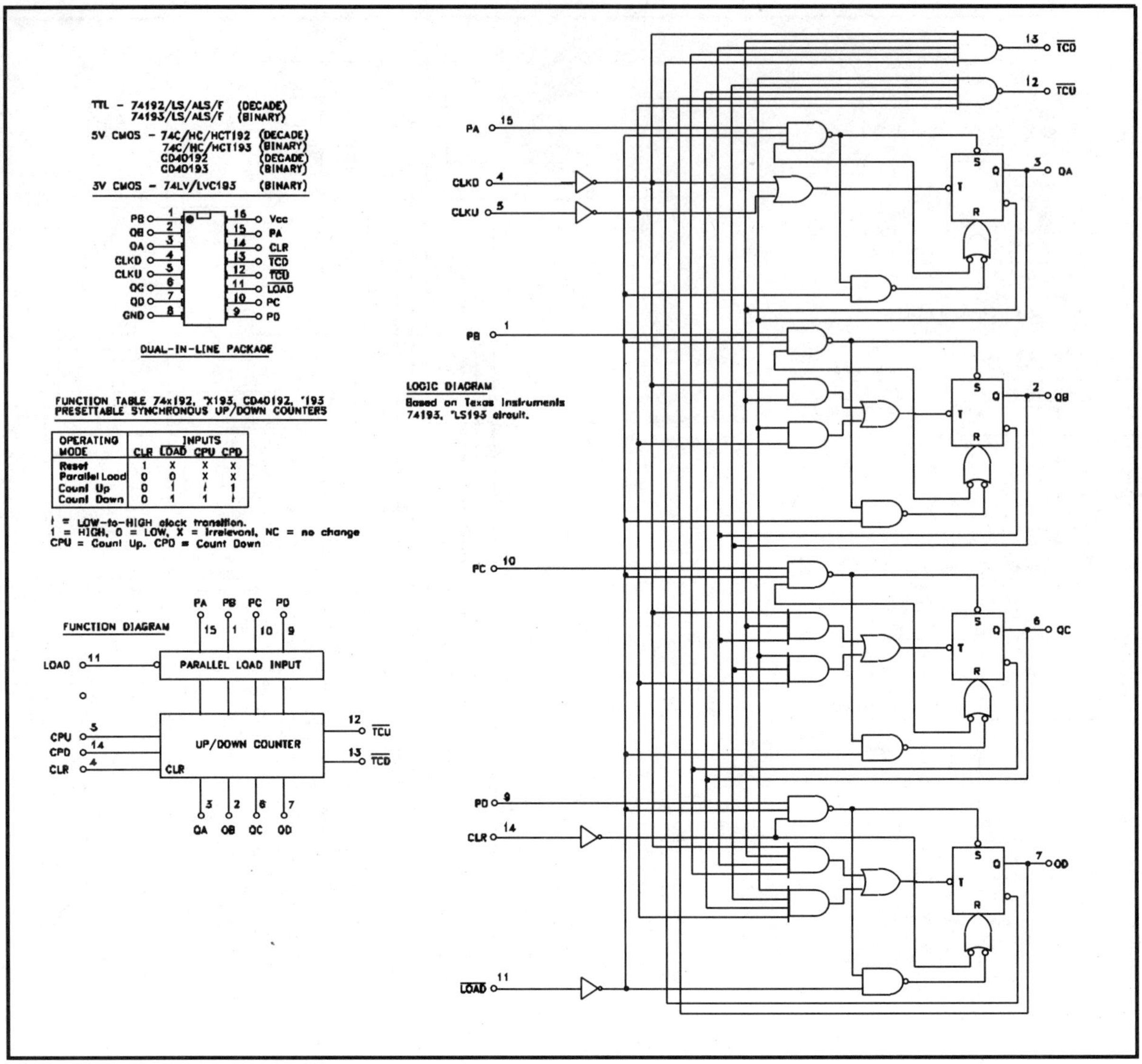

OPERATING MODE	CLR	LOAD	CPU	CPD
Reset	1	X	X	X
Parallel Load	0	0	X	X
Count Up	0	1	↑	1
Count Down	0	1	1	↑

Figure 7.15. The 74x192/74x193 are synchronous up/down BCD/binary counters, respectively, employing four master-slave flip-flops. The count direction is determined by which count input is pulsed while the other remains high. The counters are fully programmable; the outputs are asynchronously presettable to either logic level by entering a low on the load input with the data present at the four inputs. This enables utilizing the counters as modulo-n dividers by using the preset inputs to modify the count length. All inputs are buffered. Both carry and borrow outputs are available for cascading the up and down counting functions.

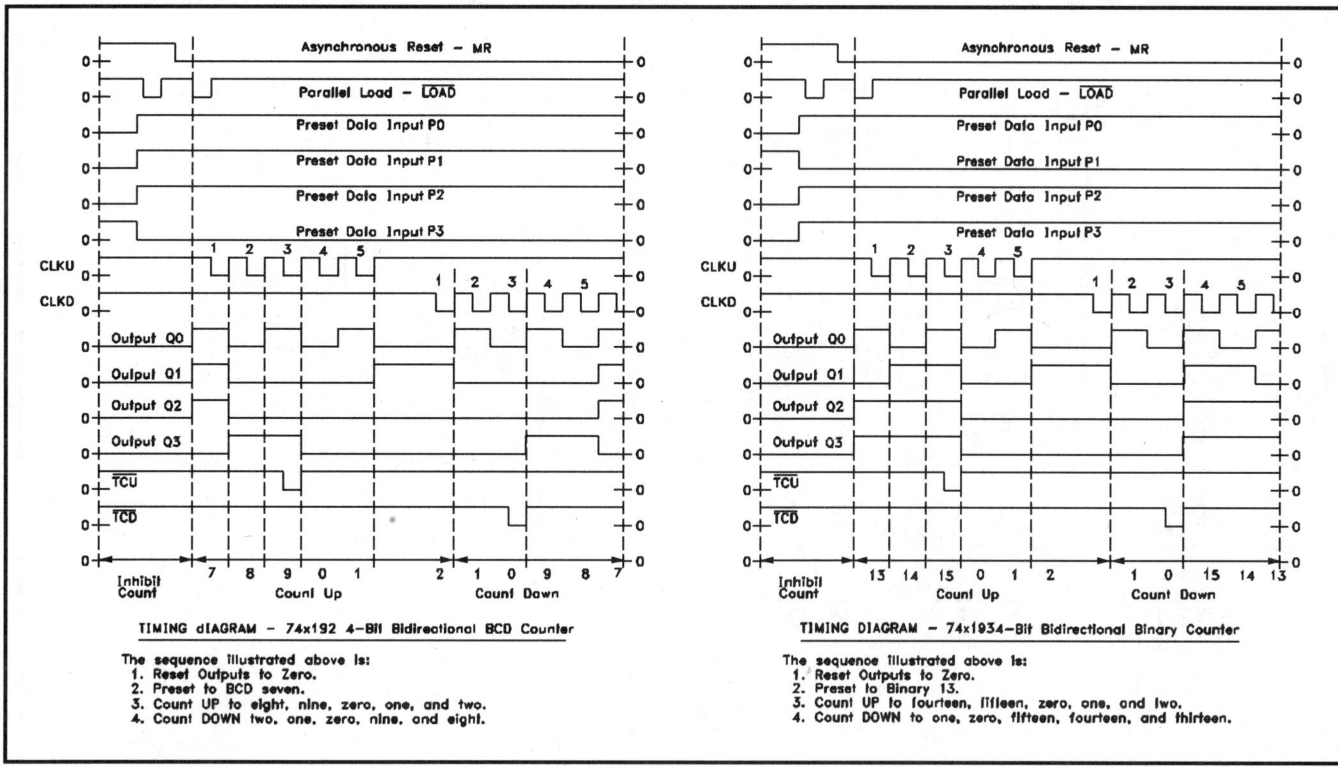

Figure 7.16. Timing diagrams for the 74x192/74x193 synchronous up/down counters.

TTL – 74F269

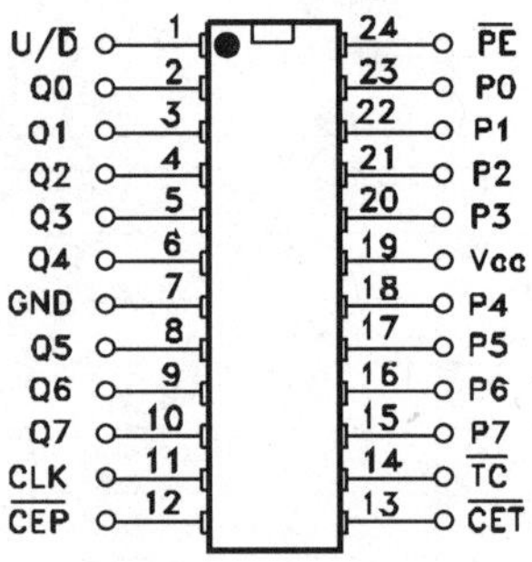

DUAL-IN-LINE PACKAGE

FUNCTION DIAGRAM

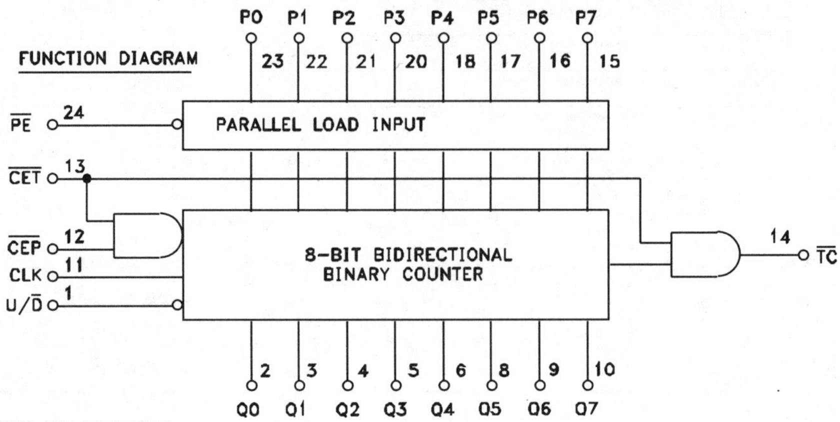

FUNCTION TABLE 74F269 8-BIT SYNCHRONOUS BIDIRECTIONAL PRESETTABLE BINARY COUNTER

OPERATING MODE	INPUTS						OUTPUTS	
	CLK	U/D̄	$\overline{CEP}$	$\overline{CET}$	$\overline{PE}$	Pn	Qn	$\overline{TC}$
Parallel Load	↑	X	X	X	l	l	0	a
	↑	X	X	X	l	h	1	a
Count Up	↑	h	l	l	h	X	C_Up	a
Count Down	↑	l	l	l	h	X	C_Dn	a
Hold	↑	X	h	X	h	X	qn	a
	↑	X	X	h	h	X	qn	H

↑ = LOW-to-HIGH clock transition.
h = HIGH voltage level one setup time prior to the ↑ clock transition.
l = LOW voltage level one setup time prior to the ↑ clock transition.
Pn, Qn = Lower case indicates state of referenced input (or output) one setup time prior to the ↑ clock transition.
(a) The TC is LOW when CET is LOW and the counter is at Terminal Count. Terminal Count Up is with all Qn HIGH; Terminal Count is Down with all Qn LOW.

Figure 7.17. The 74F269 is a fully synchronous 8-bit bidirectional binary counter. The counter is programmable, features carry look-ahead for ease of cascading, and an up/down input to select the count direction. All state changes are initiated by the rising edge of the clock input pulse. Typical count frequency is 115 MHz.

TTL - 74290/LS

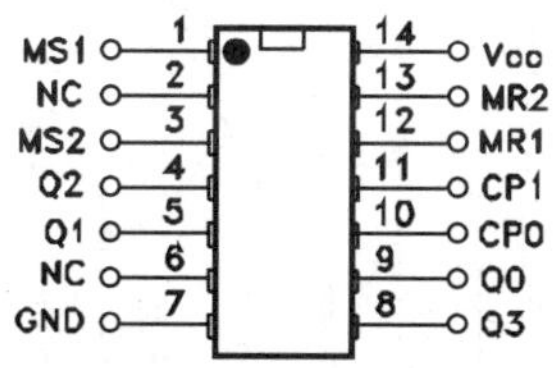

DUAL-IN-LINE PACKAGE

FUNCTION TABLE 74x290 4-BIT RIPPLE DECADE COUNTER - MODE SELECTION

RESET/SET INPUTS				OUTPUTS			
MR1	MR2	MS1	MS2	Q0	Q1	Q2	Q3
1	1	0	X	0	0	0	0
1	1	X	0	0	0	0	0
X	X	1	1	1	0	0	1
0	X	0	X	Count			
X	0	X	0	Count			
0	X	X	0	Count			
X	0	0	X	Count			

1 = HIGH voltage level
0 = LOW voltage level
x = Don't Care

NOTE: THE 74x290 IS FUNCTIONALLY IDENTICAL TO THE 7490.

LOGIC SYMBOL

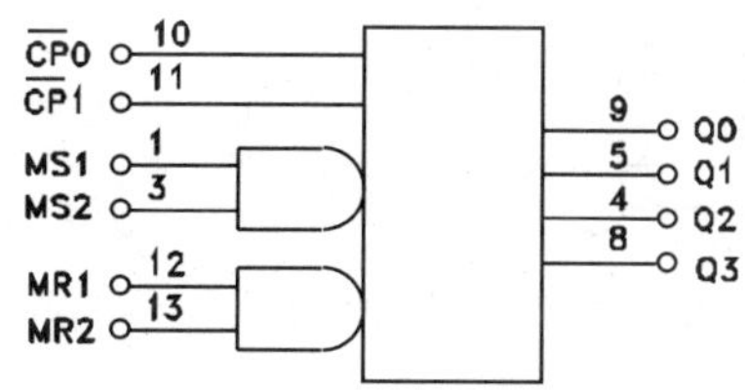

TTL - 74293/LS

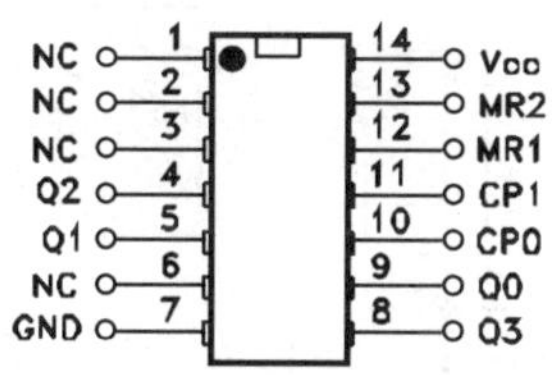

DUAL-IN-LINE PACKAGE

FUNCTION TABLE 74x293 4-BIT BINARY RIPPLE COUNTER - MODE SELECTION

RESET INPUTS		OUTPUTS			
MR1	MR2	Q0	Q1	Q2	Q3
1	1	0	0	0	0
0	1	Count			
1	0	Count			
0	0	Count			

1 = HIGH voltage level
0 = LOW voltage level
x = Don't Care

NOTE: THE 74x293 IS FUNCTIONALLY IDENTICAL TO THE 7493.

LOGIC SYMBOL

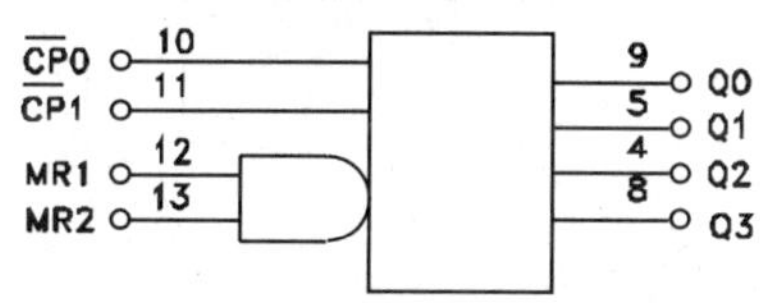

Figure 7.18. The 74x290/74x293 are electrically and functionally identical to the 7490/7493, respectively. The timing diagram of *Figure 7.6* applies to these counters. The terminal arrangement for these counters does not correspond to the prior devices, however.

TTL - 74LS/ALS390
5V CMOS - 74HC/HCT390

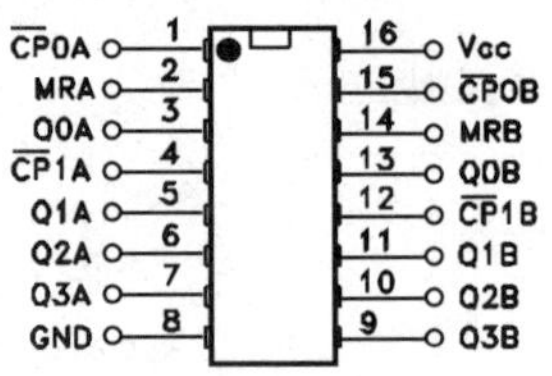

DUAL-IN-LINE PACKAGE

LOGIC SYMBOL

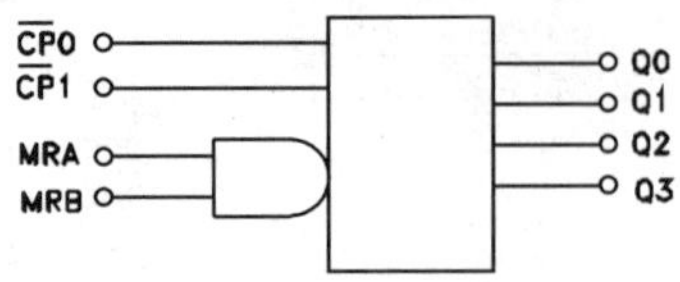

COUNT SEQUENCE TABLES 74x390 DUAL DECADE RIPPLE COUNTERS FOR 1/2 '390

BCD COUNT SEQUENCE

	OUTPUT			
COUNT	Q3	Q2	Q1	Q0
0	0	0	0	0
1	0	0	0	1
2	0	0	1	0
3	0	0	1	1
4	0	1	0	0
5	0	1	0	1
6	0	1	1	0
7	0	1	1	1
8	1	0	0	0
9	1	0	0	1

Output Q0 connected to Clock-2 input.

BIQUINARY COUNT SEQ

	OUTPUT			
COUNT	Q3	Q2	Q1	Q0
0	0	0	0	0
1	0	0	0	1
2	0	0	1	0
3	0	0	1	1
4	0	1	0	0
5	1	0	0	1
6	1	0	1	0
7	1	0	1	1
8	1	0	0	0
9	1	1	0	1

Output Q3 connected to Clock-1 input.

FUNCTION TABLE 74x390 DUAL 4-BIT DECADE COUNTER - MODE SELECTION

RESET/SET INPUTS	OUTPUTS
MR	Q0 Q1 Q2 Q3
1	0 0 0 0
0	Count

1 = HIGH voltage level
0 = LOW voltage level
x = Don't Care

LOGIC DIAGRAM 1/2'390

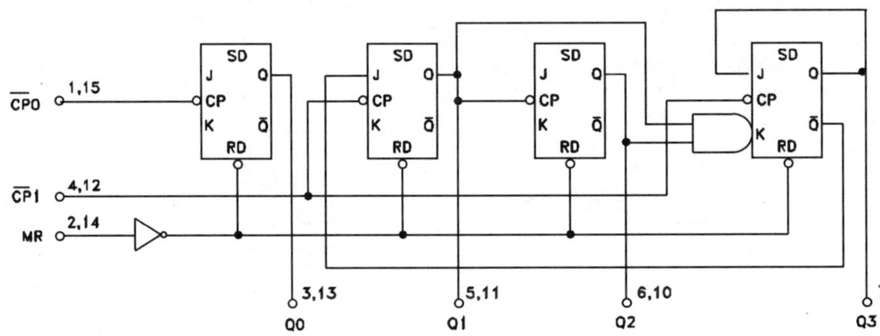

Figure 7.19. The 74x390 is a dual version of the 74x90 decade counter. Like the prior device, the '390 includes a divide-by-two and a divide-by-five feature. These can be used to provide cycle lengths equal to any whole and/or cumulative multiples of 2 and/or 5 up to divide-by-100. With the bi-quinery connection, the independent divide-by-two circuit can be used to provide a square wave output. The timing diagram of *Figure 7.6* applies to this counter.

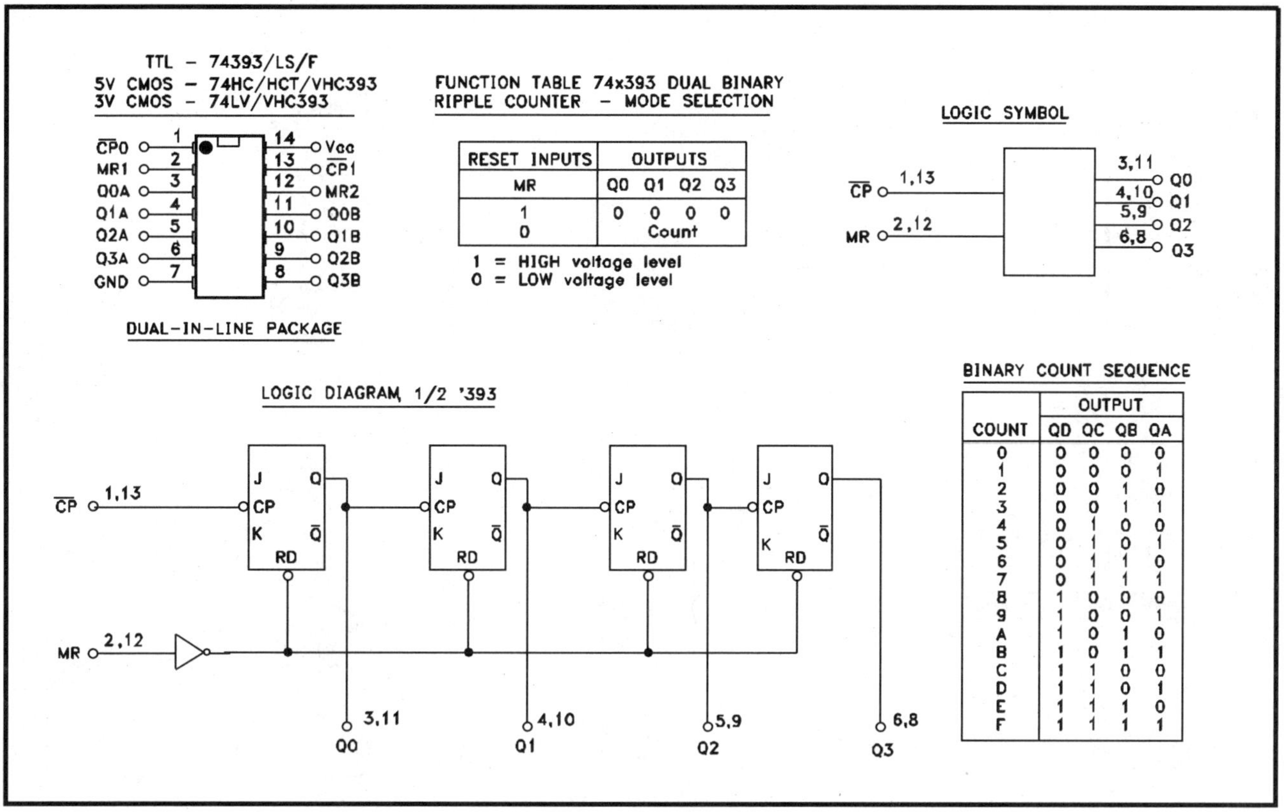

Figure 7.20. The 74x393 is a dual version of the 74x93, providing two independent binary counters with a clear and clock input. N-bit binary counters can be implemented with each counter, providing a capability of division by up to 256. The timing diagram of *Figure 7.6* applies to this counter.

TTL - 74490/LS
5V CMOS - 74HC490

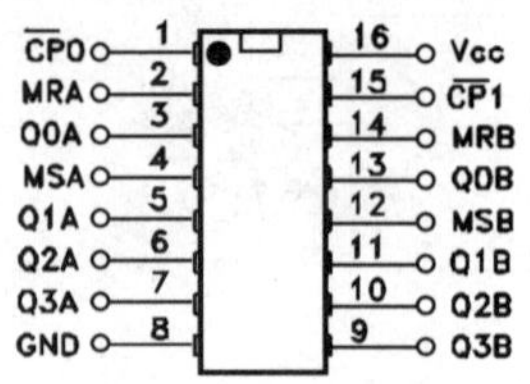

DUAL-IN-LINE PACKAGE

LOGIC SYMBOL

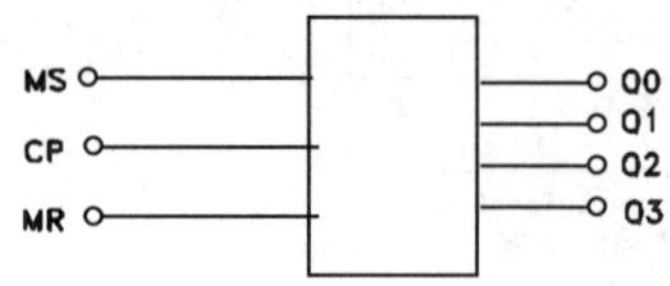

BCD COUNT SEQUENCE

COUNT	OUTPUT Q3	Q2	Q1	Q0
0	0	0	0	0
1	0	0	0	1
2	0	0	1	0
3	0	0	1	1
4	0	1	0	0
5	0	1	0	1
6	0	1	1	0
7	0	1	1	1
8	1	0	0	0
9	1	0	0	1

LOGIC DIAGRAM 1/2 '490

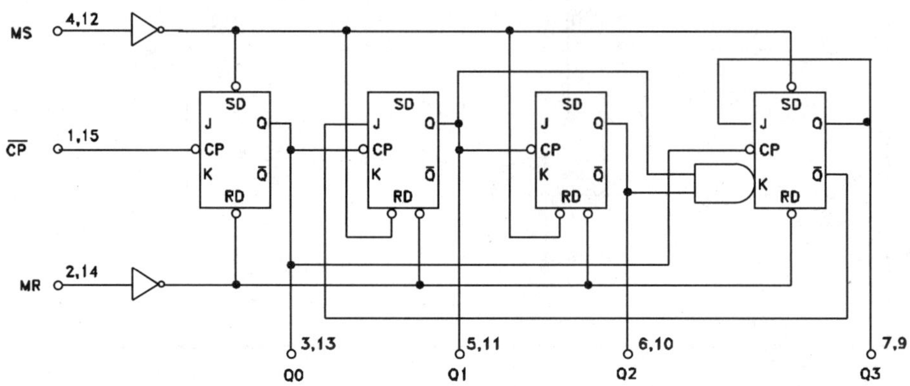

MODE SELECTION TRUTH TABLE 74x490
DUAL BCD DECADE RIPPLE COUNTER

RESET/SET INPUTS MR	MS	OUTPUTS Q0	Q1	Q2	Q3
1	X	0	0	0	0
X	1	1	0	0	1
0	0	Count			

1 = HIGH voltage level
0 = LOW voltage level
x = Don't Care

Figure 7.21. The 74LS490 is a dual version of the74LS90 and shares its versatile features. The timing diagram of *Figure 7.6* applies to this counter. The terminal arrangement is unique to this device, however.

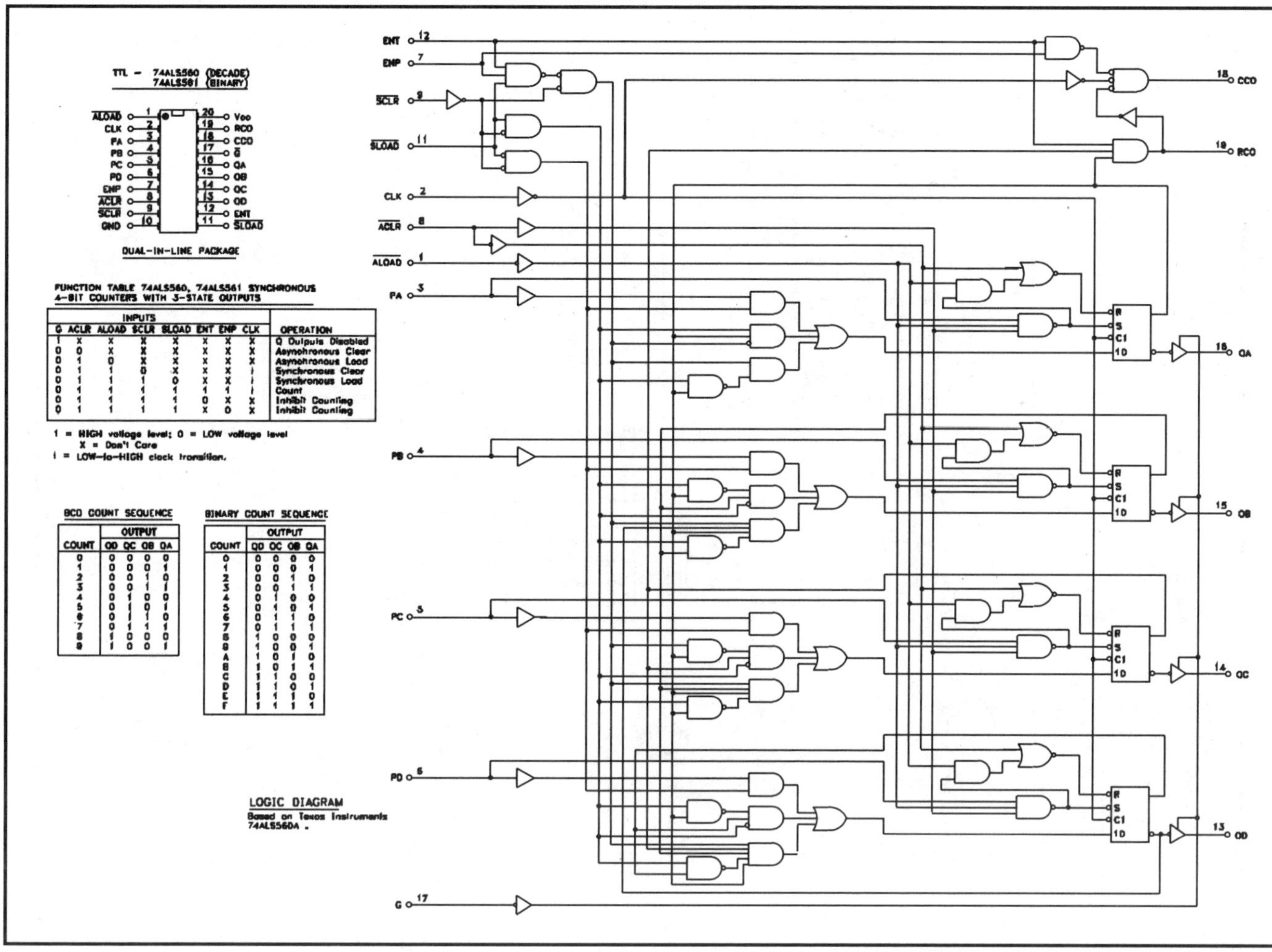

INPUTS								OPERATION
G	ACLR	ALOAD	SCLR	SLOAD	ENT	ENP	CLK	
1	X	X	X	X	X	X	X	Q Outputs Disabled
0	0	X	X	X	X	X	X	Asynchronous Clear
0	1	0	X	X	X	X	X	Asynchronous Load
0	1	1	0	X	X	X	↑	Synchronous Clear
0	1	1	1	0	X	X	↑	Synchronous Load
0	1	1	1	1	1	1	↑	Count
0	1	1	1	1	0	X	X	Inhibit Counting
0	1	1	1	1	X	0	X	Inhibit Counting

BCD COUNT SEQUENCE

COUNT	OUTPUT QD	QC	QB	QA
0	0	0	0	0
1	0	0	0	1
2	0	0	1	0
3	0	0	1	1
4	0	1	0	0
5	0	1	0	1
6	0	1	1	0
7	0	1	1	1
8	1	0	0	0
9	1	0	0	1

BINARY COUNT SEQUENCE

COUNT	OUTPUT QD	QC	QB	QA
0	0	0	0	0
1	0	0	0	1
2	0	0	1	0
3	0	0	1	1
4	0	1	0	0
5	0	1	0	1
6	0	1	1	0
7	0	1	1	1
8	1	0	0	0
9	1	0	0	1
A	1	0	1	0
B	1	0	1	1
C	1	1	0	0
D	1	1	0	1
E	1	1	1	0
F	1	1	1	1

Figure 7.22. The 74ALS560/74ALS561 are programmable synchronous decade/binary 4-bit counters with 3-state outputs. Provision for both synchronous and asynchronous loading and clearing are included. Synchronous functions are all executed on the rising (positive) edge of the clock input. Clear is initiated by applying a low to either of the two clear inputs. A high level on the output enable, /G, forces the outputs into a high impedance state. Input ENT is fed forward to enable the ripple carry output, RCO, to provide a high level output with a maximum count, 9 or 15. The clocked carry output, CCO, also produces a high level pulse of a duration equal to the low level of the clock when RCO is high and the counter enabled. For high-speed counting RCO should be used for cascading.

TTL – 74ALS/F568 (DECADE)
74ALS/F569 (BINARY)

Signal	Pin	Pin	Signal
U/$\overline{D}$	1	20	Vcc
CLK	2	19	$\overline{TC}$
D0	3	18	GC
D1	4	17	$\overline{OE}$
D2	5	16	Q0
D3	6	15	Q1
$\overline{CEP}$	7	14	Q2
$\overline{MR}$	8	13	Q3
$\overline{SR}$	9	12	$\overline{CET}$
GND	10	11	$\overline{PE}$

DUAL-IN-LINE PACKAGE

TERMINAL COUNT TRUTH TABLE 74x568

INPUTS				COUNTER STATES				OUTPUTS	
CLK	U/$\overline{D}$	$\overline{CEP}$	$\overline{CET}$	Q0	Q1	Q2	Q3	$\overline{TC}$	GC
1	0	0	0	0	0	0	0	0	1
0	0	0	0	0	0	0	0	0	0
X	0	1	0	0	0	0	0	0	1
X	0	X	1	0	0	0	0	1	1
1	1	0	0	1	X	X	1	0	1
0	1	0	0	1	X	X	1	0	0
X	1	1	0	1	X	X	1	0	1
X	1	X	1	1	X	X	1	1	1

74x569 IDENTICAL EXCEPT Q1, Q2 REPLACE X's WITH 1's.

FUNCTION TABLES 74x568, 'x569 UP/DOWN SYNCHRONOUS COUNTERS WITH 3-STATE OUTPUTS

MODE SELECT

OPERATING MODES	INPUTS								OUTPUTS			
	$\overline{MR}$	CLK	$\overline{SR}$	U/$\overline{D}$	$\overline{PE}$	$\overline{CEP}$	$\overline{CET}$	Dn	Q0	Q1	Q2	Q3
Asynchronous Reset	0	X	X	X	X	X	X	X	0	0	0	0
Synchronous Reset	1	↑	l	X	X	0	0	X	0	0	0	0
Parallel Load	1	↑	h	X	l	X	X	l	0	0	0	0
	1	↑	h	X	l	X	X	h	1	1	1	1
Count Up	1	↑	h	h	h	l	l	X	Count Up			
Count Down	1	↑	h	l	h	l	l	X	Count Down			
Hold	1	↑	h	X	h	h	X	X	No Change			
	1	↑	h	X	h	X	h	X	No Change			

0 = LOW voltage level. 1 = HIGH voltage level X = Don't Care
↑ = LOW-to-HIGH clock transition.
h = HIGH voltage level one setup time prior to the ↑ clock transition.
l = LOW voltage level one setup time prior to the ↑ clock transition.
dn, qn = Lower case indicates state of referenced input (or output) one setup time prior to the ↑ clock transition.

3-STATE BUFFER OPERATING MODES	INPUTS		OUTPUTS
	$\overline{OE}$	Qn-Counter	Q0, Q1, Q2, Q3
Read Counter	0	0	0
	0	1	1
Disable Outputs	1	0	Z
	1	1	Z

Z = High Impedance state.

BCD COUNT SEQUENCE

COUNT	OUTPUT			
	QD	QC	QB	QA
0	0	0	0	0
1	0	0	0	1
2	0	0	1	0
3	0	0	1	1
4	0	1	0	0
5	0	1	0	1
6	0	1	1	0
7	0	1	1	1
8	1	0	0	0
9	1	0	0	1

BINARY COUNT SEQUENCE

COUNT	OUTPUT			
	QD	QC	QB	QA
0	0	0	0	0
1	0	0	0	1
2	0	0	1	0
3	0	0	1	1
4	0	1	0	0
5	0	1	0	1
6	0	1	1	0
7	0	1	1	1
8	1	0	0	0
9	1	0	0	1
A	1	0	1	0
B	1	0	1	1
C	1	1	0	0
D	1	1	0	1
E	1	1	1	0
F	1	1	1	1

LOGIC DIAGRAM
Based on Philips 74F569

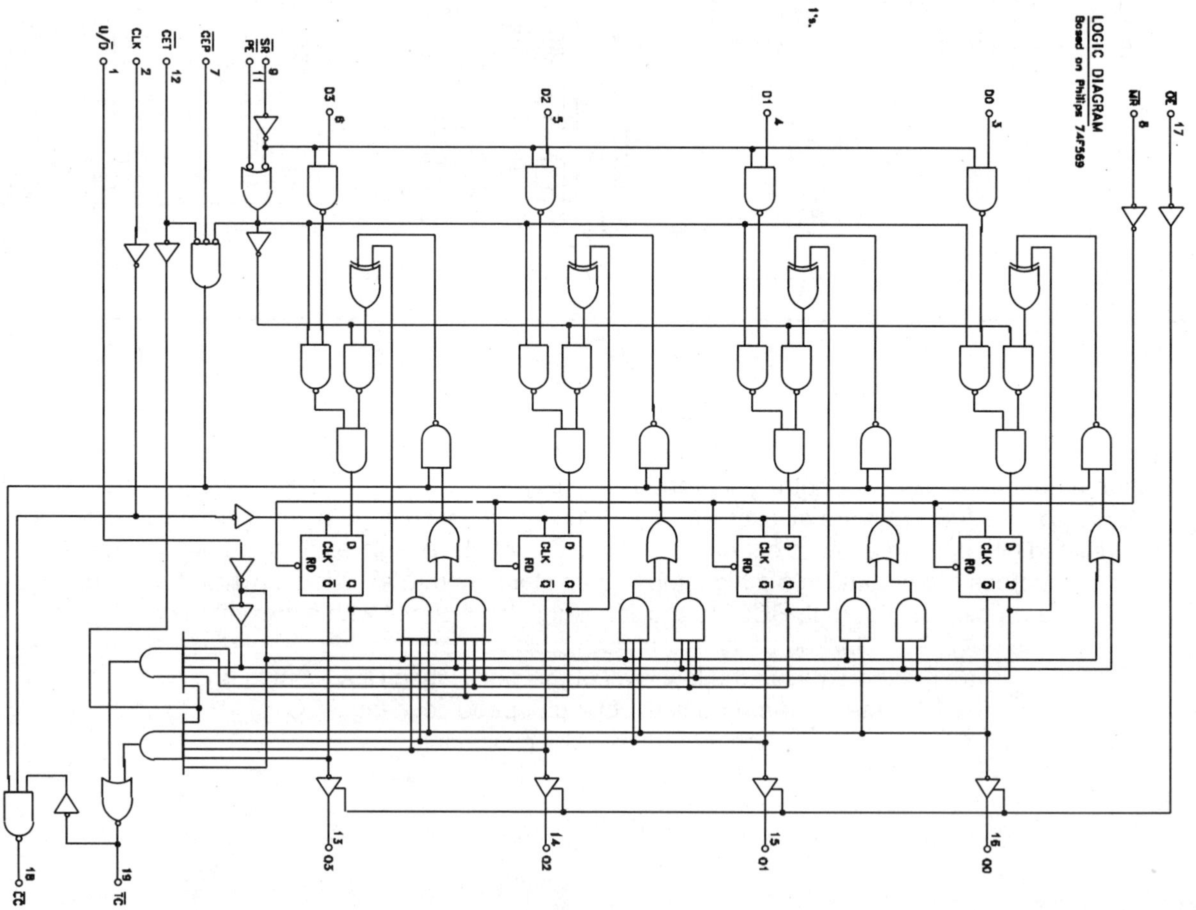

Figure 7.23. The 74ALS/F568/74ALS/F569 are programmable synchronous decade/binary 4-bit bidirectional counters with 3-state outputs. Synchronous loading and clearing and asynchronous active low reset are provided. Synchronous functions are all executed on the rising (positive) edge of the clock input. An up/down input determines the direction of counting. A high level on the output enable, /OE, forces the outputs into a high-impedance state. Internal counting, loading, and reset functions are not affected when in this state.

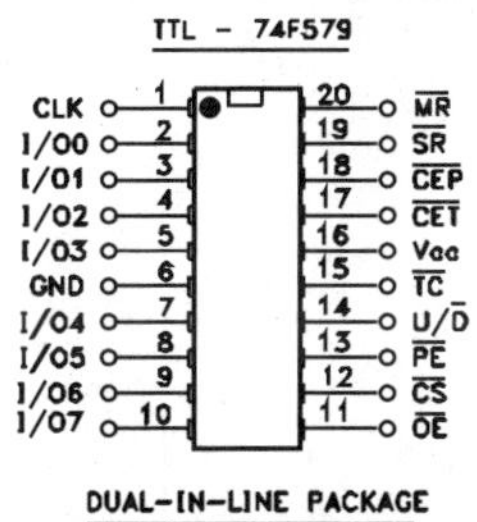

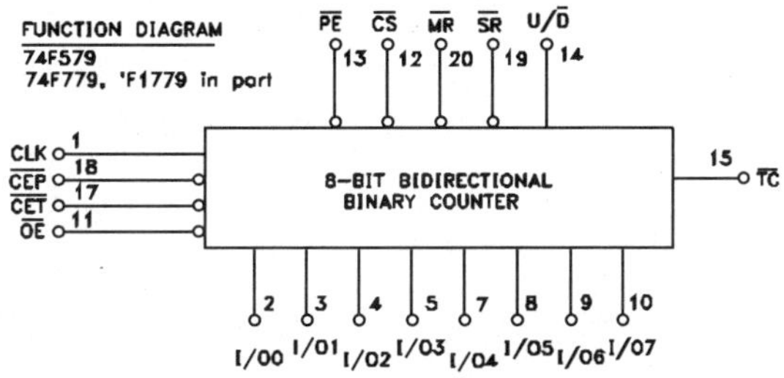

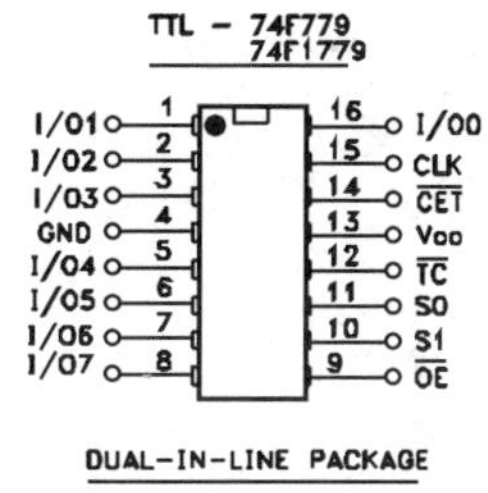

FUNCTION TABLE 74F579 8-BIT SYNCHRONOUS BIDRECTIONAL PRESETTABLE BINARY COUNTER (3-STATE)

INPUTS									OPERATING MODE
$\overline{MR}$	$\overline{SR}$	$\overline{CS}$	$\overline{PE}$	$\overline{CEP}$	$\overline{CET}$	U/$\overline{D}$	$\overline{OE}$	CLK	
X	X	1	X	X	X	X	X	X	I/O0–I/O7 in high impedance (PE disabled)
X	X	0	1	X	X	X	1	X	I/O0–I/O7 in high impedance
X	X	0	1	X	X	X	0	X	Flip-flop output appears on I/On lines
0	X	X	X	X	X	X	X	X	Asynchronous reset, all Flip-flops
1	0	X	X	X	X	X	X	↑	Synchronous reset, all Flip-flops
1	1	0	0	X	X	X	X	↑	Parallel Load, all Flip-flops
1	1	not LL		1	X	X	X	↑	Hold
1	1	not LL		X	1	X	X	↑	Hold (TC held High)
1	1	not LL		0	0	1	X	↑	Count Up
1	1	not LL		0	0	0	X	↑	Count Down

1 = HIGH voltage level, 0 = LOW voltage level, X = Don't Care
↑ = LOW-to-HIGH clock transition.
not LL = $\overline{CS}$ and $\overline{PE}$ should never be at LOW level at the same time.

FUNCTION TABLE 74F779, 'F1779 8-BIT SYNCHRONOUS BIDRECTIONAL PRESETTABLE BINARY COUNTER (3-STATE)

INPUTS					OPERATING MODE
S0	S1	$\overline{CET}$	$\overline{OE}$	CLK	
X	X	X	1	X	I/O0–I/O7 in high impedance
X	X	X	0	X	Flip-flop output appears on I/On lines
0	0	X	1	↑	Parallel Load, all Flip-flops
not LL		1	X	↑	Hold (TC held High)
1	0	0	X	↑	Count Up
0	1	0	X	↑	Count Down

1 = HIGH voltage level, 0 = LOW voltage level, X = Don't Care
↑ = LOW-to-HIGH clock transition.
not LL = $\overline{CS}$ and $\overline{PE}$ should never be at LOW level at the same time.

Figure 7.24. The 74F579 is a high speed, 115 MHz, 8-bit bidirectional binary counter with multiplexed 3-state input/output ports. Synchronous loading and clearing and asynchronous active low reset are provided. Synchronous functions are all executed on the rising (positive) edge of the clock input. An up/down input determines the direction of counting. A high level on the output enable, /OE, forces the outputs into a high-impedance state. Internal counting, loading, and reset functions are not affected when in this state. Center power pins reduce effects of package inductance.

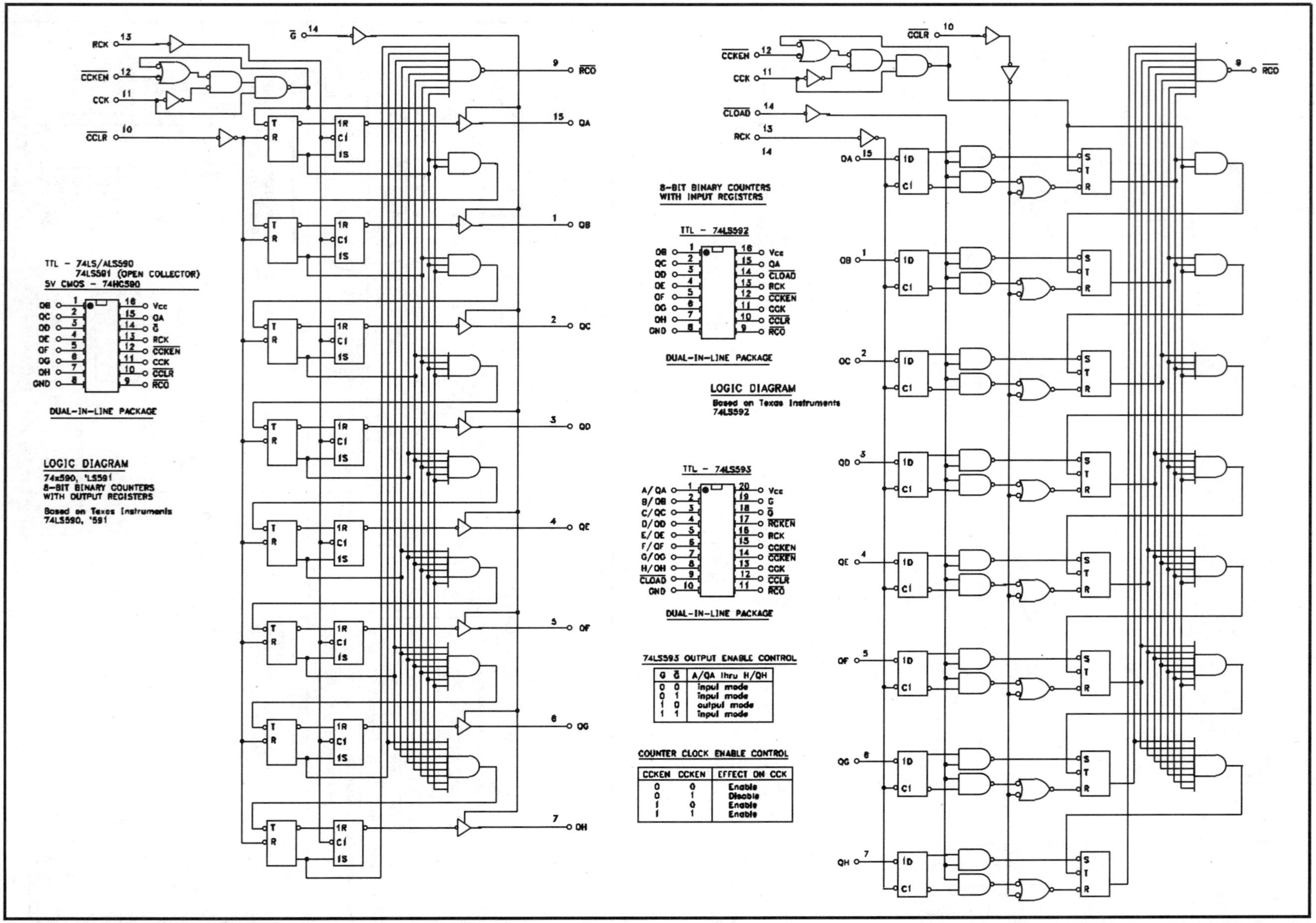

Figure 7.25. The 74LS/ALS590 and 74LS591 are 8-bit binary counters. The '590 features 3-state output registers; the '591 has open collector register outputs. Separate clocks are provided for the counter and storage registers. The counter features an active low direct clear input, /CCLR, and active low count enable input, /CCKEN. A ripple carry output, RCO, is provided for cascading. Both counter and register clocks are positive-edge triggered. If desired, the two clocks may be tied together. In this event the counter will always precede the register by one count.

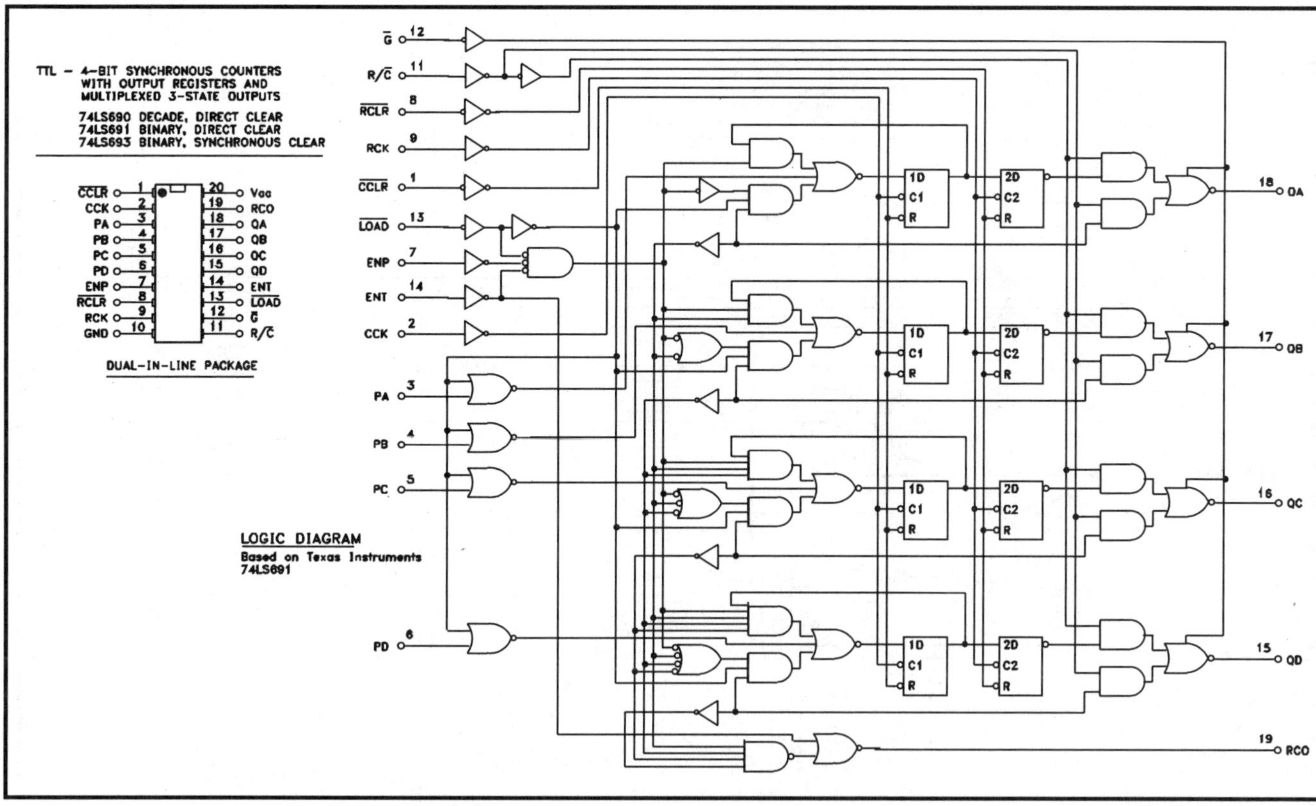

Figure 7.26. The 74LS690 is a 4-bit decade counter; the 74LS691 and 74LS693 are 4-bit binary counters. All are synchronous with 4-bit D-type flip-flops, and quadruple two-line to one-line multiplexers with 3-state outputs. The counters are programmable from data inputs with ENT and ENP enables. A ripple carry output, RCO, provides for easy expansion. The register/counter select input, R/C, selects the counter when low, the register when high for the 3-state outputs. Individual clock and clear inputs are provided for both the counter and register. Both are positive-edge triggered. Clear is active low and asynchronous on the '690 and 691; synchronous on the '693. Counter loading takes place with /LOAD low followed by a positive clock transition. Expansion is accomplished by connecting RCO to ENT of the following stage.

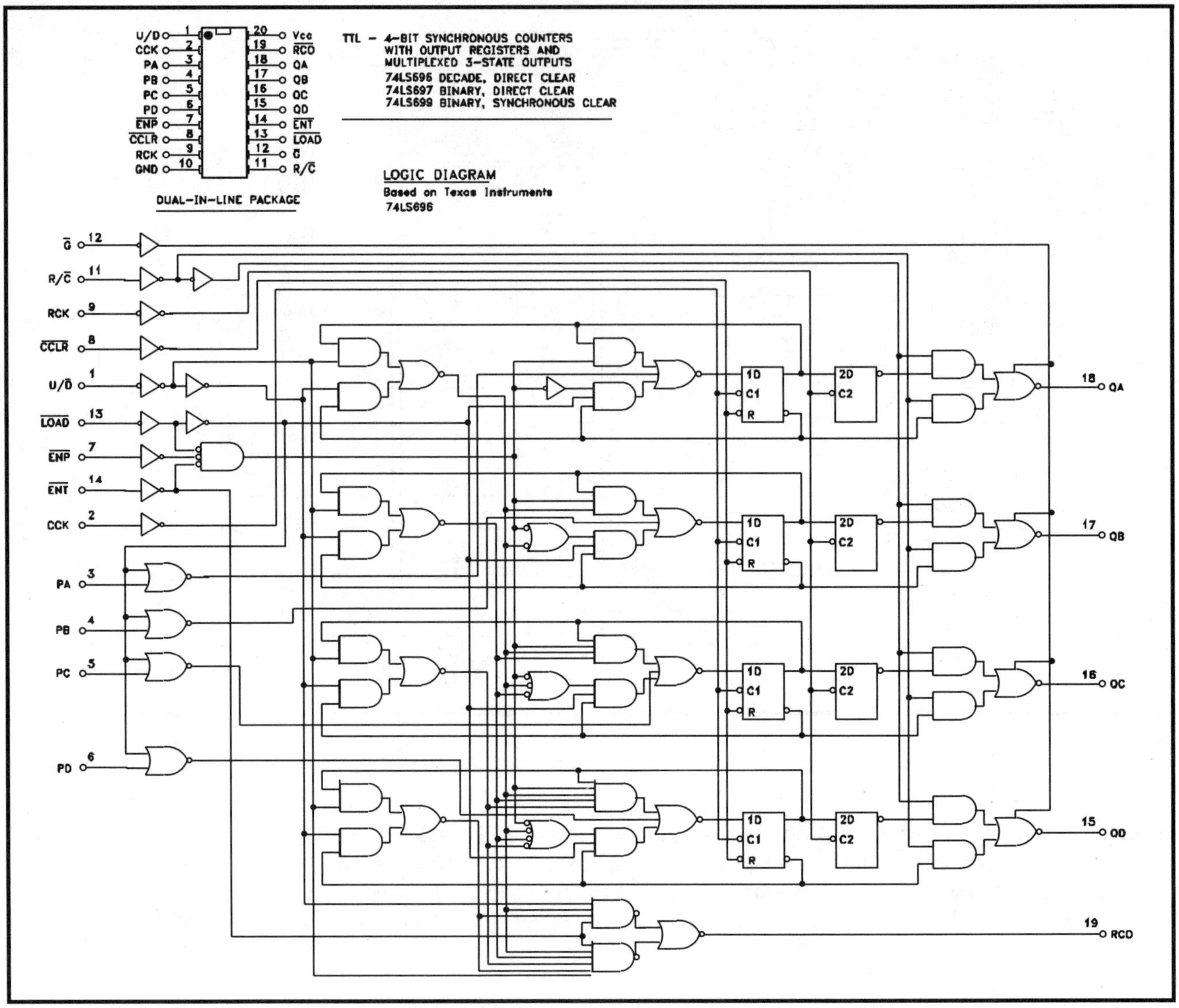

Figure 7.27. The 74LS696 is a 4-bit decade counter; the 74LS697 and 74LS699 are 4-bit binary counters. All are synchronous up/down counters with 4-bit D-type flip-flops, and quadruple two-line to one-line multiplexers with 3-state outputs. The counters are programmable from data inputs with ENT and ENP enables. A ripple carry output, RCO, provides for easy expansion. The register/counter select input, R/C, selects the counter when low, the register when high for the 3-state outputs. Individual clock and clear inputs are provided for both the counter and register. Both are positive-edge triggered. Clear is active low and asynchronous on the '696 and 697; synchronous on the '699. Counter loading takes place with /LOAD low followed by a positive clock transition. Expansion is accomplished by connecting RCO to ENT of the following stage.

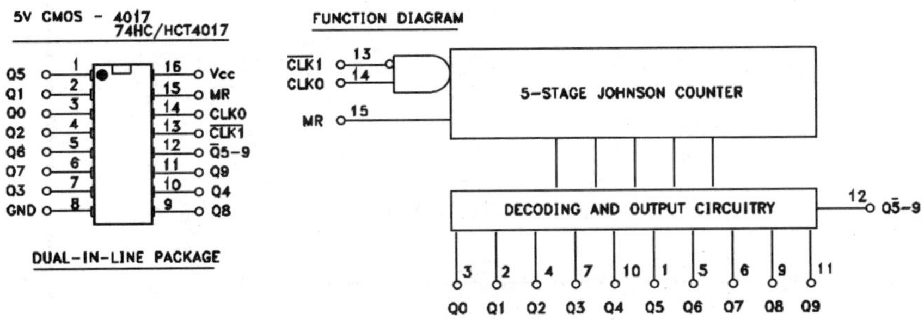

FUNCTION TABLE 5-STAGE JOHNSON COUNTER WITH 10 DECODED ACTIVE HIGH OUTPUTS

MR	CLKO	$\overline{CLK1}$	Operation
1	X	X	Q0=Q5-9=1;Q1 to Q9=0
0	1	⌐ (falling edge)	Counter advances
0	⌐ (rising edge)	0	Counter advances
0	0	X	No change
0	X	1	No change
0	1	(rising edge)	No Change
0	(falling edge)	0	No change

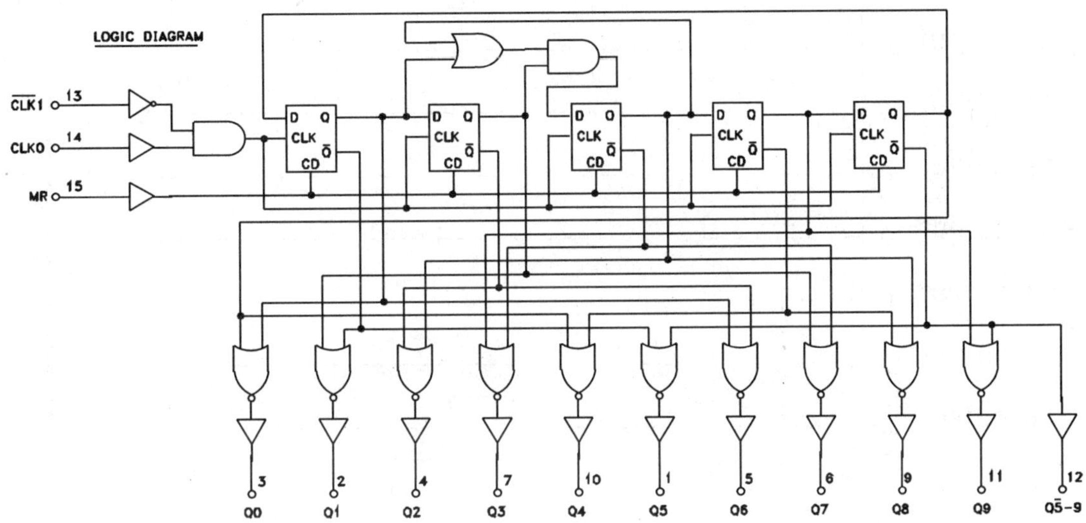

Figure 7.28. The CD4017 and 74HC/HCT4017 is a five-stage Johnson counter provided with ten spike-free decoded active-high outputs. An active low output is obtained from the most significant flip-flop. Two clock inputs, one active high, the other low and an asynchronous active high master reset, MR, are provided. The counter is advanced by a low-to-high transition to CLK0 with /CLK1 low or a high-to-low transition at /CLK1 with CLK0 high. The Q5-9 output goes high on the last count and can be used as input to the next counter. Automatic code correction takes place internally: following any illegal code the counter returns to correct operation within 11 clock pulses. *Figure 7.29* is the counter timing diagram showing the decoded outputs.

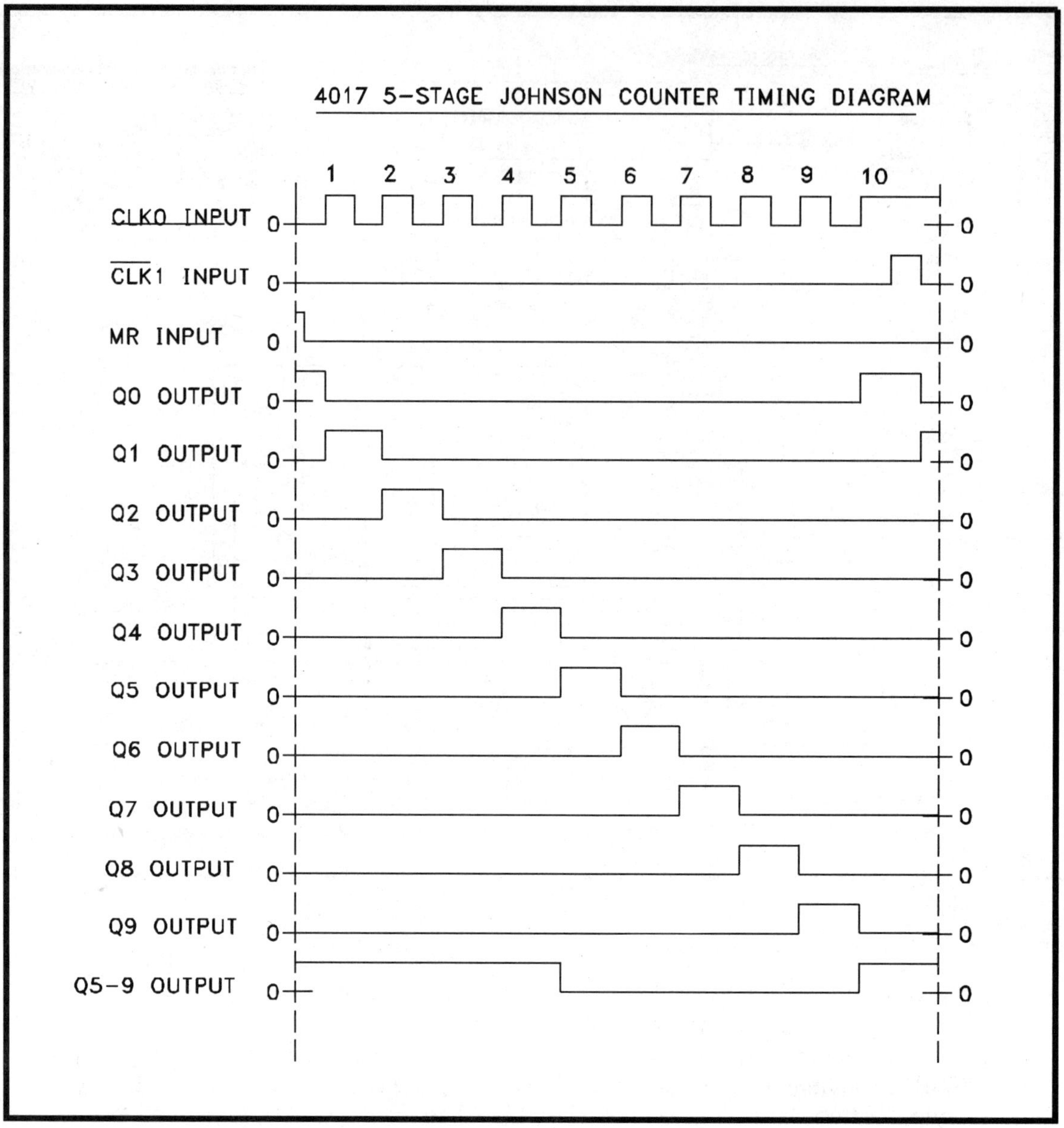

Figure 7.29. This figure is the CD4017 counter timing diagram showing the decoded outputs, clocking, reset, and Q5-9 output.

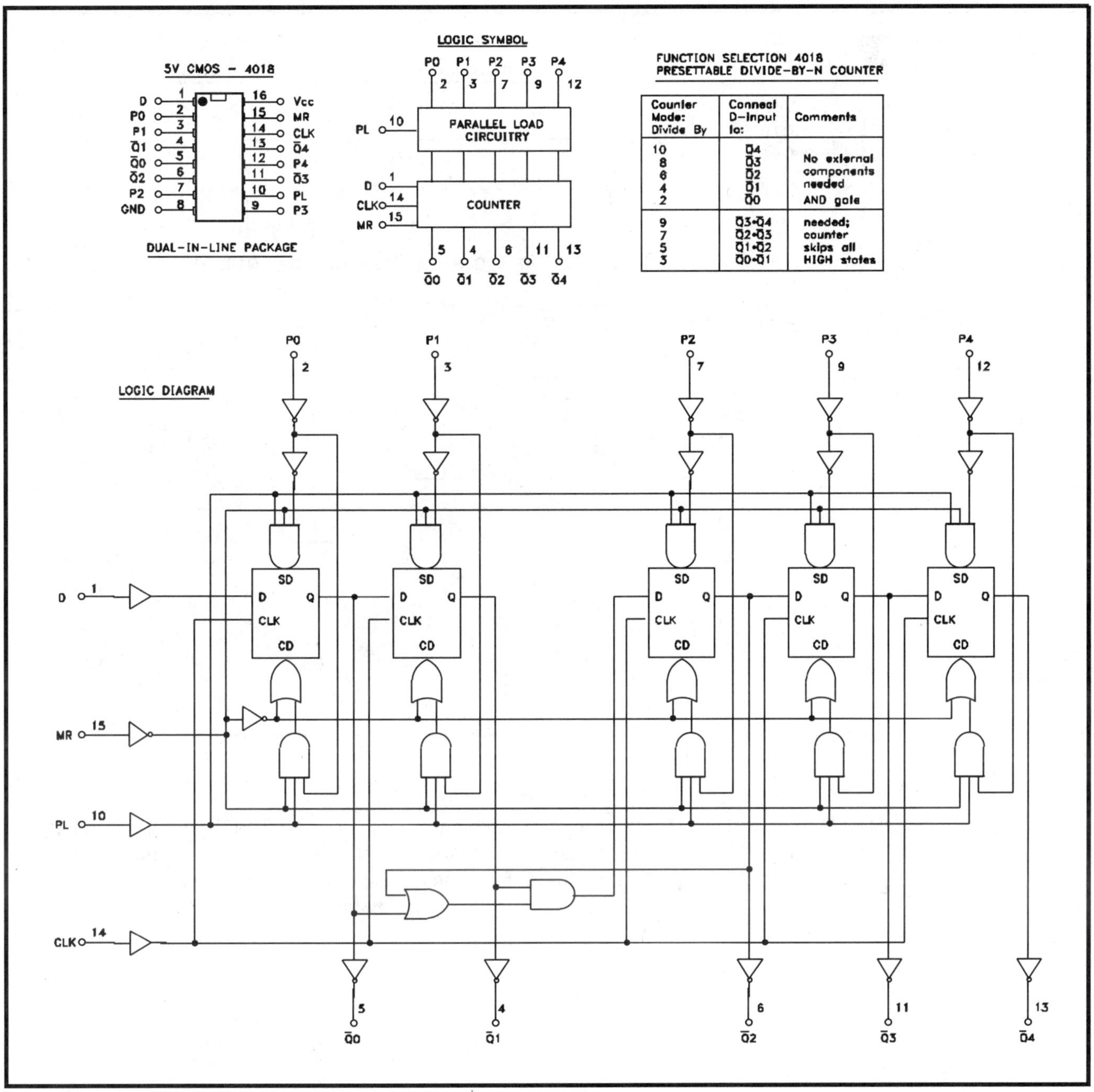

Counter Mode: Divide By	Connect D-Input to:	Comments
10	Q̄4	
8	Q̄3	No external
6	Q̄2	components
4	Q̄1	needed
2	Q̄0	AND gate
9	Q̄3•Q̄4	needed;
7	Q̄2•Q̄3	counter
5	Q̄1•Q̄2	skips all
3	Q̄0•Q̄1	HIGH states

Figure 7.30. The CD4018 is a five-stage Johnson counter having a clock input, CLK, data input, D, an asynchronous parallel load input, PL, five parallel inputs, P0-P4, five active low buffered outputs, Q0-Q4, and an asynchronous master reset, MR. Data input is asynchronously loaded when PL is high independent of the clock and D inputs. With PL low, the counter advances on the low-to-high clock transition. Connecting /Q0 to /Q4 to D causes the counter to function as a divide-by-n counter where n = 2 to 10. An internal code correction circuit provides automatic code correction from any illegal to a proper counting mode within eleven clock pulses. Reset occurs with MR high.

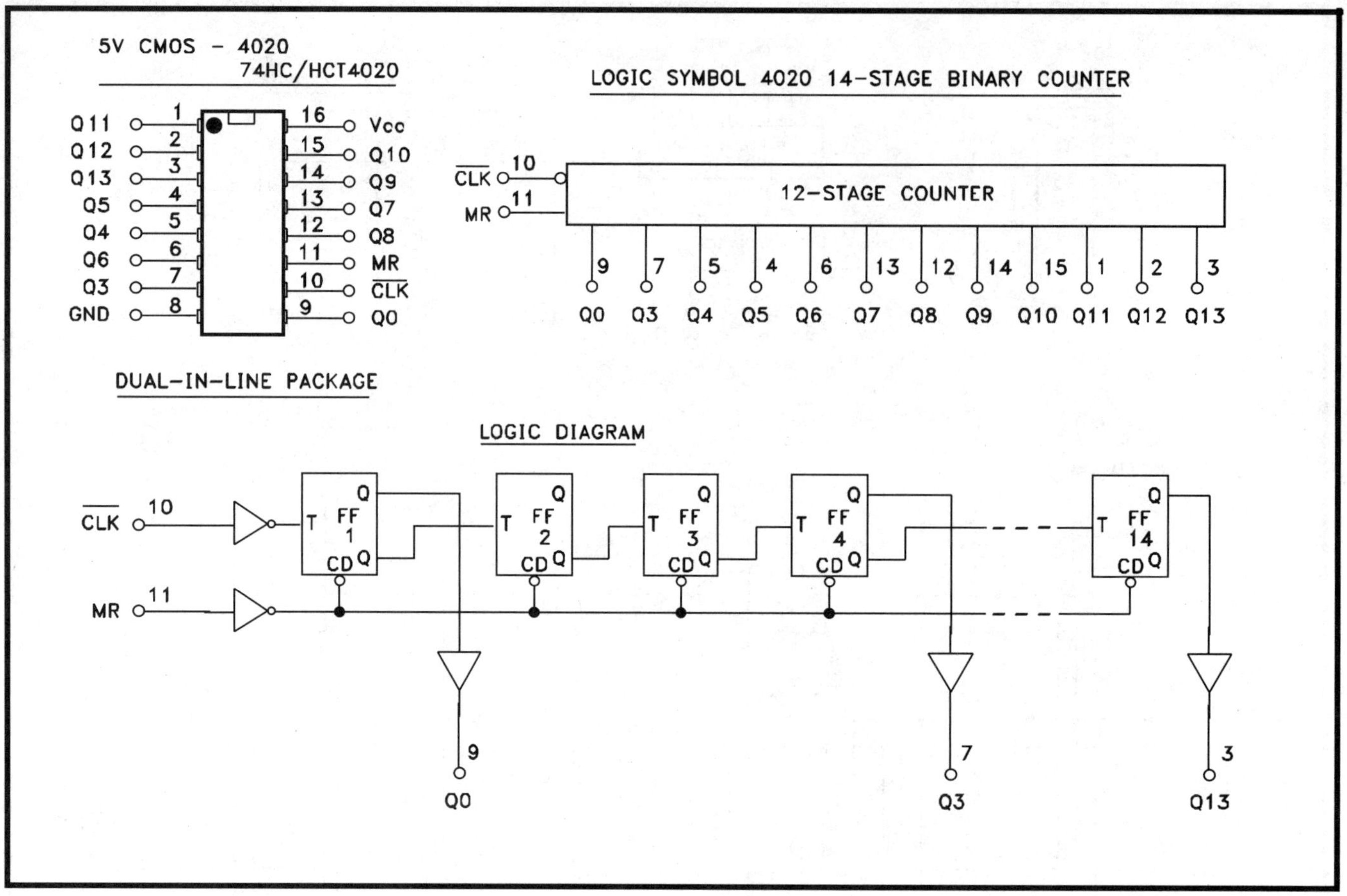

Figure 7.31. The CD4020 and 74HC/HCT4020 is a 14-stage binary ripple counter. The device has two inputs: clock, /CLK, and an active high asynchronous master reset, MR. Note that only 12 outputs are brought out, Q0, and Q3 - Q13. The counter features high speed with 35 MHz at a Vcc of 15 volts. *Figure 7.32* is the timing diagram.

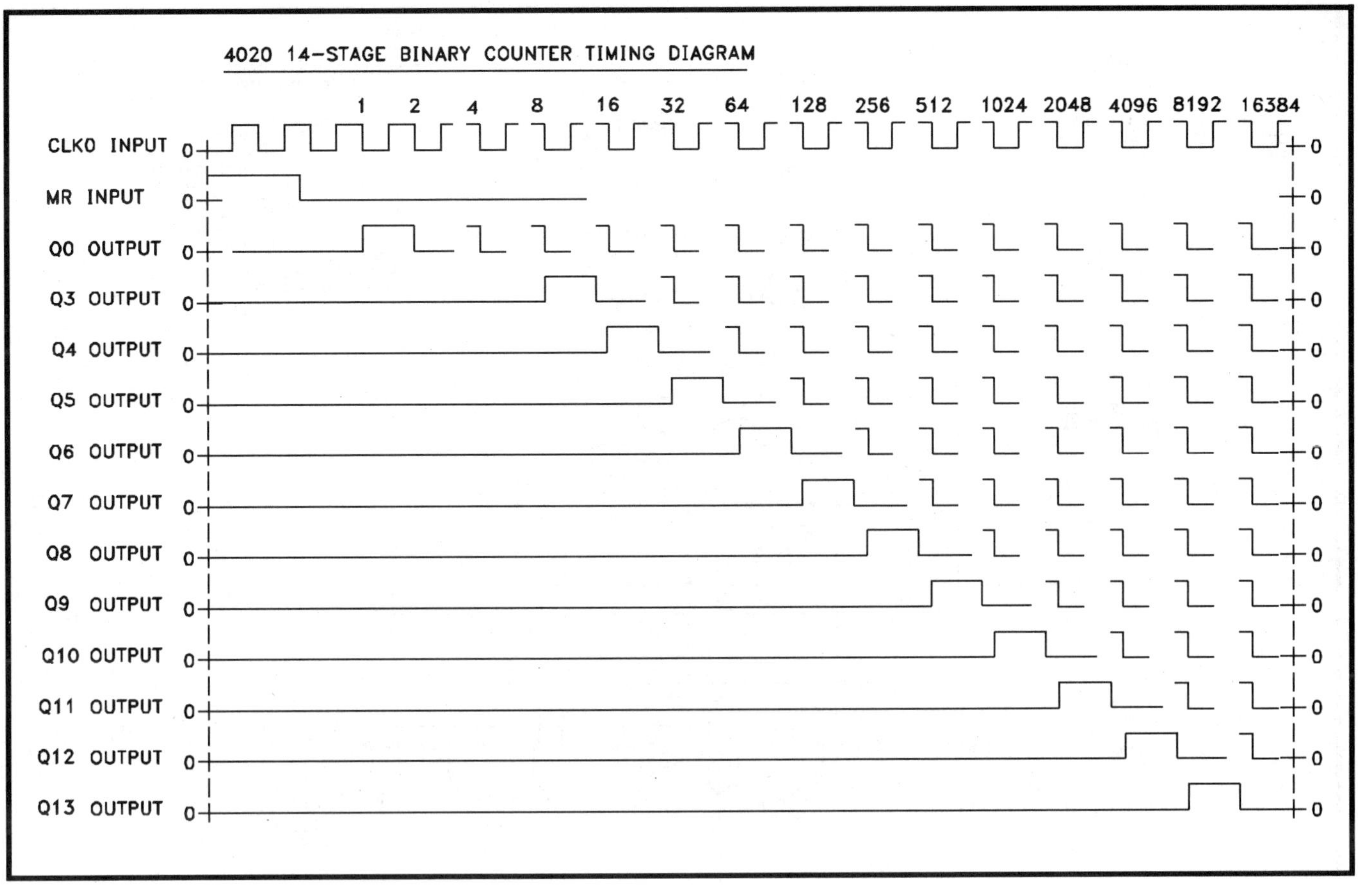

Figure 7.32. This Figure is the timing diagram for the CD4020 14-stage ripple counter. Note that each Qout is related to a specific clock pulse as shown in the clock line.

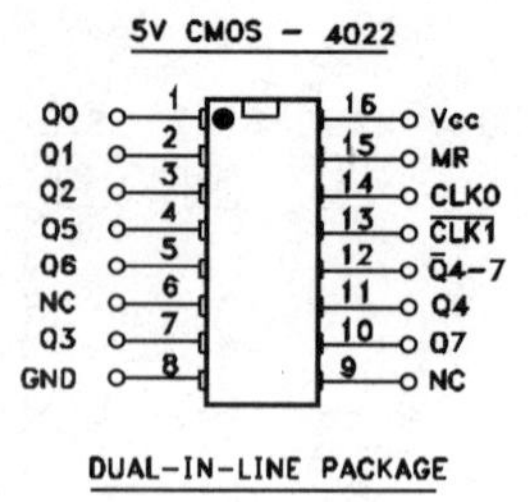

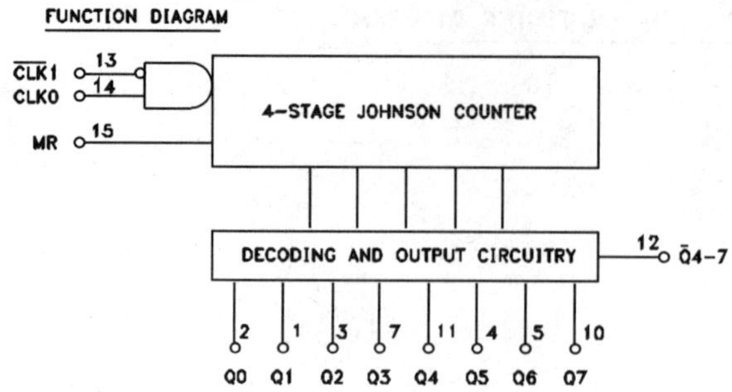

FUNCTION TABLE 4-STAGE JOHNSON COUNTER WITH 8 DECODED ACTIVE HIGH OUTPUTS

MR	CLKO	/CLK1	Operation
1	X	X	Q0=Q4-7=H; Q1-Q7=L
0	1	⌐	Counter advances
0	⌐	0	Counter advances
0	0	X	No change
0	X	1	No change
0	1	⌐	No Change
0	⌐	0	No change

H=1=HIGH voltage level
L=0=LOW voltage level

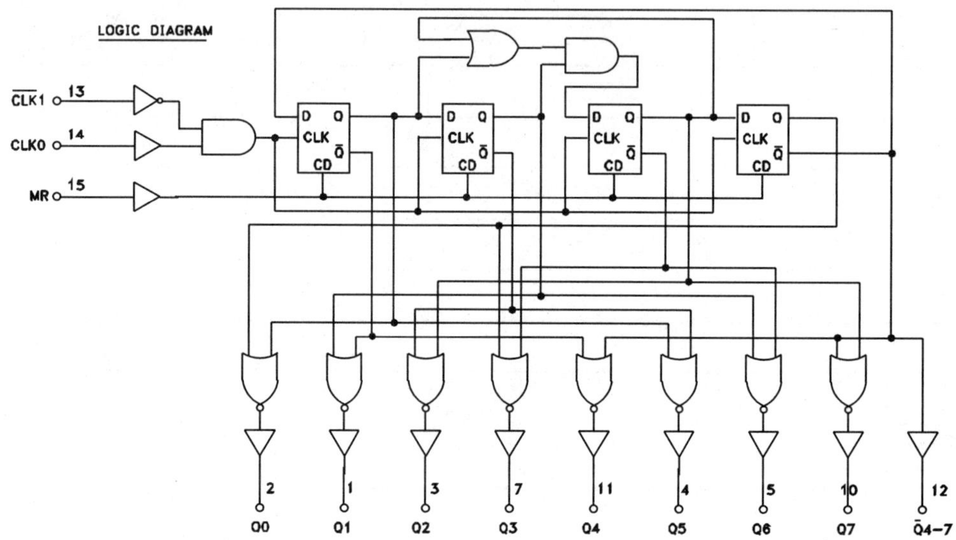

Figure 7.33. The CD4022 is a 4-stage divide-by-8 Johnson counter having eight spike-free decoded active high outputs, Q0-Q7. An active low output, /Q4-7, is obtained from the most significant flip-flop. Two clock inputs, CLKO and /CLK1, and an asynchronous master reset, MR, are the sole inputs. The counter advances on a low-to-high transition of CLKO with /CLK1 low, or a high-to-low transition on /CLK1 while CLKO is high. Either clock input may serve as the clock with the other as clock enable input. Q4-7 is used as next counter input when cascading counters. An internal code correction circuit provides automatic code correction from any illegal to a proper counting mode within eleven clock pulses. The counter outputs are similar to those shown in the timing diagram for the CD4017, *Figure 7.29*.

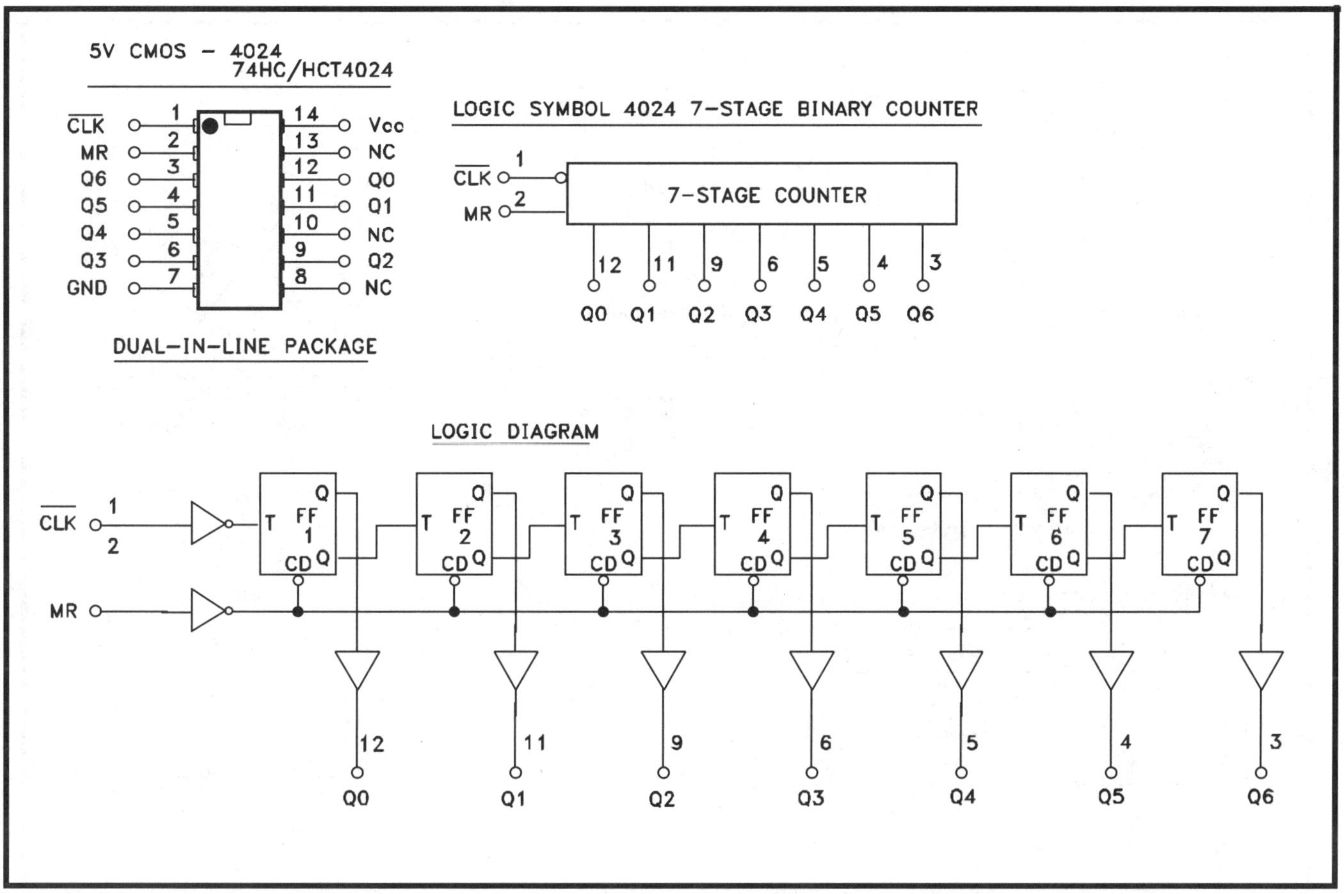

Figure 7.34. The CD4024, 74HC/HCT4024 is a 7-stage ripple counter having two inputs: a clock active with a high-to-low input transition and an active high asynchronous master reset, MR. Each counter stage is a static toggle flip-flop. All seven outputs are brought out.

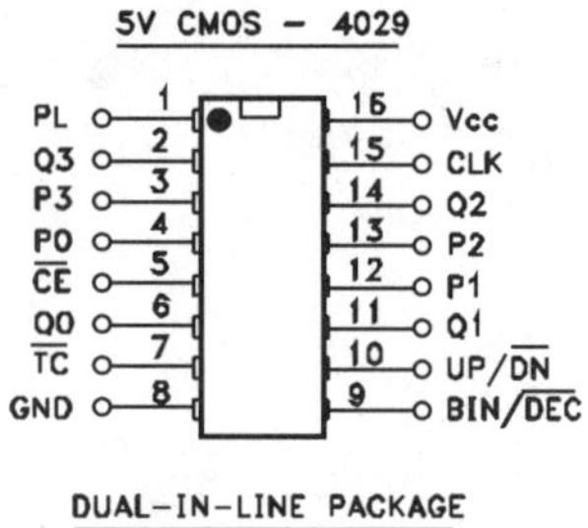

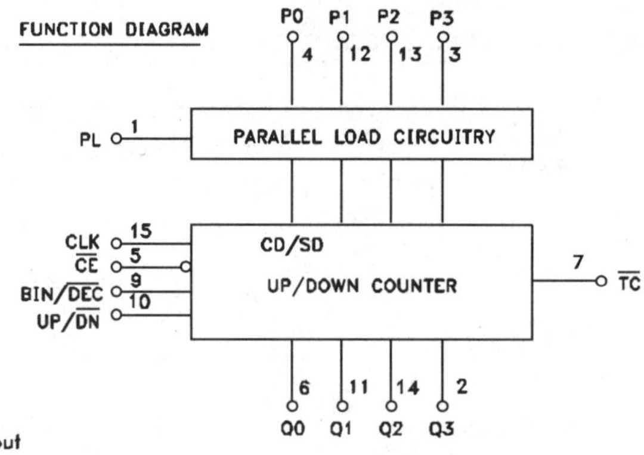

PACKAGE PINNING

PL	Parallel load input
P0–P3	Parallel data inputs
BIN/DEC	Binary/Decode control input
UP/DN	Up/Down control input
CE	Count Enable input (active LOW)
CLK	Clock input, LOW to HIGH, edge triggered
Q0–Q3	Buffered parallel outputs
TC	Terminal Count output (active LOW)

FUNCTION TABLE 4029 4-BIT BINARY/BCD SYNCHRONOUS UP/DOWN COUNTER

PL	BIN/DEC	UP/DN	CE	CLK	MODE
1	X	X	X	X	Parallel load
0	X	X	1	X	No change
0	0	0	0	ʃ	Count down, Decade
0	0	1	0	ʃ	Count up, Decade
0	1	0	0	ʃ	Count down, Binary
0	1	1	0	ʃ	Count up, Binary

H=1=HIGH voltage level
L=0=LOW voltage level
X= DON'T CARE

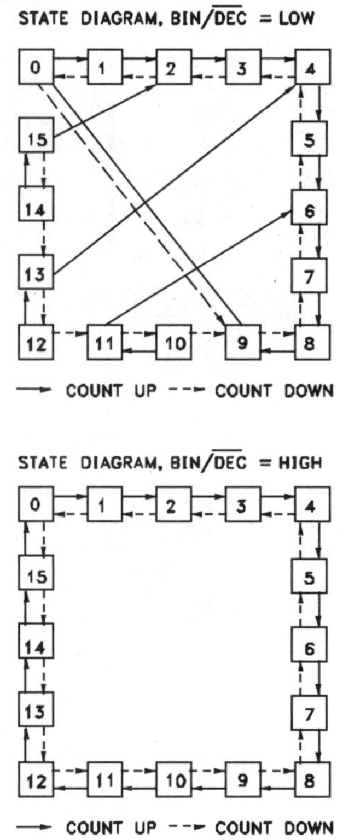

Figure 7.35. The CD4029 is a synchronous edge-triggered up/down 4-bit binary/BCD decade counter. Inputs are the clock, an active low count enable, an up/down control, a binary/decade control, an asynchronous active high parallel load, and four parallel data inputs. Four parallel data buffered outputs and an active low terminal count output are provided. Parallel input data is asynchronously loaded with the enable at high level. The counter is advanced on the low-to-high clock transition when /CE and PL are low. The /TC output is normally high, going low when the counter reaches its maximum count in the UP mode; minimum in the DOWN mode.

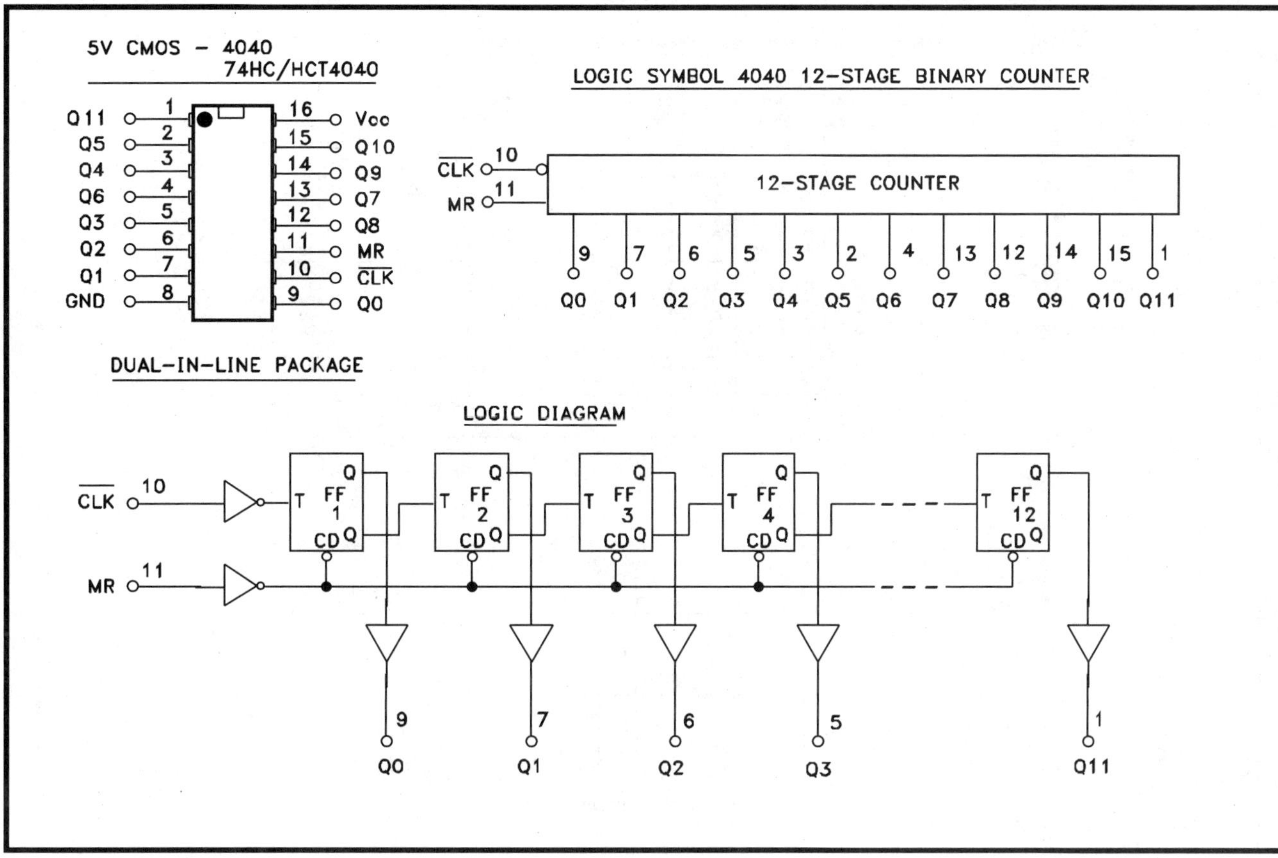

Figure 7.36. The CD4040, 74HC/HCT4040 is a 12-stage binary ripple counter having two inputs: a clock active with a high-to-low input transition and an active high asynchronous master reset, MR. Each counter stage is a static toggle flip-flop. Schmitt-trigger action in the clock input makes the circuit tolerant of slow-rise clock signals. All twelve outputs are brought out.

5V CMOS – 4059
74HC/HCT4059

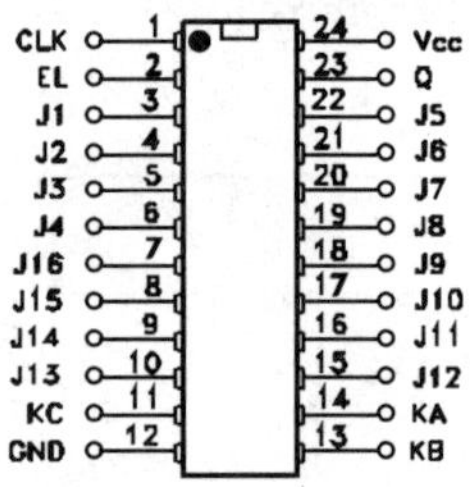

DUAL-IN-LINE PACKAGE

FUNCTION TABLE 4059 PROGRAMMABLE DIVIDE-BY-N COUNTER

Latch Enable Input	Mode Select Inputs			First counting Section Decade 1			Last counting Section Decade 5			Counter Range		
LE	KA	KB	KC	Mode	Max Preset State	Jam Inputs Used	Divide By	Max Preset State	Jam Inputs Used	BCD Max	Binary Max	Operation
1	1	1	1	2	1	J1	8	7	J2J3J4	15999	17331	Timer Mode
1	0	1	1	4	3	J1J2	4	3	J3J4	15999	18663	
1	1	0	1	5	4	J1J2J3	2	1	J4	9999	13329	
1	0	0	1	8	7	J1J2J3	2	1	J4	15999	21327	
1	1	1	0	10	9	J1J2J3J4	1	0	–	9999	16659	
0	1	1	1	2	1	J1	8	7	J2J3J4	15999	17331	Divide-by-n Mode
0	0	1	1	4	3	J1J2	4	3	J3J4	15999	18663	
0	1	0	1	5	4	J1J2J3	2	1	J4	9999	13329	
0	0	0	1	8	7	J1J2J3	2	1	J4	15999	21327	
0	1	1	0	10	9	J1J2J3J4	1	0	–	9999	16659	
1	0	1	0	10	9	J1J2J3J4	1	0	–	9999	16659	
0	0	1	0	preset inhibited			preset inhibited			fixed	–	Divide by 1000 Mode
X	X	0	0	master preset			master preset			10000 –	–	Master Preset Mode

NOTE: For LE, KA, KB, KC:
1 = HIGH voltage level; 0 = LOW voltage level; X = Don't Care

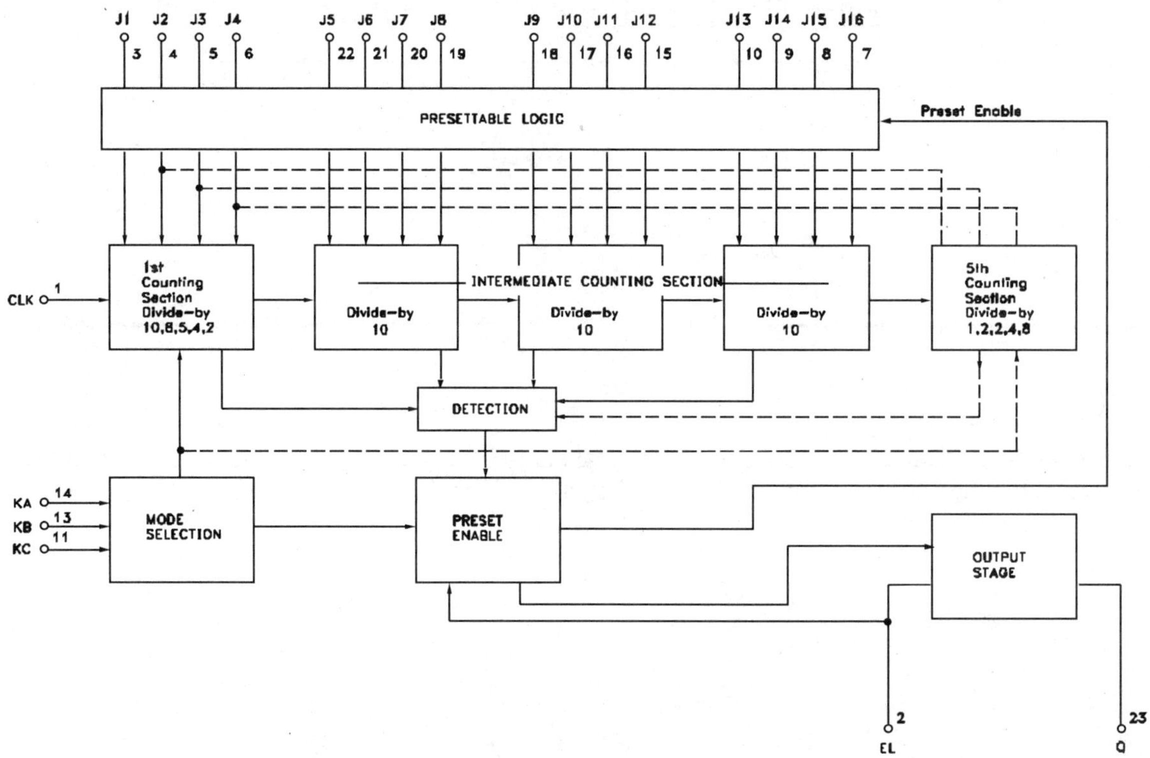

Figure 7.37. The CD4059 and 74HC/HCT4059 is a divide-by-n counter progammable for division of the input frequency by *n* over the range of 3 to 15,999. The output is a single clock cycle wide pulse occurring at a rate equal to the input frequency divided by *n*. Setup is too complex to detail here; refer to the manufacturer's data for instructions.

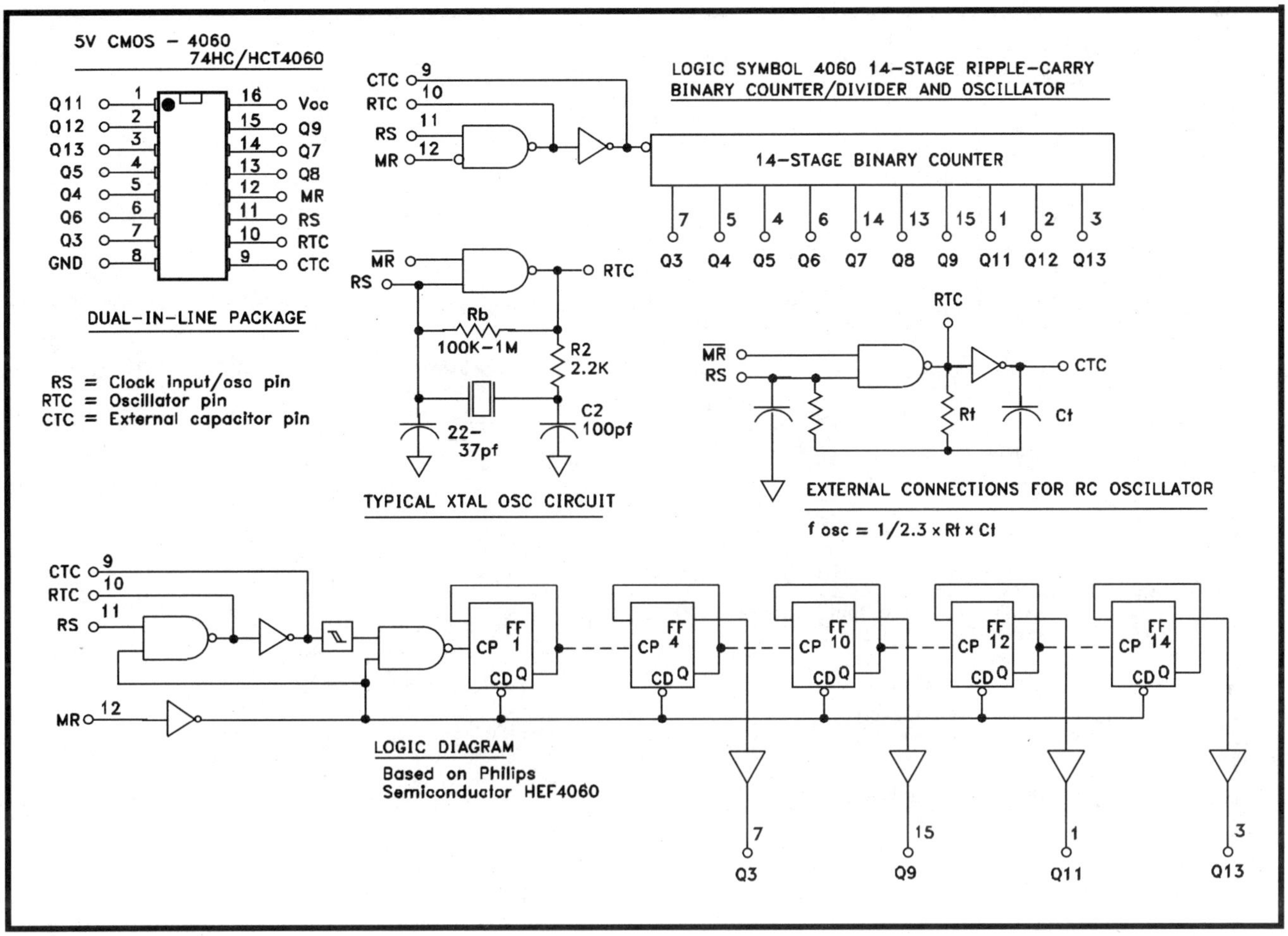

Figure 7.38. The CD4060, 74HC/HCT4060 is a 14-stage ripple-carry binary counter/ divider and oscillator. Inputs are three oscillator terminals, RS, RTC, and CTC and an active high asynchronous master reset, MR. The oscillator design provides for either RC or crystal circuitry. These may be replaced by an external oscillator to the RS input. Ten buffered outputs are provided. Schmitt-trigger action in the clock input makes the circuit tolerant of slow-rise clock signals.

5V CMOS – 4510
74HC/HCT4510

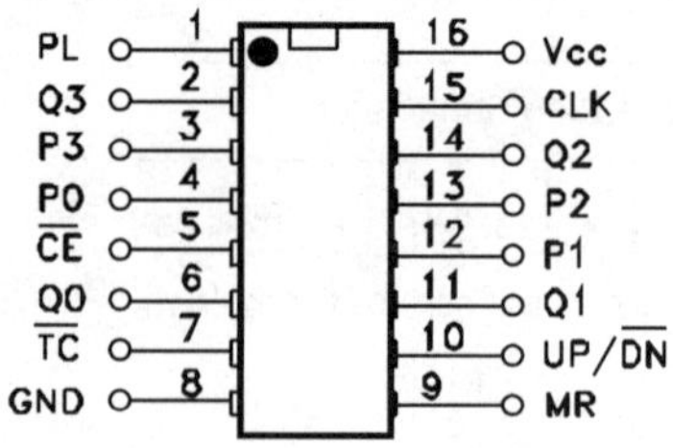

DUAL-IN-LINE PACKAGE

FUNCTION DIAGRAM

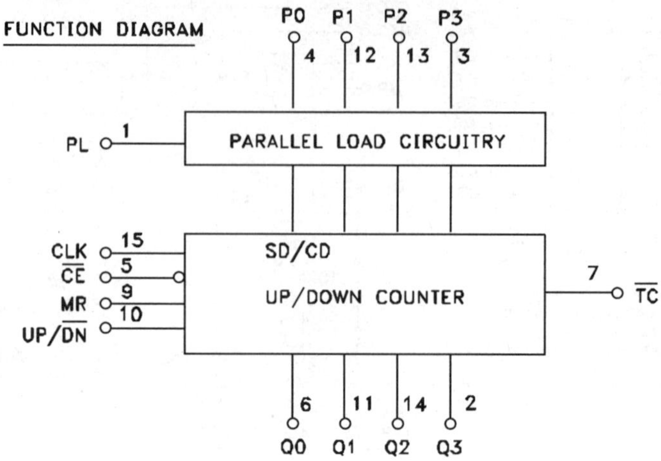

PACKAGE PINNING

PL	Parallel load input
P0–P3	Parallel data inputs
UP/DN	Up/Down control input
CE	Count Enable input (active LOW)
CLK	Clock input, LOW to HIGH, edge triggered
Q0–Q3	Buffered parallel outputs
TC	Terminal Count output (active LOW)
MR	Master Reset input

STATE DIAGRAM

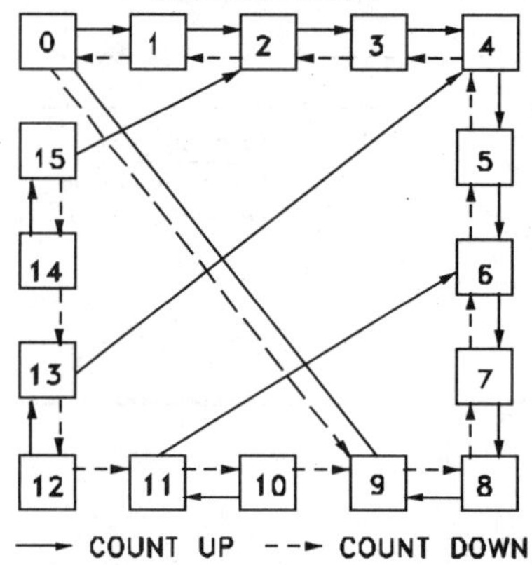

FUNCTION TABLE 4510 4–BIT BCD SYNCHRONOUS UP/DOWN COUNTER

MR	PL	UP/DN	CE	CLK	MODE
0	1	X	X	X	Parallel load
0	0	X	1	X	No change
0	0	0	0	⎍	Count down
0	0	1	0	⎍	Count Up
1	X	X	X	X	Reset

H=1=HIGH voltage level
L=0=LOW voltage level
X= DON'T CARE

Figure 7.39. The CD4510, 74HC/HCT4510 is an edge-triggered synchronous up/down BCD counter. Inputs are the clock signal, the up/down control, an active low count enable, an asynchronous high parallel load, an asynchronous high master reset, and four parallel data inputs. Four parallel outputs and an active low terminal count are provided. Data is loaded into the counter with the enable, PL, high and MR low. With PL low, the counter changes state on the low-to-high clock transition. The UP/DN input controls the count direction; high for counting up, low for down. When counting up, /TC is low when Q0 and Q3 are high. When counting down, /TC is low when Q0, Q3, and /CE are low.

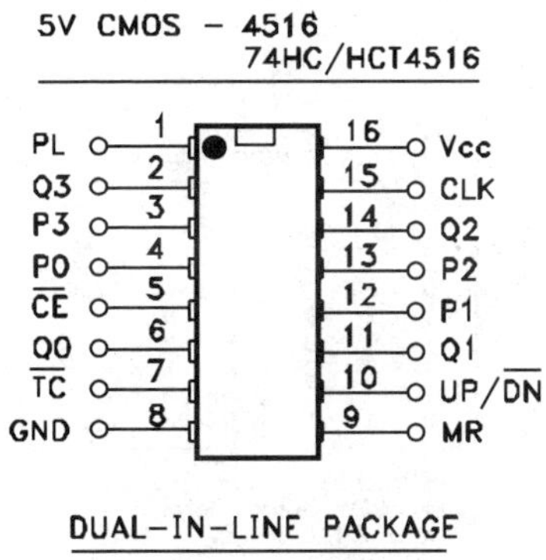

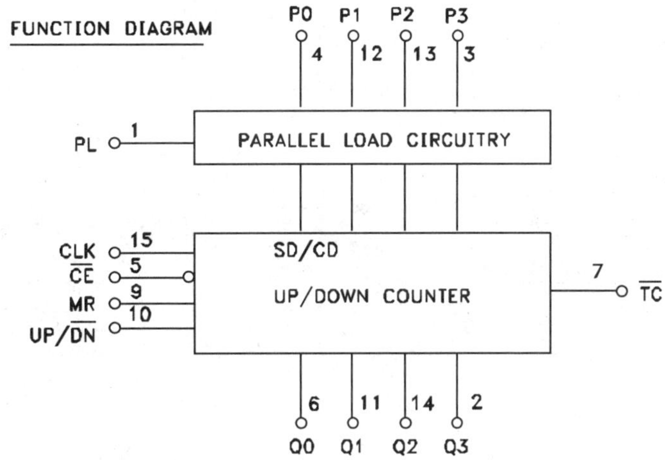

PACKAGE PINNING

PL	Parallel load input
P0–P3	Parallel data inputs
UP/DN	Up/Down control input
CE	Count Enable Input (active LOW)
CLK	Clock input, LOW to HIGH, edge triggered
Q0–Q3	Buffered parallel outputs
TC	Terminal Count output (active LOW)
MR	Master Reset input

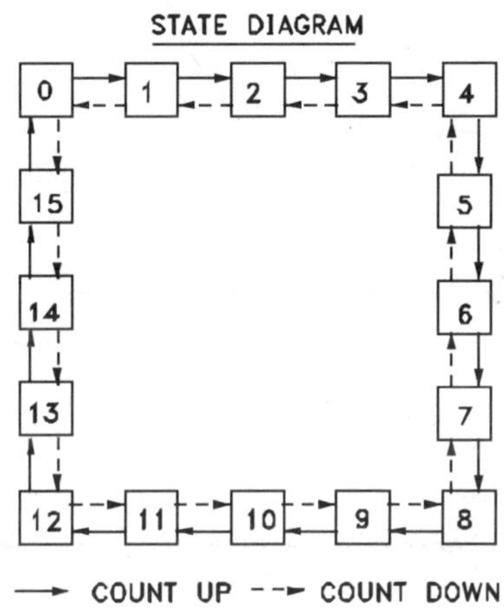

FUNCTION TABLE 4516 4-BIT BINARY SYNCHRONOUS UP/DOWN COUNTER

MR	PL	UP/DN	CE	CLK	MODE
0	1	X	X	X	Parallel load
0	0	X	1	X	No change
0	0	0	0	⨍	Count down
0	0	1	0	⨍	Count Up
1	X	X	X	X	Reset

H=1=HIGH voltage level
L=0=LOW voltage level
X= DON'T CARE

Figure 7.40. The CD4516, 74HC/HCT4516 is an edge-triggered synchronous up/down binary counter. Inputs are the clock signal, the up/down control, an active low count enable, an asynchronous high parallel load, an asynchronous high master reset, and four parallel data inputs. Four parallel outputs and an active low terminal count are provided. Data is loaded into the counter with the enable, PL, high and MR low. With PL low, the counter changes state on the low-to-high clock transition. The UP/DN input controls the count direction; high for counting up, low for down. When counting up, /TC is low when Q0 and Q3 are high. When counting down, /TC is low when Q0, Q3, and /CE are low.

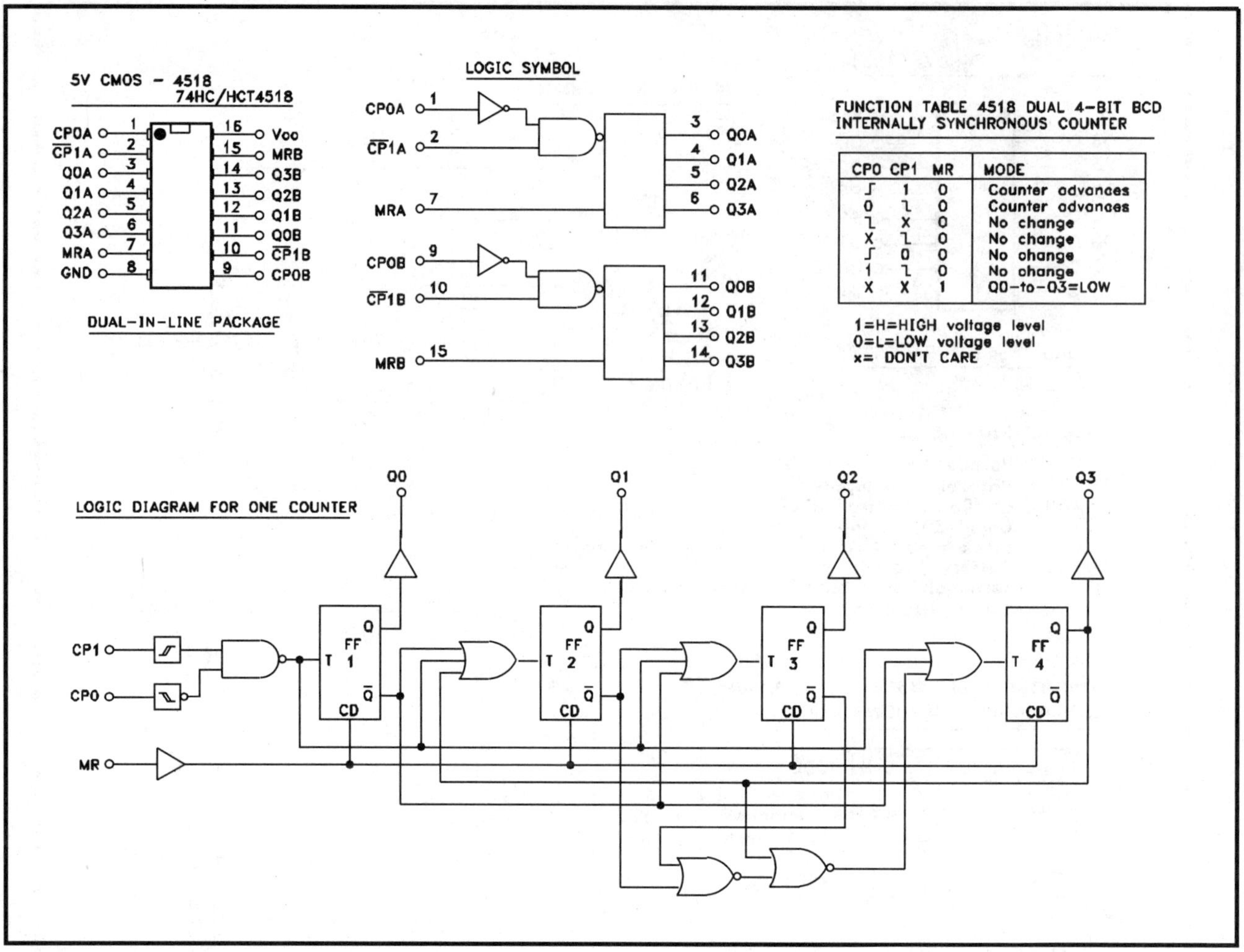

Figure 7.41. The CD4518, 74HC/HCT4518 is a dual 4-bit synchronous BCD counter. Inputs are an active high clock, CP0, an active low clock, /CP1, and an active high asynchronous master reset, MR. The counter advances on the low-to-high transition of CP0 with /CP1 high or a high-to-low transition of /CP1 with CP0 low. Either CP0 or /CP1 may be used as the clock input with the other functioning as a clock enable. Schmitt-trigger action in the clock input makes the circuit tolerant of slow-rise clock signals.

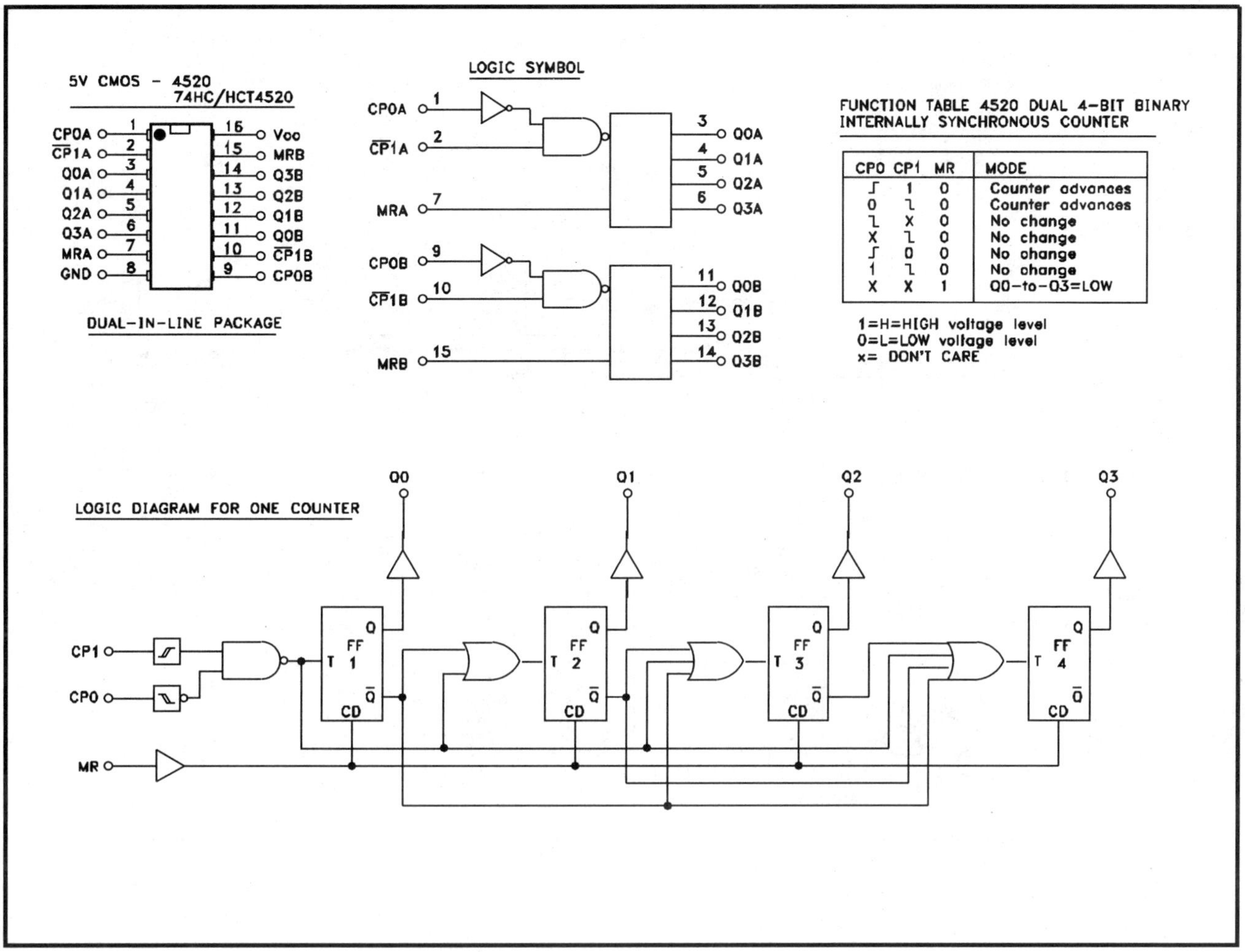

Figure 7.42. The CD4520, 74HC/HCT4520 is a dual 4-bit synchronous binary counter. Inputs are an active high clock, CP0, an active low clock, /CP1, and an active high asynchronous master reset, MR. The counter advances on the low-to-high transition of CP0 with /CP1 high or a high-to-low transition of /CP1 with CP0 low. Either CP0 or /CP1 may be used as the clock input with the other functioning as a clock enable. Schmitt-trigger action in the clock input makes the circuit tolerant of slow-rise clock signals.

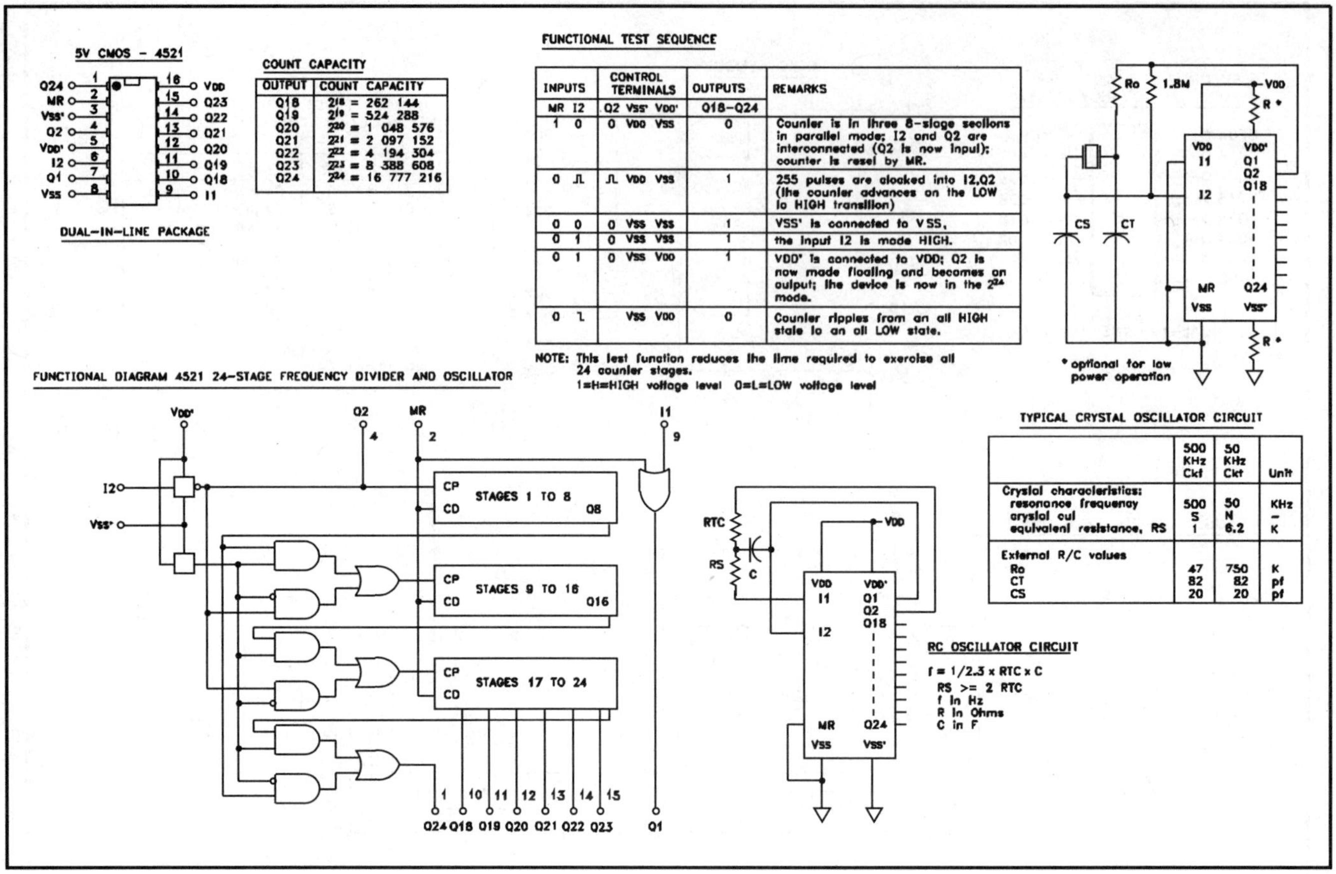

COUNT CAPACITY

OUTPUT	COUNT CAPACITY
Q18	2^{18} = 262 144
Q19	2^{19} = 524 288
Q20	2^{20} = 1 048 576
Q21	2^{21} = 2 097 152
Q22	2^{22} = 4 194 304
Q23	2^{23} = 8 388 608
Q24	2^{24} = 16 777 216

FUNCTIONAL TEST SEQUENCE

INPUTS	CONTROL TERMINALS	OUTPUTS	REMARKS
MR I2	Q2 VSS' VDD'	Q18–Q24	
1 0	0 VDD VSS	0	Counter is in three 8-stage sections in parallel mode; I2 and Q2 are interconnected (Q2 is now input); counter is reset by MR.
0 ⎍	⎍ VDD VSS	1	255 pulses are clocked into I2,Q2 (the counter advances on the LOW to HIGH transition)
0 0	0 VSS VSS	1	VSS' is connected to VSS,
0 1	0 VSS VSS	1	the input I2 is made HIGH.
0 1	0 VSS VDD	1	VDD' is connected to VDD; Q2 is now made floating and becomes an output; the device is now in the 2^{24} mode.
0 ⌉	VSS VDD	0	Counter ripples from an all HIGH state to an all LOW state.

NOTE: This test function reduces the time required to exercise all 24 counter stages.
1=H=HIGH voltage level 0=L=LOW voltage level

	500 KHz Ckt	50 KHz Ckt	Unit
Crystal characteristics:			
resonance frequency	500	50	KHz
crystal cut	S	N	–
equivalent resistance, RS	1	6.2	K
External R/C values			
Ro	47	750	K
CT	82	82	pf
CS	20	20	pf

Figure 7.43. The CD4521 is a chain of 24 toggle flip-flops with an overriding asynchronous master reset, MR. The input circuit provides for three modes of operation. A single inverting stage, I2/Q2, will function as a crystal oscillator, or in combination with I1 as an RC oscillator, or as input buffer for an external oscillator. Low power operation as a crystal oscillator is enabled by connecting external resistors to pins 3 and 5. The device will count up to 2 to the 34th power, 16,777,216. The count advances on the high-to-low clock transition. The outputs of the final seven stages are made available.

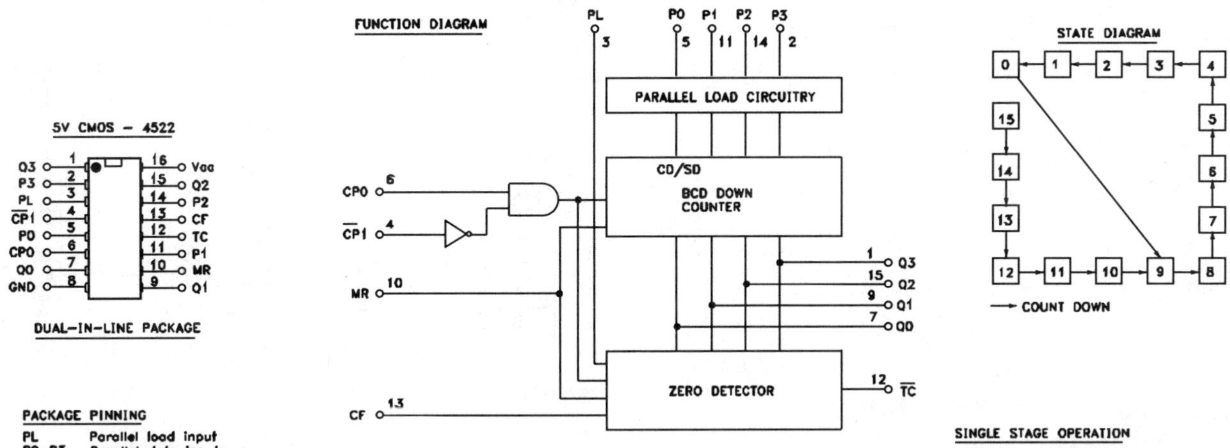

PACKAGE PINNING

PL	Parallel load input
P0–P3	Parallel data inputs
CF	Cascade Feedback input
CP0	Clock input, LOW–to–HIGH, triggered
$\overline{CP1}$	Clock input, HIGH–to–LOW, triggered
Q0–Q3	Buffered parallel outputs
TC	Terminal Count output
MR	Asynchronous Master Reset

FUNCTION TABLE 4522 PROGRAMMABLE 4-BIT BCD DOWN COUNTER

MR	PL	CP0	$\overline{CP1}$	MODE
1	X	X	X	Asynchronous reset
0	1	X	X	Asynchronous preset
0	0	⎍	1	No change
0	0	0	⎍	No change
0	0	⎍	X	No change
0	0	X	⎍	No change
0	0	⎍	0	Counter advances
0	0	1	⎍	Counter Advances

H=1=HIGH voltage level
L=0=LOW voltage level
X= DON'T CARE

COUNTING MODE:
CF=HIGH;
PL=LOW;
MR=LOW

COUNT	OUTPUTS Q3	Q2	Q1	Q0
9	1	0	0	1
8	1	0	0	0
7	0	1	1	1
6	0	1	1	0
5	0	1	0	1
4	0	1	0	0
3	0	0	1	1
2	0	0	1	0
1	0	0	0	1
0	0	0	0	0

SINGLE STAGE OPERATION
Divide-by n; MR=LOW;
CF=HIGH; $\overline{CP1}$=LOW

PL	P3	P2	P1	P0	Divide by	TC Output Pulse Width
0	X	X	X	X	10	One clock period
TC	1	1	1	1	15	Clock Pulse HIGH
TC	1	1	1	0	14	
TC	1	1	0	1	13	
TC	1	1	0	0	12	
TC	1	0	1	1	11	
TC	1	0	1	0	10	
TC	1	0	0	1	9	
TC	1	0	0	0	8	
TC	0	1	1	1	7	
TC	0	1	1	0	6	
TC	0	1	0	1	5	
TC	0	1	0	0	4	
TC	0	0	1	1	3	
TC	0	0	1	0	2	
TC	0	0	0	1	1	
TC	0	0	0	0	No Operation	

Figure 7.44. The CD4522 is a synchronous programmable 4-bit BCD down counter. Inputs are an active high clock, CP0, an active low clock, /CP1, an asynchronous parallel load, PL, four parallel data inputs, a cascade feedback, CF, and an overriding master reset, MR. Four buffered outputs and a terminal count, TC, are provided. The device is a cascadable programmable down counter with a decoded TC output for divide-by-n usage. In single stage applications the TC output is tied to PL. CF allows for divide-by-n operation with no external gates required. The four data inputs are loaded into the counter with PL high and MR low. When PL and /CP1 are low, the counter advances on low-to-high CP0 transition. With Pl low and CP0 high, the counter advances on high-to-low transition of CP1. Schmitt-trigger action in the clock input makes the circuit tolerant of slow-rise clock signals.

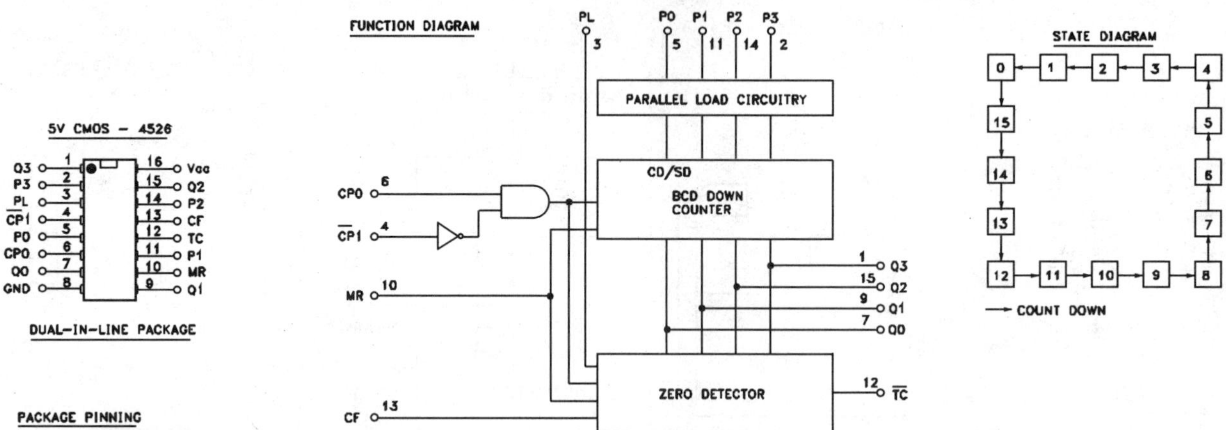

PACKAGE PINNING

PL	Parallel load input
P0–P3	Parallel data inputs
CF	Cascade Feedback input
CP0	Clock input, LOW-to-HIGH, triggered
$\overline{CP1}$	Clock input, HIGH-to-LOW, triggered
Q0–Q3	Buffered parallel outputs
TC	Terminal Count output
MR	Asynchronous Master Reset

FUNCTION TABLE 4526 PROGRAMMABLE 4-BIT BINARY DOWN COUNTER

MR	PL	CP0	$\overline{CP1}$	MODE
1	X	X	X	Asynchronous reset
0	1	X	X	Asynchronous preset
0	0	ʃ	1	No change
0	0	0	˥	No change
0	0	˥	X	No change
0	0	X	ʃ	No change
0	0	ʃ	0	Counter advances
0	0	1	˥	Counter Advances

H=1=HIGH voltage level
L=0=LOW voltage level
X= DON'T CARE

COUNTING MODE:
CF=HIGH;
PL=LOW;
MR=LOW

COUNT	OUTPUTS Q3	Q2	Q1	Q0
15	1	1	1	1
14	1	1	1	0
13	1	1	0	1
12	1	1	0	0
11	1	0	1	1
10	1	0	1	0
9	1	0	0	1
8	1	0	0	0
7	0	1	1	1
6	0	1	1	0
5	0	1	0	1
4	0	1	0	0
3	0	0	1	1
2	0	0	1	0
1	0	0	0	1
0	0	0	0	0

SINGLE STAGE OPERATION
Divide-by n; MR=LOW;
CF=HIGH; $\overline{CP1}$=LOW

PL	P3	P2	P1	P0	Divide by	TC Output Pulse Width
0	X	X	X	X	16	One clock period
TC	1	1	1	1	15	Clock Pulse HIGH
TC	1	1	1	0	14	
TC	1	1	0	1	13	
TC	1	1	0	0	12	
TC	1	0	1	1	11	
TC	1	0	1	0	10	
TC	1	0	0	1	9	
TC	1	0	0	0	8	
TC	0	1	1	1	7	
TC	0	1	1	0	6	
TC	0	1	0	1	5	
TC	0	1	0	0	4	
TC	0	0	1	1	3	
TC	0	0	1	0	2	
TC	0	0	0	1	1	
TC	0	0	0	0	No Operation	

Figure 7.45. The CD4526 is a synchronous programmable 4-bit binary down counter. Inputs are an active high clock, CP0, an active low clock, /CP1, an asynchronous parallel load, PL, four parallel data inputs, a cascade feedback, CF, and an overriding master reset, MR. Four buffered outputs and a terminal count, TC, are provided. The device is a cascadable programmable down counter with a decoded TC output for divide-by-n usage. In single-stage applications the TC output is tied to PL. CF allows for divide-by-n operation with no external gates required. The four data inputs are loaded into the counter with PL high and MR low. When PL and /CP1 are low, the counter advances on low-to-high CP0 transition. With Pl low and CP0 high, the counter advances on high-to-low transition of CP1. Schmitt-trigger action in the clock input makes the circuit tolerant of slow-rise clock signals.

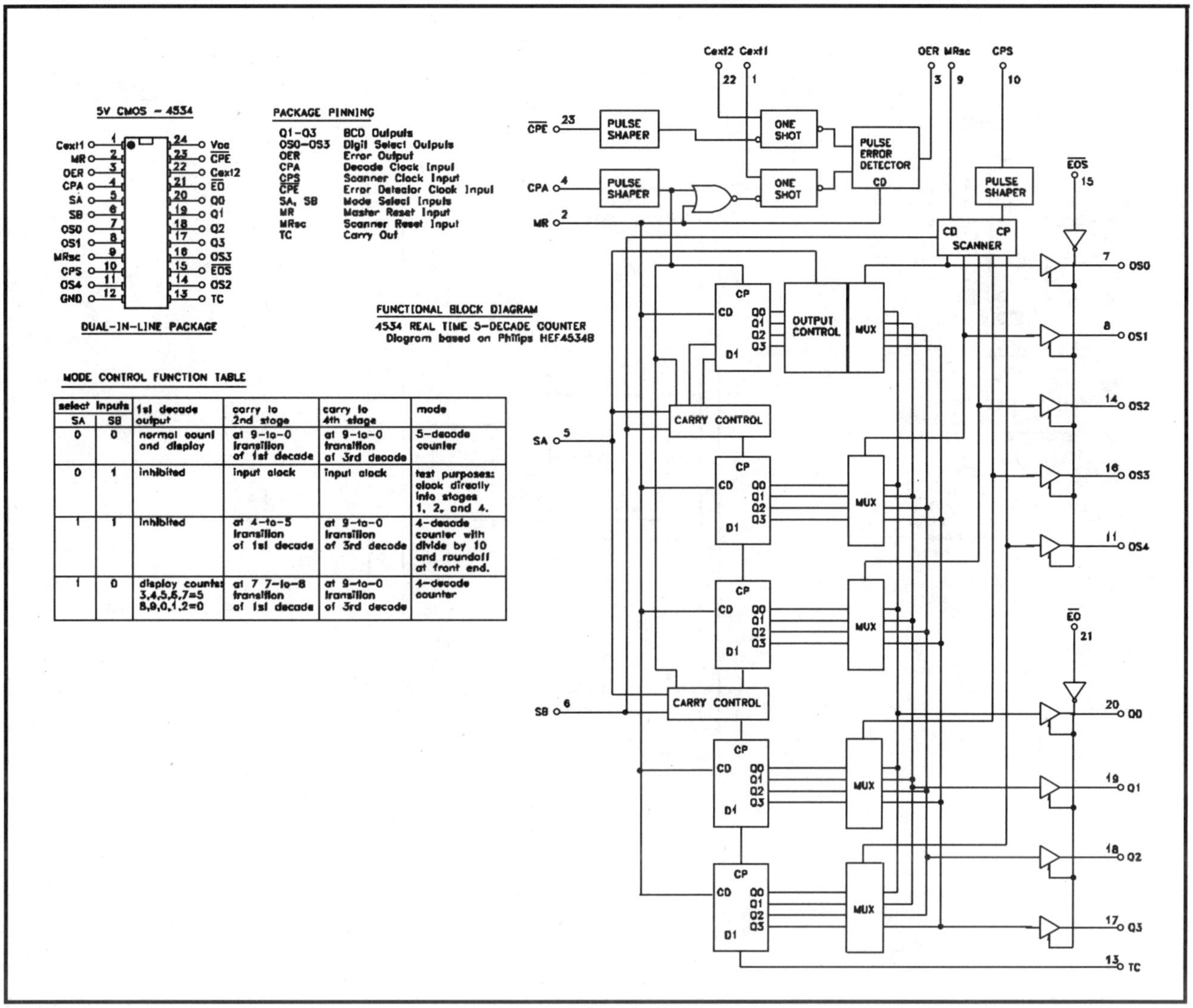

MODE CONTROL FUNCTION TABLE

select inputs SA	select inputs SB	1st decade output	carry to 2nd stage	carry to 4th stage	mode
0	0	normal count and display	at 9-to-0 transition of 1st decade	at 9-to-0 transition of 3rd decade	5-decade counter
0	1	inhibited	input clock	input clock	test purposes: clock directly into stages 1, 2, and 4.
1	1	inhibited	at 4-to-5 transition of 1st decade	at 9-to-0 transition of 3rd decade	4-decade counter with divide by 10 and roundoff at front end.
1	0	display counts 3,4,5,6,7=5 8,9,0,1,2=0	at 7 7-to-8 transition of 1st decade	at 9-to-0 transition of 3rd decade	4-decade counter

Figure 7.46. The CD4534 is a 5-decade BCD ripple counter. The binary outputs are time-multiplexed by an internal scanner on the four BCD outputs Q0-Q3. The selected decade is identified by a logic 1 on the appropriate digit select output. The binary and select outputs are 3-state controlled via enable inputs /EO and /EOS, respectively. Cascading is enabled by the carry out, TC. Triggering is by a low-to-high clock CPA transition. The scanner is triggered by a low-to-high clock CPS transition. Four counter modes are selected by inputs SA and Sb. Schmitt-trigger action in the clock input makes the circuit tolerant of slow-rise clock signals.

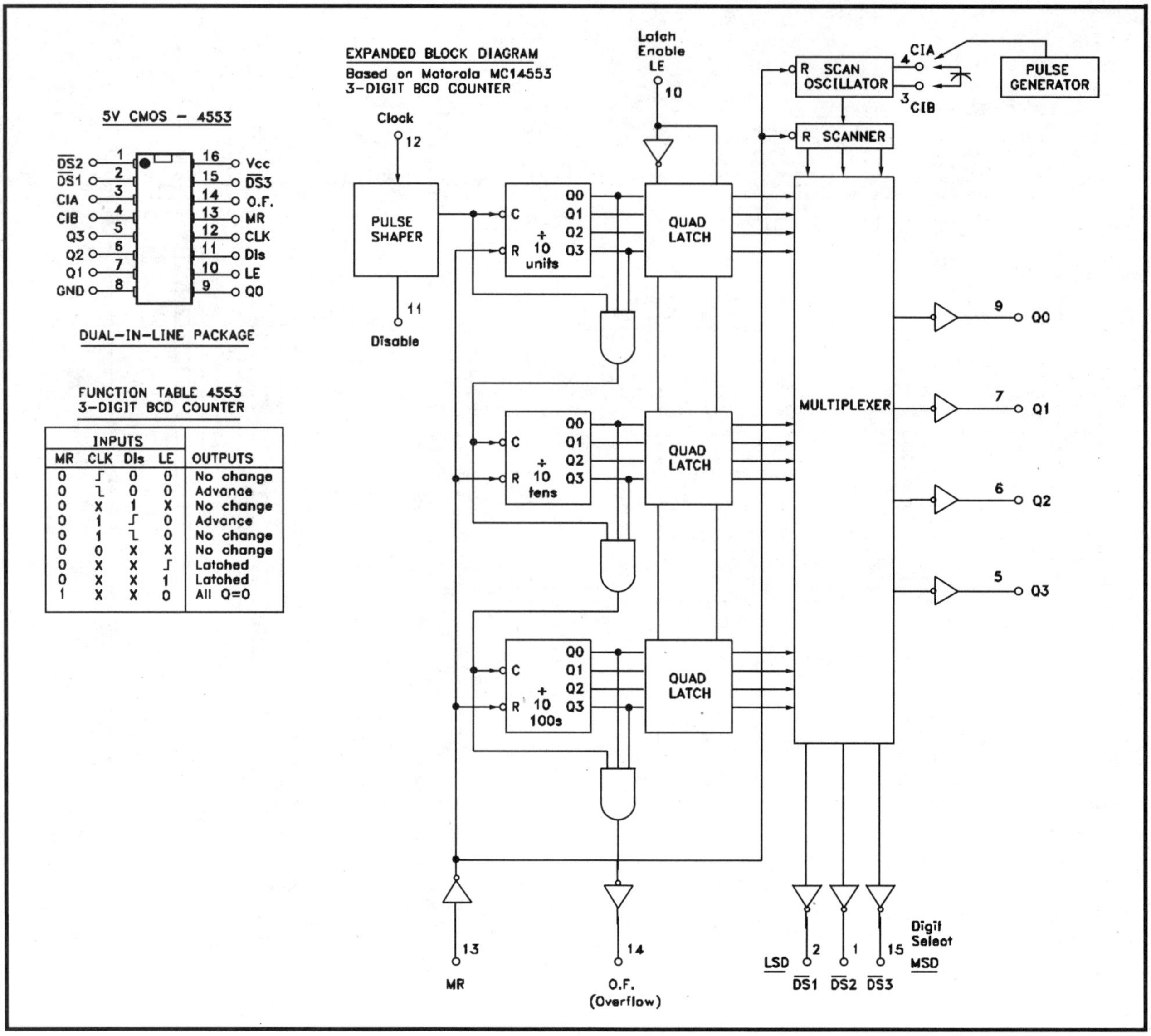

INPUTS				
MR	CLK	Dis	LE	OUTPUTS
0	⌐	0	0	No change
0	⌙	0	0	Advance
0	X	1	X	No change
0	1	⌐	0	Advance
0	1	⌙	0	No change
0	0	X	X	No change
0	X	X	⌐	Latched
0	X	X	1	Latched
1	X	X	0	All Q=0

Figure 7.47. The CD4534 is a 3-digit counter consisting of three negative-edge triggered BCD counters synchronously cascaded. A quad latch at the output of each provides for storage of any given count. The three outputs, after going through the latch, are multiplexed to provide one number or digit at a time. Active low digit selects provide display control. All outputs are TTL compatible. An on-chip oscillator functions as a slow-scan clock for the multiplexer. The frequency is controllable by an external capacitor or by an external oscillator. Cascading is by the overflow which provides one pulse for every 1000 counts. When taken high, the master reset initializes the counter and the scanning circuit. The disable input, when high, inhibits the counter while retaining the last count.

5V CMOS – 4568

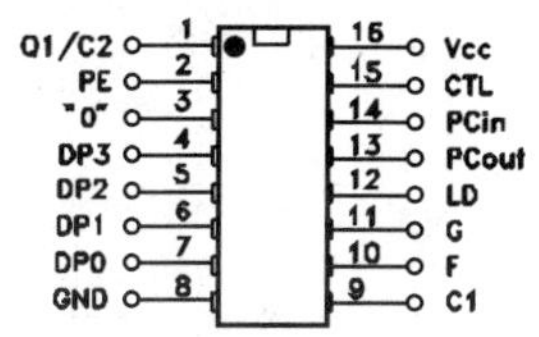

DUAL-IN-LINE PACKAGE

BLOCK DIAGRAM 4568
Diagram based on Motorola MC14568

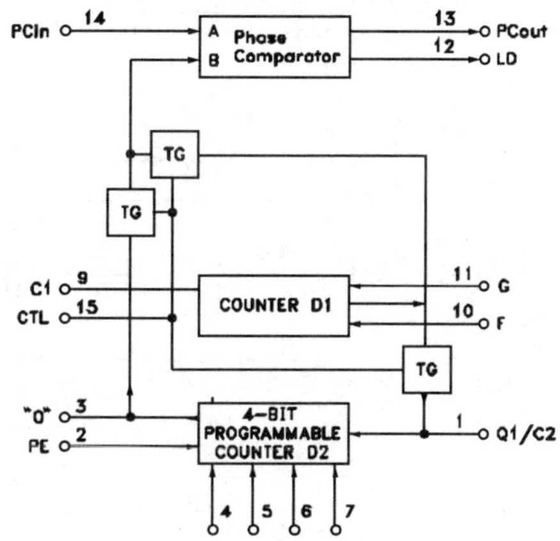

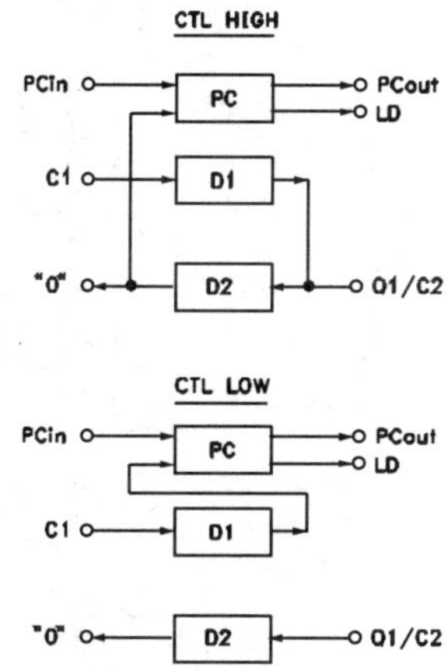

TRUTH TABLE 4568
PHASE COMPARATOR AND
PROGRAMMABLE COUNTER

F Pin 10	G Pin 11	Division Ratio of Counter D1
0	0	4
0	1	16
1	0	64
1	1	100

The divide by zero state on the programmable divide-by-N 4-bit binary counter, D2, is illegal.

5V CMOS – 4569

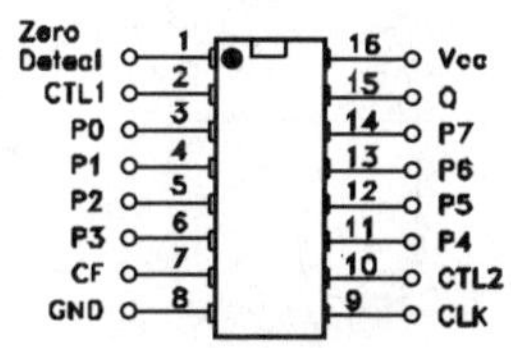

DUAL-IN-LINE PACKAGE

BLOCK DIAGRAM
Based on Motorola MC14569

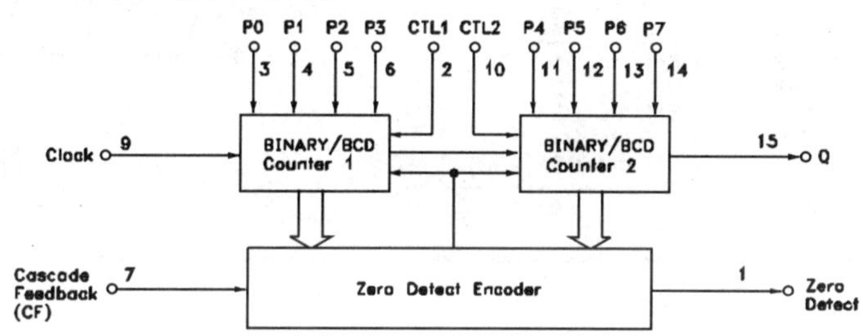

FUNCTION TABLE – MODE CONTROLS
4569 PROGRAMMABLE DIVIDE-BY-N
DUAL 4-BIT BINARY/BCD DOWN
COUNTER

Cascade Feedback, CF, = LOW

Ctr Ctrl Values		Divide Ratio	
CTL1	CTL2	Zero Detect	Q
0	0	256	256
0	1	160	160
1	0	160	160
1	1	100	100

NOTE: DATA for CF HIGH too much to include. Refer to manufacturer's data book.

TIMING DIAGRAM MC14569

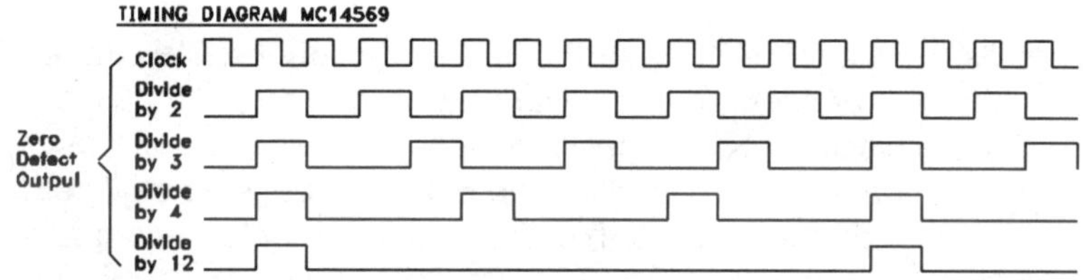

Figure 7.48. The MC14568 is comprised of a phase comparator, a divide-by-4, -16, -6 or -100 counter, and a programmable divide-by-n 4-bit counter. Triggering occurs on the positive clock edge. The device was designed for frequency synthesizer and phase-locked loop applications. The MC14569 is a programmable divide-by-n dual 4-bit binary or BCD down counter designed for use with the MC14568. Each counter can divide independently in binary or BCD. It may be cascaded with the 4568, 4522, or 4526 for frequency synthesizer applications. The clock triggering is Schmitt conditioned.

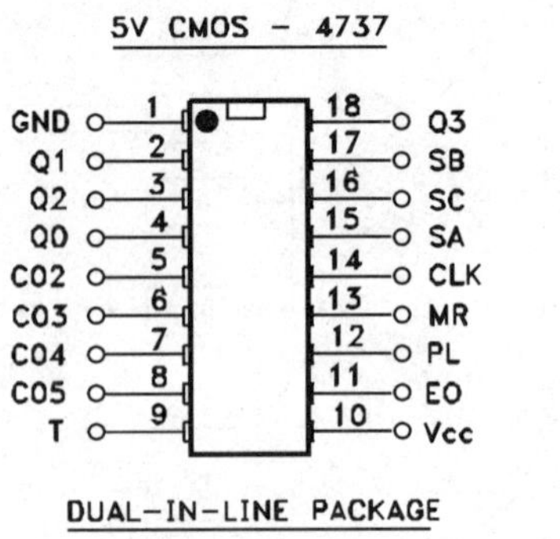

PACKAGE PINNING:

CLK	Count input
MR	Asynchronous reset input
PL	Asynchronous preset input
T	Transfer input
SA,SB,SC	Digit select inputs
EO	Output enable input
Q0–Q3	BCD outputs
CO2–CO5	Carry outputs

DIGIT SELECT INPUTS SA, SB, SC

SA	SB	SC	
0	0	0	Selects D1 (LSD)
1	0	0	Selects D2
0	1	0	Selects D3
1	1	0	Selects D4
X	X	1	Selects D5 (MSD)

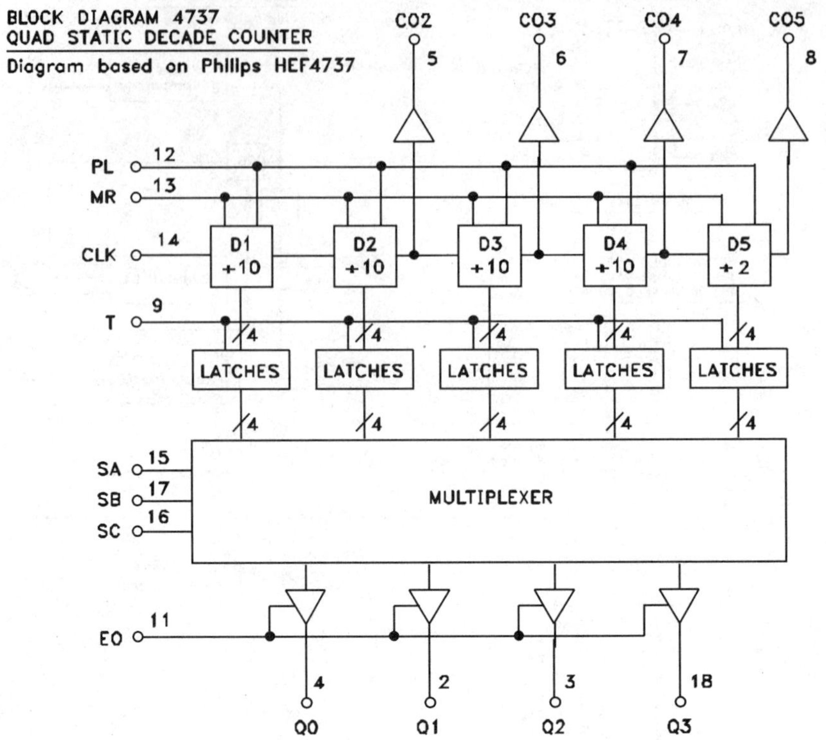

Figure 7.49. The HEF4737 quadruple decade counter is designed for frequencies of 0 to 10 MHz. The counters are provided with an overload flip-flop providing a total count capability of 19,999. Inputs are a clock, CLK; transfer, T; asynchronous reset, MR; asynchronous preset, PL; output enable, EO, controlling the BCD outputs; and digit select inputs, SA-SC, which perform selection of the latch content to the outputs. Outputs Q0-Q3 are 3-state. Carry outputs CO2-CO5 provide carry signals from all but the first decade. Schmitt triggering is provided in the inputs.

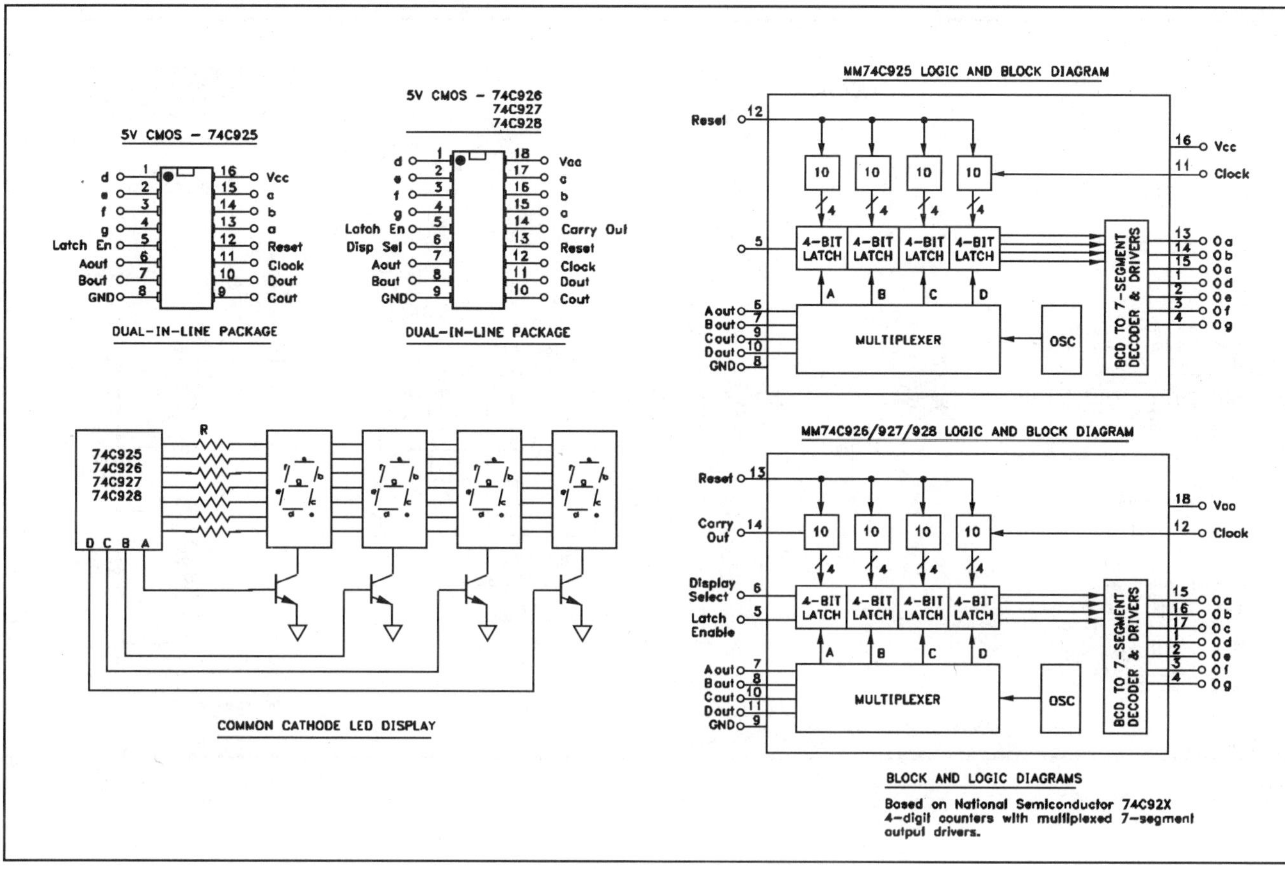

Figure 7.50. The 74C925, '926, '927, '928 are 4-digit counters with multiplexed 7-segment output drivers. Other features include an internal output latch, NPN output drivers for the display, and internal multiplexing for the drivers. The multiplexer circuit provides its own free-running oscillator: no external clock is required. The counters advance on the negative edge of the clock input. The asynchronous reset is active high. The 74C925 is a 4-decade counter with latch enable, clock and reset inputs. The 74C926 is similar but includes a display select and carry out for cascading. The carry out goes high at 6000; low at 0000. The 74C927 is similar to the '926 except the second most significant digit divides by 6 rather than 10. Given a 10 Hz clock, the display is in tenths of seconds and minutes. The 74C928 is similar to the '927 except the most significant digit divides by 2 rather than ten. The carry out is an overflow indicator—high at 2000, returning to zero only when the counter is reset; functions as a 3 1/2-digit counter.

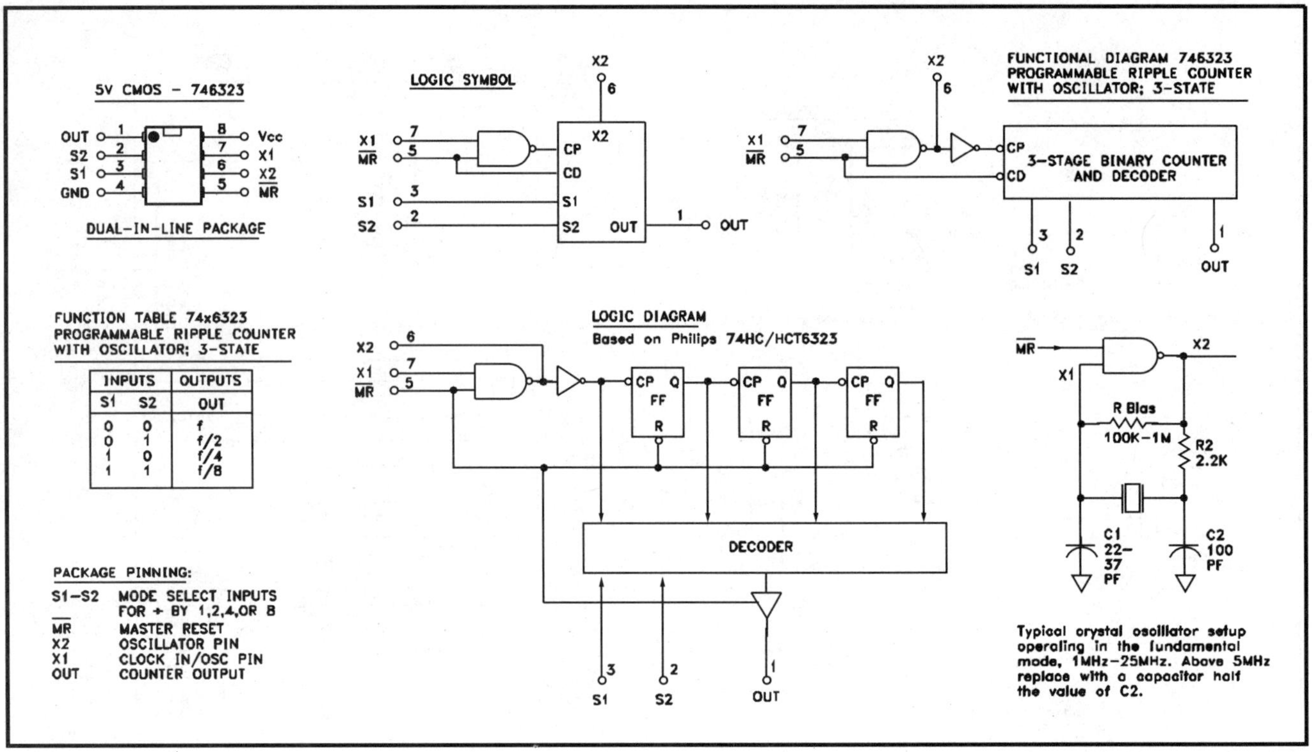

Figure 7.51. The 74HC/HCT6323 is a programmable 3-stage ripple counter with an oscillator function. It is suitable for over-tone crystal application up to 50 MHz with a five-volt Vcc. Two internal capacitors are included. The asynchronous reset is active low. It resets the counter, stops the oscillator, and sets the output buffer to 3-state mode. The two select inputs enable four modes: divide-by-1, -2, -4, or -8. If floating, divide is by 8. Operation is in the fundamental or overtone mode depending on the crystal and external components. The oscillator may be replaced by an external clock signal at input X1; input X2 must then be left floating.

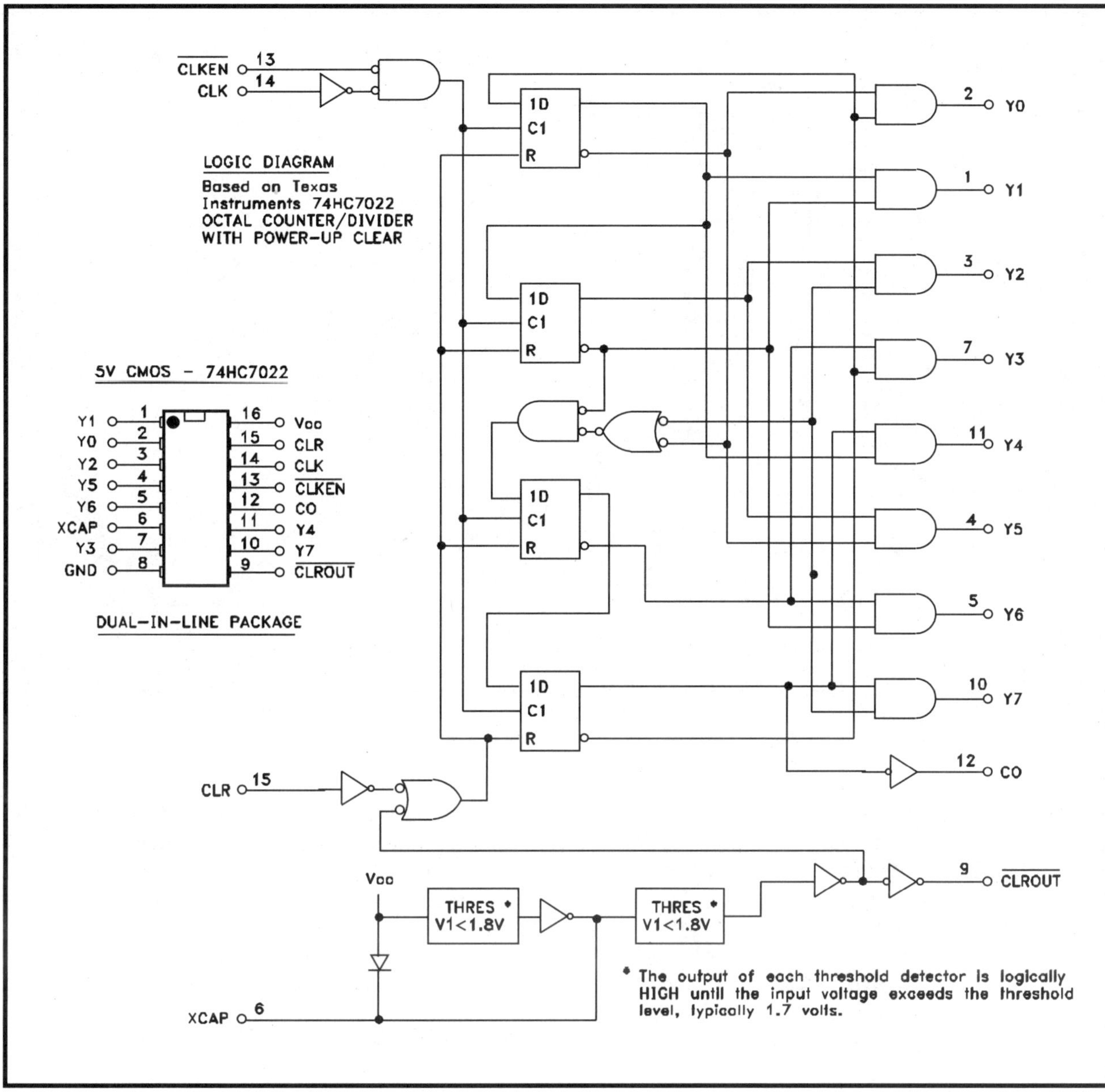

Figure 7.52. The 74HC7022 is a 4-stage divide-by-8 Johnson counter with 8 decoded outputs and a carry-out bit. The decoder outputs are normally low, going high only at their respective octal time periods. A high signal on the CLR input asynchronously clears the counter and sets the carry output and output Y0 high. With /CLKEN low, the count advances on a low-to-high transition of CLK. Alternatively, if CLK is high, the count advances on the high-to-low transition of /CLKEN. Each decoded output remains high for one clock cycle. Carry out CO is high while any of Y0-Y3 are high; then low while any of Y4-Y7 are high. A power-up circuit resets the counter on powering up. The /CLROUT is an active-low open drain output which can be used to clear or reset external circuitry.

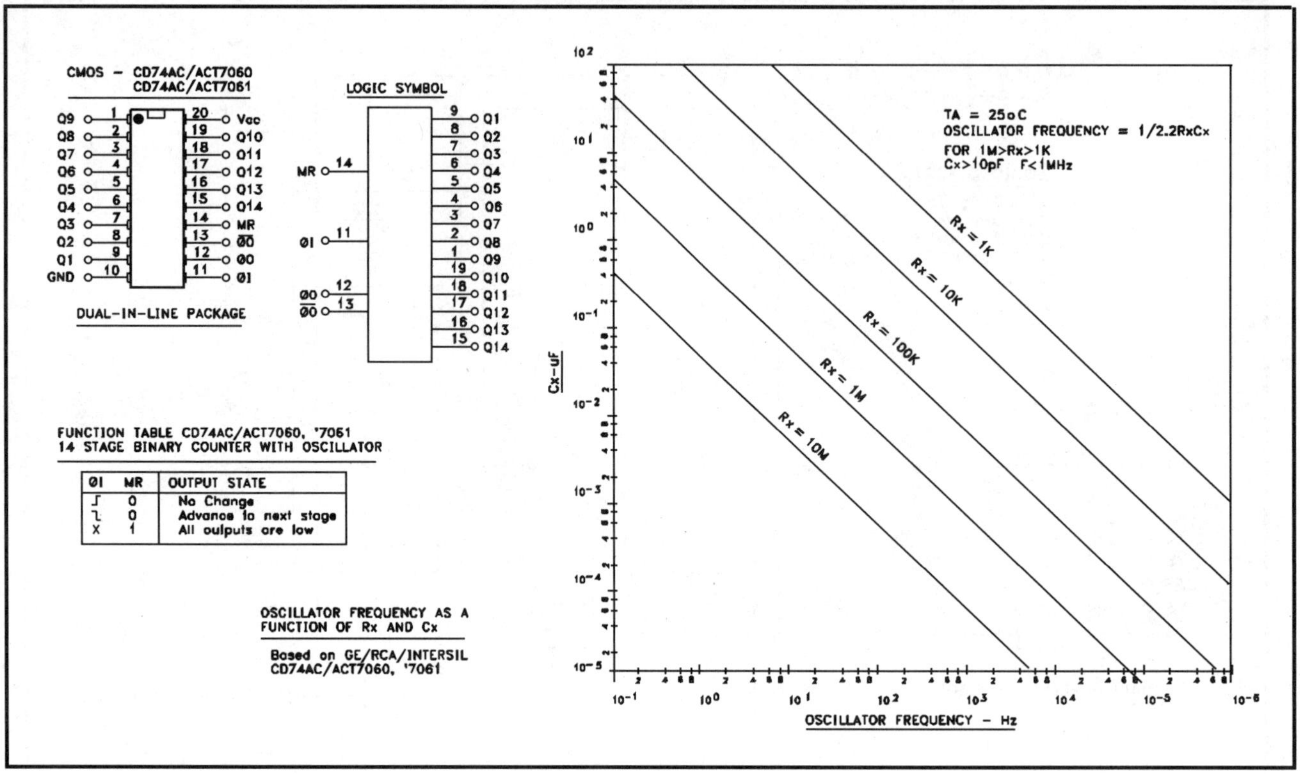

Figure 7.53. The CD74AC/ACT7060, '7061 are 14-stage binary ripple counters with oscillator. The oscillator provides for either crystal or RC input operation. An active high master reset input resets the counter to all 0's and disables the oscillator in the '7060. With the '7061 the oscillator continues to run. The counter is advanced one step on each negative transition of f0 and f1. The input pulse line features Schmitt triggering for slower rise and fall inputs. The MR input in the ACT series has TTL switching levels. The output fanout is to 15 FAST* ICs and is able to drive 50-ohm transmission lines.

* *FAST is a registered trademark of Fairchild Semiconductor Corporation.*

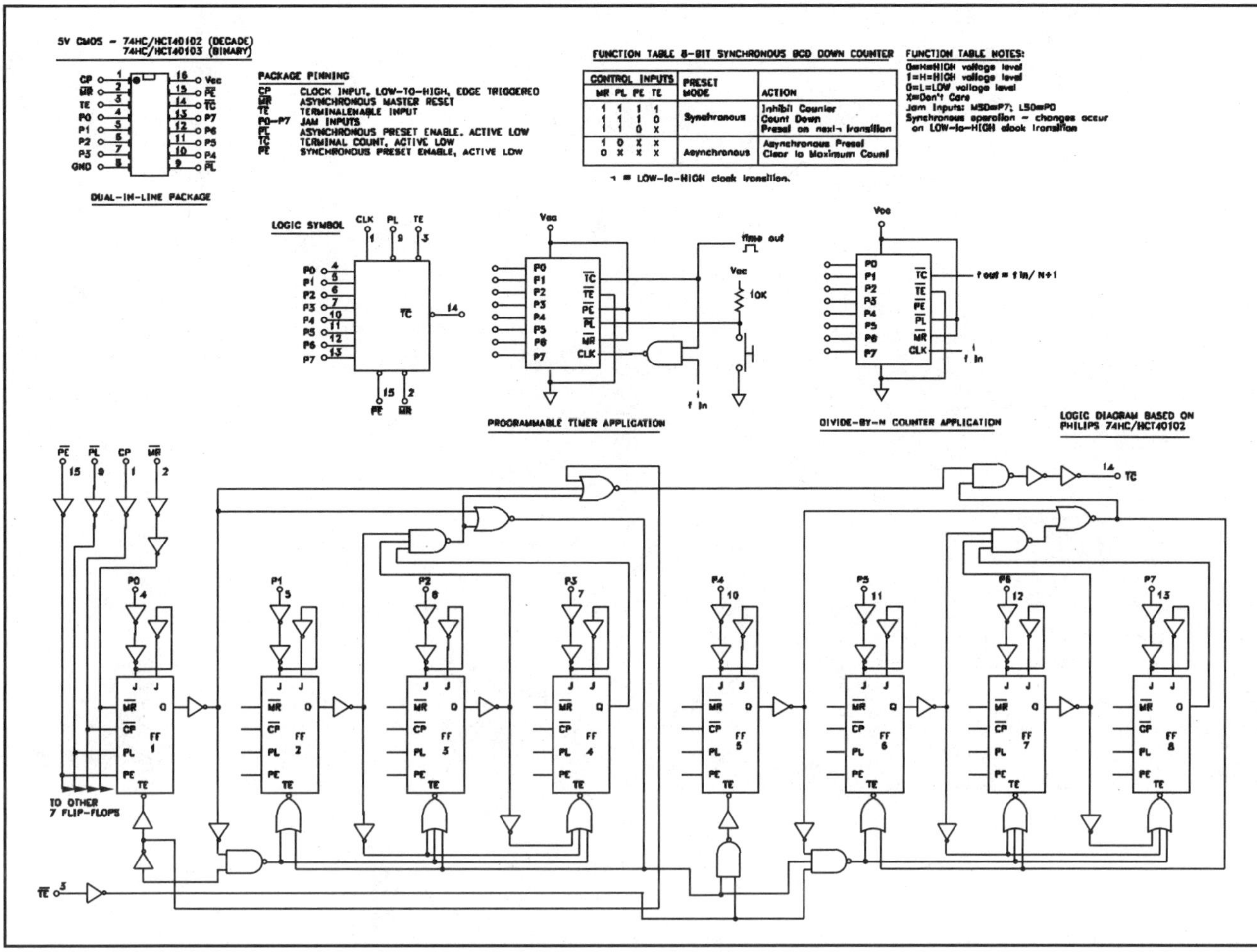

Figure 7.54. The CD40102/40103, 74HC40102/40103 are 8-bit synchronous BCD/binary down counters, respectively. These consist of an 8-bit down counter with a single output. The '40102 is configured as two cascaded 4-bit BCD counters. The '40103 has a single 8-bit binary counter. Each provide inputs for enabling or disabling the clock, for clearing the counter to its maximum value, and for either synchronously or asynchronously presetting the counter. All control inputs and the terminal count output, /TC, are active low. The counter is decremented on each positive-going clock transition. Counting is inhibited when the terminal enable input, /TE, is high. /TC goes low when the count reaches zero if /TE is low, and remains low for one full clock period.

Chapter 8

Decoders and Encoders, Multiplexers and Demultiplexers

The devices in this chapter share many features in common. They also have their own distinguishing capabilities which we need to be aware of. A primary distinction is the source of the data to be processed. With decoders/encoders we can say the data is in the address. With multiplexers/demultiplexers the address points to the data. Some devices embody both features, which is the reason for their combined presence in this chapter.

Decoders and Encoders

Decoding is a feature of virtually every digital application. With few exceptions digital counters, as an example, do not provide a decoded output. The four output lines from a decade or binary counter are used as address inputs to a decoder to select a specific decimal or octal value, as the case may be. Devices such as the 74x42 and 74x138 perform this function. A more sophisticated device combines decoding with driving. We see this in the 7-segment readout decoder/drivers such as the TTL 74x46 - 74x49 BCD to 7-segment decoder drivers. These are frequently combined with gating to time-share the displayed digits.

While BCD or 8-4-2-1 decoding is quite common, other coding requirements exist. Some examples are 2-4-2-1, Excess-3, and Excess-3 Gray. These codes have been developed to meet specific requirements. Codes such as these are called *weighted.* Weighted codes can be mathematically derived from straight binary. The Excess-3 is a weighted code, whereas the Excess-3 Gray is unweighted. An integer change from I to I+/-1 in the 10-bit Gray code, for example, changes by one bit only, a valuable feature for rotary shaft encoding. Unweighted codes require some sort of reference showing the state assignments.

Any number of bits can be decoded. There are devices in this chapter that will decode up to 16-bits. For a greater number an expansion scheme is required. Also dual and quad devices exist. Various approaches have been taken to avoid false data decoding; that is, a BCD counter output is valid only for outputs of 0 -9. Strobes and data latches will be found embedded in many devices as auxiliary functions.

Another area of importance is keyboard encoding. The National/Fairchild 74C922 and 74C923 are excellent examples of a complete encoding system. These are described in *Figure 8.52*.

The *priority encoder* is employed when the data selection must prioritized. The 74x147 and 74x148 described in *Figures 8.10 and 8.11* perform this function. With these, priority is given to the highest numbered input. This type of encoding finds application in keyboard encoders, computer interrupts, and other applications where urgency is a requirement.

Multiplexers and Demultiplexers

The multiplexer is a data selector, performing a function similar to that of a multi-position rotary switch. Thus, an 8-input multiplexer selects one of the eight inputs to appear at the output. Devices described in this chapter have application to both digital and analog data. Where analog data is to be multiplexed the device will have a negative voltage input in order to cover the positive and negative range of an AC signal. While the figure captions note a value for some of these, care must be taken with the CMOS devices. In this book reference is to "5-volt" and "3-volt" integrated circuits. However the "5-volt" devices have a typical power supply range of 2 to 18 volts, while the ranges of the low-voltage types vary considerably. Therefore, in using these with analog signals the manufacturers' data should be referred to if disaster is to be avoided.

The multiplexer also has application to the simplification of logic functions. The details are beyond the scope of this text, but can be found in texts on the design of digital systems.

Although the input maximum for a single IC is sixteen, multiplexers to any dimension can be created by interconnection in "trees." A 64-input multiplexer can be constructed with eight eight-input mux's in a first level, followed by a single 8-input mux for a second level output device. Six address lines are required: A - C for the output, and D - F for each of the eight input devices.

The opposite of the multiplexer is the demultiplexer, often also called a decoder. Here the binary address determines which one of many outputs will go low (or high with some devices). Some devices combine the functions of demultiplexing with decoding. An example is the 74154, shown in *Figure 8.16*. This device has two active low enables. With one of these held high, the IC functions as a decoder to select an output based on the four address inputs. But if the second enable is utilized as a data input, the data can be output to any of the sixteen output lines through address selection.

Chapter Organization

The first four tables identify manufacturers of the various TTL family devices, the 4000 series of CMOS, and the standard and low-voltage CMOS series. This provides a quick reference as to who makes what. While it is not all-inclusive, it provides at least one source for each device, and in many cases, multiple sources. Keep in mind that many of the devices included are no longer available from their original sources.

The remaining tables provide propagation time information. This is for comparison purposes only—it provides a speed indication that gives insight into performance, but is not sufficiently exact for design purposes.

The drawings of *Figures 8.1 to 8.57* provide Dual In-Line (DIP) package pinouts, a Function Table, and Logic Diagram. To simplify comparisons as well as conserve space, all devices with identical pinouts and functions are included in the same drawing. Some newer devices may not be available in the older DIP packages. In general, catalog descriptions include the package type.

Summary

This chapter provides tables and drawings describing many variations of decoders, encoders, multiplexers, and demultiplexers, from the early 7441 to the sophisticated multiplexed bus exchangers, devices with storage, and some having 3-state outputs. Some points to keep in mind are:

- Timing information in the tables is for comparison purposes only. There is a general lack of consistency in how these parameters are defined and not just between manufacturers. Where timing is critical consult the manufacturer's data manual.

- In general, 3-state devices are not affected internally when in the high-impedance output state.

- All timing data is shown for the industrial/commercial or recommended free-air operating temperature range. Where data is for 25°C it is given here.

- The specifications for CMOS devices may cover voltage ranges from 2 to 15 volts. Data is typically provided for VCC's of 5, 10, and 15 volts, but not always. Some manufacturers use 5.0 volts; others 4.5. There is a significant speed factor—in general, the higher the voltage the faster the performance. Again, where this is important to your application consult the data manual.

- A similar situation exists for the more recent low-voltage devices. Typical VCC ranges given are 2.7 - 3.6 volts. But values may be found for 2, 2.6, 3.0, 3.3 or 3.6 volts. (Glossary *Figures G.1 and G.2* may provide helpful insights on performance.)

DEVICE	STD		L		H		S		LS		AS		ALS		F		Figure	DEVICE
74---	NM	ST	NM	ST	NM	ST	NM	ST	NM	ST	NM	ST	NM	ST	NM	ST	No.	74---
7441	x																8.1	7441
7442	x	x x		x					x x	x x							8.2	7442
7443		x x		x													8.2	7443
7444		x x		x													8.2	7444
7445	x	x x															8.2	7445
7446	x	x x		x													8.3	7446
7447	x	x x		x					x	x							8.3	7447
7448		x x							x	x							8.4	7448
7449										x							8.5	7449
74131												x	x	x			8.6	74131
73137									x			x	x	x			8.7	73137
74138							x	x x	x x	x x		x	x	x	x	x	8.8	74138
74139							x	x x	x x	x x				x	x	x	8.9	74139
74141	x																8.1	74141
74145	x	x x							x	x x							8.2	74145
74147		x x							x	x							8.10	74147
74148	x	x x							x	x					x	x	8.11	74148
74150	x	x x															8.12	74150
74151	x	x x					x	x x	x x	x x		x	x	x	x	x x	8.13	74151
74152		x x								x							8.14	74152
74153	x	x x		x			x	x x	x x	x x		x	x	x	x	x x	8.15	74153
74154	x	x x							x							x	8.16	74154
74155	x	x x							x								8.17	74155
74156		x x							x					x			8.17	74156
74157	x	x x					x	x x	x		x	x	x	x	x	x x	8.18	74157
74158		x x					x	x x	x		x	x	x	x	x	x x	8.18	74158
74159		x															8.16	74159

Key to Manufacturer: N – National/Fairchild. M – Motorola,
S – Signetics/Philips, T – Texas Instruments

Table 8.1a. TTL Decoder, Encoder, Multiplexer, and Demultiplexer source summary by manufacturer.

DEVICE	STD		L		H		S		LS		AS		ALS		F		Figure	DEVICE
74---	NM	ST	NM	ST	NM	ST	NM	ST	NM	ST	NM	ST	NM	ST	NM	ST	No.	74---
74246		x															8.3	74246
74247		x							x	x							8.3	74247
74248									x	x							8.4	74248
74249									x								8.4	74249
74250												x					8.12	74250
74251		x x					x	x x	x x	x x			x	x	x	x x	8.13	74251
74253							x	x x	x x	x x		x	x	x	x	x x	8.15	74253
74257							x	x x	x x	x x		x	x	x	x	x x	8.18	74257
74258							x	x x	x x	x x		x	x	x	x	x x	8.18	74258
74261										x x							8.19	74261
74298		x							x	x							8.20	74298
74348									x	x							8.21	74348
74352									x x	x		x	x	x	x		8.15	74352
74353										x		x	x	x	x	x	8.15	74353
74354										x							8.22	74354
74355										x							8.22	74355
74356										x							8.22	74356
74399										x							8.23	74399
74537															x	x	8.24	74537
74538															x	x	8.25	74538
74539															x	x	8.26	74539
74748									x								8.11	74748
74848									x								8.21	74848
74850												x					8.27	74850
74851												x					8.27	74851
74857												x	x	x			8.28	74857

Key to Manufacturer: N – National/Fairchild M – Motorola,
S – Signetics/Philips, T – Texas Instruments

Table 8.1b. TTL Decoder, Encoder, Multiplexer, and Demultiplexer source summary by manufacturer. (continuation)

DEVICE	4000				4100				4500				4700				Figure No.	DEVICE
	N	M	S	T	N	M	S	T	N	M	S	T	N	M	S	T		
4016	x	x															8.29	4016
4028	x	x	x														8.30	4028
4051	x	x	x														8.31	4051
4052	x	x	x														8.32	4052
4053	x	x	x														8.33	4053
4066	x	x															8.29	4066
4067		x	x														8.34	4067
4097		x															8.35	4097
4511									x	x	x						8.36	4511
4512									x	x							8.37	4512
4513										x							8.36	4513
4514									x	x	x						8.38	4514
4515									x	x	x						8.39	4515
4529										x							8.40	4529
4532										x							8.11	4532
4539										x							8.15	4539
4543									x	x	x						8.41	4543
4544										x							8.42	4544
4547										x							8.43	4547
4551										x							8.44	4551
4555										x							8.45	4555
4556										x	x						8.45	4556
4558										x							8.46	4558

Key to Manufacturer:	Prefix
N – National/Fairchild.	CD
S – Signetics/Philips,	HEC/HEF
M – Motorola,	MC1
T – Texas Instruments	––

Table 8.2. 5V CMOS 4000 series Decoder, Encoder, Multiplexer, and Demultiplexer source summary by manufacturer.

DEVICE 74---	C NM	C ST	HC NM	HC ST	HCT NM	HCT ST	AHC (5V) NM	AHC (5V) ST	AHCT NM	AHCT ST	VHC (5V) NM	VHC (5V) ST	VHCT NM	VHCT ST	AC (5V) NM	AC (5V) ST	ACT NM	ACT ST	Figure No.	DEVICE 74---
7442	x		x x	x x		x													8.2	7442
7448	x																		8.4	7448
74137				x x		x													8.7	74137
74138				x x		x x		x		x	x		x		x		x		8.8	74138
74139				x x		x		x		x	x				x		x		8.9	74139
74147				x x		x													8.10	74147
74148				x															8.11	74148
74150	x																		8.12	74150
74151	x			x x		x									x		x		8.13	74151
74152				x											x		x		8.14	74152
74153				x x		x													8.15	74153
74154	x			x x		x													8.16	74154
74157	x			x x		x		x		x	x				x		x		8.18	74157
74158			x	x x		x		x		x					x		x		8.18	74158
74237			x	x x		x													8.47	74237
74238				x x		x													8.48	74238
74239				x															8.49	74239
74298				x x		x													8.20	74298
74251			x	x x		x									x		x		8.13	74251
74253			x	x x		x									x		x		8.15	74253
74257			x	x x		x									x		x		8.18	74257
74258				x x		x									x		x		8.18	74258
74352				x															8.15	74352
74353				x															8.15	74353
74354			x	x x		x													8.22	74354
74356				x x		x													8.22	74356
74604				x															8.50	74604
74915	x																		8.51	74915
74922	x																		8.52	74922
74923	x																		8.52	74923
744016			x	x		x													8.29	744016
744051			x	x		x													8.31	744051
744052			x	x		x													8.32	744052
744053			x	x		x													8.33	744053
744066			x	x		x													8.29	744066
744067			x	x		x													8.34	744067
744316			x	x		x													8.53	744316
744351			x	x		x													8.54	744351
744352			x	x		x													8.55	744352
744353			x	x		x													8.56	744353
744511			x	x		x													8.36	744511
744514			x	x x		x													8.38	744514
744515				x		x													8.39	744515
744543				x		x													8.41	744543

Key to Manufacturer: N – National/Fairchild M – Motorola
S – Signetics/Philips T – Texas Industries

Table 8.3. 5V CMOS 74xxx series Decoder, Encoder, Multiplexer, and Demultiplexer source summary by manufacturer.

DEVICE 74---	LV NM	LV ST	LVC NM	LVC ST	LVT NM	LVT ST	LVX NM	LVX ST	LVQ NM	LVQ ST	LCX NM	LCX ST	AC (3V) NM	AC (3V) ST	ALVCH NM	ALVCH ST	HL NM	HL ST	AHC (3V) NM	AHC (3V) ST	VHC (3V) NM	VHC (3V) ST	Figure No.	DEVICE 74---
74138		x		x x			x		x		x		x							x	x		8.8	74138
74139		x x		x x									x							x	x		8.9	74139
74151									x				x										8.13	74151
74153		x											x								x		8.15	74153
74154		x																					8.16	74154
74157		x		x x			x		x		x		x							x	x		8.18	74157
74158				x							x		x										8.18	74158
74251		x											x										8.13	74251
74253													x										8.15	74253
74257		x		x																	x		8.18	74257
744051		x											x								x		8.31	744051
744052		x																			x		8.32	744052
744053		x																			x		8.33	744053
744066		x																			x		8.29	744066
744067		x																					8.34	744067
744316																					x		8.53	744316
7416271																x							8.57	7416271
7416272																x							8.57	7416272

Key to Manufacturer: N – National/Fairchild. M – Motorola,
S – Signetics/Philips, T – Texas Instruments

Table 8.4. 5V CMOS 74xxx series Decoder, Encoder, Multiplexer, and Demultiplexer source summary by manufacturer.

DEVICE 74---	STD	L	H	S	LS	AS/A/B	ALS/A/B	F	DATA	Figure	DEVICE 74---
-7441	No Data	--	--	--	--	--	--	--	tPLH	-8.1	-7441
----	Given	--	--	--	--	--	--	--	tPHL	---	----
-7442	14-30	10-70	--	--	15-30	--	--	--	tPLH	-8.2	-7442
----	17-30	10-70	--	--	15-30	--	--	--	tPHL	---	----
-7443	35	10-70	--	--	--	--	--	--	tPLH	-8.2	-7443
----	35	10-70	--	--	--	--	--	--	tPHL	---	----
-7444	35	10-70	--	--	--	--	--	--	tPLH	-8.2	-7444
----	35	10-70	--	--	--	--	--	--	tPHL	---	----
-7445	30	--	--	--	--	--	--	--	tPLH	-8.2	-7445
----	30	--	--	--	--	--	--	--	tPHL	---	----
-7446	100	200	--	--	--	--	--	--	tPLH	-8.3	-7446
----	100	200	--	--	--	--	--	--	tPHL	---	----
-7447	100	200	--	--	100	--	--	--	tPLH	-8.3	-7447
----	100	200	--	--	100	--	--	--	tPHL	---	----
-7448	--	--	--	--	100	--	--	--	tPLH	8.4	-7448
----	--	--	--	--	100	--	--	--	tPHL	---	----
-7449	--	--	--	--	100	--	--	--	tPLH	-8.5	-7449
----	--	--	--	--	100	--	--	--	tPHL	---	----
-74131	--	--	--	--	--	2-14.5	--	--	tPLH	-8.6	-74131
----	--	--	--	--	--	2-9.5	--	--	tPHL	---	----
-73137	--	--	--	--	11-27	2-13.5	4-27	--	tPLH	-8.7	-73137
----	--	--	--	--	16-38	2-14	5-25	--	tPHL	---	----
-74138	--	--	--	--	11-27	2-10	3-22	3.0-9.0	tPLH	-8.8	-74138
----	--	--	--	--	13-39	2-10	4-18	3.0-9.0	tPHL	---	----
-74139	--	--	--	--	5-12	--	3-14	3.0-8.5	tPLH	-8.9	-74139
----	--	--	--	--	8-12	--	3-15	2.5-9.0	tPHL	---	----
-74141	No Data	--	--	--	--	--	--	--	tPLH	-8.1	-74141
----	Given	--	--	--	--	--	--	--	tPHL	---	----
-74145	50	--	--	--	50	--	--	--	tPLH	-8.2	-74145
----	50	--	--	--	50	--	--	--	tPHL	---	----
-74147	9-19	--	--	--	12-33	--	--	--	tPLH	-8.10	-74147
----	7-19	--	--	--	12-23	--	--	--	tPHL	---	----
-74148	6-30	--	--	--	7-55	--	--	2.5-10	tPLH	-8.11	-74148
----	9-30	--	--	--	9-40	--	--	2.0-12	tPHL	---	----

NOTE: For some devices td, tp, or equivalent time delays are defined instead of tPLH, tPHL.

Table 8.5a. 74xxx TTL device representative propagation delay times in nS.

DEVICE 74---	STD	L	H	S	LS	AS/A/B	ALS/A/B	F	DATA	Figure	DEVICE 74---
-74150	8.5-35	--	--	--	--	--	--	--	tPLH	-8.12	-74150
----	13-33	--	--	--	--	--	--	--	tPHL	---	----
-74151	8-38	--	--	8-16.5	13-43	1.5-14.5	3.0-24	2.0-12	tPLH	-8.13	-74151
----	8-38	--	--	8-18	12-32	1.0-15	4.0-24	1.0-10	tPHL	---	----
-74152	20-35	--	--	--	--	--	--	--	tPLH	-8.14	-74152
----	14-33	--	--	--	--	--	--	--	tPHL	---	----
-74153	18-34	24-68	--	9.0-18	15-29	2-12.5	3-21	2.5-12	tPLH	-8.15	-74153
----	23-34	30-68	--	9.018	26-38	2-11	5-21	2.5-12	tPHL	---	----
-74154	30-36	--	--	--	30-36	--	--	1.5-10.5	tPLH	-8.16	-74154
----	27-33	--	--	--	27-33	--	--	3.0-10.5	tPHL	---	----
-74155	20-32	--	--	--	15-27	--	--	--	tPLH	-8.17	-74155
----	27-32	--	--	--	27-30	--	--	--	tPHL	---	----
-74156	23-34	--	--	--	46-51	--	13-50	--	tPLH	-8.17	-74156
----	30-34	--	--	--	48-51	--	6-23	--	tPHL	---	----
-74157	14-23	--	--	--	14-23	1-11	4-24	2.5-11.5	tPLH	-8.18	-74157
----	14-27	--	--	--	14-27	1-10	2-17	1.0-10.5	tPHL	---	----
-74158	14-23	--	--	6.0-12	12-20	1-9.5	4-18	2.5-11.5	tPLH	-8.18	-74158
----	14-27	--	--	6.0-12	12-24	1-10	2-18	1.0-10.5	tPHL	---	----
-74159	15-36	--	--	--	--	--	--	--	tPLH	-8.16	-74159
----	22-36	--	--	--	--	--	--	--	tPHL	---	----
-74246	100	--	--	--	--	--	--	--	tPLH	-8.3	-74246
----	100	--	--	--	--	--	--	--	tPHL	---	----
-74247	100	--	--	--	100	--	--	--	tPLH	-8.3	-74247
----	100	--	--	--	100	--	--	--	tPHL	---	----
-74248	--	--	--	--	100	--	--	--	tPLH	-8.4	-74248
----	--	--	--	--	100	--	--	--	tPHL	---	----
-74249	--	--	--	--	100	--	--	--	tPLH	-8.5	-74249
----	--	--	--	--	100	--	--	--	tPHL	---	----
-74250	--	--	--	--	--	2-13	--	--	tPLH	-8.12	-74250
----	--	--	--	--	--	2-10.5	--	--	tPHL	---	----
-74251	10-45	--	--	7-18	10-45	--	1-38	3.0-14	tPLH	-8.13	-74251
----	9-45	--	--	7-40	9-45	--	2-34	1.5-10.5	tPHL	---	----
-74253	--	--	--	9.0-18	17-45	2.5-13.5	2-21	3.0-13.5	tPLH	-8.15	-74253
----	--	--	--	9.0-18	13-32	2.5-11.5	3-21	3.0-10	tPHL	---	----
-74257	--	--	--	7.5-15	10-21	1-11	2-18	1.5-10.5	tPLH	-8.18	-74257
----	--	--	--	6.5-15	12-21	1-10	2-22	2.0-8.0	tPHL	---	----
-74258	--	--	--	6.0-12	10-21	1-9.5	2-25	2.0-8.5	tPLH	-8.18	-74258
----	--	--	--	6.0-12	12-21	1-10	2-20	1.0-8.0	tPHL	---	----

NOTE: For some devices td, tp, or equivalent time delays are defined instead of tPLH, tPHL.

Table 8.5b. 74xxx TTL device representative propagation delay times in nS (continuation).

DEVICE 74---	STD	L	H	S	LS	AS/A/B	ALS/A/B	F	DATA	Figure	DEVICE 74---
-74261	--	--	--	--	22-42	--	--	--	tPLH	- 8.19	-74261
----	--	--	--	--	22-37	--	--	--	tPHL	---	----
-74298	18-27	--	--	--	18-27	--	--	--	tPLH	- 8.20	-74298
----	21-32	--	--	--	21-32	--	--	--	tPHL	---	----
-74348	--	--	--	--	11-55	--	--	--	tPLH	- 8.21	-74348
----	--	--	--	--	9-40	--	--	--	tPHL	---	----
-74352	--	--	--	--	13-29	2-11	3-24	2.0-12.5	tPLH	- 8.15	-74352
----	--	--	--	--	17-38	2-13	2-21	1.0-9.5	tPHL	---	----
-74353	--	--	--	--	11-45	3-9	4-24	2.0-12.5	tPLH	- 8.15	-74353
----	--	--	--	--	13-32	2-12	3-21	1.0-9.5	tPHL	---	----
-74354	--	--	--	--	18-51	--	--	--	tPLH	- 8.22	-74354
----	--	--	--	--	23-60	--	--	--	tPHL	---	----
-74355	--	--	--	--	30-68	--	--	--	tPLH	- 8.22	-74355
----	--	--	--	--	18-68	--	--	--	tPHL	---	----
-74356	--	--	--	--	18-54	--	--	--	tPLH	- 8.22	-74356
----	--	--	--	--	18-60	--	--	--	tPHL	---	----
-74399	--	--	--	--	18-27	--	--	--	tPLH	- 8.23	-74399
----	--	--	--	--	21-32	--	--	--	tPHL	---	----
-74537	--	--	--	--	--	--	--	3.5-17	tPLH	- 8.24	-74537
----	--	--	--	--	--	--	--	2.0-12.5	tPHL	---	----
-74538	--	--	--	--	--	--	--	3.5-16.5	tPLH	- 8.25	-74538
----	--	--	--	--	--	--	--	3.5-13.5	tPHL	---	----
-74539	--	--	--	--	--	--	--	3.0-15.5	tPLH	- 8.26	-74539
----	--	--	--	--	--	--	--	2.5-13	tPHL	---	----
-74748	--	--	--	--	7-55	--	--	--	tPLH	- 8.11	-74748
----	--	--	--	--	9-45	--	--	--	tPHL	---	----
-74848	--	--	--	--	11-35	--	--	--	tPLH	- 8.21	-74848
----	--	--	--	--	9-45	--	--	--	tPHL	---	----
-74850	--	--	--	--	--	3-15	--	--	tPLH	- 8.27	-74850
----	--	--	--	--	--	1-17.5	--	--	tPHL	---	----
-74851	--	--	--	--	--	3-18	--	--	tPLH	- 8.27	-74851
----	--	--	--	--	--	1-20	--	--	tPHL	---	----
-74857	--	--	--	--	--	2	4	--	tPLH	- 8.28	-74857
----	--	--	--	--	--	18	37	--	tPHL	---	----

NOTE: For some devices td, tp, or equivalent time delays are defined instead of tPLH, tPHL.

Table 8.5c. 74xxx TTL device representative propagation delay times in nS (continuation).

DEVICE CD\MC1	4000	4100	4500	4700	DATA	Figure	DEVICE
- 4016	15–45	--	--	--	t_{PLH}	- 8.29	- 4016
----	15–45	--	--	--	t_{PHL}	---	----
- 4028	300–600	--	--	--	t_{PLH}	- 8.30	- 4028
----	300–600	--	--	--	t_{PHL}	---	----
- 4051	35–90	--	--	--	t_{PLH}	- 8.31	- 4051
----	35–90	--	--	--	t_{PHL}	---	----
- 4052	30–75	--	--	--	t_{PLH}	- 8.32	- 4052
----	30–75	--	--	--	t_{PHL}	---	----
- 4053	25–65	--	--	--	t_{PLH}	- 8.33	- 4053
----	25–65	--	--	--	t_{PHL}	---	----
- 4066	20–40	--	--	--	t_{PLH}	- 8.29	- 4066
----	20–40	--	--	--	t_{PHL}	---	----
- 4067	35–90	--	--	--	t_{PLH}	- 8.34	- 4067
----	35–90	--	--	--	t_{PHL}	---	----
- 4097	25–65	--	--	--	t_{PLH}	- 8.35	- 4097
----	25–65	--	--	--	t_{PHL}	---	----
4511	--	--	640–1280	--	t_{PLH}	8.36	4511
----	--	--	720–1440	--	t_{PHL}	---	----
- 4512	--	--	330–650	--	t_{PLH}	- 8.37	- 4512
----	--	--	330–650	--	t_{PHL}	---	----
- 4513	--	--	640–1280	--	t_{PLH}	- 8.36	- 4513
----	--	--	720–1440	--	t_{PHL}	---	----
- 4514	--	--	550–1100	--	t_{PLH}	- 8.38	- 4514
----	--	--	550–1100	--	t_{PHL}	---	----
- 4515	--	--	550–1100	--	t_{PLH}	- 8.39	- 4515
----	--	--	550–1100	--	t_{PHL}	---	----
- 4529	--	--	20–40	--	t_{PLH}	- 8.40	- 4529
----	--	--	20–40	--	t_{PHL}	---	----
- 4532	--	--	205–410	--	t_{PLH}	- 8.11	- 4532
----	--	--	205–410	--	t_{PHL}	---	----
- 4539	--	--	210–450	--	t_{PLH}	- 8.15	- 4539
----	--	--	210–450	--	t_{PHL}	---	----
- 4543	--	--	605–1210	--	t_{PLH}	- 8.41	- 4543
----	--	--	505–1650	--	t_{PHL}	---	----
4544	--	--	605–1210		t_{PLH}	8.42	4544
----	--	--	505–1650	--	t_{PHL}	---	----
- 4547	--	--	750–1500	--	t_{PLH}	- 8.43	- 4547
----	--	--	750–1500	--	t_{PHL}	---	----
- 4551	--	--	350–875	--	t_{PLH}	- 8.44	- 4551
----	--	--	350–875	--	t_{PHL}	---	----
- 4555	--	--	200–440	--	t_{PLH}	- 8.45	- 4555
----	--	--	200–440	--	t_{PHL}	---	----
- 4556	--	--	300–600	--	t_{PLH}	- 8.45	- 4556
----	--	--	300–600	--	t_{PHL}	---	----
- 4558	--	--	580–1160	--	t_{PLH}	- 8.46	- 4558
----	--	--	780–1560	--	t_{PHL}	---	----

NOTE: For some devices td, tp, or equivalent time delays are provided instead of t_{PLH}, t_{PHL}

Table 8.6. 5V CMOS 4000 series device representative propagation delay times in nS.

DEVICE 74---	C	HC	HCT	AHC	AHCT	VHC(5V)	VHCT	AC (5V)	ACT	DATA	Figure	DEVICE 74---
-7442	200-300	17-38	20-44	--	--	--	--	--	--	tPLH	- 8.2	-7442
----	200-300	17-38	20-44	--	--	--	--	--	--	tPHL	---	----
-7448	450-2000	--	--	--	--	--	--	--	--	tPLH	- 8.4	-7448
----	450-2000	--	--	--	--	--	--	--	--	tPHL	---	----
-74137	--	18-48	18-55	--	--	--	--	--	--	tPLH	- 8.7	-74137
----	--	18-48	18-55	--	--	--	--	--	--	tPHL	---	----
-74138	--	17-38	18-50	1-11.5	1-13	5.6-12.8	6.6-9.1	1.5-11	1.5-12	tPLH	- 8.8	-74138
----	--	17-38	18-50	1-11.5	1-13	7.1-8.1	7.1-11.4	1.5-9.5	1.5-11.5	tPHL	---	----
-74139	--	12-29	16-43	1-16.5	1	4.4-7.2	--	3.0-8.5	1.5-10	tPLH	- 8.9	-74139
----	--	12-29	16-43	1-16.5	1	5.9-9.2	--	2.5-7.5	1.5-9.5	tPHL	---	----
-74147	--	18-32	20-44	--	--	--	--	--	--	tPLH	-8.10	-74147
----	--	18-32	20-44	--	--	--	--	--	--	tPHL	---	----
-74148	--	19-57	--	--	--	--	--	--	--	tPLH	-8.11	-74148
----	--	19-57	--	--	--	--	--	--	--	tPHL	---	----
-74150	120-650	--	--	--	--	--	--	--	--	tPLH	-8.12	-74150
----	120-650	--	--	--	--	--	--	--	--	tPHL	---	----
-74151	170-360	15-51	16-54	--	--	--	--	1.5-13	2.5-15.5	tPLH	-8.13	-74151
----	170-360	15-51	16-54	--	--	--	--	1.5-13	2.5-15.5	tPHL	---	----
-74152	--	18-77	--	--	--	--	--	--	--	tPLH	-8.14	-74152
----	--	18-77	--	--	--	--	--	--	--	tPHL	---	----
-74153	--	12-38	13-43	--	--	--	--	1.5-11	2.0-11.5	tPLH	-8.15	-74153
----	--	12-38	13-43	--	--	--	--	1.5-11	2.0-11.5	tPHL	---	----
-74154	265-400	13-38	15-44	--	--	--	--	--	--	tPLH	-8.16	-74154
----	265-400	13-38	15-44	--	--	--	--	--	--	tPHL	---	----
-74157	150-300	13-31	15-46	--	--	--	--	1.5-9.0	1.5-10.0	tPLH	-8.18	-74157
----	150-300	13-31	15-46	--	--	--	--	1.5-9.0	1.5-9.5	tPHL	---	----
-74158	--	15-36	16-44	--	--	--	--	1.5-9.5	1.5-9.5	tPLH	-8.18	-74158
----	--	15-36	16-44	--	--	--	--	1.5-9.0	1.5-95	tPHL	---	----
-74237	--	17-48	20-53	--	--	--	--	--	--	tPLH	-8.47	-74237
----	--	17-48	20-53	--	--	--	--	--	--	tPHL	---	----
-74238	--	17-40	17-48	--	--	--	--	--	--	tPLH	- 8.48	-74238
----	--	17-40	17-48	--	--	--	--	--	--	tPHL	---	----
-74239	--	8-38	--	--	--	--	--	--	--	tPLH	- 8.49	-74239
----	--	8-38	--	--	--	--	--	--	--	tPHL	---	----
-74251	--	18-51	22-55	--	--	--	--	--	--	tPLH	-8.13	-74251
----	--	18-51	22-55	--	--	--	--	--	--	tPHL	---	----
-74253	--	20-44	22-50	--	--	--	--	--	--	tPLH	-8.15	-74253
----	--	20-44	22-50	--	--	--	--	--	--	tPHL	---	----
-74257	--	13-38	16-44	--	--	--	--	--	--	tPLH	- 8.18	-74257
----	--	13-38	16-44	--	--	--	--	--	--	tPHL	---	----
-74258	--	11-35	16-43	--	--	--	--	--	--	tPLH	- 8.18	-74258
----	--	11-35	16-43	--	--	--	--	--	--	tPHL	---	----
-74298	--	16-48	--	--	--	--	--	--	--	tPLH	- 8.20	-74298
----	--	16-48	--	--	--	--	--	--	--	tPHL	---	----

NOTE: For some devices td, tp, or equivalent time delays are defined instead of tPLH, tPHL.

Table 8.7a. 5V 74xxx series CMOS device representative propagation delay times in nS.

DEVICE 74---	C	HC	HCT	AHC	AHCT	VHC(5V)	VHCT	AC (5V)	ACT	DATA	Figure	DEVICE 74---
-74352	--	10-53	--	--	--	--	--	--	--	tPLH	-8.15	-74352
----	--	10-53	--	--	--	--	--	--	--	tPHL	---	----
-74353	--	14-56	--	--	--	--	--	--	--	tPLH	- 8.15	-74353
----	--	14-56	--	--	--	--	--	--	--	tPHL	---	----
-74354	--	22-58	25-63	--	--	--	--	--	--	tPLH	- 8.22	-74354
----	--	22-58	25-63	--	--	--	--	--	--	tPHL	---	----
-74356	--	24-54	26-63	--	--	--	--	--	--	tPLH	-8.22	-74356
----	--	24-54	26-63	--	--	--	--	--	--	tPHL	---	----
-74604	--	23-43	--	--	--	--	--	--	--	tPLH	- 8.50	-74604
----	--	23-43	--	--	--	--	--	--	--	tPHL	---	----
-74915	110-200	--	--	--	--	--	--	--	--	tPLH	- 8.51	-74915
----	150-250	--	--	--	--	--	--	--	--	tPHL	---	----
-74922	80-200	--	--	--	--	--	--	--	--	tPLH	- 8.52	-74922
----	100-250	--	--	--	--	--	--	--	--	tPHL	---	----
-74923	80-200	--	--	--	--	--	--	--	--	tPLH	- 8.52	-74923
----	100-200	--	--	--	--	--	--	--	--	tPHL	---	----
-744016	--	6-12	6-12	--	--	--	--	--	--	tPLH	- 8.29	-744016
----	--	6-12	6-12	--	--	--	--	--	--	tPHL	---	----
-744051	--	5-12	4-12	--	--	--	--	--	--	tPLH	-8.31	-744051
----	--	5-12	4-12	--	--	--	--	--	--	tPHL	---	----
-744052	--	5-12	4-12	--	--	--	--	--	--	tPLH	- 8.32	-744052
----	--	5-12	4-12	--	--	--	--	--	--	tPHL	---	----
-744053	--	5-12	4-12	--	--	--	--	--	--	tPLH	- 8.33	-744053
----	--	5-12	4-12	--	--	--	--	--	--	tPHL	---	----
-744066	--	3-12	3-12	--	--	--	--	--	--	tPLH	-8.29	-744066
----	--	3-12	3-12	--	--	--	--	--	--	tPHL	---	----
-744067	--	6-15	6-15	--	--	--	--	--	--	tPLH	-8.34	-744067
----	--	6-15	6-15	--	--	--	--	--	--	tPHL	---	----
-744316	--	6-12	4-12	--	--	--	--	--	--	tPLH	- 8.53	-744316
----	--	6-12	4-12	--	--	--	--	--	--	tPHL	---	----
-744351	--	5-12	4-12	--	--	--	--	--	--	tPLH	- 8.54	-744351
----	--	5-12	4-12	--	--	--	--	--	--	tPHL	---	----
-744352	--	6-12	5-12	--	--	--	--	--	--	tPLH	-8.55	-744352
----	--	6-12	5-12	--	--	--	--	--	--	tPHL	---	----
-744353	--	5-12	4-12	--	--	--	--	--	--	tPLH	-8.56	-744353
----	--	5-12	4-12	--	--	--	--	--	--	tPHL	---	----
-744511	--	22-60	16-60	--	--	--	--	--	--	tPLH	-8.36	-744511
----	--	22-60	16-60	--	--	--	--	--	--	tPHL	---	----
-744514	--	15-46	17-55	--	--	--	--	--	--	tPLH	-8.38	-744514
----	--	15-46	17-55	--	--	--	--	--	--	tPHL	---	----
-744515	--	18-50	18-50	--	--	--	--	--	--	tPLH	-8.39	-744515
----	--	18-50	18-50	--	--	--	--	--	--	tPHL	---	----
-744543	--	20-74	24-80	--	--	--	--	--	--	tPLH	-8.41	-744543
----	--	20-74	24-80	--	--	--	--	--	--	tPHL	---	----

NOTE: For some devices td, tp, or equivalent time delays are defined instead of tPLH, tPHL.

Table 8.7b. 5V 74xxx series CMOS device representative propagation delay times in nS (continuation).

DEVICE 74---	LV	LVC	ALVT LVT/ABT	LVX	LVQ	LCX	AC(3V)	LVCH/ ALVCH	HL	AHC (3V)	VHC(3V)	DATA	Figure No.	DEVICE 74---
-74138	14-23	3.5-6.5	--	5.5-10.6	1.5-15.5	1.5-7.0	1.5-15.5	--	--	8.1-16.3	8.1-16.3	tPLH	- 8.8	-74138
----	14-23	3.5-6.5	--	8.0-14.1	1.5-13.5	1.5-7.0	1.5-9.5	--	--	8.2-16.3	8.1-16.3	tPHL	---	----
-74139	11-18	3.2-6.0	--	--	--	--	4.0-12	--	--	6.4-14.5	6.4-11.0	tPLH	-8.9	-74139
----	11-18	3.2-6.0	--	--	--	--	3.0-10	--	--	6.4-14.5	6.4-11.0	tPHL	---	----
-74151	--	--	--	--	2.5-18.0	--	2.5-18	--	--	--	--	tPLH	- 8.13	-74151
----	--	--	--	--	2.5-18.0	--	1.5-18	--	--	--	--	tPHL	---	----
-74153	11-34	--	--	--	--	--	2.5-15	--	--	--	--	tPLH	-8.15	-74153
----	11-34	--	--	--	--	--	1.5-14.5	--	--	--	--	tPHL	---	----
-74154	12-26	--	--	--	--	--	--	--	--	--	--	tPLH	-8.16	-74154
----	12-26	--	--	--	--	--	--	--	--	--	--	tPHL	---	----
-74157	12-29	3.8-8.5	--	5.1-11.5	1.5-11.5	No	1.5-11.5	--	--	6.2-17.1	6.2-13.6	tPLH	-8.18	-74157
----	12-29	3.8-8.5	--	7.6-15.0	1.5-11.5	Time	1.5-11.0	--	--	6.2-17.1	6.2-13.6	tPHL	---	----
-74158	--	1-7.5	--	--	--	Data	1.5-12.0	--	--	6.2-17.1	--	tPLH	-8.18	-74158
----	--	1-7.5	--	--	--	Given	1.5-11.5	--	--	6.2-17.1	--	tPHL	---	----
-74251	17-48	--	--	--	--	--	1.5-17.5	--	--	--	--	tPLH	-8.13	-74251
----	17-48	--	--	--	--	--	1.5-17.5	--	--	--	--	tPHL	---	----
-74253	--	--	--	--	--	--	1.5-14.5	--	--	--	--	tPLH	-8.15	-74253
----	--	--	--	--	--	--	1.5-14.5	--	--	--	--	tPHL	---	----
-74257	12-33	4.0-8.0	--	--	--	--	1.5-10.5	--	--	--	--	tPLH	-8.18	-74257
----	12-33	4.0-8.0	--	--	--	--	1.5-10.5	--	--	--	--	tPHL	---	----
-744051	5-10	--	--	--	--	--	--	--	--	--	--	tPLH	-8.31	-744051
----	5-10	--	--	--	--	--	--	--	--	--	--	tPHL	---	----
-744052	5-10	--	--	--	--	--	--	--	--	--	--	tPLH	-8.32	-744052
----	5-10	--	--	--	--	--	--	--	--	--	--	tPHL	---	----
-744053	5-10	--	--	--	--	--	--	--	--	--	--	tPLH	- 8.33	-744053
----	5-10	--	--	--	--	--	--	--	--	--	--	tPHL	---	----
-744066	3-15	--	--	--	--	--	--	--	--	--	--	tPLH	-8.29	-744066
----	3-15	--	--	--	--	--	--	--	--	--	--	tPHL	---	----
-744067	6-18	--	--	--	--	--	--	--	--	--	--	tPLH	-8.34	-744067
----	6-18	--	--	--	--	--	--	--	--	--	--	tPHL	---	----
-744316	--	--	--	--	--	--	--	--	--	--	--	tPLH	- 8.53	-744316
----	--	--	--	--	--	--	--	--	--	--	--	tPHL	---	----
-7416271	--	--	--	--	--	--	--	0-10	--	--	--	tPLH	- 8.57	-7416271
----	--	--	--	--	--	--	--	0-10	--	--	--	tPHL	---	----
-7416272	--	--	--	--	--	--	--	0-10	--	--	--	tPLH	- 8.57	-7416272
----	--	--	--	--	--	--	--	0-10	--	--	--	tPHL	---	----

NOTE: For some devices td, tp, or equivalent time delays are defined instead of tPLH, tPHL.

Table 8.8. Low voltage 74xxx series CMOS device representative propagation delay times in nS.

TTL - 7441

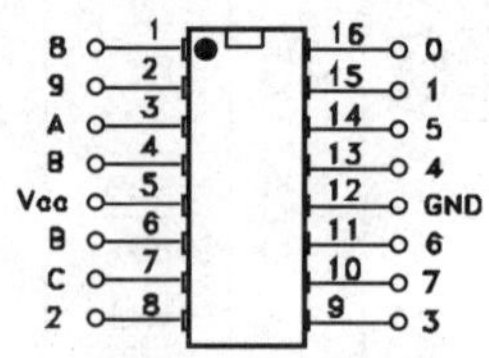

DUAL-IN-LINE PACKAGE

PACKAGE PINNING:

PIN-
1,2,8,9,10,11,13-16: OUTPUTS
3,4,6,7: INPUTS

FUNCTION TABLE 7441 BCD to DECIMAL DECODERS/DRIVERS

INPUTS D C B A	OUTPUT ON
0 0 0 0	0
0 0 0 1	1
0 0 1 0	2
0 0 1 1	3
0 1 0 0	4
0 1 0 1	5
0 1 1 0	6
0 1 1 1	7
1 0 0 0	8
1 0 0 1	9
(OVER RANGE)	
1 0 1 0	0
1 0 1 1	1
1 1 0 0	2
1 1 0 1	3
1 1 1 0	4
1 1 1 1	5

LOGIC DIAGRAM
Based on National 7441

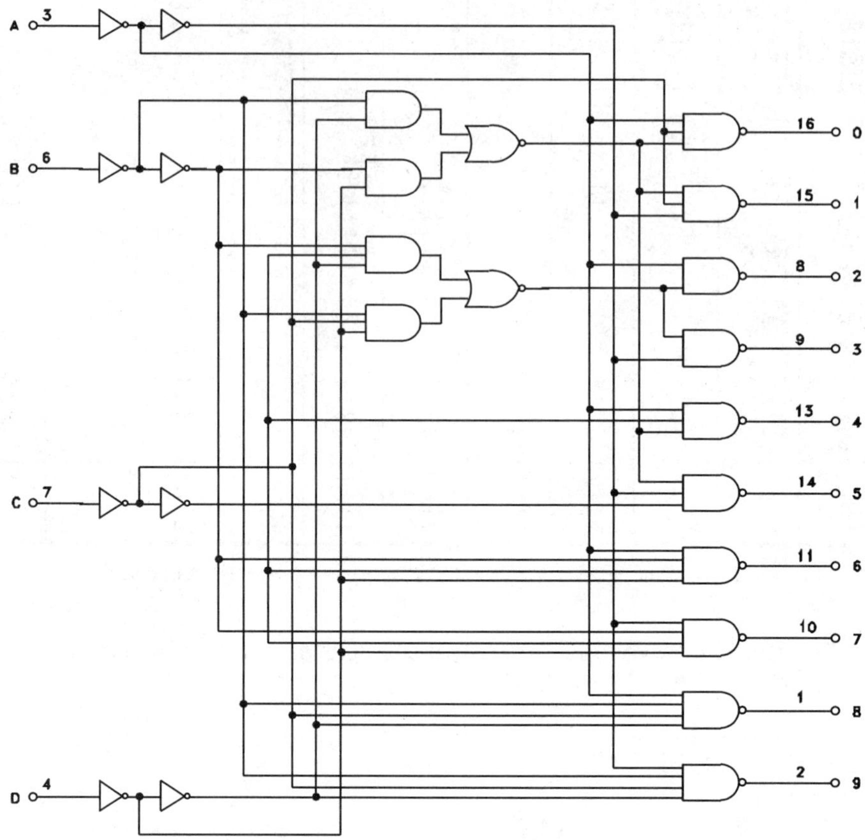

Figure 8.1. The 7441 BCD-to-decimal decoder is designed for driving gas filled NIXIE tubes. The device is also useful for driving low-current lamps and relay coils. An over-range feature provides that if binary numbers between 10 and 15 are applied on the input, the least significant bit (0 - 5) will be decoded at the output. The 74141 is designed for driving cold-cathode indicator tubes. The device differs from the 7441 in that while over-range inputs are decoded, the outputs are off.

TTL – 7442/L/LS
7443/L
7444/L
7445
74145/LS

5V CMOS – 74C/HC/HCT42

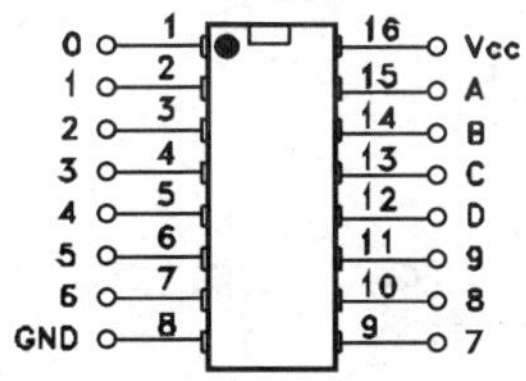

DUAL-IN-LINE PACKAGE

PACKAGE PINNING:
PIN-
1-7,9-11 OUTPUTS
12-15 INPUTS

FUNCTION TABLE 74x42, 7445, 74145/LS
BCD-to-DECIMAL DECODER
NOTE: The DECIMAL OUTPUT
TABLE IS TRUE FOR ALL DEVICES

	BCD INPUT				DECIMAL OUTPUT									
No.	D	C	B	A	0	1	2	3	4	5	6	7	8	9
0	0	0	0	0	0	1	1	1	1	1	1	1	1	1
1	0	0	0	1	1	0	1	1	1	1	1	1	1	1
2	0	0	1	0	1	1	0	1	1	1	1	1	1	1
3	0	0	1	1	1	1	1	0	1	1	1	1	1	1
4	0	1	0	0	1	1	1	1	0	1	1	1	1	1
5	0	1	0	1	1	1	1	1	1	0	1	1	1	1
6	0	1	1	0	1	1	1	1	1	1	0	1	1	1
7	0	1	1	1	1	1	1	1	1	1	1	0	1	1
8	1	0	0	0	1	1	1	1	1	1	1	1	0	1
9	1	0	0	1	1	1	1	1	1	1	1	1	1	0
I	1	0	1	0	1	1	1	1	1	1	1	1	1	1
N	1	0	1	1	1	1	1	1	1	1	1	1	1	1
V	1	1	0	0	1	1	1	1	1	1	1	1	1	1
A	1	1	0	1	1	1	1	1	1	1	1	1	1	1
L	1	1	1	0	1	1	1	1	1	1	1	1	1	1
I	1	1	1	1	1	1	1	1	1	1	1	1	1	1
D														

FUNCTION TABLE 7443/L

	EXCESS-3 INPUT			
No.	D	C	B	A
0	0	0	0	0
1	0	0	0	1
2	0	0	1	0
3	0	0	1	1
4	0	1	0	0
5	0	1	0	1
6	0	1	1	0
7	0	1	1	1
8	1	0	0	0
9	1	0	0	1
I	1	1	0	1
N	1	1	1	0
V	1	1	1	1
A	0	0	0	0
L	0	0	0	1
I	0	0	1	0
D				

FUNCTION TABLE 7444/L

	EXCESS-3 GRAY INPUT			
No.	D	C	B	A
0	0	0	0	0
1	0	0	0	1
2	0	0	1	0
3	0	0	1	1
4	0	1	0	0
5	0	1	0	1
6	0	1	1	0
7	0	1	1	1
8	1	0	0	0
9	1	0	0	1
I	1	0	1	1
N	1	0	0	1
V	1	0	0	0
A	0	0	0	0
L	0	0	0	1
I	0	0	1	1
D				

LOGIC DIAGRAM
Based on National 7442, 7445, 74145

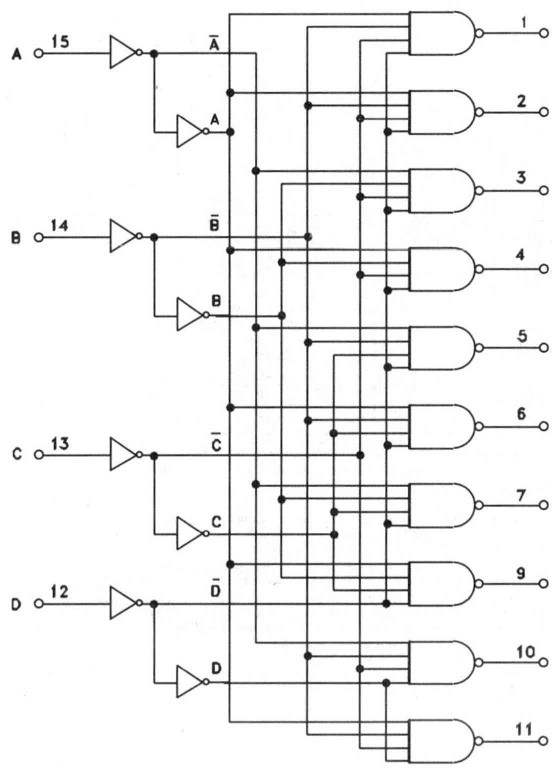

Figure 8.2. The 74x42 is a multipurpose BCD-to-decimal decoder designed to accept four BCD inputs and provide ten mutually exclusive low outputs as shown in the function table. The 74x43 is an Excess-3 to 1-of-10 decoder that accepts Excess-3 coded inputs (binary range three through 12) and generates ten mutually exclusive low outputs. This code is used in decimal arithmetic because of its self-complementing feature. The 74x44 decodes Excess-3 gray code to 1-of-10 outputs. This code finds use in decimal position encoders in making a single bit change between adjacent states, including the transition from nine to zero. The 7445 and 74x145 are open collector devices which accept BCD inputs on the A0-A3 inputs and generate ten mutually exclusive active low outputs. When an input code greater than nine appears, the outputs go off. The device can be used as a 1-of-8 decoder with A3 used as an active low enable.

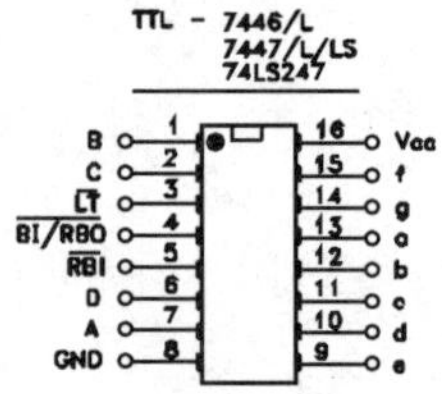

DUAL-IN-LINE PACKAGE

PACKAGE PINNING:

PIN-

1,2,6,7	SEGMENT SELECT INPUTS
3	LAMP TEST
4	BLANKING INPUT/ RIPPLE BLANKING OUTPUT
5	RIPPLE BLANKING INPUT
9-15	SEGMENT DRIVES

FUNCTION TABLE 7446, 74x47, 74LS247
BCD-TO-7-SEGMENT DECODERS/DRIVERS

DEC OR FUNC	INPUTS LT	RBI	D	C	B	A	$\overline{BI}/\overline{RBO}$†	OUTPUTS a	b	c	d	e	f	g	NOTE
0	1	X	0	0	0	0	1	on	on	on	on	on	on	off	
1	1	X	0	0	0	1	1	off	on	on	off	off	off	off	
2	1	X	0	0	1	0	1	on	on	off	on	on	off	on	
3	1	X	0	0	1	1	1	on	on	on	on	off	off	on	
4	1	X	0	1	0	0	1	off	on	on	off	off	on	on	
5	1	X	0	1	0	1	1	on	off	on	on	off	on	off	
6	1	X	0	1	1	0	1	off	off	on	on	on	on	on	
7	1	X	0	1	1	1	1	on	on	on	off	off	off	on	1
8	1	X	1	0	0	0	1	on	on	on	on	on	on	on	
9	1	X	1	0	0	1	1	on	on	on	off	off	on	on	
10	1	X	1	0	1	0	1	off	off	off	on	on	off	on	
11	1	X	1	0	1	1	1	off	off	on	on	off	off	on	
12	1	X	1	1	0	0	1	off	on	off	off	off	on	on	
13	1	X	1	1	0	1	1	on	off	off	on	off	on	on	
14	1	X	1	1	1	0	1	off	off	off	on	on	on	on	
15	1	X	1	1	1	1	1	off	off	off	off	off	off	off	
BI	X	X	X	X	X	X	0	off	off	off	off	off	off	off	2
RBI	1	0	0	0	0	0	0	off	off	off	off	off	off	off	3
LT	0	X	X	X	X	X	1	on	on	on	on	on	on	on	4

SEGMENT ID

1 = HIGH voltage level. 0 = LOW voltage level. X = Don't Care.

† $\overline{BI}/\overline{RBO}$ is wire AND logic serving as blanking input $\overline{BI}$ and /or ripple blanking output $\overline{RBO}$.

NOTES:

1. The blanking input $\overline{BI}$ must be held open or at high level when output functions 0–15 are desired. The ripple blanking input $\overline{RBI}$ must be open or high if blanking of a decimal is not desired.
2. When a low logic level is applied directly to the blanking input $\overline{BI}$, all segment outputs are low regardless of the level of any other input.
3. When ripple blanking input $\overline{RBI}$ and inputs A, B, C, and D are low with lamp test input high, all segment outputs go low and the ripple blanking output $\overline{RBO}$ goes to a low level (response condition).
4. When the blanking input/ripple output $\overline{BI}/\overline{RBO}$ is open or held high and a low is applied to the lamp test input, all segments outputs are high.

LOGIC DIAGRAM
Based on Texas Instruments 7446, 7447, 74LS47

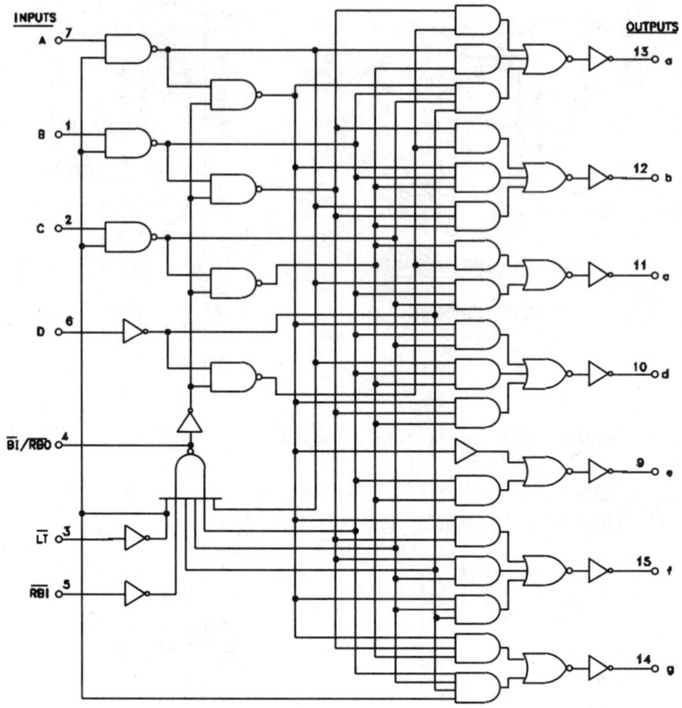

Figure 8.3. The 74x46 and 74x47 are BCD-to-7-segment decoder/drivers with open collector outputs. They accept four-bit BCD data on their inputs and provide seven active low decoded outputs to drive 7-segment or LED incandescent displays directly. Their 40 mA output sink makes them useful for driving multiplexed common-anode LED displays. Both devices feature overriding lamp test (all segments on) and ripple blanking for leading and trailing zeros. The 74246 and 74x247 offer the above features with internal pull-up resistors which eliminate the need for external resistors. These also provide a lamp intensity modulation capability.

TTL – 7448/LS
74LS248
74LS249
5V CMOS – 74C48

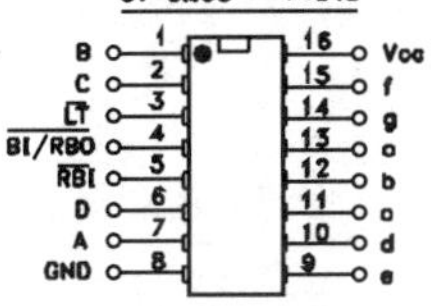

DUAL-IN-LINE PACKAGE

PACKAGE PINNING:

PIN–
1,2,6,7 SEGMENT SELECT INPUTS
3 LAMP TEST
4 BLANKING INPUT/ RIPPLE BLANKING OUTPUT
5 RIPPLE BLANKING INPUT
9–15 SEGMENT DRIVES

FUNCTION TABLE 74x48, 74LS248, 74LS249
BCD-TO-7-SEGMENT DECODERS/DRIVERS

DEC OR FUNC	INPUTS						BI/RBO†	OUTPUTS							NOTE
	LT	RBI	D	C	B	A		a	b	c	d	e	f	g	
0	1	X	0	0	0	0	1	1	1	1	1	1	1	0	1
1	1	X	0	0	0	1	1	0	1	1	0	0	0	0	
2	1	X	0	0	1	0	1	1	1	0	1	1	0	1	
3	1	X	0	0	1	1	1	1	1	1	1	0	0	1	
4	1	X	0	1	0	0	1	0	1	1	0	0	1	1	
5	1	X	0	1	0	1	1	1	0	1	1	0	1	1	
6	1	X	0	1	1	0	1	0	0	1	1	1	1	1	
7	1	X	0	1	1	1	1	1	1	1	0	0	0	0	
8	1	X	1	0	0	0	1	1	1	1	1	1	1	1	
9	1	X	1	0	0	1	1	1	1	1	0	0	1	1	
10	1	X	1	0	1	0	1	0	0	0	1	1	0	1	
11	1	X	1	0	1	1	1	0	0	1	1	0	0	1	
12	1	X	1	1	0	0	1	0	1	0	0	0	1	1	
13	1	X	1	1	0	1	1	1	0	0	1	0	1	1	
14	1	X	1	1	1	0	1	0	0	0	1	1	1	1	
15	1	X	1	1	1	1	1	0	0	0	0	0	0	0	
BI	X	X	X	X	X	X	0	0	0	0	0	0	0	0	2
RBI	1	0	0	0	0	0	0	0	0	0	0	0	0	0	3
LT	0	X	X	X	X	X	1	1	1	1	1	1	1	1	4

1 = HIGH voltage level. 0 = LOW voltage level. X = Don't Care.

† BI/RBO is wire AND logic serving as blanking input BI and /or ripple blanking output RBO.

NOTES:

1. The blanking input BI must be held open or at high level when output functions 0–15 are desired. The ripple blanking input RBI must be open or high if blanking of a decimal is not desired.
2. When a low logic level is applied directly to the blanking input BI, all segment outputs are low regardless of the level of any other input.
3. When ripple blanking input RBI and inputs A, B, C, and D are low with lamp test input high, all segment outputs go low and the ripple blanking output RBO goes to a low level (response condition).
4. When the blanking input/ripple output BI/RBO is open or held high and a low is applied to the lamp test input, all segments outputs are high.

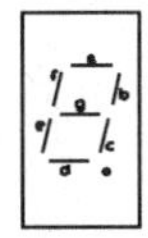

SEGMENT ID

LOGIC DIAGRAM
Based on Texas Instruments
7448, 74LS48

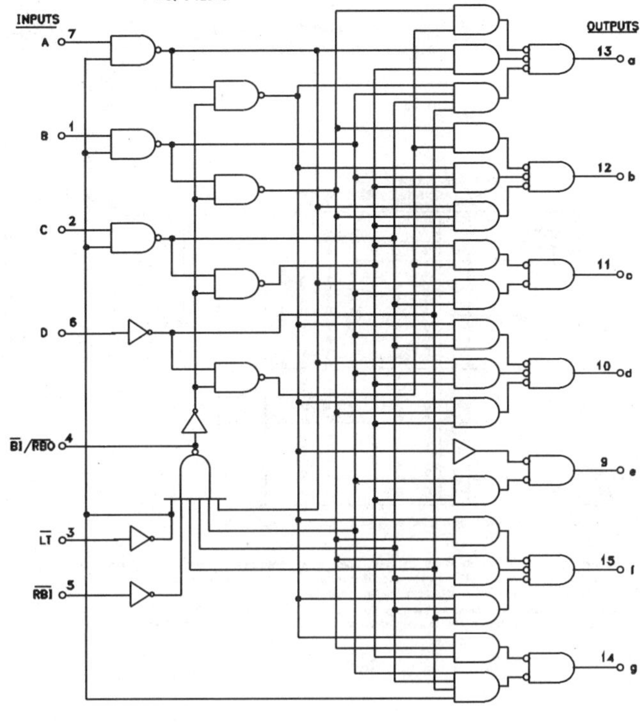

Figure 8.4. The 74x48 and 74LS248 are BCD-to-7-segment decoder/drivers. They do not provide open collector outputs. These accept four-bit BCD data on their inputs and provide seven active high decoded outputs to drive displays of various types. They are primarily used with NPN transistors or segment drivers for multiplexed displays. Both devices feature overriding lamp test (all segments on) and ripple blanking for leading and trailing zeros. Note that the 7446, '47 and '48 compose the "6" and "9" without tails; the '246, '247, 'LS247 and 'LS248 compose these with tails.

TTL – 74LS49

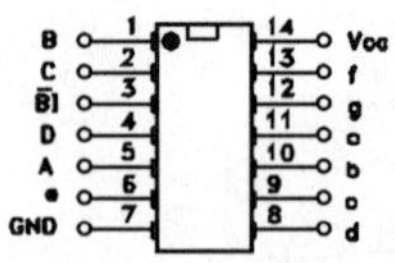

DUAL-IN-LINE PACKAGE

PACKAGE PINNING:
PIN–
1,2,4,5 SEGMENT SELECT INPUTS
3 BLANKING INPUT
6,8–13 SEGMENT DRIVES

FUNCTION TABLE 74LS49
BCD–TO–7–SEGMENT DECODERS/DRIVERS

DEC OR FUNC	INPUTS D C B A	$\overline{BI}$	OUTPUTS a	b	c	d	e	f	g	NOTE
0	0 0 0 0	1	1	1	1	1	1	1	0	1
1	0 0 0 1	1	0	1	1	0	0	0	0	
2	0 0 1 0	1	1	1	0	1	1	0	1	
3	0 0 1 1	1	1	1	1	1	0	0	1	
4	0 1 0 0	1	0	1	1	0	0	1	1	
5	0 1 0 1	1	1	0	1	1	0	1	1	
6	0 1 1 0	1	0	0	1	1	1	1	1	
7	0 1 1 1	1	1	1	1	0	0	0	0	
8	1 0 0 0	1	1	1	1	1	1	1	1	
9	1 0 0 1	1	1	1	1	0	0	1	1	
10	1 0 1 0	1	0	0	0	1	1	0	1	
11	1 0 1 1	1	0	0	1	1	0	0	1	
12	1 1 0 0	1	0	1	0	0	0	1	1	
13	1 1 0 1	1	1	0	0	1	0	1	1	
14	1 1 1 0	1	0	0	0	1	1	1	1	
15	1 1 1 1	1	0	0	0	0	0	0	0	
BI	X X X X	0	0	0	0	0	0	0	0	2

1 = HIGH voltage level. 0 = LOW voltage level. X = Don't Care.

NOTES:

1. The blanking input $\overline{BI}$ must be held open or at high level when output functions 0–15 are desired. The ripple blanking input $\overline{RBI}$ must be open or high if blanking of a decimal is not desired.
2. When a low logic level is applied directly to the blanking input $\overline{BI}$, all segment outputs are low regardless of the level of any other input.

SEGMENT ID

LOGIC DIAGRAM
Based on Texas Instruments
74LS49

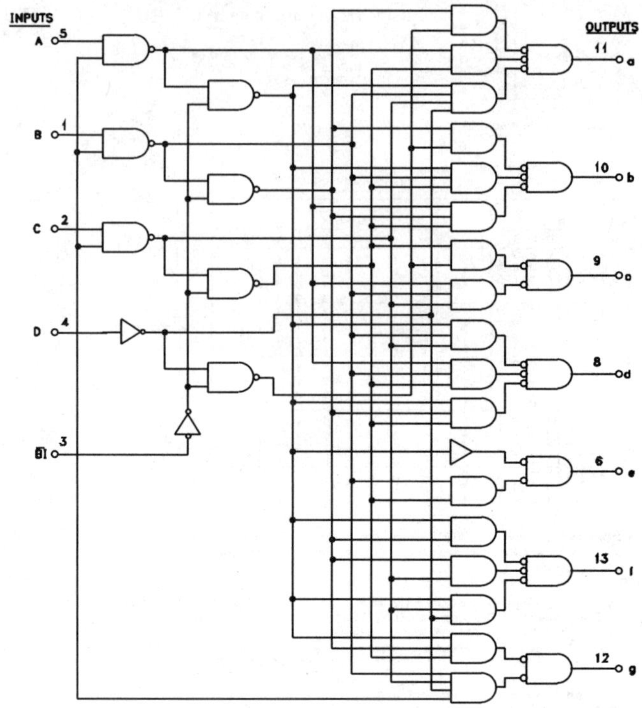

Figure 8.5. The 74LS49 is a BCD-to-7-segment decoder/driver similar to the 7448, but lacking the overriding lamp test (all segments on) and ripple blanking for leading and trailing zeros. It composes the "6" and "9" without tails. The device is provided in a 14-pin DIP. The 74LS249 is a 16-pin version of the 74LS49, providing the full functionality lacking in that device.

TTL – 74ALS131

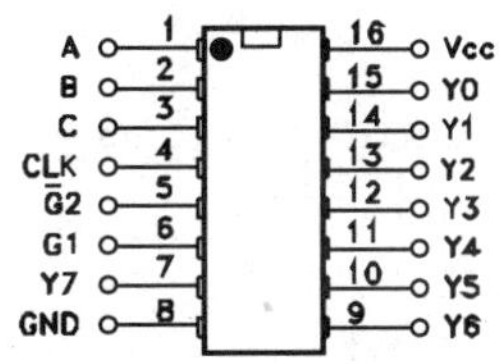

DUAL-IN-LINE PACKAGE

PACKAGE PINNING

PIN–	
1,2,3	ADDRESSING INPUTS
4	CLOCK
5,6	ENABLES
7,9–15	OUTPUTS

FUNCTION TABLE 74ALS131 3-to-8 LINE DECODER/DEMULTIPLEXER

INPUTS			SELECT			OUTPUTS							
CLK	G1	$\overline{G2}$	C	B	A	Y0	Y1	Y2	Y3	Y4	Y5	Y6	Y7
X	X	1	X	X	X	1	1	1	1	1	1	1	1
X	0	X	X	X	X	1	1	1	1	1	1	1	1
↑	1	1	0	1	0	0	1	1	1	1	1	1	1
↑	1	0	0	0	1	1	0	1	1	1	1	1	1
↑	1	0	0	1	0	1	1	0	1	1	1	1	1
↑	1	0	0	1	1	1	1	1	0	1	1	1	1
↑	1	0	1	1	0	1	1	1	1	0	1	1	1
↑	1	0	1	1	1	1	1	1	1	1	0	1	1
↑	1	0	1	1	0	1	1	1	1	1	1	0	1
↑	1	0	1	1	1	1	1	1	1	1	1	1	0
0	1	0	X	X	X	Output corresponding to stored address, 0; all others, 1.							
1	1	0	X	X	X								

1 = HIGH voltage level. 0 = LOW voltage level.
X = Don't Care

↑ = LOW-to-HIGH clock transition.

LOGIC DIAGRAM
Based on National
74ALS131

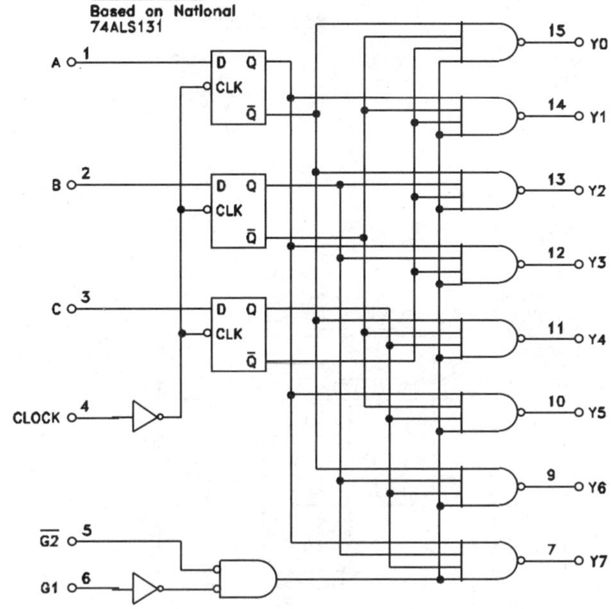

Figure 8.6. The 74AS/ALS131 are a 3-line to 8-line decoder/demultiplexer, featuring registers on the three address inputs. When the clock transits from low-to-high, the device functions as a decoder/demultiplexer. The address present at the A, B, and C inputs is stored in the registers. Further address changes are ignored until the next low-to-high clock transition. The output enable inputs control the state of the outputs independently of the select or clock inputs. The maximum clock frequency is 100 MHz with a pulse duration of 5 nS. Setup time is 3.5 nS with zero hold time.

TTL – 74LS/AS/ALS137
5V CMOS – 74HC/HCT137

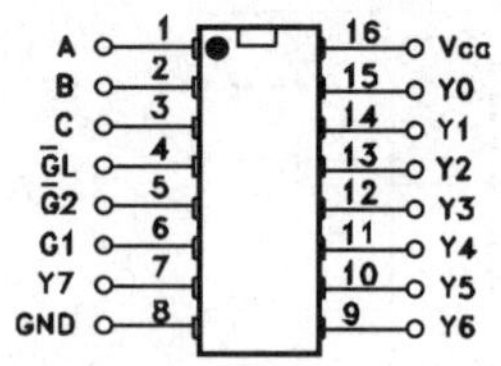

DUAL-IN-LINE PACKAGE

PACKAGE PINNING

PIN–
1,2,3 ADDRESSING INPUTS
4 LATCH ENABLE
5,6 ENABLES
7,9–15 OUTPUTS

FUNCTION TABLE 74x137 3-to-8 LINE DECODER/DEMULTIPLEXER WITH ADDRESS LATCHES

INPUTS													
ENABLES			SELECT			OUTPUTS							
$\overline{GL}$	$\overline{G1}$	G2	C	B	A	Y0	Y1	Y2	Y3	Y4	Y5	Y6	Y7
X	X	1	X	X	X	1	1	1	1	1	1	1	1
X	0	X	X	X	X	1	1	1	1	1	1	1	1
0	1	1	0	1	0	0	1	1	1	1	1	1	1
0	1	0	0	0	1	1	0	1	1	1	1	1	1
0	1	0	0	1	0	1	1	0	1	1	1	1	1
0	1	0	0	1	1	1	1	1	0	1	1	1	1
0	1	0	1	1	0	1	1	1	1	0	1	1	1
0	1	0	1	1	1	1	1	1	1	1	0	1	1
0	1	0	1	1	0	1	1	1	1	1	1	0	1
0	1	0	1	1	1	1	1	1	1	1	1	1	0
1	1	0	X	X	X	Output corresponding to stored address, 0; all others, 1.							

1 = HIGH voltage level. 0 = LOW voltage level.
X = Don't Care

LOGIC DIAGRAM
Based on National 74ALS137

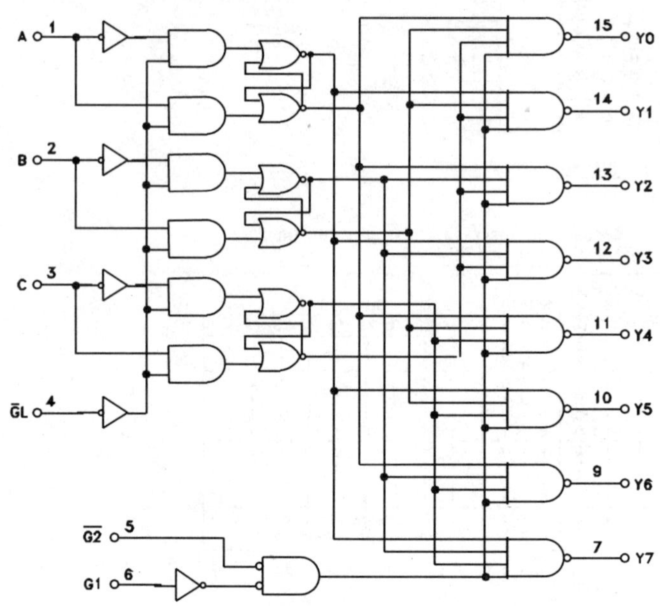

Figure 8.7. The 74AS/ALS137 and 74HC/HCT137 are a 3-line to 8-line decoder/demultiplexer featuring registers on the three address inputs. When the latch enable input /GL is low, the device functions as a decoder/demultiplexer. The address present at the A, B, and C inputs is stored in the registers. Further address changes are ignored as long as the enable remains high. The output enable inputs, G1, /G2, control the state of the outputs independently of the select inputs. Pulse duration with the latch enable low is 5 nS. Setup time is 10 nS, with 5 nS hold time.

TTL – 74S/LS/AS/ALS/F138
5V CMOS – 74HC/HCT/AHC/AHCT/VHC/VHCT/AC/ACT138
3V CMOS – 74LV/LVC/LVX/LVQ/LCX/AC/VHC138

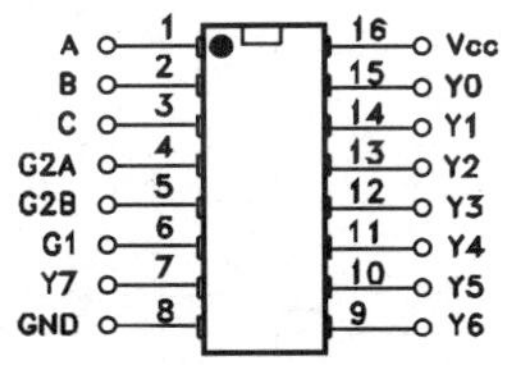

DUAL-IN-LINE PACKAGE

PACKAGE PINNING

PIN–
1,2,3 ADDRESSING INPUTS
4,5,6 LATCH ENABLES
7,9–15 OUTPUTS

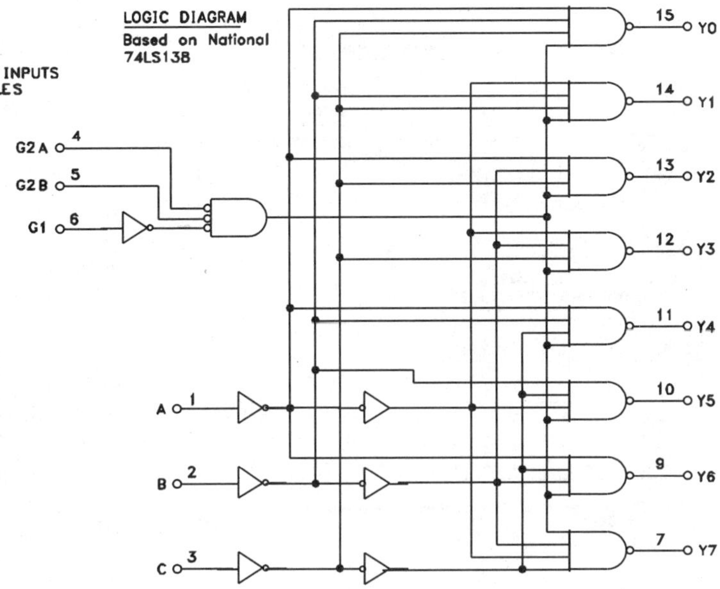

FUNCTION TABLE 74x138 DECODER/DEMULTIPLEXER

INPUTS ENABLES		SELECT			OUTPUTS							
G1	G2*	C	B	A	Y0	Y1	Y2	Y3	Y4	Y5	Y6	Y7
X	1	X	X	X	1	1	1	1	1	1	1	1
0	X	X	X	X	1	1	1	1	1	1	1	1
1	0	0	1	0	0	1	1	1	1	1	1	1
1	0	0	0	1	1	0	1	1	1	1	1	1
1	0	0	1	0	1	1	0	1	1	1	1	1
1	0	0	1	1	1	1	1	0	1	1	1	1
1	0	1	1	0	1	1	1	1	0	1	1	1
1	0	1	1	1	1	1	1	1	1	0	1	1
1	0	1	1	0	1	1	1	1	1	1	0	1
1	0	1	1	1	1	1	1	1	1	1	1	0

*G2 = G2A+G2B
1 = HIGH voltage level. 0 = LOW voltage level.
X = Don't Care

Figure 8.8. The 74AS/ALS138, and the CMOS devices shown are 3-line to 8-line decoder/demultiplexers designed for high-performance memory decoding or data routing applications requiring very short propagation delay times. Conditions at the binary select inputs and the three enable inputs select one of eight input lines. Two active low and one active high enable inputs reduce the need for external gates or inverters when expanding. A 24-line decoder can be implemented with no external logic. A 32-line decoder requires one external inverter.

TTL - 74S/LS/AS/ALS/F139
5V CMOS - 74HC/HCT/AHC/AHCT/VHC/AC/ACT139
3V CMOS - 74LV/LVC/AC/VHC139

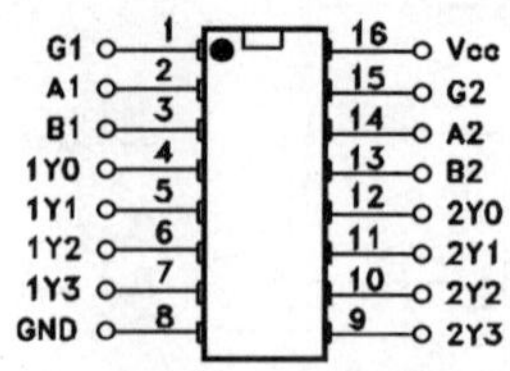

PACKAGE PINNING

PIN–
2,3,13,14 ADDRESSING INPUTS
1,15 ENABLES
4–7 SECTION 1 OUTPUTS
9–12 SECTION 2 OUTPUTS

DUAL-IN-LINE PACKAGE

FUNCTION TABLE 74x139
DUAL DECODER/DEMULTIPLEXER

INPUTS		OUTPUTS
ENABLE	SELECT	
	B A	Y0 Y1 Y2 Y3
1	X X	1 1 1 1
0	0 0	0 1 1 1
0	0 1	1 0 1 1
0	1 0	1 1 0 1
0	1 1	1 1 1 0

1 = HIGH voltage level. 0 = LOW voltage level.
X = Don't Care

LOGIC DIAGRAM
Based on National 74LS139

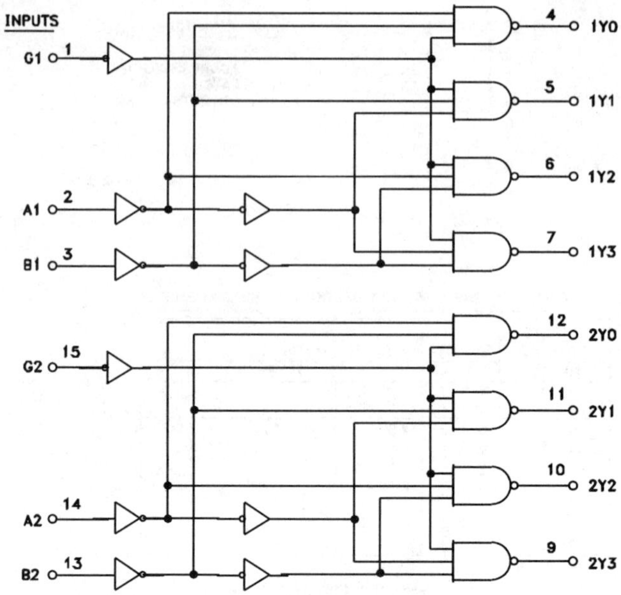

Figure 8.9. The 74AS/ALS139, and the CMOS devices shown are dual 2-line to 4-line decoder/demultiplexers designed for high-performance memory decoding or data routing applications requiring very short propagation delay times. The active low enable can be used as a data line in demultiplexing applications. Inputs are fully buffered to minimize line loading.

TTL – 74147/LS
5V CMOS – 74HC/HCT147

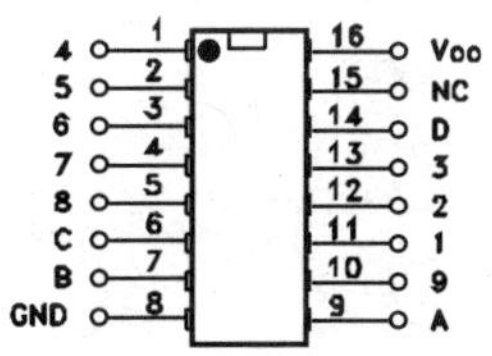

DUAL-IN-LINE PACKAGE

PACKAGE PINNING:

PIN–
1–7,10–13 DATA INPUTS
6,7,9,14 ENCODED OUTPUTS
15 NO CONNECTION

FUNCTION TABLE 74147, 'LS147 4-LINE-TO-10 LINE PRIORITY ENCODERS

INPUTS									OUTPUTS			
1	2	3	4	5	6	7	8	9	D	C	B	A
1	1	1	1	1	1	1	1	1	1	1	1	1
X	X	X	X	X	X	X	X	0	0	1	1	0
X	X	X	X	X	X	X	0	X	0	1	1	1
X	X	X	X	X	X	0	X	X	1	0	0	0
X	X	X	X	X	0	X	X	X	1	0	0	1
X	X	X	X	0	X	X	X	X	1	0	1	0
X	X	X	0	X	X	X	X	X	1	0	1	1
X	X	0	X	X	X	X	X	X	1	1	0	0
X	0	X	X	X	X	X	X	X	1	1	0	1
0	X	X	X	X	X	X	X	X	1	1	1	0

1 = HIGH voltage level. 0 = LOW voltage level.
X = Don't Care

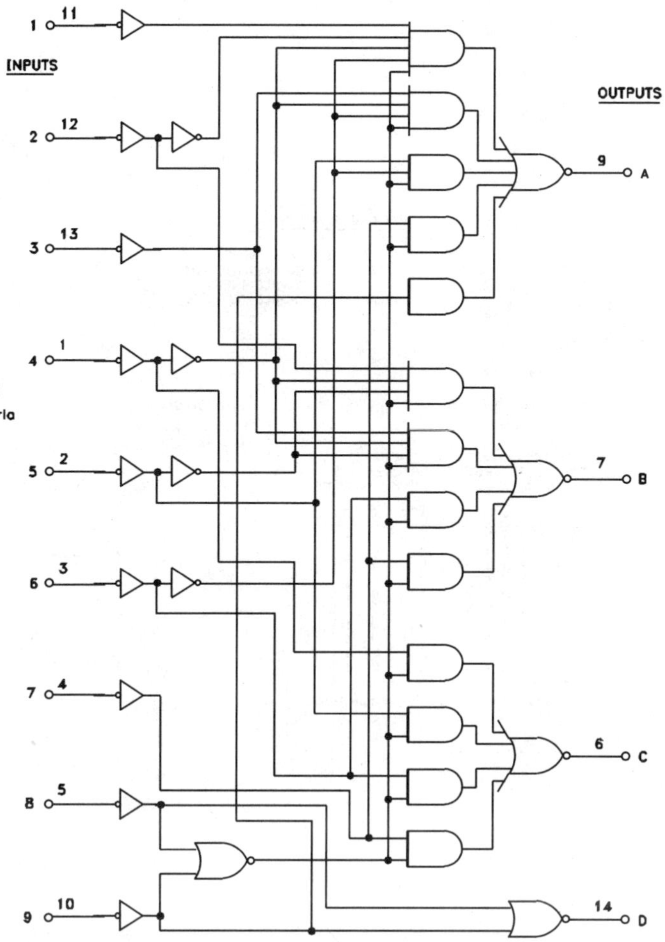

LOGIC DIAGRAM
Based on Motorola 74LS147

Figure 8.10. The 74x147 TTL and CMOS logic series are priority encoders. They provide priority decoding of the inputs to ensure that only the highest order data line is encoded. Data inputs and outputs are active at the low logic level. These devices encode nine data lines to four-line BCD (8-4-2-1). The implied decimal zero condition does not require an input condition because zero is encoded when all nine data lines are at a high logic level.

TTL – 74148/LS/F
74LS748
5V CMOS – 74HC148
CD4532

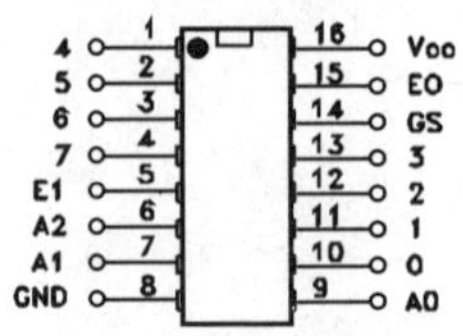

DUAL-IN-LINE PACKAGE

PACKAGE PINNING:

PIN–	
1–4,10–13	DATA INPUTS
6,7,9	ENCODED OUTPUTS
5	ENABLE INPUT
14,15	ENABLE OUTPUTS

FUNCTION TABLE 74x148, 'LS748, CD4532
8-LINE-TO-3 LINE PRIORITY ENCODERS

INPUTS									OUTPUTS				
EI	0	1	2	3	4	5	6	7	C	B	A	GS	EO
1	X	X	X	X	X	X	X	X	1	1	1	1	1
0	1	1	1	1	1	1	1	1	1	1	1	1	0
0	X	X	X	X	X	X	X	0	0	0	0	0	1
0	X	X	X	X	X	X	0	X	0	0	1	0	1
0	X	X	X	X	X	0	X	X	0	1	0	0	1
0	X	X	X	X	0	X	X	X	0	1	1	0	1
0	X	X	X	0	X	X	X	X	1	0	0	0	1
0	X	X	0	X	X	X	X	X	1	0	1	0	1
0	X	0	X	X	X	X	X	X	1	1	0	0	1
0	0	X	X	X	X	X	X	X	1	1	1	0	1

1 = HIGH voltage level. 0 = LOW voltage level.
X = Don't Care

LOGIC DIAGRAM
Based on Motorola 74LS148

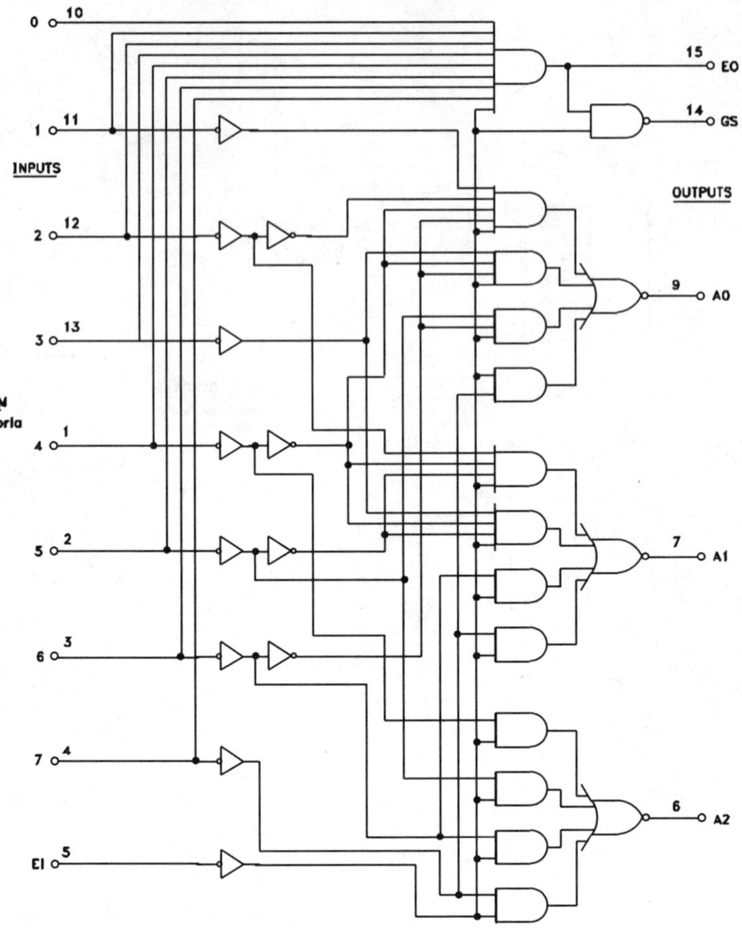

Figure 8.11. The 74x148, 74LS748 TTL and CMOS logic series are priority encoders similar to the 74x147. They provide priority decoding of the inputs to ensure that only the highest order data line is encoded. Data inputs and outputs are active at the low logic level. These devices encode eight data lines to three-line (4-2-1) binary; that is, octal. Octal expansion is possible without external logic provided.

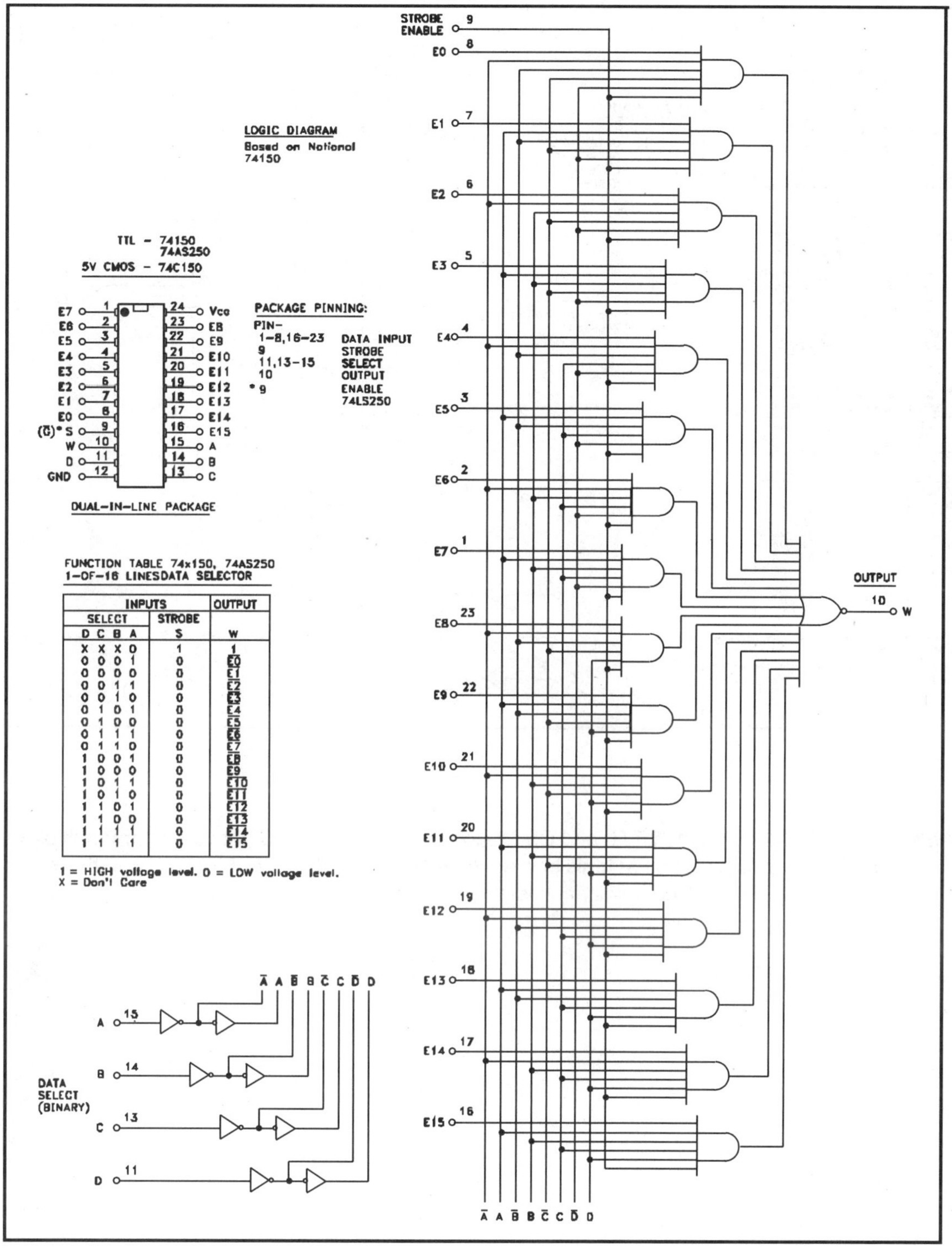

FUNCTION TABLE 74x150, 74AS250
1-OF-16 LINESDATA SELECTOR

INPUTS					OUTPUT
SELECT				STROBE	
D	C	B	A	S	W
X	X	X	0	1	1
0	0	0	1	0	$\overline{E0}$
0	0	0	0	0	$\overline{E1}$
0	0	1	1	0	$\overline{E2}$
0	0	1	0	0	$\overline{E3}$
0	1	0	1	0	$\overline{E4}$
0	1	0	0	0	$\overline{E5}$
0	1	1	1	0	$\overline{E6}$
0	1	1	0	0	$\overline{E7}$
1	0	0	1	0	$\overline{E8}$
1	0	0	0	0	E9
1	0	1	1	0	$\overline{E10}$
1	0	1	0	0	$\overline{E11}$
1	1	0	1	0	$\overline{E12}$
1	1	0	0	0	$\overline{E13}$
1	1	1	1	0	$\overline{E14}$
1	1	1	1	0	$\overline{E15}$

1 = HIGH voltage level. 0 = LOW voltage level.
X = Don't Care

Figure 8.12. The 74x150, 74AS250 TTL and 74C150 CMOS logic series are 1-of-16 data selectors/multiplexers. An enabling strobe input must be at logic low; a high level forces the output high. The device has application as a Boolean function generator.

TTL – 74151/S/LS/AS/ALS/F
74251/S/LS/ASALS/F
5V CMOS – 74C/HC/HCT/AC/ACT151
74HC/HCT/AC/ACT251
3V CMOS – 74LVX151
74LV/AC251

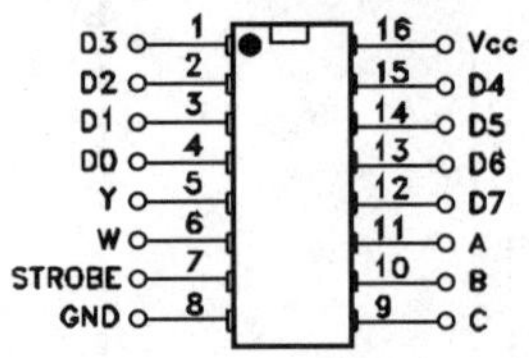

DUAL-IN-LINE PACKAGE

PACKAGE PINNING:
PIN–
1–4,12–15 DATA INPUTS
9–11 DATA SELECT
5,6 OUTPUTS
7 STROBE

LOGIC DIAGRAM
Based on Notional 74LS151

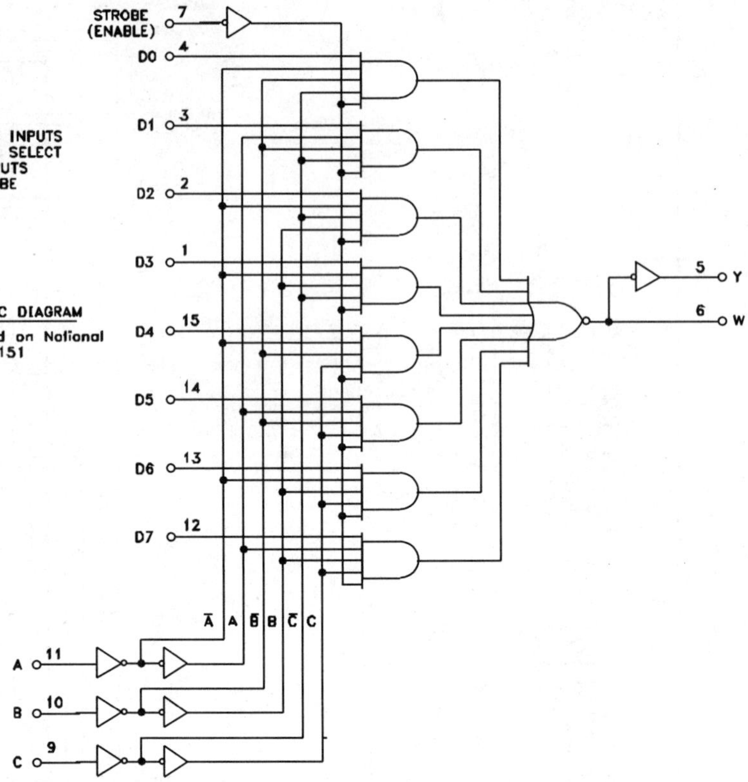

FUNCTION TABLE 74x151, 74x251
DATA SELECTOR/MULTIPLEXER

INPUTS				OUTPUTS	
SELECT			STROBE		
C	B	A	S	Y	W
X	X	X	1	0	1
0	0	0	0	D0	$\overline{D0}$
0	0	1	0	D1	$\overline{D1}$
0	1	0	0	D2	$\overline{D2}$
0	1	1	0	D3	$\overline{D3}$
1	0	0	0	D4	$\overline{D4}$
1	0	1	0	D5	$\overline{D5}$
1	1	0	0	D6	$\overline{D6}$
1	1	1	0	D7	$\overline{D7}$

1 = HIGH voltage level. 0 = LOW voltage level.
X = Don't Care
D0,D1...D7 = The level of the respective D input.

Figure 8.13. The 74X251, 74x151 TTL and CMOS logic series are 1-of-8 data selectors/multiplexers. An enabling strobe input must be at logic low; a high level forces the W output high and the Y low. The device application is similar to the 74x150/250 series.

TTL - 74152/LS
5V CMOS - 74HC/AC/ACT152
3V CMOS - 74AC152

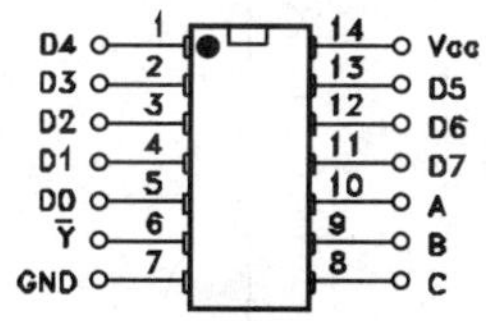

DUAL-IN-LINE PACKAGE

PACKAGE PINNING:

PIN-
1-5,11-13 DATA INPUTS
8-10 DATA SELECT
7 OUTPUT

FUNCTION TABLE 74x152 8-LINE TO 1-LINEDATA SELECTOR/MULTIPLEXER

INPUTS											OUTPUT
SELECT			DATA								
C	B	A	D0	D1	D2	D3	D4	D5	D6	D7	$\overline{Y}$
0	0	0	0	X	X	X	X	X	X	X	1
0	0	0	1	X	X	X	X	X	X	X	0
0	0	1	X	0	X	X	X	X	X	X	1
0	0	1	X	1	X	X	X	X	X	X	0
0	1	0	X	X	0	X	X	X	X	X	1
0	1	0	X	X	1	X	X	X	X	X	0
0	1	1	X	X	X	0	X	X	X	X	1
0	1	1	X	X	X	1	X	X	X	X	0
1	0	0	X	X	X	X	0	X	X	X	1
1	0	0	X	X	X	X	1	X	X	X	0
1	0	1	X	X	X	X	X	0	X	X	1
1	0	1	X	X	X	X	X	1	X	X	0
1	1	0	X	X	X	X	X	X	0	X	1
1	1	0	X	X	X	X	X	X	1	X	0
1	1	1	X	X	X	X	X	X	X	0	1
1	1	1	X	X	X	X	X	X	X	1	0

1 = HIGH voltage level. 0 = LOW voltage level.
X = Don't Care

LOGIC DIAGRAM

Based on Philips/ Signetics 74152

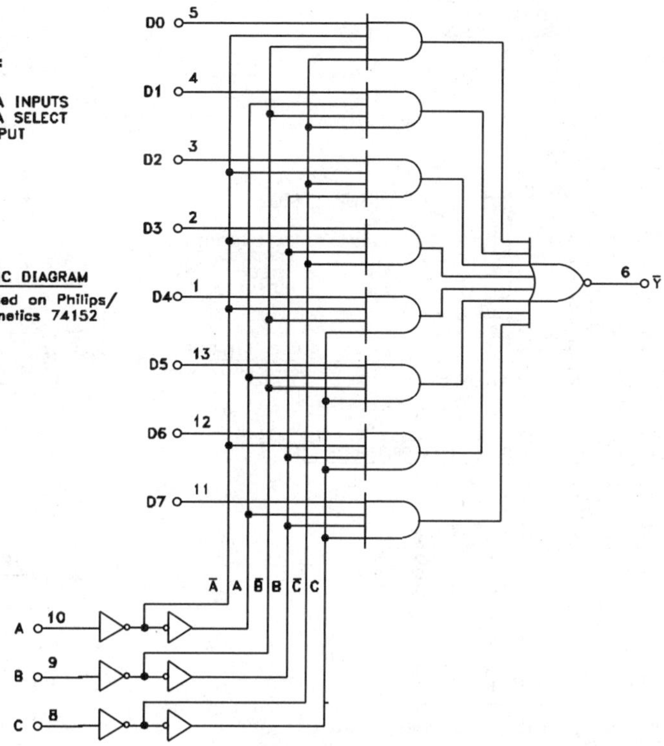

Figure 8.14. The 74x152 TTL and CMOS logic series are 1-of-8 data selectors/multiplexers similar to the series of *Figure 8.13*, without the enabling strobe. The device is the logical implementation of a single pole, 8-position rotary switch with the switch position controlled by the three select inputs.

TTL - 74153/L/S/LS/AS/ALS/F
74253/S/LS/AS/ALS/F
74LS/AS/ALS/F352
74LS/AS/ALS/F353
5V CMOS - 74HC/HCT153
74HC/HCT/AC/ACT253
74HC352
74HC353
CD4539
3V CMOS - 74LV/VHC/AC153
74AC253

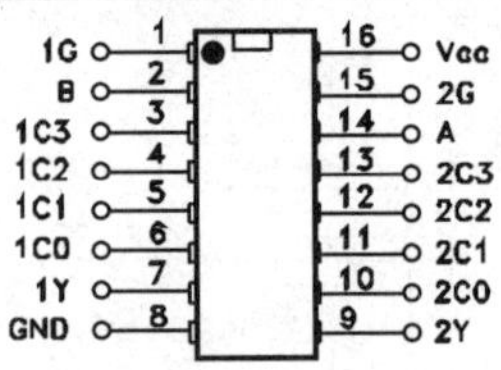

DUAL-IN-LINE PACKAGE

PACKAGE PINNING:

PIN-
3-6,10-13 DATA INPUT
1,15 STROBE
2,14 SELECT
7,9 OUTPUT

LOGIC DIAGRAM
Based on National 74LS153

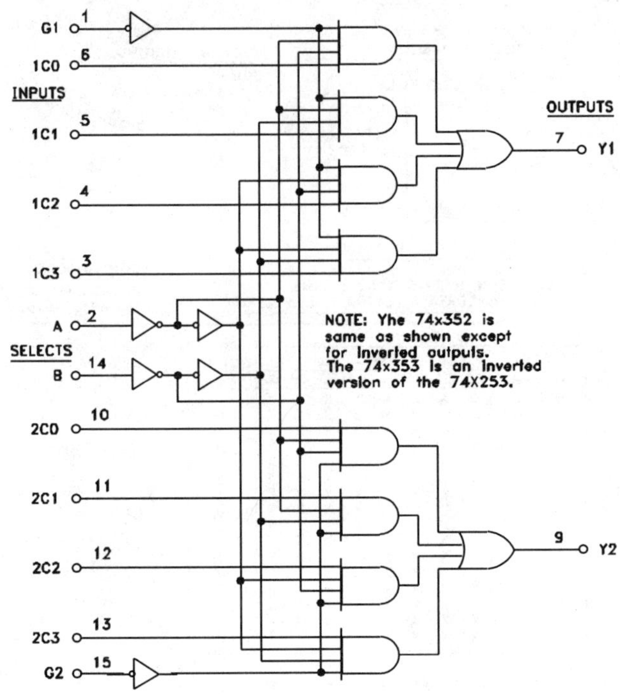

FUNCTION TABLE 74x153, 74x253, 74x352*, 74x353,** 4539
DUAL 4-LINE TO 1-LINE DATA SELECTOR/MULTIPLEXER

SELECT INPUTS		DATA INPUTS				STROBE	OUTPUT
B	A	C0	C1	C2	C3	G	Y
X	X	X	X	X	X	1	0
0	0	0	X	X	X	0	0
0	0	1	X	X	X	0	1
0	1	X	0	X	X	0	0
0	1	X	1	X	X	0	1
1	0	X	X	0	X	0	0
1	0	X	X	1	X	0	1
1	1	X	X	X	0	0	0
1	1	X	X	X	1	0	1

1 = HIGH voltage level. 0 = LOW voltage level.
X = Don't Care

* 74x352 outputs are inverted.
** 74x353 an inverted version of 74x253.

Figure 8.15. The 74x153, 'x253, 'x353 TTL and CMOS logic series are dual 4-input multiplexers with common select inputs and individual enable inputs for each section. The device is the logical implementation of a 2-pole 4-position rotary switch where the switch position is determined by the logic levels provided to the two select inputs. As a function generator the device can generate two functions of three variables.

TTL – 74154/L/LS/F
74159 (OPEN COLLECTOR)
5V CMOS – 74C/HC/HCT154
3V CMOS – 74LV154

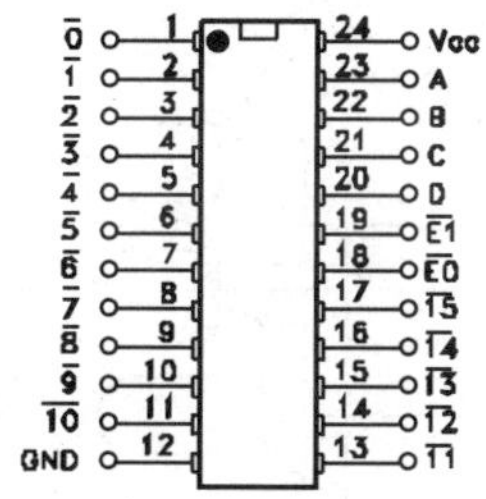

DUAL-IN-LINE PACKAGE

PACKAGE PINNING:

PIN–	
1–11,13–17	OUTPUTS
18,19	ENABLES
20–23	SELECT

FUNCTION TABLE 74x154, 74159
TO 1–LINE DATA SELECTOR/MULTIPLEXER

INPUTS						OUTPUTS *															
ENABLE		SELECT																			
$\overline{E0}$	$\overline{E1}$	D	C	B	A	$\overline{0}$	$\overline{1}$	$\overline{2}$	$\overline{3}$	$\overline{4}$	$\overline{5}$	$\overline{6}$	$\overline{7}$	$\overline{8}$	$\overline{9}$	$\overline{A}$	$\overline{B}$	$\overline{C}$	$\overline{D}$	$\overline{E}$	$\overline{F}$
0	1	X	X	X	X	1	1	1	1	1	1	1	1	1	1	1	1	1	1	1	1
1	0	X	X	X	X	1	1	1	1	1	1	1	1	1	1	1	1	1	1	1	1
1	1	X	X	X	X	1	1	1	1	1	1	1	1	1	1	1	1	1	1	1	1
0	0	0	0	0	0	0	1	1	1	1	1	1	1	1	1	1	1	1	1	1	1
0	0	0	0	0	1	1	0	1	1	1	1	1	1	1	1	1	1	1	1	1	1
0	0	0	0	1	0	1	1	0	1	1	1	1	1	1	1	1	1	1	1	1	1
0	0	0	0	1	1	1	1	1	0	1	1	1	1	1	1	1	1	1	1	1	1
0	0	0	1	0	0	1	1	1	1	0	1	1	1	1	1	1	1	1	1	1	1
0	0	0	1	0	1	1	1	1	1	1	0	1	1	1	1	1	1	1	1	1	1
0	0	0	1	1	0	1	1	1	1	1	1	0	1	1	1	1	1	1	1	1	1
0	0	0	1	1	1	1	1	1	1	1	1	1	0	1	1	1	1	1	1	1	1
0	0	1	0	0	0	1	1	1	1	1	1	1	1	0	1	1	1	1	1	1	1
0	0	1	0	0	1	1	1	1	1	1	1	1	1	1	0	1	1	1	1	1	1
0	0	1	0	1	0	1	1	1	1	1	1	1	1	1	1	0	1	1	1	1	1
0	0	1	0	1	1	1	1	1	1	1	1	1	1	1	1	1	0	1	1	1	1
0	0	1	1	0	0	1	1	1	1	1	1	1	1	1	1	1	1	0	1	1	1
0	0	1	1	0	1	1	1	1	1	1	1	1	1	1	1	1	1	1	0	1	1
0	0	1	1	1	0	1	1	1	1	1	1	1	1	1	1	1	1	1	1	0	1
0	0	1	1	1	1	1	1	1	1	1	1	1	1	1	1	1	1	1	1	1	0

1 = HIGH voltage level. 0 = LOW voltage level.
X = Don't Care
* A = 10, B = 11, C = 12, D = 13, E = 14, F = 15

LOGIC DIAGRAM
Based on Philips/Signetics 74154

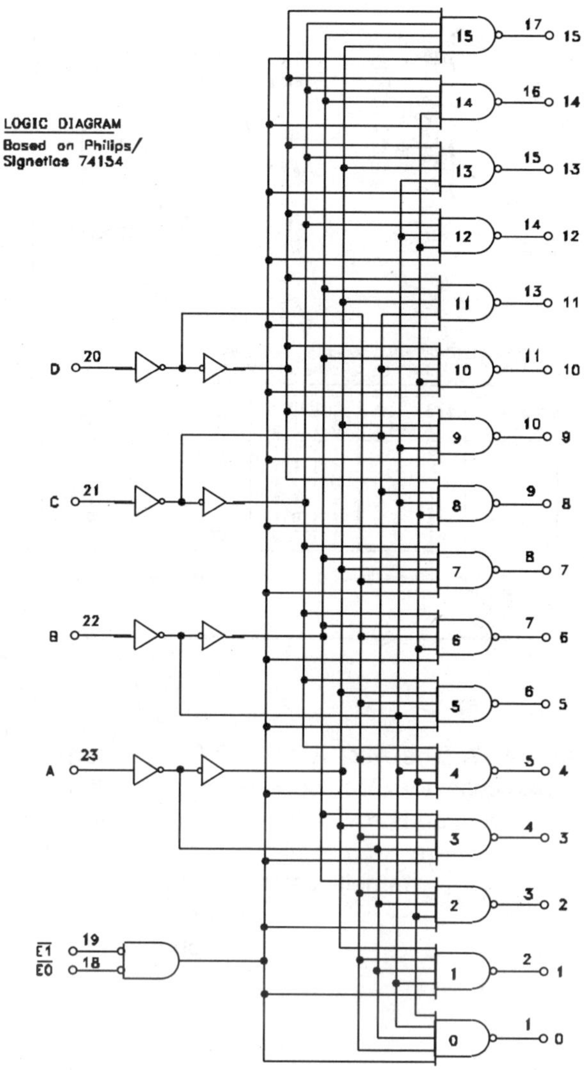

Figure 8.16. The 74159, 'x154 TTL and CMOS logic series are 4-to-16 line decoder/demultiplexers having a two-input active low enable gate. 16 mutually exclusive active low outputs are obtainable. The enable can be used as a data input to demultiplex up to 16-bits of serial data.

TTL – 74155/LS
74156/LS/ALS

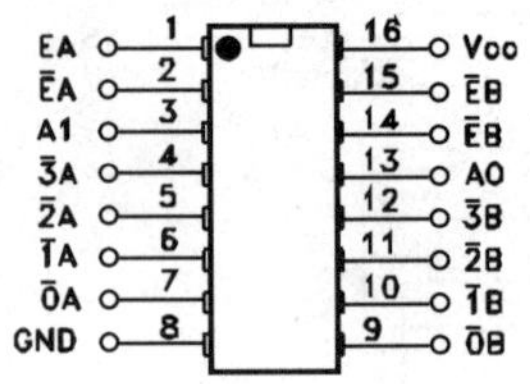

DUAL-IN-LINE PACKAGE

LOGIC SYMBOL

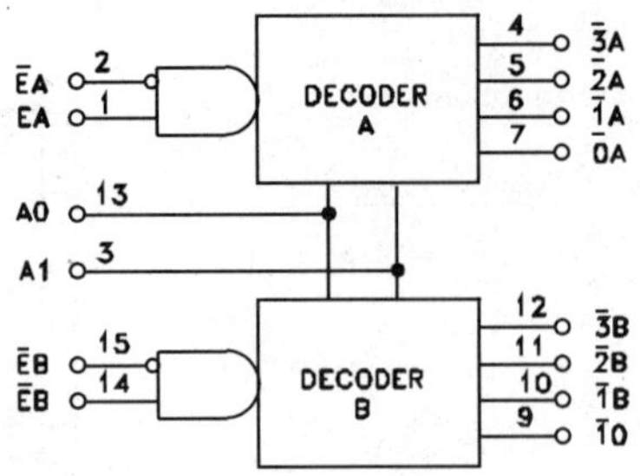

LOGIC DIAGRAM
Based on Philips/
Signetics 74x155

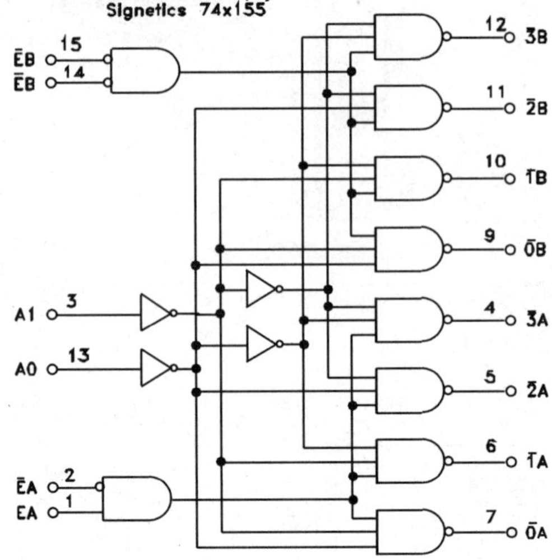

FUNCTION TABLE 74x155, 74x156
DUAL 2-LINE TO 4-LINE DECODER/DEMULTIPLEXER

ADDRESS		ENABLE A		OUTPUT A				ENABLE B		OUTPUT B			
A0	A1	EA	$\overline{EA}$	$\overline{0}$	$\overline{1}$	$\overline{2}$	$\overline{3}$	$\overline{EB}$	$\overline{EB}$	$\overline{0}$	$\overline{1}$	$\overline{2}$	$\overline{3}$
X	X	0	X	1	1	1	1	1	X	1	1	1	1
X	X	X	1	1	1	1	1	X	1	1	1	1	1
0	0	1	0	0	1	1	1	0	0	0	1	1	1
1	0	1	0	1	0	1	1	0	0	1	0	1	1
0	1	1	0	1	1	0	1	0	0	1	1	0	1
1	1	1	0	1	1	1	0	0	0	1	1	1	0

1 = HIGH voltage level. 0 = LOW voltage level.
X = Don't Care

Figure 8.17. The 74x155, 'x156 TTL logic series are dual 2-to-4 line decoder/demultiplexers. The '156 is open collector. Each decoder section, when enabled, will accept the two binary weighted address inputs and provide four mutually exclusive active low outputs. When encoder requirements are not met all outputs are high. The device can be used as a 1-of-8 decoder/demultiplexer by tying EA to /EB and relabeling the common connection as address A2. A common enable is then formed by connecting the remaining /EB to /EA.

TTL - 74157/L/LS/AS/ALS/F
74158/LS/AS/ALS/F
74S/LS/AS/ALS/F257
74S/LS/AS/ALS/F258

5V CMOS - 74C/HC/HCTAHC/AHCT/AC/ACT157
74HC/HCT/AHC/AHCT/AC/ACT158
74HC/HCT/AC/ACT257
74HC/HCT/AC/ACT258

3V CMOS - 74AHC/AC157 74LV/LVC/AC/VHC257
74AHC/AC158 74AC258

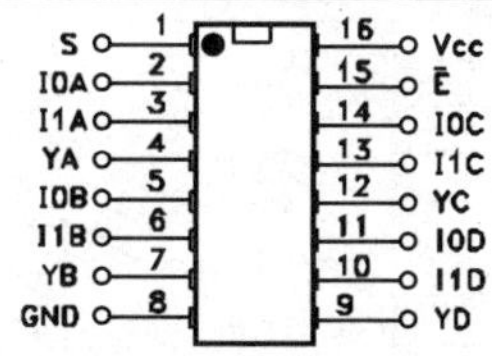

DUAL-IN-LINE PACKAGE

PACKAGE PINNING

PIN-	
1	COMMON DATA SELECT
2,5,14,11	DATA INPUTS FROM SOURCE 0
3,6,13,10	DATA INPUTS FROM SOURCE 1
4,7,12,9	MULTIPLEXER OUTPUTS
15	ENABLE

FUNCTION TABLE 74x157, 74x158, 74x257, 74x258
QUAD 2-INPUT DATA SELECTOR/MULTIPLEXER

ENABLE	SELECT INPUT	DATA INPUTS		** OUTPUT
Ē*	S	I0	I1	Y
1	X	X	X	0
0	1	X	0	0
0	1	X	1	1
0	0	0	X	0
0	0	1	X	1

1 = HIGH voltage level. 0 = LOW voltage level.
X = Don't Care

* Replace E with Output Enable OE for 74x257, '258.
** 74x258 is an inverted version of 74x257.

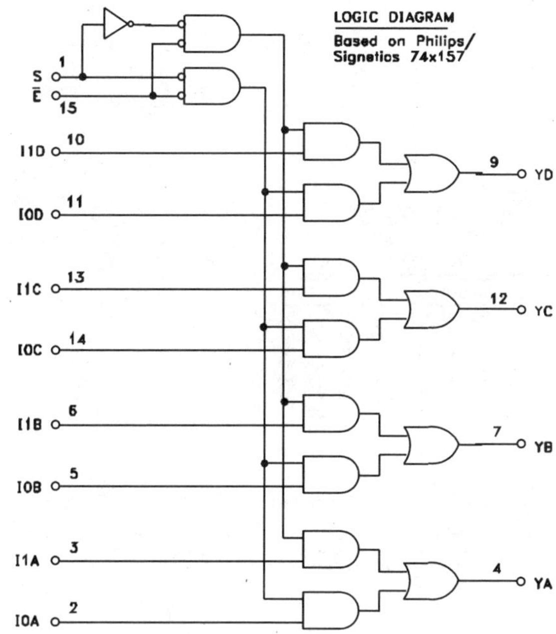

Figure 8.18. The 74x157, 'x158, 'x257, x258 TTL and CMOS logic series are quad 2-input multiplexers that can select four bits of data from two sources using the common select and enable inputs. The 'x257 and 'x258 devices, not shown in the diagram, have 3-state outputs. The '7 devices have true outputs; the '8s inverted. The devices are a logical implementation of a 4-pole, 2-position switch where the switch position is determined by the logic levels provided the two select inputs.

TTL – 74LS261

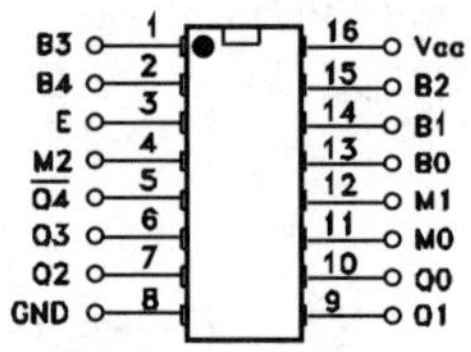

DUAL-IN-LINE PACKAGE

PACKAGE PINNING:

PIN–
1,2,13–15 DATA INPUTS
3 ENABLE
4,11,12 MULTIPLIER INPUT
5–7,9,10 OUTPUTS

FUNCTION TABLE 74LS261 MULTIPLY DECODER

INPUTS				OUTPUTS				
E	M2	M1	M0	$\overline{Q4}$	Q3	Q2	Q1	Q0
0	X	X	X	q4	q3	q2	q1	q0
1	0	0	0	1	0	0	0	0
1	0	0	1	$\overline{B4}$	B4	B3	B2	B1
1	0	1	0	$\overline{B4}$	B4	B3	B2	B1
1	0	1	1	$\overline{B4}$	B3	B2	B1	B0
1	1	0	0	B4	$\overline{B3}$	$\overline{B2}$	$\overline{B1}$	$\overline{B0}$
1	1	0	1	B4	$\overline{B4}$	$\overline{B3}$	$\overline{B2}$	$\overline{B1}$
1	1	1	0	B4	$\overline{B4}$	$\overline{B3}$	$\overline{B2}$	$\overline{B1}$
1	1	1	1	1	0	0	0	0

1 = HIGH voltage level. 0 = LOW voltage level.
X = Don't Care

q = Lower case letters indicate the state of the referenced output one setup time prior to the HIGH-to-LOW Enable transition.

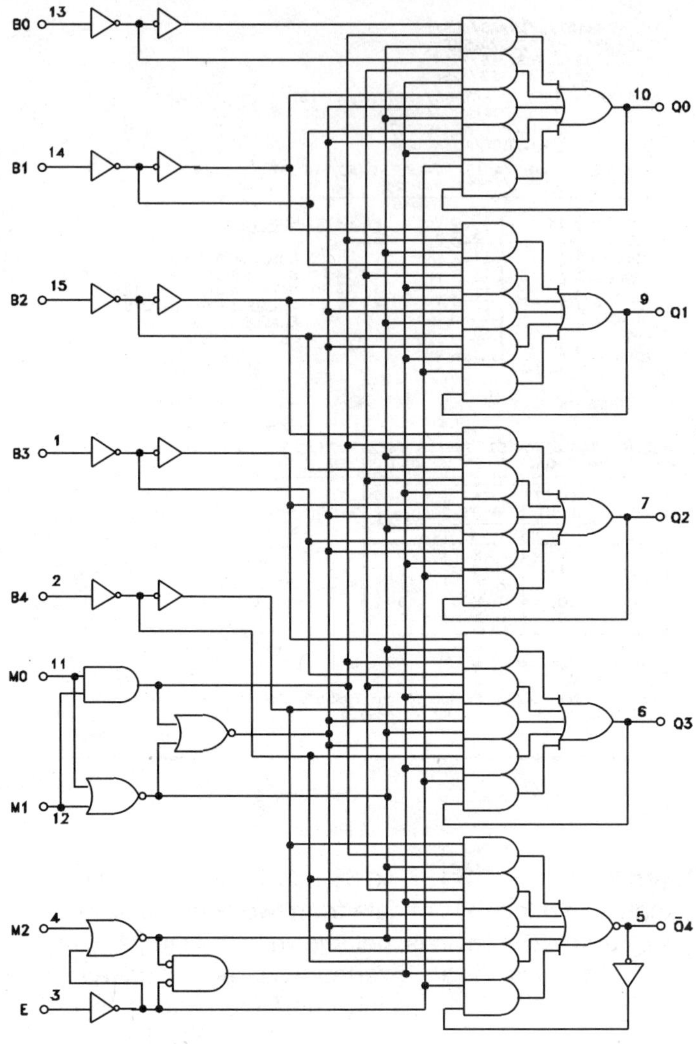

LOGIC DIAGRAM
Based on Philips/Signetic 74LS261

Figure 8.19. The 74LS261 is a multiply decoder designed for use in parallel multiplication applications. Multiplication is performed two bits at a time in two's complement form. The "M" inputs are for the multiplier; the "B" inputs for the multiplicand. The outputs represent partial products in one's complement notation. The leading, most significant, bit of the product is inverted for ease in extending the sign to left justify the partial-product bits.

TTL - 74298/LS
5V CMOS - 74C/HC298

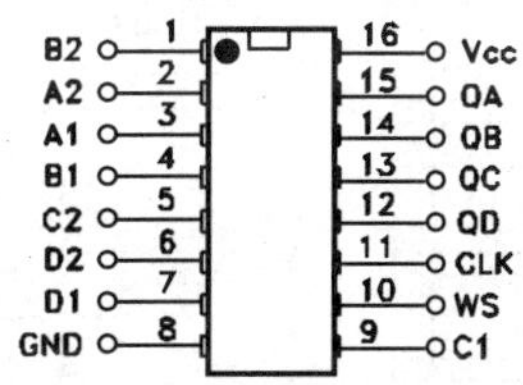

DUAL-IN-LINE PACKAGE

FUNCTION TABLE 74x298 QUAD 2-INPUT MULTIPLEXERS WITH STORAGE

INPUTS		OUTPUTS			
WORD SELECT	CLOCK	QA	QB	QC	QD
0	↓	a1	b1	c1	d1
1	↓	a2	b2	c2	d2
X	1	QA0	QB0	QC0	QD0

1 = HIGH voltage level. 0 = LOW voltage level.
X = Don't Care
↓ = HIGH-to-LOW clock transition.

LOGIC DIAGRAM
Based on Texas Instruments 74LS298

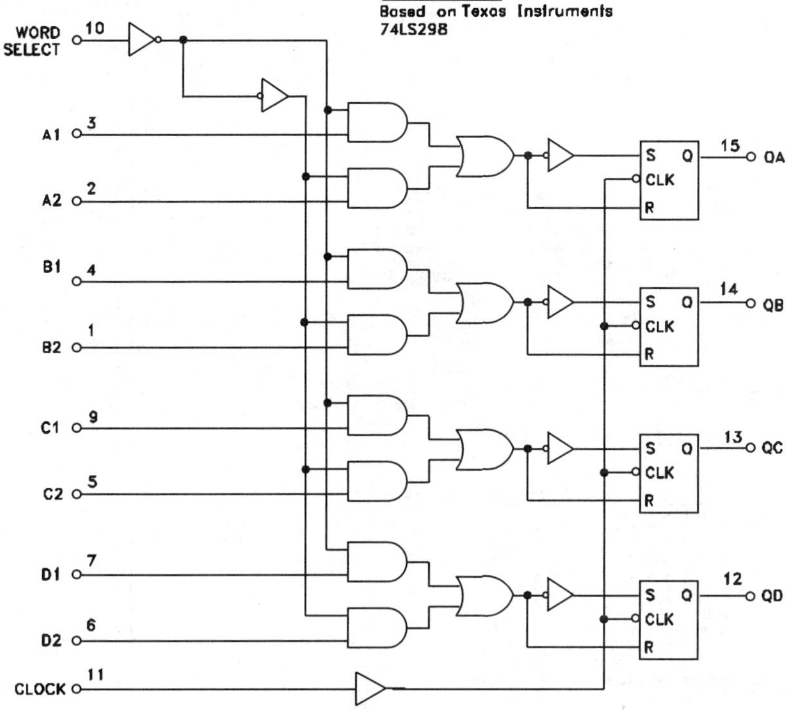

Figure 8.20. The 74x298 TTL and CMOS series are quad 2-input multiplexers with storage. The device selects one of two four-bit data sources and stores data synchronously with the clock. When the word select input is low, word 1 (A1,B1,C1,D1) is selected. With the select high, word 2 (A2,B2,C2,D2) is selected. The selected word is clocked to the output on the negative-going transition. Clock pulse, tw, is 20 nS with 25 nS select setup time and zero hold time. Data setup and hold times are 15 and 5 nS, respectively.

TTL - 74LS348
74LS848

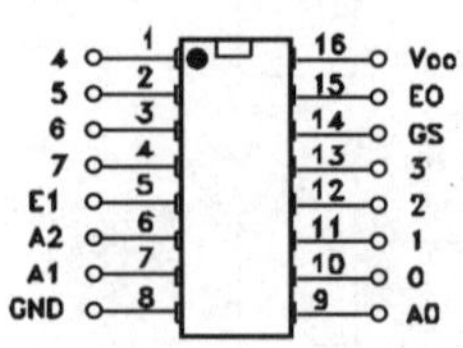

DUAL-IN-LINE PACKAGE

PACKAGE PINNING:

PIN-
1-4,10-13 DATA INPUTS
6,7,9 ENCODED OUTPUTS
5 ENABLE INPUT
14,15 ENABLE OUTPUTS

FUNCTION TABLE 74LS348, 74LS848 8-LINE TO 3-LINE PRIORITY ENCODERS

INPUTS									OUTPUTS				
EI	0	1	2	3	4	5	6	7	C	B	A	GS	EO
1	X	X	X	X	X	X	X	X	1	1	1	1	1
0	1	1	1	1	1	1	1	1	1	1	1	1	0
0	X	X	X	X	X	X	X	0	0	0	0	0	1
0	X	X	X	X	X	X	0	X	0	0	1	0	1
0	X	X	X	X	X	0	X	X	0	1	0	0	1
0	X	X	X	X	0	X	X	X	0	1	1	0	1
0	X	X	X	0	X	X	X	X	1	0	0	0	1
0	X	X	0	X	X	X	X	X	1	0	1	0	1
0	X	0	X	X	X	X	X	X	1	1	0	0	1
0	0	X	X	X	X	X	X	X	1	1	1	0	1

1 = HIGH voltage level. 0 = LOW voltage level.
X = Don't Care

LOGIC DIAGRAM
Based on Motorla 74LS348

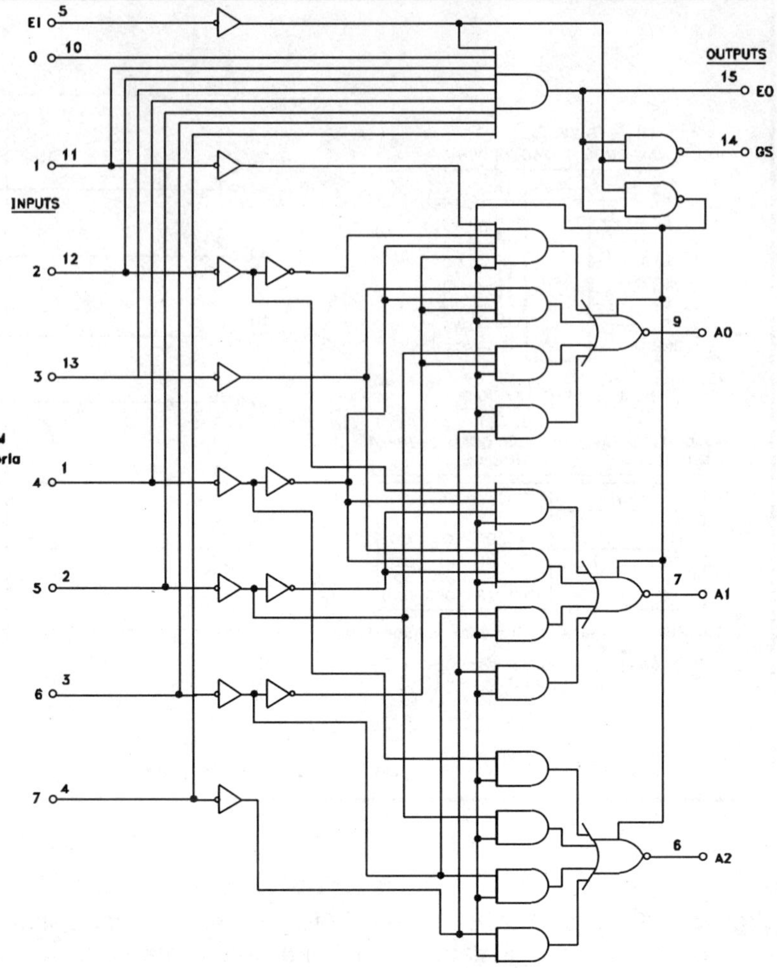

Figure 8.21. The TTL 74LS348 and 74LS848 are 8-input priority encoders with 3-state outputs. The LS848 features special internal circuitry to greatly reduce the negative-going glitch on the group signal, GS, output. The outputs and inputs are active low. The active low input having the highest priority is represented on the outputs. The group signal is active low when any of the inputs is low. It serves to indicate when an input is active. A0, A1, and A2 are the 3-state outputs.

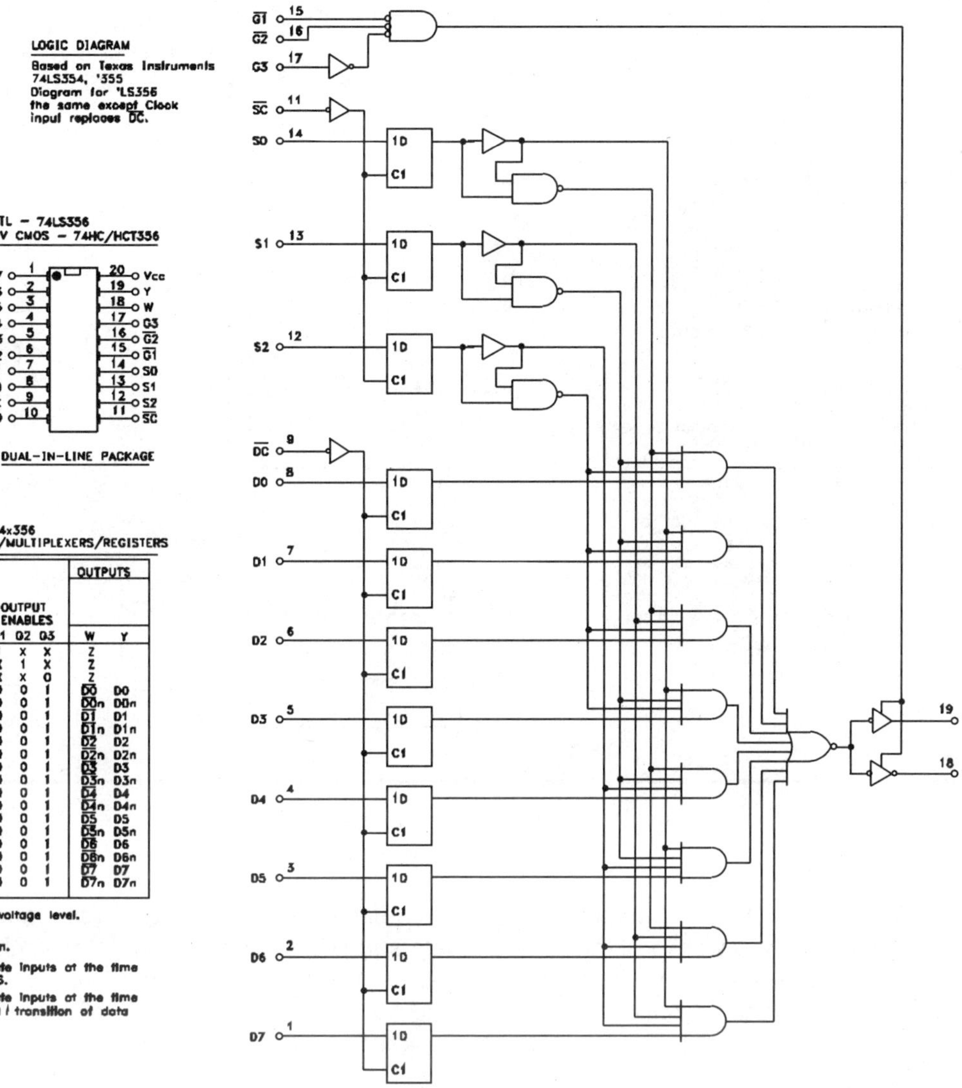

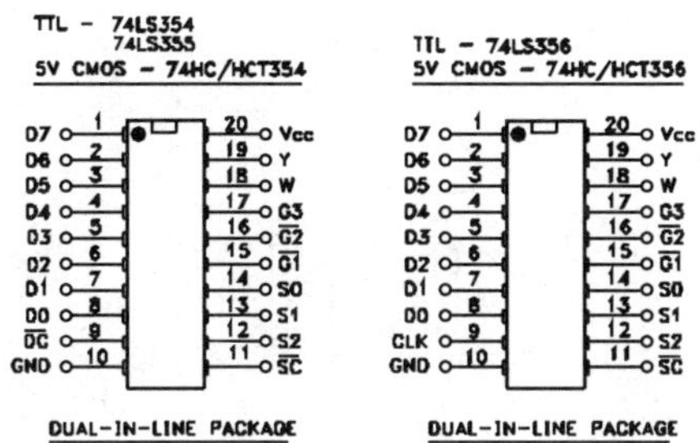

FUNCTION TABLE 74x354, 74x355, 74x356
8-LINE TO 1-LINE DATA SELECTORS/MULTIPLEXERS/REGISTERS

SELECT			INPUTS DATA CONTROL 'LS354 'LS355	CLOCK 'LS356	OUTPUT ENABLES			OUTPUTS	
S2	S1	S0			G1	G2	G3	W	Y
X	X	X	X	X	1	X	X	Z	
X	X	X	X	X	X	1	X	Z	
X	X	X	X	X	X	X	0	Z	
0	0	0	0	↑	0	0	1	$\overline{D0}$	D0
0	0	0	1	1or0	0	0	1	$\overline{D0}$n	D0n
0	0	1	0	↑	0	0	1	$\overline{D1}$	D1
0	0	1	1	1or0	0	0	1	$\overline{D1}$n	D1n
0	1	0	0	↑	0	0	1	$\overline{D2}$	D2
0	1	0	1	1or0	0	0	1	$\overline{D2}$n	D2n
0	1	1	0	↑	0	0	1	$\overline{D3}$	D3
0	1	1	1	1or0	0	0	1	$\overline{D3}$n	D3n
1	0	0	0	↑	0	0	1	$\overline{D4}$	D4
1	0	0	1	1or0	0	0	1	$\overline{D4}$n	D4n
1	0	1	0	↑	0	0	1	$\overline{D5}$	D5
1	0	1	1	1or0	0	0	1	$\overline{D5}$n	D5n
1	1	0	0	↑	0	0	1	$\overline{D6}$	D6
1	1	0	1	1or0	0	0	1	$\overline{D6}$n	D6n
1	1	1	0	↑	0	0	1	$\overline{D7}$	D7
1	1	1	1	1or0	0	0	1	$\overline{D7}$n	D7n

1 = HIGH voltage level. 0 = LOW voltage level.
X = Don't Care

↑ = LOW-to-HIGH clock transition.

D0...D7 = The level of steady state inputs at the time of the ↑ with the 'LS356.

D0n..D7n = The level of steady state inputs at the time before the most recent ↑ transition of data control or clock.

Figure 8.22. The 74x354, 74LS355 and 74x356 are 8-line to 1-line data selectors/multiplexers/registers. The '354 data select address is stored in transparent latches that are active low enabled on input /SC. The open collector '355 is similar to the '354. For these two a logic low on input /DC is also a data enable. The '356 features negative-edge triggering at the CLK input and 3-state outputs. Setup and hold times for these is 15 nS.

TTL - 74LS399

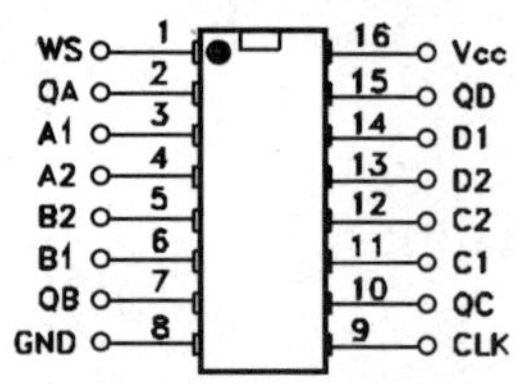

DUAL-IN-LINE PACKAGE

FUNCTION TABLE 74LS399 QUAD 2-INPUT MULTIPLEXERS WITH STORAGE

INPUTS		OUTPUTS			
WORD SELECT	CLOCK	QA	QB	QC	QD
0	↑	a1	b1	c1	d1
1	↑	a2	b2	c2	d2
X	1	QA0	QB0	QC0	QD0

1 = HIGH voltage level. 0 = LOW voltage level.
X = Don't Care
↑ = LOW-to-HIGH clock transition.

LOGIC DIAGRAM
Based on Texas Instruments 74LS399

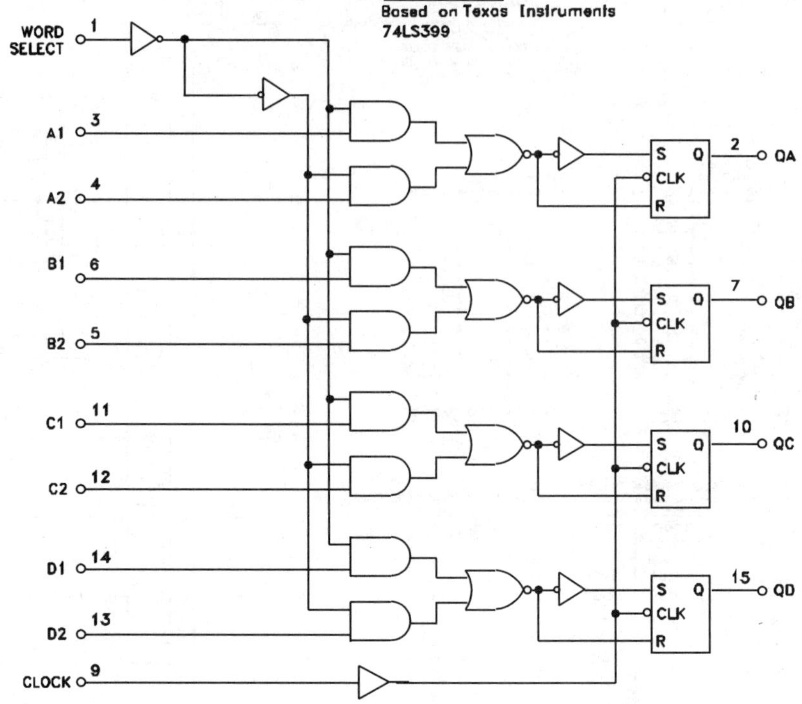

Figure 8.23. The 74LS399 is a quadruple 2-input multiplexer with storage. In operation, when the word select is low, it selects word 1 of the two 4-bit data sources and stores the data synchronously with the positive-going transition of the clock. With the word select input high, word 2 is selected. Minimum clock pulse width is 20 nS; data setup time is 25 nS; word select setup time is 45 nS. Hold time is zero.

FUNCTION TABLE 74F537 1-OF-10 DECODER WITH 3-STATE OUTPUTS

FUNCTION	INPUTS							OUTPUTS									
	OE	E1	E2	A3	A2	A1	A0	00	01	02	03	04	05	06	07	08	09
HIGH Z	1	X	X	X	X	X	X	Z	Z	Z	Z	Z	Z	Z	Z	Z	Z
Disable	0	1	X	X	X	X	X	Outputs equal P input									
	0	X	0	X	X	X	X										
Active HIGH Output (P=0)	0	0	1	0	0	0	0	1	0	0	0	0	0	0	0	0	0
	0	0	1	0	0	0	1	0	1	0	0	0	0	0	0	0	0
	0	0	1	0	0	1	0	0	0	1	0	0	0	0	0	0	0
	0	0	1	0	0	1	1	0	0	0	1	0	0	0	0	0	0
	0	0	1	0	1	0	0	0	0	0	0	1	0	0	0	0	0
	0	0	1	0	1	1	1	0	0	0	0	0	1	0	0	0	0
	0	0	1	0	1	0	0	0	0	0	0	0	0	1	0	0	0
	0	0	1	0	1	1	1	0	0	0	0	0	0	0	1	0	0
	0	0	1	1	0	0	0	0	0	0	0	0	0	0	0	1	0
	0	0	1	1	0	0	1	0	0	0	0	0	0	0	0	0	1
	0	0	1	1	X	1	X	0	0	0	0	0	0	0	0	0	0
	0	0	1	1	1	X	X	0	0	0	0	0	0	0	0	0	0
Active LOW Output (P=1)	0	0	1	0	0	0	0	0	1	1	1	1	1	1	1	1	1
	0	0	1	0	0	0	1	1	0	1	1	1	1	1	1	1	1
	0	0	1	0	0	1	0	1	1	0	1	1	1	1	1	1	1
	0	0	1	0	0	1	1	1	1	1	0	1	1	1	1	1	1
	0	0	1	0	1	0	0	1	1	1	1	0	1	1	1	1	1
	0	0	1	0	1	1	1	1	1	1	1	1	0	1	1	1	1
	0	0	1	0	1	0	0	1	1	1	1	1	1	0	1	1	1
	0	0	1	0	1	1	1	1	1	1	1	1	1	1	0	1	1
	0	0	1	1	0	0	0	1	1	1	1	1	1	1	1	0	1
	0	0	1	1	0	0	1	1	1	1	1	1	1	1	1	1	0
	0	0	1	1	X	1	X	1	1	1	1	1	1	1	1	1	1
	0	0	1	1	1	X	X	1	1	1	1	1	1	1	1	1	1

1 = HIGH voltage level. 0 = LOW voltage level.
X = Don't Care

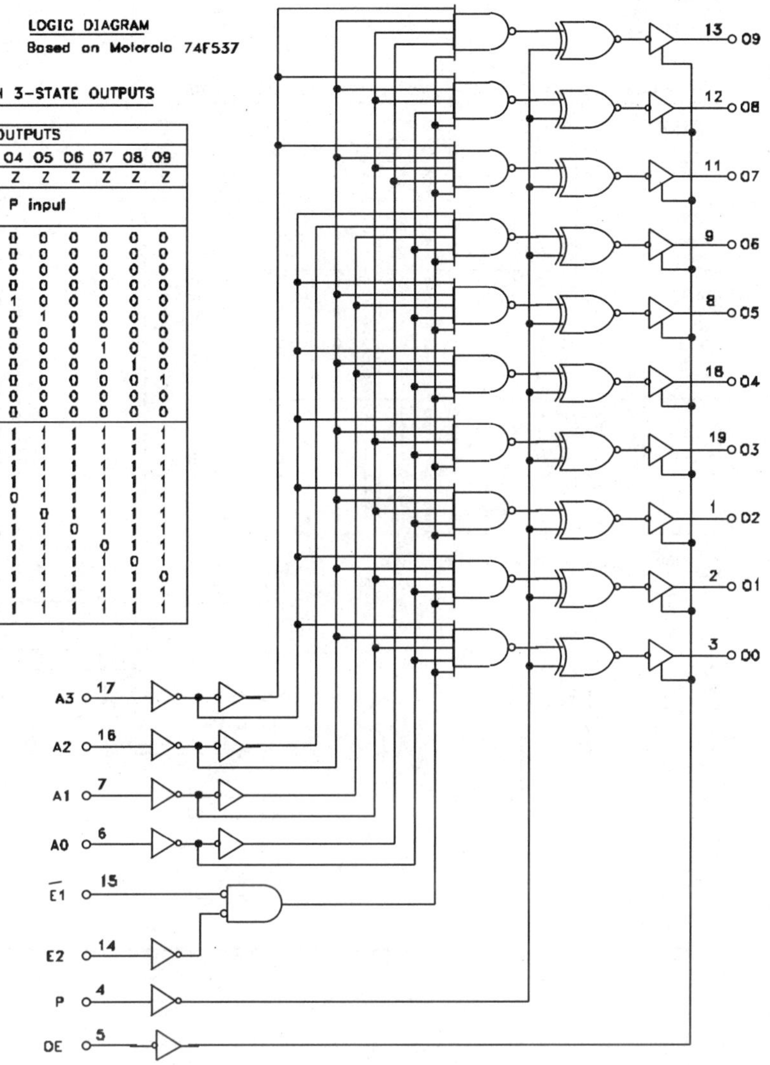

TTL - 74F537

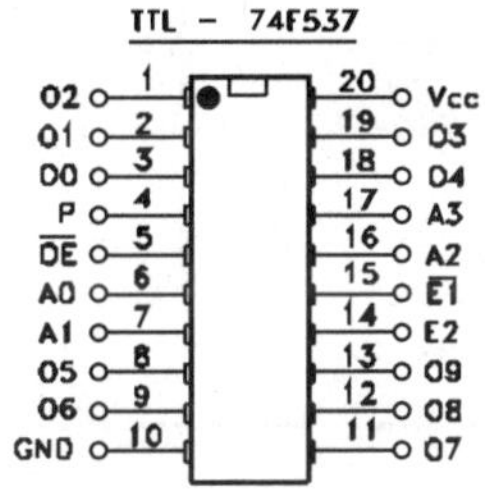

DUAL-IN-LINE PACKAGE

Figure 8.24. The 74F537 is a 1-of-10 decoder/multiplexer with 4 active high BCD inputs and 10 mutually exclusive 3-state outputs. The output enable is active low; when this input is high, the output is in a high impedance state. Two input enables provide for expansion. There is also a polarity control input. When this input is low, the decoded output polarity is high; when high, the output is low.

FUNCTION TABLE 74F538 1-OF-8 DECODER WITH 3-STATE OUTPUTS

FUNCTION	INPUTS									OUTPUTS							
	$\overline{OE1}$	$\overline{OE2}$	$\overline{E1}$	$\overline{E2}$	E3	E4	A2	A1	A0	O0	O1	O2	O3	O4	O5	O6	O7
HIGH Z	X	1	X	X	X	X	X	X	X	Z	Z	Z	Z	Z	Z	Z	Z
	1	X	X	X	X	X	X	X	X	Z	Z	Z	Z	Z	Z	Z	Z
Disable	0	0	1	X	X	X	X	X	X	Outputs equal P input							
	0	0	X	1	X	X	X	X	X								
	0	0	X	X	0	X	X	X	X								
	0	0	X	X	X	0	X	X	X								
Active HIGH Output (P=0)	0	0	0	0	1	1	0	0	0	1	0	0	0	0	0	0	0
	0	0	0	0	1	1	0	0	1	0	1	0	0	0	0	0	0
	0	0	0	0	1	1	0	1	0	0	0	1	0	0	0	0	0
	0	0	0	0	1	1	0	1	1	0	0	0	1	0	0	0	0
	0	0	0	0	1	1	1	0	0	0	0	0	0	1	0	0	0
	0	0	0	0	1	1	1	1	1	0	0	0	0	0	1	0	0
	0	0	0	0	1	1	1	0	0	0	0	0	0	0	0	1	0
	0	0	0	0	1	1	1	1	1	0	0	0	0	0	0	0	1
Active LOW Output (P=1)	0	0	0	0	1	1	0	0	0	0	1	1	1	1	1	1	1
	0	0	0	0	1	1	0	0	1	1	0	1	1	1	1	1	1
	0	0	0	0	1	1	0	1	0	1	1	0	1	1	1	1	1
	0	0	0	0	1	1	0	1	1	1	1	1	0	1	1	1	1
	0	0	0	0	1	1	1	0	0	1	1	1	1	0	1	1	1
	0	0	0	0	1	1	1	1	1	1	1	1	1	1	0	1	1
	0	0	0	0	1	1	1	0	0	1	1	1	1	1	1	0	1
	0	0	0	0	1	1	1	1	1	1	1	1	1	1	1	1	0

1 = HIGH voltage level. 0 = LOW voltage level.
X = Don't Care

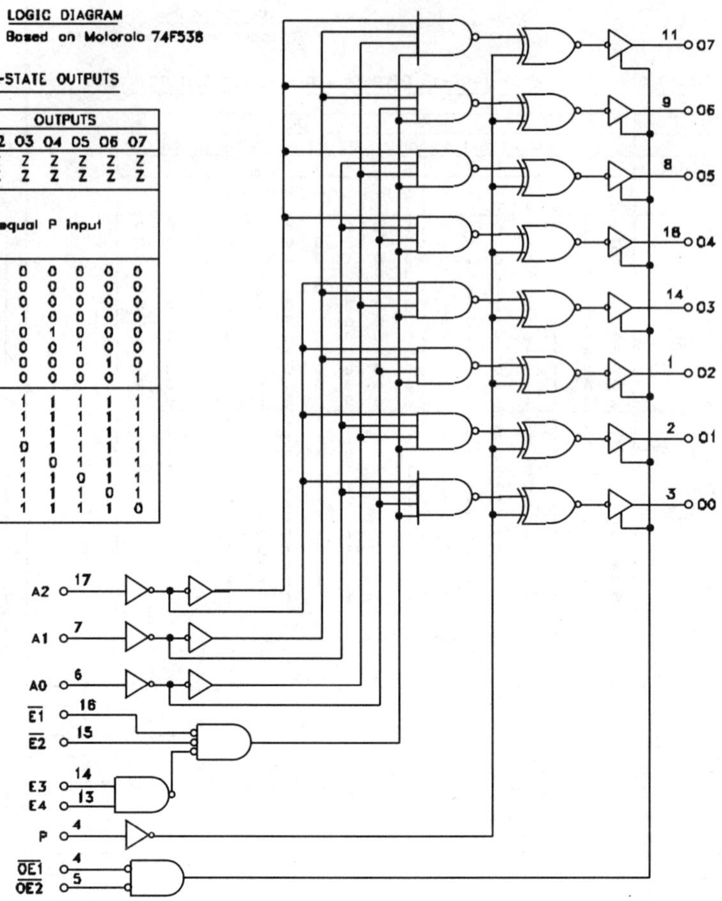

TTL - 74F538

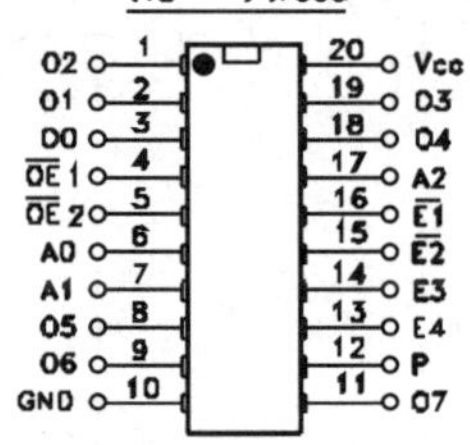

DUAL-IN-LINE PACKAGE

Figure 8.25. The 74F538 is a 1-of-8 decoder/multiplexer with 3 address inputs, A0-A2, decoded to 8 mutually exclusive 3-state outputs. A logic high on either of the two output enables forces the output to a high impedance state. The two output and four input enables provide for expansion. There is also a polarity control input. When this input is low, the decoded output polarity is high; when high, the output is low.

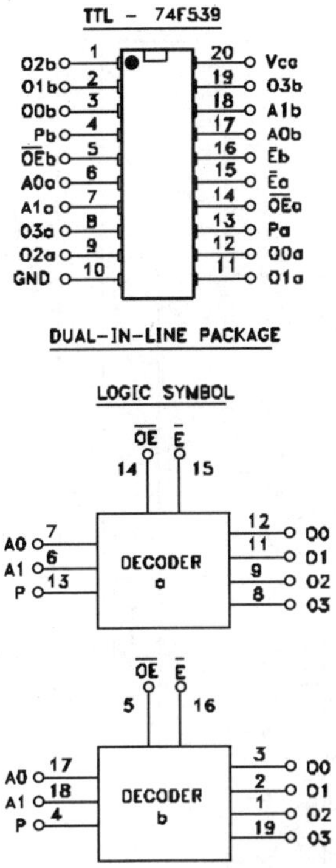

FUNCTION TABLE 74F539 DUAL 1-of-4 DECODER WITH 3-STATE OUTPUTS

FUNCTION	INPUTS				OUTPUTS			
	$\overline{OE}$	$\overline{E}$	A1	A0	O0	O1	O2	O3
HIGH Z	1	X	X	X	Z	Z	Z	Z
Disable	0	1	X	X	Outputs equal P input			
Active HIGH Output (P=0)	0	0	0	0	1	0	0	0
	0	0	0	1	0	1	0	0
	0	0	1	0	0	0	1	0
	0	0	1	1	0	0	0	1
Active LOW Output (P=1)	0	0	0	0	0	1	1	1
	0	0	0	1	1	0	1	1
	0	0	1	0	1	1	0	1
	0	0	1	1	1	1	1	0

1 = HIGH voltage level. 0 = LOW voltage level.
X = Don't Care

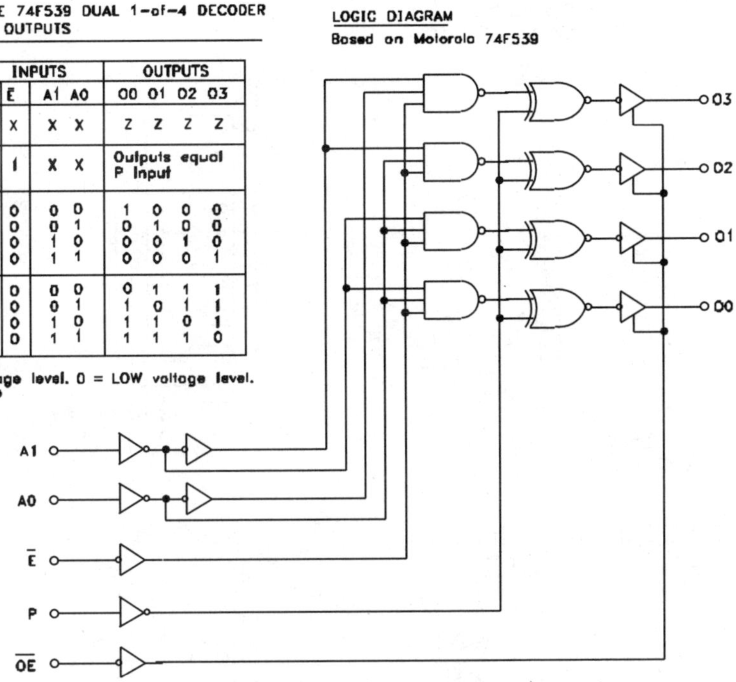

Figure 8.26. The 74F539 is a dual 1-of-4 decoder with 2 address inputs, A0, A1, decoded to select one-of-4 mutually exclusive 3-state outputs. A logic high on the output enable forces the output to a high impedance state. An input enables multiplexing; data is routed to the selected output in non-inverted form when the enable is low; inverted when high. There is also a polarity control input. When this input is low, the decoded output polarity is high; when high, the output is low.

LOGIC DIAGRAM

Based on Texas Instruments 74AS850, 'AS851

TTL – 74AS850
74AS851

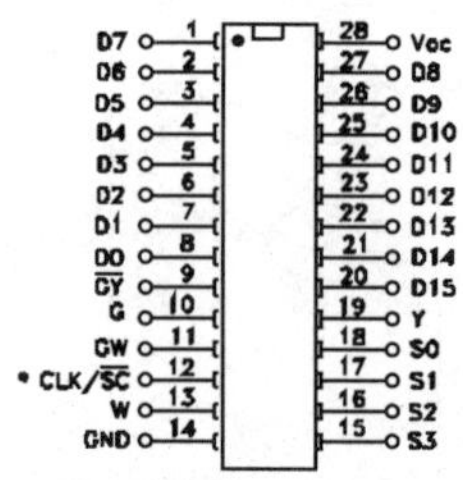

DUAL-IN-LINE PACKAGE

• CLK/$\overline{SC}$ = CLK for 'AS850
$\overline{SC}$ for 'AS851

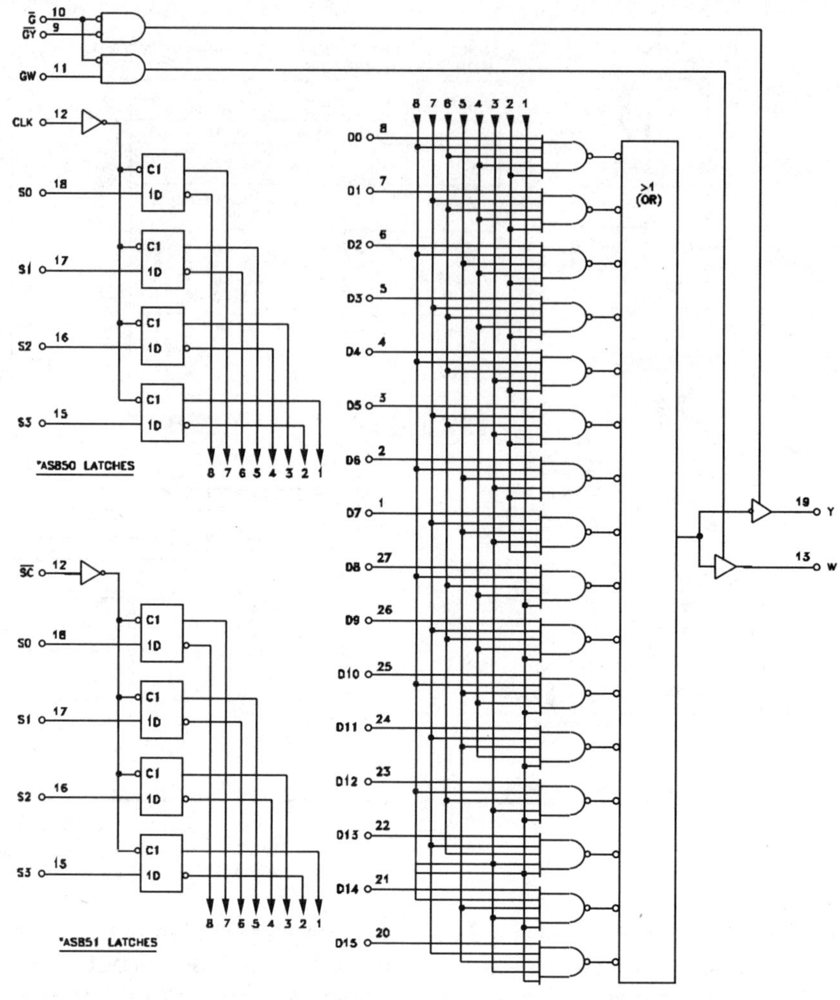

FUNCTION TABLE 74AS850, 'AS851 1-of-16 DATA SELECTORS/MULTIPLEXERS WITH 3-STATE OUTPUTS

SELECT INPUTS				'AS850	'AS851	INPUT
S3	S2	S1	S0	CLK	SC	SELECTED
0	0	0	0	↑	0	D0
0	0	0	1	↑	0	D1
0	0	1	0	↑	0	D2
0	0	1	1	↑	0	D3
0	1	0	0	↑	0	D4
0	1	0	1	↑	0	D5
0	1	1	0	↑	0	D6
0	1	1	1	↑	0	D7
1	0	0	0	↑	0	D8
1	0	0	1	↑	0	D9
1	0	1	0	↑	0	D10
1	0	1	1	↑	0	D11
1	1	0	0	↑	0	D12
1	1	0	1	↑	0	D13
1	1	1	0	↑	0	D14
1	1	1	1	↑	0	D15
X	X	X	X	1 or 0	1	Dn

1 = HIGH voltage level. 0 = LOW voltage level.
X = Don't Care

↑ = LOW-to-HIGH clock transition.

Figure 8.27. The 74AS580 'AS581 are 4-line to 1-line data selectors/multiplexers that can select one of 16 data inputs. 3-state bus driver outputs are provided. The 'AS850 offers clocked selects; the 'AS851 enable controlled selects. Each has a master output control, /G, for cascading, and individual output controls, /GY, GW, for each output. Y and W are complementary outputs. The output controls do not affect the internal operation. New data can be set up while in the high-impedance output state.

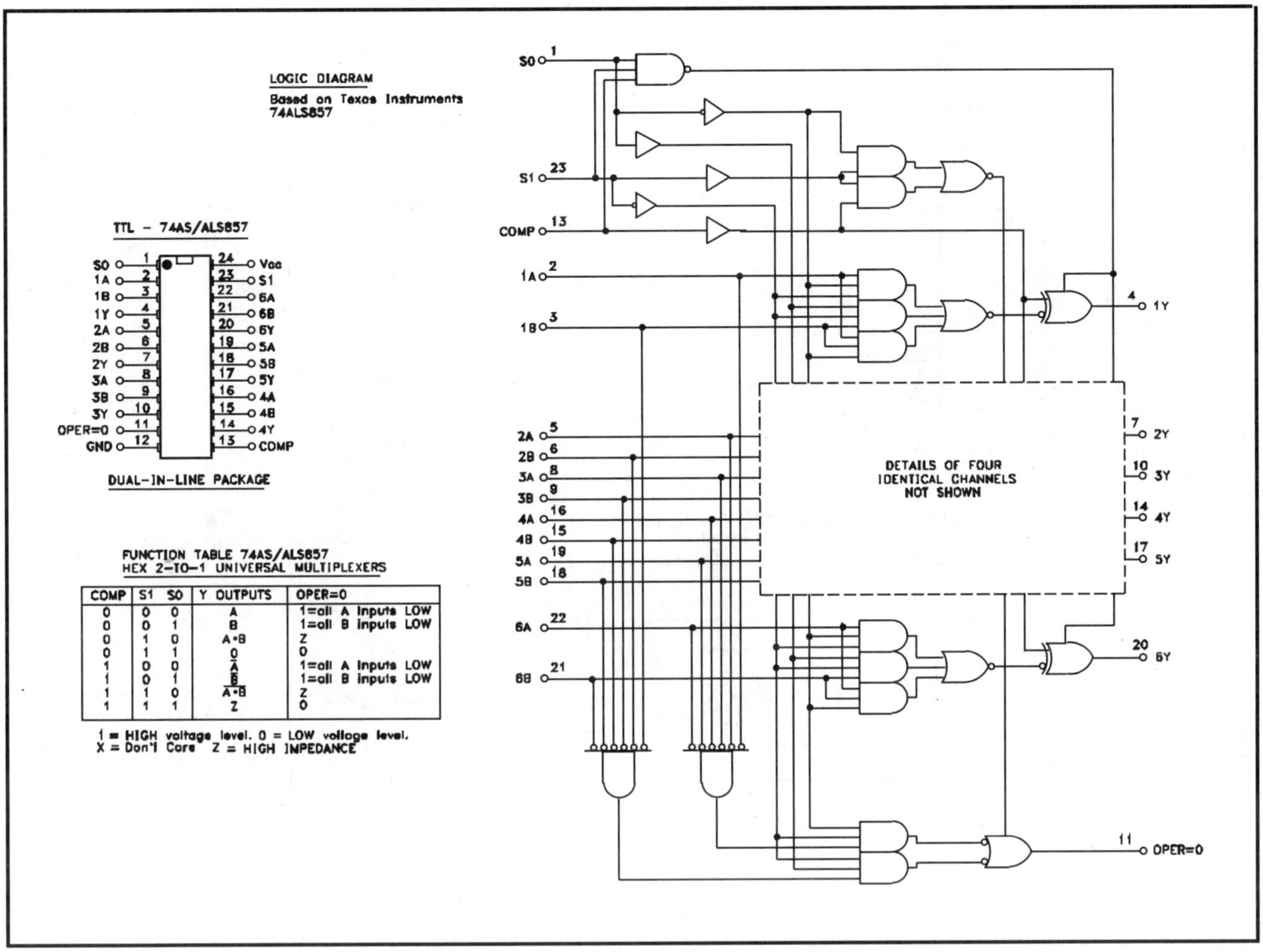

COMP	S1	S0	Y OUTPUTS	OPER=0
0	0	0	A	1=all A inputs LOW
0	0	1	B	1=all B inputs LOW
0	1	0	A•B	Z
0	1	1	0	0
1	0	0	$\overline{A}$	1=all A inputs LOW
1	0	1	$\overline{B}$	1=all B inputs LOW
1	1	0	$\overline{A \cdot B}$	Z
1	1	1	Z	0

Figure 8.28. The 74AS/ALS857 are hex 2-line to 1-line universal multiplexers with 3-state outputs. These can provide either true (COMP LOW) or inverted (COMP HIGH) data at the Y outputs. In addition, they can perform the logical AND function and clear as well. The four modes of operation are: select A data inputs; select B data inputs; AND A data with B; Clear. The OPER = 0 output is high if all the A or B inputs are low.

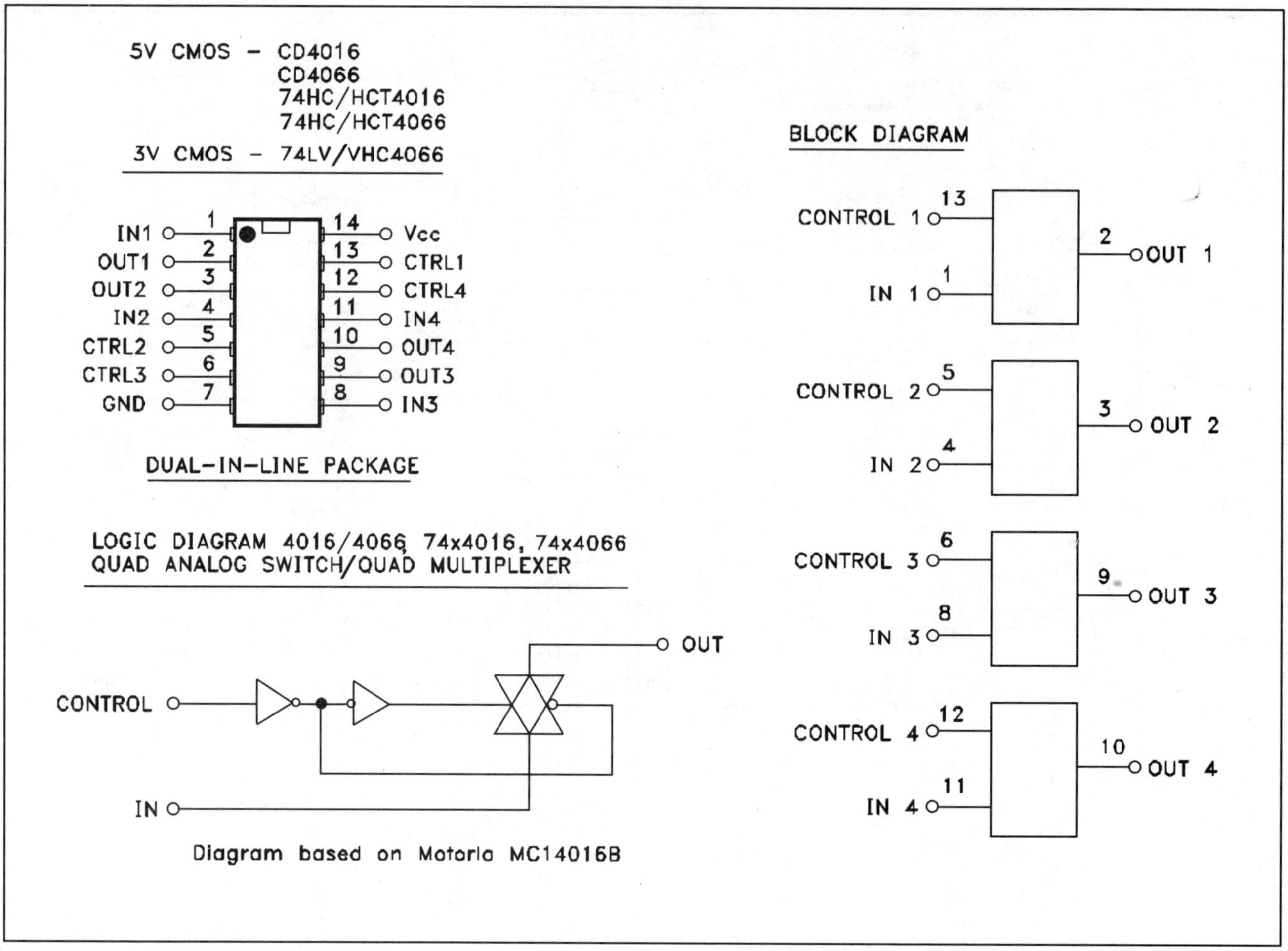

Figure 8.29. The 4016 and 4066 are a quad analog switch/quad analog multiplexer. Each consists of four independent switches capable of controlling either digital or analog signals. The 4016 and 4066 are pin compatible with the 74x4016/74x4066 high-speed and low-voltage devices. The 4066 has an improved transfer characteristic which causes more parasitic coupling capacitance than the 4016. Typical CD4016 Ron = 300 ohms @ 5 Vcc and 25°C. Typical CD4066 Ron = 250 W. Ron improves significantly with higher Vcc. 74HC4016 Ron at the same conditions = 320 W; 74HC4066 = 170 W. Keep in mind the presence of other factors, the manufacturer's data should be referred to for greater detail.

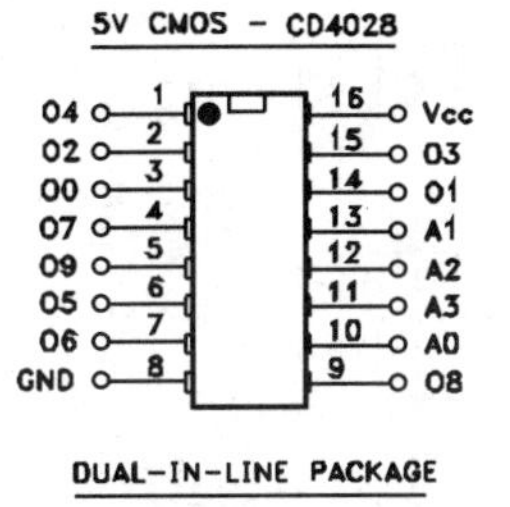

FUNCTION TABLE 4028 1-OF-10 DECODER

INPUTS				OUTPUTS									
A3	A2	A1	A0	00	01	02	03	04	05	06	07	08	09
0	0	0	0	1	0	0	0	0	0	0	0	0	0
0	0	0	1	0	1	0	0	0	0	0	0	0	0
0	0	1	0	0	0	1	0	0	0	0	0	0	0
0	0	1	1	0	0	0	1	0	0	0	0	0	0
0	1	0	0	0	0	0	0	1	0	0	0	0	0
0	1	1	1	0	0	0	0	0	1	0	0	0	0
0	1	0	0	0	0	0	0	0	0	1	0	0	0
0	1	1	1	0	0	0	0	0	0	0	1	0	0
1	0	0	0	0	0	0	0	0	0	0	0	1	0
1	0	0	1	0	0	0	0	0	0	0	0	0	1
1	0	1	0	0	0	0	0	0	0	0	0	0	0
1	0	1	1	0	0	0	0	0	0	0	0	0	0
1	1	0	0	0	0	0	0	0	0	0	0	0	0
1	1	0	1	0	0	0	0	0	0	0	0	0	0
1	1	1	0	0	0	0	0	0	0	0	0	0	0
1	1	1	1	0	0	0	0	0	0	0	0	0	0

1 = HIGH voltage level. 0 = LOW voltage level.
X = Don't Care

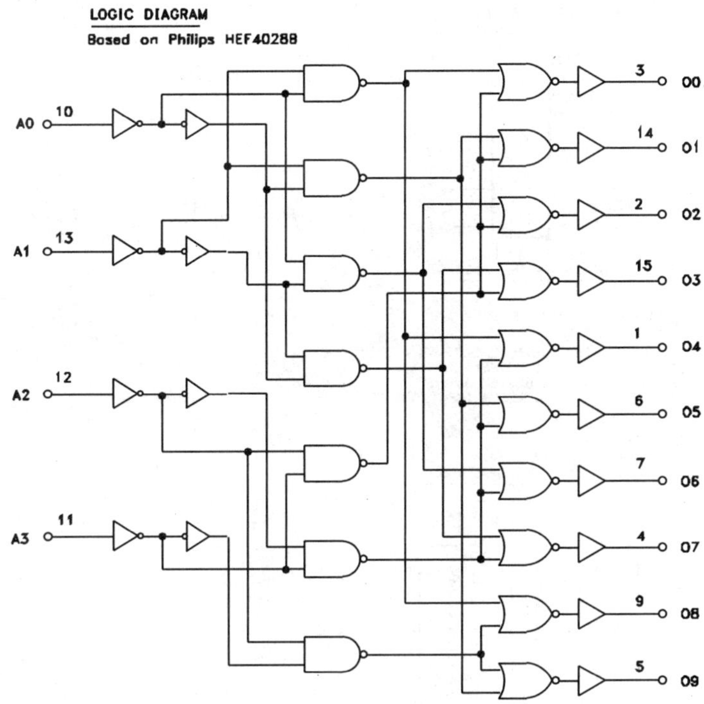

Figure 8.30. The 4028 BCD-to-decimal decoder/binary-to-octal decoder is constructed such that an 8421 BCD code on the four inputs provides a decimal (1-of-10) decoded output, while a 3-bit binary input provides a decoded octal (1-of-8) output with "D" (A3) forced to a logic zero. Expanded decoding, such as binary-to-hexadecimal (1-of-16), etc., can be achieved with additional devices.

5V CMOS - CD4051
74HC/HCT4051
3V CMOS - 74LV/VHC4051

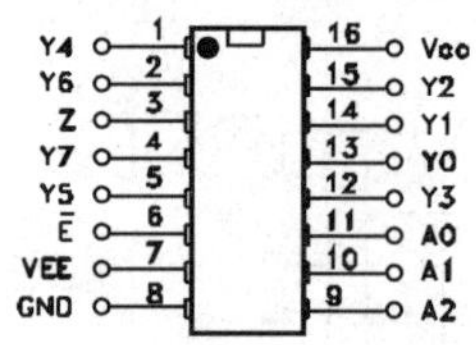

DUAL-IN-LINE PACKAGE

PACKAGE PINNING:

PIN-

1,2,4,5,12-15	INPUTS/OUTPUTS
6	INPUT ENABLE
9-11	ADDRESS INPUTS
3	COMMON INPUT/OUTPUT

FUNCTION TABLE 4051, 74x4051 8-CHANNEL ANALOG MULTIPLEXER/DEMULTIPLEXER

INPUTS				CHANNEL ON
Ē	A2	A1	A0	
0	0	0	0	Y0-Z
0	0	0	1	Y1-Z
0	0	1	0	Y2-Z
0	0	1	1	Y3-Z
0	1	0	0	Y4-Z
0	1	0	1	Y5-Z
0	1	1	0	Y6-Z
0	1	1	1	Y7-Z
1	X	X	0	None

1 = HIGH voltage level.
0 = LOW voltage level.
X = Don't Care

LOGIC DIAGRAM
Based on Philips HEF4051B

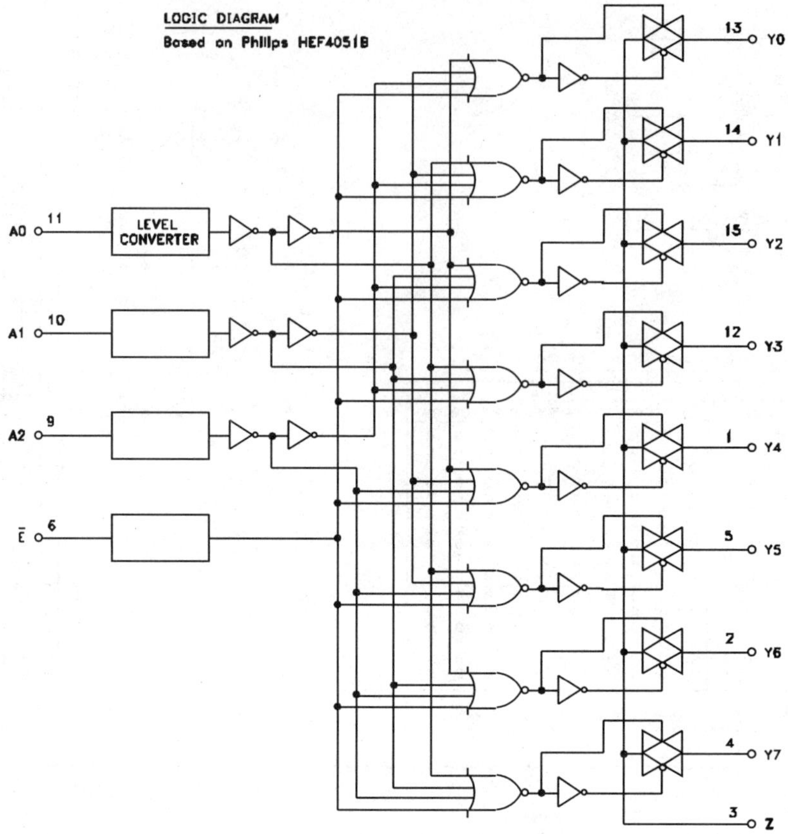

Figure 8.31. The CMOS 4051 and 74x4051 are an 8-channel analog multiplexer/demultiplexer with three address inputs (A0-A2), an active low input enable (/E), eight independent input/outputs (Y0-Y7), and a common input/output (Z). The device includes eight bidirectional analog switches with one side connected to an independent input/output and the other connected to a common input/output (Z). With /E low, one of the 8 switches is selected by A0 to A2. With /E high, all switches are in a high-impedance OFF-state. The analog inputs/outputs can swing between Vcc (or VDD) as a positive limit and VEE as the negative. Vcc-VEE may not exceed -15V, -7.5V for the low-voltage devices. With VEE connected to ground operation is as a digital multiplexer/demultiplexer.

5V CMOS – CD4052
74HC/HCT4052
3V CMOS – 74LV/VHC4052

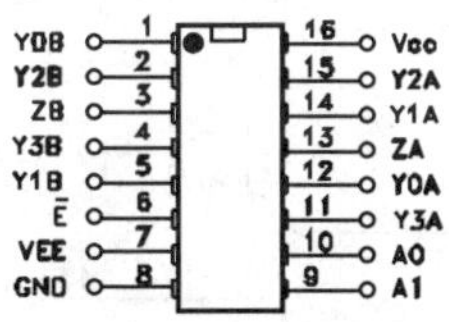

DUAL-IN-LINE PACKAGE

PACKAGE PINNING:

PIN–
1,2,4,5,11,12,14,15 INPUTS/OUTPUTS
6, INPUT ENABLE
9,10 ADDRESS INPUTS
3,13 COMMON INPUT/OUTPUT

FUNCTION TABLE 4052, 74x4052 DUAL 4-CHANNEL ANALOG MULTIPLEXER/DEMULTIPLEXER

INPUTS			CHANNEL ON
Ē	A1	A0	
0	0	0	Y0A –ZA; Y0B–ZB
0	0	1	Y1A –ZA; Y1B–ZB
0	1	0	Y2A –ZA; Y2B–ZB
0	1	1	Y3A –ZA; Y3B–ZB
1	X	0	None

1 = HIGH voltage level.
0 = LOW voltage level.
X = Don't Care

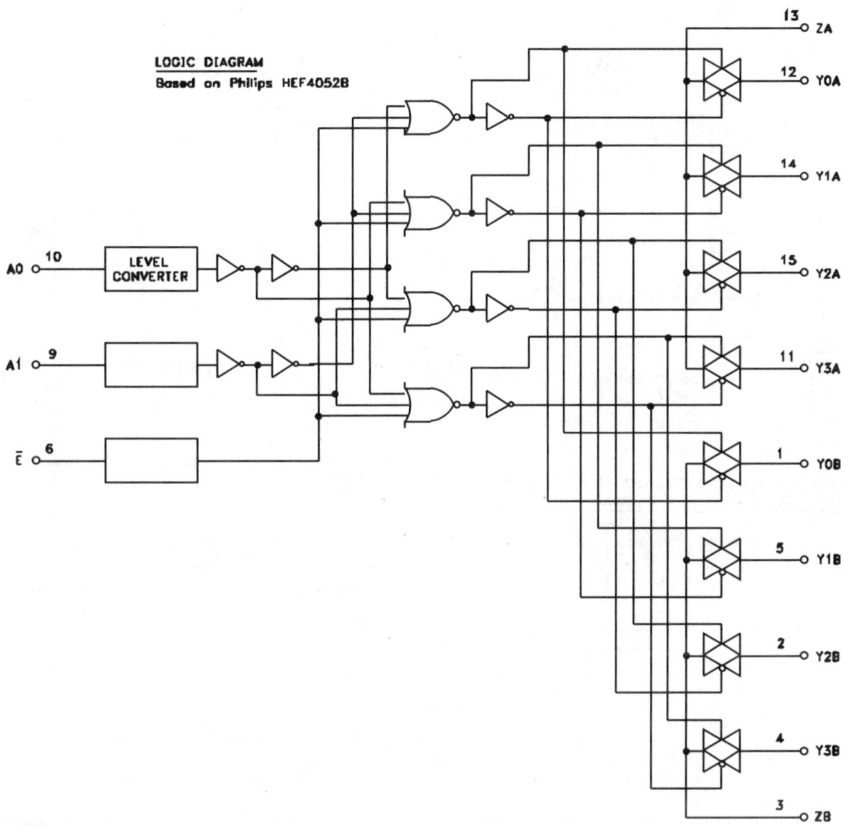

Figure 8.32. The CMOS 4052 and 74x4052 are a dual 4-channel analog multiplexer/demultiplexer with common channel select logic: two address inputs (A0,A1),and an active low input enable (/E). Each has four independent input/outputs (Y0-Y3), and a common input/output (Z). Each multiplexer/demultiplexer provides four bidirectional analog switches with one side connected to an independent input/output and the other connected to a common input/output (Z). With /E low, one of the 4 switches is selected by A0 and A1. With /E high, all switches are in a high-impedance OFF-state. The analog inputs/outputs can swing between Vcc (or VDD) as a positive limit and VEE as the negative. Vcc-VEE may not exceed -15V, -7.5V for the low-voltage devices. With VEE connected to ground operation is as a digital multiplexer/demultiplexer.

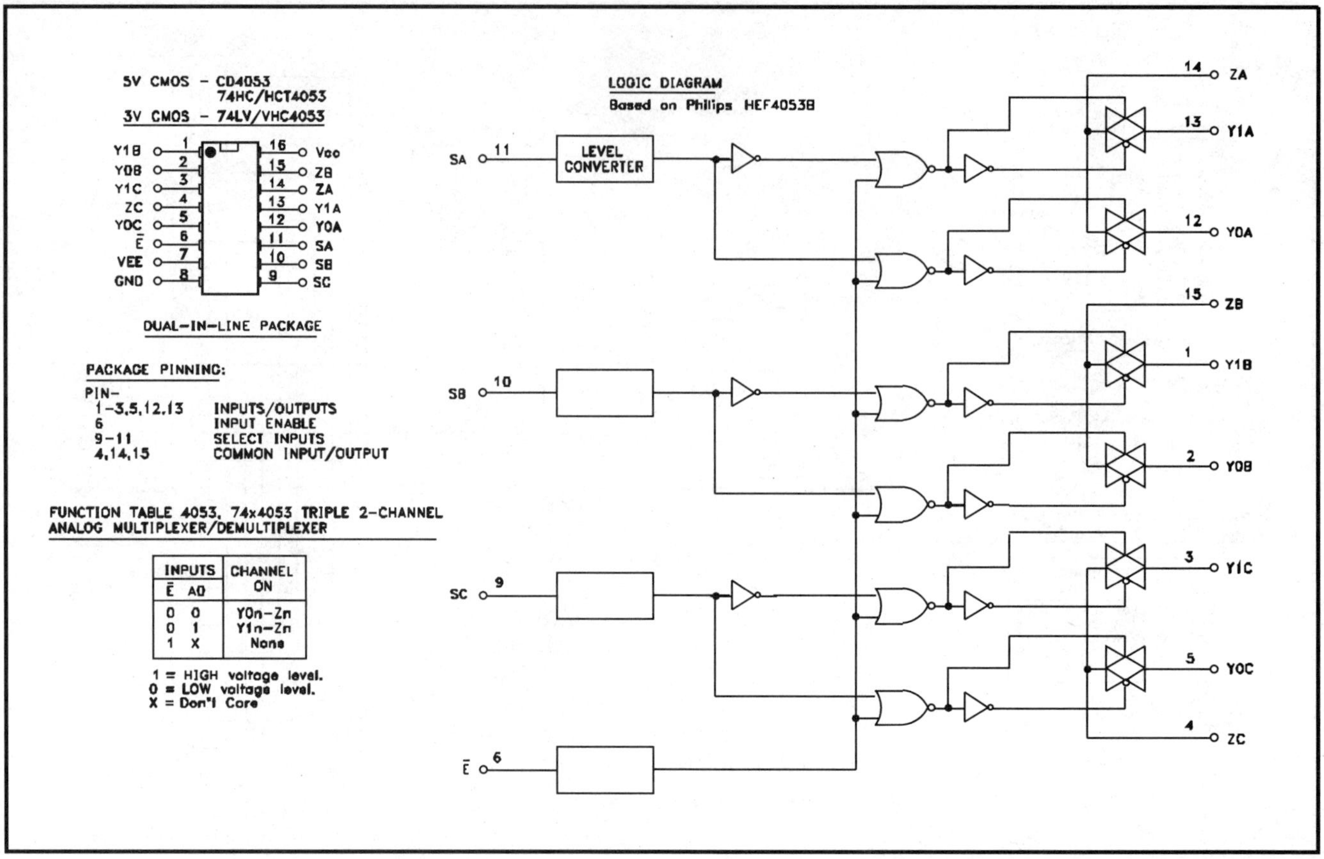

Figure 8.33. The CMOS 4053 and 74x4053 are a triple 2-channel analog multiplexer/demultiplexer with common channel select logic: three select inputs (SA-SC), and an active low input enable (/E). Each has two independent input/outputs (Y0,Y2), and a common input/output (Zn). Each multiplexer/demultiplexer provides two bidirectional analog switches with one side connected to an independent input/output and the other connected to a common input/output (Z). With /E low, one of the two switches is selected by Sn. With /E high, all switches are in a high impedance OFF-state. The analog inputs/outputs can swing between Vcc (or VDD) as a positive limit and VEE as the negative. Vcc-VEE may not exceed -15V, -7.5V for the low-voltage devices. With VEE connected to ground operation is as a digital multiplexer/demultiplexer.

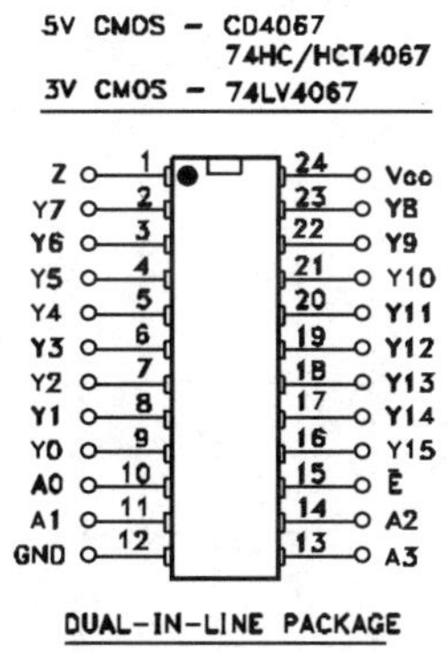

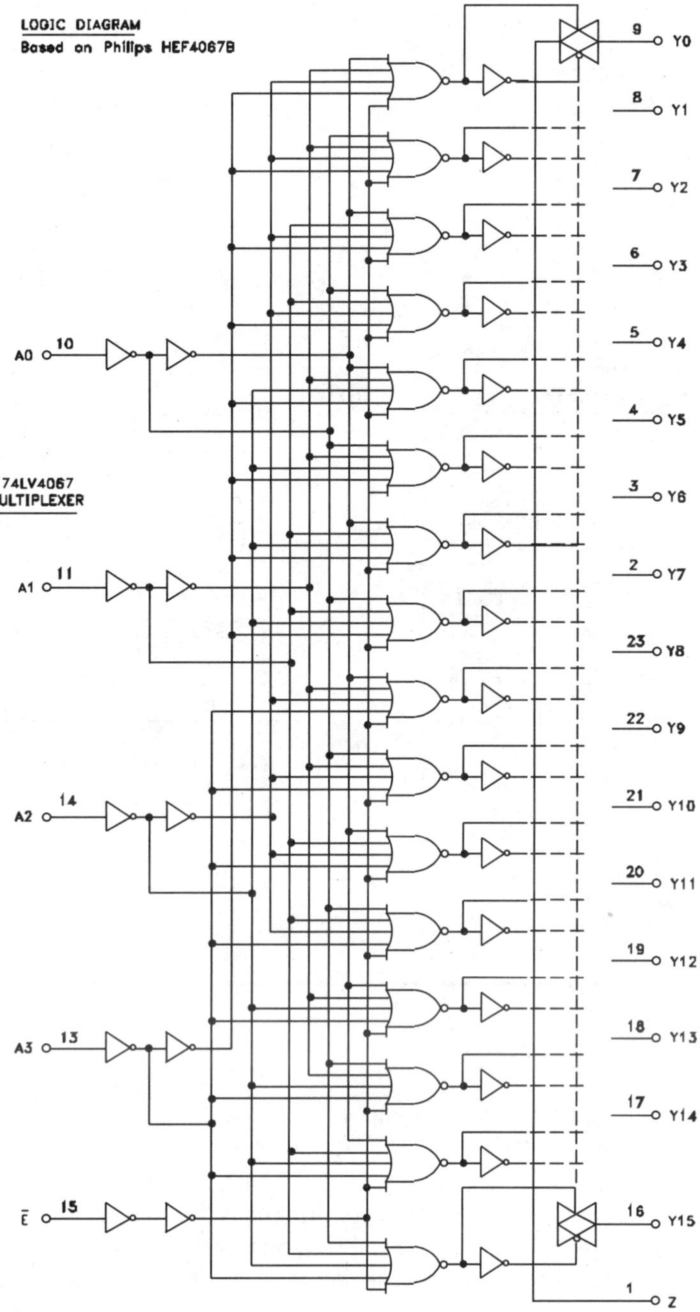

FUNCTION TABLE 4067, 74HC/HCT4067, 74LV4067
16-CHANNEL ANALOG MULTIPLEXER/DEMULTIPLEXER

INPUTS					CHANNEL ON
E1	A3	A2	A1	A0	
0	0	0	0	0	Y0-Z
0	0	0	0	1	Y1-Z
0	0	0	1	0	Y2-Z
0	0	0	1	1	Y3-Z
0	0	1	0	0	Y4-Z
0	0	1	1	1	Y5-Z
0	0	1	0	0	Y6-Z
0	0	1	1	1	Y7-Z
0	1	0	0	0	Y8-Z
0	1	0	0	1	Y9-Z
0	1	0	1	0	Y10-Z
0	1	0	1	1	Y11-Z
0	1	1	0	0	Y12-Z
0	1	1	1	1	Y13-Z
0	1	1	0	0	Y14-Z
0	1	1	1	1	Y15-Z
1	X	X	X	X	None

1 = HIGH voltage level.
0 = LOW voltage level.
X = Don't Care

Figure 8.34. The CMOS 4067 and 74x4067 are a 16-channel analog multiplexer/demultiplexer with four address inputs (A0-A3), an active low-input enable (/E), 16 independent input/outputs (Y0-Y15), and a common input/output (Z). The device provides 16 bidirectional analog switches with one side connected to an independent input/output and the other connected to a common input/output (Z). With /E low, one of the 16 switches is selected by A0-A3. With /E high, all switches are in a high-impedance OFF-state. The analog inputs/outputs can swing between Vcc (or VDD) as a positive limit and VEE as the negative. Vcc-VEE may not exceed 15V. With VEE connected to ground operation is as a digital multiplexer/demultiplexer.

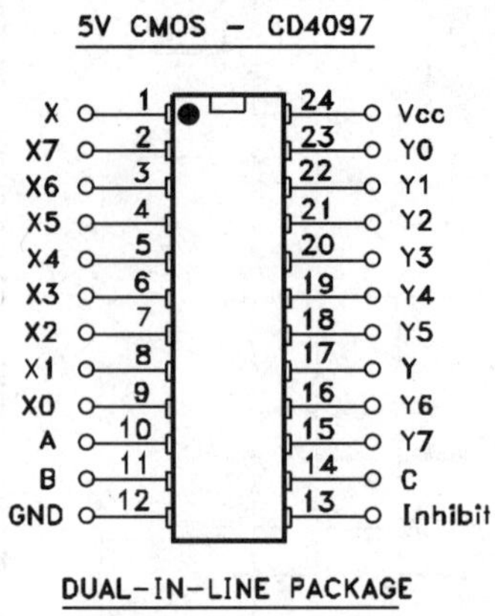

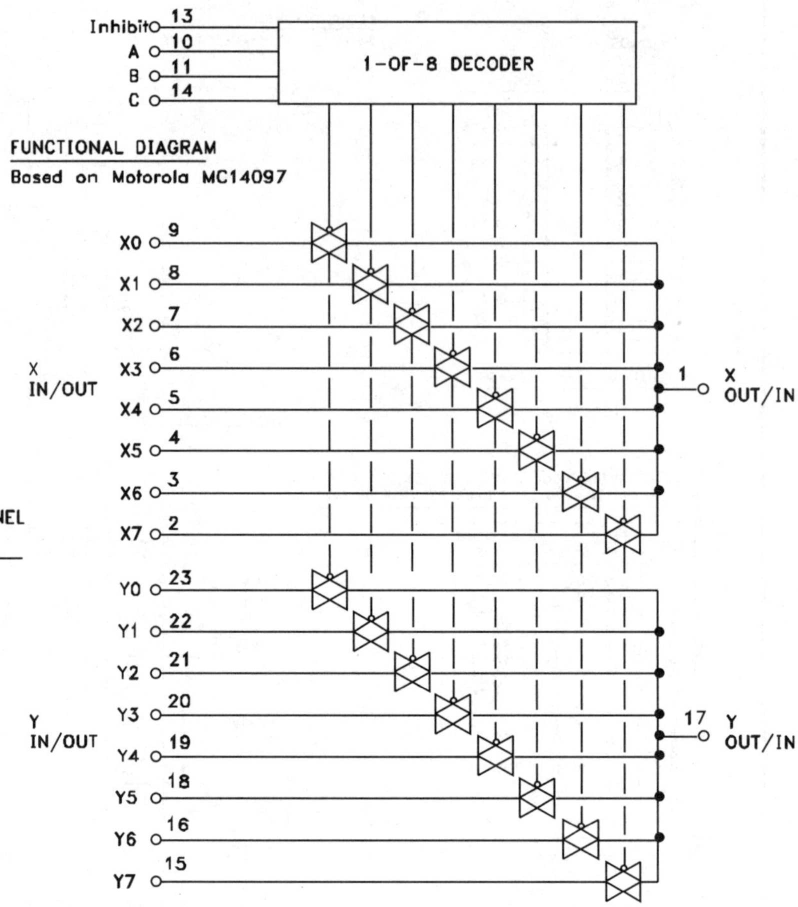

FUNCTION TABLE 4097 DUAL 8-CHANNEL ANALOG MULTIPLEXER/DEMULTIPLEXER

INPUTS				SELECTED CHANNELS	
Inh	C	B	A		
1	X	X	X	None	
0	0	0	0	X0	Y0
0	0	0	1	X1	Y1
0	0	1	0	X2	Y2
0	0	1	1	X3	Y3
0	1	0	0	X4	Y4
0	1	1	1	X5	Y5
0	1	0	0	X6	Y6
0	1	1	1	X7	Y7

1 = HIGH voltage level.
0 = LOW voltage level.
X = Don't Care

Figure 8.35. The CMOS 4097 is a differential 8-channel analog multiplexer/demultiplexer with three common address inputs (A - C), a common inhibit, dual inputs/outputs (X0-X7, Y0-Y7), and a common input/output (X, Y). The device provides 8 analog switch pairs with one side connected to an independent input/output and the other connected to a common input/output (X, Y). With the Inhibit low, one of the 8 switches in each pair is selected by A - C. The device features a 3 - 18 volt operating range with matched channel resistance and low crosstalk between channels. Typical Ron at 5V, 25°C, is 250 ohms.

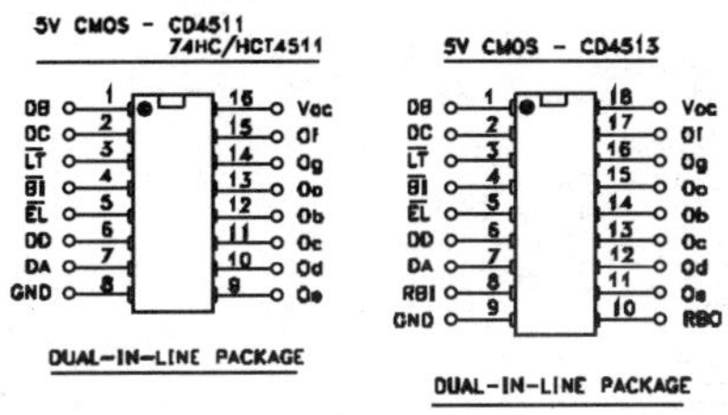

FUNCTION TABLE 4511, 74x4511, 4513 BCD to 7-SEGMENT LATCH/DECODER/DRIVER

INPUTS							OUTPUTS							
EL	BI	LT	DD	DC	DB	DA	Oa	Ob	Oc	Od	Oe	Of	Og	DISPLAY
X	X	0	X	X	X	X	1	1	1	1	1	1	1	8
X	0	1	X	X	X	X	0	0	0	0	0	0	0	blank
0	1	1	0	0	0	0	1	1	1	1	1	1	0	0
0	1	1	0	0	0	1	0	1	1	0	0	0	0	1
0	1	1	0	0	1	0	1	1	0	1	1	0	1	2
0	1	1	0	0	1	1	1	1	1	1	0	0	1	3
0	1	1	0	1	0	0	0	1	1	0	0	1	1	4
0	1	1	0	1	0	1	1	0	1	1	0	1	1	5
0	1	1	0	1	1	0	0	0	1	1	1	1	1	6
0	1	1	0	1	1	1	1	1	1	0	0	0	0	7
0	1	1	1	0	0	0	1	1	1	1	1	1	1	8
0	1	1	1	0	0	1	1	1	1	0	0	1	1	9
0	1	1	1	0	1	0	0	0	0	0	0	0	0	blank
0	1	1	1	0	1	1	0	0	0	0	0	0	0	blank
0	1	1	1	1	0	0	0	0	0	0	0	0	0	blank
0	1	1	1	1	0	1	0	0	0	0	0	0	0	blank
0	1	1	1	1	1	0	0	0	0	0	0	0	0	blank
0	1	1	1	1	1	1	0	0	0	0	0	0	0	blank
1	1	1	X	X	X	X				*				*

1 = HIGH voltage level. 0 = LOW voltage level.
X = Don't Care
* Depends upon the BCD code previously applied when LE=–.
~ RBO = RBI(DC B A) indicated by other rows of the table.

CD4513 ONLY

INPUTS		
RBI	RBO	DISPLAY
X	+	8
X	+	blank
1	1	blank
0	0	0
X	0	1
X	0	2
X	0	3
X	0	4
X	0	5
X	0	6
X	0	7
X	0	8
X	0	9
X	0	blank
X	0	blank
X	0	blank
X	0	blank
X	0	blank
X	0	blank
X	~	*

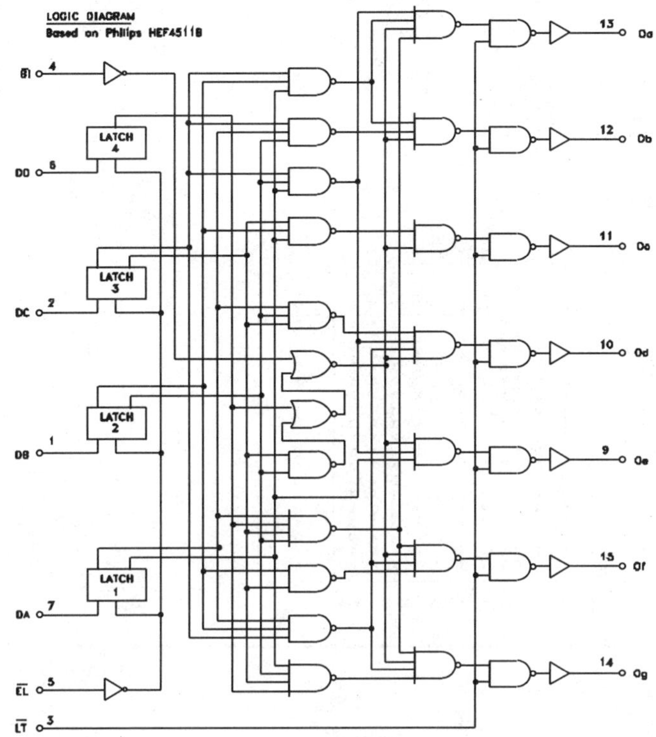

Figure 8.36. The CMOS 4511 and 74x4511 are a BCD-to-7-segment latch/decoder/driver having four address inputs (DA - DC), an active low latch enable (/EL), an active low ripple blanking input (/BI), an active low lamp test input (/LT), and seven active high NPN bipolar transistor outputs (Oa - Og). When /EL is low, the state of the segment outputs is determined by the data on DA - DD. When /EL goes high, the existing input data is stored in the latches and the output remains stable. When /LT is low, all the segment outputs are high regardless of the input data. With /LT high, a low to /BI forces all segment outputs low. Inputs /LT and /BI do not affect the latch circuit. The 4513 is similar in function to the 4511 but includes Ripple Blanking Input (RBI) and Ripple Blanking Output (RBO) for the suppression of leading and trailing zeros.

5V CMOS - CD4512

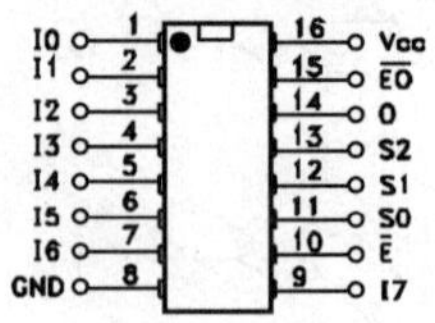

DUAL-IN-LINE PACKAGE

PACKAGE PINNING:

PIN-

1-7,9 MULTIPLEXER INPUTS
15 OUTPUT ENABLE
10 ENABLE
11-13 SELECT INPUTS
14 MULTIPLEXER OUTPUTS

FUNCTION TABLE 4512 8-INPUT MULTIPLEXER WITH 3-STATE OUTPUT

INPUTS													OUTPUT
EO	E	S2	S1	S0	I0	I1	I2	I3	I4	I5	I6	I7	O
0	1	X	X	X	X	X	X	X	X	X	X	X	0
0	0	0	0	0	0	X	X	X	X	X	X	X	0
0	0	0	0	0	1	X	X	X	X	X	X	X	1
0	0	0	0	1	X	0	X	X	X	X	X	X	0
0	0	0	0	1	X	1	X	X	X	X	X	X	1
0	0	0	1	0	X	X	0	X	X	X	X	X	0
0	0	0	1	0	X	X	1	X	X	X	X	X	1
0	0	0	1	1	X	X	X	0	X	X	X	X	0
0	0	0	1	1	X	X	X	1	X	X	X	X	1
0	0	1	0	0	X	X	X	X	0	X	X	X	0
0	0	1	0	0	X	X	X	X	1	X	X	X	1
0	0	1	0	1	X	X	X	X	X	0	X	X	0
0	0	1	0	1	X	X	X	X	X	1	X	X	1
0	0	1	1	0	X	X	X	X	X	X	0	X	0
0	0	1	1	0	X	X	X	X	X	X	1	X	1
0	0	1	1	1	X	X	X	X	X	X	X	0	0
0	0	1	1	1	X	X	X	X	X	X	X	1	1
1	X	X	X	X	X	X	X	X	X	X	X	X	X

1 = HIGH voltage level. 0 = LOW voltage level.
X = Don't Care

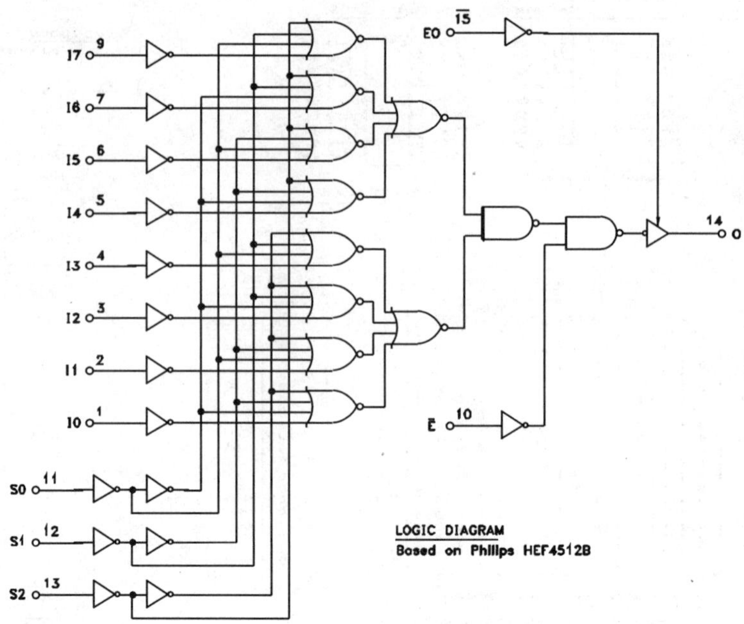

LOGIC DIAGRAM
Based on Philips HEF4512B

Figure 8.37. The CMOS 4512 is an 8-channel data selector for application to signal multiplexing. The device is provided with three input selectors (S0 - S2) and an active high 3-state output enable.

5V CMOS – CD4514
74HC/HCT4514

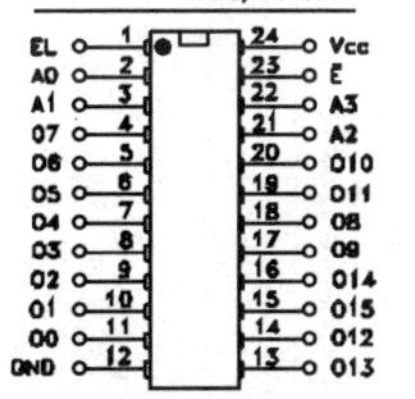

DUAL-IN-LINE PACKAGE

PACKAGE PINNING:
PIN-
2,3,21,22 ADDRESS INPUTS
23 ENABLE INPUT
1 LATCH ENABLE INPUT
4-11,13-20 DECODER OUTPUTS

LOGIC DIAGRAM
Based on Philips HEF4514B

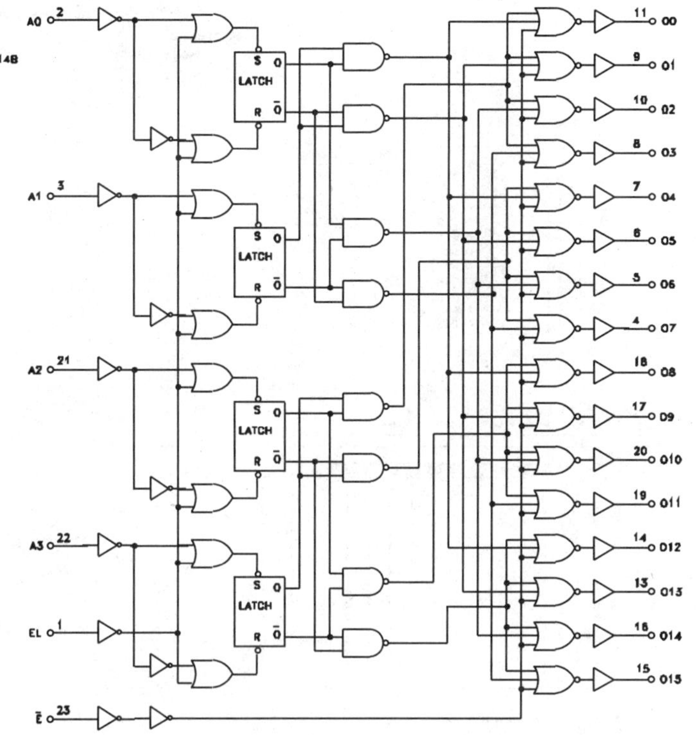

FUNCTION TABLE 4514, 74x4514 1-OF-16
DECODER/DEMULTIPLEXER WITH INPUT LATCHES

INPUTS					OUTPUTS															
E	A0	A1	A2	A3	00	01	02	03	04	05	06	07	08	09	010	011	012	013	014	015
1	X	X	X	X	0	0	0	0	0	0	0	0	0	0	0	0	0	0	0	0
0	0	0	0	0	1	0	0	0	0	0	0	0	0	0	0	0	0	0	0	0
0	1	0	0	0	0	1	0	0	0	0	0	0	0	0	0	0	0	0	0	0
0	0	1	0	0	0	0	1	0	0	0	0	0	0	0	0	0	0	0	0	0
0	1	1	0	0	0	0	0	1	0	0	0	0	0	0	0	0	0	0	0	0
0	0	0	1	0	0	0	0	0	1	0	0	0	0	0	0	0	0	0	0	0
0	1	0	1	0	0	0	0	0	0	1	0	0	0	0	0	0	0	0	0	0
0	0	1	1	0	0	0	0	0	0	0	1	0	0	0	0	0	0	0	0	0
0	1	1	1	0	0	0	0	0	0	0	0	1	0	0	0	0	0	0	0	0
0	0	0	0	1	0	0	0	0	0	0	0	0	1	0	0	0	0	0	0	0
0	1	0	0	1	0	0	0	0	0	0	0	0	0	1	0	0	0	0	0	0
0	0	1	0	1	0	0	0	0	0	0	0	0	0	0	1	0	0	0	0	0
0	1	1	0	1	0	0	0	0	0	0	0	0	0	0	0	1	0	0	0	0
0	0	0	1	1	0	0	0	0	0	0	0	0	0	0	0	0	1	0	0	0
0	1	0	1	1	0	0	0	0	0	0	0	0	0	0	0	0	0	1	0	0
0	0	1	1	1	0	0	0	0	0	0	0	0	0	0	0	0	0	0	1	0
0	1	1	1	1	0	0	0	0	0	0	0	0	0	0	0	0	0	0	0	1

EL = HIGH
1 = HIGH voltage level. 0 = LOW voltage level.
X = Don't Care

Figure 8.38. The CMOS 4514 and 74x4514 are a 1-of-16 decoder/demultiplexer with input latches having four binary weighted address inputs (A0 - A3), a latch enable input (LE), and an active low enable input (/LE). The 16 outputs, (O0 - O15), are mutually exclusive active high. When EL is high, the selected output is determined by the data on An. When /EL goes low, the last data present is stored in the latches and the outputs remain stable. When /E is low, the selected output, determined by the latch content, is high. At /E high, all outputs are low. The enable input, /E, does not affect the state of the latch. When used as a multiplexer, /E is the data input and A0-A3 are the address inputs.

LOGIC DIAGRAM
Based on Philips HEF4515B

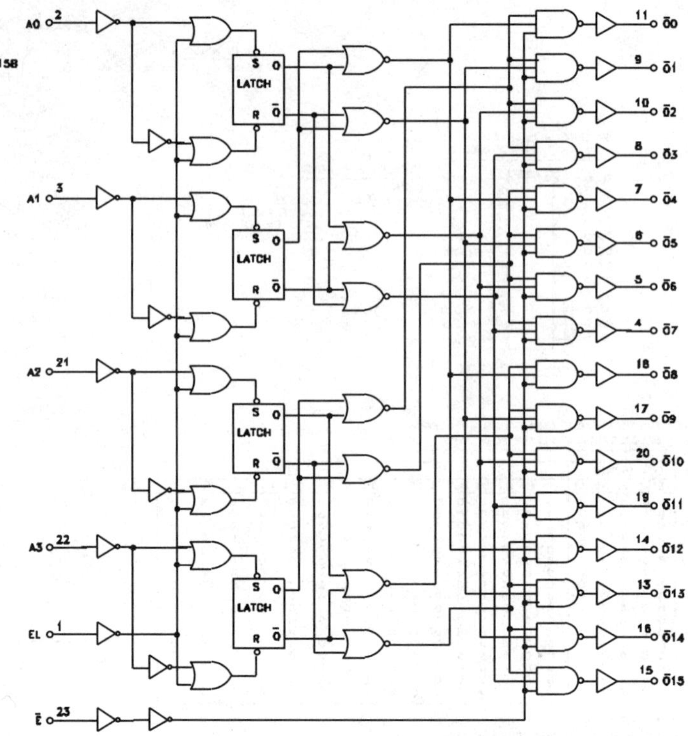

5V CMOS – CD4515
74HC/HCT4515

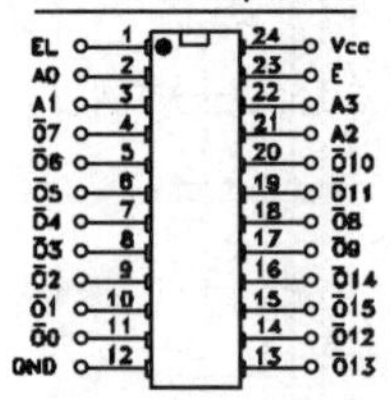

DUAL-IN-LINE PACKAGE

PACKAGE PINNING:

PIN–

2,3,21,22	ADDRESS INPUTS
23	ENABLE INPUT
1	LATCH ENABLE INPUT
4–11,13–20	DECODER OUTPUTS

FUNCTION TABLE 4515, 74x4515 1-OF-16
DECODER/DEMULTIPLEXER WITH INPUT LATCHES

INPUTS					OUTPUTS															
Ē	A0	A1	A2	A3	00	01	02	03	04	05	06	07	08	09	010	011	012	013	014	015
1	X	X	X	X	1	1	1	1	1	1	1	1	1	1	1	1	1	1	1	1
0	0	0	0	0	0	1	1	1	1	1	1	1	1	1	1	1	1	1	1	1
0	1	0	0	0	1	0	1	1	1	1	1	1	1	1	1	1	1	1	1	1
0	0	1	0	0	1	1	0	1	1	1	1	1	1	1	1	1	1	1	1	1
0	1	1	0	0	1	1	1	0	1	1	1	1	1	1	1	1	1	1	1	1
0	0	0	1	0	1	1	1	1	0	1	1	1	1	1	1	1	1	1	1	1
0	1	0	1	0	1	1	1	1	1	0	1	1	1	1	1	1	1	1	1	1
0	0	1	1	0	1	1	1	1	1	1	0	1	1	1	1	1	1	1	1	1
0	1	1	1	0	1	1	1	1	1	1	1	0	1	1	1	1	1	1	1	1
0	0	0	0	1	1	1	1	1	1	1	1	1	0	1	1	1	1	1	1	1
0	1	0	0	1	1	1	1	1	1	1	1	1	1	0	1	1	1	1	1	1
0	0	1	0	1	1	1	1	1	1	1	1	1	1	1	0	1	1	1	1	1
0	1	1	0	1	1	1	1	1	1	1	1	1	1	1	1	0	1	1	1	1
0	0	0	1	1	1	1	1	1	1	1	1	1	1	1	1	1	0	1	1	1
0	1	0	1	1	1	1	1	1	1	1	1	1	1	1	1	1	1	0	1	1
0	0	1	1	1	1	1	1	1	1	1	1	1	1	1	1	1	1	1	0	1
0	1	1	1	1	1	1	1	1	1	1	1	1	1	1	1	1	1	1	1	0

EL = HIGH
1 = HIGH voltage level. 0 = LOW voltage level.
X = Don't Care

Figure 8.39. The CMOS 4515 and 74x4515 are a 1-of-16 decoder/demultiplexer with input latches having four binary weighted address inputs (A0 -A3), a latch enable input (LE), and an active low enable input (/LE). The 16 outputs, (O0 - O15), are mutually exclusive active low. When EL is high, the selected output is determined by the data on An. When /EL goes low, the last data present is stored in the latches and the outputs remain stable. When /E is low, the selected output, determined by the latch content, is low. At /E high, all outputs are high. The enable input, /E, does not affect the state of the latch. When used as a multiplexer, /E is the data input and A0-A3 are the address inputs.

LOGIC DIAGRAM

Based on Motorola MC14529

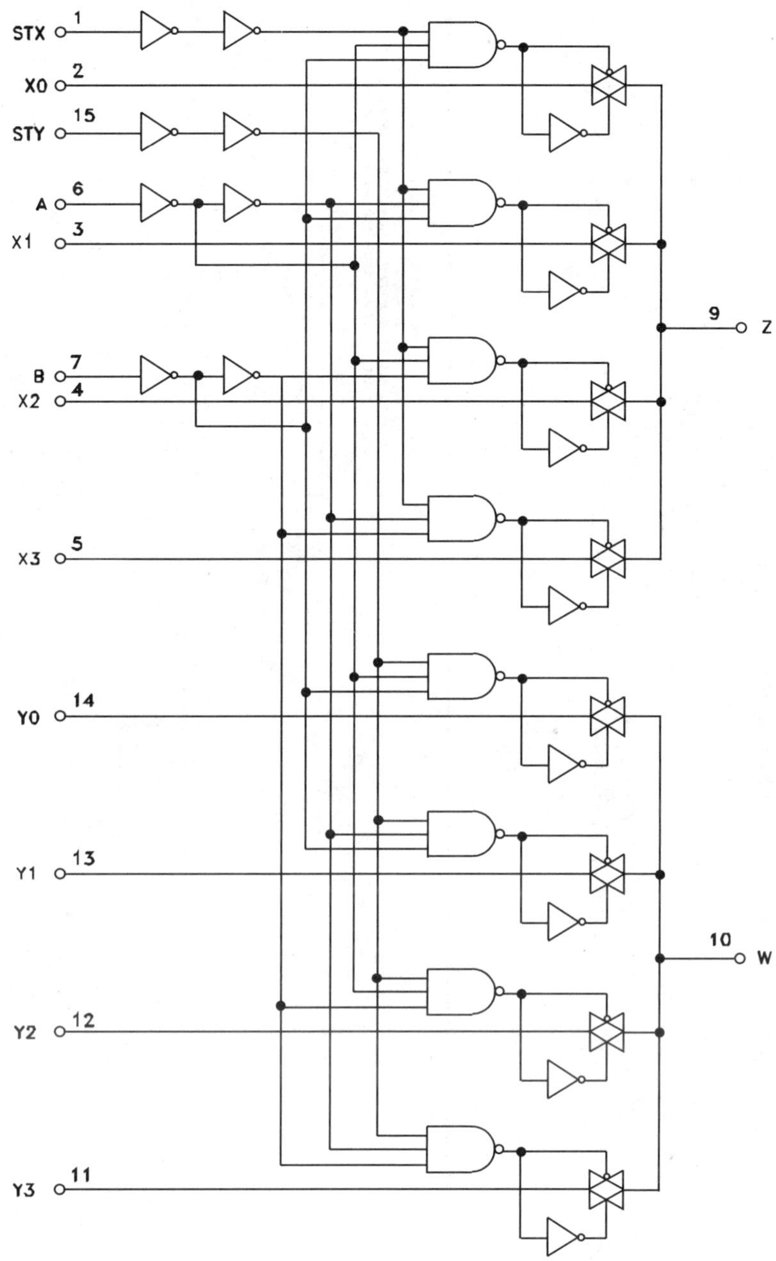

5V CMOS – CD4529

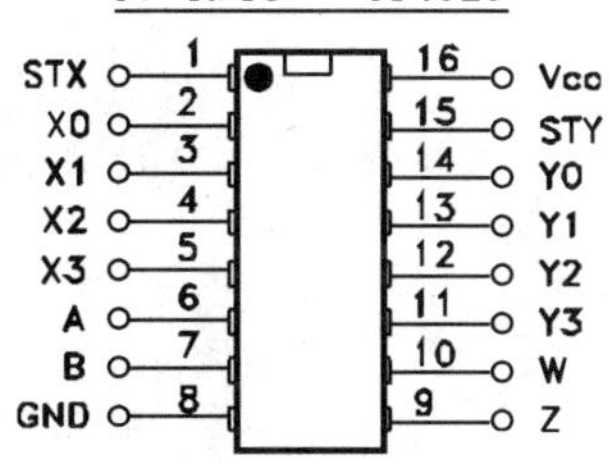

DUAL-IN-LINE PACKAGE

FUNCTION TABLE 4529 DUAL 4-CHANNEL ANALOG DATA SELECTOR

INPUTS				OUTPUTS	
STX	STY	B	A	Z	W
1	1	0	0	X0	Y0
1	1	0	1	X1	Y1
1	1	1	0	X2	Y2
1	1	1	1	X3	Y3
1	0	0	0	X0	
1	0	0	1	X1	
1	0	1	0	X2	
1	0	1	1	X3	
0	1	0	0	Y0	
0	1	0	1	Y1	
0	1	1	0	Y2	
0	1	1	1	Y3	
0	0	X	X	High Impedance	

Figure 8.40. The CMOS 4529 is a dual 4-channel analog data selector, or single 8-channel device depending on the input coding. The device has bidirectional analog capabilities which make it useful as a dual binary to 1-of-4 or a binary to 1-of-8 decoder. The outputs are 3-state. In the single channel mode, the two outputs are tied together.

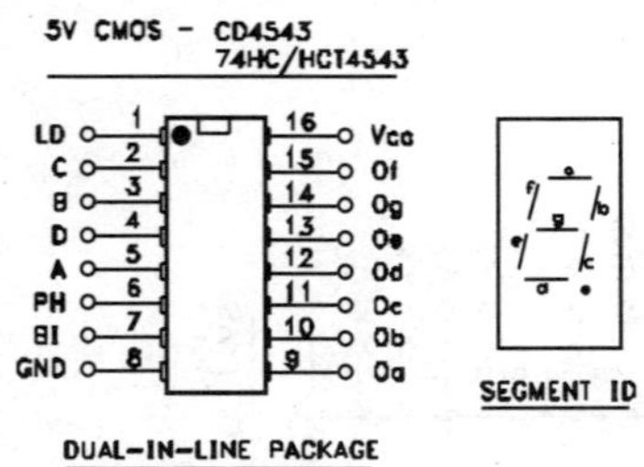

FUNCTION TABLE 4543, 74x4543 BCD-TO 7-SEGMENT LATCH/DECODER/DRIVER FOR LIQUID CRYSTALS

INPUTS							OUTPUTS							
LD	BI	PH●	DD	DC	DB	DA	Oa	Ob	Oc	Od	Oe	Of	Og	DISPLAY
X	1	0	X	X	X	X	0	0	0	0	0	0	0	blank
1	0	0	0	0	0	0	1	1	1	1	1	1	0	0
1	0	0	0	0	0	1	0	1	1	0	0	0	0	1
1	0	0	0	0	1	0	1	1	0	1	1	0	1	2
1	0	0	0	0	1	1	1	1	1	1	0	0	1	3
1	0	0	0	1	0	0	0	1	1	0	0	1	1	4
1	0	0	0	1	0	1	1	0	1	1	0	1	1	5
1	0	0	0	1	1	0	1	0	1	1	1	1	1	6
1	0	0	0	1	1	1	1	1	1	0	0	0	0	7
1	0	0	1	0	0	0	1	1	1	1	1	1	1	8
1	0	0	1	0	0	1	1	1	1	0	0	1	1	9
1	0	0	1	0	1	0	0	0	0	0	0	0	0	blank
1	0	0	1	0	1	1	0	0	0	0	0	0	0	blank
1	0	0	1	1	0	0	0	0	0	0	0	0	0	blank
1	0	0	1	1	0	1	0	0	0	0	0	0	0	blank
1	0	0	1	1	1	0	0	0	0	0	0	0	0	blank
1	0	0	1	1	1	1	0	0	0	0	0	0	0	blank
0	0	0	X	X	X	X	*							*
Above combinations apply							Inverse of the above combinations							As above

1 = HIGH voltage level. 0 = LOW voltage level.
X = Don't Care
* Depends upon the BCD code previously applied when LD=1.
● For liquid crystal readouts, apply a square wave to PH.
For Common Cathode LED readouts select PH=0.
For Common Anode LED readouts select PH=1.

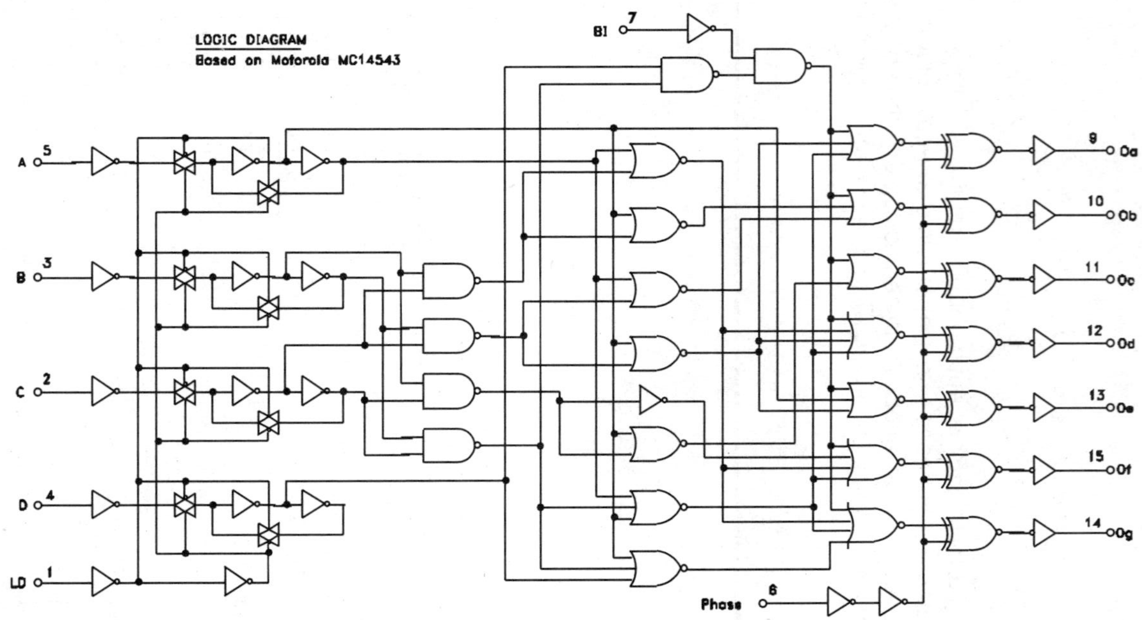

Figure 8.41. The CMOS 4543 and 74x4543 are a BCD-to-7-segment latch/decoder/driver for liquid crystal and LED displays. It has four address inputs (DA - DD), an active high latch disable input (LD), an active high blanking input (BI), an active high phase input (PH), and seven buffered segment outputs (Oa - Og). The device functions as a 4-bit storage latch and 8421 BCD-to-7-segment decoder/driver. It can invert the logic functions of the output combination using the PH input. For liquid crystal displays a square wave is applied to PH and the electrical common back-plane of the display. The device outputs are directly connected to the segments of the liquid crystal display.

5V CMOS – CD4544

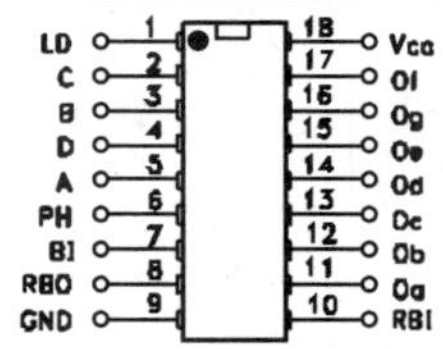

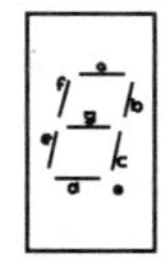

SEGMENT ID

FUNCTION TABLE 4544 BCD–TO–7–SEGMENT
LATCH/DECODER/DRIVER FOR LIQUID CRYSTALS

INPUTS									OUTPUTS							
RBI	LD	BI	PH●	DD	DC	DB	DA	RBO	Oa	Ob	Oc	Od	Oe	Of	Og	DISPLAY
X	X	1	0	X	X	X	X	#	0	0	0	0	0	0	0	blank
1	1	1	0	0	0	0	0	1	0	0	0	0	0	0	0	blank
0	1	0	0	0	0	0	0	0	1	1	1	1	1	1	0	0
X	1	0	0	0	0	0	1	0	0	1	1	0	0	0	0	1
X	1	0	0	0	0	1	0	0	1	1	0	1	1	0	1	2
X	1	0	0	0	0	1	1	0	1	1	1	1	0	0	1	3
X	1	0	0	0	1	0	0	0	0	1	1	0	0	1	1	4
X	1	0	0	0	1	0	1	0	1	0	1	1	0	1	1	5
X	1	0	0	0	1	1	0	0	1	0	1	1	1	1	1	6
X	1	0	0	0	1	1	1	0	1	1	1	0	0	0	0	7
X	1	0	0	1	0	0	0	0	1	1	1	1	1	1	1	8
X	1	0	0	1	0	0	1	0	1	1	1	0	0	1	1	9
X	1	0	0	1	0	1	0	0	0	0	0	0	0	0	0	blank
X	1	0	0	1	0	1	1	0	0	0	0	0	0	0	0	blank
X	1	0	0	1	1	0	0	0	0	0	0	0	0	0	0	blank
X	1	0	0	1	1	0	1	0	0	0	0	0	0	0	0	blank
X	1	0	0	1	1	1	0	0	0	0	0	0	0	0	0	blank
X	1	0	0	1	1	1	1	0	0	0	0	0	0	0	0	blank
X	0	0	0	X	X	X	X	#				*				*
Above combinations apply									Inverse of the above combinations							As above

1 = HIGH voltage level. 0 = LOW voltage level.
X = Don't Care
* Depends upon the BCD code previously applied when LD=1.
● For liquid crystal readouts, apply a square wave to PH.
For Common Cathode LED readouts select PH=0.
For Common Anode LED readouts select PH=1.
RBO = RBI·($\overline{A}\,\overline{B}\,\overline{C}\,\overline{D}$)

LOGIC DIAGRAM
Based on Motorola MC14544

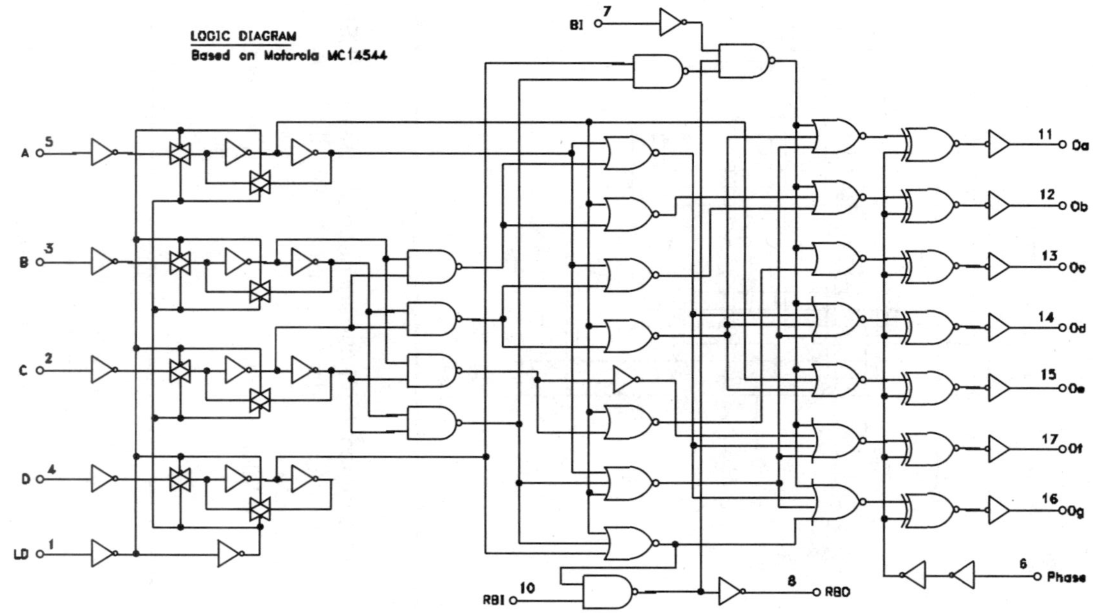

Figure 8.42. The CMOS 4544 is a BCD-to-7-segment latch/decoder/driver for liquid crystal displays. It has four address inputs (A - D), an active high latch disable input (LD), an active high blanking input (BI), an active high phase input (PH), and seven buffered segment outputs (Oa - Og). The device functions as a 4-bit storage latch and 8421 BCD-to-7-segment decoder/driver. It can invert the logic functions of the output combination using the PH input. For liquid crystal displays a square wave is applied to PH and the electrical common back-plane of the display. The device outputs are directly connected to the segments of the liquid crystal display. Ripple Blanking Input (RBI) and Ripple Blanking Output (RBO) input and output provide for leading and trailing zero suppression in the display.

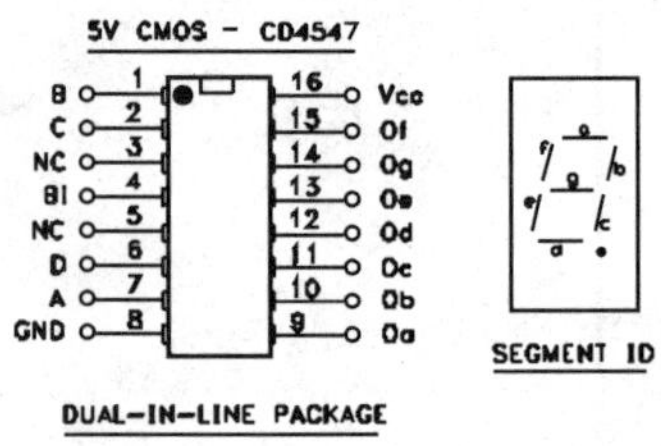

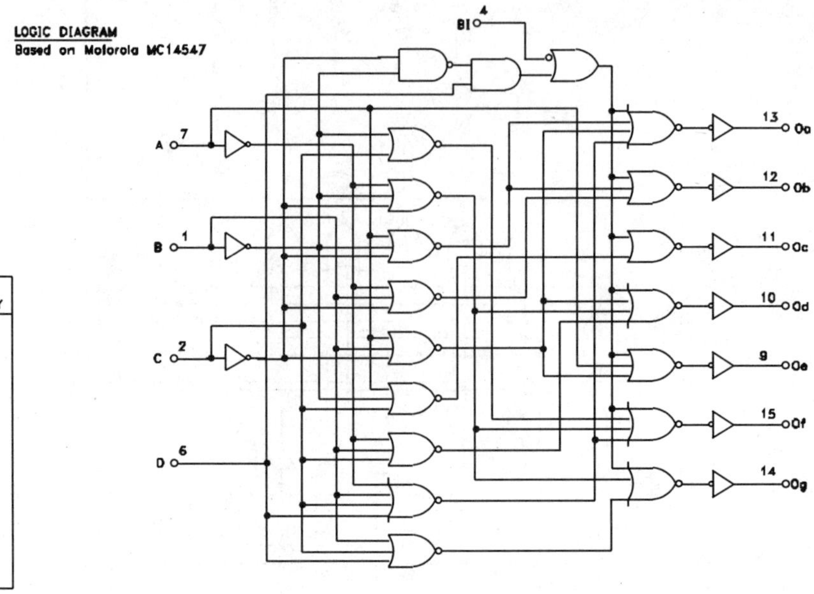

FUNCTION TABLE 4547 BCD-TO-7-SEGMENT
HIGH CURRENT DECODER/DRIVER

INPUTS					OUTPUTS							
BI	DD	DC	DB	DA	Oa	Ob	Oc	Od	Oe	Of	Og	DISPLAY
1	X	X	X	X	0	0	0	0	0	0	0	blank
0	0	0	0	0	1	1	1	1	1	1	0	0
0	0	0	0	1	0	1	1	0	0	0	0	1
0	0	0	1	0	1	1	0	1	1	0	1	2
0	0	0	1	1	1	1	1	1	0	0	1	3
0	0	1	0	0	0	1	1	0	0	1	1	4
0	0	1	0	1	1	0	1	1	0	1	1	5
0	0	1	1	0	1	0	1	1	1	1	1	6
0	0	1	1	1	1	1	1	0	0	0	0	7
0	1	0	0	0	1	1	1	1	1	1	1	8
0	1	0	0	1	1	1	1	0	0	1	1	9
0	1	0	1	0	0	0	0	0	0	0	0	blank
0	1	0	1	1	0	0	0	0	0	0	0	blank
0	1	1	0	0	0	0	0	0	0	0	0	blank
0	1	1	0	1	0	0	0	0	0	0	0	blank
0	1	1	1	0	0	0	0	0	0	0	0	blank
0	1	1	1	1	0	0	0	0	0	0	0	blank

1 = HIGH voltage level. 0 = LOW voltage level.
X = Don't Care

Figure 8.43. The CMOS 4547 is a BCD-to-7-segment high-current decoder/driver for LED, incandescent, fluorescent, or gas discharge displays. It has four address inputs (A - D), an active high blanking input (BI), and seven segment outputs (Oa - Og). The device functions as a 8421 BCD-to-7-segment decoder/driver. If a latch is required use the 4511 instead.

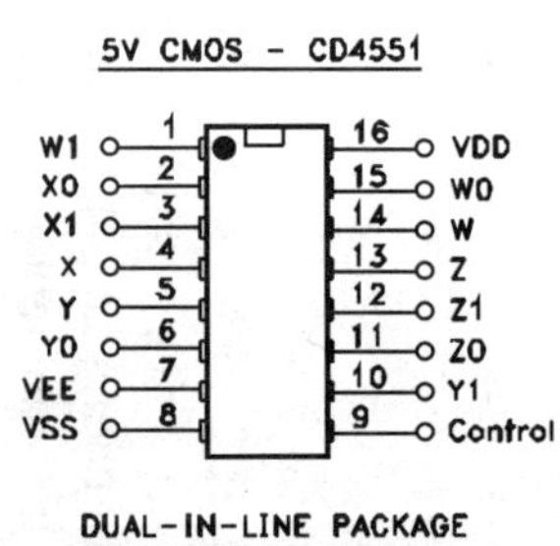

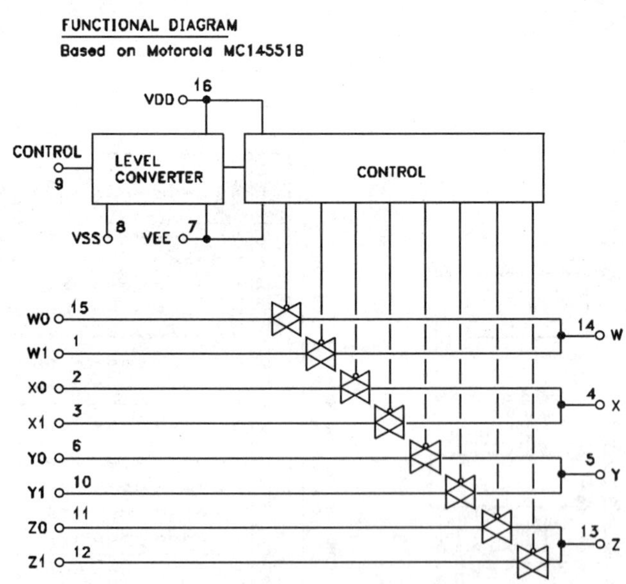

FUNCTION TABLE 4551 QUAD 2-CHANNEL
ANALOG MULTIPLEXER/DEMULTIPLEXER

CONTROL	ON
0	W0 X0 Y0 Z0
1	W1 X1 Y1 Z1

NOTES:
Control input referenced to GND.
Analog inputs and outputs referenced to VEE.
VEE must be </= VSS
1 = HIGH voltage level. 0 = LOW voltage level.

Figure 8.44. The CMOS 4551 is a quad 2-channel analog multiplexer/demultiplexer, a digitally controlled analog switch. It functions as a 4PDT switch. The analog voltage range is 3 - 18 volts. VEE must be ≤Vss.

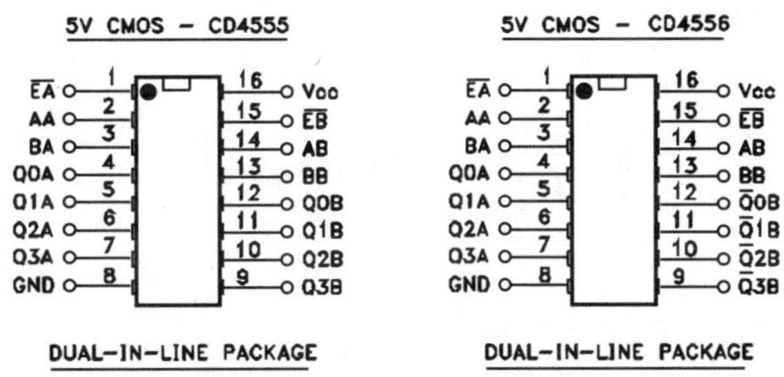

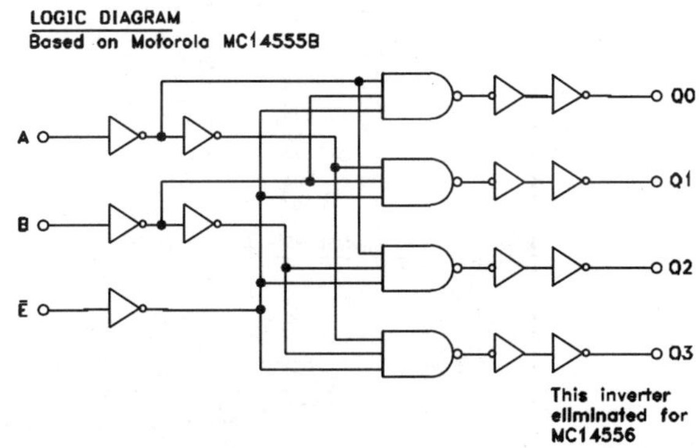

FUNCTION TABLE 4555, 4556 DUAL BINARY TO 1-OF-4 DECODER/DEMULTIPLEXER

INPUTS			OUTPUTS				OUTPUTS			
ENABLE	SELECT		4555				4556			
$\overline{E}$	B	A	Q3	Q2	Q1	Q0	$\overline{Q3}$	$\overline{Q2}$	$\overline{Q1}$	$\overline{Q0}$
0	0	0	0	0	0	1	1	1	1	0
0	0	1	0	0	1	0	1	1	0	1
0	1	0	0	1	0	0	1	0	1	1
0	1	1	1	0	0	0	0	1	1	1
1	X	X	0	0	0	0	1	1	1	1

1 = HIGH voltage level. 0 = LOW voltage level.
X = Don't Care

Figure 8.45. The CMOS 4555 and 4556 are dual binary to 1-of-4 decoder/demultiplexers. Each circuit has two select inputs, A and B, and an active low enable, /E, with four mutually exclusive outputs Q0-Q3. With the 4555, the selected outputs go to the high logic state; with the 4556, the outputs go to the low state. Expanded decoding can be achieved using additional devices.

5V CMOS – CD4558

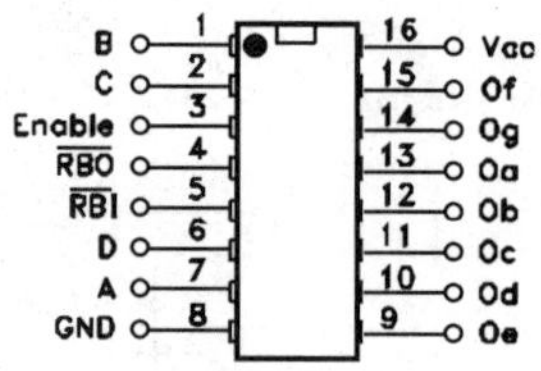

DUAL-IN-LINE PACKAGE

AUXILIARY INPUT TRUTH TABLE

Enable Pin 3	$\overline{RBI}$ Pin 5	BCD Input Code	$\overline{RBO}$ Pin 4	Function Performed
0	0	X	0	Lamp Test
0	1	X	1	Blank Segments
1	1	0	1	Display Zero
1	0	0	0	Blank Segments
1	X	1–9	1	1–9 Displayed

1 = HIGH voltage level. 0 = LOW voltage level.
X = Don't Care

FUNCTION TABLE 4558 BCD-TO-7-SEGMENT DECODER

INPUTS			OUTPUTS		
Enable Pin 3	$\overline{RBI}$ Pin 5	D C B A	a b c d e f g	$\overline{RBO}$	DISPLAY
1	1	0 0 0 0	1 1 1 1 1 1 0	1	0
1	X	0 0 0 1	0 0 0 0 1 1 0	1	1
1	X	0 0 1 0	1 1 0 1 1 0 1	1	2
1	X	0 0 1 1	1 1 1 1 0 0 1	1	3
1	X	0 1 0 0	0 1 1 0 0 1 1	1	4
1	X	0 1 0 1	1 0 1 1 0 1 1	1	5
1	X	0 1 1 0	0 0 1 1 1 1 1	1	6
1	X	0 1 1 1	1 1 1 0 0 0 0	1	7
1	X	1 0 0 0	1 1 1 1 1 1 1	1	8
1	X	1 0 0 1	1 1 1 0 0 1 1	1	9
1	0	0 0 0 0	0 0 0 0 0 0 0	0	Blank
0	0	X X X X	1 1 1 1 1 1 1	0	8
0	1	X X X X	0 0 0 0 0 0 0	1	Blank

ZERO SUPRESSION WITH LAMP TEST AND INTENSITY BLANKING

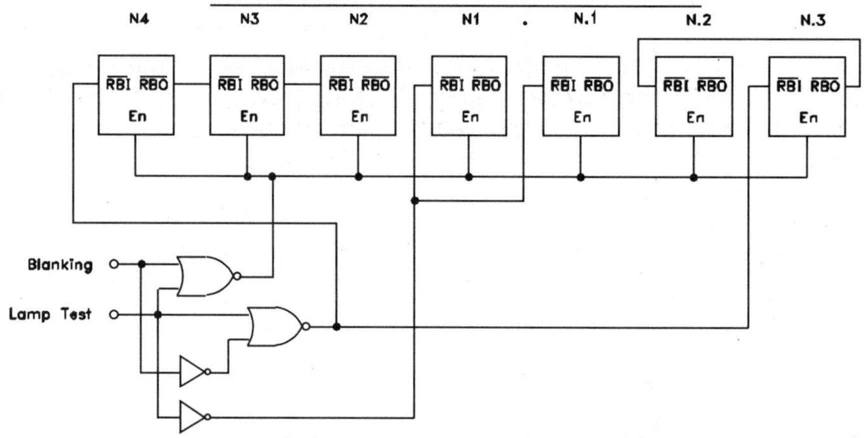

Figure 8.46. The CMOS 4558 is a BCD-to-7-segment decoder, providing an active high output depending on the status of two control inputs: an Enable and Ripple Blanking Input (RBI). A lamp test input function and Ripple Blanking Output (RBO) are included. Leading zero blanking is easily obtained with an external flip-flop in time division multiplexed systems displaying most significant decades (MSD) first.

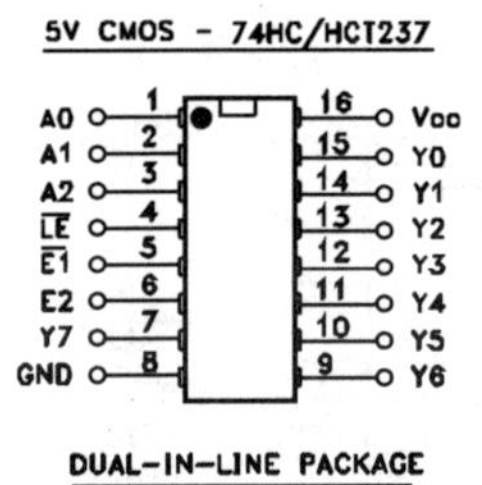

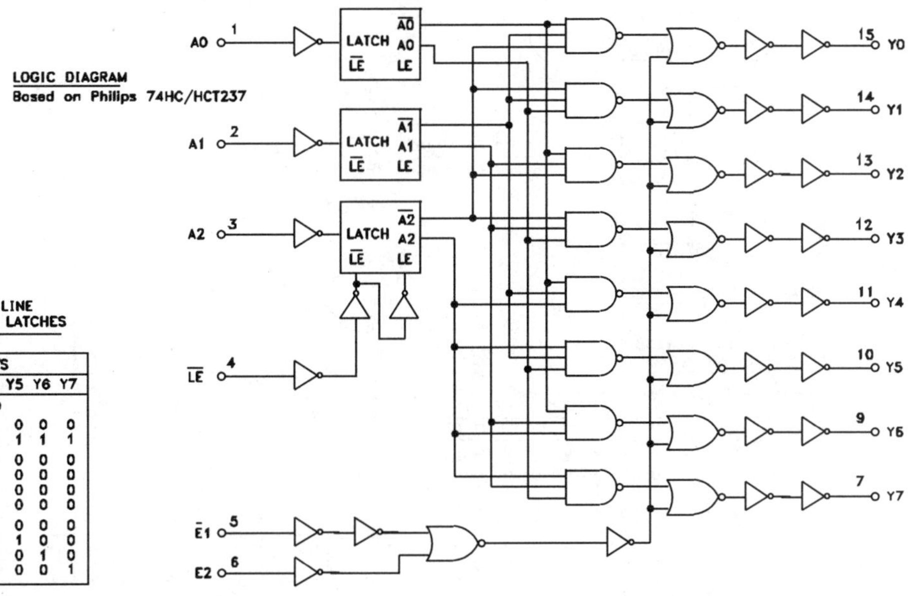

FUNCTION TABLE 74HC/HCT237 3-TO-8 LINE DECODER/DEMULTIPLEXER WITH ADDRESS LATCHES

INPUTS						OUTPUTS							
/LE	/E1	E2	A0	A1	A2	Y0	Y1	Y2	Y3	Y4	Y5	Y6	Y7
1	0	1	X	X	X	stable							
X	1	X	X	X	X	0	0	0	0	0	0	0	0
X	X	0	X	X	X	1	1	1	1	1	1	1	1
0	0	1	0	0	0	1	0	0	0	0	0	0	0
0	0	1	1	0	0	0	1	0	0	0	0	0	0
0	0	1	0	1	0	0	0	1	0	0	0	0	0
0	0	1	1	1	0	0	0	0	1	0	0	0	0
0	0	1	0	0	1	0	0	0	0	1	0	0	0
0	0	1	1	0	1	0	0	0	0	0	1	0	0
0	0	1	0	1	1	0	0	0	0	0	0	1	0
0	0	1	1	1	1	0	0	0	0	0	0	0	1

1 = HIGH voltage level. 0 = LOW voltage level.
X = Don't Care

Figure 8.47. The CMOS 74HC/HCT237 is a 3-to-8 line decoder/demultiplexer with latches at the three address inputs, An. The device combines the decoder function with a 3-bit storage latch. With the latch enabled, /LE low, the '237 is a 3-to-8 active low decoder. When the latch enable, /LE, goes from low-to-high; the last data present is stored in the latches. Further address changes are ignored as long as /LE remains high. The output enable inputs, /E1 and E2, control the state of the output independent of the address inputs or the latch operation. All outputs are high unless /E1 is low and E2 is high.

5V CMOS – 74HC/HCT238

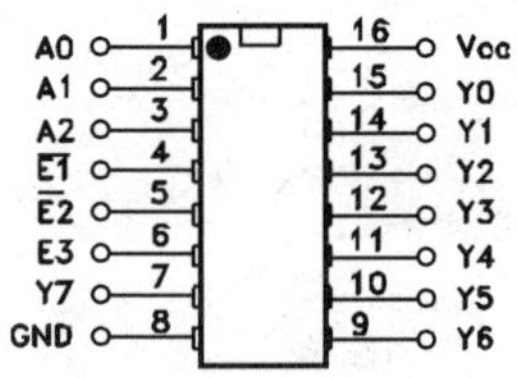

DUAL-IN-LINE PACKAGE

LOGIC DIAGRAM
Based on Philips 74HC/HCT238

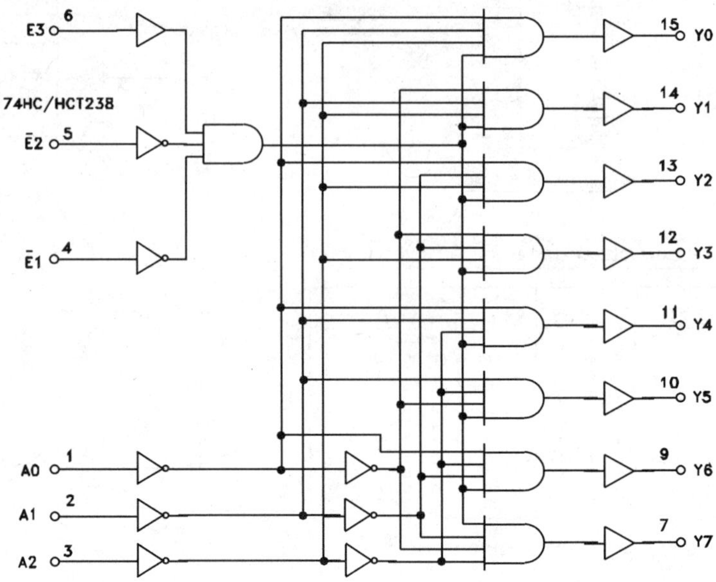

FUNCTION TABLE 74HC/HCT238 3-TO-8
LINE DECODER/DEMULTIPLEXER

INPUTS						OUTPUTS							
Ē1	Ē2	E3	A0	A1	A2	Y0	Y1	Y2	Y3	Y4	Y5	Y6	Y7
1	X	X	X	X	X	0	0	0	0	0	0	0	0
X	1	X	X	X	X	0	0	0	0	0	0	0	0
X	X	0	X	X	X	1	1	1	1	1	1	1	1
0	0	1	0	0	0	1	0	0	0	0	0	0	0
0	0	1	1	0	0	0	1	0	0	0	0	0	0
0	0	1	0	1	0	0	0	1	0	0	0	0	0
0	0	1	1	1	0	0	0	0	1	0	0	0	0
0	0	1	0	0	1	0	0	0	0	1	0	0	0
0	0	1	1	0	1	0	0	0	0	0	1	0	0
0	0	1	0	1	1	0	0	0	0	0	0	1	0
0	0	1	1	1	1	0	0	0	0	0	0	0	1

1 = HIGH voltage level. 0 = LOW voltage level.
X = Don't Care

Figure 8.48. The CMOS 74HC/HCT238 is a 3-to-8 line decoder/demultiplexer. Three binary coded address inputs, A0 - A1, when enabled, provide eight mutually exclusive active high outputs, Y0 - Y7. The '238 features three enable inputs: /E1 and /E2 are active low; E3 is active high. Every output will be low unless /E1 and /E2 are low and E3 is high. This logic arrangement provides for easy expansion to a 1-of-32 (5 lines to 32 lines) with four '238s and one inverter. The device can be used as an eight-output demultiplexer by using one of the active low enable inputs as the data input and the remaining inputs as strobes. Unused enable inputs must be permanently tied to their appropriate high or low states.

5V CMOS – 74HC239

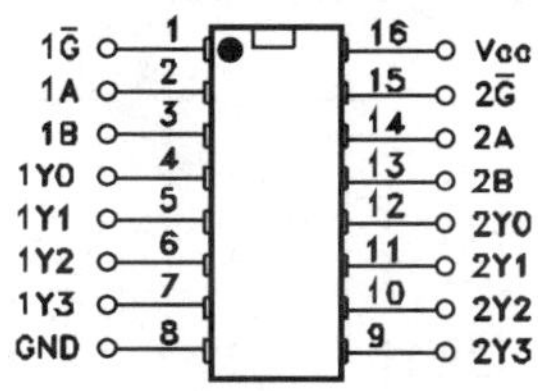

DUAL-IN-LINE PACKAGE

FUNCTION TABLE 74HC239 DUAL 2-LINE TO 4-LINE DECODER/DEMULTIPLEXER

INPUTS			OUTPUTS			
ENABLE	SELECT					
G̅	B	A	Y0	Y1	Y2	Y3
1	X	X	0	0	0	0
0	0	0	1	0	0	0
0	0	1	0	1	0	0
0	1	0	0	0	1	0
0	1	1	0	0	0	1

1 = HIGH voltage level.
0 = LOW voltage level.
X = Don't Care

LOGIC DIAGRAM

Based on Texas Instruments 74HC239

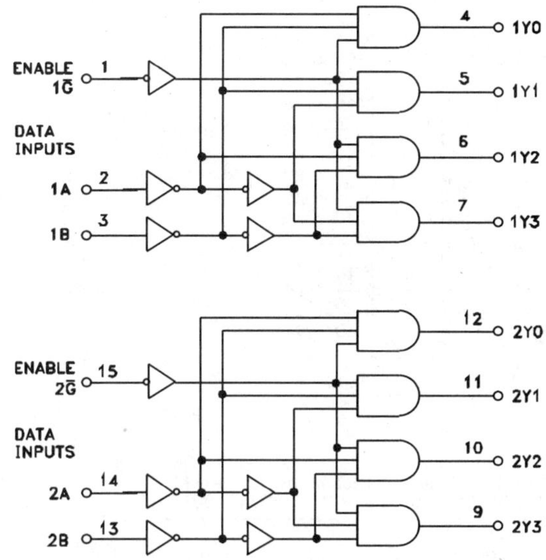

Figure 8.49. The CMOS 74HC239 is a dual 2-to-4 line decoder/demultiplexer designed for use in high-performance memory decoding or data routing applications requiring very short propagation delay times. The '239 is comprised of two individual decoders. The active low enable input can be used as a data line in demultiplexing applications.

5V CMOS - 74HC604

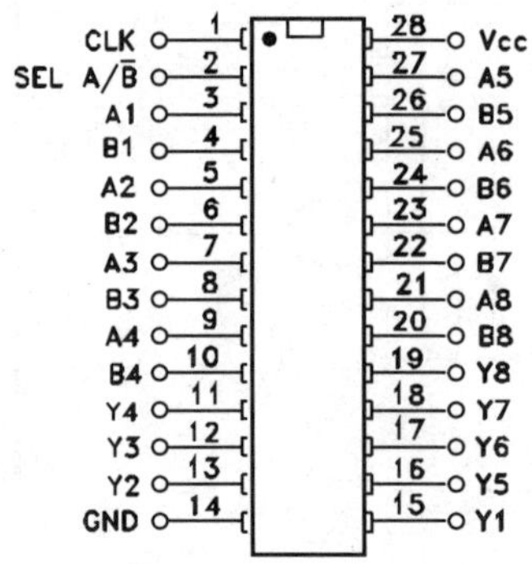

DUAL-IN-LINE PACKAGE

LOGIC DIAGRAM
Based on Texas Instruments 74HC604

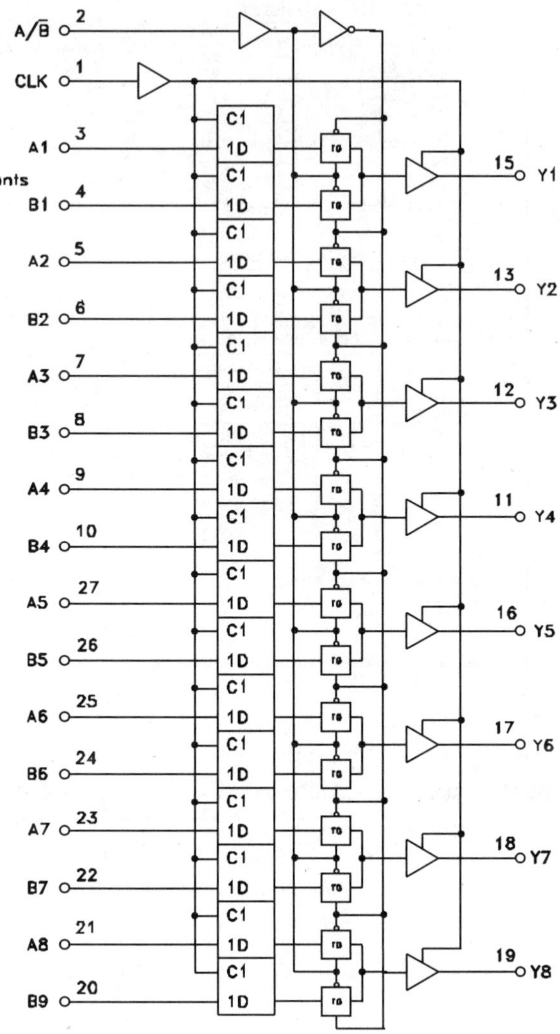

FUNCTION TABLE 74HC604 OCTAL 2-INPUT MULTIPLEXED LATCH WITH 3-STATE OUTPUTS

INPUTS				OUTPUTS
A1–A8	B1–B8	A/B̄	CLOCK	Y1–Y8
A data	B data	0	↑	B data
A data	B data	1	↑	A data
X	X	X	0	Z
X	X	0	1	B register stored data
X	X	1	1	A register stored data

1 = HIGH voltage level. 0 = LOW voltage level.
X = Don't Care

↑ = LOW-to-HIGH clock transition.

Figure 8.50. The CMOS 74HC604 is an octal 2-input multiplexed latch with 3-state outputs. The device is ideal for storing data from two input buses, A and B, and for providing the output bus with stored data from either the A or B register. Data is loaded on the positive-going clock transition. The clock pin also controls the active and high-impedance states of the outputs. When the input is low, the outputs are in the high-impedance state. When high, the outputs are enabled. These functions are ideal for interfacing from a 16-bit microprocessor to a 64K RAM board. The row and column addresses can be loaded as one word from the microprocessor and then multiplexed sequentially to the RAM.

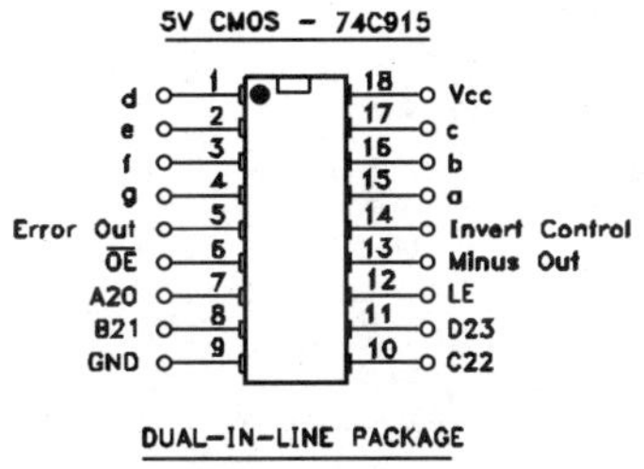

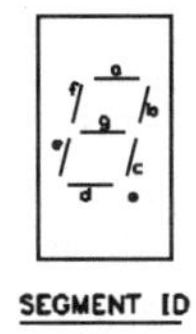

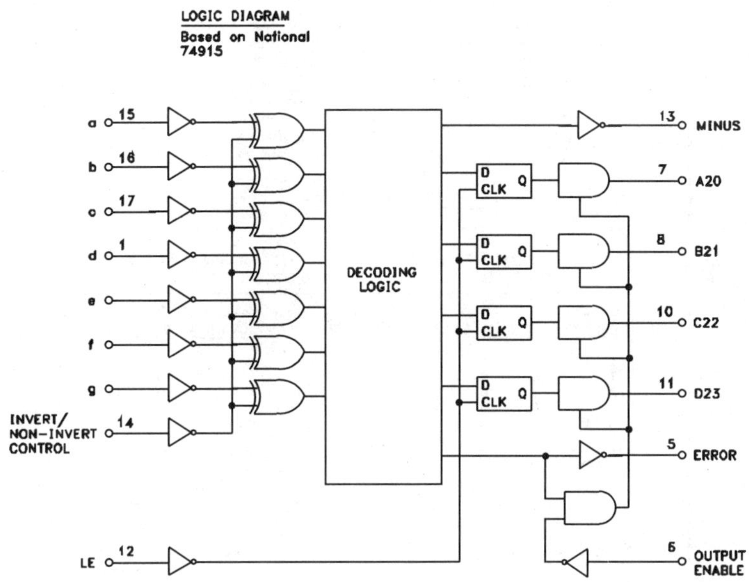

FUNCTION TABLE 74C915 7-SEGMENT TO BCD CONVERTER

CHARACTER AT SEGMENT INPUTS	BCD OUTPUTS				NON-BCD OUTPUTS	
	D23	C22	B21	A20	ERROR	MINUS
0	0	0	0	0	0	0
1 *	0	0	0	1	0	0
1	0	0	0	1	0	0
2	0	0	1	0	0	0
3	0	0	1	1	0	0
4	0	1	0	0	0	0
5	0	1	0	1	0	0
6 *	0	1	1	0	0	0
6	0	1	1	0	0	0
7	0	1	1	1	0	0
8	1	0	0	0	0	0
9 *	1	0	0	1	0	0
9	1	0	0	1	0	0
	1	1	1	1	0	0
All other	X	X	X	X	1	1
input	X	X	X	X	1	0
combinations	X	X	X	X	1	0

* Alternative character constructions.
1 = HIGH voltage level. 0 = LOW voltage level.
X = TRI-STATE condition. **

Figure 8.51. The CMOS 74C915 is a 7-segment-to-BCD converter. The circuit accepts 7-segment information and converts it to BCD. The true state of the segment inputs can be selected by use of the Invert/Non-invert control input. A logical "0" selects active high true decoding at the segment inputs. Logical "1" selects active low. In addition to four TTL compatible BCD outputs, an error output and a minus output are available. The error output goes to an active "1" whenever a nonstandard 7-segment code is input. The output is forced into a 3-state condition whenever an error is detected. This allows the user to write his own error code. A high input to the output enable forces the outputs to the 3-state condition.

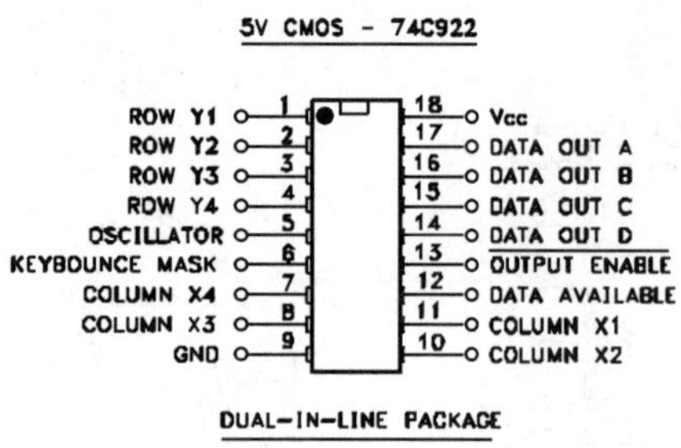

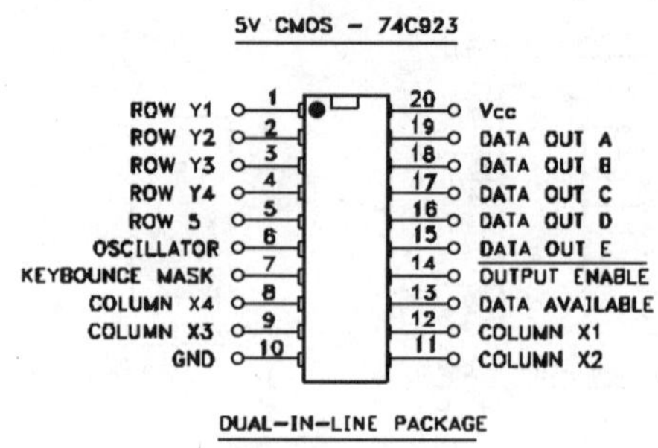

TRUTH TABLE 74C922 16-KEY, 74C923 20-KEY, ENCODERS

Switch Position	0 Y1,X1	1 Y1,X2	2 Y1,X3	3 Y1,X4	4 Y2,X1	5 Y2,X2	6 Y2,X3	7 Y2,X4	8 Y3,X1	9 Y3,X2	10 Y3,X3	11 Y3,X4	12 Y4,X1	13 Y4,X2	14 Y4,X3	15 Y4,X4	16 Y5,X1	17 Y5,X2	18 Y5,X3	19 Y5,X4
DATA OUT A	0	1	0	1	0	1	0	1	0	1	0	1	0	1	0	1	0	1	0	1
DATA OUT B	0	0	1	1	0	0	1	1	0	0	1	1	0	0	1	1	0	0	1	1
DATA OUT C	0	0	0	0	1	1	1	1	0	0	0	0	1	1	1	1	0	0	0	0
DATA OUT D	0	0	0	0	0	0	0	0	1	1	1	1	1	1	1	1	0	0	0	0
DATA OUT E*	0	0	0	0	0	0	0	0	0	0	0	0	0	0	0	0	1	1	1	1

* Omit for 74C922

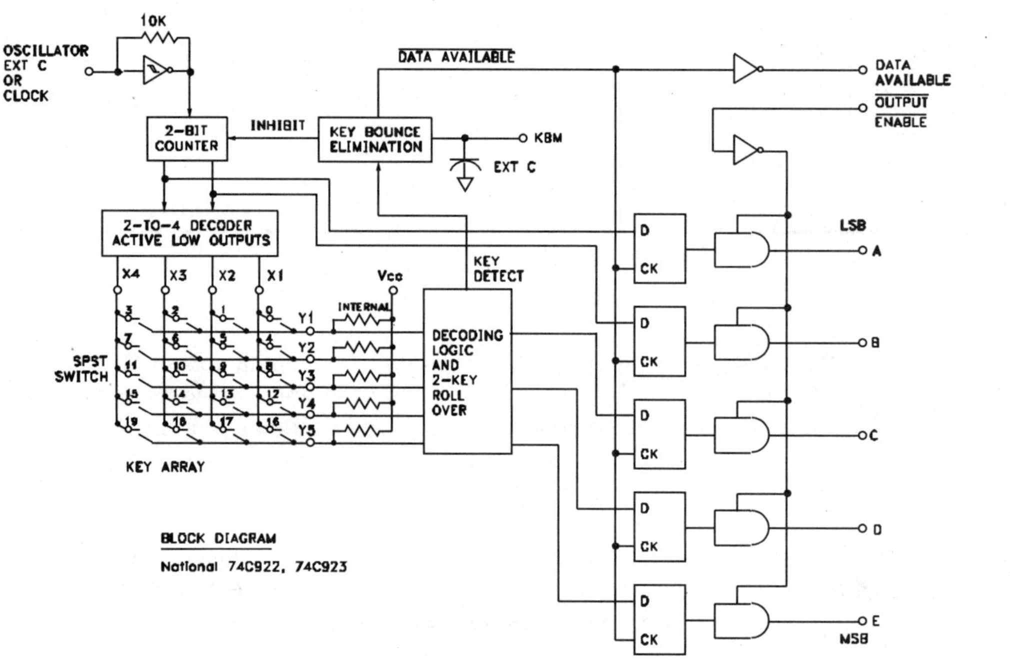

Figure 8.52. The CMOS 74C922 and 74C923 are 16-key and 20-key keypad encoders, respectively. They contain all the logic to encode an array of SPST switches. The keyboard scan can be implemented by an external clock or capacitor. On-chip pull-ups allow switches with up to 50 kW of resistance to be used. No diodes in the switch array are needed. The internal debounce needs only a single capacitor and is defeated by omitting the capacitor. A Data Available output goes to a logic "1" when a valid key entry has been made. This output goes low when the depressed key is released, even if another key has already been pressed. The Data Available will return to the high level after the debounce time. This two-key rollover functions between any two keys.

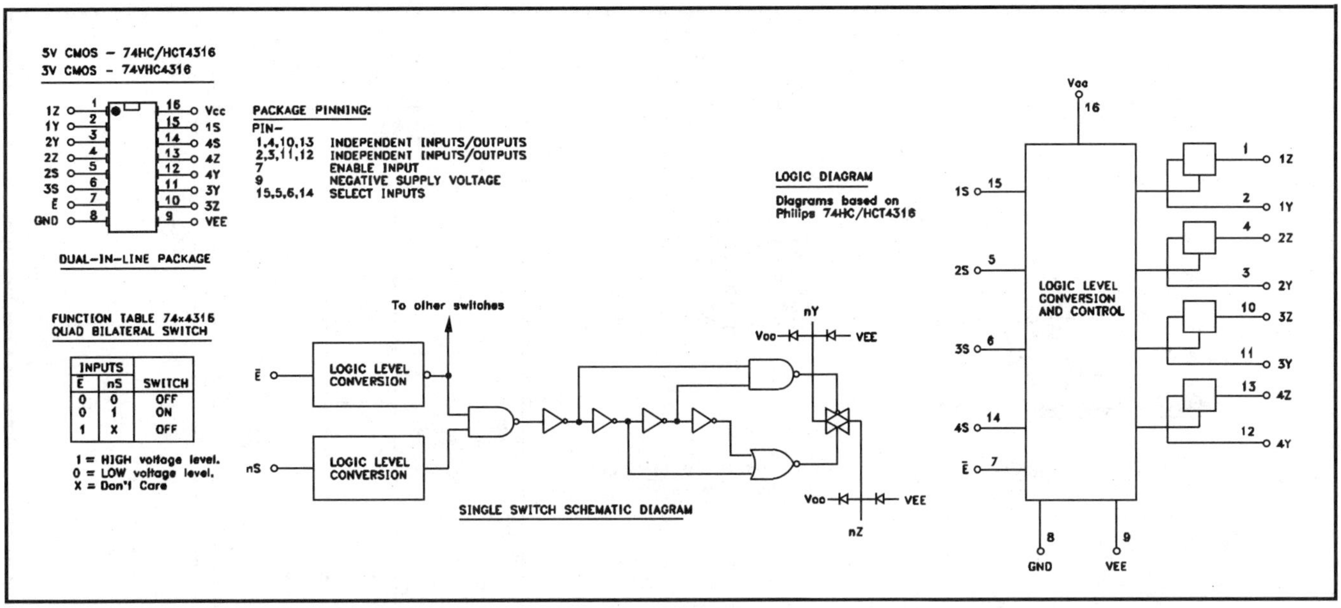

Figure 8.53. The CMOS 74x4316 quad bilateral switch contains four independent analog switches. Each switch has two input/output terminals, nY, nZ, and an active high select input, nS. When the enable input /E is high, all four switches are turned off. Vcc and ground are the supply voltage for the digital control inputs. The analog inputs can swing between Vcc as a positive limit and VEE as the negative. Vcc-VEE cannot exceed 10 volts; VEE max is -7.5 V for low-voltage devices. See the 4016 for the version without logic level translation.

5V CMOS - 74HC/HCT4351

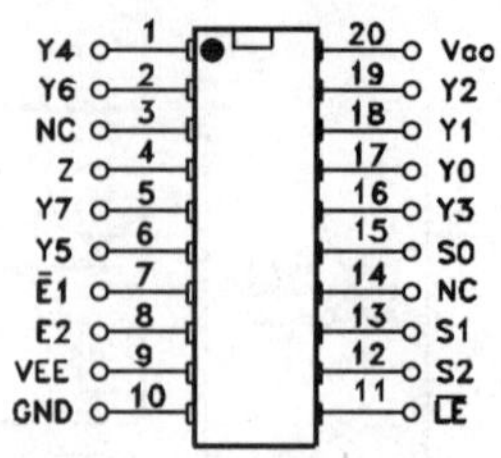

DUAL-IN-LINE PACKAGE

PACKAGE PINNING:

PIN-	
1,2,5,6,16-19	INPUTS/OUTPUTS
4	COMMON INPUT/OUTPUT
7,8	INPUT ENABLES
11	LATCH ENABLE
12,13,15	SELECT INPUTS
9	NEGATIVE VOLTAGE
3,14	NOT CONNECTED

FUNCTION TABLE 74HC/HCT4351 8-CHANNEL ANALOG MULTIPLEXER/DEMULTIPLEXER WITH LATCH

INPUTS						CHANNEL ON
Ē1	E2	L̄Ē	S2	S1	S0	
1	X	X	X	X	X	none
X	0	X	X	X	X	none
0	1	1	0	0	0	Y0
0	1	1	0	0	1	Y1
0	1	1	0	1	0	Y2
0	1	1	0	1	1	Y3
0	1	1	1	0	0	Y4
0	1	1	1	0	1	Y5
0	1	1	1	1	0	Y6
0	1	1	1	1	1	Y7
0	1	0	X	X	X	*
X	X	↓	X	X	X	**

* Last selected channel ON
** Selected channels latched
1 = HIGH voltage level. 0 = LOW voltage level.
X = Don't Care
↓ = HIGH-to-LOW L̄Ē transition.

FUNCTIONAL DIAGRAM
Based on Philips 74HC/HCT4351

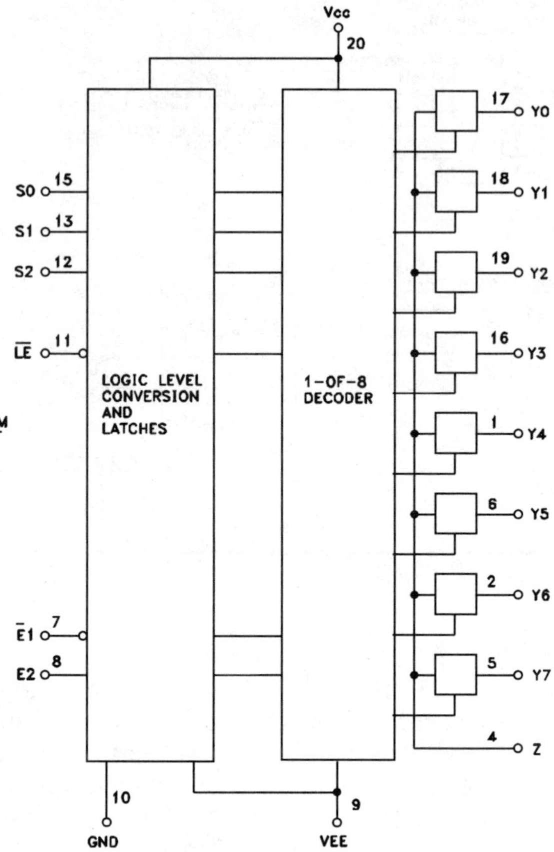

Figure 8.54. The CMOS 74HC/HCT4351 is an 8-channel analog multiplexer/demultiplexer with three select inputs, S0-S2, two enable inputs, /E1, E2, a latch enable input, /LE, eight independent inputs/outputs, Y0-Y7, and a common input/output, Z. One of the switches is selected with /E1 low and E2 high. The input data may be latched by use of the active low latch, /LE. When /LE is high, the latch is transparent. Whenever one of the two enables, /E1 or E2 is inactive, all switches are in an off state.

5V CMOS - 74HC/HCT4352

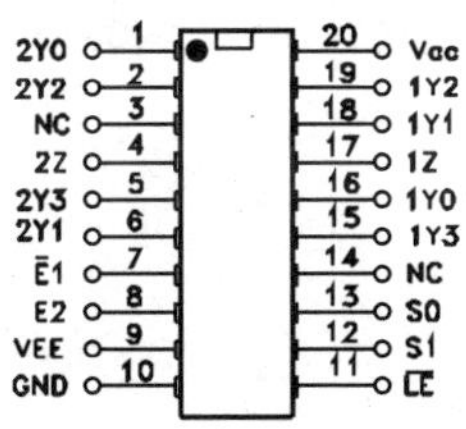

DUAL-IN-LINE PACKAGE

PACKAGE PINNING:

PIN-

1,2,5,6,15-19	INPUTS/OUTPUTS
7,8	INPUT ENABLES
11	LATCH ENABLE
12,13	SELECT INPUTS
9	NEGATIVE VOLTAGE
3,14	NOT CONNECTED
4,17	COMMON INPUTS/OUTPUTS

FUNCTION TABLE 74HC/HCT4352 DUAL 4-CHANNEL ANALOG MULTIPLEXER/DEMULTIPLEXER WITH LATCH

INPUTS					CHANNEL ON
$\overline{E1}$	E2	$\overline{LE}$	S0	S1	
1	X	X	X	X	none
X	0	X	X	X	none
0	1	1	0	0	nY0-nZ
0	1	1	0	1	nY1-nZ
0	1	1	1	0	nY2-nZ
0	1	1	1	1	nY3-nZ
0	1	0	X	X	*
X	X	↓	X	X	**

* Last selected channel ON
** Selected channels latched
1 = HIGH voltage level. 0 = LOW voltage level.
X = Don't Care
↓ = HIGH-to-LOW $\overline{LE}$ transition.

FUNCTIONAL DIAGRAM
Based on Philips 74HC/HCT4352

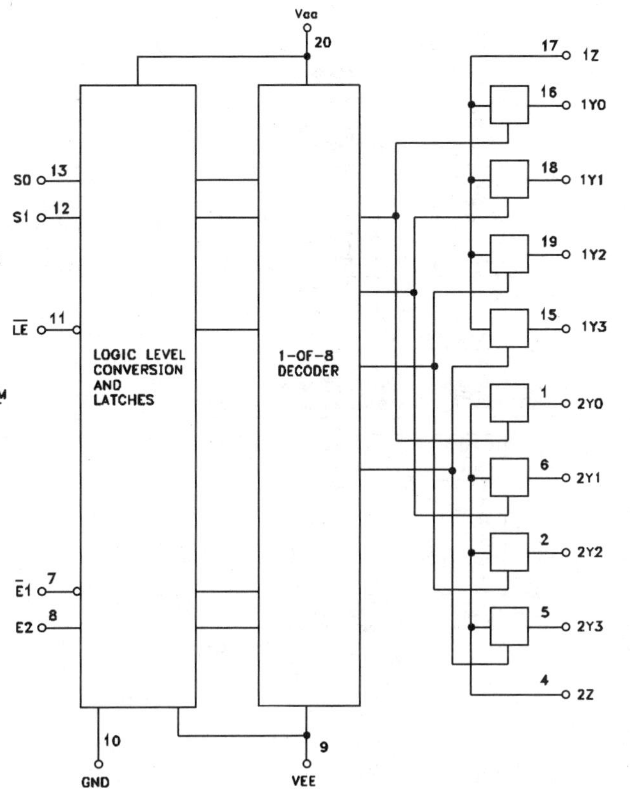

Figure 8.55. The CMOS 74HC/HCT4352 is a dual 4-channel analog multiplexer/demultiplexer with latch and common select logic. Each multiplexer has four independent inputs/outputs, nY0 - nY7, and a common input/output, nZ. The common channel select logics include two select inputs, S0, S1, two enable inputs, /E1,E2, and a latch enable input, /LE. With /E1 low and E2 high, one of the four switches is selected by S0 and S1. The data at the select inputs may be latched by using the active low /LE enable. When LE is high, the latch is transparent. When either of the two enable inputs is inactive all analog switches are in an off state.

5V CMOS – 74HC/HCT4353

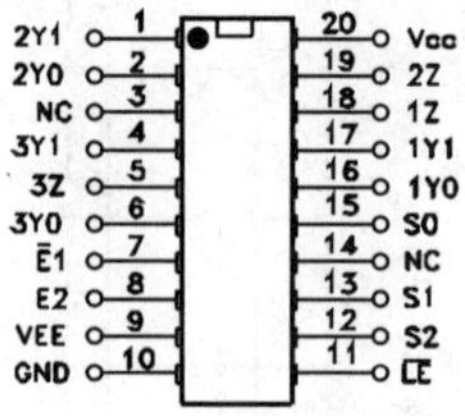

DUAL-IN-LINE PACKAGE

PACKAGE PINNING:

PIN–

1,2,4,6,16,17,19	INPUTS/OUTPUTS
7,8	INPUT ENABLES
11	LATCH ENABLE
12,13,15	SELECT INPUTS
9	NEGATIVE VOLTAGE
3,14	NOT CONNECTED
5,18,19	COMMON INPUTS/OUTPUTS

FUNCTION TABLE 74HC/HCT4353 TRIPLE 2-CHANNEL ANALOG MULTIPLEXER/DEMULTIPLEXER WITH LATCH

INPUTS				CHANNEL ON
Ē1	E2	LE	Sn	
1	X	X	X	none
X	0	X	X	none
0	1	1	0	nY0–nZ
0	1	1	0	nY1–nZ
0	1	0	X	*
X	X	↓	X	**

* Last selected channel ON
** Selected channels latched
1 = HIGH voltage level. 0 = LOW voltage level.
X = Don't Care
↓ = HIGH-to-LOW LE transition.

FUNCTIONAL DIAGRAM
Based on Philips
74HC/HCT4353

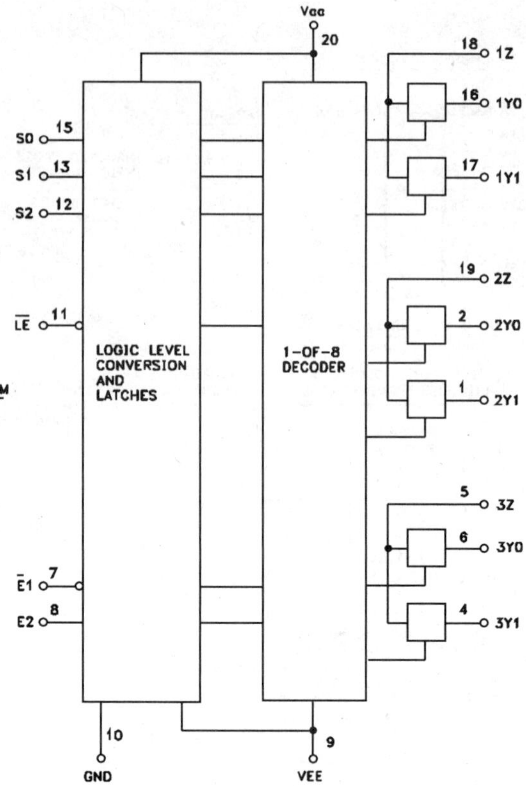

Figure 8.56. The CMOS 74HC/HCT4353 is a triple 2-channel analog multiplexer/demultiplexer with latch and common select logic. Each multiplexer has two independent inputs/outputs, nY0, nY1, and a common input/output, nZ. The common channel select logics include three select inputs, S0 - S2, two common enable inputs, /E1,E2, and a latch enable input, /LE. With /E1 low and E2 high, one of the two switches is selected by S0 to S2. The data at the select inputs may be latched by using the active low /LE enable. When LE is high, the latch is transparent. When either of the two enable inputs is inactive all analog switches are in an off state.

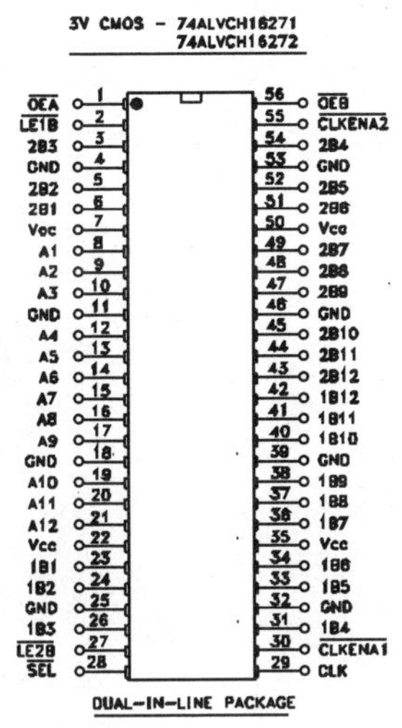

FUNCTION TABLES 74ALVCH16271, '16272
12-BIT-TO-24-BIT MULTIPLEXED BUS
EXCHANGERS WITH 3-STATE OUTPUTS

OUTPUT ENABLE

INPUTS		OUTPUTS	
$\overline{OEA}$	$\overline{OEB}$	A	1B,2B
1	1	Z	Z
1	0	Z	Active
0	1	Active	Z
0	0	Active	Active

A-TO-B STORAGE ($\overline{OEB}$=LOW)

INPUTS				OUTPUTS	
$\overline{CLKENA1}$	$\overline{CLKENA2}$	CLK	A	1B	2B
1	1	X	X	1B0†	2B0†
0	X	↑	0	0	X
0	X	↑	1	1	X
X	0	↑	0	X	0
X	0	↑	1	A0	1

† Two CLK edges are needed with the '16272 to propagate data.

B-TO-A STORAGE ($\overline{OEA}$-LOW)

INPUTS				OUTPUT
$\overline{LE}$	SEL	1B	2B	A
1	X	X	X	A0*
1	X	X	X	A0*
0	1	0	X	0
0	1	1	X	1
0	0	X	0	0
0	0	X	1	1

* Output level before the indicated steady-state input conditions were established.

1 = HIGH voltage level. 0 = LOW voltage level.
X = Don't Care

↑ = LOW-to-HIGH clock transition.

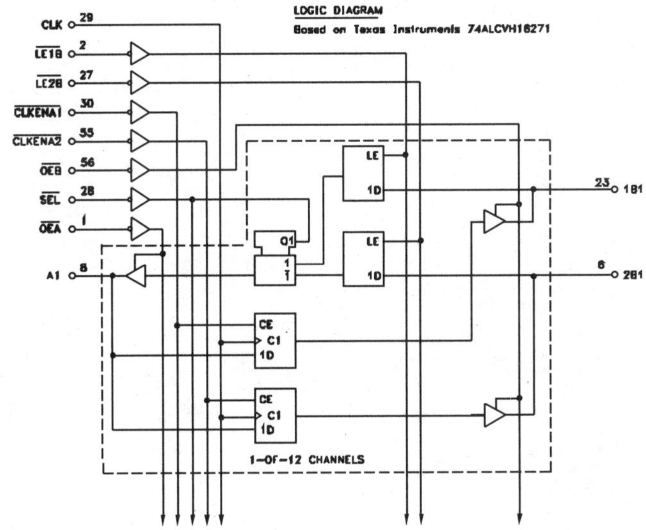

Figure 8.57. The CMOS 74ALVCH16271 and 74ALVCH16272 are 12-bit to 24-bit multiplexed bus exchangers with 3-state outputs. These devices are intended for applications where two separate datapaths must be multiplexed onto, or demultiplexed from, a single datapath. With the '271 A, data is stored in the internal A-to-B registers on the low-to-high transition of the clock, provided the /CLKENA inputs are low. Proper control of these inputs allows two sequential 12-bit words to be presented as a 24-bit word on the B port. With the '272, data from the A inputs is stored in the internal registers on the low-to-high transition of the clock input when the /CLKENA inputs are low. A two-stage pipeline is provided in each of the A-to-1B and A-to-2B paths to serve as a shallow write buffer. With both devices transparent latches are provided in the B-to-A path to allow asynchronous operation to maximize memory access throughput. These latches transfer data when the /LE enable inputs are low. The SEL line selects 1B or 2B data for the A outputs. Data flow is controlled by the active low output enables /OEA,/OEB.

Chapter 9

Arithmetic and Logical Functions

Our interests in this chapter relate to devices providing Arithmetic and Logical Functions, Parity Generation and Checking, and Digital Comparisons. Before the advent of medium scale integration (MSI) if we needed an adder or magnitude comparator we dug out the databooks and assembled our function from the small scale (SSI) gate logic available. Now we enjoy the fruits of technology advancements with a wealth of devices to choose from.

The variety of devices in this chapter is larger than most. For convenience a chapter figure listing for the functions is provided for ready reference as follows:

Device Function	**Figure Numbers**
Adder/Subtracter	9.1, .2, .3, .5, .9, .13, .20, .24, .39, .40, .44, .45
Multiplier	9.14, .19, .41, .43
Arithmetic Logic Unit	9.7, .17, .18, .31, .36
Magnitude/Identity Comparator	9.4, .21, .22, .23, .29, .30, .33, .34, .35, .38, .47, .50
Parity Generator/Checker	9.6, .11, .15, .42, .48, .49, .51, .52
Look-Ahead Carry Generator	9.8, .12, .37, .46
4-Bit Word Shifter	9.16
Error Detection/Correction	9.25, .26, .27, .28
Binary/BCD Conversion	9.10

A brief summary of each function follows in the sequence above.

Adders and Multipliers

The function of the adder is to provide the Boolean sum of two or more binary or BCD numbers. There are half adders, full adders, and serial adders. The distinction between a half and full adder is the presence of a carry input, not provided with the half adder.

A review of binary arithmetic may prove helpful. Computer arithmetic employs complement notation in its computations. The three basics are 1's complement, 2's complement, and sign and magnitude notation. For most additions and subtractions the 2's complement is simplest. The next best is the 1's complement. The sign and magnitude are the most complex of these in addition but works out better for multiplication and division. The disadvantage in 1's complement is the need for an end-around carry.

Table 9.9 provides a summary of the basic operations.

Devices range from the basic full adder such as the 7480, *Figure 9.1*, to sophisticated devices such as the 74LS385, *Figure 9.20*, a quadruple serial adder/subtracter containing four independent adder/subtracter circuits. The '385 can be used in conjunction with the 74LS384, *Figure 9.19*, a 2's complement multiplier to provide a 2's complement product without additional external logic.

A unique device is the 74H87, *Figure 9.5*. This device provides a true/complement conversion of 4-input bits depending on the logic of two control inputs. The logic may also force the outputs to be all one's or all zero's.

Arithmetic Logic Units (ALU)

The first of the TTL integrated 4-Bit ALU's was the 74181, *Figure 9.7*. The elegant circuitry in this device enables the computation of the sixteen possible logic operations plus a variety of arithmetic operations as well. The device may be expanded in 4-bit increments and includes internal look-ahead carry logic for high-speed operation. A more advanced ALU, the 74LS861, *Figure 9.31*, integrates the arithmetic logic with storage registers. The '681 performs 16 arithmetic and 16 logic operations, with full carry look-ahead. These are defined in the tables of *Figure 9.31*.

Magnitude/Identity Comparator

In essence these are identical in their function. A *Magnitude* comparator looks at two sets of inputs, A and B, and provides three outputs: $A < B$, $A = B$, $A > B$. A *Magnitude/Identity* comparator typically offers one output only: $A = B$.

The 74x85, *Figure 9.4*, are expandable 4-bit Magnitude Comparators. Later versions, such as the 74AS866 are high-speed devices capable of 8-Bit binary, BCD and 2's complement inputs.

Parity Generator/Checker

The transmission of data between devices is susceptible to error from any number of causes. Many, if not most, of these will be a change in a single bit, appearing as a reversal from a one to a zero, or the opposite. The system of parity generation and checking was initiated to detect the error and provide an opportunity to correct it.

A word in computer terminology consists of eight bits. Parity generation requires the addition of a ninth, termed the *parity bit*. The basic concept is to look at the number of 1's or 0's in the 8-bit word. We have a choice of electing to have an even number or odd number of one or the other. Suppose we choose an even number of 1's. We set up a logic scheme to look at the eight bits and count the 1's. If the total is even—that is, 2, 4, 6 or 8—there is nothing to be done, the parity bit is 0. If the number of 1's is odd—1, 3, 5 or 7—the parity bit is set to 1, thus maintaining the even bit requirement across all nine in the transmission.

At the receiving end the nine bits are checked against the even parity requirement. If all is well, no action is required. If the parity turns out to be odd, the simplest corrective action is to request a repeat of the transmission. This will often resolve the difficulty and the operations can continue.

The parity generation and checking are combined in one for the majority of devices. Although an 8-bit word has been used in this example, the scheme can be applied to words of any bit length.

Look-Ahead Carry Generator

Another term for look-ahead is *simultaneous*. In a ripple adder the least significant bits have to propagate from stage to stage to reach the output. For an adder with many stages, this represents a significant delay. The concept behind look-ahead is to determine the carry at each stage and present it to the next along with the data input.

The logic diagrams and function table of the figures allow us to trace through the device operations. The 7482, *Figure 9.2*, provides carry input and output without carry look-ahead. Contrast the logic of this circuit with that of the 7483, *Figure 9.3*, which incorporates this added feature. The 74AS282, *Figure 9.12*, is a standalone look-ahead carry generator for use with 4-bit ALU's.

4-Bit Word Shifter

The 74F350 is a shifter that shifts a 4-bit word by 0, 1, 2, or 3 places. Unlike your usual shift register, this device does not require a clock, and indeed, is not provided with a clock input. Shifting may be logical, arithmetic, or end-around. Devices may be cascaded to handle any number of bits. Applications described in the Philips databook are for an 8-bit end-around shifter utilizing four devices and for a 13-bit 2's complementer. The output is 3-state.

Error Detection and Correction

Figure 9.25 illustrates the 74ALS616 and '617. These highly sophisticated devices provide 16-bit parallel error detection and correction. *Figure 9.25* provides function tables and text describing the basic features of these devices. Four such devices, similar in capabilities and features, are described in *Figures 9.25 - 9.28*.

Binary/BCD Conversion

The 74184 and 74185, *Figure 9.10*, comprise a set for conversion between binary and BCD logic in either direction. These Motorola devices are based on two 256-bit Read-Only Memories (ROM). They illustrate the versatility of ROM-based devices in building a variety of reference-based systems. The databook contains a variety of schemes for BCD-to-Binary and Binary-to-BCD conversions.

Chapter Organization

Nine tables are provided. *Tables 9.1 - 9.4* provide a listing of devices by manufacturer. There are sources beyond those listed, but those given provide a realistic picture of what has been and is currently available.

Tables 9.5 - 9.8 provide the LOW-to-HIGH (t_{PLH}) and HIGH-to-LOW (t_{PHL}) propagation delays. For some devices, these are not defined as such, in which case a propagation or time delay, t_d or t_p, is given. Because of the manner in which manufacturers define these quantities, the data in the tables is not precise. The intent is to provide insight helpful for performance comparisons. In particular, we can see the spectacular gains in speed and reduced power dissipation, as ell as complexity, achieved with the latest devices.

Table 9.9 provides examples of basic binary arithmetic.

The drawings of *Figures 9.1 - 9.52* describe the Dual In-Line (DIP) package pinning, a Function Table, Logic Diagrams and/or Logic Symbol as available. Some diagrams are too large and/or complex to be included. To simplify comparisons, as well as to conserve space, all devices with matching features, specifically pinouts and function tables, are included in the same drawing.

Summary

This chapter provides tables and drawings describing many variations of decoders, encoders, multiplexers, and demultiplexers, from the early 7441 to the sophisticated multiplexed bus exchangers, devices with storage, and some having 3-state outputs. Some points to keep in mind are:

- Timing information in the tables is for comparison purposes only. There is a general lack of consistency in how these parameters are defined and not just between manufacturers. Where timing is critical consult the manufacturer's data manual.

- In general, 3-state devices are not affected internally when in the high-impedance output state.

- All timing data is shown for the industrial/commercial or recommended free-air operating temperature range. Where data is for 25°C it is given here.

- The specifications for CMOS devices may cover voltage ranges from 2 to 15 volts. Data is typically provided for V_{CC}'s of 5, 10, and 15 volts, but not always. Some manufacturers use 5.0 volts; others 4.5. There is a significant speed factor—in general, the higher the voltage the faster the performance. Again, where this is important to your application consult the data manual.

- A similar situation exists for the more recent low-voltage devices. Typical V_{CC} ranges given are 2.7 - 3.6 volts. But values may be found for 2, 2.6, 3.0, 3.3 or 3.6 volts. (Glossary *Figures G.1 and G.2* may provide helpful insights on performance.)

DEVICE 74---	STD NM	STD ST	L NM	L ST	H NM	H ST	S NM	S ST	LS NM	LS ST	AS NM	AS ST	ALS NM	ALS ST	F NM	F ST	Figure No.	DEVICE 74---
7480		x x															9.1	7480
7482		x															9.2	7482
7483		x x							x	x							9.3	7483
7485	x	x x		x				x	x x	x x						x	9.4	7485
7487						x											9.5	7487
74180	x	x x															9.6	74180
74181	x	x x					x	x x	x	x x		x			x	x	9.7	74181
74182		x x					x	x x							x	x	9.8	74182
74183						x				x							9.9	74183
74184	x																9.10	74184
74185	x																9.10	74185
74280							x	x x	x	x	x			x	x	x x	9.11	74280
74282											x	x					9.12	74282
74283		x					x	x	x	x x					x	x x	9.13	74283
74284		x															9.14	74284
74285		x															9.14	74285
74286											x	x					9.15	74286
74350								x							x	x x	9.16	74350
74381							x	x		x					x	x	9.17	74381
74382										x					x	x x	9.18	74382
74384										x							9.19	74384
74385										x						x	9.20	74385
74518													x	x			9.21	74518
74519													x	x			9.21	74519
74520													x	x			9.22	74520
74521													x	x			9.22	74521
74522													x	x			9.22	74522
74526														x			9.23	74526
74527														x			9.23	74527
74528														x			9.23	74528
74616														x			9.25	74616
74617														x			9.25	74617
74630										x							9.26	74630
74632												x		x			9.27	74632
74633														x			9.27	74633
74634												x		x			9.27	74634
74635														x			9.27	74635
74636										x							9.28	74636
74637										x							9.28	74637
74677														x			9.29	74677
74678														x			9.29	74678
74679														x			9.30	74679
74680														x			9.30	74680
74681										x							9.31	74681
74682										x							9.32	74682
74684										x							9.32	74684
74685										x							9.32	74685
74686										x							9.33	74686
74687										x							9.33	74687
74688										x				x			9.34	74688
74689													x	x			9.35	74689
74866														x			9.38	74866
74881												x					9.36	74881
74882											x	x					9.37	74882
74885												x					9.38	74885

Key to Manufacturer: N – National/Fairchild. M – Motorola,
S – Signetics/Philips, T – Texas Instruments

Table 9.1. TTL Arithmetic, Digital Comparator, and Parity Generator/Checker source summary by manufacturer.

DEVICE	4000		4100		4500		4700		Figure No.	DEVICE
	NM	ST	NM	ST	NM	ST	NM	ST		
4008	x	x							9.39	4008
4032	x								9.40	4032
4038	x								9.40	4038
4527					x	x			9.41	4527
4531					x	x			9.42	4531
4554					x	x			9.43	4554
4560					x				9.44	4560
4561					x				9.45	4561
4581					x				9.7	4581
4582					x				9.46	4582
4585					x	x			9.47	4585

Key to Manufacturer:	Prefix
N – National/Fairchild.	CD
S – Signetics/Philips,	HEC/HEF
M – Motorola,	MC1
T – Texas Instruments	––

Table 9.2. 5V CMOS 4000 series Arithmetic, Digital Comparator, and Parity Generator/Checker source summary by manufacturer.

DEVICE 74---	C		HC		HCT		AHC (5V)		AHCT		VHC (5V)		VHCT		AC (5V)		ACT		Figure No.	DEVICE 74---
	NM	ST	NM	ST	NM	ST	NM	ST	NM	ST	NM	ST	NM	ST	NM	ST	NM	ST		
7483	x																		9.3	7483
7485	x		x x	x x		x													9.4	7485
74180				x															9.6	74180
74181			x	x		x													9.7	74181
74182			x	x		x													9.8	74182
74280			x x	x x		x									x				9.11	74280
74283			x x	x x		x													9.13	74283
74520																			9.22	74520
74521			x		x										x		x		9.22	74521
74583				x		x													9.24	74583
74677				x															9.29	74677
74678				x															9.29	74678
74679				x															9.30	74679
74680				x															9.30	74680
74682				x															9.32	74682
74684				x															9.32	74684
74688			x x	x x	x	x													9.33	74688
747080				x		x													9.48	747080
7411286																		x	9.49	7411286
7411520																x			9.50	7411520
7416833																		x	9.51	7416833

Key to Manufacturer: N – National/Fairchild. M – Motorola,
S – Signetics/Philips, T – Texas Instruments

Table 9.3. 5V CMOS 74xxx series Arithmetic, Digital Comparator, and Parity Generator/Checker source summary by manufacturer.

DEVICE 74---	LV		LVCH LVC		ALVT LVT		LVX		LVQ		LCX		AC (3V)		ALVCH		HL		AHC (3V)		VHC (3V)		Figure No.	DEVICE 74---
	NM	ST	NM	ST	NM	ST	NM	ST	NM	ST	NM	ST	NM	ST	NM	ST	NM	ST	NM	ST	NM	ST		
74280													x										9.11	74280
74520													x										9.22	74520
74521													x										9.22	74521
74688		x																					9.33	74688
7411520														x									9.50	7411520
7416901																x							9.51	7416901

Key to Manufacturer: N – National/Fairchild. M – Motorola,
S – Signetics/Philips, T – Texas Instruments

Table 9.4. 5V CMOS 74xxx series Arithmetic, Digital Comparator, and Parity Generator/Checker source summary by manufacturer.

DEVICE 74---	STD	L	H	S	LS	AS/A/B	ALS/A/B	F	DATA	Figure	DEVICE 74---
-7480	17-70	--	--	--	--	--	--	--	tPLH	- 9.1	-7480
----	12-80	--	--	--	--	--	--	--	tPHL	---	----
-7482	19-42	--	--	--	--	--	--	--	tPLH	- 9.2	-7482
----	19-42	--	--	--	--	--	--	--	tPHL	---	----
-7483	32-60	--	--	--	--	--	--	--	tPLH	- 9.3	-7483
----	32-60	--	--	--	--	--	--	--	tPHL	---	----
-7485	31-70	75-150	--	7.5-16	20-45	--	--	2.5-11.0	tPLH	- 9.4	-7485
----	30-69	55-150	--	8.5-16.5	17-45	--	--	3.5-14.0	tPHL	---	----
-7487	--	--	13-25	--	--	--	--	--	tPLH	- 9.5	-7487
----	--	--	13-25	--	--	--	--	--	tPHL	---	----
-74180	20-60	--	--	--	--	--	--	--	tPLH	- 9.6	-74180
----	10-68	--	--	--	--	--	--	--	tPHL	---	----
-74181	18-50	--	--	10.5-23	27-50	2-21	--	3.0-19.0	tPLH	- 9.7	-74181
----	18-50	--	--	10.5-30	10-62	2-21	--	2.5-12.5	tPHL	---	----
-74182	17-22	--	--	6.5-10	--	--	--	1.5-10.0	tPLH	- 9.8	-74182
----	17-22	--	--	7.0-10	--	--	--	1.5-7.5	tPHL	---	----
-74183	--	--	10-15	--	--	--	--	--	tPLH	- 9.9	-74183
----	--	--	12-18	--	--	--	--	--	tPHL	---	----
-74184	35-35	--	--	--	--	--	--	--	tPLH	- 9.10	-74184
----	35-35	--	--	--	--	--	--	--	tPHL	---	----
-74185	35-35	--	--	--	--	--	--	--	tPLH	- 9.10	-74185
----	35-35	--	--	--	--	--	--	--	tPHL	---	----
-74280	--	--	--	21-21	23-50	3-12	3-20	4.0-10.5	tPLH	- 9.11	-74280
----	--	--	--	18-18	29-50	3-12	3-22	4.0-13.0	tPHL	---	----
-74282	--	--	--	--	--	2-14	--	--	tPLH	- 9.12	-74282
----	--	--	--	--	--	2-11	--	--	tPHL	---	----
-74283	9-24	--	--	6-18	11-24	--	--	3.5-7.0	tPLH	- 9.13	-74283
----	11-24	--	--	7.5-18	11-24	--	--	3.5-7.0	tPHL	---	----
-74284	20-60	--	--	--	--	--	--	--	tPLH	- 9.14	-74284
----	20-60	--	--	--	--	--	--	--	tPHL	---	----
-74285	20-60	--	--	--	--	--	--	--	tPLH	- 9.14	-74285
----	20-60	--	--	--	--	--	--	--	tPHL	---	----
-74286	--	--	--	--	--	3-16.5	--	--	tPLH	- 9.15	-74286
----	--	--	--	--	--	3-16.5	--	--	tPHL	---	----
-74350	--	--	--	5-17	--	--	--	3.0-10.0	tPLH	- 9.16	-74350
----	--	--	--	8-20	--	--	--	2.5-8.5	tPHL	---	----
-74381	--	--	10-30	18-53	--	--	--	2.5-20	tPLH	- 9.17	-74381
----	--	--	10-30	14-51	--	--	--	2.5-12.5	tPHL	---	----
-74382	--	--	--	10-57	--	--	--	3.0-15.0	tPLH	- 9.18	-74382
----	--	--	--	11-54	--	--	--	2.5-11.0	tPHL	---	----
-74384	--	--	--	15-23	--	--	--	--	tPLH	- 9.19	-74384
----	--	--	--	15-25	--	--	--	--	tPHL	---	----
-74385	--	--	--	14-22	--	--	--	3.0-9.5	tPLH	- 9.20	-74385
----	--	--	--	18-30	--	--	--	3.5-9.0	tPHL	---	----
-74518	--	--	--	--	--	--	15-33	--	tPLH	- 9.21	-74518
----	--	--	--	--	--	--	15-33	--	tPHL	---	----
-74519	--	--	--	--	--	--	15-33	--	tPLH	- 9.21	-74519
----	--	--	--	--	--	--	15-33	--	tPHL	---	----
-74520	--	--	--	--	--	--	2-12	--	tPLH	- 9.22	-74520
----	--	--	--	--	--	--	2-12	--	tPHL	---	----
-74521	--	--	--	--	--	--	2-12	--	tPLH	- 9.22	-74521
----	--	--	--	--	--	--	2-12	--	tPHL	---	----
-74522	--	--	--	--	--	--	8-25	--	tPLH	- 9.22	-74522
----	--	--	--	--	--	--	5-23	--	tPHL	---	----

NOTE: For some devices td, tp, or equivalent time delays are defined instead of tPLH, tPHL.

Table 9.5a. 74xxx TTL device representative propagation delay times in nS.

DEVICE 74---	STD	L	H	S	LS	AS/A/B	ALS/A/B	F	DATA	Figure	DEVICE 74---
-74526	--	--	--	--	--	--	2-15	--	tPLH ↑	- 9.23	-74526
----	--	--	--	--	--	--	2-15	--	tPHL ↓	---	----
-74527	--	--	--	--	--	--	2-15	--	tPLH ↑	- 9.23	-74527
----	--	--	--	--	--	--	2-15	--	tPHL ↓	---	----
-74528	--	--	--	--	--	--	2-15	--	tPLH ↑	- 9.23	-74528
----	--	--	--	--	--	--	2-15	--	tPHL ↓	---	----
-74616	--	--	--	--	--	--	7-55	--	tPLH ↑	- 9.25	-74616
----	--	--	--	--	--	--	7-55	--	tPHL ↓	---	----
-74617	--	--	--	--	--	--	26-40	--	tPLH ↑	- 9.25	-74617
----	--	--	--	--	--	--	26-40	--	tPHL ↓	---	----
-74630	--	--	--	--	27-65	--	--	--	tPLH ↑	- 9.26	-74630
----	--	--	--	--	45-65	--	--	--	tPHL ↓	---	----
-74632	--	--	--	--	--	--	5-55	--	tPLH ↑	- 9.27	-74632
----	--	--	--	--	--	--	5-55	--	tPHL ↓	---	----
-74633	--	--	--	--	--	--	2-60	--	tPLH ↑	- 9.27	-74633
----	--	--	--	--	--	--	2-60	--	tPHL ↓	---	----
-74634	--	--	--	--	--	--	5-55	--	tPLH ↑	- 9.27	-74634
----	--	--	--	--	--	--	5-55	--	tPHL ↓	---	----
-74635	--	--	--	--	--	--	24-40	--	tPLH ↑	- 9.27	-74635
----	--	--	--	--	--	--	24-40	--	tPHL ↓	---	----
-74636	--	--	--	--	27-45	--	--	--	tPLH ↑	- 9.28	-74636
----	--	--	--	--	20-65	--	--	--	tPHL ↓	---	----
-74637	--	--	--	--	20-55	--	--	--	tPLH ↑	- 9.28	-74637
----	--	--	--	--	28-65	--	--	--	tPHL ↓	---	----
-74677	--	--	--	--	--	--	3-25	--	tPLH ↑	- 9.29	-74677
----	--	--	--	--	--	--	5-38	--	tPHL ↓	---	----
-74678	--	--	--	--	--	--	3-22	--	tPLH ↑	- 9.29	-74678
----	--	--	--	--	--	--	5-48	--	tPHL ↓	---	----
-74679	--	--	--	--	--	--	3-25	--	tPLH ↑	- 9.30	-74679
----	--	--	--	--	--	--	5-35	--	tPHL ↓	---	----
-74680	--	--	--	--	--	--	3-22	--	tPLH ↑	- 9.30	-74680
----	--	--	--	--	--	--	5-42	--	tPHL ↓	---	----
-74681	--	--	--	--	9-65	--	--	--	tPLH ↑	- 9.31	-74681
----	--	--	--	--	9-65	--	--	--	tPHL ↓	---	----
-74682	--	--	--	--	13-30	--	--	--	tPLH ↑	- 9.32	-74682
----	--	--	--	--	15-30	--	--	--	tPHL ↓	---	----
-74684	--	--	--	--	15-30	--	--	--	tPLH ↑	- 9.32	-74684
----	--	--	--	--	15-30	--	--	--	tPHL ↓	---	----
-74685	--	--	--	--	24-45	--	--	--	tPLH ↑	- 9.32	-74685
----	--	--	--	--	16-35	--	--	--	tPHL ↓	---	----
-74686	--	--	--	--	11-30	--	--	--	tPLH ↑	- 9.33	-74686
----	--	--	--	--	15-30	--	--	--	tPHL ↓	---	----
-74687	--	--	--	--	21-35	--	--	--	tPLH ↑	- 9.33	-74687
----	--	--	--	--	15-30	--	--	--	tPHL ↓	---	----
-74688	--	--	--	--	12-18	--	3-12	--	tPLH ↑	- 9.34	-74688
----	--	--	--	--	13-23	--	5-22	--	tPHL ↓	---	----
-74689	--	--	--	--	--	--	8-25	--	tPLH ↑	- 9.35	-74689
----	--	--	--	--	--	--	5-25	--	tPHL ↓	---	----
-74866	--	--	--	--	--	1-20	--	--	tPLH ↑	- 9.38	-74866
----	--	--	--	--	--	1-16	--	--	tPHL ↓	---	----
-74881	--	--	--	--	--	2-19	--	--	tPLH ↑	- 9.36	-74881
----	--	--	--	--	--	2-19	--	--	tPHL ↓	---	----
-74882	--	--	--	--	--	2-9	--	--	tPLH ↑	- 9.37	-74882
----	--	--	--	--	--	2-14	--	--	tPHL ↓	---	----
-74885	--	--	--	--	--	5-17.5	--	--	tPLH ↑	- 9.38	-74885
----	--	--	--	--	--	5.5-15	--	--	tPHL ↓	---	----

NOTE: For some devices td, tp, or equivalent time delays are defined instead of tPLH, tPHL.

Table 9.5b. 74xxx TTL device representative propagation delay times in nS (continuation).

DEVICE	4000	4100	4500	4700	DATA	Figure	DEVICE
-4008	170-800	--	--	--	tPLH↓	- 9.39	-4008
----	170-800	--	--	--	tPHL↑	---	----
-4032	280-2400	--	--	--	tPLH↓	- 9.40	-4032
----	280-2400	--	--	--	tPHL↑	---	----
-4038	280-2400	--	--	--	tPLH↓	- 9.40	-4038
----	280-2400	--	--	--	tPHL↑	---	----
- 4527	--	--	125-800	--	tPLH↓	- 9.41	- 4527
----	--	--	125-800	--	tPHL↑	---	----
-4531	--	--	250-1320	--	tPLH↓	- 9.42	-4531
----	--	--	250-1320	--	tPHL↑	---	----
-4554	--	--	270-1700	--	tPLH↓	- 9.43	-4554
----	--	--	270-1700	--	tPHL↑	---	----
-4560	--	--	550-2250	--	tPLH↓	- 9.44	-4560
----	--	--	550-2250	--	tPHL↑	---	----
-4561	--	--	400-1000	--	tPLH↓	- 9.45	-4561
----	--	--	400-1000	--	tPHL↑	---	----
-4581	--	--	305-1910	--	tPLH↓	- 9.7	-4581
----	--	--	305-1910	--	tPHL↑	---	----
- 4582	--	--	345-690	--	tPLH↓	- 9.46	- 4582
----	--	--	345-690	--	tPHL↑	---	----
-4585	--	--	430-860	--	tPLH↓	- 9.47	- 4585
----	--	--	430-860	--	tPHL↑	---	----

NOTE: For some devices, td, tp, or equivalent time delays are defined instead of tPLH, tPHL.

Table 9.6. 5V CMOS 4000 series device representative propagation delay times in nS.

DEVICE 74---	C	HC	HCT	AHC	AHCT	VHC(5V)	VHCT	AC (5V)	ACT	DATA	Figure	DEVICE 74---
-7483	180-400	--	--	--	--	--	--	--	--	tPLH↑	-9.3	-7483
----	180-400	--	--	--	--	--	--	--	--	tPHL↓	---	----
-7485	120-550	12-36	18-44	--	--	--	--	--	--	tPLH↑	-9.4	-7485
----	120-550	12-36	18-44	--	--	--	--	--	--	tPHL↓	---	----
-74180	--	15-52	--	--	--	--	--	--	--	tPLH↑	-9.6	-74180
----	--	15-52	--	--	--	--	--	--	--	tPHL↓	---	----
-74181	--	20-51	32-56	--	--	--	--	--	--	tPLH↑	-9.7	-74181
----	--	20-51	32-56	--	--	--	--	--	--	tPHL↓	---	----
-74182	--	14-34	17-43	--	--	--	--	--	--	tPLH↑	-9.8	-74182
----	--	14-34	17-43	--	--	--	--	--	--	tPHL↓	---	----
-74280	--	20-40	21-45	--	--	--	--	3.0-13	--	tPLH↑	-9.11	-74280
----	--	20-40	21-45	--	--	--	--	3.0-13	--	tPHL↓	---	----
-74283	--	19-46	18-53	--	--	--	--	--	--	tPLH↑	-9.13	-74283
----	--	19-46	18-53	--	--	--	--	--	--	tPHL↓	---	----
-74520	--	--	--	--	--	--	--	--	--	tPLH↑	-9.22	-74520
----	--	--	--	--	--	--	--	--	--	tPHL↓	---	----
- 74521	--	14-30	10-30	--	--	--	--	2.5-6.0	2.0-9.0	tPLH↑	- 9.22	- 74521
----	--	14-30	10-30	--	--	--	--	2.0-8.5	2.5-10	tPHL↓	---	----
-74583	--	18-73	20-73	--	--	--	--	--	--	tPLH↑	-9.24	-74583
----	--	18-73	20-73	--	--	--	--	--	--	tPHL↓	---	----
-74677	--	14-125	--	--	--	--	--	--	--	tPLH↑	-9.29	-74677
----	--	14-125	--	--	--	--	--	--	--	tPHL↓	---	----
-74678	--	19-125	--	--	--	--	--	--	--	tPLH↑	- 9.29	-74678
----	--	19-125	--	--	--	--	--	--	--	tPHL↓	---	----
-74679	--	15-60	--	--	--	--	--	--	--	tPLH↑	- 9.30	-74679
----	--	15-60	--	--	--	--	--	--	--	tPHL↓	---	----
-74680	--	15-60	--	--	--	--	--	--	--	tPLH↑	- 9.30	-74680
----	--	15-60	--	--	--	--	--	--	--	tPHL↓	---	----
-74682	--	-69	--	--	--	--	--	--	--	tPLH↑	- 9.32	-74682
----	--	-69	--	--	--	--	--	--	--	tPHL↓	---	----
-74684	--	-69	--	--	--	--	--	--	--	tPLH↑	- 9.32	-74684
----	--	-69	--	--	--	--	--	--	--	tPHL↓	---	----
-74688	--	16-42	18-34	--	--	--	--	--	--	tPLH↑	- 9.33	-74688
----	--	16-42	18-34	--	--	--	--	--	--	tPHL↓	---	----
-747080	--	15-56	18-63	--	--	--	--	--	--	tPLH↑	- 9.48	-747080
----	--	15-56	18-63	--	--	--	--	--	--	tPHL↓	---	----
-7411286	--	--	--	--	--	--	--	--	2.2-9.7	tPLH↑	- 9.49	7411286
----	--	--	--	--	--	--	--	--	3.1-11.4	tPHL↓	---	----
-7411520	--	--	--	--	--	--	--	1.5-11.1	--	tPLH↑	- 9.50	7411520
----	--	--	--	--	--	--	--	1.5-10.1	--	tPHL↓	---	----
-7416833	--	--	--	--	--	--	--	--	4-12	tPLH↑	- 9.51	7416833
----	--	--	--	--	--	--	--	--	3.2-12.4	tPHL↓	---	----

NOTE: For some devices td, tp, or equivalent time delays are defined instead of tPLH, tPHL.

Table 9.7. 5V 74xxx series CMOS device representative propagation delay times in nS.

DEVICE 74---	LV	LVC	ALVT LVT/ABT	LVX	LVQ	LCX	AC(3V)	ALVCH	HL	AHC (3V)	VHC(3V)	DATA	Figure No.	DEVICE 74---
-74280	--	--	--	--	--	--	5.0-17.0	--	--	--	--	tPLH↑	- 9.11	-74280
----	--	--	--	--	--	--	5.0-17.0	--	--	--	--	tPHL↓	---	----
-74520	--	--	--	--	--	--	3.5-7.5	--	--	--	--	tPLH↑	- 9.22	-74520
----	--	--	--	--	--	--	3.5-12.0	--	--	--	--	tPHL↓	---	----
-74521	--	--	--	--	--	--	3.0-11.0	--	--	--	--	tPLH↑	- 9.22	-74521
----	--	--	--	--	--	--	3.0-11.5	--	--	--	--	tPHL↓	---	----
-74688	17	--	--	--	--	--	--	--	--	--	--	tPLH↑	- 9.33	-74688
----	17	--	--	--	--	--	--	--	--	--	--	tPHL↓	---	----
7411520	--	--	--	--	--	--	1.5-16.5	--	--	--	--	tPLH↑	- 9.50	7411520
----	--	--	--	--	--	--	1.5-14.4	--	--	--	--	tPHL↓	---	----
7416901	--	--	--	--	--	--	--	1.5-11.1	--	--	--	tPLH↑	- 9.51	7416901
----	--	--	--	--	--	--	--	1.5-11.1	--	--	--	tPHL↓	---	----

NOTE: For some devices td, tp, or equivalent time delays are defined instead of tPLH, tPHL.

Table 9.8. Low-voltage 74xxx series CMOS device representative propagation delay times in nS.

ADDITION

Of Two Positive Numbers

No Carry		With Carry	
12	0.001100	18	0.010010
33	0.100001	22	0.010110
45	0.101101	40	0.101000

Of Two Negative Numbers

No Carry		With Carry	
−21	1.010101	−11	1.001011
− 8	1.001000	−19	1.010011
−29	1.011101	−30	1.011110

SUBTRACTION

Of Two Positive Numbers

No Borrow		With Borrow	
22	0.010110	25	0.011001
18	0.010010	13	0.001101
4	0.000100	12	0.001100

Of Two Positive Numbers

No Borrow		With Borrow	
−53	1.110101	−43	1.101011
− 4	1.000100	−28	1.011100
−49	1.110001	−15	1.001111

Addition Of Oppositely Signed Numbers

No Borrow

Usual Manner

+17	0.010001
−11	1.001011
+ 6	0.000110

With End Around Carry

Using 1's Complement

0.010001	augend
1.110100	addend 1's complement
0.000001	add 1 (End Around Carry)
0.000110	sum

With Borrow

Using 2's Complement

+18	0.010010
−22	1 010110
− 4	1.111100

Complement Magnitude and add 1

	1.000011
+	0.000001
	1.000100 = −4

Multiplication Using Left Shift

11	001011
x2	010110 = 22

Division Using Right Shift

14	001110
÷2	000111 = 7

Table 9.9. Some examples of binary arithmetic with signed numbers.

TTL - 7480

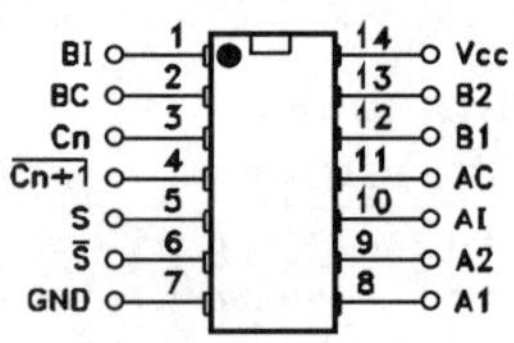

DUAL-IN-LINE PACKAGE

PACKAGE PINNING:

PIN-	
1,10	INVERTING DATA INPUTS
8,9,12,13	NON-INVERTING DATA INPUTS
2,11	CONTROL INPUTS
3	CARRY INPUT
5,6	SUM OUTPUTS
4	ACTIVE LOW CARRY OUTPUT

FUNCTION TABLE 7480 GATED FULL ADDER

INPUTS			OUTPUTS		
Cn	B	A	$\overline{Cn+1}$	$\overline{S}$	S
0	0	0	1	1	0
0	0	1	1	0	1
0	1	0	1	0	1
0	1	1	0	0	0
1	0	0	1	1	1
1	0	1	0	0	0
1	1	0	0	1	0
1	1	1	0	0	1

1 = HIGH voltage level.
0 = LOW voltage level.

$A = \overline{AI \cdot AC}$, $B = \overline{BI \cdot BC}$
where $AI = \overline{A1 \cdot A2}$, $BI = \overline{B1 \cdot B2}$

When AI or BI are used as inputs, A1 and A2 or B1 and B2, respectively must be connected to GND.

When A1 and A2 or B1 and B2 are used as inputs, AI or BI, respectively must be open or used to perform DOT-OR logic.

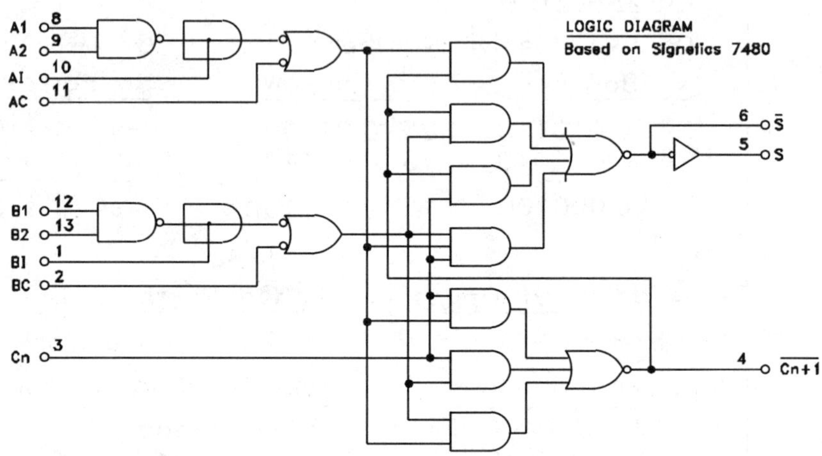

LOGIC DIAGRAM
Based on Signetics 7480

Figure 9.1. The TTL 7480 is a single-bit, high-speed, Binary Full Adder with gated complementary inputs, complementary sum (S and /S) outputs and inverted carry output. It is designed for multiple bit, parallel add/serial carry applications.

TTL - 7482

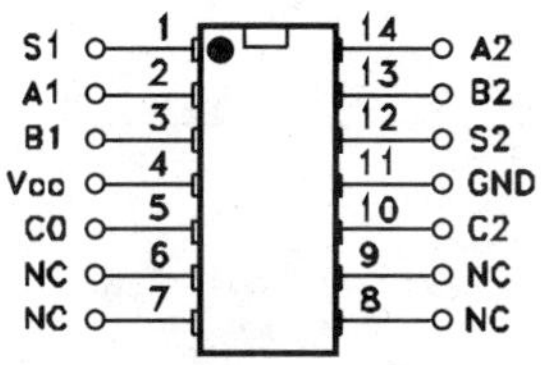

DUAL-IN-LINE PACKAGE

PACKAGE PINNING

PIN–
2,3,13,14 DATA INPUTS
1,12 SUM OUTPUTS
10 CARRY OUTPUT
5 CONTROL INPUT

FUNCTION TABLE 7482 2-BIT BINARY FULL ADDER

INPUTS				OUTPUTS					
				CO=0			CO=1		
A1	B1	A2	B2	S1	S2	C2	S1	S2	C2
0	0	0	0	0	0	0	1	0	0
1	0	0	0	1	0	0	0	1	0
0	1	0	0	1	0	0	0	1	0
1	1	0	0	0	1	0	1	1	0
0	0	1	0	0	1	0	1	1	0
1	0	1	0	1	1	0	0	0	1
0	1	1	0	1	1	0	0	0	1
1	1	1	0	0	0	1	1	0	1
0	0	0	1	0	1	0	1	1	0
1	0	0	1	1	1	0	0	0	1
0	1	0	1	1	1	0	0	0	1
1	1	0	1	0	0	1	1	0	1
0	0	1	1	0	0	1	1	0	1
1	0	1	1	1	0	1	0	1	1
0	1	1	1	1	0	1	0	1	1
1	1	1	1	0	1	1	1	1	1

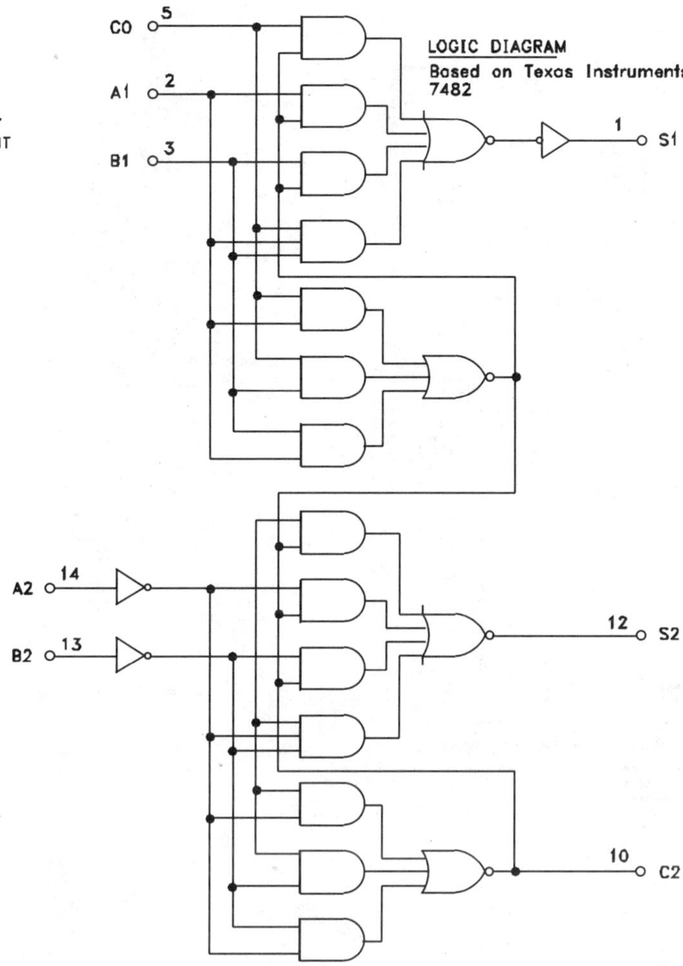

Figure 9.2. The TTL 7482 is a high-speed, 2-Bit Binary Full Adder for the addition of two 2-bit binary numbers. The sum (S) outputs are provided for each bit—the resultant carry is obtained from the second bit. It is designed for multiple-bit, parallel add/serial carry applications.

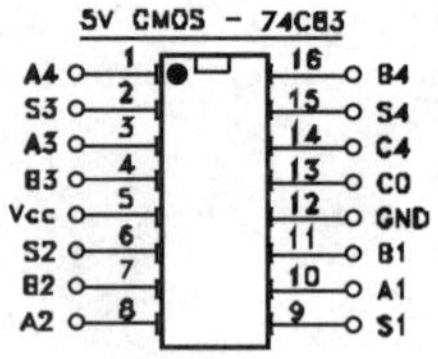

DUAL-IN-LINE PACKAGE

PACKAGE PINNING:

PIN-	
10,11,3,4	OPERAND INPUTS
88,7,1,16	OPERAND INPUTS
13	CARRY INPUT
9,6,2,15	SUM OUTPUTS
14	CARRY OUTPUT

FUNCTION TABLE 74x83 4-BIT BINARY FULL ADDERS WITH FAST CARRY

INPUT				OUTPUT WHEN C0=0 / WHEN C2=0			OUTPUT WHEN C0=1 / WHEN C2=1		
A1 / A3	B1 / B3	A2 / A4	B2 / B4	S1 / S3	S2 / C4	C2 / S3	S1 / S3	S2 / S4	C2 / C4
0	0	0	0	0	0	0	1	0	0
1	0	0	0	1	0	0	0	1	0
0	1	0	0	1	0	0	0	1	0
1	1	0	0	0	1	0	1	1	0
0	0	1	0	0	1	0	1	1	0
1	0	1	0	1	1	0	0	0	1
0	1	1	0	1	1	0	0	0	1
1	1	1	0	0	0	1	1	0	1
0	0	0	1	0	1	0	1	1	0
1	0	0	1	1	1	0	0	0	1
0	1	0	1	1	1	0	0	0	1
1	1	0	1	0	0	1	1	0	1
0	0	1	1	0	0	1	1	0	1
1	0	1	1	1	0	1	0	1	1
0	1	1	1	1	0	1	0	1	1
1	1	1	1	0	1	1	1	1	1

1 = HIGH voltage level. 0 = LOW voltage level.

Input conditions at A1, B1, A2, B2, and C0 are used to determine outputs S1 and S2 and the value of internal carry C2. The values at C2, A3, B3, A4, and B4 are then used to determine outputs S3, S4, and C4.

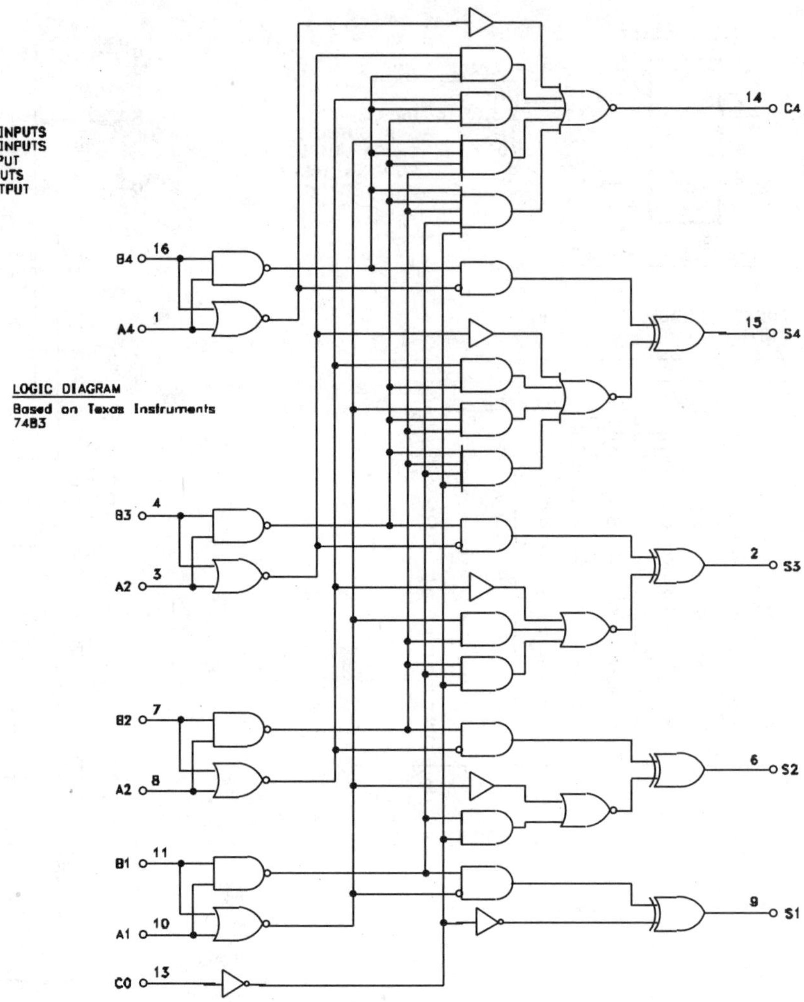

Figure 9.3. The TTL 7483 and CMOS 74C83 are high-speed, 4-Bit Binary Full Adders with internal look-ahead. These accept two 4-bit binary words and a carry input. The sum of the two words is combined with the carry input and presented at the four sum (S1 - S4) outputs and the carry output.

TTL - 7485/L/S/LS/F
5V CMOS - 74C/HC/HCT85

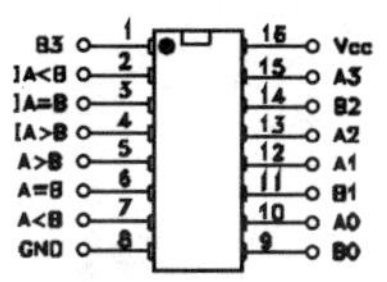

DUAL-IN-LINE PACKAGE

PACKAGE PINNING:
PIN-
1,9-15 OPERAND INPUTS
2-4 EXPANDER INPUTS
5-7 OUTPUTS

LOGIC DIAGRAM
Based on Signetics 7485

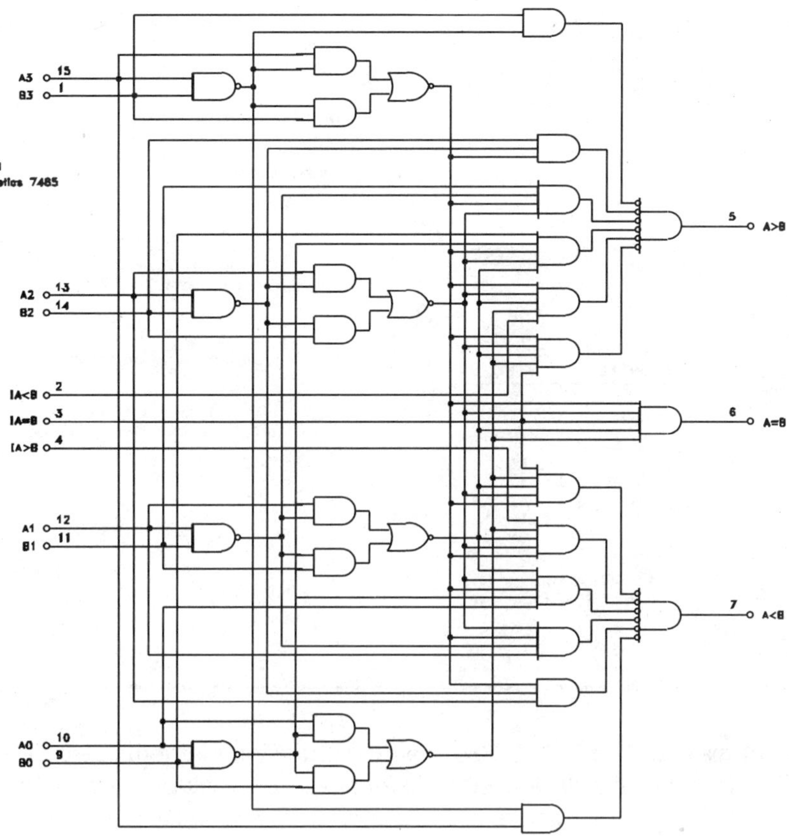

FUNCTION TABLE 74x85 4-BIT MAGNITUDE COMPARATOR

COMPARING INPUTS				CASCADING INPUTS			OUTPUTS		
A3,B3	A2,B2	A1,B1	A0,B0	IA>B	IA<B	IA=B	A>B	A<B	A=B
A3>B3	X	X	X	X	X	X	1	0	0
A3<B3	X	X	X	X	X	X	0	1	0
A3=B3	A2>B2	X	X	X	X	X	1	0	0
A3=B3	A2<B2	X	X	X	X	X	0	1	0
A3=B3	A2=B2	A1>B1	X	X	X	X	1	0	0
A3=B3	A2=B2	A1<B1	X	X	X	X	0	1	0
A3=B3	A2=B2	A1=B1	A0>B0	X	X	X	1	0	0
A3=B3	A2=B2	A1=B1	A0<B0	X	X	X	0	1	0
A3=B3	A2=B2	A1=B1	A0=B0	1	0	0	1	0	0
A3=B3	A2=B2	A1=B1	A0=B0	0	1	0	0	1	0
A3=B3	A2=B2	A1=B1	A0=B0	0	0	1	0	0	1
A3=B3	A2=B2	A1=B1	A0=B0	X	X	1	0	0	1
A3=B3	A2=B2	A1=B1	A0=B0	1	1	0	0	0	0
A3=B3	A2=B2	A1=B1	A0=B0	0	0	0	1	1	0

1 = HIGH voltage level. 0 = LOW voltage level.
X = Don't Care

The upper part of the table describes the normal operation under all conditions that will occur in a single device or in a series expansion scheme. In this part of the table the three outputs are mutually exclusive. In the lower part of the table the outputs reflect the feed-forward conditions that exist in the parallel expansion scheme.

Figure 9.4. The TTL 74x85 and CMOS 74x85 are expandable 4-Bit Magnitude Comparators. These compare two 4-bit binary weighted words (A0 - A3, B0 - B3) where A3 and B3 are the most significant bits. The expansion inputs are treated as least significant bits for comparison purposes. The three active high outputs are "A greater than B," (A > B), "A equals B," (A = B), and "A less than B" (A < B). The devices are cascadable to almost any length.

TTL – 74H87

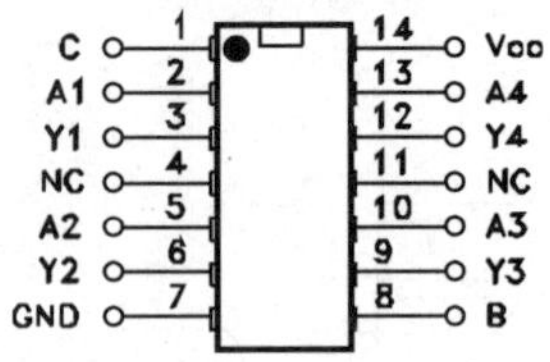

DUAL-IN-LINE PACKAGE

PACKAGE PINNING:

PIN–
2,5,10,13 DATA INPUTS
3,6,9,12 DATA OUTPUTS
1,8 CONTROL INPUTS
4,11 NO CONNECTION

LOGIC DIAGRAM

Based on Texas Instruments 74H87

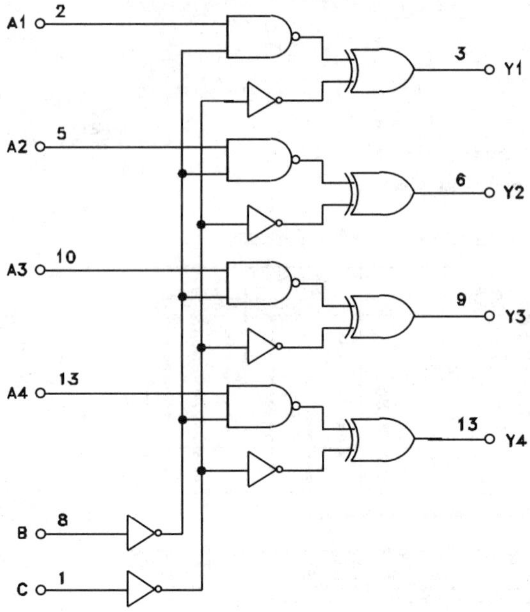

FUNCTION TABLE 74H87 4-BIT TRUE/COMPLEMENT, ZERO/ONE ELEMENT

CONTROL INPUTS		OUTPUTS			
B	C	Y1	Y2	Y3	Y4
0	0	$\overline{A1}$	$\overline{A2}$	$\overline{A3}$	$\overline{A4}$
0	1	A1	A2	A3	A4
1	0	1	1	1	1
1	1	0	0	0	0

1 = HIGH voltage level.
0 = LOW voltage level.

Figure 9.5. The TTL 74H85 is a 4-Bit True/Complement, Zero/One Element. By the use of two control lines, B and C, the device will transfer a 4-bit binary input, An, to the output Y in either true or complement form. Additionally, the control lines will also set all outputs to either logic state, 1 or 0, independent of the state of the inputs.

TTL – 74180
5V CMOS – 74HC180

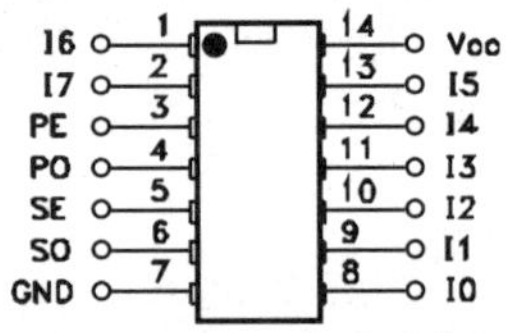

DUAL-IN-LINE PACKAGE

LOGIC DIAGRAM

Based on Signetics 74180

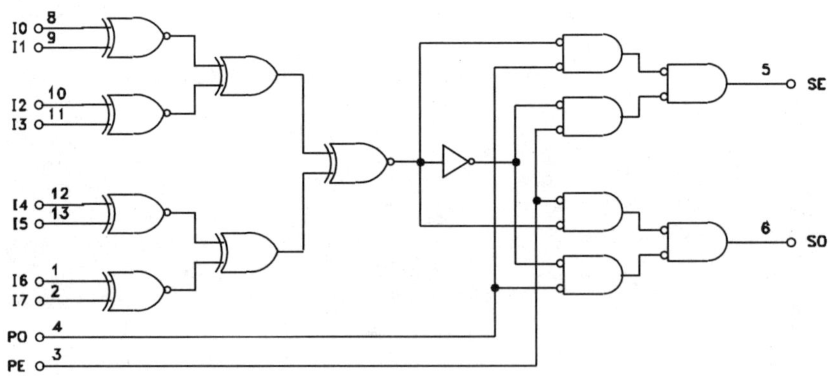

FUNCTION TABLE 74x180 9-BIT ODD/EVEN PARITY GENERATOR/CHECKER

INPUTS				OUTPUTS	
Number of HIGH Data Inputs (I0–I7)	PE	PO	SE	SO
Even	1	0	1	0
Odd	1	0	0	1
Even	0	1	0	1
Odd	0	1	1	0
X	1	1	0	0
X	0	0	1	1

1 = HIGH voltage level.
0 = LOW voltage level.
X = Don't Care

Figure 9.6. The TTL 74180 and 74HC180 are 9-Bit Odd/Even Parity Generator/Checkers. These are commonly employed to detect errors in high-speed data transmission or data retrieval systems. Both even and odd parity enable inputs and parity outputs are available for generating or checking parity on 8-Bits.

TTL - 74181\S\LS\F
5V CMOS - 74HC/HCT181
4581

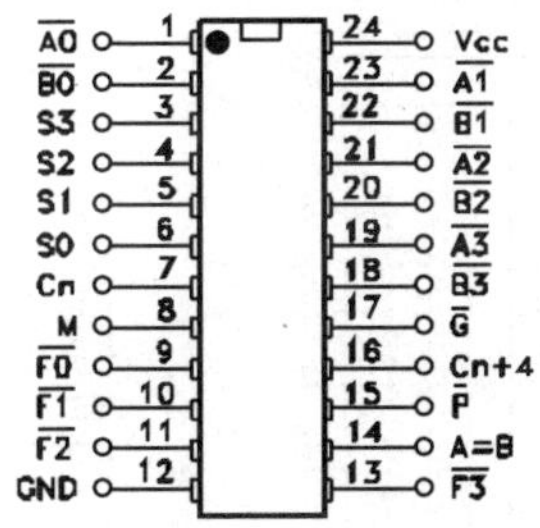

DUAL-IN-LINE PACKAGE

PACKAGE PINNING:

PIN–	
2,19,21,23	OPERAND A INPUT
1,18,20,22	OPERAND B INPUTS
3–6	FUNCTION SELECT INPUTS
8	MODE CONTROL INPUT
7	CARRY INPUT
9–11,13	FUNCTION OUTPUTS (ACTIVE LOW)
17	CARRY GEN OUTPUT (ACTIVE LOW)
15	CARRY PROPAGATE OUTPUT (ACTIVE LOW)
16	RIPPLE CARRY OUTPUT

MODE SELECT FUNCTION TABLE 74x181 4581
4-BIT ARITHMETIC LOGIC UNIT

MODE SELECT INPUTS				ACTIVE HIGH INPUTS AND OUTPUTS	
S3	S2	S1	S0	LOGIC (M=H)	ARITHMETIC" (M=L)(Cn=H)
0	0	0	0	$\bar{A}$	A
0	0	0	1	$\overline{A+B}$	A+B
0	0	1	0	$\bar{A}B$	A+B
0	0	1	1	Logical 0	Minus 1
0	1	0	0	$\overline{AB}$	A plus $A\bar{B}$
0	1	0	1	$\bar{B}$	(A+B) plus $A\bar{B}$
0	1	1	0	$A\oplus B$	A minus B minus 1
0	1	1	1	$A\bar{B}$	AB minus 1
1	0	0	0	$\bar{A}+B$	A plus AB
1	0	0	1	$\overline{A\oplus B}$	A plus B
1	0	1	0	B	$(A+\bar{B})$ plus AB
1	0	1	1	AB	AB minus 1
1	1	0	0	Logical 1	A plus A*
1	1	0	1	$A+\bar{B}$	(A+B) plus A
1	1	1	0	A+B	$(A+\bar{B})$ plus A
1	1	1	1	A	A minus 1

MODE SELECT INPUTS				ACTIVE LOW INPUTS AND OUTPUTS	
S3	S2	S1	S0	LOGIC (M=H)	ARITHMETIC" (M=L)(Cn=L)
0	0	0	0	$\bar{A}$	A minus 1
0	0	0	1	$\overline{AB}$	AB minus 1
0	0	1	0	A+B	$A\bar{B}$ minus 1
0	0	1	1	Logical 1	minus 1
0	1	0	0	$\overline{A+B}$	A plus $(A+\bar{B})$
0	1	0	1	$\bar{B}$	AB plus $(A+\bar{B})$
0	1	1	0	$\overline{A\oplus B}$	A minus B minus 1
0	1	1	1	$A+\bar{B}$	$A+\bar{B}$
1	0	0	0	$\bar{A}B$	A plus (A+B)
1	0	0	1	A+B	A plus B
1	0	1	0	B	$A\bar{B}$ plus (A+B)
1	0	1	1	A+B	A+B
1	1	0	0	Logical 0	A plus A*
1	1	0	1	$A\bar{B}$	AB plus A
1	1	1	0	AB	$A\bar{B}$ plus A
1	1	1	1	A	A

1 = HIGH voltage level. 0 = LOW voltage level.

* Each bit is shifted to the next more significant position.
" Arithmetic operations expressed in 2s complement notation.

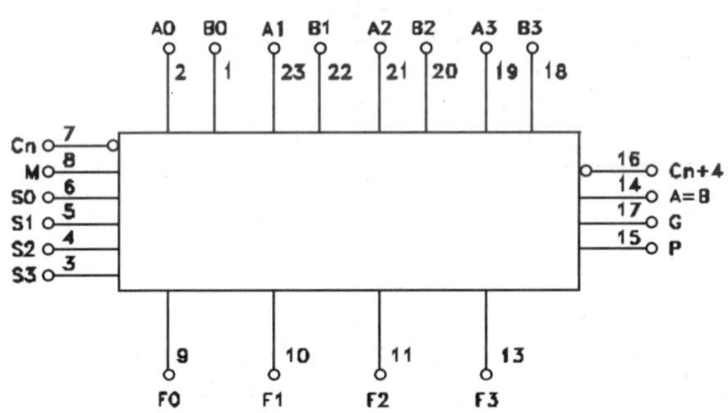

ACTIVE HIGH OPERANDS

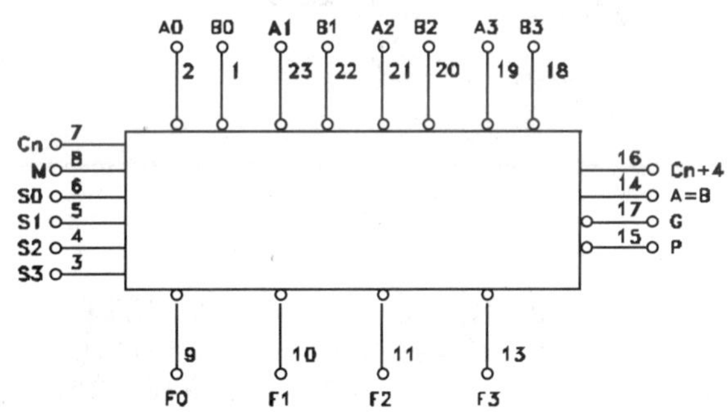

ACTIVE LOW OPERANDS

Figure 9.7. The TTL 74x181, CMOS 4581 and 74HC/HCT181 are 4-Bit Arithmetic Logic Units (ALU) which are capable of performing all of the possible 16 logic operations on two variables and a variety of arithmetic operations. The device is expandable in 4-bit increments, enabling its operation with any word length. Full look-ahead carry provides for high-speed operation on long words.

TTL – 74182/S/F

5V CMOS – 74HC/HCT182

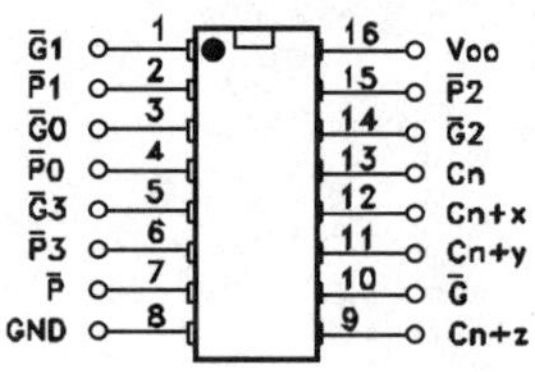

DUAL-IN-LINE PACKAGE

PACKAGE PINNING:

PIN–	
13	CARRY INPUT
2,4,6,15	CARRY PROPAGATE (ACTIVE LOW) INPUT
1,3,5,14	CARRY GENERATE (ACTIVE LOW) INPUT
9,11,12	CARRY OUTPUTS
7	CARRY PROPAGATE (ACTIVE LOW) OUTPUT
10	CARRY GENERATE (ACTIVE LOW) OUTPUT

FUNCTION TABLE 74x182 LOOKAHEAD CARRY GENERATOR

INPUTS									OUTPUTS				
Cn	$\overline{G0}$	$\overline{P0}$	$\overline{G1}$	$\overline{P1}$	$\overline{G2}$	$\overline{P2}$	$\overline{G3}$	$\overline{P3}$	Cn+x	Cn+y	Cn+z	$\overline{G}$	$\overline{P}$
X	1	1							0				
0	1	X							0				
X	0	X							1				
1	X	0							1				
X	X	X	1	1						0			
X	1	1	1	X						0			
0	1	X	1	X						0			
X	X	X	0	X						1			
X	0	X	X	0						1			
1	X	0	X	0						1			
X	X	X	X	X	1	1					0		
X	X	X	1	1	1	X					0		
X	1	1	1	X	1	X					0		
0	1	X	1	X	1	X					0		
X	X	X	X	X	0	X					1		
X	X	X	0	X	X	0					1		
X	0	X	X	0	X	0					1		
1	X	0	X	0	X	0					1		
	X		X	X	X	X	1	1				1	
	X		X	X	1	1	1	X				1	
	X		1	1	1	X	1	X				1	
	1		1	X	1	X	1	X				1	
	X		X	X	X	X	0	X				0	
	X		X	X	0	X	X	0				0	
	X		0	X	X	0	X	0				0	
	0		X	0	X	0	X	0				0	
		1		X		X		X					1
		X		1		X		X					1
		X		X		1		X					1
		X		X		X		1					1
		0		0		0		0					0

1 = HIGH voltage level. 0 = LOW voltage level. X = Don't Care

Logic equations provided at the outputs are:

$Cn+x = G0+P0Cn$

$Cn+y = G1+P1G0 = P1P0Cn$

$Cn+z = G2+P2G1+P2P1G0+P2P1P0Cn$

$G = \overline{G3+P3G2+P3P2G1+P3P2P1G0}$

$P = \overline{P3P2P1P0}$

LOGIC DIAGRAM

Based on Signetics 74182

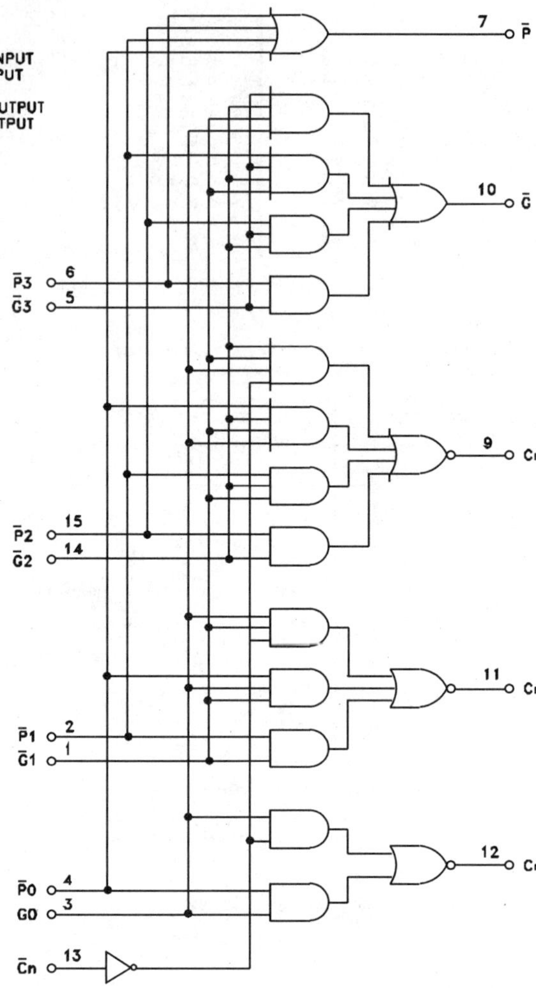

Figure 9.8. The TTL 74x182 and 74HC/HCT182 are a high-speed Look-ahead Carry Generator. These are generally used with Arithmetic Logic Units such as the 74x181. They enhance the operation with word lengths of more than four bits. The device accepts up to four sets of Carry Generate functions, Carry Propagate functions, and a carry input from four ALU's and provides the three Carry Out signals required by the ALU's. It also provides the next level of auxiliary functions needed for further levels of look-ahead.

TTL – 74LS183
74H183

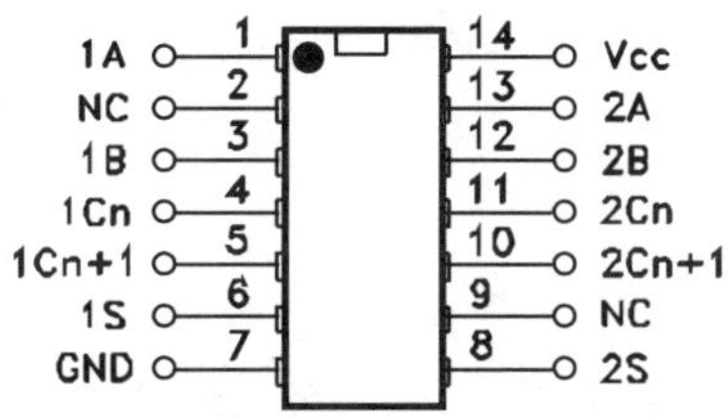

DUAL-IN-LINE PACKAGE

FUNCTION TABLE 74LS183
DUAL CARRY-SAVE FULL ADDER

INPUTS			OUTPUTS	
Cn	B	A	S	Cn+1
0	0	0	0	0
0	0	1	1	0
0	1	0	1	0
0	1	1	0	1
1	0	0	1	0
1	0	1	0	1
1	1	0	0	1
1	1	1	1	1

1 = HIGH voltage level.
0 = LOW voltage level.

LOGIC DIAGRAM

Based on Texas Instruments 74LS183
Diagram applies to both adders.

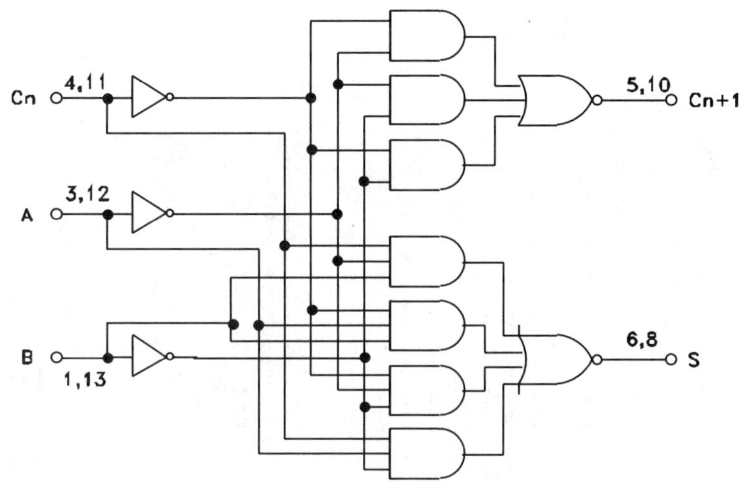

Figure 9.9. The TTL 74LS183 and 74H183 are a dual Carry-Save Full Adder featuring an individual carry from each bit for use in multiple-input, carry-save techniques to produce the true sum and true carry outputs with no more than two gage delays.

TTL - 74184 (BCD-to-BINARY)
74185 (BINARY-to-BCD)

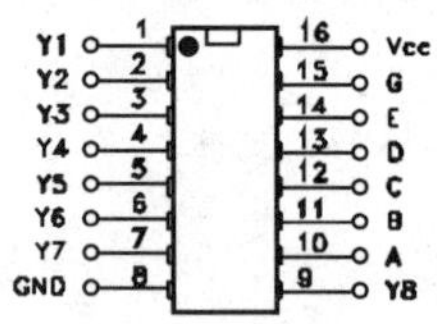

DUAL-IN-LINE PACKAGE

74185 6-BIT BINARY-to-BCD CONVERTER

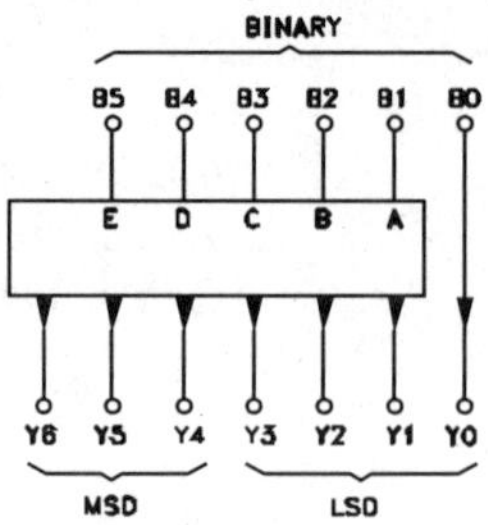

FUNCTION TABLES 74184, 74185 BCD-to-BINARY AND BINARY-to-BCD CONVERTERS

Binary-to-BCD Conversion

Binary Words	INPUTS: Binary Select E	D	C	B	A	Enable G	OUTPUTS: Y8	Y7	Y6	Y5	Y4	Y3	Y2	Y1
0 1	0	0	0	0	0	0	1	1	0	0	0	0	0	0
2 3	0	0	0	0	1	0	1	1	0	0	0	0	0	1
4 5	0	0	0	1	0	0	1	1	0	0	0	0	1	0
6 7	0	0	0	1	1	0	1	1	0	0	0	0	1	1
8 9	0	0	1	0	0	0	1	1	0	0	0	1	0	0
10 11	0	0	1	0	1	0	1	1	0	0	1	0	0	0
12 13	0	0	1	1	0	0	1	1	0	0	1	0	0	1
14 15	0	0	1	1	1	0	1	1	0	0	1	0	1	0
16 17	0	1	0	0	0	0	1	1	0	0	1	0	1	1
18 19	0	1	0	0	1	0	1	1	0	0	1	1	0	0
20 21	0	1	0	1	0	0	1	1	0	1	0	0	0	0
22 23	0	1	0	1	1	0	1	1	0	1	0	0	0	1
24 25	0	1	1	0	0	0	1	1	0	1	0	0	1	0
26 27	0	1	1	0	1	0	1	1	0	1	0	0	1	1
28 29	0	1	1	1	0	0	1	1	0	1	0	1	0	0
30 31	0	1	1	1	1	0	1	1	0	1	1	0	0	0
32 33	1	0	0	0	0	0	1	1	0	1	1	0	0	1
34 35	1	0	0	0	1	0	1	1	0	1	1	0	1	0
36 37	1	0	0	1	0	0	1	1	0	1	1	0	1	1
38 39	1	0	0	1	1	0	1	1	0	1	1	1	0	0
40 41	1	0	1	0	0	0	1	1	1	0	0	0	0	0
42 43	1	0	1	0	1	0	1	1	1	0	0	0	0	1
44 45	1	0	1	1	0	0	1	1	1	0	0	0	1	0
46 47	1	0	1	1	1	0	1	1	1	0	0	0	1	1
48 49	1	1	0	0	0	0	1	1	1	0	0	1	0	0
50 51	1	1	0	0	1	0	1	1	1	0	1	0	0	0
52 53	1	1	0	1	0	0	1	1	1	0	1	0	0	1
54 55	1	1	0	1	1	0	1	1	1	0	1	0	1	0
56 57	1	1	1	0	0	0	1	1	1	0	1	0	1	1
58 59	1	1	1	0	1	0	1	1	1	0	1	1	0	0
60 61	1	1	1	1	0	0	1	1	1	0	0	0	0	0
62 63	1	1	1	1	1	0	1	1	1	0	0	0	0	1
All	X	X	X	X	X	1	1	1	1	1	1	1	1	1

1 = HIGH voltage level. 0 = LOW voltage level.
X = Don't Care

BCD-to-Binary Conversion

BCD Words	INPUTS (See note A) E	D	C	B	A	G	OUTPUTS (See note B) Y5	Y4	Y3	Y2	Y1
0 1	0	0	0	0	0	0	0	0	0	0	0
2 3	0	0	0	0	1	0	0	0	0	0	1
4 5	0	0	0	1	0	0	0	0	0	1	0
6 7	0	0	0	1	1	0	0	0	0	1	1
8 9	0	0	1	0	0	0	0	0	1	0	0
10 11	0	1	0	0	0	0	0	0	1	0	1
12 13	0	1	0	0	1	0	0	0	1	1	0
14 15	0	1	0	1	0	0	0	0	1	1	1
16 17	0	1	0	1	1	0	0	1	0	0	0
18 19	0	1	1	0	0	0	0	1	0	0	1
20 21	1	0	0	0	0	0	0	1	0	1	0
22 23	1	0	0	0	1	0	0	1	0	1	1
24 25	1	0	0	1	0	0	0	1	1	0	0
26 27	1	0	0	1	1	0	0	1	1	0	1
28 29	1	0	1	0	0	0	0	1	1	1	0
30 31	1	1	0	0	0	0	0	1	0	1	1
32 33	1	1	0	0	1	0	1	0	0	0	0
34 35	1	1	0	1	0	0	1	0	0	0	1
36 37	1	1	0	1	1	0	1	0	0	1	0
38 39	1	1	1	0	0	0	1	0	1	1	1
Any	X	X	X	X	X	1	1	1	1	1	1

BCD 9s or BCD 10s Complement Converter

BCD Words	INPUTS (See note C) E	D	C	B	A	G	OUTPUTS (See note D) Y8	Y7	Y6
0	0	0	0	0	0	0	1	0	0
1	0	0	0	0	1	0	1	0	1
2	0	0	0	1	0	0	0	1	0
3	0	0	0	1	1	0	0	1	1
4	0	0	1	0	0	0	0	1	0
5	0	1	0	0	0	0	0	1	1
6	0	1	0	0	1	0	0	0	0
7	0	1	0	1	0	0	0	0	1
8	0	1	0	1	1	0	0	0	0
9	0	1	1	0	0	0	0	0	1
0	1	0	0	0	0	0	0	0	0
1	1	0	0	0	1	0	1	0	0
2	1	0	0	1	0	0	1	0	0
3	1	0	0	1	1	0	0	1	1
4	1	0	1	0	0	0	0	1	1
5	1	1	0	0	0	0	0	1	0
6	1	1	0	0	1	0	0	1	0
7	1	1	0	1	0	0	0	1	1
8	1	1	0	1	1	0	0	0	1
9	1	1	1	0	0	0	0	0	0
Any	X	X	X	X	X	1	1	1	1

Note A. Input conditions other than those shown produce highs at outputs Y1–Yy5
B. Outputs Y6, Y7, and Y8 are not used in BCD-to-Binary conversion.
C. Input conditions other than those shown produce highs at outputs Y6, Y7, and Y8.
D. Outputs Y1–Y5 are not used for BCD 9s or BCD 10s complement conversion. When these devices are used as complement converters, Input E is used as a mode control. With this input low, the BCD 9s complement is generated; when it is high the BCD 10s complement is generated.

74184 BCD-to-BINARY CONVERTER FOR TWO BCD DECADES

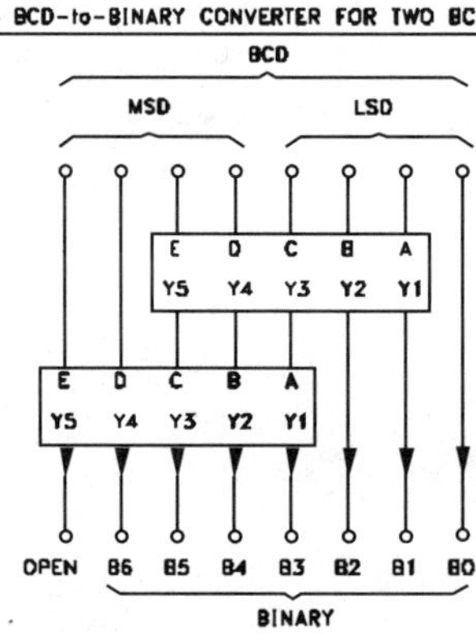

Figure 9.10. The TTL 74184 and 74185 function as a set for BCD-to-Binary and Binary-to-BCD conversions. These are derived from National's DM7488, a 256-bit memory unit. Emitter connections are made to provide direct readout of converted codes at outputs Y8 through Y1, as shown in the function tables. The drawing illustrates one possible application; the manufacturer's databook shows a variety of conversion applications.

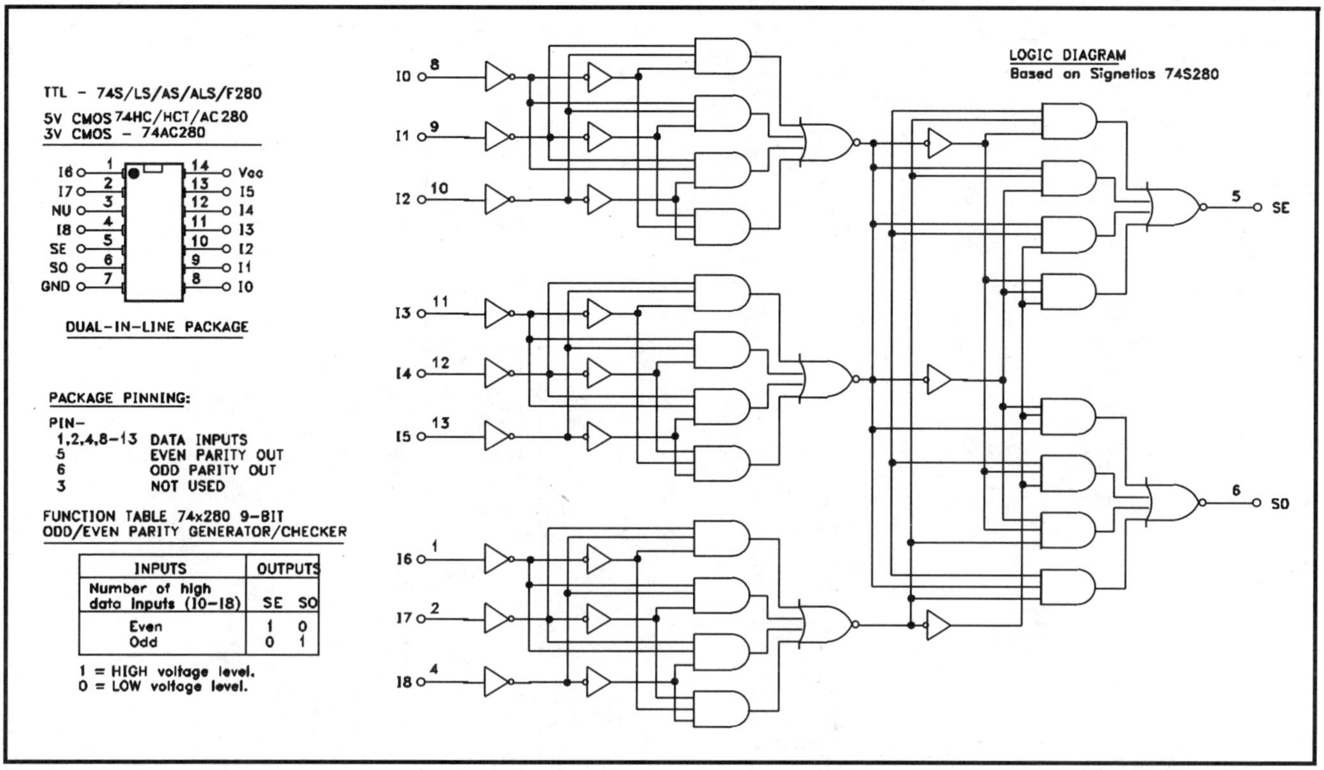

Figure 9.11. The TTL and CMOS 74x280 are a 9-Bit Odd/Even Parity Generator/Checker. The existence of odd/even outputs facilitates operation of either odd or even parity applications for nine data lines. The devices are cascadable to meet n-bit requirements. These may be used to upgrade many systems utilizing the 74x180 parity generator/checker.

TTL – 74AS282

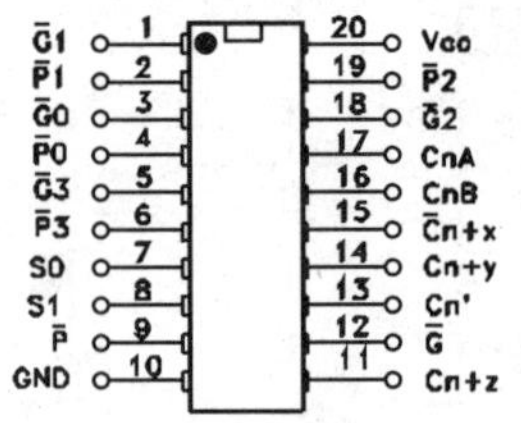

DUAL-IN-LINE PACKAGE

PACKAGE PINNING:

PIN–

26,17	CARRY INPUTS
2,4,6,19	CARRY PROPAGATE (ACTIVE LOW) INPUT
1,3,5,18	CARRY GENERATE (ACTIVE LOW) INPUT
11,14,15	CARRY OUTPUTS
9	CARRY PROPAGATE (ACTIVE LOW) OUTPUT
12	CARRY GENERATE (ACTIVE LOW) OUTPUT
7,8	CARRY SELECT INPUTS
13	SELECTED CARRY

FUNCTION TABLE 74AS282 LOOKAHEAD CARRY GENERATOR WITH SELECTABLE CARRY INPUTS

INPUTS								OUTPUT		OUTPUTS		
$\overline{G3}$	$\overline{G2}$	$\overline{G1}$	$\overline{G0}$	$\overline{P3}$	$\overline{P2}$	$\overline{P1}$	$\overline{P0}$	S1	S0	$\overline{G}$	$\overline{P}$	Cn'
0	X	X	X	X	X	X				0		
X	0	X	X	0	X	X				0		
X	X	0	X	0	0	X				0		
X	X	X	0	0	0	0				0		
All other combinations										1		
				0	0	0	0				0	
All other combinations											1	
								0	0			CnA
								0	1			$\overline{CnA}$
								1	0			CnB
								1	1			$\overline{CnB}$
$\overline{G3}$	$\overline{G2}$	$\overline{G1}$	$\overline{G0}$	$\overline{P3}$	$\overline{P2}$	$\overline{P1}$	$\overline{P0}$	Cn'		Cn+x	Cn+y	Cn+z
			0				X	X		1		
			X				0	1		1		
All other combinations										0		
		0	X			X	X	X			1	
		X	0			0	X	X			1	
		X	X			0	0	1			1	
All other combinations											0	
	0	X	X		X	X	X	X				1
	X	0	X		0	X	X	X				1
	X	X	0		0	0	X	X				1
	X	X	X		0	0	0	1				1
All other combinations												0

1 = HIGH voltage level. 0 = LOW voltage level. X = Don"t Care

Logic equations provided at the outputs are:

$$Cn+x = G0+P0Cn$$
$$Cn+y = G1+P1G0 = P1P0Cn$$
$$Cn+z = G2+P2G1+P2P1G0+P2P1P0Cn$$
$$\overline{G} = \overline{G3+P3G2+P3P2G1+P3P2P1G0}$$
$$\overline{P} = \overline{P3P2P1P0}$$

LOGIC DIAGRAM

Based on Texas Instruments 74AS282

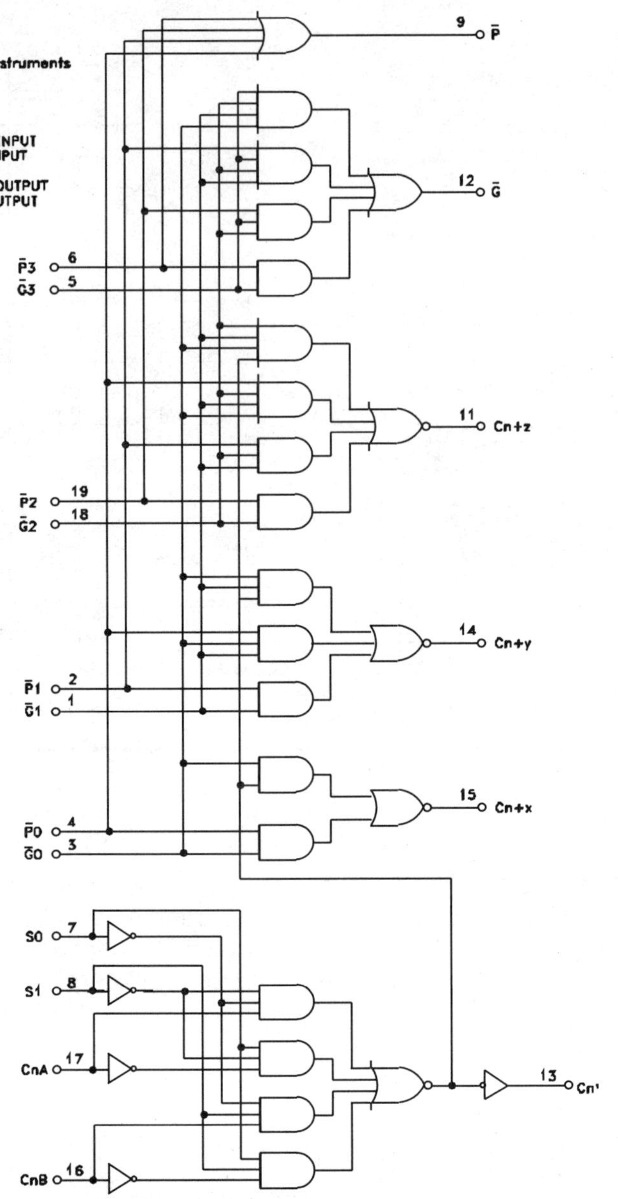

Figure 9.12. The TTL 74AS282 is a Look-Ahead Carry Generator with selectable carry inputs allowing for a double precision carry. It is capable of anticipating a carry across four binary adders or a group of adders. Functionally, the device is the same as the 74AS182 except for the provision of carry inputs CnA and CnB and their complements. The logic equations are written in terms of the selected carry, Cn. This is also available at output Cn'.

TTL – 74283
74/S/LS/AS/ALS/F283
5V CMOS – 74HC/HCT283

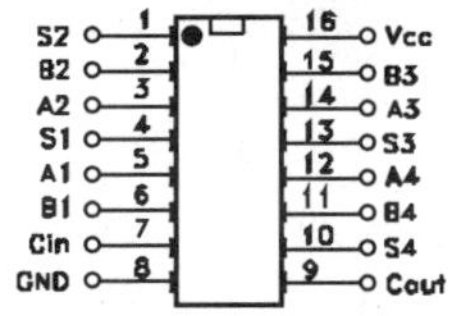

DUAL-IN-LINE PACKAGE

PACKAGE PINNING:

PIN–
3,5,12,14 OPERAND INPUTS
2,6,11,15 OPERAND INPUTS
7 CARRY INPUT
1,4,10,13 SUM OUTPUTS
9 CARRY OUTPUT

LOGIC DIAGRAM
Based on Signetics 74LS283

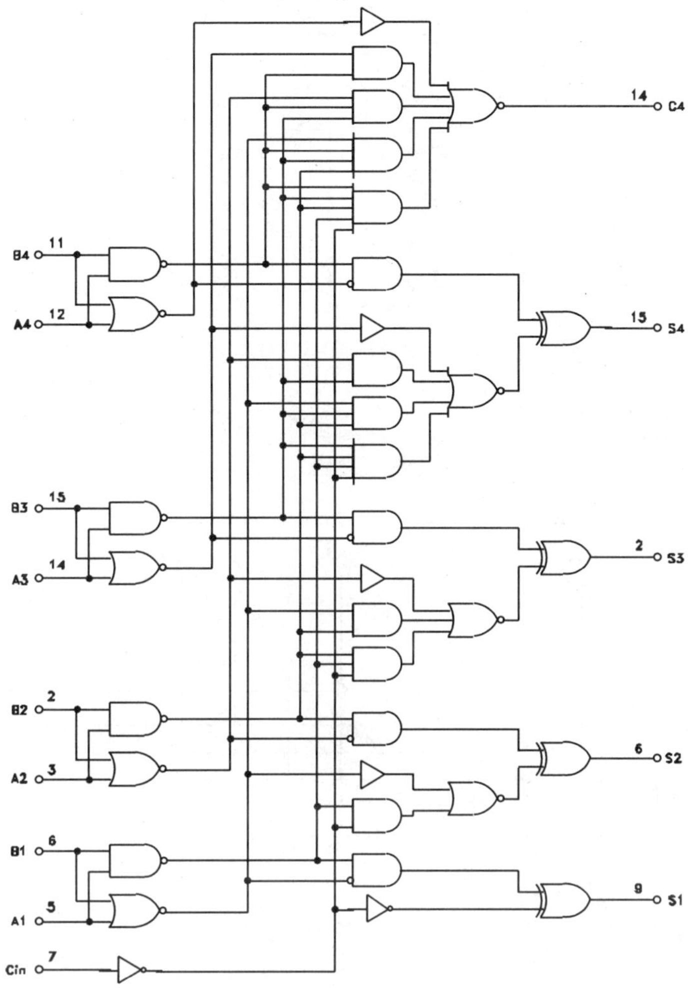

FUNCTION TABLE 74x283 4-BIT
FULL ADDERS WITH FAST CARRY

PINS	Cin	A1	A2	A3	A4	B1	B2	B3	B4	S1	S2	S3	S4	Cout
Logic Levels	L	L	H	L	H	H	L	L	H	H	H	L	L	H
Active HIGH*	0	0	1	0	1	1	0	0	1	1	1	0	0	1
Active LOW**	1	1	0	1	0	0	1	1	0	0	0	1	1	0

1 = HIGH voltage level. 0 = LOW voltage level.

* (10+9=19)
** (Carry +5+6=12)

The "283" adds two 4-bit binary words (An+Bn) plus the incoming carry. The binary sum appears on the Sum outputs (S1–S4) and the outgoing carry according to the equation:

Cin+(A1+B1)+2(A2+B2)+4(A3+B3)+8(A4+B4)
= S1+2S2+4S3+8S4+16Cout
Where (+) = plus

Figure 9.13. The TTL 74x283 and CMOS 74HC/HCT283 are a 4-bit Binary Full Adder, electrically and functionally equivalent to the 7483, but having a different package pinout. The device performs the addition of two 4-bit binary words. The sum (S) outputs are provided for each bit and the resultant carry (C4) is obtained from the 4th bit. Full internal look-ahead is provided all four bits. The adder logic, the carry included, is implemented in its true form.

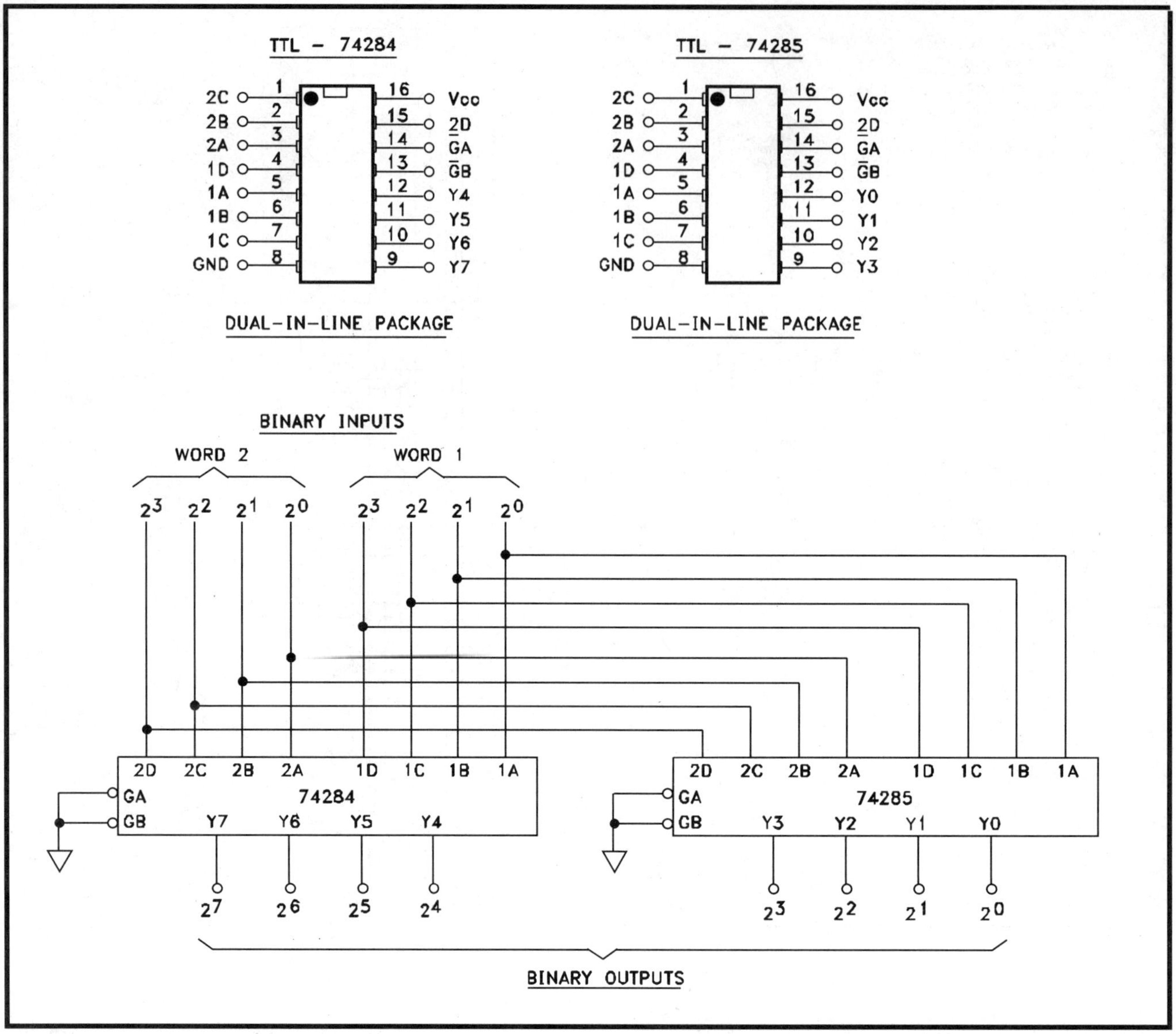

Figure 9.14. The TTL 74284 and 74285 are a set providing 4-Bit by 4-Bit Parallel Binary Multiplication yielding an 8-bit product in a typical 40 nS. When connected as shown, they perform the positive logic multiplication of two 4-bit binary words. This basic pair can be utilized as a block for the construction of larger multipliers. For these applications the '181, '182, and '183 are included to achieve a high-performance system. The scheme is expandable by employing NxM bit multipliers.

TTL - 74AS286

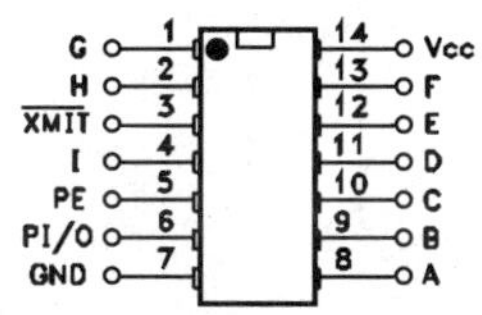

DUAL-IN-LINE PACKAGE

FUNCTION TABLE 74AS286 9-BIT PARITY GENERATOR/CHECKER WITH BUS DRIVER PARITY I/O PORT

NUMBER OF HIGH INPUTS (A THRU I)	$\overline{XMIT}$	PARITY I/O	PARITY ERROR
0,2,4,6,8	l	H	H
1,3,5,7,9	l	L	H
0,2,4,6,8	h	h	H
	h	l	L
1,3,5,7,9	h	h	L
	h	l	H

h = High level input l = Low level input
H = High level output L = Low level output

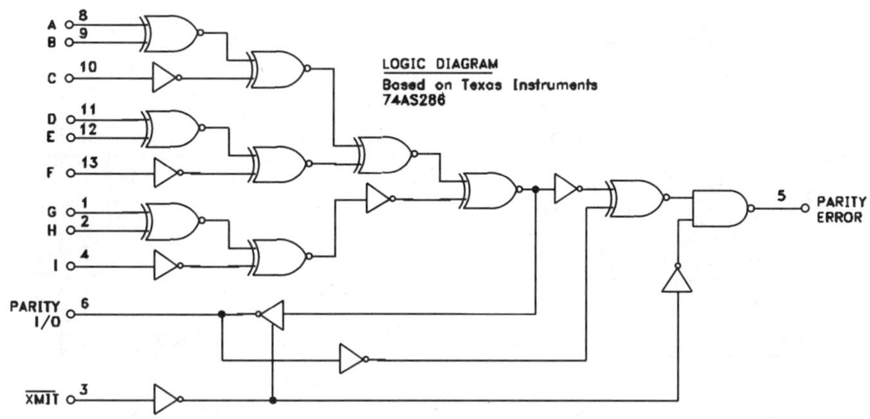

Figure 9.15. The TTL 74AS286 is a 9-Bit Parity Generator/Checker with a 48 mA bus-driving parity I/O port. The /XMIT control input is implemented specifically to enable cascading. When this input is low, the parity tree is disabled and the Parity Error (PE) output remains high regardless of the input levels. When the control is high, the parity tree is enabled. The circuit design is such that the I/O port will remain in a high-impedance state during power-up or power-down.

TTL - 74F350

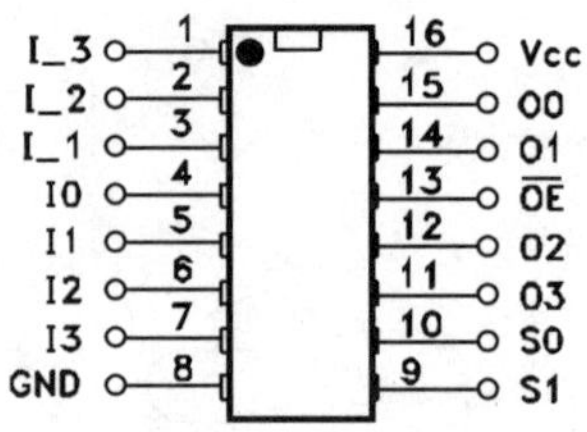

DUAL-IN-LINE PACKAGE

FUNCTION TABLE 74F350 4-BIT SHIFTER WITH 3-STATE OUTPUTS

INPUTS			OUTPUTS			
$\overline{OE}$	S1	S0	00	01	02	03
1	X	X	Z	Z	Z	Z
0	0	0	I0	I1	I2	I3
0	0	1	I_1	I0	I1	I2
0	1	0	I_2	I_1	I0	I1
0	1	1	I_3	I_2	I_1	I0

1 = HIGH voltage level. 0 = LOW voltage level.

LOGIC EQUATIONS:

$00 = \overline{S0}\,\overline{S1}\,I0 + S0\,\overline{S1}\,I\_1 + \overline{S0}\,S1\,I\_2 + S0\,S1\,I\_3$

$01 = \overline{S0}\,\overline{S1}\,I1 + S0\,\overline{S1}\,I0 + \overline{S0}\,S1\,I\_1 + S0\,S1\,I\_2$

$02 = \overline{S0}\,\overline{S1}\,I2 + S0\,\overline{S1}\,I1 + \overline{S0}\,S1\,I0 + S0\,S1\,I\_1$

$03 = \overline{S0}\,\overline{S1}\,I3 + S0\,\overline{S1}\,I2 + \overline{S0}\,S1\,I1 + S0\,S1\,I0$

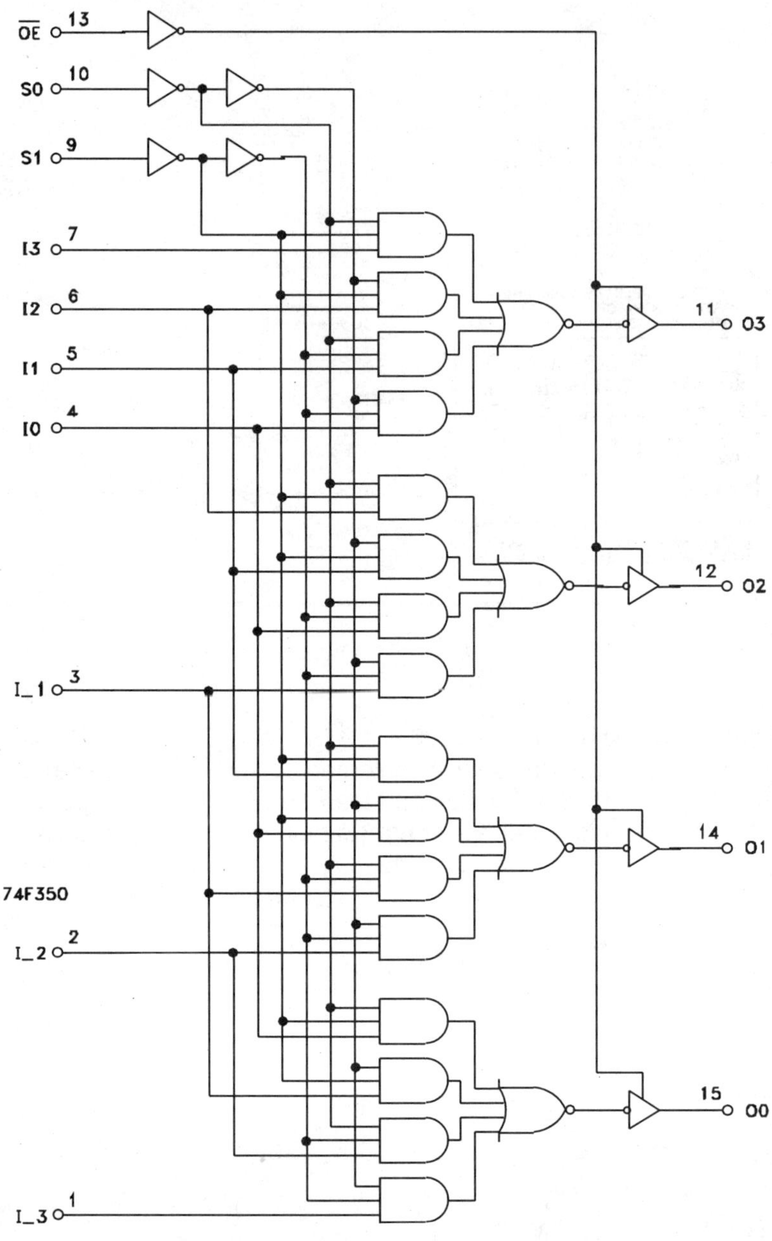

LOGIC DIAGRAM

Based on Motorola 74F350

Figure 9.16. The TTL 74F350 is a Shifter that shifts a 4-Bit word by 0,1, 2, or 3 places. No clocking is required as with shift registers. With suitable interconnection, any number of bits can be shifted any number of places up or down. Shifting may be logical, with logic zeros filled in at either end; arithmetic where the sign bit is extended during a shift down; or end-around where the data word forms a continuous loop. The output is 3-state.

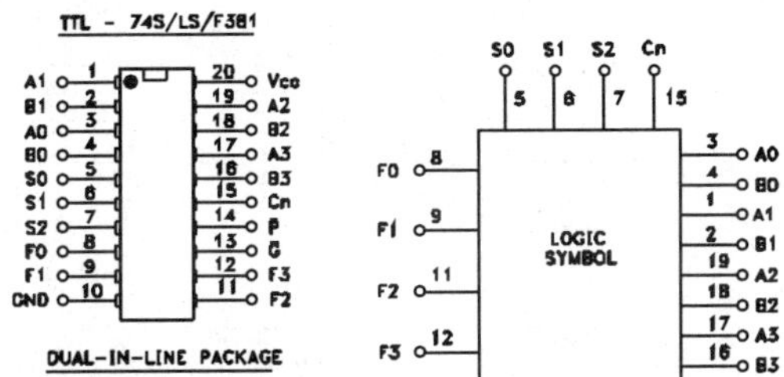

FUNCTION SELECT TABLE 74x381
4-BIT ARITHMETIC LOGIC UNIT

SELECT			
S0	S1	S2	OPERATION
0	0	0	Clear
1	0	0	B Minus A
0	1	0	A Minus B
1	1	0	A Plus B
0	0	1	A⊕B
1	0	1	A+B
0	1	1	AB
1	1	1	Preset

1 = HIGH voltage level.
0 = LOW voltage level.

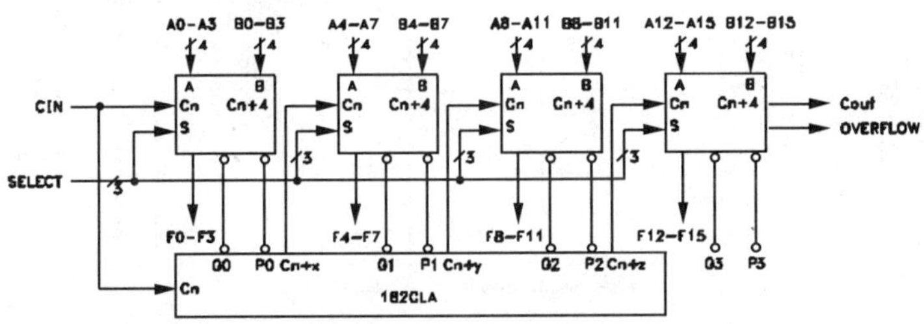

16-BIT LOOKAHEAD CARRY ALU EXPANSION

TRUTH TABLE

	INPUTS						OUTPUTS					
FUNCTION	S0	S1	S2	Cn	An	Bn	F0	F1	F2	F3	$\bar{G}$	$\bar{P}$
CLEAR	0	0	0	X	X	X	0	0	0	0	0	0
B MINUS A	1	0	0	0	0	0	1	1	1	1	1	0
				0	0	1	0	1	1	1	0	0
				0	1	0	0	0	0	0	1	1
				0	1	1	1	1	1	1	1	0
				1	0	0	0	0	0	0	1	0
				1	0	1	1	1	1	1	0	0
				1	1	0	1	0	0	0	1	1
				1	1	1	0	0	0	0	1	0
A MINUS B	0	1	0	0	0	0	1	1	1	1	1	0
				0	0	1	0	0	0	0	1	1
				0	1	0	0	1	1	1	0	0
				0	1	1	1	1	1	1	1	0
				1	0	0	0	0	0	0	1	0
				1	0	1	1	0	0	0	1	1
				1	1	0	1	1	1	1	0	0
				1	1	1	0	0	0	0	1	0
A PLUS B	1	1	0	0	0	0	0	0	0	0	1	1
				0	0	1	1	1	1	1	1	0
				0	1	0	1	1	1	1	1	0
				0	1	1	0	1	1	1	0	0
				1	0	0	1	0	0	0	1	1
				1	0	1	0	0	0	0	1	0
				1	1	0	0	0	0	0	1	0
				1	1	1	1	1	1	1	0	0
A⊕B	0	0	1	X	0	0	0	0	0	0	0	0
				X	0	1	1	1	1	1	1	1
				X	1	0	1	1	1	1	1	0
				X	1	1	0	0	0	0	0	0
A+B	1	0	1	X	0	0	0	0	0	0	0	0
				X	0	1	1	1	1	1	1	1
				X	1	0	1	1	1	1	1	1
				1	1	1	1	1	1	1	1	0
AB	0	1	1	X	0	0	0	0	0	0	0	0
				X	0	1	0	0	0	0	1	1
				X	1	0	0	0	0	0	0	0
				X	1	1	1	1	1	1	1	0
PRESET	1	1	1	X	0	0	1	1	1	1	1	1
				X	0	1	1	1	1	1	1	1
				X	1	0	1	1	1	1	1	1
				X	1	1	1	1	1	1	1	0

1 = HIGH voltage level. 0 = LOW voltage level. X = Don't Care

Figure 9.17. The TTL 74LS/F381 is an Arithmetic Logic Unit (ALU) capable of performing three arithmetic and three logical operations on two 4-bit words. Carry Propagate (/P) and Generate (/G) outputs are provided for use with the '182 Carry Look-Ahead Generator for expansion to longer word lengths. For ripple expansion refer to the '382 device data. Signals applied to the select inputs determine the operating mode as defined in the function table.

TTL – 74LS/F382

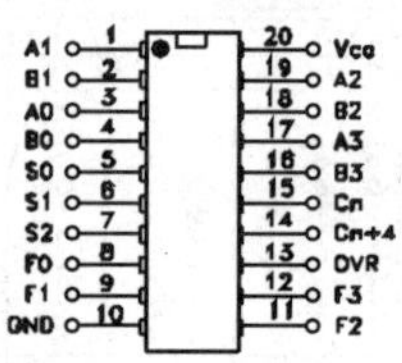

DUAL-IN-LINE PACKAGE

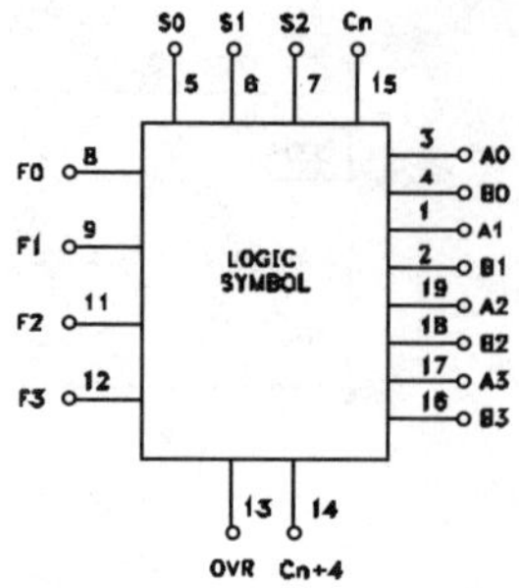

FUNCTION SELECT TABLE 74x382
4-BIT ARITHMETIC LOGIC UNIT

SELECT			
S0	S1	S2	OPERATION
0	0	0	Clear
1	0	0	B Minus A
0	1	0	A Minus B
1	1	0	A Plus B
0	0	1	A⊕B
1	0	1	A+B
0	1	1	AB
1	1	1	Preset

1 = HIGH voltage level.
0 = LOW voltage level.

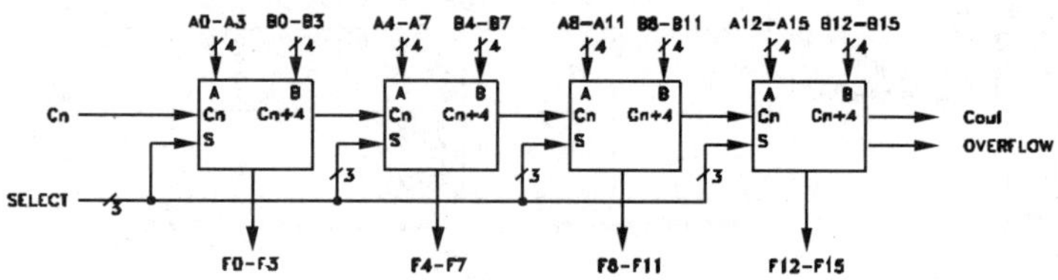

16-BIT RIPPLE CARRY ALU EXPANSION

TRUTH TABLE

FUNCTION	INPUTS S0	S1	S2	Cn	An	Bn	OUTPUTS F0	F1	F2	F3	OVR	Cn+4
CLEAR	0	0	0	0	X	X	0	0	0	0	1	1
				1	X	X	0	0	0	0	1	1
B MINUS A	1	0	0	0	0	0	1	1	1	1	0	0
				0	0	1	0	1	1	1	0	1
				0	1	0	0	0	0	0	0	0
				0	1	1	1	1	1	1	0	0
				1	0	0	0	0	0	0	0	1
				1	0	1	1	1	1	1	0	1
				1	1	0	1	0	0	0	0	0
				1	1	1	0	0	0	0	0	1
A MINUS B	0	1	0	0	0	0	1	1	1	1	0	0
				0	0	1	0	0	0	0	0	0
				0	1	0	0	1	1	1	0	1
				0	1	1	1	1	1	1	0	0
				1	0	0	0	0	0	0	0	1
				1	0	1	1	0	0	0	0	0
				1	1	0	1	1	1	1	0	1
				1	1	1	0	0	0	0	0	1
A PLUS B	1	1	0	0	0	0	0	0	0	0	0	0
				0	0	1	1	1	1	1	0	0
				0	1	0	1	1	1	1	0	0
				0	1	1	0	1	1	1	0	1
				1	0	0	1	0	0	0	0	0
				1	0	1	0	0	0	0	0	1
				1	1	0	0	0	0	0	0	1
				1	1	1	1	1	1	1	0	1
A⊕B	0	0	1	X	0	0	0	0	0	0	0	0
				X	0	1	1	1	1	1	0	0
				0	1	0	1	1	1	1	0	0
				X	1	1	0	0	0	0	1	1
				1	1	0	1	1	1	1	1	1
A+B	1	0	1	X	0	0	0	0	0	0	0	0
				X	0	1	1	1	1	1	0	0
				X	1	0	1	1	1	1	0	0
				0	1	1	1	1	1	1	0	0
				1	1	1	1	1	1	1	1	1
AB	0	1	1	X	0	0	0	0	0	0	1	1
				X	0	1	0	0	0	0	0	0
				X	1	0	0	0	0	0	1	1
				0	1	1	1	1	1	1	0	0
				1	1	1	1	1	1	1	1	1
PRESET	1	1	1	X	0	0	1	1	1	1	0	0
				X	0	1	1	1	1	1	0	0
				X	1	0	1	1	1	1	0	0
				0	1	1	1	1	1	1	0	0
				1	1	0	1	1	1	1	1	1

1 = HIGH voltage level. 0 = LOW voltage level. X = Don't Care

Figure 9.18. The TTL 74LS/F382 is an Arithmetic Logic Unit (ALU) capable of performing three arithmetic and three logical operations on two 4-bit words. Three select input codes, (S0 - S2), force the function outputs low or high. An overflow output is provided for convenience with 2's complement arithmetic. A carry output is provided for ripple expansion. Signals applied to the select inputs determine the operating mode as defined in the function table. The circuit performs the arithmetic function for either active-high or active-low operands, with output levels in the same convention.

TTL – 74LS384

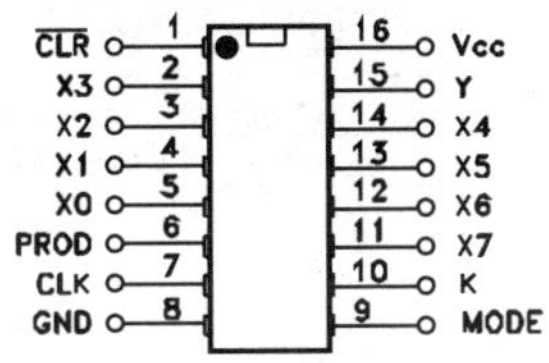

DUAL-IN-LINE PACKAGE

FUNCTION TABLE 74LS384 8-BIT BY 1-BIT TWO'S COMPLEMENT MULTIPLIER

INPUTS				INTERNAL	OUTPUT	
$\overline{CLR}$	CLK	XI	Y	Y_1	PROD	FUNCTION
0	X	Data	X	0	0	Load new multiplicand and clear internal sum and carry registers.
1	↑	X	0	0	Output	Shift sum register
1	↑	X	0	1	per	Add multiplicand to sum register and shift
1	↑	X	1	0	Booth's	Subtract multiplicand from sum register and shift
1	↑	X	1	1	Algorithm	Shift sum register

1 = HIGH voltage level. 0 = LOW voltage level.
X = Don't Care

↑ = LOW-to-HIGH clock transition.

LOGIC DIAGRAM

Based on Texas Instruments 74LS384

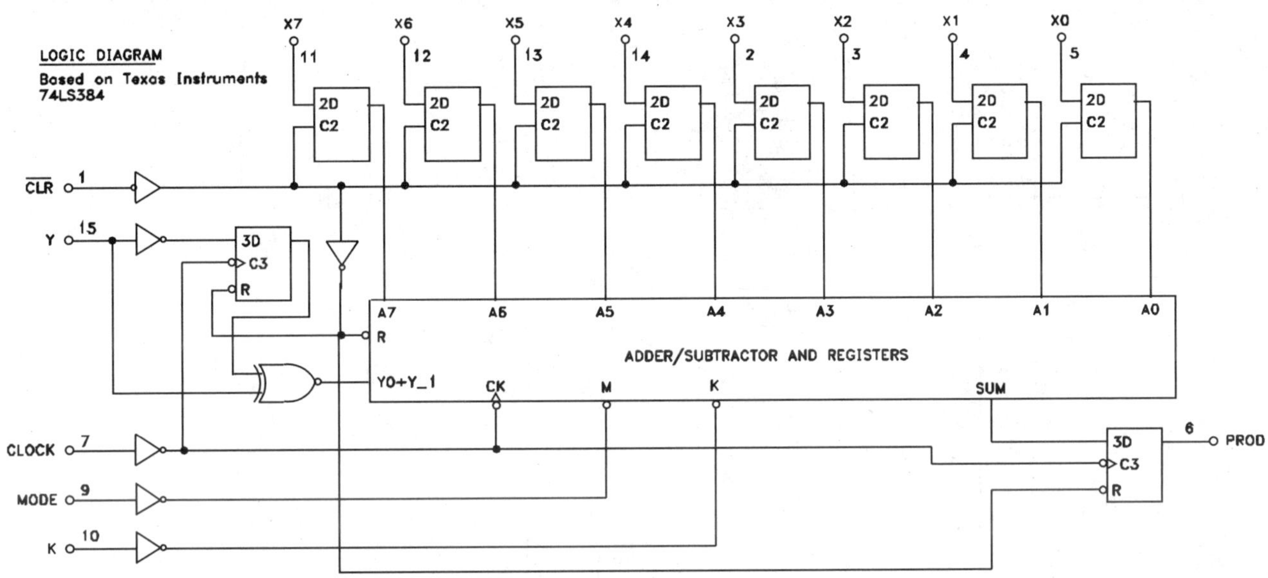

Figure 9.19. The TTL 74LS384 is an 8-Bit by 1-Bit 2's Complement Multiplier. It performs digital multiplication of two numbers presented in 2's complement form to produce a 2's complement product without external correction through internal functioning. The device accepts an 8-bit multiplicand (X input) which is stored in eight latches. These are controlled by the Clear input. When this input is low, all latches are cleared and available to accept new input. When this input is high, the latches do not respond to changes in the X variables. The multiplier word data is passed by the Y input in a serial bit stream, least significant bit first. The product is clocked out of the PROD output. Typical clock frequency is 40 MHz.

TTL – 74LS/F385

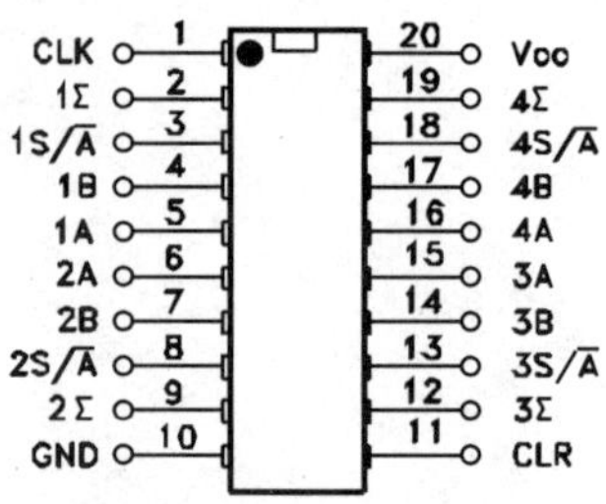

DUAL-IN-LINE PACKAGE

FUNCTION TABLE 74LS385 QUADRUPLE SERIAL ADDER/SUBTRACTOR

SELECTED FUNCTION	INPUTS					DATA IN CARRY FLIP-FLOP		OUTPUT
	CLR	S/$\overline{A}$	A	B	CLK	BEFORE ↑	AFTER ↑	AFTER ↑
CLEAR	0	0	X	X	X	0	0	0
	0	1	X	X	X	1	1	0
ADD	1	0	0	0	↑	0	0	0
	1	0	0	0	↑	1	0	1
	1	0	0	1	↑	0	0	1
	1	0	0	1	↑	1	1	0
	1	0	1	0	↑	0	0	1
	1	0	1	0	↑	1	1	0
	1	0	1	1	↑	0	1	0
	1	0	1	1	↑	1	1	1
SUBTRACT	1	1	0	0	↑	0	0	1
	1	1	0	0	↑	1	1	0
	1	1	0	1	↑	0	0	0
	1	1	0	1	↑	1	0	1
	1	1	1	0	↑	0	1	0
	1	1	1	0	↑	1	1	1
	1	1	1	1	↑	0	0	1
	1	1	1	1	↑	1	1	0

1 = HIGH voltage level. 0 = LOW voltage level.
X = Don't Care
↑ = LOW-to-HIGH clock transition.

LOGIC DIAGRAM
Based on Texas Instruments 74LS385
(ONE OF FOUR ADDER/SUBTRACTORS)

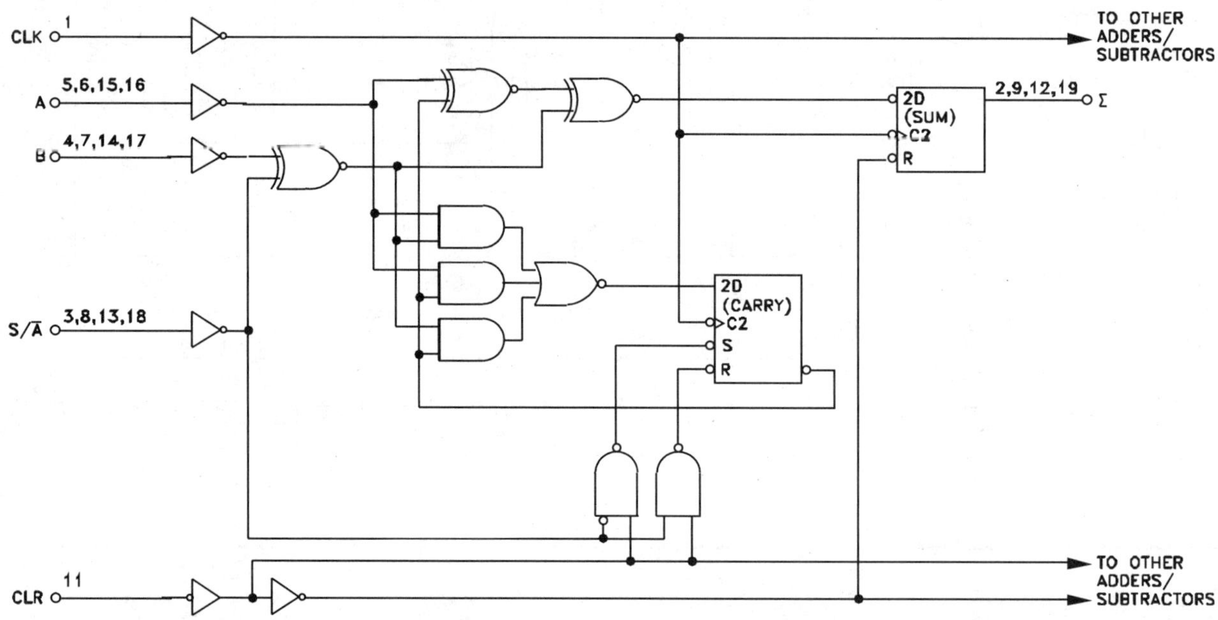

Figure 9.20. The TTL 74LS/F385 is a Quadruple Serial Adder/Subtracter. It is a useful companion to the '384 serial/parallel 2's complement multiplier. The '385 contains four independent adder/subtracters elements with common clock and clear. Each of the independent sum (S) outputs reflects its respective input as controlled by S//A control. With this control high, the S function is A minus B; when low, A plus B. The active low Clear asynchronously resets the sum and carry flip-flops: high in the subtract mode, low in the add mode. The clock is positive edge-triggered, providing control as shown in the function table.

TTL – 74ALS518
74ALS519

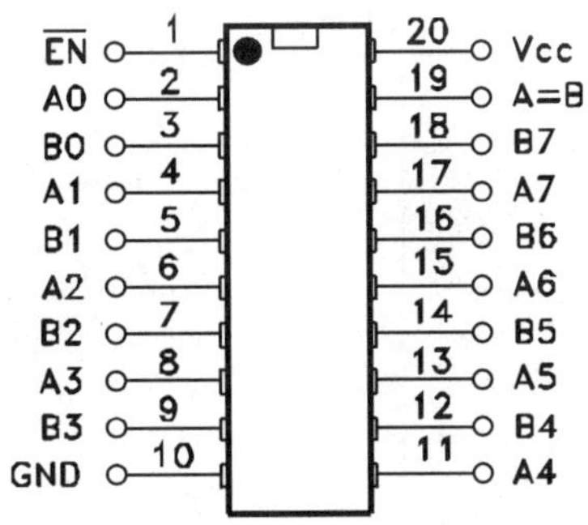

DUAL-IN-LINE PACKAGE

FUNCTION TABLE 74ALS518, '519
8-BIT COMPARATORS

INPUTS		OUTPUT
$\overline{EN}$	Data	A=B
0	A=B	1
0	A≠B	0
1	X	0

1 = HIGH voltage level.
0 = LOW voltage level.
X = Don't Care

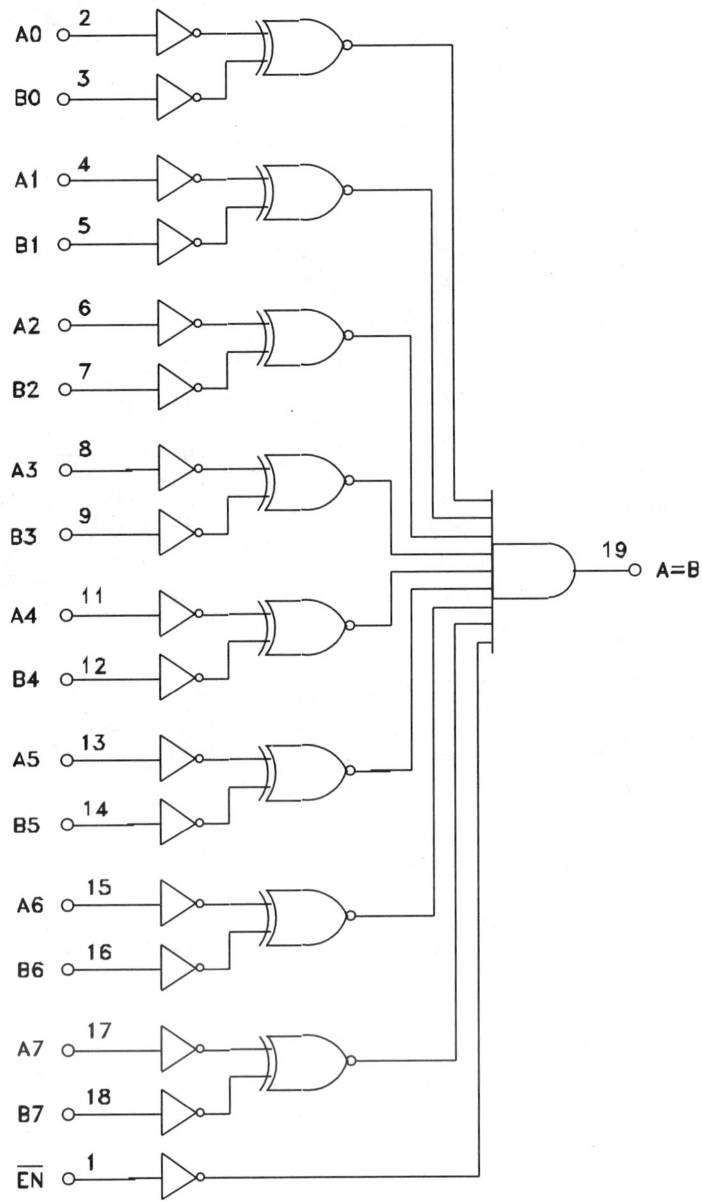

LOGIC DIAGRAM

Based on National 74ALS518

Figure 9.21. The TTL 74ALS518 and '519 are 8-Bit Identity Comparators. These perform comparisons on two eight-bit binary or BCD words. P = Q outputs are provided in contrast the '520, '521, and '522, which provide /P = Q outputs (*Figure 9.22*). These two devices feature open collector outputs. The '518 also provides a 20K pull-up termination resistor on the Q inputs for analog or switch inputs.

TTL – 74ALS520
74ALS521
74ALS522
5V CMOS – 74AC/ACT520
74HC/HCT/AC/ACT521
3V CMOS – 74AC520
74AC521

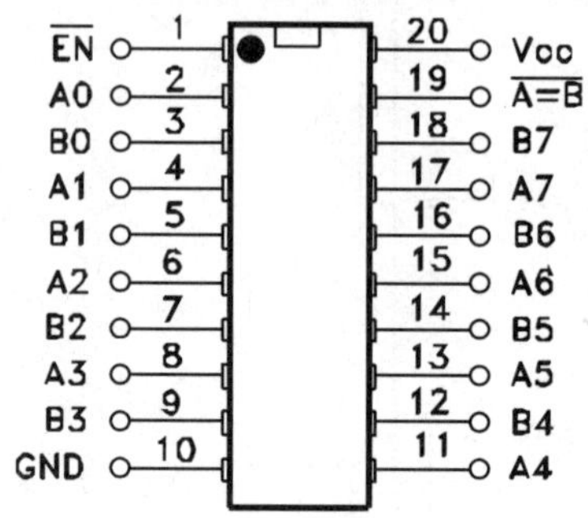

DUAL-IN-LINE PACKAGE

FUNCTION TABLE 74ALS520,'521,'522
8-BIT COMPARATORS

INPUTS		OUTPUT
$\overline{EN}$	Data	$\overline{A=B}$
0	A=B	1
0	A≠B	0
1	X	0

1 = HIGH voltage level.
0 = LOW voltage level.
X = Don't Care

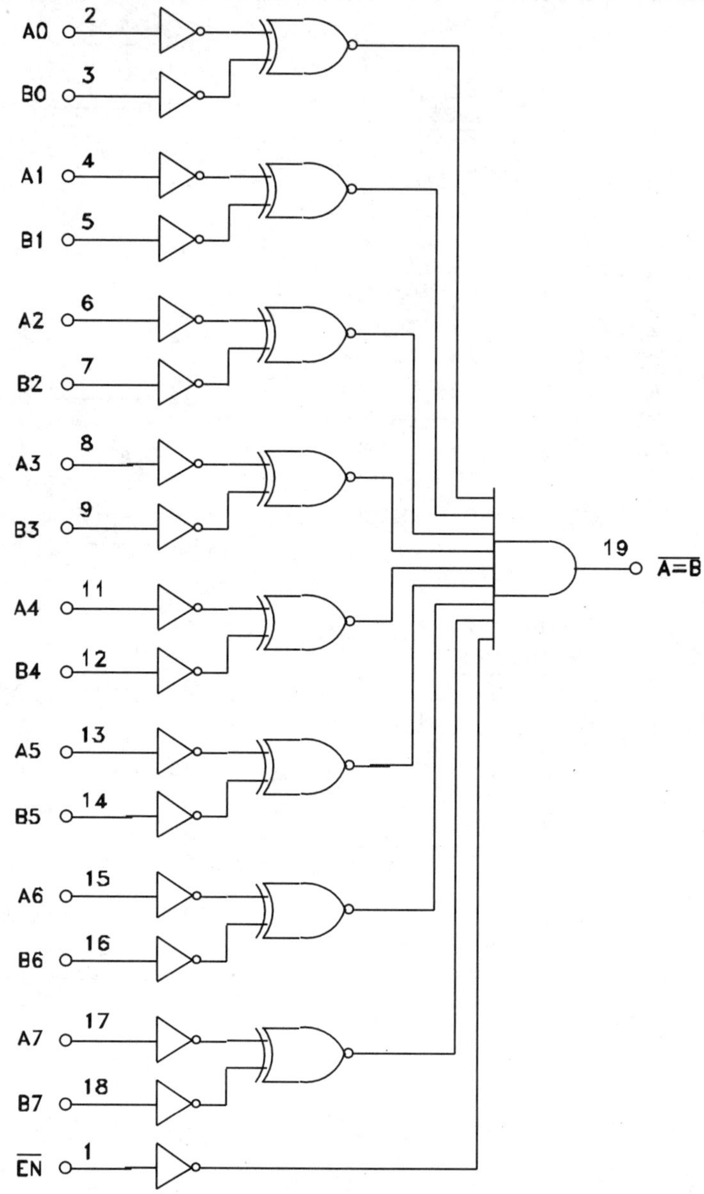

LOGIC DIAGRAM
Based on National 74ALS520

Figure 9.22. The TTL 74ALS520, '521 and '522 and their CMOS counterparts are 8-Bit Identity Comparators similar to the '518 and '519 (*Figure 9.21*) in their functions. These perform comparisons on two 8-bit binary or BCD words. /P = Q outputs are provided in contrast to the '518 and '519, which provide P = Q outputs. The 'ALS522 features open collector outputs. The 'ALS520 'ALS522 also provide a 20K pull-up termination resistor on the Q inputs for analog or switch inputs. The '521 is functionally and pin compatible to the TTL and CMOS versions of the '688 (*Figure 9.34*).

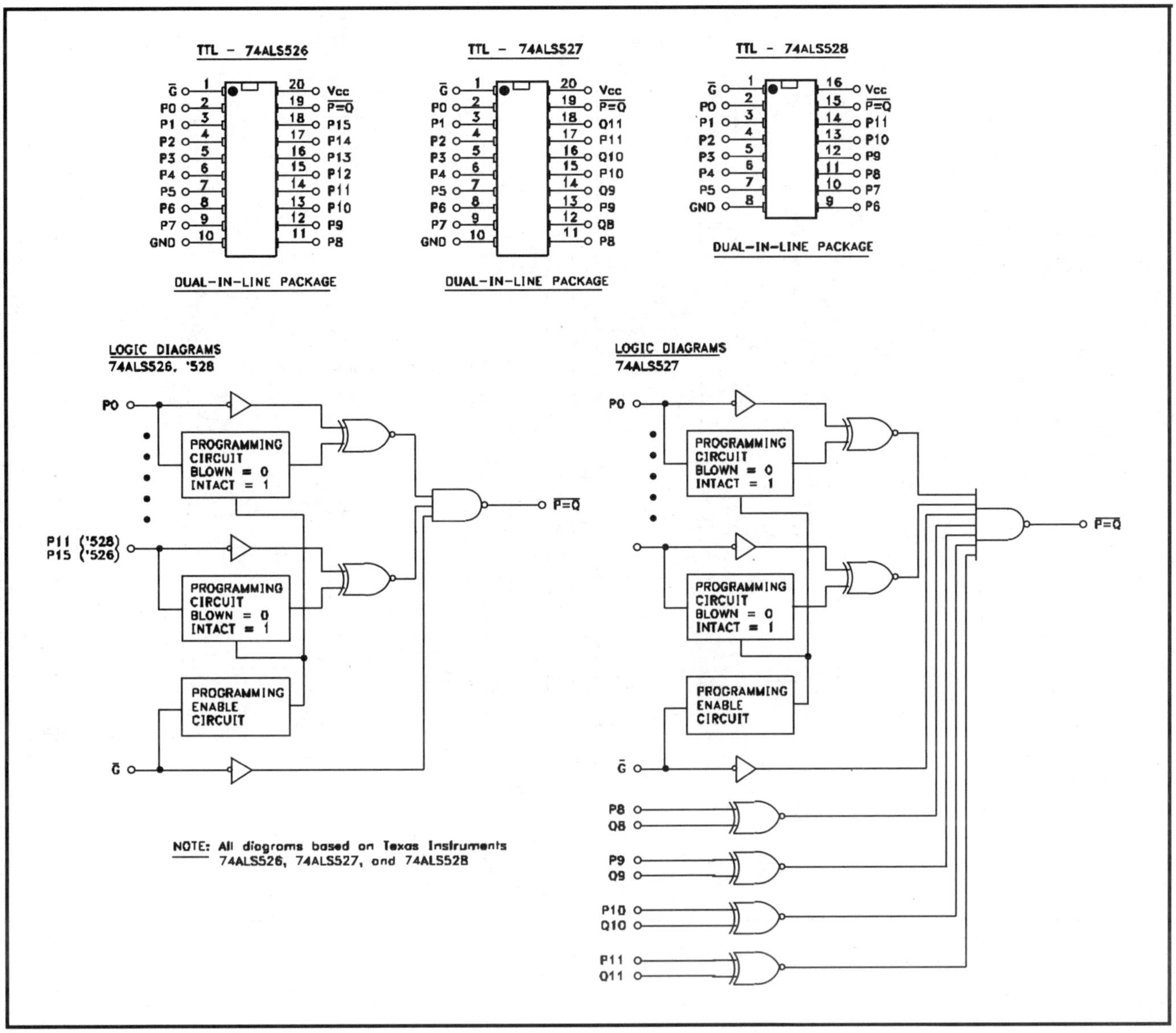

Figure 9.23. The TTL 74ALS526, '527 and '528 are Fuse Programmable Identity Comparators. The '526 is a 16-bit identity comparator; the '527 an 8-bit identity comparator and 4-bit comparator; the '528 a 12-bit identity comparator. These devices may be programmed and verified on most incoming test equipment.

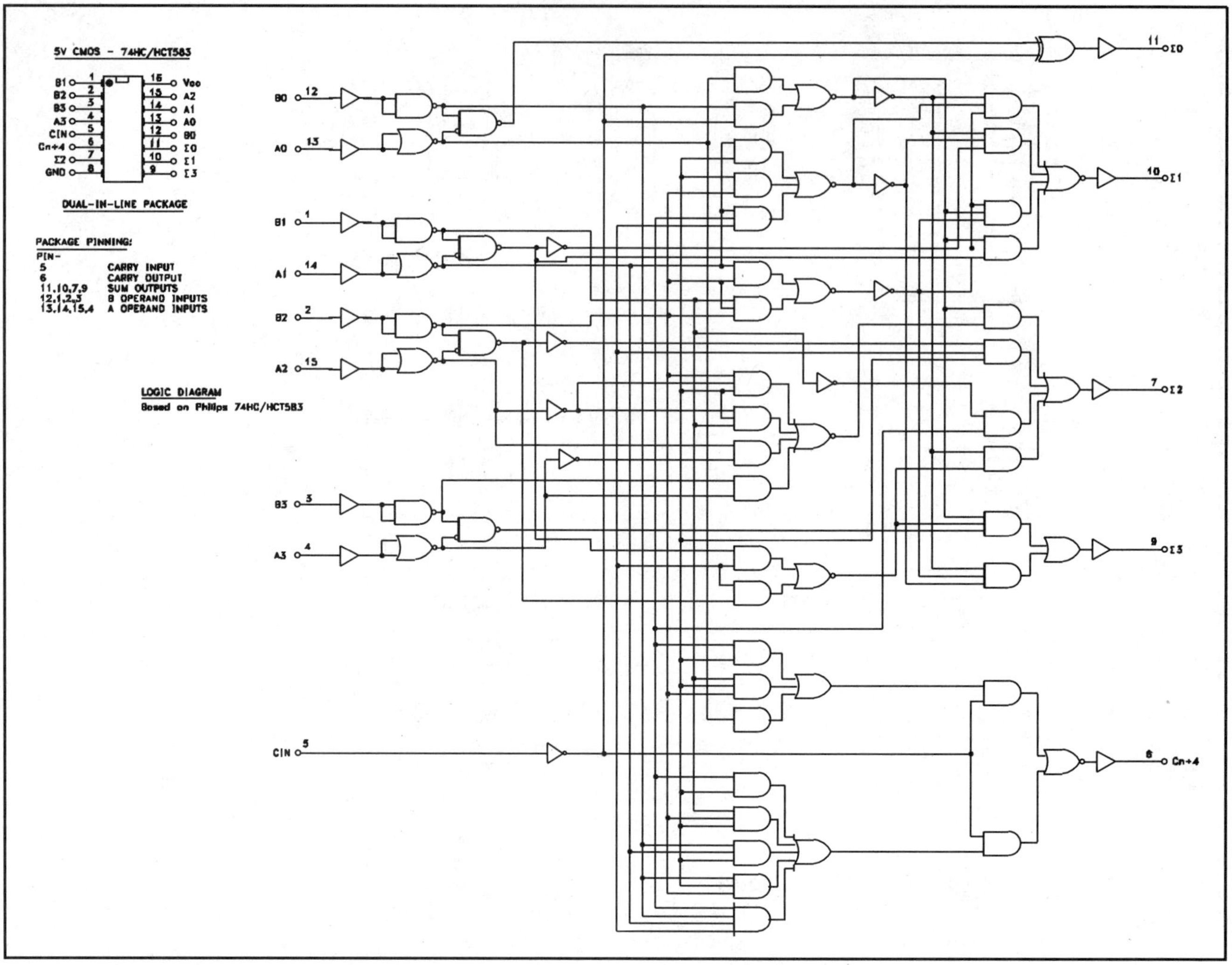

Figure 9.24. The CMOS 74HC/HCT583 are 4-Bit BCD Full Adders With Fast Carry and internal carry look-ahead. The device accepts two 4-bit decimal numbers (A0 - A3,B0 - B3) and a carry input, Cin. The output is the decimal sum, S0 - S3, and the carry output, Cn + 4, if the sum is greater than 9. For input values greater than nine, a valid BCD sum and carry result. For these, the binary value is converted to BCD. The conversion requires grounding one set of inputs while applying the four inputs to the other set. If the input is between 0 and 9, a BCD output is provided. If between 10 and 15, the output is BCD with a carry. Conversion of larger numbers may be accomplished by cascading devices.

MEMORY WRITE CYCLE:
During this cycle check bits CB0–CB5 are internally generated in the EADC by 8–input parity generators using the 16–bit data word as defined in Table 2. These are stored with the original 16–bit word.

ERROR DETECTION AND CORRECTION:
During a memory read cycle the 6–bit check word is retrieved along with the actual data. The error flags are then tested for the high level. The first case in Table 3 is the normal, no error condition. The next two cases are correctable errors. The last three generate an interrupt.
Error detection is accomplished as the 6–bit check word and the 16–bit data word from memory are applied to the internal parity generators/checkers.

If the parity of one or more of the check groups is incorrect an error has occurred and appropriate error flags are set low. The 2–bit error is not correctable as the parity tree can detect only single bit errors. Both error flags are set low on detection of a 2–bit error.

FUNCTION TABLES 74ALS616, '617 16–BIT PARALLEL ERROR DETECTION AND CORRECTION CIRCUITS

TABLE 1. WRITE CONTROL FUNCTION

MEMORY CYCLE	EDAC FUNCTION	CONTROL S1 S0	DATA I/O	DB CONTROL $\overline{OEB0}$ & $\overline{OEB1}$	DB OUTPUT LATCH $\overline{LEDBO}$	CHECK I/O	CB CONTOL $\overline{OECB}$	ERROR FLAGS $\overline{ERR}$ $\overline{MERR}$
Write	Generate Check Word	0 0	Input	1	x	Output Check Bits	0	1 1

TABLE 2. PARITY ALGORITHM

CHECK WORD BIT	15	14	13	12	11	10	9	8	7	6	5	4	3	2	1	0
	16–BIT DATA WORD															
CB0			X		X	X	X			X					X	X
CB1		X		X		X	X	X				X		X		X
CB2	X			X	X			X			X			X		X
CB3	X	X	X				X	X			X	X	X			
CB4	X	X	X	X	X	X			X	X						
CB5									X	X	X	X	X	X	X	X

TABLE 3. ERROR FUNCTION

TOTAL NUMBER OF ERRORS		ERROR FLAGS		DATA CORRECTION
16–BIT DATA WORD	6–BIT CHECK WORD	ERR	MERR	
0	0	H	H	Not Applicable
1	0	L	H	Correction
0	1	L	H	Correction
1	1	L	L	Interrupt
2	0	L	L	Interrupt
0	2	L	L	Interrupt

TABLE 4. READ, FLAG, AND CORRECT FUNCTION

MEMORY CYCLE	EDAC FUNCTION	CONTROL S1 S0	DATA I/O	DB CONTROL $\overline{OEB0}$ & $\overline{OEB1}$	DB OUTPUT LATCH $\overline{LEDBO}$	CHECK I/O	CB CONTOL $\overline{OECB}$	ERROR FLAGS ERR MERR
Read	Reas $ Flag	H L	Input	H	x	Input	H	Enabled*
Read	Latch Input Data & Check Bits	H H	Latched Input Data	H	L	Latched Input Check Word	H	Enabled*
Read	Output Corrected Data and Syndrome Bits	H H	Output Corrected Data Word	L	X	Output Syn–drome Bits	L	Enabled*

TTL – 74ALS616 3–STATE
74ALS617 OPEN–COLLECTOR

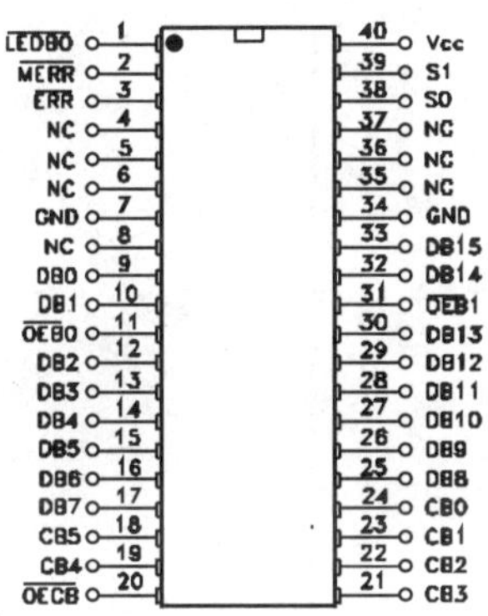

DUAL–IN–LINE PACKAGE

NOTES:
1 = HIGH voltage level.
0 = LOW voltage level.
X = Don't Care

CB X = error in check bit X
DB Y = error in data bit Y
2–bit = double–bit error
unc = uncorrectable multibit error

This drawing is based on Texas Instruments 74ALS616, '617.

TABLE 5. SYNDROME DECODING

SYNDROME BITS 5 4 3 2 1 0	ERROR
0 0 0 0 0 0	2–bit
0 0 0 0 0 1	unc
0 0 0 0 1 0	unc
0 0 0 0 1 1	2–bit
0 0 0 1 0 0	unc
0 0 0 1 0 1	2–bit
0 0 0 1 1 0	2–bit
0 0 0 1 1 1	unc
0 0 1 0 0 0	unc
0 0 1 0 0 1	2–bit
0 0 1 0 1 0	2–bit
0 0 1 0 1 1	DB7
0 0 1 1 0 0	2–bit
0 0 1 1 0 1	unc
0 0 1 1 1 0	DB6
0 0 1 1 1 1	2–bit

SYNDROME BITS 5 4 3 2 1 0	ERROR
0 1 0 0 0 0	unc
0 1 0 0 0 1	2–bit
0 1 0 0 1 0	2–bit
0 1 0 0 1 1	DB5
0 1 0 1 0 0	2–bit
0 1 0 1 0 1	DB4
0 1 0 1 1 0	DB3
0 1 0 1 1 1	2–bit
0 1 1 0 0 0	2–bit
0 1 1 0 0 1	DB2
0 1 1 0 1 0	DB1
0 1 1 0 1 1	2–bit
0 1 1 1 0 0	DB0
0 1 1 1 0 1	2–bit
0 1 1 1 1 0	2–bit
0 1 1 1 1 1	CB5

SYNDROME BITS 5 4 3 2 1 0	ERROR
1 0 0 0 0 0	unc
1 0 0 0 0 1	2–bit
1 0 0 0 1 0	2–bit
1 0 0 0 1 1	DB15
1 0 0 1 0 0	2–bit
1 0 0 1 0 1	DB14
1 0 0 1 1 0	DB13
1 0 0 1 1 1	2–bit
1 0 1 0 0 0	2–bit
1 0 1 0 0 1	DB12
1 0 1 0 1 0	DB11
1 0 1 0 1 1	2–bit
1 0 1 1 0 0	DB10
1 0 1 1 0 1	2–bit
1 0 1 1 1 0	2–bit
1 0 1 1 1 1	CB4

SYNDROME BITS 5 4 3 2 1 0	ERROR
1 1 0 0 0 0	2–bit
1 1 0 0 0 1	DB8
1 1 0 0 1 0	unc
1 1 0 0 1 1	2–bit
1 1 0 1 0 0	DB9
1 1 0 1 0 1	2–bit
1 1 0 1 1 0	2–bit
1 1 0 1 1 1	CB3
1 1 1 0 0 0	unc
1 1 1 0 0 1	2–bit
1 1 1 0 1 0	2–bit
1 1 1 0 1 1	CB2
1 1 1 1 0 0	2–bit
1 1 1 1 0 1	CB1
1 1 1 1 1 0	CB0
1 1 1 1 1 1	none

Figure 9.25. The TTL 74ALS616 and 'ALS617 are 16-Bit Parallel Error Detection and Correction devices. These use a modified Hamming code to generate a 6-bit check word from a 16-bit data word. The check word is stored along with the data word during the memory write cycle. During memory read cycles the 22-bit word is processed to determine if errors have occurred in memory. Single bit errors are corrected and flagged. The data word is unaltered. A 6-bit error syndrome pinpoints the error location. Dual errors are pinpointed but not corrected.

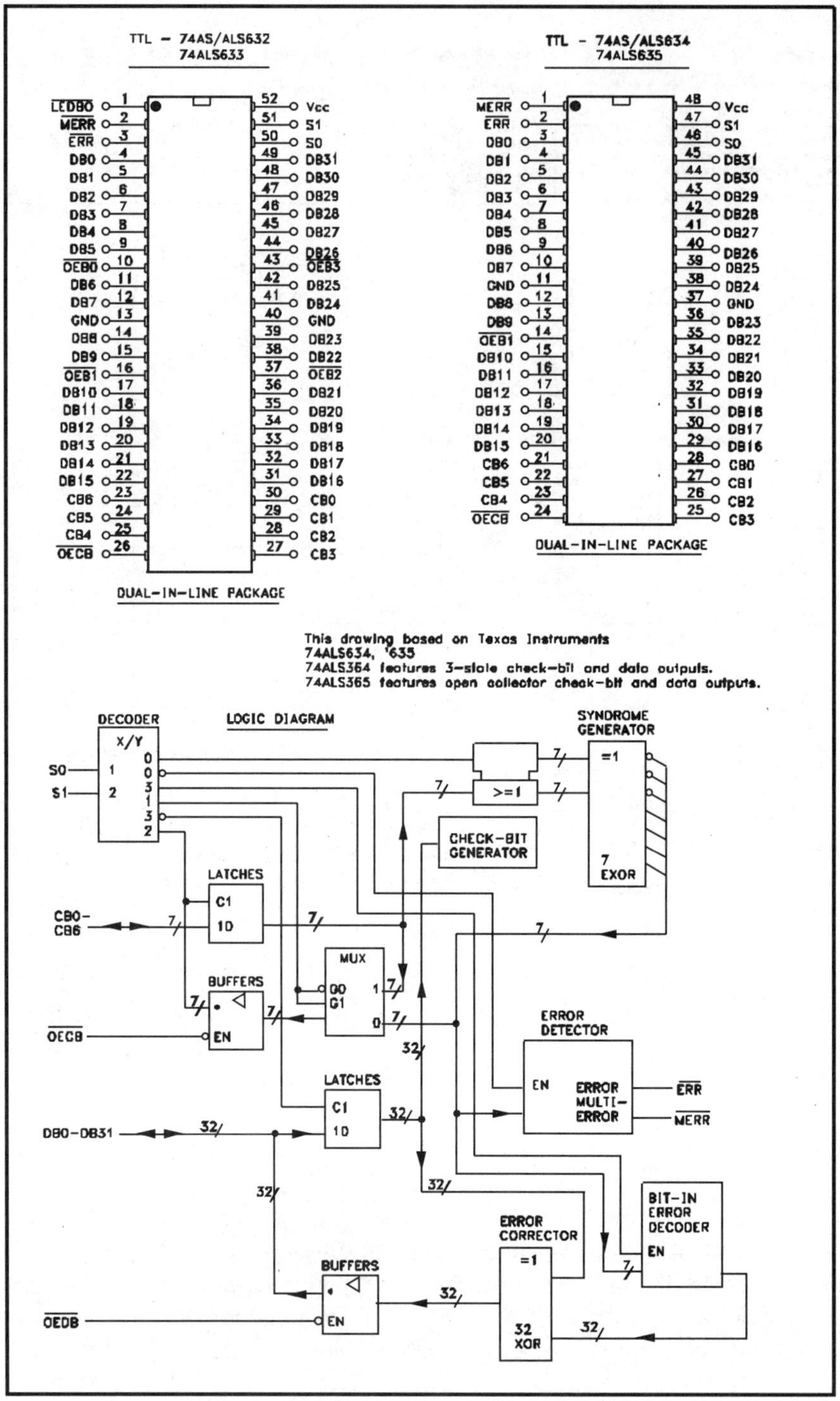

Figure 9.26. The TTL 74LS630 is a 16-Bit Parallel Error Detection and Correction device with similarities to the 'ALS616/617 (*Figure 9.25*). These use a modified Hamming code to generate a 6-bit check word from a 16-bit data word. The check word is stored along with the data word during the memory write cycle. During memory read cycles the 22-bit word is processed to determine if errors have occurred in memory. Single-bit errors in the 16-bit data word are corrected and flagged. Single-bit errors in the 6-bit check word are flagged and a correction cycle occurs even though the data word is not in error. The 16-bit data word is passed along unaltered. A 6-bit error syndrome pinpoints the error location. Dual errors are pin pointed but not corrected.

TTL – 74LS636 3-STATE
74LS637 OPEN COLLECTOR

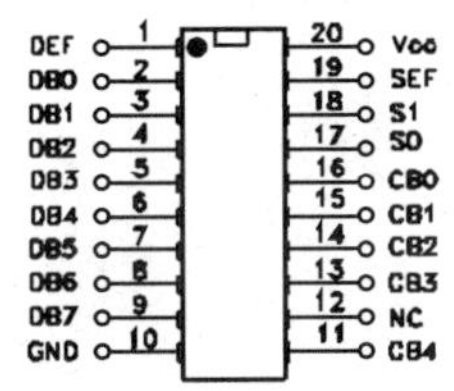

DUAL-IN-LINE PACKAGE

FUNCTION TABLES 74LS636, '637 8-BIT ERROR DETECTION AND CORRECTION CIRCUITS

ERROR FUNCTION TABLE

Total Number of Errors		Error Flags		
8-Bit Data	5-Bit Checkword	SEF	DEG	Data Correction
0	0	L	L	Not Applicable
1	0	H	L	Correction
0	1	H	L	Correction
1	1	H	H	Interrupt
2	0	H	H	Interrupt
0	2	H	H	Interrupt

CHECKWORD BIT	8-BIT DATA WORD							
	0	1	2	3	4	5	6	7
CB0	x	x		x	x			
CB1	x		x	x		x	x	
CB2		x	x		x	x		x
CB3	x	x	x				x	x
CB4				x	x	x	x	x

NOTE: The data in this drawing was obtained from the Texas Instruments 74LS636, '637 8-Bit Parallel Error Detection and Correction Circuits data.

ERROR SYNDROME TABLE

	SYNDROME ERROR CODE				
ERROR LOCATION	CB0	CB1	CB2	CB3	CB4
DB0	0	0	1	0	1
DB1	0	1	0	0	1
DB2	1	1	0	0	1
DB3	0	0	1	1	0
DB4	0	1	0	1	0
DB5	1	0	0	1	0
DB6	1	0	1	0	0
DB7	1	1	0	1	0
CB0	0	1	1	1	1
CB1	1	0	1	1	1
CB2	1	1	0	1	1
CB3	1	1	1	0	1
CB4	1	1	1	1	0
NO ERROR	1	1	1	1	1

1 = HIGH voltage level.
0 = LOW voltage level.
X = Don't Care

LOGIC DIAGRAM

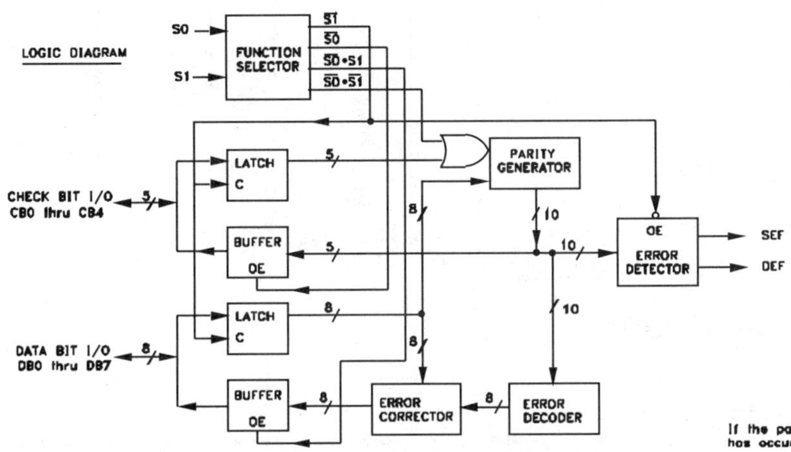

The first case in the error function table represents the normal, no error condition. The next two cases of single-bit errors require data correction. Although the EDAC can discern the single check bit error and ignore it, the error flags are identical to the single error in the 8-bit data word. The CPU will ask for data correction in both cases. An interrupt condition to the CPU results in each of the last three cases.

During a memory write cycle, five check bits, CB0–CB4, are generated by eight-input parity generators using the data bits as defined. During a memory read cycle, the 5-bit check word is retrieved along with the actual data. Error detection is accomplished as the five-bit check word and the 8-bit data word from memory are applied to internal parity generator/checkers. If the parity of all five groupings of data and check bits are correct it is assumed that no error has occurred and both error flags will be low.

If the parity of one or more of the check groups is incorrect an error has occurred and the proper error flag or flags will be set high.

Any single error in the 8-bit data word will change the sense of exactly three bits of the five-bit check word. Any single error in the six bit check word changes the sense of only that one bit. The single error flag will be set high while the dual error flag will remain low.

Any two-bit error will change the sense of an even number of check bits. The two-bit error is not correctable. Both error flags are set high.

Three or more simultaneous bit errors can fool the EDAC into believing that no error has occurred and produce erroneous results.

Error correction is achieved by identifying the bad bit and inverting it.

As the corrected word is made available on the data word I/O port, the check word I/O port presents a 5-bit syndrome error code, useable for identifying the bad memory chip.

Figure 9.27. The TTL 74ALS632, '633, '634 and '635 are 32-Bit Parallel Error Detection and Correction Circuits with similarities to those described previously. (*Figures 9.25, 26*). These use a modified Hamming code to generate a 7-bit check word from a 32-bit data word. The check word is stored along with the data word during the memory write cycle. During memory read cycles the 39-bit word is processed to determine if errors have occurred in memory. Single-bit errors in the 32-bit data word are corrected and flagged. Single-bit errors in the 7-bit check word are flagged and a correction cycle occurs even though the data word is not in error. The 32-bit data word is passed along unaltered. A 7-bit error syndrome pinpoints the error location. Dual errors are pinpointed but not corrected. The '632 and '634 outputs are 3-state; the '633 and '635 are open collector.

TTL – 74LS636 3-STATE
74LS637 OPEN COLLECTOR

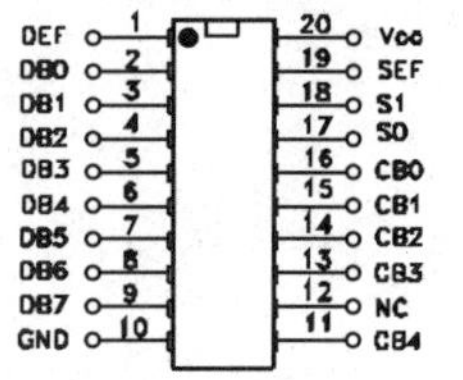

DUAL-IN-LINE PACKAGE

FUNCTION TABLES 74LS636, '637 8-BIT ERROR DETECTION AND CORRECTION CIRCUITS

ERROR FUNCTION TABLE

Total Number of Errors		Error Flags		
8-Bit Data	5-Bit Checkword	SEF	DEC	Data Correction
0	0	L	L	Not Applicable
1	0	H	L	Correction
0	1	H	L	Correction
1	1	H	H	Interrupt
2	0	H	H	Interrupt
0	2	H	H	Interrupt

NOTE: The data in this drawing was obtained from the Texas Instruments 74LS636, '637 8-Bit Parallel Error Detection and Correction Circuits data.

CHECKWORD BIT	8-BIT DATA WORD 0	1	2	3	4	5	6	7
CB0	x	x		x	x			
CB1	x		x	x		x	x	
CB2		x	x		x	x		x
CB3	x	x	x				x	x
CB4				x	x	x	x	x

ERROR SYNDROME TABLE

	SYNDROME ERROR CODE				
ERROR LOCATION	CB0	CB1	CB2	CB3	CB4
DB0	0	0	1	0	1
DB1	0	1	0	0	1
DB2	1	1	0	0	1
DB3	0	0	1	1	0
DB4	0	1	0	1	0
DB5	1	0	0	1	0
DB6	1	0	1	0	0
DB7	1	1	0	1	0
CB0	0	1	1	1	1
CB1	1	0	1	1	1
CB2	1	1	0	1	1
CB3	1	1	1	0	1
CB4	1	1	1	1	0
NO ERROR	1	1	1	1	1

1 = HIGH voltage level.
0 = LOW voltage level.
X = Don't Care

LOGIC DIAGRAM

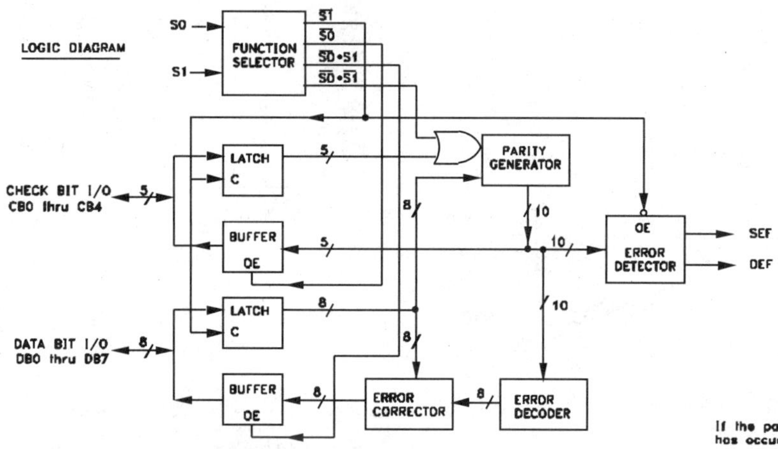

The first case in the error function table represents the normal, no error condition. The next two cases of single-bit errors require data correction. Although the EDAC can discern the single check bit error and ignore it, the error flags are identical to the single error in the 8-bit data word. The CPU will ask for data correction in both cases. An interrupt condition to the CPU results in each of the last three cases.

During a memory write cycle, five check bits, CB0–CB4, are generated by eight-input parity generators using the data bits as defined. During a memory read cycle, the 5-bit check word is retrieved along with the actual data. Error detection is accomplished as the five-bit check word and the 8-bit data word from memory are applied to internal parity generator/checkers. If the parity of all five groupings of data and check bits are correct it is assumed that no error has occurred and both error flags will be low.

If the parity of one or more of the check groups is incorrect an error has occurred and the proper error flag or flags will be set high.

Any single error in the 8-bit data word will change the sense of exactly three bits of the five-bit check word. Any single error in the six bit check word changes the sense of only that one bit. The single error flag will be set high while the dual error flag will remain low.

Any two-bit error will change the sense of an even number of check bits. The two-bit error is not correctable. Both error flags are set high.

Three or more simultaneous bit errors can fool the EDAC into believing that no error has occurred and produce erroneous results.

Error correction is achieved by identifying the bad bit and inverting it.

As the corrected word is made available on the data word I/O port, the check word I/O port presents a 5-bit syndrome error code, useable for identifying the bad memory chip.

Figure 9.28. The TTL 74LS637 and '638 are 8-Bit Parallel Error Detection and Correction Circuits with similarities to those described previously. (*Figures 9.25, 26, 27*). These use a modified Hamming code to generate a 5-bit check word from a 8-bit data word. The check word is stored along with the data word during the memory write cycle. During memory read cycles the 13-bit word is processed to determine if errors have occurred in memory. Single-bit errors in the 8-bit data word are corrected and flagged. Single-bit errors in the 5-bit check word are flagged and a correction cycle occurs even though the data word is not in error. The 8-bit data word is passed along unaltered. A 5-bit error syndrome pinpoints the error location. Dual errors are pinpointed but not corrected. The '636 output is 3-state; the '637 is open collector.

TTL - 74ALS677
5V CMOS - 74HC677

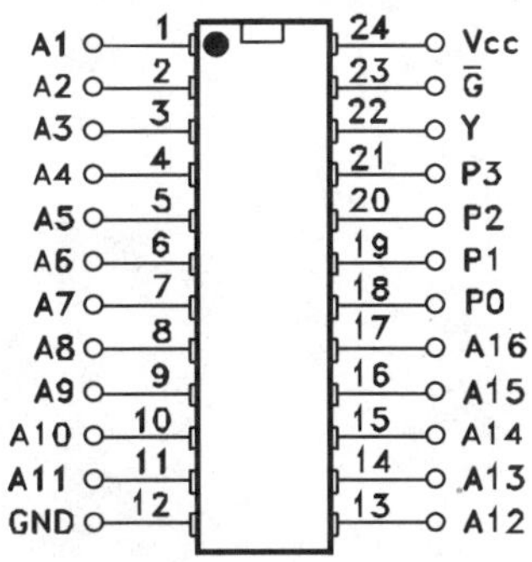

DUAL-IN-LINE PACKAGE

TTL - 74ALS678
5V CMOS - 74HC678

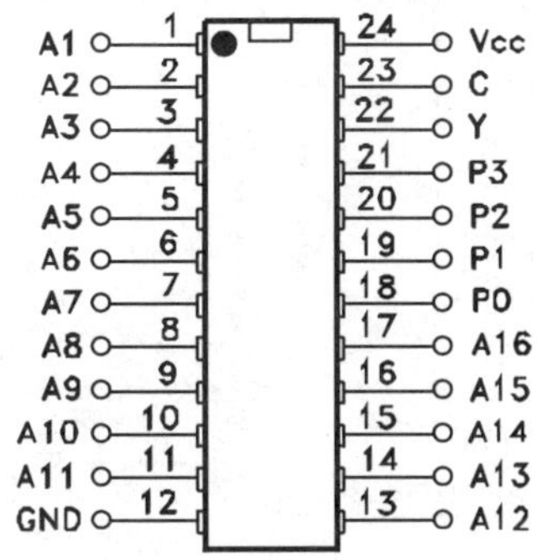

DUAL-IN-LINE PACKAGE

FUNCTION TABLE 74x677 AND 74x678
16-BIT ADDRESS COMPARATORS

'x677	'x678	INPUTS COMMON TO BOTH 'x677 AND 'x678																				OUTPUT
G	C	P3	P2	P1	P0	A1	A2	A3	A4	A5	A6	A7	A8	A9	A10	A11	A12	A13	A14	A15	A16	Y
0	1	0	0	0	0	1	1	1	1	1	1	1	1	1	1	1	1	1	1	1	1	0
0	1	0	0	0	1	0	1	1	1	1	1	1	1	1	1	1	1	1	1	1	1	0
0	1	0	0	1	0	0	0	1	1	1	1	1	1	1	1	1	1	1	1	1	1	0
0	1	0	0	1	1	0	0	0	1	1	1	1	1	1	1	1	1	1	1	1	1	0
0	1	0	1	0	0	0	0	0	0	1	1	1	1	1	1	1	1	1	1	1	1	0
0	1	0	1	0	1	0	0	0	0	0	1	1	1	1	1	1	1	1	1	1	1	0
0	1	0	1	1	0	0	0	0	0	0	0	1	1	1	1	1	1	1	1	1	1	0
0	1	0	1	1	1	0	0	0	0	0	0	0	1	1	1	1	1	1	1	1	1	0
0	1	1	0	0	0	0	0	0	0	0	0	0	0	1	1	1	1	1	1	1	1	0
0	1	1	0	0	1	0	0	0	0	0	0	0	0	0	1	1	1	1	1	1	1	0
0	1	1	0	1	0	0	0	0	0	0	0	0	0	0	0	1	1	1	1	1	1	0
0	1	1	0	1	1	0	0	0	0	0	0	0	0	0	0	0	1	1	1	1	1	0
0	1	1	1	0	0	0	0	0	0	0	0	0	0	0	0	0	0	1	1	1	1	0
0	1	1	1	0	1	0	0	0	0	0	0	0	0	0	0	0	0	0	1	1	1	0
0	1	1	1	1	0	0	0	0	0	0	0	0	0	0	0	0	0	0	0	1	1	0
0	1	1	1	1	1	0	0	0	0	0	0	0	0	0	0	0	0	0	0	0	1	0
0	1	All other combinations																				1
1		'x677: Any combination																				1
	0	'x678: Any combination																				Latched

1 = HIGH voltage level. 0 = LOW voltage level.

Figure 9.29. The TTL 74ALS677 and 'ALS678 and 74HC677 and 'HC678 are 16-Bit Address Comparators. Their usage is the simplification of memory board and peripheral addressing. The four P inputs are typically hardwired with a preprogrammed address. An internal decoder determines what input information applied to the 16 A inputs must be low or high to cause a low state at the output, Y. The '677 features an active low enable input, /G. When /G is low, the device is enabled. When high, it is disabled, with the output high regardless of the A and P inputs. The '678 features a transparent latch and latch enable C. With C high, the device mode is transparent. When low, the previous logic state of Y is latched.

TTL – 74ALS679
5V CMOS – 74HC679

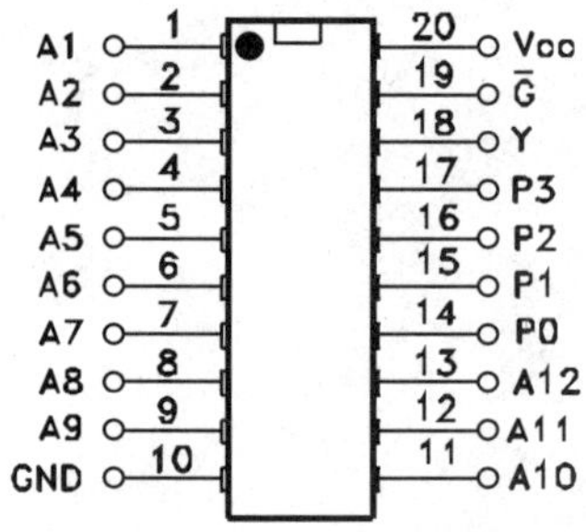

DUAL-IN-LINE PACKAGE

TTL – 74ALS680
5V CMOS – 74HC780

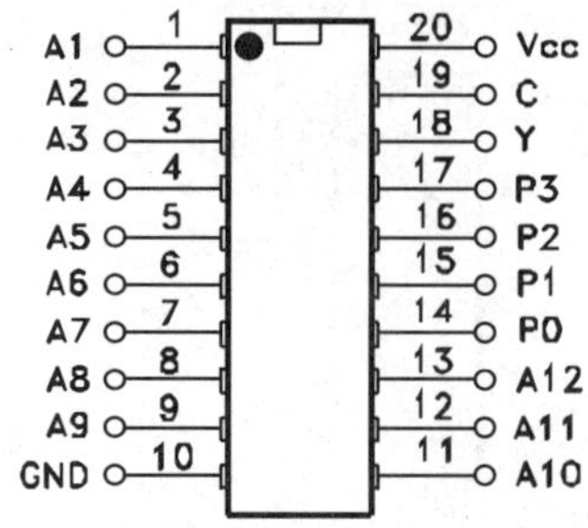

DUAL-IN-LINE PACKAGE

FUNCTION TABLE 74x679 AND 74x680
12-BIT ADDRESS COMPARATORS

'x679	'x680	INPUTS COMMON TO BOTH 'x679 AND 'x680																	OUTPUT
G	C	P3	P2	P1	P0	A1	A2	A3	A4	A5	A6	A7	A8	A9	A10	A11	A12		Y
0	1	0	0	0	0	1	1	1	1	1	1	1	1	1	1	1	1		0
0	1	0	0	0	1	0	1	1	1	1	1	1	1	1	1	1	1		0
0	1	0	0	1	0	0	0	1	1	1	1	1	1	1	1	1	1		0
0	1	0	0	1	1	0	0	0	1	1	1	1	1	1	1	1	1		0
0	1	0	1	0	0	0	0	0	0	1	1	1	1	1	1	1	1		0
0	1	0	1	0	1	0	0	0	0	0	1	1	1	1	1	1	1		0
0	1	0	1	1	0	0	0	0	0	0	0	1	1	1	1	1	1		0
0	1	0	1	1	1	0	0	0	0	0	0	0	1	1	1	1	1		0
0	1	1	0	0	0	0	0	0	0	0	0	0	0	1	1	1	1		0
0	1	1	0	0	1	0	0	0	0	0	0	0	0	0	1	1	1		0
0	1	1	0	1	0	0	0	0	0	0	0	0	0	0	0	1	1		0
0	1	1	0	1	1	0	0	0	0	0	0	0	0	0	0	0	1		0
0	1	1	1	0	0	0	0	0	0	0	0	0	0	1	1	1	0	} See note below	0
0	1	1	1	0	1	0	0	0	0	0	0	0	0	0	1	1	0	} See note below	0
0	1	1	1	1	0	0	0	0	0	0	0	0	0	0	0	1	0	} See note below	0
0	1	1	1	1	1	0	0	0	0	0	0	0	0	0	0	0	0		0
0	1	All other combinations																	1
1		'x679: Any combination																	1
	0	'x680: Any combination																	Latched

} NOTE: Refer to Manufacturer's Data Book for procedures with which data in these rows is valid.

1 = HIGH voltage level. 0 = LOW voltage level.

Figure 9.30. The TTL 74ALS679 and 'ALS680 and 74HC679 and 'HC680 are 12-Bit Address Comparators. Their usage is the simplification of memory board and peripheral addressing. The four P inputs are typically hardwired with a preprogrammed address. An internal decoder determines what input information applied to the 12 A inputs must be low or high to cause a low state at the output, Y. The '679 features an active low enable input, /G. When /G is low, the device is enabled. When high, it is disabled, with the output high regardless of the A and P inputs. The '680 features a transparent latch and latch enable C. With C high, the device mode is transparent. When low, the previous logic state of Y is latched.

TTL - 74LS681

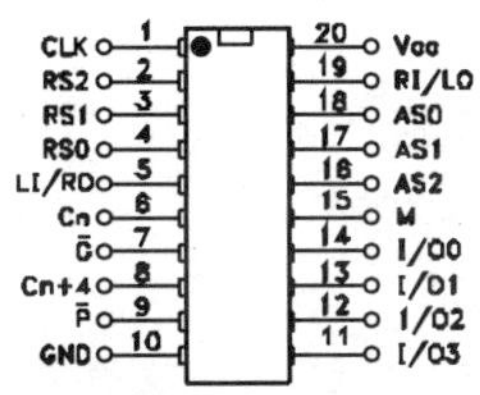

DUAL-IN-LINE PACKAGE

THIS DRAWING BASED ON TEXAS INSTRUMENTS 74LS681

FUNCTIONAL BLOCK DIAGRAM

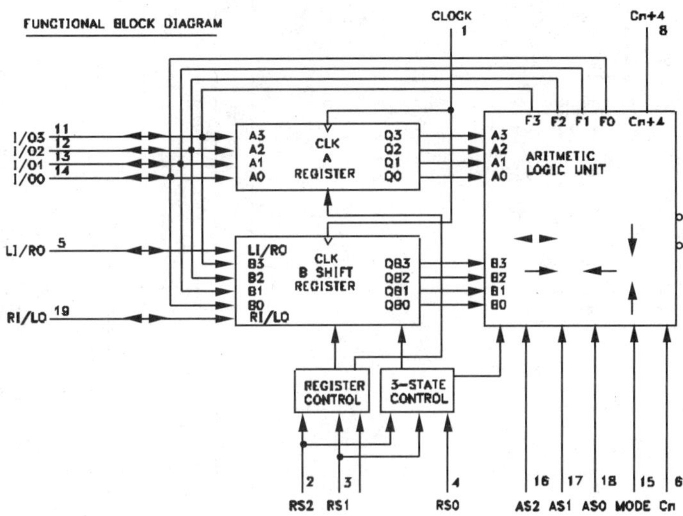

FUNCTION TABLES 74LS681 4-BIT PARALLEL BINARY ACCUMULATOR

TABLE 1. ARITHMETIC FUNCTIONS — Mode Control (M) = Low

ALU SELECTION			ACTIVE HIGH DATA	
AS2	AS1	AS0	Cn=H (With Carry)	Cn=L (No Carry)
0	0	0	Fi=L	Fi=H
0	0	1	F=B Minus A	F=B Minus A Minus 1
0	1	0	F=A Minus B	F=A Minus B Minus 1
0	1	1	F=A Plus B Plus 1	F=A Plus B
1	0	0	F=B Plus 1	Fi=Bi
1	0	1	F=$\bar{B}$ Plus 1	Fi=$\bar{B}$i
1	1	0	F=A Plus 1	Fi=Ai
1	1	1	F=$\bar{A}$ Plus 1	Fi=$\bar{A}$i

TABLE 2. LOGIC FUNCTIONS — Mode Control (M) = High

ALU SELECTION			ACTIVE HIGH DATA	
AS2	AS1	AS0	Cn=H (With Carry)	Cn=L (No Carry)
0	0	0	F0=H,F1=F2=F3=L	Fi=L
0	0	1	Fi=Ai⊕Bi+1	Fi=Ai⊕Bi
0	1	0	Fi=$\overline{Ai \oplus Bi}$+1	Fi=$\overline{Ai \oplus Bi}$
0	1	1	Fi=L	Fi=H
1	0	0	Fi=AiBi Plus 1	Fi=AiBi
1	0	1	Fi=$\overline{Ai+Bi}$ Plus 1	Fi=$\overline{Ai+Bi}$
1	1	0	Fi=$\overline{AiBi}$ Plus 1	Fi=$\overline{AiBi}$
1	1	1	Fi=Ai+Bi Plus 1	Fi=Ai+Bi

TABLE 3. REGISTER FUNCTIONS

	INPUTS BEFORE L-TO-H CLOCK TRANSITION									INTERNAL OUTPUTS AFTER LTO-H CLOCK TRANSITION													
	REGISTER SELECTION			DATA INPUTS						A REGISTER				B SHIFT REGISTER						ALU			
FUNCTION	RS2	RS1	RS0	LI/RO	I/O3	I/O2	I/O1	I/O0	RI/LO	QA3	QA2	QA1	QA0	LI/RO	QB3	QB2	QB1	QB0	RI/LO	F3	F2	F1	F0
ACCUM	L	L	L	Z	F3	F2	F1	F0	Z	QA30	QA20	QA10	QA00	Z	F3n	F2n	F1n	F0n	Z	F3	F2	F1	F0
LOAD B	L	H	H	Z	b3	b2	b1	b0	Z	QA30	QA20	QA10	QA00	Z	b3	b2	b1	b0	z	Z	Z	Z	Z
LEFT Shift LOGICAL	L	H	L	li	F3	F2	F1	F0	QB0	QA30	QA20	QA10	QA00	li	li	QB3n	QB2n	QB1n	QB1n	F3	F2	F1	F0
LEFT Shift ARITH	L	H	H	li	F3	F2	F1	F0	QB0	QA30	QA20	QA10	QA00	li	QB3n	li	QB2n	QB1n	QB1n	F3	F2	F1	F0
RIGHT Shift LOGICAL	H	L	L	QB3	F3	F2	F1	F0	ri	QA30	QA20	QA10	QA00	QB2n	QB2n	QB1n	QB0n	ri	ri	F3	F2	F1	F0
RIGHT Shift ARITH	H	L	H	QB2	F3	F2	F1	F0	ri	QA30	QA20	QA10	QA00	QB1n	QB3n	QB1n	QB0n	ri	ri	F3	F2	F1	F0
HOLD	H	H	L	Z	F3	F2	F1	F0	Z	QA30	QA20	QA10	QA00	Z	QB30	QB20	QB10	QB00	Z	F30	F20	F10	F00
LOAD A	H	H	H	Z	a3	a2	a1	a0	Z	a3	a2	a1	a0	Z	QB30	QB20	QB10	QB00	Z	Z	Z	Z	Z

1 = HIGH voltage level. 0 = LOW voltage level. STEADY STATE.
Z = HIGH Inmpedance (Output off)
a0...a3, b0...b3 = the level of steady state condition at I/O0 thru I/O3, respectively as A or B input data established.
F0...F3 = Internal ALU results
QA00...QB00, F00...F30 = the level of I/O0 thru I/O3 and F0-F3, respectively, before the indicated steady state input conditions were established.
QA0n...QB3n = the level of QA0 thru QB3 before the most recent ↑ transition of the clock.
ri, li = the level of steady-state conditions at RI/LO or LI/RO, respectively.

Figure 9.31. The TTL 74LS681 is a 4-Bit Parallel Binary Accumulator. The device integrates a high-speed arithmetic logic unit (ALU) with registers for words A and B. The ALU performs 16 arithmetic and 16 logic functions. Full carry look-ahead provides fast carry of the 4-bit words. The carry input, Cn, and propagate and generate outputs, /P, /G, enable direct use with 74S182 carry look-ahead generators for optimum performance with longer words. The A and B registers have three control inputs, RS0 - RS2, which define eight register modes The A register provides simple storage; the B is a combination storage/shift/accumulator register.

TTL – 74LS682
74LS684
74LS685

5V CMOS – 74HC682
74HC684

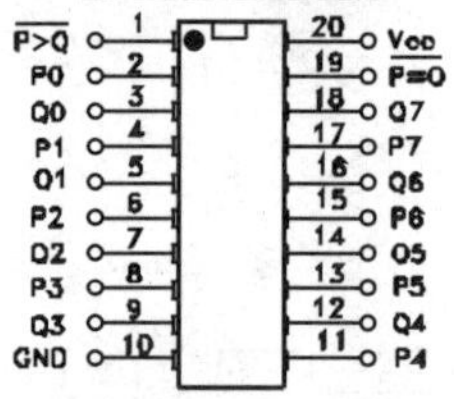

DUAL-IN-LINE PACKAGE

FUNCTION TABLE 74x682, 'x684, 'LS685
8-BIT MAGNITUDE/IDENTITY COMPARATORS

INPUTS DATA P,Q	OUTPUTS $\overline{P=Q}$	$\overline{P>Q}$
P=Q	0	1
P>Q	1	0
P<Q	1	1
P=Q	1	1
P>Q	1	1
X	1	1

1 = HIGH voltage level.
0 = LOW voltage level.
X = Don't Care

LOGIC DIAGRAM
Based on Texas Instruments 74LS682, 'LS684, 'LS685

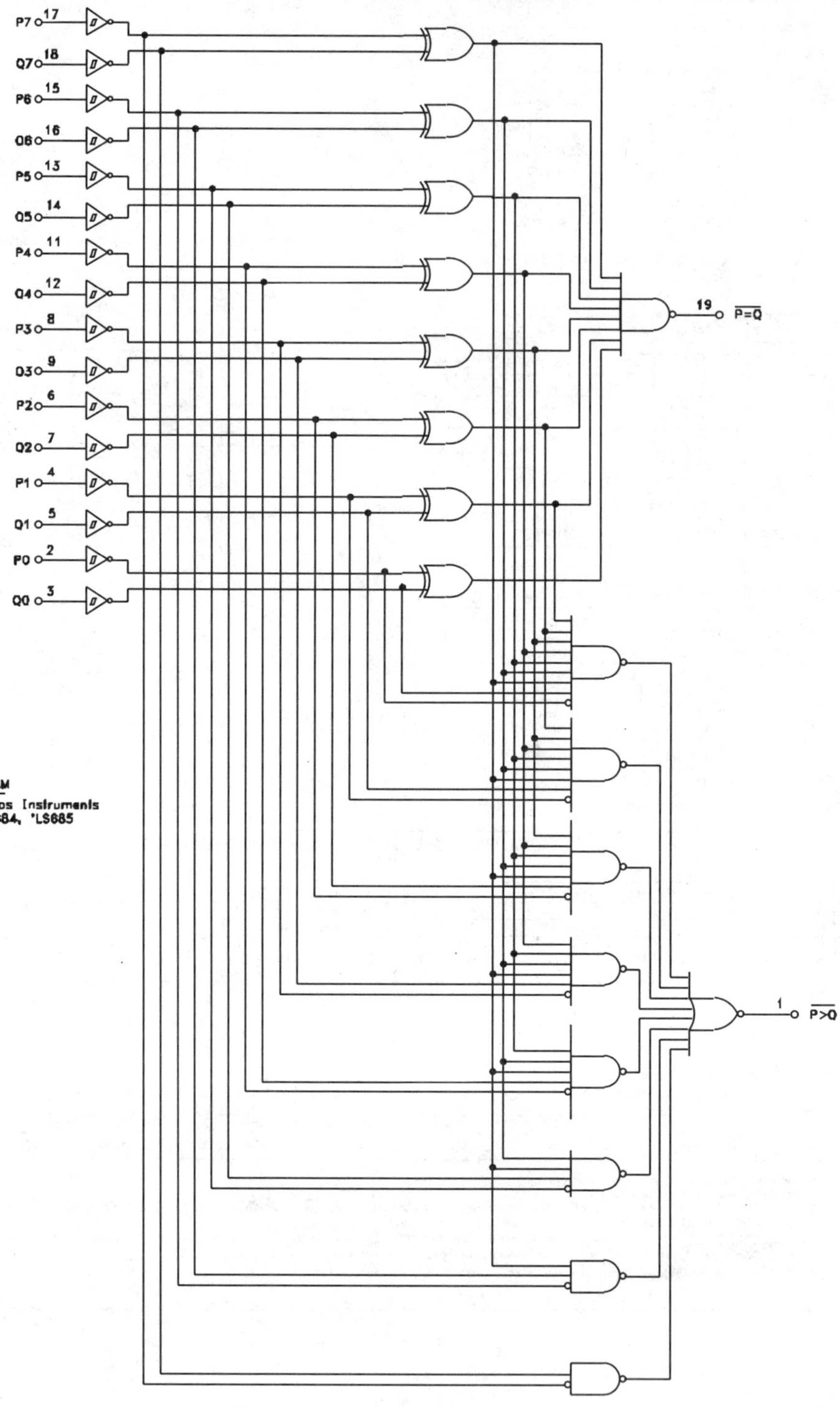

Figure 9.32. The TTL 74LS682, 'LS684, LS685 and CMOS counterparts 74HC682, and '684 are 8-Bit Magnitude/Identity Comparators. These perform comparisons of two 8-bit binary or BCD words. Two outputs are provided: /P = Q and /P > Q. The 'LS682 and 'LS684 feature totem pole outputs; the 'LS685 open collector. Also, the 'LS682 provides 20K pull-ups termination resistors on the Q inputs for analog or switch data.

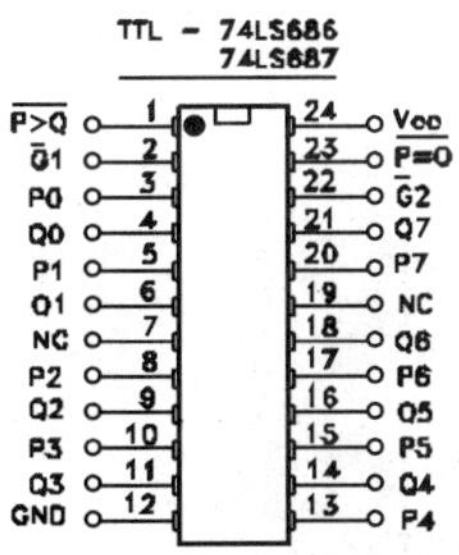

DUAL-IN-LINE PACKAGE

FUNCTION TABLE 74LS686, 'LS687
8-BIT MAGNITUDE/IDENTITY COMPARATORS

INPUTS			OUTPUTS	
DATA	ENABLES			
P,Q	G1	G2	$\overline{P=Q}$	$\overline{P>Q}$
P=Q	0	X	0	1
P>Q	X	0	1	0
P<Q	X	X	1	1
P=Q	1	X	1	1
P>Q	X	1	1	1
X	1	1	1	1

1 = HIGH voltage level.
0 = LOW voltage level.
X = Don't Care

LOGIC DIAGRAM
Based on Texas Instruments
74LS686, '687

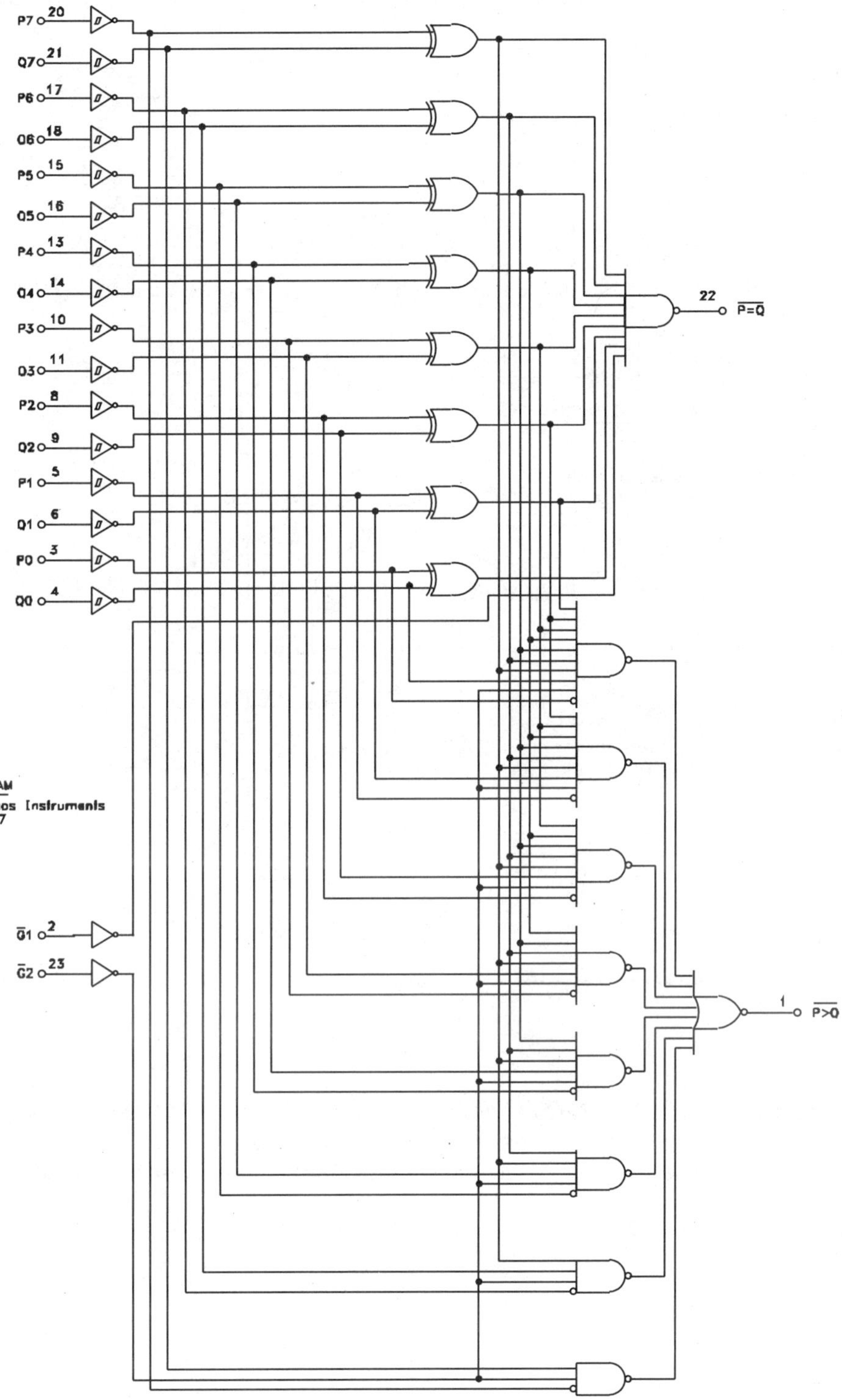

Figure 9.33. The TTL 74LS686 and 'LS687 are 8-Bit Magnitude/Identity Comparators. These perform comparisons of two 8-bit binary or BCD words. Two outputs are provided: /P = Q and /P > Q. The 'LS686 features totem pole outputs; the 'LS687 open collector.

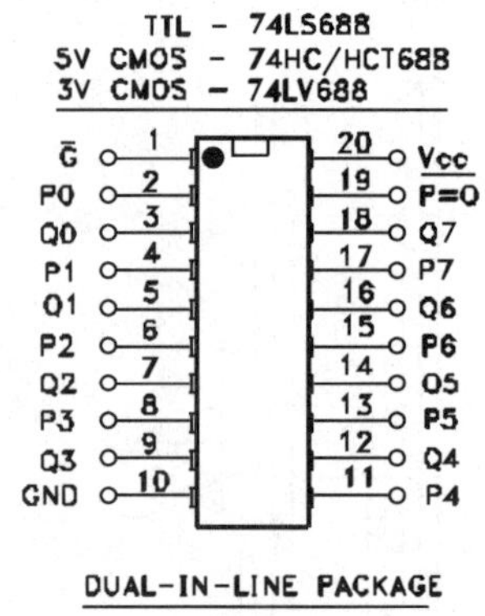

FUNCTION TABLE 74x688
8-BIT MAGNITUDE/IDENTITY COMPARATORS

INPUTS DATA P,Q	ENABLE G	OUTPUTS $\overline{P=Q}$	$\overline{P>Q}$
P=Q	0	0	1
P>Q	X	1	0
P<Q	X	1	1
P=Q	1	1	1
P>Q	X1	1	1
X		1	1

1 = HIGH voltage level.
0 = LOW voltage level.
X = Don't Care

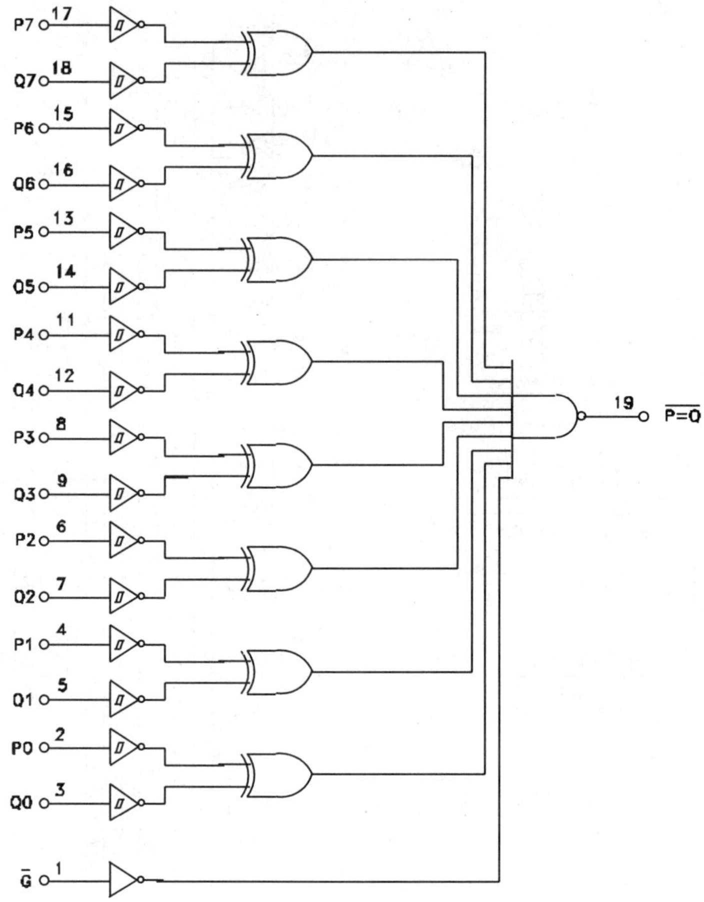

LOGIC DIAGRAM
Based on Texas Instruments 74LS688

Figure 9.34. The TTL 74LS/ALS688 and its CMOS counterparts are 8-Bit Magnitude/Identity Comparators. These perform comparisons of two 8-bit binary or BCD words. A single output is provided: /P = Q. The 'LS688 features totem pole outputs.

TTL – 74ALS689

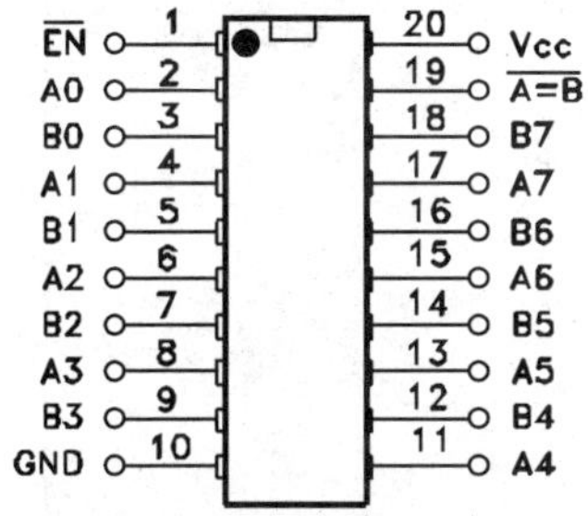

DUAL-IN-LINE PACKAGE

FUNCTION TABLE 74ALS689
8-BIT COMPARATOR

INPUTS		OUTPUT
EN	DATA	A=B
0	A=B	0
0	A≠B	1
1	X	1

1 = HIGH voltage level.
0 = LOW voltage level.
X = Don't Care

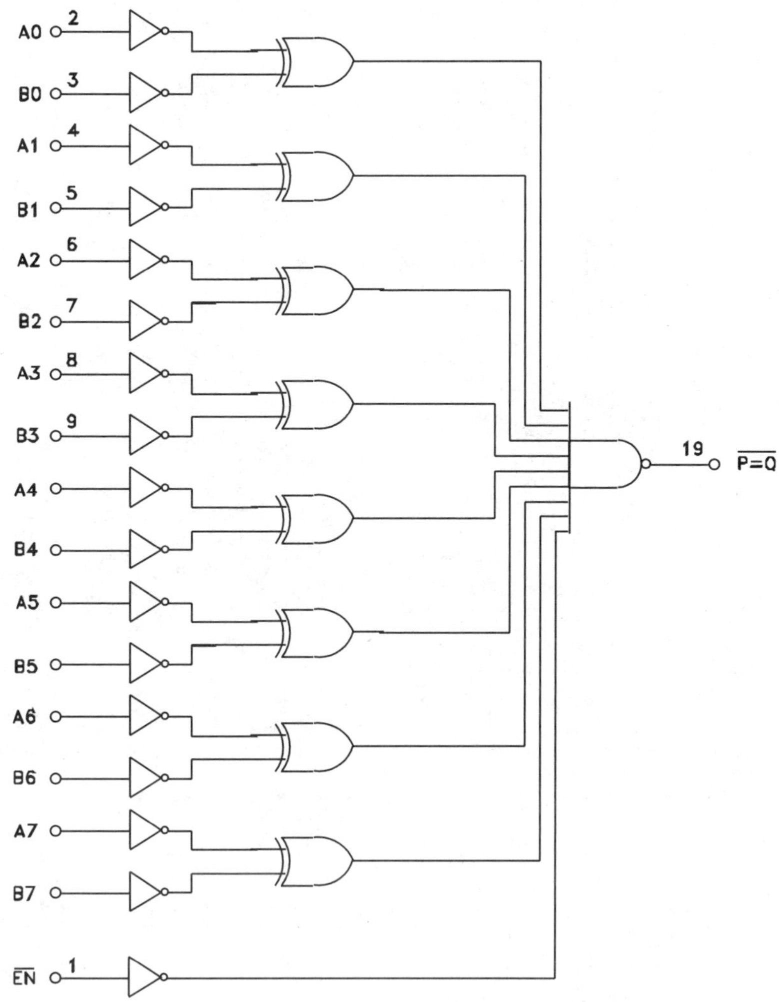

LOGIC DIAGRAM
Based on National/Fairchild 74ALS689

Figure 9.35. The TTL 74ALS689 is an 8-Bit Magnitude/Identity Comparator. It performs comparisons of two 8-bit binary or BCD words. A single output is provided: /P = Q. The 'LS689 features open collector outputs.

TTL – 74AS881

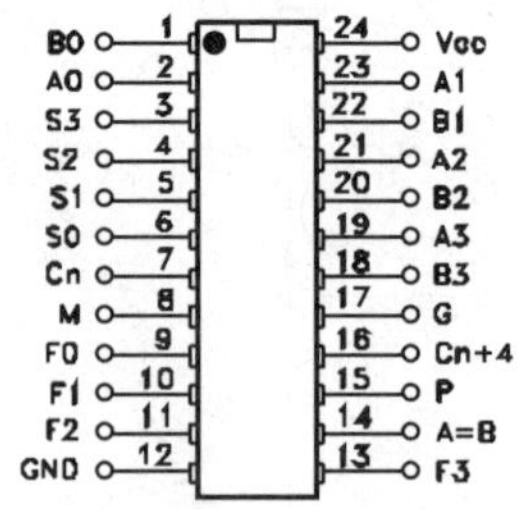

DUAL-IN-LINE PACKAGE

PACKAGE PINNING:

PIN-	
19,21,23,2	WORD A INPUTS
18,20,22,1	WORD B INPUTS
3,4,5,6	FUNCTION SELECT INPUTS
7	INV. CARRY INPUT
8	MODE CONTROL INPUT
13,11,10,9	FUNCTION OUTPUTS
14	COMPARATOR OUTPUT
15	CARRY PROPAGATE OUTPUT
16	INV. CARRY OUTPUT
17	CARRY GENERATE OUTPUT

FUNCTION TABLES 74AS881 4-BT ARITHMETIC LOGIC UNIT/FUNCTION GENERATOR

INPUT Cn	OUTPUT Cn+4	ACTIVE-LOW DATA Table 1	ACTIVE-HIGH DATA Table 2
1	1	A≥B	A≤B
1	0	A≤B	A≥B
0	1	A≥B	A≤B
0	0	A≤B	A≥B

TABLE 1. ACTIVE-LOW DATA

SELECTION				M=H LOGIC	M=L; ARITHMETIC OPERATIONS	
S3	S2	S1	S0	FUNCTIONS	Cn=L (No Carry)	Cn=H (With Carry)
0	0	0	0	F= $\overline{A}$	F= A Minus 1	F= A
0	0	0	1	F= $\overline{AB}$	F= AB Minus 1	F= AB
0	0	1	0	F= $\overline{A}$+B	F= A$\overline{B}$ Minus 1	F= A$\overline{B}$
0	0	1	1	F= 1	F= Minus 1 (2's Comp)	F= Zero
0	1	0	0	F= $\overline{A+B}$	F= A Plus(A+$\overline{B}$)	F= A Plus(A+$\overline{B}$) Plus 1
0	1	0	1	F= B	F= AB Plus(A+$\overline{B}$)	F= AB Plus(A+$\overline{B}$) Plus 1
0	1	1	0	F= $\overline{A \oplus B}$	F= A Minus B Minus 1	F= A Minus B
0	1	1	1	F= A+B	F= A+$\overline{B}$	F= (A+$\overline{B}$) Plus 1
1	0	0	0	F= $\overline{A}$B	F= A Plus(A+B)	F= A Plus(A+B) Plus 1
1	0	0	1	F= A⊕B	F= A Plus B	F= A Plus B Plus 1
1	0	1	0	F= B	F= $\overline{A}$B Plus(A+B)	F= A$\overline{B}$ Plus(A+B) Plus 1
1	0	1	1	F= A+B	F= A+B	F= (A+B) Plus 1
1	1	0	0	F= 0	F= A+A*	F= A Plus A Plus 1
1	1	0	1	F= A$\overline{B}$	F= AB Plus A	F= AB Plus A Plus 1
1	1	1	0	F= AB	F= A$\overline{B}$ PLus A	F= A$\overline{B}$ Plus A Plus 1
1	1	1	1	F= A	F= A	F= A Plus 1

TABLE 2. ACTIVE-HIGH DATA

SELECTION				M=H LOGIC	M=L; ARITHMETIC OPERATIONS	
S3	S2	S1	S0	FUNCTIONS	Cn=H (No Carry)	Cn=L (With Carry)
0	0	0	0	F= $\overline{A}$	F= A	F= A Plus 1
0	0	0	1	F= $\overline{A+B}$	F= A+B	F= (A+B) Plus 1
0	0	1	0	F= $\overline{A}$B	F= A+B	F= (A+$\overline{B}$) Plus 1
0	0	1	1	F= 0	F= Minus 1 (2's Comp)	F= Zero
0	1	0	0	F= $\overline{AB}$	F= A Plus AB	F= A Plus A$\overline{B}$ Plus 1
0	1	0	1	F= $\overline{B}$	F= (A+$\overline{B}$) Plus A$\overline{B}$	F= (A+B) Plus A$\overline{B}$ Plus 1
0	1	1	0	F= A⊕B	F= A Minus B Minus 1	F= A Minus B
0	1	1	1	F= A$\overline{B}$	F= A$\overline{B}$ Minus 1	F= A$\overline{B}$
1	0	0	0	F= $\overline{A}$+B	F= A Plus AB	F= A Plus AB Plus 1
1	0	0	1	F= $\overline{A \oplus B}$	F= A Plus B	F= A Plus B Plus 1
1	0	1	0	F= B	F= (A+$\overline{B}$) Plus AB	F= (A+B) Plus AB Plus 1
1	0	1	1	F= AB	F= AB Minus 1	F= AB
1	1	0	0	F= 1	F= A Plus A*	F= A Plus A Plus 1
1	1	0	1	F= A+$\overline{B}$	F= (A+B) Plus A	F= (A+B) Plus A Plus 1
1	1	1	0	F= A+B	F= (A+$\overline{B}$) Plus A	F= (A+$\overline{B}$) Plus A Plus 1
1	1	1	1	F= A	F= A Minus 1	F= A

* Each bit is shifted to the next more significant position.

The 74AS881 will accomodate active-high or active-low data if the pin designations are interpreted as below.

Pin Number	2	1	23	22	21	20	19	18	9	10	11	13	7	16	15	17
Active-Low Data (Table 1)	$\overline{A0}$	$\overline{B0}$	$\overline{A1}$	$\overline{B1}$	$\overline{A2}$	$\overline{B2}$	$\overline{A3}$	$\overline{B3}$	$\overline{F0}$	$\overline{F1}$	$\overline{F2}$	$\overline{F3}$	Cn	Cn+4	$\overline{P}$	$\overline{G}$
Active-High Data (Table 2)	A0	B0	A1	B1	A2	B2	A3	B3	F0	F1	F2	F3	Cn	Cn+4	P	G

NOTE: This drawing is based on the National/Fairchild 74AS881

Subtraction is accomplished by 1's complement addition where the 1's complement of the subtrahend is generated internally. The resultant ouput is A–B–1, which requires an end-around or forced carry to provide A–B.

The 74AS881 can also be utilized as a comparator. The A=B output is internally decoded from the function outputs (F0,F1,F2,F3) so that when two words of equal magnitude are applied at the A and B inputs, it will assume a high level to indicate equality (A=B). The ALU must be in the subtract mode with Cn=H when performing this comparison.

The A=B output is open collector so that it can be wire-ANDed to give a comparison for more than 4 bits.

The carry output (Cn+4) can also be used to supply relative magnitude information. Again the ALU must be in the subract mode by placing the function select inputs S3,S2,S1,S0 at o,1,1,0 respectively.

1 = HIGH voltage level. 0 = LOW voltage level. X = Don't Care

Figure 9.36. The TTL 74AS881 is an Arithmetic Logic Unit (ALU)/Function Generator with full look-ahead for high-speed operation on long words. Arithmetic operating modes are addition, subtraction, shift operand A one position, magnitude comparison, plus 12 other operations. Logic functions are exclusive-OR, comparator, AND, NAND, OR, NOR, register checks, plus ten other functions.

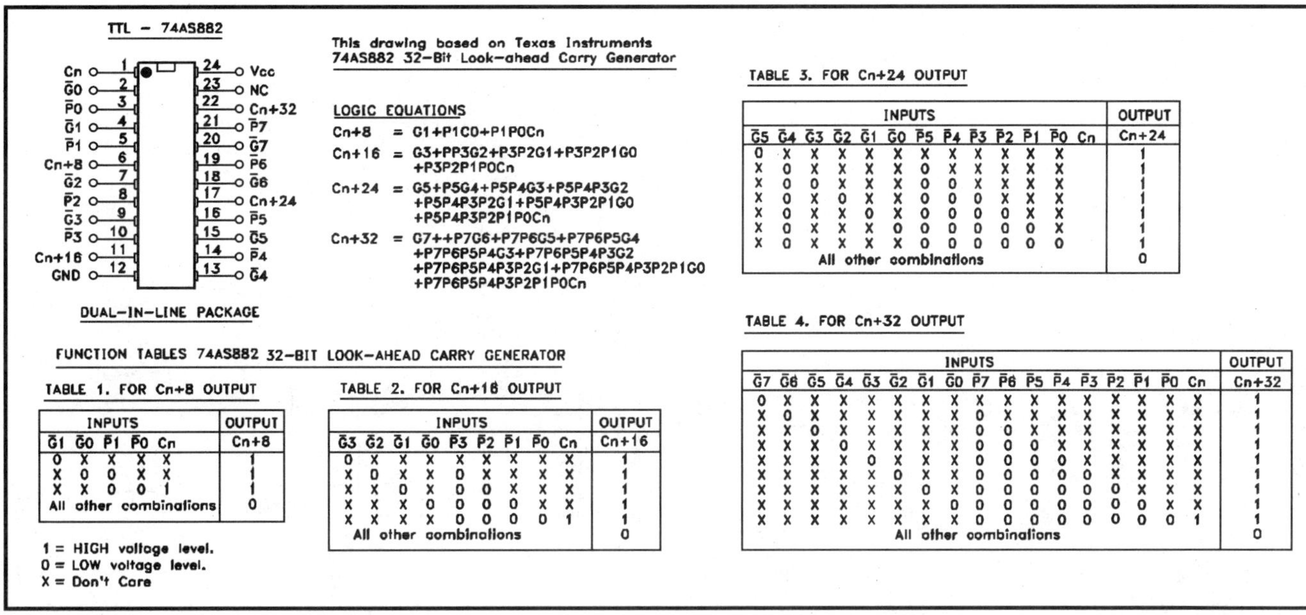

Figure 9.37. The TTL 74AS882 is a 32-Bit Look-Ahead Carry Generator capable of anticipating the carry across a group of eight 4-bit adders, permitting the designer to implement look-ahead for a 32-bit ALU. The device is directly compatible with the 'AS181 and the 'AS881 ALU's.

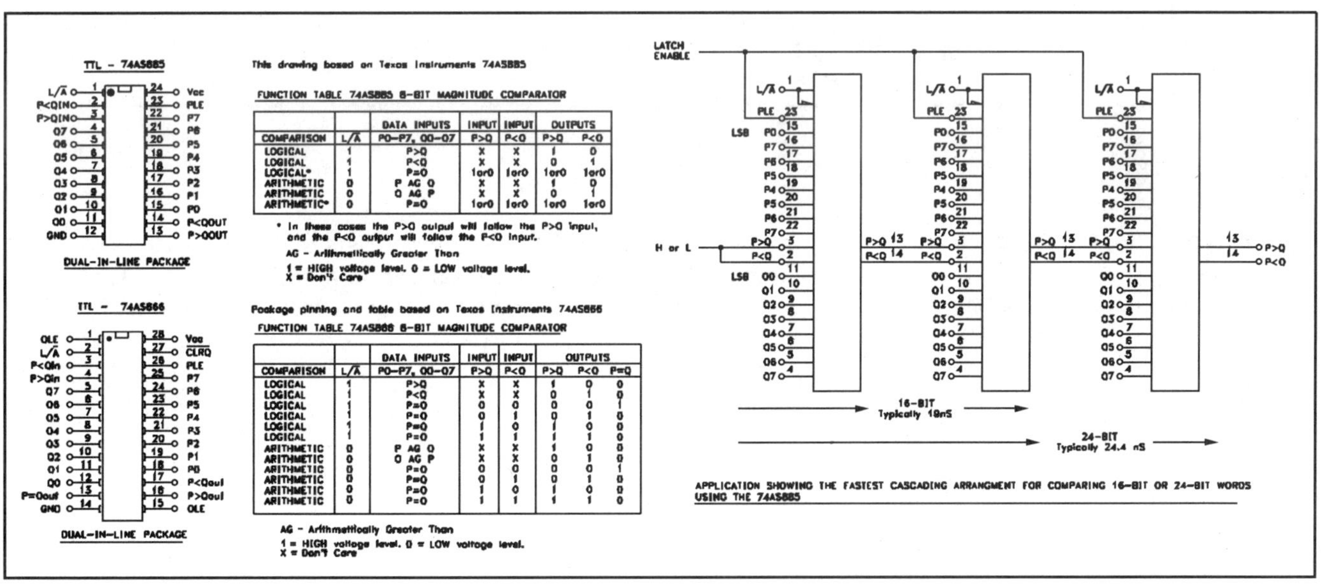

Figure 9.38. The TTL 74AS866 and 'AS885 are 8-Bit Magnitude Comparators capable of high-speed arithmetic or magnitude comparisons on two 8-bit binary or 2's complement words. Three fully decoded words are available at the outputs of the 'AS866; two at the 'AS885. In other respects the two devices carry out the same functions.

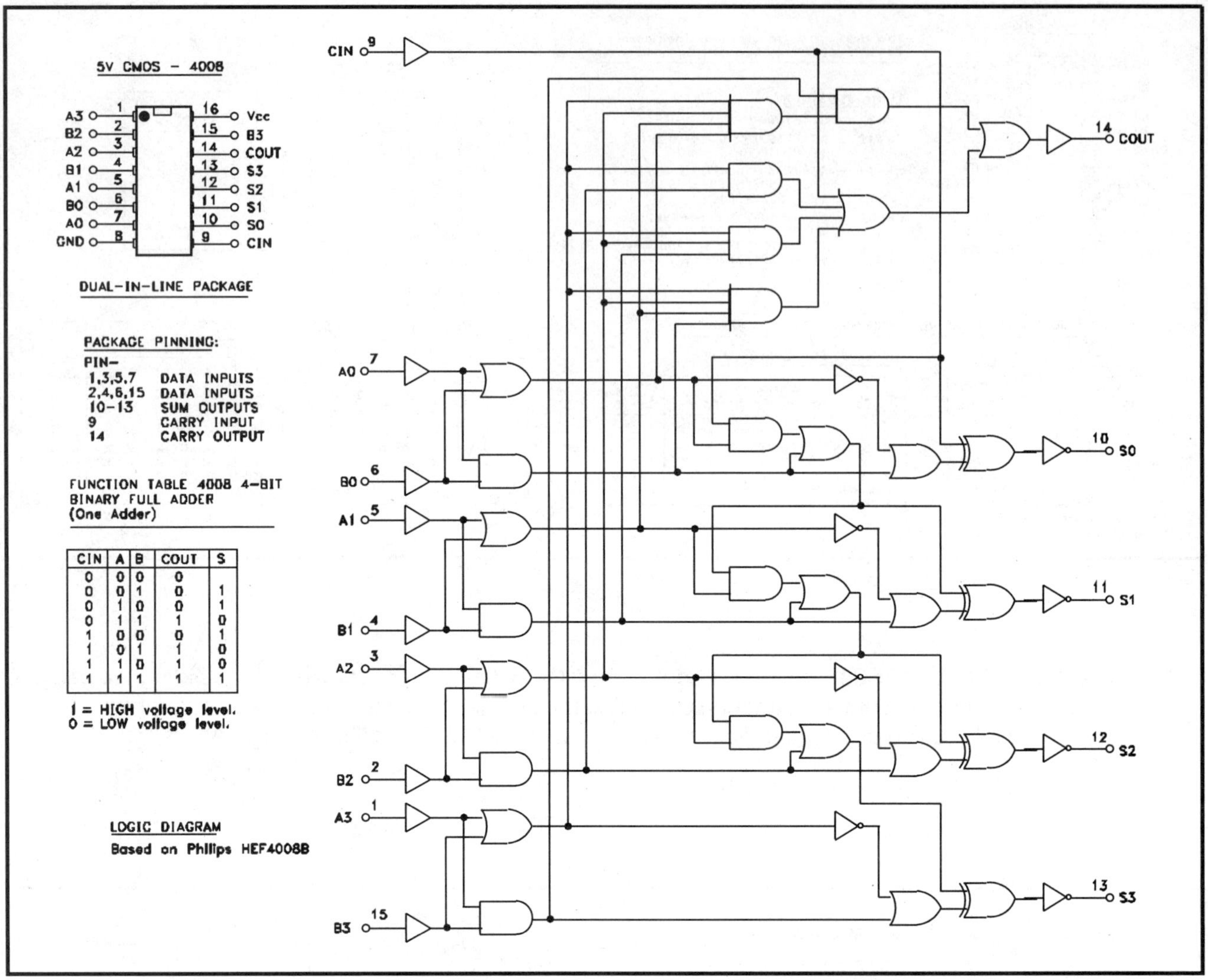

Figure 9.39. The CMOS 4000 series 4008 is a 4-Bit Full Adder. The device consists of four full adders with fast internal look-ahead carry output. It is useful in binary addition and other arithmetic operations. The fast carry output bit allows high-speed operation when used with other adders in a system.

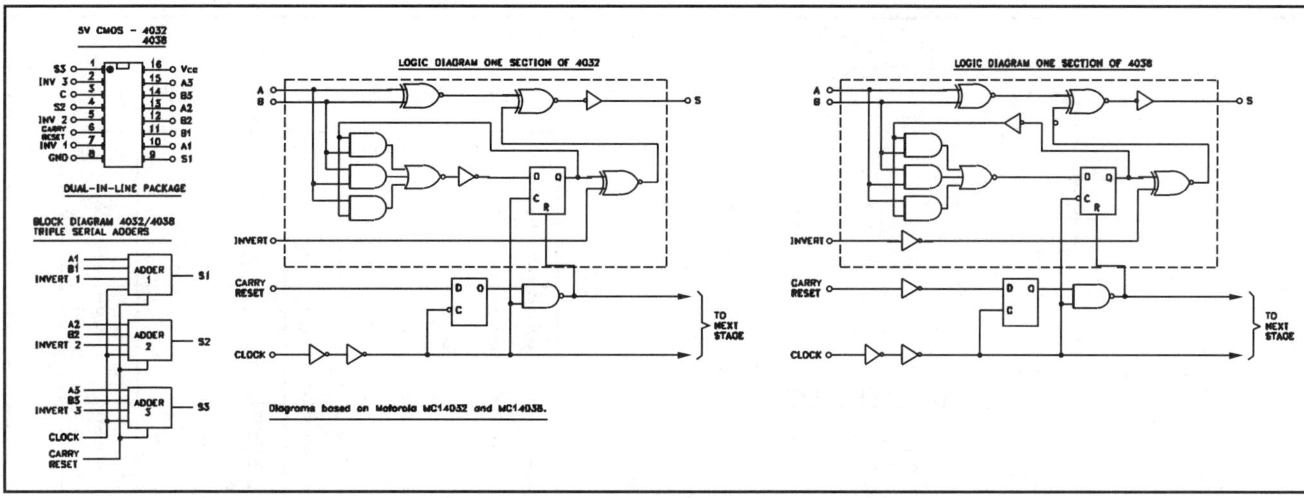

Figure 9.40. The CMOS 4000 series 4032 and 4038 are Triple Serial Adders having the clock and carry reset inputs common to all three adders. The carry is added on the positive-going clock transition for the 4032; on the negative-going for the 4038. In other respects the two devices are identical. Typical applications include serial arithmetic units, digital correlators, digital servo control systems, datalink computers, and flight control computers.

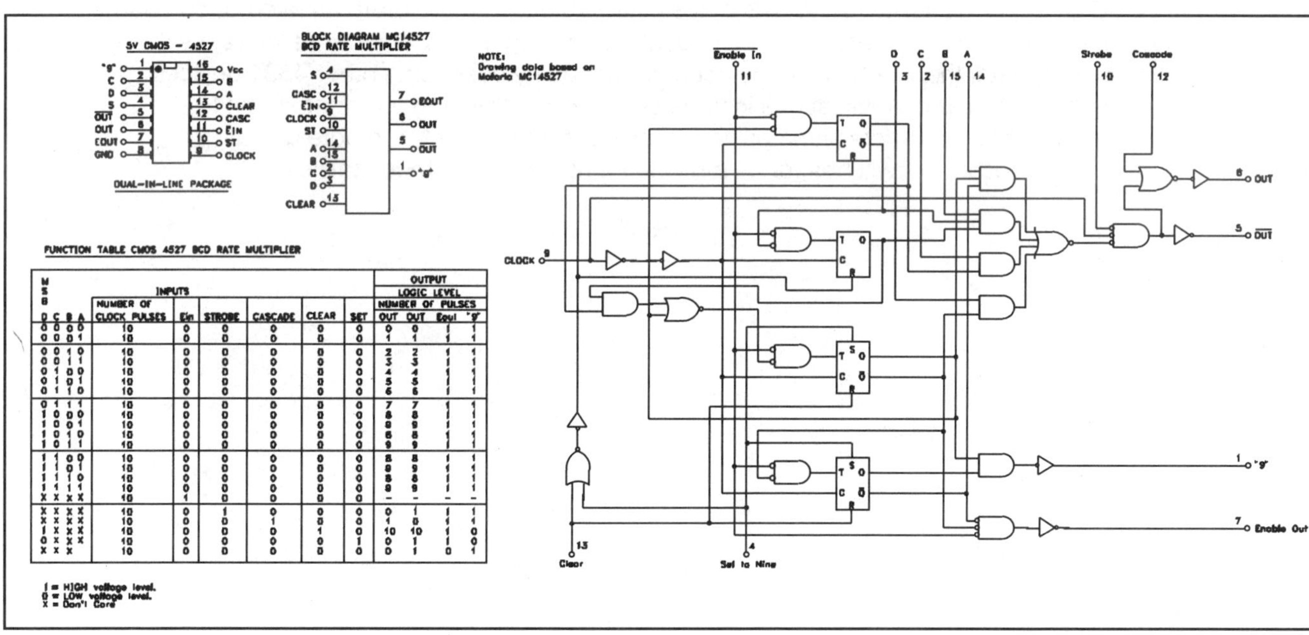

MSB D C B A	NUMBER OF CLOCK PULSES	Ein	STROBE	CASCADE	CLEAR	SET	OUT	OUT	Eout	"9"
0 0 0 0	10	0	0	0	0	0	0	0	1	1
0 0 0 1	10	0	0	0	0	0	1	1	1	1
0 0 1 0	10	0	0	0	0	0	2	2	1	1
0 0 1 1	10	0	0	0	0	0	3	3	1	1
0 1 0 0	10	0	0	0	0	0	4	4	1	1
0 1 0 1	10	0	0	0	0	0	5	5	1	1
0 1 1 0	10	0	0	0	0	0	6	6	1	1
0 1 1 1	10	0	0	0	0	0	7	7	1	1
1 0 0 0	10	0	0	0	0	0	8	8	1	1
1 0 0 1	10	0	0	0	0	0	9	9	1	1
1 0 1 0	10	0	0	0	0	0	8	8	1	1
1 0 1 1	10	0	0	0	0	0	9	9	1	1
1 1 0 0	10	0	0	0	0	0	8	8	1	1
1 1 0 1	10	0	0	0	0	0	9	9	1	1
1 1 1 0	10	0	0	0	0	0	8	8	1	1
1 1 1 1	10	0	0	0	0	0	9	9	1	1
X X X X	10	1	0	0	0	0	–	–	–	–
X X X X	10	0	1	0	0	0	0	1	1	1
X X X X	10	0	0	1	0	0	1	0	1	1
1 X X X	10	0	0	0	1	0	10	10	1	0
0 X X X	10	0	0	0	0	1	0	1	1	0
X X X	10	0	0	0	0	0	0	1	0	1

1 = HIGH voltage level.
0 = LOW voltage level.
X = Don't Care

Figure 9.41. The CMOS 4000 series 4527 is a BCD Rate Multiplier providing an output pulse rate based upon the BCD input number. To illustrate, if six is the input number, the device will output six pulses for every ten input pulses. The device has application to arithmetic operations of multiplication and division. Typical uses include digital filters, motor speed control, and frequency synthesizers.

5V CMOS 4531

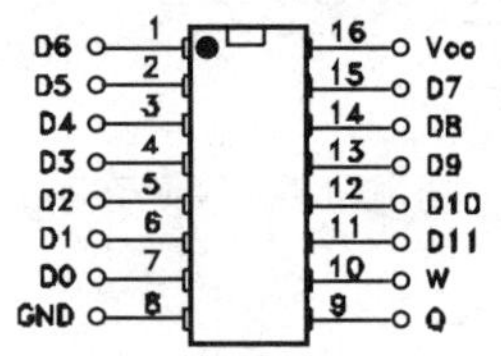

DUAL-IN-LINE PACKAGE

NOTE:
Drawing data based on Motorola MC14531

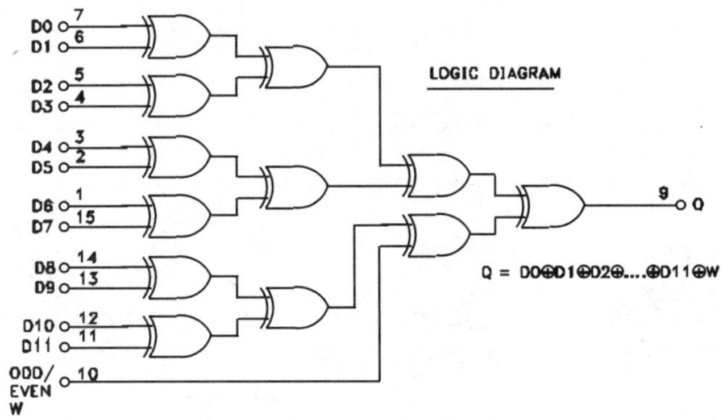

FUNCTION TABLE CMOS 4531 12-BIT PARITY TREE

INPUTS								OUTPUT
W	D11	D10	...	D2	D1	D0	DECIMAL (OCTAL) EQUIVALENT	Q*
0	0	0	...	0	0	0	0 (0)	0
0	0	0	...	0	0	1	1 (1)	1
0	0	0	...	0	1	0	2 (2)	1
0	0	0	...	0	1	1	3 (3)	0
0	0	0	...	1	0	0	4 (4)	1
0	0	0	...	1	0	1	5 (5)	0
0	0	0	...	1	1	0	6 (6)	0
0	0	0	...	1	1	1	7 (7)	1
.	.	.	...	.	.	.	. .	.
.	.	.	...	.	.	.	. .	.
.	.	.	...	.	.	.	. .	.
1	1	1	...	0	0	0	8184 (17770)	0
1	1	1	...	0	0	1	8185 (17771)	1
1	1	1	...	0	1	0	8186 (17772)	1
1	1	1	...	0	1	1	8187 (17773)	0
1	1	1	...	1	0	0	8188 (17774)	1
1	1	1	...	1	0	1	8189 (17775)	0
1	1	1	...	1	1	0	8190 (17776)	0
1	1	1	...	1	1	1	8191 (17777)	1

* 0 = EVEN PARITY
1 = ODD PARITY

NOTE: May redefine to suit applications by manipulating W and/or other available D's.

1 = HIGH voltage level.
0 = LOW voltage level.

Figure 9.42. The CMOS 4000 series 4531 is a 12-Bit Parity Tree. The circuit consists of 12 data-bit inputs, (D0 - D11), and an even/odd parity select input, (W) and an output,(Q). The parity selection bit can be considered an additional parity bit. Words of less than 13 bits can generate an even or odd parity output if the remaining inputs are selected to contain an even or odd number of 1s, respectively. Words of greater than 12 bits can be accommodated by cascading other 4531 devices using the W input. Applications include checking or including a redundant bit to a word for error detection/correction systems, controller for remote digital sensors or switches, or as multiple input summer without carries.

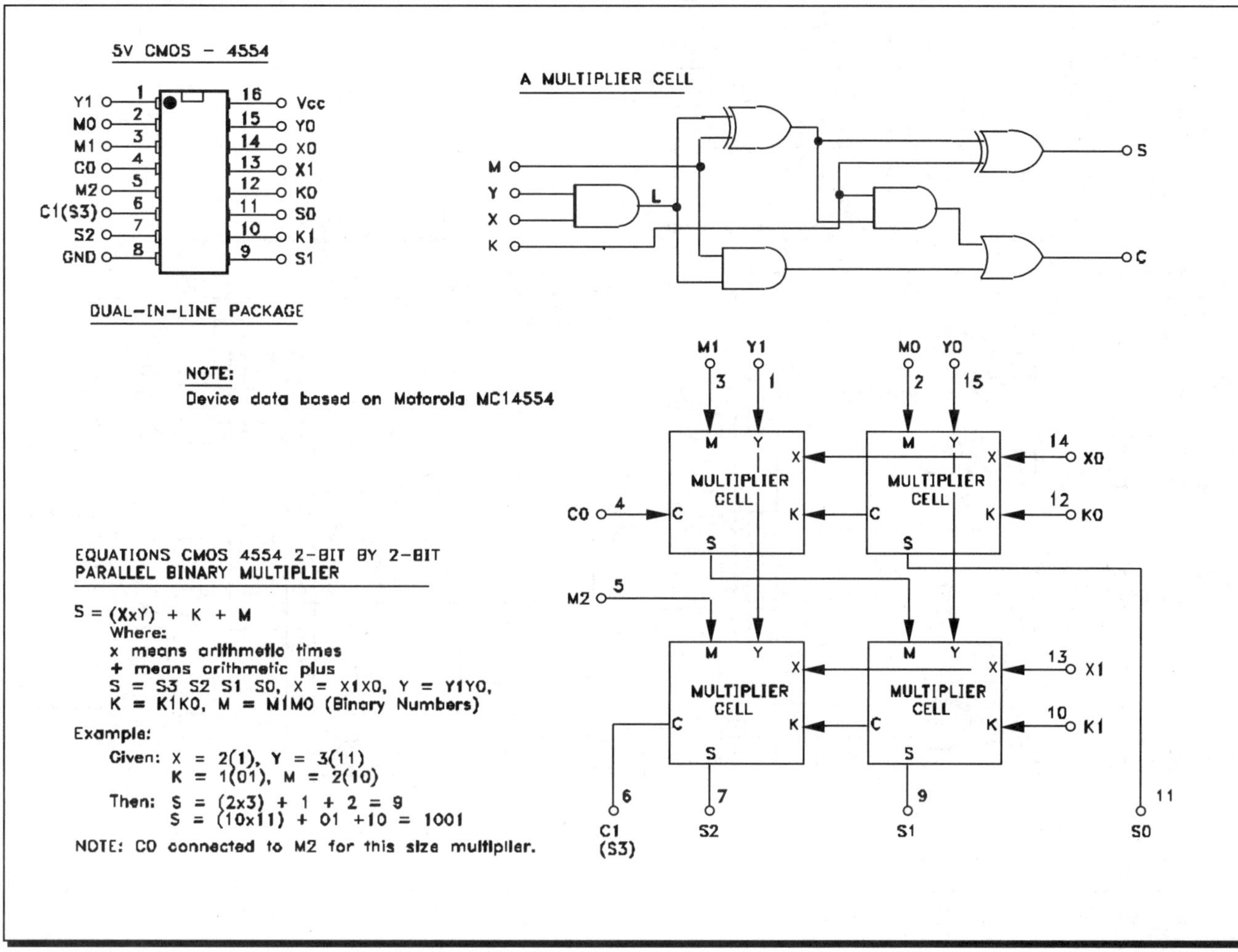

Figure 9.43. The CMOS 4000 series 4554 is a 2-Bit by 2-Bit Parallel Binary Multiplier. The device can perform the multiplication of two binary numbers and simultaneously add two other binary numbers to the product. There are two multiplicand inputs, X0, X1, two multiplier inputs, Y0, Y1, five cascading or adding inputs, K0, K1, M0 - M2, and five sum and carry outputs, S0, S1, S2, C1 (S3), and C0. The basic multiplier can be expanded into a straightforward n-bit parallel multiplier without additional logic elements. Applications include arithmetic processing and process and machine controls.

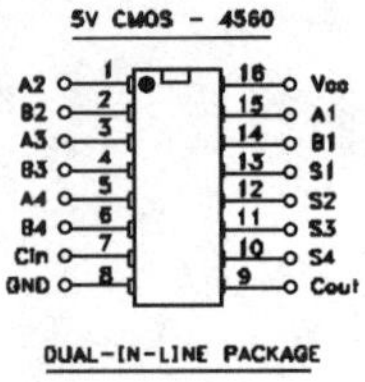

FUNCTION TABLE CMOS 4560 NATURAL BINARY CODE (NBCD) ADDER

INPUT										OUTPUT			
A4	A3	A2	A1	B4	B3	B2	B1	CIN	COUT	S4	S3	S2	S1
0	0	0	0	0	0	0	0	0	0	0	0	0	0
0	0	0	0	0	0	0	0	1	0	0	0	0	1
0	1	0	0	0	0	1	1	0	0	0	1	1	1
0	1	0	0	0	0	1	1	1	0	1	0	0	0
0	1	1	1	0	1	0	0	0	1	0	0	0	1
0	1	1	1	0	1	0	0	1	1	0	0	1	0
1	0	0	0	0	1	0	1	0	1	0	0	1	1
0	1	1	0	1	0	0	0	0	1	0	1	0	0
1	0	0	1	1	0	0	1	1	1	1	0	0	1

This is a partial truth table to show logic operation for representative input values.

NOTE:
Device data shown based on Motorola MC14560

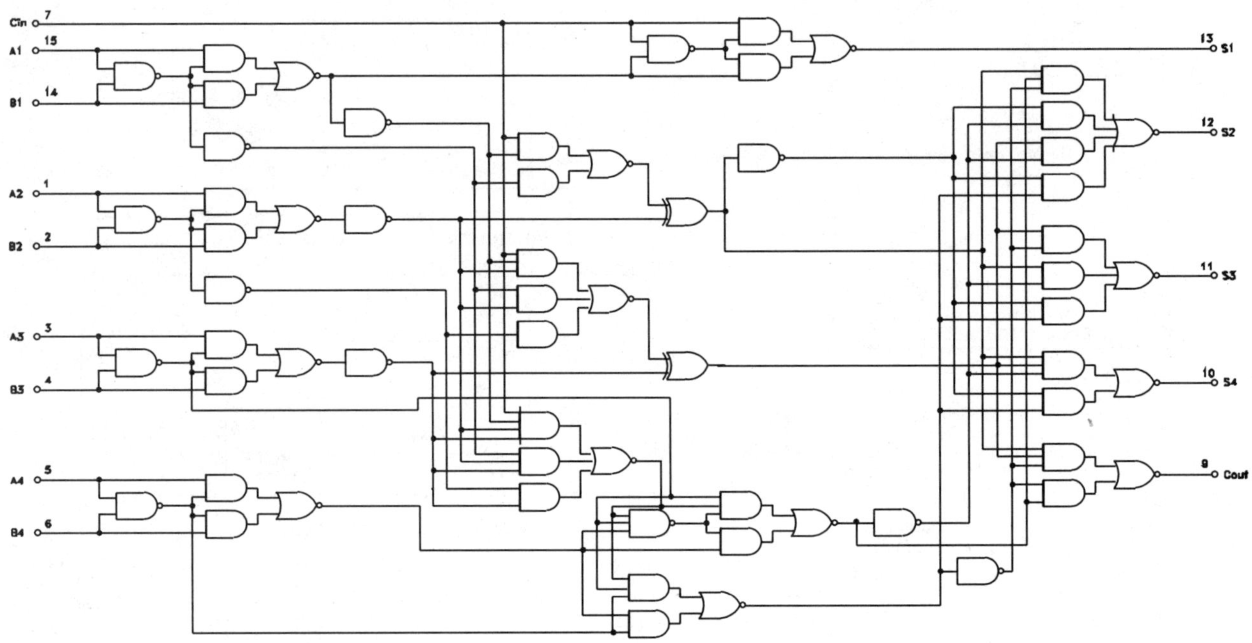

Figure 9.44. The CMOS 4000 series 4560 is a NBCD (Natural Binary Coded Decimal) Adder. The device adds two 4-bit numbers in NBCD format, resulting in sum and carry outputs, also in NBCD code. The device can subtract when one set of inputs is complemented with a 9's complementer (the 4561). All inputs and outputs are active high. The carry input for the least significant bit is tied to ground for no carry in.

5V CMOS - 4561

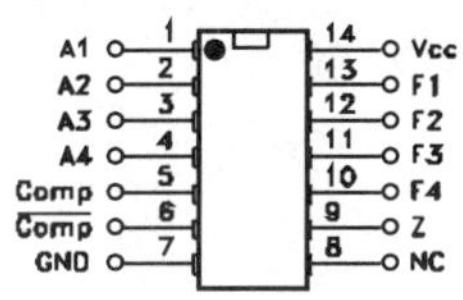

DUAL-IN-LINE PACKAGE

FUNCTION TABLE CMOS 4561 9's COMPLEMENTER

Z	Comp	$\overline{\text{Comp}}$	F1	F2	F3	F4	MODE
0	0	0	A1	A2	A3	A4	Straight Thru
0 0 0	0 1 1	1 1 0	A1	A2	A2A3 A2A3	A2A3A4	Complement
1	X	X	0	0	0	0	Zero

1 = HIGH voltage level.
0 = LOW voltage level.
X = Don't Care

NOTE:
Device details shown based on Motorola MC14561

TRUTH TABLE - COMPLEMENT MODE

	Decimal Equivalent Input	INPUTS A4	A3	A2	A1	Decimal Equivalent Output	OUTPUTS F4	F3	F2	F1
	0	0	0	0	0	9	1	0	0	1
	1	0	0	0	1	8	1	0	0	0
	2	0	0	1	0	7	0	1	1	1
	3	0	0	1	1	6	0	1	1	0
	4	0	1	0	0	5	0	1	0	1
	5	0	1	0	1	4	0	1	0	0
	6	0	1	1	0	3	0	0	1	1
	7	0	1	1	1	2	0	0	1	0
	8	1	0	0	0	1	0	0	0	1
	9	1	0	0	1	0	0	0	0	0
Illegal BCD Input Codes	10	1	0	1	0	7	0	1	1	1
	11	1	0	1	1	6	0	1	1	0
	12	1	1	0	0	5	0	1	0	1
	13	1	1	0	1	4	0	1	0	0
	14	1	1	1	0	3	0	0	1	1
	15	1	1	1	1	2	0	0	1	0

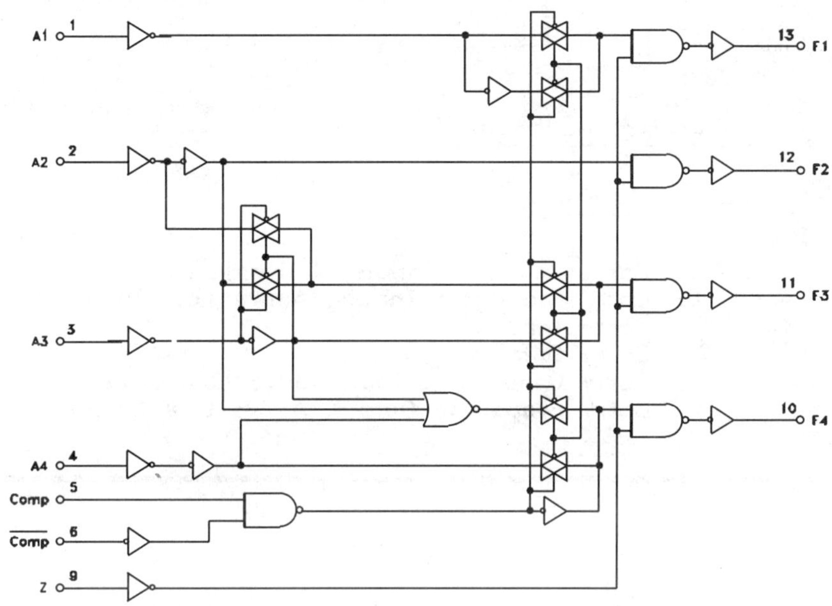

Figure 9.45. The CMOS 4000 series 4561 is a 9's Complementer. It functions as a companion to the 4560 NBCD Adder to allow for subtraction. A BCD number code (8-4-2-1) is applied to the inputs. If the complement control is low, the BCD number appears at the outputs unmodified. The complement disable, /Comp, allows the complement control to be gated, or an inverted control signal to be used. If the complement input is high and the disable input low, the 9's complement of the number is displayed at the outputs. The zero control, Z, when high, forces outputs low, regardless of the state of the other inputs. When used to perform subtraction in conjunction with the 4560 NBCD Adder, the complement control becomes another add/subtract control.

5V CMOS – 4582

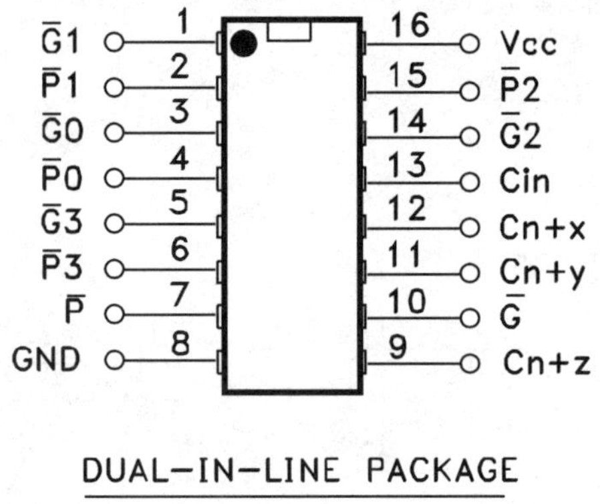

LOGIC EQUATIONS

$$Cn{+}x = \bar{G}0+(\bar{P}0\cdot Cn)$$
$$Cn{+}y = \bar{G}1+(\bar{P}1\cdot \bar{G}0)+(\bar{P}1\cdot \bar{P}0\cdot Cn)$$
$$Cn{+}z = \bar{G}2+(\bar{P}2\cdot \bar{G}1)+(\bar{P}2\cdot \bar{P}1\cdot \bar{G}0)+(\bar{P}2\cdot \bar{P}1\cdot \bar{P}0\cdot Cn)$$
$$\bar{G} = \bar{G}3+(\bar{P}3\cdot \bar{G}2)+(\bar{P}3\cdot \bar{P}2\cdot \bar{G}1)+(\bar{P}1\cdot \bar{P}2\cdot \bar{P}3\cdot \bar{G}0)$$
$$\bar{P} = \bar{P}3\cdot \bar{P}2\cdot \bar{P}1\cdot \bar{P}0$$

NOTE:
Device data shown based on Motorola MC14582

PACKAGE PINNING:

PIN–	
3,1,14,5	Carry Generate Inputs, Active Low
4,2,15,6,	Carry Propagate Inputs, Active Low
13	Carry Input
12,11,9	Carry Outputs
10	Carry Generate Output, Active Low Group
7	Carry Propagate Output, Active Low Group

Figure 9.46. The CMOS 4000 series 4582 is a Look-Ahead Carry Block. It is capable of anticipating a carry across four binary adders or groups of adders. It may be cascaded to anticipate a carry across a cluster of n-bit adders. Carry, generate-carry, and propagate carry functions are provided as shown in the package pinning and logic equations of the drawing.

5V CMOS - 4585

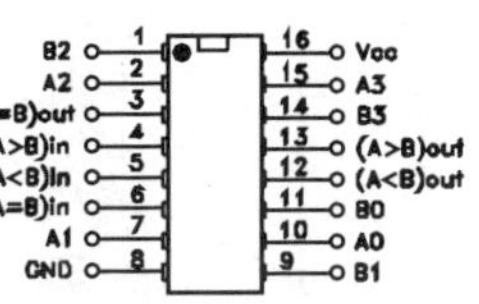

DUAL-IN-LINE PACKAGE

NOTE:
Device data shown based on Motorola MC14585

FUNCTION TABLE CMOS 4585 4-BIT MAGNITUDE COMPARATOR

COMPARING INPUTS				CASCADING INPUTS			OUTPUTS		
A3,B3	A2,B2	A1,B1	A0,B0	A<B	A=B	A>B	A<B	A=B	A>B
A3>B3	X	X	X	X	X	X	0	0	1
A3=B3	A2>B2	X	X	X	X	X	0	0	1
A3=B3	A2=B2	A1>B1	X	X	X	X	0	0	1
A3=B3	A2=B2	A1=B1	A0>B0	X	X	X	0	0	1
A3=B3	A2=B2	A1=B1	A0=B0	0	0	X	0	0	1
A3=B3	A2=B2	A1=B1	A0=B0	0	1	X	0	1	0
A3=B3	A2=B2	A1=B1	A0=B0	1	0	X	1	0	0
A3=B3	A2=B2	A1=B1	A0=B0	1	1	X	1	1	0
A3=B3	A2=B2	A1=B1	A0<B0	X	X	X	1	0	0
A3=B3	A2=B2	A1<B1	X	X	X	X	1	0	0
A3=B3	A2<B2	X	X	X	X	X	1	0	0
A3<B3	X	X	X	X	X	X	1	0	0

1 = HIGH voltage level. 0 = LOW voltage level.
X = Don't Care

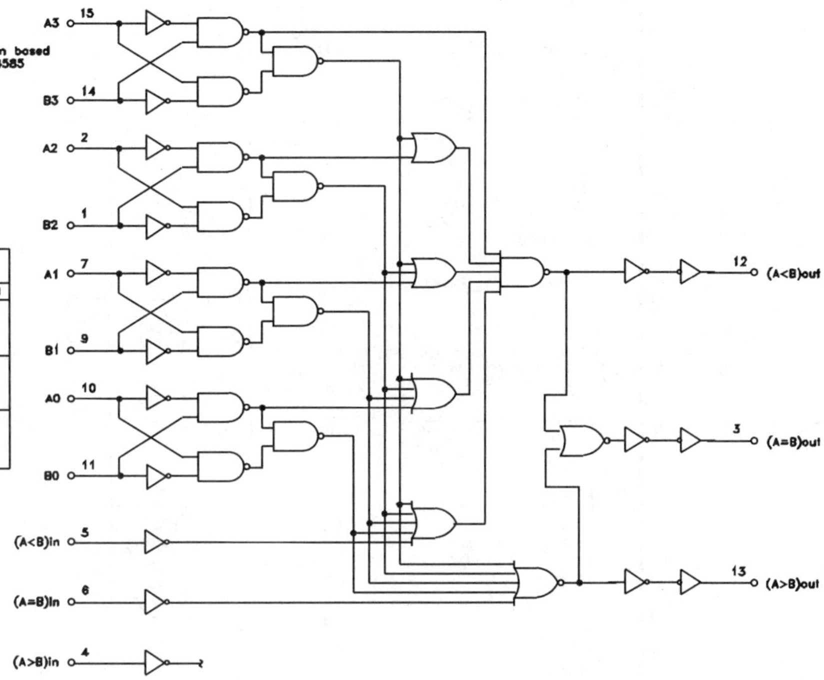

Figure 9.47. The CMOS 4000 series 4585 is a 4-Bit Magnitude Comparator. The circuit has eight comparing inputs (A3, B3, A2, B2, A1, B1, A0, B0), three cascading inputs, (A < B, A = B, A > B), and three outputs, (A < B, A = B, A > B). The device compares two 4-bit words and determines whether they are "less-than," "equal to," or "greater than" by a high level on the appropriate output. For words exceeding 4-bits devices can be cascaded by connecting outputs to the corresponding inputs of the next significant comparator. Inputs A < B, A = B, A > B on the least significant (first) comparator are connected to a low, a high, and a low respectively.

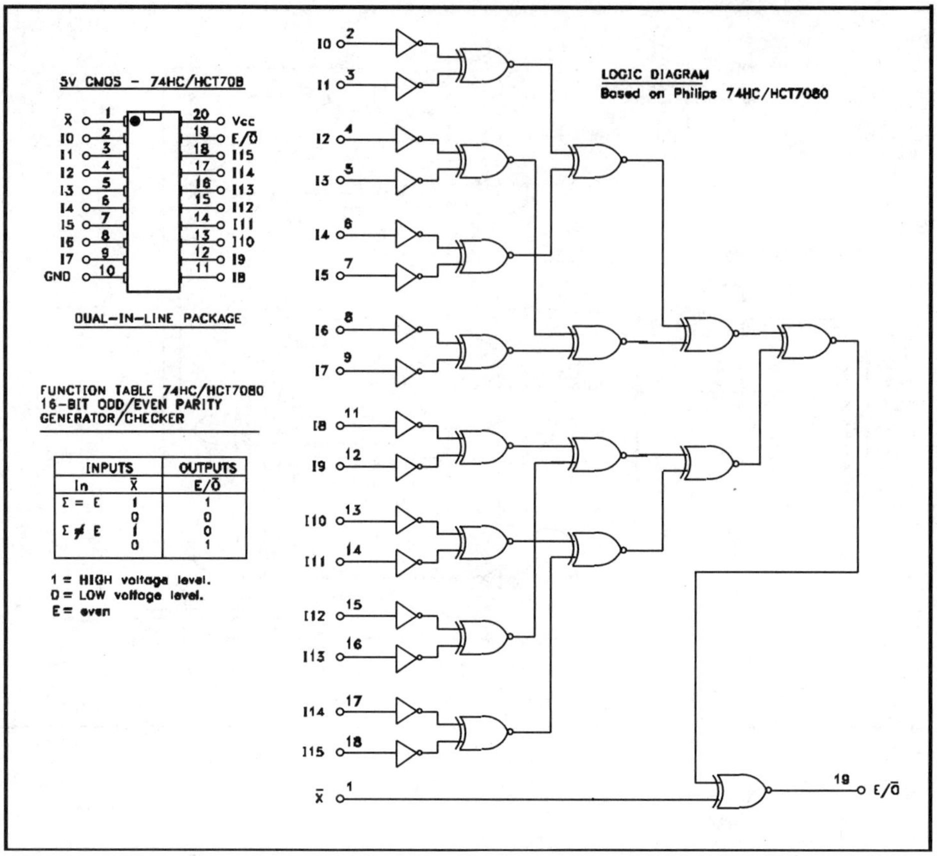

Figure 9.48. The CMOS 74HC/HCT7080 is a 16-Bit Even/Odd Parity Generator/Checker. The even/odd parity output is available for generating or checking even/odd parity up to 16-bits. The even/odd parity output, E//O, is high when an even number of data inputs are high and the cascade/even-odd-changing input, /X, is high. Expansion to larger word sizes is accomplished by connecting the even/odd parity output, E//O, to the cascade/even-odd-changing input /X to the final stage.

5V CMOS – 74ACT11286

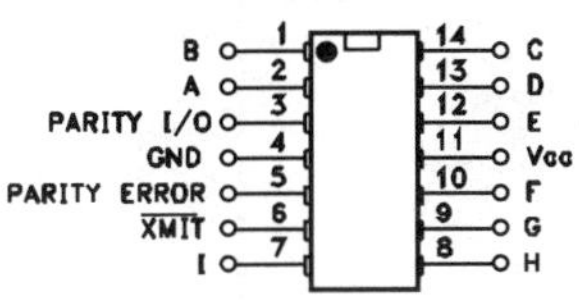

DUAL-IN-LINE PACKAGE

FUNCTION TABLE 74ACT11286
9-BIT PARITY GENERATOR/CHECKER
WITH BUS DRIVER PARITY I/O PORTS

NUMBER OF INPUTS (A–I) THAT ARE HIGH	$\overline{XMIT}$ INPUT	PARITY I/O	PARITY ERROR OUTPUT
0,2,4,6,8	l	H	H
1,3,5,7,9	l	L	H
0,2,4,6,8	h h	h l	H L
1,3,5,7,9	h h	h l	L H

h = high level input
l = low input level
H = high output level
L = low output level

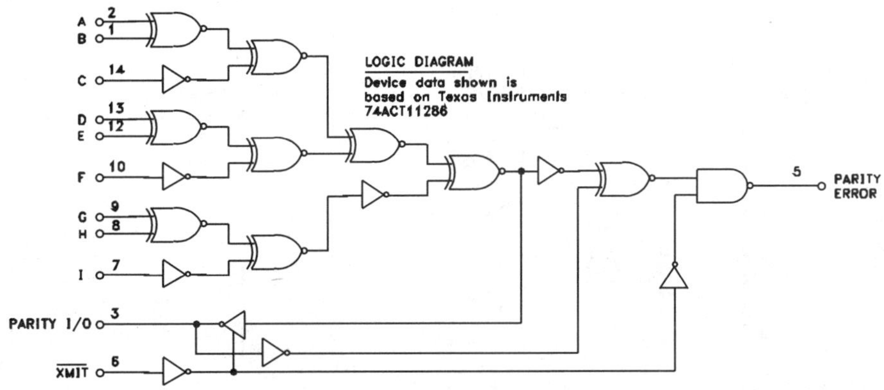

Figure 9.49. The CMOS 74ACT11286 is a 9-Bit Parity Generator/Checker with Bus-Driven Parity I/O Ports. The device features a local output for parity checking and a bus driving parity I/O port for parity generation/checking. The word length capability is expanded by cascading. The /XMIT input is provided for cascading. When /XMIT is low, the parity tree is disabled, the parity output error output remains at a high level regardless of the input levels. When /XMIT is high, the parity tree is enabled. Parity Error indicates an error when either an even number of inputs are high and parity I/O is forced low or when an odd number of inputs are high and parity I/O is forced high. The I/O circuitry is designed to maintain the I/O port in the high-impedance state during power up or power down.

5V CMOS – 74AC11520
3V CMOS – 74AC11520

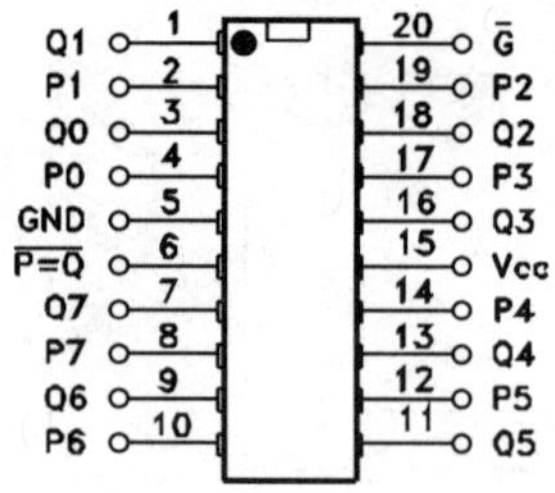

DUAL-IN-LINE PACKAGE

FUNCTION TABLE 74AC11520
8-BIT IDENTITY COMPARATOR

INPUTS		OUTPUTS
DATA P,Q	ENABLE $\overline{G}$	$\overline{P=Q}$
P=Q	0	0
P>Q	0	1
P<Q	0	1
X	1	1

1 = HIGH voltage level.
0 = LOW voltage level.
E = even

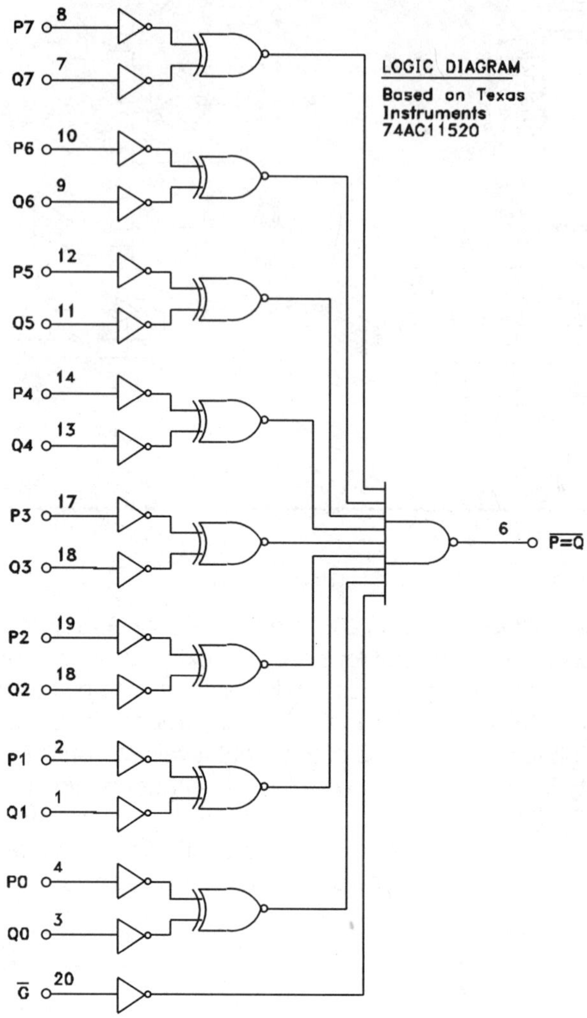

Figure 9.50. The CMOS 74ACT11520 is an 8-Bit Identity Comparator. The device performs comparisons on two 8-bit binary of BCD words. Features include 20K pull-up termination resistors on the Q inputs for analog or switch data input and a /P = Q totem-pole output. The package pinning is a center-pin Vcc and Ground configuration to minimize high-speed switching noise. Performance specifications are provided for two Vcc voltage levels: 5 ±0.5 and 3.3 ±0.3 volts.

5V CMOS - 74ACT16833

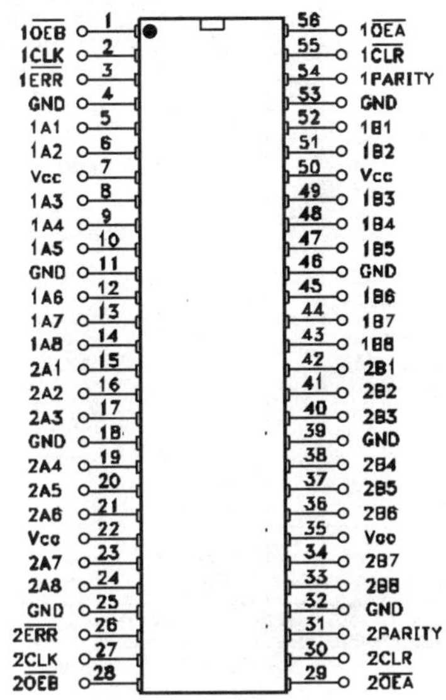

DUAL-IN-LINE PACKAGE

NOTE:
Device data shown is for Texas Instruments 74ACT16833

FUNCTION TABLE 74ACT16883 DUAL 8-BIT TO 9-BIT PARITY BUS TRANCEIVERS

INPUTS						OUTPUT AND I/O				FUNCTION
$\overline{OEB}$	$\overline{OEA}$	$\overline{CLR}$	CLK	Ai Σ OF H	Bi* Σ OF H	A	B	PARITY	$\overline{ERR}$+	
0	1	X	X	Odd Even	NA	NA	A	0 1	NA	A data to B bus and generate parity
1	0	1	↑	NA	Odd Even	B	NA	NA	1 0	B data to A bus and generate parity
X	X	0	X	X	X	X	NA	NA	1	Check error-flag register
1	1	1 0 1 1	Not ↑ Not ↑ ↑ ↑	X X Odd Even	X	Z	Z	Z	NC 1 1 0	Isolation%
0	0	X	X	Odd Even	NA	NA	A	1 0	NA	A data to B bus and generate inverted parity

1 = HIGH voltage level. 0 = LOW voltage level.
NA = Not Applicable. NC = No Change. X = Don't Care
* = Summation of high level inputs includes PARITY along with Bi inputs.
+ = Output states shown assume ERR was previously high.
% = In this mode ERR, when clocked, shows inverted parity of the A bus.
↑ = LOW-to-HIGH clock transition.

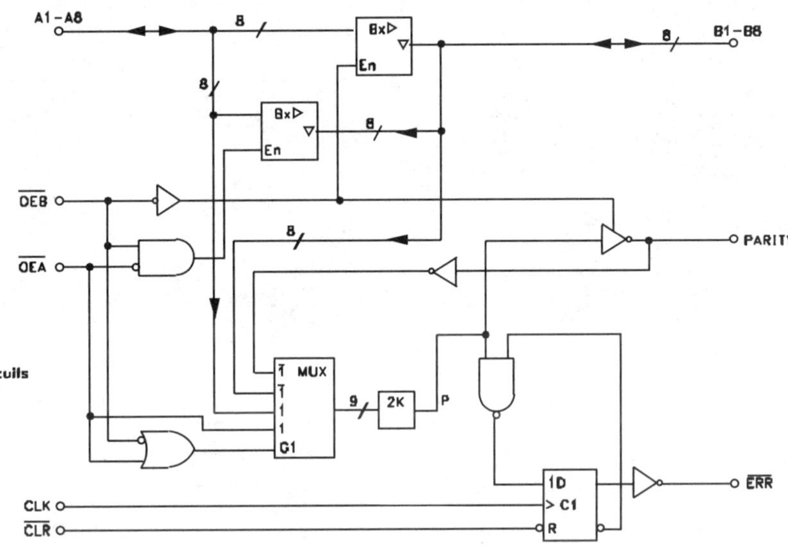

LOGIC DIAGRAM
One of two dual circuits

Figure 9.51. The CMOS 74ACT16833 is a dual 8-Bit to 9-Bit Parity Bus Transceiver. The device consists of two noninverting 8-bit to 9-bit parity bus transceivers designed for communication between data buses. For each transceiver, when data is transmitted from the A bus to the B bus an odd parity bit is generated and outputted on the parity I/O pin (1PARITY or 2PARITY). When data is transmitted from the B bus to the A bus, 1PARITY or 2PARITY is configured as an input and combined with the 8-bit data to generate an active-low error flag if odd parity is not detected. The /1ERR, /2ERR output is configured as an open-collector output. Refer to the *Function Table* for explicit details on parity operations.

3V CMOS – 74ALVCH16901

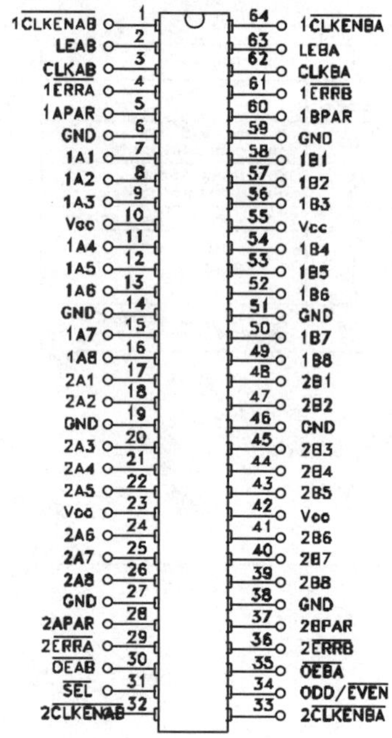

DGG PACKAGE

PARITY FUNCTION TABLE

INPUTS								OUTPUTS			
SEL	OEBA	OEAB	ODD/EVEN	Σ OF INPUTS A1–A8=H	Σ OF INPUTS B1–B8=H	APAR	BPAR	APAR	ERRA	BPAR	ERRB
0	1	0	0	0,2,4,6,8	N/A	0	N/A	N/A	1	0	Z
0	1	0	0	1,3,5,7	N/A	0	N/A	N/A	0	1	Z
0	1	0	0	0,2,4,6,8	N/A	1	N/A	N/A	0	0	Z
0	1	0	0	1,3,5,7	N/A	1	N/A	N/A	1	1	Z
0	0	1	0	N/A	0,2,4,6,8	N/A	0	0	Z	N/A	1
0	0	1	0	N/A	1,3,5,7	N/A	0	1	Z	N/A	0
0	0	1	0	N/A	0,2,4,6,8	N/A	1	0	Z	N/A	0
0	0	1	0	N/A	1,3,5,7	N/A	1	1	Z	N/A	1
0	1	0	1	0,2,4,6,8	N/A	0	N/A	N/A	0	1	Z
0	1	0	1	1,3,5,7	N/A	0	N/A	N/A	1	0	Z
0	1	0	1	0,2,4,6,8	N/A	1	N/A	N/A	1	1	Z
0	1	0	1	1,3,5,7	N/A	1	N/A	N/A	0	0	Z
0	0	1	1	N/A	0,2,4,6,8	N/A	0	1	Z	N/A	0
0	0	1	1	N/A	1,3,5,7	N/A	0	0	Z	N/A	1
0	0	1	1	N/A	0,2,4,6,8	N/A	1	1	Z	N/A	1
0	0	1	1	N/A	1,3,5,7	N/A	1	0	Z	N/A	0
0	1	1	0	0,2,4,6,8	0,2,4,6,8	0	0	Z	1	Z	1
0	1	1	0	1,3,5,7	1,3,5,7	0	0	Z	0	Z	0
0	1	1	0	0,2,4,6,8	0,2,4,6,8	1	1	Z	0	Z	0
0	1	1	0	1,3,5,7	1,3,5,7	1	1	Z	1	Z	1
0	1	1	1	0,2,4,6,8	0,2,4,6,8	0	0	Z	0	Z	0
0	1	1	1	1,3,5,7	1,3,5,7	0	0	Z	1	Z	1
0	1	1	1	0,2,4,6,8	0,2,4,6,8	1	1	Z	1	Z	1
0	1	1	1	1,3,5,7	1,3,5,7	1	1	Z	0	Z	0
0	0	0	0	N/A	N/A	N/A	N/A	PE@	Z	PE@	Z
0	0	0	1	N/A	N/A	N/A	N/A	PE#	Z	PE#	Z

@ = Parity output is set to the level so that the specific bus side set to even parity
\# = Parity output is set to the level so that the specific bus side set to odd parity

NOTE:
Device data shown is based on Texas Instruments 74ALVCH16901

FUNCTION TABLE 74ALVCH1690 18-BIT UNIVERSAL BUS TRANSCEIVER WITH PARITY GERNERATOR/CHECKER

INPUTS					OUTPUT
CLKENAB	OEAB	LEAB	CLKAB	A	B
X	1	X	X	X	Z
X	0	1	X	0	0
X	0	1	X	1	1
1	0	0	X	X	B0*
0	0	0	↑	0	0
0	0	0	↑	1	1
0	0	0	0	X	B0*
0	0	0	1	X	B0%

* = A-to-B data flow is shown; B-to-A flow is similar, but uses OEBA, LEBA, and CLKENBA.
% = Output level before the indicated steady-state input conditions were established, provided that CLKAB was low before LEAB went low.
1 = HIGH voltage level.
0 = LOW voltage level.
X = Don't Care
↑ = LOW-to-HIGH clock transition.

BLOCK DIAGRAM

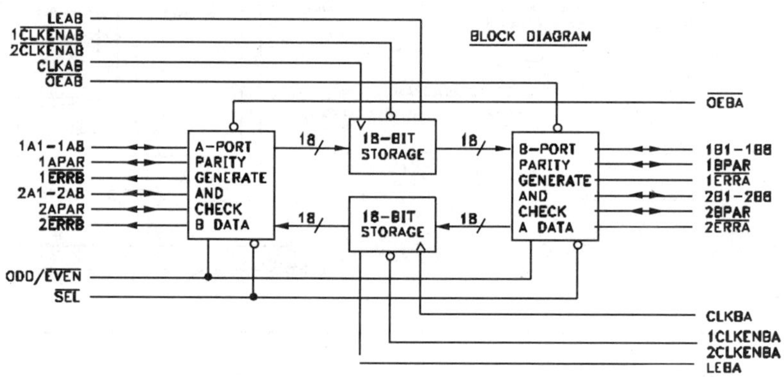

PARITY ENABLE FUNCTION TABLE

INPUTS			OPERATION OR FUNCTION	
SEL	OEBA	OEAB		
0	1	0	Parity is checked on port A and is generated on Port B	
0	0	1	Parity is checked on port B and is generated on port A	
0	1	1	Parity is checked on port A and port B	
0	0	0	Parity is generated on port A and port B if device is in FF mode	
1	0	0	Parity functions are disabled; device acts as a standard 18-bit registered transceiver	QA data to B, QB data to A
1	0	1		QB data to A
1	1	0		QA data to B
1	1	1		Isolation

Figure 9.52. The CMOS 74ALVCH16901 is an 18-Bit Universal Bus Transceiver with Parity Generator/Checkers designed for 2.3 to 3.6 volt operation. The '901 is a dual 9-bit to dual 9-bit parity transceiver with registers. The device can operate as a feedthrough transceiver or it can generate/check parity from the two 8-bit data buses in either direction. Features include an independent clock, latch and dual 9-bit clock enable inputs, parity enable and parity select inputs, and separate error-signal outputs for checking parity. The direction of data flow is controlled by /OEAB and /OEBA. When /SEL is low, the parity functions are enabled. When /SEL is high, the parity functions are disabled and the device acts as an 18-bit registered transceiver.

Chapter 10

Buffers and Line Drivers/Receivers

In this chapter we find descriptions of devices intended to lend protection and/or oomph to the transmission of data. The earliest devices, such as the 74125 and 74126, are little more than versions of the 7404 hex inverter in a quad format and the addition of a 3-state output enable. The sink current rating for the TTL 7404 and 74125 reads the same: 16 mA. Advances in technology raised the ante on expectations of bus width, speed, and current drive, resulting in devices such as the low-voltage CMOS 74ALVCH162827 operating with a Vcc range of 2.3 to 3.6 volts, propagation delays of less than 5 nS, a 20-bit bus width, and ESD protection exceeding 2,000 volts. Its sink current rating of 12 mA may appear less impressive in contrast to the TTL FAST 75F655A octal buffer sink rating of 64 mA.

But the diversity explains the need for a broad range of devices. As we browse through the chapter figures we will see many devices with duplicate features, but with an added touch to meet requirements not previously available.

And so it goes when the name of the game is progress.

Chapter Organization

Eight tables are provided. *Tables 10.1 - 10.4a, 4b* provide a listing of devices by manufacturer. There are sources beyond those listed but those given provide a realistic picture of what has been and is currently available. Unlike previous chapters, there is no listing for 4000 series of CMOS devices. These were included in Chapter 2 with the inverters and combinational gate logic.

Tables 10.5 - 10.6a, 6b provide the LOW-to-HIGH (t_{PLH}) and HIGH-to-LOW (t_{PHL}) propagation delays. For some low-voltage CMOS devices these are not defined as such, in which case a propagation or time delay, t_d or t_p, is given. Because of the manner in which manufacturers define these quantities, the data in the tables is not precise. The intent is to provide insight helpful for performance comparisons. In particular, we can see the spectacular gains in speed and reduced power dissipation, as well as complexity, achieved with the latest devices.

The drawings of *Figures 10.1 - 10.23* describe the Dual In-Line (DIP) package pinning, a Truth or Function Table, and Logic Diagrams. To simplify comparisons as well as to conserve space, all devices with matching features, specifically pinouts and function tables, are included in the same drawing.

Summary

This chapter provides tables and drawings describing varieties of integrated circuit buffers and line drivers/ receivers. While of simpler logic than the sometimes exotic devices of earlier chapters, many devices display an impressive array of performance abilities. Some points to keep in mind are:

- Timing information in the tables is for comparison purposes only. There is a general lack of consistency in how these parameters are defined and not just between manufacturers. Where timing is critical consult the manufacturer's data manual.

- All timing data is shown for the industrial/commercial or recommended free-air operating temperature range. Where data is for 25°C it is given here.

- The specifications for CMOS devices may cover voltage ranges from 2 to 15 volts. Data is typically provided for VCC's of 5, 10, and 15 volts, but not always. Some manufacturers use 5.0 volts; others 4.5. There is a significant speed factor—in general, the higher the voltage the faster the performance. Again, where this is important to your application consult the data manual.

- A similar situation exists for the more recent low-voltage devices. Typical VCC ranges given are 2.7 - 3.6 volts. But values may be found for 2, 2.6, 3.0, 3.3 or 3.6 volts. (Glossary *Figures G.1 and G.2* may provide helpful insights on performance.)

DEVICE	STD		L		H		S		LS		AS		ALS		F		Figure	DEVICE
74---	NM	ST	NM	ST	NM	ST	NM	ST	NM	ST	NM	ST	NM	ST	NM	ST	No.	74---
74125	x	x x							x x	x x						x x	10.1	74125
74126		x x							x x	x x						x x	10.1	74126
74140							x										10.1	74140
74230											x	x		x			10.2	74230
74231											x	x		x			10.2	74231
74240							x	x	x x	x x	x	x	x	x		x x	10.3	74240
74241							x	x	x x	x x	x	x	x	x		x x	10.3	74241
74244							x	x	x x	x x	x	x	x	x		x x	10.3	74244
74365	x	x x							x x	x x					x	x	10.4	74365
74366		x x							x x	x x					x	x	10.4	74366
74367	x	x x							x x	x x					x	x	10.4	74367
74368	x	x x							x x	x x					x	x	10.4	74368
74455																x	10.5	74455
74456																x	10.5	74456
74465									x	x			x	x			10.6	74465
74466									x	x			x	x			10.6	74466
74467									x	x			x	x			10.6	74467
74468									x	x			x	x			10.6	74468
74540									x	x			x	x		x	10.7	74540
74541									x	x			x	x		x x	10.7	74541
74655																x	10.8	74655
74656																x	10.8	74656
74746														x			10.7	74746
74747														x			10.7	74747
74756												x		x		x	10.9	74756
74757												x				x	10.9	74757
74760												x		x			10.10	74760
74762												x		x			10.10	74762
74763												x		x			10.10	74763
74795									x								10.6	74795
74796									x								10.6	74796
74797									x								10.6	74797
74798									x								10.6	74798
74827																x	10.11	74827
74828																x	10.11	74828
74940							x										10.12	74940
74941							x										10.12	74941
741240													x	x		x	10.3	741240
741241													x			x	10.3	741241
741244													x	x		x	10.3	741244
742240														x		x	10.3	742240
742241																x x	10.3	742241
742244																x	10.3	742244
74540														x			10.7	742540
742541													x	x			10.7	742541
743037																x	10.13	743037
743038																x	10.13	743038
743040																x	10.13	743040
7429827														x			10.14	7429827
7429828														x			10.14	7429828
7430240																x	10.15	7430240
7430244																x	10.15	7430244

Key to Manufacturer: N – National/Fairchild. M – Motorola,
S – Signetics/Philips, T – Texas Instruments

Table 10.1. TTL Buffer and Line Driver/Receiver source summary by manufacturer.

DEVICE 74---	C NM	C ST	HC NM	HC ST	HCT NM	HCT ST	AHC (5V) NM	AHC (5V) ST	AHCT NM	AHCT ST	VHC (5V) NM	VHC (5V) ST	VHCT NM	VHCT ST	AC (5V) NM	AC (5V) ST	ACT NM	ACT ST	Figure No.	DEVICE 74---
74125			x x	x x		x		x		x	x x				x		x		10.1	74125
74126			x x	x x		x		x		x	x x								10.1	74126
74240			x x	x x	x x	x x		x		x	x x		x		x	x	xq	x	10.3	74240
74241			x x	x x	x x	x x									x	x	xq	x	10.3	74241
74244			x x	x x	x x	x x		x		x	x x		x		x	x	xq	x	10.3	74244
74365			x x	x x		x													10.4	74365
74366			x x	x x		x													10.4	74366
74367			x x	x x		x													10.4	74367
74368			x x	x x		x													10.4	74368
74540			x x	x x	x x	x x		x		x	x				x				10.7	74540
74541			x x	x x	x x	x x		x		x	x x				x				10.7	74541
747540				x		x													10.16	747540
747541				x		x													10.16	747541
749014				x		x													10.17	749014
749015				x		x													10.17	749015
749114				x		x													10.17	749114
749115				x		x													10.17	749115
7411240																x		x	10.18	7411240
7411244																x		x	10.18	7411244
7416240								x		x						x	q	x	10.19	7416240
7416241																		x	10.19	7416241
7416244								x		x							q		10.19	7416244
7416540																	q	x	10.20	7416540
7416541																	q	x	10.20	7416541
7416825																		x	10.21	7416825
7416827																		x	10.21	7416827
7418825																	q		10.21	7418825
7418826																	q		10.21	7418826

Key to Manufacturer: N – National/Fairchild. M – Motorola,
S – Signetics/Philips, T – Texas Instruments
Devices shown as "q" are National/Fairchild 74ACTQ series

Table 10.2. 5V CMOS 74xxx series Buffer and Line Driver/Receiver source summary by manufacturer.

DEVICE 74---	LV NM	LV ST	LVCH LVC NM	LVCH LVC ST	ALVT LVT NM	ALVT LVT ST	LVX NM	LVX ST	LVQ NM	LVQ ST	LCX NM	LCX ST	AC (3V) NM	AC (3V) ST	ALVCH NM	ALVCH ST	HL NM	HL ST	AHC (3V) NM	AHC (3V) ST	VHC (3V) NM	VHC (3V) ST	Figure No.	DEVICE 74---
74125		x x		x x	x	x x	x		x		x		x							x	x x		10.1	74125
74126		x		x		x														x	x x		10.1	74126
74240		x x		x x	x	x x	x		x		x x		x							x	x x		10.3	74240
74241		x		x x	x	x x			x				x										10.3	74241
74244		x x		x x	x	x x	x		x		x x		x							x	x x		10.3	74244
74365		x																					10.4	74365
74366																							10.4	74366
74367		x																					10.4	74367
74368		x																					10.4	74368
74540				x							x									x			10.7	74540
74541		x		x							x									x			10.7	74541
74827				x x																			10.11	74827
74828				x																			10.11	74828
742240				x	x	x																	10.3	742240
742241				x		x																	10.3	742241
742244				x	x	x																	10.3	742244
742827				x																			10.21	742827
7411240													x										10.18	7411240
7411244													x										10.18	7411244
7416240				x	x	x					x					x x				x			10.19	7416240
7416241				x		x																	10.19	7416241
7416244				x	x	x					x					x x				x			10.19	7416244
7416540				x																			10.20	7416540
7416541				x		x																	10.20	7416541
7416825																x							10.21	7416825
7416827						x										x							10.21	7416827
7416835																x							10.22	7416835
74162240				x		x																	10.19	74162240
74162241				x																			10.19	74162241
74162244				x		x										x							10.19	74162244
74162540				x																			10.20	74162540
74162541																							10.20	74162541
74162827						x										x							10.21	74162827
7433240																		x					10.23	7433240
7433241																		x					10.23	7433241
7433244																		x					10.23	7433244

Key to Manufacturer: N – National/Fairchild. M – Motorola,
S – Signetics/Philips, T – Texas Instruments

Table 10.3. Low-Voltage CMOS 74xxx series Buffer and Line Driver/Receiver source summary by manufacturer.

DEVICE 74---	STD	L	H	S	LS	AS/A/B	ALS/A/B	F	DATA	Figure	DEVICE 74---
-74125	13	--	--	--	18	--	--	2.0-6.5	t_{PLH}↑	-10.1	-74125
----	18	--	--	--	15	--	--	3.0-8.0	t_{PHL}↓	---	----
-74126	13	--	--	--	18	--	--	2.0-6.5	t_{PLH}↑	-10.1	-74126
----	18	--	--	--	15	--	--	3.0-8.0	t_{PHL}↓	---	----
-74140	--	--	--	2-9	--	--	--	--	t_{PLH}↑	-10.1	-74140
----	--	--	--	2-9	--	--	--	--	t_{PHL}↓	---	----
-74230	--	--	--	--	--	2.5-6.5	----	--	t_{PLH}↑	-10.2	-74230
----	--	--	--	--	--	2-6.2	----	--	t_{PHL}↓	---	----
-74231	--	--	--	--	--	2-6.5	2-9	--	t_{PLH}↑	-10.2	-74231
----	--	--	--	--	--	2-8.5	2-9	--	t_{PHL}↓	---	----
-74240	--	--	--	2-10	14	2-6.5	2-9	2.5-6.5	t_{PLH}↑	-10.3	-74240
----	--	--	--	2-12	18	2-5.7	2-9	2.0-5.8	t_{PHL}↓	---	----
-74241	--	--	--	2-12	18	2-6.2	3-11	2.5-6.5	t_{PLH}↑	-10.3	-74241
----	--	--	--	2-12	18	2-6.2	3-10	2.0-5.8	t_{PHL}↓	---	----
-74244	--	--	--	2-12	18	3-10	3-10	2.5-5.7	t_{PLH}↑	-10.3	-74244
----	--	--	--	2-12	18	3-10	3-10	2.5-5.2	t_{PHL}↓	---	----
-74365	16	--	--	--	15	--	--	2.5-6.5	t_{PLH}↑	-10.4	-74365
----	22	--	--	--	12	--	--	2.0-5.5	t_{PHL}↓	---	----
-74366	17	--	--	--	10	--	--	2.5-6.5	t_{PLH}↑	-10.4	-74366
----	16	--	--	--	12	--	--	2.0-5.5	t_{PHL}↓	---	----
-74367	16	--	--	--	15	--	--	2.5-6.5	t_{PLH}↑	-10.4	-74367
----	22	--	--	--	12	--	--	2.0-5.5	t_{PHL}↓	---	----
-74368	17	--	--	--	10	--	--	2.5-6.5	t_{PLH}↑	-10.4	-74368
----	16	--	--	--	12	--	--	2.0-5.5	t_{PHL}↓	---	----
-74455	--	--	--	--	--	--	--	2.0-13	t_{PLH}↑	-10.5	-74455
----	--	--	--	--	--	--	--	1.0-14.5	t_{PHL}↓	---	----
-74456	--	--	--	--	--	--	--	2.0-13	t_{PLH}↑	-10.5	-74456
----	--	--	--	--	--	--	--	1.0-14.5	t_{PHL}↓	---	----
-74465	--	--	--	--	9-15	--	2-13	--	t_{PLH}↑	-10.6	-74465
----	--	--	--	--	12-18	--	4-12	--	t_{PHL}↓	---	----
-74466	--	--	--	--	7-12	--	3-12	--	t_{PLH}↑	-10.6	-74466
----	--	--	--	--	9-15	--	2-9	--	t_{PHL}↓	---	----
-74467	--	--	--	--	9-15	--	2-13	--	t_{PLH}↑	-10.6	-74467
----	--	--	--	--	12-18	--	4-12	--	t_{PHL}↓	---	----
-74468	--	--	--	--	7-12	--	2-12	--	t_{PLH}↑	-10.6	-74468
----	--	--	--	--	9-15	--	2-9	--	t_{PHL}↓	---	----
-74540	--	--	--	--	9-15	--	2-12	2.5-6.5	t_{PLH}↑	-10.7	-74540
----	--	--	--	--	9-15	--	2-9	1.5-7.0	t_{PHL}↓	---	----
-74541	--	--	--	--	9-15	--	4-14	2.5-6.5	t_{PLH}↑	-10.7	-74541
----	--	--	--	--	10-18	--	2-10	2.5-6.5	t_{PHL}↓	---	----
-74655	--	--	--	--	--	--	--	2.0-13	t_{PLH}↑	-10.8	-74655
----	--	--	--	--	--	--	--	1.0-14.5	t_{PHL}↓	---	----
-74656	--	--	--	--	--	--	--	2.0-13	t_{PLH}↑	-10.8	-74656
----	--	--	--	--	--	--	--	1.0-14.5	t_{PHL}↓	---	----
-74746	--	--	--	--	--	--	3-12	--	t_{PLH}↑	-10.7	-74746
----	--	--	--	--	--	--	2-9	--	t_{PHL}↓	---	----
-74747	--	--	--	--	--	--	4-14	--	t_{PLH}↑	-10.7	-74747
----	--	--	--	--	--	--	2-10	--	t_{PHL}↓	---	----
-74756	--	--	--	--	--	3-19.5	8-24	--	t_{PLH}↑	-10.9	-74756
----	--	--	--	--	--	1-7.5	2-20	--	t_{PHL}↓	---	----
-74757	--	--	--	--	--	3-21		--	t_{PLH}↑	-10.9	-74757
----	--	--	--	--	--	1-7.5		--	t_{PHL}↓	---	----

Table 10.4a. 74xxx TTL device representative propagation delay times in nS.

DEVICE 74---	STD	L	H	S	LS	AS/A/B	ALS/A/B	F	DATA	Figure	DEVICE 74---
-74760	--	--	--	--	--	3-18.5	22-25	--	t_{PLH}↑	-10.10	-74760
----	--	--	--	--	--	1-8	13-24	--	t_{PHL}↓	---	----
-74762	--	--	--	--	--	3-19.5	14-17	--	t_{PLH}↑	-10.10	-74762
----	--	--	--	--	--	1-7.5	6-18	--	t_{PHL}↓	---	----
-74763	--	--	--	--	--	3-20	16-18	--	t_{PLH}↑	-10.10	-74763
----	--	--	--	--	--	1-8	5-13	--	t_{PHL}↓	---	----
-74795	--	--	--	--	11-16	--	--	--	t_{PLH}↑	-10.6	-74795
----	--	--	--	--	15-22	--	--	--	t_{PHL}↓	---	----
-74796	--	--	--	--	6.0-10	--	--	--	t_{PLH}↑	-10.6	-74796
----	--	--	--	--	13-17	--	--	--	t_{PHL}↓	---	----
-74797	--	--	--	--	11-16	--	--	--	t_{PLH}↑	-10.6	-74797
----	--	--	--	--	15-22	--	--	--	t_{PHL}↓	---	----
-74798	--	--	--	--	6.0-10	--	--	--	t_{PLH}↑	-10.6	-74798
----	--	--	--	--	13-17	--	--	--	t_{PHL}↓	---	----
-74827	--	--	--	--	--	--	--	2.0-17	t_{PLH}↑	-10.11	-74827
----	--	--	--	--	--	--	--	1.0-17	t_{PHL}↓	---	----
-74828	--	--	--	--	--	--	--	2.0-17	t_{PLH}↑	-10.11	-74828
----	--	--	--	--	--	--	--	1.0-17	t_{PHL}↓	---	----
-74940	--	--	--	2-10	--	--	--	--	t_{PLH}↑	-10.12	-74940
----	--	--	--	2-12	--	--	--	--	t_{PHL}↓	---	----
-74941	--	--	--	2-10	--	--	--	--	t_{PLH}↑	-10.12	-74941
----	--	--	--	2-12	--	--	--	--	t_{PHL}↓	---	----
-741240	--	--	--	--	--	--	2-13	2.5-6.5	t_{PLH}↑	-10.3	-741240
----	--	--	--	--	--	--	2-13	1.5-6.5	t_{PHL}↓	---	----
-741241	--	--	--	--	--	--	2-13	2.5-6.5	t_{PLH}↑	-10.3	-741241
----	--	--	--	--	--	--	2-13	1.5-6.5	t_{PHL}↓	---	----
-741244	--	--	--	--	--	--	3-14	2.5-6.0	t_{PLH}↑	-10.3	-741244
----	--	--	--	--	--	--	3-14	2.0-7.5	t_{PHL}↓	---	----
-742240	--	--	--	--	--	3-10	--	3.0-7.0	t_{PLH}↑	-10.3	-742240
----	--	--	--	--	--	3-10	--	2.0-6.5	t_{PHL}↓	---	----
-742241	--	--	--	--	--	--	--	3.0-7.0	t_{PLH}↑	-10.3	-742241
----	--	--	--	--	--	--	--	2.0-6.5	t_{PHL}↓	---	----
-742244	--	--	--	--	--	--	--	3.0-7.0	t_{PLH}↑	-10.3	-742244
----	--	--	--	--	--	--	--	2.5-7.0	t_{PHL}↓	---	----
-742540	--	--	--	--	--	--	2-12	--	t_{PLH}↑	-10.7	-742540
----	--	--	--	--	--	--	2-11	--	t_{PHL}↓	---	----
-742541	--	--	--	--	--	--	2-15	--	t_{PLH}↑	-10.7	-742541
----	--	--	--	--	--	--	2-12	--	t_{PHL}↓	---	----
-743037	--	--	--	--	--	--	--	1.0-5.0	t_{PLH}↑	-10.13	-743037
----	--	--	--	--	--	--	--	1.0-4.5	t_{PHL}↓	---	----
-743038	--	--	--	--	--	--	--	6.0-11.5	t_{PLH}↑	-10.13	-743038
----	--	--	--	--	--	--	--	1.0-5.0	t_{PHL}↓	---	----
-743040	--	--	--	--	--	--	--	1.0-5.0	t_{PLH}↑	-10.13	-743040
----	--	--	--	--	--	--	--	1.0-4.5	t_{PHL}↓	---	----
7429827	--	--	--	--	--	--	3.5-11	--	t_{PLH}↑	-10.14	7429827
----	--	--	--	--	--	--	6.5-16	--	t_{PHL}↓	---	----
7429828	--	--	--	--	--	--	4-10	--	t_{PLH}↑	-10.14	7429828
----	--	--	--	--	--	--	3-12.9	--	t_{PHL}↓	---	----
7430240	--	--	--	--	--	--	--	4.0-14.5	t_{PLH}↑	-10.15	7430240
----	--	--	--	--	--	--	--	1.0-9.0	t_{PHL}↓	---	----
7430244	--	--	--	--	--	--	--	4.0-14.5	t_{PLH}↑	-10.15	7430244
----	--	--	--	--	--	--	--	1.0-9.0	t_{PHL}↓	---	----

Table 10.4b. 74xxx TTL device representative propagation delay times in nS (continuation).

DEVICE 74---	C	HC	HCT	AHC	AHCT	VHC(5V)	VHCT	AC (5V)	ACT	DATA	Figure	DEVICE 74---
-74125	--	11-25	15-31	3.8-7.5	3.8-7.5	3.8-6.5	--	1.0-7.0	1.0-9.0	t_{PLH}↑	-10.1	-74125
----	--	11-25	15-31	3.8-7.5	3.8-7.5	5.3-8.5	--	1.0-7.0	1.0-9.0	t_{PHL}↓	---	----
-74126	--	11-25	14-30	3.8-7.5	3.8-7.5	----	--	--	--	t_{PLH}↑	-10.1	-74126
----	--	11-25	14-30	3.8-7.5	3.8-7.5	----	--	--	--	t_{PHL}↓	---	----
-74240	--	11-25	11-25	3.6-85	5.3-11	3.6-6.5	5.4-8.5	1.5-6.5	1.5-8.5	t_{PLH}↑	-10.3	-74240
----	--	11-25	11-25	3.6-8.5	5.3-11	5.1-8.5	5.9-9.5	1.5-6.0	1.5-7.5	t_{PHL}↓	---	----
-74241	--	9-25	13-28	--	--	--	--	1.5-7.0	1.5-9.0	t_{PLH}↑	-10.3	-74241
----	--	9-25	13-28	--	--	--	--	1.5-7.0	1.5-9.0	t_{PHL}↓	---	----
-74244	--	11-28	13-28	3.9-7.5	5.4-8.4	3.9-6.5	5.4-8.5	1.5-7.0	2.0-9.0	t_{PLH}↑	-10.3	-74244
----	--	11-28	13-28	3.9-7.5	5.4-8.4	5.4-8.5	5.9-9.5	1.5-7.0	2.0-9.0	t_{PHL}↓	---	----
-74365	--	11-24	14-31	--	--	--	--	--	--	t_{PLH}↑	-10.4	-74365
----	--	11-24	14-31	--	--	--	--	--	--	t_{PHL}↓	---	----
-74366	--	12-25	13-30	--	--	--	--	--	--	t_{PLH}↑	-10.4	-74366
----	--	12-25	13-30	--	--	--	--	--	--	t_{PHL}↓	---	----
-74367	--	10-24	14-31	--	--	--	--	--	--	t_{PLH}↑	-10.4	-74367
----	--	10-24	14-31	--	--	--	--	--	--	t_{PHL}↓	---	----
-74368	--	11-24	13-30	--	--	--	--	--	--	t_{PLH}↑	-10.4	-74368
----	--	11-24	13-30	--	--	--	--	--	--	t_{PHL}↓	---	----
-74540	--	11-25	13-30	3.7-7	4-8.5	3.7-6.0	--	1.5-6.0	--	t_{PLH}↑	-10.7	-74540
----	--	11-25	13-30	3.7-7	4-8.5	5.2-8.0	--	1.5-5.5	--	t_{PHL}↓	---	----
-74541	--	12-29	15-35	3.5-7	4.1-8.5	3.5-6.0	--	1.5-6.0	--	t_{PLH}↑	-10.7	-74541
----	--	12-29	15-35	3.5-7	3.7-8.5	5.0-8.0	--	1.5-6.0	--	t_{PHL}↓	---	----
-747540	--	14-30	19-40	--	--	--	--	--	--	t_{PLH}↑	-10.16	-747540
----	--	14-30	19-40	--	--	--	--	--	--	t_{PHL}↓	---	----
-747541	--	14-30	19-40	--	--	--	--	--	--	t_{PLH}↑	-10.16	-747541
----	--	14-30	19-40	--	--	--	--	--	--	t_{PHL}↓	---	----
-749014	--	12-26	19-40	--	--	--	--	--	--	t_{PLH}↑	-10.17	-749014
----	--	12-26	19-40	--	--	--	--	--	--	t_{PHL}↓	---	----
-749015	--	12-26	18-40	--	--	--	--	--	--	t_{PLH}↑	-10.17	-749015
----	--	12-26	18-40	--	--	--	--	--	--	t_{PHL}↓	---	----
-749114	--	13-28	17-39	--	--	--	--	--	--	t_{PLH}↑	-10.17	-749114
----	--	13-28	17-39	--	--	--	--	--	--	t_{PHL}↓	---	----
-749115	--	13-28	18-39	--	--	--	--	--	--	t_{PLH}↑	-10.17	-749115
----	--	13-28	18-39	--	--	--	--	--	--	t_{PHL}↓	---	----
-7411240	--	--	--	--	--	--	--	1.5-7.5	1.5-9.9	t_{PLH}↑	-10.18	-7411240
----	--	--	--	--	--	--	--	1.5-6.6	1.5-8.0	t_{PHL}↓	---	----
-7411244	--	--	--	--	--	--	--	1.5-6.7	1.5-8.9	t_{PLH}↑	-10.18	-7411244
----	--	--	--	--	--	--	--	1.5-6.4	1.5-8.6	t_{PHL}↓	---	----
-7416240	--	--	--	3.6-8.5	5.4-9.5	--	--	1.3-5.3	2.3-7.7	t_{PLH}↑	-10.19	-7416240
----	--	--	--	3.6-8.5	5.4-9.5	--	--	1.9-6.5	4.1-9.2	t_{PHL}↓	---	----
-7416241	--	--	--	--	--	--	--	--	3.3-8.4	t_{PLH}↑	-10.19	-7416241
----	--	--	--	--	--	--	--	--	2.3-8.2	t_{PHL}↓	---	----
-7416244	--	--	--	3.9-8.5	5.4-9.5	--	--	--	3.0-7.3	t_{PLH}↑	-10.19	-7416244
----	--	--	--	3.9-8.5	5.4-9.5	--	--	--	2.5-6.8	t_{PHL}↓	---	----
-7416540	--	--	--	--	--	--	--	--	2.7-7.3	t_{PLH}↑	-10.20	-7416540
----	--	--	--	--	--	--	--	--	3.0-7.3	t_{PHL}↓	---	----
-7416541	--	--	--	--	--	--	--	--	3.0-7.3	t_{PLH}↑	-10.20	-7416541
----	--	--	--	--	--	--	--	--	2.5-7.3	t_{PHL}↓	---	----
-7416825	--	--	--	--	--	--	--	--	4.1-9.3	t_{PLH}↑	-10.21	-7416825
----	--	--	--	--	--	--	--	--	3.1-9.6	t_{PHL}↓	---	----
-7416827	--	--	--	--	--	--	--	--	3.6-9.8	t_{PLH}↑	-10.21	-7416827
----	--	--	--	--	--	--	--	--	2.8-9.8	t_{PHL}↓	---	----
-7418825	--	--	--	--	--	--	--	--	2.0-8.4	t_{PLH}↑	-10.21	-7418825
----	--	--	--	--	--	--	--	--	2.0-8.7	t_{PHL}↓	---	----
-7418826	--	--	--	--	--	--	--	--	2.0-8.2	t_{PLH}↑	-10.21	-7418826
----	--	--	--	--	--	--	--	--	2.0-8.9	t_{PHL}↓	---	----

Table 10.5. 5V 74xxx series CMOS device representative propagation delay times in nS.

DEVICE 74---	LV	LVCH/ LVC	ALVT LVT/ABT	LVX	LVQ	LCX	AC(3V)	LVCH/ ALVCH	HL	AHC (3V)	VHC(3V)	DATA	Figure No.	DEVICE 74---
-74125	10-14	6	1.0-3.9	4.4-8.5	1.0-9.0	1.5-6.5	1.0-9.0	--	--	5.6-13	5.6-9.5	tPLH	-10.1	-74125
-----	10-14	6	1.0-3.9	6.9-12	1.0-9.0	1.5-6.5	1.0-9.0	--	--	5.6-13	8.1-13	tPHL	---	-----
-74126	10-14	-----	1.0-3.9	--	--	--	--	--	--	5.6-13	5.6-9.5	tPLH	-10.1	-74126
-----	10-14	-----	1.0-3.9	--	--	--	--	--	--	5.6-13	8.1-13	tPHL	---	-----
-74240	10-14	1.5-8.0	1.0-4.3	4.3-7.5	2.0-10	1.5-7.5	1.5-8.0	--	--	5.3-11	5.3-9.0	tPLH	-10.3	-74240
-----	10-14	1.5-8.0	1.0-4.3	6.8-11	2.0-10	1.5-7.5	1.5-8.0	--	--	5.3-11	7.8-12.5	tPHL	---	-----
-74241	11-23	1.5-6.5	1.0-4.3	--	2.0-9.0	--	1.5-9.0	--	--	--	--	tPLH	-10.3	-74241
-----	11-23	1.5-6.5	1.0-4.3	--	2.0-9.0	--	1.5-9.0	--	--	--	--	tPHL	---	-----
-74244	9-14	1.5-6.5	1.0-4.1	4.7-8.5	2.0-9.0	1.5-7.5	2.0-9.0	--	--	5.8-11.9	5.8-10.0	tPLH	-10.3	-74244
-----	9-14	1.5-6.5	1.0-4.1	7.2-12	2.0-9.0	1.5-7.5	2.0-9.0	--	--	5.8-11.9	8.3-13.5	tPHL	---	-----
-74365	14-26	--	--	--	--	--	--	--	--	--	--	tPLH	-10.4	-74365
-----	14-26	--	--	--	--	--	--	--	--	--	--	tPHL	---	-----
-74367	13-24	--	--	--	--	--	--	--	--	--	--	tPLH	-10.4	-74367
-----	13-24	--	--	--	--	--	--	--	--	--	--	tPHL	---	-----
-74368	14-26	--	--	--	--	--	--	--	--	--	--	tPLH	-10.4	-74368
-----	14-26	--	--	--	--	--	--	--	--	--	--	tPHL	---	-----
-74540	--	1.5-7.5	--	--	--	--	--	--	--	4.8-10.5	--	tPLH	-10.7	-74540
-----	--	1.5-7.5	--	--	--	--	--	--	--	4.8-10.5	--	tPHL	---	-----
-74541	--	1.5-7	--	--	--	--	--	--	--	5.0-10.5	--	tPLH	-10.7	-74541
-----	--	1.5-7	--	--	--	--	--	--	--	5.0-10.5	--	tPHL	---	-----
-74827	--	1.5-7	--	--	--	--	--	--	--	--	--	tPLH	-10.11	-74827
-----	--	1.5-7	--	--	--	--	--	--	--	--	--	tPHL	---	-----
-74828	--	1.5-8	--	--	--	--	--	--	--	--	--	tPLH	-10.11	-74828
-----	--	1.5-8	--	--	--	--	--	--	--	--	--	tPHL	---	-----
-742240	--	1.5-8.0	--	--	--	--	--	--	--	--	--	tPLH	-10.3	-742240
-----	--	1.5-8.0	--	--	--	--	--	--	--	--	--	tPHL	---	-----
-742241	--	1.5-8.0	--	--	--	--	--	--	--	--	--	tPLH	-10.3	-742241
-----	--	1.5-8.0	--	--	--	--	--	--	--	--	--	tPHL	---	-----
-742244	--	1.5-8.0	--	--	--	--	--	--	--	--	--	tPLH	-10.3	-742244
-----	--	1.5-8.0	--	--	--	--	--	--	--	--	--	tPHL	---	-----
-742827	--	---8.0	--	--	--	--	--	--	--	--	--	tPLH	-10.21	-742827
-----	--	---8.0	--	--	--	--	--	--	--	--	--	tPHL	---	-----
-7411240	--	--	--	--	--	--	1.5-10.5	--	--	--	--	tPLH	-10.18	-7411240
-----	--	--	--	--	--	--	1.5-8.6	--	--	--	--	tPHL	---	-----

Table 10.6a. Low-Voltage 74xxx series CMOS device representative propagation delay times in nS.

DEVICE 74---	LV	LVC	ALVT LVT/ABT	LVX	LVQ	LCX	AC(3V)	LVCH/ ALVCH	HL	AHC (3V)	VHC(3V)	DATA	Figure No.	DEVICE 74---
-7411244	--	--	--	--	--	--	1.5-9.3	--	--	--	--	tPLH	-10.18	-7411244
-----	--	--	--	--	--	--	1.5-9.3	--	--	--	--	tPHL	---	-----
-7416240	--	--5.2	0.5-2.4	--	--	1.5-4.9	--	1.0-4.2	--	1-12.5	--	tPLH	-10.19	-7416240
-----	--	--5.2	0.5-2.4	--	--	1.5-4.9	--	1.0-4.2	--	1-12.5	--	tPHL	---	-----
-7416241	--	2.8-5.0	0.5-2.4	--	--	--	--	--	--	--	--	tPLH	-10.19	-7416241
-----	--	2.8-5.0	0.5-2.4	--	--	--	--	--	--	--	--	tPHL	---	-----
-7416244	--	2.8-5.0	1.0-2.4	--	--	1.5-5.2	--	1.0-4.0	--	1-13.5	--	tPLH	-10.19	-7416244
-----	--	2.8-5.0	1.0-2.4	--	--	1.5-5.2	--	1.0-4.0	--	1-13.5	--	tPHL	---	-----
-7416540	--	2.8-5.0	--	--	--	--	--	--	--	--	--	tPLH	-10.20	-7416540
-----	--	2.8-5.0	--	--	--	--	--	--	--	--	--	tPHL	---	-----
-7416541	--	--5.2	0.5-2.4	--	--	--	--	--	--	--	--	tPLH	-10.20	-7416541
-----	--	--5.2	0.5-2.4	--	--	--	--	--	--	--	--	tPHL	---	-----
-7416825	--	--	--	--	--	--		1-3.9	--	--	--	tPLH	-10.21	-7416825
-----	--	--	--	--	--	--		1-3.9	--	--	--	tPHL	---	-----
-7416827	--	--	1.0-3.5	--	--	--	--	1-3.9	--	--	--	tPLH	-10.21	-7416827
-----	--	--	1.0-3.5	--	--	--	--	1-3.9	--	--	--	tPHL	---	-----
-7416835	--	--	--	--	--	--	--	1-5.6	--	--	--	tPLH	-10.22	-7416835
-----	--	--	--	--	--	--	--	1-5.6	--	--	--	tPHL	---	-----
-74162240	--	--7.0	----	--	--	--	--	--	--	--	--	tPLH	-10.19	-74162240
-----	--	--7.0	----	--	--	--	--	--	--	--	--	tPHL	---	-----
-74162241	--	--5.2	--	--	--	--	--	--	--	--	--	tPLH	-10.19	-74162241
-----	--	--5.2	--	--	--	--	--	--	--	--	--	tPHL	---	-----
-74162244	--	2.8-5.0	1.5-3.5	--	--	--	--	2.8-4.6	--	--	--	tPLH	-10.19	-74162244
-----	--	2.8-5.0	1.5-3.5	--	--	--	--	2.8-4.6	--	--	--	tPHL	---	-----
-74162540	--	--6.5	--	--	--	--	--	--	--	--	--	tPLH	-10.20	-74162540
-----	--	--6.5	--	--	--	--	--	--	--	--	--	tPHL	---	-----
-74162541	--	--6.8	--	--	--	--	--	--	--	--	--	tPLH	-10.20	-74162541
-----	--	--6.8	--	--	--	--	--	--	--	--	--	tPHL	---	-----
-74162827	--	--	1.5-3.5	--	--	--	--	1.5-4.6	--	--	--	tPLH	-10.21	-74162827
-----	--	--	1.5-3.5	--	--	--	--	1.5-4.6	--	--	--	tPHL	---	-----
-7433240	--	--	--	--	--	--	--	--	3.5-4.0	--	--	tPLH	-10.23	-7433240
-----	--	--	--	--	--	--	--	--	3.5-4.0	--	--	tPHL	---	-----
-7433241	--	--	--	--	--	--	--	--	3.5-4.0	--	--	tPLH	-10.23	-7433241
-----	--	--	--	--	--	--	--	--	3.5-4.0	--	--	tPHL	---	-----
-7433244	--	--	--	--	--	--	--	--	3.5-4.0	--	--	tPLH	-10.23	-7433244
-----	--	--	--	--	--	--	--	--	3.5-4.0	--	--	tPHL	---	-----

NOTE: For some devices td, tp, or equivalent time delays are defined instead of tPLH, tPHL.

Table 10.6b. Low-Voltage 74xxx series CMOS device representative propagation delay times in nS. (continuation).

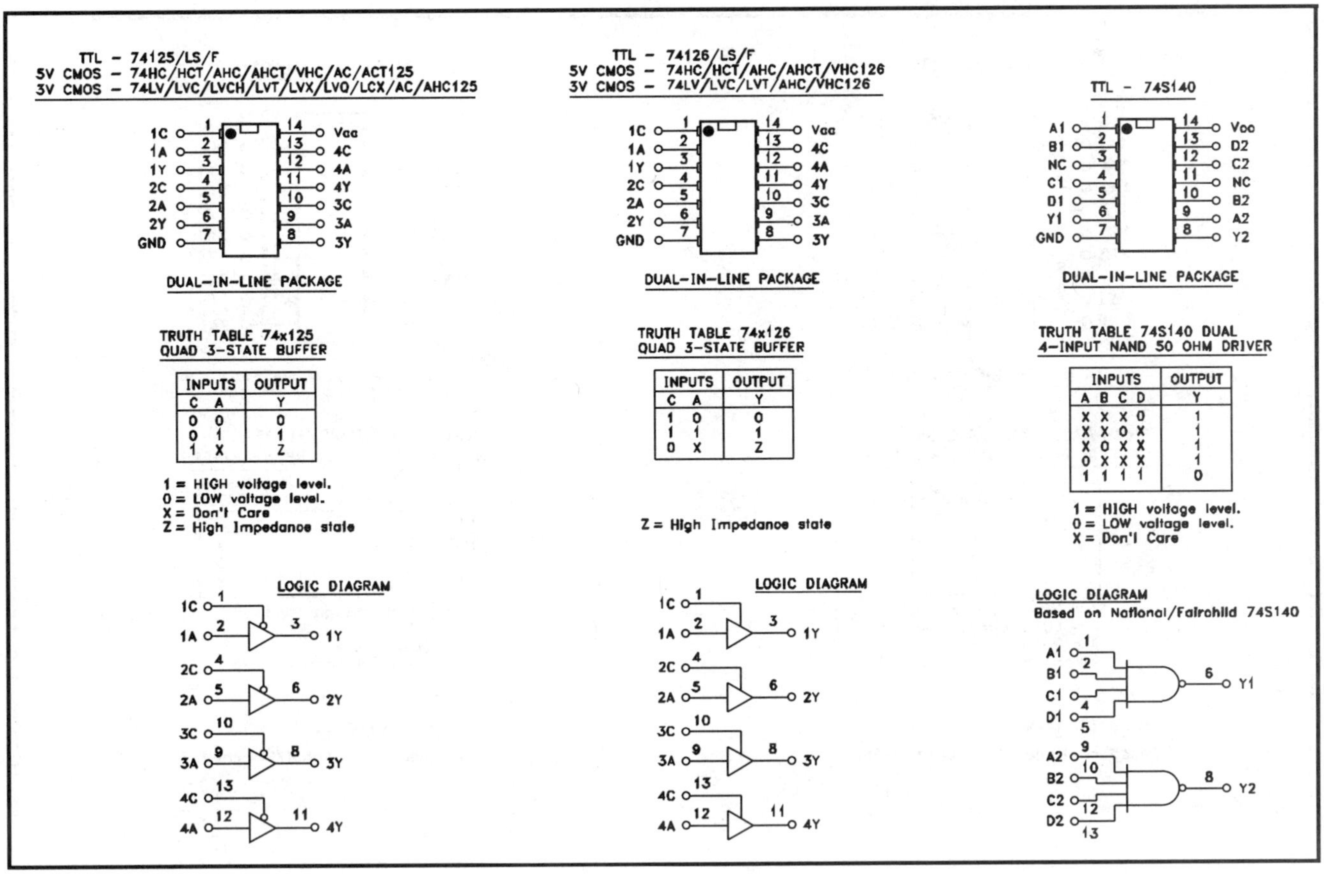

Figure 10.1. The 74x125 are quad inverting buffers provided with 3-state outputs. The 74x126 are non-inverting, but similar in other respects. The 74S140 contains two independent quad NAND 50W driver gates.

TTL - 74AS/ALS230

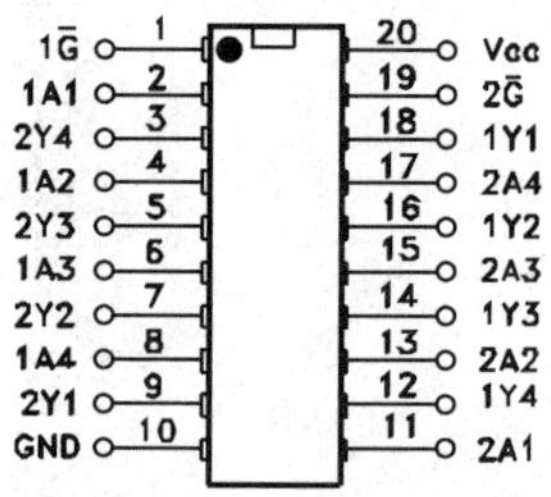

DUAL-IN-LINE PACKAGE

TRUTH TABLE 74AS/ALS230
3-STATE BUS DRIVERS/RECEIVERS

INPUTS		OUTPUTS	
G̅	A	1Y	2Y
0	0	1	0
0	1	0	1
1	X	Z	Z

1 = HIGH voltage level.
0 = LOW voltage level.
X = Don't Care
Z = High Impedance state

LOGIC DIAGRAM
Based on National/Fairchild 74AS230

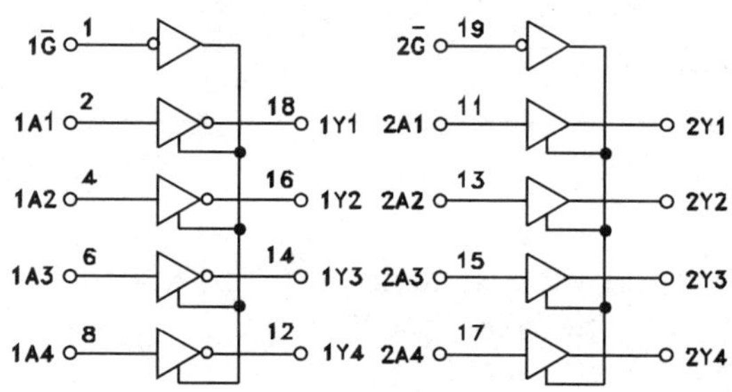

TTL - 74AS/ALS231

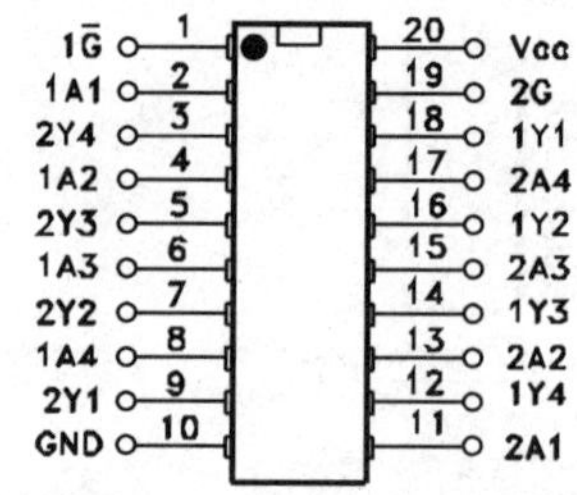

DUAL-IN-LINE PACKAGE

TRUTH TABLE 74AS/ALS231
3-STATE BUS DRIVERS/RECEIVERS

INPUTS				OUTPUTS	
1G̅	1A	2G	2A	1Y	2Y
0	0	1	0	1	1
0	1	1	1	0	0
1	X	0	X	Z	Z

1 = HIGH voltage level.
0 = LOW voltage level.
X = Don't Care
Z = High Impedance state

LOGIC DIAGRAM
Based on National/Fairchild 74AS231

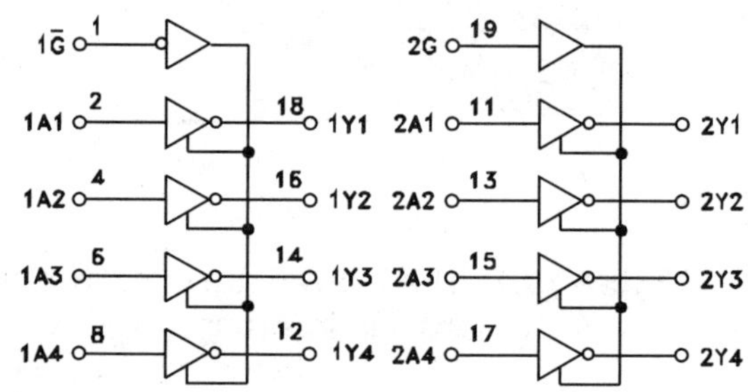

Figure 10.2. The TTL 74AS/ALS230 and '231 are designed for either bidirectional or unidirectional buffer interfacing in memory, microprocessor, or communications systems. The output circuits are low impedance for driving loads as low as 133W. The inputs feature high impedance to minimize line loading.

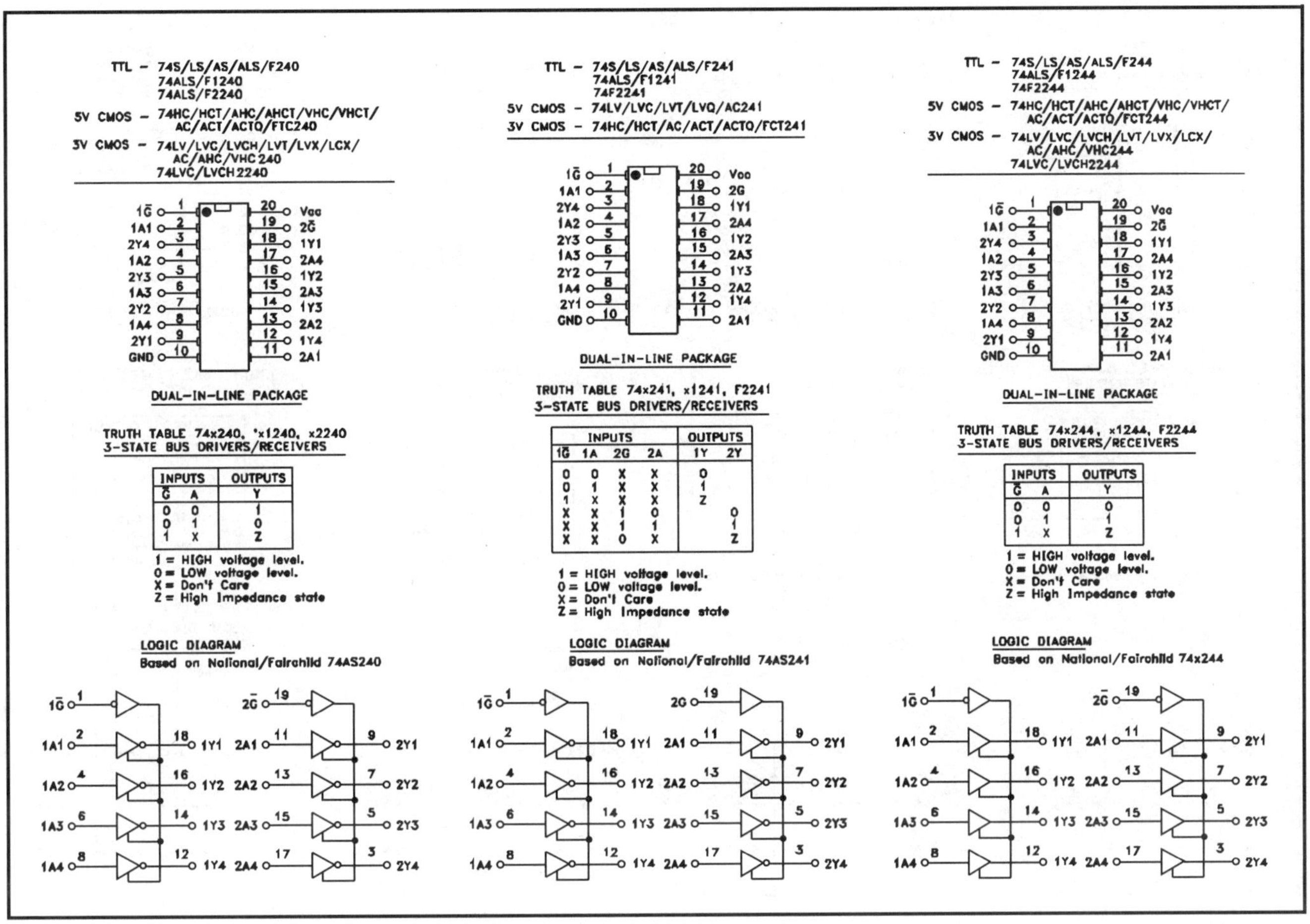

TRUTH TABLE 74x240, 'x1240, x2240

INPUTS		OUTPUTS
Ḡ	A	Y
0	0	1
0	1	0
1	X	Z

TRUTH TABLE 74x241, x1241, F2241

INPUTS				OUTPUTS	
1Ḡ	1A	2G	2A	1Y	2Y
0	0	X	X	0	
0	1	X	X	1	
1	X	X	X	Z	
X	X	1	0		0
X	X	1	1		1
X	X	0	X		Z

TRUTH TABLE 74x244, x1244, F2244

INPUTS		OUTPUTS
Ḡ	A	Y
0	0	0
0	1	1
1	X	Z

Figure 10.3. The 74x240, 'x241 and 'x244 are a popular series of bus driver/receivers featuring a versatile selection of logic options and 3-state output circuitry. These are designed for either bidirectional or unidirectional buffer interfacing in memory, microprocessor, or communications systems. The output circuits are constructed for driving low-impedance loads. The inputs feature high impedance to minimize line loading.

TTL - 74365/LS/F
5V CMOS - 74HC/HCT365
3V CMOS - 74LV365

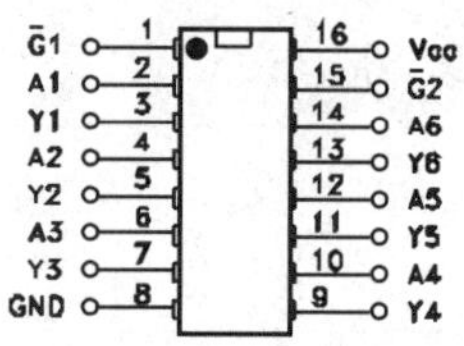

DUAL-IN-LINE PACKAGE

TRUTH TABLE 74x365
HEX 3-STATE BUFFERS

INPUT			OUTPUT
G̅1	G̅2	A	Y
1	X	X	HI-Z
X	1	X	HI-Z
0	0	1	1
0	0	0	0

1 = HIGH voltage level.
0 = LOW voltage level.
X = Don't Care
Z = High Impedance state

LOGIC DIAGRAM
Based on National/Fairchild 74LS365

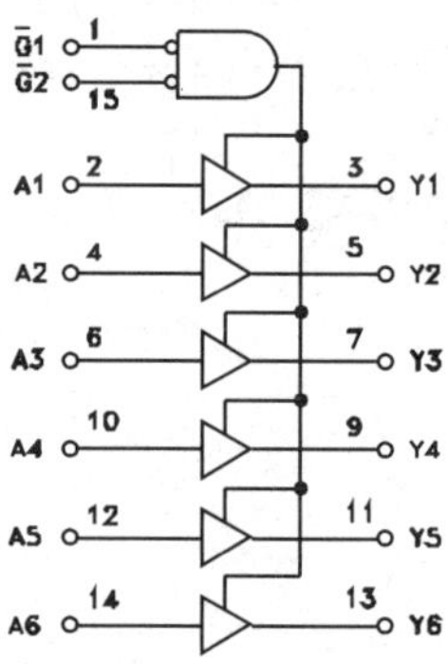

TTL - 74366/LS/F
5V CMOS - 74HC/HCT366

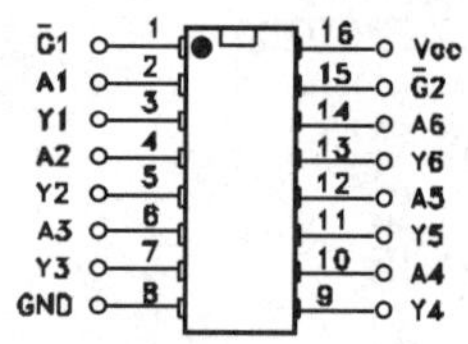

DUAL-IN-LINE PACKAGE

TRUTH TABLE 74x366
INVERTING HEX 3-STATE BUFFERS

INPUT			OUTPUT
G̅1	G̅2	A	Y
1	X	X	HI-Z
X	1	X	HI-Z
0	0	0	1
0	0	1	0

1 = HIGH voltage level.
0 = LOW voltage level.
X = Don't Care
Z = High Impedance state

LOGIC DIAGRAM
Based on National/Fairchild 74LS366

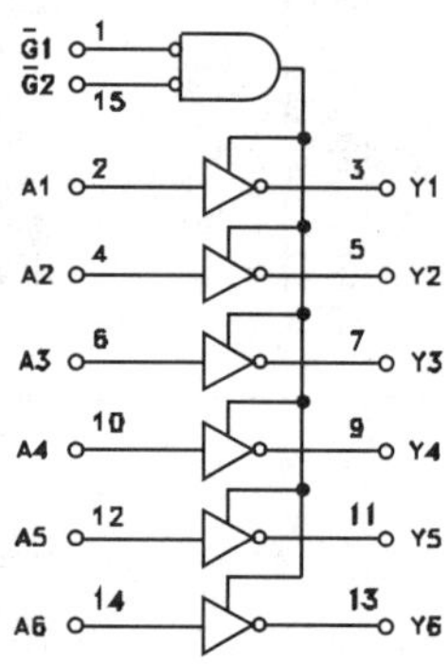

TTL - 74367/LS/F
5V CMOS - 74HC/HCT367
3V CMOS - 74LV367

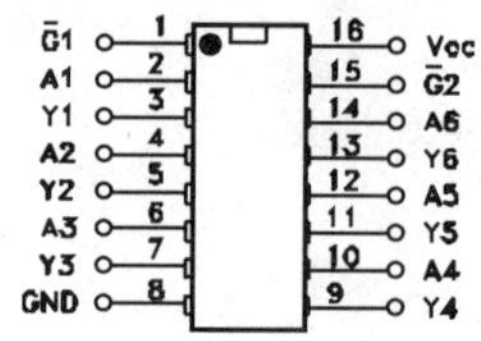

DUAL-IN-LINE PACKAGE

TRUTH TABLE 74x367
HEX 3-STATE BUFFERS

INPUT		OUTPUT
G̅	A	Y
0	0	0
0	1	1
1	X	HI-Z

1 = HIGH voltage level.
0 = LOW voltage level.
X = Don't Care
Z = High Impedance state

LOGIC DIAGRAM
Based on National/Fairchild 74LS367

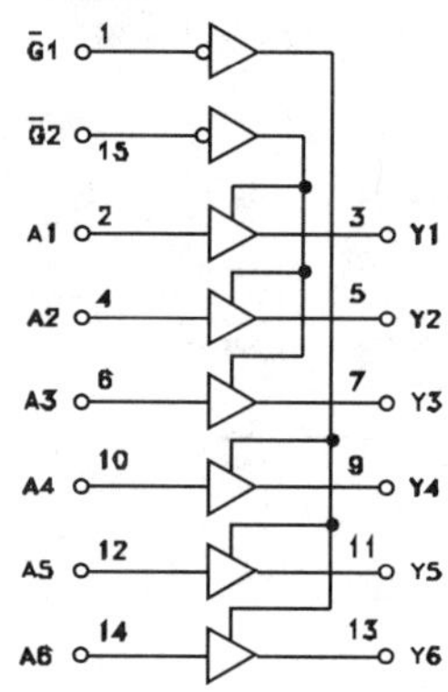

TTL - 74368/LS/F
5V CMOS - 74HC/HCT368
3V CMOS - 74LV368

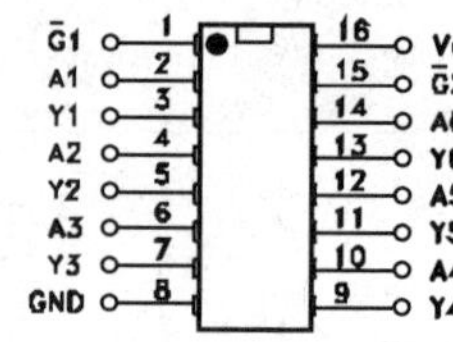

DUAL-IN-LINE PACKAGE

TRUTH TABLE 74x368
INVERTING HEX 3-STATE BUFFERS

INPUT		OUTPUT
G̅	A	Y
0	0	1
0	1	0
1	X	HI-Z

1 = HIGH voltage level.
0 = LOW voltage level.
X = Don't Care
Z = High Impedance state

LOGIC DIAGRAM
Based on National/Fairchild 74LS368

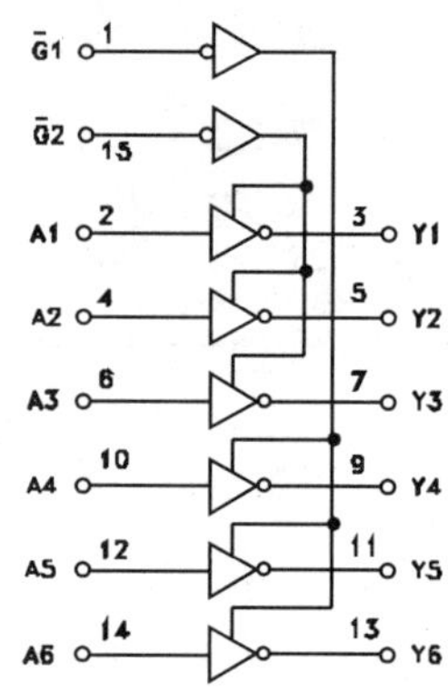

Figure 10.4. The 74x365, 'x366, 'x367 and 'x368 are a popular series of hex bus driver/receivers featuring a versatile selection of true and inverted logic with 3-state output provisions. The output circuits are constructed for driving low-impedance loads. The inputs feature high impedance to minimize line loading.

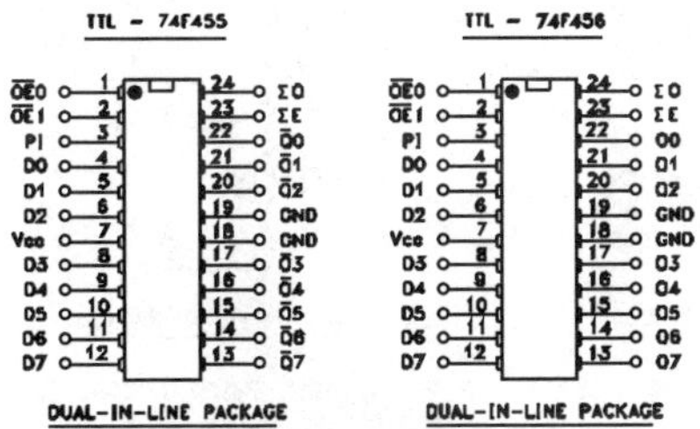

FUNCTION TABLE 74F455, 'F456
OCTAL BUFFER/DRIVER, WITH PARITY
3-STATE INVERTING ('455)
3-STATE NONINVERTING ('456)

INPUTS			OUTPUTS	
			'F455	'F456
$\overline{OE0}$	$\overline{OE1}$	Dn	$\overline{Qn}$	On
0	0	0	1	0
0	0	1	0	1
1	X	X	Z	Z
X	1	X	Z	Z

1 = HIGH voltage level.
0 = LOW voltage level.
X = Don't Care
Z = High Impedance state

FUNCTION TABLE FOR PARITY OUTPUTS

INPUTS	OUTPUTS	
Number of inputs High (PI,D0-D7)	ΣE	ΣO
Even - 0,2,4,6,8	1	0
Odd - 1,3,5,7,9	0	1
Any OEn = High	Z	Z

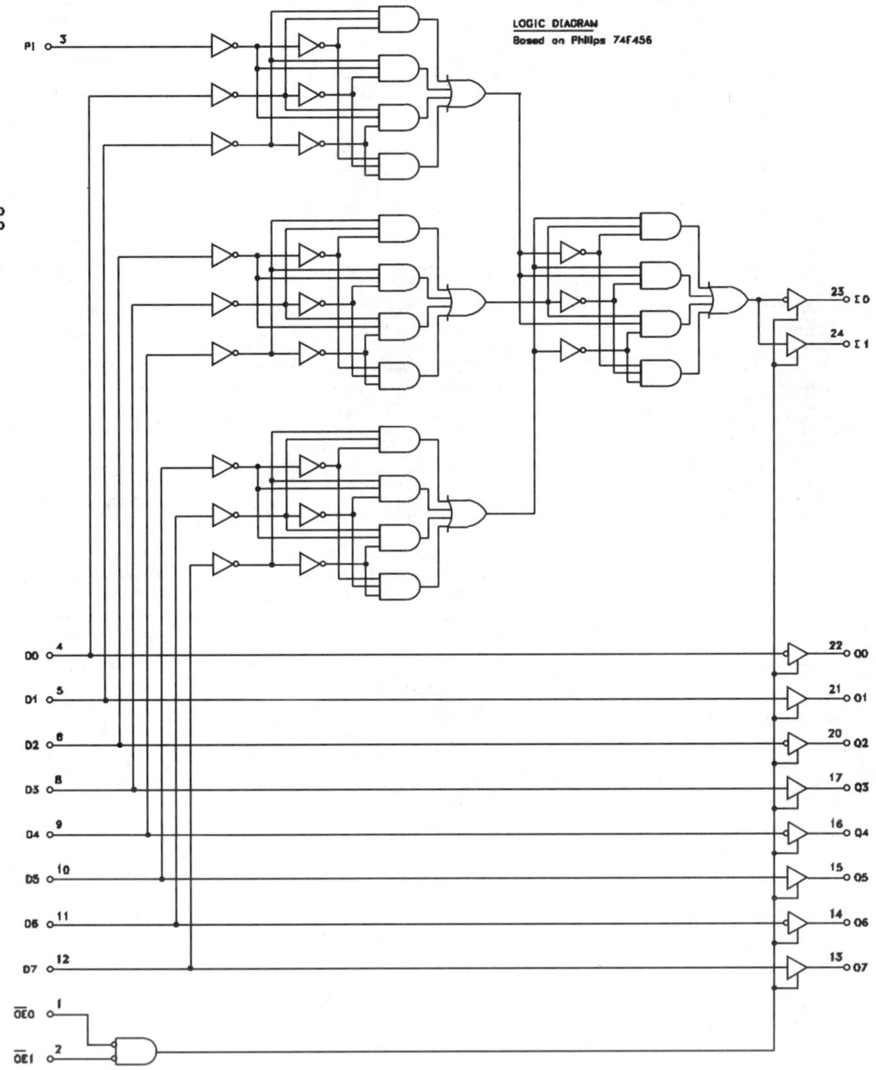

Figure 10.5. The 74F455 and 'F456 are octal buffers/drivers with parity featuring high-impedance NPN base inputs and 3-state outputs capable of sinking 64 mA and sourcing 15 mA. The 'F455 combines 'F240 functions and 'F280 parity generation and checking features in a single package. The 'F456 provides similarly for the 'F244 and 'F280.

TTL – 74LS/ALS465
74LS795

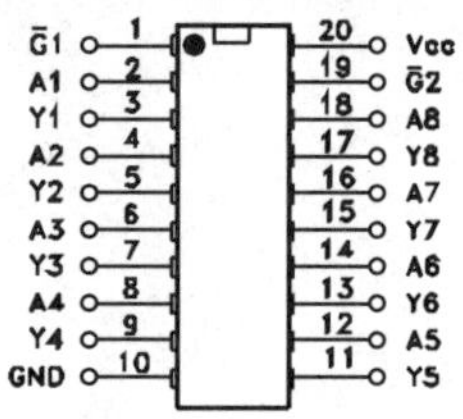

DUAL-IN-LINE PACKAGE

TRUTH TABLE 74LS/ALS465, 'LS795
3-STATE OCTAL BUFFERS

INPUTS			OUTPUTS
Ḡ1	Ḡ2	A	Y
1	X	X	Hi-Z
X	1	X	Hi-Z
0	0	1	1
0	0	0	0

1 = HIGH voltage level.
0 = LOW voltage level.
X = Don't Care
Z = High Impedance state

LOGIC DIAGRAM
Based on National/Fairchild 74LS465

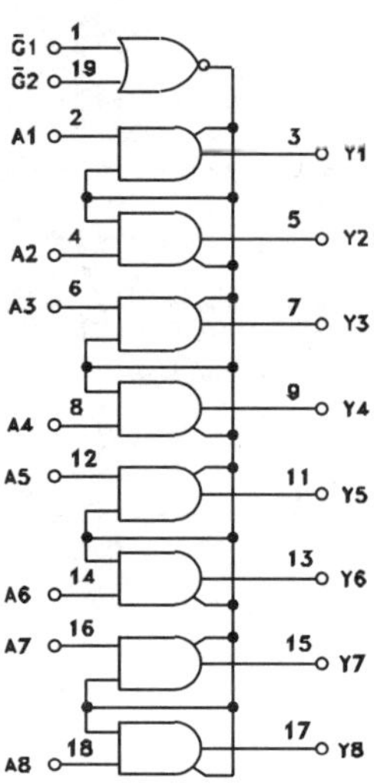

TTL – 74LS/ALS466
74LS796

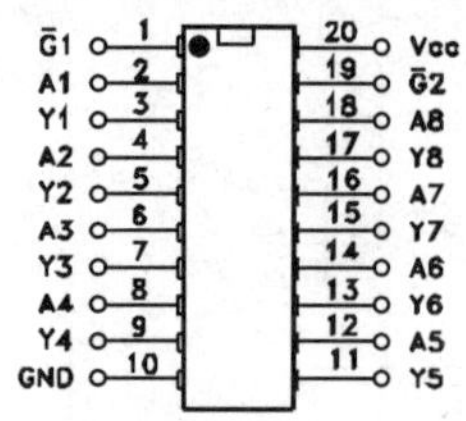

DUAL-IN-LINE PACKAGE

TRUTH TABLE 74LS/ALS466, 'LS796
3-STATE OCTAL BUFFERS

INPUTS			OUTPUTS
Ḡ1	Ḡ2	A	Y
1	X	X	Hi-Z
X	1	X	Hi-Z
0	0	1	0
0	0	0	1

1 = HIGH voltage level.
0 = LOW voltage level.
X = Don't Care
Z = High Impedance state

LOGIC DIAGRAM
Based on National/Fairchild 74LS466

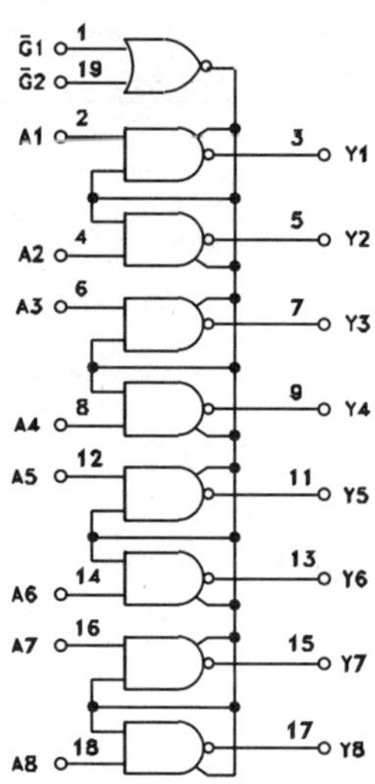

TTL – 74LS/ALS467
74LS797

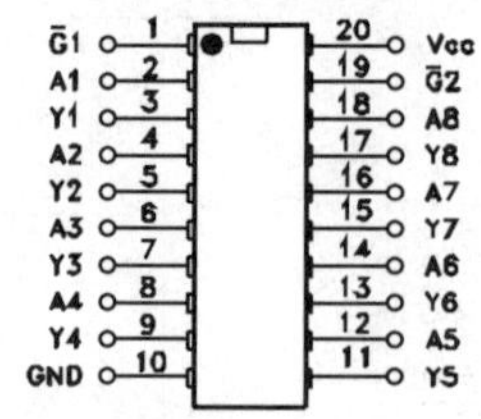

DUAL-IN-LINE PACKAGE

TRUTH TABLE 74LS/ALS467, 'LS797
3-STATE OCTAL BUFFERS

INPUTS		OUTPUTS
Ḡ	A	Y
1	X	Hi-Z
0	1	1
0	0	0

1 = HIGH voltage level.
0 = LOW voltage level.
X = Don't Care
Z = High Impedance state

LOGIC DIAGRAM
Based on National/Fairchild 74LS467

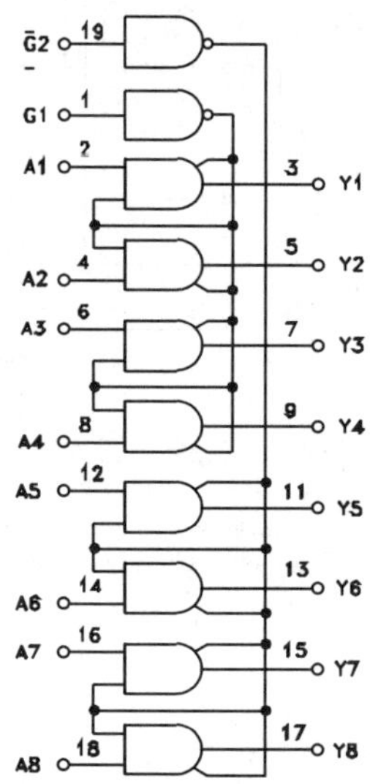

TTL – 74LS/ALS468
74LS798

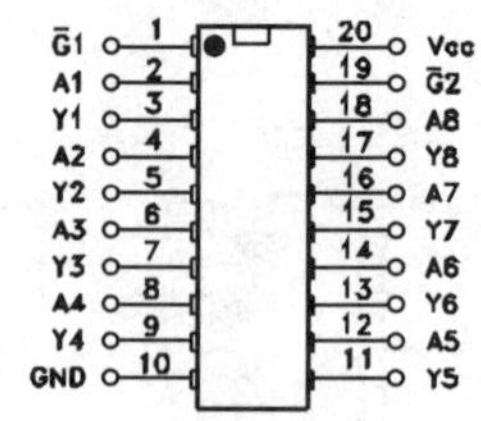

DUAL-IN-LINE PACKAGE

TRUTH TABLE 74LS/ALS468, 'LS798
3-STATE OCTAL BUFFERS

INPUTS		OUTPUTS
Ḡ	A	Y
1	X	Hi-Z
0	1	0
0	0	1

1 = HIGH voltage level.
0 = LOW voltage level.
X = Don't Care
Z = High Impedance state

LOGIC DIAGRAM
Based on National/Fairchild 74LS468

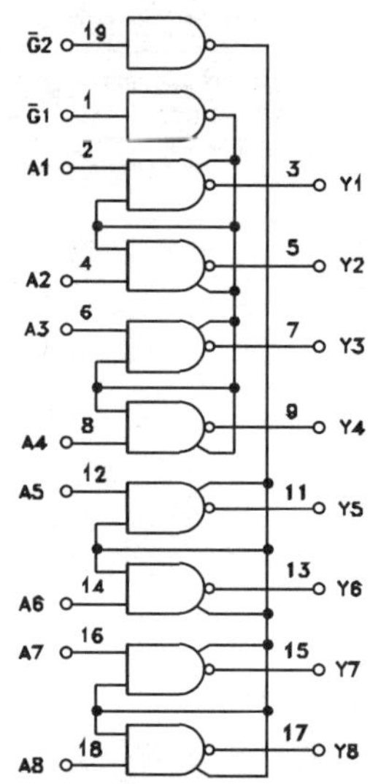

Figure 10.6. The 74LS/ALS465, '466, '467' and '468 provide eight 2-input buffers in a single package. One of the two control lines is used in a control mode to gate the output into a high-impedance state while the other passes data through the buffer. The '465 and '467 pass true data at their outputs; the '466 and '468 are inverting. Note the '467 and '468 can be configured for either single octal or dual quad operation.

TTL – 74LS/ALS/F540
74ALS746
74ALS2540
5V CMOS – 74HC/HCT/AHC/AHCT/AC540
3V CMOS – 74LV/LVC/LCX/AHC540

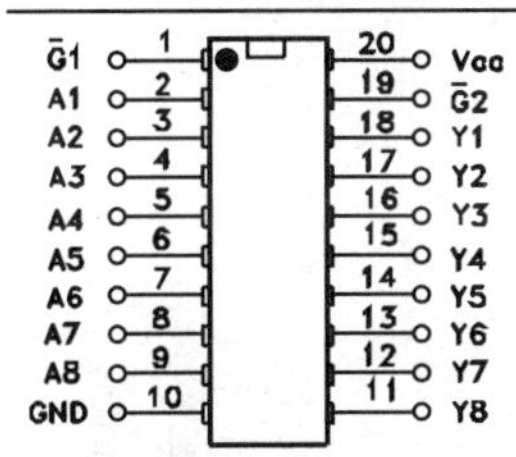

DUAL-IN-LINE PACKAGE

NOTE: The 74ALS746 has 20 KOhm input pullup resistors.

TRUTH TABLE 74x540, 'ALS746, 'ALS2540
OCTAL INVERTING BUFFER/LINE DRIVERS WITH 3-STATE OUTPUTS

INPUT			OUTPUT
Ḡ1	Ḡ2	A	Y
1	X	X	HI-Z
X	1	X	HI-Z
0	0	0	1
0	0	1	0

1 = HIGH voltage level.
0 = LOW voltage level.
X = Don't Care
Z = High Impedance state

LOGIC DIAGRAM
Based on National/Fairchild 74ALS540

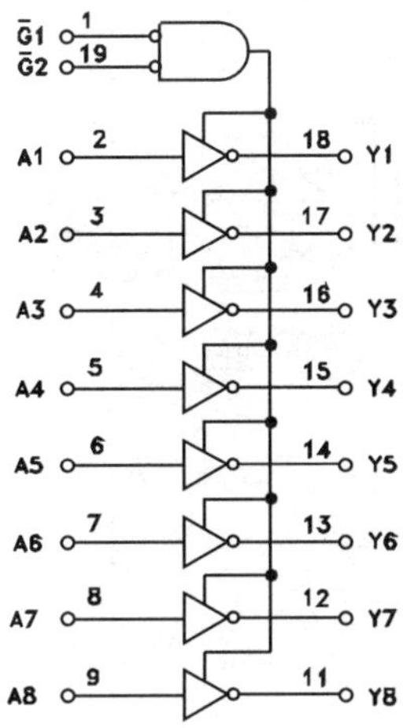

TTL – 74LS/ALS/F541
74ALS747
74ALS2541
5V CMOS – 74HC/HCT/AHC/AHCTAC541
3V CMOS – 74LV/LVC/LCX/AHC541

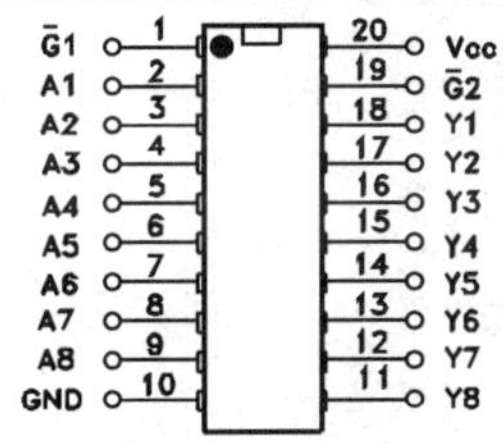

DUAL-IN-LINE PACKAGE

NOTE: The 74ALS747 has 20 KOhm input pullup resistors.

TRUTH TABLE 74x541, 'ALS747, 'ALS2541
OCTAL BUFFERS/LINE DRIVERS WITH 3-STATE OUTPUTS

INPUT			OUTPUT
Ḡ1	Ḡ2	A	Y
1	X	X	HI-Z
X	1	X	HI-Z
0	0	0	0
0	0	1	1

1 = HIGH voltage level.
0 = LOW voltage level.
X = Don't Care
Z = High Impedance state

LOGIC DIAGRAM
Based on National/Fairchild 74ALS541

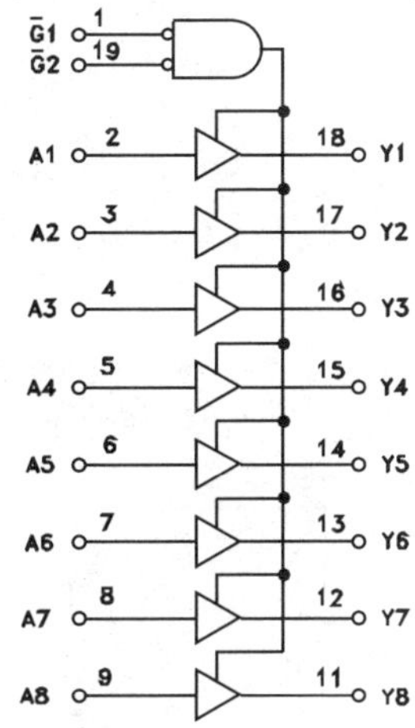

Figure 10.7. The 74x540 and 'x541 are octal buffers/line drivers designed to provide the performance of the popular 'x240 series. The package pinouts have been revised to position inputs and outputs on opposite sides to facilitate printed circuit board layout. The 3-state control is a NOR function such that with either control input high all outputs are in the high-impedance state.

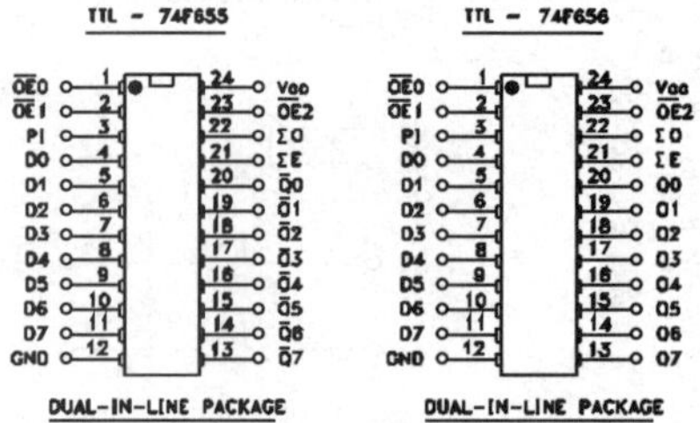

FUNCTION TABLE 74F655, 'F656
OCTAL BUFFER/DRIVER, WITH PARITY
3-STATE INVERTING ('F655)
3-STATE NONINVERTING ('F656)

INPUTS				OUTPUTS 'F655	OUTPUTS 'F656
$\overline{OE0}$	$\overline{OE1}$	$\overline{OE2}$	Dn	$\overline{Qn}$	Qn
0	0	0	0	1	0
0	0	0	1	0	1
1	X	X	X	Z	Z
X	1	X	X	Z	Z
X	X	1	X	Z	Z

1 = HIGH voltage level.
0 = LOW voltage level.
X = Don't Care
Z = High Impedance state

FUNCTION TABLE FOR PARITY OUTPUTS

INPUTS	OUTPUTS	
Number of inputs High (PI,D0-D7)	ΣE	ΣO
Even - 0,2,4,6,8	1	0
Odd - 1,3,5,7,9	0	1
Any OEn = High	Z	Z

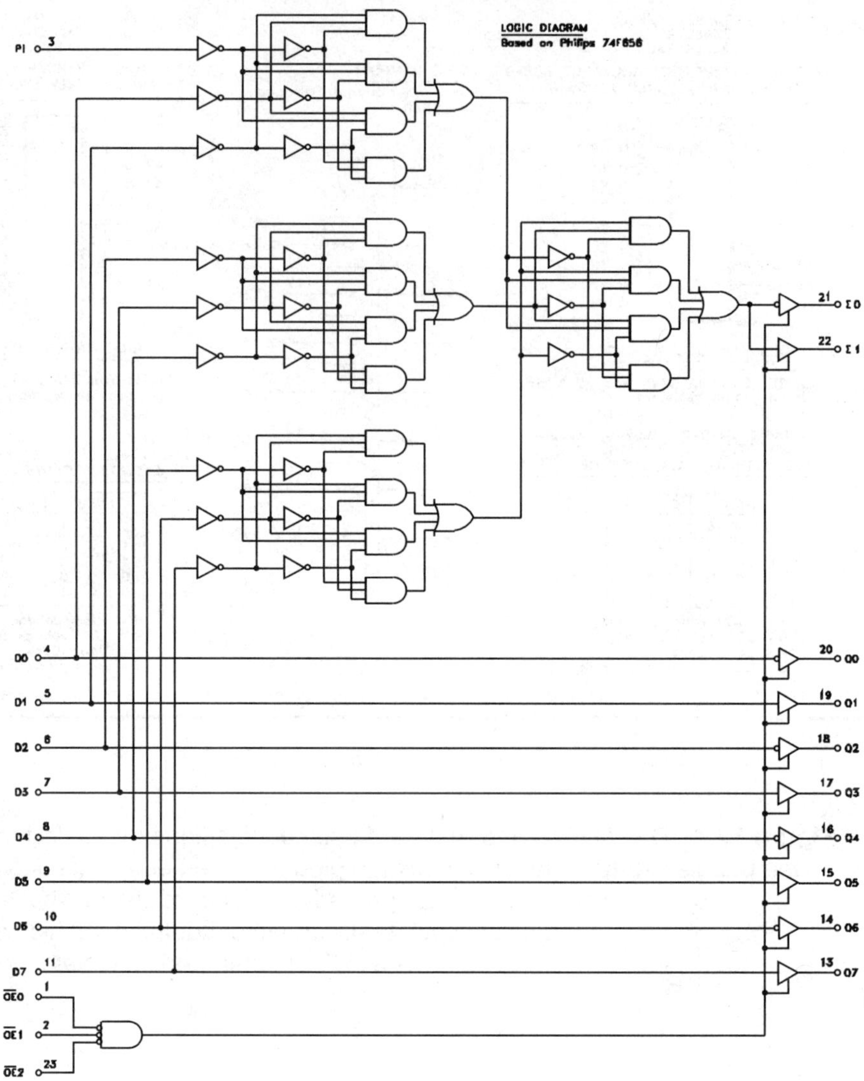

Figure 10.8. The 74F655 and 'F656 are equal in function to the 'F455, 'F456 of *Figure 10.5* except for the change in package power pinouts. These devices also provide the parity functions of the 'x280 series.

TTL – 74F756
74AS/ALS756

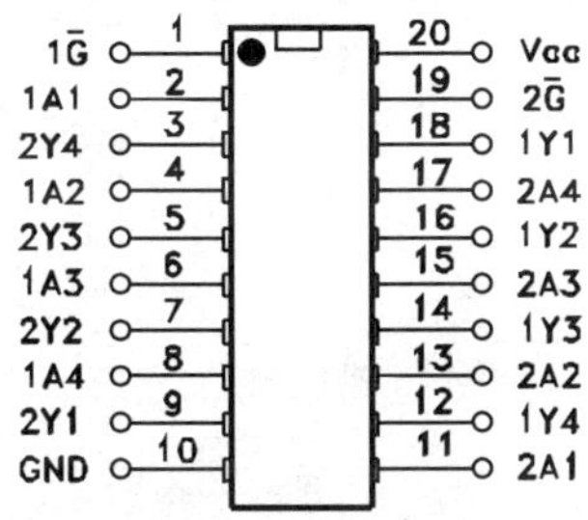

DUAL-IN-LINE PACKAGE

TTL – 74F757
74AS757

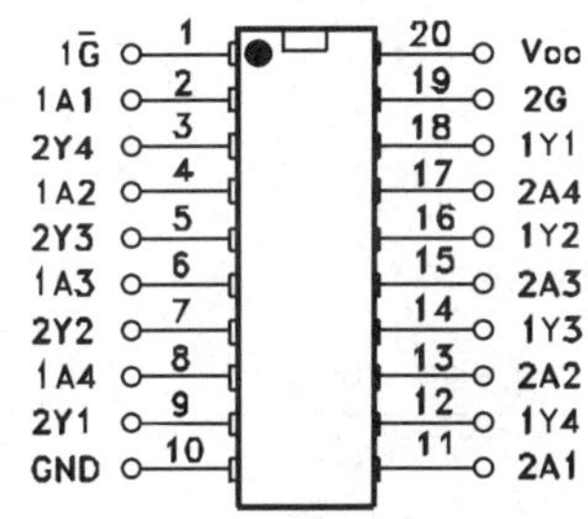

DUAL-IN-LINE PACKAGE

TRUTH TABLE 74x756 OCTAL INVERTER BUFFER (OPEN COLLECTOR)

INPUTS				OUTPUTS	
$1\overline{G}$	1A	$2\overline{G}$	2A	1Y	2Y
0	0	0	0	1	1
0	1	0	1	0	0
1	X	1	X	Z	Z

1 = HIGH voltage level.
0 = LOW voltage level.
X = Don't Care
Z = High Impedance state

TRUTH TABLE 74x757 OCTAL BUFFER (OPEN COLLECTOR)

INPUTS				OUTPUTS	
$1\overline{G}$	1A	2G	2A	1Y	2Y
0	0	1	0	0	0
0	1	1	1	1	1
1	X	0	X	Z	Z

1 = HIGH voltage level.
0 = LOW voltage level.
X = Don't Care
Z = High Impedance state

LOGIC DIAGRAM
Based on Texas Instruments 74AS/ALS756

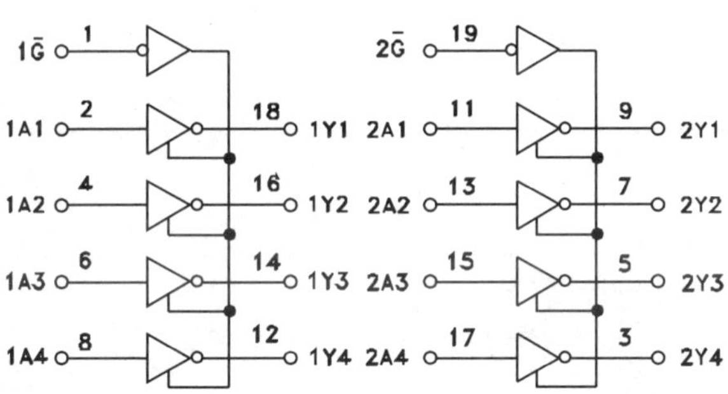

LOGIC DIAGRAM
Based on Texas Instruments 74AS757

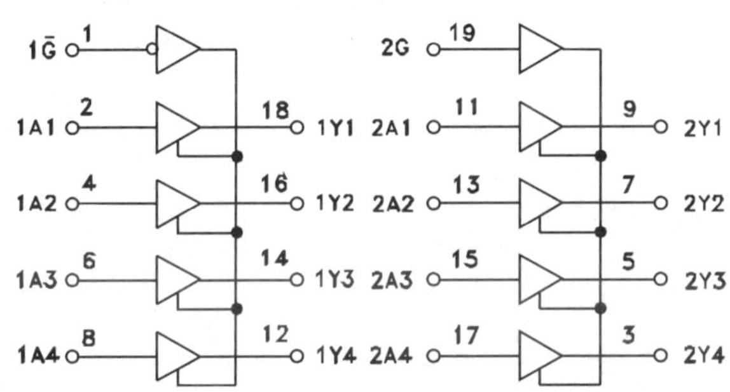

Figure 10.9. The 74x756 and 'x757 are open collector versions of the 'x240 and x241 devices equal in function to similar devices depicted in *Figure 10.3.*

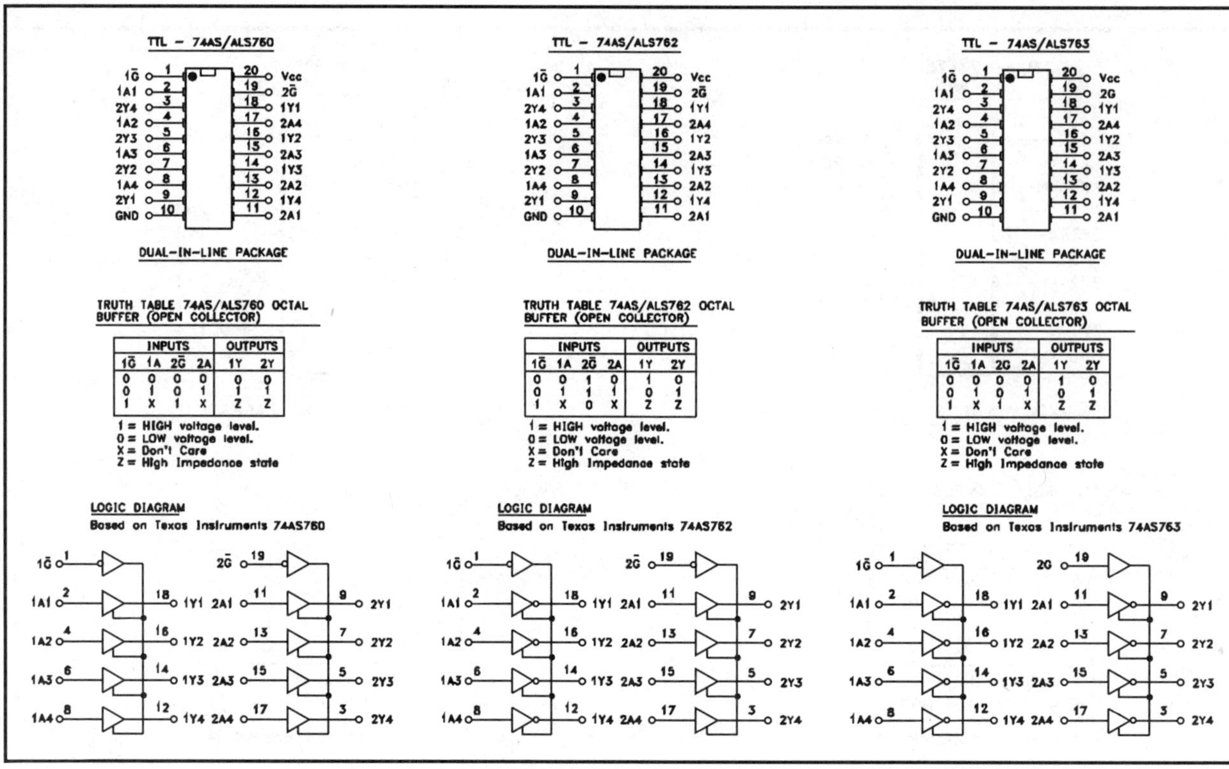

TRUTH TABLE 74AS/ALS760 OCTAL BUFFER (OPEN COLLECTOR)

INPUTS				OUTPUTS	
$\overline{1G}$	1A	$\overline{2G}$	2A	1Y	2Y
0	0	0	0	0	0
0	1	0	1	1	1
1	X	1	X	Z	Z

1 = HIGH voltage level.
0 = LOW voltage level.
X = Don't Care
Z = High Impedance state

TRUTH TABLE 74AS/ALS762 OCTAL BUFFER (OPEN COLLECTOR)

INPUTS				OUTPUTS	
$\overline{1G}$	1A	$\overline{2G}$	2A	1Y	2Y
0	0	1	0	1	0
0	1	1	1	0	1
1	X	0	X	Z	Z

1 = HIGH voltage level.
0 = LOW voltage level.
X = Don't Care
Z = High Impedance state

TRUTH TABLE 74AS/ALS763 OCTAL BUFFER (OPEN COLLECTOR)

INPUTS				OUTPUTS	
$\overline{1G}$	1A	2G	2A	1Y	2Y
0	0	0	0	1	0
0	1	0	1	0	1
1	X	1	X	Z	Z

1 = HIGH voltage level.
0 = LOW voltage level.
X = Don't Care
Z = High Impedance state

Figure 10.10. The 74F760 is an open collector version of the 'F244. The 74AS/ALS760, '762 and 'x763 are similarly open collector versions of the 'x240, 'x241, equal in function to similar devices depicted in *Figure 10.3*.

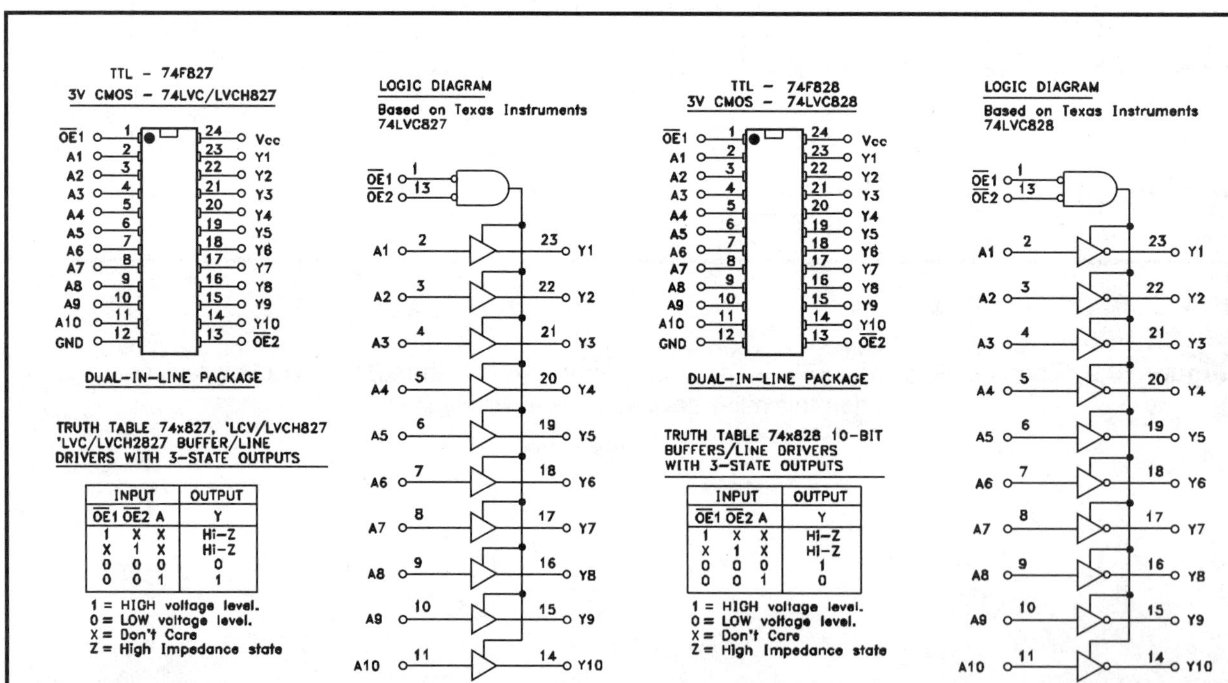

TRUTH TABLE 74x827, 'LCV/LVCH827 'LVC/LVCH2827 BUFFER/LINE DRIVERS WITH 3-STATE OUTPUTS

INPUT			OUTPUT
$\overline{OE1}$	$\overline{OE2}$	A	Y
1	X	X	Hi-Z
X	1	X	Hi-Z
0	0	0	0
0	0	1	1

1 = HIGH voltage level.
0 = LOW voltage level.
X = Don't Care
Z = High Impedance state

TRUTH TABLE 74x828 10-BIT BUFFERS/LINE DRIVERS WITH 3-STATE OUTPUTS

INPUT			OUTPUT
$\overline{OE1}$	$\overline{OE2}$	A	Y
1	X	X	Hi-Z
X	1	X	Hi-Z
0	0	0	1
0	0	1	0

1 = HIGH voltage level.
0 = LOW voltage level.
X = Don't Care
Z = High Impedance state

Figure 10.11. The 74x827 and 'x828 are ten-bit buffers/line drivers with non-inverting and inverting outputs, respectively. The TTL device offers an NPN high-impedance input. The low-voltage CMOS versions accept either 3.3 or 5 volts at their inputs; in 3-state operation they can provide five volts at their outputs.

TTL - 74S940

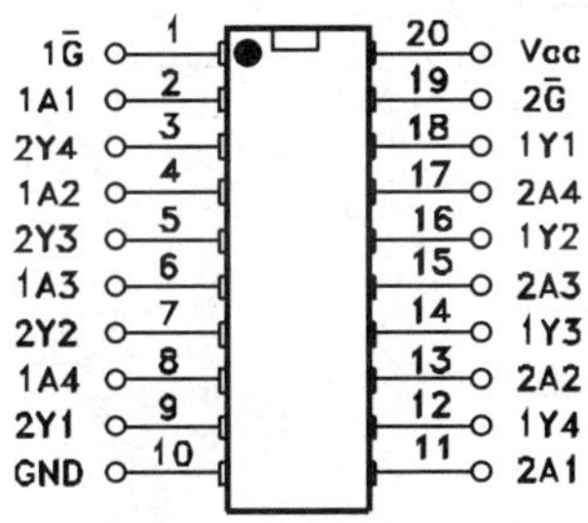

DUAL-IN-LINE PACKAGE

TRUTH TABLE 74S940 3-STATE BUFFER/LINE DRIVER LINE RECEIVER

INPUTS				OUTPUTS	
1Ḡ	1A	2Ḡ	2A	1Y	2Y
0	0	0	0	1	1
0	1	0	1	0	0
1	X	1	X	Z	Z

1 = HIGH voltage level.
0 = LOW voltage level.
X = Don't Care
Z = High Impedance state

LOGIC DIAGRAM
Based on National/Fairchild 74S940

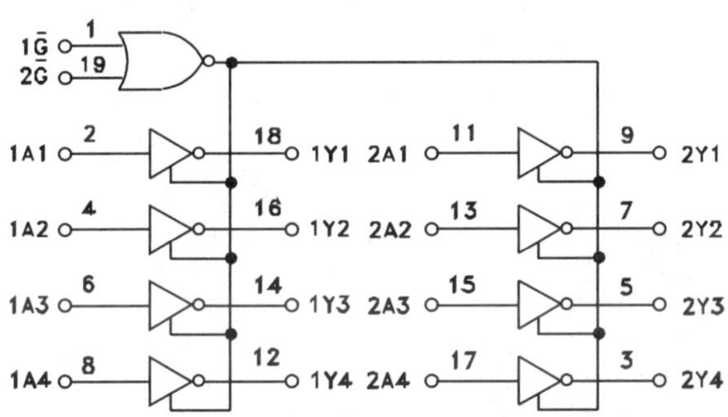

TTL - 74S941

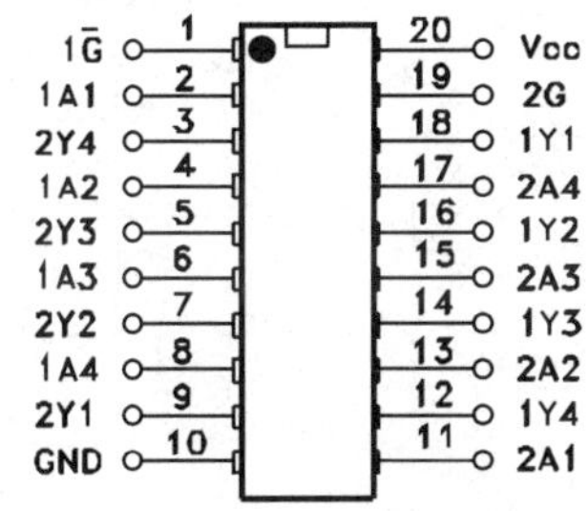

DUAL-IN-LINE PACKAGE

TRUTH TABLE 74S941 3-STATE BUFFER/LINE DRIVER LINE RECEIVER

INPUTS				OUTPUTS	
1Ḡ	1A	2G	2A	1Y	2Y
0	0	1	0	0	0
0	1	1	1	1	1
1	X	0	X	Z	Z

1 = HIGH voltage level.
0 = LOW voltage level.
X = Don't Care
Z = High Impedance state

LOGIC DIAGRAM
Based on National/Fairchild 74S941

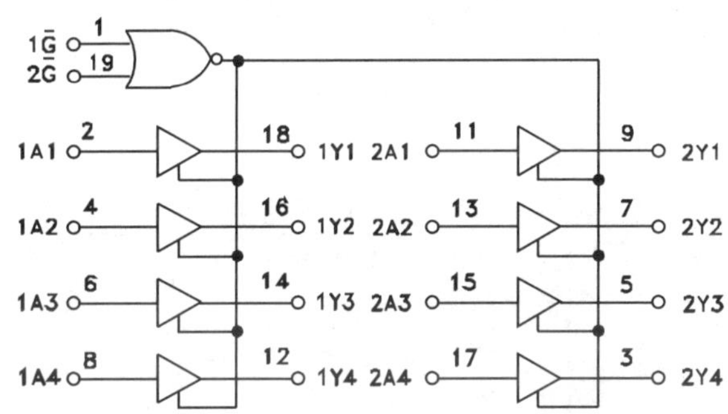

Figure 10.12. The 74S940/941 are octal 3-state buffer/line drivers/line receivers. Their design improves both performance and printed-circuit board density. The design features 400 mV of hysteresis at each low-current PNP input. The outputs can drive terminated lines down to 133 W.

TTL - 74F3037
74F3038

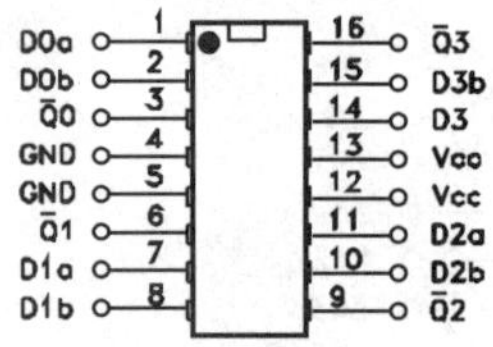

DUAL-IN-LINE PACKAGE

LOGIC DIAGRAM
Based on Philips 74F3037, 'F3038

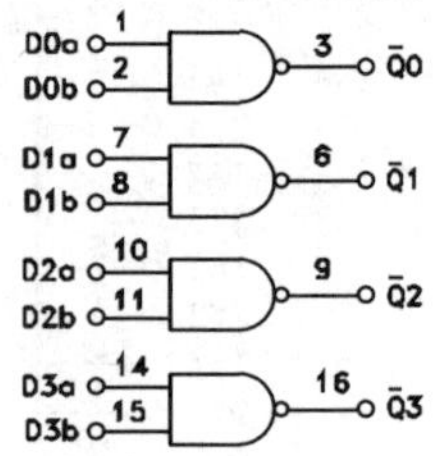

NOTE: The 74F3038 is an open collector version of the 74F3037.

TRUTH TABLE 74F3037, '3038 QUAD 2-INPUT NAND 30 OHM DRIVER

INPUTS		OUTPUTS
Dna	Dnb	Q̄n
0	0	1
0	1	1
1	0	1
1	1	0

1 = HIGH voltage level.
0 = LOW voltage level.

TTL - 74F3040

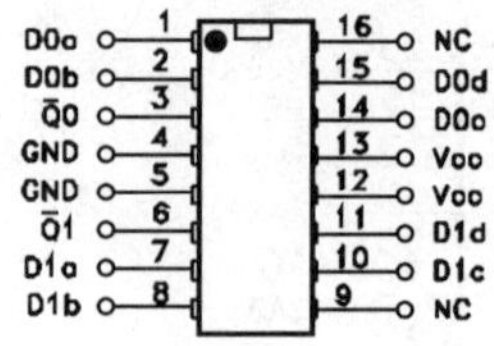

DUAL-IN-LINE PACKAGE

LOGIC DIAGRAM
Based on Philips 74F3040

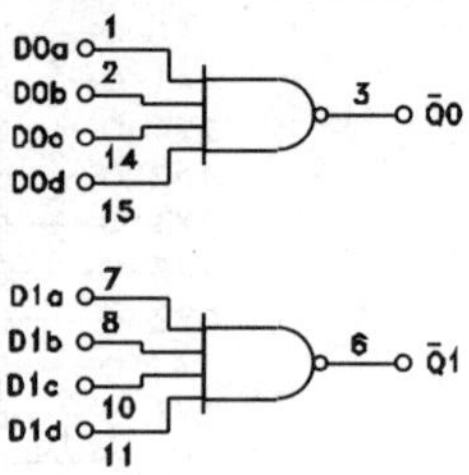

TRUTH TABLE 74F3040 DUAL 4-INPUT NAND 30 OHM LINE DRIVER

INPUTS				OUTPUTS
Dna	Dnb	Dnc	Dnd	Q̄n
0	X	X	X	1
X	0	X	X	1
X	X	0	X	1
X	X	X	0	1
1	1	1	1	0

1 = HIGH voltage level.
0 = LOW voltage level.
X = Don't Care

Figure 10.13. The 74F3037/3038 are quad 2-input NAND high current 30W line drivers. The 74F3040 dual 4-input NAND is also a high-current 30W line driver. The package pinning has been designed to deal with the transmission line effects of printed-circuit boards. The devices will sink up to 160 mA with a Vcc of 4.5 volts.

TTL - 74ALS29827

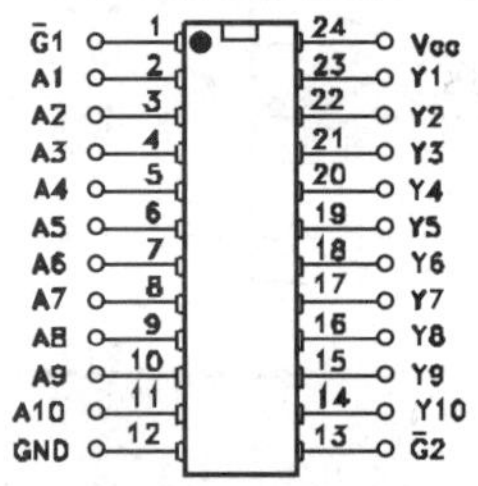

DUAL-IN-LINE PACKAGE

TRUTH TABLE 74ALS29827 10-BUFFER AND BUS DRIVER WITH 3-STATE OUTPUTS

INPUT			OUTPUT
Ḡ1	Ḡ2	An	Yn
1	X	X	Hi-Z
X	1	X	Hi-Z
0	0	1	1
0	0	0	0

1 = HIGH voltage level.
0 = LOW voltage level.
X = Don't Care
Z = High Impedance state

LOGIC DIAGRAM
Based on Texas Instruments 74ALS29827

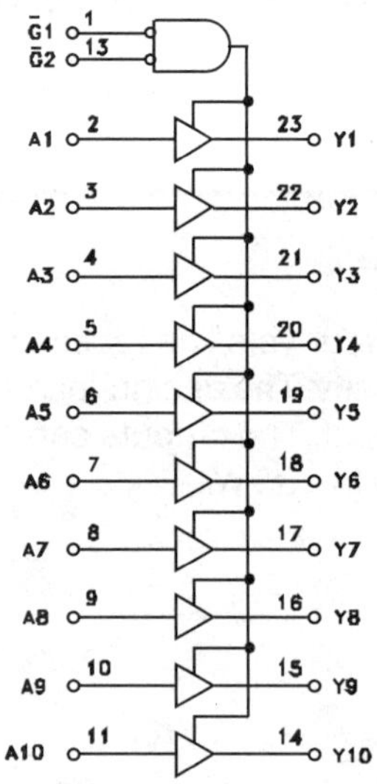

TTL - 74ALS29828

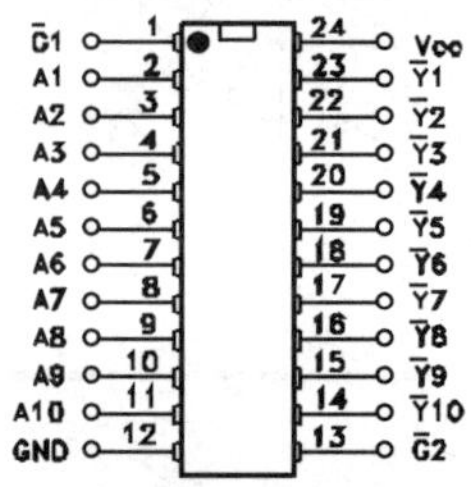

DUAL-IN-LINE PACKAGE

TRUTH TABLE 74ALS29828 10-BUFFER AND BUS DRIVER WITH 3-STATE OUTPUTS

INPUT			OUTPUT
Ḡ1	Ḡ2	An	Ȳn
1	X	X	Hi-Z
X	1	X	Hi-Z
0	0	0	0
0	0	1	1

1 = HIGH voltage level.
0 = LOW voltage level.
X = Don't Care
Z = High Impedance state

LOGIC DIAGRAM
Based on Texas Instruments 74ALS29828

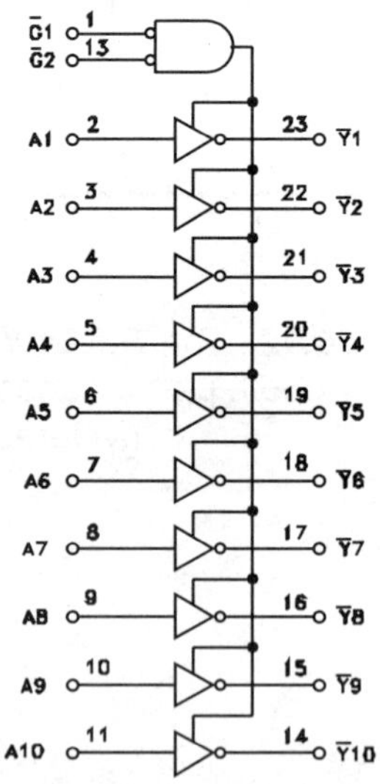

Figure 10.14. The 74ALS29827/29828 are 10-bit buffers and line drivers with 3-state outputs. The devices are non-inverting and inverting, respectively. Package pinout features data inputs and outputs on opposite sides to facilitate printed-circuit board layout. If either control input is high, the output is in the high-impedance state.

TTL – 74F30240

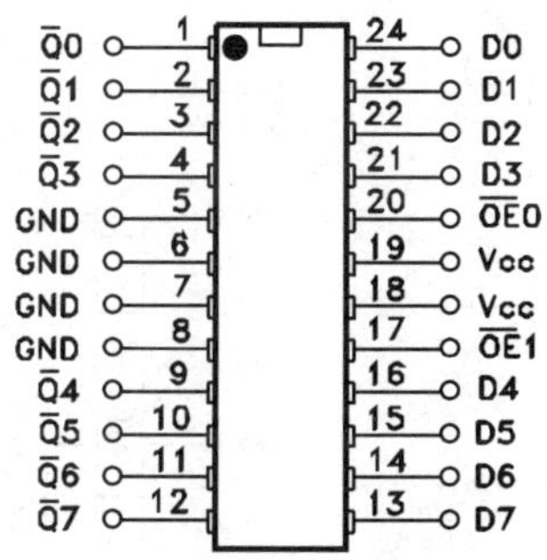

DUAL-IN-LINE PACKAGE

TTL – 74F30244

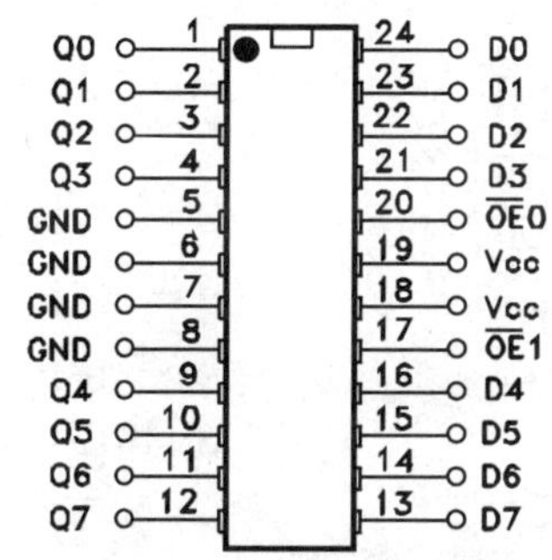

DUAL-IN-LINE PACKAGE

TRUTH TABLE 74F30240 OCTAL
30 OHM LINE DRIVER, OPEN COLLECTOR
INVERTING

INPUTS		OUTPUTS
$\overline{OEn}$	Dn	$\overline{Qn}$
0	0	1
0	1	0
1	X	OFF

1 = HIGH voltage level.
0 = LOW voltage level.
X = Don't Care

TRUTH TABLE 74F30244 OCTAL
30 OHM LINE DRIVER, OPEN COLLECTOR
NON-INVERTING

INPUTS		OUTPUTS
$\overline{OEn}$	Dn	Qn
0	0	0
0	1	1
1	X	OFF

1 = HIGH voltage level.
0 = LOW voltage level.
X = Don't Care

LOGIC DIAGRAM
Based on Philips 74f30240

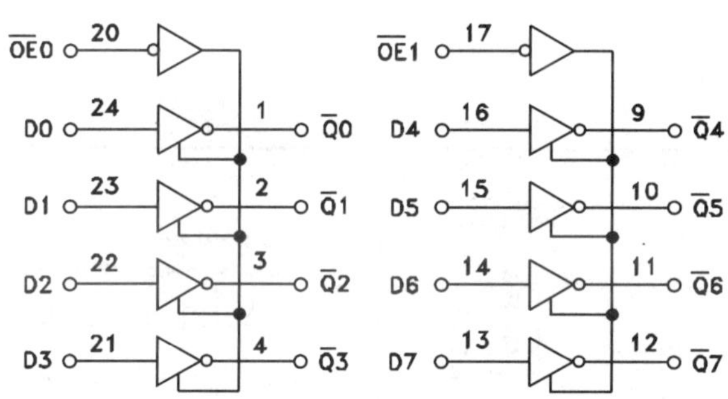

LOGIC DIAGRAM
Based on Philips 74F30244

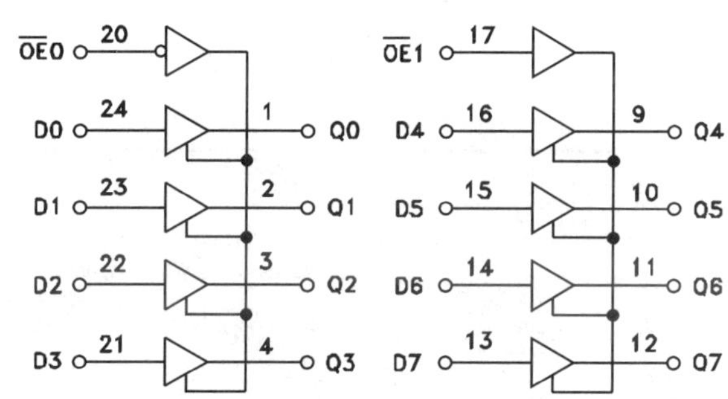

Figure 10.15. The 74F30240/30244 are octal open collector 30W line drivers. The 'F30240 is inverting, the 'F30244 non-inverting. The package pinning features center power pins for lead inductance reduction providing improved speed and noise immunity.

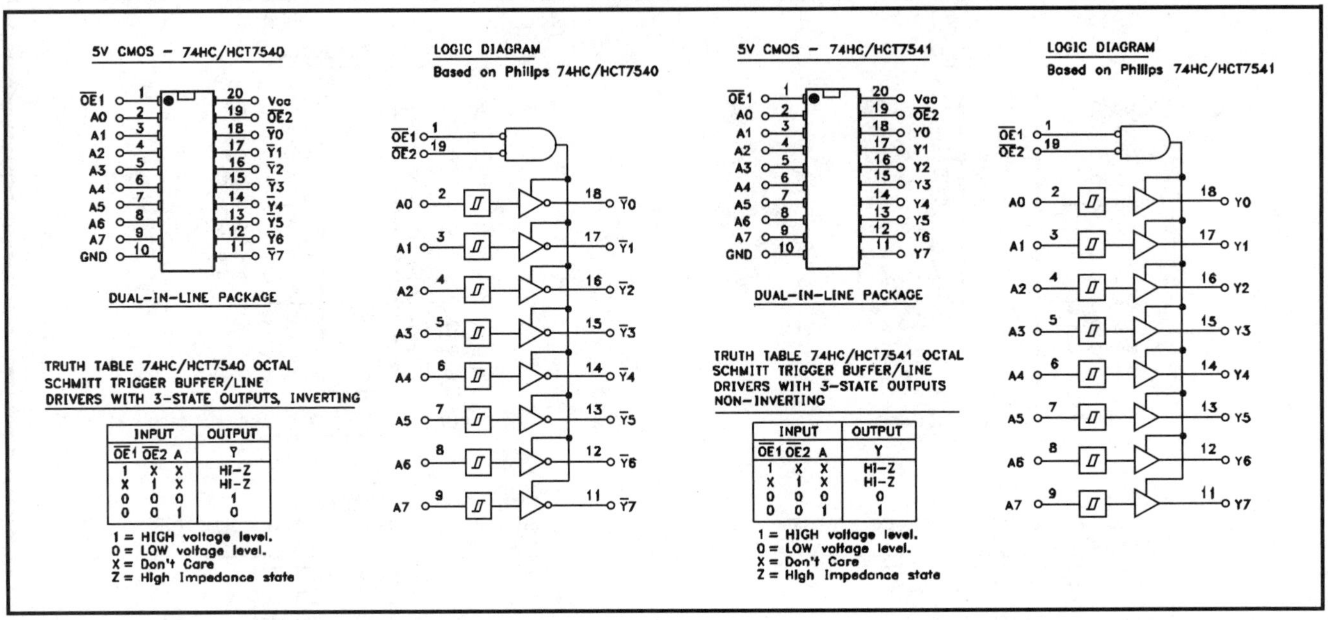

74HC/HCT7540:

INPUT			OUTPUT
$\overline{OE1}$	$\overline{OE2}$	A	$\overline{Y}$
1	X	X	HI-Z
X	1	X	HI-Z
0	0	0	1
0	0	1	0

74HC/HCT7541:

INPUT			OUTPUT
$\overline{OE1}$	$\overline{OE2}$	A	Y
1	X	X	HI-Z
X	1	X	HI-Z
0	0	0	0
0	0	1	1

Figure 10.16. The 74HC/HCT7540/7541 are octal Schmitt trigger buffers/line drivers with 3-state outputs. The '7540 is inverting; the '7541 non-inverting. A logic high on either control input results in a high-impedance output state.

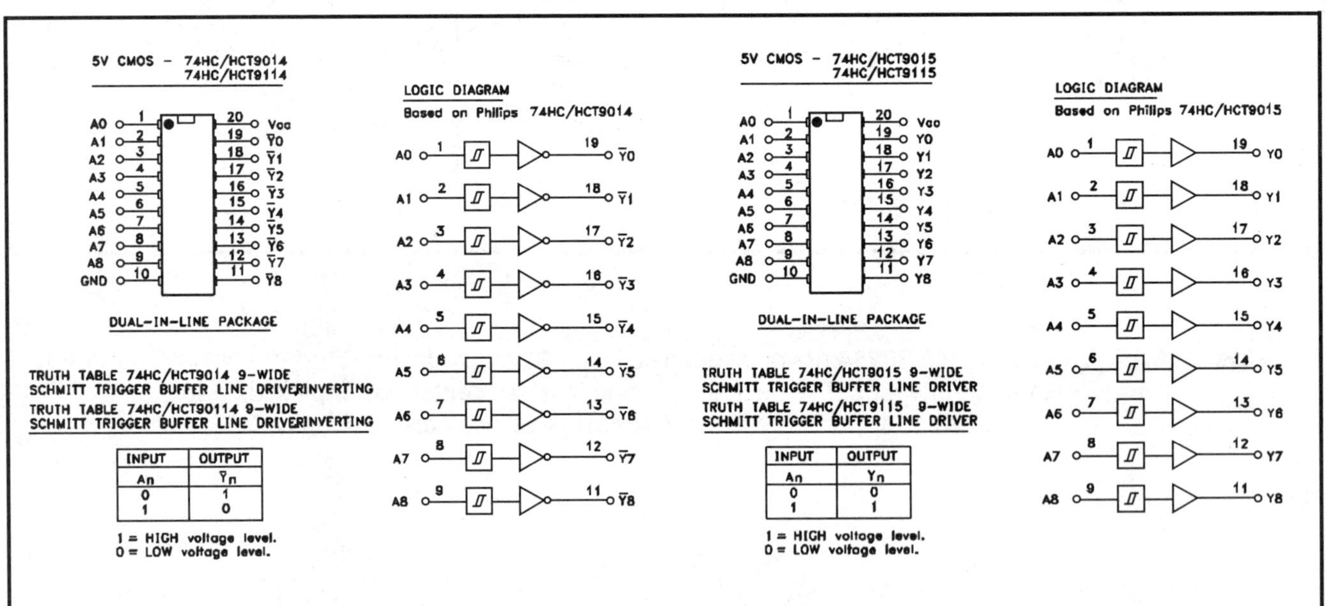

74HC/HCT9014 / 9114:

INPUT	OUTPUT
An	$\overline{Y}n$
0	1
1	0

74HC/HCT9015 / 9115:

INPUT	OUTPUT
An	Yn
0	0
1	1

Figure 10.17. The 74HC/HCT9014/9015/9114/9115 are nine-wide Schmitt trigger buffers/line drivers. The '9014, '9114 are inverting; the '9015 and '9115 non-inverting. The '9114/9115 feature open drain outputs. The inputs transform slowly changing signals into sharply defined jitter-free output signals.

5V CMOS – 74AC/ACT11240
3V CMOS – 74AC11240

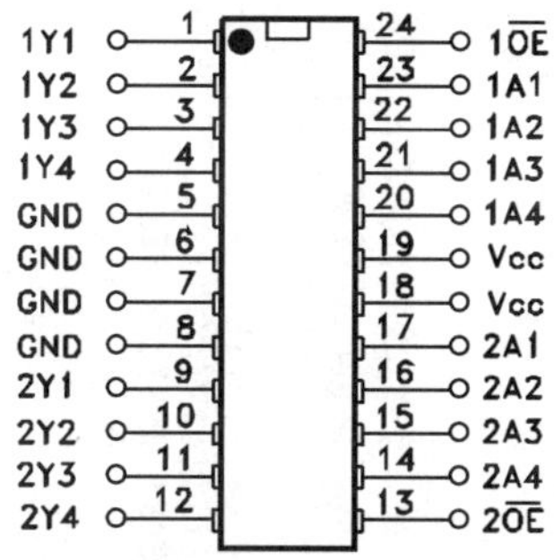

DUAL–IN–LINE PACKAGE

TRUTH TABLE 74AC/ACT11240 OCTAL BUFFER/LINE DRIVER WITH 3-STATE OUTPUTS INVERTING

INPUTS		OUTPUTS
nOE	nA	nY
0	0	1
0	1	0
1	X	Z

1 = HIGH voltage level.
0 = LOW voltage level.
X = Don't Care
Z = High Impedance

LOGIC DIAGRAM
Based on Texas Instruments 74AC/ACT11240

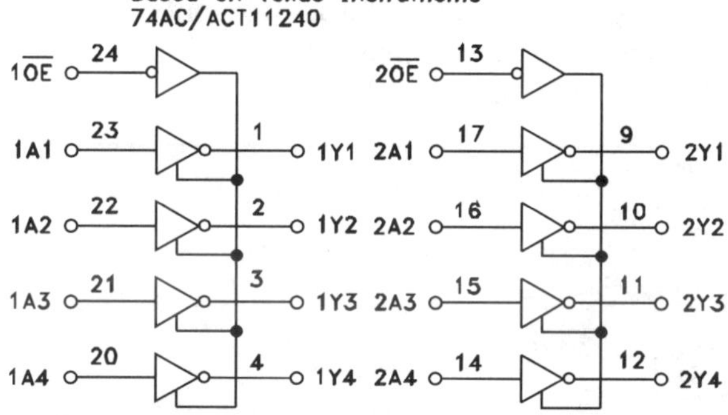

5V CMOS – 74AC/ACT11244
3V CMOS – 74AC11244

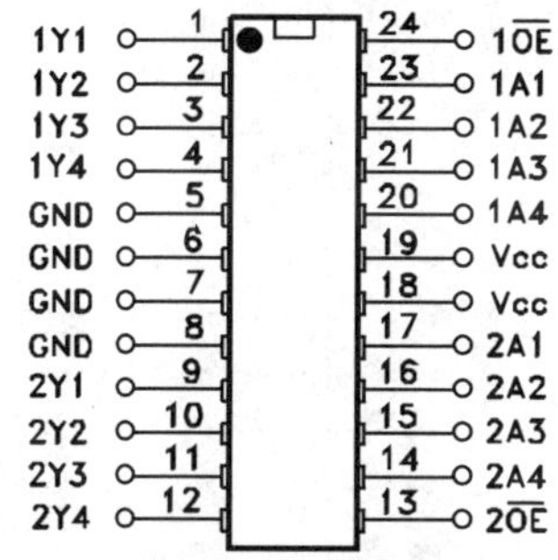

DUAL–IN–LINE PACKAGE

TRUTH TABLE 74AC/ACT11244 OCTAL BUFFER/LINE DRIVER WITH 3-STATE OUTPUTS NON-INVERTING

INPUTS		OUTPUTS
nOE	nA	nY
0	0	0
0	1	1
1	X	Z

1 = HIGH voltage level.
0 = LOW voltage level.
X = Don't Care
Z = High Impedance

LOGIC DIAGRAM
Based on Texas Instruments 74AC/ACT11244

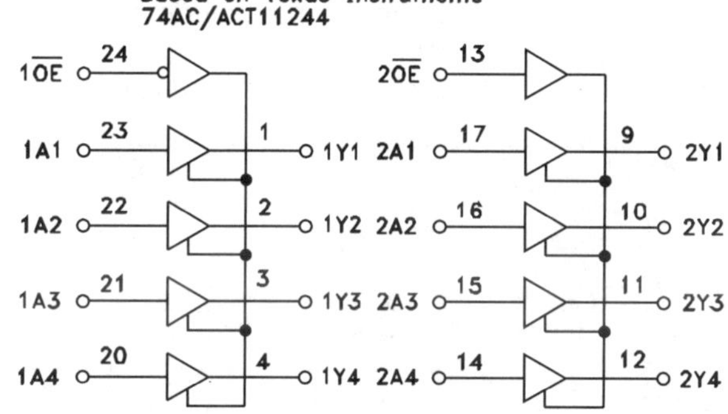

Figure 10.18. The 7AC/ACT11240, '11244 are octal buffers/line drivers with 3-state outputs. The '11240 is inverting; the '11244 non-inverting. Package center pin power connections minimize high-speed switching noise; the flow-through layout improves the printed-circuit artwork. Typical latch-up immunity is 500 mA at 125 °C.

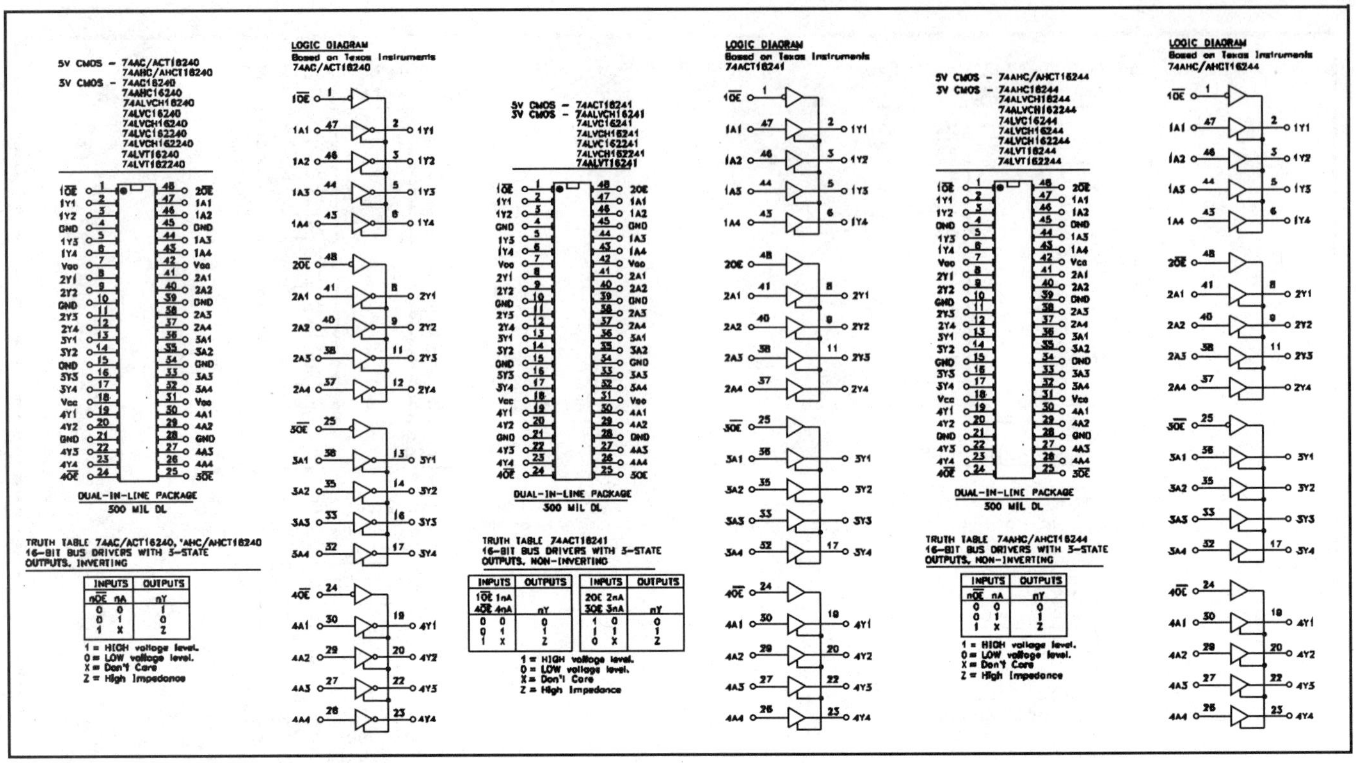

Figure 10.19. The 74x16240, 'x16241 and 'x16244 are 16-bit buffers/drivers with 3-state outputs. The 'x16240 is inverting; the 'x16241 non-inverting with mixed selection logic; the '16244 non-inverting. Package power pin connections minimize high-speed switching noise; the flow-through layout improves the printed-circuit artwork. Typical latch-up immunity for the AC/ACT devices is 500 mA at 125 °C.

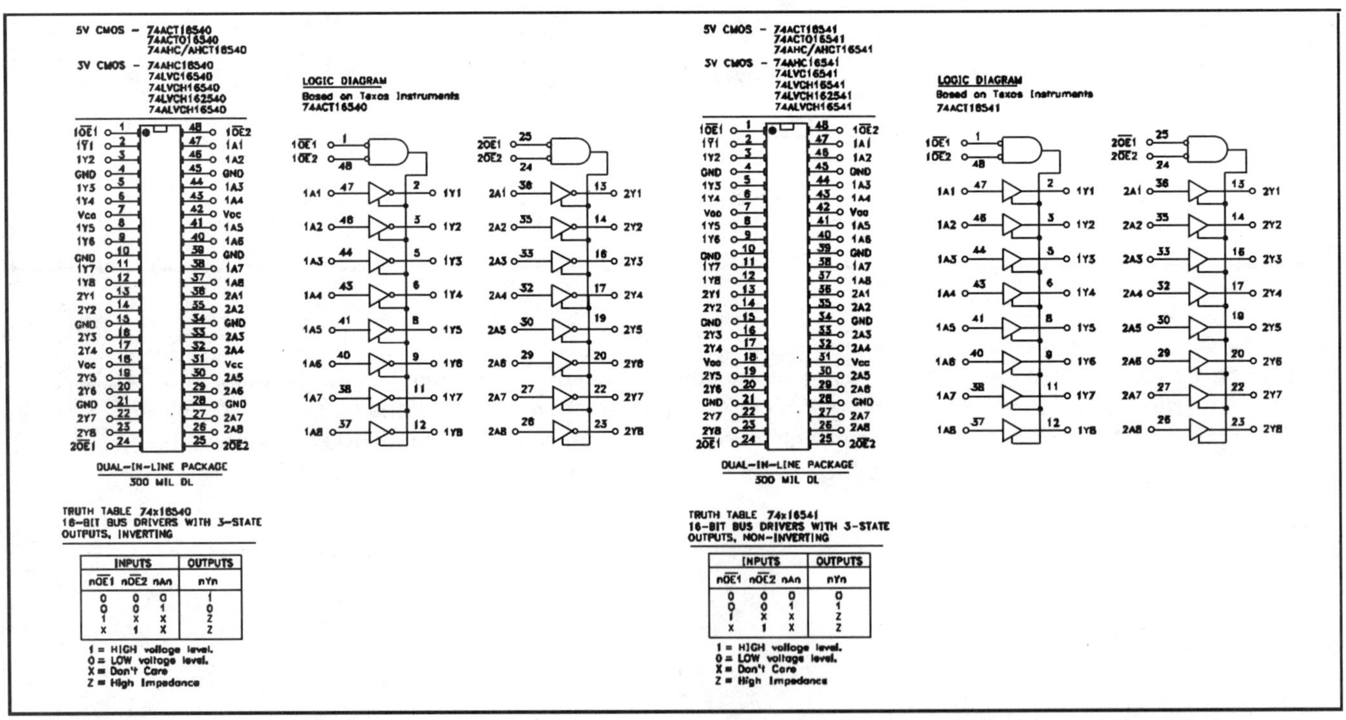

Figure 10.20. The 74x16540, and 'x16541 are 16-bit buffers/drivers with 3-state outputs. The 'x16540 is inverting; the '16541 non-inverting. Package power pin connections minimize high-speed switching noise; the flow-through layout improves the printed-circuit artwork. Typical latch-up immunity for the AC/ACT devices is 500 mA at 125 °C.

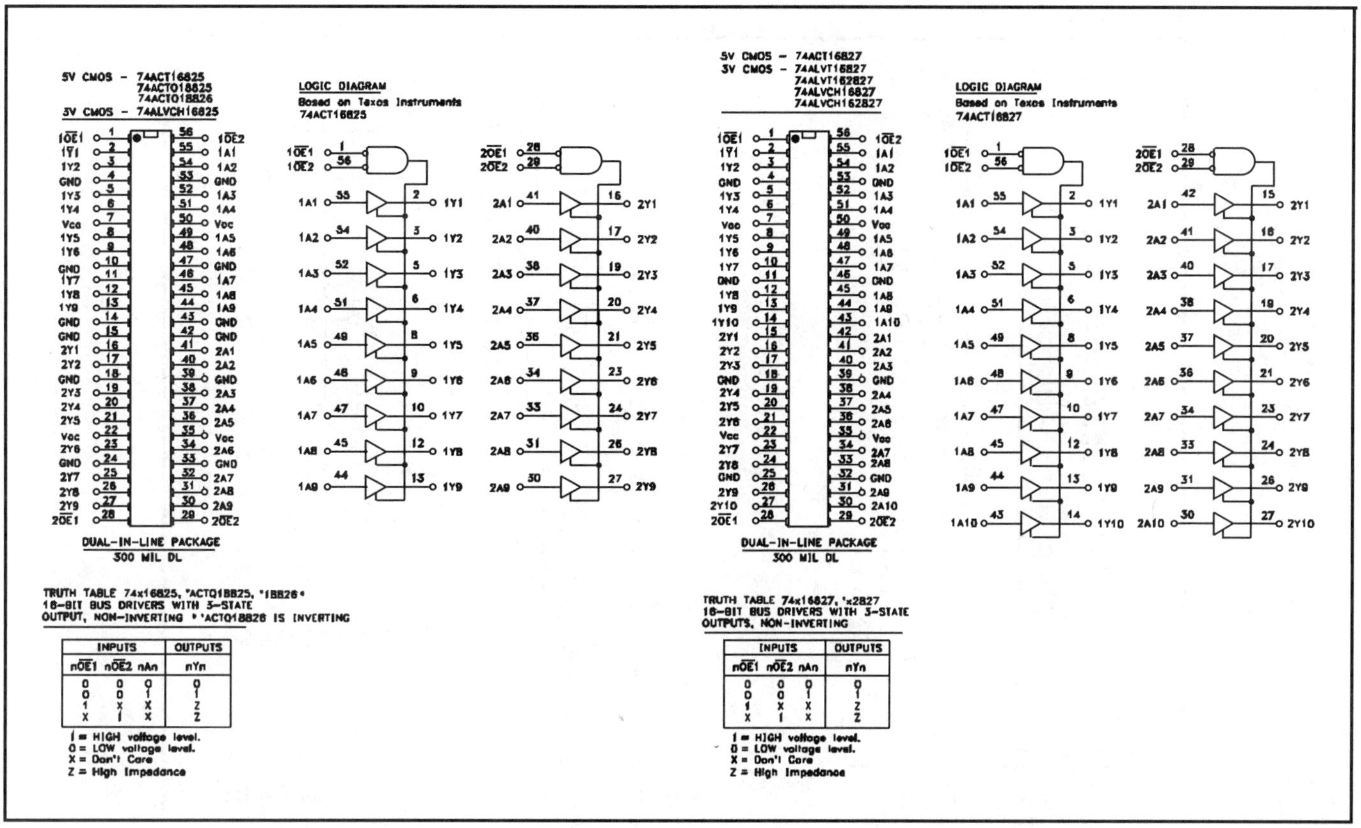

Figure 10.21. The 74x16825, and 'ACTQ16826 are 18-bit non-inverting buffers/drivers with 3-state outputs. The 'x16827 are 20-bit versions. Package power pin connections minimize high-speed switching noise; the flow-through layout improves the printed-circuit artwork.

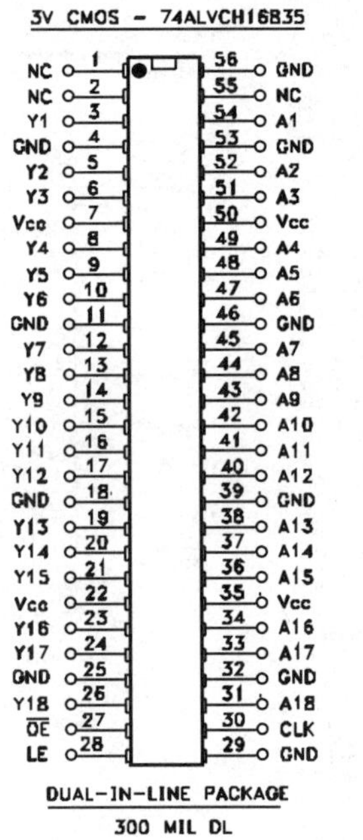

FUNCTION TABLE 74ALVCH16835 18-BIT UNIVERSAL BUS DRIVER WITH 3-STATE OUTPUTS

INPUTS				OUTPUT
$\overline{OE}$	LE	CLK	A	Y
1	X	X	X	Z
0	1	X	0	0
0	1	X	1	1
0	0	↑	0	0
0	0	↑	1	1
0	0	1	X	Y0 *
0	0	0	X	Y0 **

* = Output level before the indicated state-state input conditions were established, provided that CLK is high before LE goes low.
** = Output level before the indicated state-state input conditions were established.
↑ = LOW-to-HIGH clock transition.
1 = HIGH voltage level.
0 = LOW voltage level.
X = Don't Care
Z = High Impedance state

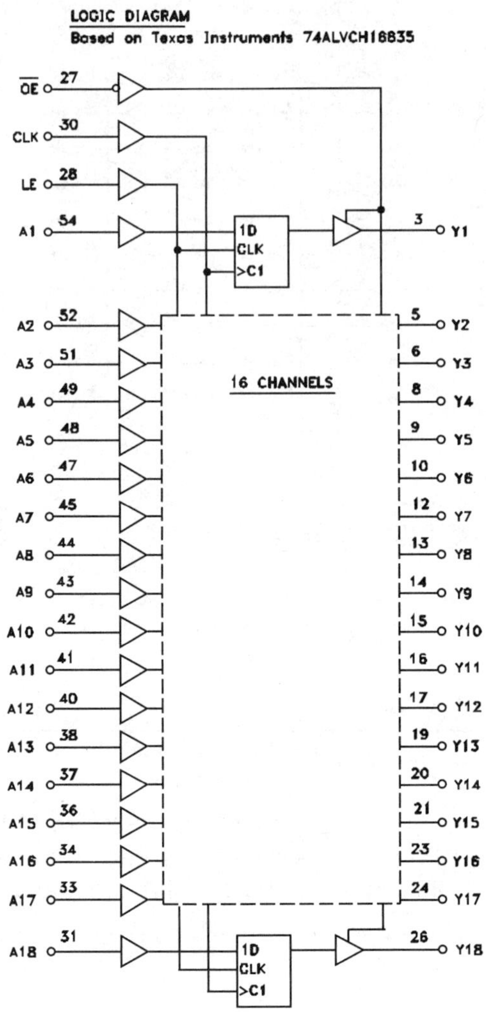

Figure 10.22. The 74ALVCH16835 is an 18-bit non-inverting universal bus/driver with 3-state outputs. Package power pin connections minimize high-speed switching noise; the flow through layout improves the printed-circuit artwork. The device operates in the transparent mode when LE is high. The A data is latched with the clock at logic low or high if LE is low. The bus hold on the data inputs eliminates the need for external pullup/pulldown resistors. To ensure the high-impedance state during power-up, /OE should be tied to Vcc through a pull-up resistor. The maximum clock frequency is given as 150 MHz.

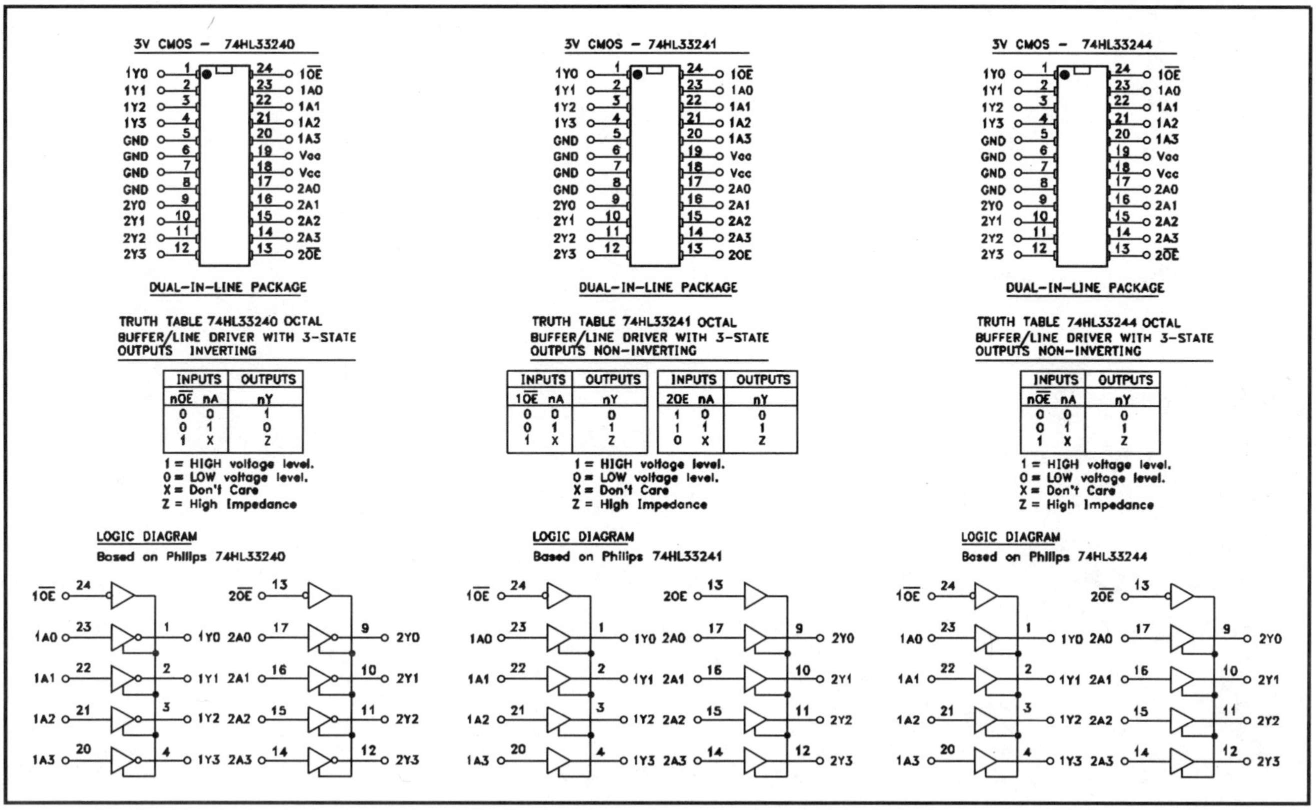

Truth table 74HL33240:

INPUTS		OUTPUTS
nOE	nA	nY
0	0	1
0	1	0
1	X	Z

Truth table 74HL33241:

INPUTS		OUTPUTS	INPUTS		OUTPUTS
1OE	nA	nY	2OE	nA	nY
0	0	0	1	0	0
0	1	1	1	1	1
1	X	Z	0	X	Z

Truth table 74HL33244:

INPUTS		OUTPUTS
nOE	nA	nY
0	0	0
0	1	1
1	X	Z

Figure 10.23. The 74HL33240, '33241, and '33244 octal buffers/line drivers with 3-state outputs replicate the features of the popular 74x240 series shown in *Figure 10.3*, with an important difference of package center pin connections for improved high-speed performance. These low-voltage CMOS devices provide direct interface to TTL levels with 5 volt to 3.3 volt level shifting.

Chapter 11

Bus Transceivers

We cannot fail but note that this chapter is significantly larger than any that precede it. And by a pretty good margin. Beyond that the distribution is unique. As with Chapter 10, there are no 4000-series devices represented. When we look at *Tables 11.1a and 11.1b* we see no standard TTL devices. The emphasis is on the more complex devices that appear as the TTL number sequence increases. With respect to the high-speed CMOS we see these clustered in the HC/HCT grouping, with only a scattering of the more complex devices. A much greater range of low-voltage devices is provided than in any previous chapter. This device spread suggests the influence of requirements for ever increasing speed with reduced operating voltages and the popularity of laptop computers.

The basic Transceiver is a fairly elementary circuit, as illustrated by the 74x242, *Figure 11.1.* In the late '70s I created a similar circuit using the LM339 analog comparator to interface the I/O ports of a KIM_1 single-board computer. The ports were bidirectional, but I wanted a buffer to isolate the ports from the applications I was working with. While it sufficed for the needs at the time, such a circuit would be woefully inadequate now.

Advances in technology led to the need for more sophisticated transceivers, as evidenced by the versatile assortment of features now available. These include greater width for wide bus applications, distributed power pins for noise and crosstalk reduction, built-in resistors to eliminate pull-ups, storage registers and parity generators/checkers.

Chapter Organization

Eleven tables are provided. *Tables 11.1a, 1b - 11.3* provide a listing of devices by manufacturer. There are sources beyond those listed, but those given provide a realistic picture of what has been and is currently available.

Tables 11.4a, 4b, 4c - 11.6a, 6b provide the LOW-to-HIGH (tPLH) and HIGH-to-LOW (tPHL) propagation delays. For some devices these are not defined as such, in which case a propagation or time delay, td or tp, is given. Because of the manner in which manufacturers define these quantities, the data in the tables is not precise. The intent is to provide insight helpful for performance comparisons. In particular, we can see the spectacular gains in speed and reduced power dissipation, as well as complexity, achieved with the latest devices.

The drawings of *Figures 11.1 - 11.65* describe the Dual In-Line (DIP) package pinning, a Truth or Function Table, and Logic Diagrams. To simplify comparisons as well as to conserve space, all devices with matching features, specifically pinouts and function tables, are included in the same drawing.

Summary

This chapter provides tables and drawings describing varieties of integrated circuit buffers and line drivers/ receivers. While of simpler logic than the sometimes exotic devices of earlier chapters, many devices display an impressive array of performance abilities. Some points to keep in mind are:

- Timing information in the tables is for comparison purposes only. There is a general lack of consistency in how these parameters are defined, and not just between manufacturers. Where timing is critical consult the manufacturer's data manual.
- All timing data is shown for the industrial/commercial or recommended free-air operating temperature range. Where data is for 25°C it is given here.
- The specifications for CMOS devices may cover voltage ranges from 2 to 15 volts. Data is typically provided for VCC's of 5, 10, and 15 volts, but not always. Some manufacturers use 5.0 volts; others 4.5. There is a significant speed factor—in general, the higher the voltage the faster the performance. Again, where this is important to your application consult the data manual.
- A similar situation exists for the more recent low-voltage devices. Typical VCC ranges given are 2.7 - 3.6 volts. But values may be found for 2, 2.6, 3.0, 3.3 or 3.6 volts. (Glossary *Figures G.1 and G.2* may provide helpful insights on performance.)

DEVICE	STD		L		H		S		LS		AS		ALS		F		Figure	DEVICE
74---	NM	ST	NM	ST	NM	ST	NM	ST	NM	ST	NM	ST	NM	ST	NM	ST	No.	74---
74242									x	x x	x	x	x	x	x	x	11.1	74242
74243									x x	x x	x	x	x	x	x	x	11.1	74243
74245									x x	x x	x	x		x	x	x x	11.2	74245
74440										x							11.3	74440
74441										x							11.3	74441
74442										x							11.3	74442
74444										x							11.3	74444
74446										x							11.4	74446
74449										x							11.4	74449
74543															x	x x	11.5	74543
74544															x	x x	11.5	74544
74545																x	11.6	74545
74552																x	11.7	74552
74614														x			11.8	74614
74615														x			11.8	74615
74620										x	x	x		x	x	x x	11.9	74620
74621										x		x		x		x x	11.10	74621
74622												x		x		x x	11.10	74622
74623										x	x	x		x		x x	11.9	74623
74638										x							11.11	74638
74639										x							11.11	74639
74640									x	x	x	x		x	x	x	11.12	74640
74641									x	x		x		x		x	11.13	74641
74642									x	x		x		x		x	11.13	74642
74643												x		x			11.12	74643
74644										x		x		x			11.13	74644
74645									x	x	x	x		x			11.12	74645
74646									x	x	x	x	x	x		x	11.14	74646
74647										x		x		x		x	11.14	74647
74648										x	x	x	x	x		x	11.14	74648
74649										x				x		x	11.14	74649
74651										x		x	x	x		x	11.15	74651
74652										x	x	x	x	x		x	11.15	74652
74653										x				x		x	11.15	74653
74654										x				x		x	11.15	74654
74657															x	x x	11.16	74657

Key to Manufacturer: N – National/Fairchild. M – Motorola,
S – Signetics/Philips, T – Texas Instruments

Table 11.1a. TTL Bus Transceiver source summary by manufacturer.

DEVICE 74---	STD NM	STD ST	L NM	L ST	H NM	H ST	S NM	S ST	LS NM	LS ST	AS NM	AS ST	ALS NM	ALS ST	F NM	F ST	Figure No.	DEVICE 74---
74758												x		x			11.1	74758
74759												x					11.1	74759
74776																x	11.19	74776
74777																x	11.20	74777
74807																x	11.21	74807
74852												x					11.22	74852
74856												x					11.23	74856
74861																x	11.24	74861
74862																x	11.24	74862
74863																x	11.24	74863
74864																x	11.24	74864
75877												x					11.25	74877
741242														x		x	11.26	741242
741243																x	11.26	741243
741245													x	x	x	x	11.27	741245
741640														x			11.12	741640
741645													x	x			11.12	741645
742242														x			11.12	742242
742620												x					11.1	742620
742623												x					11.10	742623
742640												x					11.10	742640
742643												x					11.12	742643
742645											x	x	x				11.12	742645
742952																x	11.28	742952
742953																x	11.28	742953
743893																x	11.30	743893
745245													x				11.2	745245
745620													x				11.2	745620
748960																x	11.32	748960
748961																x	11.32	748961
748962																x	11.33	748962
748963																x	11.33	748963
748965																x	11.34	748965
748966																x	11.34	748966
7429833														x			11.57	7429833
7429854														x			11.58	7429854
7429861														x			11.59	7429861
7429862														x			11.59	7429862
7429863														x			11.60	7429863
7429864														x			11.60	7429864

Key to Manufacturer: N – National/Fairchild. M – Motorola,
S – Signetics/Philips, T – Texas Instruments

Table 11.1b. TTL Bus Transceiver source summary by manufacturer (Continuation).

DEVICE 74---	C NM	C ST	HC NM	HC ST	HCT NM	HCT ST	AHC (5V) NM	AHC (5V) ST	AHCT NM	AHCT ST	VHC(5V) NM	VHC(5V) ST	VHCT NM	VHCT ST	AC (5V) NM	AC (5V) ST	ACT NM	ACT ST	Figure No.	DEVICE 74---
74242			x x	x x		x x													11.1	74242
74243			x	x x	x	x x													11.1	74243
74245			x x	x x	x x	x x		x		x	x x		x x		xq	x	xq	x	11.2	74245
74543															q		q		11.5	74543
74544															q		q		11.5	74544
74620			x x	x	x	x													11.9	74620
74623			x x	x	x	x													11.9	74623
74640			x x	x x	x x	x x													11.12	74640
74643			x x	x	x	x x													11.12	74643
74645				x		x													11.12	74645
74646			x x	x x	x	x x									xq		xq		11.14	74646
74648			x x	x x		x x									x				11.14	74648
74651				x		x													11.15	74651
74652				x x		x x											q		11.15	74652
74657															q		q		11.16	74657
74658				x		x													11.17	74658
74659				x		x													11.17	74659
74664				x		x													11.18	74664
74665				x		x													11.18	74665
7411245																x		x	11.35	7411245
7411652																x		x	11.36	7411652
7416245								x		x						x	q	x	11.37	7416245
7416470																		x	11.42	7416470
7416472																x			11.43	7416472
7416474																		x	11.43	7416474
7416475																		x	11.43	7416475
7416543																x	q	x	11.46	7416543
7416544																		x	11.46	7416544
7416620																x		x	11.48	7416620
7416623																		x	11.48	7416623
7416640																x		x	11.49	7416640
7416646																x	q	x	11.50	7416646
7416648																		x	11.50	7416648
7416651																		x	11.51	7416651
7416652																x		x	11.51	7416652
7416657																		x	11.52	7416657
7416861																		x	11.53	7416861
7416863																		x	11.54	7416863
7416864																		x	11.55	7416864
7416952																		x	11.56	7416952

Key to Manufacturer: N – National/Fairchild, M – Motorola,
S – Signetics/Philips, T – Texas Instruments

NOTE: "q" identifies 74ACQ/QCTQnnnnn CMOS series.

Table 11.2. 5V CMOS 74xxx series Bus Transceiver source summary by manufacturer.

DEVICE 74---	LV NM	LV ST	LVCH LVC NM	LVCH LVC ST	ALVT LVT NM	ALVT LVT ST	LVX NM	LVX ST	LVQ NM	LVQ ST	LCX NM	LCX ST	AC (3V) NM	AC (3V) ST	ALVCH NM	ALVCH ST	HL NM	HL ST	AHC (3V) NM	AHC (3V) ST	VHC (3V) NM	VHC (3V) ST	Figure No.	DEVICE 74---
74245		X X		X X	X	X X	X		X		X X			X						X	X X		11.2	74245
74543				X X		X X					X		q										11.5	74543
74544				X X									q										11.5	74544
74623				X																			11.9	74623
74640																							11.12	74640
74646				X X	X	X X					X X		X										11.14	74646
74648													X										11.14	74648
74652				X X	X	X X					X X												11.15	74652
74657																							11.16	74657
74861				X																			11.24	74861
74863				X																			11.24	74863
742245				X																			11.2	742245
742952				X X		X X					X												11.28	742952
743245							X X																11.29	743245
744245				X X			X X																11.29	744245
7411245														X									11.35	7411245
7411652														X									11.36	7411652
7416245				X		X										X X				X			11.37	7416245
7416269																X							11.39	7416269
7416270																X							11.40	7416270
7416409																X							11.41	7416409
7416472														X									11.43	7416472
7416500						X					X					X X							11.44	7416500
7416501						X					X					X X							11.44	7416501
7416525																X X							11.45	7416525
7416543						X					X			X		X X							11.46	7416543
7416600						X										X X							11.47	7416600
7416601						X										X X							11.47	7416601
7416620														X									11.48	7416620
7416623				X												X							11.48	7416623
7416640				X										X									11.49	7416640
7416646					X	X					X X			X		X X							11.50	7416646
7416652					X	X					X X			X		X X							11.51	7416652
7416952																X X							11.56	7416952
74162245				X		X																	11.37	74162245
74162268																X							11.38	74162268
74162269																X							11.39	74162269
74162409																X							11.41	74162409
74162525				X												X							11.45	74162525
74162601																X							11.47	74162601
74162623				X																			11.48	74162623
74162640				X																			11.49	74162640
74164245																X							11.31	74164245
7422952				X																			11.28	7422952
7433245																		x					11.61	7433245
7433620																		x					11.62	7433620
7433623																		x					11.62	7433623
7433640																		x					11.61	7433640
7433646																		x					11.63	7433646
7433648																		x					11.63	7433648
7433651																		x					11.64	7433651
7433652																		x					11.64	7433652
7433952																		x					11.65	7433952

Key to Manufacturer: N – National/Fairchild. M – Motorola,
S – Signetics/Philips, T – Texas Instruments

Table 11.3. Low-Voltage CMOS 74xxx series Bus Transceiver source summary by manufacturer.

DEVICE 74---	STD	L	H	S	LS	AS/A/B	ALS/A/B	F	DATA	Figure	DEVICE 74---
-74242	--	--	--	--	14	2-6.5	2-11	2.5-6.0	tPLH↑	-11.1	-74242
----	--	--	--	--	18	2-5.7	2-10	2.0-4.0	tPHL↓	---	----
-74243	--	--	--	--	18	3-7.5	4-11	2.5-6.0	tPLH↑	-11.1	-74243
----	--	--	--	--	18	3-6.5	4-11	2.0-4.0	tPHL↓	---	----
-74245	--	--	--	--	12	2-7.5	3-10	2.5-6.0	tPLH↑	-11.2	-74245
----	--	--	--	--	12	2-7.0	3-10	2.5-6.0	tPHL↓	---	----
-74440	--	--	--	--	24-35	--	--	--	tPLH↑	-11.3	-74440
----	--	--	--	--	20-30	--	--	--	tPHL↓	---	----
-74441	--	--	--	--	21-30	--	--	--	tPLH↑	-11.3	-74441
----	--	--	--	--	9-15	--	--	--	tPHL↓	---	----
-74442	--	--	--	--	10-14	--	--	--	tPLH↑	-11.3	-74442
----	--	--	--	--	13-20	--	--	--	tPHL↓	---	----
-74444	--	--	--	--	9-14	--	--	--	tPLH↑	-11.3	-74444
----	--	--	--	--	7-20	--	--	--	tPHL↓	---	----
-74446	--	--	--	--	8-13	--	--	--	tPLH↑	-11.4	-74446
----	--	--	--	--	7-12	--	--	--	tPHL↓	---	----
-74449	--	--	--	--	10-15	--	--	--	tPLH↑	-11.4	-74449
----	--	--	--	--	11-17	--	--	--	tPHL↓	---	----
-74543	--	--	--	--	--	--	--	2.5-11.5	tPLH↑	-11.5	-74543
----	--	--	--	--	--	--	--	2.5-9.5	tPHL↓	---	----
-74544	--	--	--	--	--	--	--	3.0-11.5	tPLH↑	-11.5	-74544
----	--	--	--	--	--	--	--	3.0-9.5	tPHL↓	---	----
-74545	--	--	--	--	--	--	--	1.5-5.5	tPLH↑	-11.6	-74545
----	--	--	--	--	--	--	--	2.5-6.5	tPHL↓	---	----
-74552	--	--	--	--	--	--	--	3.0-16.5	tPLH↑	-11.7	-74552
----	--	--	--	--	--	--	--	4.0-17.0	tPHL↓	---	----
-74614	--	--	--	--	--	--	9-64	--	tPLH↑	-11.8	-74614
----	--	--	--	--	--	--	2-22	--	tPHL↓	---	----
-74615	--	--	--	--	--	--	6-64	--	tPLH↑	-11.8	-74615
----	--	--	--	--	--	--	4-25	--	tPHL↓	---	----
-74620	--	--	--	--	6-15	1-7	2-10	2.5-6.5	tPLH↑	-11.9	-74620
----	--	--	--	--	6-15	2-6	2-10	1.0-4.5	tPHL↓	---	----
-74621	--	--	--	--	17-40	5-24	10-39	6.0-15.0	tPLH↑	-11.10	-74621
----	--	--	--	--	16-50	1-10	5-35	3.5-10.5	tPHL↓	---	----
-74622	--	--	--	--	--	5-25	8-38	7.5-14.5	tPLH↑	-11.10	-74622
----	--	--	--	--	--	1-10.5	5-35	1.5-10.0	tPHL↓	---	----
-74623	--	--	--	--	8-15	1-9	2-13	2.0-5.5	tPLH↑	-11.9	-74623
----	--	--	--	--	8-15	1-8.5	3-11	2.5-7.0	tPHL↓	---	----
-74638	--	--	--	--	6-40	--	--	--	tPLH↑	-11.11	-74638
----	--	--	--	--	8-60	--	--	--	tPHL↓	---	----
-74639	--	--	--	--	8-40	--	--	--	tPLH↑	-11.11	-74639
----	--	--	--	--	11-50	--	--	--	tPHL↓	---	----
-74640	--	--	--	--	6-15	2-7	2-11	2.0-7.0	tPLH↑	-11.12	-74640
----	--	--	--	--	6-15	2-6	2-10	1.0-5.0	tPHL↓	---	----
-74641	--	--	--	--	17-40	5-22	5-32	6.0-10.5	tPLH↑	-11.13	-74641
----	--	--	--	--	16-50	1-10	3-32	4.0-7.5	tPHL↓	---	----
-74642	--	--	--	--	19-40	5-24	10-30	7.5-12.5	tPLH↑	-11.13	-74642
----	--	--	--	--	14-60	1-11.5	5-38	1.5-10.5	tPHL↓	---	----
-74643	--	--	--	--	--	2-10	2-13	--	tPLH↑	-11.12	-74643
----	--	--	--	--	--	2-9	2-13	--	tPHL↓	---	----
-74644	--	--	--	--	17-40	5-24	8-30	--	tPLH↑	-11.13	-74644
----	--	--	--	--	14-60	1-10	5-35	--	tPHL↓	---	----
-74645	--	--	--	--	8-15	2-9.5	3-10	--	tPLH↑	-11.12	-74645
----	--	--	--	--	11-15	2-9	3-10	--	tPHL↓	---	----

NOTE: For some devices td, tp, or equivalent time delays are defined instead of tPLH, tPHL.

Table 11.4a. 74xxx TTL device representative propagation delay times in nS.

DEVICE 74---	STD	L	H	S	LS	AS/A/B	ALS/A/B	F	DATA	Figure	DEVICE 74---
-74646	--	--	--	--	12-50	2-11	5-35	4.0-10.0	tPLH↑	-11.14	-74646
----	--	--	--	--	13-35	1-9	3-20	4.0-10.0	tPHL↓	---	----
-74647	--	--	--	--	17-60	--	9-60	7.0-16.0	tPLH↑	-11.14	-74647
----	--	--	--	--	18-50	--	2-22	4.0-18.0	tPHL↓	---	----
-74648	--	--	--	--	12-55	2-11	3-39	3.0-9.5	tPLH↑	-11.14	-74648
----	--	--	--	--	15-40	1-9	2-22	4.0-9.5	tPHL↓	---	----
-74649	--	--	--	--	15-55	--	8-62	7.0-16.0	tPLH↑	-11.14	-74649
----	--	--	--	--	20-50	--	2-22	4.0-18.0	tPHL↓	---	----
-74651	--	--	--	--	9-47	2-11	4-38	3.0-10.5	tPLH↑	-11.15	-74651
----	--	--	--	--	20-35	1-9	2-21	3.0-11.0	tPHL↓	---	----
-74652	--	--	--	--	12-50	2-11	5-35	3.0-10.5	tPLH↑	-11.15	-74652
----	--	--	--	--	13-36	1-9	3-20	3.0-11.0	tPHL↓	---	----
-74653	--	--	--	--	10-57	--	5-64	4.0-20.0	tPLH↑	-11.15	-74653
----	--	--	--	--	16-55	--	2-25	4.5-12.5	tPHL↓	---	----
-74654	--	--	--	--	--	--	5-64	4.0-20.0	tPLH↑	-11.15	-74654
----	--	--	--	--	--	--	2-25	4.5-12.5	tPHL↓	---	----
-74657	--	--	--	--	--	--	--	2.5-20.5	tPLH↑	-11.16	-74657
----	--	--	--	--	--	--	--	3.0-20.5	tPHL↓	---	----
-74758	--	--	--	--	--	3-21	10-28	--	tPLH↑	-11.1	-74758
----	--	--	--	--	--	1-7.5	2-21	--	tPHL↓	---	----
-74759	--	--	--	--	--	3-21	--	--	tPLH↑	-11.1	-74759
----	--	--	--	--	--	1-7.5	--	--	tPHL↓	---	----
-74776	--	--	--	--	--	--	--	3.0-12.0	tPLH↑	-11.19	-74776
----	--	--	--	--	--	--	--	3.0-9.0	tPHL↓	---	----
-74777	--	--	--	--	--	--	--	3.0-13.0	tPLH↑	-11.20	-74777
----	--	--	--	--	--	--	--	5.0-12.0	tPHL↓	---	----
-74807	--	--	--	--	--	--	--	6.5-24.0	tPLH↑	-11.21	-74807
----	--	--	--	--	--	--	--	5.0-25.5	tPHL↓	---	----
-74852	--	--	--	--	--	2-10	--	--	tPLH↑	-11.22	-74852
----	--	--	--	--	--	3-12.5	--	--	tPHL↓	---	----
-74856	--	--	--	--	--	2-7.5	--	--	tPLH↑	-11.23	-74856
----	--	--	--	--	--	2-19	--	--	tPHL↓	---	----
-74861	--	--	--	--	--	--	--	4.0-9.0	tPLH↑	-11.24	-74861
----	--	--	--	--	--	--	--	3.0-8.0	tPHL↓	---	----
-74862	--	--	--	--	--	--	--	4.0-9.0	tPLH↑	-11.24	-74862
----	--	--	--	--	--	--	--	1.5-6.5	tPHL↓	---	----
-74863	--	--	--	--	--	--	--	4.0-9.0	tPLH↑	-11.24	-74863
----	--	--	--	--	--	--	--	3.0-8.0	tPHL↓	---	----
-74864	--	--	--	--	--	--	--	4.0-9.0	tPLH↑	-11.24	-74864
----	--	--	--	--	--	--	--	1.5-6.5	tPHL↓	---	----
-74877	--	--	--	--	--	2-10	--	--	tPLH↑	-11.25	-74877
----	--	--	--	--	--	3-11.5	--	--	tPHL↓	---	----
-741242	--	--	--	--	--	--	2-12	3.0-6.0	tPLH↑	-11.26	-741242
----	--	--	--	--	--	--	2-10	1.5-4.0	tPHL↓	---	----
-741243	--	--	--	--	--	--	--	3.0-6.0	tPLH↑	-11.26	-741243
----	--	--	--	--	--	--	--	1.5-4.0	tPHL↓	---	----
-741245	--	--	--	--	--	--	2-13	2.0-6.5	tPLH↑	-11.27	-741245
----	--	--	--	--	--	--	2-13	2.5-7.5	tPHL↓	---	----
-741640	--	--	--	--	--	--	5-15	--	tPLH↑	-11.12	-741640
----	--	--	--	--	--	--	2-10	--	tPHL↓	---	----
-741645	--	--	--	--	--	--	2-13	--	tPLH↑	-11.12	-741645
----	--	--	--	--	--	--	2-13	--	tPHL↓	---	----

NOTE: For some devices td, tp, or equivalent time delays are defined instead of tPLH, tPHL.

Table 11.4b. 74xxx TTL device representative propagation delay times in nS (continuation).

DEVICE 74---	STD	L	H	S	LS	AS/A/B	ALS/A/B	F	DATA	Figure	DEVICE 74---
-742242	--	--	--	--	--	--	2-11	--	tPLH	-11.12	-742242
----	--	--	--	--	--	--	2-10	--	tPHL	---	----
-742620	--	--	--	--	--	1-8	--	--	tPLH	-11.1	-742620
----	--	--	--	--	--	1-6.5	--	--	tPHL	---	----
-742623	--	--	--	--	--	1-8.5	--	--	tPLH	-11.10	-742623
----	--	--	--	--	--	1-7.5	--	--	tPHL	---	----
-742640	--	--	--	--	--	1-7.5	--	--	tPLH	-11.10	-742640
----	--	--	--	--	--	1-6.5	--	--	tPHL	---	----
-742643	--	--	--	--	--	1-10	--	--	tPLH	-11.12	-742643
----	--	--	--	--	--	1-9.5	--	--	tPHL	---	----
-742645	--	--	--	--	--	1-10	--	--	tPLH	-11.12	-742645
----	--	--	--	--	--	1-9.5	--	--	tPHL	---	----
-742952	--	--	--	--	--	--	--	3.0-7.5	tPLH	-11.28	-742952
----	--	--	--	--	--	--	--	3.5-8.5	tPHL	---	----
-742953	--	--	--	--	--	--	--	3.0-7.5	tPLH	-11.28	-742953
----	--	--	--	--	--	--	--	3.5-8.5	tPHL	---	----
-743893	--	--	--	--	--	--	--	1.0-5.0	tPLH	-11.30	-743893
----	--	--	--	--	--	--	--	1.5-6.0	tPHL	---	----
-745245	--	--	--	--	--	--	3-10	--	tPLH	-11.2	-745245
----	--	--	--	--	--	--	3-10	--	tPHL	---	----
-745620	--	--	--	--	--	--	2-10	--	tPLH	-11.2	-745620
----	--	--	--	--	--	--	2-10	--	tPHL	---	----
-748960	--	--	--	--	--	--	--	2.5-8.5	tPLH	-11.32	-748960
----	--	--	--	--	--	--	--	3.5-13.5	tPHL	---	----
-748961	--	--	--	--	--	--	--	3.0-12.0	tPLH	-11.32	-748961
----	--	--	--	--	--	--	--	3.0-9.0	tPHL	---	----
-748962	--	--	--	--	--	--	--	3.5-10.0	tPLH	-11.33	-748962
----	--	--	--	--	--	--	--	3.0-10.5	tPHL	---	----
-748963	--	--	--	--	--	--	--	2.0-10.0	tPLH	-11.33	-748963
----	--	--	--	--	--	--	--	2.0-8.5	tPHL	---	----
-748965	--	--	--	--	--	--	--	1.0-10.0	tPLH	-11.34	-748965
----	--	--	--	--	--	--	--	1.5-9.5	tPHL	---	----
-748966	--	--	--	--	--	--	--	1.0-10.0	tPLH	-11.34	-748966
----	--	--	--	--	--	--	--	1.5-9.5	tPHL	---	----
-7429833	--	--	--	--	--	--	8-22	--	tPLH	-11.57	-7429833
----	--	--	--	--	--	--	8-24	--	tPHL	---	----
-7429854	--	--	--	--	--	--	8-22	--	tPLH	-11.58	-7429854
----	--	--	--	--	--	--	8-22	--	tPHL	---	----
-7429861	--	--	--	--	--	--	8-15	--	tPLH	-11.59	-7429861
----	--	--	--	--	--	--	8-15	--	tPHL	---	----
-7429862	--	--	--	--	--	--	7-14	--	tPLH	-11.59	-7429862
----	--	--	--	--	--	--	7.5-14	--	tPHL	---	----
-7429863	--	--	--	--	--	--	8-15	--	tPLH	-11.60	-7429863
----	--	--	--	--	--	--	8-15	--	tPHL	---	----
-7429864	--	--	--	--	--	--	8-14	--	tPLH	-11.60	-7429864
----	--	--	--	--	--	--	7.5-14	--	tPHL	---	----

NOTE: For some devices td, tp, or equivalent time delays are defined instead of tPLH, tPHL.

Table 11.4c. 74xxx TTL device representative propagation delay times in nS (continuation).

DEVICE 74---	C	HC	HCT	AHC	AHCT	VHC(5V)	VHCT	AC (5V)	ACT	DATA	Figure	DEVICE 74---
-74242	--	9-23	12-25	--	--	--	--	--	--	t_{PLH}↑	-11.1	-74242
----	--	9-23	12-25	--	--	--	--	--	--	t_{PHL}↓	---	----
-74243	--	8-23	13-28	--	--	--	--	--	--	t_{PLH}↑	-11.1	-74243
----	--	8-23	13-28	--	--	--	--	--	--	t_{PHL}↓	---	----
-74245	--	9-23	12-28	1-13.5	1-11	40-5.5	4.9-7.7	1-9	1-8	t_{PLH}↑	-11.2	-74245
----	--	9-23	12-28	1-13.5	1-11	5.5-7.5	5.4-8.7	1-9	1-9	t_{PHL}↓	---	----
-74543	--	--	--	--	--	--	--	1.5-8.5	1.5-9.0	t_{PLH}↑	-11.5	-74543
----	--	--	--	--	--	--	--	1.5-8.5	1.5-9.0	t_{PHL}↓	---	----
-74544	--	--	--	--	--	--	--	1.5-8.5	1.5-9.0	t_{PLH}↑	-11.5	-74544
----	--	--	--	--	--	--	--	1.5-8.5	1.5-9.0	t_{PHL}↓	---	----
-74620	--	26-34	25-47	--	--	--	--	--	--	t_{PLH}↑	-11.9	-74620
----	--	26-34	25-47	--	--	--	--	--	--	t_{PHL}↓	---	----
-74623	--	26-34	25-47	--	--	--	--	--	--	t_{PLH}↑	-11.9	-74623
----	--	26-34	25-47	--	--	--	--	--	--	t_{PHL}↓	---	----
-74640	--	26-47	23-34	--	--	--	--	--	--	t_{PLH}↑	-11.12	-74640
----	--	26-47	23-34	--	--	--	--	--	--	t_{PHL}↓	---	----
-74643	--	28-49	23-34	--	--	--	--	--	--	t_{PLH}↑	-11.12	-74643
----	--	28-49	23-34	--	--	--	--	--	--	t_{PHL}↓	---	----
-74645	--	26-34	25-38	--	--	--	--	--	--	t_{PLH}↑	-11.12	-74645
----	--	26-34	25-38	--	--	--	--	--	--	t_{PHL}↓	---	----
-74646	--	34-48	31-69	--	--	--	--	2.5-9.5	2.0-10.5	t_{PLH}↑	-11.14	-74646
----	--	34-48	31-69	--	--	--	--	2.5-9.5	2.0-10.5	t_{PHL}↓	---	----
-74648	--	34-48	31-69	--	--	--	--	1.5-11	--	t_{PLH}↑	-11.14	-74648
----	--	34-48	31-69	--	--	--	--	1.5-10.5	--	t_{PHL}↓	---	----
-74651	--	34-48	31-69	--	--	--	--	--	--	t_{PLH}↑	-11.15	-74651
----	--	34-48	31-69	--	--	--	--	--	--	t_{PHL}↓	---	----
-74652	--	34-48	31-69	--	--	--	--	--	2.0-10.5	t_{PLH}↑	-11.15	-74652
----	--	34-48	31-69	--	--	--	--	--	2.0-10.5	t_{PHL}↓	---	----
-74657	--	--	--	--	--	--	--	1.5-10.5	1.5-11.0	t_{PLH}↑	-11.16	-74657
----	--	--	--	--	--	--	--	1.5-10.5	1.5-11.0	t_{PHL}↓	---	----
-74658	--	38-79	34-79	--	--	--	--	--	--	t_{PLH}↑	-11.17	-74658
----	--	38-79	34-79	--	--	--	--	--	--	t_{PHL}↓	---	----
-74659	--	35-79	35-79	--	--	--	--	--	--	t_{PLH}↑	-11.17	-74659
----	--	35-79	35-79	--	--	--	--	--	--	t_{PHL}↓	---	----
-74664	--	38-79	34-79	--	--	--	--	--	--	t_{PLH}↑	-11.18	-74664
----	--	38-79	34-79	--	--	--	--	--	--	t_{PHL}↓	---	----
-74665	--	35-79	35-79	--	--	--	--	--	--	t_{PLH}↑	-11.18	-74665
----	--	35-79	35-79	--	--	--	--	--	--	t_{PHL}↓	---	----

NOTE: For some devices t_d, t_p, or equivalent time delays are defined instead of t_{PLH}, t_{PHL}.

Table 11.5a. 5-volt 74xxx series CMOS device representative propagation delay times in nS.

DEVICE 74---	C	HC	HCT	AHC	AHCT	VHC(5V)	VHCT	AC (5V)	ACT	DATA	Figure	DEVICE 74---
-7411245	--	--	--	--	--	--	--	1.5-9.5	1.5-10	tPLH↑	-11.35	-7411245
----	--	--	--	--	--	--	--	1.5-6.9	1.5-9.1	tPHL↓	---	----
-7411652	--	--	--	--	--	--	--	2.4-10.7	2.8-13.1	tPLH↑	-11.36	-7411652
----	--	--	--	--	--	--	--	3.1-12	3.4-14.4	tPHL↓	---	----
-7416245	--	--	--	1-8.5	1-9.5	--	--	2-7.9	3.2-10.5	tPLH↑	-11.37	-7416245
----	--	--	--	1-8.5	1-9.5	--	--	2.5-8.9	2.6-10.2	tPHL↓	---	----
-7416470	--	--	--	--	--	--	--	--	3.9-11.8	tPLH↑	-11.42	-7416470
----	--	--	--	--	--	--	--	--	3.8-11.7	tPHL↓	---	----
-7416472	--	--	--	--	--	--	--	2.9-11.7	--	tPLH↑	-11.43	-7416472
----	--	--	--	--	--	--	--	3.1-10.9	--	tPHL↓	---	----
-7416474	--	--	--	--	--	--	--	--	4-11.5	tPLH↑	-11.43	-7416474
----	--	--	--	--	--	--	--	--	4.2-11.4	tPHL↓	---	----
-7416475	--	--	--	--	--	--	--	--	3.8-12.5	tPLH↑	-11.43	-7416475
----	--	--	--	--	--	--	--	--	4.2-12.6	tPHL↓	---	----
-7416543	--	--	--	--	--	--	--	2.7-11.5	3.5-13.8	tPLH↑	-11.46	-7416543
----	--	--	--	--	--	--	--	2.9-10.9	3.1-13.5	tPHL↓	---	----
-7416544	--	--	--	--	--	--	--	--	2.8-14	tPLH↑	-11.46	-7416544
----	--	--	--	--	--	--	--	--	2.8-13.5	tPHL↓	---	----
-7416620	--	--	--	--	--	--	--	2.1-6.8	2-8.5	tPLH↑	-11.48	-7416620
----	--	--	--	--	--	--	--	3.1-8.2	4-10.5	tPHL↓	---	----
-7416623	--	--	--	--	--	--	--	--	4.2-10.4	tPLH↑	-11.48	-7416623
----	--	--	--	--	--	--	--	--	3.1-10.3	tPHL↓	---	----
-7416640	--	--	--	--	--	--	--	1.8-7.3	2.2-9.1	tPLH↑	-11.49	-7416640
----	--	--	--	--	--	--	--	2.6-8.6	4.1-10.5	tPHL↓	---	----
-7416646	--	--	--	--	--	--	--	2.9-12.5	3.9-13.8	tPLH↑	-11.50	-7416646
----	--	--	--	--	--	--	--	2.9-13.1	3.4-14.9	tPHL↓	---	----
-7416648	--	--	--	--	--	--	--	--	2.4-15.8	tPLH↑	-11.50	-7416648
----	--	--	--	--	--	--	--	--	3.8-16.6	tPHL↓	---	----
-7416651	--	--	--	--	--	--	--	--	3-17.3	tPLH↑	-11.51	-7416651
----	--	--	--	--	--	--	--	--	4.6-17.8	tPHL↓	---	----
-7416652	--	--	--	--	--	--	--	2.7-13.8	3.7-12.3	tPLH↑	-11.51	-7416652
----	--	--	--	--	--	--	--	3-13.8	3-16.9	tPHL↓	---	----
-7416657	--	--	--	--	--	--	--	--	4-14.6	tPLH↑	-11.52	-7416657
----	--	--	--	--	--	--	--	--	3.2-14.7	tPHL↓	---	----
-7416861	--	--	--	--	--	--	--	--	3.1-10.4	tPLH↑	-11.53	-7416861
----	--	--	--	--	--	--	--	--	2.9-11.1	tPHL↓	---	----
-7416863	--	--	--	--	--	--	--	--	4.1-11.1	tPLH↑	-11.54	-7416863
----	--	--	--	--	--	--	--	--	3.1-11.8	tPHL↓	---	----
-7416864	--	--	--	--	--	--	--	--	1.6-10.2	tPLH↑	-11.55	-7416864
----	--	--	--	--	--	--	--	--	3.7-11.3	tPHL↓	---	----
-7416952								--	4.7-11.8	tPLH↑	-11.56	-7416952
----								--	4.9-11.7	tPHL↓	---	----

NOTE: For some devices td, tp, or equivalent time delays are defined instead of tPLH, tPHL.

Table 11.5b. 5-volt 74xxx series CMOS device representative propagation

DEVICE 74---	LV	ALVC LVC	ALVT LVT/ABT	LVX	LVQ	LCX	AC(3V)	LVCH/ ALVCH	HL	AHC (3V)	VHC(3V)	DATA	Figure No.	DEVICE 74---
-74245	9-14	1.5-7.0	1.0-4.0	1.0-8.5	2.0-10.0	1.5-7.0	1-9	--	--	1-10	5.8-8.4	tPLH	-11.2	-74245
----	9-14	1.5-7.0	1.0-4.0	1.0-8.5	2.0-10.0	1.5-8.0	1-9	--	--	1-10	8.3-11.9	tPHL	---	----
-74543	--	1.5-10	1.0-3.6	--	--	1.5-8.5	1.5-12.5	--	--	--	--	tPLH	-11.5	-74543
----	--	1.5-10	1.0-4.2	--	--	1.5-8.5	1.5-8.0	--	--	--	--	tPHL	---	----
-74544	--	1.5-10	--	--	--	--	1.5-12.0	--	--	--	--	tPLH	-11.5	-74544
----	--	1.5-10	--	--	--	--	1.5-8.0	--	--	--	--	tPHL	---	----
-74623	--	1.5-6.5	--	--	--	--	--	--	--	--	--	tPLH	-11.9	-74623
----	--	1.5-6.5	--	--	--	--	--	--	--	--	--	tPHL	---	----
-74640	--	--	--	--	--	--	--	--	--	--	--	tPLH	-11.12	-74640
----	--	--	--	--	--	--	--	--	--	--	--	tPHL	---	----
-74646	--	1.5-8.9	1.3-6.2	--	--	1.5-8.5	2.0-16.5	--	--	--	--	tPLH	-11.14	-74646
----	--	1.5-8.9	1.0-6.2	--	--	1.5-9.5	1.5-14.5	--	--	--	--	tPHL	---	----
-74648	--	--	--	--	--	--	1.5-15.5	--	--	--	--	tPLH	-11.14	-74648
----	--	--	--	--	--	--	1.5-13.5	--	--	--	--	tPHL	---	----
-74652	--	1.5-8.9	1.2-6.4	--	--	1.5-8.5	--	--	--	--	--	tPLH	-11.15	-74652
----	--	1.5-8.9	1.0-6.2	--	--	1.5-9.5	--	--	--	--	--	tPHL	---	----
-74657	--	--	--	--	--	--	--	--	--	--	--	tPLH	-11.16	-74657
----	--	--	--	--	--	--	--	--	--	--	--	tPHL	---	----
-74861	--	1.5-7	--	--	--	--	--	--	--	--	--	tPLH	-11.24	-74861
----	--	1.5-7	--	--	--	--	--	--	--	--	--	tPHL	---	----
-74863	--	1.5-7	--	--	--	--	--	--	--	--	--	tPLH	-11.24	-74863
----	--	1.5-7	--	--	--	--	--	--	--	--	--	tPHL	---	----
-742245	--	1.5-8.0	--	--	--	--	--	--	--	--	--	tPLH	-11.2	-742245
----	--	1.5-8.0	--	--	--	--	--	--	--	--	--	tPHL	---	----
-742952	--	--8.5	1.3-6.1	--	--	--	--	--	--	--	--	tPLH	-11.28	-742952
----	--	--8.5	1.8-6.0	--	--	--	--	--	--	--	--	tPHL	---	----
-743245	--	--	--	1.0-7.5	--	--	--	--	--	--	--	tPLH	-11.29	-743245
----	--	--	--	1.0-8.0	--	--	--	--	--	--	--	tPHL	---	----
-744245	--	1.5-6.5	--	1.0-8.5	--	--	--	--	--	--	--	tPLH	-11.29	-744245
----	--	1.5-6.5	--	1.0-8.5	--	--	--	--	--	--	--	tPHL	---	----
-7411245	--	--	--	--	--	--	1.5-11.2	--	--	--	--	tPLH	-11.35	-7411245
----	--	--	--	--	--	--	1.5-8.5	--	--	--	--	tPHL	---	----
-7411652	--	--	--	--	--	--	2.9-14.3	--	--	--	--	tPLH	-11.36	-7411652
----	--	--	--	--	--	--	3.9-16.2	--	--	--	--	tPHL	---	----
-7416245	--	--5.2	1.0-3.3	--	--	--	--	1-5.2	--	5.8-8.4	--	tPLH	-11.37	-7416245
----	--	--5.2	1.0-3.3	--	--	--	--	1-5.2	--	5.8-8.4	--	tPHL	---	----
-7416269	--	--	--	--	--	--	--	1-6.2	--	--	--	tPLH	-11.39	-7416269
----	--	--	--	--	--	--	--	1-6.2	--	--	--	tPHL	---	----
-7416270	--	--	--	--	--	--	--	1-5.5	--	--	--	tPLH	-11.40	-7416270
----	--	--	--	--	--	--	--	1-5.5	--	--	--	tPHL	---	----
-7416409	--	--	--	--	--	--	--	1.5-5.1	--	--	--	tPLH	-11.41	-7416409
----	--	--	--	--	--	--	--	1.5-5.1	--	--	--	tPHL	---	----
-7416500	--	--	0.5-5.4	--	--	1.5-7.0	--	NA	--	--	--	tPLH	-11.44	-7416500
----	--	--	0.5-5.4	--	--	1.5-7.0	--	NA	--	--	--	tPHL	---	----
-7416501	--	--	0.5-5.4	--	--	1.5-7.0	--	1-4.9	--	--	--	tPLH	-11.44	-7416501
----	--	--	0.5-5.4	--	--	1.5-7.0	--	1-4.9	--	--	--	tPHL	---	----
-7416525	--	--	--	--	--	--	--	1-4.2	--	--	--	tPLH	-11.45	-7416525
----	--	--	--	--	--	--	--	1-4.2	--	--	--	tPHL	---	----
-7416543	--	--	1.0-4.8	--	--	1.5-6.5	--	NA	--	--	--	tPLH	-11.46	-7416543
----	--	--	1.0-4.8	--	--	1.5-6.5	--	NA	--	--	--	tPHL	---	----

NOTE: For some devices td, tp, or equivalent time delays are defined instead of tPLH, tPHL.

Table 11.6a. Low-Voltage 74xxx series CMOS device representative propagation delay times in nS.

DEVICE 74---	LV	LVC	ALVT LVT/ABT	LVX	LVQ	LCX	AC(3V)	ALVCH	HL	AHC (3V)	VHC(3V)	DATA	Figure No.	DEVICE 74---
-7416600	--	--	1.5–3.8	--	--	--	--	1–5.7	--	--	--	tPLH	-11.47	-7416600
----	--	--	1.5–4.0	--	--	--	--	1–5.7	--	--	--	tPHL	---	----
-7416601	--	--	1.0–2.6	--	--	--	--	1.2–5.0	--	--	--	tPLH	-11.47	-7416601
----	--	--	1.0–3.2	--	--	--	--	1.2–5.0	--	--	--	tPHL	---	----
-7416620	--	--	--	--	--	--	2.7–8.7	--	--	--	--	tPLH	-11.48	-7416620
----	--	--	--	--	--	--	3.9–10.6	--	--	--	--	tPHL	---	----
-7416623	--	--5.2	--	--	--	--	--	1.5–3.6	--	--	--	tPLH	-11.48	-7416623
----	--	--5.2	--	--	--	--	--	1.5–3.6	--	--	--	tPHL	---	----
-7416640	--	--5.2	--	--	--	--	2.2–9.1	--	--	--	--	tPLH	-11.49	-7416640
----	--	--5.2	--	--	--	--	3–11	--	--	--	--	tPHL	---	----
-7416646	--	--	1.0–4.0	--	--	1.5–6.0	3.4–13.2	3.0–6.7	--	--	--	tPLH	-11.50	-7416646
----	--	--	1.0–3.6	--	--	1.5–6.0	3.6–13.4	3.0–6.7	--	--	--	tPHL	---	----
-7416652	--	--	1.0–2.4	--	--	1.5–6.5	3.6–18.7	3.0–6.7	--	--	--	tPLH	-11.51	-7416652
----	--	--	1.0–2.1	--	--	1.5–6.5	4.1–18.3	3.0–6.7	--	--	--	tPHL	---	----
-7416952	--	--	--	--	--	--	--	3.2–6.8	--	--	--	tPLH	-11.56	-7416952
----	--	--	--	--	--	--	--	3.2–6.8	--	--	--	tPHL	---	----
-74162245	--	--7.0	0.5–3.6	--	--	--	--	--	--	--	--	tPLH	-11.37	-74162245
----	--	--7.0	0.5–3.1	--	--	--	--	--	--	--	--	tPHL	---	----
-74162268	--	--	--	--	--	--	--	1.8–5.4	--	--	--	tPLH	-11.38	-74162268
----	--	--	--	--	--	--	--	1.8–5.4	--	--	--	tPHL	---	----
-74162269	--	--	--	--	--	--	--	1.6–6.7	--	--	--	tPLH	-11.39	-74162269
----	--	--	--	--	--	--	--	1.6–6.7	--	--	--	tPHL	---	----
-74162409	--	--	--	--	--	--	--	1.5–6.2	--	--	--	tPLH	-11.41	-74162409
----	--	--	--	--	--	--	--	1.5–6.2	--	--	--	tPHL	---	----
-74162525	--	1–4.7	--	--	--	--	--	1–4.7	--	--	--	tPLH	-11.45	-74162525
----	--	1–4.7	--	--	--	--	--	1–4.7	--	--	--	tPHL	---	----
-74162601	--	--	--	--	--	--	--	1–5.5	--	--	--	tPLH	-11.47	-74162601
----	--	--	--	--	--	--	--	1–5.5	--	--	--	tPHL	---	----
-74162623	--	--6.2	--	--	--	--	--	--	--	--	--	tPLH	-11.48	-74162623
----	--	--6.2	--	--	--	--	--	--	--	--	--	tPHL	---	----
-74162640	--	--6.5	--	--	--	--	--	--	--	--	--	tPLH	-11.49	-74162640
----	--	--6.5	--	--	--	--	--	--	--	--	--	tPHL	---	----
-74164245	--	--	--	--	--	--	--	3.6–5.5	--	--	--	tPLH	-11.31	-74164245
----	--	--	--	--	--	--	--	3.6–5.5	--	--	--	tPHL	---	----
-7422952	--	--9.5	--	--	--	--	--	--	--	--	--	tPLH	-11.28	-7422952
----	--	--9.5	--	--	--	--	--	--	--	--	--	tPHL	---	----
-7433245	--	--	--	--	--	--	--	--	3.5–4.0	--	--	tPLH	-11.61	-7433245
----	--	--	--	--	--	--	--	--	3.5–4.0	--	--	tPHL	---	----
-7433620	--	--	--	--	--	--	--	--	3.5–4.0	--	--	tPLH	-11.62	-7433620
----	--	--	--	--	--	--	--	--	3.5–4.0	--	--	tPHL	---	----
-7433623	--	--	--	--	--	--	--	--	3.5–4.0	--	--	tPLH	-11.62	-7433623
----	--	--	--	--	--	--	--	--	3.5–4.0	--	--	tPHL	---	----
-7433640	--	--	--	--	--	--	--	--	3.5–4.0	--	--	tPLH	-11.61	-7433640
----	--	--	--	--	--	--	--	--	3.5–4.0	--	--	tPHL	---	----
-7433646	--	--	--	--	--	--	--	--	4.8–7.0	--	--	tPLH	-11.63	-7433646
----	--	--	--	--	--	--	--	--	4.8–7.0	--	--	tPHL	---	----
-7433648	--	--	--	--	--	--	--	--	4.9–7.0	--	--	tPLH	-11.63	-7433648
----	--	--	--	--	--	--	--	--	4.9–7.0	--	--	tPHL	---	----
-7433651	--	--	--	--	--	--	--	--	4.9–7.0	--	--	tPLH	-11.64	-7433651
----	--	--	--	--	--	--	--	--	4.9–7.0	--	--	tPHL	---	----
-7433652	--	--	--	--	--	--	--	--	4.8–7.0	--	--	tPLH	-11.64	-7433652
----	--	--	--	--	--	--	--	--	4.8–7.0	--	--	tPHL	---	----
-7433952	--	--	--	--	--	--	--	--	5.9–6.8	--	--	tPLH	-11.65	-7433952
----	--	--	--	--	--	--	--	--	5.9–6.8	--	--	tPHL	---	----

NOTE: For some devices td, tp, or equivalent time delays are defined instead of tPLH, tPHL.

Table 11.6b. Low-Voltage 74xxx series CMOS device representative propagation delay times in nS (continuation).

TTL – 74LS/AS/ALS/F242
74LS/AS/ALS/F243
74AS/ALS758 (OC)
74AS759 (OC)
74ALS2242

5V CMOS – 74HC/HCT242
74HC/HCT243

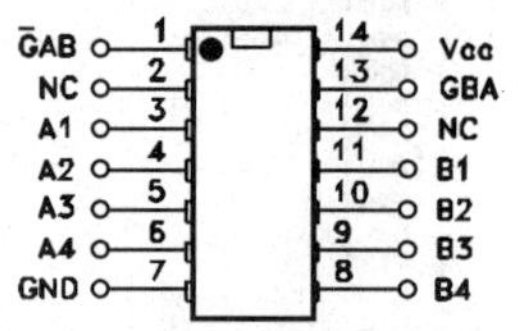

DUAL-IN-LINE PACKAGE

FUNCTION TABLE 74x242, 'x243
74x758, 'AS759 OPEN COLLECTOR
74ALS2242 TRANSCEIVER/MOS DRIVER
QUAD BUS TRANSCEIVERS
(EACH TRANSCEIVER)

INPUTS		'x242	x243
$\overline{GAB}$	GBA		
0	0	A to B	A to B
1	1	B to A	B to A
1	0	isolation	isolation
0	1	Latch A and B (A=B)	Latch A and B (A=B)

1 = HIGH voltage level.
0 = LOW voltage level.

LOGIC DIAGRAM
TBased on Texas Instruments 74LS242

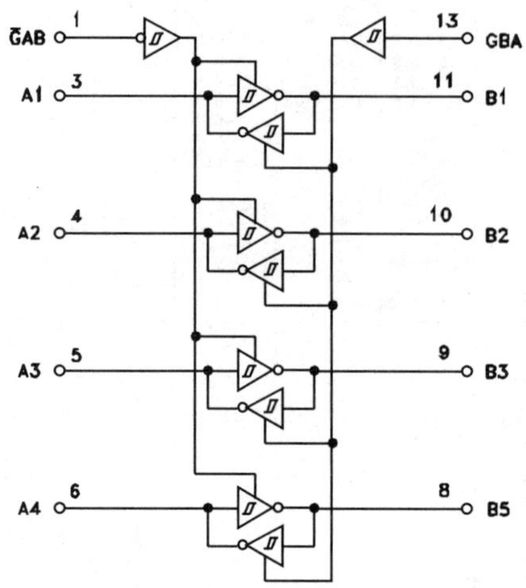

LOGIC DIAGRAM
TBased on Texas Instruments 74LS243

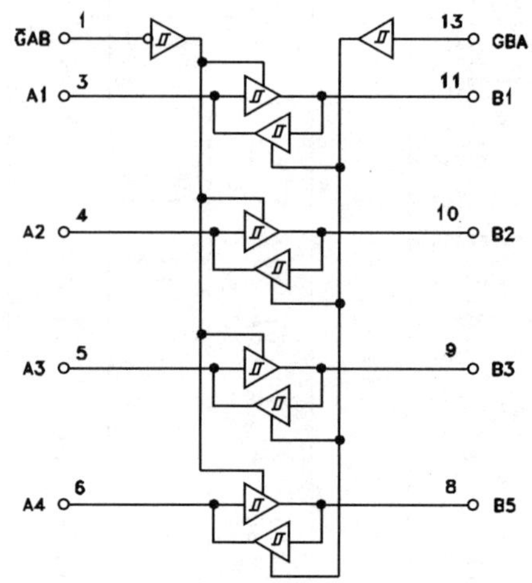

Figure 11.1. The 74x242 and 'x243 series of quadruple bus transceivers provide two-way asynchronous communication between data buses. The '243 is a non-inverting version of the '242. The TTL devices feature PNP inputs to minimize loading. The LS versions can drive loads as low as 133 ohms. The 74AS/ALS758 and 74AS759 are open collector versions. The output enable inputs provide a means for isolating buses.

TTL – 74LS/AS/ALS/F245
74ALS5245
74ALS5620
5V CMOS – 74HC/HCT/AHC/AHCTVHC/VHCT/
AC/ACQ/ACT/ACTQ245
3V CMOS – 74LV/LVC/LVCH/LVT/LCX/LVX/
LVQ/LVT ABT/VHC/AC/ACQ/AHC/245
74LVC/LVCH2245

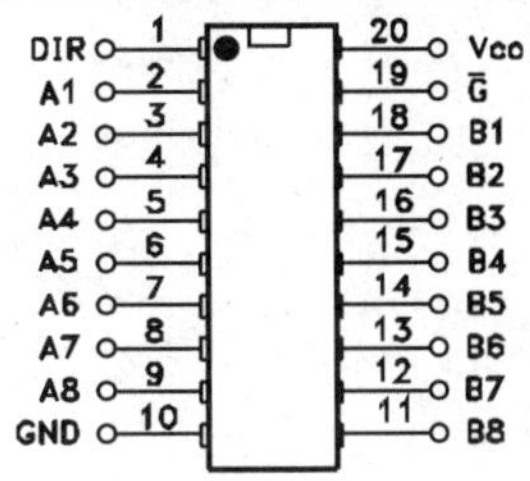

NOTE:
The SSOP and TSSOP Packages follow the DIP Package pinning.

DUAL-IN-LINE PACKAGE

FUNCTION TABLE 74x245, 'ALS5245
OCTAL BUS TRANSCEIVERS WITH 3-STATE OUTPUTS

ENABLE G̅	DIRECTION CONTROL DIR	OPERATION
0	0	B data to A bus
0	1	A data to B bus
1	X	Isolation

1 = HIGH voltage level.
0 = LOW voltage level.
X = Don't Care

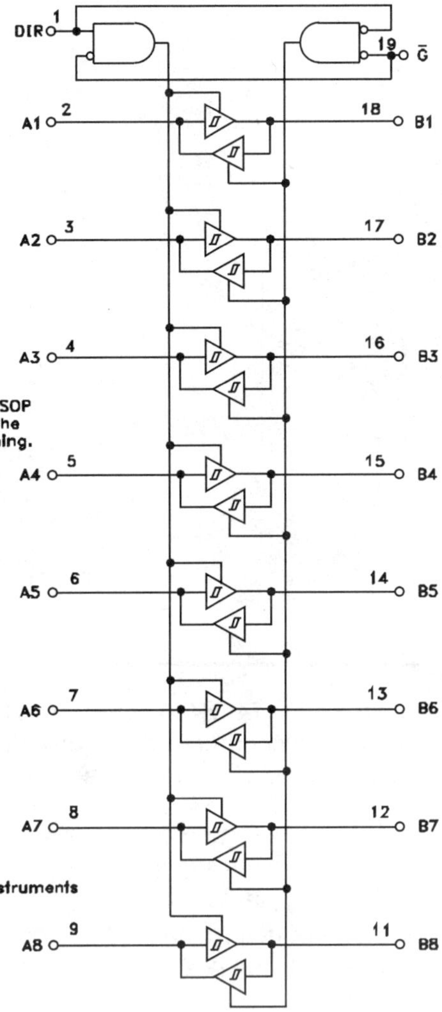

LOGIC DIAGRAM
Based on Texas Instruments 74LS245

Figure 11.2. The 74x245 series of TTL and CMOS devices are noninverting octal bus transceivers providing two-way asynchronous communication between data buses. The TTL devices feature hysteresis with PNP inputs to minimize loading. 3-state output enable inputs provide a means for isolating buses in both directions. The 74ALS5245 is comparable to the TTL versions of the '245 series. The 74ALS5620 is an inverting version.

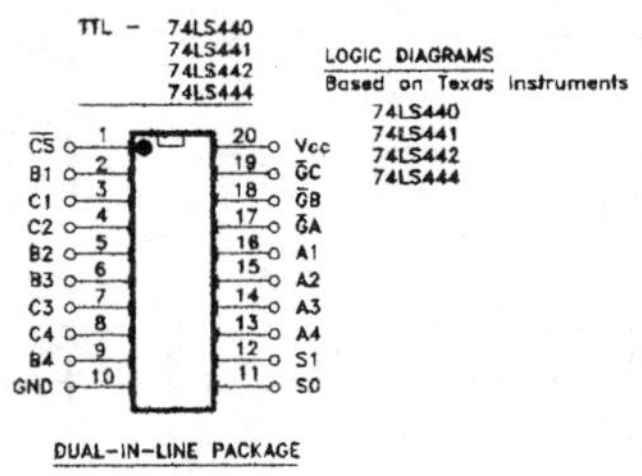

FUNCTION TABLE 74LS440, '441, '442 '444
QUAD TRIDIRECTIONAL BUS TRANCEIVERS

INPUTS						TRANSFERS BETWEEN BUSES		
$\overline{CS}$	S1	S0	$\overline{GA}$	$\overline{GB}$	$\overline{GC}$	'LS440 'LS442	'LS441	'LS444
1	X	X	X	X	X	None	None	None
X	1	1	X	X	X	None	None	None
X	X	X	1	1	1	None	None	None
X	0	0	X	1	1	None	None	None
X	0	1	1	X	1	None	None	None
X	1	0	1	1	X	None	None	None
0	0	0	X	0	0	A←B, A←C	$\overline{A}$←B, $\overline{A}$←C	$\overline{A}$←B, $\overline{A}$←C
0	0	1	0	X	0	B←C, B←A	$\overline{B}$←C, $\overline{B}$←A	B←C, $\overline{B}$←A
0	1	0	0	0	X	C←A, B←B	$\overline{C}$←A, $\overline{C}$←B	$\overline{C}$←A, C←B
0	0	0	X	0	1	A←B	$\overline{A}$←B	$\overline{A}$←B
0	0	1	1	X	0	B←C	$\overline{B}$←C	B←C
0	1	0	0	1	X	C←A	$\overline{C}$←A	$\overline{C}$←A
0	0	0	X	1	0	A←C	$\overline{A}$←C	$\overline{A}$←C
0	0	1	0	X	1	B←A	$\overline{B}$←A	$\overline{B}$←A
0	1	0	1	0	X	C←B	$\overline{C}$←B	C←B

1 = HIGH voltage level.
0 = LOW voltage level.
X = Don't Care

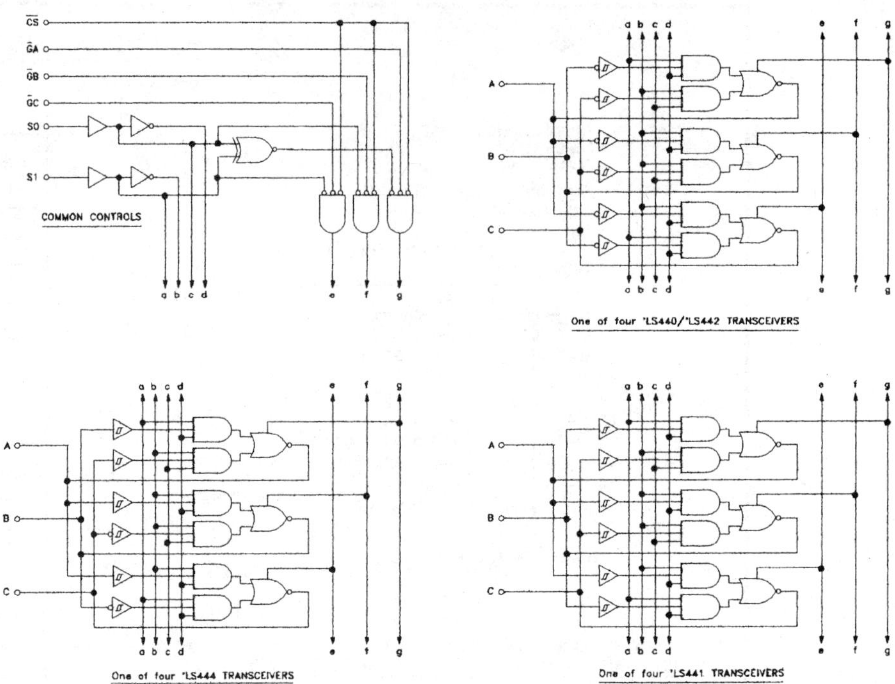

Figure 11.3. The 74LS440 - 74LS444 series of TTL devices are quadruple tridirectional bus transceivers providing three-way asynchronous communication between data buses. Input hysteresis improves noise margins. The series offer a choice of open collector or 3-state outputs.
74LS440 open collector, true output
74LS441 open collector, inverting output
74LS442 3-state, true output
74LS444 3-state. inverting output
The select inputs determine the bus from which data will be transferred; the /G inputs enable the bus, or buses, to which the data will be transferred.

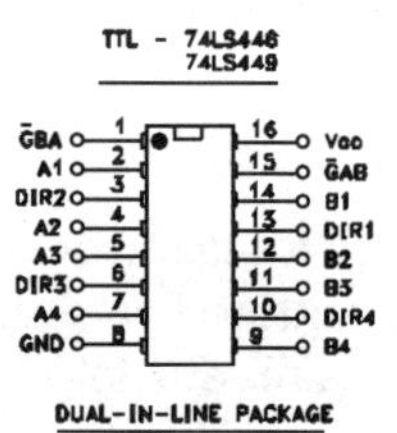

FUNCTION TABLE 74LS446, 'LS449 QUAD BUS TRANSCEIVERS WITH INDIVIDUAL DIRECTION CONTROLS

ENABLE $\overline{GBA}$	ENABLE $\overline{GAB}$	DIRECTION DIR	OPERATION 'LS446	OPERATION 'LS449
1	1	X	Isolation	Isolation
X	0	1	$\overline{A}$ data to B bus	A data to B bus
0	X	0	$\overline{B}$ data to A bus	B data to A bus
X	1	1	Isolation	Isolation
1	X	0	Isolation	Isolation

1 = HIGH voltage level.
0 = LOW voltage level.
X = Don't Care

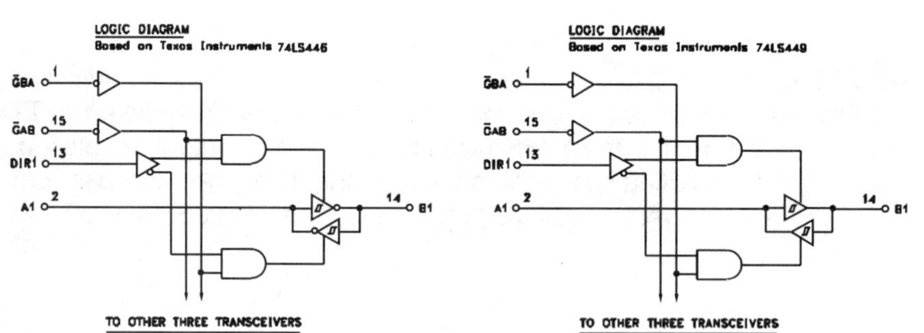

Figure 11.4. The 74LS446 and 74LS449 TTL devices are quadruple bus transceivers providing data transmission from individual lines of the A bus to individual lines of the B bus, depending on the logic levels at the direction control inputs, DIR1 - DIR4. This is shown in the function table. Hysteresis at the bus inputs improves noise margins, while PNP inputs reduce loading. The package pinning provides for flow-through data, B bus opposite A bus. The pair offer a choice of true ('LS449) and inverting ('LS446) outputs.

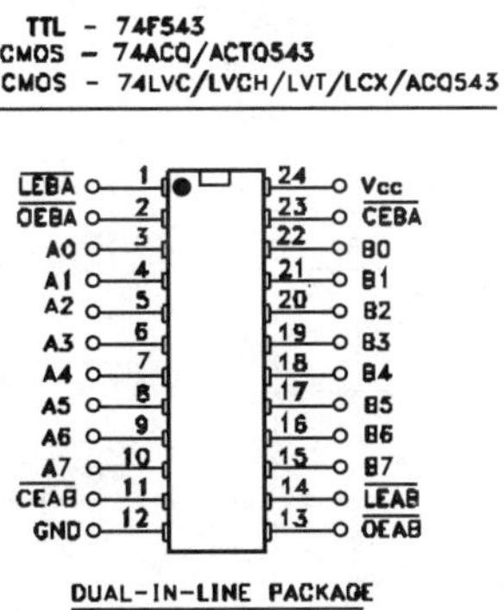

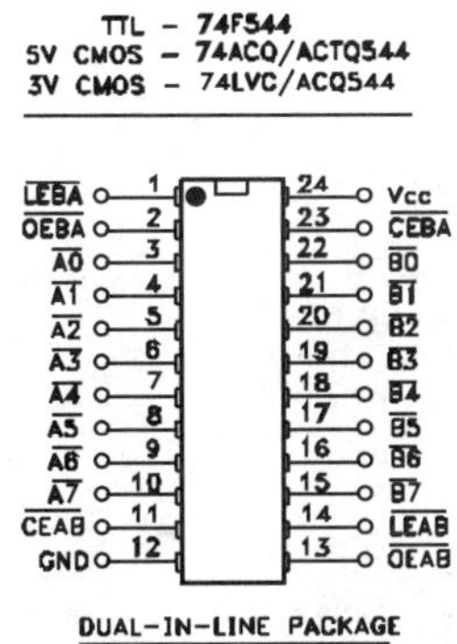

NOTE:
The SSOP and TSSOP Packages follow the DIP Package pinning.

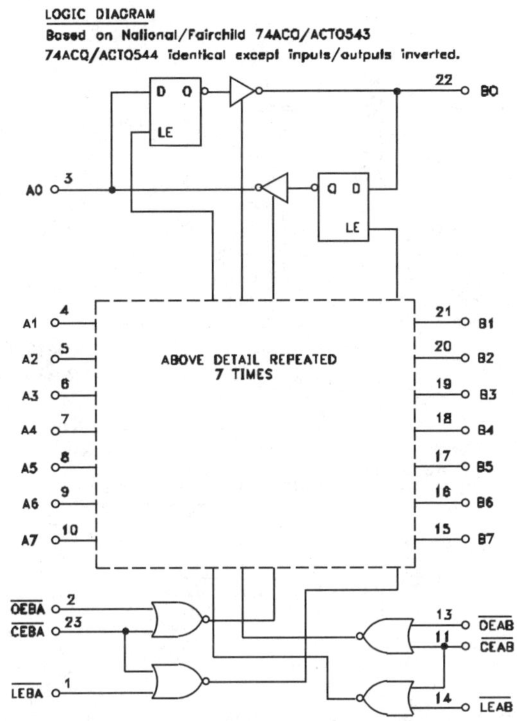

DATA I/O CONTROL TABLE 74x543, 'x544
OCTAL REGISTERED TRANSCEIVER
WITH 3-STATE OUTPUTS

INPUTS				
$\overline{CEAB}$	$\overline{LEAB}$	$\overline{OEAB}$	LATCH STATUS	OUTPUT BUFFERS
1	X	X	Latched	High Z
X	1	X	Latched	---
0	0	X	Transparent	---
X	X	1	---	High Z
0	X	0	---	Driving

1 = HIGH voltage level.
0 = LOW voltage level.
X = Don't Care
Z = High Impedance state

PIN NAMES	DESCRIPTION
$\overline{OEAB}$	A-to-B Output Enable Input (Active Low)
$\overline{OEBA}$	B-to-A Output Enable Input (Active Low)
$\overline{CEAB}$	A-to-B Enable Input (Active Low)
$\overline{CEBA}$	B-to-A Enable Input (Active Low)
$\overline{LEAB}$	A-to-B Latch Enable Input (Active Low)
$\overline{LEBA}$	B-to-A Latch Enable Input (Active Low)
A0-A7	A-to-B Data Inputs or B-to-A 3-State Outputs
B0-B7	B-to-A Data Inputs or A-to-B 3-State Outputs

Figure 11.5. The 74x543 and 74x544 TTL and CMOS devices are octal registered transceivers containing two sets of D-type registers for temporary storage of data flowing in either direction. Separate Latch and Output enables are provided for each register for independent input and output control. The pair offer a choice of true ('x543) and inverting ('x544) outputs.

TTL – 74F545

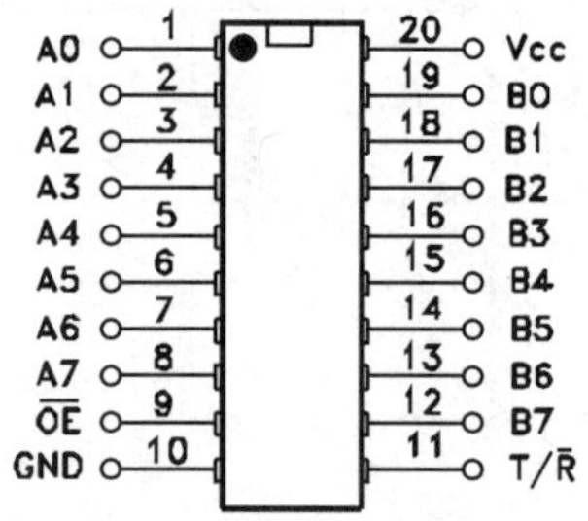

DUAL-IN-LINE PACKAGE

PACKAGE PINNING

PINS–
1–7 DATA INPUT/OUTPUT
12–19 DATA INPUT/OUTPUT
9 OUTPUT ENABLE INPUT
11 TRANSMIT/RECEIVE INPUT

FUNCTION TABLE 74F545
OCTAL BIDIRECTIONAL TRANSCEIVER
WITH 3-STATE INPUTS/OUTPUTS

INPUTS		
OE	T/R̄	OUTPUTS
0	0	Bus B data to Bus A
0	1	Bus A data to Bus B
1	X	Z

1 = HIGH voltage level.
0 = LOW voltage level.
X = Don't Care
Z = High Impedance state

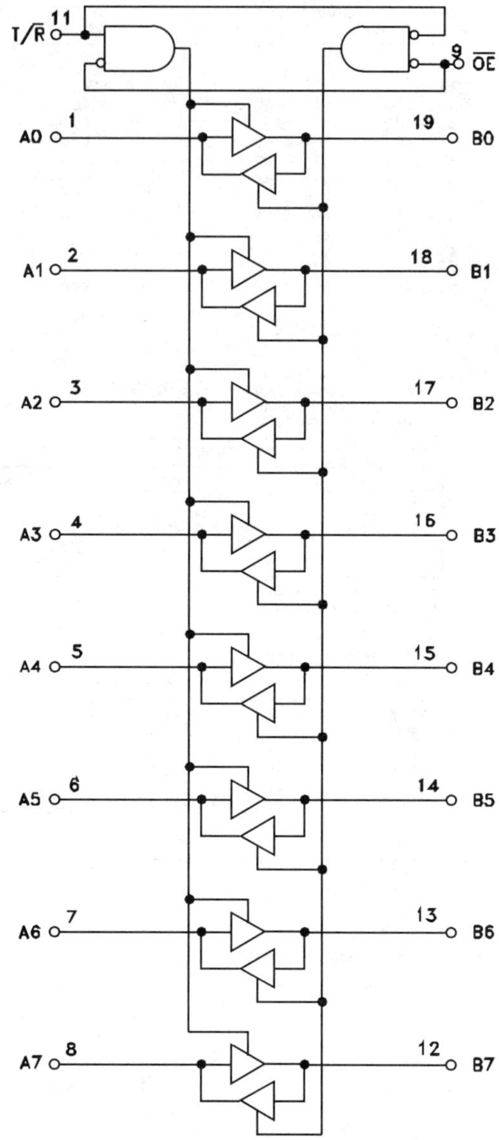

LOGIC DIAGRAM
Based on Philips 74F545

Figure 11.6. The TTL 74F545 is an octal bidirectional transceiver featuring 3-state inputs and outputs. Straight through bidirectional transceivers are featured with 24 mA bus drive capability on the A ports and 64 mA on the B ports. The Transmit//Receive (T//R) input determines the direction of logic signals through the transceivers. Transmit enables data from A to B ports; Receive from B to A ports. The output enable disables both ports by placing them in a 3-state condition. The 'F545 performs the same functions as the 'F245, the only difference being pin placement.

TTL – 74F552

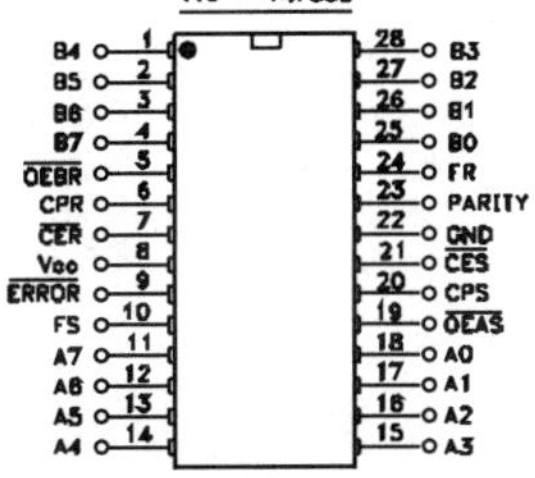

DUAL-IN-LINE PACKAGE

PACKAGE PINNING:

PINS–	
15–18,11–14	A Data Inputs
1–4,25–28	B Data Inputs
5	A-to-B Output Enable and Clear FS Output
6	R Registers Clock Input
7	R Registers Clock Enable Input
9	Parity Check Output
19	B-to-A Output Enable and Clear Fr Output
20	S Registers Clock Input
21	S Registers Clock Enable Input
23	Parity Bit Transceiver I/O
10	B-to-A Status Flag Output
24	A-to-B Status Flag Output

FUNCTION TABLES 74F552 OCTAL REGISTERED TRANSCEIVER WITH PARITY AND FLAGS, 3-STATE

R OR S REGISTER FUNCTION TABLE

INPUTS			OUTPUTS	
An or Bn	CPx	$\overline{CEx}$	Internal Q	OPERATING MODE
X	X	1	NC	Hold Data
0	↑	0	0	Load Data
1	↑	0	1	Load Data
X	↑	0	NC	Keep Old Data

↑ = LOW-to-HIGH clock transition.
↓ = HIGH-to-LOW clock transition.
1 = HIGH voltage level.
0 = LOW voltage level.
X = Don't Care
Z = High Impedance state

OUTPUT CONTROL TABLE

INPUTS	OUTPUTS		
$\overline{OExx}$	Internal Q	An or Bn	OPERATING MODE
1	X	Z	Disable Outputs
0	0	0	Enable Outputs
0	1	1	Enable Ouputs

R OR S FLAG FUNCTION TABLE

INPUTS			OUTPUTS	
CEx	$\overline{OExx}$	CPx	FR or FS	OPERATING MODE
X	↑	1	NC	Hold Flag
0	↑	0	1	Set Flag
1	↑	0	0	Clear Flag

PARITY GENERATION FUNCTION TABLE

INPUTS		OUTPUTS		
$\overline{OEBR}$	CPR	No. of Highs In the R Registe Q Outputs	PARITY	OPERATING MODE
1	X	X	Z	Hold Data
0	↑	0,2,4,6,8	1	Load Data
0	↑	1,3,5,7	0	Load Data

PARITY CHECK FUNCTION TABLE

INPUTS			OUTPUTS		
$\overline{OEAS}$	CPS	PARITY	No. of Highs In the R Registe Q Outputs	$\overline{ERROR}$	OPERATING MODE
1	X	X	X	1	Parity Check
0	↑	0	0,2,4,6,8	0	
0	↑	0	1,3,5,7	1	
0	↑	1	0,2,4,6,8	1	
0	↑	1	1,3,5,7	0	

Figure 11.7. The TTL 74F552 is an octal registered transceiver featuring parity and flags with 3-state outputs. The device contains two 8-bit registers for temporary storage of data flowing in either direction. Each register has its own clock and clock enable inputs as well as a flag flip-flop that is automatically set as the register is loaded. The flag output is reset when the output enable returns to High after reading the output port. Each register has its own output enable for its 3-state buffer. The individual clocks, flags, and enables provide considerable flexibility as I/O ports for demand response data transfer. When data is transferred from the A port to the B a parity bit is generated. On the other hand, when the transfer is from the B to the A port the parity of the B0-B7 data is checked.

TTL – 74ALS614
74ALS615

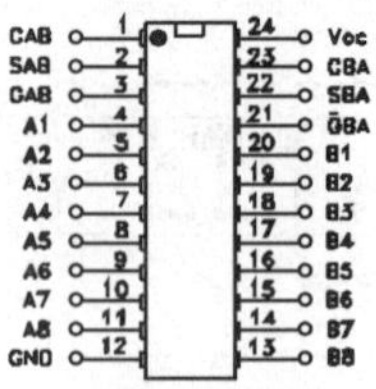

DUAL-IN-LINE PACKAGE

FUNCTION TABLE 74ALS614, 'ALS615 OCTAL BUS TRANCEIVERS AND REGISTERS, OPEN COLLECTOR OUTPUTS

INPUTS						DATA I/O		OPERATION OR FUNCTION	
GAB	GBA	CAB	CBA	SAB	SBA	A1 – A8	B1 – B8	'ALS614	'ALS615
0	1	1 or 0	1 or 0	X	X	Input	Input	Isolation	Isolation
0	1	↑	↑	X	X	Input	Input	Store A and B Data	Store A and B Data
X	1	↑	1 or 0	X	X	Input	Unspecified	Store A, Hold B	Store A, Hold B
1	1	↑	↑	X	X	Input	Output	Store A in both registers	Store A in both registers
0	X	1 or 0	↑	X	X	Unspecified	Input	Hold A, Store B	Hold A, Store B
0	0	↑	↑	X	X	Output	Input	Store B in both registers	Store B in both registers
0	0	X	X	X	0	Output	Input	Real-time B̄ Data to A Bus	Real-time B Data to A Bus
0	0	X	1 or 0	X	1	Output	Output	Stored B̄ Data to A Bus	Stored B Data to A Bus
1	1	X	X	0	X	Input	Output	Real-time Ā Data to B Bus	Real-time A Data to B Bus
1	1	1 or 0	X	1	X	Input	Output	Stored Ā Data to B Bus	Stored A Data to B Bus
1	0	1 or 0	1 or 0	1	1	Output	Output	Stored Ā Data to B Bus and Stored B̄ Data to A Bus	Stored A Data to B Bus and Stored B Data to A Bus

↑ = LOW-to-HIGH clock transition.
1 = HIGH voltage level.
0 = LOW voltage level.
X = Don't Care
Z = High Impedance state

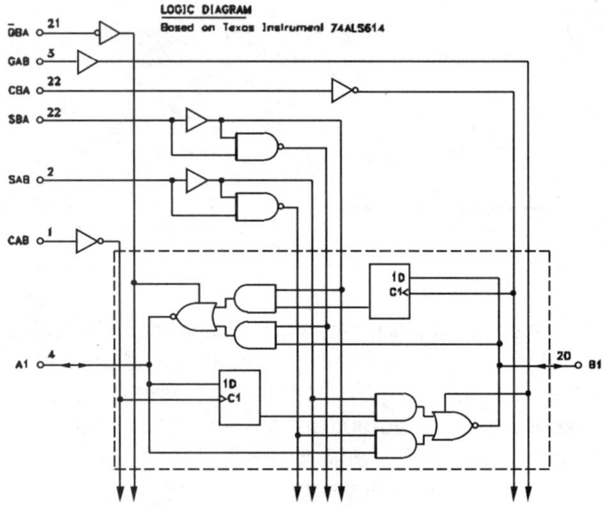

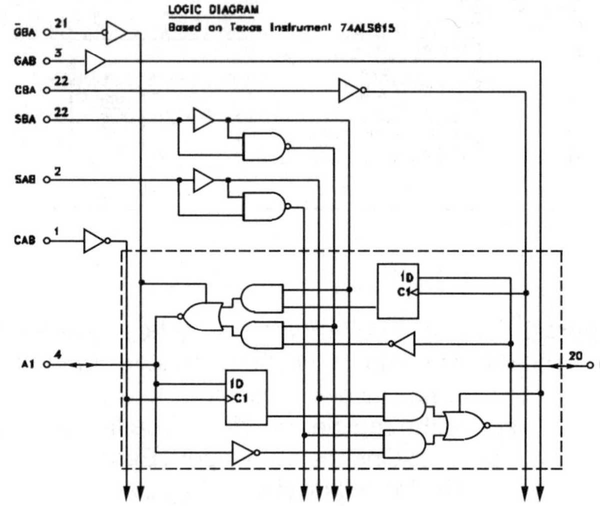

Figure 11.8. The TTL 74ALS614 and 74ALS615 are octal bus transceivers and registers with open collector outputs. The devices consist of bus transceiver circuitry, D-type flip-flops and control circuitry configured for multiplexed data transmission directly from the data bus or from the internal storage registers. Enables GAB and /GBA provide control for the transceiver functions. SAB and SBA control inputs select whether real time or stored data is transferred. The circuitry is designed to eliminate the decoding glitch that often occurs in a multiplexer during the transition. Data on either or both data buses can be stored in the internal flip-flops.

TTL - 74AS/ALS/F620
74LS/AS/ALS/F623
74AS2620
74AS2623
74ALS5620
5V CMOS - 74HC/HCT620
74HC/HCT623
3V CMOS - 74LVC/LVCH623

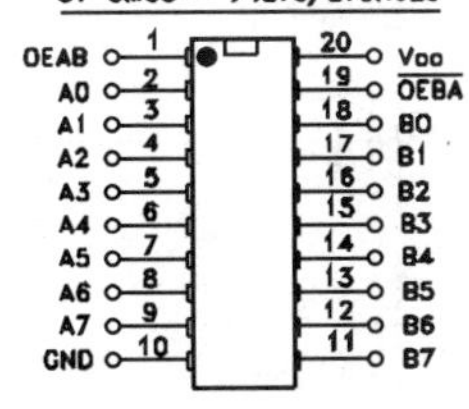

DUAL-IN-LINE PACKAGE

PACKAGE PINNING

PINS-
2-9 DATA INPUT/OUTPUT
11-18 DATA INPUT/OUTPUT
1,19 OUTPUT ENABLE INPUTS

NOTE:
The SSOP and TSSOP Packages follow the DIP Package pinning.

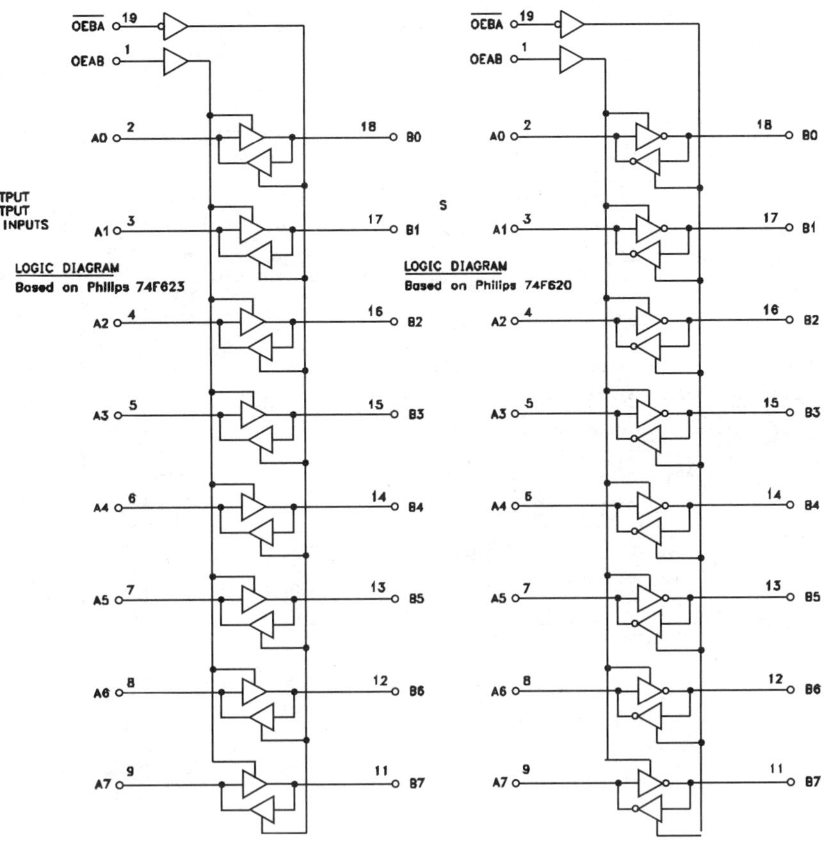

FUNCTION TABLE 74x620, 'x623, 'AS2620, 'AS2623, 'ALS5620 OCTAL BIDIRECTIONAL TRANSCEIVER WITH 3-STATE INPUTS/OUTPUTS

INPUTS		OPERATING MODES	
OEBA	$\overline{OEAB}$	'ALS5620 'x620 'AS2620	'x623 'AS2623
0	0	$\overline{B}$ Data to A Bus	B Data to A Bus
1	1	$\overline{A}$ Data to B Bus	A Data to B Bus
1	0	Z	Z
0	1	$\overline{B}$ Data to A Bus	B Data to A Bus
0	1	$\overline{A}$ Data to B Bus	A Data to B Bus

1 = HIGH voltage level.
0 = LOW voltage level.
X = Don't Care
Z = High Impedance state

Figure 11.9. The 74x620 and 74x623 are octal bus transceivers with 3-state outputs providing asynchronous two-way communication between two data buses. The dual-enable feature gives the device the ability to store data by simultaneous operation of the two control inputs. Each output reinforces its input in this configuration. The F version features high-impedance NPN inputs to minimize bus loading. The 'x620 is inverting; the 'x623 true outputs.

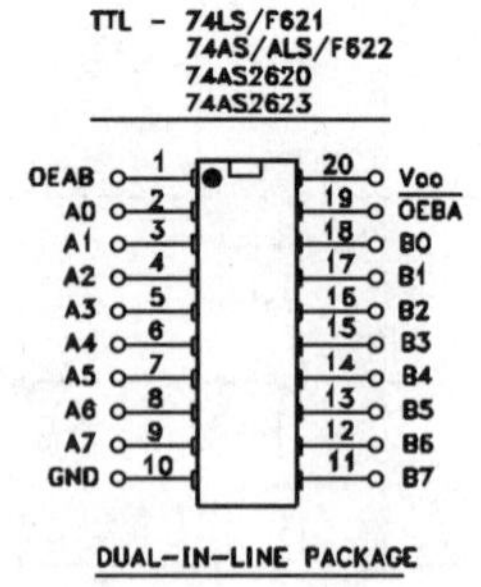

PACKAGE PINNING
PINS–
2–9 DATA INPUT/OUTPUT
11–18 DATA INPUT/OUTPUT
1,19 OUTPUT ENABLE INPUTS

FUNCTION TABLE 74x621, 'x622
74AS2620, AS2623
OCTAL BIDIRECTIONAL TRANSCEIVES

INPUTS		OPERATING MODES	
OEBA	$\overline{OEAB}$	'x620	'x623
0	0	$\overline{B}$ Data to A Bus	B Data to A Bus
1	1	$\overline{A}$ Data to B Bus	A Data to B Bus
1	0	OFF	
0	1	$\overline{B}$ Data to A Bus	B Data to A Bus
0	1	$\overline{A}$ Data to B Bus	A Data to B Bus

1 = HIGH voltage level.
0 = LOW voltage level.
X = Don't Care
OFF = HIGH if pullup resistor connected to OC output
74AS2620, 'AS2623 are MOS drivers with 25 ohm I/O port series resistors; no external resistors required.

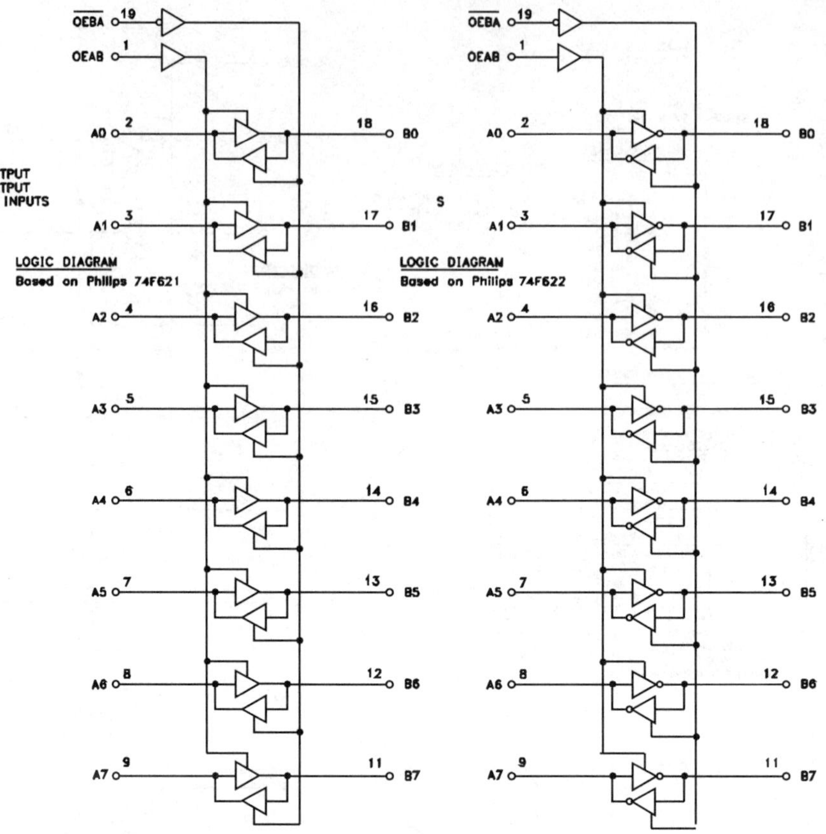

Figure 11.10. The 74x621 and 74x622 are TTL octal bus transceivers with open collector outputs providing asynchronous two-way communication between two data buses. The dual-enable feature gives the device the ability to store data by simultaneous operation of the two control inputs. Each output reinforces its input in this configuration. The F version features high-impedance NPN inputs to minimize bus loading. The 'x622 is inverting; the 'x621 true outputs.

TTL - 74LS638
74LS639

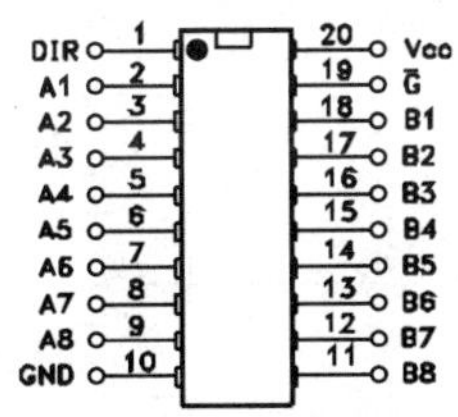

DUAL-IN-LINE PACKAGE

FUNCTION TABLE 74LS638, 'LS639 OCTAL BUS TRANSCEIVERS WITH OC 3-STATE OUTPUTS

ENABLE	DIRECTION CONTROL	OPERATION	
$\bar{G}$	DIR	'LS638	LS639
0	0	$\bar{B}$ data to A bus	B data to A bus
0	1	$\bar{A}$ data to B bus	A data to B bus
1	X	Isolation	Isolation

1 = HIGH voltage level.
0 = LOW voltage level.
X = Don't Care

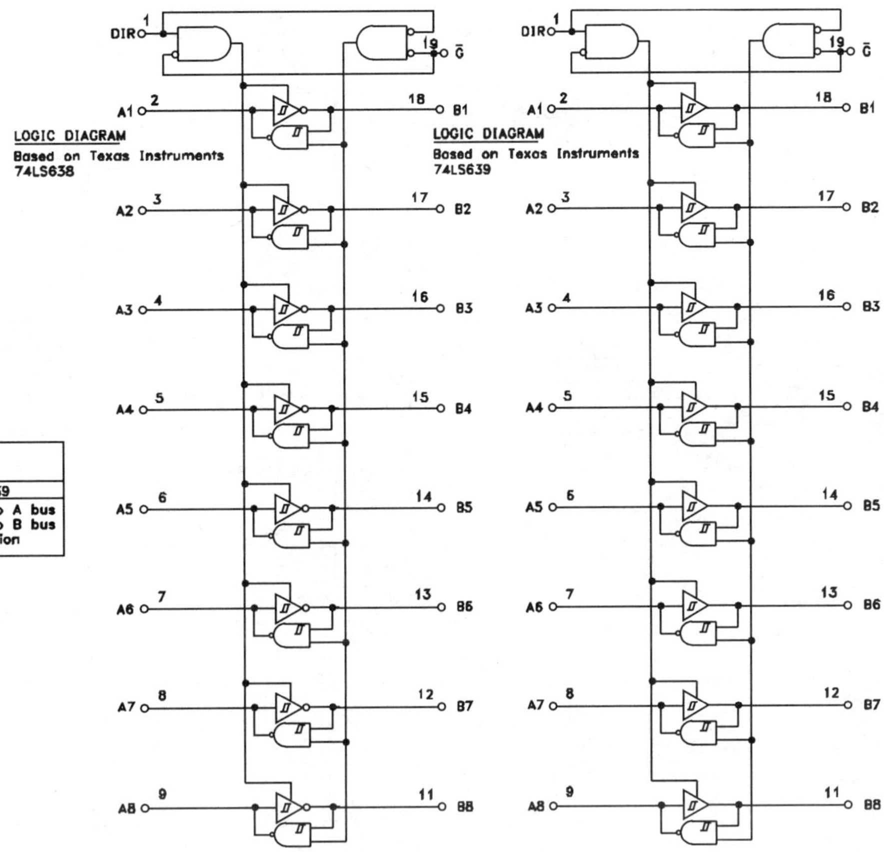

Figure 11.11. The 74LS638 and 74LS639 are TTL bidirectional octal bus transceivers providing asynchronous two-way communication between two data buses. The A bus outputs are open collector; the B bus are 3-state. The direction of data transmission between buses is determined by the logic level of the DIR input as seen in the function table. The 'x638 is inverting; the 'x639 provides true outputs.

TTL - 74LS/AS/ALS/F640
74AS/ALS643
74LS/AS/ALS645
74ALS1640, 'AS2640
74ALS1645, 'AS2643
74AS2645

5V CMOS - 74HC/HCT640
74HC/HCT643
74HC/HCT645

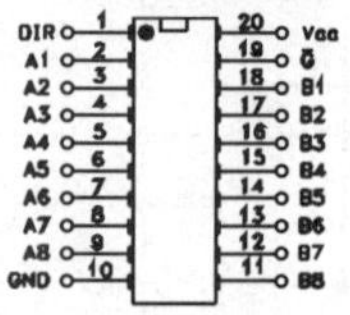

DUAL-IN-LINE PACKAGE

FUNCTION TABLE 74x640,'x643,'x645,'ALS1640.'ALS1645
OCTAL BUS TRANSCEIVERS WITH 3-STATE CONTROLS
74AS2640, 'AS2643, '2645 TRANSCEIVER/MOS DRIVER

ENABLE	DIRECTION CONTROL	OPERATION	
$\overline{G}$	DIR	'2640 'x640 'ALS1640	'x643 '2643
0	0	$\overline{B}$ data to A bus	$\overline{B}$ data to A bus
0	1	$\overline{A}$ data to B bus	$\overline{A}$ data to B bus
1	X	Isolation	Isolation

ENABLE	DIRECTION CONTROL	OPERATION
$\overline{G}$	DIR	'AS2645 'x645 'ALS1645
0	0	B data to A bus
0	1	A data to B bus
1	X	Isolation

1 = HIGH voltage level.
0 = LOW voltage level.
X = Don't Care

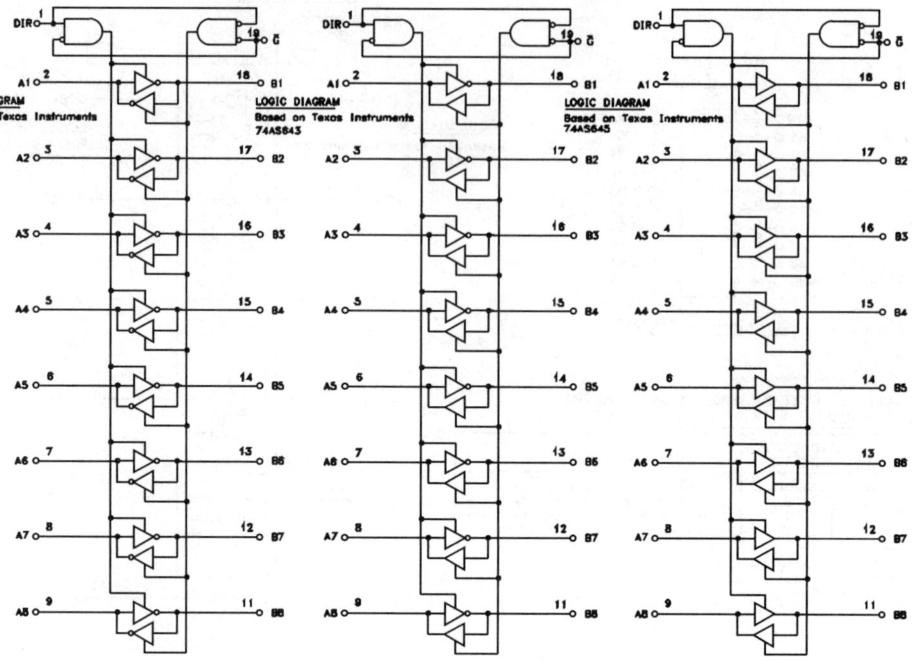

Figure 11.12. The 74x640, 'x643, and 74x645 are TTL and CMOS bidirectional octal bus transceivers providing asynchronous two-way communication between two data buses. The device outputs are 3-state. The direction of data transmission between buses is determined by the logic level of the DIR input, as seen in the function tables. The 'x640 is inverting; the 'x645 provides true outputs; the 'x643 provides both true and inverting outputs.

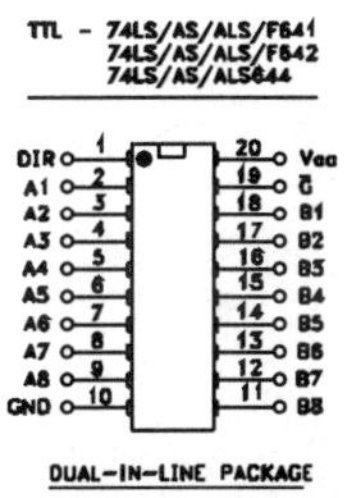

FUNCTION TABLE 74x641, 'x642, 'x644 OCTAL TRANSCEIVERS WITH OPEN COLLECTOR OUTPUTS

ENABLE	DIRECTION CONTROL	OPERATION	
$\bar{G}$	DIR	'x642	'x644
0	0	$\bar{B}$ data to A bus	$\bar{B}$ data to A bus
0	1	$\bar{A}$ data to B bus	$\bar{A}$ data to B bus
1	X	Isolation	Isolation

ENABLE	DIRECTION CONTROL	OPERATION
$\bar{G}$	DIR	'x641
0	0	B data to A bus
0	1	A data to B bus
1	X	Isolation

1 = HIGH voltage level.
0 = LOW voltage level.
X = Don't Care

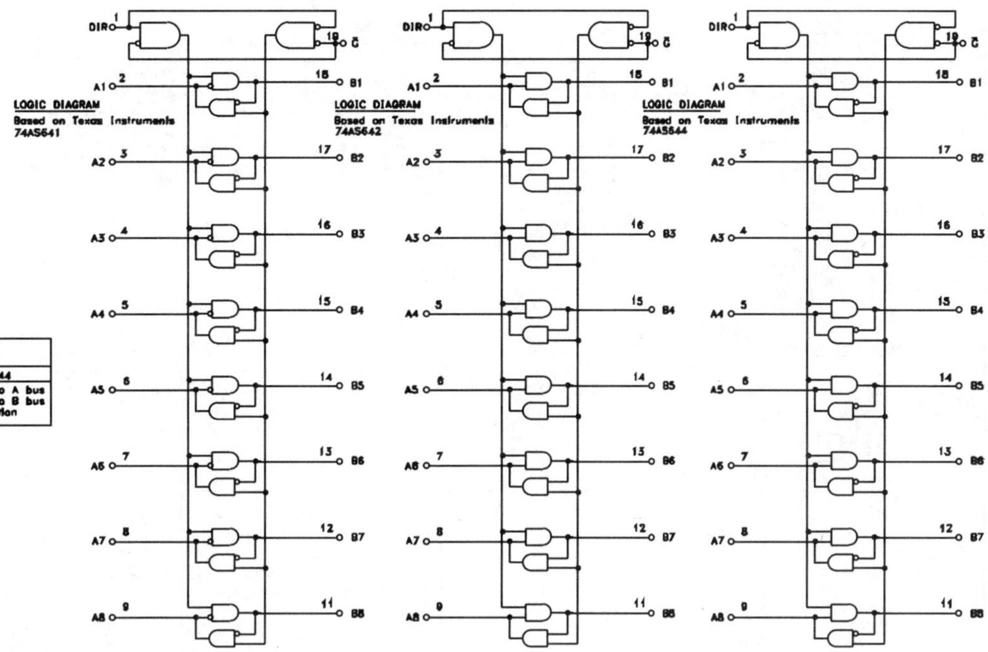

Figure 11.13. The 74ALS/AS641, 'ALS/AS642, and 'ALS/AS644 are TTL bidirectional octal bus transceivers providing asynchronous two-way communication between two data buses. The device outputs are open collector. The direction of data transmission between buses is determined by the logic level of the DIR input, as seen in the function tables. The 'x642 is inverting; the 'x641 provides true outputs; the 'x644 provides both true and inverting outputs.

TTL – 74LS/AS/ALS/F646
74LS/AS/ALS/F647
74LS/AS/ALS/F648
74LS/ALS/F649
5V CMOS – 74HC/HCTAC/ACO/
ACT/ACTQ646
74HC/HCT/AC648
3V CMOS – 74LVC/LCVH/LVT/LCX/
LVT ABT/AC/ACO/
ALVCH646
74AC648

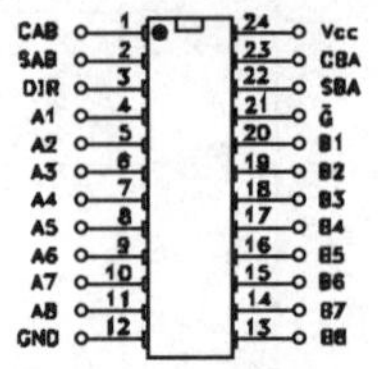

DUAL-IN-LINE PACKAGE

NOTE:
The SSOP and TSSOP Packages follow the DIP Package pinning.

FUNCTION TABLE 74x646, 'x647, 'x648, 'x649 OCTAL BUS TRANSCEIVERS AND REGISTERS. 'x646,'x648 3-STATE, 'x647,'x649 OPEN COLLECTOR

INPUTS						DATA I/O		OPERATION OR FUNCTION	
G	DIR	CAB	CBA	SAB	SBA	A1–A8	B1–B8	'x646, x'647	'x648, 'x649
X	X	↑	X	X	X	Input	Unspecified	Store A, B Unspecified	Store A, B Unspecified
X	X	X	↑	X	X	Unspecified	Input	Store B, A Unspecified	Store B, A Unspecified
1	X	↑	↑	X	X	Input	Input	Store A and B Data	Store A and B Data
1	X	1or0	1or0	X	X	Input	Input	Isolation, Hold Storage	Isolation, Hold Storage
0	0	X	X	X	0	Output	Input	Real-time B Data to A Bus	Real-time $\bar{B}$ Data to A Bus
0	0	X	1or0	X	1	Output	Input	Stored B Data to A Bus	Stored $\bar{B}$ Data to A Bus
0	1	X	X	0	X	Input	Output	Real-time A Data to B Bus	Real-time $\bar{A}$ Data to B Bus
0	1	1or0	X	1	X	Input	Output	Stored A Data to B Bus	Stored $\bar{A}$ Data to B Bus

↑ = LOW-to-HIGH clock transition.
1 = HIGH voltage level.
0 = LOW voltage level.
X = Don't Care
Z = High Impedance state

LOGIC DIAGRAM
Based on Texas Instrument 74ALS648, 'ALS649

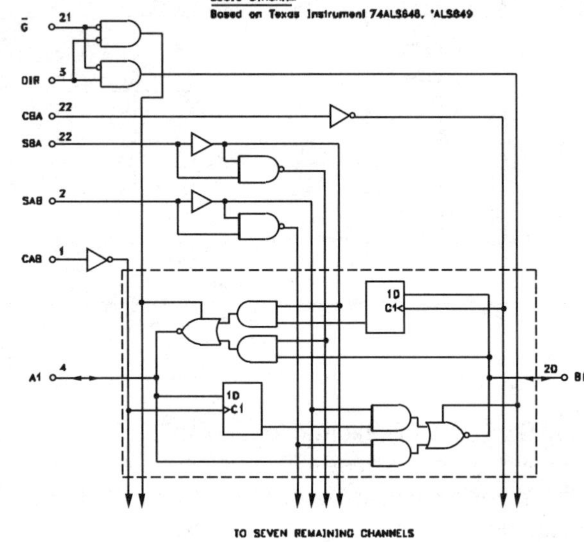

LOGIC DIAGRAM
Based on Texas Instrument 74ALS646, 'ALS647

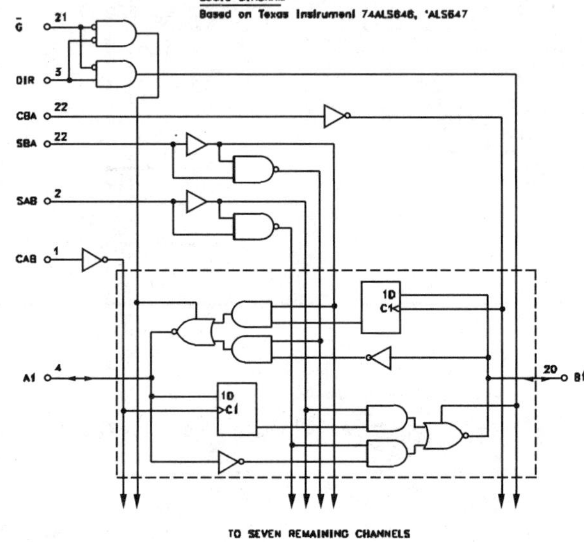

Figure 11.14. The 74ALS/AS/F646, 'ALS/F647, 'ALS/AS/F648 and 'ALS/F649 are TTL bidirectional octal bus transceivers and registers providing multiplexed transmission of data directly from the data bus or from internal storage registers. The '646 and '648 include CMOS versions of these. The direction of data transmission between buses is determined by the logic level of the DIR input, as seen in the function tables. The 'x648 and 'x649 are inverting; the 'x646 and 'x647 provide true outputs. The TTL '646 and '648 are 3-state; the 'x647 and 'x649 are open collector. TTL clock speeds range from 35 to 50 MHz for the AS/ASL devices to as high as 185 MHz for the F series. Maximum frequency for the HC/HCT is 85 MHz; for the low-voltage types as high as 150 MHz.

TTL - 74LS/AS/ALS/F651
74LS/AS/ALS/F652
74LS/ALS/F653
74ALS/F654
5V CMOS - 74HC/HCT651
74HC/HCT/ACTQ852
3V CMOS - 74LVC/LVT/LCX/LVTABT 852

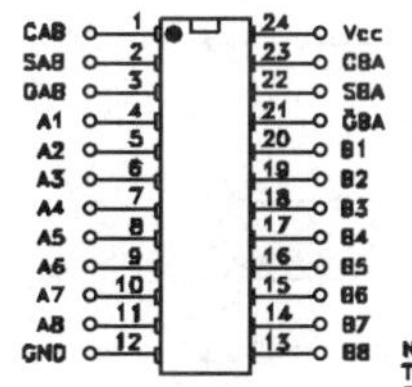

DUAL-IN-LINE PACKAGE

NOTE:
The SSOP and TSSOP Packages follow the DIP Package pinning.

FUNCTION TABLE 74x651, 'x652 OCTAL BUS TRANSCEIVERS AND REGISTERS, 3-STATE OUTPUTS
FUNCTION TABLE 74x653, 'x654 OCTAL BUS TRANSCEIVERS AND REGISTERS, TTL OPEN COLLECTOR, 'ACTO OUTPUT ENABLE CONTROL

INPUTS						DATA I/O		OPERATION OR FUNCTION	
GAB	ḠBA	CAB	CBA	SAB	SBA	A1–A8	B1–B8	'ALS614	'ALS615
0	1	1 or 0	1 or 0	X	X	Input	Input	Isolation	Isolation
0	1	↑	↑	X	X	Input	Input	Store A and B Data	Store A and B Data
X	1	↑	1 or 0	X	X	Input	Unspecified	Store A, Hold B	Store A, Hold B
1	1	↑	↑	X	X	Input	Output	Store A in both registers	Store A in both registers
0	X	1 or 0	↑	X	X	Unspecified	Input	Hold A, Store B	Hold A, Store B
0	0	↑	↑	X	X	Output	Input	Store B in both registers	Store B in both registers
0	0	X	X	X	0	Output	Input	Real-time B̄ Data to A Bus	Real-time B Data to A Bus
0	0	X	1 or 0	X	1	Output	Output	Stored B̄ Data to A Bus	Stored B Data to A Bus
1	1	X	X	0	X	Input	Output	Real-time Ā Data to B Bus	Real-time A Data to B Bus
1	1	1 or 0	X	1	X	Input	Output	Stored Ā Data to B Bus	Stored A Data to B Bus
1	0	1 or 0	1 or 0	1	1	Output	Output	Stored Ā Data to B Bus and Stored B̄ Data to A Bus	Stored A Data to B Bus and Stored B Data to A Bus

↑ = LOW-to-HIGH clock transition.
1 = HIGH voltage level.
0 = LOW voltage level.
X = Don't Care
Z = High Impedance state

LOGIC DIAGRAM
Based on Texas Instrument 74ALS651, 'ALS653

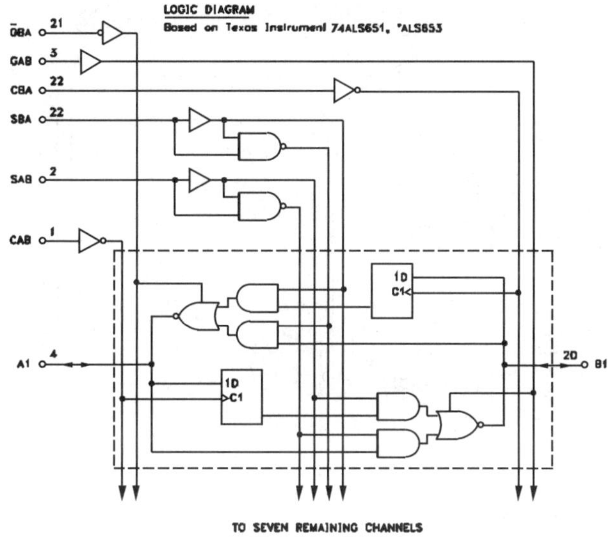

LOGIC DIAGRAM
Based on Texas Instrument 74ALS652, 'ALS654

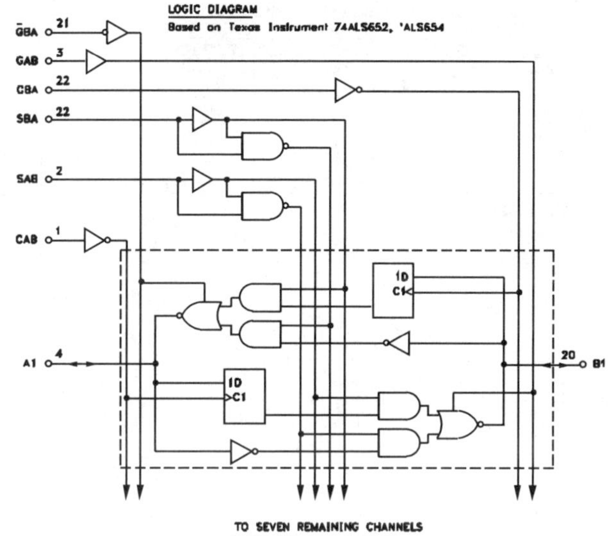

Figure 11.15. The 74LS/ALS/AS/F651, 'LS/ALS/AS/F652, 'LS/ALS/F653 and 'ALS/F654 are TTL bidirectional octal bus transceivers and registers providing multiplexed transmission of data directly from the data bus or from internal storage registers. The '651 and '652 include CMOS versions of these. Data on the A or B buses, or both, will be stored in the internal registers at the appropriate clock inputs, as seen in the function tables. The 'x651 and 'x653 are inverting; the 'x652 and 'x654 provide true outputs. The TTL '651 and '652 are 3-state; the 'x653 and 'x654 are open collector.

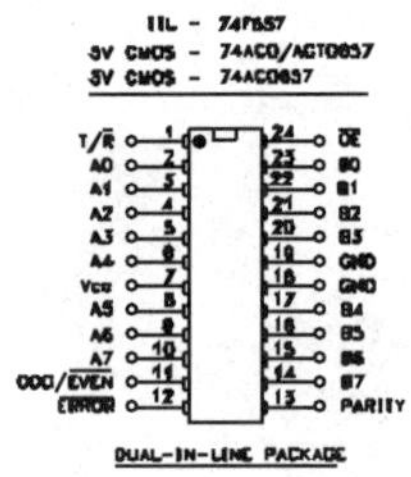

FUNCTION TABLE 74x657 OCTAL TRANSCEIVER WITH 8-BIT PARITY GENERATOR

	INPUTS			INPUT/OUTPUT	OUTPUTS	
Number of Outputs That Are High	$\overline{OE}$	T/$\overline{R}$	ODD/$\overline{EVEN}$	PARITY	$\overline{ERROR}$	OUTPUTS MODE
0,2,4,6,8	0	1	1	1	Z	Transmit
	0	1	0	0	Z	Transmit
	0	0	1	1	1	Receive
	0	0	1	0	0	Receive
	0	0	0	1	0	Receive
	0	0	0	0	1	Receive
1,3,5,7,	0	1	1	0	Z	Transmit
	0	1	0	1	Z	Transmit
	0	0	1	1	0	Receive
	0	0	1	0	1	Receive
	0	0	0	1	1	Receive
	0	0	0	0	0	Receive
	1	X	Z	Z	Z	Z

1 = HIGH voltage level.
0 = LOW voltage level.
X = Don't Care
Z = High Impedance state

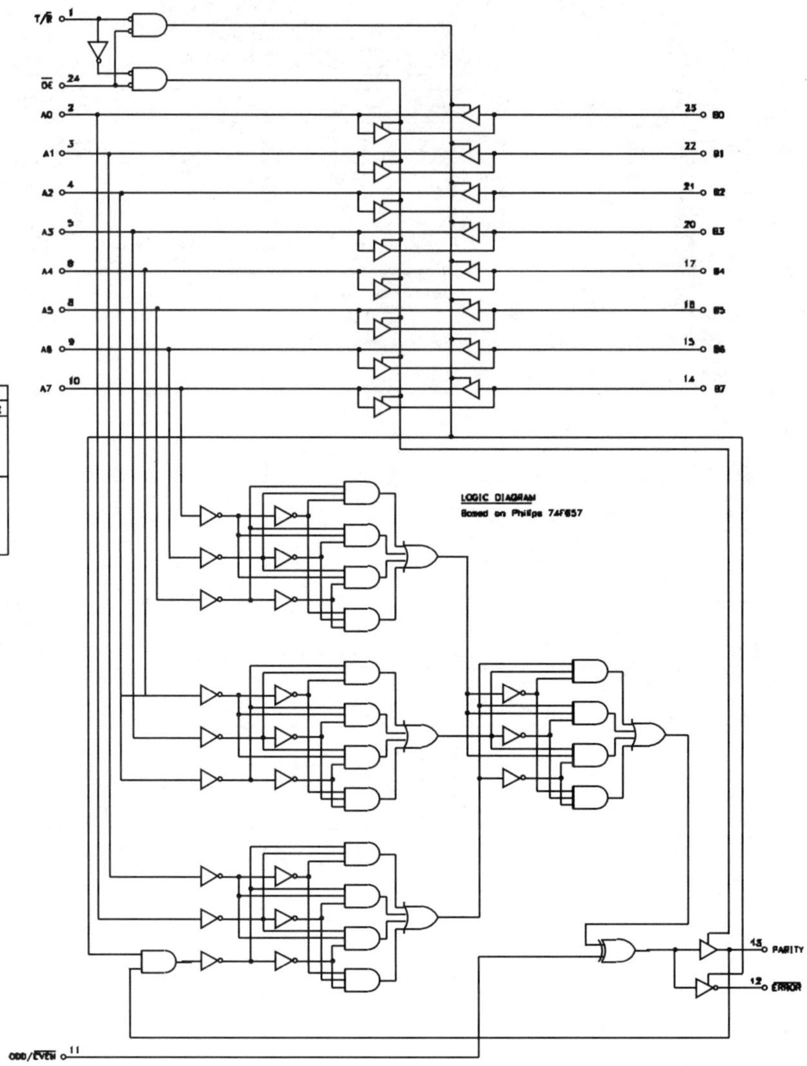

Figure 11.16. The 74x657 are octal transceivers with 8-bit parity generator/checker features. The device combines functions of the 74x245 and the 74x280. The device Transmit//Receive (T//R) input determines the direction of the data flow through the bidirectional receivers. In the transmitting mode the parity generator detects the number of bits on the A port and compares them with the parity select. With the Receiver input low, the parity select and the number of high inputs on port B are compared to the condition of the parity input. The /ERROR output is low in event of an error.

5V CMOS - 74HC/HCT658
74HC/HCT659

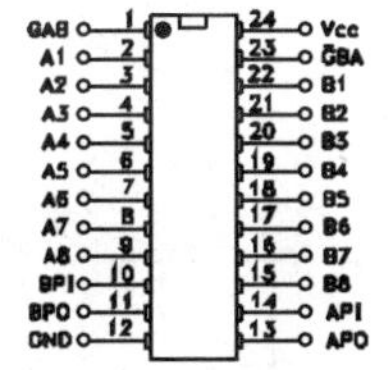

DUAL-IN-LINE PACKAGE

FUNCTION TABLE 74HC/HCT658, 'HC/HCT659
OCTAL TRANSCEIVERS WITH PARITY

CONTROL INPUTS $\overline{G}$BA	CONTROL INPUTS GAB	NUMBER OF HIGH INPUTS ON A BUS AND API	NUMBER OF HIGH INPUTS ON B BUS AND BPI
0	0	X X	0,2,4,6,8 1,3,5,7
1	1	0,2,4,6,8 1,3,5,7	X X
1	0	X	X
0	1	X X 0,2,4,6,8 1,3,5,7	0,2,4,6,8 1,3,5,7 X X

OUTPUTS APO	OUTPUTS BPO	OPERATION 'x658	OPERATION 'x659
Z Z	1 0	$\overline{B}$ Data to A Bus	B Data to A Bus
1 0	Z Z	$\overline{A}$ Data to B Bus	A Data to B Bus
Z	Z	Isolation	Isolation
  1 0	1 0	$\overline{B}$ Data to A Bus $\overline{A}$ Data to B Bus	B Data to A Bus A Data to B Bus

1 = HIGH voltage level.
0 = LOW voltage level.
X = Don't Care
Z = High Impedance state

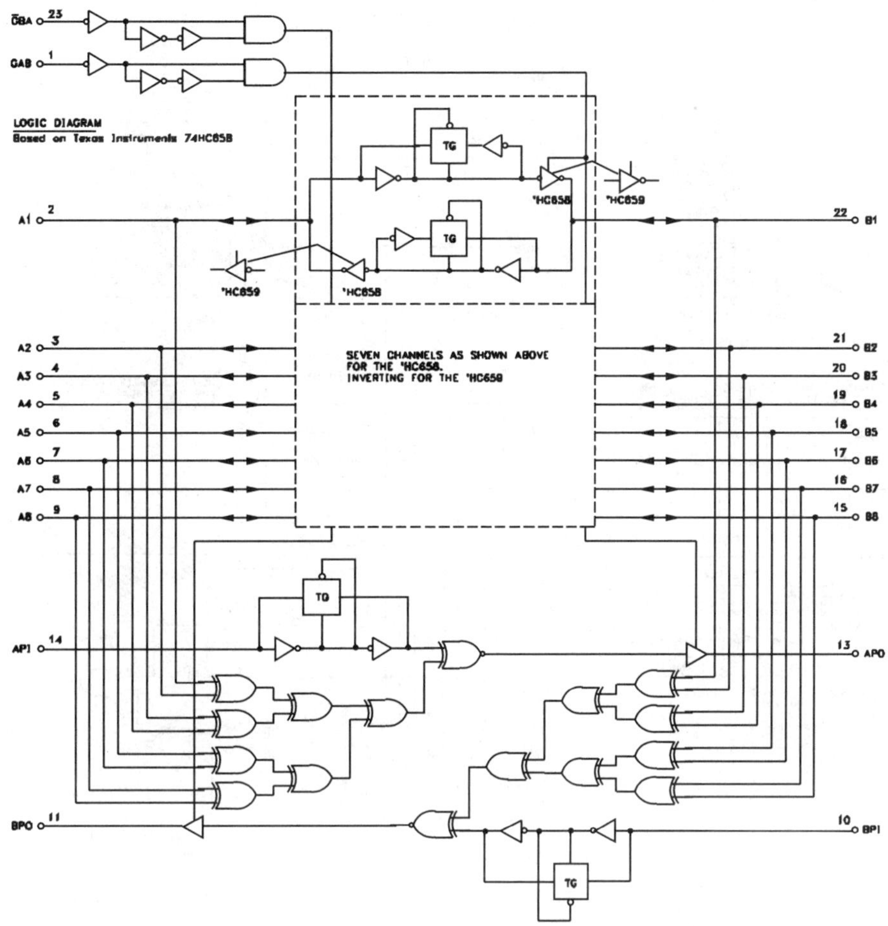

Figure 11.17. The 74HC/HCT658 and 'HC/HCT659 are octal transceivers with 8-bit parity generator ability. The I/O ports include active circuits on the input stage such that when the output shared by that stage is disabled will maintain the input in the last state assumed by the output. This state is maintained until changed by activity on the bus. The advantage is that when all outputs on the bus are disabled, the inputs will be prevented from floating. There is no need for external pull-up or pull-down resistors. The 'x658 has inverting outputs; the 'x659 true.

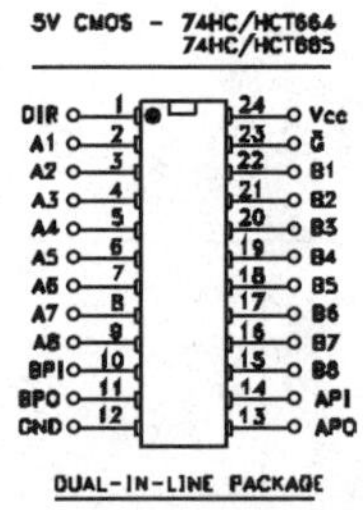

FUNCTION TABLE 74HC/HCT664, 'HC/HCT665
OCTAL BUS TRANSCEIVERS WITH PARITY

CONTROL INPUTS		NUMBER OF HIGH INPUTS ON A BUS AND API	NUMBER OF HIGH INPUTS ON B BUS AND BPI
$\bar{G}$	DIR		
0	0	X X	0,2,4,6,8 1,3,5,7
1	1	0,2,4,6,8 1,3,5,7	X X
1	X	X	X

OUTPUTS		OPERATION	
APO	BPO	'x644	'x665
Z Z	1 0	$\bar{B}$ Data to A Bus	B Data to A Bus
1 0	Z Z	$\bar{A}$ Data to B Bus	A Data to B Bus
Z	Z	Isolation	Isolation

1 = HIGH voltage level.
0 = LOW voltage level.
X = Don't Care
Z = High Impedance state

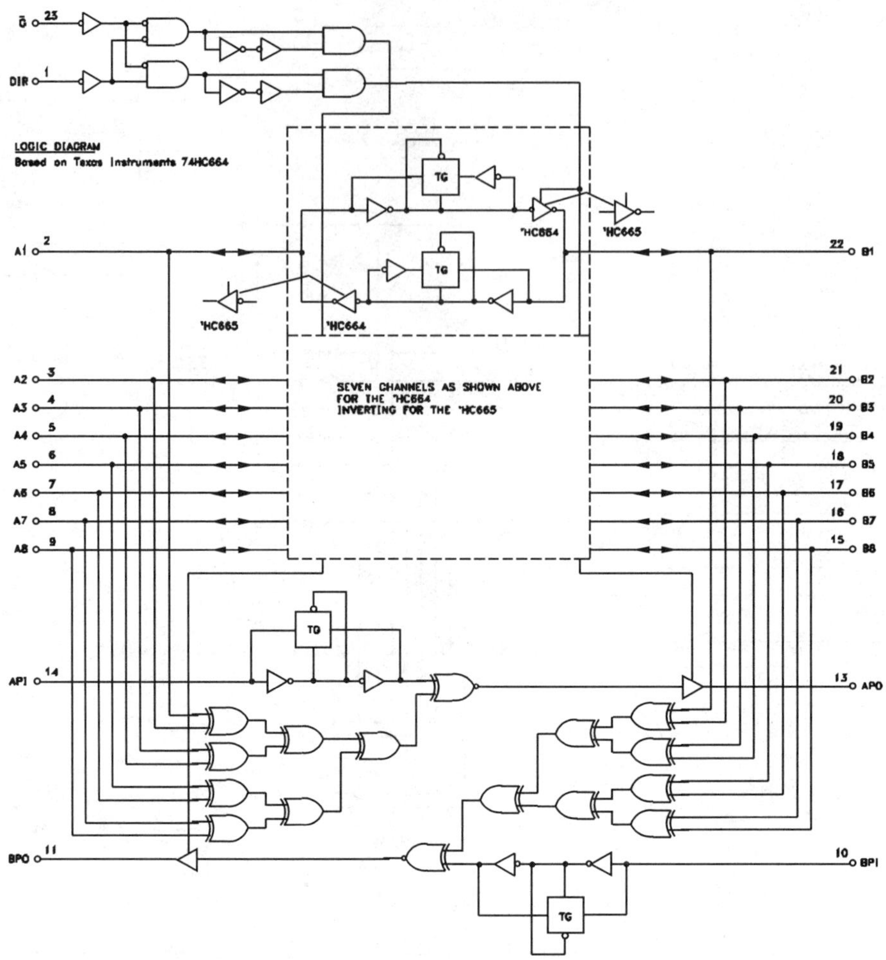

Figure 11.18. The 74HC/HCT664 and 'HC/HCT665 are octal transceivers with 8-bit parity generator ability. The DIR control input determines the direction of the transmission. The I/O ports include active circuits on the input stage such that when the output shared by that stage is disabled will maintain the input in the last state assumed by the output. This state is maintained until changed by activity on the bus. The advantage is that when all outputs on the bus are disabled, the inputs will be prevented from floating. There is no need for external pull-up or pull-down resistors. The 'x664 has inverting outputs, the 'x665 true.

TTL – 74F776

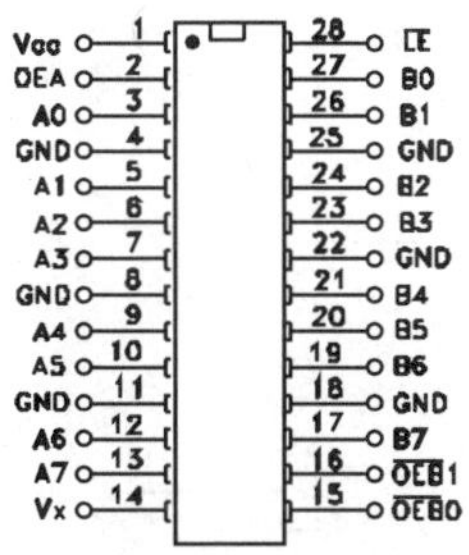

DUAL-IN-LINE PACKAGE

PACKAGE PINNING

PIN	FUNCTION
A0–A7	PNP Latched input/3-state output (with Vx control option)
B0–B7	Data input with special threshold circuitry to reject noise/open collector output, high current drive.
$\overline{OEB0}$ $\overline{OEB1}$	Enables the B outputs when both pins are low.
OEA	Enables the A outputs when high
$\overline{LE}$	Latched when high (a special feature is built in for proper enabling times.
Vx	Clamping voltage keeping VOH from rising above Vx(Vx=Vcc for normal use).

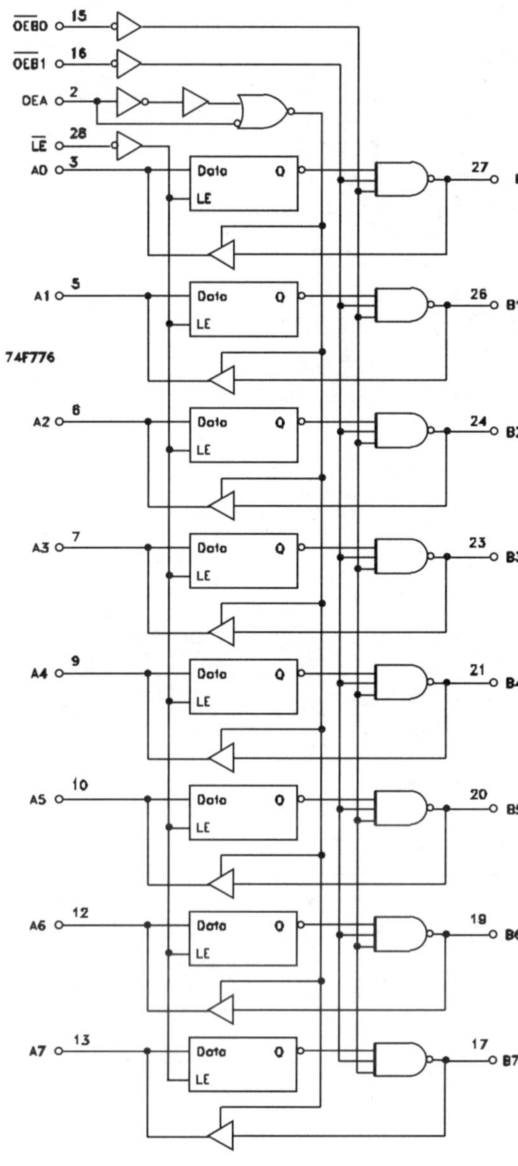

LOGIC DIAGRAM
Based on Philips 74F776

FUNCTION TABLE 74F776
PI-BUS TRANSCEIVER

INPUTS						LATCH	OUTPUTS		
An	Bn*	$\overline{LE}$	OEA	$\overline{OEB0}$	$\overline{OEB1}$	STATE	An	Bn	OPERATING MODE
1	X	0	0	0	0	1	Z	Z	A 3-state, data from A to B
0	X	0	0	0	0	0	Z	0	
X	X	1	0	0	0	Qn	Z	Qn	A 3-state, latched data to B
–	–	0	1	0	0	(1)	(1)	(1)	Feedback, A to B, B to A
–	1	1	1	0	0	H(2)	1	Z(2)	Preconditioned latchenabline data
–	0	1	1	0	0	H(2)	0	Z(2)	transfer from A to B
									Latch state to A and B
–	–	1	1	0	0	Qn	Qn	Qn	
1	X	0	0	1	X	1	Z	Z	B and A 3-state
0	X	0	0	1	X	0	Z	Z	
X	X	1	0	1	X	Qn	Z	Z	
–	1	0	1	1	X	1	1	Z	B 3-state, data from B to A
–	0	0	1	1	X	0	0	Z	
–	1	1	1	1	X	Qn	1	Z	
–	0	1	1	1	X	Qn	0	Z	
1	X	0	0	X	1	1	Z	Z	B and A 3-state
0	X	0	0	X	1	0	Z	Z	
X	X	1	0	X	1	Qn	Z	Z	
–	1	0	1	X	1	1	1	Z	B 3-state, data from B to A
–	0	0	1	X	1	0	0	Z	
–	1	1	1	X	1	Qn	1	Z	
–	0	1	1	X	1	Qn	0	Z	

– = Input not externally driven.
Qn = High or Low voltage level one setup time prior to the low-to-high $\overline{LE}$ transition.
(1) = Condition will cause a feedback loop path: A-to-B and B-to-A.
(2) = The Latch must be preconditioned such that B inputs may assume a high or low level while $\overline{OEB0}$ and $\overline{OEB1}$ are low and $\overline{LE}$ is high.
B* = Precaution should be taken to insure the B inputs do not float. If they do they are equal to low state.
1 = HIGH voltage level.
0 = LOW voltage level.
X = Don't Care
Z = High Impedance state

Figure 11.19. The 74F776 is an octal latched transceiver designed to drive heavily loaded backplanes with equivalent load impedances down to 10 ohms. The device is compatible with Pi-bus and IEEE 896 Futurebus standards. The B port features a reduced voltage swing of 1-2V and open collector outputs with 100 mA drive. A controlled output ramp and multiple ground pins minimize ground bounce. The A port has 3-state output drivers and receivers with a latch function. The high-level control voltage, Vx, limits the A side output level to a given voltage level, such as 3.3 V. For five-volt systems, Vx is simply tied to Vcc.

TTL – 74F777

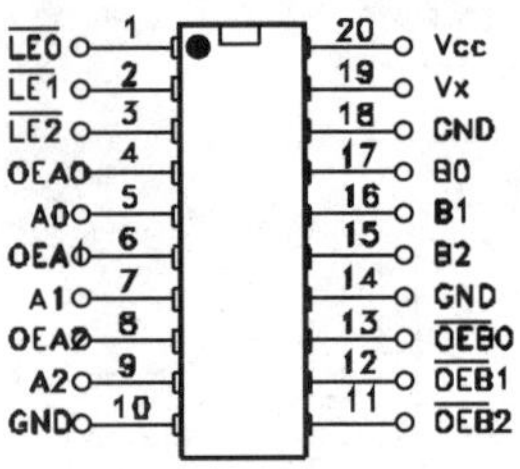

DUAL-IN-LINE PACKAGE

PACKAGE PINNING:

PIN	FUNCTION
A0–A2	PNP Latched Inputs
B0–B2	Data inputs with threshold circuitry
OEA0–OEA2	A output enable inputs (active high)
OEB0–OEB2	B output enable inputs (active low)
LE0–LE2	Latch enable inputs (active low)
A0–A2	3-state outputs
B0–B2	Open collector outputs

LOGIC DIAGRAM
Based on Philips 74F777

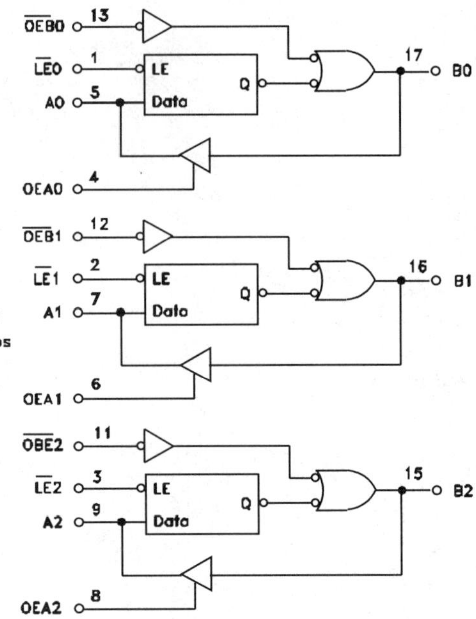

FUNCTION TABLE 74F777 TRIPLE BIDIRECTIONAL LATCHED BUS TRANSCEIVER (3-STATE+OPEN COLLECTOR)

INPUTS					LATCH	OUTPUTS		
An	Bn*	$\overline{\text{LEn}}$	$\overline{\text{OEAn}}$	$\overline{\text{OEBn}}$	STATE	An	Bn	OPERATING MODE
1	X	0	0	0	1	Z	Z	A 3-state, data from A to B
0	X	0	0	0	0	Z	0	
X	X	1	0	0	Qn	Z	Qn	A 3-state, latched data to B
–	–	0	0	0	(1)	(1)	(1)	Feedback, A to B, B to A
–	1	1	0	0	H(2)	1	Z(2)	Preconditioned latchenabline data transfer from A to B
–	0	1	0	0	H(2)	0	Z(2)	
–	–	1	0	0	Qn	Qn	Qn	Latch state to A and B
1	X	0	1	X	1	Z	Z	B and A 3-state
0	X	0	1	X	0	Z	Z	
X	X	1	1	X	Qn	Z	Z	
–	1	0	1	X	1	1	Z	B 3-state, data from B to A
–	0	0	1	X	0	0	Z	
–	1	1	1	X	Qn	1	Z	
–	0	1	1	X	Qn	0	Z	

FUNCTION TABLE NOTES:

- – = Input not externally driven.
- Qn = High or Low voltage level one setup time prior to the low-to-high $\overline{\text{LE}}$ transition.
- (1) = Condition will cause a feedback loop path: A-to-B and B-to-A.
- (2) = The Latch must be precondtioned such that B inputs may assume a high or low level while $\overline{\text{OEB0}}$ and $\overline{\text{OEB1}}$ are low and $\overline{\text{LE}}$ is high.
- B* = Precaution should be taken to insure the B inputs do not float. If they do they are equal to low state.
- 1 = HIGH voltage level.
- 0 = LOW voltage level.
- X = Don't Care
- Z = High Impedance state

Figure 11.20. The 74F777 is a triple latched bidirectional latched bus transceiver designed to drive high-performance wired-OR bus with a characteristic impedance range of 20 to 50 ohms. The device is compatible with Test Mode (TM) bus specifications. The open collector B port output driver is designed to sink 100 mA from 2 volts to minimize crosstalk and ringing. It is terminated with a 30 to 40 ohm resistor at each end. The A port has 3-state output drivers with a latch function. The high-level control voltage, Vx, limits the A side output level to a given voltage level, such as 3.3V. For five-volt systems, Vx is simply tied to Vcc.

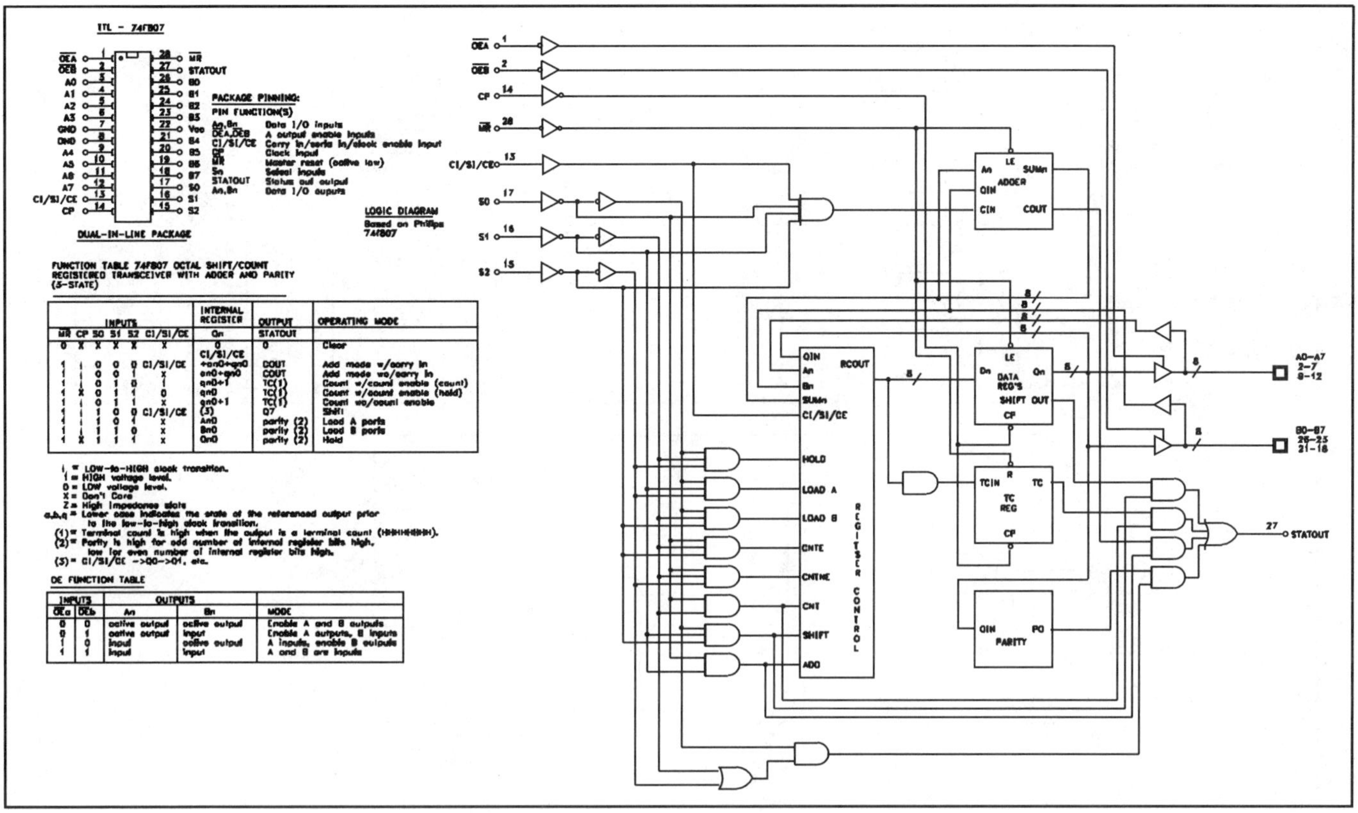

Figure 11.21. The 74F807 is an octal shift/count registered transceiver with the added capability to perform count, shift, and add functions. It also can generate a parity output. The /MR control is an overriding reset asynchronous switch. The two output enable inputs have no influence on the device's internal operations, only on the output states. These are enabled low. All starting modes other than 3-state, clear, and hold are on the rising edge of the clock. Typical clock frequency is 115 MHz.

TTL – 74AS852

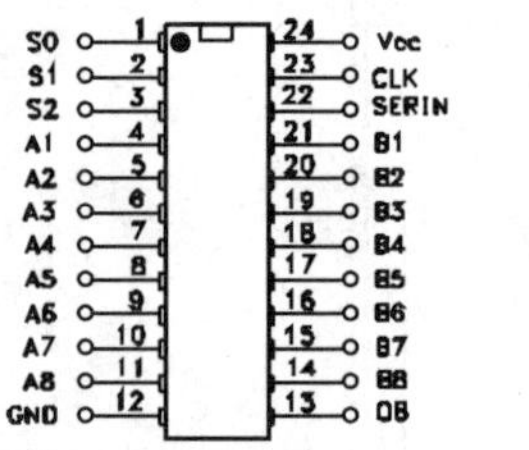

DUAL-IN-LINE PACKAGE

1 = HIGH voltage level.
0 = LOW voltage level.
X = Don't Care
Z = High Impedance state
↑ = LOW-to-HIGH clock transition.

FUNCTION TABLE 74AS852 8-BIT UNIVERSAL TRANSCEIVER PORT CONTROLLER

MODE S2 S1 S0	CLOCK	SERIN	A1 Q1 B1	A2 Q2 B2	A3 Q3 B3	A4 Q4 B4	A5 Q5 B5	A6 Q6 B6	A7 Q7 B7	A8 Q8 B8	PORT FUNCTION
0 0 0	1 or 0	X	Z Qn A1	Z Qn A2	Z Qn A3	Z Qn A4	Z Qn A5	Z Qn A6	Z Qn A7	Z Qn A8	A TO B
0 0 0	↑	X	Z A1 A1	Z A2 A2	A A3 A3	Z A4 A4	Z A5 A5	Z A6 A6	Z A7 A7	Z A8 A8	
0 0 1	1 or 0	X	B1 Qn Z	B2 Qn Z	B3 Qn Z	B4 Qn Z	B5 Qn Z	B6 Qn Z	B7 Qn Z	B8 Qn Z	B TO A
0 0 1	↑	X	B1 B1 Z	B2 B2 Z	B3 B3 Z	B4 B4 Z	B5 B5 Z	B6 B6 Z	B7 B7 Z	B8 B8 Z	
0 1 0	1 or 0	X	X Qn Q1	X Qn Q2	X Qn Q3	X Qn Q4	X Qn Q5	X Qn Q6	X Qn Q7	X Qn Q8	QN TO BN
0 1 0	↑	X	Z A1 A1	Z A2 A2	Z A3 A3	Z A4 A4	Z A5 A5	Z A6 A6	Z A7 A7	Z A8 A8	
0 1 1	1 or 0	X	Q1 Qn X	Q2 Qn X	Q3 Qn X	Q4 Qn X	Q5 Qn X	Q6 Qn X	Qn Q7 X	Q8 Qn X	QN TO AN
0 1 1	↑	X	B1 B1 Z	B2 B2 Z	B3 B3 Z	B4 B4 Z	B5 B5 Z	B6 B6 Z	B7 B7 Z	B8 B8 Z	
1 0 0	1 or 0	X	Z Qn A1	Z Qn A2	Z Qn A3	Z Qn A4	Z Qn A5	Z Qn A6	Z Qn A7	Z Qn A8	SHIFT AND A TO B
1 0 0	↑	1	Z H A1	Z Q1 A2	Z Q2 A3	Z Q3 A4	Z Q4 A5	Z Q5 A6	Z Q6 A7	Z Q7 A8	
1 0 0	↑	0	Z L A1	Z Q1 A2	Z Q2 A3	Z Q3 A4	Z Q4 A5	Z Q5 A6	Z Q6 A7	Z Q7 A8	
1 0 1	1 or 0	X	B1 Qn Z	B2 Qn Z	B3 Qn Z	B4 Qn Z	B5 Qn Z	B6 Qn Z	B7 Qn Z	B8 Qn Z	SHIFT AND B TO A
1 0 1	↑	1	B1 H Z	B2 Q1 Z	B3 Q2 Z	B4 Q3 Z	B5 Q4 Z	B6 Q5 Z	B7 Q6 Z	B8 Q7 Z	
1 0 1	↑	0	B1 L Z	B2 Q1 Z	B3 Q2 Z	B4 Q3 Z	B5 Q4 Z	B6 Q5 Z	B7 Q6 B	B8 Q7 Z	
1 1 0	1 or 0	X	Z Qn Z	Z Qn Z	Z Qn Z	Z Qn Z	Z Qn Z	Z Qn Z	Z QN Z	Z Qn Z	SHIFT
1 1 0	↑	1	Z H Z	Z Q1 Z	Z Q2 Z	Z Q3 Z	Z Q4 Z	Z Q5 Z	Z Q6 Z	Z Q7 Z	
1 1 0	↑	0	Z L Z	Z Q1 Z	Z Q2 Z	Z Q3 Z	Z Q4 Z	Z Q5 Z	Z Q6 Z	Z Q7 Z	
1 1 1	1 or 0	X	Z Qn Z	Z Qn	Z Qn Z	Z Qn Z	Z Qn Z	Z Qn Z	Z Qn Z	Z Qn Z	CLEAR
1 1 1	↑	X	Z L Z	Z L Z	Z L Z	Z L Z	Z L Z	Z L Z	Z L Z	Z L Z	

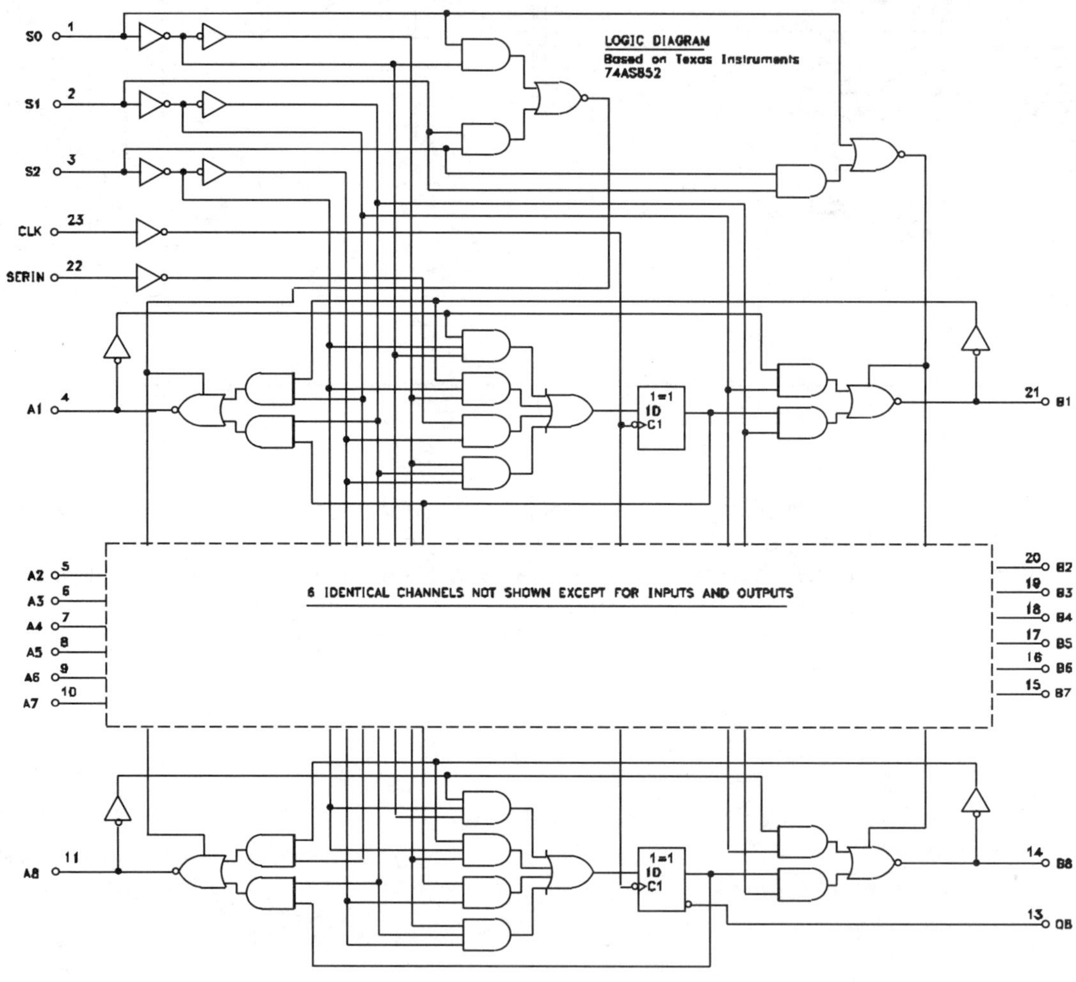

Figure 11.22. The 74AS852 is an 8-bit universal port controller with buffered 3-state outputs for driving bus lines directly. The device features two 8-bit I/O ports and 8-bit parallel-load, serial-in/parallel out shift register and control logic. These allow for performing eight selectable transceiver or port functions depending on the state of the three select inputs as shown in the function table. Data may be transferred between ports, from the register to either port, serial shifting of data from either port to the other, performing line shifts, and clearing the register. Synchronous parallel loading of the internal register can be accomplished from either port on the positive clock transition. Maximum clock frequency is 50 MHz.

TTL - 74AS856

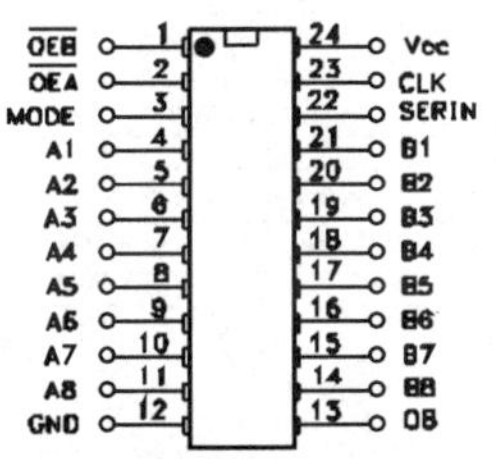

DUAL-IN-LINE PACKAGE

1 = HIGH voltage level.
0 = LOW voltage level.
X = Don't Care
Z = High Impedance state
↑ = LOW-to-HIGH clock transition.

FUNCTION TABLE 74AS856 8-BIT UNIVERSAL TRANSCEIVER PORT CONTROLLER

MODE	$\overline{OEA}$	$\overline{OEB}$	CLOCK	SERIN	A1 Q1 B1	A2 Q2 B2	A3 Q3 B3	A4 Q4 B4	A5 Q5 B5	A6 Q6 B6	A7 Q7 B7	A8 Q8 B8	PORT FUNCTION
0	0	0	1 or 0	X	Q1 Q1 Q1	Q2 Q2 Q2	Q3 Q3 Q3	Q4 Q4 Q4	Q5 Q5 Q5	Q6 Q6 Q6	Q7 Q7 Q7	Q8 Q8 Q8	FEEDBACK
0	0	0	↑	X	Q1 Q1 Q1	Q2 Q2 Q2	Q3 Q3 Q3	Q4 Q4 Q4	Q5 Q5 Q5	Q6 Q6 Q6	Q7 Q7 Q7	Q8 Q8 Q8	
0	0	1	1 or 0	X	B1 Q1 Z	B82 Q2 Z	B3 Q3 Z	B4 Q4 Z	B5 Q5 Z	B6 Q6 Z	B7 Q7 Z	B8 Q8 Z	B TO A
0	0	1	↑	X	B1 B1 Z	B2 B2 Z	B3 B3 Z	B4 B4 Z	B5 B5 Z	B6 B6 Z	B7 B7 Z	B8 B8 Z	A TO Q
0	1	0	1 or 0	X	Z Q1 Q1	Z Q2 Q2	Z Q3 Q3	Z Q4 Q4	Z Q5 Q5	Z Q6 Q6	Z Q7 Q7	Z Q8 Q8	A TO Q
0	1	0	↑	X	Z A1 A1	Z A2 A2	Z A3 A3	Z A4 A4	Z A5 A5	Z A6 A6	Z A7 A7	Z A8 A8	Q TO B
0	1	1	1 or 0	X	Z Q1 Z	Z Q2 Z	Z Q3 Z	Z Q4 Z	Z Q5 Z	Z Q6 Z	Z Q7 Z	Z Q8 Z	A TO Q
0	1	1	↑	X	Z A1 Z	Z A2 Z	Z A3 Z	Z A4 Z	Z A5 Z	Z A6 Z	Z A7 Z	Z A8 Z	
1	0	0	1 or 0	X	Q1 Qn Q1	Q2 Qn Q2	Q3 Qn Q3	Q4 Qn Q4	Q5 Qn Q5	Q6 Qn Q6	Q7 Qn Q7	Q8 Qn Q8	SHIFT TO A and B
1	0	0	↑	1	H H H	Q1 Q1 Q1	Q2 Q2 Q2	Q3 Q3 Q3	Q4 Q4 Q4	Q5 Q5 Q5	Q6 Q6 Q6	Q7 Q7 Q7	
1	0	0	↑	0	L L L	Q1 Q1 Q1	Q2 Q2 Q2	Q3 Q3 Q3	Q4 Q4 Q4	Q5 Q5 Q5	Q6 Q6 Q6	Q7 Q7 Q7	
1	0	1	1 or 0	X	Q1 Qn Z	Q2 Qn Z	Q3 Qn Z	Q4 Qn Z	Q5 Qn Z	Q6 Qn Z	Q7 Qn Z	Q8 Qn Z	SHIFT TO A
1	0	1	↑	1	H H Z	Q1 Q1 Z	Q2 Q2 Z	Q3 Q3 Z	Q4 Q4 Z	Q5 Q5 Z	Q6 Q6 Z	Q7 Q7 Z	
1	0	1	↑	0	L L Z	Q1 Q1 Z	Q2 Q2 Z	Q3 Q3 Z	Q4 Q4 Z	Q5 Q5 Z	Q6 Q6 Z	Q7 Q7 Z	
1	1	0	1 or 0	X	Z Qn Q1	Z Qn Q2	Z Qn Q3	Z Qn Q4	Z Qn Q5	Z Qn Q6	Z Qn Q7	Z Qn Q8	SHIFT TO B
1	1	0	↑	1	Z H H	Z Q1 Q1	Z Q2 Q2	Z Q3 Q3	Z Q4 Q4	Z Q5 Q5	Z Q6 Q6	Z Q7 Q7	
1	1	0	↑	0	Z L L	Z Q1 Q1	Z Q2 Q2	Z Q3 Q3	Z Q4 Q4	Z Q5 Q5	Z Q6 Q6	Z Q7 Q7	
1	1	1	1 or 0	X	Z Qn Z	Z Qn Z	Z Qn Z	Z Qn Z	Z Qn Z	Z Qn Z	Z Qn Z	Z Qn Z	SHIFT
1	1	1	↑	1	Z H Z	Z Q1 Z	Z Q2 Z	Z Q3 Z	Z Q4 Z	Z Q5 Z	Z Q6 Z	Z Q7 Z	
1	1	1	↑	0	Z L H	Z Q1 Z	Z Q2 Z	Z Q3 Z	Z Q4 Z	Z Q5 Z	Z Q6 Z	Z Q7 Z	

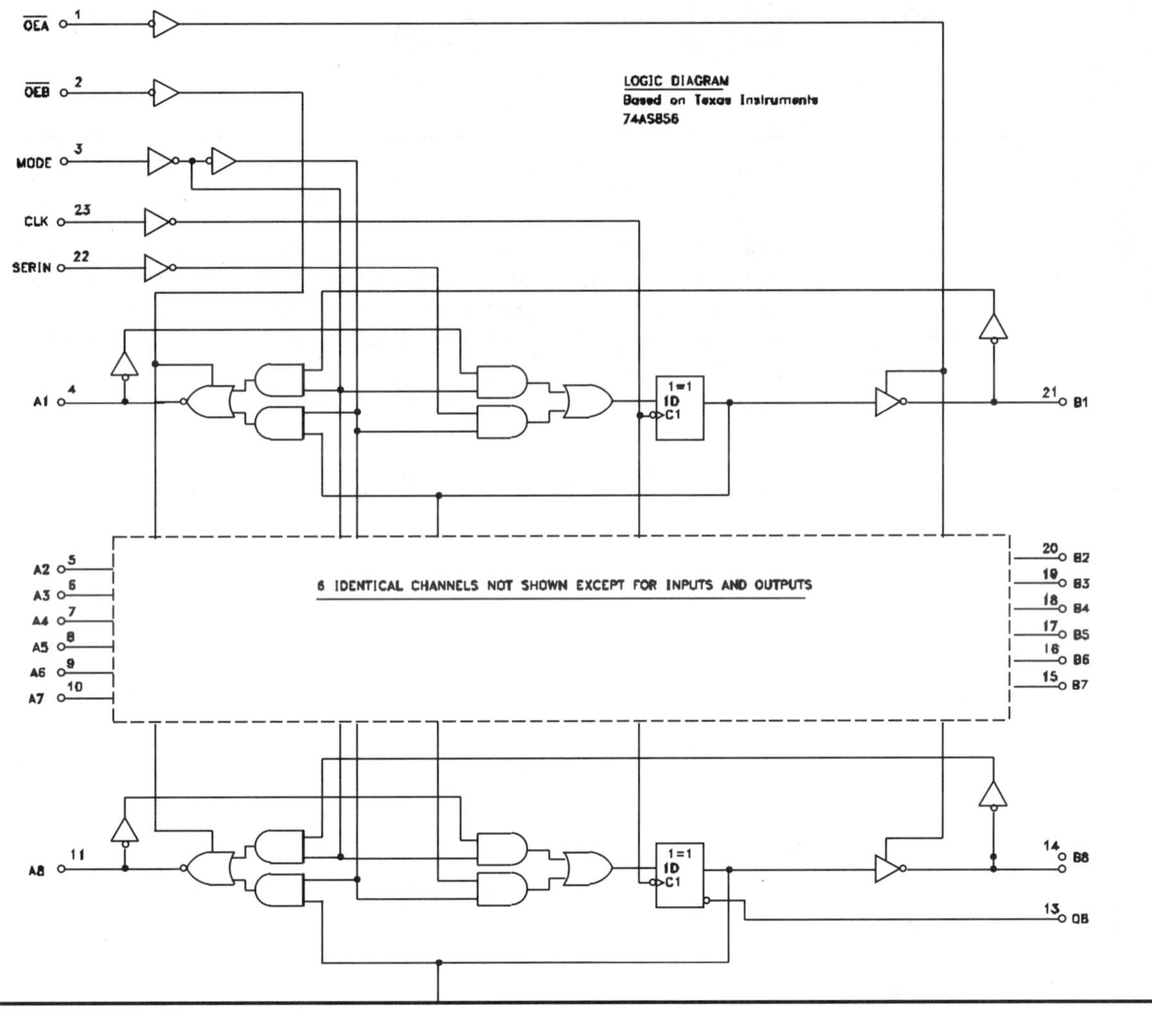

Figure 11.23. The 74AS856 is an 8-bit universal port controller with buffered 3-state outputs for driving bus lines directly. Q8 is a totem pole output. The device features two 8-bit I/O ports and 8-bit parallel-load, serial-in/parallel out shift register and control logic. These allow for performing eight selectable transceiver or port functions, depending on the state of the three control inputs, as shown in the function table. Data may be transferred between ports, from the register to either port, serial shifting of data from either port to the other, performing line shifts, and clearing the register. Synchronous parallel loading of the internal register can be accomplished from either port while serially shifting data in via the SERIN input on the positive clock transition. Maximum clock frequency is 50 MHz.

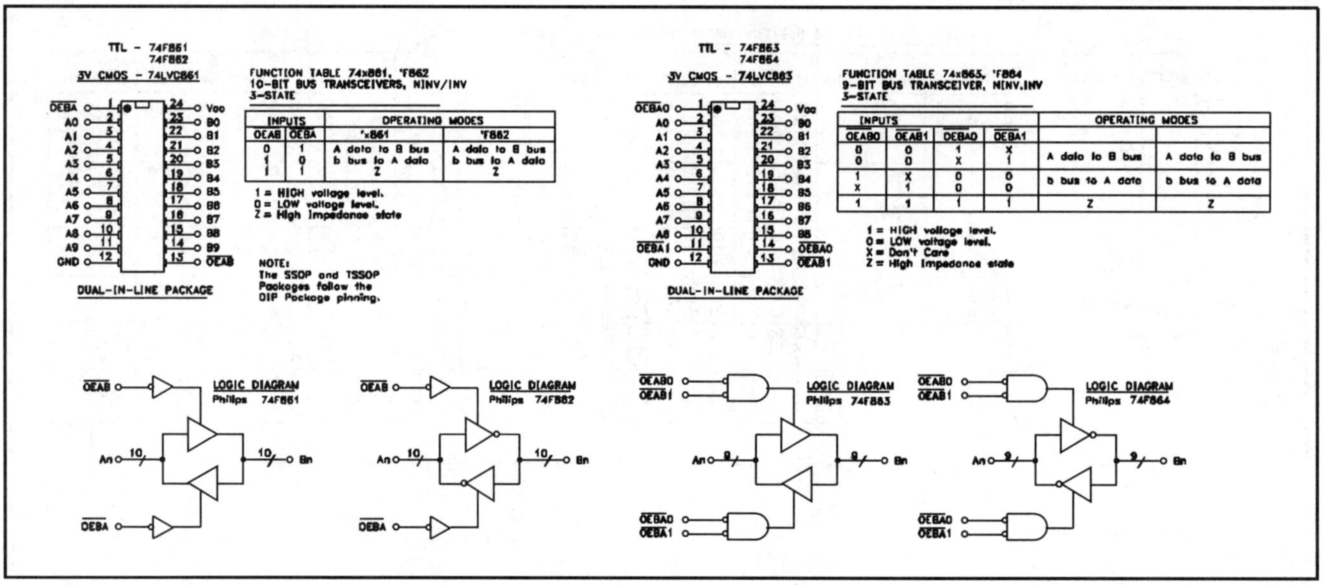

FUNCTION TABLE 74x861, 'F862
10-BIT BUS TRANSCEIVERS, NINV/INV
3-STATE

INPUTS		OPERATING MODES	
$\overline{OEAB}$	$\overline{OEBA}$	'x861	'F862
0	1	A data to B bus	A data to B bus
1	0	b bus to A data	b bus to A data
1	1	Z	Z

1 = HIGH voltage level.
0 = LOW voltage level.
Z = High Impedance state

FUNCTION TABLE 74x863, 'F864
9-BIT BUS TRANSCEIVER, NINV,INV
3-STATE

INPUTS				OPERATING MODES	
$\overline{OEAB0}$	$\overline{OEAB1}$	$\overline{OEBA0}$	$\overline{OEBA1}$		
0 0	0 0	1 X	X 1	A data to B bus	A data to B bus
1 X	X 1	0 0	0 0	b bus to A data	b bus to A data
1	1	1	1	Z	Z

1 = HIGH voltage level.
0 = LOW voltage level.
X = Don't Care
Z = High Impedance state

Figure 11.24. The 74F861, 'LVC861 and 'F862 are 10-bit bus transceivers providing high-performance bus interface buffering for wide data/address paths of buses carrying parity. The F devices have buffered control inputs for light loading or increased fan-in as required for MOS microprocessors. High-impedance NPN inputs provide reduced loading. To ensure the high-impedance state during power-up or -down the output enable should be tied to Vcc through a pull-up resistor. The 74F863, 'LVC863, and 'F864 are 9-bit counterparts sharing the same features. Outputs are 3-state.

TTL - 74AS877

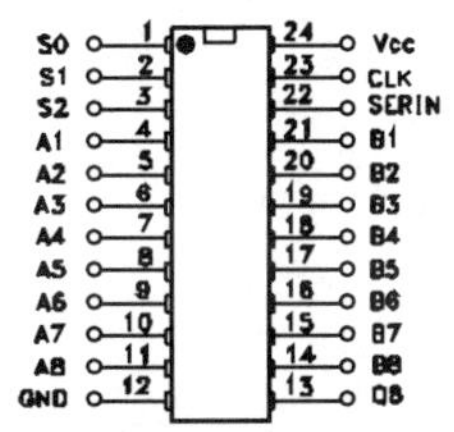

DUAL-IN-LINE PACKAGE

1 = HIGH voltage level.
0 = LOW voltage level.
X = Don't Care
Z = High impedance state
↑ = LOW-to-HIGH clock transition.

FUNCTION TABLE 74AS877 8-BIT UNIVERSAL TRANSCEIVER PORT CONTROLLER

MODE S2 S1 S0	CLOCK	SERIN	A1 Q1 B1	A2 Q2 B2	A3 Q3 B3	A4 Q4 B4	A5 Q5 B5	A6 Q6 B6	A7 Q7 B7	A8 Q8 B8	PORT FUNCTION
0 0 0	1 or 0	X	Z Qn A1	Z Qn A2	Z Qn A3	Z Qn A4	Z Qn A5	Z Qn A6	Z Qn A7	Z Qn A8	A TO B
0 0 0	↑	X	Z A1 A1	Z A2 A2	A A3 A3	Z A4 A4	Z A5 A5	Z A6 A6	Z A7 A7	Z A8 A8	
0 0 1	1 or 0	X	B1 Qn Z	B2 Qn Z	B3 Qn Z	B4 Qn Z	B5 Qn Z	B6 Qn Z	B7 Qn Z	B8 Qn Z	B TO A
0 0 1	↑	X	B1 B1 Z	B2 B2 Z	B3 B3 Z	B4 B4 Z	B5 B5 Z	B6 B6 Z	B7 B7 Z	B8 B8 Z	
0 1 0	1 or 0	X	X Qn Q1	X Qn Q2	X Qn Q3	X Qn Q4	X Qn Q5	X Qn Q6	X Qn Q7	X Qn Q8	QN TO BN
0 1 0	↑	X	Z A1 A1	Z A2 A2	Z A3 A3	Z A4 A4	Z A5 A5	Z A6 A6	Z A7 A7	Z A8 A8	
0 1 1	1 or 0	X	Q1 Qn X	Q2 Qn X	Q3 Qn X	Q4 Qn X	Q5 Qn X	Q6 Qn X	Qn Q7 X	Q8 Qn X	QN TO AN
0 1 1	↑	X	B1 B1 Z	B2 B2 Z	B3 B3 Z	B4 B4 Z	B5 B5 Z	B6 B6 Z	B7 B7 Z	B8 B8 Z	
1 0 0	1 or 0	X	Z Qn Q1	Z Qn Q2	Z QN Q3	Z Qn Q4	Z Qn Q5	Z Qn Q6	Z Qn Q7	Z Qn Q8	SHIFT TO B
1 0 0	↑	1	Z H H	Z Q1 Q1	Z Q2 Q2	Z Q3 Q3	Z Q4 Q4	Z Q5 Q5	Z Q6 Q6	Z Q7 Q7	
1 0 0	↑	0	Z L L	Z Q1 Q1	Z Q2 Q2	Z Q3 Q3	Z Q4 Q4	Z Q5 Q5	Z Q6 Q6	Z Q7 Q7	
1 0 1	1 or 0	X	Q1 Qn Z	Q2 Qn Z	Q3 Qn Z	Q4 Qn Z	Q5 Qn Z	Q6 Qn Z	Q7 Q7 Z	Q8 Qn Z	SHIFT TO A
1 0 1	↑	1	H H Z	Q1 Q1 Z	Q2 Q2 Z	Q3 Q3 Z	Q4 Q4 Z	Q5 Q5 Z	Q6 Q6 Z	Q7 Q7 Z	
1 0 1	↑	0	L L Z	Q1 Q1 Z	Q2 Q2 Z	Q3 Q3 Z	Q4 Q4 Z	Q5 Q5 Z	Q6 Q6 Z	Q7 Q7 Z	
1 1 0	1 or 0	X	Z Qn Z	Z Qn Z	Z Qn Z	Z Qn Z	Z Qn Z	Z Qn Z	Z QN Z	Z Qn Z	SHIFT
1 1 0	↑	1	Z H Z	Z Q1 Z	Z Q2 Z	Z Q3 Z	Z Q4 Z	Z Q5 Z	Z Q6 Z	Z Q7 Z	
1 1 0	↑	0	Z L Z	Z Q1 Z	Z Q2 Z	Z Q3 Z	Z Q4 Z	Z Q5 Z	Z Q6 Z	Z Q7 Z	
1 1 1	1 or 0	X	Z Qn Z	Z Qn	Z Qn Z	Z Qn Z	Z Qn Z	Z Qn Z	Z Qn Z	Z Qn Z	CLEAR
1 1 1	↑	X	Z L Z	Z L Z	Z L Z	Z L Z	Z L Z	Z L Z	Z L Z	Z L Z	

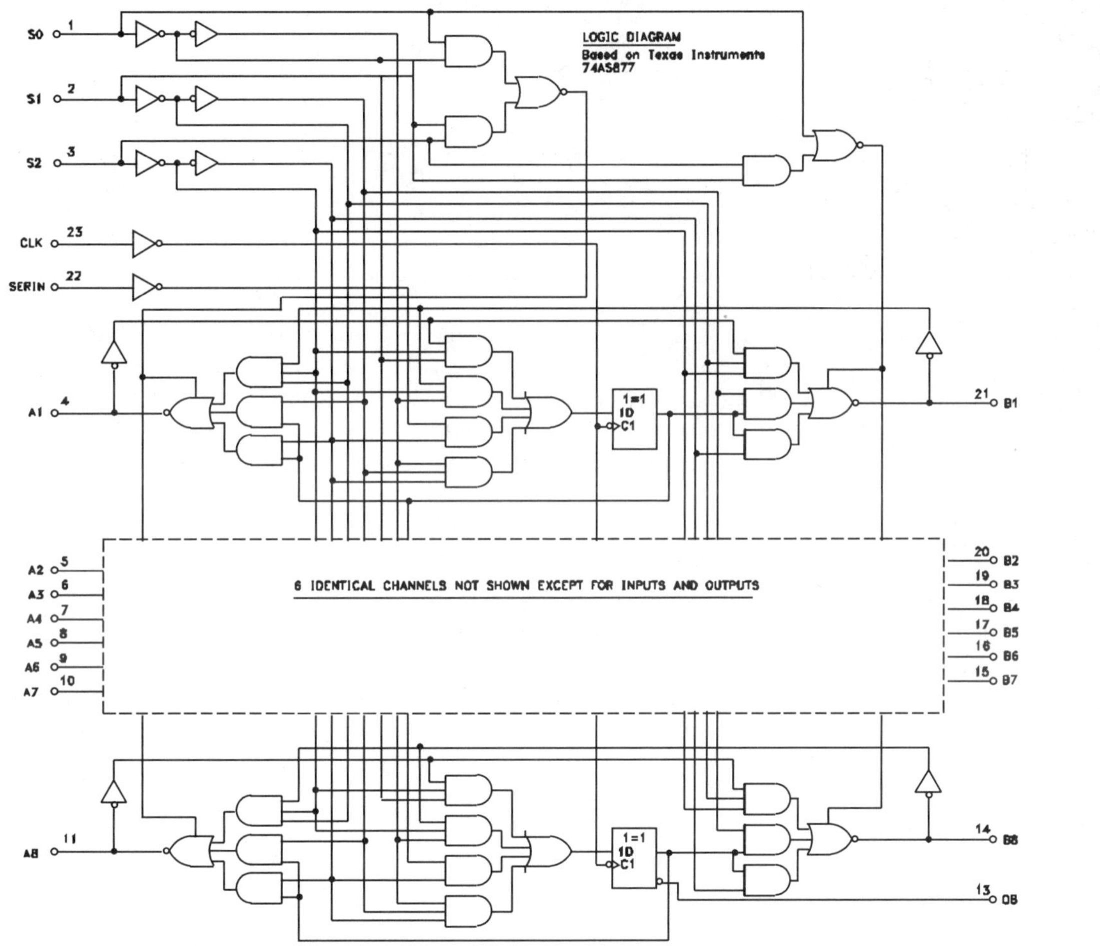

Figure 11.25. The 74AS877 is an 8-bit universal transceiver port controller featuring two 8-bit I/O ports, an 8-bit parallel load, serial-in, parallel-out shift register, and control logic. With these features the device is able to perform eight selectable transceiver or port functions, depending on the logic state of the three select input lines. Synchronous parallel loading can be accomplished from either port on the positive transition of the clock while serially shifting data in via the SERIN input. All serial data is shifted right. All outputs are buffer type designed specifically to drive bus lines directly. All are 3-state, except Q8, which is totem pole.

TTL – 74ALS/F1242
74F1243

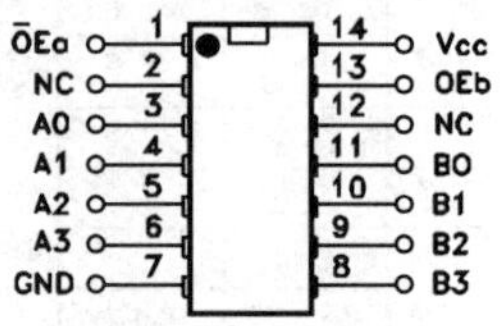

DUAL-IN-LINE PACKAGE

FUNCTION TABLE 74x1242
QUAD BUS TRANSCEIVERS
(EACH TRANSCEIVER)

INPUTS		OUTPUTS	
$\overline{OEa}$	OEb	An	Bn
0	0	INPUT	$B=\overline{A}$
1	0	Z	Z
0	1	a	a
1	1	$A=\overline{B}$	INPUT

FUNCTION TABLE 74F143

INPUTS		OUTPUTS	
$\overline{OEa}$	OEb	An	Bn
0	0	INPUT	B=A
1	0	Z	Z
0	1	a	a
1	1	A=B	INPUT

1 = HIGH voltage level.
0 = LOW voltage level.
Z = High Impedance
a = Not allowed, excessive currents

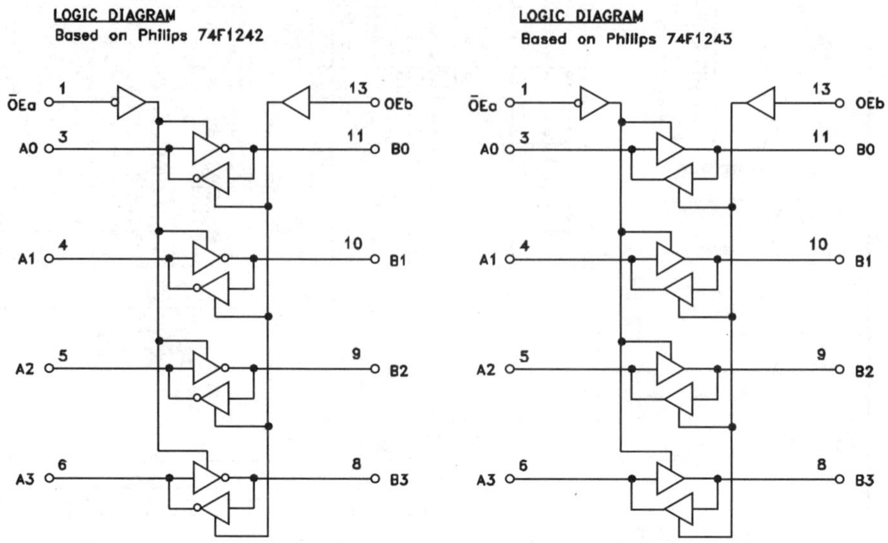

Figure 11.26. The 74ALS1242 is a low-power version of the 'ALS242. The 74F1242 and 74F1243 are light-load FAST versions of the 'F242 and 'F243, respectively. Refer to *Figure 11.1* for information on these devices.

TTL – 74ALS/F1245

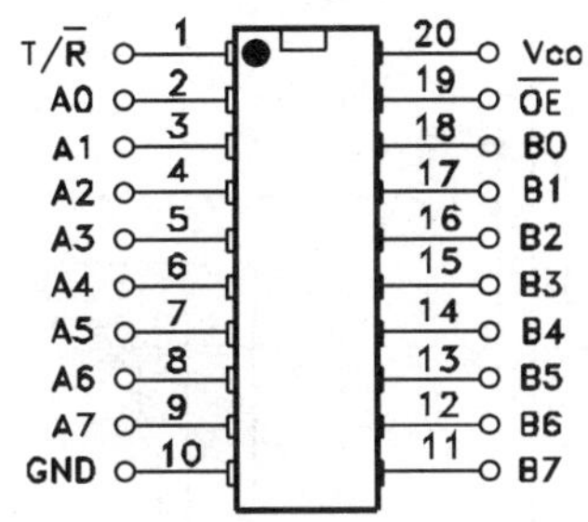

DUAL-IN-LINE PACKAGE

LOGIC DIAGRAM
Based on Philips 74F1245

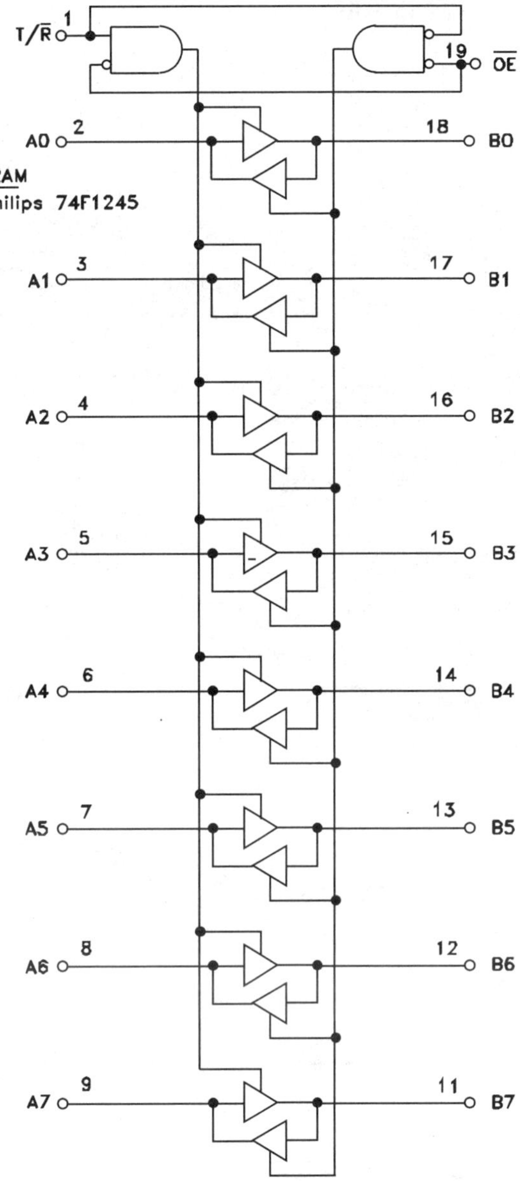

FUNCTION TABLE 74x1245 OCTAL BUS TRANCEIVERS WITH 3-STATE OUTPUTS

INPUTS		INPUTS/OUTPUTS	
$\overline{OE}$	T/$\overline{R}$	An	Bn
0	0	A=B	INPUTS
0	1	INPUTS	B=A
1	X	Z	Z

1 = HIGH voltage level.
0 = LOW voltage level.
X = Don't Care
Z = High Impedance state

Figure 11.27. The 74F1245 is a light-load FAST version of the 'F245, featuring the same functions and pinouts of the 'F245. Refer to *Figure 11.2* for functional details on this device.

TTL – 74F2952
74F2953

3V CMOS – 74LVC/LVCH2952
74LVC/LVCH22952
74LVT2952
74LCX2952

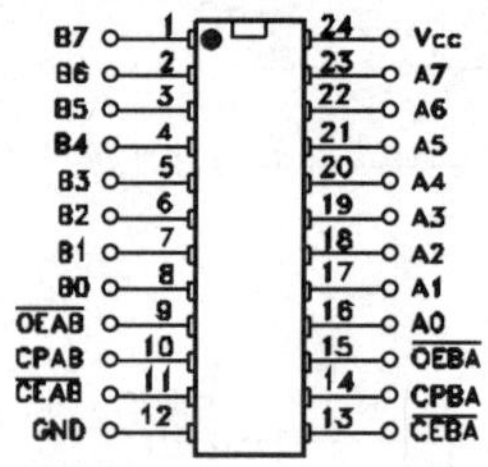

DUAL-IN-LINE PACKAGE

NOTE:
The SSOP and TSSOP Packages follow the DIP Package pinning.

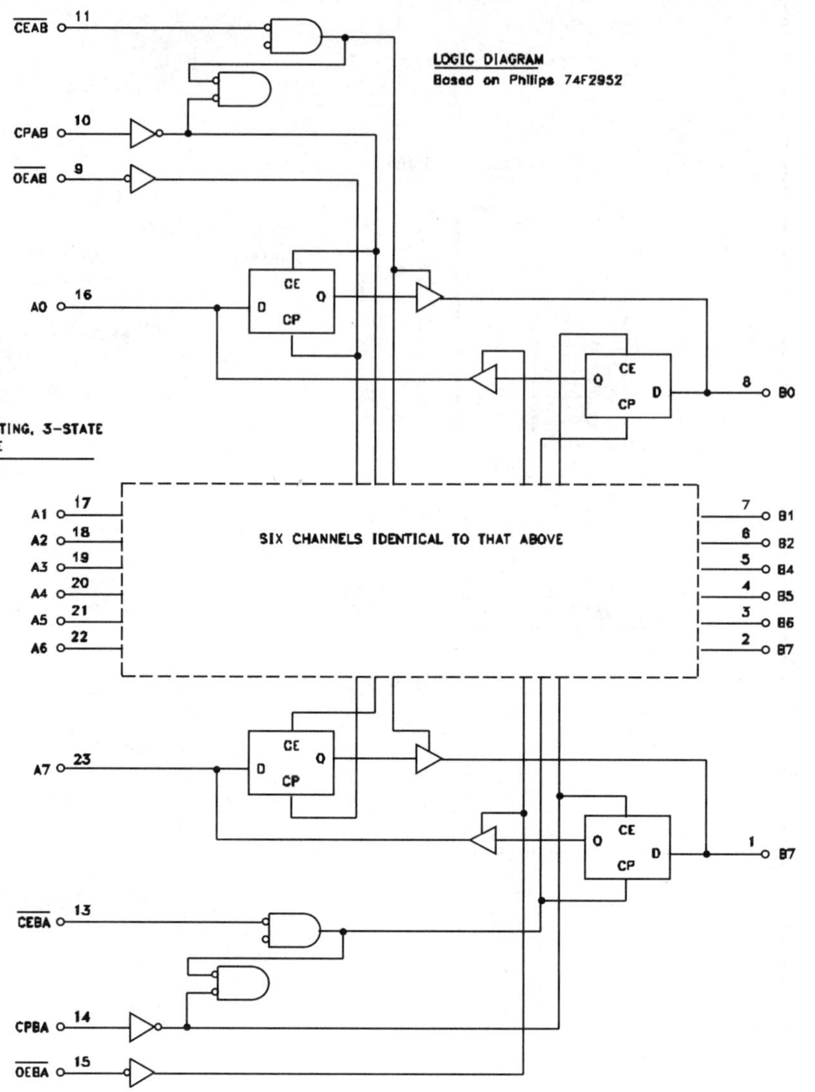

74x2952,'x22952 REGISTERED TRANSCEIVER, NON-INVERTING, 3-STATE
74F2953 REGISTERED TRANSCEIVER, INVERTING, 3-STATE

FUNCTION TABLE FOR REGISTER An OR Bn

INPUTS			INTERNAL	
An/Bn	CPXX	CEXX	Q	OPERATING MODE
X	X	1	NC	Hold Data
0	↑	0	0	Load Data
1	↑	0	1	Load Data

FUNCTION TABLE FOR OUTPUT ENABLE

INPUTS	INTERNAL	An/Bn OUTPUTS		
OEXX	Q	'F2952	'F2953	OPERATING MODE
1	X	Z	Z	Disable Outputs
0	0	0	1	Enable Outputs
0	1	1	0	Enable Outputs

↑ = LOW-to-HIGH clock transition.
1 = HIGH voltage level.
0 = LOW voltage level.
X = Don't Care
Z = High Impedance state

Figure 11.28. The 74x2952 and 74F2953 are 8-bit registered transceivers. Two 8-bit back-to-back registers store data flowing in both directions between two bidirectional buses. Data applied at the inputs is entered and stored on the rising edge of the clock, provided the clock enable is low. The data is then available at the 3-state output buffers when the output enable is low. The '2952 outputs are noninverting; the '2953 inverting.

3V CMOS – 74LVX/LVXC3245
74LVX/LVXC4245

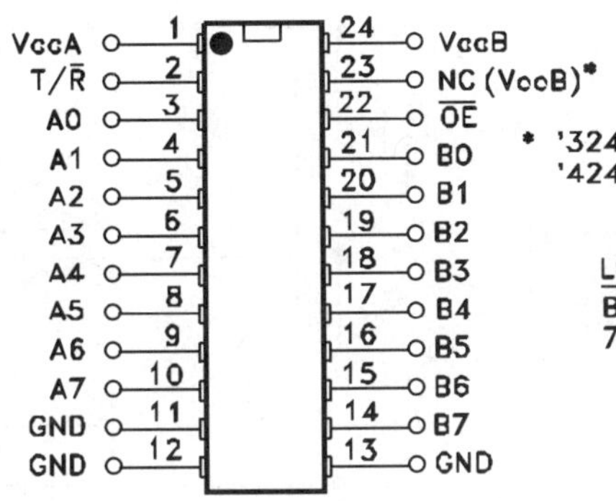

SOIC/TSSOP CASE

* '3245 = NC
'4245 = VccB

LOGIC DIAGRAM
Based on Motorola 74LVXC3245

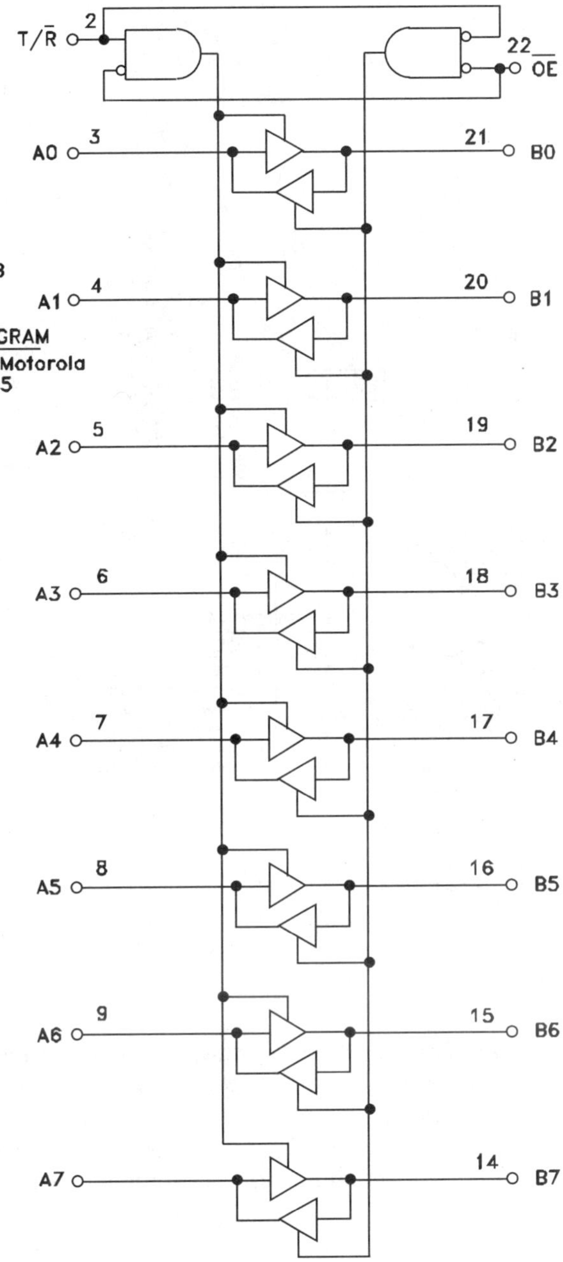

FUNCTION TABLE 74x245, 'ALS5245
OCTAL BUS TRANSCEIVERS WITH 3-STATE OUTPUTS

ENABLE OE̅	DIRECTION CONTROL T/R̄	OPERATION
0	0	B data to A bus
0	1	A data to B bus
1	X	Z

1 = HIGH voltage level.
0 = LOW voltage level.
X = Don't Care
Z = High Impedance state

Figure 11.29. The 74LVX/LVXC3245 and 74LVX/LVCX4245 are 8-bit dual supply translating transceivers with 3-state outputs. These devices provide a bidirectional interface between 3V and 5V buses. Inputs are compatible with TTL levels. With the 'LVX3245, 3V data flow is at the A port, 5V at the B port. With the 'LVX4245 these are reversed, with the 5V at port A and the 3V at port B. With the LVCX3245, the A port is dedicated to 3V; the B port tracks the Vcc B supply. These roles are reversed for the 'LVXC4245. Functionally, these are compatible with the 74 series 245 devices of *Figure 11.1*.

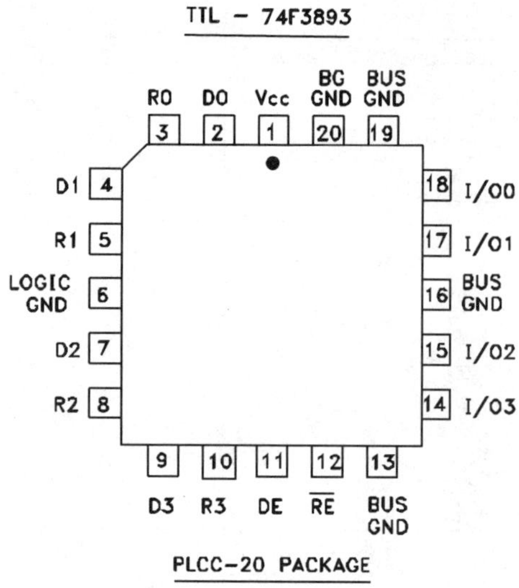

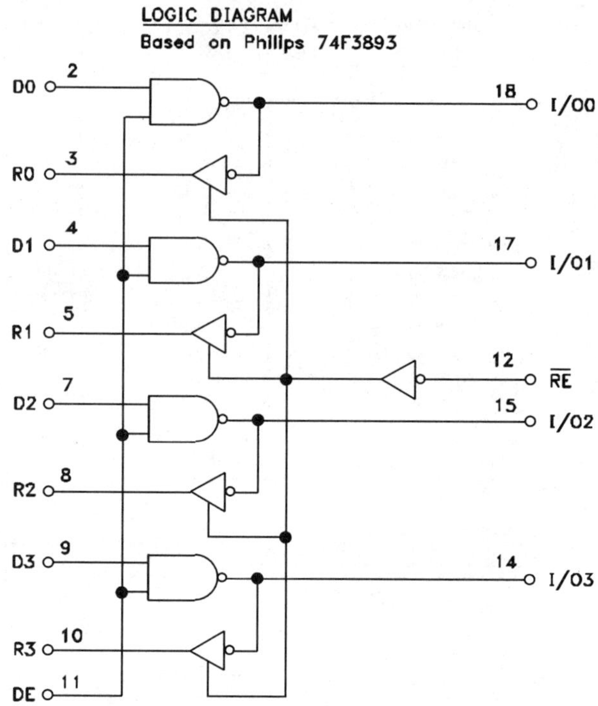

FUNCTION TABLE 74F3893 QUAD FUTUREBUS BACKPLANE TRANSCEIVER

INPUT			INPUT/OUTPUT	OUTPUT	OPERATING
DE	$\overline{RE}$	Dn	I/On	Rn	MODE
1	0	0	1	0	Transmit to Bus
1	0	1	0	1	Transmit to Bus
1	1	Dn	Dn	Z	Receiver 3-state
0	1	X	1	Z	Transmit to Bus
0	0	X	1	0	Receive, I/On=Inputs
0	0	X	0	1	Receive, I/On=Inputs

1 = HIGH voltage level.
0 = LOW voltage level.
X = Don't Care
Z = High Impedance state

Figure 11.30. The 74F3893 is a quad backplane transceiver intended for use in very high-speed bus systems. It will drive heavily loaded backplanes with equivalent load impedances down to 10 ohms. The device is compatible with IEEE 896 and IEEE 1194.1 Futurebus standards. Futurebus drivers sink 100 mA. A reduced voltage swing of 1V produces less noise and reduces power consumption. All outputs are designed to be glitch free on power-up.

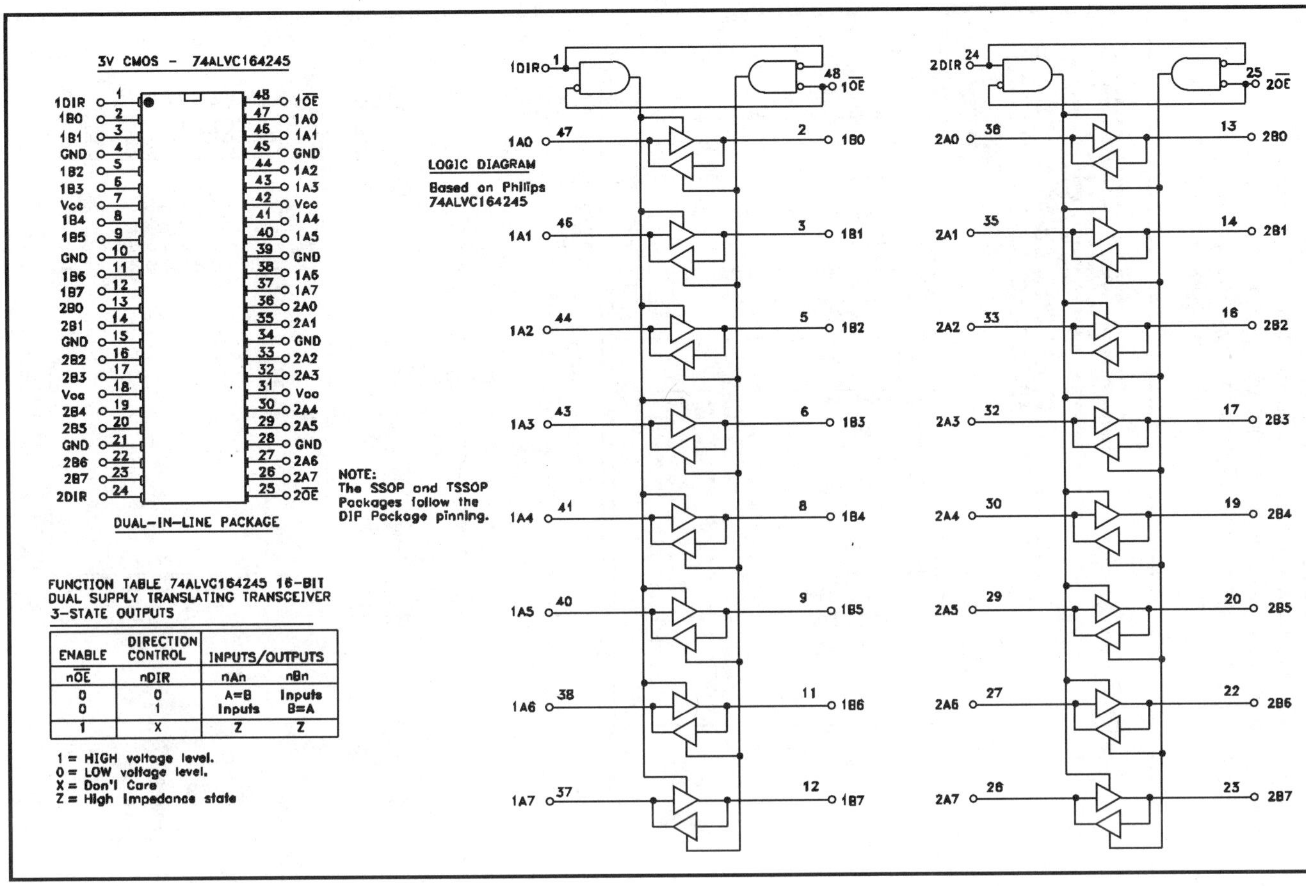

Figure 11.31. The 74ALVC164245 is a 16-bit (dual octal) backplane translating receiver designed to interface between a 5V bus and a 3V bus in a mixed 3V/5V supply environment. The direction control inputs determine the direction of the data flow. Voltage range of the A port is 1.2 to 3.6 volts; the B port 1.2 to 5.5 volts. Control inputs voltage range is 2.7 to 5.5 volts. The device is in accordance with JEDEC standard 8-1A.

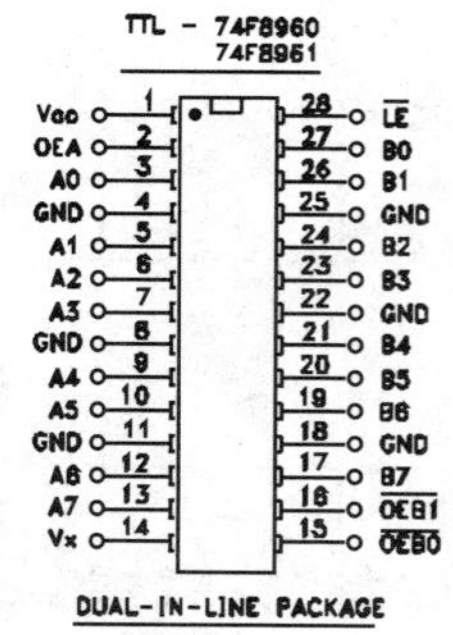

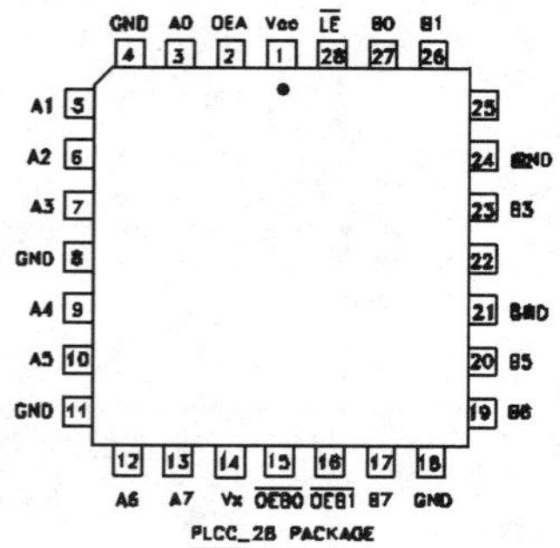

PACKAGE PINNING:

FUNCTION

A0–A7	I/O, PNP latched input/3–state output (with vX control option)
B0–B7	I/O, Data input w/special threshold circuitry to reject noise open collector output, high current drive
$\overline{OEB0}$	Input, Enables the B outputs when both pins are low
$\overline{OEB1}$	Input, Enable the A outputs when high
$\overline{LE}$	Input, Latched when high (a special feature built in for proper enabling times.
Vx	Input, Clamping voltage keeping VOH from rising above Vx (Vx = Vcc for normal use)

FUNCTION TABLE 74F8960 OCTAL LATCHED BIDIRECTIONAL TRANSCEIVER, INVERTING
3–STATE + OPEN COLLECTOR
748961 OCTAL LATCHED BIDIRECTIONAL TRANSCEIVER, NON–INVERTING
3–STATE + OPEN COLLECTOR

INPUTS						LATCH	OUTPUTS		
An	Bn*	$\overline{LE}$	OEA	$\overline{OEB0}$	$\overline{OEB1}$	STATE	An	Bn	OPERATING MODE
1	X	0	0	0	0	1	Z	0	A 3–state, data from A to B
0	X	0	0	0	0	0	Z	H**	A 3–state, data from A to B
X	X	1	0	0	0	On	Z	$\overline{On}$	A 3–state, latched data to B
–	–	0	1	0	0	(1)	(1)	(1)	Feedback: A to B, B to A
–	1	1	1	0	0	H(2)	H	Z(2)	Preconditioned latch enabling data transfer B>A
–	0	1	1	0	0	H(2)	L	Z(2)	Preconditioned latch enabling data transfer B>A
–	–	1	1	0	0	On	$\overline{On}$	$\overline{On}$	Latch state to A and B
1	X	0	0	1	X	H	Z	Z	B and A 3–state
0	X	0	0	1	X	L	Z	Z	B and A 3–state
X	X	1	0	1	X	On	Z	Z	B and A 3–state
–	1	0	1	1	X	H	H	Z	B 3–state
–	0	0	1	1	1	L	L	Z	B 3–state
–	1	1	1	1	1	On	H	Z	B 3–state
–	0	1	1	1	1	On	L	Z	B 3–state
1	X	0	0	X	1	H	Z	Z	B and A 3–state
0	X	0	0	X	1	L	Z	Z	B and A 3–state
X	X	1	0	X	1	On	Z	Z	B and A 3–state
–	1	0	1	X	1	H	H	Z	B 3–state, data from B to A
–	0	0	1	X	1	L	L	Z	B 3–state, data from B to A
–	1	1	1	X	1	On	H	Z	B 3–state, data from B to A
–	0	1	1	X	1	On	L	Z	B 3–state, data from B to A

1 = HIGH voltage level.
0 = LOW voltage level.
X = Don't Care
Z = High Impedance state
– = Input not externally driven
On = High or low voltage level one setup time prior to the low–to–high LE transition
(1) = Condition will cause a feedback loop path: A>B, and B>A
(2) = The latch must be preconditioned such that B inputs may assume a high or low level while OEB0 and OEB1 are low and LE is high
H** = Goes to level of pullup voltage
B* = Precaution should be taken to insure the B inputs do not float. If they do they are equal to low state.
Function table shown is for 74F8960. Same as for 74F8961 except where On are shown inverted these are non–inverted for the 74F8961

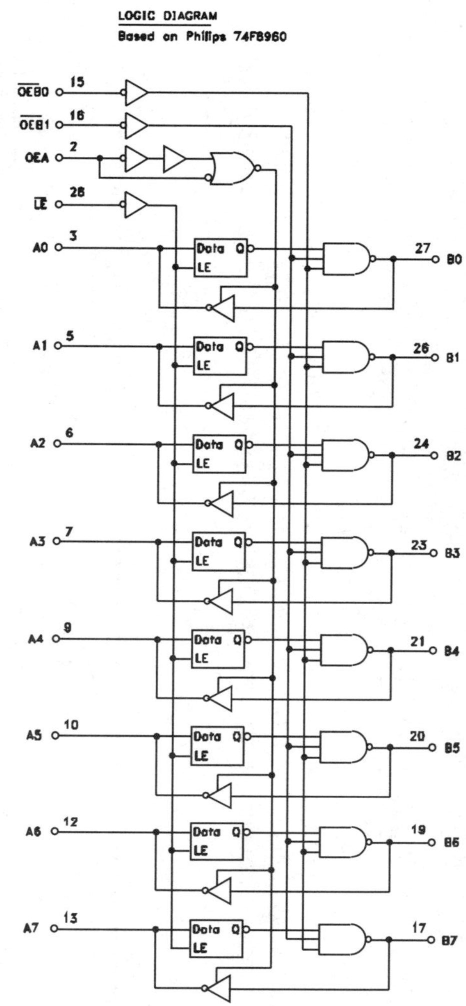

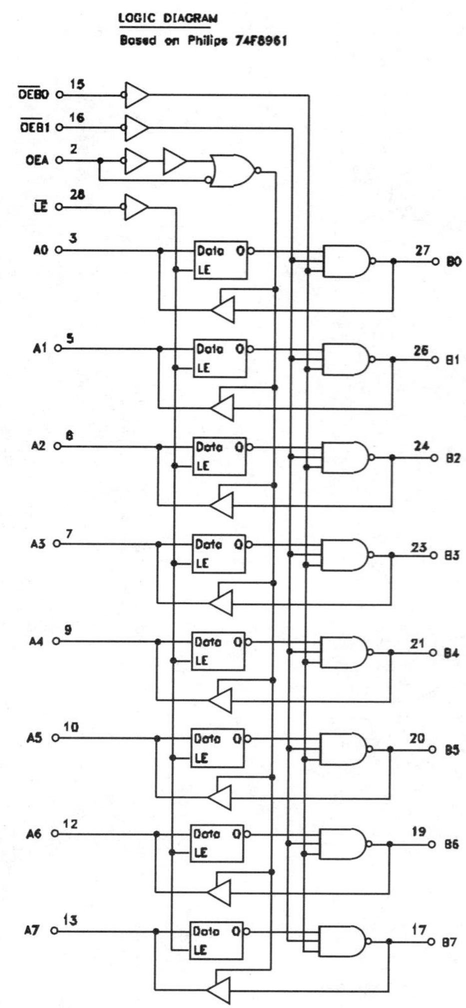

Figure 11.32. The 74F8960 and 74F8961 are octal latched bidirectional futurebus transceivers designed to provide the electrical interface to a high-performance wired-OR bus. These will drive heavily loaded backplanes with equivalent load impedances down to 10 ohms. The devices are compatible with IEEE switching standards. Open collector drivers sink 100 mA on the B port. A reduced voltage swing of 1V produces less noise and reduces power consumption. Acontrolled output ramp and multiple ground pins minimize ground bounce. All outputs are designed to be glitch-free on power-up and -down. Both devices feature 3-state plus open collector outputs. The 74F8960 is inverting; the 'F8961 noninverting.

TTL - 74F8962
74F8963

ALL PRODUCT DATA BASED ON PHILIPS 74F8962 AND 74F8963

THE PACKAGE FOR THESE IS THE 44-PIN PLCC
PIN DESCRIPTION

SYMBOL	PINS	TYPE	NAME AND FUNCTION
AI0–AI8	2,4,7,9,11,14,16,19,21	INPUT	PNP latched inputs
B0–B8	42,40,38,36,34,32,30,28,26	I/O	Data input/open collector output, high current drives
$\overline{OEAB}$	25	INPUT	Output enable input. Enables the B outputs when low.
$\overline{OEBA}$	44	INPUT	Output enable input. Enables the A outputs when high.
$\overline{LEAB}$	24	INPUT	Latch enable input. Enables the AB latches low
$\overline{LEBA}$	43	INPUT	Latch enable input. Enables the BA latches low
AO0–AO8	3,5,8,10,13,15,17,20,22	OUTPUT	TTL 3-state outputs.
GND	6,12,18,27,29,31,33,35,37,39,41	GROUND	Grounds
Vcc	1,23	POWER	Positive supply voltages

FUNCTION TABLE 74F8962, 74F8963
9-BIT LATCHED BIDIRECTIONAL FUTUREBUS TRANSCEIVERS
74F8962 INVERTING, OPEN COLLECTOR
74F8963 NON-INVERTING OPEN COLLECTOR

INPUTS						LATCH STATES		OUTPUTS		
AIn	Bn*	$\overline{LEAB}$	$\overline{LEBA}$	$\overline{OEAB}$	$\overline{OEBA}$	AB	BA	AOn	Bn	OPERATING MODES
1	1	0	0	1	1	1	1	Z	X	B and AO disabled
0	0	0	0	1	1	0	0	Z	X	B and AO disabled
X	X	1	1	1	1	Qn*	Qn*	Z	X	B and AO disabled
1	–	0	X	0	1	1	Qn*	Z	0	AO 3-state, transparent data A>B
0	–	0	X	0	1	0	Qn*	Z	H**	AO 3-state, transparent data A>B
X	1	X	0	1	0	Qn*	1	0	X	B disabled, transparent data B>AO
X	0	X	0	1	0	Qn*	0	1	X	B disabled, transparent data B>AO
X	X	1	X	0	1	Qn*	Qn*	Z	$\overline{Qn}$	AO 3-state, latched data to B
X	X	X	1	1	0	Qn*	Qn*	$\overline{Qn}$~	X	B disabled, latched to AO
X	X	1	1	0	0	Qn*	Qn*	$\overline{Qn}$~	$\overline{Qn}$~	Latched state to AO and B
1	–	0	0	0	0	1	0	1	0	Read back from AI to B to AO
0	–	0	0	0	0	0	1	0	H**	(both latches transparent)

1 = HIGH voltage level.
0 = LOW voltage level.
X = Don't Care
Z = High Impedance state
– = Input not externally driven
Qn = High or low voltage value one setup time prior to the low-to-high LEXX transition.
H** = Goes to level of pullup voltage
B* = Precaution should be taken to insure B inputs do not float. If they do, they are equal to low state.
Qn* = Inverted for 74F8963
Qn~ = Not inverted for 74F8963

LOGIC DIAGRAM 74F8962

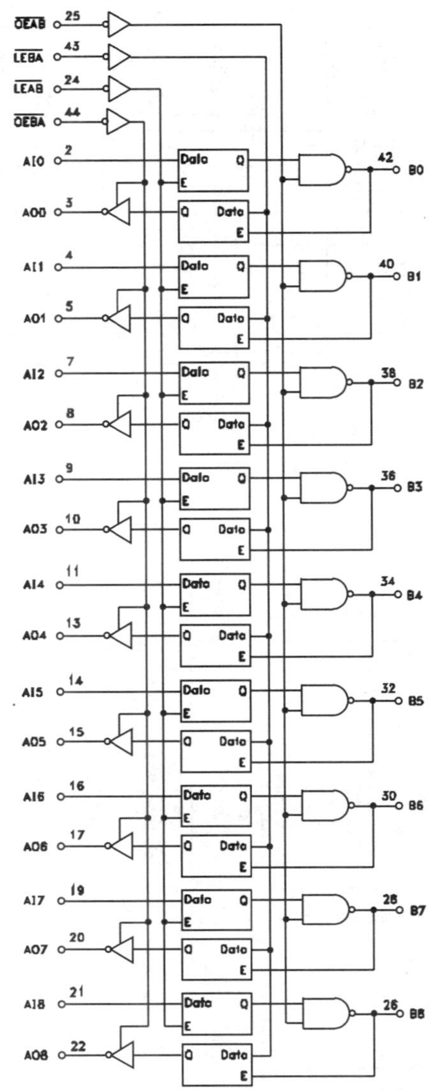

LOGIC DIAGRAM 74F8963

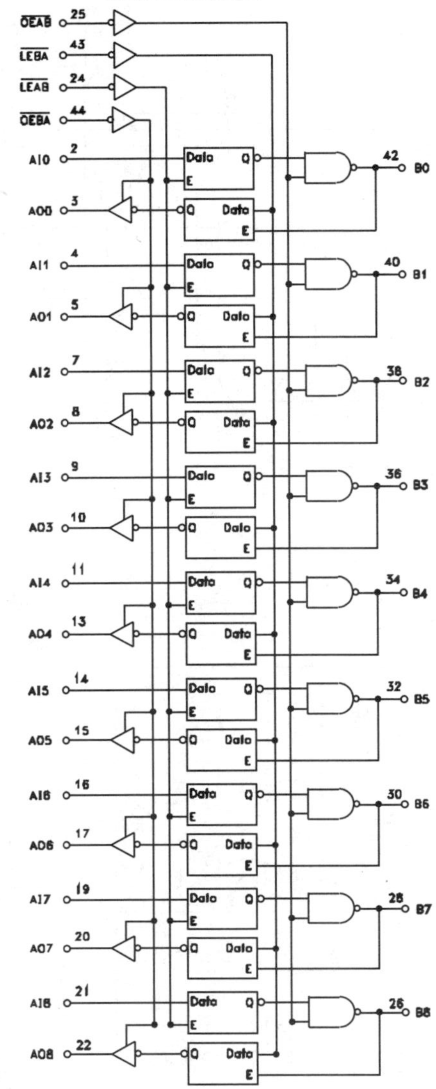

Figure 11.33. The 74F8962 and 74F8963 are 9-bit latched bidirectional futurebus transceivers designed to provide the electrical interface to a high-performance wired-OR bus. These will drive heavily loaded backplanes with equivalent load impedances down to 10 ohms. The devices are compatible with IEEE 896 futurebus standards. Open collector drivers sink 100 mA from 2V on the B port. A reduced voltage swing of 1-2V produces less noise and reduces power consumption. The A ports have 3-state drivers and TTL receivers with a latch function. Multiple ground pins minimize ground bounce. All outputs are designed to be glitch-free on power-up and -down. The 74F8962 is inverting; the 'F8963 noninverting.

TTL – 74F8965
74F8966

THE PACKAGE FOR THESE
IS THE 44-PIN PLCC
PIN DESCRIPTION

SYMBOL	PINS	TYPE	NAME AND FUNCTION
A0–A8	4,6,8,10,12,13,15,17,19	I/O	Data inputs/TTL 3-state outputs
B0–B8	41,39,37,35,33,31,29,27,25	I/O	Data input/open collector output, high current drives
OEB0	1	INPUT	Output enable input Enables the B outputs when high
$\overline{OEB1}$	44	INPUT	Output enable input Enables the B outputs when low
OEA	2	INPUT	Output enable input Enables the A outputs when high
$\overline{LE}$	22	INPUT	Latch enable input Enables latch when low
$\overline{LS}$	23	INPUT	Latch select input Selects latch when low ('F8965)
$\overline{IAREQ}$	23	INPUT	Idle arbitration request ('F8966)
$\overline{IAMC}$	20	OUTPUT	Idle arbitration/multiple competitors output (OC output) ('F8966)
Bus GND	26,28,30,32,34,36,38,40,42	GROUND	Bus Ground (0V)
Logic GND	5,7,9,11,14,16,18,20('F8965)	GROUND	Logic Ground (0V)
Bus Vcc	24,43	POWER	Positive supply voltages
Logic Vcc	3,21	POWER	Positive supply voltages

ALL PRODUCT DATA BASED ON PHILIPS
74F8965 AND 74F8966
LOGIC DIAGRAM 74F8965

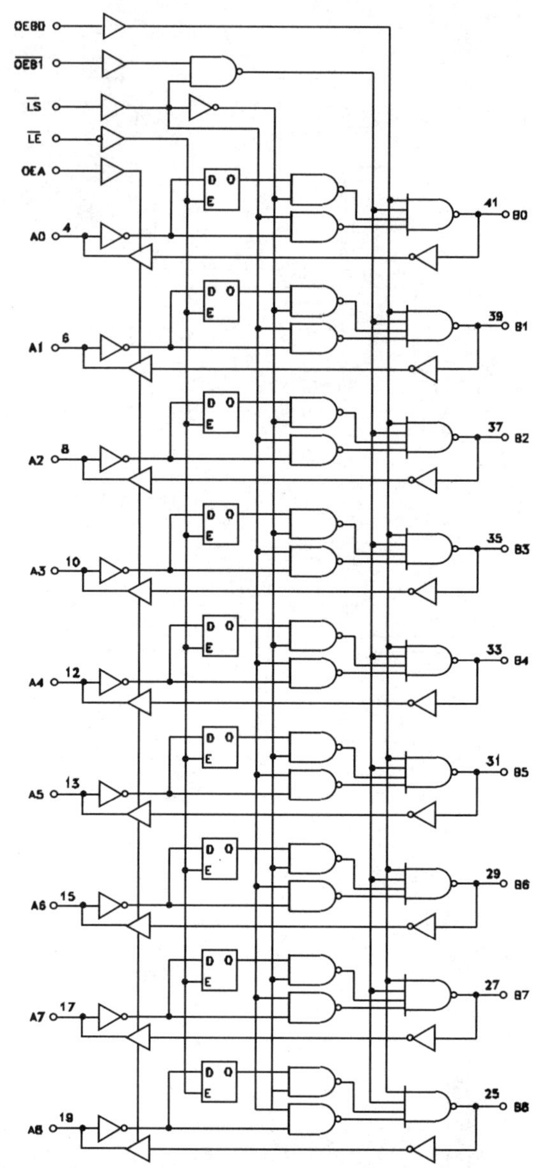

FUNCTION TABLE 74F8965
ADDRESS/DATA FUTUREBUS TRANSCEIVER, ADT

INPUTS							LATCH	OUTPUTS		
AIn	Bn*	$\overline{OEB0}$	$\overline{OEB1}$	$\overline{LS}$	OEA	$\overline{LE}$	STATE	An	Bn	OPERATING MODES
0	–	1	0	1	0	X	X	Input	H**	An to $\overline{B}$n bypass latch
1	–	1	0	1	0	X	X	Input	L	
0	–	1	0	0	0	0	1	Input	H**	An to $\overline{B}$n transparent latch
1	–	1	0	0	0	0	0	Input	L	
l	–	1	0	0	0	↑	1	Input	H**	An to $\overline{B}$n latch and read
h	–	1	0	0	0	↑	0	Input	L	
–	–	1	0	0	1	1	1	L	H**	An to $\overline{B}$n outputs latched and read (preconditioned latch)
–	–	1	0	0	1	1	0	H	L	
X	–	1	0	0	0	1	NC	Input	L	An to $\overline{B}$n hold
X	X	0	X	X	X	X	X	X	H**	Disable $\overline{B}$n outputs
X	X	X	1	1	X	X	X	X	H**	
–	0	0	1	1	1	X	X	H	Input	$\overline{B}$n to An
–	1	0	1	1	1	X	X	L	Input	
–	X	X	X	X	0	X	X	Z	x	Disable A outputs

FUNCTION TABLE 74F8966
ADDRESS/DATA FUTUREBUS TRANSCEIVER, ADT

INPUTS								LATCH	OUTPUTS			
AIn	Bn*	$\overline{OEB0}$	$\overline{OEB1}$	$\overline{IAREQ}$	$\overline{LS}$	OEA	$\overline{LE}$	STATE	An	Bn	IAMC	OPERATING MODES
0	–	1	0	0	1	0	X	X	Input	H**	H**	An to $\overline{B}$n bypass latch
1	–	1	0	0	1	0	X	X	Input	L	H**	
0	–	1	0	0	0	0	0	1	Input	H**	H**	An to $\overline{B}$n transparent latch
1	–	1	0	0	0	0	0	0	Input	L	H**	
l	–	1	0	0	0	0	↑	1	Input	H**	H**	An to $\overline{B}$n latch and read
h	–	1	0	0	0	0	↑	0	Input	L	H**	
–	–	1	0	0	0	1	1	1	L	H**	H**	An to $\overline{B}$n outputs latched and read (preconditioned latch)
–	–	1	0	0	0	1	1	0	H	L	H**	
X	–	1	0	0	0	0	1	NC	Input	L	H**	An to $\overline{B}$n hold
X	X	0	X	X	X	X	X	X	X	H**	H*	Disable $\overline{B}$n outputs
X	X	X	1	1	1	X	X	X	X	H**	H**	
–	0	0	1	1	1	1	X	X	H	Input	H**	$\overline{B}$n to An
–	1	0	1	1	1	1	X	X	L	Input	H**	
–	Bn	0	1	1	↓*	1	1	Bn	Z	Bn	L	Latch Bn data idle arbitration request (preconditioned latch)
–	Bn	0	1	1	↓*	1	1	Bn	Z	Bn	H**	
–	X	X	X	X	X	0	X	X	Z		X	Disable A outputs

↑ = Low-to-high transition
↓* = High-to-low transition, latch must be preconditioned before $\overline{IAREQ}$
1 = HIGH voltage level.
0 = LOW voltage level.
X = Don't Care
Z = High Impedance state
– = Input not externally driven
h = High voltage level one setup time prior to the low-to-high $\overline{LE}$ transition
l = Low voltage level one set up time prior to the low-to-high $\overline{LE}$ transition
H** = Goes to level of pullup voltage
B* = Precaution should be taken to insure B inputs do not float. If they do, they are equal to low state.
Qn* = Inverted for 74F8963
Qn~ = Not inverted for 74F8963

Figure 11.34. The 74F8965 and 74F8966 are 9-bit latchable bidirectional futurebus transceivers designed to provide the electrical interface to a high-performance wired-OR bus. These will drive heavily loaded backplanes with equivalent load impedances down to 10 ohms. The devices are compatible with IEEE 896 and IEEE1194 Backplane Transistor Logic (BTL) standards. Controlled ramp open collector inverting drivers sink 100 mA from 2V on the B port. A reduced voltage swing of 1-2V produces less noise and reduces power consumption. The A ports have 3-state drivers and TTL receivers with a latch function. Multiple ground pins minimize ground bounce. All outputs are designed to be glitch-free on power-up and -down. The 74F8962 is inverting; the 'F8963 noninverting. The 'F8966 includes an idle arbitrator/multiple competitors output.

5V CMOS – 74AC/ACT11245
3V CMOS – 74AC11245

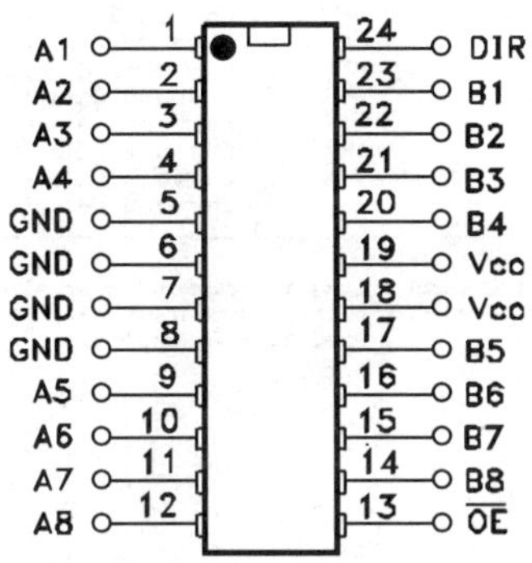

DUAL–IN–LINE PACKAGE

LOGIC DIAGRAM
Based on Texas Instruments 74AC11245

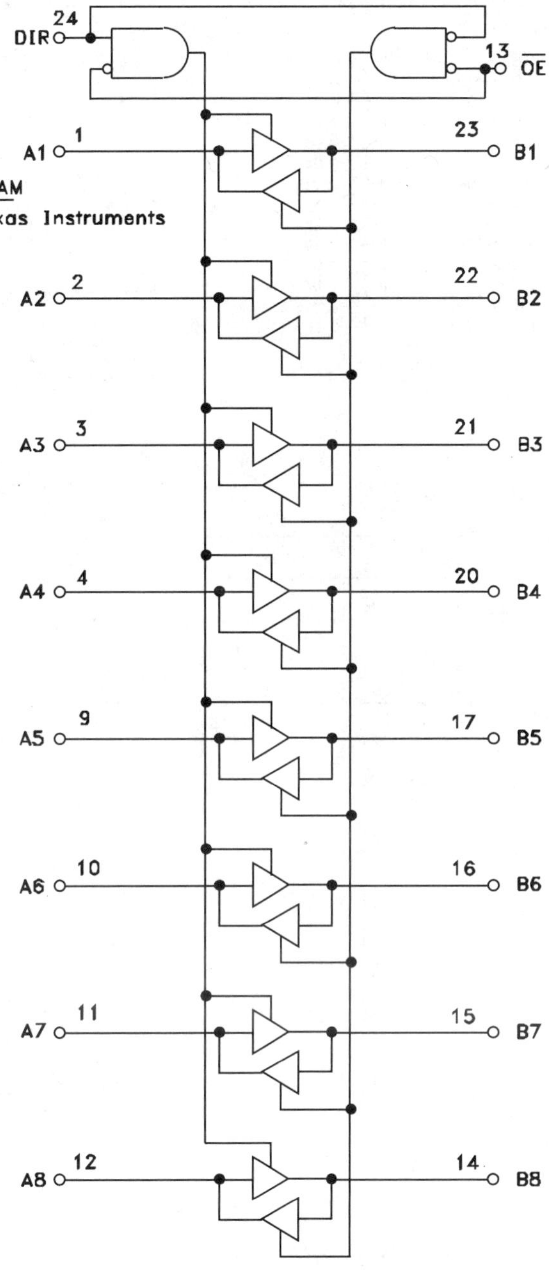

FUNCTION TABLE 74x245, 'ALS5245
OCTAL BUS TRANSCEIVERS WITH 3-STATE OUTPUTS

ENABLE $\overline{OE}$	DIRECTION CONTROL DIR	OPERATION
0	0	B data to A bus
0	1	A data to B bus
1	X	Isolation

1 = HIGH voltage level.
0 = LOW voltage level.
X = Don't Care

Figure 11.35. The 74AC/ACT11245 are octal bus transceivers with 3-state outputs for driving bus lines directly. The devices provide for asynchronous two-way communication between data buses. Direction of the noninverted data transmission between the two buses is determined by the DIR control input logic state. The output enable control can be used to disable the device so the buses are effectively isolated.

5V CMOS – 74AC/ACT11652
3V CMOS – 74AC11652

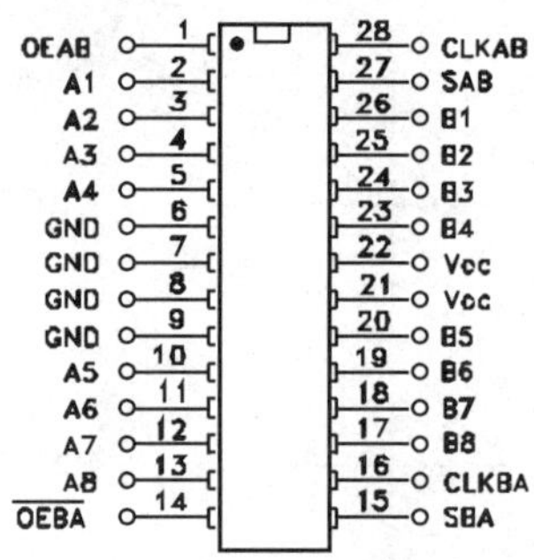

DUAL-IN-LINE PACKAGE

FUNCTION TABLE 74AC/ACT11652 OCTAL BUS TRANSCEIVER/REGISTERS W/3-STATE OUTPUTS

INPUTS						DATA I/O		
OEAB	$\overline{OEBA}$	CLKAB	CLKBA	SAB	SBA	A1–A8	B1–B8	OPERATION OR FUNCTION
0	1	0	0	X	X	Input	Input	Isolation
0	1	↑	↑	X	X	Input	Input	Store A and B Data
X	1	↑	0	X	X	Input	Unspecified	Store A, Hold B
1	1	↑	↑	X*	X	Input	Output	Store A in both registers
0	X	0	↑	X	X	Unspecified	Input	Hold A, Store B
0	0	↑	↑	X	X*	Output	Input	Store B in both registers
0	0	X	X	X	0	Output	Input	Real-time B Data to A Bus
0	0	X	0	X	1	Output	Output	Stored B Data to A Bus
1	1	X	X	0	X	Input	Output	Real-time A Data to B Bus
1	1	0	X	1	X	Input	Output	Stored A Data to B Bus
1	0	0	0	1	1	Output	Output	Stored A Data to B Bus and Stored B Data to A Bus

↑ = LOW-to-HIGH clock transition.
1 = HIGH voltage level.
0 = LOW voltage level.
X = Don't Care
Z = High Impedance state

* Select Control = Low; clocks can occur simultaenously
Select Control = High; clocks must be staggered to load both registers.

LOGIC DIAGRAM
Based on Texas Instrument 74AC11652

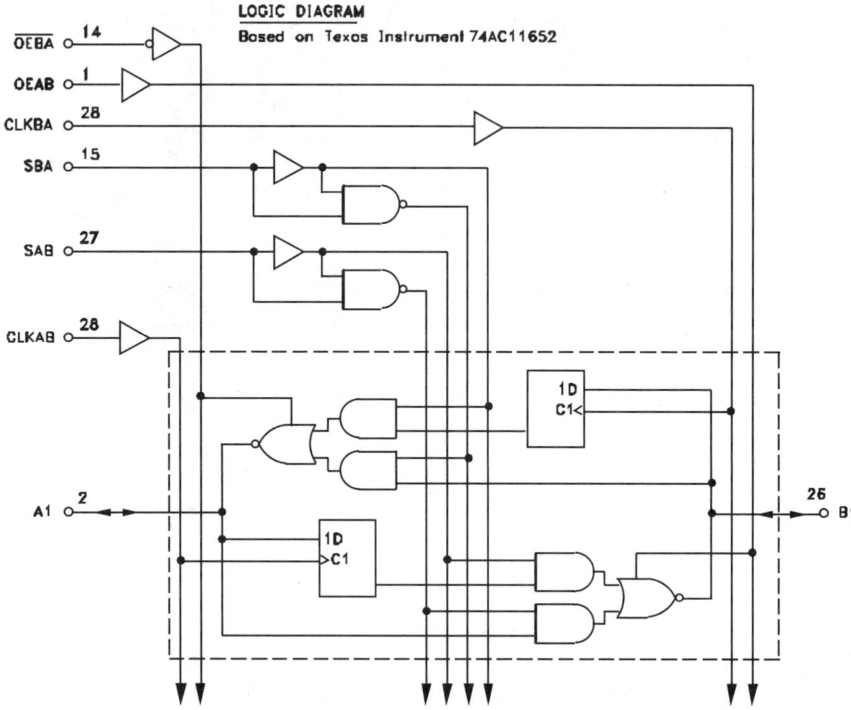

TO SEVEN REMAINING CHANNELS

Figure 11.36. The 74AC/ACT11652 are octal bus transceivers and registers with 3-state outputs. The devices are comprised of bus transceiver circuits, D-type flip-flops, and control circuits for multiplexed transmission of data directly from data bus to data bus or from the internal storage registers. Output enable inputs control the transceiver functions. The select control inputs are employed to determine whether real-time or stored data is transferred. Bus data is stored on the low-to-high clock transition. Center pin power connections minimize high-speed switching noise. Flow-through architecture optimizes PCB layout. Maximum clock frequencies are 65 MHz for 3.3 V; 105 MHz for 5.5 volt operation.

5V CMOS – 74AC/ACT16245
74AHC/AHCT16245

3V CMOS – 74AC/ALVCH/LVC/LVCH/LVT/
ALVT/AHC/AHCT16245
74LVC/LVCH162245

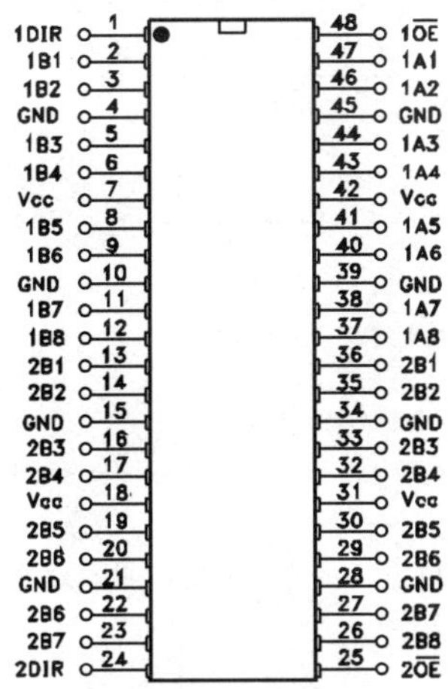

DUAL–IN–LINE PACKAGE
DGG OR DL

FUNCTION TABLE 74x16245 16–BIT BUS TRANSCEIVERS WITH 3–STATE OUTPUTS

ENABLE $\overline{OE}$	DIRECTION CONTROL DIR	OPERATION
0	0	B data to A bus
0	1	A data to B bus
1	X	Isolation

1 = HIGH voltage level.
0 = LOW voltage level.
X = Don't Care

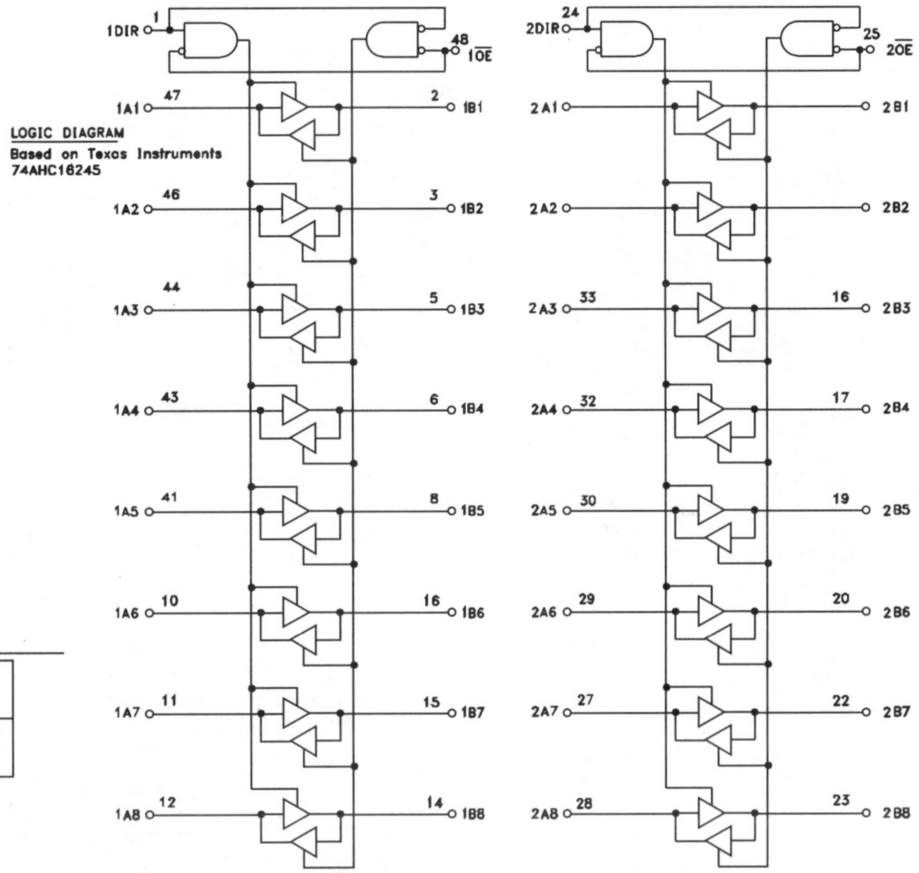

Figure 11.37. The 74x16245 are 16-bit (dual octal) non-inverting bus transceivers with 3-state outputs. The devices provide for asynchronous two-way communication between data buses. Direction of the noninverted data transmission between the two buses is determined by the DIR control input logic state. The output enable control can be used to disable the device so the buses are effectively isolated. Distributed power connections minimize high-speed switching noise. Flow through architecture optimizes PCB layout.

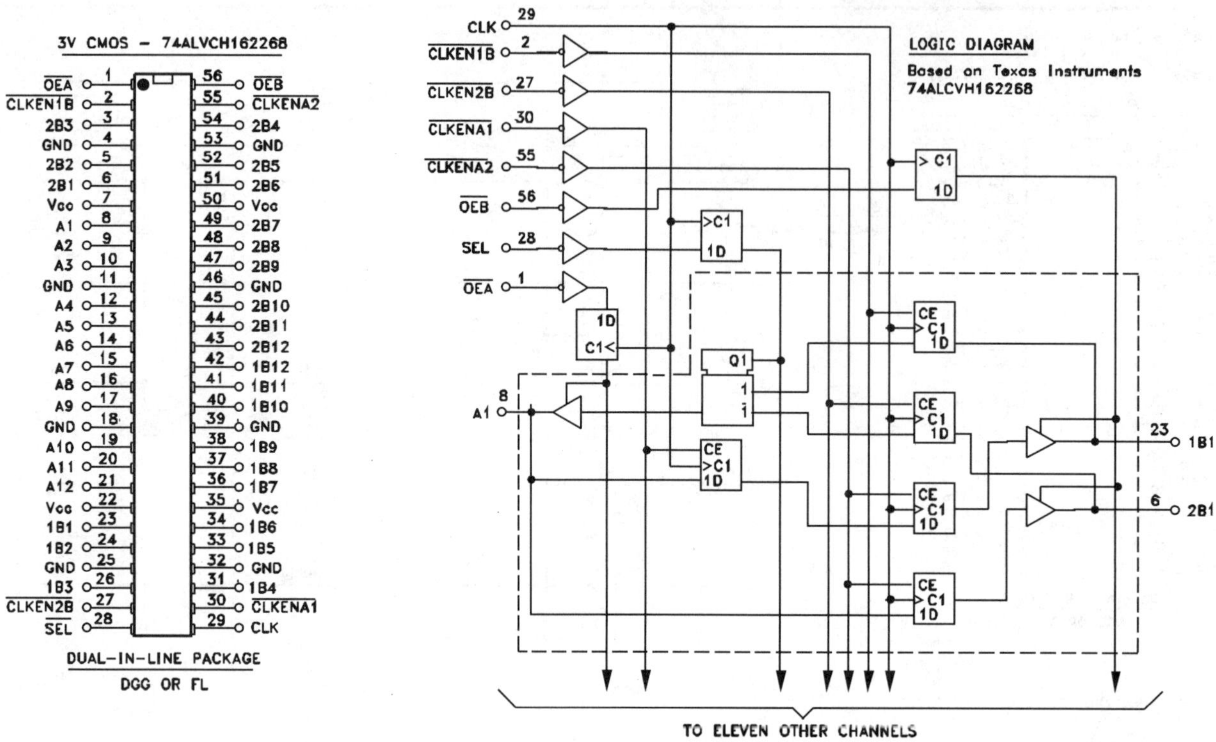

FUNCTION TABLE 74ALVCH162268 12-BIT TO 24-BIT REGISTERED BUS TRANSCEIVER WITH 3-STATE OUTPUTS

B-TO-A STORAGE ($\overline{\text{OEA}}$=L)

INPUTS						OUTPUT
$\overline{\text{CLKEN1B}}$	$\overline{\text{CLKEN2B}}$	CLK	$\overline{\text{SEL}}$	1B	2B	A
1	X	X	1	X	X	A0+
X	1	X	0	X	X	A0+
0	X	↑	1	0	X	0
0	X	↑	1	1	X	1
X	0	↑	0	X	0	0
X	0	↑	0	X	1	1

A-TO-B STORAGE ($\overline{\text{OEB}}$=L)

INPUTS				OUTPUTS	
$\overline{\text{CLKENA1}}$	$\overline{\text{CLKENA2}}$	CLK	A	1B	2B
1	1	X	X	1B0+	2B0+
0	X	↑	0	0%	X
0	X	↑	1	1%	X
X	0	↑	0	X	0
X	0	↑	1	X	1

OUTPUT ENABLE

INPUTS			OUTPUTS	
CLK	$\overline{\text{OEA}}$	$\overline{\text{OEB}}$	A	1B,2B
↑	1	1	Z	Z
↑	1	0	Z	Active
↑	0	1	Active	Z
↑	0	0	Active	Active

1 = HIGH voltage level. 0 = LOW voltage level. ↑ = LOW-to-HIGH clock transition.
X = Don"t Care Z = HIGH Impedance % = Two CLK edges needed to propagate data.
+ = Output level before the indicated steady-state input conditions were established.

Figure 11.38. The 74ALVCH162268 is a 12-bit to 24-bit registered bus exchanger with 3-state outputs. The device has application where data must be transferred from a narrow high-speed bus to a wide, lower frequency bus. The data exchange is synchronous between the two ports. Data is stored in the internal registers on the low-to-high transition of the clock when the appropriate CLKEN input is low. The select line is synchronous with the clock and selects 1B or 2B inputs for the A outputs. For data transfer in the A > B direction a two-stage pipeline is provided in the A > 1B path with a single storage register in the A > 2B path. Control of these inputs allows two sequential 12-bit words to be presented synchronously as a 24-bit word on the B port. Data flow is controlled by the active low output enables. The B outputs contain 26 ohm resistors to reduce over- and undershoot. /OE should be tied to Vcc through a pull-up to assure the high-impedance state during power-up or -down. Active bus-hold circuitry maintains unused or floating data at a valid logic level.

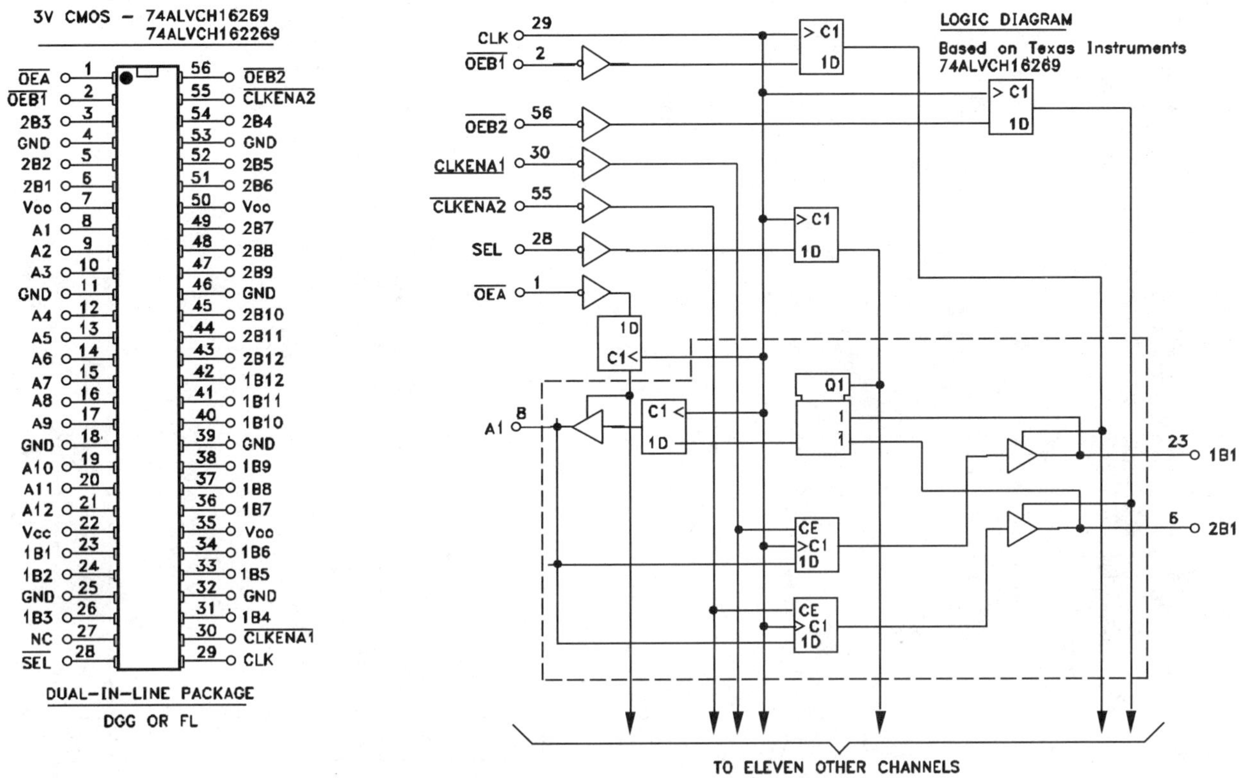

FUNCTION TABLE 74ALVCH16269 12-BIT TO 24-BIT REGISTERED BUS TRANSCEIVER WITH 3-STATE OUTPUTS

B-TO-A STORAGE ($\overline{OEA}$=L)

INPUTS				OUTPUT
CLK	$\overline{SEL}$	1B	2B	A
X	1	X	X	AO+
X	0	X	X	AO+
↑	1	0	X	0
↑	1	1	X	1
↑	0	X	0	0
↑	0	X	1	1

A-TO-B STORAGE ($\overline{OEB}$=L)

INPUTS				OUTPUT	
$\overline{CLKENA1}$	$\overline{CLKENA2}$	CLK	A	1B	2B
1	1	X	X	1BO+	2BO+
0	X	↑	0	0	X
0	X	↑	1	1	X
X	0	↑	0	X	0
X	0	↑	1	X	1

OUTPUT ENABLE

INPUTS			OUTPUTS	
CLK	$\overline{OEA}$	$\overline{OEB}$	A	1B,2B
↑	1	1	Z	Z
↑	1	0	Z	Active
↑	0	1	Active	Z
↑	0	0	Active	Active

1 = HIGH voltage level. 0 = LOW voltage level. ↑ = LOW-to-HIGH clock transition.
X = Don't Care Z = HIGH impedance
+ = Output level before the indicated steady-state input conditions were established.

Figure 11.39. The 74ALVCH16269 and 74ALVCH162269 are a 12-bit to 24-bit registered bus transceiver with 3-state outputs. The device has application where two separate ports must be multiplexed or demultiplexed from a single port. The data exchange is synchronous between the two ports. It is useful as an interface between synchronous DRAMS and high-speed microprocessors. Data is stored in the internal registers on the low-to-high transition of the clock when the appropriate CLKEN input is low. Control of these inputs allows two sequential 12-bit words to be presented synchronously as a 24-bit word on the B port. For data transfer in the B > A direction a single storage register is provided. The select line selects 1B or 2B data transfer in the B > A direction for the A outputs. The register on the A output permits the fastest possible data transfer. The control terminals are registered so that all transactions are synchronous with the clock. /OE should be tied to Vcc through a pull-up to assure the high-impedance state during power-up or -down. Active bus-hold circuitry maintains unused or floating data at a valid logic level. The '162269 outputs include 26 ohm resistors to reduce over- and undershoot.

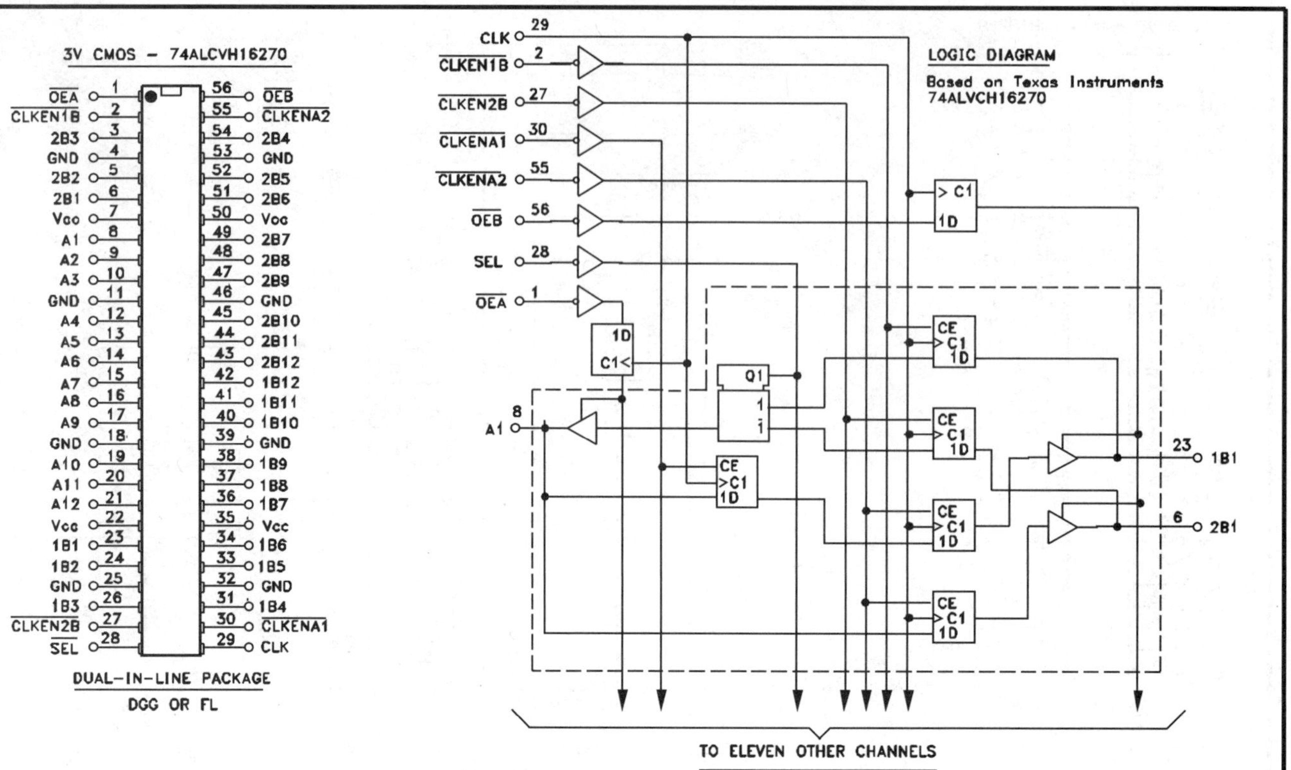

FUNCTION TABLE 74ALVCH16270 12-BIT TO 24-BIT REGISTERED BUS TRANSCEIVER WITH 3-STATE OUTPUTS

B-TO-A STORAGE (OEA=L)

INPUTS						OUTPUT
CLKEN1B	CLKEN2B	CLK	SEL	1B	2B	A
1	X	X	1	X	X	A0+
X	1	X	0	X	X	A0+
0	X	↑	1	0	X	0
0	X	↑	1	1	X	1
X	0	↑	0	X	0	0
X	0	↑	0	X	1	1

A-TO-B STORAGE (OEB=L)

INPUTS				OUTPUTS	
CLKENA1	CLKENA2	CLK	A	1B	2B
0	1	↑	0	0%	2B0+
0	1	↑	1	1%	2B0+
0	0	↑	0	0%	0
0	0	↑	1	1%	1
1	0	↑	0	1B0+	0
1	0	↑	1	1B0+	1
1	1	X	X	1B0+	2B0+

OUTPUT ENABLE

INPUTS			OUTPUTS	
CLK	OEA	OEB	A	1B,2B
↑	1	1	Z	Z
↑	1	0	Z	Active
↑	0	1	Active	Z
↑	0	0	Active	Active

1 = HIGH voltage level. 0 = LOW voltage level. ↑ = LOW-to-HIGH clock transition.
X = Don't Care Z = HIGH impedance % = Two CLK edges needed to propagate data.
+ = Output level before the indicated steady-state input conditions were established.

Figure 11.40. The 74ALVCH16270 is a 12-bit to 24-bit registered bus exchanger with 3-state outputs. The device has application where data must be transferred from a narrow high-speed bus to a wide, lower frequency bus. The data exchange is synchronous between the two ports. Data is stored in the internal registers on the low-to-high transition of the clock when the appropriate CLKEN input is low. The select line selects 1B or 2B inputs for the A outputs. For data transfer in the A > B direction a two-stage pipeline is provided in the A > 1B path with a single storage register in the A > 2B path. Control of these inputs allows two sequential 12-bit words to be presented synchronously as a 24-bit word on the B port. Data flow is controlled by the active low output enables. The control terminals are registered to synchronize the bus direction changes with the clock. /OE should be tied to Vcc through a pull-up to assure the high-impedance state during power-up or -down. Active bus-hold circuitry maintains unused or floating data at a valid logic level.

3V CMOS – 74ALVCH16409
74ALVCH162409

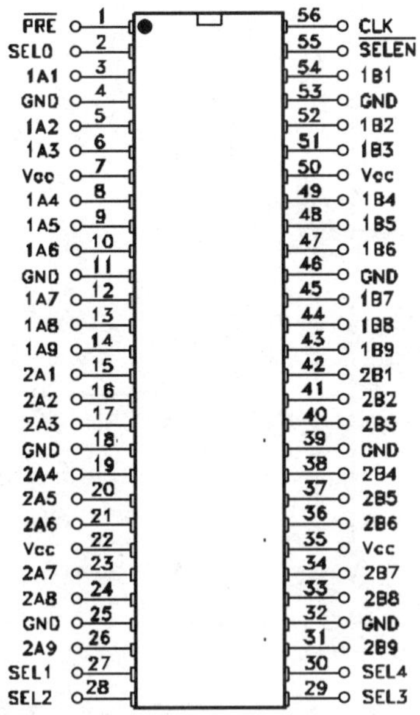

DUAL-IN-LINE PACKAGE
DGG OR DL

FUNCTION TABLE 74ALVCH16409 9-BIT, 4-PORT UNIVERSAL BUS EXCHANGER WITH 3-STATE OUTPUTS

DATA-FLOW CONTROL

INPUTS							
SELEN	CLK	SEL0	SEL1	SEL2	SEL3	SEL4	DATA FLOW
1	↑	X	X	X	X	X	No change
0	↑	0	0	0	0	0	None, all I/Os off
0	↑	0	0	0	0	1	Not used
0	↑	0	0	0	1	0	Not used
0	↑	0	0	0	1	1	Not used
0	↑	0	0	1	0	0	Not used
0	↑	0	0	1	0	1	Not used
0	↑	0	0	1	1	0	Not used
0	↑	0	0	1	1	1	Not used
0	↑	0	1	0	0	0	2A>1A and 1B>2b
0	↑	0	1	0	0	1	2A>1A
0	↑	0	1	0	1	0	2B>1B
0	↑	0	1	0	1	1	2A>1A and 2B>1B
0	↑	0	1	1	0	0	1A>2A and 1B>2B
0	↑	0	1	1	0	1	1A>2A
0	↑	0	1	1	1	0	1B>2B
0	↑	0	1	1	1	1	1A>2A and 2B>1B
0	↑	1	0	0	0	0	1A>1B and 2B>2A
0	↑	1	0	0	0	1	1A>1B
0	↑	1	0	0	1	0	2A>2B
0	↑	1	0	0	1	1	1A>1B and 2A>2B
0	↑	1	0	1	0	0	1B>1A and 2A>2B
0	↑	1	0	1	0	1	1B>1A
0	↑	1	0	1	1	0	2B>2A
0	↑	1	0	1	1	1	1B>1A and 2B>2A
0	↑	1	1	0	0	0	2B>1A and 2A>1B
0	↑	1	1	0	0	1	1B>2A
0	↑	1	1	0	1	0	2B>1A
0	↑	1	1	0	1	1	2B>1A and 1B>2A
0	↑	1	1	1	0	0	1A>2B and 1B>2A
0	↑	1	1	1	0	1	1A>2B
0	↑	1	1	1	1	0	2A>1B
0	↑	1	1	1	1	1	1A>2B and 2A>1B

1 = HIGH voltage level. 0 = LOW voltage level.
X = Don't Care
↑ = LOW-to-HIGH clock transition.

FUNCTION TABLE

INPUTS		OUTPUT
CLK	SEND PORT	RECEIVE PORT
X	X	BO+
X	0	0
X	1	1
↑	0	0
↑	1	1
1	X	BO+
0	X	BO+

BO+ = Output level before the indicated steady-state input conditions were established.

LOGIC DIAGRAM
Based on Texas Instruments 74ALVCH16409

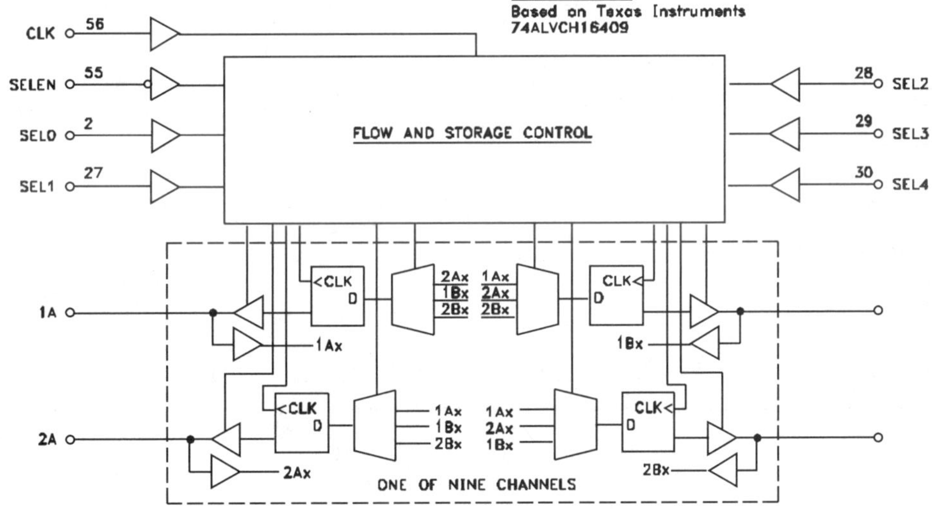

Figure 11.41. The 74ALVCH16409 and 74ALVCH162409 are 9-bit, 4-port universal bus exchangers with 3-state outputs. These allow synchronous data transfers between four different buses. Data flow is controlled by the select inputs, SEL0-SEL4. A data flow state is stored on the rising edge of the clock input if the select enable input is low. Once a data flow state has been established, data is stored in the flip-flop on the rising clock edge if /SELEN is high. The data flow control logic is designed for glitch-free transmission. To ensure the high-impedance state during power-up or -down, /SELEN should be tied to Vcc through a pull-up resistor. Active bus-hold circuitry holds unused or floating data inputs at a valid logic level. Maximum clock speed at 3.3V is 120 MHz. The 74ALVCH162409 B-output ports are provided with 26 ohm series resistors.

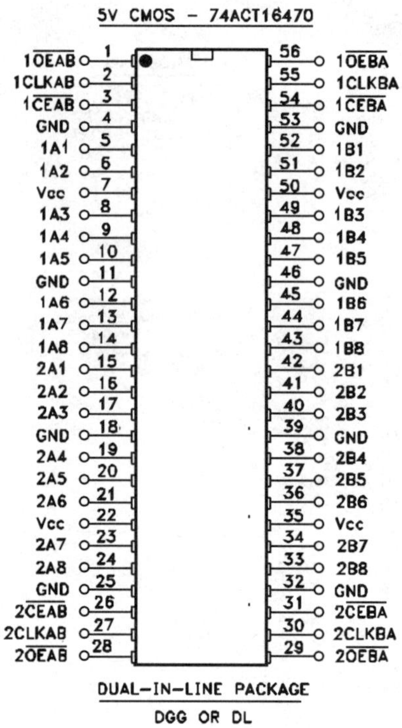

DEVICE DATA SHOWN IS BASED ON THE
Texas Instruments 74ACT16470

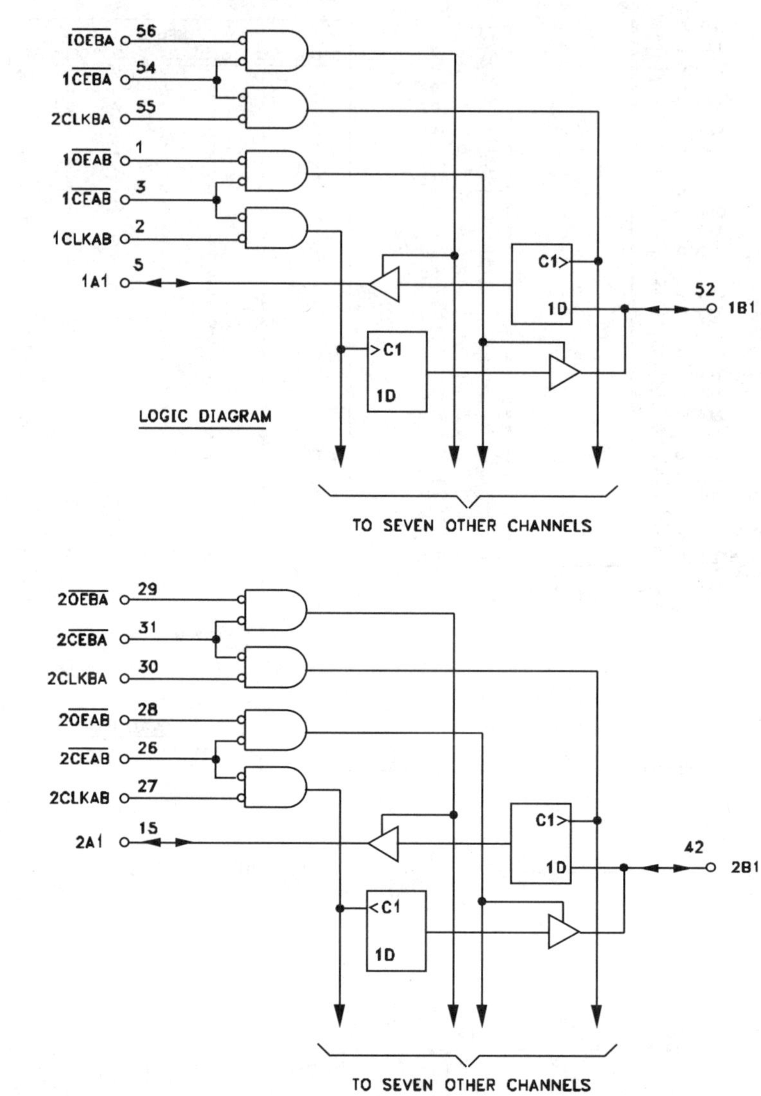

FUNCTION TABLE 74ACT16470%
16-BIT REGISTERED TRANSCEIVER
WITH 3-STATE OUTPUTS

INPUTS				OUTPUT
CEAB	CLKBA	OEAB	A	B
1	X	X	X	Z
X	X	1	X	Z
0	1	0	X	B0+
0	↑	0	0	0
0	↑	0	1	1

% = A-to-B data flow is shown; B-to-A flow control is the same except it uses CEBA, CLKBA, and OEBA.
+ = Output level before the indicated steady-state input conditins were established
1 = HIGH voltage level.
0 = LOW voltage level.
X = Don't Care
Z = High Impedance state
↑ = LOW-to-HIGH clock transition.

Figure 11.42. The 74ACT16470 is a 16-bit registered transceiver with 3-state outputs. These contain two sets of D-type flip-flops for temporary storage of data flowing in either direction. They may be used as two 8-bit transceivers or a single 16-bit device. Separate clocks and output enable inputs are provided for each register to permit independent control in either direction of data flow. The A-to-B enable, /CEAB, must be low to enter data from A or to output data to B. If both /CEAB and CLKAB are low, B port will have the level of A port prior to the most recent low-to-high transition of CLKAB. Data flow from B to A is similar, but requires the use of /CEBA, CLKBA, and /OEBA inputs. To avoid false clocking of the flip-flops, /CE should not be switched from high-to-low while CLK is high. Distributed power pins minimizes high-speed switching noise. Flow-through architecture optimizes PCB layout.

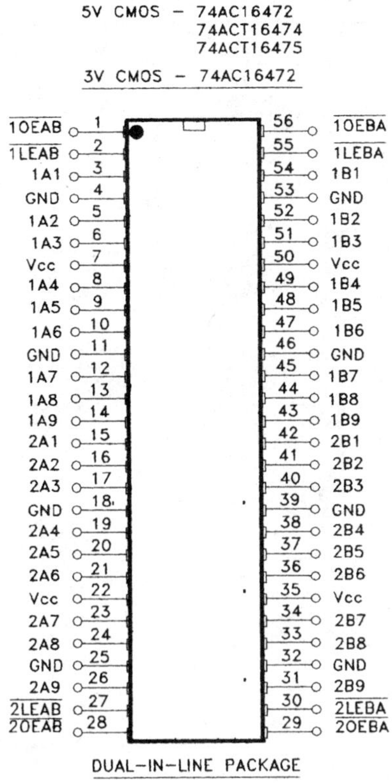

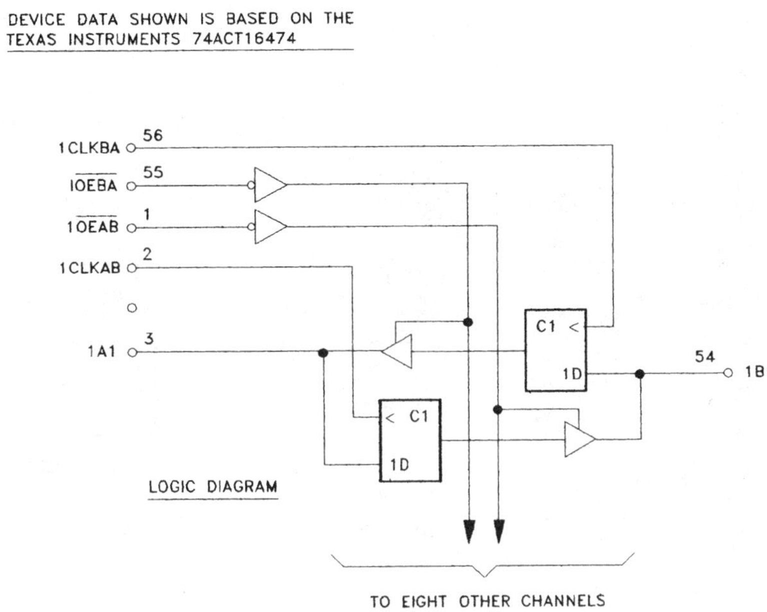

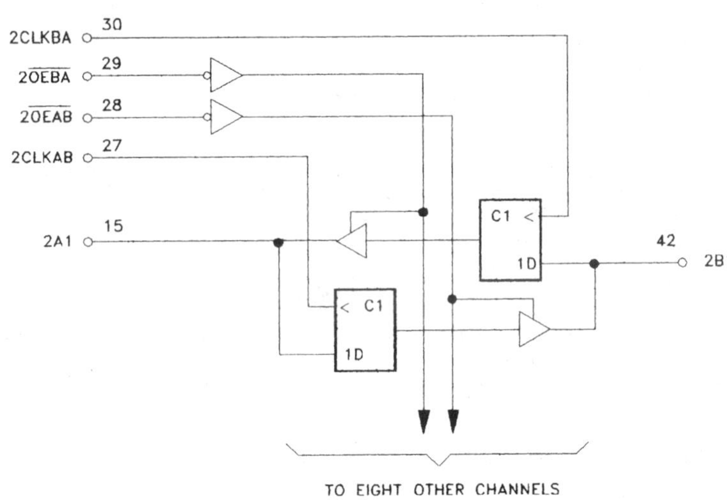

FUNCTION TABLE 16AC16472,'ACT16474, 'ACT16475 18-BIT REGISTERED BUS TRANSCEIVERS WITH 3-STATE OUTPUTS %

INPUTS			OUTPUT
CLKAB	OEAB	A	B
X	1	X	Z
0	0	X	B0+
↑	0	1	1
↑	0	0	0

% = A-to-B data flow shown: B-to-A flow similar, but uses CLKBA and OEBA.
+ = Output level before the indicated steady-state input conditions were established.
↑ = LOW-to-HIGH clock transition.
1 = HIGH voltage level.
0 = LOW voltage level.
X = Don't Care
Z = High Impedance state

Figure 11.43. The 74AC16472, 'ACT16474 and 'ACT16475 are 18-bit registered transceivers with 3-state outputs. These contain two sets of D-type latches for temporary storage of data flowing in either direction. They may be used as two 9-bit transceivers or a single 18-bit device. Separate latch and output enable inputs are provided for each register to permit independent control in either direction of data flow. When /OEAB and /LEAB are low, the A > B latches are transparent; a subsequent low-to-high transition of /LEAB puts the A latches in the storage mode. With /OEAB low, the B outputs are active and reflect the data present at the output of the A latches. Data flow from B to A is similar but requires the use of the /LEBA nd /OEBA enables. Distributed power pins minimizes high-speed switching noise. Flow-through architecture optimizes PCB layout.

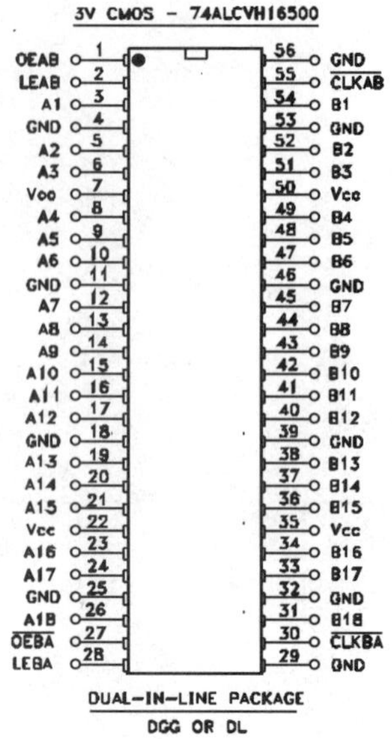

FUNCTION TABLE 74ALVCH16500
18-BIT UNIVERSAL BUS TRANSCEIVER
WITH 3-STATE OUTPUTS%

INPUTS				OUTPUT
OEAB	LEAB	/CLKAB	A	B
0	X	X	X	Z
1	1	X	0	0
1	1	X	1	1
1	0	↓	0	0
1	0	↓	1	1
1	0	1	X	B0+
1	0	0	X	B0*

% = A-to-B data flow shown: B-to-A flow similar, but uses /OEBA, LEBA, and /CLKBA
+ = Output level before the indicated steady-state input conditions were established.
* = Output level before the indicated steady-state input conditions were established, provided that /CLKAB was low before LEAB went low.
↓ = HIGH-to-LOW clock transition.
1 = HIGH voltage level.
0 = LOW voltage level.
X = Don't Care
Z = High Impedance state

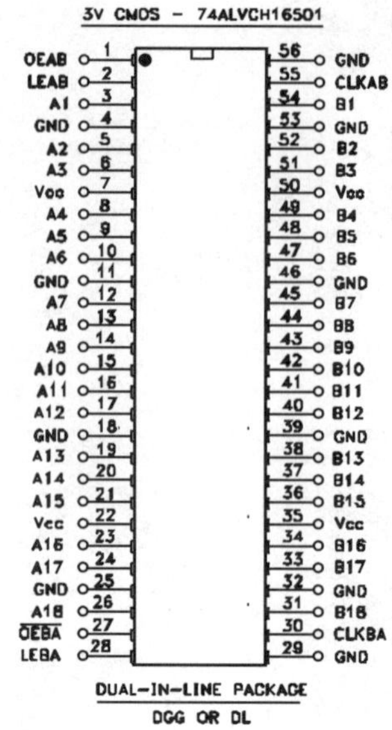

FUNCTION TABLE 74ALVCH16501
18-BIT UNIVERSAL BUS TRANSCEIVER
WITH 3-STATE OUTPUTS%

INPUTS				OUTPUT
OEAB	LEAB	CLKAB	A	B
0	X	X	X	Z
1	1	X	0	0
1	1	X	1	1
1	0	↑	0	0
1	0	↑	1	1
1	0	1	X	B0+
1	0	0	X	B0*

% = A-to-B data flow shown: B-to-A flow similar, but uses /OEBA, LEBA, and CLKBA
+ = Output level before the indicated steady-state input conditions were established.
* = Output level before the indicated steady-state input conditions were established, provided that CLKAB was low before LEAB went low.
↑ = LOW-to-HIGH clock transition.
1 = HIGH voltage level.
0 = LOW voltage level.
X = Don't Care
Z = High Impedance state

DEVICE DATA SHOWN BASED ON
TEXAS INSTRUMENTS 74ALVCH16500

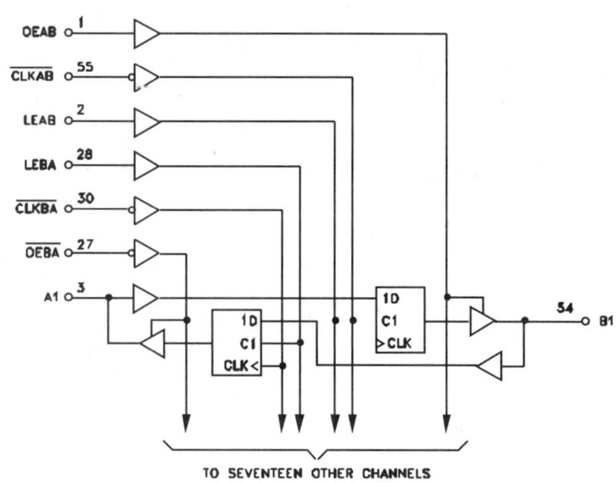

DEVICE DATA SHOWN BASED ON
TEXAS INSTRUMENTS 74ALVCH16501

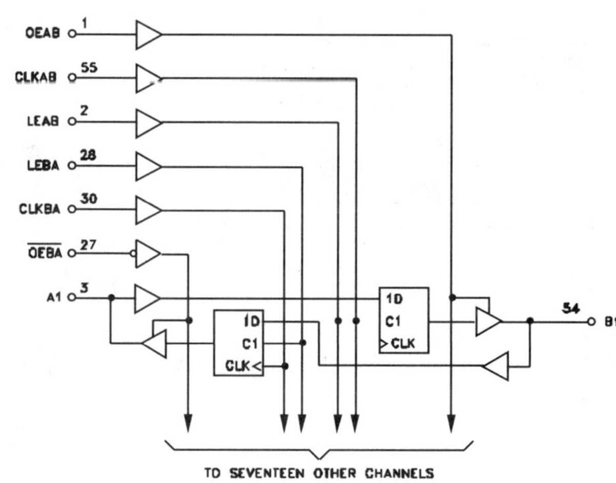

Figure 11.44. The 74ALVCH16500 and 74ALVCH16501 are 18-bit universal bus transceivers with 3-state outputs. These combine D-type latches and D-type flip-flops for operation in transparent, latched, or clock mode. Dataflow in either direction is controlled by output-enable (OEAB and /OEBA), latch-enable (LEAB and /LEBA), and clock (/CLKAB, /CLKBA, '16500; CLKAB, CLKBA, '16501). For A > B data flow, the device operates in the transparent mode when LEAB is high. When LEAB is low, the A data is latched if the /CLKAB, or CLKAB, is held at a high or low logic level. If LEAB is low, the A-bus data is stored in the latch/flip-flop on the clock transition. When OEAB is high, the B-port outputs are active. Data flow from B to A is similar but requires the use of the LEBA, /OEBA and clock BA functions. Distributed power pins minimizes high-speed switching noise. Bus hold on data inputs eliminates the need for external pull-up/pull-down resistors. Maximum clock frequency is 150 MHz.

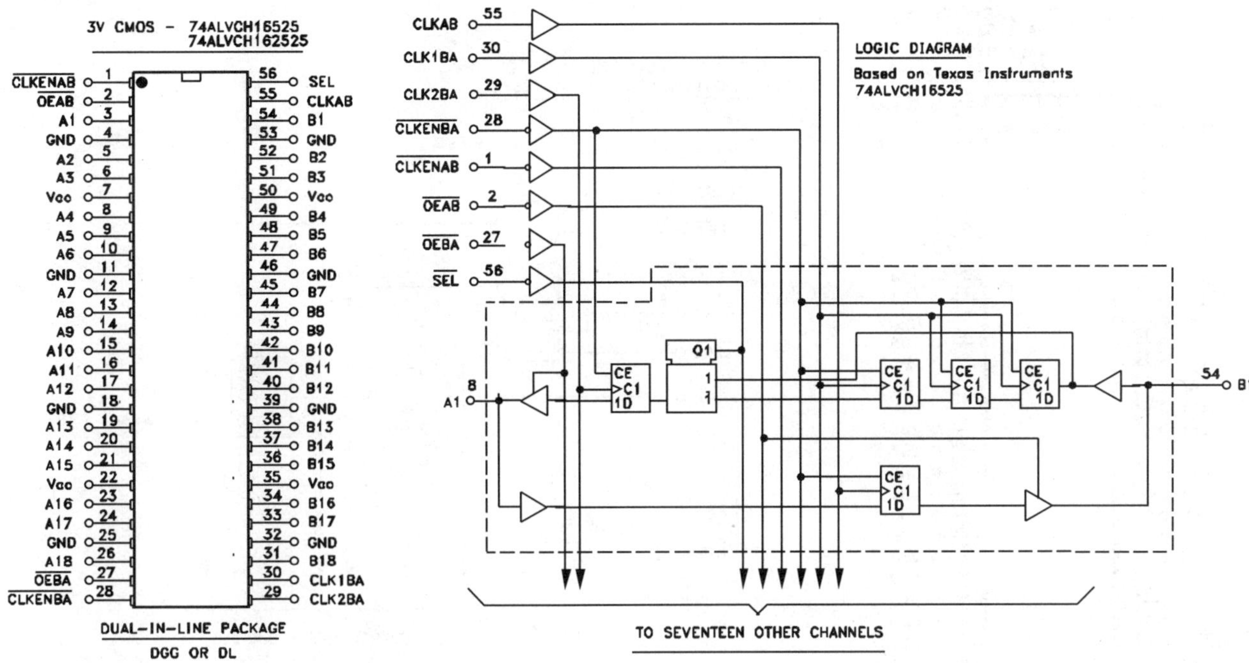

FUNCTION TABLE 74ALVCH16525, 'ALVCH162525 18-BIT REGISTERED BUS TRANSCEIVER WITH 3-STATE OUTPUTS

B-TO-A STORAGE (OEBA=L)

INPUTS					OUTPUT
CLKENBA	CLK2BA	CLK1BA	SEL	B	A
1	X	X	X	X	AO+
0	↑	X	1	0	0
0	↑	X	1	1	1
0	↑	↑	0	0	0%
0	↑	↑	0	1	1%

A-TO-B STORAGE (OEAB=L)

INPUTS			OUTPUT
CLKENAB	CLKAB	A	B
1	X	X	BO+
0	↑	0	0
0	↑	1	1

1 = HIGH voltage level. 0 = LOW voltage level. ↑ = LOW-to-HIGH clock transition.
X = Don't Care Z = HIGH Impedance
% = Three CLK1BA edges and one CLK2BA edge are needed to propagate data from B-to-A when SEL is low
+ = Output level before the indicated steady-state input conditions were established.

Figure 11.45. The 74ALVCH16525 and 74ALVCH162525 are 18-bit registered bus transceivers with 3-state outputs. Dataflow in either direction is controlled by output (OEAB and /OEBA) and clock (/CLKENAB, /CLKENBA) enables. For A > B data flow, the data flows through a single register. The B > A data can flow through a four-stage pipeline register or a single register path depending on the state of /SEL. Data is stored in the internal registers on the low-to-high transition of the CLK input, provided that the appropriate CLKEN inputs are low. The A > B data transfer is synchronized to the CLKAB input; the B > A is synchronized with the CLK1BA and CLK2BA inputs. Distributed power pins minimizes high-speed switching noise. Bus hold on data inputs eliminates the need for external pull-up/pull-down resistors. The '162525 B-port outputs have equivalent 26-ohm resistors so no external resistors are required. Maximum clock frequency is 150 MHz.

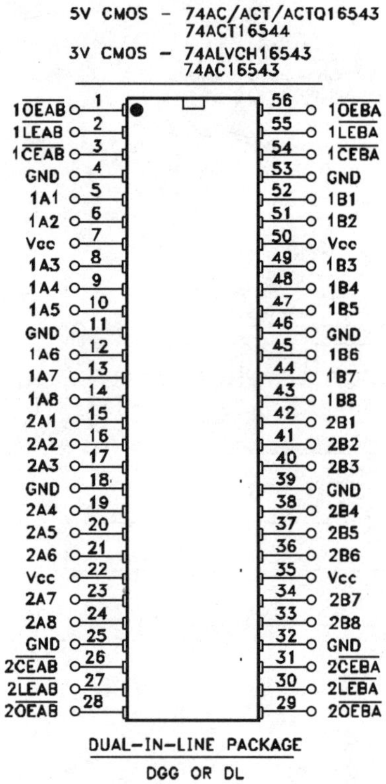

FUNCTION TABLE 74ALVCH16543%
16-BIT REGISTERED TRANSCEIVER
WITH 3-STATE OUTPUTS

INPUTS				OUTPUT
CEAB	LEAB	OEAB	A	B
1	X	X	X	Z
X	X	1	X	Z
0	1	0	X	B0+
0	0	0	0	0
0	0	0	1	1

% = A-to-B data flow is shown; B-to-A flow control is the same except it uses CEBA, LEBA and OEBA

+ = Output level before the indicated steady-state input conditins were established

1 = HIGH voltage level.
0 = LOW voltage level.
X = Don't Care
Z = High Impedance state

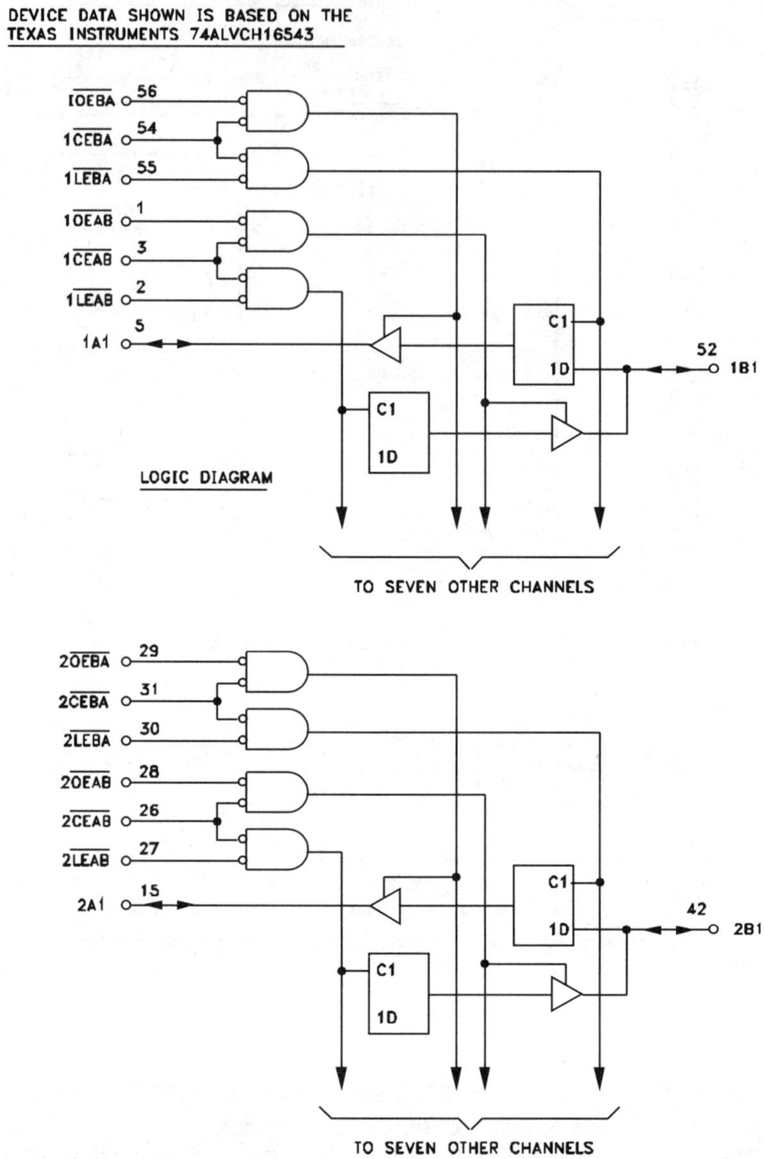

Figure 11.46. The 74AC/ACT16543, 74ACT16544 and 74ALVCH16543 are 16-bit registered bus transceivers with 3-state outputs. These contain two sets of D-type latches for temporary data storage flowing in either direction. They can be used as two 8-bit or a single 16-bit device. Separate latch (/LEAB, /LEBA) and output (/OEAB, /OEBA) enables provided for each register permit independent control in either direction of data flow. The A > B enable /CEAB must be low to enter data from A or to output data to B. Having /CEAB and /LEAB low makes the A-to-B latches transparent; a subsequent low-to-high transition at /LEAB puts the latches in the storage mode. Data flow from B-to-A is similar but requires using /CEBA, /LEBA, and /OEBA inputs. Distributed power pins minimizes high-speed switching noise.

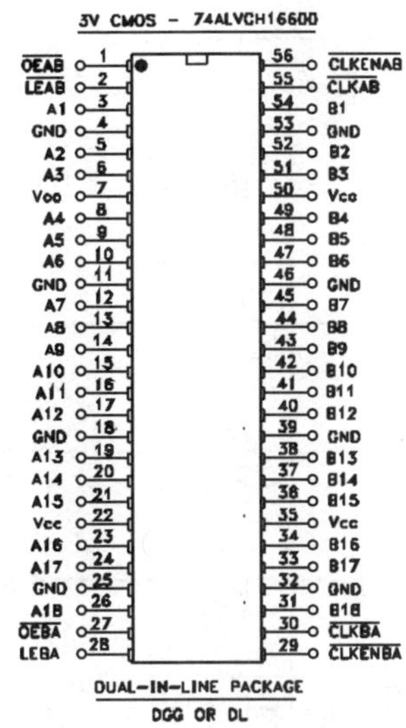

FUNCTION TABLE 74ALVCH16600
18-BIT UNIVERSAL BUS TRANSCEIVER
WITH 3-STATE OUTPUTS%

INPUTS					OUTPUT
/CLKENAB	/OEAB	LEAB	/CLKAB	A	B
X	1	X	X	X	Z
X	0	1	X	0	0
X	0	1	X	1	1
1	0	0	1	X	B0+
1	0	0	0	X	B0+
0	0	0	↓	0	0
0	0	0	↓	1	1
0	0	0	1	X	B0+
0	0	0	0	X	B0*

% = A-to-B data flow shown: B-to-A flow similar, but uses /OEBA, LEBA, /CLKBA and /CLKENBA
\+ = Output level before the indicated steady-state input conditions were established.
* = Output level before the indicated steady-state input conditions were established, provided that /CLKAB was low before LEAB went low.
↓ = HIGH-to-LOW clock transition.
1 = HIGH voltage level.
0 = LOW voltage level.
X = Don't Care
Z = High Impedance state

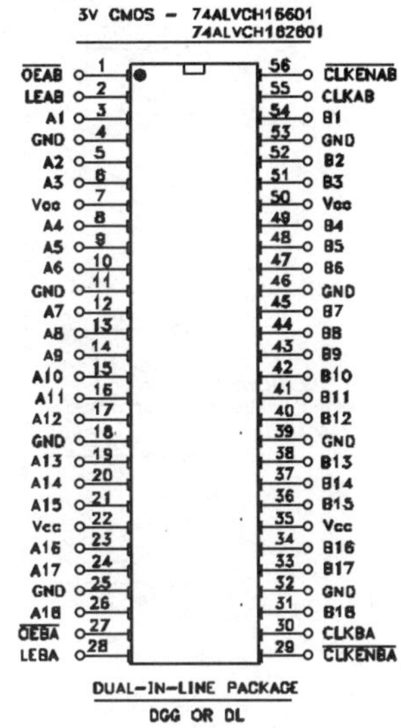

FUNCTION TABLE 74ALVCH16601, 74ALVCH162601
18-BIT UNIVERSAL BUS TRANSCEIVER
WITH 3-STATE OUTPUTS%

INPUTS					OUTPUT
/CLKENAB	/OEAB	LEAB	CLKAB	A	B
X	1	X	X	X	Z
X	0	1	X	0	0
X	0	1	X	1	1
1	0	0	1	X	B0+
1	0	0	0	X	B0+
0	0	0	↑	0	0
0	0	0	↑	1	1
0	0	0	1	X	B0+
0	0	0	0	X	B0*

% = A-to-B data flow shown: B-to-A flow similar, but uses /OEBA, LEBA, /CLKBA and /CLKENBA
\+ = Output level before the indicated steady-state input conditions were established.
* = Output level before the indicated steady-state input conditions were established, provided that CLKAB was low before LEAB went low.
↑ = LOW-to-HIGH clock transition.
1 = HIGH voltage level.
0 = LOW voltage level.
X = Don't Care
Z = High Impedance state

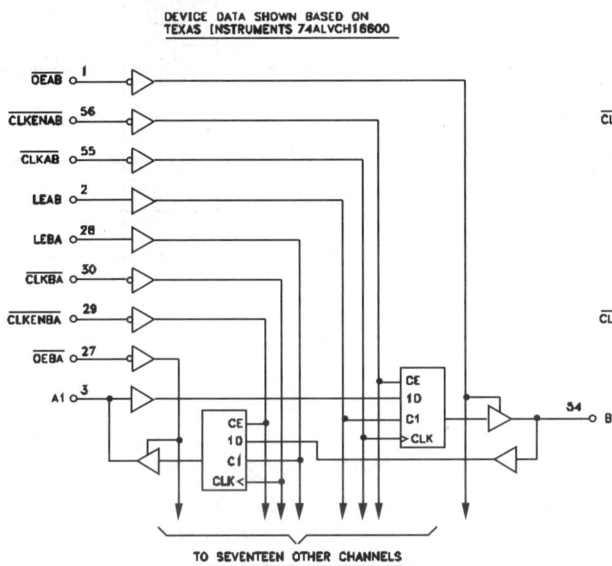

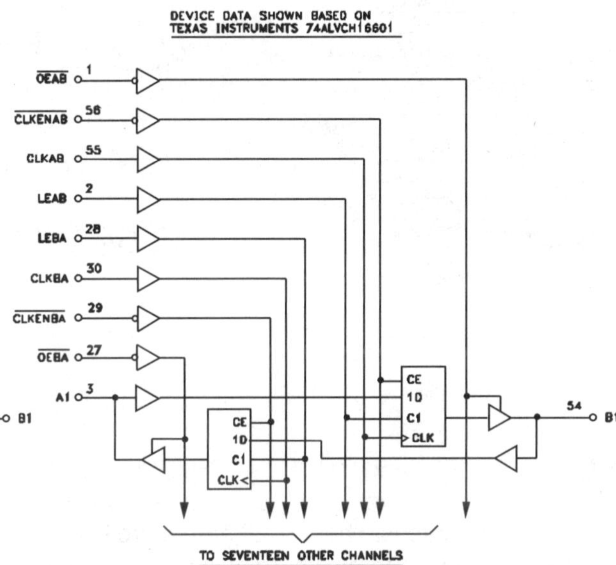

Figure 11.47. The 74ALVCH16600, 74ALVCH16601 and 74ALVCH162601 are 18-bit universal bus transceivers with 3-state outputs. These combine D-type latches and D-type flip-flops for operation in transparent, latched, or clock mode. Dataflow in either direction is controlled by output-enable (OEAB and /OEBA), latch-enable (LEAB, LEBA), and clock (/CLKAB, /CLKBA) input. Clock control is by the /CLKAB,/CLKBA, '16600; CLKAB, CLKBA, '16606,'162601. For A > B data flow, the device operates in the transparent mode when LEAB is high. When LEAB is low, the A data is latched if the /CLKAB is held at a high or low logic level. If LEAB is low, the A-bus data is stored in the latch/flip-flop on the clock transition. When /OEAB is low, the outputs are active. Data flow from B to A is similar but requires the use of the LEBA, /OEBA and clock BA functions. Distributed power pins minimizes high-speed switching noise. Bus hold on data inputs eliminates the need for external pull-up/pull-down resistors. The '162601 outputs have equivalent 26-ohm resistors so no external resistors are required. Maximum clock frequency is 150 MHz.

5V CMOS – 74AC/ACT16620
74ACT16623
3V CMOS – 74AC16620
74LVC/LVCH16623
74LVC/LVCH162623

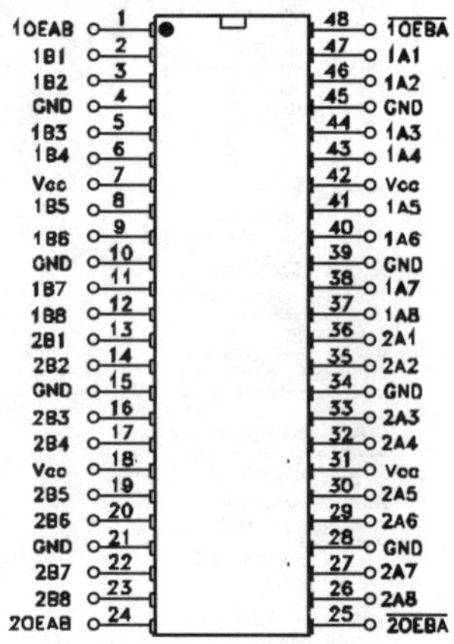

DUAL-IN-LINE PACKAGE
DL

FUNCTION TABLE 74x16620,'x16623,'x162623
16-BIT BUS TRANSCEIVERS
WITH 3-STATE OUTPUTS

INPUTS		OPERATING MODES	
$\overline{OEBA}$	OEAB	'x16620	'x16623,'x162623
0	0	$\overline{B}$ Data to A Bus	B Data to A Bus
0 0	1 1	$\overline{B}$ Data to A Bus $\overline{A}$ Data to B Bus	B Data to A Bus A Data to B Bus
1	0	Z	Z
1	1	$\overline{A}$ Data to B Bus	A Data to B Bus

1 = HIGH voltage level.
0 = LOW voltage level.
X = Don't Care
Z = High Impedance state

LOGIC DIAGRAM
Based on Texas Instruments
74AC16620

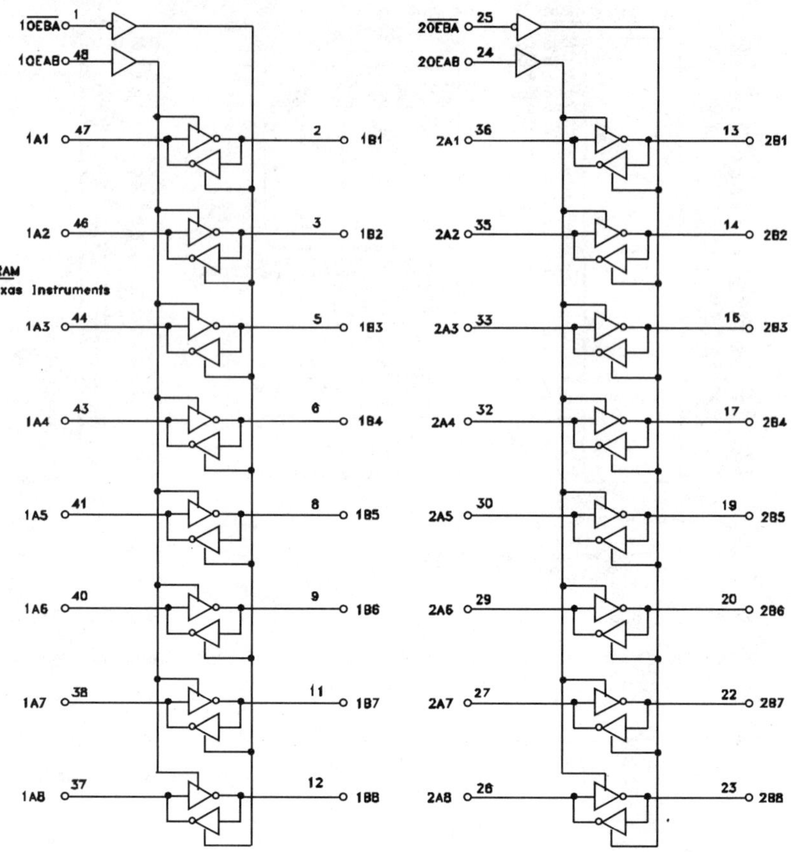

Figure 11.48. The TTL and CMOS 74AC/ACT16620, 74ACT16623, 74LVC/LVCH16623 and 74ILVC/LVCH162623 are 16-bit bus transceivers with dual enables and 3-state outputs designed for asynchronous communication between data buses. The '16620 series are inverting; the '16623 and '162623 noninverting. The control function allows for maximum flexibility in timing. These may be used as two 8-bit or a single 16-bit transceiver. The direction of data transmission between buses is determined by the logic levels at the complementary output-enables, as shown in the function table. Distributed power pins minimizes high-speed switching noise. Flow-through architecture optimizes PCB layout. The '162623 outputs have integrated 30-ohm resistors so no external resistors are required.

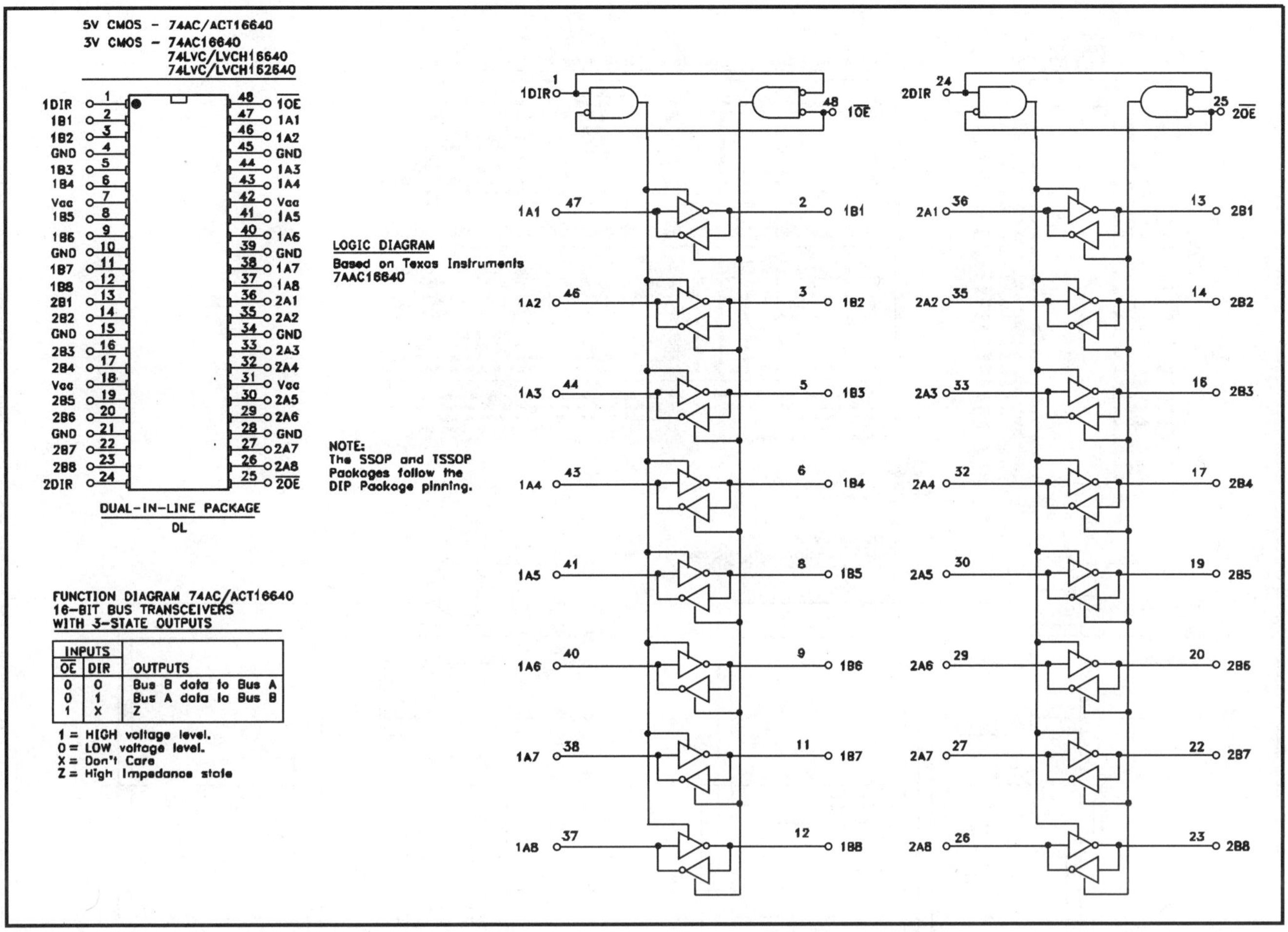

Figure 11.49. The CMOS 74AC/ACT16640 are 16-bit inverting bus transceivers with 3-state outputs designed for asynchronous communication between data buses. These may be used as two 8-bit or a single 16-bit transceiver. The direction of data transmission between buses is determined by the logic levels at the direction control inputs, 1DIR, 2DIR, as shown in the function table. Distributed power pins minimize high-speed switching noise. Flow-through architecture optimizes PCB layout.

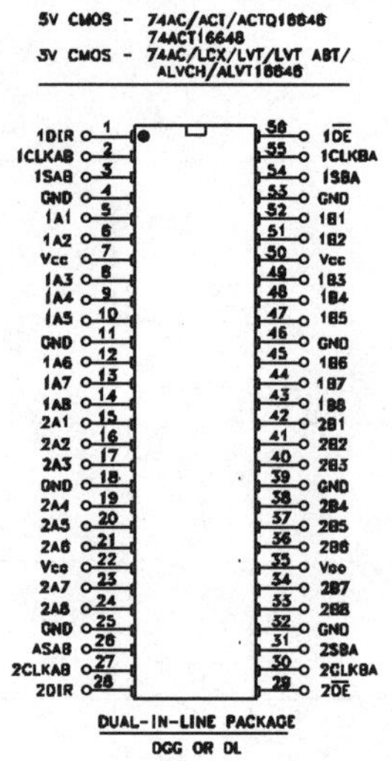

FUNCTION TABLE 74x16646 16-BIT BUS TRANSCEIVERS AND REGISTERS W/3-STATE OUTPUTS

INPUTS						DATA I/O		
$\overline{OE}$	DIR	CLKAB	CLKBA	SAB	SBA	A1–A8	B1–B8	OPERATION OR FUNCTION
X	X	↑	X	X	X	Input	Unspecified	Store A, B unspecified+
X	X	X	↑	X	X	Unspecified	Input	Store B, A unspecified+
1	X	↑	↑	X	X	Input	Input	Store A and B Data
1	X	1 or 0	1 or 0	X	X	Input	Input	Isolation, hold storage
0	1	X	X	X	0	Output	Input	Real-time B Data to A Bus
0	1	X	1 or 0	X	1	Output	Input	Stored B Data to A Bus
0	0	X	X	0	X	Input	Output	Real-time A Data to B Bus
0	0	1 or 0	X	1	X	Input	Output	Stored A Data to Bus

FUNCTION TABLE 74ACT16648 16-BIT BUS TRANSCEIVER AND REGISTER W/3-STATE OUTPUTS

INPUTS						DATA I/O		
$\overline{OE}$	DIR	CLKAB	CLKBA	SAB	SBA	A1–A8	B1–B8	OPERATION OR FUNCTION
X	X	↑	X	X	X	Input	Unspecified	Store A, B unspecified+
X	X	X	↑	X	X	Unspecified	Input	Store B, A unspecified+
1	X	↑	↑	X	X	Input	Input	Store A and B Data
1	X	0	0	X	X	Input Disabled	Input Disabled	Isolation, hold storage
0	0	X	X	X	0	Output	Input	Real-time $\bar{B}$ Data to A Bus
0	0	X	0	X	1	Output	Input	Stored $\bar{B}$ Data to A Bus
0	1	X	X	0	X	Input	Output	Real-time $\bar{A}$ Data to B Bus
0	1	0	X	1	X	Input	Output	Stored $\bar{A}$ Data to Bus

↑ = LOW-to-HIGH clock transition.
1 = HIGH voltage level.
0 = LOW voltage level.
X = Don't Care
Z = High Impedance state

+ = The data-output functions may be enabled or disbled by various signals at OE or DIR. Data input functions are always enabled, i.e., data at the bus terminals is stored on every low-to-high transition of the clock inputs.

NOTE:
The SSOP and TSSOP Packages follow the DIP Package pinning.

LOGIC DIAGRAM
Based on National/Fairchild 74AC16646 and Texas Instruments 74AC16646

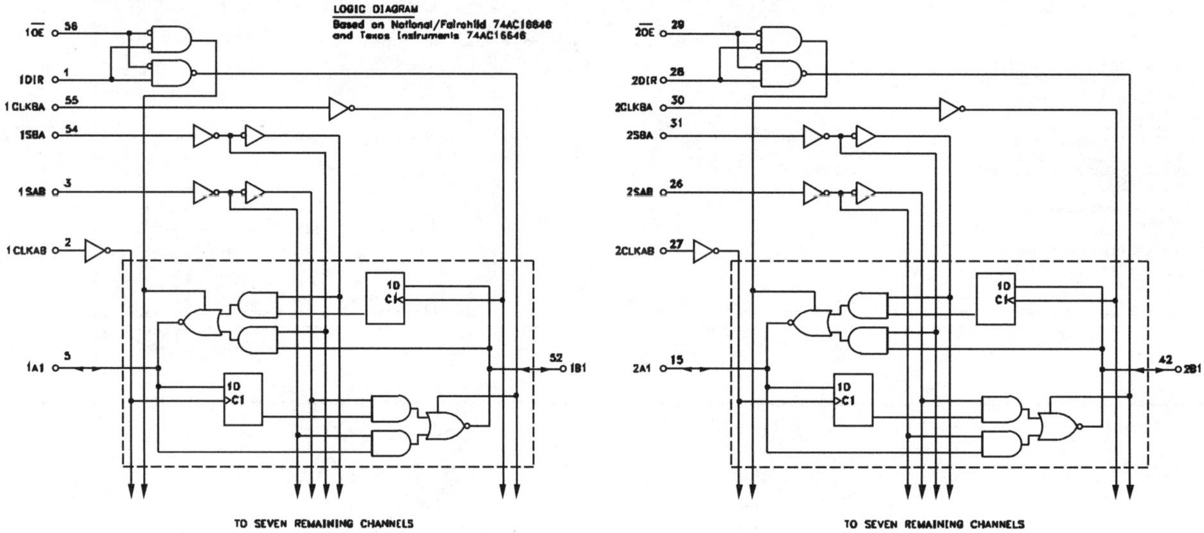

Figure 11.50. The CMOS 74x16646 and 74ACT16648 are 16-bit bus transceivers and registers with 3-state outputs configured for multiplexed transmission of data directly from the data bus or from the storage registers. These may be used as two 8-bit or a single 16-bit transceiver. Data on the A or B bus is clocked into the registers on the low-to-high transition of the appropriate clock. Four fundamental bus-management functions can be performed with the bus transceivers and registers using the output-enable, /OE and the direction-control, DIR, inputs. In the transceiver mode, data present at the high-impedance port may be stored in either register or both. The select controls can multiplex stored and real-time (transparent mode) data. With /OE input high, A data may be stored in one register and/or B data may be stored in the other register. Distributed power pins minimize high-speed switching noise. Flow-through architecture optimizes PCB layout.

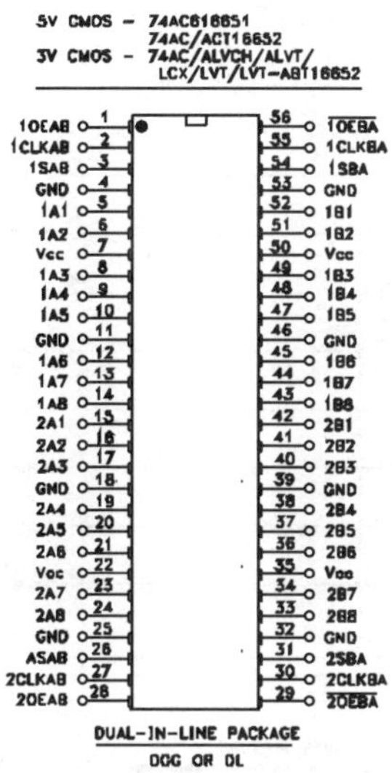

FUNCTION TABLE 74x16651,'x16652 16-BIT TRANSCEIVERS AND REGISTERS WITH 3-STATE OUTPUTS

INPUTS						DATA I/O%		OPERATION OR FUNCTION
OEAB	$\overline{OEBA}$	CLKAB	CLKBA	SAB	SBA	A1–A8	B1–B8	
0	1	0	0	X	X	Input	Input	Isolation
0	1	↑	↑	X	X	Input	Input	Store A and B Data
X	1	↑	0	X	X	Input	Unspecified	Store A, Hold B
1	1	↑	↑	X*	X	Input	Output	Store A in both registers
0	X	0	↑	X	X	Unspecified	Input	Hold A, Store B
0	0	↑	↑	X	X*	Output	Input	Store B in both registers
0	0	X	X	X	0	Output	Input	Real-time B Data to A Bus
0	0	X	0	X	1	Output	Output	Stored B Data to A Bus
1	1	X	X	0	X	Input	Output	Real-time A Data to B Bus
1	1	0	X	1	X	Input	Output	Stored A Data to B Bus
1	0	0	0	1	1	Output	Output	Stored A Data to B Bus and Stored B Data to A Bus

↑ = LOW-to-HIGH clock transition.
1 = HIGH voltage level.
0 = LOW voltage level.
X = Don't Care
Z = High Impedance state

* Select Control = Low; clocks can occur simultaneously
Select Control = High; clocks must be staggered to load both registers.

% = The data output functions may be enabled or disabled by a variety of level combinations of OEAB or OEBA. Data input functions are always enabled; i.e., data at the bus terminals is stored on every low-to-high transition of the clock inputs.

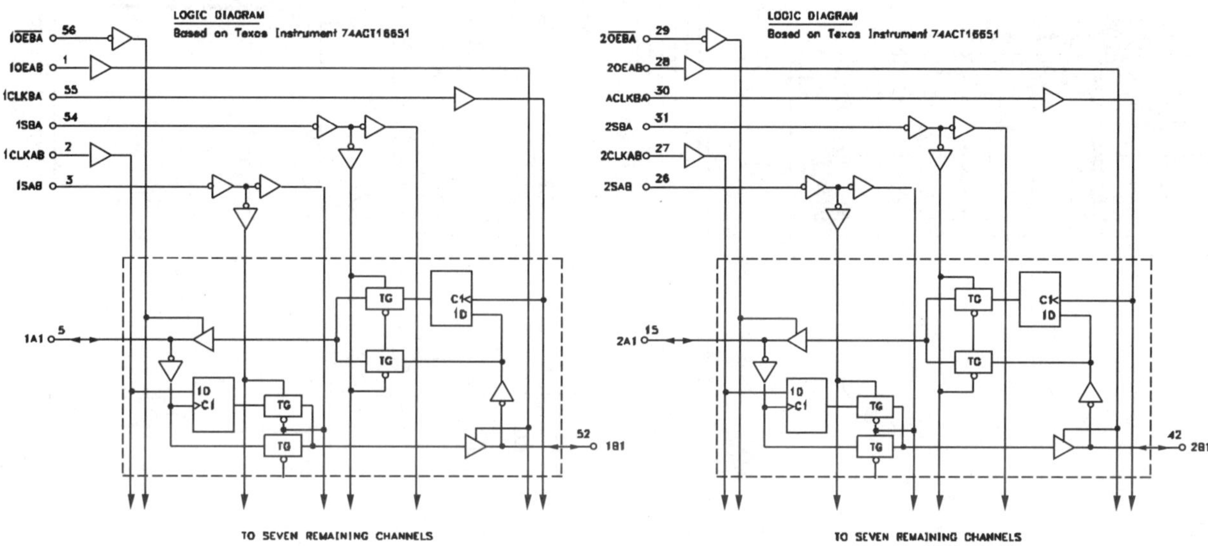

Figure 11.51. The CMOS 74AC16651 and 74x16652 are 16-bit transceivers and registers with 3-state outputs consisting of D-type flip-flops and control circuitry for multiplexed transmission of data directly from the data bus or from the storage registers. These may be used as two 8-bit or a single 16-bit transceiver. Complementary output-enable inputs, OEAB, /OEBA control the transceiver functions. Select controls, SAB, SBA, determine whether real-time or stored data is transferred. Four fundamental bus-management functions can be performed with the bus transceivers and registers. Distributed power pins minimize high-speed switching noise. Flow-through architecture optimizes PCB layout.

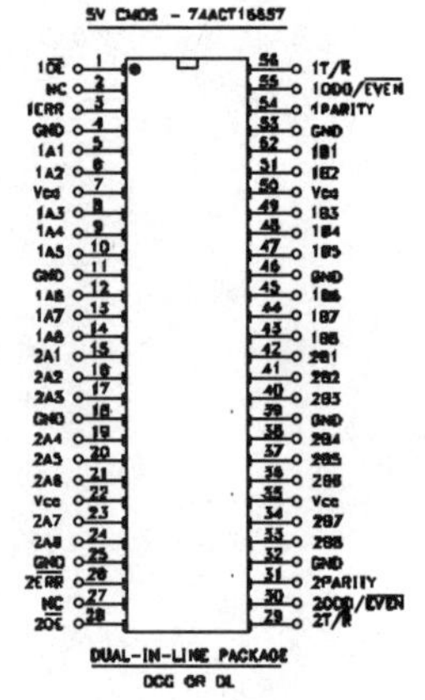

FUNCTION TABLE 74ACT16657 16-BIT TRANSCEIVER WITHPARITY GENERATOR/ CHECKER AND 3-STATE OUTPUT

NUMBER OF A or B INPUTS HIGH	INPUTS			INPUT/OUTPUT	OUTPUTS	
	$\overline{OE}$	T/$\overline{R}$	ODD/$\overline{EVEN}$	PARITY	$\overline{ERR}$	OUTPUTS MODE
0,2,4,6,8	0	1	1	1	Z	Transmit
	0	1	0	0	Z	Transmit
	0	0	1	1	1	Receive
	0	0	1	0	0	Receive
	0	0	0	1	0	Receive
	0	0	0	0	1	Receive
1,3,5,7.	0	1	1	0	Z	Transmit
	0	1	0	1	Z	Transmit
	0	0	1	1	0	Receive
	0	0	1	0	1	Receive
	0	0	0	1	1	Receive
	0	0	0	0	0	Receive
	1	X	Z	Z	Z	Z

1 = HIGH voltage level.
0 = LOW voltage level.
X = Don't Care
Z = High Impedance state

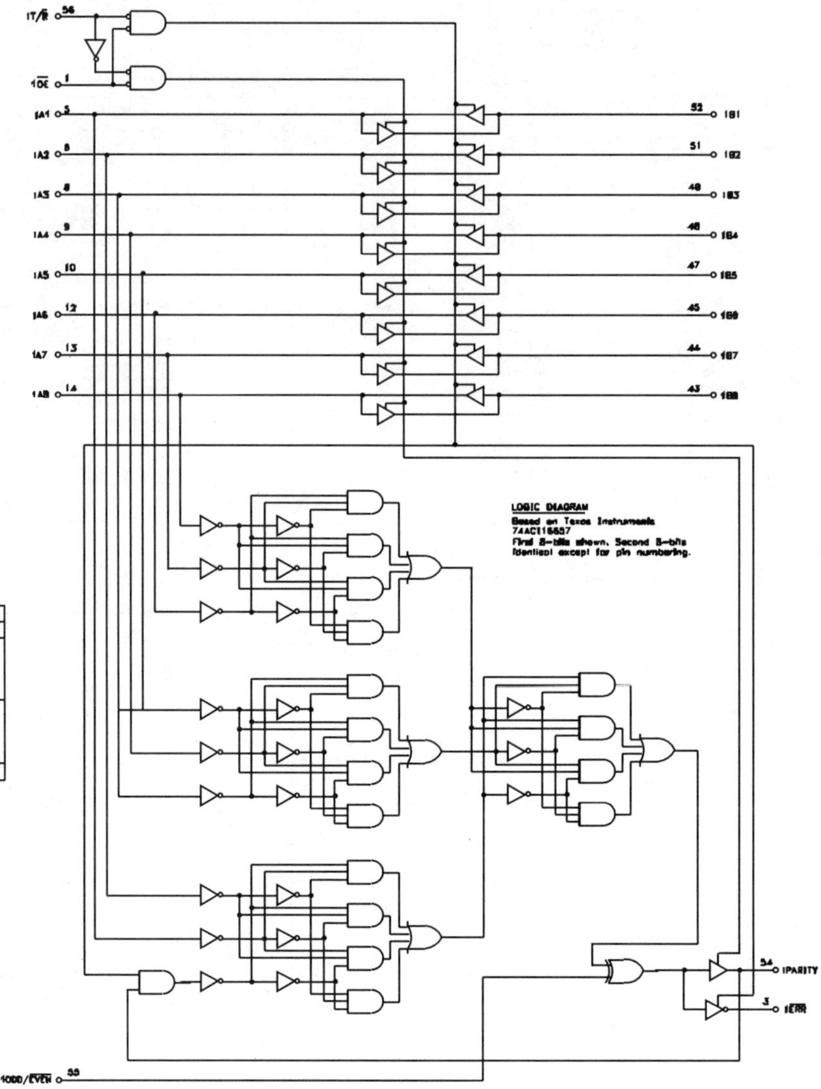

Figure 11.52. The CMOS 74ACT16657 is a 16-bit transceiver (two non-inverting octal transceiver sections), with parity generator/checkers and 3-state outputs. For either section the Transmit//Receive, T//R, input determines the direction of data flow. With either, high data flows from its respective A port to the matching B port. When low, the data flow is opposite in direction. When the output enables are high, both ports are in a high impedance state. Odd or even parity is selected by a logic high or low level, respectively, on the ODD//EVEN input. PARITY carries the respective parity bit value. In the transmit mode, it is an output from the generator/checker; an input in the receive mode. Distributed power pins minimize high-speed switching noise. Flow-through architecture optimizes PCB layout.

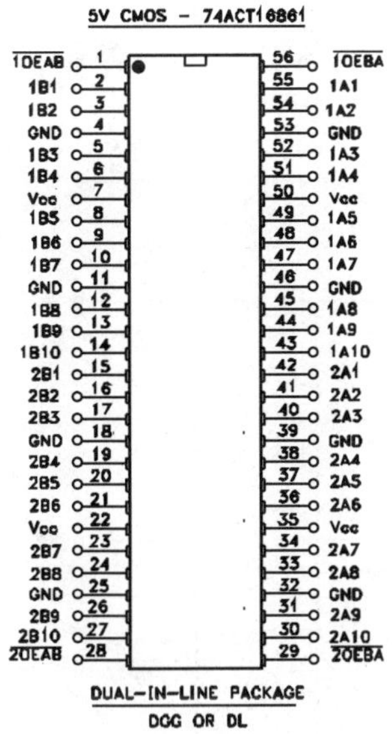

FUNCTION TABLE 74ACT16861 20-BIT BUS TRANSCEIVER WITH 3-STATE OUTPUTS

INPUTS		
OEAB	OEBA	OPERATING MODES
0	0	LATCH A and B (A=B)
0	1	A Data to B Bus
1	0	B Data to A Bus
1	1	Z

1 = HIGH voltage level.
0 = LOW voltage level.
X = Don't Care
Z = High Impedance state

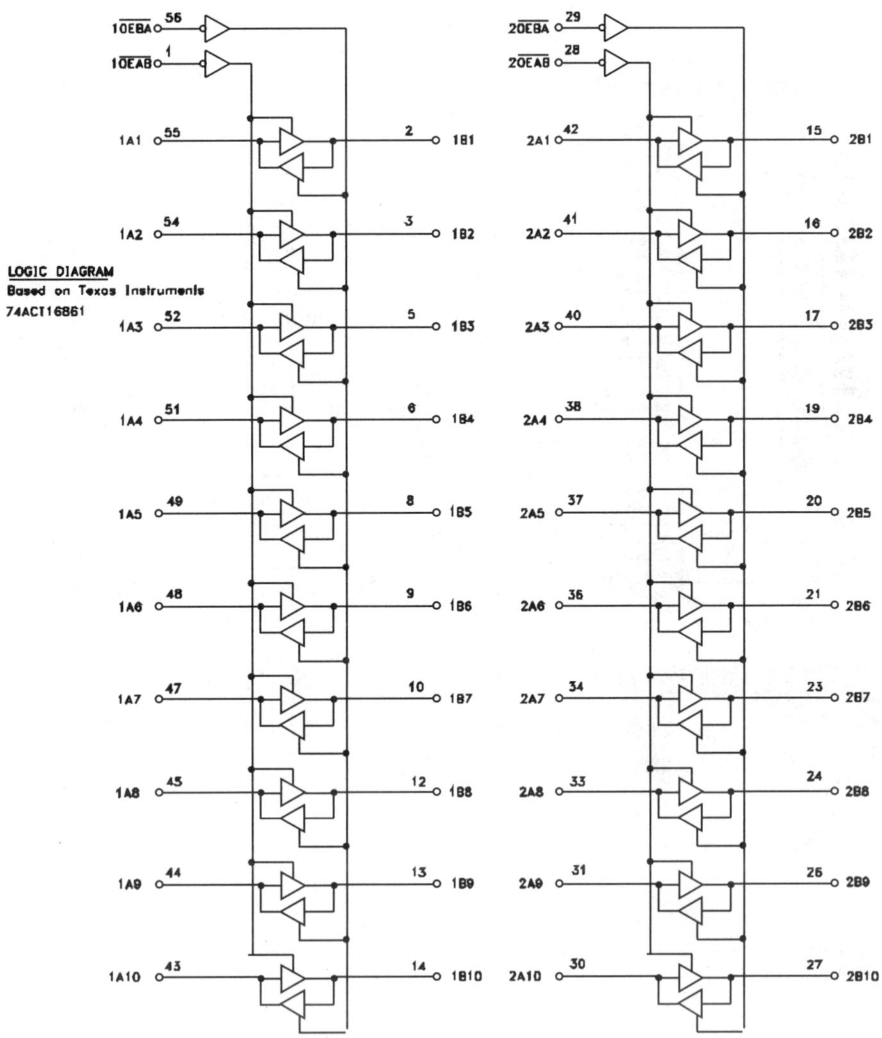

Figure 11.53. The CMOS 74ACT16861 is a 20-bit noninverting bus transceiver with 3-state outputs designed for asynchronous communication between data buses. This may be used as two 10-bit or a single 20-bit transceiver. The direction of data transmission between buses is determined by the logic levels at the output enable inputs, /OEAB, /OEBA, as shown in the function table. Distributed power pins minimize high-speed switching noise. Flow-through architecture optimizes PCB layout.

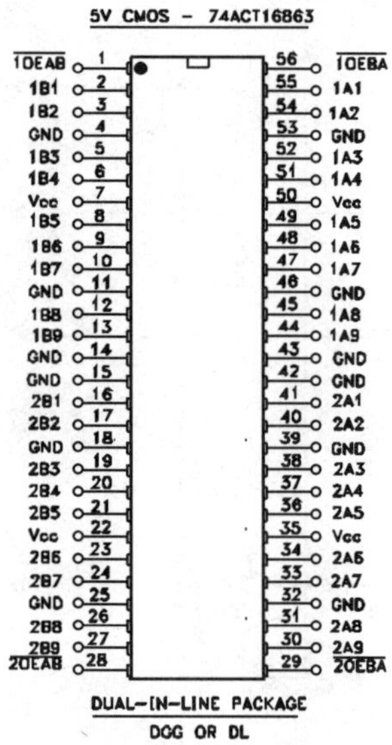

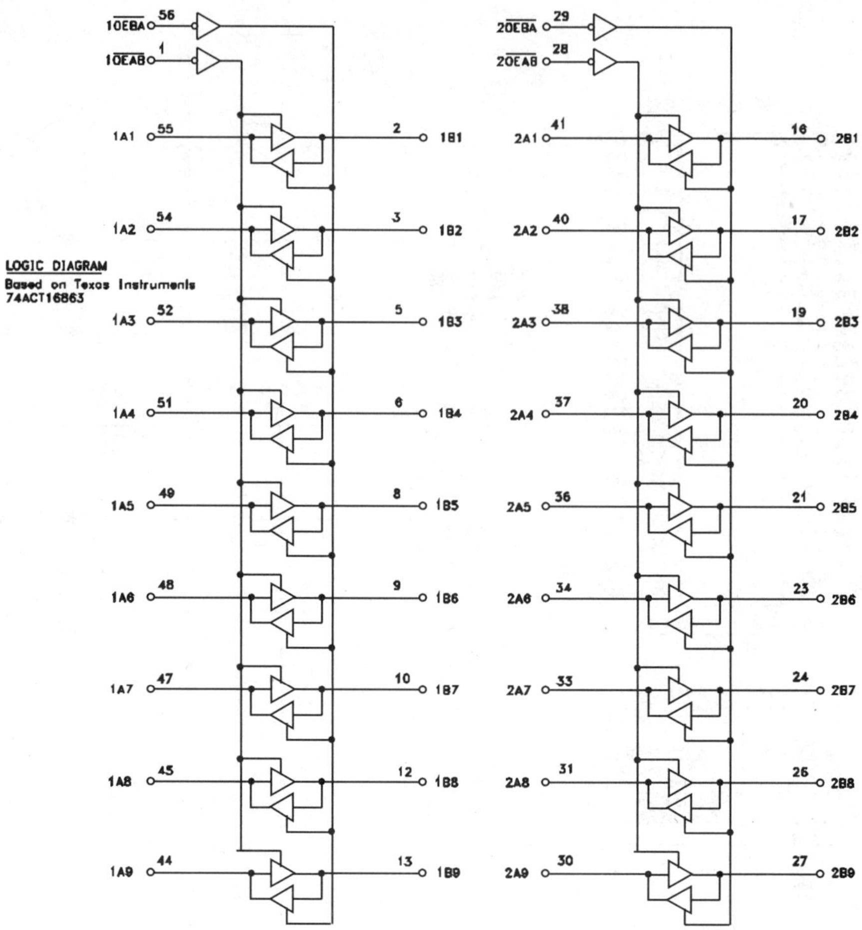

FUNCTION TABLE 74ACT16863 18-BIT BUS TRANSCEIVER WITH 3-STATE OUTPUTS

INPUTS		
$\overline{OEAB}$	$\overline{OEBA}$	OPERATING MODES
1	0	B Data to A Bus
0	1	A Data to B Bus
1	1	Z

1 = HIGH voltage level.
0 = LOW voltage level.
X = Don't Care
Z = High Impedance slole

Figure 11.54. The CMOS 74ACT16863 is an 18-bit noninverting bus transceiver with 3-state outputs designed for asynchronous communication between data buses. This may be used as two 9-bit or a single 18-bit transceiver. The direction of data transmission between buses is determined by the logic levels at the output enable inputs, /OEAB, /OEBA, as shown in the function table. Distributed power pins minimize high-speed switching noise. Flow-through architecture optimizes PCB layout.

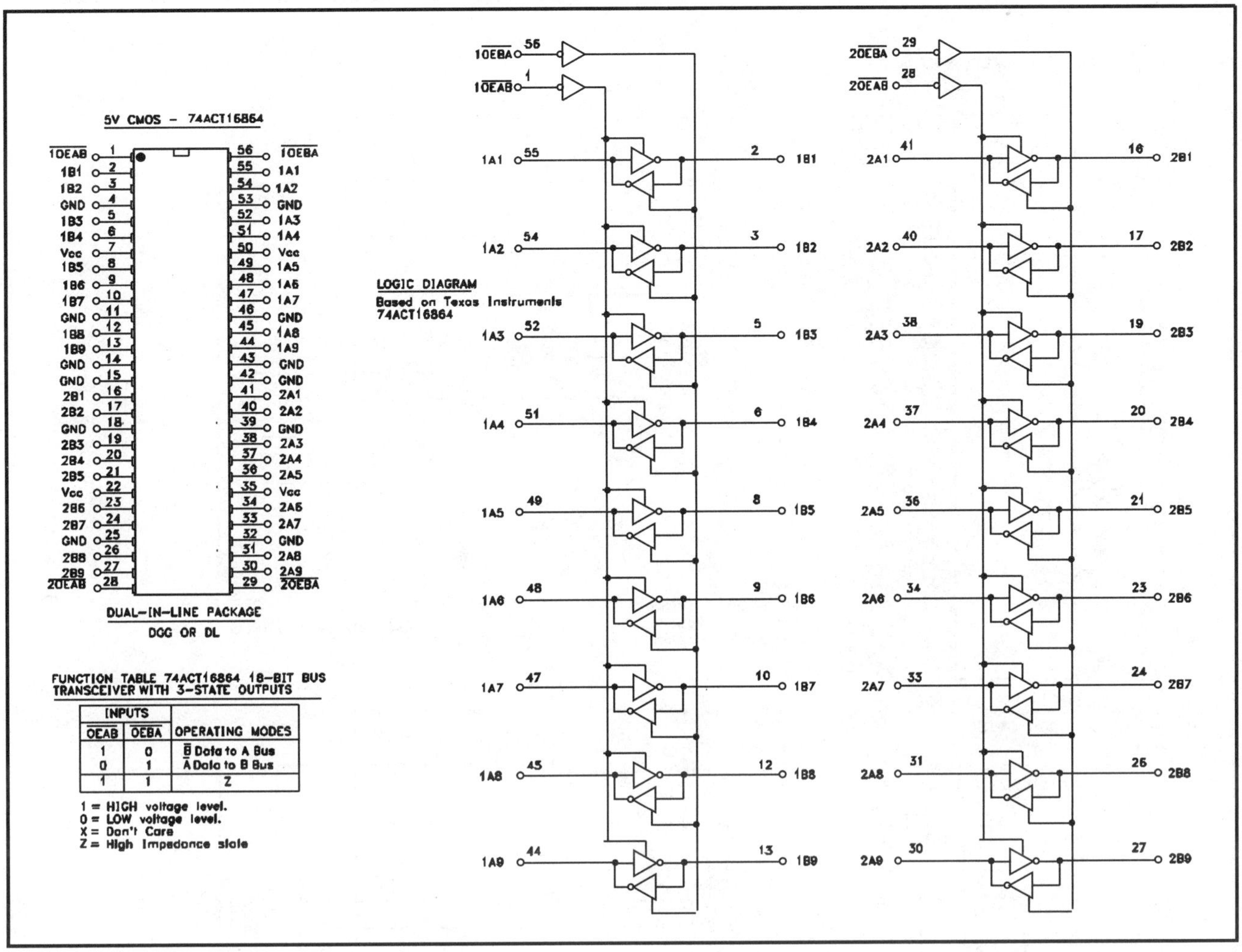

INPUTS		
OEAB	OEBA	OPERATING MODES
1	0	B Data to A Bus
0	1	A Data to B Bus
1	1	Z

Figure 11.55. The CMOS 74ACT16864 is an 18-bit inverting bus transceiver with 3-state outputs designed for asynchronous communication between data buses. This may be used as two 9-bit or a single 18-bit transceiver. The direction of data transmission between buses is determined by the logic levels at the output enable inputs,/OEAB,/OEBA, as shown in the function table. Distributed power pins minimize high-speed switching noise. Flow-through architecture optimizes PCB layout.

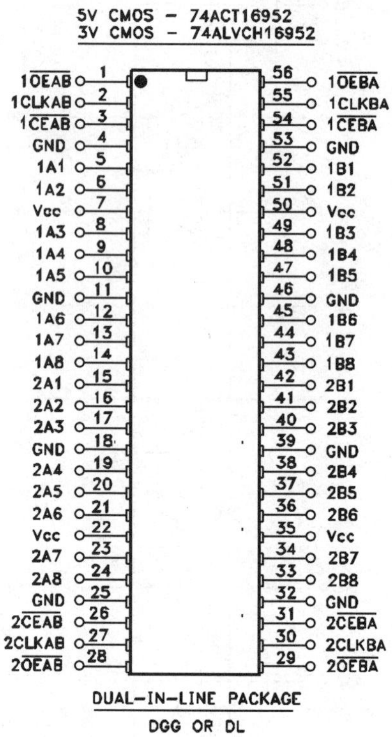

DEVICE DATA SHOWN IS BASED ON THE Texas Instruments 74ACT16952

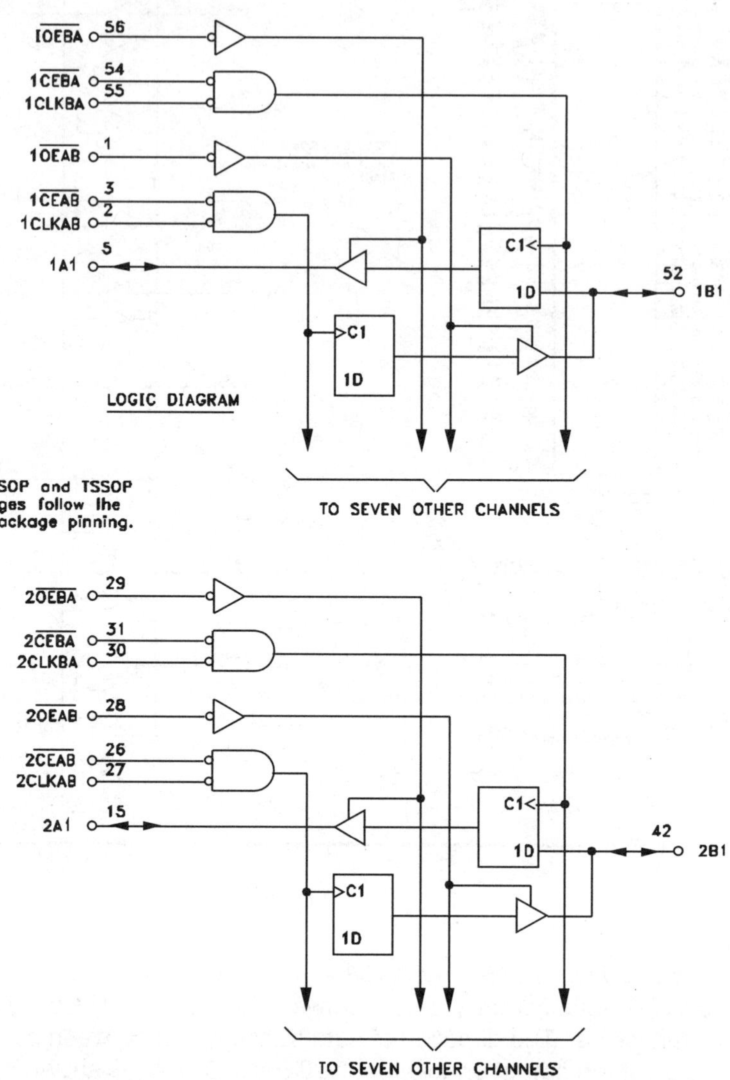

NOTE: The SSOP and TSSOP Packages follow the DIP Package pinning.

FUNCTION TABLE 74ACT16952%
16-BIT REGISTERED TRANSCEIVER WITH 3-STATE OUTPUTS

INPUTS				OUTPUT
$\overline{CEAB}$	CLKAB	$\overline{OEAB}$	A	B
1	X	0	X	B0+
X	1	0	X	B0+
0	↑	0	0	0
0	↑	0	1	1
X	X	1	X	Z

% = A-to-B data flow is shown; B-to-A flow control is the same except it uses $\overline{CEBA}$, CLKBA, and $\overline{OEBA}$.
+ = Output level before the indicated steady-state input conditions were established
1 = HIGH voltage level.
0 = LOW voltage level.
X = Don't Care
Z = High Impedance state
↑ = LOW-to-HIGH clock transition.

Figure 11.56. The CMOS 74ACT16952 and 74ALVCH16952 are 16-bit registered transceivers with noninverting 3-state outputs provided with two sets of D-type flip-flops for temporary storage of data flowing in either direction. These may be used as two 8-bit or a single 16-bit transceiver. Data on the A or B bus is stored on registers on the low-to-high transitions of the clock input provided that the clock enable, /CEAB, /CEBA, is low. Taking the output enable, /OEAB, /OEBA, low accesses the data on either port. To avoid false clocking of the flip-flops, /CEAB, /CEBA, should not be switched from high-to-low with CLKAB or CLKBA low. Distributed power pins minimize high-speed switching noise. Flow-through architecture optimizes PCB layout.

TTL – 74ALS29833

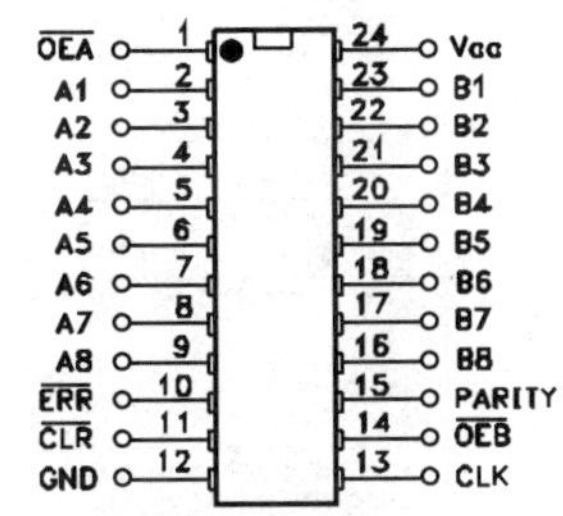

DUAL-IN-LINE PACKAGE

FUNCTION TABLE 74ALS29833 8-BIT TO 9-BIT PARITY BUS TRANSCEIVER

INPUTS						OUTPUT AND I/O				FUNCTION
$\overline{OEB}$	$\overline{OEA}$	$\overline{CLR}$	CLK	Ai Σ OF H	Bi* Σ OF H	A	B	PARITY	$\overline{ERR}$+	
0	1	X	X	Odd Even	NA	NA	A	0 1	NA	A data to B bus and generate parity
1	0	1	↑	NA	Odd Even	B	NA	NA	1 0	B data to A bus and generate parity
X	X	0	X	X	X	X	NA	NA	1	Check error–flag register
1	1	1 0 1 1	Not ↑ Not ↑ ↑ ↑	X X Odd Even	X	Z	Z	Z	NC 1 1 0	Isolation%
0	0	X	X	Odd Even	NA	NA	A	1 0	NA	A data to B bus and generate inverted parity

1 = HIGH voltage level. 0 = LOW voltage level.
NA = Not Applicable. NC = No Change. X = Don't Care
• = Summation of high level inputs includes PARITY along with Bi inputs.
+ = Output states shown assume ERR was previously high.
% = In this mode ERR, when clocked, shows inverted parity of the A bus.
↑ = LOW–to–HIGH clock transition.

ERROR-FLAG FUNCTIONS

INPUTS		INTERNAL TO DEVICE	OUTPUT PRESTATE	OUTPUT	
CLR	CLK	POINT P	ERRn–1+	ERR	FUNCTION
1 1 1	↑ ↑ ↑	1 X 0	1 0 X	1 0 0	Sample
0	X	X	X	1	Clear

ERRn–1 represents the state of ERR before any changes at CLR, CLK or POINT P.

NOTE:
Device data shown is for Texas Instruments 74ALS29833

LOGIC DIAGRAM
74ALS29833

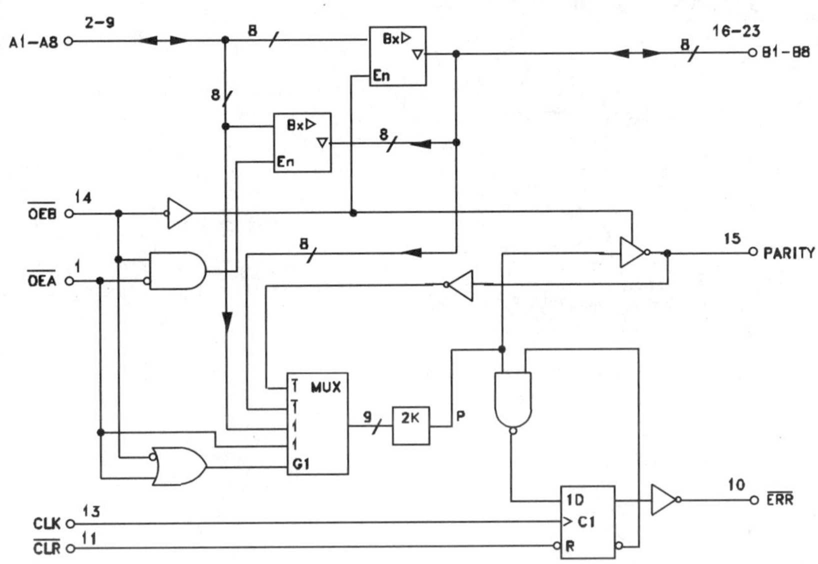

Figure 11.57. The TTL 74ALS29833 is an 8-bit to 9-bit high-speed bus transceiver with parity generator/checker designed for two-way communication between data buses. When data is transmitted from the A bus to the B bus a parity bit is generated. When data is transmitted from the B bus to the A bus with its corresponding parity bit the parity error, /ERR, output indicates whether or not an error has occurred on the B bus. The output enable inputs, /OEA, /OEB, can be used to effectively isolate the buses. A 9-bit parity generator/checker monitors the I/O port parity with an open collector error flag, /ERR. /ERR is clocked into the register on the rising edge of the CLK input. When both /OEA,/OEB are low, data is transferred from the A bus to the B bus and inverted parity is generated. Inverted parity is a forced error that gives the designer more system diagnostic ability. Minimum clock pulse duration is 10 nS.

TTL – 74ALS29854

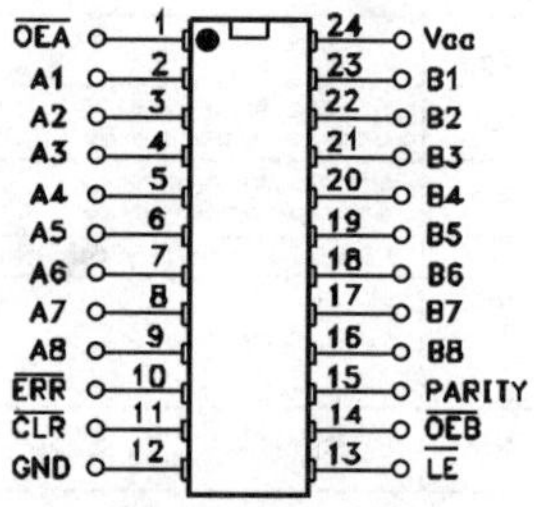

DUAL-IN-LINE PACKAGE

FUNCTION TABLE 74ALS29854 8-BIT TO 9-BIT PARITY BUS TRANSCEIVER

INPUTS						OUTPUT AND I/O				FUNCTION
OEB	OEA	CLR	LE	Ai Σ OF H	Bi* Σ OF H	A	B	PARITY	ERR+	
0	1	X	X	Odd Even	NA	NA	A	0 1	NA	A data to B bus and generate parity
1	0	1	0	NA	Odd Even	B	NA	NA	1 0	B data to A bus and generate parity
X	X	0	1	X	X	X	NA	NA	1	Check error-flag register
1	1	1 0 1 1	1 1 0 0	X X Odd Even	X	Z	Z	Z	NC 1 1 0	Isolation%
0	0	X	X	Odd Even	NA	NA	A	1 0	NA	A data to B bus and generate inverted parity

1 = HIGH voltage level. 0 = LOW voltage level.
NA = Not Applicable. NC = No Change. X = Don't Care
* = Summation of high level inputs includes PARITY along with Bi inputs.
\+ = Output states shown assume ERR was previously high.
% = In this mode ERR, when enabled, shows inverted parity of the A bus.

ERROR-FLAG FUNCTIONS

INPUTS		INTERNAL TO DEVICE	OUTPUT PRESTATE	OUTPUT	
LE	CLR	POINT P	ERRn-1+	ERR	FUNCTION
0 0	0 1	0 1	X X	0 1	Pass
0 0 0	1 1 1	0 X 1	X 0 1	0 0 1	Sample
0	0	X	X	1	Clear
1 1	1 1	X X	0 1	0 1	Store

ERRn-1 represents the state of ERR before any changes at CLR, LE, or POINT P

NOTE:
Device data shown is for Texas Instruments 74ALS29854

LOGIC DIAGRAM
74ALS29854

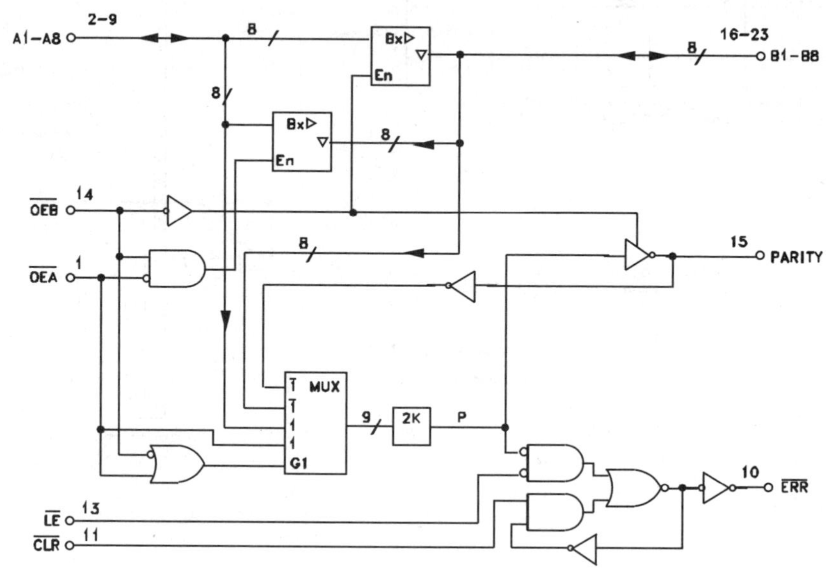

Figure 11.58. The TTL 74ALS29854 is an 8-bit to 9-bit high-speed bus transceiver with parity generator/checker designed for two-way communication between data buses. When data is transmitted from the A bus to the B bus, a parity bit is generated. When data is transmitted from the B bus to the A bus with its corresponding parity bit, the parity error, /ERR, output indicates whether or not an error has occurred on the B bus. The output enable inputs, /OEA, /OEB, can be used to effectively isolate the buses. A 9-bit parity generator/checker monitors the I/O port parity with an open collector error flag, /ERR. /ERR can be either passed, sampled, stored, or cleared from the latch using the latch enable, /LE, and clear, /CLR, inputs. When both /OEA, /OEB are low, data is transferred from the A bus to the B bus and inverted parity is generated. Inverted parity is a forced error that gives the designer more system diagnostic ability. Minimum /LE and CLR pulse duration is 10 nS.

TTL – 74ALS29861
74ALS29862

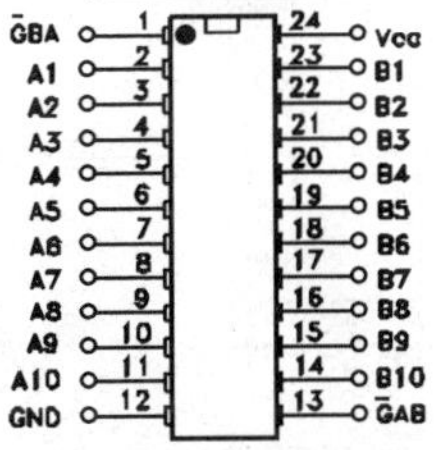

DUAL-IN-LINE PACKAGE

PACKAGE PINNING

PINS–
2–11 DATA INPUT/OUTPUT
14–23 DATA INPUT/OUTPUT
1,13 OUTPUT ENABLE INPUTS

LOGIC DIAGRAM
Based on Texas Instruments
74ALS29861

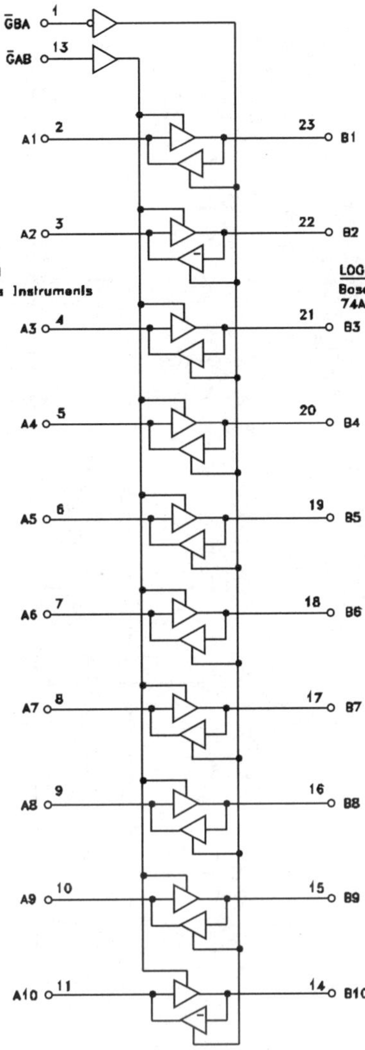

LOGIC DIAGRAM
Based on Texas Instruments
74ALS29862

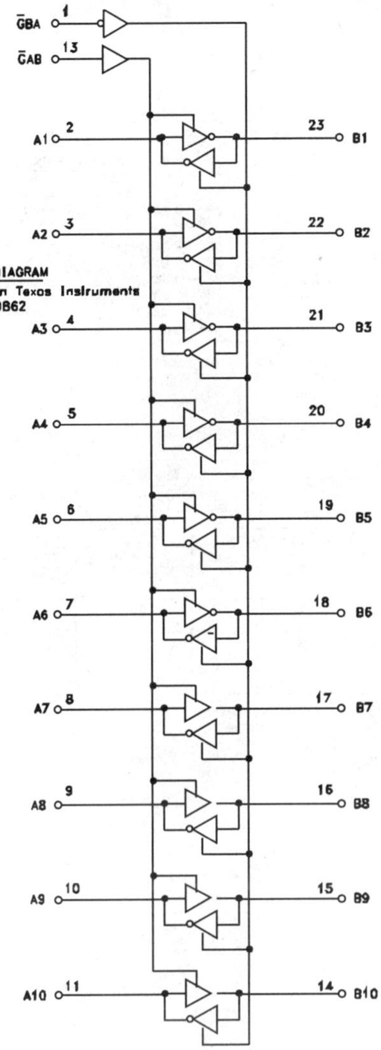

FUNCTION TABLE 74ALS29861, 'ALS29862
10-BIT BUS TRANSCEIVERS WITH 3-STATE PUTPUTS

INPUTS		OPERATING MODES	
$\overline{G}$AB	$\overline{G}$BA	'29861	'29862
0	1	B Data to A Bus	$\overline{B}$ Data to A Bus
1	0	A Data to B Bus	$\overline{A}$ Data to B Bus
1	1	Z	Z
0	0	Latch A and B (A=B)	Latch A and B (A=B)

1 = HIGH voltage level.
0 = LOW voltage level.
X = Don't Care
Z = High Impedance state

Figure 11.59. The TTL 74ALS29861 and 74ALS29862 are 10-bit bus tranceivers with 3-state outputs. The 'ALS29861 provides true outputs; the 'ALS29862 inverting. These devices are designed for asynchronous two-way communication between data buses. The control function implementation allows for maximum flexibility in timing. Data may be transmitted between buses in either direction, depending on the logic levels at the enable inputs, as shown in the function table.

TTL – 74ALS29863
74ALS29864

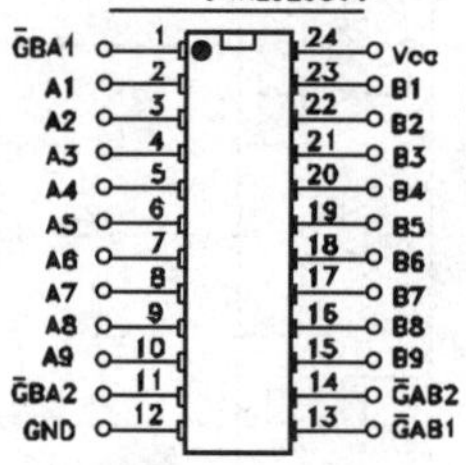

DUAL-IN-LINE PACKAGE

PACKAGE PINNING
PINS–
2–10 DATA INPUT/OUTPUT
15–23 DATA INPUT/OUTPUT
1,11,13,14 OUTPUT ENABLE INPUT

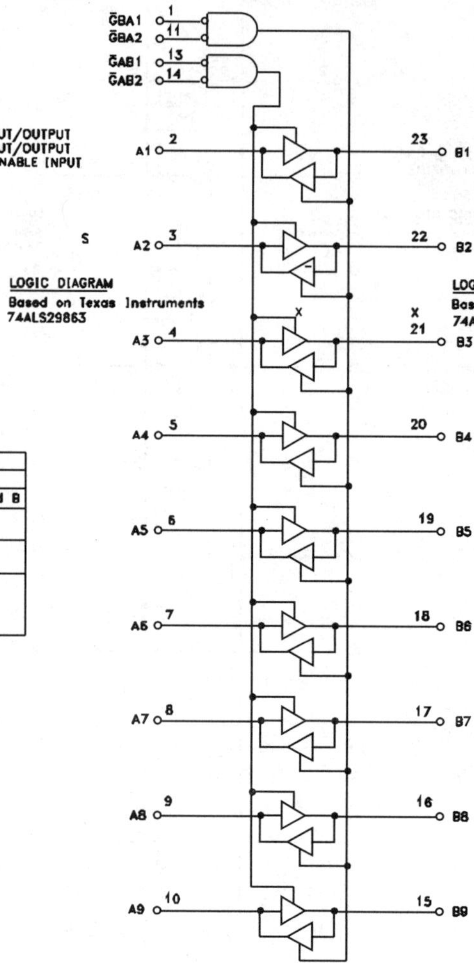

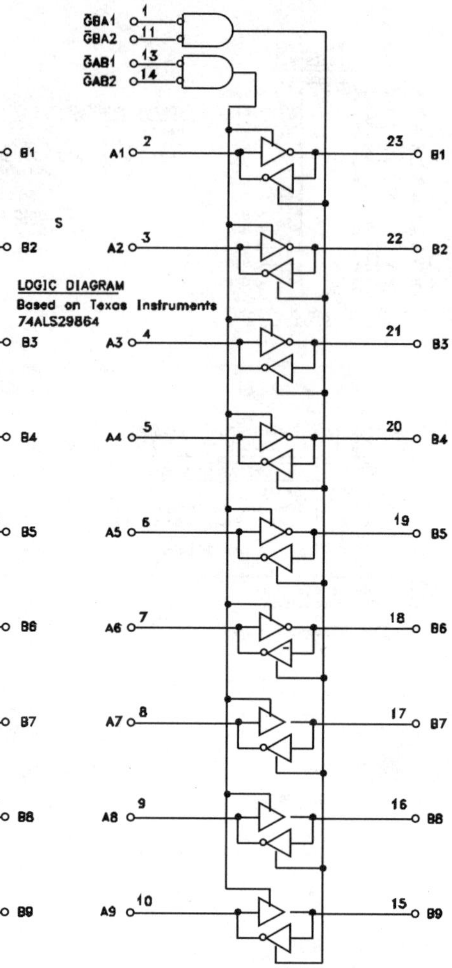

FUNCTION TABLE 74ALS29863, 'ALS29864
9-BIT BUS TRANSCEIVERS WITH 3-STATE OUTPUTS

ENABLE INPUTS				OPERATING MODES	
ḠAB1	ḠAB2	ḠBA1	ḠBA2	'ALS29863	'ALS29864
0	0	0	0	LATCH A and B	LATCH A and B
0 0	0 0	1 X	X 1	A TO B	A TO $\overline{B}$
1 X	X 1	0 0	0 0	B TO A	B TO $\overline{A}$
1 1 X X	X X 1 1	1 X X 1	X 1 1 X	Z	Z

1 = HIGH voltage level.
0 = LOW voltage level.
X = Don't Care
Z = High Impedance state

Figure 11.60. The TTL 74ALS29863 and 74ALS29864 are 9-bit bus tranceivers with 3-state outputs. The 'ALS29863 provides true outputs; the 'ALS29864 inverting. These devices are designed for asynchronous two-way communication between data buses. The control function implementation allows for maximum flexibility in timing. Data may be transmitted between buses in either direction, depending on the logic levels at the enable inputs, as shown in the function table.

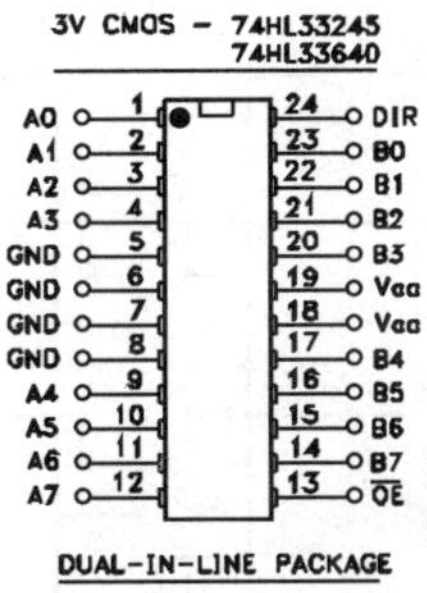

NOTE:
The SSOP and TSSOP Packages follow the DIP Package pinning.

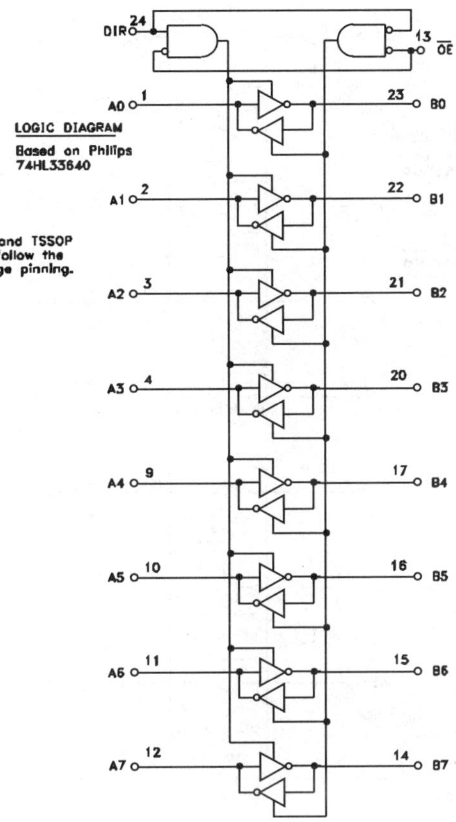

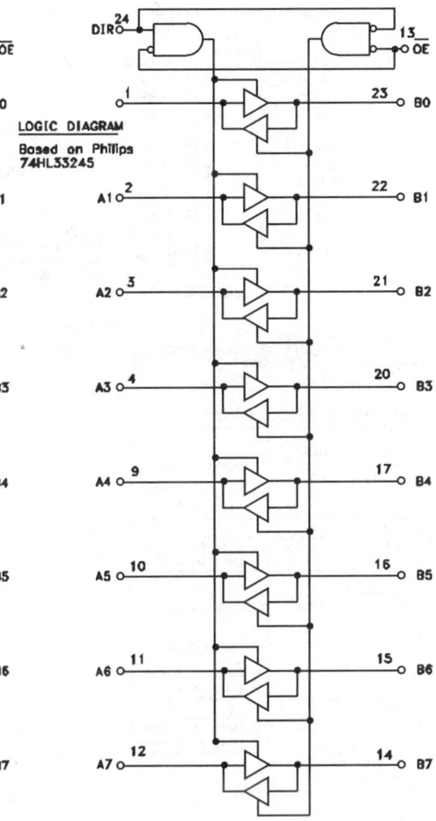

FUNCTION TABLE OCTAL TRANSCEIVERS WITH DIRECTION INPUT, 3-STATE OUTPUTS

74HL33245 - NON-INVERTING

ENABLE	DIRECTION CONTROL	INPUTS/OUTPUTS	
$\overline{OE}$	DIR	An	Bn
0	0	A=B	Inputs
0	1	Inputs	B=A
1	X	Z	Z

74HL33640 - INVERTING

ENABLE	DIRECTION CONTROL	INPUTS/OUTPUTS	
$\overline{OE}$	DIR	An	Bn
0	0	$A=\bar{B}$	Inputs
0	1	Inputs	$B=\bar{A}$
1	X	Z	Z

1 = HIGH voltage level.
0 = LOW voltage level.
X = Don't Care
Z = High Impedance state

Figure 11.61. The low-voltage CMOS 74HL33246 and 74HL33640 are octal bus transceivers with 3-state outputs. The 'HL33245 provides true outputs; the 'HL33640 inverting. These devices are designed for asynchronous two-way communication between data buses. The /OE output enable provides for easy cascading. Data may be transmitted between buses in either direction, depending on the DIR logic level at the enable input, as shown in the function table.

3V CMOS – 74HL33620
74HL33623

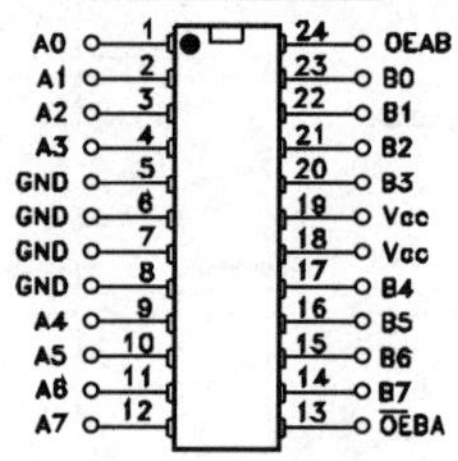

DUAL-IN-LINE PACKAGE

NOTE:
The SSOP and TSSOP Packages follow the DIP Package pinning.

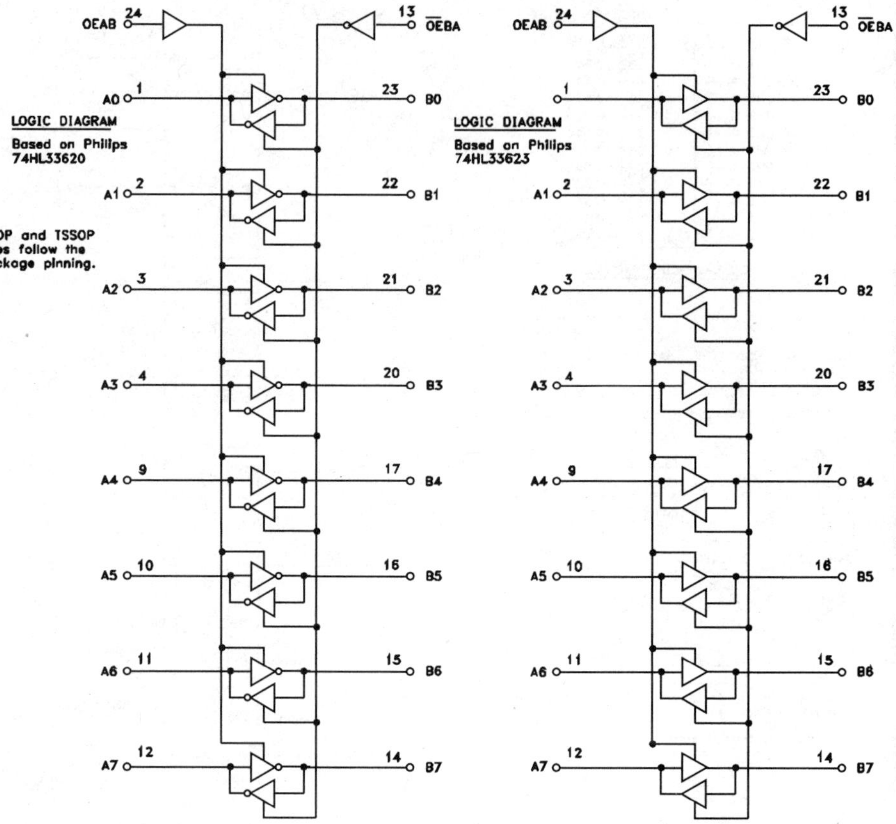

FUNCTION TABLE OCTAL TRANSCEIVERS WITH DUAL ENABLES AND 3-STATE OUTPUTS

74HL33623 – NON-INVERTING

ENABLE INPUTS		OPERATION
OEAB	$\overline{OEBA}$	
0 1	0 1	B Data to A Bus A Data to B Bus
0	1	Z Z
1	0	B Data to A Bus A Data to B Bus

74HL33620 – INVERTING

ENABLE INPUTS		OPERATION
OEAB	$\overline{OEBA}$	
0 1	0 1	$\overline{B}$ Data to A Bus $\overline{A}$ Data to B Bus
0	1	Z Z
1	0	$\overline{B}$ Data to A Bus $\overline{A}$ Data to B Bus

1 = HIGH voltage level.
0 = LOW voltage level.
Z = High Impedance state

Figure 11.62. The low-voltage CMOS 74HL33620 and 74HL33623 are octal bus transceivers with dual enables and 3-state outputs in both send and receive directions. The ‘HL33623 provides true outputs; the ‘HL33620 inverting. These devices are designed for asynchronous two-way communication between data buses. Data may be transmitted between buses in either direction, depending on the logic levels at the enable inputs, /OEAB, /OEBA, as shown in the function tables.

5V CMOS - 74HL33646

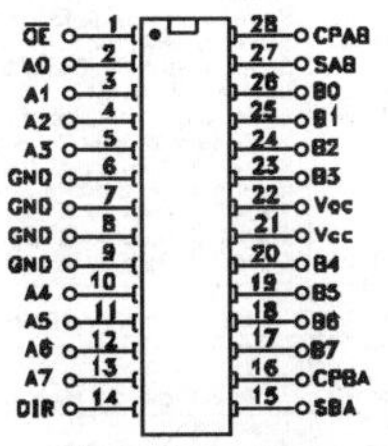

DUAL-IN-LINE PACKAGE

NOTE:
The SSOP and TSSOP Packages follow the DIP Package pinning.

5V CMOS - 74HL33648

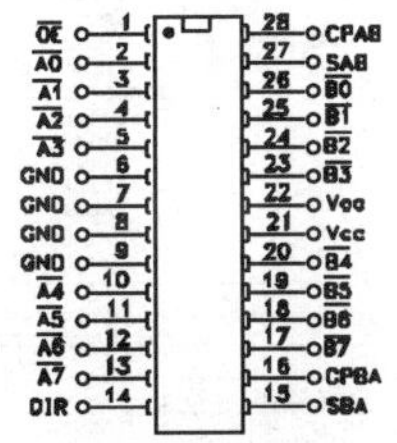

DUAL-IN-LINE PACKAGE

NOTE:
The SSOP and TSSOP Packages follow the DIP Package pinning.

FUNCTION TABLE 74HL33646 OCTAL BUS TRANSCEIVER/REGISTER WITH 3-STATE OUTPUTS

INPUTS						DATA I/O		
$\overline{OE}$	DIR	CPAB	CPBA	SAB	SBA	A1–A8	B1–B8	OPERATION OR FUNCTION
X	X	↑	X	X	X	Input	Unspecified	Store A, B unspecified+
X	X	X	↑	X	X	Unspecified	Input	Store B, A unspecified+
1	X	↑	↑	X	X	Input	Input	Store A and B Data
1	X	1 or 0	1 or 0	X	X	Input	Input	Isolation, hold storage
0	0	X	X	X	0	Output	Input	Real-time B Data to A Bus
0	0	X	1 or 0	X	1	Output	Input	Stored B Data to A Bus
0	1	X	X	0	X	Input	Output	Real-time A Data to B Bus
0	1	1 or 0	X	1	X	Input	Output	Stored A Data to Bus

FUNCTION TABLE 74HL33648 OCTAL BUS TRANSCEIVER/REGISTER WITH 3-STATE OUTPUTS

INPUTS						DATA I/O		
$\overline{OE}$	DIR	CPAB	CPBA	SAB	SBA	$\overline{A1}$–$\overline{A8}$	$\overline{B1}$–$\overline{B8}$	OPERATION OR FUNCTION
X	X	↑	X	X	X	Input	Unspecified	Store A, B unspecified+
X	X	X	↑	X	X	Unspecified	Input	Store B, A unspecified+
1	X	↑	↑	X	X	Input	Input	Store A and B Data
1	X	1 or 0	1 or 0	X	X	Input	Input	Isolation, hold storage
0	0	X	X	X	0	Output	Input	Real-time $\overline{B}$ Data to A Bus
0	0	X	1 or 0	X	1	Output	Input	Stored $\overline{B}$ Data to A Bus
0	1	X	X	0	X	Input	Output	Real-time $\overline{A}$ Data to B Bus
0	1	1 or 0	X	1	X	Input	Output	Stored $\overline{A}$ Data to Bus

↑ = LOW-to-HIGH clock transition.
1 = HIGH voltage level.
0 = LOW voltage level.
X = Don't Care
Z = High Impedance state

+ = The data-output functions may be enabled or disbled by various signals at OE or DIR. Data Input functions are always enabled, i.e., data at the bus terminals is stored on every low-to-high transition of the clock inputs.

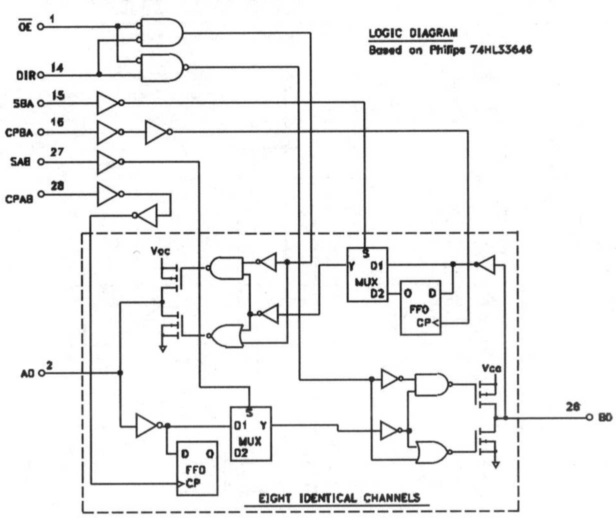

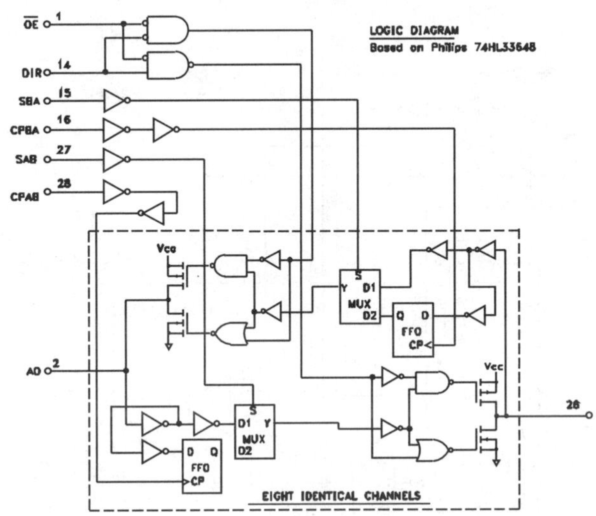

Figure 11.63. The low-voltage CMOS 74HL33646 and 74HL33648 are octal bus transceiver/registers with 3-state outputs. The 'HL33646 provides true outputs; the 'HL33648 inverting. These devices are designed for multiplexed transmission of data directly from the internal registers. Data on the A or B bus will be clocked in the internal registers as the appropriate clock, CPAB, CPBA, goes to a high logic level. Output enable, /OE and direction, DIR, inputs control the transceiver function. In the transceiver mode, data present at the high-impedance port may be stored in either the A or B register or in both. The select source inputs, SAB,SBA, can multiplex stored and real-time data. The DIR input determines which bus will receive data when /OE is low. A data may be stored in the B registers and/or B data may be stored in the A register, as shown in the function tables.

3V CMOS - 74HL33651

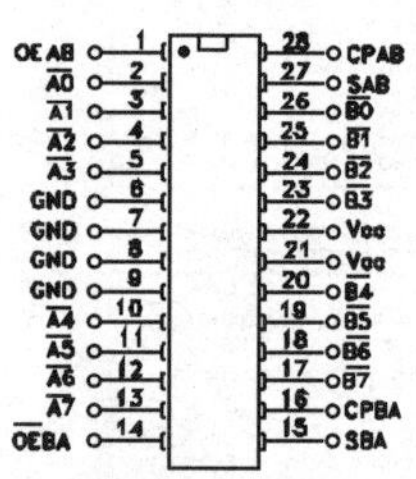

DUAL-IN-LINE PACKAGE

NOTE:
The SSOP and TSSOP Packages follow the DIP Package pinning.

3V CMOS - 74HL33652

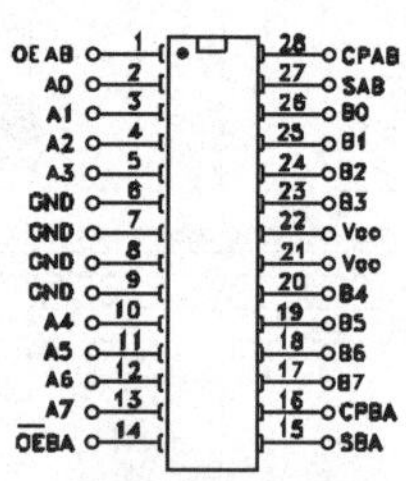

DUAL-IN-LINE PACKAGE

NOTE:
The SSOP and TSSOP Packages follow the DIP Package pinning.

FUNCTION TABLE 74HL33652 OCTAL TRANSCEIVER/REGISTER WITH DUAL ENABLE,NON-INVERTING

INPUTS						DATA I/O+		
OEAB	/OEBA	CPAB	CPBA	SAB	SBA	A1-A8	B1-B8	OPERATION OR FUNCTION
0	1	1 or 0	1 or 0	X	X	Input	Input	Isolation
0	1	↑	↑	X	X	Input	Input	Store A and B Data
X	1	↑	1 or 0	X	X	Input	Unspecified	Store A, Hold B
1	1	↑	↑	0	X	Input	Output	Store A in both Registers
0	X	1 or 0	↑	X	X	Unspecified	Input	Hold A, Store B
0	0	↑	↑	X	0	Output	Input	Store B in both Registers
0	0	X	X	X	0	Output	Input	Real-time B Data to A Bus
0	0	X	1 or 0	X	1	Output	Input	Stored B Data to A Bus
1	1	X	X	0	X	Input	Output	Real-time A Data to B Bus
1	1	1 or 0	X	1	X	Input	Output	Stored A Data to B Bus
1	0	1 or 0	1 or 0	1	1	Output	Output	Stored A Data to B Bus and
1	0	1 or 0	1 or 0	1	1	Output	Output	Stored B Data to A Bus

FUNCTION TABLE 74HL33651 OCTAL TRANSCEIVER/REGISTER WITH DUAL ENABLE,INVERTING

INPUTS						DATA I/O+		
OEAB	/OEBA	CPAB	CPBA	SAB	SBA	A1-A8	B1-B8	OPERATION OR FUNCTION
0	1	1 or 0	1 or 0	X	X	Input	Input	Isolation
0	1	↑	↑	X	X	Input	Input	Store /A and /B Data
X	1	↑	1 or 0	X	X	Input	Unspecified	Store A, Hold B
1	1	↑	↑	0	X	Input	Output	Store A in both Registers
0	X	1 or 0	↑	X	X	Unspecified	Input	Hold A, Store B
0	0	↑	↑	X	0	Output	Input	Store B in both Registers
0	0	X	X	X	0	Output	Input	Real-time /B Data to A Bus
0	0	X	1 or 0	X	1	Output	Input	Stored /B Data to A Bus
1	1	X	X	0	X	Input	Output	Real-time /A Data to B Bus
1	1	1 or 0	X	1	X	Input	Output	Stored /A Data to B Bus
1	0	1 or 0	1 or 0	1	1	Output	Output	Stored /A Data to B Bus and
1	0	1 or 0	1 or 0	1	1	Output	Output	Stored /B Data to A Bus

↑ = LOW-to-HIGH clock transition.
1 = HIGH voltage level.
0 = LOW voltage level.
X = Don't Care
Z = High Impedance state

+ = The data-output functions may be enabled or disabled by various signals at OEAB or /OEBA. Data input functions are always enabled, i.e., data at the bus terminals is stored on every low-to-high transition of the clock inputs.

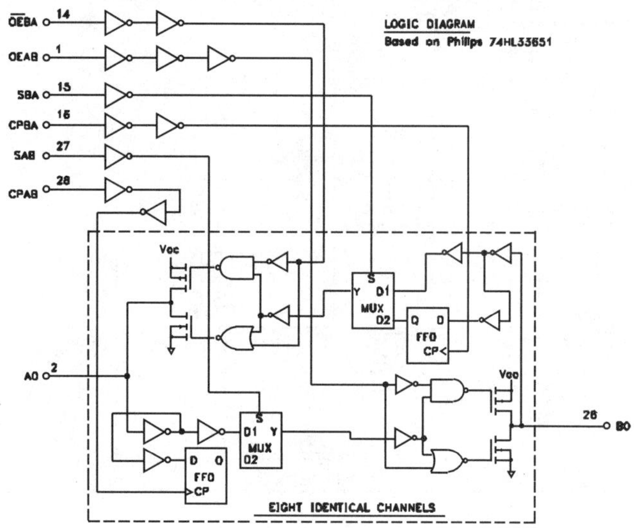

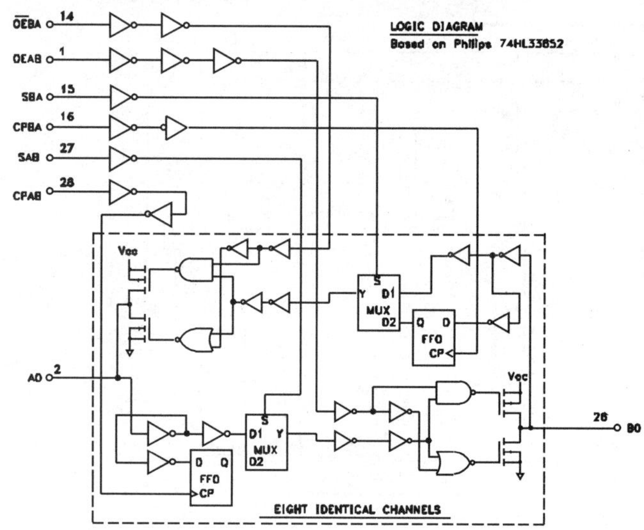

Figure 11.64. The low-voltage CMOS 74HL33651 and 74HL33652 are octal bus transceiver/registers with dual enables and 3-state outputs. The 'HL33652 provides true outputs; the 'HL33651 inverting. These devices are designed for multiplexed transmission of data directly from the internal registers. Data on the A or B bus will be stored in the internal registers at the appropriate clock inputs, CPAB, CPBA, regardless of the select inputs, SAB, SBA, or output enable OEAB, /OEBA. Depending on the select inputs, data can go directly from input to output (real time mode) or can be controlled by the clock (storage mode). The output enable inputs control determine the transceiver operation mode, as shown in the function tables.

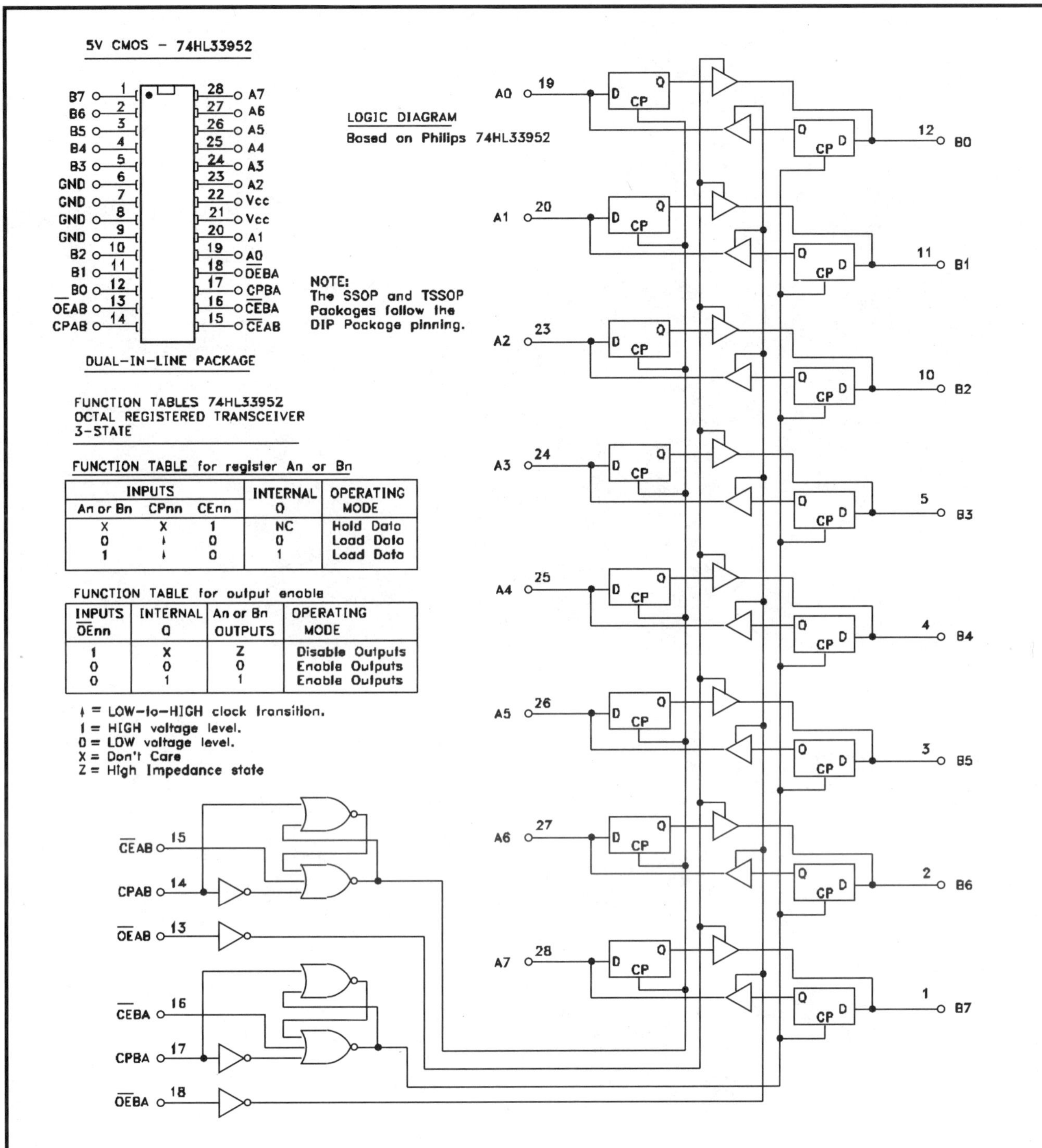

FUNCTION TABLE for register An or Bn

INPUTS			INTERNAL	OPERATING
An or Bn	CPnn	CEnn	Q	MODE
X	X	1	NC	Hold Data
0	↑	0	0	Load Data
1	↑	0	1	Load Data

FUNCTION TABLE for output enable

INPUTS OEnn	INTERNAL Q	An or Bn OUTPUTS	OPERATING MODE
1	X	Z	Disable Outputs
0	0	0	Enable Outputs
0	1	1	Enable Outputs

Figure 11.65. The low-voltage CMOS 74HL33952 is an octal non-inverting registered transceiver, consisting of two 8-bit back-to-back data registers, with data flowing in both directions between two bidirectional data buses. Outputs are 3-state. Data applied to the inputs is entered and stored on the rising edge of the clock, provided that the clock enable is low. The data is then present at the 3-state output buffers, but is accessible only with the output enable low. Data flow from the A inputs to the B outputs is identical, as for the B inputs to the A outputs.

Appendix

Manufacturers' Credits by Chapter

Introduction

Table G.1. A glossary of Logic Terms.
Signetics Corporation, *Application Memos,* September 1969.
Glossary of Logic Terms, pp 1-3 to 1-3.19

Table G.2. Symbols commonly employed with function and truth tables.
Texas Instruments *High-Speed CMOS Logic Data Book,* 1988, p1-13.

Table G.3. Letter symbols, terms, and definitions.
Based on the following reference sources:
Texas Instruments *ALS/AS Logic Data Book,* 1986, Appendix B, p4-47.
Texas Instruments *High-Speed CMOS Logic*, 1988, Glossary, pp1-15 -1-17.

Table G.4. Low voltage logic definitions that differ or add to those defined in Table G.3.1
National Semiconductor *CROSSVOLT Low Voltage Logic Series Databook,* 1994, pp2-3, -4.

Figure G.1. (a) A tabulation of minimum rise and fall time rates for the currently available logic families.
(b) Graphical display of pulse timing definitions.
Texas Instruments Inc., *AHC/AHCT Advanced High-Speed CMOS Logic Data Book,* 1997 (a) "Transition Times," p5-55, Table 1., (b) p5-54, Figure 15.

Figure G.2. Logic voltage definitions for the currently available logic families.
Texas Instruments Inc., *LVC Low-Voltage CMOS Logic Data Book,* 1997, p6-12, Figure 1.

Chapter 1

Figure 1.1. Point contact transistor construction examples.

Lloyd P. Hunter, Editor, *Handbook of Semiconductor Electronics, 3rd Ed., McGraw-Hill Book, 1970 Co.,* p1-4, Figure 1.1.

Figure 1.2. (a) Typical characteristics, N-type point contact transistor.[1]
(b) Typical PN junction characteristics.[1]
(c) NPN transistor characteristics.[2]

1. Lloyd P. Hunter, Editor, *Handbook of Semiconductor Electronics, 3rd Ed., McGraw-Hill Book, 1970 Co. (a)* p1-6, Figure 1.3. (b) p1.7, Figure 1.4
2. Clement S. Pepper, *Build Your Own Home Lab, Howard W. Sams, Prompt Publications, 1997, p133, Figure 5.1.6.1*

Figure 1.3. Bipolar transistor basic concepts. *Though the NPN only is shown the PNP model is identical with diode polarities and voltages reversed.*

Clement S. Pepper, *Build Your Own Home Lab,* Howard W. Sams, Prompt Publications, 1997, p130, Figure 5.1.3.

Figure 1.4. JFET constant-voltage biased amplifier, output characteristics and transfer curve.

Clement S. Pepper, *Build Your Own Home Lab,* Howard W. Sams, Prompt Publications, 1997, p138, Figure 5.1.11.

Figure 1.5. DMOS FET switch. (a) Electrical symbol. (b) Application as a ±10 volt analog switch.

Clement S. Pepper, *Build Your Own Home Lab,* Howard W. Sams, Prompt Publications, 1997, p143, Figure 5.1.15.

Figure 1.6. Resistor-Transistor Logic (RTL), Emitter-Coupled Logic, and Diode-Transistor Logic paved the way to Transistor-Transistor Logic (TTL).

Robert L. Morris, John R. Miller, Editors, Texas Instruments, Inc., *Designing With TTL Integrated Circuits*, McGraw-Hill Book Co., 1971, pp 8-12, Figures 1.10, 1.12, 1.13, 1.14.

Figure 1.7. Logic level thresholds and DC margins.

Signetics Corporation, *Application Memos,* September 1969, p 1-3.13

Figure 1.8. 7400 series Standard, Low-Power (L), Low-Power Schottky (LS), and Schottky (S) TTL gate circuit examples.

National Semiconductor Corporation, *ALS/AS Databook,* "Guide To Bipolar Logic Device Families," 1987, pp 1-3, 1-4, Figures 1, 2, 3, and 4.

Figure 1.9. Derivation of the Schottky clamped transistor.

Texas Instruments, *ALS/AS Logic Data Book,* 1986, p 4-11, Figure 4.5.

Figure 1.10. 7400ALS/AS gate circuit examples.

National Semiconductor Corporation, *ALS/AS Databook,* "Guide To Bipolar Logic Device Families," 1987, p 1-5, Figures 5, and 6.

Figure 1.11. 7400 FAST series input circuit options.
Signetics Corporation, *FAST Data Manual,* 1987, "Circuit Characteristics," pp 3-3, 3-4, Figures 1 and 3.

Figure 1.12. 7400 FAST series output circuit options.
Signetics Corporation, *FAST Data Manual,* 1987, "Circuit Characteristics," p 3-8, Figures 11 and 12.

Figure 1.13. (a) The MC14001 Buffered NOR gate (one of four).
(b) Two examples of Electrostatic Breakdown (ESD) Protection.
Motorola Incorporated, *CMOS Logic Data*, Series C, 1991. (a) p 6-10, MC14001B. (b) p 5-2, Figures 1A, 1B.

Figure 1.14 (a) Unbuffered NAND gate. (b) Buffered NAND gate.
National Semiconductor Corporation, *CMOS Logic Databook,* 1988 Edition, p 2-84, Figures 1(a) and 1(b).

Figure 1.15. (a) CMOS inverter with bipolar parasetics revealed.
(b) The CMOS inverter layup cross section with the latchup model circuit.
National Semiconductor Corporation, *VHC Advanced CMOS Logic Databook,* 1996 Edition. (a) p 3-15, Figure 3-29. (b) p 3-14, Figure 3-28.

Figure 1.16. 74HC CMOS circuit construction examples.
Texas Instruments, *High-Speed CMOS Logic Databook,* 1988, pp 4-5 to 4-7, Figures 1 - 7.

Figure 1.17. Some 74HC/HCT LSTTL/HC-CMOS and standard CMOS/HC-CMOS circuit interfacing examples.
National Semiconductor Corporation, *VHC Advanced CMOS Logic Databook,* 1996, "Interfacing to MM54HC/MM74HC High-Speed CMOS logic," pp 5-89 to 5-95, Figures 2, 4, and 11.

Figure 1.18. Input and output circuit ESD protection for advance high-speed CMOS.
Texas Instruments, *AHC/AHCT Advanced High-Speed CMOS Logic Data Book,* 1996, "Advanced High-Speed CMOS (AHC) Protection Circuitry," pp 5-8,9 Figures 1 and 2.

Figure 1.19. Comparison of CMOS AC/ACT output circuitry with TTL FAST.
GE Corporation, GE Solid State, *RCA Advanced CMOS Logic Integrated Circuits AC/ACT,* 1988, p 19, Figures 13, 14; p 20, Figure 16, p 23, Figure 20.

Figure 1.20. Comparison of 3.3V and 5V interfaces.
Texas Instruments, *Low-Voltage Logic, ALB,ALVC,LV,LVT,LVTZ,GTL Families Data Book,* 1996, p 12-9, Figure 2.

Figure 1.21. Comparison of CMOS LVT output circuitry with TTL FAST.
1. (a) GE Corporation, GE Solid State, *RCA Advanced CMOS Logic Integrated Circuits AC/ACT,* 1988, p 19, Figure 13.
2. (b) National Semiconductor Corporation, *Crossvolt Low Voltage Series Databook,* 1994 Edition, p 1-14, Figure 7.
3. (c) Texas Instruments, *Low-Voltage Logic, ALB,ALVC,LV,LVT,LVTZ,GTL Families Data Book,* 1996, p 12-11, Figure 3.

Chapter 2

The quantity and variety of logic gates illustrated in the figures of this chapter are too numerous to credit individually. The data shown includes package pinning, logic symbol, and truth tables. These are composites of all the various manufacturers.

The following manufacturer's data books are represented:

Motorola Incorporated, *CMOS Logic Data, DL131/D,* Series C, 1991.
Motorola Incorporated, *High-Speed CMOS Logic Data,* DL129, Rev 3, 1988
Motorola Incorporated, *High-Speed CMOS Logic Data,* DL129, Rev 6, 1996.
Motorola Incorporated, *FAST and LSTTL Data,* DL121, Rev. 5, 1992.
Motorola Incorporated, *Advanced High-Speed CMOS Data,* DL203/D, Rev 1, 03/97.
Motorola Incorporated, *LCX Data Low-Voltage CMOS Logic,* BR1339/D, Rev 3, 04/96.

National Semiconductor Corporation, *ALS/AS Databook,* "Guide To Bipolar Logic Device Families," 1987.
National Semiconductor Corporation, *CMOS Logic Databook,* 1988 Edition.
National Semiconductor Corporation, *Crossvolt Low Voltage Logic Series Databook,* 1994 Edition.
National Semiconductor Corporation, *VHC Advanced CMOS Logic Databook,* 1996 Edition.
National Semiconductor Corporation, *LS/S/TTL Logic Databook,* 1987.

Philips Semiconductors, *HE4000B Logic Family CMOS,* Data Handbook IC04, 1997.
Philips Semiconductors, *High Speed CMOS Logic Family,* Data Handbook IC06, 1996.
Philips Semiconductors, *FAST TTL Logic Series,* Data Book IC15, 1992.
Philips Semiconductors, *Low Voltage CMOS & BiCMOS Logic,* Data Handbook IC24, 1996.

Signetics Corporation, *Signetics Logic - TTL - Specifications, Military Summary, Data Manual,* 1978.
Signetics Company, a division of North American Philips Corporation, *FAST Data Manual,* 1987.

Texas Instruments, *The Integrated Circuits Catalog For Design Engineers,* 1971.
Texas Instruments, *TTL Logic Standard TTL, Schottky, Low-Power Schottky Data Book,* 1988.
Texas Instruments, *ALS/AS Logic Data Book,* 1986.
Texas Instruments, *ALS/AS Logic Data Book,* 1995.
Texas Instruments, *High-Speed CMOS Logic Data Book,* 1988.
Texas Instruments, *AC/ACT CMOS Logic Data Book,* 1997.
Texas Instruments, *AHC/AHCT Advanced High-Speed CMOS Logic Data Book,* 1997.
Texas Instruments, *Low-Voltage Logic, ALB, ALVC, LV, LVT, LVTZ, GTL Families Data Book,* 1996.
Texas Instruments, *LVC Low-Voltage CMOS Logic Data Book,* 1997.
Texas Instruments, *F Logic (SN54/74F),* 1994.

Chapter 3

Table 3.1. Motorola Inc., *CMOS Logic Data,* DL131/D, Rev. 3.
National Semiconductor, *CMOS Logic Databook,* 1988 Edition.
National Semiconductor, *VHC Advanced CMOS Logic Databook,* 1996 Edition.
National Semiconductor, *LS/S/TTL Logic Databook,* 1987 Edition.
Signetics Company, *High-Speed CMOS Data Manual,* 1988.

Figure 3.2. The S9602 Dual Retriggerable Resettable Monostable Multivibrator
Signetics Corporation, *Digital, Linear, MOS Data Book,* 1974.

Figure 3.3. The 74121 Retriggerable Monostable Multivibrator
Signetics Corporation, *Signetics Logic - TTL - Specifications, Military Summary, Data Manual,* 1978.
74122/LS122 Retriggerable Monostable Multivibrator
Texas Instruments, *TTL Logic Standard TTL, Schottky, Low-Power Schottky Data Book,* 1988.

Figure 3.4. TTL and CMOS dual monostables. These share many similarities with identical DIP package terminations.

Figure 3.5. The MC14528 Dual Monostable Multivibrator
The MC14538 Dual Precision Monostable Multivibrator
The MC14536B Programmable Timer
The MC14541B Programmable Oscillator/Timer
Motorola Incorporated, *CMOS Logic Data,*DL131/D 04/91.

Figure 3.6. The CD4047 Low-Power Monostable/Astable Multivibrator.
National Semiconductor, *CMOS Logic Databook,* 1988 Edition.

Figure 3.7. A performance selection guide for Motorola CMOS monostables.
Motorola Incorporated, *CMOS Logic Data,*DL131/D 04/91, p 6-365.

Figure 3.9. The 558 Quad IC monostable/astable timer.
Signetics Company, a Division of North American Philips Corporation, *High-Speed CMOS Data Manual,* 1988.

Figure 3.10. The HC/HCT5555 Programmable Delay Timer with Oscillator.
Signetics Company, a Division of North American Philips Corporation, *High-Speed CMOS Data Manual,* 1988.

Figure 3.12. The 74S124 Dual-Voltage Controlled Oscillator.
Texas Instruments, *TTL Logic Standard TTL, Schottky, Low-Power Schottky Data Book,* 1988.

Figure 3.13. The HEF4753 Universal Timer Module.
Philips Semiconductors, *HE4000B Logic Family CMOS,* Data Handbook IC04, 1997.

Chapter 4

Figure 4.4. 74H101/H102 Negative Edge Triggered Flip-Flops
Texas Instruments, *The Integrated Circuits Catalog For Design Engineers,* 1971.

Figure 4.5. The 74110/74111 Gated J-K Master-Slave Flip-Flops with Data Lockout.
Texas Instruments, *The Integrated Circuits Catalog For Design Engineers,* 1971.

Figure 4.6. The 7473 and 74H103 Dual K-K Master Slave Flip-Flops
The 7476 and 74H106 Dual K-K Master Slave Flip-Flops with Preset and Clear
Texas Instruments, *The Integrated Circuits Catalog For Design Engineers,* 1971.

Figure 4.7. The 74LS78A Dual J-K Flip-Flops with Common Clock, and Common Clear.
Texas Instruments, *TTL Logic Standard TTL, Schottky, Low-Power Schottky Data Book,* 1988.
The 74H108 Dual J-K Negative Edge Triggered Flip-Flops
Texas Instruments, *The Integrated Circuits Catalog For Design Engineers,* 1971.

Figure 4.8. The 74LS107 Dual J-K Flip-Flops with Clear.
The 74LS109 Dual J-/K Positive-Edge-Triggered Flip-Flops with Preset and Clear.
Texas Instruments, *TTL Logic Standard TTL, Schottky, Low-Power Schottky Data Book,* 1988.

Figure 4.9. The 74LS112 Dual J-K Negative-Flip-Flops with Preset and Clear.
Texas Instruments, *TTL Logic Standard TTL, Schottky, Low-Power Schottky Data Book,* 1988.
The MC14027 Dual J-K Flip-Flop
Motorola Incorporated, *CMOS Logic Data, DL131/D,* Series C, 1991.

Figure 4.10. The 74F173 Quad D-type Flip-Flop
Philips Semiconductors, *FAST TTL Logic Series,* Data Book IC15, 1992.

Figure 4.11. The 74LS174 Hex/Quadruple D-type Flip-Flop with Clear
Texas Instruments, *TTL Logic Standard TTL, Schottky, Low-Power Schottky Data Book,* 1988.

Figure 4.12. The 74LS175 Hex/Quadruple D-type Flip-Flop with Clear
Texas Instruments, *TTL Logic Standard TTL, Schottky, Low-Power Schottky Data Book,* 1988.

Figure 4.13. The 74LS273 Octal D-type Flip-Flop with Clear.
Texas Instruments, *TTL Logic Standard TTL, Schottky, Low-Power Schottky Data Book,* 1988.

Figure 4.14. 74276 Quadruple J-/K Flip-Flops
Texas Instruments, *TTL Logic Standard TTL, Schottky, Low-Power Schottky Data Book,* 1988.

Figure 4.15. 74LS364 Octal D-type Flip-Flop with 3-State Outputs
Signetics Corporation, *Signetics Logic - TTL - Specifications, Military Summary, Data Manual,* 1978.

Figure 4.16. 74LS374 Octal D-type Edge Triggered Flip-Flop
Texas Instruments, *TTL Logic Standard TTL, Schottky, Low-Power Schottky Data Book,* 1988.

Figure 4.17. 74376 Quadruple J-/K Flip-Flops
Texas Instruments, *TTL Logic Standard TTL, Schottky, Low-Power Schottky Data Book,* 1988.

Figure 4.18. 74LS377 Octal D-type Flip-Flop with Enable.
Texas Instruments, *TTL Logic Standard TTL, Schottky, Low-Power Schottky Data Book,* 1988.

Figure 4.19. 74LS378 Hex D-type Flip-Flop with Enable.
Texas Instruments, *TTL Logic Standard TTL, Schottky, Low-Power Schottky Data Book,* 1988.

Figure 4.20. 74LS379 Quad D-type Flip-Flop with Enable.
Texas Instruments, *TTL Logic Standard TTL, Schottky, Low-Power Schottky Data Book,* 1988.

Figure 4.21. 74AS/ALS534 Octal D-type Edge Triggered Flip-Flop with 3-State Output.
Texas Instruments, *ALS/AS Logic Data Book,* 1986.

Figure 4.22. The 74AS/ALS574 Octal D-type Edge Triggered Flip-Flop with 3-State Output.
Texas Instruments, *ALS/AS Logic Data Book,* 1986.

Figure 4.23. The 74AS/ALS575 Octal D-type Edge Triggered Flip-Flop with 3-State Output.
Texas Instruments, *ALS/AS Logic Data Book,* 1986.

Figure 4.24. The 74AS821 and 74AS822 10-bit Bus Interface Flip-Flop with 3-State Output.
Texas Instruments, *ALS/AS Logic Data Book,* 1986.

Figure 4.25. The 74AS823 and 74AS824 9-bit Bus Interface Flip-Flop with 3-State Output.
Texas Instruments, *ALS/AS Logic Data Book,* 1986.

Figure 4.26. The 74AS825 and 74AS826 8-bit Bus Interface Flip-Flop with 3-State Output.
Texas Instruments, *ALS/AS Logic Data Book,* 1986.

Figure 4.27. The 74ALS874B and 74AS874A Dual 4-bit Edge-Triggered D-type Flip-Flops.
Texas Instruments, *ALS/AS Logic Data Book,* 1986.

Figure 4.28. The 74ALS876A and 74AS876 Dual 4-bit Edge-Triggered D-type Flip-Flops.
Texas Instruments, *ALS/AS Logic Data Book,* 1986.

Figure 4.29. The 74ALS878A/879A and 74AS878/879 Dual 4-bit Edge-Triggered D-type Flip-Flops with 3-State Outputs.
Texas Instruments, *ALS/AS Logic Data Book,* 1986.

Figure 4.30. The 74AS4374B Octal Edge-Triggered D-type Dual Rank Flip-Flop with 3-State Outputs
Texas Instruments, *ALS/AS Logic Data Book,* 1995.

Figure 4.31. The 74HC7074 6-Section Multifunction (NAND, INVERT, NOR, FLIP-FLOP) Circuit.
Texas Instruments, *High-Speed CMOS Logic Data Book,* 1988.

Figure 4.32. The 74HC7075 6-Section Multifunction (NAND, INVERT, FLIP-FLOP) Circuit.
Texas Instruments, *High-Speed CMOS Logic Data Book,* 1988.

Figure 4.33. The 74HC7076 6-Section Multifunction (INVERT, FLIP-FLOP) Circuit.
Texas Instruments, *High-Speed CMOS Logic Data Book,* 1988.

Figure 4.34. The 74AC/ACT11074 Dual Positive-Edge Triggered D-type Flip-Flop with Clear and Preset.
Texas Instruments, *AC/ACT CMOS Logic Data Book,* 1997.

Figure 4.35. The 74AC/ACT11374 Octal D-type Edge-TriggeredFlip-Flop with 3-State Outputs
Texas Instruments, *AC/ACT CMOS Logic Data Book,* 1997.

Figure 4.36. The 74AC/ACT16374 16-BIT D-type Edge-TriggeredFlip-Flop with 3-State Outputs
Texas Instruments, *AC/ACT CMOS Logic Data Book,* 1997.

Figure 4.37. The 74ALVCH16721 20-bit Flip-Flop with 3-State Outputs.
Texas Instruments, *Low-Voltage Logic, ALB, ALVC, LV, LVT, LVTZ, GTL Families Data Book,* 1996.

Figure 4.38. The 74ALVCH16820/162820 10-bit Flip-Flop with Dual Outputs.
Texas Instruments, *Low-Voltage Logic, ALB, ALVC, LV, LVT, LVTZ, GTL Families Data Book,* 1996.

Figure 4.39. The 74ALVCH16821/162821 20-bit Bus Interface Flip-Flop with 3-State Outputs.
Texas Instruments, *Low-Voltage Logic, ALB, ALVC, LV, LVT, LVTZ, GTL Families Data Book,* 1996.

Figure 4.40. The 74ALVCH16823 18-bit Bus Interface Flip-Flop with 3-State Outputs.
Texas Instruments, *Low-Voltage Logic, ALB, ALVC, LV, LVT, LVTZ, GTL Families Data Book,* 1996.

Figure 4.41. The 74GTL16921 20-bit Flip-Flop with GTL I/O Levels.
Texas Instruments, *Low-Voltage Logic, ALB, ALVC, LV, LVT, LVTZ, GTL Families Data Book,* 1996.

Chapter 5

Figure 5.1. The 7475, 7477 Dual 2-Bit Transparent Latch
Signetics Corporation, *Signetics Logic - TTL - Specifications, Military Summary, Data Manual,* 1978.

Figure 5.2. The 74100 Dual 4-Bit Transparent Latch
Signetics Corporation, *Signetics Logic - TTL - Specifications, Military Summary, Data Manual,* 1978.

Figure 5.3. The 74116 Dual 4-Bit Transparent Latch
Signetics Corporation, *Signetics Logic - TTL - Specifications, Military Summary, Data Manual,* 1978.

Figure 5.4. The 74256 Dual 4-Input Addressable Latch
Signetics Corporation, *Signetics Logic - TTL - Specifications, Military Summary, Data Manual,* 1978.

Figure 5.5. The 74259 8-Bit Addressable Latch
Signetics Corporation, *Signetics Logic - TTL - Specifications, Military Summary, Data Manual,* 1978.

Figure 5.6. The 74279 Quad Set-Reset Latch
Signetics Corporation, *Signetics Logic - TTL - Specifications, Military Summary, Data Manual,* 1978.

Figure 5.7. The 74LS363 Octal Transparent Latch with 3-State Outputs.
Signetics Corporation, *Signetics Logic - TTL - Specifications, Military Summary, Data Manual,* 1978.

Figure 5.8. The 74LS373 Octal Transparent Latch with 3-State Outputs.
Signetics Corporation, *Signetics Logic - TTL - Specifications, Military Summary, Data Manual,* 1978.

Figure 5.9. The 74LS375 Dual 2-Bit Transparent Latch.
Signetics Corporation, *Signetics Logic - TTL - Specifications, Military Summary, Data Manual,* 1978.

Figure 5.10. The 74F412/432 Multi-Mode Buffered Latches.
Signetics Company, a division of North American Philips Corporation, *FAST Data Manual,* 1987.

Figure 5.11. The 74AS/ALS533 Octal D-Type Transparent Latches with 3-State Outputs.
Texas Instruments, *ALS/AS Logic Data Book,* 1986.

Figure 5.12. The 74ALS563 Octal D-Type Transparent Latch with 3-State Outputs
Texas Instruments, *ALS/AS Logic Data Book,* 1986.

Figure 5.13. The 74LS604 Octal 2-Input Multiplexed Latches
Texas Instruments, *TTL Logic Standard TTL, Schottky, Low-Power Schottky Data Book,* 1988.

Figure 5.14. The 74ALS666, 74ALS667 8-Bit D-Type Transparent Read-Back Latches with 3-State Outputs
Texas Instruments, *ALS/AS Logic Data Book,* 1986.

Figure 5.15. The 74AS/ALS841/842 10-Bit Bus Interface D-Type Latches with 3-State Outputs
Texas Instruments, *ALS/AS Logic Data Book,* 1986.

Figure 5.16. The 74AS/ALS843/844 9-Bit Bus Interface D-Type Latches with 3-State Outputs
Texas Instruments, *ALS/AS Logic Data Book,* 1986.

Figure 5.17. The 74AS/ALS845/846 8-Bit Bus Interface D-Type Latches with 3-State Outputs.
Texas Instruments, *ALS/AS Logic Data Book,* 1986.

Figure 5.18. The 74ALS/AS873 Dual 4-Bit D-Type Latches with 3-State Outputs.
Texas Instruments, *ALS/AS Logic Data Book,* 1986.

Figure 5.19. The 74ALS/AS880 Dual 4-Bit D-Type Latches with 3-State Outputs.
Texas Instruments, *ALS/AS Logic Data Book,* 1986.

Figure 5.20. The 74ALS990/991 8-Bit D-Type Transparent Read-Back Latches
Texas Instruments, *ALS/AS Logic Data Book,* 1986.

Figure 5.21. The 74ALS992/993 9-Bit D-Type Transparent Read-Back Latches with 3-State Outputs
Texas Instruments, *ALS/AS Logic Data Book,* 1986.

Figure 5.22. The 74ALS994/995 10-Bit D-type Transparent Read-Back Latches.
Texas Instruments, *ALS/AS Logic Data Book,* 1986.

Figure 5.23. The 74ALS996 8-Bit D-type Edge-Triggered Read-Back Latch.
Texas Instruments, *ALS/AS Logic Data Book,* 1986.

Figure 5.24. The MC14042 Quad Transparent Latch.
Motorola Incorporated, *Logic Data, DL131/D,* Series C, 1991.

Figure 5.25. The MC14043 Quad R-S Latch.
Motorola Incorporated, *Logic Data, DL131/D,* Series C, 1991.

Figure 5.26. The MC14044 Quad R-S Latch.
Motorola Incorporated, *Logic Data, DL131/D,* Series C, 1991.

Figure 5.27. The MC14076 4-Bit D-Type Register with 3-State Outputs
Motorola Incorporated, *Logic Data, DL131/D,* Series C, 1991.

Figure 5.28. The MC14099 8-Bit Addressable Latch.
Motorola Incorporated, *Logic Data, DL131/D,* Series C, 1991.

Figure 5.29. The MC14508 Dual 4-Bit Latch.
Motorola Incorporated, *Logic Data, DL131/D,* Series C, 1991.

Figure 5.30. The MC14597 8-Bit Bus-Compatible Latch
Motorola Incorporated, *Logic Data, DL131/D,* Series C, 1991.

Figure 5.31. The MC14598 8-Bit Bus-Compatible Latch.
Motorola Incorporated, *Logic Data, DL131/D,* Series C, 1991.

Figure 5.32. The MC14599 8-Bit Addressable Latch.
Motorola Incorporated, *Logic Data, DL131/D,* Series C, 1991.

Figure 5.33. The CD4723, CD4724 Dual 4-Bit, 8-Bit Addressable Latches.
National Semiconductor Corporation, *CMOS Logic Databook,* 1988 Edition.

Figure 5.34. The 74AC/ACT11373 Octal Transparent D-type Latch with 3-State Outputs.
Texas Instruments, *AC/ACT CMOS Logic Data Book,* 1997.

Figure 5.35. The 74ACT16373 16-Bit D-type Transparent Latch with 3-State Outputs.
Texas Instruments, *AC/ACT CMOS Logic Data Book,* 1997.

Figure 5.36. The 74ALVCH16260 12-Bit to 24-Bit Multiplexed D-Type Latch with 3-State Outputs.
Texas Instruments, *Low-Voltage Logic, ALB, ALVC, LV, LVT, LVTZ, GTL Families Data Book,* 1996.

Figure 5.37. The 74ALVCH16841 20-Bit Bus Interface D-Type Latch with 3-State Outputs.
Texas Instruments, *Low-Voltage Logic, ALB, ALVC, LV, LVT, LVTZ, GTL Families Data Book,* 1996.

Figure 5.38. The 74ALVCH16843 18-Bit Bus Interface D-Type Latch with 3-State Outputs.
Texas Instruments, *Low-Voltage Logic, ALB, ALVC, LV, LVT, LVTZ, GTL Families Data Book,* 1996.

Figure 5.39. The 74HL33373, 74HL33533 Octal D-Type Transparent Latches with 3-State Outputs.
Philips Semiconductors, *Low Voltage CMOS & BiCMOS Logic,* Data Handbook IC24, 1996.

Chapter 6

Figure 6.1. The 7491 8-Bit Shift Register.
Signetics Corporation, *Signetics Logic - TTL - Specifications, Military Summary, Data Manual,* 1978.

Figure 6.2. The 7494 4-Bit Shift Register.
Signetics Corporation, *Signetics Logic - TTL - Specifications, Military Summary, Data Manual,* 1978.

Figure 6.3. The 7495 4-bit Shift Register.
Signetics Corporation, *Signetics Logic - TTL - Specifications, Military Summary, Data Manual,* 1978.

Figure 6.4. The 7496 5-Bit Shift Register.
Signetics Corporation, *Signetics Logic - TTL - Specifications, Military Summary, Data Manual,* 1978.

Figure 6.5. The 74L98 4-Bit Data Selector/Storage Register.
Texas Instruments, *The Integrated Circuits Catalog For Design Engineers,* 1971.

Figure 6.6. The 74164 8-Bit Serial-In, Parallel-Our Shift Register.
Signetics Corporation, *Signetics Logic - TTL - Specifications, Military Summary, Data Manual,* 1978.

Figure 6.7. The 74165 8-bit Serial/Parallel-In, Serial-Out Shift Register.
Signetics Corporation, *Signetics Logic - TTL - Specifications, Military Summary, Data Manual,* 1978.

Figure 6.8. The 74166 Serial/Parallel-In, Serial-Out 8-bit Shift Register.
Signetics Corporation, *Signetics Logic - TTL - Specifications, Military Summary, Data Manual,* 1978.

Figure 6.9. The 74170 4x4 Register File.
Signetics Corporation, *Signetics Logic - TTL - Specifications, Military Summary, Data Manual,* 1978.

Figure 6.10. The 74S172 16-Bit Multiple Port Register File.
Signetics Corporation, *Signetics Logic - TTL - Specifications, Military Summary, Data Manual,* 1978.

Figure 6.11. The 74194 4-Bit Bidirectional Universal Shift Register.
Signetics Corporation, *Signetics Logic - TTL - Specifications, Military Summary, Data Manual,* 1978.

Figure 6.12. The 74LS194 4-Bit Bidirectional Universal Shift Register.
Motorola Incorporated, *FAST and LSTTL Data,* DL121, Rev. 5, 1992.

Figure 6.13. The 74195 Universal 4-Bit Shift Register.
Motorola Incorporated, *FAST and LSTTL Data,* DL121, Rev. 5, 1992.

Figure 6.14. The 74198 8-bit Bidirectional Universal Shift Register.
Signetics Corporation, *Signetics Logic - TTL - Specifications, Military Summary, Data Manual,* 1978.

Figure 6.15 The 74199 8-bit Parallel Access Shift Register.
Signetics Corporation, *Signetics Logic - TTL - Specifications, Military Summary, Data Manual,* 1978.

Figure 6.16. The 74LS295 4-bit Shift Register with 3-State Outputs.
Signetics Corporation, *Signetics Logic - TTL - Specifications, Military Summary, Data Manual,* 1978.

Figure 6.17. The 74298 Quadruple 2-Port Register.
Signetics Corporation, *Signetics Logic - TTL - Specifications, Military Summary, Data Manual,* 1978.

Figure 6.18. The 74299 8-bit Universal Shift/Storage Register.
Signetics Corporation, *Signetics Logic - TTL - Specifications, Military Summary, Data Manual,* 1978.

Figure 6.19. The 74F322 8-bit Shift Registers with Sign Extend.
Texas Instruments, *TTL Logic Standard TTL, Schottky, Low-Power Schottky Data Book,* 1988.

Figure 6.20. The 74F350 4-bit Shifter.
Philips Semiconductors, *FAST TTL Logic Series,* Data Book IC15, 1992.

Figure 6.21. The 74F395 4-bit Cascadable Shift Register (3-state).
Philips Semiconductors, *FAST TTL Logic Series,* Data Book IC15, 1992.

Figure 6.22. The 74LS396 Octal Register with 3-State Outputs.
Texas Instruments, *TTL Logic Standard TTL, Schottky, Low-Power Schottky Data Book,* 1988.

Figure 6.23. The 74LS398 Quad 2-Port Registers.
Signetics Corporation, *Signetics Logic - TTL - Specifications, Military Summary, Data Manual,* 1978.

Figure 6.24. The 74LS399 Quad 2-Port Registers.
Signetics Corporation, *Signetics Logic - TTL - Specifications, Military Summary, Data Manual,* 1978.

Figure 6.25. The 74HC589 8-Bit Shift Register with Output Latches.
National Semiconductor Corporation, *CMOS Logic Databook,* 1988 Edition.

Figure 6.26. The 74LS594 8-Bit Shift Register with Output Latches.
Texas Instruments, *TTL Logic Standard TTL, Schottky, Low-Power Schottky Data Book,* 1988.

Figure 6.27. The 74VHC595 8-bit Shift Register (3-State).
Motorola Incorporated, *High-Speed CMOS Logic Data,* DL129, Rev 6, 1996.

Figure 6.28. The 74HC597 8-bit Serial or Parallel-Input Serial-Output Shift Register.
Motorola Incorporated, *High-Speed CMOS Logic Data,* DL129, Rev 6, 1996.

Figure 6.29. The 74F670 4x4 Register File.
Philips Semiconductors, *FAST TTL Logic Series,* Data Book IC15, 1992.

Figure 6.30. The 74LS671 4-bit Universal Shift Register/Latche.
Texas Instruments, *TTL Logic Standard TTL, Schottky, Low-Power Schottky Data Book,* 1988.

Figure 6.31. The 74LS673 16-bit Shift Register.
Texas Instruments, *TTL Logic Standard TTL, Schottky, Low-Power Schottky Data Book,* 1988.

Figure 6.32. The 74LS674 16-bit Shift Register.
Texas Instruments, *TTL Logic Standard TTL, Schottky, Low-Power Schottky Data Book,* 1988.

Figure 6.33. The 74F676 16-bit Serial/Parallel-in, Serial-Out Shift Register (3-state)
Philips Semiconductors, *FAST TTL Logic Series,* Data Book IC15, 1992.

Figure 6.34. The 74F835 8-bit Shift Register with 2:1 Mux-in; Latched "B" Inputs, Serial Out.
Philips Semiconductors, *FAST TTL Logic Series,* Data Book IC15, 1992.

Figure 6.35. The 74AS/ALS870, 74AS/ALS871 Dual 16-by-4 Register Files.
Texas Instruments, *ALS/AS Logic Data Book,* 1986.

Figure 6.36. The 74ALS963 Dual-Rank 8-bit Shift Register with 3-State Outputs.
Texas Instruments, *ALS/AS Logic Data Book,* 1986.

Figure 6.37. The 74ALS964 Dual-Rank 8-bit Shift Register with 3-State Outputs.
Texas Instruments, *ALS/AS Logic Data Book,* 1986.

Figure 6.38. The MC14006 18-bit Static Shift Register.
Motorola Incorporated, *CMOS Logic Data, DL131/D,* Series C, 1991.

Figure 6.39. The MC14014 and 4021 8-bit Static Shift Register.
Motorola Incorporated, *CMOS Logic Data, DL131/D,* Series C, 1991.

Figure 6.40. The MC14015 Dual 4-bit Static Shift Register.
Motorola Incorporated, *CMOS Logic Data, DL131/D,* Series C, 1991.

Figure 6.41. The HEF4031 64 Stage Static Shift Register.
Philips Semiconductors, *HE4000B Logic Family CMOS,* Data Handbook IC04, 1997.

Figure 6.42. The CD4034 8-BIT 8-Stage TRI-State Bidirectional Parallel/Serial Input/Output Bus Register.
National Semiconductor Corporation, *CMOS Logic Databook,* 1988 Edition.
TRI-State is a Registered Trademark of National Semiconductor Corporation.

Figure 6.43. The HEF4035 4-bit Universal Shift Register.
Philips Semiconductors, *HE4000B Logic Family CMOS,* Data Handbook IC04, 1997.

Figure 6.44. The HEF4094 8-stage Shift-and-Store Bus Register.
Philips Semiconductors, *HE4000B Logic Family CMOS,* Data Handbook IC04, 1997.

Figure 6.45. The MC14517 64-bit Static Shift Register.
Motorola Incorporated, *CMOS Logic Data, DL131/D,* Series C, 1991.

Figure 6.46. The HEF4557 1-to-64-bit Variable Length Shift Register.
Philips Semiconductors, *HE4000B Logic Family CMOS,* Data Handbook IC04, 1997.

Figure 6.47. The MC14562 128-bit Static Shift Register.
Motorola Incorporated, *CMOS Logic Data, DL131/D,* Series C, 1991.

Figure 6.48. The HEF4731 Quadruple 64-bit Static Shift Register.
Philips Semiconductors, *HE4000B Logic Family CMOS,* Data Handbook IC04, 1997.

Figure 6.49. The 74HC/HCT7030 9-bit by 64 Word FIFO register; 3-state.
Philips Semiconductors, *High Speed CMOS Logic Family,* Data Handbook IC06, 1996.

Figure 6.50. The 74HC/HCT7403 4-bit by 64 Word FIFO register; 3-state.
Philips Semiconductors, *High Speed CMOS Logic Family,* Data Handbook IC06, 1996.

Figure 6.51. The 74HC/HCT7404 5-bit by 64 Word FIFO register; 3-state.
Philips Semiconductors, *High Speed CMOS Logic Family,* Data Handbook IC06, 1996.

Figure 6.52. The 74HC/HCT7597 8-bit Shift Register with Input Latches.
Philips Semiconductors, *High Speed CMOS Logic Family,* Data Handbook IC06, 1996.

Figure 6.53. The 74HC/HCT7731 Quad 64-bit Static Shift Register.
Philips Semiconductors, *High Speed CMOS Logic Family,* Data Handbook IC06, 1996.

Figure 6.54. The 74HC/HCT40104 4-bit Bidirectional Universal Shift Register; 3-State
Philips Semiconductors, *High Speed CMOS Logic Family,* Data Handbook IC06, 1996.

Figure 6.55. The 74HC/HCT40105 4-bit by 16-word FIFO Register.
Philips Semiconductors, *High Speed CMOS Logic Family,* Data Handbook IC06, 1996.

Chapter 7

Figure 7.2. The 74LS68 Dual Decade and 74LS69 Dual Binary Counters.
Texas Instruments, *TTL Logic Standard TTL, Schottky, Low-Power Schottky Data Book,* 1988.

Figure 7.3. The 7490 TTL Decade Counter.
Signetics Corporation, *Signetics Logic - TTL - Specifications, Military Summary, Data Manual,* 1978.

Figure 7.4. The 7492 and 74LS92 Divide-by-Twelve Counter.
Signetics Corporation, *Signetics Logic - TTL - Specifications, Military Summary, Data Manual,* 1978.

Figure 7.5. The 7493 Binary Counter.
Signetics Corporation, *Signetics Logic - TTL - Specifications, Military Summary, Data Manual,* 1978.

Figure 7.7. The 7497 Six-Bit Synchronous Binary Multiplier/Counter.
Texas Instruments, *TTL Logic Standard TTL, Schottky, Low-Power Schottky Data Book,* 1988.

Figure 7.8. The 74LS160 and 74LS162 Synchronous Presettable Decade Counters.
Texas Instruments, *TTL Logic Standard TTL, Schottky, Low-Power Schottky Data Book,* 1988.

Figure 7.9. The HEF40161 Synchronous Presettable Binary Counters.
Philips Semiconductors, *HE4000B Logic Family CMOS,* Data Handbook IC04, 1997.

Figure 7.11. The 74167 Synchronous Decade Rate Multiplier/Counter.
Texas Instruments, *TTL Logic Standard TTL, Schottky, Low-Power Schottky Data Book,* 1988.

Figure 7.12. The 74S169 Synchronous 4-bit Up/Down Binary Counter.
Texas Instruments, *TTL Logic Standard TTL, Schottky, Low-Power Schottky Data Book,* 1988.

Figure 7.13. The 74176 35 MHz Presettable Decade/Binary Counter/Latche.
Texas Instruments, *TTL Logic Standard TTL, Schottky, Low-Power Schottky Data Book,* 1988.

Figure 7.14. The 74190,74LS190 Synchronous Up/Down Decade counters.
Texas Instruments, *TTL Logic Standard TTL, Schottky, Low-Power Schottky Data Book,* 1988.

Figure 7.15. The 7419,74LS193 Synchronous Up/Down Binary Counters.
Texas Instruments, *TTL Logic Standard TTL, Schottky, Low-Power Schottky Data Book,* 1988.

Figure 7.17. The 74F269 Synchronous 8-bit Bidirectional Binary Counter.
Philips Semiconductors, *FAST TTL Logic Series,* Data Book IC15, 1992.

Figure 7.19. The 74LS390 Dual Decade Ripple Counter.
Signetics Corporation, *Signetics Logic - TTL - Specifications, Military Summary, Data Manual,* 1978.

Figure 7.20. The 74LS393 Dual 4-bit Binary Counter.
Signetics Corporation, *Signetics Logic - TTL - Specifications, Military Summary, Data Manual,* 1978.

Figure 7.21. The 74LS490 Dual BCD Decade Ripple Counter.
Signetics Corporation, *Signetics Logic - TTL - Specifications, Military Summary, Data Manual,* 1978.

Figure 7.22. The 74ALS560 Programmable Synchronous Decade Counter with 3-state outputs.
Texas Instruments, *ALS/AS Logic Data Book,* 1986.

Figure 7.23. The 74F569 4-bit Bidirectional Binary Synchronous Counter.
Philips Semiconductors, *FAST TTL Logic Series,* Data Book IC15, 1992.

Figure 7.24. The 74F579 8-bit Bidirectional Binary Counter (3-state).
Philips Semiconductors, *FAST TTL Logic Series,* Data Book IC15, 1992.

Figure 7.25. The 74LS590, 74LS592 8-bit Binary Counters with Output Registers.
Texas Instruments, *TTL Logic Standard TTL, Schottky, Low-Power Schottky Data Book,* 1988.

Figure 7.26. The 74LS691 4-bit Synchronous Binary Counter.
Texas Instruments, *TTL Logic Standard TTL, Schottky, Low-Power Schottky Data Book,* 1988.

Figure 7.27. The 74LS696 4-bit Synchronous Decade Up/Down Counter.
Texas Instruments, *TTL Logic Standard TTL, Schottky, Low-Power Schottky Data Book,* 1988.

Figure 7.28. The HEF4017 Five-Stage Johnson Counter.
Philips Semiconductors, *HE4000B Logic Family CMOS,* Data Handbook IC04, 1997.

Figure 7.29. The HEF4017 Counter Timing Diagram.
Philips Semiconductors, *HE4000B Logic Family CMOS,* Data Handbook IC04, 1997.

Figure 7.30. The HEF4018 Presettable Divide-by-N Counter.
Philips Semiconductors, *HE4000B Logic Family CMOS,* Data Handbook IC04, 1997.

Figure 7.31. The HEF4020 14-stage Binary Ripple Counter.
Philips Semiconductors, *HE4000B Logic Family CMOS,* Data Handbook IC04, 1997.

Figure 7.32. The timing diagram for the HEF4020 14-stage Ripple Counter.
Philips Semiconductors, *HE4000B Logic Family CMOS,* Data Handbook IC04, 1997.

Figure 7.33. The HEF4022 4-stage Divide-by-8 Johnson Counter.
Philips Semiconductors, *HE4000B Logic Family CMOS,* Data Handbook IC04, 1997.

Figure 7.34. The HEF4024 7-Stage Binary Ripple Counter.
Philips Semiconductors, *HE4000B Logic Family CMOS,* Data Handbook IC04, 1997.

Figure 7.35. The HEF4029 Synchronous Up/Down 4-bit Binary/BCD Counter.
Philips Semiconductors, *HE4000B Logic Family CMOS,* Data Handbook IC04, 1997.

Figure 7.36. The HEF4040 12-stage Binary Ripple Counter.
Philips Semiconductors, *HE4000B Logic Family CMOS,* Data Handbook IC04, 1997.

Figure 7.37. The HEF4059 Programmable Divide-by-N Counter.
Philips Semiconductors, *HE4000B Logic Family CMOS,* Data Handbook IC04, 1997.

Figure 7.38. The HEF4060 14-stage Ripple-Carry Binary Counter/Divider and Oscillator.
Philips Semiconductors, *HE4000B Logic Family CMOS,* Data Handbook IC04, 1997.

Figure 7.39. The HEF4510 BCD Up/Down Counter.
Philips Semiconductors, *HE4000B Logic Family CMOS,* Data Handbook IC04, 1997.

Figure 7.40. The HEF4516 Binary Up/Down Counter.
Philips Semiconductors, *HE4000B Logic Family CMOS,* Data Handbook IC04, 1997.

Figure 7.41. The HEF4518 Dual BCD Counter.
Philips Semiconductors, *HE4000B Logic Family CMOS,* Data Handbook IC04, 1997.

Figure 7.42. The HEF4520 Dual Binary Counter.
Philips Semiconductors, *HE4000B Logic Family CMOS,* Data Handbook IC04, 1997.

Figure 7.43. The HEF4521 24-Stage Frequency Divider and Oscillator
Philips Semiconductors, *HE4000B Logic Family CMOS,* Data Handbook IC04, 1997.

Figure 7.44. The HEF4522 Programmable 4-bit BCD Down Counter.
Philips Semiconductors, *HE4000B Logic Family CMOS,* Data Handbook IC04, 1997.

Figure 7.45. The HEF4526 Programmable 4-bit Binary Down Counter.
Philips Semiconductors, *HE4000B Logic Family CMOS,* Data Handbook IC04, 1997.

Figure 7.46. The HEF4534B Real-Time 5-Decade Counter.
Philips Semiconductors, *HE4000B Logic Family CMOS,* Data Handbook IC04, 1997.

Figure 7.47. The MC14553 3-digit BCD Counter.
Motorola Incorporated, *CMOS Logic Data, DL131/D,* Series C, 1991.

Figure 7.48. The MC14568 Phase Comparator and Programmable Counters.
Motorola Incorporated, *CMOS Logic Data, DL131/D,* Series C, 1991.

Figure 7.48. The MC14569 Programmable Divide-by-N Dual 4-bit Binary/BCD Down Counter.
Motorola Incorporated, *CMOS Logic Data, DL131/D,* Series C, 1991.

Figure 7.49. The HEF4737 Quadruple Static Decade Counter.
Philips Semiconductors, *HE4000B Logic Family CMOS,* Data Handbook IC04, 1997.

Figure 7.50. The MM74C925, '926, '927, '928 4-Digit Counters with Multiplexed 7-segment Output Drivers.
National Semiconductor Corporation, *CMOS Logic Databook,* 1988 Edition.

Figure 7.51. The 74HC/HCT6323 Programmable Ripple Counter with Oscillator; 3-State.
Philips Semiconductors, *High Speed CMOS Logic Family,* Data Handbook IC06, 1996.

Figure 7.52. The 74HC7022 Octal Counter/Divider with Power-Up Clear.
Texas Instruments, *High-Speed CMOS Logic Data Book,* 1988.

Figure 7.53. The CD74AC/ACT7060, '7061 14-stage Binary Counters with oscillator.
GE/RCA/Intersil, *Advanced CMOS Logic ICs,* SSD-283A, 1988.

Figure 7.54. The 74HC/HCT40102 8-bit Synchronous BCD Down Counters.
Philips Semiconductors, *High Speed CMOS Logic Family,* Data Handbook IC06, 1996.

Chapter 8

Figure 8.1. The 7441 BCD-to-Decimal Decoder.
National Semiconductor Corporation, *LS/S/TTL Logic Databook,* 1987.

Figure 8.2. The 7442 BCD-to-Decimal Decoder.
National Semiconductor Corporation, *LS/S/TTL Logic Databook,* 1987.

Figure 8.3. The 7446 and 7447 BCD-to-7-Segment Decoder/Drivers.
Texas Instruments, *TTL Logic Standard TTL, Schottky, Low-Power Schottky Data Book,* 1988.

Figure 8.4. The 7448 and 74LS48 BCD-to-7-Segment Decoder/Drivers.
Texas Instruments, *TTL Logic Standard TTL, Schottky, Low-Power Schottky Data Book,* 1988.

Figure 8.5. The 74LS49 BCD-to-7-Segment Decoder/Driver.
Texas Instruments, *TTL Logic Standard TTL, Schottky, Low-Power Schottky Data Book,* 1988.

Figure 8.6. The 74AS/ALS131 3-Line to 8-Line Decoder/Demultiplexer.
National Semiconductor Corporation, *ALS/AS Databook,* "Guide To Bipolar Logic Device Families," 1987.

Figure 8.7. The 74AS/ALS137 3-Line to 8-Line Decoder/Demultiplexer.
National Semiconductor Corporation, *ALS/AS Databook,* "Guide To Bipolar Logic Device Families," 1987.

Figure 8.8. The 74AS/ALS138 3-Line to 8-Line Decoder/Demultiplexers.
National Semiconductor Corporation, *ALS/AS Databook,* "Guide To Bipolar Logic Device Families," 1987.

Figure 8.9. The 74LS139 Dual 2-Line to 4-Line Decoder/Demultiplexers.
National Semiconductor Corporation, *LS/S/TTL Logic Databook,* 1987.

Figure 8.10. The 74LS147 Priority Encoders.
Motorola Incorporated, *FAST and LSTTL Data,* DL121, Rev. 5, 1992.

Figure 8.11. The 74LS148 Priority Encoder.
Motorola Incorporated, *FAST and LSTTL Data,* DL121, Rev. 5, 1992.

Figure 8.12. The 74150 1-of-16 Data Selectors/Multiplexers.
National Semiconductor Corporation, *LS/S/TTL Logic Databook,* 1987.

Figure 8.13. The 74LS151 1-of-8 Data Selectors/Multiplexers.
National Semiconductor Corporation, *LS/S/TTL Logic Databook,* 1987.

Figure 8.14. The 74152 1-of-8 Data Selectors/Multiplexer.
Signetics Corporation, *Signetics Logic - TTL - Specifications, Military Summary, Data Manual,* 1978.

Figure 8.15. The 74LS153 Dual 4-Input Multiplexers.
National Semiconductor Corporation, *LS/S/TTL Logic Databook,* 1987.

Figure 8.16. The 74159 4-to-16 Line Decoder/Demultiplexers.
Signetics Corporation, *Signetics Logic - TTL - Specifications, Military Summary, Data Manual,* 1978.

Figure 8.17. The 74155 Dual 2-to-4 Line Decoder/Demultiplexers.
Signetics Corporation, *Signetics Logic - TTL - Specifications, Military Summary, Data Manual,* 1978.

Figure 8.18. The 74157 quad 2-Input Multiplexers.
Signetics Corporation, *Signetics Logic - TTL - Specifications, Military Summary, Data Manual,* 1978.

Figure 8.19. The 74LS261 Multiply Decoder.
Signetics Corporation, *Signetics Logic - TTL - Specifications, Military Summary, Data Manual,* 1978.

Figure 8.20. The 74LS298 Quad 2-Input Multiplexers.
Texas Instruments, *TTL Logic Standard TTL, Schottky, Low-Power Schottky Data Book,* 1988.

Figure 8.21. The 74LS348 8-Input Priority Encoders with 3-State Outputs.
Motorola Incorporated, *FAST and LSTTL Data,* DL121, Rev. 5, 1992.

Figure 8.22. The 74LS354, 74LS355 and 74LS356 8-Line to 1-Line Data Selectors/Multiplexers/Registers.
Texas Instruments, *TTL Logic Standard TTL, Schottky, Low-Power Schottky Data Book,* 1988.

Figure 8.23. The 74LS399 Quadruple 2-Input Multiplexer with Storage.
Texas Instruments, *TTL Logic Standard TTL, Schottky, Low-Power Schottky Data Book,* 1988.

Figure 8.24. The 74F537 1-of-10 Decoder/Multiplexer.
Motorola Incorporated, *FAST and LSTTL Data,* DL121, Rev. 5, 1992.

Figure 8.25. The 74F538 1-of-8 Decoder/Multiplexer.
Motorola Incorporated, *FAST and LSTTL Data,* DL121, Rev. 5, 1992.

Figure 8.26. The 74F539 Dual 1-of-4 Decoder.
Motorola Incorporated, *FAST and LSTTL Data,* DL121, Rev. 5, 1992.

Figure 8.27. The 74AS580 and 74AS581 4-Line to 1-Line Data Selectors/Multiplexers.
Texas Instruments, *ALS/AS Logic Data Book,* 1986.

Figure 8.28. The 74AS/ALS857 Hex 2-Line to 1-Line Universal Multiplexers.
Texas Instruments, *ALS/AS Logic Data Book,* 1986.

Figure 8.29. The MC14016 and MC14066 Quad Analog Switch/Quad Analog Multiplexer.
Motorola Incorporated, *CMOS Logic Data, DL131/D,* Series C, 1991.

Figure 8.30. The HEF4028 BCD-to-Decimal Decoder/Binary-to-Octal Decoder.
Philips Semiconductors, *HE4000B Logic Family CMOS,* Data Handbook IC04, 1997.

Figure 8.31. The HEF4051 8-Channel Analog Multiplexer/Demultiplexer.
Philips Semiconductors, *HE4000B Logic Family CMOS,* Data Handbook IC04, 1997.

Figure 8.32. The HEF4052 Dual 4-Channel Analog Multiplexer/Demultiplexer.
Philips Semiconductors, *HE4000B Logic Family CMOS,* Data Handbook IC04, 1997.

Figure 8.33. The HEF4053 Triple 2-Channel Analog Multiplexer/Demultiplexer.
Philips Semiconductors, *HE4000B Logic Family CMOS,* Data Handbook IC04, 1997.

Figure 8.34. The HEF4067 16-Channel Analog Multiplexer/Demultiplexer.
Philips Semiconductors, *HE4000B Logic Family CMOS,* Data Handbook IC04, 1997.

Figure 8.35. The MC1409 Differential 8-Channel Analog Multiplexer/Demultiplexer.
Motorola Incorporated, *CMOS Logic Data, DL131/D,* Series C, 1991.

Figure 8.36. The HEF4511 BCD to 7-Segment Latch/Decoder/Driver.
Philips Semiconductors, *HE4000B Logic Family CMOS,* Data Handbook IC04, 1997.

Figure 8.37. The HEF4512 8-Channel Data Selector.
Philips Semiconductors, *HE4000B Logic Family CMOS,* Data Handbook IC04, 1997.

Figure 8.38. The HEF4514 1-of-16 Decoder/Demultiplexer.
Philips Semiconductors, *HE4000B Logic Family CMOS,* Data Handbook IC04, 1997.

Figure 8.39. The HEF4515 1-of-16 Decoder/Demultiplexer.
Philips Semiconductors, *HE4000B Logic Family CMOS,* Data Handbook IC04, 1997.

Figure 8.40. The MC14529 Dual 4-Channel Analog Data Selector.
Motorola Incorporated, *CMOS Logic Data, DL131/D,* Series C, 1991.

Figure 8.41. The MC14543 BCD to 7-Segment Latch/Decoder/Driver.
Motorola Incorporated, *CMOS Logic Data, DL131/D,* Series C, 1991.

Figure 8.42. The MC14544 BCD to 7-Segment Latch/Decoder/Driver.
Motorola Incorporated, *CMOS Logic Data, DL131/D,* Series C, 1991.

Figure 8.43. The MC14547 BCD to 7-Segment High Current Decoder/Driver.
Motorola Incorporated, *CMOS Logic Data, DL131/D,* Series C, 1991.

Figure 8.44. The MC14551 Quad 2-Channel Analog Multiplexer/Demultiplexer.
Motorola Incorporated, *CMOS Logic Data, DL131/D,* Series C, 1991.

Figure 8.45. The MC14555 Dual Binary to 1-of 4 Decoder/Demultiplexers.
Motorola Incorporated, *CMOS Logic Data, DL131/D,* Series C, 1991.

Figure 8.46. The MC14558 BCD to 7-Segment Decoder.
Motorola Incorporated, *CMOS Logic Data, DL131/D,* Series C, 1991.

Figure 8.47. The 74HC/HCT237 3-to-8 Line Decoder/Demultiplexer.
Philips Semiconductors, *High Speed CMOS Logic Family,* Data Handbook IC06, 1996.

Figure 8.48. The 74HC/HCT238 3-to-8 Line Decoder/Demultiplexer.
Philips Semiconductors, *High Speed CMOS Logic Family,* Data Handbook IC06, 1996.

Figure 8.49. The 74HC239 Dual 2-to-4 Line Decoder/Demultiplexer.
Texas Instruments, *High-Speed CMOS Logic Data Book,* 1988.

Figure 8.50. The 74HC604 Octal 2-Input Multiplexed Latch with 3-State Outputs.
Texas Instruments, *High-Speed CMOS Logic Data Book,* 1988.

Figure 8.51. The 74C915 7-segment to BCD Converter.
National Semiconductor Corporation, *CMOS Logic Databook,* 1988 Edition.

Figure 8.52. The 74C922 and 74C923 16-Key and 20-Key Keypad Encoders.
National Semiconductor Corporation, *CMOS Logic Databook,* 1988 Edition.

Figure 8.53. The 74HC/HCT4316 Quad Bilateral Switch.
Philips Semiconductors, *High Speed CMOS Logic Family,* Data Handbook IC06, 1996.

Figure 8.54. The 74HC/HCT4351 8-Channel Analog Multiplexer/Demultiplexer.
Philips Semiconductors, *High Speed CMOS Logic Family,* Data Handbook IC06, 1996.

Figure 8.55. The 74HC/HCT4352 Dual 4-Channel Analog Multiplexer/Demultiplexer.
Philips Semiconductors, *High Speed CMOS Logic Family,* Data Handbook IC06, 1996.

Figure 8.56. The 74HC/HCT4353 Triple 2-Channel Analog Multiplexer/Demultiplexer.
Philips Semiconductors, *High Speed CMOS Logic Family,* Data Handbook IC06, 1996.

Figure 8.57. The 74ALVCH16271 and 74ALVCH16272 12-bit to 24-bit Multiplexed Bus Exchangers.
Texas Instruments, *Low-Voltage Logic, ALB, ALVC, LV, LVT, LVTZ, GTL Families Data Book,* 1996.

Chapter 9

Figure 9.1. The 7480 High Speed Binary Full Adder.
Signetics Corporation, *Signetics Logic - TTL - Specifications, Military Summary, Data Manual,* 1978.

Figure 9.2. The 7482 High Speed 2-Bit Binary Full Adder.
Texas Instruments, *TTL Logic Standard TTL, Schottky, Low-Power Schottky Data Book,* 1988.

Figure 9.3. The 7483 High Speed 4-Bit Binary Full Adder.
Texas Instruments, *TTL Logic Standard TTL, Schottky, Low-Power Schottky Data Book,* 1988.

Figure 9.4. The TTL 7485 Expandable 4-Bit Magnitude Comparator.
Signetics Corporation, *Signetics Logic - TTL - Specifications, Military Summary, Data Manual,* 1978.

Figure 9.5. The 74H87 4-Bit True/Complement, Zero/One Element.
Texas Instruments, *The Integrated Circuits Catalog For Design Engineers,* 1971.

Figure 9.6. The TTL 74180 9-Bit Odd/Even Parity Generator/Checkers.
Signetics Corporation, *Signetics Logic - TTL - Specifications, Military Summary, Data Manual,* 1978.

Figure 9.7. The 74181 4-Bit Arithmetic Logic Units.
Texas Instruments, *TTL Logic Standard TTL, Schottky, Low-Power Schottky Data Book,* 1988.

Figure 9.8. The 74182 High Speed Lookahead Carry Generator.
Signetics Corporation, *Signetics Logic - TTL - Specifications, Military Summary, Data Manual,* 1978.

Figure 9.9. The 74LS183 Dual Carry-Save Full Adder.
Texas Instruments, *TTL Logic Standard TTL, Schottky, Low-Power Schottky Data Book,* 1988.

Figure 9.10. The 74184 and 74185 BCD-to-Binary and Binary-to-BCD Conversions.
National Semiconductor Corporation, *LS/S/TTL Logic Databook,* 1987.

Figure 9.11. The 74S280 9-Bit Odd/Even Parity Generator/Checker.
Signetics Corporation, *Signetics Logic - TTL - Specifications, Military Summary, Data Manual,* 1978.

Figure 9.12. The 74AS282 Look-Ahead Carry Generator.
Texas Instruments, *ALS/AS Logic Data Book,* 1995.

Figure 9.13. The TTL 74LS283 4-bit Binary Full Adder.
Signetics Corporation, *Signetics Logic - TTL - Specifications, Military Summary, Data Manual,* 1978.

Figure 9.14. The 74284 and 74285 4-Bit by 4-Bit Parallel Binary Multiplication.
Texas Instruments, *TTL Logic Standard TTL, Schottky, Low-Power Schottky Data Book,* 1988.

Figure 9.15. The 74AS286 9-Bit Parity Generator/Checker.
Texas Instruments, *ALS/AS Logic Data Book,* 1995.

Figure 9.16. The 74F350 Shifter.
Motorola Incorporated, *FAST and LSTTL Data,* DL121, Rev. 5, 1992.

Figure 9.17. The 74F381 Arithmetic Logic Unit (ALU).
Philips Semiconductors, *FAST TTL Logic Series,* Data Book IC15, 1992.

Figure 9.18. The 74F382 Arithmetic Logic Unit (ALU).
Philips Semiconductors, *FAST TTL Logic Series,* Data Book IC15, 1992.

Figure 9.19. The 74LS384 8-Bit by 1-Bit Two's Complement Multiplier.
Texas Instruments, *TTL Logic Standard TTL, Schottky, Low-Power Schottky Data Book,* 1988.

Figure 9.20. The 74LS385 Quadruple Serial Adder/Subtractor.
Texas Instruments, *TTL Logic Standard TTL, Schottky, Low-Power Schottky Data Book,* 1988.

Figure 9.21. The 74ALS518 and 74ALS519 8-Bit Identity Comparators.
National Semiconductor Corporation, *ALS/AS Databook,* "Guide To Bipolar Logic Device Families," 1987.

Figure 9.22. The 74ALS520 8-Bit Identity Comparator.
National Semiconductor Corporation, *ALS/AS Databook,* "Guide To Bipolar Logic Device Families," 1987.

Figure 9.23. The 74ALS526, 74ALS527 and 74ALS528 Fuse Programmable Identity Comparators.
Texas Instruments, *ALS/AS Logic Data Book,* 1995.

Figure 9.24. The 74HC/HCT583 4-Bit BCD Full Adders With Fast Carry.
Philips Semiconductors, *High Speed CMOS Logic Family,* Data Handbook IC06, 1996.

Figure 9.25. The 74ALS616 and 74ALS617 16-Bit Parallel Error Detection and Correction Circuits.
Texas Instruments, *ALS/AS Logic Data Book,* 1995.

Figure 9.26. The 74LS630 16-Bit Parallel Error Detection and Correction Circuits.
Texas Instruments, *TTL Logic Standard TTL, Schottky, Low-Power Schottky Data Book,* 1988.

Figure 9.27. The 74ALS632 - 74ALS635 32-Bit Parallel Error Detection and Correction CircuitS.
Texas Instruments, *ALS/AS Logic Data Book,* 1995.

Figure 9.28. The 74LS637 and 74LS638 8-Bit Parallel Error Detection and Correction Circuits.
Texas Instruments, *TTL Logic Standard TTL, Schottky, Low-Power Schottky Data Book,* 1988.

Figure 9.29. The 74ALS677 and 74ALS678 16-Bit Address Comparators.
Texas Instruments, *ALS/AS Logic Data Book,* 1995.

Figure 9.30. The 74ALS679 and 74ALS680 12-Bit Address Comparators.
Texas Instruments, *ALS/AS Logic Data Book,* 1995.

Figure 9.31. The 74LS681 4-Bit Parallel Binary Accumulator.
Texas Instruments, *TTL Logic Standard TTL, Schottky, Low-Power Schottky Data Book,* 1988.

Figure 9.32. The 74LS682, 74LS684, and 74LS685 8-Bit Magnitude/Identity Comparators.
Texas Instruments, *TTL Logic Standard TTL, Schottky, Low-Power Schottky Data Book,* 1988.

Figure 9.33. The 74LS686 and 74LS687 8-Bit Magnitude/Identity Comparators.
Texas Instruments, *TTL Logic Standard TTL, Schottky, Low-Power Schottky Data Book,* 1988.

Figure 9.34. The 74LS688 8-Bit Magnitude/Identity Comparators.
Texas Instruments, *TTL Logic Standard TTL, Schottky, Low-Power Schottky Data Book,* 1988.

Figure 9.35. The 74ALS689 8-Bit Magnitude/Identity Comparator.
National Semiconductor Corporation, *ALS/AS Databook,* "Guide To Bipolar Logic Device Families," 1987.

Figure 9.36. The 74AS881 Arithmetic Logic Unit (ALU)/Function Generator.
National Semiconductor Corporation, *ALS/AS Databook,* "Guide To Bipolar Logic Device Families," 1987.

Figure 9.37. The 74AS882 32-Bit Look-Ahead Carry Generator.
Texas Instruments, *ALS/AS Logic Data Book,* 1995.

Figure 9.38. The 74AS866 and 74AS885 8-Bit Magnitude Comparators.
Texas Instruments, *ALS/AS Logic Data Book,* 1995.

Figure 9.39. The HEF4008 4-Bit Full Adder.
Philips Semiconductors, *HE4000B Logic Family CMOS,* Data Handbook IC04, 1997.

Figure 9.40. The MC14032 and MC14038 Triple Serial Adders.
Motorola Incorporated, *CMOS Logic Data, DL131/D,* Series C, 1991.

Figure 9.41. The MC14527 BCD Rate Multiplier.
Motorola Incorporated, *CMOS Logic Data, DL131/D,* Series C, 1991.

Figure 9.42. The MC14531 12-Bit Parity Tree.
Motorola Incorporated, *CMOS Logic Data, DL131/D,* Series C, 1991.

Figure 9.43. The MC14554 2-Bit by 2-Bit Parallel Binary Multiplier.
Motorola Incorporated, *CMOS Logic Data, DL131/D,* Series C, 1991.

Figure 9.44. The MC14560 NBCD (Natural Binary Coded Decimal) Adder.
Motorola Incorporated, *CMOS Logic Data, DL131/D,* Series C, 1991.

Figure 9.45. The MC14561 9's Complementer.
Motorola Incorporated, *CMOS Logic Data, DL131/D,* Series C, 1991.

Figure 9.46. The MC14582 Look-Ahead Carry Block.
Motorola Incorporated, *CMOS Logic Data, DL131/D,* Series C, 1991.

Figure 9.47. The MC14585 4-Bit Magnitude Comparator.
Motorola Incorporated, *CMOS Logic Data, DL131/D,* Series C, 1991.

Figure 9.48. The 74HC/HCT7080 16-Bit Even/Odd Parity Generator/Checker.
Philips Semiconductors, *High Speed CMOS Logic Family,* Data Handbook IC06, 1996.

Figure 9.49. The 74ACT11286 9-Bit Parity Generator/Checker with Bus-Driven Parity I/O Ports.
Texas Instruments, *AC/ACT CMOS Logic Data Book,* 1997.

Figure 9.50. The 74ACT11520 8-Bit Identity Comparator.
Texas Instruments, *AC/ACT CMOS Logic Data Book,* 1997.

Figure 9.51. The 74ACT16833 Dual 8-Bit to 9-Bit Parity Bus Transceiver.
Texas Instruments, *AC/ACT CMOS Logic Data Book,* 1997.

Figure 9.52. The 74ALVCH16901 18-Bit Universal Bus Transceiver with Parity Generator/Checker.
Texas Instruments, *Low-Voltage Logic, ALB, ALVC, LV, LVT, LVTZ, GTL Families Data Book,* 1996.

Chapter 10

Figure 10.1. The 74S140 Dual 4-Input NAND 50W driver gates.
National Semiconductor Corporation, *LS/S/TTL Logic Databook,* 1987.

Figure 10.2. The 74AS/ALS230 and 74AS/ALS231 3-state Bus Drivers/Receivers
National Semiconductor Corporation, *ALS/AS Databook,* "Guide To Bipolar Logic Device Families," 1987.

Figure 10.3. The 74AS240, 74AS241 and 74AS244 3-State Bus Drivers/Receivers.
National Semiconductor Corporation, *ALS/AS Databook,* "Guide To Bipolar Logic Device Families," 1987.

Figure 10.4. The 74LS365, 74LS366, 74LS367 and 74LS368 HEX 3-State Buffers
National Semiconductor Corporation, *LS/S/TTL Logic Databook,* 1987.

Figure 10.5. The 74F455 and 74F456 Octal Buffers/Drivers with Parity
Philips Semiconductors, *FAST TTL Logic Series,* Data Book IC15, 1992.

Figure 10.6. The 74LS465, 74LS466, 74LS467 and 74LS468 3-State Octal Buffers.
National Semiconductor Corporation, *LS/S/TTL Logic Databook,* 1987.

Figure 10.7. The 74ALS540 and 74ALS541 Octal Buffers/Line Drivers.
National Semiconductor Corporation, *ALS/AS Databook,* "Guide To Bipolar Logic Device Families," 1987.

Figure 10.8. The 74F655 and 74F656 Octal Buffers/Line Drivers
Philips Semiconductors, *FAST TTL Logic Series,* Data Book IC15, 1992.

Figure 10.9. The 74AS756 and 74AS757 Octal Buffer
Texas Instruments, *ALS/AS Logic Data Book,* 1995.

Figure 10.10. The 74AS760, 74AS762 and 74AS763 Octal Buffers.
Texas Instruments, *ALS/AS Logic Data Book,* 1995.

Figure 10.11. The 74LVC827 and 74LVC828 10-bit Buffers/Line Drivers.
Texas Instruments, *Low-Voltage Logic, ALB, ALVC, LV, LVT, LVTZ, GTL Families Data Book,* 1996.

Figure 10.12. The 74S940 and 74S941 Octal 3-state Buffer/Line Drivers/Line Receivers.
National Semiconductor Corporation, *LS/S/TTL Logic Databook,* 1987.

Figure 10.13. The 74F3037 and 74F3038 Quad 2-input NAND High Current 30 W Line Drivers.
Philips Semiconductors, *FAST TTL Logic Series,* Data Book IC15, 1992.

Figure 10.14. The 74ALS29827 and 74ALS29828 10-bit Buffers and Line Drivers.
Texas Instruments, *ALS/AS Logic Data Book,* 1995.

Figure 10.15. The 74F30240 and 74F30244 Octal Open Collector 30 W Line Drivers.
Philips Semiconductors, *FAST TTL Logic Series,* Data Book IC15, 1992.

Figure 10.16. The 74HC7540 and 74HC7541 Octal Schmitt Trigger Buffers/Line Drivers.
Philips Semiconductors, *High Speed CMOS Logic Family,* Data Handbook IC06, 1996.

Figure 10.17. The 74HC/HCT9014/9015/9114/9115 Nine Wide Schmitt Trigger Buffers/Line Drivers.
Philips Semiconductors, *High Speed CMOS Logic Family,* Data Handbook IC06, 1996.

Figure 10.18. The 7AC/ACT11240, 74AC/ACT11244 Octal Buffers/Line Drivers.
Texas Instruments, *AC/ACT CMOS Logic Data Book,* 1997.

Figure 10.19. The 74AC/ACT16240, 74ACT16241 and 74AHC/AHCT16244 16-bit Buffers/Drivers.
Texas Instruments, *AC/ACT CMOS Logic Data Book,* 1997.
Texas Instruments, *AHC/AHCT Advanced High-Speed CMOS Logic Data Book,* 1997.

Figure 10.20. The 74ACT16540 and 74ACT16541 16-bit Buffers/Drivers.
Texas Instruments, *AC/ACT CMOS Logic Data Book,* 1997.

Figure 10.21. The 74ACT16825 and 74ACT16826 18-bit Buffers/Drivers.
Texas Instruments, *AC/ACT CMOS Logic Data Book,* 1997.

Figure 10.22. The 74ALVCH16835 18-bit Non-Inverting Universal Bus/Driver.
Texas Instruments, *Low-Voltage Logic, ALB, ALVC, LV, LVT, LVTZ, GTL Families Data Book,* 1996.

Figure 10.23. The 74HL33240, 74hl33241, and 74hl33244 Octal Buffers/Line Drivers.
Philips Semiconductors, *Low Voltage CMOS & BiCMOS Logic,* Data Handbook IC24, 1996.

Chapter 11

Figure 11.1. The 74ALS/AS242, 74ALS243 Quadruple Bus Transceivers.
Texas Instruments, *TTL Logic Standard TTL, Schottky, Low-Power Schottky Data Book,* 1988.

Figure 11.2. The 74LS245 Octal Bus Transceivers.
Texas Instruments, *TTL Logic Standard TTL, Schottky, Low-Power Schottky Data Book,* 1988.

Figure 11.3. The 74LS440 - 74LS444 Quadruple Tridirectional Bus Transceivers.
Texas Instruments, *TTL Logic Standard TTL, Schottky, Low-Power Schottky Data Book,* 1988.

Figure 11.4. The 74LS446 and 74LS449 Quadruple Bus Transceivers.
Texas Instruments, *TTL Logic Standard TTL, Schottky, Low-Power Schottky Data Book,* 1988.

Figure 11.5. The 74ACQ543 Octal Registered Transceivers.
National Semiconductor Corporation, *FACT Advanced CMOS Logic Databook,* 400022, 1993.

Figure 11.6. The 74F545 Octal Bidirectional Transceiver.
Philips Semiconductors, *FAST TTL Logic Series,* Data Book IC15, 1992.

Figure 11.7. The 74F552 Octal Registered Transceiver.
Philips Semiconductors, *FAST TTL Logic Series,* Data Book IC15, 1992.

Figure 11.8. The 74ALS614 and 74ALS615 Octal Bus Transceivers and Registers.
Texas Instruments, *ALS/AS Logic Data Book,* 1986.

Figure 11.9. The 74F620 and 74F623 Octal Bus Transceivers.
Philips Semiconductors, *FAST TTL Logic Series,* Data Book IC15, 1992.

Figure 11.10. The 74F621 and 74F622 Octal Bus Transceivers.
Philips Semiconductors, *FAST TTL Logic Series,* Data Book IC15, 1992.

Figure 11.11. The 74LS638 and 74LS639 Bidirectional Octal Bus Transceivers
Texas Instruments, *TTL Logic Standard TTL, Schottky, Low-Power Schottky Data Book,* 1988.

Figure 11.12. The 74AS640, 74AS643, and 74AS645 Bidirectional Octal Bus Transceivers.
Texas Instruments, *ALS/AS Logic Data Book,* 1986.

Figure 11.13. The 74ALS/AS641, 74ALS/AS642, and 74ALS/AS644 Bidirectional Octal Bus Transceivers.
Texas Instruments, *ALS/AS Logic Data Book,* 1995.

Figure 11.14. The 74ALS646, 74ALS647, 74ALS648 and 74ALS649 Bidirectional Octal Bus Transceivers.
Texas Instruments, *ALS/AS Logic Data Book,* 1995.

Figure 11.15. The 74ALS651, 74ALS652, 74ALS653 and 74ALS654 Bidirectional Octal Bus Transceivers.
Texas Instruments, *ALS/AS Logic Data Book,* 1995.

Figure 11.16. The 74F657 Octal Transceivers with 8-bit Parity Generator/Checker.
Philips Semiconductors, *FAST TTL Logic Series,* Data Book IC15, 1992.

Figure 11.17. The 74HC658 and 74HC659 Octal Transceivers with 8-bit Parity Generator.
Texas Instruments, *High-Speed CMOS Logic Data Book,* 1988.

Figure 11.18. The 74HC664 and 74HC665 Octal Transceivers with 8-bit Parity Generator.
Texas Instruments, *High-Speed CMOS Logic Data Book,* 1988.

Figure 11.19. The 74F776 Octal Latched Transceiver.
Philips Semiconductors, *FAST TTL Logic Series,* Data Book IC15, 1992.

Figure 11.20. The 74F777 Triple Latched Bidirectional Bus Transceiver.
Philips Semiconductors, *FAST TTL Logic Series,* Data Book IC15, 1992.

Figure 11.21. The 74F807 Octal Shift/Count Registered Transceiver.
Philips Semiconductors, *FAST TTL Logic Series,* Data Book IC15, 1992.

Figure 11.22. The 74AS852 8-bit Universal Port Controller.
Texas Instruments, *ALS/AS Logic Data Book,* 1995.

Figure 11.23. The 74AS856 8-bit Universal Port Controller.
Texas Instruments, *ALS/AS Logic Data Book,* 1995.

Figure 11.24. The 74F861, 74F862, 74F863, and 74F864 10-bit Bus Transceivers.
Philips Semiconductors, *FAST TTL Logic Series,* Data Book IC15, 1992.

Figure 11.25. The 74AS877 8-bit Universal Transceiver Port Controller.
Texas Instruments, *ALS/AS Logic Data Book,* 1995.

Figure 11.26. The 74F1242 and 74F1243 Quad Bus Tranceivers.
Philips Semiconductors, *FAST TTL Logic Series,* Data Book IC15, 1992.

Figure 11.27. The 74F1245 Octal Bus Transceiver.
Philips Semiconductors, *FAST TTL Logic Series,* Data Book IC15, 1992.

Figure 11.28. The 74F2952 8-bit Registered Transceivers.
Philips Semiconductors, *FAST TTL Logic Series,* Data Book IC15, 1992.

Figure 11.29. The 74LVXC3245 8-bit Dual Supply Translating Transceiver.
Motorola Incorporated, *LCX Data Low-Voltage CMOS Logic,* BR1339/D, Rev 3, 04/96.

Figure 11.30. The 74F3893 Quad Backplane Transceiver.
Philips Semiconductors, *FAST TTL Logic Series,* Data Book IC15, 1992.

Figure 11.31. The 74ALVC164245 16-bit (Dual Octal) Backplane Translating Receiver.
Philips Semiconductors, *Low Voltage CMOS & BiCMOS Logic,* Data Handbook IC24, 1996.

Figure 11.32. The 74F8960 and 74F8961 Octal Latched Bidirectional Futurebus Transceivers.
Philips Semiconductors, *FAST TTL Logic Series,* Data Book IC15, 1992.

Figure 11.33. The 74F8962 and 74F8963 9-bit Latched Bidirectional Futurebus Transceivers.
Philips Semiconductors, *FAST TTL Logic Series,* Data Book IC15, 1992.

Figure 11.34. The 74F8965 and 74F8966 9-bit Latchable Bidirectional Futurebus Transceivers.
Philips Semiconductors, *FAST TTL Logic Series,* Data Book IC15, 1992.

Figure 11.35. The 74AC11245 Octal Bus Transceiver.
Texas Instruments, *AC/ACT CMOS Logic Data Book,* 1997.

Figure 11.36. The 74AC11652 Octal Bus Transceivers and Registers.
Texas Instruments, *AC/ACT CMOS Logic Data Book,* 1997.

Figure 11.37. The 74AHC16245 16-bit (Dual Octal) Non-inverting Bus Transceivers.
Texas Instruments, *AHC/AHCT Advanced High-Speed CMOS Logic Data Book,* 1997.

Figure 11.38. The 74ALVCH162268 12-bit to 24-bit Registered Bus Exchanger.
Texas Instruments, *Low-Voltage Logic, ALB, ALVC, LV, LVT, LVTZ, GTL Families Data Book,* 1996.

Figure 11.39. The 74ALVCH16269 and 74ALVCH162269 12-bit to 24-bit Registered Bus Transceiver.
Texas Instruments, *Low-Voltage Logic, ALB, ALVC, LV, LVT, LVTZ, GTL Families Data Book,* 1996.

Figure 11.40. The 74ALVCH16270 12-bit to 24-bit Registered Bus Exchanger.
Texas Instruments, *Low-Voltage Logic, ALB, ALVC, LV, LVT, LVTZ, GTL Families Data Book,* 1996.

Figure 11.41. The 74ALVCH16409 and 74ALVCH162409 9-bit, 4-port Universal Bus Exchangers.
Texas Instruments, *Low-Voltage Logic, ALB, ALVC, LV, LVT, LVTZ, GTL Families Data Book,* 1996.

Figure 11.42. The 74ACT16470 16-bit Registered Transceiver.
Texas Instruments, *AC/ACT CMOS Logic Data Book,* 1997.

Figure 11.43. The 74AC16472, 74ACT16474 and 74ACT16475 18-bit Registered Transceivers.
Texas Instruments, *AC/ACT CMOS Logic Data Book,* 1997.

Figure 11.44. The 74ALVCH16500 and 74ALVCH16501 18-bit Universal Bus Transceivers.
Texas Instruments, *Low-Voltage Logic, ALB, ALVC, LV, LVT, LVTZ, GTL Families Data Book,* 1996.

Figure 11.45. The 74ALVCH16525 and 74ALVCH162525 18-bit Registered Bus Transceivers.
Texas Instruments, *Low-Voltage Logic, ALB, ALVC, LV, LVT, LVTZ, GTL Families Data Book,* 1996.

Figure 11.46. The 74ALVCH16543 are 16-bit Registered Bus Transceivers.
Texas Instruments, *Low-Voltage Logic, ALB, ALVC, LV, LVT, LVTZ, GTL Families Data Book,* 1996.

Figure 11.47. The 74ALVCH16600, 74ALVCH16601 and 74ALVCH162601 18-bit Universal bus Transceivers.
Texas Instruments, *Low-Voltage Logic, ALB, ALVC, LV, LVT, LVTZ, GTL Families Data Book,* 1996.

Figure 11.48. The 74AC16620 16-bit Bus Transceivers.
Texas Instruments, *AC/ACT CMOS Logic Data Book,* 1997.

Figure 11.49. The 74AC16640 16-bit Inverting Bus Transceiver.
Texas Instruments, *AC/ACT CMOS Logic Data Book,* 1997.

Figure 11.50. The 74AC16646 16-bit Bus Transceivers and Registers.
Texas Instruments, *AC/ACT CMOS Logic Data Book,* 1997.

Figure 11.51. The 74AC16651 16-bit Transceivers and Registers.
Texas Instruments, *AC/ACT CMOS Logic Data Book,* 1997.

Figure 11.52. The 74ACT16657 16-bit Transceiver with Parity Generator/Checker.
Texas Instruments, *AC/ACT CMOS Logic Data Book,* 1997.

Figure 11.53. The 74ACT16861 20-bit Non-inverting Bus Transceiver.
Texas Instruments, *AC/ACT CMOS Logic Data Book,* 1997.

Figure 11.54. The 74ACT16863 18-bit Non-inverting Bus Transceiver.
Texas Instruments, *AC/ACT CMOS Logic Data Book,* 1997.

Figure 11.55. The 74ACT16864 18-bit Inverting Bus Transceiver.
Texas Instruments, *AC/ACT CMOS Logic Data Book,* 1997.

Figure 11.56. The 74ACT16952 16-bit Registered Transceiver.
Texas Instruments, *AC/ACT CMOS Logic Data Book,* 1997.

Figure 11.57. The 74ALS29833 8-bit to 9-bit High Speed Bus Transceiver.
Texas Instruments, *ALS/AS Logic Data Book,* 1995.

Figure 11.58. The 74ALS29854 8-bit to 9-bit High Speed Bus Transceiver.
Texas Instruments, *ALS/AS Logic Data Book,* 1995.

Figure 11.59. The 74ALS29861 and 74ALS29862 10-bit Bus Tranceivers.
Texas Instruments, *ALS/AS Logic Data Book,* 1995.

Figure 11.60. The 74ALS29863 and 74ALS29864 9-bit Bus Tranceivers.
Texas Instruments, *ALS/AS Logic Data Book,* 1995.

Figure 11.61. The 74HL33246 and 74HL33640 Octal Bus Transceivers.
Philips Semiconductors, *Low Voltage CMOS & BiCMOS Logic,* Data Handbook IC24, 1996.

Figure 11.62. The 74HL33620 and 74HL33623 Octal Bus Transceivers
Philips Semiconductors, *Low Voltage CMOS & BiCMOS Logic,* Data Handbook IC24, 1996.

Figure 11.63. The 74HL33646 and 74HL33648 Transceiver/Registers.
Philips Semiconductors, *Low Voltage CMOS & BiCMOS Logic,* Data Handbook IC24, 1996.

Figure 11.64. The 74HL33651 and 74HL33652 Octal Bus Transceiver/Registers.
Philips Semiconductors, *Low Voltage CMOS & BiCMOS Logic,* Data Handbook IC24, 1996.

Figure 11.65. The 74HL33952 Octal Non-inverting Registered Transceiver.
Philips Semiconductors, *Low Voltage CMOS & BiCMOS Logic,* Data Handbook IC24, 1996.

PROMPT®
PUBLICATIONS

PC Hardware Projects, Volume 3

by James "J.J." Barbarello

PC Hardware Projects, Volume 3 shows you how to construct ComponentLAB, a device that connects to the digital I/O card of previous volumes and gives you the capability to measure and record DC voltage, resistance and capacitance, and test digital ICs. ComponentLAB also contains an 8-bit A/D converter that you can use to capture any analog signal ranging from a few hundred millivolts all the way up to 5 volts.

This book also explains how to export the data you store into other applications for your own data processing needs, and provides you with all the inside information on how to use the digital I/O card for other purposes.

PC Hardware Projects, Volume 3 comes with a disk containing the application programs for ComponentLAB, the A/D converter, and the digital I/O card. It also contains definition files for 95 of the most common TTL and CMOS ICs, and various sample data files.

Projects
204 pages • paperback • 7-3/8 x 9-1/4"
ISBN: 0-7906-1151-1 • Sams: 61151
$29.95

Real-World Interfacing With Your PC

Second Edition

by James "J.J." Barbarello

As the computer becomes increasingly prevalent in society, its functions and applications continue to expand. Modern software allows users to do everything from balance a checkbook to create a family tree.

Interfacing, however, is truly the wave of the future for those who want to use their computer for things other than manipulating text, data, and graphics.

Real-World Interfacing With Your PC provides all the information necessary to use a PC's parallel port as a gateway to electronic interfacing. In addition to hardware fundamentals, this book provides a basic understanding of how to write software to control hardware.

While the book is geared toward electronics hobbyists, it includes a chapter on project design and construction techniques, a checklist for easy reference, and a recommended inventory of starter electronic parts to which readers at every level can relate.

Projects
120 pages • paperback • 7-3/8 x 9-1/4"
ISBN: 0-7906-1145-7 • Sams: 61145
$29.95

CALL 1-800-428-7267 TODAY FOR THE NAME OF YOUR NEAREST PROMPT PUBLICATIONS DISTRIBUTOR

Prices subject to change.

PROMPT® PUBLICATIONS

Optoelectronics, *Volume 1*

Vaughn D. Martin

Welcome to the field of optoelectronics - an exciting technology which is useful, important, rapidly emerging, and constantly evolving. *Optoelectronics, Vol. 1* is a comprehensive and practical coverage of optoelectronics for readers of all levels of experience.

Optoelectronics, Vol. 1 walks you through the field of optoelectronics at your own pace and allows you to verify your progress. This book gives you a thorough understanding of optoelectronics, and bridges the gap between theories and more practical aspects and applications. Equations are used only when no other means of explanation can clearly illustrate a point.

Topics covered in *Optoelectronics, Vol. 1* include: Terminology and Concepts, Measuring and Testing, Visible Light-Emitting Sources, Photocells, Photodiodes and Photomultipliers, LED Secondary Optics, Basic Optical Math, AND MUCH, MUCH MORE!

Professional Reference
264 pages • paperback • 8-1/2 x 11"
ISBN: 0-7906-1091-4 • Sams: 61091
$34.95

Optoelectronics, *Volume 2*

Vaughn D. Martin

Written for experienced technicians and electronics students who want to broaden their knowledge of optoelectronics, *Optoelectronics, Vol. 2* presents you with comprehensive information on radiometrics, color CRTs, and much more. This book also contains fascinating and easy-to-follow projects that will show you how to put your newly acquired optoelectronics knowledge to practical use.

Topics covered in *Optoelectronics, Vol. 2* include:

Ultraviolet (UV) Theory and Applications
IR Diodes
Bar Codes and IR Optical Sensing Mechanisms
Optocoupler Applications
Phototransistor and Optointerrupters
Optical Triac Drivers
Modern Optoelectric ICs
5 Optoelectronic Projects, including a UV Radiation Monitor and a Car Burglar Alarm

Professional Reference
258 pages • paperback • 8-1/2 x 11"
ISBN: 0-7906-1110-4 • Sams: 61110
$34.95

CALL 1-800-428-7267 TODAY FOR THE NAME OF YOUR NEAREST PROMPT PUBLICATIONS DISTRIBUTOR

Prices subject to change.

Electronic Circuit
Guidebook
SENSORS
Joseph
J. Carr
PROMPT

Electronic Circuit
Guidebook
IC TIMERS
Joseph
J. Carr
PROMPT

PROMPT®
PUBLICATIONS

Electronic Circuit Guidebook, Volume 3: Op Amps

Joseph Carr

Third in a series covering electronic instrumentation and circuitry, *Electronic Circuit Guidebook, Volume 3: Op Amps* is designed to give some insight into how practical linear IC amplifiers work in actual, real-life circuits. Because of their widespread popularity, operational amplifiers figure heavily in this book, though other types of amplifiers are not overlooked. Volume 3 allows you to design and configure your own circuits, and is intended to be a practical workshop aid.

Topics covered: Linear IC Amplifiers, The Ideal Op Amp, Inverting Follower Op Amp Circuits, Noninverting Follower Op Amp Circuits, Ideal vs. Non-Ideal Op Amps, DC Differential Op Amp circuits, Instrumentation Amps, Isolation Amps, Operational Transconductance Amps (OTAs), Current Difference Amps (CDAs), Solid-State Audio Circuits, Wideband and Video Linear IC Amps, Active Analog Filter Circuits, Waveform Generators and Timer Circuits, Instrumentation and Other Circuits.

Electronics Technology
273 pages • paperback • 7-3/8 x 9-1/4"
ISBN: 0-7906-1131-7 • Sams: 61131
$29.95

Electronic Circuit Guidebook, Volume 4: Electro-Optics

Joseph Carr

Volume 4 is mostly about E-O sensors – those electronic transducers that convert light waves into a proportional voltage, current or resistance. The coverage of the sensors is wide enough to allow you to understand the physics behind the theory of operation of the device, and also the circuits used to make these sensors into useful devices. This book examines the photoelectric effect, photoconductivity, photovoltaics, and PN junction photodiodes and phototransistors. Also examined is the operation of lenses, mirrors, prisms, and other optical elements keyed to light physics.

After discussing the basic sensors and optical elements, the book examines other devices including optoisolators, optical fibers, X-ray sensors, TV sensors and some additional optical instruments such as telescopes and scientific instruments. The circuits for using sensors in E-O instruments, and for interfacing them with computers, are also discussed in sufficient depth to allow you to apply them to practical situations.

Electronics Technology
405 pages • paperback • 7-3/8 x 9-1/4"
ISBN: 0-7906-1132-5 • Sams: 61132
$29.95

CALL 1-800-428-7267 TODAY FOR THE NAME OF YOUR NEAREST PROMPT PUBLICATIONS DISTRIBUTOR

Prices subject to change.